Molecular Biology

Molecular Biology
Principles and Practice

Second Edition

Michael M. Cox
University of Wisconsin–Madison

Jennifer A. Doudna
University of California, Berkeley

Michael O'Donnell
The Rockefeller University

W. H. FREEMAN
& COMPANY

A Macmillan Education Imprint
New York

Publisher: Kate Ahr Parker
Senior Acquisitions Editor: Lauren Schultz
Developmental Editors: Anna Bristow, Erica Pantages Frost, Lisa Samols
Media Editors: Anna Bristow, Erica Pantages Frost
Assistant Editor: Tue Tran
Art Director: Diana Blume
Cover and text designer: Diana Andrews, DreamIt, Inc.
Senior Project Editor: Elizabeth Geller
Manuscript Editors: Linda Strange, Brook Soltvedt
Illustrations: H. Adam Steinberg, Dragonfly Media Group
Photo Editor: Jennifer Atkins
Photo Researcher: Teri Stratford
Illustration Coordinator: Janice Donnola
Production Coordinator: Paul Rohloff
Marketing Director: John Britch
Marketing Assistant: Bailey James
Composition: MPS Limited
Printing and Binding: Quad Versailles

Front cover image: Courtesy Illumina, Inc.

Throughout the text, a number of illustrations have been adapted from *Lehninger Principles of Biochemistry*, Sixth Edition, by David L. Nelson and Michael M. Cox.

Library of Congress Control Number: 2015930844

ISBN-13: 978-1-4641-2614-7
ISBN-10: 1-4641-2614-3

Printed in the United States of America

First Printing

W. H. Freeman and Company
41 Madison Avenue
New York, NY 10010
Houndmills, Basingstoke RG21 6XS, England

www.whfreeman.com

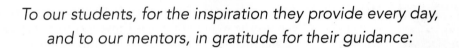

*To our students, for the inspiration they provide every day,
and to our mentors, in gratitude for their guidance:*

Tom Cech
Fred Grieman
Bill Jencks
Arthur Kornberg
Bob Lehman
Sharon Panasenko
David Sheppard
Jack Szostak
Hal White
Charles Williams

About the Authors

MICHAEL M. COX was born in Wilmington, Delaware. After graduating from the University of Delaware, he went to Brandeis University to do his doctoral work with William P. Jencks, and then to Stanford for postdoctoral study with I. Robert Lehman. He is currently Professor of Biochemistry at the University of Wisconsin–Madison. His research focuses on recombinational DNA repair processes. Cox has received awards for both teaching and research, including the 1989 Eli Lilly Award in Biological Chemistry from the American Chemical Society and two major teaching awards from the University of Wisconsin. He has coauthored five editions of *Lehninger Principles of Biochemistry*.

JENNIFER A. DOUDNA grew up on the Big Island of Hawaii and became interested in chemistry and biochemistry in high school. She received her B.A. in biochemistry from Pomona College and her Ph.D. from Harvard University, working in the laboratory of Jack Szostak, with whom she also did postdoctoral research. She then went to the University of Colorado as a Lucille P. Markey scholar and postdoctoral fellow with Thomas Cech. Doudna is currently Professor of Molecular and Cell Biology and Professor of Chemistry at the University of California, Berkeley, and an Investigator of the Howard Hughes Medical Institute. She is a member of the National Academy of Sciences, the American Academy of Arts and Sciences, and the Institute of Medicine. She is also a Fellow of the American Association for the Advancement of Science.

MICHAEL O'DONNELL grew up in a neighborhood on the banks of the Columbia River outside Vancouver, Washington. He had several inspirational teachers at Hudson Bay High School who led him into science. He received his B.A. in biochemistry from the University of Portland and his Ph.D. from the University of Michigan, where he worked under Charles Williams, Jr., on electron transfer in the flavoprotein thioredoxin reductase. He performed postdoctoral work on *E. coli* replication with Arthur Kornberg and then on herpes simplex virus replication with I. Robert Lehman in the Biochemistry Department at Stanford University. O'Donnell is currently Professor of Biochemistry and Structural Biology at The Rockefeller University and an Investigator of the Howard Hughes Medical Institute. He is a member of the National Academy of Sciences.

Contents in Brief

Contents

II NUCLEIC ACID STRUCTURE AND METHODS

Preface

As teachers, we know that undergraduate science education is evolving. Simply conveying facts does not produce a scientifically literate student, a long-held perception now reinforced by numerous studies. Students of science need more: a better window on what science is and how it is done, a clear presentation of key concepts that rises above the recitation of details, an articulation of the philosophical underpinnings of the scientific discipline at hand, exercises that demand analysis of real data, and an appreciation for the contributions of science to the well-being of humans throughout the world. As undergraduate science educators rise to these challenges, we are faced with both higher numbers of students and declining resources. How can we all do more with less?

Textbooks are an important part of the equation. A good textbook must now be more than a guide to the information that defines a discipline. For instructors, a textbook must organize information, incorporate assessment tools, and provide resources to help bring a discipline to life. For students, a textbook must relate science to everyday experience, highlight the key concepts, and show each student the process that generated those key concepts.

This book had its genesis at a meeting of the authors in Napa Valley in January 2006. From the outset, we set ambitious goals designed to address the key challenges we face as teachers.

Students see science as a set of facts rather than an active human endeavor. Molecular biology has a wealth of important stories to tell. We wanted to convey the excitement that drives modern molecular biology, the creativity at the bench, and the genuine wonder that takes hold as the workings of a new biological process are revealed. This theme is set in the first chapter, dedicated in large measure to an introduction to the scientific process. Every chapter then begins with a *Moment of Discovery,* highlighting a researcher's own description of a memorable moment in his or her career. After Chapter 1, every chapter ends with a *How We Know* section, with stories relating the often circuitous path to a new insight. Additional anecdotes—scientists in action—are woven into the text and the accompanying *Highlights.* As students read the text, the laboratories and the people behind the discoveries will never be far away.

This second edition is an update, and much more. It has allowed us to refine the initial vision we had when we started this project and to augment that vision with unparalleled resources that will bring the subject to life for students and educators alike.

◀ MOMENT OF DISCOVERY

Scientific breakthroughs represent the exhilarating culmination of a lot of hard work. Each chapter opens with a description of a significant breakthrough in molecular biology, told by the scientist who made the discovery. The scientists featured in the Moments of Discovery are David Allis, Norm Arnheim, Bonnie Bassler, Steve Benner, James Berger, Carlos Bustamante,

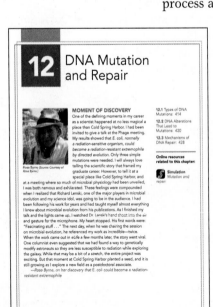

Rose Byrne, Jamie Cate, Joe DeRisi, Roxana Georgescu, Lin He, Tracy Johnson, Melissa Jurica, Judith Kimble, Robert Lehman, Steve Mayo, Harry Noller, Smita Patel, Lorraine Symington, Jack Szostak, Robert Tjian, and Wei Yang.

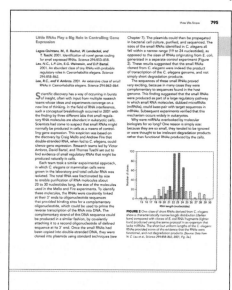

◀ HOW WE KNOW

Each chapter ends with a How We Know section that combines fascinating stories of research and researchers with experimental data for students to analyze.

Students often view science as a completed story. The reality is far different. Data can take a researcher in unexpected directions. An experiment designed to test one hypothesis can end up revealing something quite different. The analysis of real data is a fundamental skill to be honed by every student of science. We have tried to address this need aggressively. Each chapter in this text features a challenging set of problems, including at least one requiring the analysis of data from the scientific literature. Many of these are linked to the discoveries described in the How We Know sections. Each chapter also ends with some *Unanswered Questions*, providing just a sampling of the endless challenges that remain for those with the motivation to tackle them.

UNANSWERED QUESTIONS ▶

A short section at the end of each chapter describes important areas still open to discovery, showing students that even well-covered subjects, such as nucleic acid structure and DNA replication, are far from fully explored.

? UNANSWERED QUESTIONS

The study of protein function is, arguably, the oldest subdiscipline in biochemistry and molecular biology. But there is still much to learn. The relatively young science of genomics keeps pointing to genes that encode proteins about which we know little or nothing at all. Some shortcuts to functional discovery are discussed in later chapters.

1. How does protein structure relate to function? This is an old but still very relevant question for every scientist who studies proteins. Advanced methods of structural analysis are providing more information than ever before, but many of these structural pictures are static. A clear picture of a complete binding or catalytic cycle can require a detailed knowledge of the structure of multiple protein conformations. Certain structural motifs and domains (e.g., the OB fold of single-stranded DNA–binding proteins and other proteins, the AAA$^+$ ATPase domain, and simple β-barrel structures) appear in proteins that often have seemingly unrelated functions. The manner in which particular structures are adapted to different functions is an ongoing area of investigation.

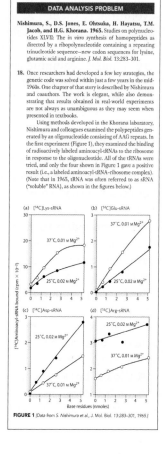

FIGURE 1 (Data from S. Nishimura et al., *J. Mol. Biol.* 13:283–301, 1965.)

◀ END-OF-CHAPTER PROBLEMS

Extensive problem sets at the end of each chapter give students the opportunity to think about and work with the chapter's key ideas. New problems have been added in each chapter for this second edition. Each problem set concludes with a Data Analysis Problem, giving students the critical experience of interpreting real research data. Solutions to all problems can be found at the back of the book.

Students get lost in the details. Presenting the major concepts clearly, in the text as well as in the illustrations, is crucial to teaching students how science is done. We have worked to use straightforward language and a conversational writing style to draw students in to the material. We have collaborated closely with our illustrator, Adam Steinberg, to create clean, focused figures. Featured *Key Conventions* highlight the implicit but often unstated conventions used when sequences and structures are displayed and in naming biological molecules.

KEY CONVENTION

DNA and RNA are defined by the type of sugar in the polynucleotide backbone (deoxyribose or ribose), not by the presence of thymine or uracil.

◀ **KEY CONVENTIONS**

In brief paragraphs, the Key Conventions clearly lay out for students some fundamental principles often glossed over.

◀ **ILLUSTRATIONS**

Good figures should speak for themselves. We have worked to keep our figures simple and the figure legends as brief as possible. The illustrations in the text are the product of close collaboration with our colleague Adam Steinberg. Together with the talented artists at Dragonfly Media Group, Adam has helped to hone and implement our vision.

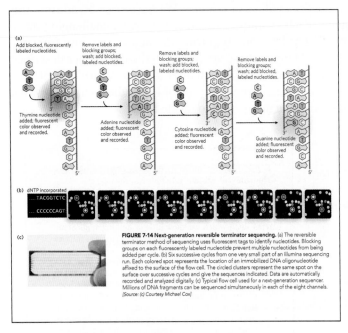

FIGURE 7-14 **Next-generation reversible terminator sequencing.** (a) The reversible terminator method of sequencing uses fluorescent tags to identify nucleotides. Blocking groups on each fluorescently labeled nucleotide prevent multiple nucleotides from being added per cycle. (b) Six successive cycles from one very small part of an Illumina sequencing run. Each colored spot represents the location of an immobilized DNA oligonucleotide affixed to the surface of the flow cell. The circled clusters represent the same spot on the surface over successive cycles and give the sequences indicated. Data are automatically recorded and analyzed digitally. (c) Typical flow cell used for a next-generation sequencer. Millions of DNA fragments can be sequenced simultaneously in each of the eight channels. [Source: (c) Courtesy Michael Cox]

Students see evolution as an abstract theory. Every time a molecular biologist studies a developmental pathway in nematodes, identifies key parts of an enzyme active site by determining what parts are conserved among species, or searches for the gene underlying a human genetic disease, he or she is relying on evolutionary theory. *Evolution is a foundational concept, upon which every discipline in the biological sciences is built.* In this text, evolution is a theme that pervades every chapter, beginning with a major section in Chapter 1 and continuing as the topic of many Highlights and chapter segments.

◀ **HIGHLIGHTS**

These discussions are designed to enhance students' understanding and appreciation of the relevance of each chapter's material. There are four categories of Highlights:

- **Medicine** explores diseases that arise from defects in biochemical pathways, and how concepts uncovered in molecular biology have contributed to drug therapies and other treatments.
- **Technology** focuses on cutting-edge molecular biology methods.
- **Evolution** reveals the role of molecular biology research in understanding key biological processes and the connections among organisms.
- **A Closer Look** examines a wide variety of additional, intriguing topics.

HIGHLIGHT 8-2   EVOLUTION

Phylogenetics Solves a Crime

In the summer of 1994, a nurse in Lafayette, Louisiana, broke off a 10-year affair with a physician. The physician had been giving the nurse vitamin shots for fatigue. He gave her one more of these shots in August 1994, after the breakup. The nurse had donated blood to a local blood bank on several occasions; she was tested and found negative for HIV in October 1992, May 1993, and April 1994. In late 1994, the nurse became ill and tested positive for both HIV-1 and hepatitis C, although she had no history of contacts that could have led to the infections. The nurse accused the doctor of infecting her with HIV.

Investigators found records indicating that the doctor had treated and drawn blood from his only HIV-infected patient and a hepatitis C-infected patient just before giving the nurse her vitamin injection in August 1994. But how does one link the patients' blood to the nurse-victim in this case? The subsequent trial of the doctor was the first to use phylogenetics in a court case. The investigation focused on the HIV infection.

Once HIV begins to replicate in a new host, the virus mutates rapidly, an evolution that occurs within one infected individual. Samples taken from a person with HIV years after infection can be used to build a phylogenetic tree that can trace the evolution of the virus in that individual. Blood samples were collected from the doctor's HIV-infected patient and from the nurse. Control samples were collected from 30 different HIV-positive patients selected at random in the Lafayette area. The HIV in the samples was sequenced and analyzed independently by two different laboratories at Baylor University and the University of Michigan. Both analyses yielded the same result. The phylogenetic analysis of the victim's HIV strains showed that they were most closely related to, and nested within, the strains from the doctor's patient (Figure 1).

With this and other evidence, the doctor was convicted of attempted second-degree murder in 1998. The verdict was upheld by a Louisiana appeals court in 2000, and the U.S. Supreme Court refused to hear the case in 2002, ending court proceedings. The doctor is now serving a 50-year prison sentence. The same methodology has since been used in rape and child abuse cases.

FIGURE 1 A phylogenetic tree reveals the diversity of HIV samples in the Lafayette, Louisiana, area. The part of the tree derived from the doctor's HIV-positive patient is highlighted, with the nurse-victim's DNA clearly nested within this set of sequences. [Source: Data from M. L. Metzker et al., Proc. Natl. Acad. Sci. USA 99:14,292–14,297, 2002.]

EXPERIMENTAL TECHNIQUES

As researchers, we know that it is critical to understand the benefits and limitations of experimental techniques. We strive to give students a sense of how an experiment is designed and what makes the use of a particular technique or model organism appropriate. The techniques covered in this book are:

NEW AND UPDATED CONTENT

The second edition addresses recent discoveries and advances, corresponding to our ever-changing understanding of molecular biology. In addition to the text updates listed here, there are numerous new figures and photos, along with significantly updated figures in every chapter. There are also new end-of-chapter problems for every chapter and many new Unanswered Questions.

Chapter 1

• Updated discussions of evolution and the scientific method

Chapter 2
• Updated discussion of the central dogma
• Updated and expanded discussion of the types of RNA

Chapter 3
• New Moment of Discovery
• Expanded discussion of nucleosides
• Revised and expanded section: The Hydrophobic Effect Brings Together Nonpolar Molecules
• New section: Electronic Interactions between Bases in Nucleic Acids

Chapter 4
• Expanded section: Amino Acids Are Categorized by Chemical Properties
• Significantly expanded discussion of protein purification, including Highlight 4-1
• New section: Intrinsically Unstructured Proteins Have Versatile Binding Properties
• Expanded section on protein families
• Significantly expanded section on protein folding and computational biology

Chapter 5
• New Moment of Discovery

Chapter 6
• Expanded discussion of the instability of RNA
• New Highlight 6-1: DNA Nanotechnology
• New discussion of riboswitches

Chapter 7
• Expanded discussion on obtaining DNA fragments to clone
• Thoroughly updated section on next-gen and other modern DNA sequencing technologies.
• New section: Genomic Sequencing Is Aided by New Generations of DNA Sequencing Methods, incorporating the exciting new advances with programmable nucleases

Chapter 8
• Expanded Highlight 8-1, now including discussion of the microbiome
• Updated section on noncoding DNA
• Expanded section on mass spectrometry

Chapter 10
• New Moment of Discovery
• Significantly expanded discussion of histone modifications, including a new table

Chapter 11
• Expanded discussion of the β sliding clamp
• Expanded discussion of the Pol III holoenzyme
• Updated and expanded discussion of eukaryotic replication forks
• Updated and expanded section: Eukaryotic Origins "Fire" Only Once per Cell Cycle
• New section: Telomeres and Telomerase Solve the End Replication Problem in Eukaryotes
• New Highlight 11-2: Short Telomeres Portend Aging Diseases

Chapter 12
• New Moment of Discovery
• New table presenting overview of DNA repair processes

Chapter 13
- Updated and expanded sections on double-strand break repair and reconstruction of replication forks
- Updated section on meiotic recombination

Chapter 14
- Updated and expanded introductory section on transposable elements and site-specific recombination
- Updated and expanded section: Precise DNA Rearrangements Are Promoted by Site-Specific Recombinases
- Reorganized section on the use of site-specific recombination systems in biotechnology
- Updated and expanded sections on transposition

Chapter 15
- Updated section on transcription elongation
- Updated and expanded discussion of the role of transcription factors
- Updated and expanded discussion of termination mechanisms among RNA polymerases

Chapter 16
- Streamlined chapter organization
- Expanded discussion of P bodies

Chapter 18
- Streamlined chapter organization
- Updated discussion of protein release factors
- Updated discussion of nuclear export signals

Chapter 19
- Updated section: Gene Expression Is Regulated through Feedback Loops, now including inducer exclusion

Chapter 22
- Expanded section on alternative splicing, including ESEs and ESSs
- Updated section on RNA interference
- New section: RNAs Regulate a Wide Range of Cellular Processes
- Updated section on the developmental potential of stem cells

MEDIA

Simulations

One of our central goals in tackling the revision of this textbook was to provide special resources to engage students (and educators) in molecular biology. New to the second edition are simulations that cover core molecular biology concepts and techniques. Created using the art from the text, the simulations reinforce students' understanding by allowing them to interact with the structures and processes they have encountered. A game-like format guides students through the simulations, unlocking them in order, and multiple-choice questions after each simulation ensure that instructors can assess whether students have thoroughly understood each topic. These simulations are the product of many days of meetings among the authors, editors, and media developers. From storyboarding to the finished product, these simulations were one of the most challenging as well as stimulating efforts associated with preparing the second edition. We are excited to present this new approach to learning key concepts.

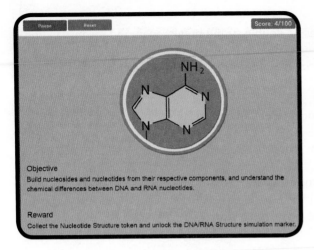

Objective
Build nucleosides and nucleotides from their respective components, and understand the chemical differences between DNA and RNA nucleotides.

Reward
Collect the Nucleotide Structure token and unlock the DNA/RNA Structure simulation marker.

Nucleotide Structure (Chapter 3)
DNA/RNA Structure (Chapter 6)
PCR (Chapter 7)
Sanger Sequencing (Chapter 7)
CRISPR (Chapters 7 and 19)
DNA Replication (Chapter 11)
DNA Polymerase (Chapter 11)
Mutation and Repair (Chapter 12)
Transcription (Chapter 15)
mRNA Processing (Chapter 16)
Translation (Chapter 18)

Nature Articles with Assessment

These articles engage students in reading about primary research and encourage critical thinking. Specifically selected for both alignment with the text coverage and exploration of identified difficult topics, the *Nature* articles include assessment questions that can be automatically graded. Also included are open-ended questions that are suitable for use in flipped classrooms and active learning discussions either in class or online.

The simulations and *Nature* articles for *Molecular Biology: Principles and Practice* are available in our LaunchPad system, along with many additional resources.

This dynamic, fully integrated learning environment brings together all of our teaching and learning resources in one place. It also contains the fully interactive **e-Book** and other newly updated resources for students and instructors, including the following:

New Clicker Questions allow instructors to integrate active learning in the classroom and to assess students' understanding of key concepts during lectures.

Updated Test Bank contains at least 40 multiple-choice and short-answer questions for each chapter.

LearningCurve allows students to test their comprehension of the chapter concepts. The system adapts to students' individual level of preparedness by giving them questions at varying levels of difficulty, depending on whether they answer a question without help, or they need help but eventually get the question right, or they are unable to answer the question. Links to the appropriate e-Book section, hints, and feedback help students realize where they need more practice on a topic.

Key Term Flashcards allow students to review the definitions of all the glossary terms and quiz themselves.

Textbook Images and Tables are offered as high-resolution JPEG files. Each image has been fully optimized to increase type size and adjust color saturation. These images have been tested in a large lecture hall to ensure maximum clarity and visibility.

Acknowledgments

This text represents our best effort to synthesize a complex and ever-shifting field and to contribute to the broadening requirements of twenty-first-century education in molecular biology. We welcome your comments and suggestions. We thank our many colleagues whose input has helped shape this book:

Steven Ackerman, *University of Massachusetts*
Ravi Allada, *Northwestern University*
Rick Amasino, *University of Wisconsin-Madison*
Andrew Andres, *University of Nevada, Las Vegas*
Brian Ashburner, *University of Toledo*
Matthew Bahamonde, *Farmingdale State College*
Kenneth Belanger, *Colgate University*
Joel Belasco, *NYU School of Medicine*
Morgan Benowitz-Fredericks, *Bucknell University*
Bradford Berges, *Brigham Young University*
Xin Bi, *University of Rochester*
Robert Borgon, *University of Central Florida*
David Bourgaize, *Whittier College*
Nicole Bournias-Vardiabasis, *California State University, San Bernardino*
John Boyle, *Mississippi State University*
Jeremy Bruenn, *University at Buffalo*
Douglas Burks, *Wilmington College*
Aaron Cassill, *University of Texas, San Antonio*
Karl Chai, *University of Central Florida*
Davis Cheng, *California State University, Fresno*
William Cody, *University of Dallas*
Mary Connell, *Appalachian State University*
Scott Covey, *University of British Columbia*
Fred Cross, *Rockefeller University*
Cristina Cummings, *Providence College*
Rodney Dale, *Loyola University Chicago*
Susan DiBartolomeis, *Millersville University*
Dessislava Dimova, *Rutgers University, New Brunswick*
David Donze, *Louisiana State University, Baton Rouge*
Arri Eisen, *Emory University*
Danielle Ellis, *California State University, Pomona*
R. Paul Evans, *Brigham Young University*
Nicholas Ewing, *California State University, Sacramento*
Jason Fitzgerald, *Southeastern Illinois College*
Gerald Frenkel, *Rutgers University, Newark*
Louise Glass, *University of California, Berkeley*

Ann Grens, *Indiana University South Bend*
Theresa Grove, *Valdosta State University*
Nancy Guild, *University of Colorado Boulder*
Immo Hansen, *New Mexico State University*
Daniel Herman, *University of Wisconsin-Eau Claire*
Margaret Hollingsworth, *University at Buffalo*
Stan Ivey, *Delaware State University*
Russell Johnson, *Colby College*
Jason Kahn, *University of Maryland*
Mijung Kim, *Chicago State University*
Timothy Lane, *University of California, Los Angeles*
Curtis Loer, *University of San Diego*
Charles Mallery, *University of Miami, College of Arts and Sciences*
Kathryn McMenimen, *Mount Holyoke College*
Mitch McVey, *Tufts University*
Thomas Mennella, *Bay Path College*
Karl Miletti, *Delaware State University*
Yuko Miyamoto, *Elon University*
Evangelos Moudrianakis, *Johns Hopkins University*
Arunachalam Muthaiyan, *Chicago State University*
Hao Nguyen, *California State University, Sacramento*
Brent Nielsen, *Brigham Young University*
James Ntambi, *University of Wisconsin-Madison*
Greg Odorizzi, *University of Colorado*
James Olesen, *Ball State University*
Harold Olivey, *Indiana University Northwest*
Anthony Otsuka, *Illinois State University*
Rekha Patel, *University of South Carolina*
Bruce Patterson, *University of Arizona, Tucson*
Brian Poole, *Brigham Young University*
Megan Porter, *University of South Dakota*
Ted Powers, *University of California, Davis*
April Pyle, *University of California, Los Angeles*
Brian Ring, *Valdosta State University*
Herve Roy, *University of Central Florida*
Edmund Rucker, *University of Kentucky*
Ivan Sadowski, *University of British Columbia*
Steven Sandler, *University of Massachusetts Amherst*
Brian Sato, *University of California, Irvine*
Mary Schuler, *University of Illinois Urbana-Champaign*
William Scovell, *Bowling Green State University*
Andrei Seluanov, *University of Rochester*
Konstantin Severinov, *Rutgers University, New Brunswick*

Xueyan Shan, *Mississippi State University*
Elaine Sia, *University of Rochester*
Ron Siu, *University of California, Los Angeles, Extension*
Agnes Southgate, *College of Charleston*
Daniel Stoebel, *Harvey Mudd College*
Derek Tan, *Sloan-Kettering Institute for Cancer Research*
Ignatius Tan, *New York University*
Lloyd Turniten, *University of Wisconsin–Eau Claire*
Jill Wildonger, *University of Wisconsin–Madison*
Bruce Wolff, *University of Waterloo*
Michael Yu, *University at Buffalo*

This book would not have been possible without the support of our publishers at W. H. Freeman. A book of this sort is an undertaking measured not just in hours but in sleepless nights, almost-met deadlines, conference calls, and occasional levity. It is an enterprise in which teachers sometimes become students. The needed guidance has been provided by an exceptionally talented team of editors and copy editors. Kate Ahr Parker has overseen the effort from the beginning. Few human beings are as gifted in the art of articulating urgency with grace.

Anna Bristow, Erica Frost, and Lisa Samols have been our development editors. Guided by their capable hands, first-draft chapters have been created, reworked, broken up, and sometimes merged. They provided encouragement and pointed out deficiencies. They have been our partners throughout, scrutinizing every word we produced. As the project progressed, the work was honed and the chapters integrated with the help of Brook Soltvedt and Linda Strange. Both Brook and Linda are long-time veterans of the *Lehninger Biochemistry* series, and their expertise added immeasurably to the final product in your hands. In the end, they have managed the impressive feat of merging three voices into one. We are extraordinarily grateful to all of the editors for their dedication to this project. We are fortunate to have had the benefit of their insights and expertise.

The artwork for this book was a labor of love handled by Adam Steinberg and the artists at Dragonfly Media. Adam is also a *Lehninger Principles of Biochemistry* veteran, and his experience and skill are evident on almost every page of this book. He worked hand in hand with the authors to create illustrations that convey concepts concisely and in a unified style.

Our thanks also go to the consummate professionals who ensured the high quality in the production of the book: art director Diana Blume, project editor Elizabeth Geller, production coordinator Paul Rohloff, illustration coordinator Janice Donnola, photo editor Jennifer Atkins, and photo researcher Teri Stratford. We greatly appreciated their flexibility and creativity in working with complex material and ever-shifting schedules.

We express our appreciation to our colleagues, friends, and families for their patience and support.

Last, but certainly not least, we are grateful to the Moment of Discovery authors, who shared some of their favorite scientific career moments with us. Each of them provided valuable time and effort for this project and helped us add a personal touch to every chapter.

Michael M. Cox
Jennifer A. Doudna
Michael O'Donnell

December 2014

Mike Cox, Jennifer Doudna, and Mike O'Donnell

Courtesy Mike Cox

1 Evolution, Science, and Molecular Biology

MOMENT OF DISCOVERY

Jack Szostak [Source: © Jim Sugar/Corbis.]

A big question in the origin of life concerns how primitive cells might have evolved. My own approach to this question involved lots of discussions with Irene Chen and others in my lab about how lipid vesicles containing RNA, which might mimic a simple self-replicating life form, could be capable of dividing. In other words, as the amount of genetic material (here, RNA) increased by making more copies of itself, *how would the increased RNA content affect the physical properties of the vesicle?* We envisioned that osmotic pressure might make vesicles grow by extracting lipids from neighboring vesicles, ultimately leading to division by rupture and resealing. This idea seemed pretty far out, though, until Irene began doing experiments with vesicles containing lipids bearing fluorescent dyes. We could encapsulate RNA inside the vesicles and watch the vesicles change in size (or not) under different conditions by following the level of fluorescence as a function of vesicle surface area. Irene found that empty vesicles or vesicles "swollen" with RNA were stable over time, but when she mixed them together, the swollen vesicles started to grow by stealing lipid molecules from neighboring empty vesicles! So the system worked exactly as we had imagined, demonstrating that vesicle growth and division is a process that can occur spontaneously.

More recently, we found that vesicles loaded with RNA can also take up nucleotides (the building blocks of RNA and DNA) from the environment, disproving an old idea that it would be hard for primitive cells to survive by taking up small molecules, including negatively charged nucleotides, from their surroundings. It has been very exciting to find that each potential roadblock to primitive cellular replication that we have explored so far can be overcome, often without requiring specialized catalysts or input energy.

—*Jack Szostak, on his discovery of self-dividing vesicles that mimic growing cells*

Online resources related to this chapter:

***Nature* exercise** Genome dynamics during experimental evolution

Born in the second half of the twentieth century, molecular biology has only recently come of age. Broadly speaking, **molecular biology** is the study of essential cellular macromolecules, including DNA, RNA, and proteins, and the biological pathways between them. Over the decades, molecular biology has become firmly associated with the structure, function, and regulation of information pathways at the molecular level. All of the processes required to reliably pass genetic information from one generation to another and from DNA to RNA to protein are included in this area of study. Of the requirements for life, it is the information in our genetic material that links all organisms to each other and documents their intertwined history. The biological information pathways that maintain, use, and transmit that information are the focus of this book.

Molecular biology may have a relatively short history, but its impact on the human experience is already considerable. Medicine, modern agriculture, forensic science, and many other endeavors rely on technologies developed by molecular biologists. Our current understanding of information pathways has given rise to diagnostic tests for genetic diseases, forensic DNA analysis, crops with improved yields and resistance to disease, new cancer therapies, an unprecedented ability to track pandemics, new wastewater treatment methods, new approaches to the generation of energy, and much more. Many of these advances are chronicled throughout this textbook.

This first chapter introduces three of the most important themes that link the book's topics. The first theme concerns the two key requirements for life: **biological information**, the genetic instructions that shape every living cell and virus, and **catalysis**, a capacity to accelerate critical molecular processes. Molecular biology deals with both, and much of the discipline focuses on the interplay between information-containing polymers (nucleic acids and proteins) and the enzymes that catalyze and regulate their synthesis, modification, function, and degradation.

The second theme is **evolution**. Many of the processes we will consider can be traced back billions of years, and a few can be traced to the last universal common ancestor. Genetic information is a kind of molecular clock that can help define ancestral relationships among species. Shared information pathways connect humans to every other living organism on Earth and to all the organisms that came before.

The third theme in this book is how we look at molecular biology as a scientific endeavor. Any scientific discipline is a construct not only of the knowledge it has generated but also of the human processes behind that knowledge. Molecular biology has both an inspirational history and a promising future, to be forged by contributors as yet unnamed. Breakthroughs rely on more than technology and ideas: they require an understanding of the scientific process and are informed by the struggles of the past.

1.1 THE EVOLUTION OF LIFE ON EARTH

All organisms on Earth are connected by an evolutionary journey spanning more than 3 billion years. The diversity of life we see around us is the sum of a limitless number of **mutations**, changes in genetic information that are usually subtle but sometimes dramatic. When Charles Darwin proposed that natural selection acts on variation in populations, he had no knowledge of the mechanisms that give rise to that variation. Such mechanisms lie at the heart of modern molecular biology.

What Is Life?

Almost anyone can distinguish a living organism from an inanimate object. However, a rigorous scientific description of life is harder to achieve. Life differs from non-life in identifiable ways, as summarized in **Figure 1-1**. Organisms move, reproduce, grow, and alter their environment in ways that inanimate objects cannot. But such characteristics alone provide an unsatisfying definition of life, particularly when a few of them may be shared by inanimate substances. In 1994, the United States National Aeronautics and Space Administration (NASA) convened a panel to consider the question, "What is life?" A simple definition resulted: *Life is a chemical system capable of Darwinian evolution.* The importance of evolutionary theory to all biological sciences gains full expression in this concise statement.

Every living system we know about has several requirements for its existence. Two of these—raw materials and energy—are supplied by a home planet endowed with an abundance of both. Molecules in Earth's life forms are made up largely of the elements hydrogen, oxygen, nitrogen, and carbon. These are the smallest and most abundant atoms that can make, respectively, one, two, three, and four covalent bonds with other atoms. The molecules formed by these elements tend to be quite stable and can be very complex. The energy required for life is derived from the sun. Plants and photosynthetic microorganisms collect and store the energy derived from sunlight in the chemical bonds of complex biomolecules.

A third requirement for a living system is an envelope, creating a barrier between the living and inanimate worlds and establishing a means of selective interaction between a cell and its environment. The work of Jack Szostak, chronicled in this chapter's Moment of Discovery, may be replicating some key evolutionary moments that led to enveloped living systems (**Figure 1-2** on p. 4).

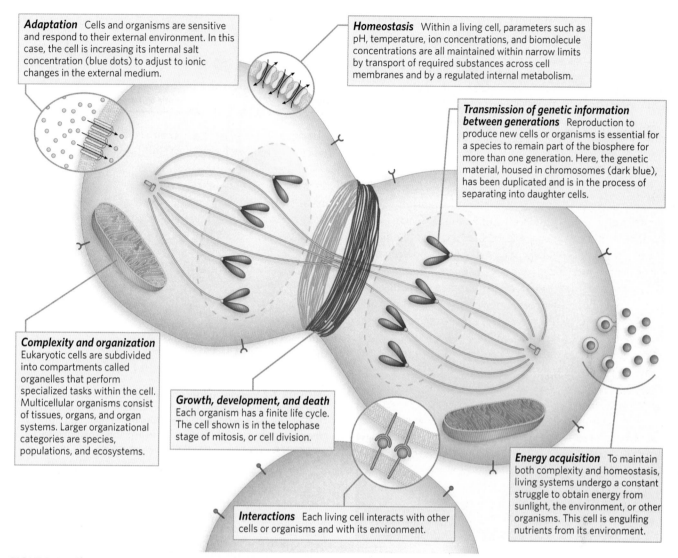

Adaptation Cells and organisms are sensitive and respond to their external environment. In this case, the cell is increasing its internal salt concentration (blue dots) to adjust to ionic changes in the external medium.

Homeostasis Within a living cell, parameters such as pH, temperature, ion concentrations, and biomolecule concentrations are all maintained within narrow limits by transport of required substances across cell membranes and by a regulated internal metabolism.

Transmission of genetic information between generations Reproduction to produce new cells or organisms is essential for a species to remain part of the biosphere for more than one generation. Here, the genetic material, housed in chromosomes (dark blue), has been duplicated and is in the process of separating into daughter cells.

Complexity and organization Eukaryotic cells are subdivided into compartments called organelles that perform specialized tasks within the cell. Multicellular organisms consist of tissues, organs, and organ systems. Larger organizational categories are species, populations, and ecosystems.

Growth, development, and death Each organism has a finite life cycle. The cell shown is in the telophase stage of mitosis, or cell division.

Energy acquisition To maintain both complexity and homeostasis, living systems undergo a constant struggle to obtain energy from sunlight, the environment, or other organisms. This cell is engulfing nutrients from its environment.

Interactions Each living cell interacts with other cells or organisms and with its environment.

FIGURE 1-1 Characteristics of living systems. Each characteristic distinguishes living organisms from inanimate matter.

The final two requirements—catalysis and biological information—are particularly important, truly distinguishing a living organism from an inanimate object. These requirements are the domain of molecular biology. The energy transactions that support homeostasis (the maintenance of parameters such as pH and biomolecule concentrations within the narrow range needed to support life) and enable the transmission of genetic information from one generation to the next are initiated by powerful catalysts called **enzymes**. Enzymes are highly specific, and each enzyme accelerates only one or a small number of chemical reactions. Most enzymes are proteins, although a few catalytic RNA molecules play important roles in cells. The catalysts that a particular organism possesses define which reactions can occur in that organism. Enzymes determine what a cell takes in for nourishment, how fast the cell grows, how it discards wastes, how it constructs its cellular membranes, how it responds to other cells, and how it reproduces.

The presence of enzymes in a cell depends on the faithful transmission of the genetic information that encodes them from one generation to the next. Enzymes, as well as the myriad other proteins and RNA molecules that regulate their synthesis and function, are the actual molecular targets of evolution. When a cell acquires a new function, it generally reflects the presence of a new enzyme or set of enzymes, or an alteration in the regulation or function of an existing enzyme or process. The new functions arise through changes in genes—changes that are shaped by evolutionary processes. In the biosphere of today, DNA is the standard macromolecule for the long-term storage and transmission of biological information. It is exquisitely adapted to that function (**Figure 1-3** on p. 5). However, as we shall see, there were probably stages in the evolution of life when DNA did not serve as the primary genetic library in living systems.

(a)

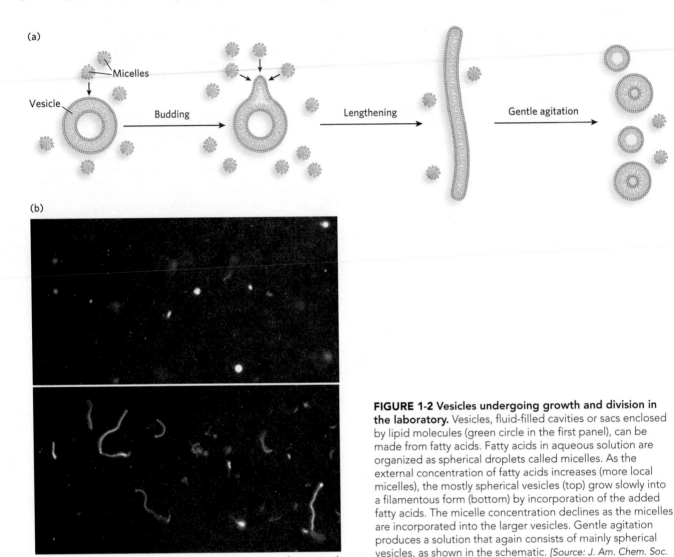

(b)

30 μm

FIGURE 1-2 Vesicles undergoing growth and division in the laboratory. Vesicles, fluid-filled cavities or sacs enclosed by lipid molecules (green circle in the first panel), can be made from fatty acids. Fatty acids in aqueous solution are organized as spherical droplets called micelles. As the external concentration of fatty acids increases (more local micelles), the mostly spherical vesicles (top) grow slowly into a filamentous form (bottom) by incorporation of the added fatty acids. The micelle concentration declines as the micelles are incorporated into the larger vesicles. Gentle agitation produces a solution that again consists of mainly spherical vesicles, as shown in the schematic. [Source: J. Am. Chem. Soc. 134(51):20812–20819, 2012, Fig. 7.]

Evolution Underpins Molecular Biology

In 1973, the geneticist Theodosius Dobzhansky published an article in the professional journal *The American Biology Teacher* entitled "Nothing in Biology Makes Sense Except in the Light of Evolution." This sentiment has special meaning in molecular biology, because the pathways and processes in living systems give rise to the genetic variation on which natural selection acts (**Figure 1-4**). They also inform the ongoing investigations into how life arose on Earth.

Evolution relies on spontaneous and generally random changes in an organism's genomic material, called mutations. In spite of the elaborate cellular mechanisms we consider in this book, all of which help ensure accurate transmission of genetic information from one generation to the next, mutations regularly occur. Mutations can be as simple as a change in a single base pair of DNA or base of RNA or as substantial as the inversion, deletion,

or insertion of large segments of genetic material. As we will be discussing in detail, errors can arise during replication (Chapter 11), and DNA damage can lead to permanent mutation when repair systems (Chapter 12) go awry. Larger chromosomal changes can arise from recombination (Chapter 13) or transposition (Chapter 14). Some mutations affect genes directly; others affect the ways in which DNA is transcribed into RNA or RNA is processed or translated (Chapters 15–18). Relatively minor changes in genes involved in regulatory processes (Chapters 19–22) can give rise to dramatic changes in the organism; this realization has created a new field, essentially a modern merger of the fields of evolutionary and developmental biology, dubbed "evo-devo" (described in Chapter 22). All the processes that contribute to information transfer are highly, but not perfectly, accurate, and the slow accumulation of alterations is inevitable. Many organisms even have mechanisms to speed up the pace of mutational change, which they draw upon in times of stress.

(a)

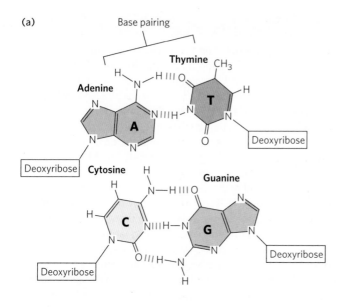

(b)

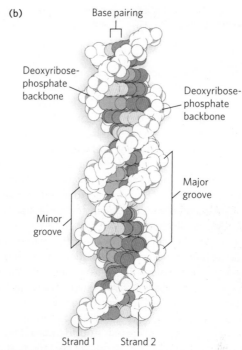

FIGURE 1-3 DNA structure. Because of its structural properties, DNA is well suited for long-term information storage. Genomic DNA almost always consists of two complementary strands of deoxyribonucleic acid. Each strand has a backbone consisting of deoxyribose residues connected by phosphate groups, and a base is attached to each ribose. Strand complementarity is enforced by specific interactions between the bases in each strand. The interactions create base pairs. (a) The G≡C and A═T base pairs are similarly sized, giving the DNA double helix a uniform width and allowing base pairs, in any sequence, to stack. Complementary base pairing facilitates replication and transmission from one generation to the next. (b) The double-helical structure and base stacking confer stability. Major and minor helical grooves in the structure provide access to genetic information for a wide range of DNA-binding proteins. The uniform structure of the DNA backbone allows the synthesis of very long polymers.

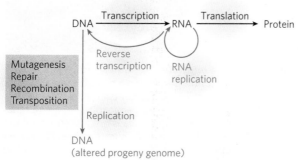

FIGURE 1-4 Pathways of biological information flow. In almost all living systems, information is stored in DNA, then transcribed into RNA, which is processed and translated into protein. DNA is replicated to prepare for cell division. The transfer and maintenance of genetic information are regulated at each of these stages. Exceptions to this general pattern are found in certain viruses (RNA viruses and retroviruses) that store their genetic information in RNA. Viruses with RNA genomes make use of additional pathways (denoted by the red arrows)—RNA replication and reverse transcription (creation of DNA from RNA, instead of the other way around)—to maintain their genomes. The yellow highlighting represents points of regulation. Processes in the gray shaded box, along with occasional errors in replication, reverse transcription, and RNA replication, give rise to genomic alterations (mutations) that fuel evolution.

An understanding of these processes has also given us insights into the origins of life and the process of evolution. Continuing explorations of RNA structure (Chapter 6) and metabolism (Chapters 15 and 16) have informed new theories of prebiotic evolution. The genetic code (Chapter 17) provides a particularly vivid look at the shared history of every organism on Earth.

Molecular biology has provided the enzymes that make most of the methods of biotechnology possible (Chapter 7). These increasingly powerful methods for studying the genes of many different organisms allow us to trace their evolution. Through modern genomics (Chapter 8), molecular biology is opening a window onto evolution that Charles Darwin would marvel at.

The interrelationship of molecular biology and evolution is of more than academic interest. Human beings exist in a world where every organism continues to evolve. Microorganisms, with their short life cycles, evolve most rapidly (**Highlight 1-1**). Of special concern are human pathogens, as well as the microorganisms, fungi, insects, and other organisms that affect our food crops, livestock, and water supply. Molecular biology

Observing Evolution in the Laboratory

The bacterium *Deinococcus radiodurans* has a remarkable capacity to survive the effects of ionizing radiation (IR, or gamma rays). A human being would be killed by exposure to 2 Gy (1 Gy (gray) = 100 rads) of IR, but cultures of *Deinococcus* routinely survive 5,000 Gy with no lethality. *Deinococcus* is a desert dweller, and this characteristic reflects its adaptation to the effects of desiccation. In dry conditions, the bacterium cannot grow and its cellular metabolism shuts down. Spontaneous damage to the cellular DNA accumulates, including strand breaks. DNA repair processes, which require ATP generated by cellular metabolism, do not take place. However, the bacterium can repair its genome quickly when conditions favorable for growth return. Like desiccation, IR also generates numerous DNA strand breaks, and that same extraordinary capacity for DNA repair is put to use after exposure to IR.

How long does it take for a bacterium to evolve extreme resistance to IR? A recent study demonstrated that *Escherichia coli*, the common laboratory bacterium, can acquire this resistance by directed evolution. Twenty cycles of exposure to enough IR to kill more than 99% of the cells, with each cycle followed by the outgrowth of survivors, produced an *E. coli* population with a radiation resistance approaching that of *Deinococcus*. The entire selection process can be achieved in less than a month. Complete genomic sequencing of cells isolated from the evolved populations typically reveals 40 to 80 mutations. The answer to survival varies from cell to cell, with different cells displaying different arrays of mutations, even when they come from the same evolved population. In just a single, small bacterial culture, evolution can take many paths, and a variety of solutions are found that lead to acquisition of a new trait.

This is just one of many experiments demonstrating that dramatic changes in microorganisms can be readily generated and observed in the laboratory within short periods of time. The same kind of evolutionary processes are occurring constantly in microorganisms in our environment, including human pathogens. When AIDS appeared as a new threat to human health in the early 1980s, the power of evolutionary theory was quickly on display. The causative agent, HIV, was soon isolated and its genomic sequence determined.

Characterizing this novel and very dangerous virus from scratch would have delayed treatments for years. But scientists had a shortcut at hand. A deep reservoir of information about viruses and their evolutionary relationships had already been built up over decades of research. The small HIV genome thus held all the clues that science needed for a rapid understanding of its infection cycle and the development of a medical response. Its genome revealed that HIV is a type of RNA virus called a retrovirus, with clear evolutionary relationships to other viruses that were already known and understood (Figure 1). It was immediately evident which HIV genes encode the enzymes essential to the virus life cycle, and these enzymes rapidly became drug targets. One result was the development of highly effective treatments at an unprecedented rate, ranging from AZT to protease inhibitors (see Highlights 5-2 and 14-3 for more detailed descriptions of the retrovirus life cycle). Millions of lives have been saved, in large measure because all biological and medical research is carried out in the context of evolutionary theory.

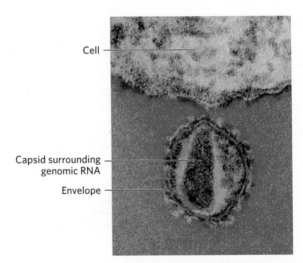

FIGURE 1 HIV is a retrovirus. Like other retroviruses, it has an RNA genome condensed within a proteinaceous capsid. The capsid is surrounded by a spherical lipid envelope derived from its host cell's cytoplasmic (plasma) membrane. Its relationship to other retroviruses is not just structural but embedded in definable ways in its chromosome. [*Source: Hans Gelderblom / Getty Images.*]

---

provides essential tools for use in tracking pandemics, investigating new microbial pathogens, identifying the genes underlying human genetic diseases, solving crimes, tracing the origin of diseases, treating cancer, and engineering microorganisms for new purposes in bioremediation and bioenergy. All of these efforts rely heavily on the concepts of evolutionary biology. Indeed, modern society relies on countless innovations in medicine and agriculture that would not exist but for Darwin's great insight.

Life on Earth Probably Began with RNA

About 4.6 billion years ago, the sun and Earth and the other planets and asteroids of our solar system were formed. Within the first billion years of our planet's existence, life appeared on its surface. How did this happen, and how likely is it that this has happened on other, similar worlds? Modern geologists, paleontologists, and molecular biologists are slowly piecing together the history of life on Earth from the rich trove of clues in the

>3.5 billion years ago

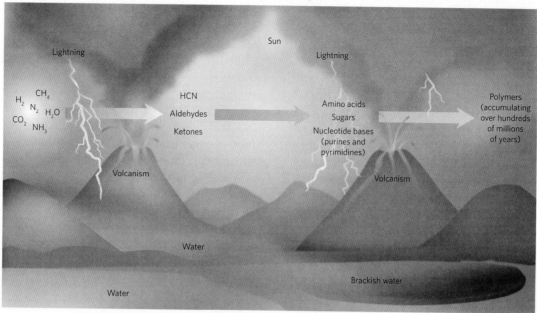

FIGURE 1-5 Prebiotic chemistry. Over hundreds of millions of years, and with constant energy input from solar radiation, volcanism, and other sources, the molecular constituents of Earth's early atmosphere were converted from simple molecules such as water, methane, ammonia, hydrogen, nitrogen, and carbon dioxide into a range of more complex organic molecules and polymers. The resulting tarry substance may have coated the planet's surface and turned bodies of water into concentrated and complex solutions.

geologic, fossil, and genomic records. A plausible sequence emerges, providing a wide range of hypotheses that can be tested using modern chemical and physical methods.

The first few hundred million years were a time of prebiotic chemistry (**Figure 1-5**). No life was present, but chemical reactions were happening everywhere. The atmosphere contained primarily water, methane, ammonia, hydrogen, nitrogen, and carbon dioxide. Reactions driven by the constant stream of energy coming from the sun were slowly yielding more complex molecules such as simple sugars, amino acids, and nucleotide bases. And the accumulation of organic material was supplemented by materials from a multitude of collisions between early Earth and meteors laden with organic matter. Prebiotic chemistry is being studied by a large community of researchers. A small sampling of their work is presented in the How We Know section at the end of this chapter.

Over a period of millions of years, the accumulation of reaction products yielded a soup containing molecules and polymers. As they grew increasingly complex, particular polymers acquired the capacity to duplicate themselves. The first self-replicating polymer, possessing two of the key requirements for life—catalysis and biological information—might be considered the first life form.

We do not know what this first "living" polymer was. However, modern molecular biology has given us many reasons to think that RNA either was the first self-replicator or arose as a much-improved descendant of that first self-replicator. RNA differs from DNA only in that it uses ribose instead of deoxyribose in its backbone. That single additional hydroxyl group in each monomeric unit of the polymer allows RNA to take up a plethora of complex structures that are inaccessible to DNA. The structural malleability of RNA gives it a capacity for both catalysis and information storage that has made it indispensable for life, from its beginnings to the present time.

The **RNA world hypothesis** was first proposed as a stage in evolution by molecular biologists Carl Woese, Francis Crick, and Leslie Orgel, in separate papers published in the late 1960s. The hypothesis describes a living system (or set of living systems) based on RNA. In this system, a variety of RNA enzymes could catalyze all of the reactions needed to synthesize the molecules required for life from simpler molecules available in the environment. The RNA enzymes would include replicators to duplicate all of the RNA catalysts. The "RNA organism," out of equilibrium with its surroundings, would have to be defined by a boundary. The experiments of Szostak and colleagues show one way in which lipid-enclosed RNA systems can arise (see the How We Know section at the end of this chapter).

Four more-recent lines of evidence have added much breadth and depth to the RNA world proposal. The first was the discovery by Thomas Cech and Sidney Altman, in the early 1980s, of **catalytic RNAs**, or **ribozymes**—enzymes that are made of RNA instead of protein. Thus we learned that some extant RNA molecules catalyze reactions and so possess both of the key conditions for life—biological information and catalysis. In modern organisms, ribozymes catalyze a relatively narrow range of reactions, such as the cleavage and ligation of other RNA molecules—a range insufficient to support an RNA world.

What is the real catalytic potential of RNA? The second line of supportive research demonstrated that RNA molecules generated in the laboratory can catalyze almost any imaginable reaction needed in a living system—certainly a range of reactions much broader than those attributable to ribozymes existing today. Early RNA molecules could clearly have catalyzed all of the reactions required to set up a primordial cellular metabolism.

The third and fourth discoveries have further broadened our perspectives on RNA function. We now know that in ribosomes, the large ribonucleoprotein complexes that translate RNA into protein, the RNA is the active component with the capacity to catalyze protein synthesis (**Figure 1-6**; see also the Moment of Discovery for Chapter 18). Finally, and most recently, RNA sequences capable of simple forms of self-replication have been discovered (discussed in Chapter 16).

Ongoing research thus makes it possible to visualize a highly plausible sequence of events unfolding on the pathway from prebiotic soup to living systems. Arising from a myriad random primordial polymers, an RNA world came into being and gradually became more complex. An RNA capable of reliable self-replication may have been the first living entity. Self-replicators would have diversified to synthesize other ribozymes, leading to an RNA-based metabolism capable of providing a greater supply of needed RNA

precursors. Ribozyme groupings became enclosed within lipid membranes. Particular groupings were successful, resulting in the first cells and a capacity to maintain a metabolic state out of equilibrium with the surroundings. As the RNA molecules in those cells increased in size and structural complexity, a need for stabilization and auxiliary functions arose. Peptides (proteins) were synthesized to neutralize the negative charges of the phosphates in the RNA backbone, to stabilize RNA structure in other ways, and to augment early metabolism. As more peptides were synthesized, some with catalytic activities arose. Proteins gradually supplanted RNA as catalysts, because the greater catalytic potential of proteins yielded an advantage. The protein world emerged, but not without retaining important vestiges of the RNA world (ribosomes and some other RNA catalysts), as we find them today.

The Last Universal Common Ancestor Is the Root of the Tree of Life

Countless nascent life forms probably arose from the primordial soup, along with many biological advances that improved their fitness. Successful combinations of RNA catalysts gave way to systems based on protein catalysts. Improvements in catalytic efficiency appeared, along with systematized genetic codes to link genetic information in RNA and DNA to protein sequences. Additional changes facilitated cellular metabolism and reproduction. Protein synthesis was systematized through the evolution of an efficient ribosome machine. RNA became more specialized for information storage and transmission. Cell membranes became more structured and specialized, eventually including mechanisms to selectively transport materials into and out of the cell as needed. And some processes became regulated. In this way, a variety of primitive cells may have evolved—each of them a viable living system. Organisms living today exhibit shared properties, telling us that one of these early experimental cells won out over the others. This cell, sometimes called **LUCA (last universal common ancestor)** (**Figure 1-7**), ultimately gave rise to all life now present on Earth.

LUCA is a special source of fascination for molecular biologists. Although LUCA probably lived more than 3 billion years ago, our speculation about what this cell was like is informed by experiment. One approach is to determine the minimum protein and genetic requirements for life. Attempts to create a minimal life form, either by reconstituting basic components or by taking bacteria and stripping them of all unnecessary parts, are underway in laboratories around the world. These experiments are not only defining properties that must have been present in LUCA; they are also setting the stage for the laboratory generation of engineered bacterial cells that can be used to manufacture chemicals for bioenergy, agriculture, and medicine.

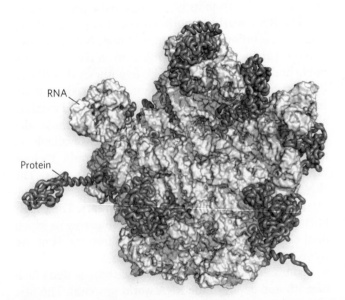

FIGURE 1-6 The 50S subunit of a bacterial ribosome. The gray parts of the subunit are RNA and the blue parts are protein. The structure is a huge ribozyme that evolved for the synthesis of protein. *[Source: PDB ID 1VSA.]*

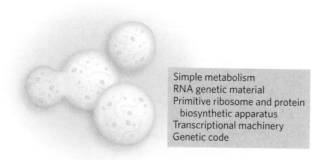

Simple metabolism
RNA genetic material
Primitive ribosome and protein
   biosynthetic apparatus
Transcriptional machinery
Genetic code

FIGURE 1-7 The last universal common ancestor. LUCA and its immediate descendants probably had a simple metabolism and a form of transcriptional machinery to replicate their RNA genome. A primitive ribosome and protein-biosynthetic apparatus would have used the same universal genetic code found in all modern organisms.

Another approach to understanding LUCA is to survey all types of living systems on Earth to determine which genes or characteristics are universal. The only genes that are truly universal in living systems are those encoding the cellular machinery for protein synthesis and some components of RNA transcription. All organisms also share (with very minor modifications discussed in Chapter 17) the same genetic code. That same code must have been present in LUCA. To support protein synthesis and RNA synthesis, a simple metabolism must have been present that allowed the uptake of chemical energy and its use to synthesize amino acids, nucleotides, and whatever lipids existed in the cell membrane

from precursors available in the environment. The study of LUCA is described in more detail in Chapter 8.

The appearance of LUCA signaled the beginning of biological evolution on Earth. New types of cells gradually appeared, and new environments were exploited. The first cells were capable of taking up organic molecules from their surroundings and converting them to the molecules needed to support protein and RNA synthesis. Cellular complexity resulted in ever-increasing requirements for cellular genomic information. DNA, with a more uniform structure and some stability advantages relative to RNA, may first have appeared in viruses. It then gradually supplanted RNA as the most stable platform for the long-term storage and transmission of genetic information, and DNA replication and systems for the segregation of replicated DNA chromosomes into daughter cells evolved.

The early single-celled organisms derived from LUCA diversified to inhabit all niches in the ecosystem of this early Earth. The diversification eventually generated the three major groups of organisms that we recognize today: **bacteria**, **archaea**, and **eukaryotes** (**Figure 1-8**).

Many additional events helped shape the life we see around us. Notably, photosynthesis appeared about 2.5 billion years ago, as evidenced by the sudden rise in the concentration of atmospheric oxygen documented in the geologic record. As cells engulfed other cells, some endosymbiotic relationships developed and became permanent. The engulfed cells became organelles within their hosts more than 1 billion years ago, and we see these organelles today as chloroplasts and mitochondria. Loose

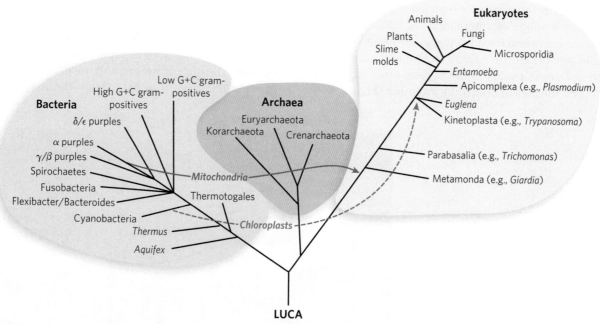

FIGURE 1-8 The universal tree of life. A current version of the tree is shown here, with branches for the three main groups of known organisms: bacteria, archaea, and eukaryotes. Particular types of bacteria, engulfed by other cells, gave rise to mitochondria and chloroplasts. [Source: Data from J. R. Brown, "Universal tree of life," in Encyclopedia of Life Sciences, Wiley InterScience (online), 2005.]

clusters of unicellular organisms led to cell specialization, and more permanent assemblies produced multicellular organisms. Diversification of body plans became more rapid about 600 million years ago, eventually generating all the major types of organisms we observe today.

Evolution by Natural Selection Requires Variation and Competition

Charles Darwin (1809–1882) was one of the most influential thinkers in history, and his name is forever associated with the concept of evolution. In his book *On the Origin of Species*, published in 1859, Darwin developed several general observations and ideas, laying out the evidence he had collected during and after his now famous voyage on the *Beagle*. He documented the variation among individuals in a population and inheritance of the variations by offspring. He noted that individuals in a population compete for resources. He argued that those individuals best adapted to exploit the prevailing resources are the ones most likely to survive and reproduce. These ideas together constitute a mechanism for evolution that can be described by the term **natural selection**.

Charles Robert Darwin, 1809–1882 *[Source: © Stapleton Collection/Corbis.]*

The Origin of Species had a tremendous and immediate influence on scientific thought, due in part to the huge volume of work it described and in part to the story it told. Darwin contributed a detailed body of evidence to support a range of interconnected ideas—some his own, some borrowed from his predecessors and contemporaries (see the How We Know section at the end of this chapter). Darwin's study of finches and other organisms introduced the idea of branching evolution (**Figure 1-9**),

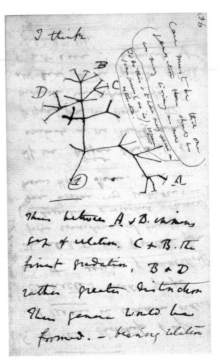

FIGURE 1-9 An evolutionary tree as sketched by Darwin in his 1837 notebook. *[Source: Reproduced by kind permission of the Syndics of Cambridge University Library.]*

which ultimately led to the idea that all life on Earth has a common ancestor. For natural selection to work, evolution must be gradual, with no discontinuities. All of these ideas coalesced, in *The Origin of Species*, into an internally consistent and compelling story describing the development of life in its many forms. Darwin's definition of the mechanism by which all of this occurred—natural selection—was the crowning achievement.

Natural selection depends on two characteristics of a population: variation and competition (**Figure 1-10**). However, the source of a population's variation eluded Darwin. The genetic program that exists in every organism was unknown to him, as were the mechanisms by

FIGURE 1-10 Variation and competition. On the plains of Africa, predation eliminates the weakest individuals from a population. *[Source: Gary Dublanko / Alamy.]*

which it is handed down from one generation to the next. Darwin was unaware that the work that would eventually reveal these mechanisms had been begun by one of his contemporaries, Gregor Mendel (see Chapter 2). Mendel's work was little appreciated during his lifetime.

Darwin's ideas have been expanded and developed into a modern synthesis of the theory of evolution, a direct outgrowth of the development of genetics in the early twentieth century—a time when Mendel's work was rediscovered, giving rise to the term *Mendelian genetics*. The concept of the gene was developed by influential geneticists such as Thomas Hunt Morgan, J. B. S. Haldane, and Theodosius Dobzhansky, providing the necessary mechanism of inheritance. As our understanding of genetic mechanisms emerged and matured, the concept of organismal variation as observed by Darwin and Mendel slowly evolved into the modern concept of mutation—a chemically definable change in a gene. By the 1950s, the theory of evolution could be stated in more detail and was bolstered by more evidence on mechanisms than was conceivable in Darwin's day. Populations, as we now know, contain inherent genetic variation generated by random mutation and genetic recombination. The frequency of different forms of genes in a population changes from generation to generation as a result of several processes (discussed in Chapters 8 and 14). Organismal changes can occur as the result of inherited mutations in a gene. Whether the change remains in a population depends to a large degree on whether the change confers an advantage to the organism. If the advantage is small or nonexistent, random genetic drift—changes in the frequency of a particular form of a gene—can occur, especially in small, isolated populations. Organisms can also acquire new genes through gene flow from other species in a process called **horizontal gene transfer.** Several mechanisms by which horizontal gene transfer can occur in bacteria are outlined in **Figure 1-11** (see Chapters 8 and 14).

Darwin's theory of natural selection provides a mechanism that responds directly to the environment. Most of the genetic changes that do not kill an organism produce only small changes in protein function or expression, resulting in a small change in the whole organism. As a result, evolutionary change is usually gradual. Sufficient diversification leads to new species and, with time, to new genera and phyla.

On the most practical level, our connectivity with other species through the tree of life has a critical effect on the study of molecular biology: we can learn about ourselves by studying other organisms (as described in detail in the Model Organisms Appendix). Even the simplest organisms have much to teach us about the inner workings of our own cells. As we will see throughout this book, the processes involved in the flow of biological information,

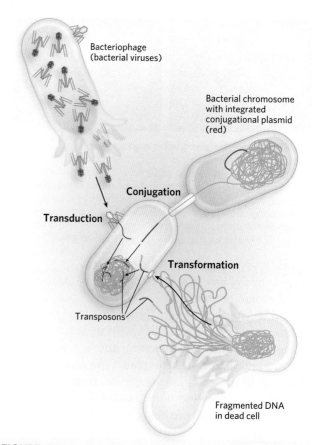

FIGURE 1-11 Horizontal gene transfer. Genetic material is transferred between organisms, especially bacteria, by several mechanisms. New DNA can be introduced by bacterial viruses (bacteriophages) via transduction. In some cases, genes are passed purposefully from one bacterium to another by a kind of bacterial sexual exchange called conjugation. Segments of DNA (released from broken cells) may be taken up from the environment in a process called transformation. In all cases, the new DNA may be incorporated into the chromosome by recombination. The movement of genetic elements called transposons (sometimes referred to colloquially as "hopping genes") can augment the effects of all these processes, if transposons are part of the introduced DNA. These processes are described in more detail in Chapter 14.

though common to all organisms, are often much more complex in eukaryotes than in bacteria. Much of our understanding of these processes is due to groundbreaking research on bacteria or yeast, followed by further research on more complex model organisms such as worms, insects, or mice (**Figure 1-12**). In this way, the elucidation of gene functions in a relatively simple organism such as yeast can lead to cures for human disease. Discoveries made in bacteria can generate improvements in agriculture. Fruit flies instruct us about the intricacies of human cognition and the complexities of fetal development. Pandemics of the future can be predicted and tamed by studying the pathogens of the past. Each investigation into the molecular biology of an organism is made more

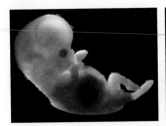 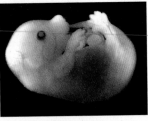

FIGURE 1-12 Similarities among organisms during development. As an example, a human embryo (left) is compared with a mouse embryo (right). Although the adult forms differ greatly in appearance, the embryos reveal similarities in body plan and development. These similarities, and many more that exist on a molecular level, allow us to learn about ourselves through the analysis of model organisms. *[Source: (left) © Last Refuge, Ltd. / Phototake; (right) CB2/ZOB/WENN.com/Newscom.]*

valuable by the fact that all species are related through a shared evolutionary history.

As Darwin remarked in *The Origin of Species*, "There is grandeur in this view of life, with its several powers, having been originally breathed into a few forms or into one; and that, whilst this planet has gone cycling on according to the fixed law of gravity, from so simple a beginning endless forms most beautiful and most wonderful have been, and are being, evolved."

SECTION 1.1 SUMMARY

- Living systems have definable characteristics and requirements. Catalysis and biological information are particularly important requirements for any life form.
- The first molecule that fulfilled the requirements of catalysis and biological information may have been a self-replicating RNA, according to the RNA world hypothesis.
- LUCA, the last universal common ancestor of all life now present on Earth, can be studied by identifying the common characteristics of living organisms and defining the minimal complement of genes necessary to support a living cell.
- Evolution by natural selection is a result of genetic variation within a population and competition between individuals for limited resources. Darwin's theory of evolution by natural selection has been strengthened by modern studies that reveal the sources of genetic variation, mutation, and recombination.
- A common evolutionary heritage links all organisms, allowing the study of model organisms to aid in our understanding of ourselves.

1.2 HOW SCIENTISTS DO SCIENCE

Science provides a story about the natural universe around us. It is a story that inspires wonder and, at the same time, has enormous practical implications for every

aspect of human existence. As is the case for any scientific discipline, success in molecular biology is defined to a large extent by the contributions a scientist makes to the larger scientific story. That ongoing story provides a context and community within which scientists frame every experiment.

The history and philosophical underpinnings of science have brought about guidelines and rules for new experimentation at every level of this intertwined enterprise. The scientific community is highly interactive, with ongoing discussions supplying both constraints and insights that can stimulate progress. The collective discussion is carried out in informal conversations, at scientific meetings, and, most importantly, in the peer-reviewed scientific literature. A successful contribution is usually one that eventually appears in this literature.

Molecular biology is a scientific enterprise, giving rise both to information, as conveyed in the chapters of this text, and to future advances. An understanding of the scientific enterprise can help in our assimilation of existing information and accelerate success in any scientific effort.

Science Is a Path to Understanding the Natural Universe

Science is both a body of knowledge and a process for generating that knowledge. In the scientific community, scholars attempt to answer questions about the natural universe. Since the dawn of humanity, people have employed many approaches to understanding the world around them, guiding human development for millennia. The modern version of the scientific method, relying on careful observation and experimentation, has been widely applied for only about four centuries.

The scientific approach to discovery has at least three characteristics. First, science focuses only on the natural universe. The realm of science is thus limited to what we can observe and measure. Second, science relies on ideas that can be tested by experiments and on observations that can be reproduced. Finally, the experiments are carried out within an ever-expanding web of scientific theories that provide guidance and insight along the way.

Science makes one philosophical assumption: that forces and phenomena existing in the universe are not subject to capricious or arbitrary change. They can be understood by applying a systematic process of inquiry—the scientific method. The French biochemist Jacques Monod referred to this basic underlying assumption as the **postulate of objectivity**. Stated more simply, science cannot succeed in a universe that plays tricks on us. This assumption is made every time a scientist performs an experiment. If an experiment is repeated and the results are different, there is always a reason;

something must have been different during the second experiment. When faced with seemingly contradictory results, every scientist is trained to figure out why. Often, inconsistencies have a trivial cause, such as a degraded or inactive reagent; but sometimes they lead to great insights.

In modern science, an initial idea is called a **hypothesis**—a proposal that provides a reasonable explanation for one or more observations but is not yet substantiated by sufficient experimental tests to stand up to rigorous critical examination. A **scientific theory** is much more than a hypothesis; it provides an explanation for a body of experimental observations and is thus a firm basis for further inquiry. When a scientific theory has been repeatedly tested and validated by many types of experiments, it can be considered a fact.

Scientific hypotheses and theories can also be defined by their presence in the peer-reviewed scientific literature. Papers are accepted or rejected for inclusion in that literature based on professional reviews by other working scientists. According to the Publishers Association of the United Kingdom, there are more than 16,000 peer-reviewed scientific journals worldwide, publishing some 1.4 million papers each year. This continuing rich harvest of information is ultimately the foundation of scientific progress.

The Scientific Method Underlies Scientific Progress

The first recorded efforts to systematize scientific inquiry have been credited to the classical Greeks, who were clearly influenced by other civilizations in Babylon, Assyria, Egypt, Persia, and elsewhere (**Figure 1-13**). Key thinkers replaced trial and error, chance discoveries, and appeals to the supernatural with a more formalized system of reasoning. Euclid, Aristotle, Plato, and others made significant advances by using inductive reasoning, making a broad conclusion from specific facts (e.g., my stove is hot, my brother's stove is hot, so all stoves are hot), and by deductive reasoning, forming a conclusion from a

Hypatia, 370–415 *[Source: © Bettmann / Corbis.]*

premise (e.g., premise: all flying organisms are birds; observation: a crow flies; conclusion: a crow is a bird). Experiments played a minor role. The process worked reasonably well for mathematics and engineering, less well for most other pursuits. Scientific progress declined in the West with the decline of the Roman Empire, symbolized by the assassination in 415 CE of the last director of the great library of Alexandria in Egypt, the accomplished Greek mathematician and philosopher Hypatia.

A reawakening began with the emergence of Islamic science during the Middle Ages, building on Greek and Roman ideas with rational arguments combined more liberally with observations and experiments. The physicist Abū ʿAlīal-Ḥasan ibn al-Hasan ibn al-Haytham (965–1040), known as Alhazen in the West, is regarded as the father of modern optics. Alhazen was particularly influential in refocusing the scientific process on reproducible experimentation. Latin translations of Arabic and Greek texts filtered into Europe, helping to stimulate the Renaissance. The Franciscan friar Roger Bacon (1214–1294) was inspired by these texts. He was among the first to describe a method of inquiry involving observation, hypothesis, and experiment, while also stressing independent repetition to verify results.

By the beginning of the seventeenth century, the flaws in Aristotelian philosophy had become clear to many scholars. René Descartes (1596–1650), Galileo Galilei (1564–1642), and Francis Bacon (1561–1626) led a revolution in scientific thinking. Frustrated with the impediments inherent in the Aristotelian system (e.g., if one encounters a bat, the premise that all flying organisms are birds is falsified), they laid the groundwork for the modern scientific method.

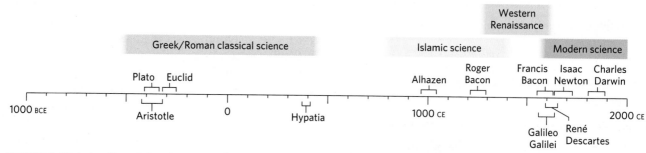

FIGURE 1-13 A timeline of development of the scientific method. The life spans of a few of the major contributors are shown.

In 1619, Descartes set down four rules for inquiry, which we paraphrase here: (1) never accept anything as true unless that truth can be clearly demonstrated; (2) reduce a problem to its parts; (3) begin with the simplest part and gradually work to the more complex; and (4) be thorough. Francis Bacon divided inquiry into a somewhat similar series of steps, first gathering relevant observations, and then deriving increasingly complex conclusions from them. Galileo combined careful observation and rational argument to develop new fields of inquiry.

The **scientific method** had matured by the beginning of the nineteenth century. In its most basic form, it begins with a question about nature. Relevant observations are made, and a hypothesis is crafted to explain them. Assumptions that underlie the hypothesis are defined, and each assumption is carefully tested by experimentation. Experimental results lead to acceptance, rejection, or modification of the hypothesis. Additional experiments can lead to the acceptance of a new theory or fact about the natural world.

In modern molecular biology, this now classical version of the scientific method underlies huge numbers of advances. One example is the discovery of the DNA-synthesizing enzyme DNA polymerase by Arthur Kornberg and his coworkers in the 1950s (discussed in Chapter 11). With the elucidation of DNA structure by James Watson and Francis Crick, Kornberg and others hypothesized that an enzyme must exist to synthesize this polymer. They assumed that the enzyme could be purified and studied, and its need for a template determined. To test these assumptions, they lysed (broke open) bacterial cells, fractionated the extracts to separate the many proteins and enzymes, and developed assays to measure the activity of the hypothetical enzyme in the presence of all plausible substrates and precursors. The effort paid off in 1956, with an assay that detected the incorporation of nucleotides into DNA polymers. DNA polymerase was subsequently purified, named, and thoroughly characterized by the same team, substantiating their original hypothesis; the existence of a DNA polymerase became fact.

The Scientific Method Is a Versatile Instrument of Discovery

The path to scientific discovery is rarely as rigid, linear, and empirical as implied in the preceding paragraphs. Molecular biology is certainly replete with

Galileo Galilei, 1564–1642
[Source: © Pictorial Press Ltd. / Alamy.]

examples of the purposeful application of the scientific method, but it is also full of surprises, excitement, and occasional messiness. Major discoveries are often characterized by hard work, extraordinary perseverance (a surprising number of discoveries seem to be consummated in the middle of the night), and more than a little innovative thinking. Many such examples are chronicled throughout this book.

To understand how science is done, we need to expand the concept of the scientific method to include the contribution of scientific context. Scientific inquiries are not carried out in a vacuum. The discovery of DNA polymerase, described above, was much more than "hypothesize the existence of this enzyme and then go get it." The search was carried out in the context of the ideas and facts that prevailed at the time. These included information about the structure of DNA, the chemistry and thermodynamics of enzyme catalysis, cellular physiology and metabolism, the properties of proteins, and the recent finding of a related enzyme, ribonucleotide phosphorylase, by Severo Ochoa. The entire discovery process was guided by the interconnected web of theories and information within which the DNA polymerase hypothesis was developed and tested.

To understand the scientific method, it is also instructive to examine other examples of how important ideas were conceived and tested. The examples below are meant to demonstrate the variety of ways in which science is advanced, and they underscore the fact that science is a very human experience, despite the rigor of the discipline.

Hypothesis and Discovery This classical path to scientific discovery led Kornberg and colleagues to DNA polymerase. It was also used by Jack Szostak in his exploration of prebiotic chemistry (see Moment of Discovery at the opening of this chapter), as well as by countless other scientists. Bob Lehman describes the quite similar path to his discovery of DNA ligase, an enzyme that joins segments of DNA (see the Moment of Discovery for Chapter 11). These discoveries started with hypotheses and ideas reasonably based on the knowledge of the time, and the hypotheses were carefully tested. In addition, the discoveries were made within a broader context of constraints imposed by theories related to the geologic history of our planet, basic chemistry, nucleic acid chemistry, structural biology, the properties of enzymes, and many other areas of knowledge.

Model Building and Calculation As we will see in Chapter 6, Watson and Crick relied on intuition and important experimental clues from other researchers to build a model of DNA structure so clearly fitting the data that its acceptance was both shocking and immediate. A similar approach can be seen in the work of Steve Mayo in his efforts to deduce the three-dimensional structure of proteins from their amino acid sequence (see the Moment of Discovery for Chapter 4). In this variant of the scientific method, a researcher draws on experiments performed by others and mines the unrealized implications of that existing body of knowledge to bring about a significant advance in understanding. Insights obtained in this way are almost always subjected to further experimental examination to confirm them. In both of the cases described here, later solution of the three-dimensional structures of the macromolecules in question confirmed the results of the original investigations.

Hypothesis and Deduction Soon after the structure of DNA was determined, it became apparent that the information in DNA is converted into RNA, and then into protein. However, how amino acids could bind to RNA molecules to guide assembly of the amino acids into proteins remained unclear. Then, in 1955, Crick wrote an influential note to some colleagues that laid out his "adaptor hypothesis," proposing that there was an as yet undiscovered molecule that linked each amino acid to sequence information in the RNA. This insight, never published but widely disseminated, was not an endpoint. The hypothesis was subjected to experimental confirmation, and the adaptor—transfer RNA (tRNA)—was discovered within a year by Paul Zamecnik and Mahlon Hoagland (see the How We Know section in Chapter 17).

Exploration and Observation Darwin's voyage on the *Beagle* was not designed to answer the questions that his work eventually addressed, and no hypotheses were offered when the ship set sail. Thousands of individual observations (and context provided by the work of many predecessors and contemporaries) simply gelled in a great mind to create a transformational theory. Molecular biologists routinely embark on similar voyages, with computers and new technologies as vessels of discovery. Rather than focusing on one or a few enzymes and reactions, we can now approach issues related to entire systems. Cells, organisms, and even ecosystems are being explored. Many of these technologies are described in Chapter 8. The new technologies enable us to explore broadly, asking: what's out there that we have missed? These efforts provide a rich context for new ideas and hypotheses.

Inspiration The polymerase chain reaction, often abbreviated PCR, is a convenient process for amplifying a DNA sample. PCR is now so integral to biotechnology and forensic science that it is hard to imagine doing molecular biology without it. PCR was invented by Kary Mullis in a flash of inspiration that came to him while driving along a northern California highway in 1983. Reflecting on one problem, Mullis stumbled on a solution to an even greater one: a method to produce almost unlimited copies of a DNA sequence in a test tube. (This story is presented in more detail in the How We Know section in Chapter 7.) Although this advance was not precipitated by an orderly application of the conventional "question, hypothesis, experiment" approach, the inspiration was a logical and probably inevitable outgrowth of the growing body of knowledge that constituted the molecular biology of that moment.

Serendipity Accidental discoveries are, by definition, unplanned and unexpected, and they can change the way we think about living systems. Such discoveries are not preceded by a carefully orchestrated quest organized around a posed question. They just happen—and they happen often. One example involves the discovery of RNA catalysis. In the early 1980s, Thomas Cech was investigating the mechanism by which segments of RNA (introns) are removed from a messenger RNA in a reaction known as RNA splicing (discussed in Chapter 16). Searching for the protein enzymes that catalyzed the process, Cech and his coworkers were startled when the splicing reaction still occurred in a necessary control trial, in which the protein extract was left out of the reaction mixture. After long and careful experimentation to eliminate any possible source of protein contamination, they were able to report the paradigm-shifting discovery of an RNA-catalyzed reaction.

These many pathways vary in detail, but they have characteristics in common that identify them as part of a modern and adaptable scientific method. They all focus exclusively on questions and properties related to the natural universe. All the ideas, insights, and experimental facts that arise from these endeavors rely on reproducible observation and/or experiment. Ultimately, the information from the various approaches fits into the web of scientific knowledge. The experimental efforts and results can be used by other scientists anywhere in the world to build new hypotheses and make new discoveries. An attempt to construct a flow chart for a realistic application of the scientific method is shown in **Figure 1-14**.

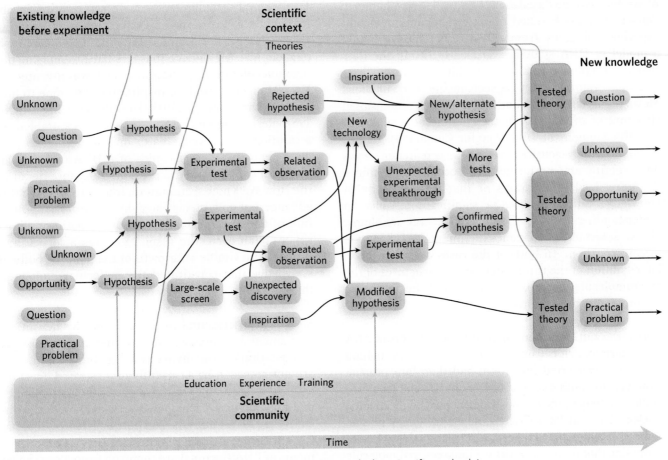

FIGURE 1-14 A flow chart of the scientific method. When scientists apply the scientific method, it is a nonlinear process with many and varied inputs.

Scientists Work within a Community of Scholars

Any individual who uses the scientific method to answer questions about the natural universe is a scientist—from a high school student examining pond water with a microscope to an investigator using high-tech equipment at a major research institute. A specific academic degree is not a prerequisite for making a scientific contribution. However, success in science is strongly correlated with training and experience.

Scientific training is much more than lessons about experimental methods; it is an education in how to think about and approach scientific problems. Progress often demands a willingness to give up a favorite hypothesis. In his book *Advice to a Young Scientist* (1979), the Nobel Prize winner Peter Medawar expressed it thus: "I cannot give any scientist of any age better advice than this: the intensity of the conviction that a hypothesis is true has no bearing on whether it is true or not."

If any idea about the natural universe does not align with reproducible observations, a scientist must challenge it. Except for the postulate of objectivity, no assumption is inviolate. The ideas that a scientist accepts must be based on evidence that can be observed, measured, and reproduced. The standard of reproducibility, coupled with the continual advance of new experimentation, provides a robust system for weeding out the inevitable errors and for guiding insightful reinterpretations.

The many decisions along any path to discovery, such as determining the number and types of tests to use in examining a hypothesis, are guided by training and experience. There is no prescribed formula for these decisions, but they are much less subjective than they might at first appear. The context of scientific theory and information plays a major role in ensuring rigor. When a DNA-synthesizing enzyme was first purified from bacterial crude extracts by Kornberg and his colleagues, specific tests had to be performed to establish that the enzyme was involved in normal chromosomal replication. These tests were dictated by the broader scientific context in which the discovery was made. The experiments were conceived, in part, by the need to fit this new enzyme into the existing web of scientific information.

When a result contradicts the original hypothesis, a scientist must decide whether to reject or revise. This decision is similarly rooted in both the context of scientific information known at the time and the experience of the investigator. When Bob Lehman and colleagues detected the action of DNA ligase (in the bacterium *Escherichia coli*) for the first time, they anticipated that a high-energy source, probably ATP, would function as a necessary cofactor for the reaction. This expectation was rooted in a great deal of precedent from similar reactions. But when ATP failed to stimulate the ligation reaction, the DNA ligase hypothesis was not abandoned. Instead, the investigators questioned their assumptions about ATP. A search eventually identified another cofactor, NAD^+, as the critical energy source for the *E. coli* DNA ligase, revealing a new chemical function for this important cellular molecule.

In science, the work is not done by individuals in isolation, and scientific decisions are influenced and affected by the scientific community at large. It is not enough to demonstrate a new idea to oneself. To integrate an advance into the web of scientific information, a scientist must make a case that is compelling to the broader community. A continual discussion within this community fosters progress and helps ensure the integrity of the information generated. A few years after Kornberg first isolated DNA polymerase, the involvement of this enzyme in chromosomal replication was questioned. The rate at which it synthesized DNA seemed insufficient to match the observed rate of chromosomal duplication in cells, and cells somehow survived when the gene encoding the enzyme was eliminated. These observations led to discovery of another DNA polymerase in the same bacteria, now called DNA polymerase III, that serves as the primary enzyme in chromosomal replication. The original DNA polymerase (now called DNA polymerase I) is more abundant than DNA polymerase III, hence its initial detection. DNA polymerase I has a variety of auxiliary functions in DNA replication and DNA repair, as we will see in Chapters 11 and 12.

When an idea is tested, the most compelling case is made when the idea passes multiple tests involving two or more different techniques. Collaboration among scientists often strengthens the case. Different individuals bring different perspectives and expertise to bear on a problem. The cross-talk can correct errors, provide novel insights, and make connections that one individual might miss. The case becomes even more compelling when tests are replicated in multiple laboratories. To gain entry to the scientific literature, the work must pass the inspection of fellow scientists. Peer reviews of scientific papers, and of grant applications, are sometimes challenging and even painful experiences for the authors, but they are critical for ensuring quality and rigor. A conscientious participation in this peer-review system is a shared duty of every working scientist.

Ultimately, this literature documents the scientific story about the natural universe. That story is ever expanding, never complete, and always ready for new contributors.

SECTION 1.2 SUMMARY

- Science is both a body of knowledge and a process of inquiry by a community of scholars. Science is a path to knowledge about the natural universe, relying on reproducible observation and experiment. Science relies on one foundational assumption: the postulate of objectivity, that the universe is governed by immutable, and therefore discoverable, laws.
- The classic form of the modern scientific method poses questions, generates explanatory hypotheses, and tests the hypotheses with experiments. Hypotheses are accepted, rejected, or modified depending on the results of the experiments.
- The scientific method is malleable, and there are multiple paths to scientific knowledge. All pathways are linked by a reliance on reproducible observation or experiment and by their capacity to generate knowledge consistent with the broader context of scientific information.
- Scientists are individuals who apply the scientific method to answer questions about the natural universe. Their decisions are guided by training, their knowledge of the scientific context of their work, and continuing interactions with the worldwide scientific community.

? UNANSWERED QUESTIONS

The questions that are relevant to this introductory chapter include some of the most fundamental and far-reaching issues in molecular biology. Many of them relate to discussions in chapters throughout the book.

1. What is the minimal set of genes required in a living cell? Efforts to define a cell at its most basic level are well underway. Scientists may create an engineered cell in the near future. The research has practical implications, because such a cell could be a living scaffold for the engineering of cells to bioremediate toxic waste or generate biofuels. The work should also tell us something about LUCA. To this end, a completely synthetic version of the genome of the bacterium *Mycoplasma mycoides* has been constructed and transplanted into recipient cells, where it displaced the resident DNA. A more

complete engineering of an artificial cell remains a goal.

2. Can we reconstruct the entire tree of life? Developing a complete tree of life is an ongoing effort of evolutionary biology, making use of the tools of molecular biology and complementary information from the fossil and geologic records (see Chapter 8).

3. What is the full set of mechanisms that drive evolution, and how much does each mechanism contribute? Genetic variation in a species results from spontaneous DNA damage, replication errors, transposition, DNA repair processes gone awry, genetic recombination, and many other processes. Molecular biology provides the roadmap to a comprehensive listing and understanding. Many mechanisms are increasingly understood, but surprises are still common.

4. Can evolution be controlled? When bacteria are subjected to stress, specific sets of genes turn on in programmed responses. One part of that response in *E. coli* involves the production of higher levels of mutagenesis. Can the development of resistance of pathogenic bacteria to antibiotics—an increasing problem in medical treatments—be slowed by inhibiting bacterial stress responses to DNA damage? Some academic and biotechnology industry researchers around the world are betting that it can.

HOW WE KNOW

Adenine Could Be Synthesized with Prebiotic Chemistry

Ferris, J.P., and L.E. Orgel. 1966. An unusual photochemical rearrangement in the synthesis of adenine from hydrogen cyanide. *J. Am. Chem. Soc.* 88:1074.

Oro, J., and A.P. Kimball. 1962. Synthesis of purines under possible primitive Earth conditions. II. Purine intermediates from hydrogen cyanide. *Arch. Biochem. Biophys.* 96:293–313.

Few biomolecules are more important than adenine. It is the base component not only of the adenine nucleotides in RNA and DNA but also of a wide range of enzyme cofactors, including ATP (adenosine

triphosphate), an important energy source that drives many cellular reactions. Thus, over the course of evolution, adenine took on a larger role than other nucleotide bases in the biochemistry of living systems, presumably reflecting some particular aspect of the evolution of life. Experiments carried out by Juan Oro in the 1950s and early 1960s, and continued by Leslie Orgel and colleagues, demonstrated that there are good chemical reasons to think that adenine may have been present in much higher concentrations than other nucleotide bases in the prebiotic soup. This suggests that adenine's current place in the scheme of life reflects the conditions when life was formed: adenine was there!

One of the most common reagents generated from the molecules thought to predominate in the atmosphere of early Earth is hydrogen cyanide (HCN). HCN can be converted into a variety of organic molecules, one of the most abundant products being adenine. Adenine can be synthesized from HCN in several ways, but the scheme shown in Figure 1 is the

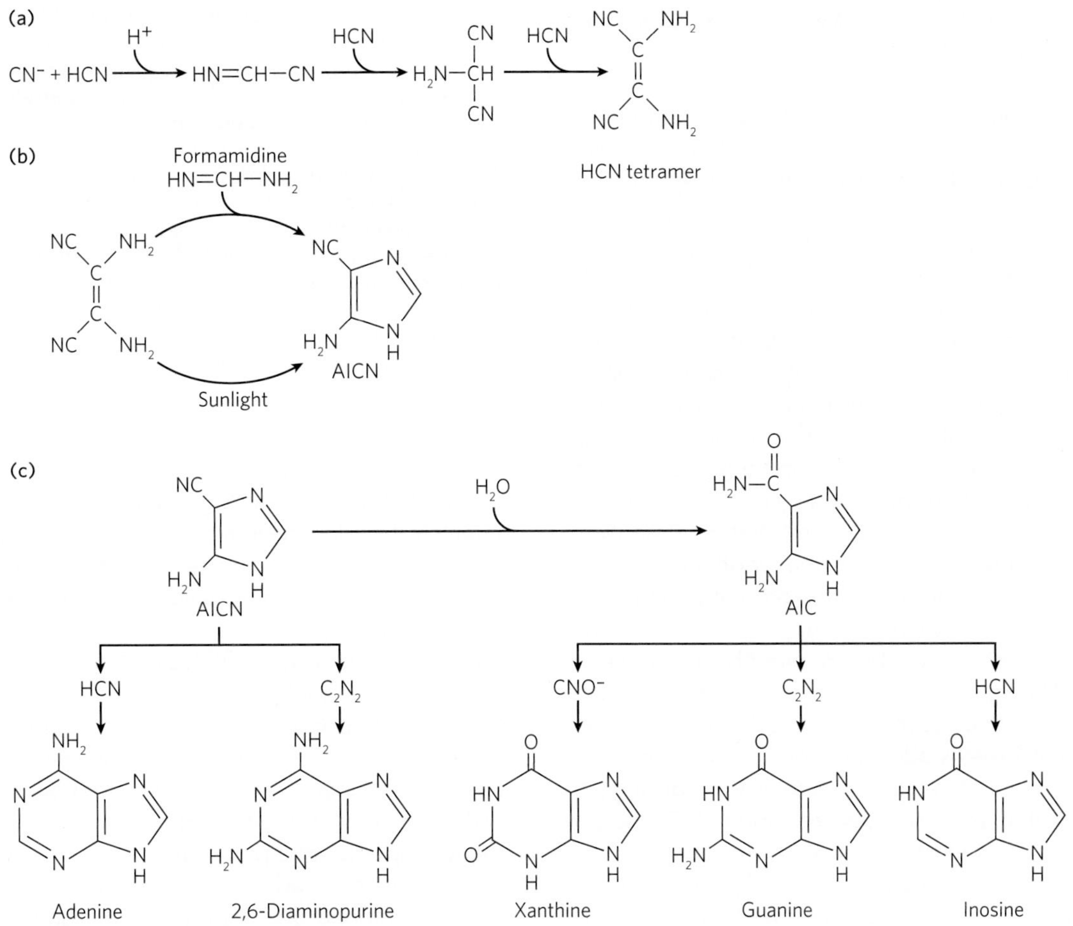

FIGURE 1 HCN can be chemically converted into adenine and related molecules. [Source: Data from L. E. Orgel, Crit. Rev. Biochem. Mol. Biol. 39:99, 2004.]

HOW WE KNOW

most plausible under the conditions believed to have existed on early Earth. An HCN tetramer forms in the first step (Figure 1a), and this reacts with formamidine (also formed from HCN) to produce 4-amino-5-cyanoimidazole (AICN) (Figure 1b). Adenine is formed from AICN. Some related molecules are formed from AICN's hydrolysis product, 4-aminoimidazole-5-carboxamide (AIC), providing a primordial source of some other nucleotide bases (Figure 1c).

Clay Had a Role in Prebiotic Evolution

Hanczyc, M.M., S.M. Fujikawa, and J.W. Szostak. 2003. Experimental models of primitive cellular compartments: Encapsulation, growth, and division. *Science* 302:618–622.

Huang, W.H., and J.P. Ferris. 2003. Synthesis of 35–40 mers of RNA oligomers from unblocked monomers: A simple approach to the RNA world. *Chem. Commun.* 12:1458–1459.

Sodium montmorillonite, first mined at Fort Benton, Wyoming, and commonly known as bentonite, is a type of clay used commercially. It generally contains various mineral impurities (10% to 20%) that give it a layered aluminosilicate structure. The clay readily expands to permit large molecules to enter the interlayers. Products making use of montmorillonite clays include lubricating grease, paints, copy paper, dynamite, plaster, cat litter, matches, cement tiles, shoe polish, concrete, cleaning agents, wall boards, crayons, and bleaching agents. Some forms of montmorillonite are claimed to have health-giving properties and are used in edible preparations marketed by health spas and health food companies. The claims that montmorillonite clays are "living clays" may not be far off the mark.

Any nucleotides formed in the prebiotic soup would have had to be joined into polymers before the properties of RNA could be exploited by evolution. Wenhua Huang and James Ferris demonstrated that a phosphoramadite-activated nucleotide, based on the structure of 1-methyladenine, is readily polymerized in the presence of montmorillonite clay, producing polymers of up to 40 nucleotides (Figure 2). These clays have also been shown to facilitate a variety of other reactions leading to the production of polynucleotides and other complex organic molecules that may have been part of the prebiotic soup and key ingredients in the cauldron that generated the first living organisms.

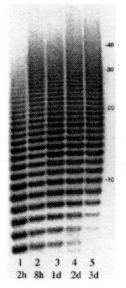

FIGURE 2 Gel electrophoresis is a technique (described in Chapters 4 and 6) that can be used to separate macromolecules by size. Molecules migrate through a gel (here, from top to bottom) in response to an electric current, with small molecules migrating faster than larger ones. In the experiment shown here, the polymerization of 1-methyladenine nucleotides is shown by the increasing size of DNA fragments with time. The numbers on the right indicate the length of the polymer in number of nucleotides; the labels below the lane numbers indicate the duration of polymerization of each sample in hours (h) and days (d). [Source: W. H. Huang and J. P. Ferris, Chem. Commun. 12:1458–1459, 2003, Fig. 2. Photo reproduced by permission of The Royal Society of Chemistry/Courtesy of James Ferris.]

Jack Szostak's research group has long been interested in identifying reactions that might explain how early replicating polymers could have become enclosed in lipid envelopes (see this chapter's Moment of Discovery). Szostak and his coworkers found that montmorillonite clay facilitates the conversion of lipids into vesicles. In their experiments, clay particles often became entrapped in the vesicles, bringing with them any RNA embedded in the particle surfaces. As more lipids were added, the vesicles would grow and bud off, without dilution of their contents. This system models many properties of a primordial cell.

We have included here just a small sampling of studies exploring the properties of this unusual clay material. A wide array of reactions are facilitated by this clay, many of them of interest to evolutionary biologists.

Darwin's World Helped Him Connect the Dots

Mayr, E. 2000. Darwin's influence on modern thought. *Sci. Am.* 283(July):78–83.

Padian, K. 2008. Darwin's enduring legacy. *Nature* 451:632–634.

Darwin was not the first to come up with the idea of evolution. The sense that the world changes over time and that the organisms inhabiting it also change can be traced back to classical Greece. By the end of the eighteenth century, the study of geology had impressed on the scientists of the day that Earth is much older than previously thought and that the environment changes over time. The discovery of and growing interest in fossils had similarly shown that there were animals on Earth in past eras that were no longer present. They had either become extinct or undergone great changes in appearance.

One of the first influential efforts to explain all these observations came from the French biologist Jean-Baptiste Lamarck (1744–1829). Lamarck was the first scientist to use the term *biology*, coined in 1802 to encompass the sciences of botany and zoology. Lamarckian evolution is linear, not branched. He also proposed that changes in the environment triggered adaptive changes in individuals that could be passed on to succeeding generations, an idea known as the inheritance of acquired characteristics.

Lamarck's ideas were hotly debated for many decades, and a variety of competing ideas about the transmutation or transformation of animals and plants appeared in contemporary writings. The basic idea that organisms change, or evolve, over time was familiar, if not well accepted, before Darwin presented his theory of evolution in *The Origin of Species*. However, until he set pen to paper, no plausible mechanism of evolution had been proposed.

Drawing on his experiences during the 1831–1836 voyage of H.M.S. *Beagle* (Figure 3), Darwin formulated his own new ideas over a period of years. Observations he had accumulated during a five-week stay in the Galápagos Islands had a particularly strong influence. The species on the islands differed from those on the mainland, consisting only of representatives of groups that could have managed a trip over a large expanse of ocean. Patterns of divergence were evident in certain groups, such as the multiple finch species on the islands. Darwin's observations indicated that a few individuals of one finch species had originally colonized the islands. That species had then diversified to exploit the various island environments.

Darwin was influenced by the works of Thomas Malthus (1766–1834), who had argued that human reproduction, if not controlled, would lead to a population that would outstrip the food supply. This would lead to an inevitable struggle to survive. A competition between organisms for resources was built into Darwin's ideas about natural selection.

Darwin's great work was published in 1859, after he discovered that similar ideas had been developed by Alfred Russel Wallace (1823–1913). Wallace deserves credit as a codiscoverer of natural selection, but Darwin conceived the idea earlier and developed it more fully.

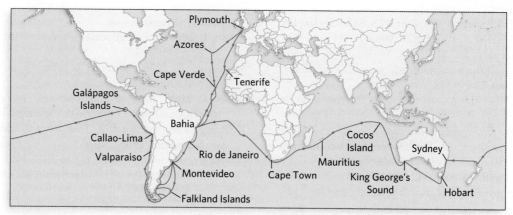

FIGURE 3 The route taken and the lands visited during the 1831–1836 voyage of H.M.S. *Beagle.* [*Source: Charles Darwin, The Voyage of the Beagle, 1839.*]

molecular biology, 2

biological information, 2

catalysis, 2

evolution, 2

mutation, 2

enzyme, 3

RNA world hypothesis, 7

catalytic RNA, 7

ribozyme, 7

LUCA (last universal
   common ancestor), 8

bacteria, 9

archaea, 9

eukaryotes, 9

natural selection, 10

horizontal gene transfer,
   11

postulate of objectivity,
   12

hypothesis, 13

scientific theory, 13

scientific method, 14

ADDITIONAL READING

General

Carroll, S.B. 2005. *Endless Forms Most Beautiful: The New Science of Evo Devo and the Making of the Animal Kingdom.* New York: W. W. Norton & Company. An accessible exploration of evolutionary developmental biology by one of the leaders in the field.

Dobzhansky, T. 1973. Nothing in biology makes sense except in the light of evolution. *Am. Biol. Teach.* 35:125–129.

Wilson, E.O., ed. 2005. *From So Simple a Beginning: Darwin's Four Great Books.* New York: W. W. Norton & Company. A one-volume compendium of the four great works of Charles Darwin: *Voyage of the H.M.S. Beagle* (1845), *The Origin of Species* (1859), *The Descent of Man* (1871), and *The Expression of Emotions in Man and Animals* (1872).

The Evolution of Life on Earth

Barrick, J.E., and R.E. Lenski. 2013. Genome dynamics during experimental evolution. *Nat. Rev. Genet.* 14:827–839.

Byrne, R.T., A.J. Klingele, E.L. Cabot, W.S. Schackwitz, J.A. Martin, J. Martin, Z. Wang, E.A. Wood, C. Pennacchio, L.A. Pennacchio, et al. 2014. Evolution of extreme resistance to ionizing radiation via genetic adaptation of DNA repair. *eLife* 3:e01322.

Dawkins, R. 2004. *The Ancestor's Tale: A Pilgrimage to the Dawn of Evolution.* New York: Houghton Mifflin Company.

Gould, S.J. 2002. *The Structure of Evolutionary Theory.* Boston: Harvard University Press. The last contribution from one of the great popularizers of evolutionary biology.

Hanczyc, M.M., and J.W. Szostak. 2004. Replicating vesicles as models of primitive cell growth and division. *Curr. Opin.*

Chem. Biol. 8:660–664. A summary of some current efforts to re-create a protocell in the laboratory.

Kirschner, M.W., and J.C. Gerhart. 2005. *The Plausibility of Life: Resolving Darwin's Dilemma.* New Haven, CT: Yale University Press.

Lazcano, A., and S.L. Miller. 1996. The origin and early evolution of life: Prebiotic chemistry, the pre-RNA world, and time. *Cell* 85:793–798.

Lincoln, T.A., and G.F. Joyce. 2009. Self-sustained replication of an RNA enzyme. *Science* 323:1229–1232. Yes, RNA molecules are capable of self-replication.

Mayr, E. 2000. Darwin's influence on modern thought. *Sci. Am.* 283(July):78–83. A terrific explanation of how Darwin has affected the way in which all of us approach our world.

Orgel, L.E. 2004. Prebiotic chemistry and the origin of the RNA world. *Crit. Rev. Biochem. Mol. Biol.* 39:99–123. An excellent summary of the field exploring prebiotic chemistry.

Woese, C.R. 2004. A new biology for a new century. *Microbiol. Mol. Biol. Rev.* 68:173–186. A thoughtful appeal for some new approaches in biology.

Zimmer, C. 2001. *Evolution: The Triumph of an Idea.* New York: HarperCollins Books. The companion volume to the PBS (WGBH) series on evolution.

Zuckerkandl, E., and L. Pauling. 1965. Molecules as documents of evolutionary history. *J. Theor. Biol.* 8:357–366. One of the first articles to recognize that much of evolutionary history is documented in the genomes of existing organisms.

How Scientists Do Science

Bryson, B. 2004. *A Short History of Nearly Everything.* New York: Broadway Books. Easy to read, this book provides an excellent lay description of how scientists approach problems.

Dennett, D.C. 1995. *Darwin's Dangerous Idea: Evolution and the Meanings of Life.* New York: Touchstone.

Feynman, R. 1999. *The Pleasure of Finding Things Out: The Best Short Works of Richard Feynman.* New York: Helix Books / Perseus Books.

Kosso, P. 2009. The large-scale structure of scientific method. *Sci. Educ.* 18:33–42.

Medawar, P. 1979. *Advice to a Young Scientist.* New York: HarperCollins.

Sober, E. 2008. *Evidence and Evolution: The Logic behind the Science.* Cambridge: Cambridge University Press.

Working Group on Teaching Evolution, National Academy of Sciences. 1998. *Teaching about Evolution and the Nature of Science.* Washington, DC: National Academy Press. www.nap.edu. This excellent resource is available as a free download.

2 DNA: The Repository of Biological Information

James Berger
[Source: Courtesy James Berger.]

MOMENT OF DISCOVERY

The first time I had an "Aha!" moment in science was when I was a graduate student. The question that intrigued me was related to the mechanism proposed for topoisomerases, which are essential enzymes that coil or uncoil DNA during DNA synthesis in all cells. Topoisomerase II–type enzymes (called Topo II) pass DNA strands through each other by cutting and rejoining DNA without marking or changing the genome in any way. In textbooks, the enzyme was shown as a sphere that bound to one segment of DNA, cut it, and then split in half to pass a second DNA segment through the split. But what held the DNA ends together during the passage of the DNA duplex through the double-stranded break? There had to be something else going on.

Francis Crick once said that you can't understand how an enzyme works unless you see its structure, and *I wanted to see the structure of Topo II.* I spent the next couple of years trying to crystallize the enzyme with no success, and eventually reached the point where I wondered if my project would ever work, and whether I had what it took be a scientist. I made one last preparation of the enzyme, and after working overnight in the lab, I put the purified enzyme on ice and went home to bed. When I came back the next day, the protein in the tube had turned white, and I was crushed, thinking it had precipitated into a useless aggregate. But when I looked at a sample under the microscope, I saw crystals growing in the tube! At that moment I knew I had a project. I spent the next nine months solving the molecular structure of the enzyme, and I'll never forget the thrill of seeing the structure for the first time.

It was instantly clear how Topo II must work. The enzyme has two jaws, one of which grabs and cleaves the DNA duplex and holds it while the other jaw passes a different segment of DNA through the gap. I experienced the intense joy of discovering this fundamental mechanism of DNA metabolism, and of knowing that at that moment I was the first person in the world to have this understanding of the natural world.

—James Berger, on his discovery of the structure and mechanism of topoisomerase II

enetics is the science of heredity and the variation of inherited characteristics. Today, we know that biological information is stored and transmitted from generation to generation by deoxyribonucleic acid, or DNA, but this understanding arose only gradually. DNA was not widely accepted as the chemical of heredity until the 1940s, and its structure was not determined until 1953, when James Watson and Francis Crick introduced the world to the DNA double helix. (The structure of DNA is described in Chapter 6.) Our knowledge of the beautiful double-helical DNA structure has transformed the way that science is performed, to the extent that it is tempting to think of the field of genetics in terms of before and after DNA structure. But genetics has a wonderfully rich and varied history, every bit as exciting in the decades before the double helix as afterward.

The beginnings of modern genetics can be traced to the 1850s, when Gregor Mendel studied the inheritance of traits in the garden pea. He deduced that organisms contain particles of heredity (what we now call genes) that exist in pairs and that the paired particles split up when **gamete cells** (sex cells, the ovum and pollen in peas) are formed; pairs of hereditary particles are reformed on the union of two gametes during fertilization. Mendel was absolutely correct, but decades ahead of his time. His marvelous work went unnoticed for more than 30 years, until well after his death.

In contrast, a contemporary of Mendel's, Charles Darwin, was exceedingly famous in his lifetime. Darwin's theory of evolution started an awakening, one that continues to this day (see Chapter 1). For evolutionary theory to work, there must be diversity among individuals within a species, and variants more suited to the environment are selected and survive to produce offspring. Darwin's evolutionary theory, as wonderful as it is, completely lacks an explanation for how this diversity is produced. In fact, Darwin spent considerable time pondering this problem. He espoused the theory of *pangenesis*, first proposed by the ancient Greeks, in which genetic traits are shaped by life experience and transferred by "pangenes" to gamete cells, via the blood, enabling the traits to be inherited. In principle, the mistaken pangenesis theory is a variation of Jean-Baptiste Lamarck's theory of inheritance of acquired characteristics (see the How We Know section at the end of Chapter 1).

Darwin's theory of evolution became widely known, but Mendel's work fell into obscurity. During the late 1800s, advances in microscopy pushed the optical limits, enabling scientists to visualize subcellular structures. Of particular interest to geneticists were chromosomes, structures found in the nuclei of cells. A rash of intense studies documented chromosome behavior during cell division, fertilization, and the formation of gamete cells.

New discoveries revealed that the number of chromosomes in **somatic cells** (all cells in a multicellular organism other than sex cells) is constant for a given species and that the total number of chromosomes is halved to form gametes. When Mendel's work was rediscovered in 1900, his principles of heredity and particles of inheritance fit nicely with the behavior of chromosomes observed under the microscope.

Proof that genes reside on chromosomes soon followed, from a series of wonderful studies on fruit flies started in 1908 by Thomas Hunt Morgan. Central to Morgan's work were mutants, flies displaying physical traits not found in the average fly. The variety of mutant flies accumulated by Morgan's lab during 15 years of study—generations of flies reared in milk bottles—was amazing, including flies with bodies of different shapes and sizes, a variety of wing patterns, legs of different sizes, and a whole spectrum of eye colors. These fly mutants simply appeared spontaneously over generations of growth in Morgan's lab. Here was the answer to the variation required to make Darwin's theory of evolution work. Spontaneous mutants are infrequent, but given the expanse of evolutionary time, sufficient numbers and types of mutants are produced for nature to select and mold new species.

Genes and mutations explain heredity and illuminate evolutionary theory. But what are genes made of, and how is the information within them translated into the physical traits of an organism? Chromosomes were known to consist of both DNA and protein—but which of these is the genetic material? Several elegant and now classic experiments identified DNA as the molecule of heredity and found that DNA contains a code to direct the synthesis of RNA and proteins. The structure of the DNA double helix intuited by Watson and Crick revealed an architecture more beautiful than anyone could have imagined. The DNA molecule consists of two long strands twisted about each other, each chain a series of repeating units called nucleotides.

Watson and Crick immediately realized that the cell must have a mechanism to untwist the two strands in order to duplicate the DNA molecule and pass the genetic information to the next generation. Indeed, as we shall see in Chapter 9, the cell contains a complex arsenal of enzymes devoted to untwisting and altering the topology of DNA. These enzymes, called topoisomerases, are important targets of anticancer drugs, and their mechanisms of action are still actively investigated by James Berger and many other researchers. By extension, it is also critical to have a thorough understanding of the DNA molecule itself, and that is the subject of this and several other chapters in this textbook.

Despite the intrinsic beauty of the newly discovered double helix, the process by which a sequence of nucleotides could code for a sequence of amino acids in a protein remained a mystery. In a rapid series of advances in the 10 years following Watson and Crick's breakthrough, mRNA, tRNA, and rRNA were discovered and the workings of the directional flow of biological information, DNA→RNA→protein, were understood.

The field of molecular biology developed from these great discoveries. It seeks a detailed explanation of how biological information brings order to living processes. Molecular biology—the subject of this book—is a rapidly evolving scientific pursuit. The fundamental discoveries described in this chapter provided the groundwork for all subsequent studies in the field.

2.1 MENDELIAN GENETICS

Gregor Mendel was a monk at the monastery of St. Thomas at Brünn (now Brno in the Czech Republic). Renowned for teaching the sciences, the monastery sent Mendel to the University of Vienna in 1851, to obtain his teaching credentials. Although Mendel failed his final exams, he returned to the monastery and began a 10-year program of experiments that were so well conceived and executed that his results form the cornerstone of modern genetics.

In Mendel's time, plant hybrids were highly desired for their unique ornamental varieties. But the inheritance of colorful hybrid flower patterns was perplexing and unpredictable. The inability to decipher general principles of inheritance was not for lack of trying. Many well-known scientists performed extensive plant-breeding experiments, but no fundamental principles of inheritance could be formulated from these endeavors.

Gregor Mendel, 1822–1884
[Source: Pictorial Press Ltd/Alamy.]

Mendel's success where others failed can be attributed to his sound scientific approach to the problem. For his studies Mendel picked the garden pea, *Pisum sativum*, an excellent choice for several reasons. Because the pea was economically important, many varieties were available from seed merchants. The pea plant is also small, so that many plants can be grown in a confined space; and it grows quickly, reaching maturity in one growing season. But perhaps more important than Mendel's choice of experimental organism was his approach to studying it. Others before him, in studying plant breeding, had looked at the plant as a whole, dooming a study of heredity from the start because many genes control the overall appearance of an organism. Mendel focused instead on separate features of the plant, carefully observing isolated characteristics of the seeds, flowers, stem, and seed pods (**Figure 2-1**).

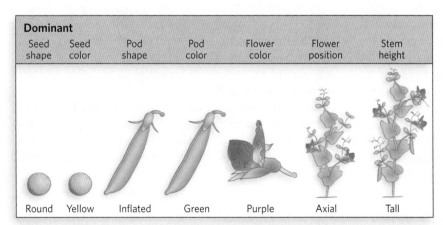

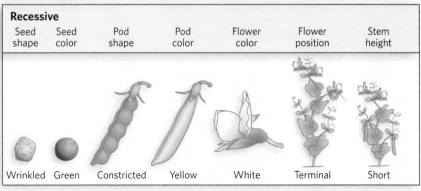

FIGURE 2-1 Traits of the garden pea examined by Mendel. Mendel picked seven pairs of traits to study: seed shape, seed color, seed pod shape, seed pod color, flower color, flower position along the stem, and stem height. The dominant and recessive forms of the traits are shown.

Mendel's First Law: Allele Pairs Segregate during Gamete Formation

Mendel spent the first two years growing different varieties of peas to ensure that each was a true-breeding, or **purebred**, with the offspring produced by crossing two plants of the same variety always having the same appearance as the parents. Then he carefully selected seven different pairs of traits and cross-pollinated plants with contrasting traits. For example, plants with round seeds were crossed with plants having wrinkled seeds. The parental plants are referred to as the **P generation** (*P* for parental). The **hybrid** offspring are called the **F_1 generation** (*F* for filial, from the Latin for "son or daughter"; F_1 indicating first filial). This first generation produced only round seeds; the wrinkled-seed trait seemed to have disappeared. Mendel observed a similar result in crosses for all seven pairs of traits (**Table 2-1**). He referred to the trait that appears in the F_1 generation as the **dominant** trait, and the trait that disappears as the **recessive** trait (see Figure 2-1).

Mendel's finding that one trait is dominant and the other is recessive was novel and completely contrary to the prevailing view that parental traits blend together in the offspring. Other experimenters might have stopped at this new and dramatic discovery, but not Mendel. In his next experiment, he allowed F_1 plants to self-pollinate and produce the **F_2 generation** (second filial generation). Surprisingly, the F_2 generation was a mixture: most plants produced round seeds (the dominant trait), but some had wrinkled seeds. The recessive wrinkled-seed trait that disappeared in the F_1 generation reappeared in the F_2 generation! In contrast to the view that parental traits blend in the offspring, Mendel did not observe a blending of the traits he studied. For example, there were no partly wrinkled seeds, and seed colors were either yellow or green, not yellow-green.

Unlike other scientists before him, Mendel kept close track of the numbers of offspring with the dominant and recessive traits. He counted 5,474 dominant round seeds and 1,850 recessive wrinkled seeds in the F_2 generation, for a ratio of 2.96 dominant to 1 recessive trait. His experiments examining seed color produced a similar result. He observed 6,022 dominant yellow seeds and 2,001 recessive green seeds, for a ratio of 3.01:1. The other pairs of traits also appeared in a 3:1 ratio of dominant to recessive offspring in the F_2 generation, as summarized in Table 2-1.

Mendel's interpretation of these results was brilliant. Reappearance of the recessive wrinkled-seed trait in the F_2 generation suggested that traits do not really disappear. Mendel therefore proposed that traits are "hereditary particles"—now called **genes**—and that they come in pairs. Organisms that carry two copies of each gene are **diploid**, and the different variants of a given gene are called **alleles**. In other words, a diploid parental plant has two alleles of the seed-shape gene (for example, one for smooth seeds and one for wrinkled seeds). Furthermore, Mendel proposed that one allele could mask the appearance of the other. This explained why traits could disappear but then reappear in future generations. Even though the F_1 plant carried one dominant allele (round seed) and one recessive allele (wrinkled seed), only the dominant round-seed allele is evident in the outward appearance, or **phenotype**, of the F_1 plant. Mendel also reasoned that each parent contributes only one copy of each gene to the offspring; that is, the gamete cells are **haploid**, having only one allele of each gene. When two F_1 generation gametes combine and each carries the recessive allele for seed shape, the resulting F_2 plant will produce wrinkled seeds.

We have introduced several genetic terms in the preceding paragraphs. These terms, and others that follow, are part of the scientific language of genetics and are defined in **Table 2-2**.

Mendel was well-trained in mathematics, which helped him make sense of his results. To explain the ratios of phenotypes in the offspring in mathematical terms, he referred to the dominant allele with a capital

TABLE 2-1				
Results of Mendel's Single-Factor Crosses				
Characteristic Isolated for Study	*Parental Cross*	*F_1 Phenotype (dominant trait)*	*F_2 Phenotypes (dominant and recessive)*	*F_2 Ratio (dominant:recessive)*
Seed shape	Round × wrinkled	All round	5,474 round and 1,850 wrinkled	2.96:1
Seed color	Yellow × green	All yellow	6,022 yellow and 2,001 green	3.01:1
Pod shape	Inflated × constricted	All inflated	882 inflated and 299 constricted	2.95:1
Pod color	Green × yellow	All green	428 green and 152 yellow	2.82:1
Flower color	Purple × white	All purple	705 purple and 224 white	3.15:1
Flower position	Axial × terminal	All axial	651 axial and 207 terminal	3.14:1
Stem height	Tall × short	All tall	787 tall and 277 short	2.84:1

TABLE 2-2

Commonly Used Terms in Genetics

Term	Definition
P generation	Parents used in a cross
F_1 generation	Progeny resulting from a cross in the P generation
F_2 generation	Progeny resulting from a cross in the F_1 generation (succeeding generations are F_3, F_4, etc.)
Purebred	Individual homozygous for a given trait or set of traits
Hybrid	Progeny resulting from a cross of parents with different genotypes
Gene	Section of DNA encoding a protein or functional RNA
Allele	Variant of the gene encoding a trait (e.g., seed color: yellow or green)
Phenotype	Outward appearance of an organism
Genotype	Alleles contained in an organism
Homozygous	Having two identical copies of an allele for a gene
Heterozygous	Having two different alleles for a gene
Dominant allele	Allele expressed in the phenotype of a heterozygous organism
Recessive allele	Allele masked in the phenotype of a heterozygous organism

letter (e.g., *R* for round) and to the recessive allele with the lowercase version of the same letter (*r* for wrinkled). A lowercase *w* might seem more fitting for the wrinkled-seed allele, but using different letters would make it harder to keep track of allele pairs of the same gene. A purebred round-seed plant has two *R* alleles, *RR*, and a purebred wrinkled-seed plant has two recessive alleles, *rr*. *RR* plants exhibit the dominant *R* trait (round seeds), and *rr* plants exhibit the recessive *r* trait (wrinkled seeds). This double-letter nomenclature, representing the allelic makeup of an organism, is a way of denoting the organism's **genotype**.

In a cross of purebred parents, round (*RR*) and wrinkled (*rr*), all F_1 progeny receive one allele from each parent and are thus *Rr*. The *R* allele is dominant to the *r* allele, so all F_1 *Rr* hybrid plants have round seeds. The F_1 plant produces *R* and *r* gametes in equal amounts, and therefore self-pollination of F_1 plants produces three different diploid genotypes: *RR*, *Rr*, and *rr*. A convenient way of displaying the genes that come together during a cross such as this is a **Punnett square** analysis (**Figure 2-2**). In this analysis, the gamete genotypes of one parent are written along the top of the Punnett square, and those of the other parent are written along the left side of the square. The various combinations in which the alleles can come together during pollination are entered in the grid. The results yield the three different F_2 genotypes in the following ratios: 1 *RR*, 2 *Rr*, and 1 *rr*. These genotypes, together with the concept of dominant and recessive alleles, explain the ratio of phenotypes that Mendel observed: 3 dominant (1 *RR* + 2 *Rr*)

to 1 recessive (1 *rr*). We now refer to an organism with identical alleles for a given gene as **homozygous** for that gene (such as *RR* or *rr*). An organism with two different alleles, such as an *Rr* plant, is characterized as **heterozygous**.

To determine whether the F_2 plants really were of three genotypes in a 1:2:1 ratio, Mendel analyzed the offspring of self-fertilized F_2 plants (the F_3 generation). The F_2 recessive wrinkled-seed plants (25% of the total F_2 plants) all bred true, giving only wrinkled-seed F_3 progeny, and thus were homozygous *rr*. The dominant round-seed F_2 plants were of two types. One-third (25% of the total) bred true; their offspring always produced round seeds, and therefore these F_2 plants were homozygous *RR*. The remaining two-thirds of the round-seed F_2 plants (50% of the total) produced F_3 plants with round and wrinkled seeds in a 3:1 ratio, and therefore these F_2 plants were heterozygous *Rr*. The analysis fit the 1:2:1 ratio for the F_2 genotype exactly: 1*RR*:2*Rr*:1*rr* (see Figure 2-2).

In summary, Mendel hypothesized that traits are carried by particulate genes, that somatic cells contain two copies (two alleles) of each gene, and that gamete cells obtain only one allele for each gene during gamete formation. When two gametes fuse at fertilization, allele pairs are restored, producing the diploid genotype of the offspring. The general principle summarizing this proposal is often referred to as Mendel's first law, or the **law of segregation**, which states that equal and independent segregation of alleles occurs during formation of gamete cells.

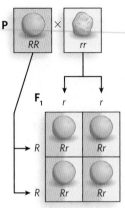

Genotypes: all *Rr*
Phenotypes: all round

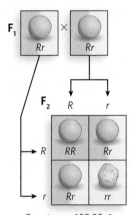

Genotypes: 1*RR*:2*Rr*:1*rr*
Phenotypes: 3 round:1 wrinkled

FIGURE 2-2 An example of Mendel's first law. Alleles of the same gene segregate independently into gametes. Plants that are homozygous for dominant round-seed shape (*RR*) were cross-pollinated with homozygous recessive wrinkled-seed plants (*rr*) to produce F₁ progeny. The Punnett square analysis shows the gametes of each parent along the top and left side of the grid, from which—if the gametes are formed in equal amounts—one can predict the possible progeny and their frequency. Punnett square analysis is a common way of illustrating genetic crosses today, but was developed after Mendel's work. In the F₁ generation, the only progeny that can be produced are *Rr* hybrids. F₁ plants were self-fertilized, and as the Punnett square analysis for the F₂ generation predicts (based on the assumption that the different alleles (*R* and *r*) segregate independently into F₁ generation gametes), round seeds and wrinkled seeds were produced in a 3:1 ratio.

Mendel's Second Law: Different Genes Assort Independently during Gamete Formation

Mendel's results demonstrated that two alleles for one gene separate during gamete formation, but how do alleles for two *different* genes behave? There are two possibilities. Alleles for two different genes could separate during gamete formation, assorting randomly into the gametes. Alternatively, they could remain associated, traveling together into the same gamete cells. These two scenarios have distinct outcomes. For example, if particular alleles for seed shape and seed color stay

together during the formation of gamete cells, future offspring will retain both the same seed shape and the same seed color as one parent or the other. But if alleles for the two genes separate during gamete formation, some of the F₂ offspring will exhibit new combinations of seed shape and color, distinct from those of either parent.

To test these hypotheses, Mendel's next experiments were two-factor crosses, analyzing the transmission of two different genes in each cross. He began by cross-pollinating purebred plants having round, yellow seeds with plants having wrinkled, green seeds. First let's consider the genotype of the two plants. We already know that the round-seed allele (*R*) is dominant to the wrinkled-seed allele (*r*). The genotype of the purebred plant with dominant yellow seeds is *YY*, and the purebred plant with green seeds is homozygous for the recessive green-seed allele, *yy*. Thus, the genotype of a purebred round, yellow-seed plant is *RRYY*, and the genotype of a plant with wrinkled, green seeds is *rryy*. A cross between these plants yields F₁ progeny of genotype *RrYy*, phenotypically round and yellow. If the four alleles for the two genes separate and assort randomly during gamete formation, F₁ plants will produce four different gametes (*Ry*, *rY*, *RY*, and *ry*), and all combinations of seed shape and color will be observed in F₂ plants (**Figure 2-3**).

Mendel's observations of F₂ phenotypes are shown in **Table 2-3**. All possible combinations of traits occurred, and therefore the alleles of the two different genes are not physically connected; instead, they separate during the formation of gamete cells. The analysis of the four possible genotypes, as shown in Figure 2-3, predicts a 9:3:3:1 ratio, close to the observed result. Mendel performed many two-factor crosses analyzing different gene combinations, and the results were always consistent with the random assortment of genes during gamete formation. Mendel's second law, or the **law of independent assortment**, states that different genes assort into gametes independently of one another.

Mendel studied garden peas, but his basic principles hold true for sex-based inheritance in all animals and plants. Indeed, many human genetic diseases that can be traced through a family pedigree follow Mendel's simple rules of inheritance.

There Are Exceptions to Mendel's Laws

The transmission of dominant and recessive traits documented by Mendel is sometimes referred to as "Mendelian behavior." However, not all genes behave in such an ideal fashion. There are many exceptions to Mendel's principles of heredity, a few of which we review here.

Incomplete Dominance Some alleles of a gene are neither dominant nor recessive. Instead, hybrid progeny display

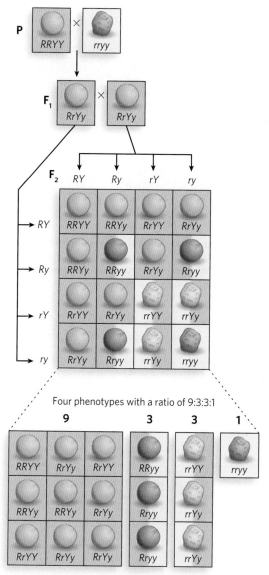

a phenotype intermediate between those of the two parents. This type of non-Mendelian behavior is called **incomplete dominance**. An example of incomplete dominance can be seen in the gene for flower color in four o'clock plants (**Figure 2-4**). Homozygotes are either red (*RR*) or white (*R′R′*; primed capitals are used so as not to confuse this case with recessive alleles), but the F₁ heterozygote (*RR′*) is neither red nor white; it is pink. The molecular explanation for the pink heterozygote is the production of sufficient red color from the single *R* allele in the heterozygote to yield a pink coloration.

Interestingly, an example of incomplete dominance can also be found in Mendel's published work. He studied different alleles for the gene controlling the pea plant's flowering time. The F₁ progeny had a flowering time that was intermediate between the flowering times of the two parents.

Codominance Recessive alleles often produce nonfunctional proteins, or none at all. However, there are many examples of two alleles of a gene that produce two different functional proteins, neither of which is dominant to the other. This non-Mendelian behavior is known as **codominance**. An example of codominance is human blood type (**Figure 2-5**). The allele for A-type blood, *I^A*, results in a cell surface glycoprotein different from the glycoprotein encoded by the allele for B-type blood, *I^B*. People with A-type blood are homozygous *I^A I^A*, and those with B-type blood are homozygous *I^B I^B*. AB-type individuals are *I^A I^B* heterozygotes. O-type individuals lack both varieties of surface glycoprotein; they are homozygous for the recessive *i* allele.

Linked Genes The most common non-Mendelian behavior is seen in **linked genes**, in which alleles for two different genes assort together in the gametes, rather than assorting independently. We now know that genes are located on chromosomes, and diploid organisms have two copies of each chromosome, known as **homologous chromosomes**, or homologs. During gamete formation, whole chromosomes, not individual genes, assort into gametes. Genes that are close together on one chromosome are inherited together, contrary to Mendel's second law.

FIGURE 2-3 An example of Mendel's second law. Different genes assort independently into gamete cells. The parental cross (round, yellow seeds × wrinkled, green seeds) yields uniform F₁ progeny with the dominant phenotype (round, yellow seeds) and genotype *RrYy*. The Punnett square analysis assumes random assortment of the different alleles into the gametes formed by the F₁ plant. All possible gamete genotypes are written across the top and left side of the grid. The predicted outcome for independent assortment is seeds of four different phenotypes in a 9:3:3:1 ratio, as illustrated below the Punnett square.

TABLE 2-3				
Mendel's Results from a Two-Factor Cross				
Characteristic Isolated for Study	*Parental Cross*	*F₁ Phenotype*	*F₂ Phenotype*	*F₂ Ratio*
Seed shape and seed color	Round, yellow × wrinkled, green	All round, yellow	315 round, yellow	8.3
			101 wrinkled, yellow	2.7
			108 round, green	2.8
			38 wrinkled, green	1.0

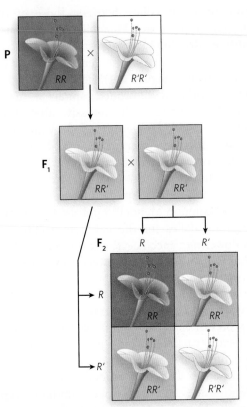

FIGURE 2-4 Non-Mendelian behavior: incomplete dominance in four o'clock plants. Cross-pollination of red- and white-flowered plants yields an F$_1$ plant with flowers of intermediate color (pink). Therefore, neither parental allele is completely dominant. The single *R* allele gives rise to sufficient red pigment to produce a pink coloration. Genotypes are given below each flower.

Assortment of linked genes into gamete cells is shown in **Figure 2-6** and is discussed in detail later in the chapter.

Mendel picked traits whose genes assorted independently. However, some of the genes that he studied are on the same chromosome. How could Mendel have observed independent assortment of genes on the same chromosome? As we describe later in the chapter, homologous chromosomes associate together during the cell divisions that lead to gametes. At that time, there is often an exchange of genetic material between the chromosome pair, resulting in some alleles previously found on one chromosome now being found on the other. For the traits Mendel selected for study, genes on the same chromosome are spaced far apart, and this swapping of genetic material occurs frequently between them. Therefore, the genes assort as though they are on different chromosomes.

Many other types of non-Mendelian behavior exist besides those described here. These include traits determined by multiple genes, traits derived from interactions between different genes (epistasis), the inheritance of traits encoded by organelle genes (cytoplasmic inheritance), and traits that depend on whether the gene is inherited from the male or female parent (genomic imprinting).

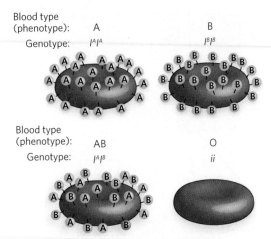

FIGURE 2-5 Non-Mendelian behavior: codominance in human blood types. The cell surface glycoprotein antigens on red blood cells (erythrocytes) determine human blood type. Two different alleles encode two variants of the enzyme glycosylase (an enzyme involved in formation of glycoproteins) and produce different cell surface glycoproteins, A (allele *I^A*, yellow circles) and B (allele *I^B*, blue circles). Both alleles are expressed in the heterozygote (AB blood type, genotype *I^{A}I^B*). Because both alleles produce functional surface glycoproteins, neither allele is dominant to the other. Individuals with O-type blood have two null alleles (*ii*) and thus produce no A or B surface antigens on their red blood cells.

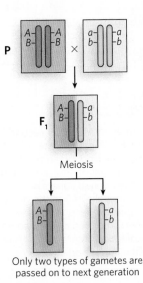

Only two types of gametes are passed on to next generation

FIGURE 2-6 Non-Mendelian behavior: linked genes. Genes *A* and *B* are located on the same chromosome. Shown here is a cross between a homozygous dominant parent and a homozygous recessive parent that results in F$_1$ hybrid progeny, all *AaBb*. Because the alleles for the two genes are close together on the same chromosome (*A, B* and *a, b*), they cannot separate during the formation of gametes. Each gamete receives one copy of this chromosome, and thus either the *A* and *B* alleles or the *a* and *b* alleles; no gametes containing *A* and *b* or *a* and *B* are formed. The two genes *A* and *B* are linked and do not assort independently during meiosis in the formation of gamete cells.

- Mendel's studies on the garden pea revealed an underlying mathematical pattern in inheritance.
- Mendel postulated that genetic traits are carried by hereditary particles, now called genes. Diploid organisms contain two copies, or alleles, of each gene and produce haploid gametes that contain one allele for each gene.
- Individuals homozygous for a particular gene have two identical alleles for that gene. In heterozygous individuals, the two alleles for a gene are different. The allelic makeup of an organism is its genotype.
- The different alleles for a gene may be dominant or recessive. In a heterozygote, the dominant allele masks the recessive allele in the outward appearance, or phenotype, of the organism.
- Mendel's first law, the law of segregation, states that the two alleles for each gene segregate independently into haploid gamete cells.
- Mendel's second law, the law of independent assortment, states that alleles for different genes assort into gametes randomly. However, we now know that genes reside on chromosomes and that chromosomes, not genes, assort randomly into gametes.
- There are exceptions to Mendel's laws. For example, a gene exhibits incomplete dominance when the phenotype of heterozygous progeny is intermediate between those of the two homozygous parents. Gene alleles exhibit codominance when both produce functional proteins and neither is dominant to the other, as in human blood types. Alleles for two genes close together on the same chromosome are linked and do not assort independently into gametes.

2.2 CYTOGENETICS: CHROMOSOME MOVEMENTS DURING MITOSIS AND MEIOSIS

In 1865, Mendel presented his findings on inheritance in two lectures to the Brünn Society for the Study of the Natural Sciences, and they were then published in an obscure journal. Only about 150 copies of this journal were printed, and Mendel's findings lay dormant for decades, to be resurrected only after his death. However, in his lifetime, Mendel was well appreciated at his monastery, was elected abbot, and managed one of the wealthiest cloisters in the land. His claim to fame was an incident in which he refused to pay a new tax imposed on the monastery by the Habsburg Empire. Mendel met the sheriff at the gate and dared him to take the keys from his pocket before he'd pay another pfennig! Of course, this is not what we know Mendel for today.

The years between 1880 and 1900 saw amazing discoveries in **cytology**, the study of cells, which intersected with the rediscovery of Mendel's work. Microscopes had become more advanced, and chromosomes could be stained and visualized in the cell nucleus. Cytologists observed that, unlike other cellular components, chromosomes were meticulously divided between the two new cells during cell division. The diploid nature of somatic cells and the haploid nature of gametes were also discovered around this time. It was in this scientific environment of explosive growth in **cytogenetics** that, in 1900, Mendel's principles of heredity were confirmed experimentally and rediscovered independently by three scientists: Hugo de Vries, Carl Correns, and Erich von Tschermak. The behavior of chromosomes was seen to remarkably mirror the behavior of Mendel's hereditary particles, and the idea that the nucleus, and perhaps the chromosomes themselves, formed the basis of heredity was bandied about.

In this section, we describe the architecture of the cell and the chromosome movements that occur during somatic cell division and gamete formation, setting the stage for the chromosome theory of inheritance (the subject of Section 2.3).

Cells Contain Chromosomes and Other Internal Structures

Robert Hooke was the first to notice the cellular composition of a biological specimen, during his microscopic examination of cork in 1665 (**Figure 2-7**). In his famous book *Micrographia*, Hooke described a multitude of tiny boxes in the cork sample and coined the word **cell** (Latin *cellula*, "small compartment"). By the early 1800s, it became clear that plants are made up of cells. In 1833, Robert Brown identified the nucleus, the first subcellular structure to be discovered. In 1839, Theodor Schwann realized that animal tissue contains nuclei throughout the cells, and he proposed the **cell theory**, which states that all animals and plants consist of large assemblages of cells.

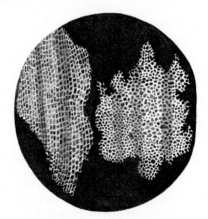

FIGURE 2-7 Hooke's microscopic examination of cork. Robert Hooke used a compound microscope to visualize cork cells and catalogued his work with meticulous drawings. This drawing is from his book *Micrographia*. His studies provided the first clue that organisms are cell-based. [*Source: U.S. National Library of Medicine.*]

Microscopic studies of chromosomes within nuclei were first made in plant cells by Karl Wilhelm von Nägeli and Wilhelm Hofmeister between 1842 and 1849. **Chromosomes** were named (from the Greek for "colored body") for their property of taking up large amounts of colored dye. The term was coined by Heinrich von Waldeyer in 1888. **Figure 2-8** shows rapidly dividing cells of an onion, stained to show the chromosomes. Development of the electron microscope in 1931 eventually brought into view the detailed structure of the cell as we know it today. Each cell is bounded by the **cytoplasmic membrane**, encasing the cytoplasm and its variety of subcellular structures called **organelles**. **Figure 2-9** is a schematic depiction of a typical animal cell, a eukaryotic cell; the caption describes each organelle and its function.

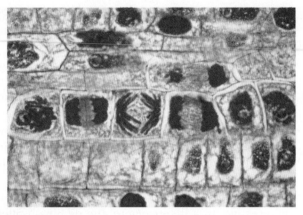

FIGURE 2-8 Plant cell chromosomes. A cross section of a rapidly dividing root tip of the onion (*Allium cepa*) shows the chromosomes as darkly stained bodies. Cells in different stages of division (mitosis) are apparent. *[Source: Manfred Kage/Science Source.]*

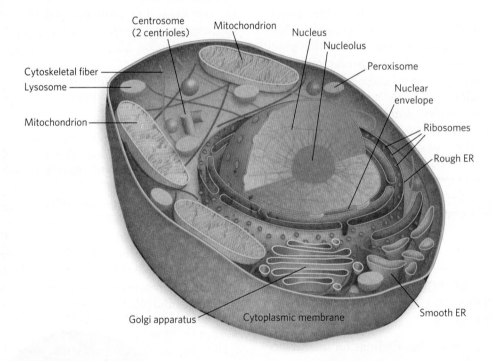

FIGURE 2-9 Animal cell structure. Eukaryotic cells are bounded by a cytoplasmic membrane. Chromosomes (not visible as individual structures in a nondividing cell) are located in the nucleus, and the nuclear envelope (nuclear membrane) is a double membrane with large pores through which RNA and proteins move. The nucleolus, a substructure within the nucleus, is the site of rRNA synthesis. All other organelles are in the cytoplasm. The centrosome consists of two centrioles, perpendicular cylinder-shaped protein complexes. Centrosomes organize microtubule spindles that attach to the cytoskeleton. During cell division, the centrosome duplicates and the two centrosomes migrate to opposite poles of the cell. The centrosomes then organize the spindle apparatus, in which microtubules connect chromosomes to the centrosomes for partitioning of the chromosomes into daughter cells. Mitochondria, the energy factories of animal cells, oxidize fuels to produce ATP. Lysosomes, containing degradative enzymes, aid in digestion of intracellular debris and recycle certain components. Peroxisomes help detoxify chemicals and degrade fatty acids. The smooth endoplasmic reticulum (ER) is the site of lipid synthesis and drug detoxification. Ribosomes, composed of both RNA and protein, act as protein-synthesizing factories; many attach to the ER, giving it a rough appearance. The rough ER sorts proteins destined for the cytoplasmic membrane or for other organelles; it is continuous with the outer membrane of the nuclear envelope. The Golgi apparatus, a membranous network, receives proteins from the ER and modifies and directs them to their proper compartments. Cytoskeletal fibers are a network of structural proteins that give shape to the cell and aid in cell movement.

Mitosis: Cells Evenly Divide Chromosomes between New Cells

For biological information to be faithfully transmitted to daughter cells, it must be duplicated and then each complete information packet correctly partitioned into its own cell. As cells grow, they proceed through four phases, collectively known as the **cell cycle**: G_1, S, G_2, and M (**Figure 2-10**). In G_1 **phase**, cells are diploid, containing two copies of each chromosome. The cellular chromosome content in G_1 cells is represented as $2n$, where n is the number of unique chromosomes of that species. G_1 is also called the first gap, because it represents a gap in time before S phase.

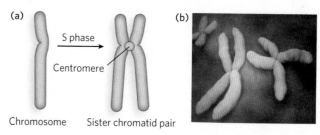

FIGURE 2-11 Formation of sister chromatid pairs by chromosome duplication. (a) Chromosomes are duplicated as cells proceed through S phase. Each resulting sister chromatid pair is held together at the centromere. (b) Three sister chromatid pairs, as seen by electron microscopy. [*Source: (b) MedicalRF / The Medical File / Peter Arnold Inc.*]

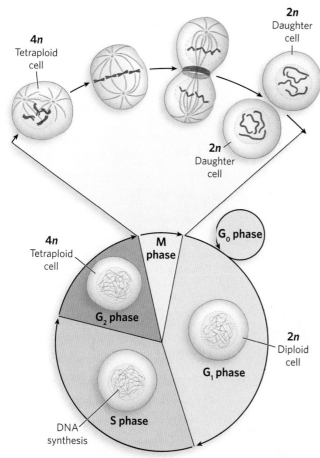

FIGURE 2-10 The eukaryotic cell cycle. Cells start in G_1 phase and progress to S phase, in which chromosomes are duplicated. In G_1, cells are diploid ($2n$, where n is the species' chromosome number). After S phase, cells are tetraploid ($4n$) and enter G_2 phase (chromosomes are not visible in interphase; they begin to condense and become visible in the microscope at the beginning of M phase). In M phase, the duplicated chromosomes are equally divided, and the cell splits (by cytokinesis) into two daughter cells, each $2n$. These cells can enter a quiescent phase, G_0, which removes them from the cell cycle, or can undergo further division. The duration of each phase varies with species and cell type. A typical human cell in tissue culture has a cell cycle of about 24 hours: G_1 phase, 6–12 hours; S phase, 6–8 hours; G_2 phase, 3–4 hours; and M phase, 1 hour.

During **S phase** (*S* for synthesis), each chromosome is duplicated, and the two identical chromosomes remain together as a **sister chromatid pair**. The point where the sister chromatids are joined is called the **centromere** (**Figure 2-11**). At the end of S phase, each homologous chromosome exists as a sister chromatid pair, and thus the cell now contains four copies of each chromosome in the form of two sister chromatid pairs (i.e., the cell is $4n$, or tetraploid). The cell next enters G_2 **phase**, or the second gap in time, after S phase.

The final phase of the cell cycle is **M phase**, or **mitosis**, in which the duplicated chromosomes separate completely and the cell divides into two daughter cells, each $2n$. The two new cells reenter G_1 phase and then either continue through another division or cease to divide, entering a quiescent phase (G_0) that may last hours, days, or the lifetime of the cell (see Figure 2-10). Differentiated cells such as hepatocytes (liver cells) or adipocytes (fat cells) have acquired their specialized function and thereafter remain in G_0 phase.

Many scientists contributed to the description of events during mitosis, and Walther Flemming figures most prominently among them. By the late 1870s, the quality of microscopes included such developments as the oil immersion lens and the substage condenser. These advances made possible Flemming's detailed observations of dividing cells, published in 1878 and 1882, revealing the stages of mitosis as we know them today. The steps of mitosis are illustrated on the left side of **Figure 2-12** and summarized here.

Interphase. Cells not in mitosis are in interphase (which comprises G_1, S, and G_2). The cell is metabolically active and growing, and the chromosomes are duplicated (in S phase) in preparation for mitosis. The chromosomes are decondensed and not yet visible in the microscope.

Prophase. Prophase is the first stage of mitosis, following G_2 phase. As cells enter prophase, the sister

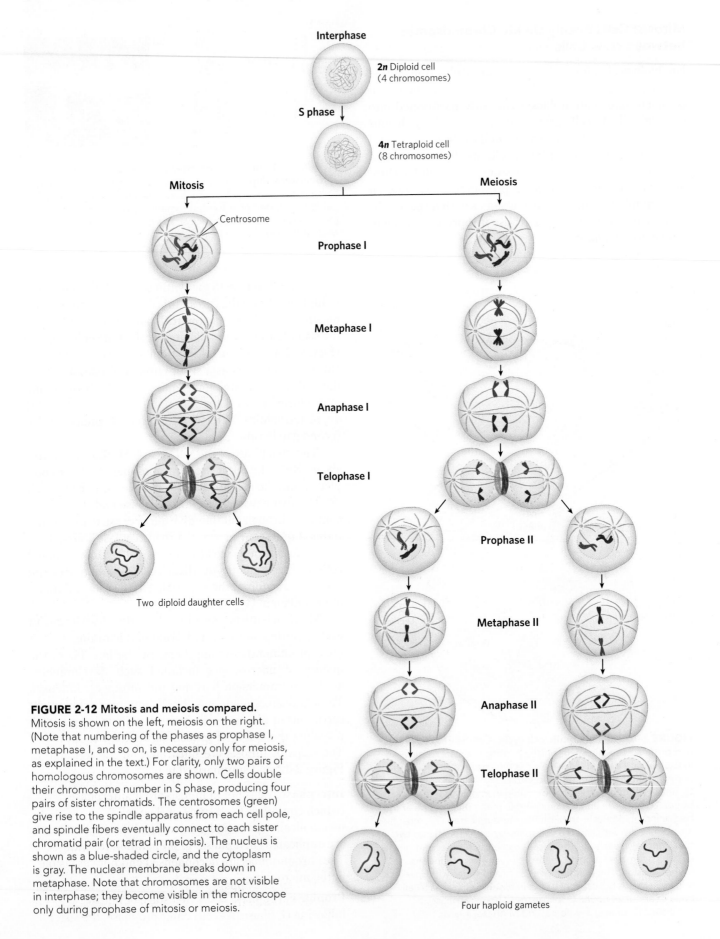

FIGURE 2-12 Mitosis and meiosis compared.
Mitosis is shown on the left, meiosis on the right.
(Note that numbering of the phases as prophase I,
metaphase I, and so on, is necessary only for meiosis,
as explained in the text.) For clarity, only two pairs of
homologous chromosomes are shown. Cells double
their chromosome number in S phase, producing four
pairs of sister chromatids. The centrosomes (green)
give rise to the spindle apparatus from each cell pole,
and spindle fibers eventually connect to each sister
chromatid pair (or tetrad in meiosis). The nucleus is
shown as a blue-shaded circle, and the cytoplasm
is gray. The nuclear membrane breaks down in
metaphase. Note that chromosomes are not visible
in interphase; they become visible in the microscope
only during prophase of mitosis or meiosis.

chromatid pairs condense and become visible. Two organelles, called **centrosomes**, move to opposite poles of the cell. There they give rise to the spindle apparatus, an organized structure of protein fibers, which also becomes visible during prophase.

Metaphase. In metaphase, the membrane surrounding the nucleus dissolves. The spindle apparatus becomes fully developed and attaches to the centromeres of the sister chromatid pairs, directing them to align in the equatorial plane of the cell, a site also known as the **metaphase plate**.

Anaphase. Each sister chromatid pair separates at the centromere, becoming two separate chromosomes. The spindle apparatus moves the separated chromosomes toward opposite poles of the cell.

Telophase. The two chromosome sets reach opposite cell poles, nuclear membranes (nuclear envelopes) re-form, and the chromosomes become less distinct as they decondense. Telophase ends with **cytokinesis**, the physical splitting of the cytoplasmic membrane and cell contents to form two daughter cells.

Meiosis: Chromosome Number Is Halved during Gamete Formation

Studies of fertilization in the late 1800s revealed that gamete cells contain only half the number of chromosomes found in somatic cells, and the union of two gametes reestablishes the diploid chromosome number; mitosis keeps this chromosome number constant during somatic cell division. These findings nicely explained how chromosome number is established and maintained in an organism, but they posed a new riddle: how are haploid gamete cells formed? The answer came in the 1880s from studies by Edouard van Beneden, Oskar Hertwig, and Theodor Boveri. Their studies of the ovary cells of a parasitic worm, *Ascaris*, revealed that the haploid female gamete, the egg (ovum), is formed by two consecutive cell divisions, in a process known as **meiosis** (see Figure 2-12, right). Later studies revealed that male gametes are also formed by meiotic cell divisions.

The most commonly studied organisms (such as *Ascaris*, sea urchin, or salamander larvae) had small, similar-looking chromosomes, and thus it was unclear whether chromosomes had unique identities. As far as anyone could tell, the cell simply divided an amorphous pool of chromosomes into equal parts to form daughter cells or gametes, rather than teasing apart two exact sets of different chromosomes. The unique nature of chromosomes, and the true precision with which the cell deals with them, was to come from work by Walter Sutton, in studies of the grasshopper. The grasshopper has chromosomes with unique morphologies that allow

individual chromosomes to be observed during cell division. Sutton's observations demonstrated that during mitosis, one complete set of chromosomes is partitioned into each daughter cell (see the How We Know section at the end of this chapter). Chromosome segregation during meiosis proved to be just as precise, yielding one haploid set of chromosomes per gamete cell.

The Process of Meiosis Meiosis involves a halving of chromosome number to form haploid (n) gametes. One might expect that a diploid cell simply divides once to form two haploid gamete cells, but this is not so. Meiosis involves two successive cell divisions, and four haploid gametes are formed from one diploid cell. The meiotic cell first goes through S phase, just as in mitosis, thereby increasing the number of each chromosome to four copies per cell ($4n$). In sharp contrast to mitosis, however, the homologous chromosomes—each of which is a pair of sister chromatids—find each other in meiosis and physically associate to form a **tetrad**. In the first meiotic cell division, the tetrad splits and the two sister chromatid pairs segregate into two new cells, each $2n$. This differs from mitosis, in which each sister chromatid pair splits at the centromere, resulting in two chromosomes that segregate into the two daughter cells.

Whereas mitosis involves only one cell division, the daughter cells from this first meiotic division divide a second time, but without an intervening S phase (no additional chromosome duplication). This second cell division closely resembles mitosis, except that the cells are diploid ($2n$) going into the second meiotic division (rather than $4n$, as in mitosis), so the second division reduces the diploid chromosome number by half, to form haploid gametes (n). In other words, the second meiotic cell division resembles mitosis in that sister chromatids separate, but in meiosis, for each chromosome, there is only one sister chromatid pair to split apart, whereas in mitosis the sister chromatid pairs of both homologous chromosomes are present at the metaphase plate, and each pair splits apart.

The phases of meiosis are summarized (and contrasted with mitosis) here and illustrated in Figure 2-12.

Interphase. Chromosomes are duplicated to form sister chromatid pairs; no obvious difference from mitosis.

Prophase I. Sister chromatid pairs become visible and the spindle apparatus forms. The difference from mitosis is that two homologous sister chromatid pairs find and associate with each other, forming a tetrad. In mitosis, two homologous sister chromatid pairs remain independent and do not associate with each other.

Metaphase I. The nuclear membrane breaks down, and the spindle apparatus moves the four homologous chromosomes to the metaphase plate as a tetrad, rather than moving two homologous but independent sister chromatid pairs as in mitosis.

Anaphase I. Centromeres stay intact, and sister chromatids do not separate. Instead, the tetrad splits and the two sets of sister chromatid pairs move to opposite poles. By contrast, in mitosis, sister chromatids split at the centromere and individual chromosomes move apart.

Telophase I. Telophase occurs as in mitosis. The nuclear membrane re-forms and the cell divides.

The second meiotic cell division is a lot like mitosis, but there is no S phase between divisions and the cell is diploid going into the second cell division.

Prophase II. As in mitosis, sister chromatid pairs are visible, but there are half as many as in mitosis because the homologous sister chromatid pair is no longer present (it is in the other daughter cell formed from the first division).

Metaphase II. As in mitosis, the nuclear membrane breaks down and sister chromatid pairs align in the equatorial plane.

Anaphase II. As in mitosis, the centromere splits and the two separated chromosomes move to opposite poles of the cell.

Telophase II. As in mitosis, cytokinesis results in two cells and the two nuclear membranes form. Unlike mitosis, where the resulting daughter cells are diploid ($2n$), the second meiotic division produces daughter cells that are haploid (n).

In overview, two cells enter meiosis II, and the end result of meiosis is four n cells from a single $2n$ cell. This is different from mitosis in two ways: the number of daughter cells formed and the n value of each daughter cell.

Sex Determination Cytological studies in many types of cells documented the existence of one or two chromosomes that behaved strangely in meiosis during the formation of male gametes. These were called accessory or X chromosomes; they either pair with a morphologically distinct partner chromosome or do not pair at all. Meiotic divisions therefore produce two types of sperm that differ in the accessory chromosome they contain. In 1905, Edmund B. Wilson and Nettie Stevens identified these accessory chromosomes in insects as the determinants of male and female sex and referred to them as X and Y chromosomes, or **sex chromosomes**. All other chromosomes are called **autosomes**.

Sex can be determined in many different ways, depending on the type of organism. For example, in mammals, a common way is the XY system (**Figure 2-13**). In XY determination, the female is XX and the male is XY; the male gametes are of two varieties, carrying either the

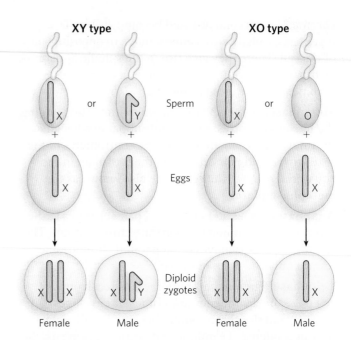

FIGURE 2-13 The chromosomal basis of sex determination. In the XY type of sex determination, meiosis produces sperm with either an X or a Y chromosome, in a 50:50 ratio (autosomes are not shown). All eggs have one X. Fertilization results in either an XX (female) or XY (male) zygote. The Y chromosome confers maleness. In XO determination, meiosis produces two types of sperm in a 50:50 ratio, either with or without an X chromosome (X or O). All eggs have one X. Fertilization results in either an XX (female) or an XO (male) zygote. The number of copies of X determines the sex.

X or Y chromosome. In many insects, sex is determined by the XO system, in which females are XX and males have one X and no other sex chromosome. The male gametes contain either an X chromosome or no sex chromosome. In both XY and XO determination, the union of male and female gametes is equally likely to produce male or female offspring. In birds, some insects, and other organisms, the ZW system determines sex. It is like the XY system but in reverse: males have two of the same chromosome (ZZ), whereas females have one copy each of the Z and W chromosomes.

Edmund B. Wilson, 1856–1939
[*Source: Courtesy of University Archives, Columbia University in the City of New York.*]

Nettie Stevens, 1861–1912
[*Source: Courtesy of University Archives, Columbia University in the City of New York.*]

SECTION 2.2 SUMMARY

- Organisms are composed of cells, which have intricate intracellular structures, including chromosomes located in the nucleus.
- Cells that are not actively dividing contain two complete sets of unique chromosomes in the nucleus; they are diploid, or $2n$ (except gametes, which are haploid, or n).
- The cell cycle consists of four stages; G_1 phase, S phase (synthesis), G_2 phase, and M phase (mitosis). The chromosomes of a diploid cell are duplicated in S phase (during interphase) and then carefully segregated into two daughter cells during mitosis, which proceeds through four stages: prophase, metaphase, anaphase, and telophase. The resulting daughter cells are also diploid.
- Meiosis is a specialized type of cell division that halves the diploid chromosome number ($2n$) to produce haploid gametes (n), each containing one complete set of chromosomes.
- Haploid gametes unite during fertilization to reestablish the diploid state of the organism.
- Sex is determined by an accessory chromosome that is paired either with a similar chromosome or with a distinct, differently shaped chromosome, or has no partner at all. These special chromosomes are called sex chromosomes; all other chromosomes are autosomes.

2.3 THE CHROMOSOME THEORY OF INHERITANCE

Walter Sutton's studies on chromosomes were performed just as Mendel's work was being rediscovered. Sutton found himself at a remarkable intersection of two fields: cytology and genetics. He made the connection between chromosomes and Mendel's particles of heredity in his classic 1903 paper "The Chromosomes in Heredity" (see the How We Know section at the end of this chapter). He proposed that chromosomes contain Mendel's particles of heredity and that the particles come in pairs: chromosomes exist as homologous pairs in diploid cells. Mendel's particles—gene pairs—separate and assort independently into gamete cells; homologous chromosome pairs also separate and assort into haploid gamete cells.

Sutton's hypothesis that genes are located on chromosomes received much attention and became known as the **chromosome theory of inheritance**. But there was still no proof that genes were actually on chromosomes. This would be left for Thomas Hunt Morgan and his students to establish in their classic studies of fruit flies. Interestingly, Morgan did not initially believe in the chromosome theory of inheritance. But his experiments would inevitably lead him to this conclusion, and his name would become as linked to the chromosome theory as are genes themselves.

Sex-Linked Genes in the Fruit Fly Reveal That Genes Are on Chromosomes

In 1908, Morgan initiated his studies of the fruit fly, *Drosophila melanogaster* (see the Model Organisms Appendix). In those days, it was essential to keep costs to a minimum, as funding for science was scarce. The fruit fly is small; it could be grown in large numbers and was inexpensive to maintain. Flies also have an array of phenotypic features suitable for genetic studies, and they have just four homologous pairs of chromosomes that can be visualized under the microscope. Most important of all, the generation time of the fly is less than 2 weeks, and each female can lay hundreds of eggs. These features made fruit flies far superior to other model organisms of the day.

Thomas Hunt Morgan, 1866–1945 [*Source: Courtesy of the Archives, California Institute of Technology.*]

Morgan's famous Fly Room at Columbia University was small and cramped, but conducive to science; several of his students became famous for their discoveries in genetics. In 1910, Morgan noticed a male fly with white eyes that spontaneously appeared in a bottle of red-eyed flies. He immediately set up crosses to determine whether the white-eye trait was inheritable. It was, but in an unusual way. **Figure 2-14** shows Morgan's first experiment. Normal flies have red eyes, straight wings, and a gray body, referred to as **wild-type** traits.

Morgan crossed a red-eyed female with the mutant white-eyed male. All the F_1 hybrid progeny had red eyes, the wild-type phenotype. This told Morgan that the allele for red eyes is dominant to the allele for white eyes. His next cross, an F_1 female with an F_1 male, produced some F_2 progeny with white eyes—the expected result for a typical recessive allele. But surprisingly, all the white-eyed flies were male. All the F_2 females had red eyes, and about half the F_2 males had red eyes. It seemed that the trait for white eyes was somehow connected to sex. Morgan performed a variety of additional crosses and found, again to his surprise, that the white-eye trait mirrors the segregation behavior of the X chromosome

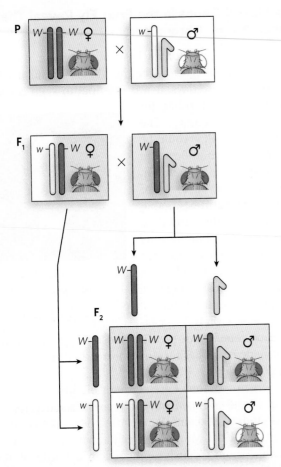

FIGURE 2-14 X-linkage of the white-eye allele. Results of a cross between a red-eyed female and a white-eyed male fly. F₁ flies are red-eyed. The white-eye trait reappears in the F₂ generation. Only males in the F₂ generation have white eyes. Morgan realized that these results make sense if the white-eye allele is located on the X chromosome. [*Source: Adapted from T. H. Morgan et al.,* Mechanism of Mendelian Heredity, *Henry Holt, 1915.*]

produce visible abnormalities in the chromosomes themselves. Bridges crossed white-eyed female flies (X^wX^w) with red-eyed males (X^WY) that he had in his fly collection. Most progeny were the expected white-eyed males and red-eyed females. However, Bridges noticed a few rare (<0.1%) white-eyed females and red-eyed males, which he called "primary exceptionals." Bridges made the unusual prediction that if genes are truly on chromosomes, then primary exceptional flies will have an abnormal chromosome number. He reasoned that the primary exceptional phenotype might be explained by defective meiosis in the female parent, in which X chromosomes did not separate, so producing an egg with two X chromosomes and an egg with no X chromosome (**Figure 2-15a**). Thus, exceptional white-eyed females, which must have two X^w chromosomes, received them from the abnormal X^wX^w egg, plus a Y chromosome from the sperm, for a genotype of X^wX^wY (note that an X^W sperm would bring in a dominant red-eye gene) (**Figure 2-15b**). By similar reasoning, the exceptional red-eyed male originated from fertilization of the abnormal egg having no X chromosome by a sperm containing a single X^W chromosome, for a genotype of X^WO. Note that although flies (like mammals) have X and Y chromosomes, sex in *D. melanogaster* is determined by the number of copies of the X chromosome, not by the presence or absence of the Y chromosome. Thus, an XXY fly is female and an XO fly is male.

When Bridges examined the chromosomes of primary exceptional flies, the results followed his predictions precisely (**Figure 2-15c**). His study was an impressive demonstration that genes are located on chromosomes, because, to explain the genetic results, he had hypothesized highly unusual outcomes that could be verified by examining the chromosomes directly. This abnormal assortment of chromosomes during meiosis is called **nondisjunction**.

Linked Genes Do Not Segregate Independently

Chromosomes, not individual genes, segregate into gamete cells, so one might expect two different genes on the same chromosome to stay together during meiosis and thus to be inherited together (i.e., they would not obey Mendel's second law). Take, for example, two genes, *A* and *B*, on the same chromosome. A cross of *AABB* and *aabb* parents will produce the *AaBb* F₁ hybrid, but particular combinations of alleles (*A, B* and *a, b*) are linked on the same chromosome. Therefore, the F₁ hybrid can produce only two types of gametes, *AB* and *ab*, rather than all four possible gametes produced if the genes separated and assorted randomly—*AB, Ab, aB, ab*.

To determine the genotype of an F₁ hybrid experimentally, it is crossed with a strain that is homozygous recessive (*aabb*), and the progeny reveal both the recessive and dominant alleles of the F₁ gametes. Such a cross is

(see Figure 2-14). Morgan's findings, linking a genetic trait to a particular chromosome, were convincing evidence that genes are located on chromosomes.

The alleles for this eye-color gene can be represented as X^W for red-eyed and X^w for white-eyed. In this genetic nomenclature, the *X* represents the X chromosome, and a superscript *W* is used for the dominant red-eye allele and a superscript *w* for the recessive white-eye allele. The letters *R* and *r* (for red and white, respectively), which might be expected from the convention introduced earlier in the chapter, are not used in this case because there are many different mutant alleles that affect eye color. There is only one wild-type color, so the wild-type and different mutant alleles are named according to the different mutant colors.

Further evidence that genes are located on chromosomes came from Calvin Bridges, an associate in Morgan's laboratory. Bridges hypothesized that if genes are located on chromosomes, then some genetic anomalies should also

(a)

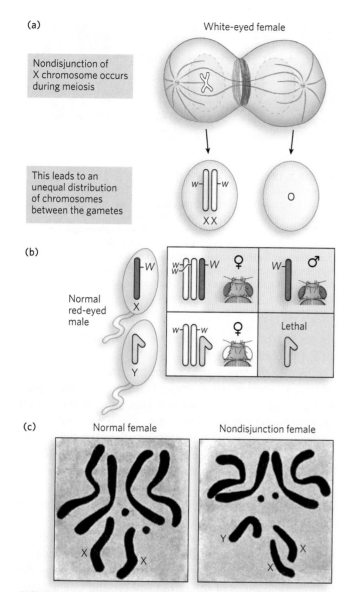

(b)

(c)

FIGURE 2-15 Nondisjunction. A rare occurrence during meiosis, nondisjunction produces gametes that have either one extra chromosome or one fewer chromosome. (a) Nondisjunction of the X chromosome is shown here; the white-eye allele is on the X chromosome. (b) Fertilization with normal sperm produces adult flies with odd numbers of chromosomes, as illustrated in the Punnett square. (c) Bridges predicted the occurrence of nondisjunction to explain rare progeny phenotypes, and cytologic examination of chromosomes in the rare progeny confirmed the predicted extra Y chromosome in a white-eyed female. [Source: (c) C. Bridges, Genetics 1:107–163, 1916.]

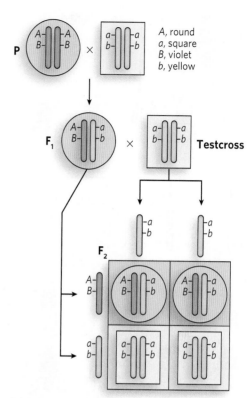

FIGURE 2-16 The inheritance of linked genes. Linked genes segregate together because they are on the same chromosome—that is, they are part of the same DNA molecule. In this hypothetical example, the dominant and recessive genes and their phenotypes are: A, round; a, square; B, violet; b, yellow. The cross between homozygous dominant and homozygous recessive parents produces F_1 AaBb progeny with linked alleles A, B and a, b. A testcross with a double-recessive homozygous individual (aabb) reveals the genotypes of the gametes produced by the F_1 progeny. The Punnett square shows the expected results for completely linked genes. The F_1 generation can produce only AB and ab gametes, and thus only two types of F_2 progeny are observed; they have the same phenotype as the original P generation.

known as a **testcross**. If the two genes separate in the gametes of the F_1 hybrid, the F_2 generation will exhibit all four possible phenotypes. If the two genes are linked, the F_1 hybrid will produce only two types of gametes (AB and ab) and the F_2 generation will display only the two parental phenotypes (**Figure 2-16**).

An example of linked genes in *Drosophila* is illustrated in **Figure 2-17**, for a body-color gene with alleles b (black body) and B (gray body), and a wing-shape gene with alleles v (vestigial wings) and V (long wings). Consider the parental cross BBvv (gray body, vestigial wings) × bbVV (black body, long wings). All F_1 progeny (BbVv) have a gray body and long wings. To determine whether the two genes are linked, a testcross is performed between an F_1 fly and a double-recessive bbvv fly. The F_2 progeny are mainly of two types and exhibit the same characteristics as the P generation (gray body, vestigial wings; black body, long wings). Thus, the two genes are linked. Had the genes assorted completely independently, mixed phenotypes would have been observed in the F_2 generation (black body, vestigial wings; gray body, long wings) in amounts equal to the parental phenotypes.

The results of the experiment, however, do not show complete gene linkage. There are some F_2 generation flies with mixed phenotypes, indicating that linked genes sometimes unlink. How can this happen—how do linked genes become unlinked?

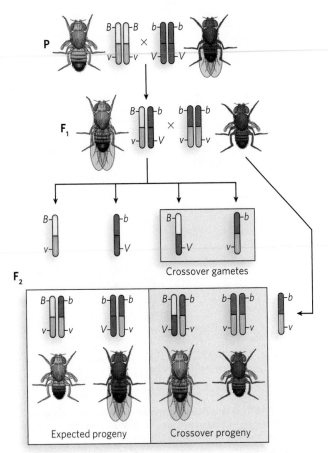

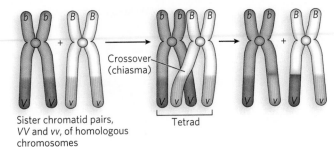

FIGURE 2-18 Crossing over in the tetrad. Two homologous chromosomes are duplicated in S phase to produce two sister chromatid pairs (left), one homozygous *VV* and the other homozygous *vv*. The sister chromatid pairs are homologous to each other, and they pair to form a tetrad in prophase I of meiosis, before the first cell division (middle). Recombination—as evidenced by the exchange of *V* and *v* alleles—occurs at crossovers, or chiasmata, the sites where chromosomes intertwine, resulting in genetic exchange between the two chromosomes of the sister chromatid pairs (right).

FIGURE 2-17 Unlinking genes by crossing over. Linked genes can become unlinked by chromosome recombination, or crossing over, during meiosis. The chromosomes containing the linked genes are illustrated in diploid cells and in gametes. To analyze gametes produced by F$_1$ flies, the F$_1$ hybrid is crossed with a double-recessive fly of genotype *bbvv*. In the F$_1$ gametes, *B* and *v* are linked, and *b* and *V* are linked, so all F$_2$ progeny are expected to contain these same two combinations. The double-recessive fly always contributes a *bv* gamete. But, in fact, four types of F$_2$ progeny are observed: two are the expected phenotypes; the other two contain *b*, *v* and *B*, *V*, resulting from gametes in which the linked alleles were unlinked by recombination during meiosis. The two crossover phenotypes are produced at equal frequency (17% each). [Source: Adapted from T. H. Morgan et al., Mechanism of Mendelian Heredity, Henry Holt, 1915.]

Recombination Unlinks Alleles

Morgan noticed that linked genes do not always stay linked, but instead show a low, though reproducible, frequency of separating. Take, for example, the cross of flies with linked genes shown in Figure 2-17. Linked genes should give only parental phenotypes in the F$_2$ progeny, yet a low frequency of mixed-phenotype F$_2$ progeny was observed. These **recombinant** flies could be produced only if the linked genes were unlinked and separated during gamete formation. Both possible types of mixed-phenotype recombinant flies were produced (black body, vestigial wings; gray body, long wings) and

appeared with equal frequency: each was 17% of the total F$_2$ population.

To explain how linked genes become unlinked, and why they produce equal amounts of the two mixed phenotypes, Morgan hypothesized that one of the linked alleles on one chromosome (e.g., the long-wing allele) trades places with the homologous allele (vestigial-wing allele) on the homologous chromosome (**Figure 2-18**). In other words, genes hop from one homologous chromosome to the other and do so in a reciprocal fashion. This reciprocal exchange of alleles between chromosomes is called **recombination**, or **crossing over**.

The idea that chromosomes exchange genetic material had been suggested earlier, in cytological studies by F. A. Janssens in 1909. Janssens noticed that during meiosis, the four chromosomes of the tetrad coil around one another and form cross-shaped junctions, which he called **chiasmata** (see Figure 2-18). He proposed that, as the mechanical forces pull the sister chromatid pairs apart during the first division of meiosis, the intertwined chromosomes break at the same place and then rejoin, but with the opposite chromosome. The first experimental proof that genetic crossing over is mediated by physical recombination between two chromosomes came from a study of corn by Barbara McClintock and Harriet Creighton (see the How We Know section at the end of this chapter).

We now know that recombination events are mediated by specialized proteins that catalyze DNA breakage and rejoining within homologous chromosomes of the tetrad. Crossing over is a frequent event during meiosis (**Figure 2-19**), occurring at least once in each tetrad. It is thought that meiotic recombination was selected for during evolution because it helps generate diversity within a species. Homologous recombination is discussed in detail in Chapter 13.

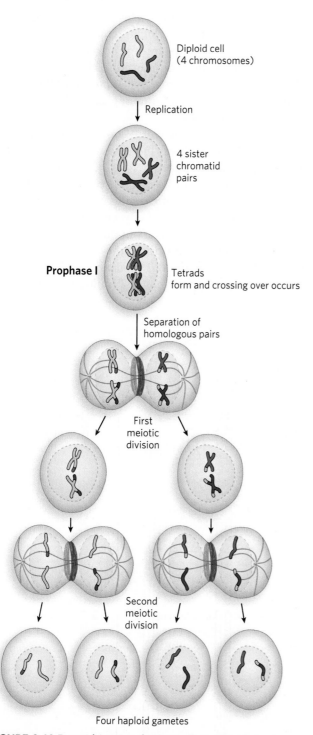

Prophase I

FIGURE 2-19 Recombination during meiosis. The chromosomes of a diploid cell (four chromosomes, two homologous pairs, are shown here) replicate, and each pair is held together at the centromere, forming four sister chromatid pairs. In prophase I, at the start of the first meiotic division, the two homologous sets of sister chromatid pairs align to form tetrads. Crossovers occur within the tetrads. In the first meiotic division, homologous pairs of chromosomes segregate into daughter cells. Each sister chromatid pair then lines up in preparation for the second meiotic division, which produces four haploid gamete cells. Each gamete has two chromosomes, half the number of the diploid cell. *[Source: Adapted from D. L. Nelson and M. M. Cox,* Lehninger Principles of Biochemistry, *5th ed., W. H. Freeman, 2008, Fig. 25-31.]*

Recombination Frequency Can Be Used to Map Genes along Chromosomes

Different pairs of linked genes exhibit different frequencies of crossing over, but the frequency is constant for a given pair of genes. Alfred Sturtevant, a student of Morgan's, rationalized this observation by assuming that the frequency of crossing over corresponds to the distance between the two linked genes. He reasoned that the greater the distance, the more room there is for recombination to occur, thereby allowing linked genes to separate with greater frequency. With this logic, he used the frequency of crossing over to map the relative positions of pairs of linked genes along *Drosophila* chromosomes (**Figure 2-20**). Genetic map units, calculated from the

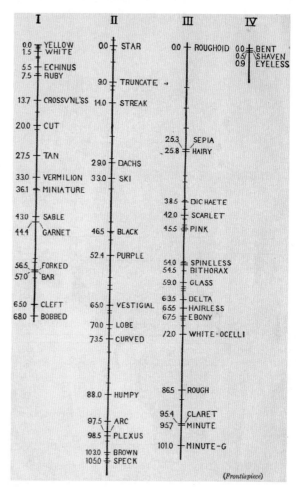

FIGURE 2-20 Using recombination frequency to create genetic maps. Sturtevant created genetic maps showing the positions of genes along the four *Drosophila* chromosomes, based on the frequency of crossing over between many pairs of linked genes. Linked genes fall into four groups, corresponding to the four different chromosomes in *Drosophila*, represented here by the vertical lines. Numbers on the left side of each chromosome are genetic map units, in centimorgans. Along the right side are the names of mutant alleles used in the crosses. *[Source: From The Mechanism of Mendelian Heredity by T. H. Morgan, A. H. Sturtevant, H. J. Muller, C. B. Bridges, 1915, 1922 Henry Holt and Company, NY.]*

frequency of crossing over, are called centimorgans (cM) in honor of Thomas Hunt Morgan; however, they do not necessarily reflect accurate physical distances between genes. Some regions of chromosomes tend to promote recombination, giving the impression that genes are farther apart than they really are; conversely, other regions repress crossing over, and genes seem to be closer than they are. The accuracy of genetic map distances is also limited by one crossing-over event interfering with another. However, recombination frequencies do provide useful genetic maps, because the data reveal the linear order of genes along a chromosome and provide a first approximation of the distance between them.

An example of **recombination mapping** is illustrated in **Figure 2-21** for linked genes A, B, and C. Consider the frequency of crossing over of the linked gene pair A and B in fruit flies (Figure 2-21a). A parent

homozygous for dominant alleles is crossed with a double-recessive fly ($AABB \times aabb$), and the frequency of crossing over (the frequency of production of Ab and aB gametes by the F_1 flies) is determined from the percentage of recombinant F_2 progeny ($Aabb$ and $aaBb$). This is repeated for the A, C pair and B, C pair. The results are shown in Figure 2-21b.

The greater frequency of recombinants for the A, B pair than for the A, C pair indicates that genes A and B are farther apart than genes A and C. However, gene C could be between A and B or on the opposite side of A from B. The frequency of crossing over of the B, C pair resolves the ambiguity: C is between A and B.

The frequency of recombinants for the A, B pair (26%) is somewhat less than the added frequencies of recombinants for the B, C pair and C, A pair (28%). This is because the probability of multiple crossing-over

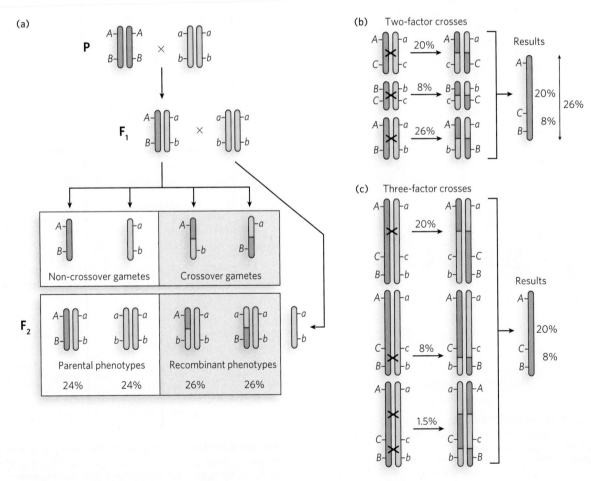

FIGURE 2-21 Recombination mapping. (a) The procedure for analyzing a two-factor cross. Diploid cells and gametes are illustrated to show the origin of recombinant F_2 progeny. Chromosomes with dominant linked genes are purple; chromosomes with recessive linked genes are yellow. Crossing over results in purple-and-yellow hybrid chromosomes. F_1 progeny are crossed with a homozygous double-recessive fly to analyze the genotypes of gametes produced by the F_1 hybrid. (b) Analysis of linked genes using three two-factor crosses. Crossing over is indicated by × between two chromosomes. (c) Analysis of linked genes using three three-factor crosses. Two- and three-factor crosses lead to the same conclusion about gene order (ACB).

events between linked genes is higher the farther they are apart. For example, a single crossover unlinks the genes, but a second crossover links them again. Therefore, an odd number of crossovers will unlink genes, and an even number of crossovers relinks them, resulting in a maximum frequency of recombination of 50%. The frequency of independent assortment of genes on different chromosomes is also 50%, because there is a 50:50 chance that two chromosomes will segregate together into the same gamete. Because crossing over is a frequent occurrence in meiosis, genes on the same chromosome often assort independently. Therefore, recombination mapping is accurate only for pairs of linked genes that are close together.

Analysis of three genes in a single experiment, known as a three-factor cross, provides a convenient method to identify or confirm their order along the chromosome. To illustrate this, consider a three-factor cross between genes *A*, *B*, and *C* (Figure 2-21c). A fly that is homozygous dominant for three linked genes is crossed with a fly that is double recessive for all three genes. The F_1 progeny (*AaBbCc*) are then crossed with a fly that is double recessive for all three genes. Most F_2 progeny exhibit the parental phenotypes, but crossing over will produce six possible recombinants: three recombinants containing two dominant traits and three reciprocal recombinants containing one dominant trait. If the gene order is *ACB*, generation of the *AcB* and *aCb* recombinants requires two crossover events—one between *A* and *C*, and another between *C* and *B*. The *aCB* (and *Acb*) or *ACb* (and *acB*) recombinants each require only one crossover. Because a double crossover is much less frequent than a single crossover, the far lower frequency of the double crossover (1.5% in this example, yielding *AcB* and *aCb*) reveals which gene (*C* in this case) is between the other two.

SECTION 2.3 SUMMARY
- Direct evidence that genes are located on chromosomes came from intensive studies of the fruit fly, Drosophila melanogaster, by Thomas Hunt Morgan. Segregation of the white-eye mutant allele with the X chromosome suggested that genes are associated with chromosomes.
- Calvin Bridges's correlation of mutant genes with chromosome abnormalities showed definitively that genes are located on chromosomes.
- Linked genes, genes on the same chromosome, violate Mendel's second law and assort together into gametes. However, linked genes must be close together on the chromosome to stay linked. The farther apart they are, the more likely they are to be separated by recombination during meiosis.
- Recombination frequency can be used to map the relative positions of genes along a chromosome.

2.4 FOUNDATIONS OF MOLECULAR GENETICS

The union of genetics and cytology in the early 1900s was an exceedingly productive time. Heredity was based in genes, which were located on chromosomes. But what are genes made of? To some scientists, genes were almost unreal, a mental construct to explain real phenomena. We now know, of course, that genes are made of DNA. In fact, DNA was discovered decades before its significance was understood. The recognition of DNA as the genetic material, and the solution of its chemical and three-dimensional structure, brought genetics out of the realm of imagination and into the realm of chemistry. These discoveries sparked the fusion of chemistry and genetics to give us an entirely new scientific discipline: **molecular genetics** or, more generally, **molecular biology**. In this section, we outline some discoveries that led to our current understanding of DNA as the repository of biological information. We also describe how the information in DNA is translated into functional RNAs and proteins, and how this knowledge furthers our understanding of human health and disease.

DNA Is the Chemical of Heredity

Deoxyribonucleic acid (DNA), as we have noted, was identified long before its importance was recognized. The history begins with Friedrich Miescher, who carried out the first systematic chemical studies of cell nuclei in 1868. Miescher obtained white blood cells from pus that he collected from discarded surgical bandages. He carefully isolated the nuclei and then ruptured the nuclear membranes, releasing an acidic, phosphorus-containing substance that he called *nuclein*. Nuclein, a **nucleic acid**, was a new type of chemical polymer, different from all others previously identified. Around the turn of the century, Albrecht Kossel investigated the chemical structure of nucleic acids—both DNA and a similar molecule called **ribonucleic acid (RNA)**—and found that they contain nitrogenous bases, or nitrogen-containing basic compounds. Kossel identified five types of nitrogenous bases: adenine (A), guanine (G), cytosine (C), thymine (T), and uracil (U) (described in Chapters 3 and 6). By 1910, other investigators had determined that nitrogenous bases were but one component of a larger unit called a **nucleotide**, which consists of a phosphate group, a pentose sugar, and a nitrogenous base. DNA is composed of nucleotides that contain the bases A, G, C, or T, and RNA is composed of nucleotides with the bases A, G, C, or U.

By the 1920s, the chemical basis of heredity was thought to lie in chromosomes, but chromosomes are composed of both DNA and protein. Which one is the

Oswald Avery, 1877–1955
[Source: Science Source.]

Frederick Griffith, 1879–1941
[Source: Science Source.]

an observation made in 1928 by English microbiologist Frederick Griffith, who studied the pneumonia-causing bacterium *Streptococcus pneumoniae*. This pneumococcus exists as two types, virulent (disease-causing) and nonvirulent. Griffith noticed that virulent bacteria produced smooth colonies when grown on Petri plates, but colonies of a nonvirulent strain appeared rough. The difference in appearance lies in a polysaccharide capsule coat present only on the virulent strains. Griffith found that heat-killed virulent bacteria transformed live nonvirulent bacteria into live virulent bacteria (**Figure 2-22a–d**). The nonvirulent bacteria somehow acquired the smooth-colony trait from the heat-killed bacteria. The results suggested that the genetic material coding for capsules remained intact even after the virulent (smooth-colony) bacteria were killed, and this material could enter another cell and recombine with its genetic material.

hereditary chemical? DNA was initially ruled out because it was believed to be too simple—just a repeating polymer of four different nucleotides. Surely such a monotonous molecule lacked the complexity to code for the working apparatus of a living cell! Attention turned to the other biopolymer, protein, for a chemical explanation of heredity. Only after biochemical studies conducted in the 1940s pointed to DNA as the genetic molecule was attention refocused on the structure and function of this molecule.

DNA was shown to be the chemical of heredity in the 1940s, by Oswald T. Avery and his colleagues at the Rockefeller Institute in New York City. Their starting point was

Avery and his colleagues reproduced Griffith's results, and they analyzed the heat-killed virulent bacterial extract for the chemical nature of the transforming factor. They selectively removed either DNA, RNA, or protein from the heat-killed virulent bacterial extract by treatment with DNase, RNase, or proteases (enzymes that specifically break down one of these cellular components, leaving the others intact). The DNase-treated extract lost the capacity to transform nonvirulent rough-colony cells into a virulent smooth-colony strain. The researchers then extracted DNA from virulent bacteria, purified it of contaminating proteins and RNA, and

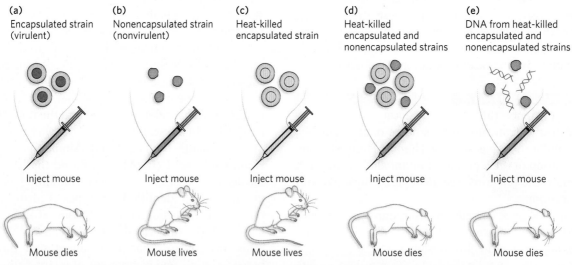

(a) Encapsulated strain (virulent)

(b) Nonencapsulated strain (nonvirulent)

(c) Heat-killed encapsulated strain

(d) Heat-killed encapsulated and nonencapsulated strains

(e) DNA from heat-killed encapsulated and nonencapsulated strains

Inject mouse   Inject mouse   Inject mouse   Inject mouse   Inject mouse

Mouse dies   Mouse lives   Mouse lives   Mouse dies   Mouse dies

FIGURE 2-22 Transformation of nonvirulent bacteria to virulent bacteria by DNA. When injected into mice, (a) the encapsulated strain of pneumococcus (*Streptococcus pneumoniae*), producing smooth colonies, is lethal, whereas (b) the nonencapsulated strain, producing rough colonies, and (c) the heat-killed encapsulated strain are harmless. (d) Griffith's research had shown that adding heat-killed virulent bacteria to a live nonvirulent strain (each harmless to mice on their own) permanently transformed the live strain into lethal, virulent, encapsulated bacteria. (e) Avery and his colleagues extracted the DNA from heat-killed virulent pneumococci, removing RNA and protein as completely as possible, and added this DNA to nonvirulent bacteria, which were permanently transformed into a virulent strain.

showed that this pure DNA was still capable of transforming nonvirulent bacteria into the virulent strain (**Figure 2-22e**). In 1944, Avery and colleagues reported their surprising conclusion that DNA was the carrier of genetic information. Another classic experiment, by Alfred Hershey and Martha Chase, supported this conclusion that DNA is the chemical of heredity (see the How We Know section at the end of this chapter).

Genes Encode Polypeptides and Functional RNAs

DNA and protein are chemically very different, and it was puzzling how a DNA sequence could code for a protein sequence. Regardless of the details, however, it now became easy to understand that mutations in a gene could lead to altered enzymes. In fact, even before the DNA structure was solved, the relationship between genes, mutations, and enzymes was well understood.

In 1902, the physician Archibald Garrod studied patients with alkaptonuria, a disease of little consequence for the patients, except that they excreted urine that turned black. Mendel's work had recently been rediscovered, and by noticing how alkaptonuria was inherited, Garrod realized that this disorder behaved as a recessive trait. It was already known that the synthesis and breakdown of biomolecules occur in multistep pathways, each step requiring a different **enzyme**—a protein catalyst that facilitates the reaction. Garrod hypothesized that alkaptonuria was caused by a mutation that inactivated a gene required for the production of one enzyme in a metabolic pathway. Without this functional enzyme, the pathway was blocked, resulting in the build-up of an intermediate compound, which was excreted in the urine and turned black. Garrod's reasoning drew the connection between a mutation in a gene and a mutation in an enzyme.

Formal proof that genes encode enzymes came from a series of elegant experiments in the 1940s by George Beadle and Edward L. Tatum. They introduced a new microorganism into the study of genetics: the bread mold, *Neurospora crassa* (see the Model Organisms Appendix). This haploid organism can grow on a simple, defined medium, called minimal medium. Minimal medium contains sugar, nitrogen, inorganic salts, and biotin, and the cell must make all the rest of the biochemicals that it needs to live from these simple starting compounds. Beadle and Tatum irradiated *Neurospora* spores to intentionally produce mutations, then germinated individual spores on a complete medium (i.e., one made with cell extracts that have all the necessary amino acids, nucleotides, and vitamins) to obtain genetically pure colonies and their spores. These spores were then tested for their ability to germinate on the minimal medium.

George Beadle, 1903–1989
[Source: Courtesy of the Archives, California Institute of Technology.]

Edward Tatum, 1909–1975
[Source: American Stock Archives / Getty Images.]

An inability to grow on minimal medium indicates a mutation in one of the metabolic pathways required for growth. These mutants are called **auxotrophs**. Spores of different auxotrophs were then analyzed for growth on a range of minimal media supplemented with selected compounds, to identify the defective metabolic pathways and the steps affected. An example of one such study is illustrated in **Figure 2-23**, for auxotrophs of arginine metabolism.

Beadle and Tatum had a collection of *Neurospora* mutants that were auxotrophic for the amino acid arginine. The arginine synthetic pathway was known to include the intermediate compounds ornithine and citrulline, so they tested their arginine auxotrophs for growth on minimal media containing ornithine, citrulline, or arginine. The arginine auxotrophs fell into three classes, depending on which intermediate(s) they required for growth (see Figure 2-23). Beadle and Tatum also mapped the mutant genes and found that mutants in a particular class of auxotrophs mapped to the same chromosomal location. They concluded that each class of mutant was caused by a single defective gene. Their findings also held true for genes in other metabolic pathways.

On the basis of these experiments, Beadle and Tatum proposed the *one gene, one enzyme* hypothesis, which stated that each gene codes for one enzyme. We now know that some enzymes are composed of multiple subunits encoded by different genes; furthermore, not all proteins are enzymes. So, the hypothesis was later revised to *one gene, one polypeptide*. A **polypeptide** is a chain of amino acids, and a functional protein can be composed of a single polypeptide or multiple polypeptide subunits. For a large number of genes, *one gene, one polypeptide* holds true. But as we will see throughout this textbook, even this hypothesis is not entirely accurate. Some genes code for functional RNAs rather than for protein. And through a process called alternative splicing (see Chapter 16), some genes code for more than one type of polypeptide.

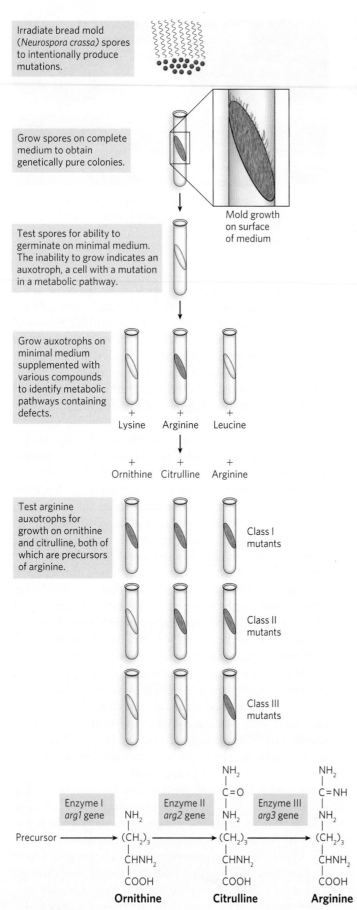

Irradiate bread mold (*Neurospora crassa*) spores to intentionally produce mutations.

Grow spores on complete medium to obtain genetically pure colonies.

Mold growth on surface of medium

Test spores for ability to germinate on minimal medium. The inability to grow indicates an auxotroph, a cell with a mutation in a metabolic pathway.

Grow auxotrophs on minimal medium supplemented with various compounds to identify metabolic pathways containing defects.

+ Lysine   + Arginine   + Leucine

+ Ornithine   + Citrulline   + Arginine

Test arginine auxotrophs for growth on ornithine and citrulline, both of which are precursors of arginine.

Class I mutants

Class II mutants

Class III mutants

Enzyme I *arg1* gene   Enzyme II *arg2* gene   Enzyme III *arg3* gene

Precursor

$$NH_2$$
$$|$$
$$(CH_2)_3$$
$$|$$
$$CHNH_2$$
$$|$$
$$COOH$$
Ornithine

$$NH_2$$
$$|$$
$$C=O$$
$$|$$
$$NH_2$$
$$|$$
$$(CH_2)_3$$
$$|$$
$$CHNH_2$$
$$|$$
$$COOH$$
Citrulline

$$NH_2$$
$$|$$
$$C=NH$$
$$|$$
$$NH_2$$
$$|$$
$$(CH_2)_3$$
$$|$$
$$CHNH_2$$
$$|$$
$$COOH$$
Arginine

The Central Dogma: Information Flows from DNA to RNA to Protein—Usually

Watson and Crick's determination of DNA structure was a turning point in understanding how information flows in biological systems. Their model of DNA structure, which they reasoned from data collected by other scientists— most notably Rosalind Franklin—consists of two strands of DNA wound about one another in a spiral, double helix. Each strand is composed of a long string of the four nucleotides containing the bases adenine (A), guanine (G), cytosine (C), and thymine (T). The nucleotides in one strand pair with those in the other. Because A pairs only with T, and G pairs only with C, the sequence of each strand contains information about the sequence of the other, and the two strands are said to be **complementary**. The A–T and G–C pairs are referred to as **base pairs**. The detailed structure of DNA and the nucleotide bases, how the nucleotides base-pair in a specific way, and the critical contributions Rosalind Franklin made to Watson and Crick's structure are described in Chapter 6.

The double-helical DNA structure immediately suggested a mechanism for the transmission of genetic information. The essential feature of the model is the complementarity of the two DNA strands. As Watson and Crick realized well before confirmatory data became available, DNA could logically be replicated by separating the two strands and using each as a template to synthesize a new, complementary strand, thereby generating two new DNA duplexes that are identical to each other and to the original double-stranded DNA.

With discovery of the DNA structure, genetics could now be described in chemical terms. Both DNA and proteins are linear polymers, so the sequence of nucleotides in DNA must somehow be converted to a sequence of amino acids. But DNA is located in the nucleus, whereas proteins are synthesized in the cytoplasm. Therefore, there must be an intermediary molecule to shuttle information between

FIGURE 2-23 "One gene, one polypeptide" analysis of a *Neurospora crassa* auxotroph. Beadle and Tatum identified mutant *Neurospora* that were unable to synthesize the amino acid arginine. To investigate the metabolic pathway of arginine synthesis, they analyzed arginine auxotrophs for growth on minimal medium plus ornithine or citrulline, both precursors of arginine (or on minimal medium plus arginine, to be sure that the mutant grew when supplied with arginine). They found that class I mutants grow when supplied with any of the three compounds, so these mutants lack an enzyme that is upstream of these three compounds (i.e., an enzyme catalyzing an earlier reaction) in the synthetic pathway. Class II mutants do not grow on ornithine, and thus lack an enzyme downstream of this intermediate but upstream of citrulline. Class III mutants grow only on arginine and therefore lack an enzyme involved in the conversion of citrulline to arginine.

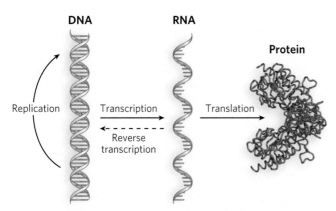

FIGURE 2-24 Crick's central dogma of information flow: DNA→RNA→protein. The information to replicate DNA is inherent in its structure (curved arrow). Information flows from DNA to RNA by transcription. Information flows from RNA to protein by translation. In some instances, information can also flow backward, from RNA to DNA (reverse transcription), and some viruses encode enzymes that produce RNA from RNA (not shown in the figure). Other types of information flow exist that do not fall within Crick's central dogma.

the two locations. RNA was believed to play a role in this, and the similarities between DNA and RNA made it a simple matter to understand how an RNA molecule could be made from a DNA template. Crick proposed that biological information flows in the direction DNA→RNA→protein and that DNA acts as a template for its own synthesis (DNA→DNA) (**Figure 2-24**). Crick's proposal is known as the **central dogma** of information flow.

Over the years, it has become obvious that the nice and tidy linear flow of information in Crick's central dogma is not really all that simple after all. Several different paths of information flow are now known to exist. Among these different pathways are the ability of some enzymes to synthesize DNA from RNA (RNA→DNA) and the ability of some viruses to use RNA as a template to make more RNA (RNA→RNA). But the most profound change in what we know about information flow is the finding that the cell makes a huge amount of RNA that is not translated into protein, and it is not just tRNA and rRNA (whose functions we describe below). Indeed, much of the mammalian genome is transcribed into RNA that does not code for protein. We discuss this topic briefly at the end of this section.

RNA was widely expected to be the molecule that mediates the transfer of information from DNA in the nucleus to the site of protein synthesis in the cytoplasm. However, no one imagined that three different types of RNA would be required for the process.

Ribosomal RNA In the early 1950s, Paul Zamecnik and his colleagues identified the site of protein synthesis as particles in the cytoplasm called **ribosomes**. Ribosomes

are large structures composed of both protein and RNA. The RNA component is called **ribosomal RNA (rRNA)**. In bacteria and eukaryotes, ribosomes consist of a large subunit and a small subunit.

Messenger RNA The combined findings that ribosomes are the site of protein synthesis and that rRNA is the most abundant RNA (>80%) in the cell led most researchers to believe that rRNA was the carrier of information from DNA to protein. However, some features of rRNA are incompatible with its function as an information carrier. For example, rRNA is an integral part of the ribosome, so there would have to be specific ribosomes to make each specific protein. Further, the nucleotide composition of rRNAs from different organisms was relatively constant, whereas the nucleotide composition of chromosomal DNA varied considerably from one organism to the next.

Studies by Sydney Brenner, Jacques Monod, and Matthew Meselson in the early 1960s, using *Escherichia coli* (see the Model Organisms Appendix), suggested that another type of RNA carries the message from DNA to protein. They discovered a class of RNA that targets preexisting ribosomes, and the nucleotide composition of this RNA was more similar to chromosomal DNA than was rRNA. These properties are exactly those expected for a true messenger between DNA and protein. The investigators called this RNA **messenger RNA (mRNA)** and concluded that ribosomes are protein-synthesizing factories that use mRNA as a template to direct construction of the protein sequence.

RNA synthesis is carried out by the enzyme **RNA polymerase**, which synthesizes RNA by reading one strand of the duplex DNA, pairing RNA bases to the bases in the DNA strand, to synthesize a single-stranded RNA molecule that has a sequence directed by the DNA sequence (**Figure 2-25**). This process of making single-stranded RNA copies of a DNA strand is known as **transcription**.

Transfer RNA The discovery of mRNA was a crucial piece of the information puzzle. But a problem remained: how is a sequence of nucleotides in mRNA converted to a sequence of amino acids in protein? Furthermore, DNA and RNA each consist of only four different nucleotides, whereas proteins have 20 different amino acids. Hence, one must assume the existence of a code that uses combinations of nucleotides to specify amino acids. Combinations of two nucleotides yield only 16 permutations (4^2). Combinations of three nucleotides yield 64 permutations (4^3), more than enough to specify a code for 20 amino acids.

In 1955, Crick hypothesized the existence of an adaptor molecule, perhaps a small RNA, that could read three nucleotides and also carry amino acids. It was not

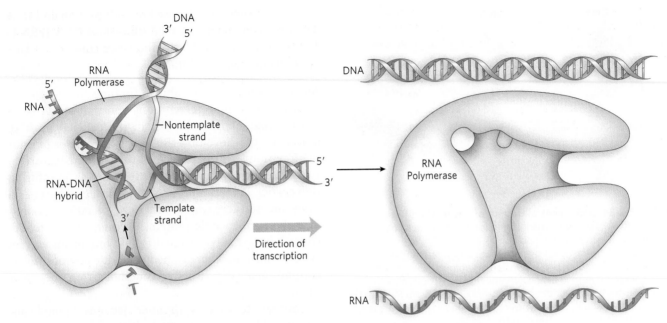

FIGURE 2-25 The process of transcription. RNA polymerase opens the DNA duplex and uses one strand as a template for RNA synthesis. The polymerase matches incoming nucleotides to the DNA template strand by base pairing and joins them together to form an RNA chain. As RNA polymerase advances along the template strand, the two DNA strands reassociate behind it to re-form the double helix. When the gene has been completely transcribed, the polymerase dissociates from DNA, releasing the completed RNA transcript. The 3′ and 5′ ends of the DNA are labeled.

long after Crick's adaptor hypothesis (see Chapter 17) that Paul Zamecnik and Mahlon Hoagland discovered a small RNA to which amino acids could attach. This small RNA, later called **transfer RNA (tRNA)**, was the adaptor between nucleic acid and protein.

The discovery of tRNA, combined with the idea of a three-letter code, suggested how the DNA sequence could be converted to an amino acid sequence. Three bases in the tRNA form base pairs with a triplet sequence in the mRNA. When two amino acid–linked tRNAs align side-by-side on the mRNA by base-pairing to adjacent triplets, the amino acids attached to the tRNAs can be joined together. By continuing this process over the length of an mRNA strand, amino acids carried to the mRNA by tRNAs become connected together in a linear order specified by the mRNA sequence. These connections occur as the mRNA-tRNA complexes thread through the ribosome. The overall process of protein synthesis, involving three different types of RNA molecules, is known as **translation (Figure 2-26)**. (Translation is covered in detail in Chapter 18.)

Functional RNAs All RNAs, whether they code for protein or not, are transcribed from DNA genes. Messenger RNA is needed only transiently, to instruct the synthesis of proteins. But the end products of tRNA and rRNA genes are the RNA molecules themselves. These **functional RNAs** fold into specific three-dimensional shapes and constitute the majority of the RNA in a cell.

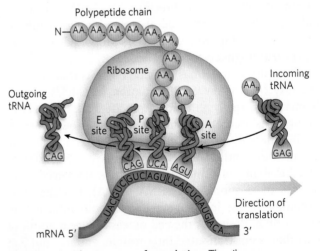

FIGURE 2-26 The process of translation. The ribosome, composed of a large and a small "subunit" (each consisting of many proteins and several rRNAs), mediates protein synthesis in cells. It associates with both mRNA and tRNAs as it synthesizes polypeptide chains. The ribosome has three major sites for binding tRNA molecules, the P site, the A site, and the E site. Two tRNAs form base pairs with their respective, adjacent, matching triplets on the mRNA: the tRNA in the P site carries the growing polypeptide chain, and the tRNA in the A site carries an amino acid (AA). The ribosome catalyzes the transfer of the polypeptide attached to the P-site tRNA to the amino acid on the A-site tRNA. The ribosome then shifts relative to mRNA so that the A-site tRNA, now holding the polypeptide, moves into the P site and the tRNA previously at the P site moves to the E site, from which it will depart after the next cycle. The next tRNA carrying an amino acid then binds to the vacated A site to continue extending the polypeptide chain. (N indicates the amino-terminal end of the polypeptide.)

There is an abundance of other functional RNAs besides tRNA and rRNA, some of which have known functions. For example, some small nuclear RNAs (<150 nucleotides) associate with protein to form ribonucleoprotein particles that process the introns from mRNAs (see Chapters 16 and 22). Other types of small RNA, the microRNAs (**miRNA**) and the short interfering RNAs (**siRNA**), have important gene regulatory functions. MicroRNAs anneal to particular mRNAs, usually causing their degradation and thereby effectively turning off, or silencing, the gene (see Chapter 22). Perhaps the most mysterious of the non-protein-coding RNAs are the "long noncoding RNAs." These RNAs (>200 nucleotides) are not translated and have no known function, yet they are more abundant than translated mRNA. There is accumulating evidence that at least some members of this abundant class of RNA may be needed for proper cell function. Identification and understanding of the function of these new RNAs is a fast-paced field, and new types of RNA are almost certain to be discovered in the near future.

There are yet other types of information flow, besides the use of RNAs, that fall outside the classic central dogma. Notable among these is the epigenetic control of gene regulation, based in specific chemical modifications of particular nucleotides and of the proteins that package DNA (in structures called nucleosomes). Combinations of these chemical changes can program the transcriptional control of a cell and can be inherited in cell divisions during an organism's development (see Chapter 10). This *epigenetic inheritance* falls outside the domain of DNA sequence. There are also many other types of protein modifications that transduce the flow of information within the cell and from one cell to another. Suffice it to say that the new dogma is that there is no simple "central dogma." The flow of information is so vital to life and evolution that it takes many forms, some hard to recognize, and scientists have no shortage of work ahead of them to elucidate these important mechanisms.

Mutations in DNA Give Rise to Phenotypic Change

Most cellular functions are carried out by proteins. The precise sequence of amino acids in each protein molecule and the specific rules governing the timing and quantity of its production are programmed into an organism's DNA. When changes in the DNA sequence occur, cellular function can be altered. Mutations in DNA can be beneficial or harmful to an organism, or can have no effect at all. For example, if the mutation does not change the sequence of a protein or how the protein is regulated, the mutation has no effect and is said to be silent. Evolution depends on mutations that are beneficial, and these usually alter the sequence or regulation of a protein in a way that enhances its function or confers a new, beneficial function that increases the viability of the organism. However, most mutations that change a protein sequence are harmful, because they lead to altered proteins with decreased function or new, detrimental function, and give rise to various diseases. When these DNA mutations occur in germ-line cells (cells that give rise to gametes), the disease can be inherited. There are many examples of inherited diseases, some of which have altered the course of history. One such disease is hemophilia.

Hemophilia afflicted the interrelated royal families of England, Russia, Spain, and Prussia in the 1800s. At the root of this malady is an inability of the blood to clot, resulting in excessive bleeding from even the slightest injury. The disease typically results in death at an early age. Tracing hemophilia through the royal families of Europe indicates that it originated with Queen Victoria (**Figure 2-27a**). It is interesting that none of the current family members are carriers, presumably the result of natural selection against this trait.

Hemophilia is about 10,000 times more common in males than in females. This is because the blood-clotting factor involved in 90% of cases of this disease is factor VIII, encoded by a gene on the X chromosome. A mutant recessive allele of the factor VIII gene is responsible for hemophilia A, the most common form of the disease. Males have only one copy of the X chromosome, and the recessive allele, when inherited, is always expressed. Females have two X chromosomes, and if one X contains a wild-type allele, it masks the expression of the mutant recessive allele. A female with only one copy of the recessive allele is called a carrier, because she is phenotypically normal but may pass on this allele to her offspring (**Figure 2-27b**).

Many other inherited diseases have been mapped to their particular genes. One of the first to be identified was the gene involved in Huntington disease (**Figure 2-28 on p. 51**). The gene, *HTT*, is located on chromosome 4. The disease is associated with a region of the *HTT* gene that can have a variable number of repeats of the triplet nucleotide sequence CAG (encoding the amino acid glutamine). The *HTT* gene in healthy individuals has about 27 or fewer of these repeats, but when the number exceeds 36, it is often associated with disease. The likelihood of having Huntington disease increases with the number of trinucleotide repeats in the *HTT* gene. The function of the protein encoded by *HTT* is unknown, but the disease results in the degeneration of neurons in areas of the brain that affect motor coordination, memory, and cognitive function.

(a)

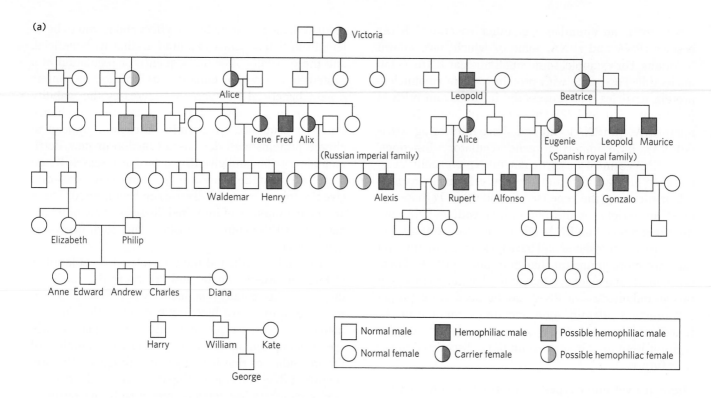

(b)

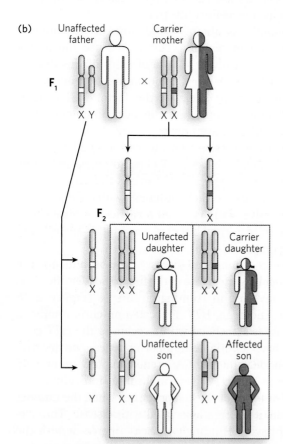

FIGURE 2-27 The inheritance of hemophilia. (a) The hereditary pattern of hemophilia in the royal families of Europe reveals that it is a recessive X-linked disease. (b) Because females have two copies of the X chromosome, they can carry one copy of the mutant gene for hemophilia without exhibiting the disease; they have hemophilia only if both X chromosomes carry the mutant gene. Male offspring, having only one X chromosome, are more likely to have the disease; hemophilia occurs about 10,000 times more frequently in males than in females.

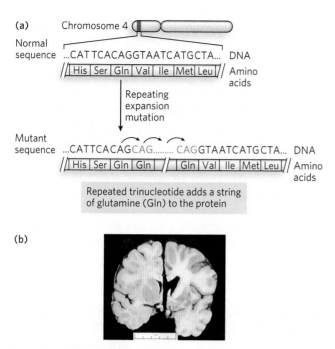

(a)

Chromosome 4

Normal sequence ...CATTCACAGGTAATCATGCTA... DNA

His | Ser | Gln | Val | Ile | Met | Leu / Amino acids

Repeating expansion mutation

Mutant sequence ...CATTCACAGCAG......... CAGGTAATCATGCTA... DNA

His | Ser | Gln | Gln / Gln | Val | Ile | Met | Leu / Amino acids

Repeated trinucleotide adds a string of glutamine (Gln) to the protein

(b)

FIGURE 2-28 Huntington disease. Huntington disease is an inherited autosomal dominant neurological disease. (a) The gene for Huntington disease (*HTT*) is located on the short arm of chromosome 4, and the disease is associated with CAG repeats (CAG encodes glutamine) in this gene. When the number of CAG repeats increases above 36 copies, disease may occur in midlife. (b) Huntington disease affects the brain by causing degeneration. [*Source: (b) Biophoto Associates/Science Source.*]

The number of triplet repeats in *HTT* can increase during gamete production, resulting in earlier onset and increased severity of the disease over successive generations. This is thought to occur by template slippage (the same segment of DNA replicated more than once) during DNA synthesis due to the repetitive nature of the sequence. Other diseases caused by "triplet expansion" of this type have now been identified. These include Kennedy disease, spinocerebellar ataxia, and Machado-Joseph disease, all caused by an increase in CAG repeats. The CGG repeat is associated with fragile X syndrome, a neurological disorder; expansion of the CTG repeat is associated with myotonic dystrophy, a muscular wasting disease. In these triplet repeat mutations, it is not the sequence of the repeat that matters; rather, it is the disruptive effect of the iterative amino acid they encode within the sequence of the expressed protein that causes the disease.

Cystic fibrosis is another genetic disease that has been identified at the molecular level. The gene (*CFTR*) is on chromosome 7 and encodes a chloride channel protein, the cystic fibrosis transmembrane conductance regulator (M_r 168,173). The protein contains five domains: two domains that span the cytoplasmic membrane for chloride transport; two domains that bind and

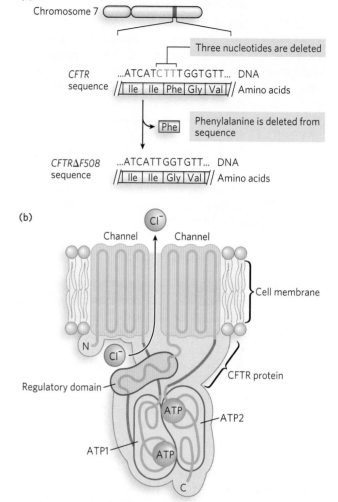

(a)

Chromosome 7

Three nucleotides are deleted

CFTR sequence ...ATCATCTTTGGTGTT... DNA

Ile | Ile | Phe | Gly | Val / Amino acids

Phe   Phenylalanine is deleted from sequence

CFTRΔF508 sequence ...ATCATTGGTGTT... DNA

Ile | Ile | Gly | Val / Amino acids

(b)

Cl⁻

Channel   Channel

Cell membrane

N   Cl⁻

CFTR protein

Regulatory domain

ATP

ATP2

ATP1

ATP

C

FIGURE 2-29 Cystic fibrosis. Cystic fibrosis is caused by a mutation that affects the function of a chloride ion channel. (a) The *CFTR* gene is on chromosome 7. It encodes a channel protein that transports chloride ions. The most common *CFTR* mutation leading to cystic fibrosis is a deletion of three nucleotides that results in the omission of phenylalanine (Phe) at position 508. The isoleucine (Ile) at position 507 remains the same, because both ATC and ATT code for an Ile residue. The omission of Phe[508] prevents proper protein folding. (b) The chloride ion channel consists of five domains: two domains that form the channel across the cytoplasmic membrane (red), two domains that bind and use ATP as an energy source (ATP1 and ATP2), and a regulatory domain. Phe[508] is in the ATP1 domain.

use ATP, the energy that fuels transport of the chloride ions; and a regulatory domain (**Figure 2-29**). The most common mutation (occurring in about 60% of cases) is *CFTRΔF508*, in which three nucleotides are deleted (denoted by Δ), resulting in the deletion of phenylalanine (denoted by F) at position 508 in the amino acid sequence. This amino acid residue is located in the first of the ATP-binding domains, and its deletion prevents proper folding of the protein. Many other mutations in *CFTR* have also been discovered. *CFTR* mutations are most prevalent in Caucasians from Northern Europe.

KEY CONVENTION

The molecular mass of a molecule is commonly expressed in one of two ways, and these are used interchangeably in this book. The first is molecular weight, also called relative molecular mass (M_r). The relative molecular mass of a molecule is the ratio of the mass of that molecule to one-twelfth the mass of carbon-12. Because it is a ratio, M_r is dimensionless and has no units. The second common method is molecular mass (*m*), which is the molar mass of the substance divided by Avogadro's number and is expressed in daltons (Da) or kilodaltons (kDa). One dalton is equal to one-twelfth the mass of carbon-12. For example, the molecular mass of a protein that is 1,000 times the mass of carbon-12 can be expressed by M_r = 12,000, *m* = 12,000 Da, or *m* = 12 kDa.

The *ΔF508* mutation in *CFTR* is autosomal recessive, and therefore an individual must inherit two copies of the mutant allele to develop cystic fibrosis, one from each parent. Without functional CFTR chloride channels, individuals with cystic fibrosis develop abnormally high sweat and mucus production, and a major complication is the buildup of mucus in the lungs. Patients experience breathing difficulties and often have pneumonia. Individuals with cystic fibrosis have typically had an average life span of about 30 years, but as new treatments are developed, survival is increasing greatly.

Although many mutations are detrimental, other mutations can be beneficial. For example, the protein CCR5 is a coreceptor for HIV, the AIDS virus. There is

HIGHLIGHT 2-1 **MEDICINE**

The Molecular Biology of Sickle-Cell Anemia, a Recessive Genetic Disease of Hemoglobin

Genetics, molecular biology, and evolution by natural selection all converge in a striking fashion in sickle-cell anemia, a human hereditary disease. Sickle-cell anemia is a disease of the blood caused by a mutation in the hemoglobin protein. Hemoglobin, the oxygen-carrying protein of red blood cells (erythrocytes), is composed of four subunits, two α chains and two β chains. The sickle-cell mutation occurs in the β chain, and the mutant hemoglobin is called hemoglobin S. Normal hemoglobin is called hemoglobin A. Humans are diploid and thus contain two alleles of the β-chain gene. The two alleles are sometimes slightly different. About 50 genetic variants of hemoglobin are known, usually due to a single amino acid change, and most of these are quite rare. Although the effects on hemoglobin structure and function are often negligible, they can sometimes be extraordinary.

Sickle-cell anemia is a recessive genetic disease in which an individual inherits two copies of the β-chain allele for sickle-cell hemoglobin S (i.e., the sickle-cell allele). A heterozygous individual, with one sickle-cell allele and one normal allele, has nearly normal blood. Two heterozygous parents can potentially have a child who is homozygous for the recessive sickle-cell allele (Figure 1).

The nucleotide sequence of the sickle-cell allele usually contains a thymine (T) in place of an adenine (A), thereby changing one nucleotide triplet from GAG to GTG (Figure 2). This single base change results in hemoglobin S, which contains a hydrophobic (water-fearing) valine residue at one position in the β chain, instead of a hydrophilic (water-loving) glutamic acid residue. This amino acid change causes deoxygenated hemoglobin S molecules to stick together, forming insoluble fibers inside erythrocytes and

$\beta^A\beta^A$
$\beta^A\beta^S$
$\beta^A\beta^A$
$\beta^A\beta^S$
$\beta^A\beta^A$
$\beta^A\beta^S$
$\beta^A\beta^A$
$\beta^A\beta^S$
$\beta^A\beta^S$
$\beta^A\beta^A$
$\beta^A\beta^S$
$\beta^A\beta^A$
$\beta^A\beta^A$
$\beta^A\beta^S$
$\beta^A\beta^S$
$\beta^S\beta^S$

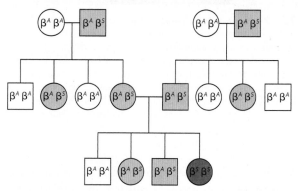

FIGURE 1 A family pedigree for sickle-cell anemia shows the genotypes for the hemoglobin β chain (circles, females; squares, males). The alleles are β^A for wild-type hemoglobin A and β^S for sickle-cell hemoglobin S. Heterozygous individuals are shaded in pink, individuals homozygous for β^S in red.

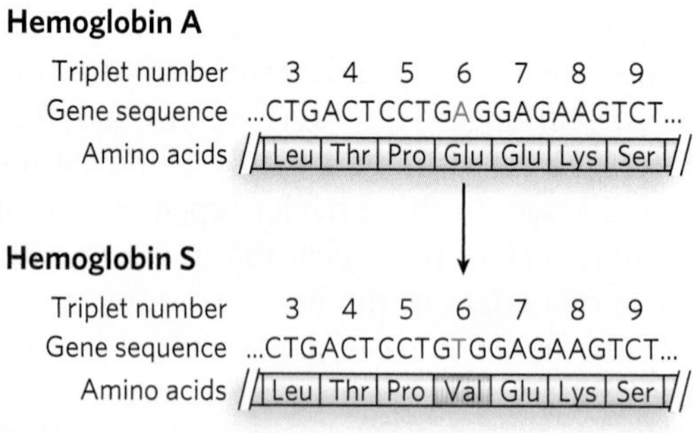

Hemoglobin A
Triplet number 3 4 5 6 7 8 9
Gene sequence ...CTGACTCCTGAGGAGAAGTCT...
Amino acids Leu Thr Pro Glu Glu Lys Ser
Hemoglobin S
Triplet number 3 4 5 6 7 8 9
Gene sequence ...CTGACTCCTGTGGAGAAGTCT...
Amino acids Leu Thr Pro Val Glu Lys Ser

FIGURE 2 A single nucleotide change in the sickle-cell allele alters the hemoglobin β chain. In hemoglobin A, triplet 6 is GAG, which codes for glutamic acid (Glu). In hemoglobin S, the most common substitution is a T for the A in triplet 6 to form a GTG triplet, which codes for valine (Val).

much speculation about a 32 amino acid deletion mutation of CCR5 (due to the *CCR5Δ32* mutation), which is widely dispersed among people of European descent (an occurrence of 5% to 14% in these groups), although much rarer among Asians and Africans. Researchers speculate that this mutation may have conferred resistance to the bubonic plague or smallpox, thereby becoming enriched in the population, by natural selection, in endemic areas. Although the allele has a negative effect on T-cell (a type of immune cell) function, it seems to provide protection against HIV infection, as well as smallpox.

Another example of a mutation that confers some benefit is the one that causes sickle-cell anemia (**Highlight 2-1**), a mutation of hemoglobin. When the mutation is inherited from both parents, the result is misshapen red blood cells that can get stuck in capillaries and impede blood flow, with possibly fatal results. However, people who are heterozygous for this mutation have enhanced resistance to malaria. Geographic areas where this mutation is prevalent in the population correlate with locations that are plagued by malaria.

Discovery of DNA as the hereditary material, and the understanding of how it is transcribed and translated into RNA and protein, is a most fascinating story in science. Darwin's theory of the origin of species through evolution by natural selection was compelling, but the mechanism that drove the variation on which natural selection could act remained a mystery in his lifetime. Yet, the

---

(a) (b)

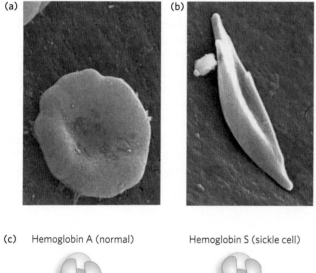

(c)   Hemoglobin A (normal)   Hemoglobin S (sickle cell)

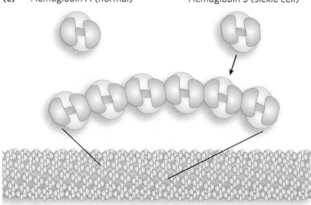

Insoluble fiber

FIGURE 3 A comparison of (a) a normal, uniform, cup-shaped erythrocyte with (b) a sickle-shaped erythrocyte seen in sickle-cell anemia. (c) The shape change in the hemoglobin molecule, due to the substitution of a Val for a Glu residue in the β chain, allows the molecules to aggregate into insoluble fibers within the erythrocytes. [*Sources: (a) and (b) CDC/Sickle Cell Foundation of Georgia. Photo by Janice Haney.*]

changing the shape of the cells. The blood of individuals with sickle-cell anemia contains many long, thin, crescent-shaped erythrocytes that look like the blade of a sickle (Figure 3). The sickle shape occurs only in veins, after the blood has become deoxygenated. Sickled cells are fragile and rupture easily, resulting in anemia (from the Greek for "lack of blood").

When capillaries become blocked, the condition is much more serious. Capillary blockage causes pain and interferes with organ function—often the cause of early death. Without medical treatment, people with sickle-cell anemia usually die in childhood. Nevertheless, the sickle-cell allele is surprisingly common in certain parts of Africa. Investigation into the persistence of an allele that is so obviously deleterious in homozygous individuals led to the finding that in heterozygous individuals, the allele confers a small but significant resistance to lethal forms of malaria. Heterozygous individuals experience a milder form of sickle-cell disease called sickle-cell trait; only about 1% of their erythrocytes become sickled on deoxygenation. These individuals can live normal lives by avoiding vigorous exercise and other stresses on the circulatory system. Natural selection has thus resulted in an allele that balances the deleterious effects of the homozygous sickle-cell condition against the resistance to malaria conferred by the heterozygous condition.

key to understanding this mystery had already been discovered by Mendel. Mutations create the natural variation needed for the forces of natural selection to mold new species. It seems almost ludicrous that Mendel and Darwin were alive at the same time and separately uncovered secrets that together explained the diversity of planetary life. Lack of a robust means of communication kept these two vital pieces of information segregated for decades—an improbable situation today, given the rapid pace of global communication. Although most mutations are deleterious, the rare mutation that carries a beneficial change eventually enters the population through natural selection over the expanse of evolutionary time. Natural selection still drives change and the evolution of new species today.

SECTION 2.4 SUMMARY

- Nucleic acids (DNA and RNA) are composed of repeating units called nucleotides. Each nucleotide contains a phosphate group, a ribose sugar, and a nitrogenous base. Four different bases are found in DNA: adenine, guanine, cytosine, and thymine. RNA also contains adenine (A), guanine (G), and cytosine (C), but uracil (U) instead of thymine (T). Information is encoded by the specificity of pairing of G with C, and of A with T (or U).

- Identification of DNA as the chemical of heredity was determined in experiments using virulent and nonvirulent bacteria. The DNA of virulent bacteria transforms nonvirulent bacteria into a virulent form.

- Even before the DNA structure was solved, studies of mutants drew the connection between genes and enzymes, as in the investigations of defective enzymes in the biosynthetic pathways of auxotrophic mutants of *Neurospora crassa*.

- Information flow in the direction DNA→RNA→ protein is known as the central dogma. RNA is synthesized from a DNA template in the process of transcription. In translation, the RNA sequence is converted to protein. The duplication of DNA is replication. Exceptions to the central dogma exist (RNA→DNA, and RNA→RNA).

- Three types of RNA are required for DNA→ RNA→protein. Ribosomal RNA combines with proteins to form ribosomes, which are factories for protein synthesis. Transfer RNAs are small adaptor RNAs to which amino acids become attached. Messenger RNAs encode proteins and are read by tRNAs in groups of three nucleotides, each of which specifies an amino acid.

- Functional RNAs are RNA sequences that are not translated into protein. Rather, the RNA sequences themselves perform functions in the cell. Both rRNA and tRNA are functional RNAs.

- Mutations are changes in DNA sequence. When a mutation affects the function of a protein or functional RNA, it results in a phenotypic change. Changes in the DNA sequence of germ-line cells underlie inherited human diseases, including hemophilia, Huntington disease, cystic fibrosis, and sickle-cell anemia. Mutations are not always deleterious—sometimes they can be beneficial and, indeed, are vital in creating the diversity needed for the evolution of new species.

HOW WE KNOW

Chromosome Pairs Segregate during Gamete Formation in a Way That Mirrors the Mendelian Behavior of Genes

Walter Sutton, 1877–1916
[Source: University of Kansas Medical Center Archives.]

Boveri, T. 1902. Ueber mehropolige Mitosen als Mittel zur Analyse des Zellkerns. *Verh. Phys. Med. Ges. Wurzburg* 35:67–90.

Sutton, W. S. 1902. On the morphology of the chromosome group in *Brachystola magna. Biol. Bull.* 4:24–39.

Sutton, W. S. 1903. The chromosomes in heredity. *Biol. Bull.* 4:231–251.

Walter Sutton was only a graduate student when, in 1902, he made observations that led to some of the most profound conclusions in biology. At the time of Sutton's studies at Columbia University, Mendel's laws had just been rediscovered by genetic methods similar to the ones used by Mendel 35 years earlier. But now there were new ways of looking at organisms—namely, observing individual cells under the microscope. Sutton was particularly interested in the process of gamete production, in which one cell undergoes two divisions; in the second division, the chromosome number is halved relative to that of the parent. This process fascinated him. Others who studied these cell divisions used organisms with chromosomes that were too small to allow the observer to discern their individual identity. But Sutton studied the great lubber grasshopper (*Brachystola magna*), which had large chromosomes with distinctive shapes (Figure 1). This allowed him to see that in meiosis, each chromosome paired with a look-alike partner (a homologous chromosome), and the homologous chromosomes (each a pair of sister chromatids) separated from each other in the first meiotic cell division. During the second cell division, the two sister chromatids of each duplicated chromosome assorted into different daughter cells. On the union of sperm and egg, the homologous pairs of chromosomes were reestablished.

The behavior of chromosomes mimicked the Mendelian behavior of segregation of traits, but on a subcellular level. Sutton hypothesized that paternal and maternal chromosomes exist in pairs and separate into gametes during meiosis, explaining the diploid particles of heredity in Mendel's laws.

Today, a scientist making a groundbreaking discovery of this caliber would have established a solid reputation in science. But in Sutton's day, there were no graduate student stipends or regular sources of scientific funding. So Sutton became a physician and went back to his hometown in Kansas to practice medicine.

Theodor Boveri, a talented German scientist, worked completely independent of Sutton yet reached similar conclusions in the same year as Sutton. Boveri studied the behavior of chromosomes in sea urchin eggs. Although sea urchin chromosomes are small and thus cannot be distinguished by their shape, their number can be observed during fertilization and cell division. Boveri's studies reached the same conclusions as Sutton's, linking chromosomes with the particles of Mendelian inheritance. He also observed that eggs from which the nucleus was removed could be fertilized and then develop into normal—albeit haploid—larvae, and that normal larvae could develop from eggs with only the female set of chromosomes in the nucleus (also haploid). He concluded that each chromosome set, contributed by either parent, had a complete set of instructions for development of the organism. The findings of the two scientists became known as the Sutton-Boveri chromosome theory of inheritance.

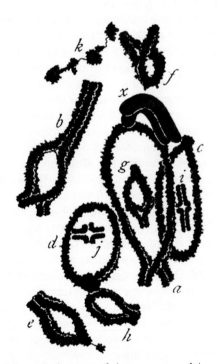

FIGURE 1 Sutton's drawings of chromosomes of the grasshopper *Brachystola magna*, showing their unique shapes and sizes. The pairs of chromosomes are labeled *a* through *k* and *x*. [Source: W. S. Sutton, Biol. Bull. 4:24–39, 1902.]

HOW WE KNOW

Corn Crosses Uncover the Molecular Mechanism of Crossing Over

Creighton, H., and B. McClintock. 1931. A correlation of cytological and genetical crossing-over in *Zea mays. Proc. Natl. Acad. Sci. USA* 17:492–497.

Barbara McClintock, 1902–1992 (left); Harriet Creighton, 1909–2004 (right) *[Source: Karl Maramorosch/ Courtesy Cold Spring Harbor Laboratory Archives.]*

Fruit flies have taught us how our body plan is determined. Who would have guessed that fruit flies would teach us so much? Among the many fruitful (pun intended) discoveries by Thomas Hunt Morgan, who developed the fly as a model for genetic study, was the finding that genes cross over between chromosomes. Although researchers presumed that genetic recombination occurred through material exchange between homologous chromosomes, there was no proof that this was indeed the case. Direct proof came in 1931 from a now classic study in corn (maize) by Harriet Creighton and Barbara McClintock.

The insightful experiments of Creighton and McClintock combined genetics and cytologic methods. To visualize crossing over between two homologous chromosomes, one first needs to find two homologous chromosomes that look different—no easy task. Creighton and McClintock searched until they found a plant with an odd-shaped chromosome; in this plant, chromosome 9 had a knob on one end and an extension on the other. Next, they showed that this plant could be crossed with a plant having a normal-shaped chromosome 9 to produce offspring having a homologous chromosome 9 pair that did not look alike. They then mapped two alleles on chromosome 9 to follow recombination genetically. These alleles were seed color—*C* (colored) and *c* (colorless); and seed texture—*Wx* (starchy) and *wx* (waxy). Creighton and McClintock crossed the two plants represented at the top of Figure 2 and looked for colorless, waxy progeny (i.e., progeny that produce colorless, waxy seeds). Genetic crossing over between the misshapen chromosome 9 and its homolog is required to produce a colorless, waxy plant of genotype *ccwxwx*. If genetic crossing over results from physical recombination between the two chromosomes, then the chromosomes of the colorless, waxy progeny

should contain chromosome pairs with the misshapen chromosome 9 having only one abnormality—either a knob or an extension at one end (see Figure 2). Indeed, chromosomes of the rare colorless, waxy offspring looked exactly as predicted, confirming that genetic recombination occurs through the physical exchange of material between homologous chromosomes.

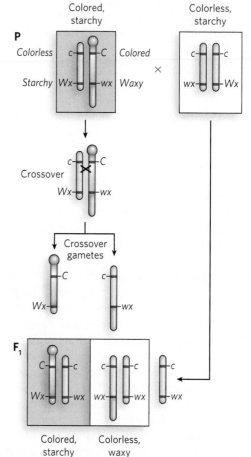

FIGURE 2 The gametes at the top left represent a corn plant with colored, starchy seeds that is heterozygous for these seed-color and seed-texture genes (*CcWxwx*). One chromosome 9 in the diploid has abnormal extremities. This plant was crossed with a corn plant (top right) having colorless, starchy seeds (*ccWxwx*). Genetic crossing over in the colored, starchy plant produced colorless, waxy progeny of genotype *ccwxwx*. Microscopic examination confirmed that genetic crossing over involves physical recombination of chromosomes: one end of the abnormal chromosome 9 was replaced with a normal end, containing the colorless-seed gene.

Hershey and Chase Settle the Matter: DNA Is the Genetic Material

Hershey, A.D. and M. Chase, 1952.
Independent functions of viral protein and nucleic acid in growth of bacteriophage. *J. Gen. Physiol.* 36:39–56.

Martha Chase, 1927–2003 (left); Alfred Hershey, 1908–1997 (right) *[Source: Karl Maramorosch/Courtesy Cold Spring Harbor Laboratory Archives.]*

In 1952, Martha Chase and Alfred Hershey performed a now classic experiment, the results of which would convince the world that DNA is the genetic material. They used a bacterial virus, mainly composed of protein and DNA, and set out to determine which of these components carries the hereditary material. Bacteriophage T2, or T2 phage, like other bacterial viruses, consists of a protein coat and a DNA core. Hershey and Chase took advantage of a key chemical difference between these two macromolecules. Using the fact that sulfur is found in proteins but not in DNA, and that phosphorus is found in DNA but not in proteins, they prepared radiolabeled T2 phage using either ^{35}S (only protein is radioactively labeled) or ^{32}P (only DNA is radioactively labeled). The two T2 phage samples were allowed to attach to their bacterial host, *Escherichia coli*, in two separate flasks (Figure 3). After infection, the bacteria were transferred to a kitchen blender and agitated to strip away any T2 phage material from the outside of the bacterial cell walls. Cells were collected by centrifugation, leaving unattached phage in the supernatant. The results were clear: ^{32}P-labeled DNA had transferred into the cells, while ^{35}S-labeled protein remained in the supernatant. Therefore, it is the DNA that carries out the genetic program of the phage. In addition, the progeny phage produced in the infected cells contained ^{32}P and no ^{35}S, further proof that the DNA is the genetic material.

Although earlier experiments by Avery had suggested that DNA was the genetic material, the Hershey and Chase experiment finalized this important conclusion and inspired Watson and Crick in their quest to determine the structure of DNA. Hershey shared the 1969 Nobel Prize in Physiology or Medicine with Max Delbrück and Salvador E. Luria for their discoveries on the replication mechanism of viruses.

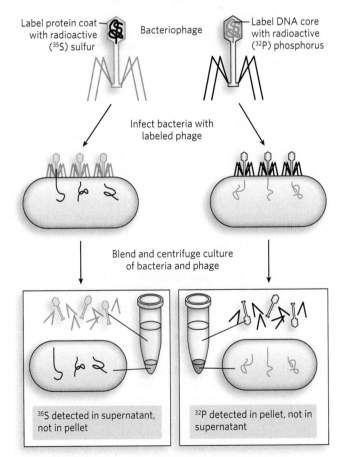

FIGURE 3 Bacterial cells infected with ^{32}P-labeled phage contained ^{32}P after blender treatment, indicating that viral ^{32}P-DNA had entered the cells. Cells infected with ^{35}S-labeled phage had no radioactivity after blender treatment. Progeny virus particles contained ^{32}P-DNA acquired from the cells infected with ^{32}P-labeled phage.

KEY TERMS

gamete cell, 24
somatic cell, 24
gene, 26
diploid, 26
allele, 26
phenotype, 26
haploid, 26
genotype, 27
homozygous, 27
heterozygous, 27
law of segregation, 27
law of independent
   assortment, 28
chromosome, 32
G_1 phase, 33
S phase, 33

centromere, 33
G_2 phase, 33
M phase, 33
mitosis, 33
cytokinesis, 35
meiosis, 35
tetrad, 35
crossing over, 40
deoxyribonucleic acid
   (DNA), 43
ribonucleic acid
   (RNA), 43
ribosome, 47
RNA polymerase, 47
transcription, 47
translation, 48

PROBLEMS

1. Two purebred pea plants are crossed. One strain has dominant round seeds; the other has recessive wrinkled seeds. (a) What phenotypes would be seen in the F_1 generation plants, and in what proportions? (b) What phenotypes would be seen in the F_2 generation plants, and in what proportions? (c) If an F_1 generation plant is crossed with a plant producing wrinkled seeds, what phenotypes are seen in the progeny, and in what proportions?

2. Two pea plants with round seeds are crossed. In the F_1 generation, all the plants have round seeds. What can you say about the genotype of the parental plants?

3. The F_1 plants from the cross in Problem 2 are next crossed at random. There are 129 plants in the F_2 generation. The majority, 121 plants, produce round seeds. However, there are 8 plants that produce wrinkled seeds. From this information, what were the genotypes of the original parental plants?

4. Purebred white-eyed male fruit flies are crossed with wild-type red-eyed females. If the progeny are crossed with each other repeatedly, which generation will be the first to contain white-eyed female flies?

5. Purebred wild-type male flies are crossed with purebred white-eyed female flies. If the progeny are crossed with each other repeatedly, which generation will be the first to contain white-eyed male flies?

6. A new species of fruit fly is found on an uncharted island. The flies are brightly colored, with blue and green

bodies. After studying these insects for a year or two, researchers find one male with an all-black body. When this male is crossed with wild-type females, all of the male progeny in the F_1 generation are black, and all of the female progeny have the blue and green coloring. This same pattern (all black males and colored females) is repeated in the F_2, F_3, and F_4 generations. Explain these observations.

7. Two purebred flowering plants are crossed. One has red flowers and small leaves (*RRll*) and the other has white flowers and large leaves (*rrLL*). Using a Punnett square analysis, and assuming that the genes are unlinked, predict the type and frequency of phenotypes in the F_2 generation.

8. If the F_1 plants in Problem 7 had genes for red flower color that exhibited incomplete dominance, the heterozygous *Rr* flower color would be pink. In that instance, what percentage of the F_1 plants in Problem 7 would have pink flowers? What percentage of the F_2 generation would have pink flowers?

9. Two purebred fruit flies are crossed. The male has white eyes and vestigial wings. The female has red eyes and normal wings. All F_1 flies have red eyes and vestigial wings. Using a Punnett square analysis, predict the percentage of F_2 generation males that will have red eyes and normal wings. Assume that the wing trait is not sex-linked.

10. A new and exotic species of fly is found, with green eyes (*G*) and striped wings (*S*). A mutant fly of the same species is found that has orange eyes (*g*) and clear (unstriped) wings (*s*). The mutant is cultured for many generations to obtain a purebred strain with the double-mutant phenotype. A *ggss* female fly is mated with a wild-type *GGSS* male. The F_1 progeny all have green eyes and striped wings, as expected. An F_1 male is mated with an F_1 female. Among the F_2 progeny of this cross, only two kinds of flies are observed: 75% with green eyes and striped wings, and 25% with orange eyes and clear wings. Some expected F_2 progeny (such as flies with green eyes and clear wings) are absent. Explain this result.

11. Both meiosis and mitosis are initiated with a complete replication of the cell's chromosomes in S phase. The replication of each chromosome produces a pair of sister chromatids. During the cell division immediately following replication, how are the chromosomes in the sister chromatid pairs distributed to daughter cells in mitosis and meiosis?

12. Two purebred plants, with genotypes *AABBCCDDEEFF* and *aabbccddeeff*, are crossed. In the F_1 generation, all individuals are heterozygous for all traits. Geneticists probe the linkage of these various genes by doing a series of crosses, examining two traits in each cross. When all the

crosses are done and the data are tabulated, the researchers find that in the F_1 plants, meiosis produces gametes that contain the following combinations of alleles at the indicated frequencies (which correspond to crossover frequencies):

$A + b$	9%
$A + e$	13%
$A + d$	50%
$B + c$	6%
$C + f$	50%
$C + e$	10%
$D + f$	16%

With these data, determine how the genes are distributed along the chromosomes. Draw a map, using the crossover frequencies as distances.

13. On one chromosome there are three linked genes designated *M*, *N*, and *O*. If crossing over occurs between *M* and *O* 5% of the time, and between *N* and *O* 8% of the time, what are the possible arrangements of the genes on the chromosome?

14. In the central dogma developed by Francis Crick and others, three kinds of RNA play important roles: rRNAs, tRNAs, and mRNAs. Describe two features that are characteristic of each type of RNA.

15. In the classic Hershey-Chase experiment (see the How We Know section for this chapter), the T2 phage was labeled with either ^{35}S or ^{32}P before using it to infect a bacterial host. In this experiment, would it have been possible for these researchers to label one batch of T2 phage with both ^{35}S and ^{32}P and still get a definitive result? Why or why not? What would the results of the experiment be?

16. In the bacterium *Escherichia coli*, the amino acid tryptophan is synthesized in a multistep pathway, beginning with an organic precursor called chorismate:

$$\text{Chorismate} \rightarrow X \rightarrow Y \rightarrow \text{Tryptophan}$$
$$\qquad\quad 1 \quad\ 2 \quad\ 3$$

The numbers denote steps in the pathway, each catalyzed by an enzyme. There are five bacterial genes that encode polypeptides associated with these enzymes, called *trpA*, *trpB*, *trpC*, and so on. A researcher isolates a series of mutations that eliminate the capacity of the cells to synthesize tryptophan, each mutation affecting one of the five genes. Each mutant cell is tested for its ability to grow on a medium containing either tryptophan, chorismate, or intermediate X or Y. The following results are obtained (+ means growth; – means absence of growth; WT means wild type):

Molecule in growth medium	trpA	trpB	trpC	trpD	trpE	WT
Tryptophan	+	+	+	+	+	+
X	–	–	–	–	–	+
Y	–	–	–	+	+	+
Chorismate	+	+	+	+	+	+

Which genes encode the enzymes involved in steps 1, 2, and 3 of the tryptophan biosynthetic pathway? Given this information, what can you say about the structure of these enzymes?

17. In mitosis and meiosis, all cellular chromosomes are replicated before cell division. In a diploid cell, there are two copies of each autosomal chromosome. For a typical diploid eukaryotic cell and a hypothetical chromosome A, the two copies of the chromosome can be labeled A1 and A2. During replication, each chromosome (including A1 and A2) is converted into two new chromosome copies that are transiently held together. The immediate tethered products of a replicated chromosome are called sister chromatids. The two sets of sister chromatids resulting from replication of chromosomes A1 and A2 can be labeled A1/A1* and A2/A2*. Thus, in all, four chromosomes are held in two pairs of sister chromatids. For both mitosis and meiosis, replication is eventually followed by a cell division (the first of two cell divisions in the case of meiosis). In that cell division, how are the four chromosomes segregated into the daughter cells in mitosis? How are the four chromosomes segregated into daughter cells in the first cell division of meiosis?

18. Food for thought: As described in the chapter, substantial segments of the genome of many organisms are transcribed into RNA that does not encode protein and does not correspond to either rRNA or tRNA. Without studying several additional chapters in this book, you may not be familiar with this RNA. However, considering some potential functions for this RNA is a useful prelude to your continued study of molecular biology. Based on the discussion of the RNA world hypothesis in Chapter 1 and other knowledge you may have about RNA, suggest one or two potential functions for this non-protein-coding RNA.

ADDITIONAL READING

Many of the books and papers listed here are available online, free of charge, from Electronic Scholarly Publishing, a collection of source material on the foundations of classical genetics (www.esp.org/foundations/genetics/classical).

General

Dolan DNA Learning Center, Cold Spring Harbor Laboratory. DNA Interactive: Timeline. www.dnai.org/timeline/index.html. This website offers an overview of the major advances in molecular genetics from 1865 to 2000.

Watson, J.D. 1968. *The Double Helix: A Personal Account of the Discovery of the Structure of DNA.* New York: Atheneum Publishers. Watson's original account of his adventures; a quick and very interesting read, warts and all.

Mendelian Genetics

Bateson, W. 1909. *Mendel's Principles of Heredity.* Cambridge: Cambridge University Press.

Mendel, G. 1866. Versuche über Pflanzen-Hybriden. In *Verhandlungen des naturforschenden Vereines* (*Proceedings of the Natural History Society*). Brünn. Fewer than 150 copies were produced; Darwin owned one of them, but the evidence from examining his copy indicates that he did not open it to Mendel's work.

Mendel Museum, Masaryk University. Gregor Mendel. www.mendel-museum.com. This original location of Mendel's work is undergoing extensive restoration as a museum.

Cytogenetics: Chromosome Movements during Mitosis and Meiosis

Hooke, R. 1664. *Micrographia: Some Physiological Descriptions of Minute Bodies Made by Magnifying Glasses with Observations and Inquiries Thereupon.* London. Available at www.gutenberg.org/etext/15491.

Sutton, W.S. 1903. The chromosomes in heredity. *Biol. Bull.* 4:231–251. An outline of the rationale behind Sutton's proposal that chromosomes carry the material of heredity.

The Chromosome Theory of Inheritance

Creighton, H.B., and B. McClintock. 1931. A correlation of cytological and genetical crossing-over in *Zea mays. Proc. Natl. Acad. Sci. USA* 17:492–497.

Morgan, T.H., A.H. Sturtevant, H.J. Muller, and C.B. Bridges. 1915. *Mechanism of Mendelian Heredity.* New York: Henry Holt and Company. This book contains the classic work from Morgan's lab and illustrates the way in which scientific discoveries were published before scientific journals became the norm.

Foundations of Molecular Genetics

Avery, O.T., C.M. MacLeod, and M. McCarty. 1944. Studies on the chemical nature of the substance inducing transformation of pneumococcal types. *J. Exp. Med.* 79:137–158.

Beadle, G.W., and E.L. Tatum. 1941. Genetic control of biochemical reactions in *Neurospora. Proc. Natl. Acad. Sci. USA* 27:499–506.

Crick, F. 1970. Central dogma of molecular biology. *Nature* 227:561–563. The classic paper in which Crick proposes the central dogma of information flow in biology.

Hershey, A.D. and M. Chase. 1952. Independent functions of viral protein and nucleic acid in growth of bacteriophage. *J. Gen. Physiol.* 36:39–56.

3

Chemical Basis of Information Molecules

Roxana Georgescu [Source: Courtesy Roxana Georgescu.]

MOMENT OF DISCOVERY

I was always interested to see a crystal structure of DNA going through the circular beta processivity clamp (a ring-shaped protein that associates with DNA during replication). But it was thought that beta only slides on DNA and doesn't stick to it, so no one tried to make a DNA-beta cocrystal. However, my studies suggested that beta might stick to DNA, and I just knew it would bind DNA in an interesting fashion. The problem was, how could I make a protein crystal that I was sure contained DNA, without solving the whole structure by x-ray crystallography? I decided to use a colored DNA that should give a colored beta crystal only if DNA is in the crystal. I ordered an oligonucleotide with a fluorescent blue dye, hybridized it to another DNA strand to make a duplex, and then mixed it with beta and waited for protein crystals to form. When I took the tray out a few days later, I was shocked to see the experiment actually worked—some crystals were brilliant blue! That was one of my "Aha" moments, and I had goose bumps, because I knew I had crystals of beta protein with DNA in them. Then I realized that all I had to do next was to solve this structure. Just one problem—I didn't know how to do x-ray crystallography.

I called Dr. Xiang-Peng Kong at NYU, who first solved the structure of this protein a decade earlier. He was more than willing to teach me, and he was such a delight to work with. But victory was not certain, because even though we knew DNA was present in the crystals, it would be completely invisible if it bound to the beta clamp protein in more than one unique way. When we collected sufficient data and the first electron density maps came into focus on the computer, we were spellbound— the electron density of helical DNA was right there in the middle of the protein, and it had a completely unexpected sharp tilt that would make the clamp spin while sliding. We would publish this for everyone to see. But for that moment, we were the only ones on Earth who knew this wonderful secret of nature, and that is a very special feeling.

—Roxana Georgescu, on her discovery of how beta processivity clamps bind DNA

Online resources related to this chapter:

Simulation Nucleotide structure

Molecular biology involves the study of molecules that store and process genetic information. The chemical properties of these molecules and the principles that govern their behavior are central to understanding the maintenance and transfer of that information. Key to the storage and use of genetically encoded information are nucleic acids and proteins, macromolecules that are major constituents of all cells. These high-molecular-weight polymers are assembled from relatively simple precursors and can form three-dimensional structures that mediate a wide variety of biological activities.

The functions of nucleic acids and proteins stem from their chemical properties. Shape, electrical charge distribution, propensity to form weak or strong chemical bonds, and preference for hydrophobic (water-fearing, or water-excluding) or hydrophilic (water-loving, or water-including) interactions—all of these contribute to the ability of DNA to function as the primary repository of genetic information and the ability of proteins to enhance biochemical reaction rates. Proteins also play important structural roles in cells, and they enable cells to communicate and respond to their environment. RNA, a chemical cousin of DNA, shares the information-bearing properties of DNA and also has some structural and functional similarities to proteins. As we shall see throughout this book, new biological activities and functions for RNA are still being discovered. RNA is an important and, until recently, underappreciated controller of gene expression in all cells.

The flow of biological information in cells and organisms makes sense only in the context of the underlying chemical behavior of these biomolecules. Molecular biologists often say that a protein "recognizes" a fragment of DNA, or that an RNA molecule "binds" to a protein, or that several proteins "assemble" into a multisubunit complex, but what do these phrases really mean? A more quantitative framework for understanding cellular function requires a familiarity with the chemical principles by which molecules fold, react, and interact.

In this chapter we discuss these chemical principles, many of which developed from concepts originally drawn from the study of small molecules. We begin with the chemical building blocks of nucleic acids and proteins and a discussion of the kinds of chemical bonds that hold them together. We then discuss constraints on the behavior of biomolecules stemming from their stereochemistry, their ionization properties within the cellular environment, and their propensities to react and interact with other large and small molecules. Understanding the chemistry that governs protein and nucleic acid function provides a strong foundation for exploring the many facets of biological behavior described throughout this book.

3.1 CHEMICAL BUILDING BLOCKS OF NUCLEIC ACIDS AND PROTEINS

We start by focusing on the underlying chemical properties that control the behavior of nucleic acids and proteins. Nucleic acid structures are discussed in more depth in Chapter 6, and amino acids are discussed in Chapter 4.

Nucleic Acids Are Long Chains of Nucleotides

Deoxyribonucleic acid (DNA) and **ribonucleic acid (RNA)** store and transmit genetic information, in part by coding for proteins. In addition, some RNA molecules function catalytically or structurally within larger, multi-molecular complexes. Both DNA and RNA are composed of building blocks (monomers) called **nucleotides,** which are linked together by **phosphodiester bonds** to form long, unbranched chains. Nucleic acids can reach chain lengths of up to many millions of nucleotides and molecular masses of up to several billion daltons. A nucleotide molecule has three components: a nitrogenous base, a five-carbon (pentose) sugar, and a phosphate group. A base and sugar without the phosphate group is referred to as a **nucleoside.** In DNA or RNA molecules, the sugars and phosphates of adjacent, individual nucleotides are chemically linked to form a sugar–phosphate backbone that has a characteristic directionality. This directionality is defined by the chemical convention for numbering carbon atoms in the nucleotide sugar ring, giving rise to a 5′ end and a 3′ end (see **Figure 3-1**). The directionality of DNA and RNA is discussed in detail in Chapter 6.

The nucleotides that make up DNA polymers are **deoxyribonucleotides** (Figure 3-1a, left), named for the type of pentose sugar found in DNA: deoxyribose. Whereas the phosphate group and the type of pentose remain constant, each deoxyribonucleotide contains one of four different nitrogenous bases: **thymine (T), adenine (A), guanine (G),** or **cytosine (C)** (Figure 3-1b, left). The type of base establishes the identity of each individual deoxyribonucleotide. Thus, the information in DNA is written in a four-letter alphabet.

Chemically, RNA is very similar to DNA. Like DNA, it is a long, unbranched polymer of nucleotides. And like DNA nucleotides, all RNA nucleotides contain a pentose and a phosphate group, and one of four different nitrogenous bases. Two small differences in their chemical components, however, give rise to important distinctions between the structures and functions of RNA and DNA. The first is the type of pentose present. RNA nucleotides contain ribose and thus are named **ribonucleotides** (see Figure 3-1a, right). Ribose has one more hydroxyl (—OH) group on the sugar ring than does deoxyribose, which

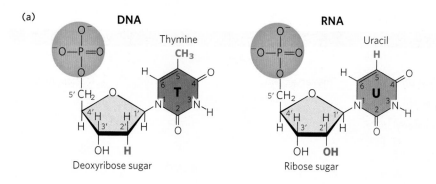

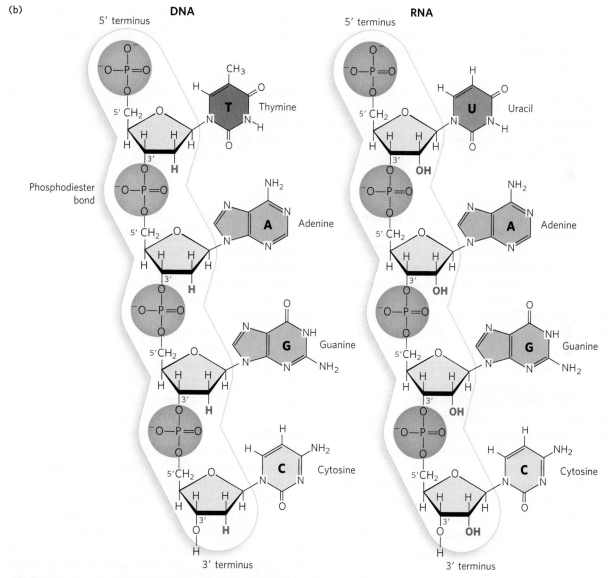

FIGURE 3-1 Chemical building blocks of DNA and RNA. (a) Chemical differences between the major nucleotides of DNA (left) and RNA (right). Atoms in the base ring are numbered sequentially, starting with the nitrogen that is bound to the sugar; carbon numbering in the sugar ring is denoted by a prime ('). DNA nucleotides contain the sugar deoxyribose, whereas RNA nucleotides contain ribose (light gray). The difference between these two sugars is a single hydroxyl group on the 2'carbon (red). In addition, DNA contains the base thymine, rather than the uracil found in RNA; these bases differ by a methyl group on carbon 5 of the base (red). (b) Segments of deoxyribonucleotide (DNA, left) and ribonucleotide (RNA, right) chains. Phosphodiester bonds connect the individual nucleotide units; the sugar–phosphate backbone (outlined) is polar and directional, with free 5' and 3' ends. The 2' hydrogens of DNA and the 2'-hydroxyl groups of RNA are highlighted in red. The color coding of bases, here and throughout the figures in this book, is as follows: cytosine, yellow; adenine, light blue; guanine, green; thymine, dark blue; uracil, purple.

TABLE 3-1
Nucleic Acid Nomenclature

| Base | Nucleoside | Nucleotide | Nucleotide Abbreviation | | Nucleic Acid |
			One-letter	Three-letter	
Adenine	Adenosine	Adenylate	A	AMP	RNA
	Deoxyadenosine	Deoxyadenylate	dA	dAMP	DNA
Guanine	Guanosine	Guanylate	G	GMP	RNA
	Deoxyguanosine	Deoxyguanylate	dG	dGMP	DNA
Cytosine	Cytidine	Cytidylate	C	CMP	RNA
	Deoxycytidine	Deoxycytidylate	dC	dCMP	DNA
Thymine	Thymidine or deoxythymidine	Thymidylate or deoxythymidylate	T or dT	TMP or dTMP	DNA
Uracil	Uridine	Uridylate	U	UMP	RNA

Note: *Nucleoside* and *nucleotide* are generic terms that include both ribo- and deoxyribo- forms. Also, ribonucleosides and ribonucleotides are here designated simply as nucleosides and nucleotides (e.g., riboadenosine as adenosine), and deoxyribonucleosides and deoxyribonucleotides as deoxynucleosides and deoxynucleotides (e.g., deoxyriboadenosine as deoxyadenosine). Both forms of naming are acceptable, but the shortened names are more commonly used. Thymine is an exception; *ribothymidine* is used to describe its unusual occurrence in RNA.

defines the RNA polynucleotide as *ribo*nucleic acid rather than *deoxyribo*nucleic acid. The second distinction is the assortment of nitrogenous bases found in RNA. Ribonucleotides contain three of the same bases found in DNA—adenine, guanine, and cytosine—but instead of thymine, the fourth base in RNA is **uracil (U)** (see Figure 3-1b, right). Uracil is structurally identical to thymine except for the absence of the methyl ($-CH_3$) group. The nucleotides of DNA and RNA are represented by both three-letter and one-letter abbreviations (**Table 3-1**).

KEY CONVENTION

DNA and RNA are defined by the type of sugar in the polynucleotide backbone (deoxyribose or ribose), not by the presence of thymine or uracil.

Even with just four types of nucleotides each, the number of possible DNA and RNA sequences (4^n, where n is the number of nucleotides in the sequence) is enormous for even the shortest molecules. Thus, an almost infinite number of distinct genetic messages can exist.

Proteins Are Long Polymers of Amino Acids

Proteins, like nucleic acids, are unbranched polymers. The building blocks of protein chains are **amino acids** (**Figure 3-2a**). When amino acids are joined together by a **peptide bond** between the amino group of one amino acid and the carboxyl group of another, a peptide is formed. Longer chains of amino acids are called

polypeptides (**Figure 3-2b**). Polypeptides have a characteristic directionality defined by the free amino group of the amino acid at one end of the polymer (the amino terminus, or N-terminus) and the free carboxyl group at the other end (the carboxyl terminus, or C-terminus). Once incorporated into a polypeptide chain, the individual amino acids are referred to as amino acid residues (see Chapter 4). A functional protein may be formed from one polypeptide chain or from several interacting polypeptides. Proteins are abundant in all cells. They perform many functions, including catalyzing biochemical reactions, serving structural roles, receiving and transmitting chemical signals within and among cells, and transporting specific ions and molecules across cellular membranes. Most proteins found in cells and viruses are composed of just 20 different amino acids.

All 20 common amino acids have a similar structure: a central carbon atom, the **alpha carbon atom** (**α carbon**, or C_α), bonded to four different chemical groups. For this reason, they are called α-amino acids. The α-amino acids have a carboxyl ($-COOH$) group, an amino ($-NH_2$) group, and a hydrogen atom, all bonded to the α carbon. Each amino acid also has a unique side chain, or **R group**, bonded to the α carbon atom (see Figure 3-2). The R groups vary in structure, size, electrical charge, and hydrophobicity. The diverse chemical properties of R groups are what give proteins the ability to form many different three-dimensional structures and to perform many different kinds of activities in biological systems. The 20 common amino acids are represented by both

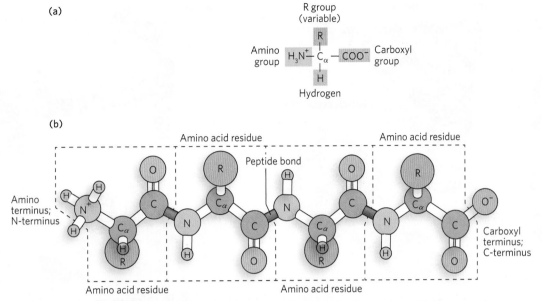

FIGURE 3-2 Chemical building blocks of proteins. (a) The structure of an amino acid. The central carbon atom (C_α) bonds to an amino group (blue), a carboxyl group (pink), a hydrogen, and a side chain (R, purple). The amino and carboxyl groups are shown in the ionized forms found in solution at physiological pH. (b) A segment of a polypeptide chain. Note that polypeptide chains have directionality, with a free amino group at one end (the amino terminus, or N-terminus) and a free carboxyl group at the other (carboxyl terminus, or C-terminus). Peptide bonds connect the amino acid residues in the polypeptide chain.

three-letter and one-letter abbreviations, which are used to indicate the composition and amino acid sequence of proteins. (The functional diversity of R groups and the nomenclature for the 20 common amino acids are discussed in detail in Chapter 4.) There are also many less-common amino acids, found both in proteins and as cellular constituents not incorporated into proteins. Note that with 20 different amino acid building blocks, the number of possible protein sequences (20^n, where n is the number of amino acids in the sequence) is vast.

Chemical Composition Helps Determine Nucleic Acid and Protein Structure

The fact that some of the crucial requirements for life are met by polymeric molecules makes good sense, from a biosynthetic standpoint. As we have noted, a huge variety of nucleic acids and proteins can be produced by varying the sequence of nucleotide or amino acid monomers in the chains. DNA molecules are typically many millions of nucleotides long, but they form relatively uniform overall structures in which the nucleotide bases in two strands pair up along their length to produce a double helix (**Figure 3-3a**). RNA molecules, except for those that store the genetic information of viruses, are much shorter and more structurally diverse than DNA. A single strand of RNA can fold back on itself to form short helices that

come together in a three-dimensional shape (**Figure 3-3b**). These differences between DNA and RNA structure stem from the role of RNA's 2′-hydroxyl groups in altering the shape and chemical properties of the sugar–phosphate backbone (which we explore in Chapter 6).

Of all biological polymers, proteins have the greatest variety of three-dimensional structures and range of functional groups, resulting from the different types of amino acid side chains. This variety underlies the role of proteins as the primary catalysts of chemical reactions (**Figure 3-4**). Of course, proteins also perform many other, noncatalytic cellular functions, made possible by the chemical diversity of their amino acid building blocks.

Chemical Composition Can Be Altered by Postsynthetic Changes

Chemical modifications of nucleotides and amino acids often occur after a DNA, RNA, or protein molecule has been synthesized. Sometimes these modifications are required for the molecule to attain its biologically active structure or to bind other molecules.

The primary modification of DNA nucleotides is the addition of methyl ($-CH_3$) groups to the C, A, and G bases (**Figure 3-5a**). DNA base methylation is critical for accurate DNA replication and, in bacteria, for the protection of DNA from degradative enzymes; in human

(a) **DNA**   (b)   **tRNA**

Self-cleaving RNA

Self-splicing intron

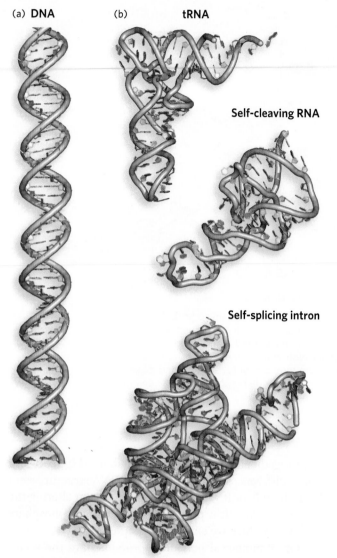

FIGURE 3-3 The helical structure of DNA and RNA. In these representations, the sugar–phosphate backbone is shown as a solid bar with the bases extending away from the backbone and available for base pairing. (a) Ribbon model of a DNA double helix, consisting of two strands of DNA. Base pairs in the helix twist around the central axis. (b) Ribbon models of three RNA molecules, each consisting of a single strand of RNA: phenylalanine-tRNA from yeast, a self-cleaving RNA from the hepatitis delta virus (HDV), and a self-splicing intron from *Tetrahymena*. Each RNA includes short stretches of helical structure that fold into a three-dimensional shape. As discussed in the text, the differences in chemical structure between DNA and RNA are the basis for the differences in the three-dimensional structures that they form. *[Sources: (b) PDB ID 1TRA (top), PDB ID 1DRZ (middle), PDB ID 1U6B (bottom).]*

(a) **Calmodulin**

(b) **Dicer**

(c) **Hemoglobin**

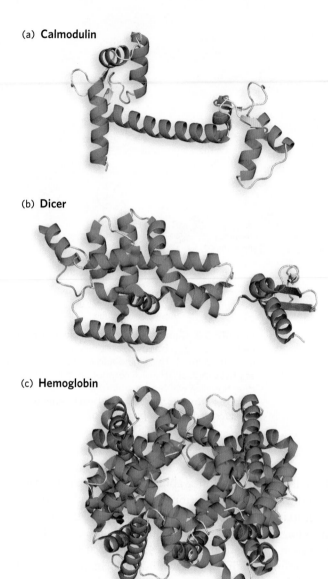

FIGURE 3-4 Examples of protein structures. Proteins can form a wide range of three-dimensional structures due to the variety of chemical properties of the 20 common amino acids. Shown here are (a) calmodulin, a Ca^{2+}-binding protein; (b) Dicer, an enzyme that cleaves double-stranded RNA; and (c) hemoglobin, the oxygen carrier in red blood cells. See Section 4.3 and Figure 4-10 for an explanation of how the molecular structures of proteins are represented throughout this book. *[Sources: (a) PDB ID 1CLL. (b) PDB ID 3C4B. (c) PDB ID 1HGA.]*

and other eukaryotic cells, it is essential for activating and silencing gene expression. RNA molecules can be modified in a greater variety of ways, including the addition of methyl groups to the nucleotide bases or to the 2′-hydroxyl group of the ribose and the substitution of less-common bases for the usual A, C, G, or U (**Figure 3-5b**). Such chemical changes affect the ability

of RNA molecules to fold into their correct three-dimensional structure and to interact with proteins.

Proteins are often modified by the addition of chemical groups to specific amino acid residues within a polypeptide chain. More than 300 types of amino acid modifications are known to occur in proteins; a few are particularly common. For example, addition of phosphate groups to hydroxyl groups in the side chains of serine, tyrosine, and threonine can dramatically change a protein's shape and function. The phosphorylation and

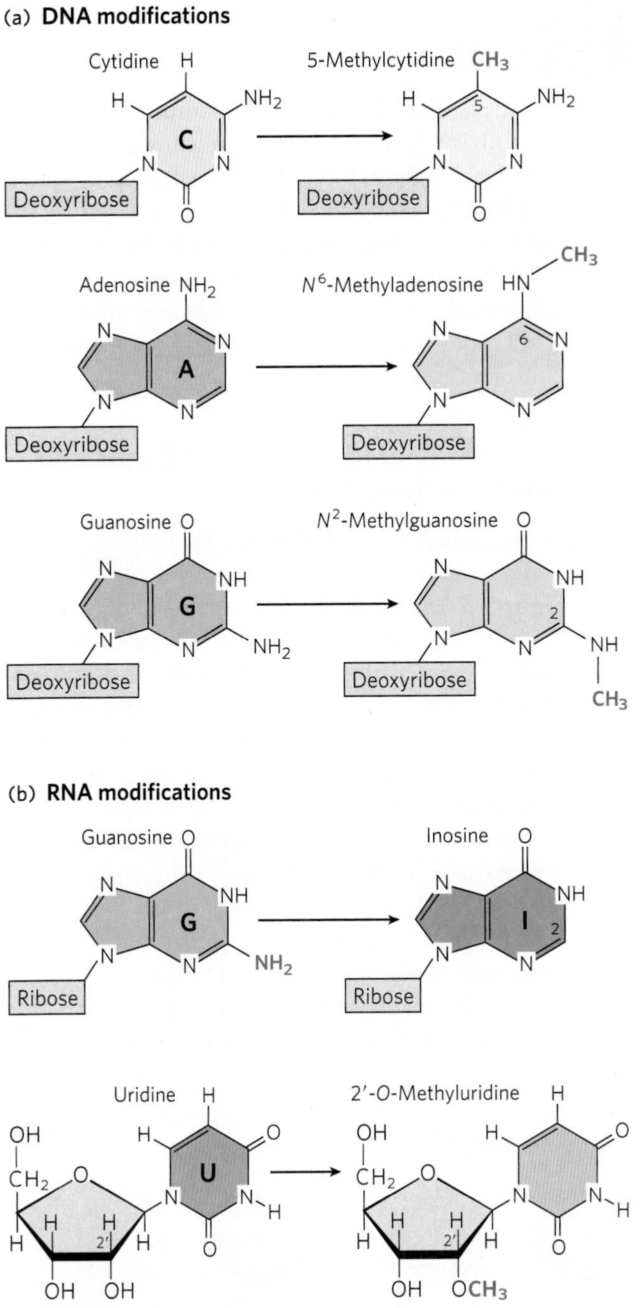

(a) DNA modifications

(b) RNA modifications

FIGURE 3-5 Chemical modification of nucleotide sugars and bases. (a) Examples of methylation modifications in DNA nucleotides. The extra methyl group on 5-methylcytidine, N^6-methyladenosine, and N^2-methylguanosine is highlighted in red. (b) Examples of modifications in RNA nucleotides. Loss of the amino group attached to the guanine ring (the exocyclic amine) gives rise to inosine; uridine can be methylated at the 2′ position to produce 2′-O-methyluridine. Modification sites are indicated in red.

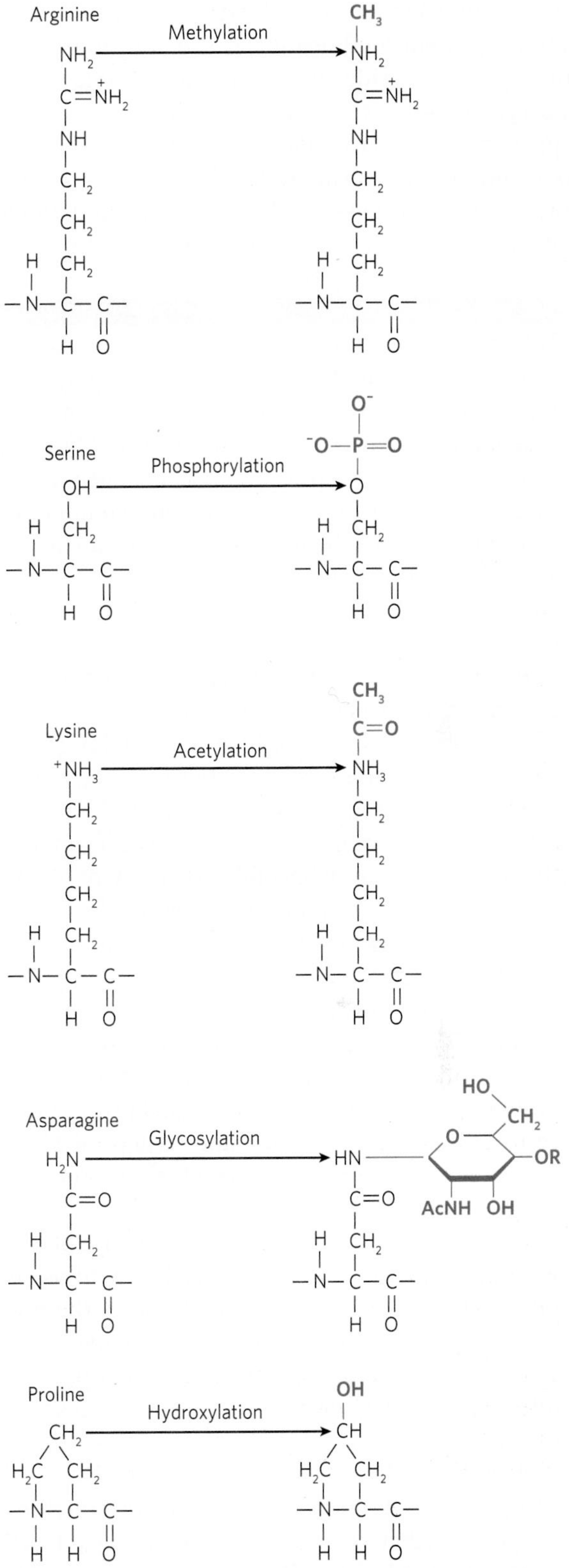

FIGURE 3-6 Chemical modification of some amino acid residues. Arginine can be methylated; the hydroxyl (—OH) group of serine is a frequent site of phosphorylation, as are those of tyrosine and threonine (not shown); lysine can be acetylated; asparagine can be glycosylated; and proline is sometimes hydroxylated. In each case, the modification alters the behavior of the protein containing the changed amino acid residue. Modification sites are indicated in red.

dephosphorylation of proteins is an important mechanism by which signals are transmitted within and among cells. Proteins are sometimes modified by the addition of sugars (glycosylation) or methyl groups, with functional consequences (**Figure 3-6**). For example, glycosylated proteins provide chemical signatures on the surfaces of cells that help distinguish "self" from "nonself." Another common protein modification is the addition of acetyl groups to lysine side chains. Lysine acetylation—and deacetylation—plays a central role in the production of proteins from particular genes. All these chemical modifications can substantially change the behavior of proteins, as we discuss in Chapters 5, 18, and 19 when taking a closer look at the varied functions of proteins.

- Polymeric molecules play crucial roles in all organisms.
- The nucleic acids, DNA and RNA, are polymers of nucleotides. Each nucleotide has three components: a deoxyribose (in DNA) or ribose (in RNA) pentose sugar, a phosphate group, and a nitrogenous base. The four bases in DNA are adenine, guanine, cytosine, and thymine; the four bases in RNA are adenine, guanine, cytosine, and uracil.
- DNA and RNA are chemically similar, with two small differences that have significant functional consequences, including different helical geometries, three-dimensional shapes, and protein-binding abilities. The ribose of RNA has a hydroxyl (—OH) group on the 2′ carbon of the sugar ring, but the deoxyribose of DNA does not; and, instead of thymine, RNA nucleotides contain uracil, an unmethylated form of the thymine base. The nucleotides of DNA or RNA are linked into chains by phosphodiester bonds.
- Proteins are polymers of amino acids. Twenty amino acid building blocks are commonly found in proteins, each consisting of a central α-carbon atom bonded to four different groups: a carboxyl group, an amino group, an R group, and a hydrogen atom. The R groups, or side chains, have chemical properties that contribute to the functional and structural diversity of proteins. The amino acid residues in proteins are linked by peptide bonds.
- DNA molecules form a two-stranded double helix, whereas RNA molecules are mainly found as single polynucleotide strands that fold back on themselves to create various three-dimensional shapes. Protein structures are even more diverse, due in part to the different chemical properties of the amino acid side chains.
- Postsynthetic chemical modifications of DNA, RNA, and proteins can dramatically affect the structure and biological activity of these macromolecules. In DNA, methylation of bases A, C, and G is common and leads to changes in gene expression. In RNA, modifications are more varied and include methylation of bases and/or ribose and other, more substantial alterations of bases. Protein modifications include the addition of phosphate, sugar, methyl, acetyl, and hydroxyl groups to specific amino acid side chains.

3.2 CHEMICAL BONDS

All molecules, whether table salt (sodium chloride) or a segment of DNA, are atomic aggregates in which the atoms are held together by attractive forces known as **chemical bonds**. Some chemical bonds are strong and can hold two atoms together indefinitely, but others are relatively weak and transient. In this section we discuss strong chemical bonds—covalent bonds and ionic bonds. Section 3.3 focuses on weaker chemical bonds and interatomic interactions.

Electrons Are Shared in Covalent Bonds and Transferred in Ionic Bonds

A **covalent bond** is formed when two atoms share a pair of electrons between their positively charged nuclei (**Figure 3-7a**). Atoms joined by a covalent bond tend to share electrons such that their outer electron shells are filled. Another type of chemical bond is an **ionic bond**, which, in contrast to covalent bonds, involves the complete transfer of one or more electrons from one atom to another. When electrons are transferred, the atoms are converted into ions—one having a positive charge and the other having a negative charge. The electrostatic attraction between the positive and negative ions holds them together (**Figure 3-7b**).

In the 1930s, Linus Pauling demonstrated that covalent bonds and ionic bonds represent opposite ends of a continuum; in reality, most chemical bonds lie somewhere in between. To determine whether two atoms in

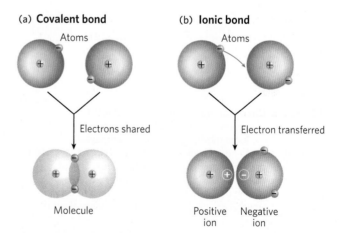

FIGURE 3-7 Covalent bonds and ionic bonds compared. (a) A covalent bond forms when two atoms share electrons so that their outer electron shells are filled. (b) An ionic bond forms when one or more electrons are completely transferred from one atom to another, such that one atom bears a positive charge, and the other a negative charge. Note the space between the atoms paired in an ionic bond; they are not as close together as atoms joined by a covalent bond.

Linus Pauling, 1901–1994
[Source: © Bettmann/CORBIS.]

a molecule are more likely to form a covalent or an ionic bond, Pauling introduced the concept of **electronegativity**, the propensity of an atom within a molecule to attract electrons to itself. Unequal electron sharing reflects different affinities of the bonded atoms for electrons. Atoms with a tendency to gain electrons are referred to as **electronegative atoms**, those with a propensity to lose electrons as **electropositive atoms**. In general, as atomic radius decreases, electronegativity increases and the atom has a greater likelihood of forming an ionic rather than a covalent bond. Ionic bonds often form between a metal and a nonmetal; the metal atom donates one or more electrons to the nonmetal atom to form a salt, such as sodium chloride. When Pauling formulated the concept of electronegativity, he developed a scale of dimensionless values ranging from 0.7 to 3.98 and assigned a value to each atom (e.g., hydrogen = 2.20). With this Pauling scale, the difference in electronegativity of two atoms can be used to predict the type of bonding between them (**Figure 3-8**). When this difference is zero or very small, the bond is purely covalent; when the difference is greater than zero but less than 1.67, the bond is considered to be **polar covalent**, meaning that

the electrons are shared between the atoms but biased toward one "pole" of the two-atom bond. A difference in electronegativity of greater than 1.67 gives rise to an ionic bond.

Although typically weaker than covalent bonds, ionic bonds do not restrict the relative orientations of the bonded atoms; thus, they are very useful in macromolecules. For example, ionic bonds—also called salt bridges—can form between pairs of oppositely charged amino acid side chains, such as arginine and glutamic acid (glutamate), to stabilize protein structure (**Figure 3-9**). In highly charged molecules such as nucleic acids, metal ions form ionic bonds with phosphate groups that help stabilize three-dimensional structure. RNA molecules require ionic bonding with magnesium ions to form their complex three-dimensional structures, which involve close packing of the negatively charged sugar–phosphate backbone (**Figure 3-10**).

The strength of an ionic bond can vary with the salt concentration and hydrophobicity of the environment. Some ionic bonds are very strong indeed. Sodium chloride is a good example of a molecule with a single strong ionic bond. Like other metals, sodium tends to form ionic bonds because it loses an electron, producing a positive ion that is strongly attracted to the negatively charged chloride ion. However, many ionic bonds found in biological molecules are weaker than the strong ionic or covalent bonds.

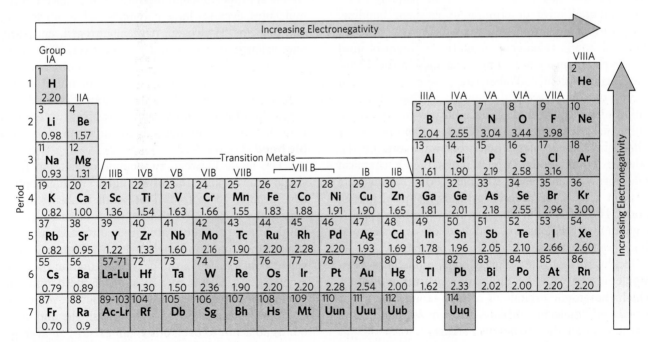

FIGURE 3-8 A periodic table of electronegativity, using the Pauling scale. The numbers below each element are the electronegativity values; low values = low electronegativity, and high values = high electronegativity. The electronegativity of an atom is affected by both its atomic weight and the distance of its outer electrons from its positively charged nucleus (i.e., its atomic radius). Electronegativity is affected by the molecular environment of an atom, and hence the electronegativities shown are average values.

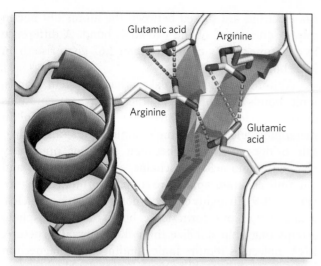

FIGURE 3-9 Salt bridges between oppositely charged amino acid side chains. In this representation of the three-dimensional structure of a polypeptide, the white tube represents the peptide backbone; dashed lines indicate salt bridges between the positively charged amino groups of an arginine side chain (nitrogens shown in blue) and the negatively charged carboxyl group of a glutamic acid side chain (oxygens shown in red). Only the side chains of the four residues involved in the salt bridges are shown. [*Source: PDB ID 1HGD.*]

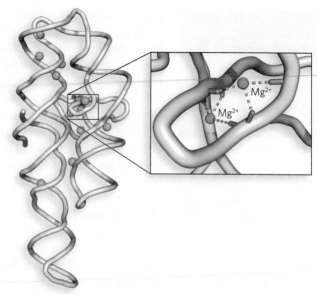

FIGURE 3-10 Stabilization of RNA structures by magnesium ions. The three-dimensional structure of this domain of a catalytic RNA molecule is critical to its biological function. Several magnesium ions (green) participate in salt bridges in the center of its folded form. The enlarged view shows magnesium ions stabilizing the RNA structure, with dashed lines indicating ionic bonds to the —O^- (red) groups of adjacent phosphates in the RNA backbone. [*Source: PDB ID 1GID.*]

Chemical Bonds Are Explainable in Quantum Mechanical Terms

Although the idea of shared electron pairs provides a useful qualitative description of covalent bonding, the nature of the strong and weak forces that produce chemical bonds remained unknown to chemists until the development of quantum mechanics in the 1920s. The German scientists Walter Heitler and Fritz London offered the first successful quantum mechanical explanation of molecular hydrogen in 1927, laying the foundation for predicting the structures and properties of other simple molecules. Their work was based on the **valence bond model**, which posits that a chemical bond forms when there is suitable overlap between the electron clouds, or **atomic orbitals**, of participating atoms. These atomic orbitals are known to have specific angular interrelationships, and thus the valence bond model can predict the bond angles observed in simple molecules. Today, the valence bond model has been supplemented with the **molecular orbital model**, in which the atomic orbitals of bonded atoms interact to form hybrid molecular orbitals. These molecular orbitals extend between the two bonding atoms.

Each element forms a characteristic number of bonds necessary to give it a complete outer shell of electrons. Because a complete outer shell, for most atoms, contains eight electrons, this is known as the octet rule. The maximum number of covalent bonds a particular atom can

form is called its **valence**. The valence of the atoms commonly found in biological molecules dictates the shape, chemical properties, and ultimately the behavior of these molecules, even for large polymers such as nucleic acids and proteins. Hydrogen, oxygen, nitrogen, and carbon have valences of 1, 2, 3, and 4, respectively. Thus, hydrogen can form just one covalent bond, and O, N, and C can form any combination of single or multiple bonds to make up the total allowable number (**Figure 3-11**). A **single bond** between two atoms involves two electrons. Four shared electrons between two atoms produce a **double bond** (**Figure 3-12**).

The angle between two bonds originating from a single atom is called the **bond angle**. The angle between two specific types of covalent bonds is always approximately the same. For example, the four single covalent bonds of a carbon atom are directed toward the corners of a tetrahedron (bond angle = 109.5°) (**Figure 3-13**). Covalent bonds differ in the degree of rotation they allow. Single bonds permit free rotation of the bound atoms around the bond, whereas double bonds are more rigid.

Many molecules that contain single and double bonds adjacent to each other can exist as an average of multiple structures, a phenomenon called **resonance**. A **resonance hybrid** is a molecule that exists in an average of two possible forms. A classic example of this in biology is the molecular structure around the peptide bond that links together two amino acids (**Figure 3-14a**).

Atom	Outer electrons	Usual number of covalent bonds	Bond geometry
Hydrogen	Ḣ	1	
Oxygen	·Ö·	2	
Nitrogen	·N̈·	3	
Carbon	·Ċ·	4	

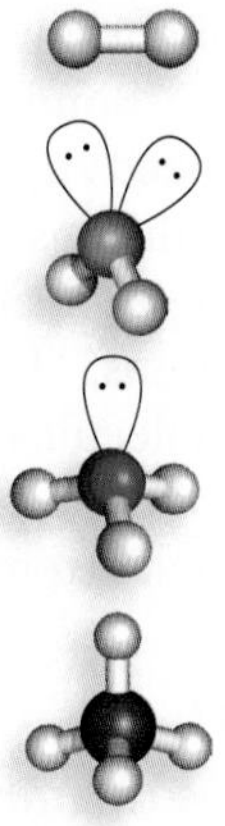

FIGURE 3-11 Valences of atoms that are common in biological molecules. Conventional representations of atoms known as Lewis structures (named for the chemist Gilbert Lewis) show the lone pairs of electrons—valence electrons—in the outer shell of an atom that are available for chemical bonding (second column). Examples of the resulting bond geometry are shown in the fourth column. [*Source: Data from H. Lodish et al.,* Molecular Cell Biology, *5th ed., W. H. Freeman, Table 2-1.*]

The peptide bond links the carboxyl group of one amino acid to the amino group of another. The carbonyl (C=O) and imino (C=N) bonds each have both double- and single-bond properties (**Figure 3-14b**). As a result, chemical groups bound together by the peptide bond in proteins must be located in the same plane, because the partial double-bond character of the carbonyl and imino bonds restricts rotation about these positions. As we shall see, this has profound consequences for the structure and function of proteins (see the How We Know section at the end of this chapter; also see Chapters 4 and 5).

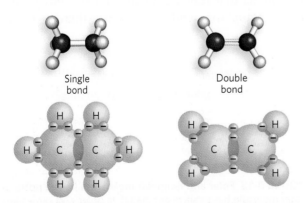

FIGURE 3-12 Shared electrons in single and double covalent bonds. Two electrons are shared between two atoms in a single covalent bond; four electrons are shared in a double covalent bond.

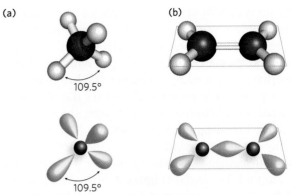

FIGURE 3-13 Geometry of single- and double-bonded carbon. (a) The four covalent bonds of a single-bonded carbon point to the corners of a tetrahedron, as shown in these representations of molecular structure (ball-and-stick model, top) and molecular orbitals (bottom). (b) The covalent bonds of double-bonded carbons lie in a plane.

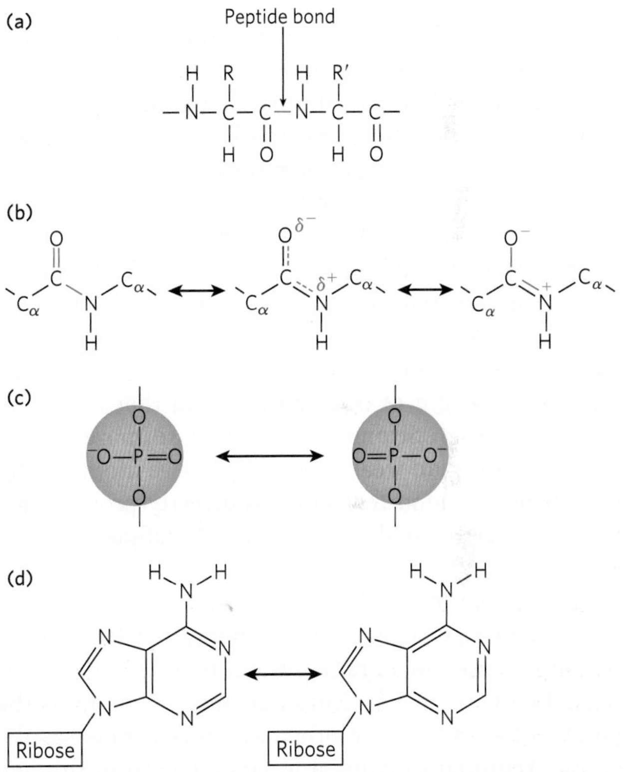

FIGURE 3-14 Resonance in peptides and nucleic acids. (a) A peptide bond covalently links the carboxyl group and amino group of adjacent amino acid residues in a protein. (b) Resonance between the resulting carbonyl and imino bonds gives each the properties of both a single and a double bond. Rotation is restricted about these bonds, and thus the attached chemical groups must lie in the same plane. Although the N atom in a peptide bond is often represented with a partial positive charge, as here, a careful consideration of bond orbitals and quantum mechanics indicates that the N has a net charge that is neutral or slightly negative. The partial positive and negative charges are represented by δ^+ and δ^-. (c) Resonance in the phosphate group (of the phosphodiester bond) of nucleic acids. (d) Resonance in the bases of nucleic acids; the adenosine nucleoside is shown here. See Chapter 6 for resonance structures of other bases.

Resonance also affects the behavior of nucleic acids, in multiple ways. The phosphodiester bonds that link the individual nucleotides of DNA and RNA have a tetrahedral geometry and include two bonds to oxygen atoms that are related by resonance (**Figure 3-14c**). As a consequence, the negative charge on the phosphate group can shift between the two oxygen atoms that are not bonded to sugars in the backbone. Also, the bases are conjugated ring systems—that is, they have alternating double and single bonds—giving rise to shared electrons around the ring(s) (**Figure 3-14d**). As we discuss in Chapter 6, the accuracy of base pairing between two DNA strands, A with T and C with G, results from the dominance of particular resonance structures of the bases.

Forming and Breaking Chemical Bonds Involves Energy Transfer

For a chemical bond to form, the total energy of the system—defined as the molecule and its environment—must be lower in the bonded state than in the nonbonded state. Therefore, bonding is an **exothermic** process, releasing energy when the bond is formed. The strength of a covalent bond increases with decreasing bond length; thus, two atoms connected by a single strong covalent bond, or by a double bond, are always closer together than identical atoms held together by a single weak covalent bond. As mentioned above, the electronegativity of an atom can be used to predict the type and strength of bonding between that atom and any other atom. Stronger bonds release more energy on formation than weaker bonds. For two atoms A and B, the rate of bond formation is directly proportional to the frequency with which A and B collide. In other words, A and B are more likely to bond if they bump into each other more often.

A calorie is the amount of energy needed to raise the temperature of 1 gram of water by 1 degree Celsius, from 14.5°C to 15.5°C. (Another unit of energy is the joule, equal to 0.239 calories, which is defined as the energy required to apply a force of 1 newton through a distance of 1 meter.) Energy changes in chemical reactions are typically expressed in kilocalories per mole (kcal/mol), because thousands of calories are involved in forming or breaking a **mole** of (that is, 6.02×10^{23}) chemical bonds. If energy is given off when two atoms combine to form a covalent bond, then the separated atoms must have had more total energy than the molecule. The amount of energy required to break a chemical bond exactly equals the amount that was released on its formation. This equivalence follows from the first law of thermodynamics, which states that energy, except where interconvertible with mass, can be neither created nor destroyed (see Section 3.6). This, then, is what holds atoms together in covalent bonds: they cannot separate unless they are given the required amount of energy.

Bonds frequently break on heating, because heat speeds up molecular motions, leading to intermolecular collisions in which some of the kinetic energy of a moving molecule is released as it pushes apart two bonded atoms. Higher temperatures produce faster-moving molecules and hence a greater chance that collisions will break bonds. Therefore, molecules are less stable at higher temperatures.

Electron Distribution between Bonded Atoms Determines Molecular Behavior

All chemical bonds, whether strong or weak, are the result of attractions between electrical charges. For example, the hydrogen molecule (H:H) has a symmetric distribution of electrons between its two hydrogen atoms, so both atoms are uncharged and the bond they share is purely covalent. In contrast, the polar covalent bonds of a water molecule (H:O:H) have a nonuniform distribution of charge. In water, the bonding electrons are unevenly shared due to the different electronegativity of H and O atoms (see Figure 3-8). In this case, the oxygen atom holds the bonding electrons more strongly and thus has a considerable negative charge, whereas the two hydrogen atoms together have an equal amount of positive charge (**Figure 3-15a**). Such a combination of separated positive and negative charges is called an **electric dipole moment**.

Molecules such as water that have a dipole moment are referred to as **polar molecules**. **Nonpolar molecules** are those with no effective dipole moment; an example is methane (**Figure 3-15b**). The large size of proteins and nucleic acids allows polar and nonpolar regions to exist within the same molecule. For example, the outer surfaces of proteins that function in the aqueous environment of the cytoplasm tend to be polar, thereby favoring

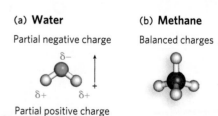

(a) Water
Partial negative charge
$\delta-$
$\delta+$ $\delta+$
Partial positive charge

(b) Methane
Balanced charges

FIGURE 3-15 Polar and nonpolar molecules. (a) The polar water molecule has positive and negative poles and carries an electric dipole moment. Dipole moment is a vector quantity and is represented by a small arrow pointing from the positive charge toward the negative charge. (b) A nonpolar methane molecule has no separation of charges.

interactions with polar water molecules. In contrast, proteins that function in the nonpolar environment of cellular membranes tend to have nonpolar surfaces, thus fostering contacts with the nonpolar fatty acid chains of the membrane.

SECTION 3.2 SUMMARY

- All molecules consist of atoms linked together by strong and/or weak chemical bonds.
- Covalent bonds share electrons equally between two atoms, whereas ionic bonds have electrons that are completely transferred from one atom to another, such that the charged atoms (ions) are drawn together by electrostatic forces. Electronegativity, a measure of how strongly an atom attracts electrons to itself, can be used to predict the type of bonding between two atoms.
- Valence is the maximum number of covalent bonds an atom can form, and the valence of the atoms in biological molecules dictates the shape of these molecules. Carbon, with a valence of 4, forms four single bonds to neighboring atoms arranged in a tetrahedral geometry.
- Resonance is an aspect of valence bond theory used to graphically represent and mathematically model molecules for which no single, conventional model can satisfactorily represent the observed molecular structure or explain its behavior. Such molecules are considered to be an intermediate or average, or resonance hybrid, of several conventional models that differ only in the placement of the valence electrons.
- Single bonds, in which a pair of atoms share two electrons, give rise to variable geometries, whereas double bonds, involving four shared electrons, give rise to planar molecular geometries.
- Exothermic bond formation is energetically favorable, and the total energy of the system is decreased in the process. The amount of energy released when a chemical bond breaks is the same as that required for its formation. Energy can be expressed in units of calories (cal) or kilocalories (kcal), or in joules. Molecular biologists typically describe energy changes in chemical reactions in terms of kilocalories per mole (kcal/mol).
- Attractions between electrical charges in atoms lead to chemical bond formation. Two bonded atoms are uncharged when the bonding electrons are positioned equally between them. When one atom holds the bonding electrons more tightly than the other, due to differences in the atoms' electronegativities, an unequal charge distribution results. One end of the molecule carries a net negative charge, the other a net positive charge. The molecule is said to have an electric dipole moment.
- Polar molecules are those with a dipole moment, such as water. Some molecules, such as methane, lack a dipole moment and are nonpolar. The polarity of biomolecules governs their locations within cells and their interactions with other molecules.

3.3 WEAK CHEMICAL INTERACTIONS

The macromolecules of most interest to molecular biologists—nucleic acids and proteins—are formed by the covalent joining of their constituent atoms. Because covalent bonds are relatively strong, stable, and not subject to spontaneous breakage under physiological conditions, they were once thought to be solely responsible for holding together the atoms in molecules. In contrast, weak chemical interactions, sometimes called **weak chemical bonds**, involve greater distances between atoms, are easily broken, and, individually, are transient. These properties, however, can be useful in biological systems, where transient chemical interactions are an essential part of cellular functions. For example, weak bonds mediate the interactions of proteins with small molecules, DNA, and/or other proteins. Such weak bonds enable countless critical processes in every cell, such as reversible binding of the hormone insulin to its receptor protein to regulate blood glucose levels or the interactions between adrenaline and receptor proteins that trigger the fight-or-flight response.

When arranged in ordered groups, weak bonds can persist for a long time and can thus play central roles in the formation and stability of the active three-dimensional shapes of macromolecules. DNA is a case in point: although DNA is a linear polymer of covalently linked nucleotides, its shape and its ability to encode genetic information are determined by the stable yet dynamic double-helical structure that it adopts. This structure is defined by a large number of individually weak contacts between nucleotides that are not covalently bonded. Likewise, protein structures are largely determined by weak interactions between amino acid residues that are not necessarily adjacent in the polypeptide sequence. Therefore, although they are not strong enough individually to effectively bind two atoms together, weak chemical interactions play central roles in the structure and behavior of biological macromolecules.

Three kinds of weak chemical interactions are important in biological systems: van der Waals forces, hydrophobic interactions, and hydrogen bonds. Most macromolecules also use weak ionic interactions, along with these weak chemical interactions, to bind other molecules and to form three-dimensional structures.

Van der Waals Forces Are Nonspecific Contacts between Atoms

The Dutch chemist Johannes van der Waals was the first to document the intermolecular forces that result from the polarization of atoms. When two atoms approach each other, as they get closer, induced fluctuating charges between them cause a weak, nonspecific attractive interaction. This **van der Waals interaction** depends heavily on the distance between the interacting atoms. As the distance decreases below a certain point, a more powerful van der Waals repulsive force is caused by overlap of the atoms' outer electron shells. The **van der Waals radius** of an atom is defined as the distance at which these attractive and repulsive forces are balanced and is characteristic for each atom (**Figure 3-16**).

The van der Waals bonding energy between two atoms that are separated by the sum of their van der Waals radii increases with the size of the atoms, but on average is only about 1 kcal/mol—just slightly above the average thermal energy of molecules at room temperature. This means that van der Waals forces are an effective binding force under physiological conditions only when they involve several atoms in each of the two interacting molecules. For several atoms to interact effectively in this way, the intermolecular fit must be exact, because the distance between any two interacting atoms must not be much different from the sum of their van der Waals radii.

The strongest kind of van der Waals contact arises when a macromolecule contains a surface that precisely fits the shape of the molecule that it binds. This is the case for antibodies, proteins that recognize antigens—specific molecules of viruses, bacteria, or other foreign particles that enter the body. Antibodies contain clefts with the same shape as the antigen they bind, enabling

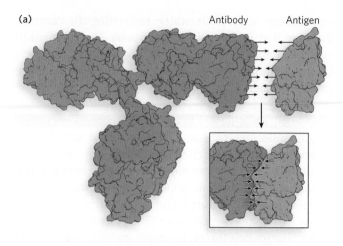

(a)

Antibody   Antigen

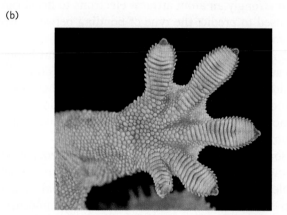

(b)

FIGURE 3-17 Examples of van der Waals interactions in nature. (a) The binding pockets of antibodies and the molecules they recognize, called antigens, typically bind through van der Waals interactions. (b) It is also van der Waals interactions that enable a gecko to climb vertical surfaces, through the enormous number of interactions between its foot pads and the molecules of the surface material, such as the glass shown here. [Sources: (a) Data from H. Lodish et al., Molecular Cell Biology, 6th ed., W. H. Freeman, Fig. 24-13; PDB ID 1IGT and PDB ID 3HFM. (b) Volker Steger/Science Source.]

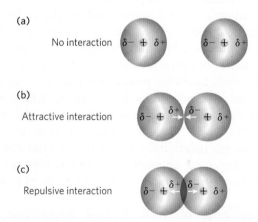

(a)

No interaction δ^- + δ^+ δ^- + δ^+

(b)

Attractive interaction δ^- + δ^+ → ← δ^+ + δ^+

(c)

Repulsive interaction δ^- + δ^+ δ^- → + δ^+

FIGURE 3-16 Attractive and repulsive van der Waals interactions. (a) Atoms that are farther apart than the sum of their van der Waals radii do not experience van der Waals forces. (b) Attractive interactions arise when atoms are separated by a distance equal to their combined van der Waals radii. (c) Repulsive interactions arise when atoms are closer than their combined van der Waals radii.

van der Waals contacts along the length of the bound antigen (**Figure 3-17a**). The additive effects of many van der Waals forces can be exceptionally strong; for instance, they are responsible for a gecko's ability to climb vertically on a glass surface and hang there by a single toe (**Figure 3-17b**).

The Hydrophobic Effect Brings Together Nonpolar Molecules

The **hydrophobic effect** arises from the strong tendency of water to exclude nonpolar groups, forcing these groups to aggregate in contact with one another. The word *hydrophobic* means "water-fearing," which describes the apparent behavior of nonpolar molecules in water. For example, when drops of oil are added to water, they combine to form a larger drop. This happens because water molecules are attracted to one another, due to their polarity, whereas the nonpolar oil molecules have no charged regions to repel or attract other molecules. The attractive forces between water molecules result in an unfavorable organization of water molecules in the vicinity of hydrophobic molecules, decreasing the disorder, or entropy (see Section 3.6), of the nearby aqueous environment. As the oil drops aggregate, minimizing the surface area in contact with water, the net disorder of the surrounding water increases. Nonpolar molecules are sometimes said to undergo "hydrophobic interactions," but their proximity is largely enforced by the effects of entropy in the surrounding aqueous medium. Such hydrophobic effects are common in biological molecules. For example, they are generally the dominant factor in stabilizing protein structures, and the energy required to unfold proteins goes mainly toward disrupting the stabilizing hydrophobic effect. The hydrophobic effect is critical to many other cellular functions as well, including insertion of protein molecules into membranes and secretion of hormones and other signaling molecules.

Adjacent Bases in Nucleic Acids Participate in Noncovalent Interactions

The helical structures in DNA and RNA are stabilized by electronic interactions that arise from the stacked arrangement of base pairs. In accordance with the molecular orbital model introduced earlier, the electrons of the atoms in a base's aromatic ring(s) are found in a decentralized cloud above and below the ring(s). **Pi stacking** is defined as the attractive, noncovalent interactions resulting from overlap between electrons of neighboring aromatic rings. This overlap of pi electrons of adjacent stacked nucleotides contributes to the stability of the nucleic acid's helical structure (**Figure 3-18**).

Hydrogen Bonds Are a Special Kind of Noncovalent Bond

A **hydrogen bond** is an attractive intermolecular force between two partial electrical charges of opposite polarity. As the name implies, one partner in the bond is a hydrogen, which must be covalently bonded to a strongly electronegative atom such as oxygen, nitrogen, or fluorine; this hydrogen is the hydrogen-bond donor. The electronegative atom attracts the electron cloud from around the hydrogen nucleus and, by decentralizing the cloud, leaves the hydrogen atom with a partial positive charge. This partial charge represents a large charge density that can attract the lone pair of electrons on another, nonhydrogen atom, which becomes the hydrogen-bond acceptor (**Figure 3-19**). Although other types of atoms can similarly acquire a partial positive charge when bonded to an electronegative element, only hydrogen is small enough to approach another atom or molecule close enough to undergo an energetically significant interaction.

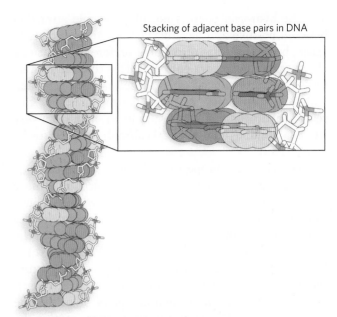

FIGURE 3-18 Electronic interactions in DNA. Stacking of the rings in adjacent bases of DNA involves favorable overlap of decentralized electrons above and below the plane of each base. Such molecular orbital overlap between bases (pi stacking) stabilizes the double helix. The DNA backbone is shown in white, with the phosphorus atoms of the phosphate groups in orange. Bases are colored as in Figure 3-3. The spheres indicate the van der Waals radii of the atoms in the bases.

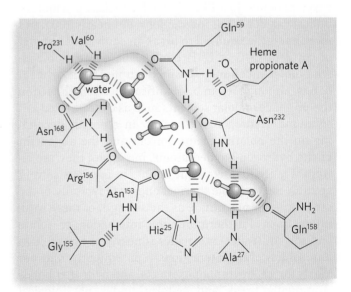

FIGURE 3-19 Hydrogen bonds. Water can donate or accept a hydrogen atom to form a hydrogen bond with another water molecule or another kind of atom. For example, shown here is a region of cytochrome f (a photosynthetic protein) in which water molecules form hydrogen bonds with one another and with amino acid residues of the protein. Each hydrogen bond is represented by three vertical blue lines; this convention is used in figures throughout the book. [*Source: Data from P. Nicolls, Cell. Mol. Life Sci. 57:987–92, 2000, Fig. 6a (redrawn from information in the PDB and a Kinemage file published by S. E. Martinez et al., Protein Sci. 5:1081–92, 1996).*]

The hydrogen bond is not like a simple attraction between positive and negative charges at two points in space, but instead has directional preference and some characteristics of a covalent bond. This covalent character is more pronounced when acceptors hydrogen-bond with donors on more-electronegative atoms. For this reason, hydrogen bonds can vary in strength from very weak (a bonding energy of 2 kcal/mol) to fairly strong (7 kcal/mol).

The chemical properties of water are mostly due to hydrogen bonds that cause water molecules to have strong attractions for one another, but hydrogen bonds can and do form among many other kinds of molecules as well. In proteins, hydrogen bonds occur between the backbone carbonyl and imino groups to stabilize α helices and β sheets, the basic motifs of protein structure that we discuss in Chapter 4. In nucleic acids, hydrogen bonding between complementary pairs of nucleotide bases links one DNA strand to its partner, forming the double helix (see Figure 6-11).

Combined Effects of Weak Chemical Interactions Stabilize Macromolecular Structures

The noncovalent interactions we have discussed (van der Waals forces, hydrophobic effects, pi bonds/stacking, and hydrogen bonds) are substantially weaker than covalent bonds. Just 1 kcal is needed to disrupt a mole of typical van der Waals interactions, but nearly 100 times more energy is required to break an equivalent number of covalent C–C or C–H bonds. The hydrophobic effect, too, is much more easily disrupted than covalent bonds, although it can be significantly strengthened in the presence of polar solvents such as concentrated salt solutions. Hydrogen bonds vary in strength depending on the polarity of the solvent and the alignment of the hydrogen-bonded atoms, but again, they are always weaker than covalent bonds. In aqueous solvent at 25°C, the available thermal energy is typically of the same order of magnitude as the strength of these weak interactions. Furthermore, the interaction of solute and solvent (water) molecules is nearly as favorable as any solute-solute interactions. Consequently, van der Waals forces, hydrophobic effects, and hydrogen bonds continually break and re-form under physiological conditions.

As mentioned earlier, although these types of interactions are individually weak relative to covalent bonds, the cumulative effect of many such interactions can be significant. Macromolecules, including DNA, RNA, and proteins, contain so many sites of possible van der Waals or hydrophobic contacts and hydrogen bonding that the combined effect of these small

binding forces is largely responsible for their molecular structure. For macromolecules, the most stable structure is usually that in which weak interactions are maximized. This principle determines the folding of a single polypeptide or polynucleotide chain into its three-dimensional shape. Complete unfolding of the structure requires the removal of all these interactions at the same time. And because these contacts are breaking and re-forming rapidly and randomly, such synchronized disruptions are very unlikely. The molecular stability conferred by many weak interactions is therefore much greater than one might expect intuitively, based on a simple summation of many small binding energies.

A special class of weak interactions in macromolecules involves the water molecules that are invariably bound to surface and interior sites by hydrogen bonds. Sometimes these water molecules are so well positioned that they behave as though they are an integral part of the macromolecule. In many cases, bound water molecules are essential to macromolecular function. Certain DNA-binding proteins, for example, use water molecules that are integral to their structure to help recognize specific DNA sequences. In RNA structures, water molecules bridge nucleotide bases that are involved in nontraditional base pairing and in interactions involving three nucleotides (base triples), thereby enabling unique three-dimensional structures to form (**Figure 3-20**).

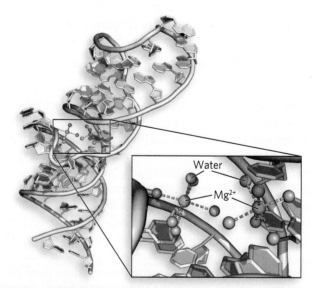

FIGURE 3-20 Ordered water molecules in an RNA molecule. Hydrogen bonding between water molecules and nucleotides is essential to the three-dimensional structure of an RNA molecule. Magnesium ions form hydrogen bonds with water molecules and with a phosphate oxygen (red) in the RNA. *[Source: PDB ID 1DUL.]*

Weak Chemical Bonds Also Facilitate Macromolecular Interactions

Interactions among DNA, RNA, and protein molecules, and interactions between these macromolecules and small organic molecules, are also mediated by weak, noncovalent chemical interactions. Such contacts involving information-carrying macromolecules govern how and when cells replicate, repair and recombine their DNA, synthesize RNA and proteins, detect and respond to chemical signals, and conduct all the other activities essential for life.

For example, the noncovalent binding of a protein to another protein, to a small molecule (such as a hormone), or to another macromolecule (such as a nucleic acid) may involve several hydrogen bonds and one or more ionic, hydrophobic, and/or van der Waals interactions (**Figure 3-21**). These weak contacts contribute, overall, to an energetically favorable interaction. It is worth noting, however, that because each hydrogen bond between two groups in a protein or nucleic acid forms at the expense of *two* hydrogen bonds between the same groups and two water molecules, the net stabilization is not as great as it might seem. The large size of nucleic acids and proteins provides extensive molecular surfaces with many opportunities for weak interactions with other molecules. The energetic favorability of such interactions reflects the molecular surface complementarity and the resulting large numbers of weak interactions of polar, charged, and hydrophobic groups on the surfaces of these molecules.

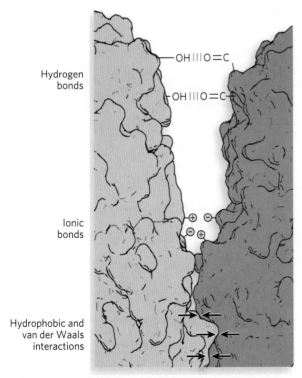

FIGURE 3-21 Stabilization of macromolecular structure by weak interactions. The combined effects of multiple noncovalent forces, including hydrogen bonds, weak ionic bonds, hydrophobic effects, and van der Waals interactions, allow the specific and stable association of molecules—in this case, two proteins. [*Source: Data from H. Lodish et al.,* Molecular Cell Biology, *6th ed., W. H. Freeman, Fig. 2-12.*]

SECTION 3.3 SUMMARY

- Weak chemical bonds differ from covalent or ionic bonds in several ways: they involve greater distances between atoms, they are easily broken, and they are often transient. These properties are useful when short-lived chemical interactions are required in cells. Weak bonds often mediate the interactions of proteins with small molecules, DNA, hormones, or other proteins.

- Van der Waals forces, hydrophobic effects, and hydrogen bonds are the three most important kinds of weak chemical interactions found in macromolecules that facilitate binding to other molecules and the formation of three-dimensional structures. Weak ionic interactions are also used.

- Van der Waals forces occur when two atoms closely approach each other, inducing fluctuating charges that cause a weak, nonspecific, attractive interaction between them. Each type of atom has its own van der Waals radius, the distance at which attractive and repulsive forces with neighboring atoms are balanced.

- The hydrophobic effect, giving rise to "hydrophobic interactions," arises from the strong tendency of water to exclude nonpolar groups, forcing these groups into contact with one another. Hydrophobic contacts stabilize protein structures and the stacking of bases in DNA and RNA helices.

- Overlap of pi electrons of adjacent stacked nucleotides contributes to the stability of the helical structure of nucleic acids.

- Hydrogen bonds occur between two atoms with partial electrical charges of opposite polarity, one of which is a hydrogen atom. Other types of atoms can also acquire partial positive charge, but only hydrogen atoms are small enough to approach another atom or molecule close enough for an energetically useful interaction.

- Although weak chemical interactions have only minor attractive or repulsive effects individually, the cumulative effects can be significant. DNA, RNA, and proteins contain so many sites of possible van der Waals or hydrophobic contacts and hydrogen bonding that the combined effect of many small binding forces is largely responsible for their molecular structure.

3.4 STEREOCHEMISTRY

The concept of **stereochemistry**, the spatial arrangement of the atoms within a molecule, is critical for understanding the structures and activities of biological molecules. As we shall see, the orientation of the chemical bonds of nucleic acids and proteins influences how these molecules fold in three dimensions, bind to other molecules, and catalyze reactions.

Three-Dimensional Atomic Arrangements Define Molecules

Our understanding of stereochemistry stems from a discovery by the French physicist Dominique Arago in 1811. Arago found that plane-polarized light, which vibrates in just one plane, rotates when it is sent through a piece of quartz crystal. Other scientists subsequently showed that many, but not all, substances share the ability to rotate the plane of plane-polarized light; a substance with this ability is said to be **optically active**. In 1848, 26-year-old chemist and crystallographer Louis Pasteur proposed that if a molecule is not superposable on its mirror image, it will be optically active; otherwise, it is optically inactive. This bold prediction subsequently proved to be correct. In fact, all objects can be classified as those that can be superposed on their mirror image, such as golf balls and champagne glasses, and those that cannot, such as hands. Objects that can be superposed on their mirror image are said to be **achiral**; those that cannot are **chiral** (**Figure 3-22**). All optically active chemical compounds are chiral, and all optically inactive compounds are achiral.

This is an important idea in molecular biology, because many biologically relevant molecules are chiral. For example, any carbon atom connected to four different groups—such as the α-carbon atom in an amino acid or certain carbons in the pentose sugars of nucleotides—is an asymmetric carbon atom, or **chiral center**. Thus, all nucleic acids and proteins are chiral, because they contain carbon atoms that are chiral centers, as we shall see shortly.

(a) **Chiral molecule:**
Rotated molecule *cannot* be superimposed on its mirror image

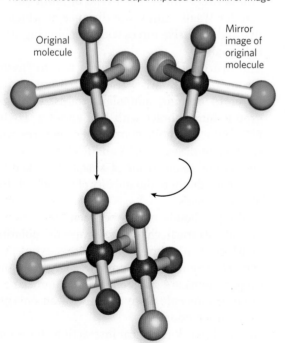

Original molecule

Mirror image of original molecule

(b) **Achiral molecule:**
Rotated molecule *can* be superimposed on its mirror image

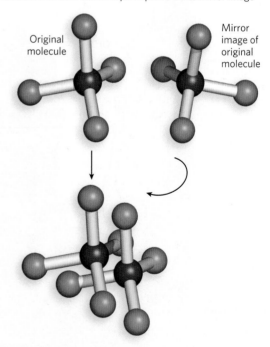

Original molecule

Mirror image of original molecule

FIGURE 3-22 Chiral versus achiral molecules. (a) A molecule is chiral if it is not identical to its mirror image (i.e., it cannot be superposed on its image). When bonded to four atoms or functional groups that are not identical to each other, a carbon atom is chiral, because the four atoms or functional groups can be arranged in two different ways that are not superposable mirror images. All four of the bonded groups must be different from each other for the carbon center to be chiral. Thus, the two forms of the molecule have the same chemical formula but different chemical behavior. (b) In contrast, a carbon atom bonded to four identical atoms is achiral, because only one configuration is possible and any arrangement of the four bound atoms is superposable on its mirror image.

Two different arrangements of atoms in a molecule that cannot be interconverted without breaking and re-forming one or more covalent bonds are said to have different **configurations**. Two different arrangements of atoms that can be interconverted without breaking and re-forming a covalent bond, such as by rotation about a single bond, are said to have different **conformations**.

Biological Molecules and Processes Selectively Use One Stereoisomer

Two molecules that have the same chemical and structural formula but differ in the arrangement of their atoms in space (i.e., are not superposable) are called **stereoisomers**. A pair of stereoisomers that are also mirror images of each other are called **enantiomers**. All physical and chemical properties of enantiomers are identical, except for two. First, the two enantiomers rotate the plane of plane-polarized light in opposite directions. Second, two enantiomers react at the same rate with any achiral compound but at different rates with any chiral compound.

For particular reference molecules selected by early chemists, including Pasteur, the one that rotates light to the right is called the (+) or d form (*dextrorotatory*), and the one that rotates light to the left is called the (−) or l form (*levorotatory*). Note that these "d" and "l" prefixes are distinct from the "D" and "L" prefixes (in small capitals) that refer to the actual configuration of each enantiomer (regardless of their light-rotating properties). Many chiral biological molecules are defined as D or L configurations based on their chemical relationship to the two forms of a reference compound, whether or not a pure solution of one form of that molecule is levorotatory or dextrorotatory. For amino acids, the reference compound is glyceraldehyde, and the version synthesized from (+) glyceraldehyde is considered to be the D form. Nine of the nineteen L-amino acids commonly found in proteins (glycine is achiral), however, are actually dextrorotatory!

Many biological molecules are chiral, and living organisms usually use only one of the two possible enantiomers. Thus, biochemical reactions have evolved to recognize and favor one stereoisomer over the other. For example, the simple sugar glucose is chiral. D-Glucose can be broken down by digestive enzymes, which are also chiral, at a rate commensurate with its use as an energy source. Because digestive enzymes are chiral, they do not recognize or digest glucose that has the opposite chirality (L-glucose), even though the two sugar molecules are identical in chemical and structural formulas.

Proteins and Nucleic Acids Are Chiral

Proteins are chiral because nearly all of the common amino acids contain an α carbon bonded to four different groups: a carboxyl group, an amino group, an R group, and a hydrogen atom. (The amino acid glycine is the exception, because its R group is simply a hydrogen atom.) The α carbon is therefore a chiral center, and the tetrahedral arrangement of the bonding orbitals around the α-carbon atom enables the four different functional groups to occupy two different possible spatial arrangements. This means that each amino acid, except glycine, has two possible stereoisomers. Because these are nonsuperposable mirror images of each other, the two forms are enantiomers (**Figure 3-23a**). Like all molecules with a chiral center, amino acids are optically active and rotate plane-polarized light. Nature uses only one of these enantiomeric forms in proteins, and it is the same one for each amino acid—the L form. D-Amino acids are almost never found in nature—a phenomenon that has been cleverly harnessed by researchers to design new therapeutics (**Highlight 3-1**).

Nucleic acids, too, are chiral molecules. Nucleotides contain several chiral centers in the ribose or deoxyribose ring—all of the carbon atoms in ribose except for C-5′ and all carbons in deoxyribose except for C-2′ or C-5′ are asymmetric (**Figure 3-23b**). Again, nature uses one enantiomeric form of the sugar—D-ribose or D-deoxyribose—in the building blocks for RNA and DNA. Enzymes that act on DNA or RNA have evolved to recognize this chiral arrangement of the nucleic acids.

SECTION 3.4 SUMMARY

- A molecule with a structure that cannot be superposed on its mirror image is chiral.
- All nucleic acids and proteins are chiral, because they contain carbon atoms bonded to four different atoms or groups. Because the four bonds of carbon point to the four corners of a tetrahedron, and the four functional groups can be spatially arranged in two different ways, these carbon atoms are chiral centers.
- Stereoisomers are pairs of molecules that have the same chemical and structural formulas but are not superposable on each other, because they differ in the spatial arrangement of their atoms. Biochemical reactions have evolved to recognize and favor one stereoisomer over the other.
- Enantiomers are pairs of stereoisomers that are mirror images of each other. Enantiomers rotate the plane of plane-polarized light in opposite directions. The D and L forms of major biological molecules are defined not by the direction in which they rotate plane-polarized light but by their structural relationship to a reference compound—glyceraldehyde in the case of α-amino acids.

(a)

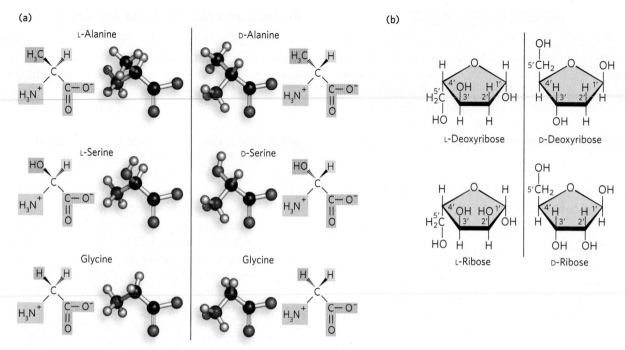

(b)

FIGURE 3-23 Enantiomers of amino acids and nucleotides. (a) Each amino acid, except glycine, has two possible stereoisomers. These nonsuperposable mirror images (enantiomers) are known as L and D forms. Only the L forms of amino acids are found in natural proteins. (b) All of the carbon atoms in ribose except C-5′ and all in deoxyribose except C-2′ and C-5′ are chiral centers. Nature uses only the D-enantiomeric form of the sugars—D-deoxyribose or D-ribose—as the building blocks for DNA and RNA.

HIGHLIGHT 3-1   MEDICINE

The Behavior of a Peptide Made of D-Amino Acids

L-Peptides capable of binding and blocking the function of important therapeutic targets—such as viral enzymes or receptor proteins—have turned out to be virtually useless for treating patients, because once introduced into the body, they are quickly destroyed. This is because cells and blood serum contain proteases, enzymes that bind to normal, L-amino acid–containing peptides and catalyze their rapid degradation. To get around this problem, Peter Kim, then an MIT professor and Howard Hughes Medical Institute investigator, wondered whether unnatural peptides made of D-amino acids would be useful as therapeutics in cells or serum.

Kim and his coworkers synthesized a portion of CD8, the human immunodeficiency virus (HIV) protein required for infection, in the unnatural, D-amino acid form, and they used it to identify L-amino acid peptides that bind specifically to this protein (Figure 1, left). These L-amino acid peptides were then resynthesized, but this time from D-amino acids. Because D- and L-amino acids are mirror images of each other, the new D-amino acid peptides bound to the target CD8 protein of the natural (L) form (Figure 1, right). Thus, the enantiomeric D-amino acid peptides could be used

therapeutically to block the activity of the natural viral protein, while avoiding degradation by human enzymes that act only on natural peptides. This clever approach was subsequently used to make D-peptide–based drugs that were tested in clinical trials for inhibiting the entry of HIV into cells. Although these peptides were not effective as HIV therapeutics, in part due to difficulties in getting them into infected cells, the strategy of using biologically inert enantiomers as drugs remains attractive.

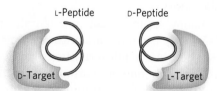

FIGURE 1 Kim's experiment to isolate L-amino acid peptides that bind to an unnatural target protein and then synthesize D-peptides that will bind to the natural protein. Because of the mirror-image relationship between L- and D-amino acids, L-peptides that bind to the D form of a target protein have the same sequence as D-peptides that bind to the L form of the target protein. [*Source: Data from T. N. Schumacher et al., Science 271:1854–1857, 1996, Fig. 1.*]

- Only the L enantiomers of amino acids are found in natural proteins. Like all molecules with a chiral center, amino acids are optically active and rotate plane-polarized light.
- Only the D enantiomer of the pentose sugar—D-ribose or D-deoxyribose—is used in the building blocks for RNA and DNA.

3.5 ## THE ROLE OF pH AND IONIZATION

Although pH and the ionization potential of molecules may seem like esoteric topics for biology, they are central to the function of biological molecules. This is because most biological processes occur within a narrow pH range, and cells and their internal compartments carefully regulate pH to suit their needs. For example, foreign molecules that enter cells are degraded in lysosomes, organelles that maintain an interior pH of 4.5, an acidic environment that is optimal for the function of the lysosome's digestive enzymes. The pH of blood must be maintained near neutral pH—neither acidic nor basic. Blood with a pH lower than 7.35 is too acidic to carry oxygen efficiently, and blood with a pH above 7.45 is too basic (alkaline). In most cases, the activities of proteins and nucleic acids depend on carefully controlled pH, and various mechanisms have evolved for regulating the pH in cells and organelles.

The Hydrogen Ion Concentration of a Solution Is Measured by pH

Many of the macromolecules and chemical reactions of central importance in molecular biology naturally occur in **aqueous solutions**, which are mixtures of molecules dissolved in water. Such solutions can be acidic, neutral, or basic, depending on the concentration of positively charged water molecules, called hydronium ions (H_3O^+). For example, a solution with $[H_3O^+] = 4 \times 10^{-3}$ mol/L is more acidic and less basic than one with $[H_3O^+] = 5 \times 10^{-4}$ mol/L. Note that we consider hydronium ions rather than hydrogen ions (H^+), because any H^+ ions (protons) in an aqueous environment quickly bind to water molecules to form H_3O^+. Because small numbers such as 4×10^{-3} or 5×10^{-4} are difficult to work with, Sören Sörenson devised another way to express $[H_3O^+]$. In 1909, he defined the quantity **pH** as the negative logarithm of the hydronium ion concentration:

$$pH = -\log [H_3O^+] \qquad (3\text{-}1)$$

It is simple to convert $[H_3O^+]$ to pH, and vice versa.

Similarly, we can define **pOH** as $-\log [OH^-]$. We can easily express the acidity or basicity of an aqueous solution by using pH and pOH; by common practice, however, pH is used rather than pOH. In aqueous solutions at equilibrium, the product $[H_3O^+][OH^-] = 10^{-14}$, and pH + pOH must equal 14. Thus, a solution with a pH of 4.6 has a pOH of 9.4. A solution of pH 7 is neutral, because $[H_3O^+] = [OH^-]$. In a solution with a pH lower than 7, $[H_3O^+] > [OH^-]$ and the solution is acidic; in one with a pH greater than 7, $[H_3O^+] < [OH^-]$ and the solution is basic. The lower the pH, the more acidic the solution; the higher the pH, the more basic the solution. Because pH values are on a logarithmic scale, a change in pH of one unit corresponds to a tenfold change in hydronium ion concentration.

In the laboratory, the pH of an aqueous solution can be determined by using chemical compounds (pH indicators) that change color over a narrow pH range, or with an instrument called a pH meter. Many foods and other common household substances are acidic or basic aqueous solutions. **Figure 3-24** lists approximate pH values of some foods and common materials, as well as some body fluids.

Buffers Prevent Dramatic Changes in pH

Most macromolecules found in biological systems have evolved to function at approximately neutral pH, because most cells and body fluids (other than stomach acid) are neither acidic nor basic. For example, the pH of normal human blood is 7.4, a value that is critical for the health of an individual. If the pH of blood were to rise to 7.8 or drop to 7.0, serious illness or death could result, because the hemoglobin protein in red blood cells would no longer be able to bind and release oxygen efficiently.

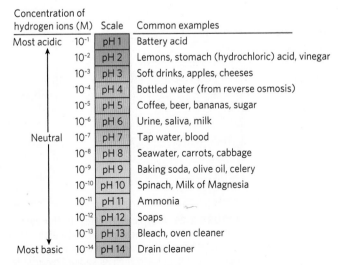

FIGURE 3-24 The pH scale. Shown here are pH values for some common substances and body fluids.

The body is able to maintain the correct pH of blood, despite the constant influx of nutrients with much higher or lower pH, because blood is buffered.

A **buffer solution**, a solution with a pH that does not change very much when H_3O^+ or OH^- ions are added, contains approximately equal amounts of a weak acid—that is, an acid that does not release all of its hydrogen atoms in solution—and its conjugate base, which is formed by the removal of a proton from the acid molecule. The dissociation of a weak acid (HA) in aqueous solution can be written as follows:

$$HA + H_2O \rightleftharpoons A^- + H_3O^+ \qquad (3\text{-}2)$$

Buffers are present in biological fluids, and they can also be prepared in the laboratory. A typical buffer solution might consist of 0.1 mol of acetic acid (CH_3COOH) and 0.1 mol of sodium acetate (CH_3COONa) dissolved in 1 L of water. Not much of the acetic acid, a weak acid, loses a proton, but just about all of the sodium acetate, an ionic compound, will dissociate into the separate ions, generating the conjugate base of acetic acid. Such a solution will have equal concentrations of the weak acid acetic acid (CH_3COOH) and its conjugate base, the acetate ion (CH_3COO^-). This solution acts as a buffer because it neutralizes any H_3O^+ or OH^- that may be added to the solution by accepting or donating a proton, respectively—up to a certain limit (**Figure 3-25**). Any added H_3O^+ ions interact with the negatively charged acetate ions and are

neutralized, whereas added OH^- ions are neutralized by the acetic acid molecules.

Every buffer solution has two major properties: its pH value and its buffering capacity. The buffer solution has a characteristic value called its **acid dissociation constant** (K_a), which equals the concentration of conjugate base multiplied by the concentration of H_3O^+, divided by the concentration of weak acid. We can write this as follows:

$$K_a = [A^-][H_3O^+]/[HA] \qquad (3\text{-}3)$$

where HA is the weak acid and A^- is its conjugate base. Just as we use pH to describe the H_3O^+ concentration on a log scale, we can define $pK_a = -\log K_a$. Each acid has a characteristic **pK_a** value that describes the ratio of charged (A^-) to neutral acid (HA) molecules at equilibrium in water. A pK_a is therefore a measure of the tendency for an acid to lose its proton in aqueous solution. The lower the pK_a, the stronger the acid and the stronger its tendency to give up its proton. According to Equation 3-3, the pH of a buffer solution prepared by dissolving equal numbers of molecules (moles) of a weak acid and its conjugate base in water equals the pK_a of the weak acid. If $[HA] = [A^-]$, then $[H_3O^+] = K_a$, and $pH = pK_a$. Buffers work well within a pH range that is within one pH unit above or below their pK_a. Thus, buffer solutions can be prepared for almost any desired pH by selecting the appropriate weak acid. Molecular biologists often use buffers in the pH range of 6 to 8 to work with biological

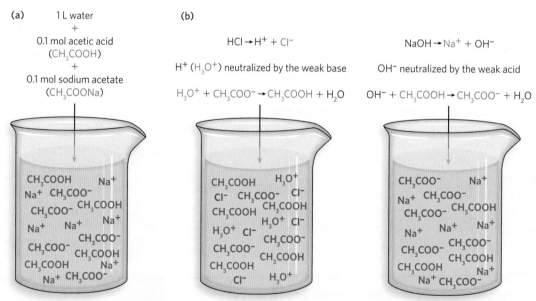

FIGURE 3-25 A typical chemical buffer system. (a) A buffer prepared with 0.1 mol of acetic acid and 0.1 mol of sodium acetate dissolved in 1 L of water has equal concentrations of the weak acid (CH_3COOH) and its conjugate base (CH_3COO^-). (b) When a strong acid such as hydrochloric acid (left) or a strong base such as sodium hydroxide (right) is added to the solution, an excess of H_3O^+ or OH^- ions is introduced. The solution acts as a buffer by neutralizing the incoming H_3O^+ or OH^- ions, keeping pH constant.

molecules that fold and function near physiological (neutral) pH.

Note that a buffer solution has a limit to its ability to neutralize acid or base, beyond which its buffering action is overwhelmed. This **buffering capacity** depends on the total concentrations, rather than the ratio, of HA and A^-. For example, consider a buffer solution Y containing 10 times more molecules of acetic acid and its conjugate base than does buffer solution Z. Both solutions have the same pH (~5), but solution Y has 10 times the buffering capacity of buffer Z because it has 10 times more molecules available to neutralize H_3O^+ and OH^- ions.

The Henderson-Hasselbalch Equation Estimates the pH of a Buffered Solution

A defined relationship exists between the pH of a solution and the concentration of a weak acid dissolved in it. The relationship is defined by the Henderson-Hasselbalch equation:

$$pH = pK_a + \log [A^-]/[HA] \qquad (3\text{-}4)$$

In other words, the pH of a solution containing a buffering weak acid equals the pK_a of that weak acid plus the log of the ratio of the concentration of its basic and acidic forms. Using this equation, it is possible to calculate the pH of a buffered solution. Or, if the solution pH and the concentrations of the basic and acidic forms of the weak acid are known, the pK_a of the weak acid can be determined.

This information can be very useful for working with biological samples, such as blood or proteins, where the pH of the solution must be carefully controlled to avoid destruction of the sample. Molecular biologists use the Henderson-Hasselbalch equation to prepare buffered solutions of a specific pH in the laboratory.

▌ SECTION 3.5 SUMMARY

- Aqueous solutions can be acidic, neutral, or basic, depending on the concentration of hydronium ions (H_3O^+) present.
- The pH value is defined as the negative logarithm of the hydronium ion concentration:

$$pH = -\log [H_3O^+]$$

- A solution with a pH lower than 7 is acidic, and $[H_3O^+] > [OH^-]$; a solution with a pH greater than 7 is basic, and $[H_3O^+] < [OH^-]$. Because pH values are on a logarithmic scale, a change in pH of one unit corresponds to a tenfold change in hydronium ion concentration.
- A buffer solution is a solution with a pH that does not change very much when H_3O^+ or OH^- ions are added.

It contains approximately equal amounts of a weak acid and its conjugate base.

- Each acid has a characteristic pK_a value, defined as $pK_a = -\log K_a$. The pK_a describes the ratio of charged (A^-) to neutral acid (HA) molecules at equilibrium in the solution.
- The relationship between the pH of a solution and the concentration of a weak acid (HA) dissolved in it is described by the Henderson-Hasselbalch equation:

$$pH = pK_a + \log [A^-]/[HA]$$

▌ 3.6 CHEMICAL REACTIONS IN BIOLOGY

Life is possible because molecules in biological systems frequently undergo chemical reactions, enabling organisms to replicate DNA, synthesize RNA and protein molecules, pump small molecules into and out of cells, and use energy in the form of food or light. In this section we discuss the physical principles governing chemical reactions. We review the fundamental laws of thermodynamics and the roles of catalysts in accelerating the rates of reactions between biomolecules. Finally, we describe how energy, in the form of high-energy phosphate compounds, is harnessed to drive certain chemical reactions that would otherwise occur too rarely or too slowly to be useful to living systems.

The Mechanism and Speed of Chemical Transformation Define Chemical Reactions

Chemical reactions involve the breakage of covalent bonds and the formation of new bonds. Typically, chemical reactions are written with the **reactants**, or starting molecules, on the left and the **products** on the right, connected by an arrow indicating the direction of the reaction. For example, the expression

$$ATP + H_2O \rightarrow ADP + P_i \qquad (3\text{-}5)$$

describes a very common reaction in biology: the breakage of a phosphorus–oxygen bond in the nucleotide adenosine triphosphate (ATP) to produce adenosine diphosphate (ADP) and inorganic phosphate (P_i). A more detailed representation of this reaction can be drawn to indicate, with curved arrows, the direction in which electrons are moving (**Figure 3-26**).

Reactions of this type involve the attack of a **nucleophile**, a strongly electronegative atom, such as oxygen or nitrogen, on a less electronegative atom, such as phosphorus or carbon. When the nucleophile initiating the reaction is part of a water molecule, as in Figure 3-26, the reaction is known as **hydrolysis**.

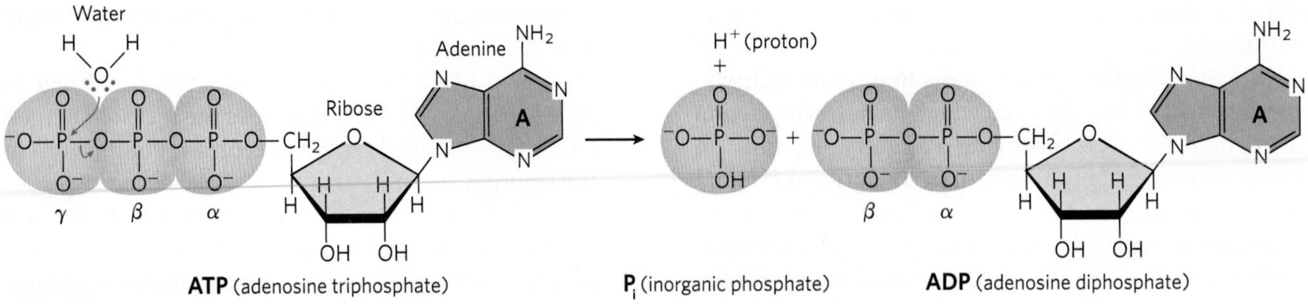

FIGURE 3-26 The hydrolysis of ATP. A phosphorus–oxygen bond in the nucleotide adenosine triphosphate (ATP) reacts with water to produce adenosine diphosphate (ADP) and inorganic phosphate (P_i).

Most chemical reactions that take place in biological systems are not spontaneous; if they were, they would be impossible to control, and life as we know it could not exist. Instead, the starting and ending points of chemical reactions are bridged by an energy barrier, called the **activation energy**, that separates the reactants from the products (**Figure 3-27**). As reactants come together and bonds are breaking and forming, the reacting species progress through a high-energy state called a **transition state**. Once past this state, the reaction proceeds spontaneously because it is energetically favorable. Any chemical reaction can, in principle, proceed in the forward or reverse direction—toward products or reactants. In practice, however, most reactions characteristically tend to proceed more rapidly in one direction than the other, due to differences in the relative energetic stability of products versus reactants. In Figure 3-27, the difference in energy between the products and the transition state is greater than the difference in energy between the reactants and the transition state. Hence, at equilibrium, there will be more products than reactants, because the energetic barrier to the reverse reaction is higher. (This is further discussed later in the chapter.)

A **reaction mechanism** is the sequence of individual steps that take place during the conversion of reactants to products. It shows which bonds are breaking and forming and which species are **reaction intermediates**—substances that form and exist for an extremely short time before being converted to other intermediates or to reaction products. Understanding the mechanisms of chemical reactions that occur in living systems is important in molecular biology because it helps us understand how these reactions are used and controlled during such processes as cell growth and responses to chemical stimuli.

Although it is often difficult to determine reaction mechanisms, understanding the **reaction kinetics**, the rates at which the reaction proceeds in the forward and reverse directions, can provide important clues. Reaction rates are a function of the concentration of reactants (when the reaction is catalyzed by an enzyme, reactants are called substrates) and their tendency to react. Chemical reactions require collisions between molecules, and collisions occur more frequently when there are more molecules per unit volume. It is important to note that reaction rates also depend on the individual steps that make up the reaction. Usually, one step is slower than the others, and this slowest step, the **rate-limiting step**, governs the overall reaction rate. For example, suppose the reaction $A + 2B \rightarrow 2C$ is found, by experiment, to occur with a reaction rate proportional to the concentration of A multiplied by the concentration of B. We can write this as:

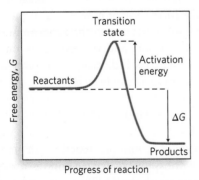

FIGURE 3-27 The activation barrier between reaction substrates and products. The starting and ending points of chemical reactions are bridged by an energy barrier, called the activation energy, that separates the reactants from the products. If the difference in energy between the products and the transition state is greater than the difference in energy between the reactants and the transition state, then at equilibrium the amount of product will be greater than the amount of reactant, because the energy barrier to the reverse reaction is higher. The difference in free energy (ΔG) between the reactants and products is negative in the forward reaction, indicating that this reaction is energetically favorable. This type of diagram is called a reaction coordinate diagram (see Chapter 5).

$$\text{Reaction rate} = k[A][B] \qquad (3\text{-}6)$$

where k is a value called the **rate constant**, a property of the overall reaction that quantitatively describes the tendency to react. A possible mechanism for this reaction might be:

$$(1) \ A + B \rightarrow D \ \text{(slow)} \qquad (3\text{-}7)$$

$$(2) \ D + B \rightarrow 2C \ \text{(fast)} \qquad (3\text{-}8)$$

Step 1, the rate-limiting step, produces an intermediate, D, that rapidly reacts with B in step 2 to yield product C. This reaction mechanism is not the only one consistent with the observed reaction kinetics, but it provides a hypothesis that could be tested experimentally.

Many biologically important reactions involve not just two but many individual steps. As we shall see, the identification of so many individual steps can be extremely challenging.

Biological Systems Follow the Laws of Thermodynamics

Living systems demand an almost constant input of energy, and as a result, organisms devote considerable molecular machinery to obtaining and using it. Biology obeys the physical laws of thermodynamics, which are the foundation for understanding energy and its effects on matter. In thermodynamics, a **system** is defined as a container or organism or other portion of the universe that is under study, and the rest of the universe, outside the system of interest, is called the **surroundings**.

The **first law of thermodynamics** states that energy can never be created or destroyed; in other words, the energy of a system is conserved. In a thermodynamic system, all readily occurring (spontaneous) processes take place without the input of additional energy from outside. The first law of thermodynamics cannot predict whether a process is spontaneous, however. For example, heat spontaneously transfers from a warmer object to a cooler one, never the reverse. Yet transfer in either direction is consistent with the first law of thermodynamics, because the total energy of the system remains unchanged in either case. To determine which direction of a process or reaction is spontaneous, we need additional criteria.

According to the **second law of thermodynamics**, all spontaneous processes take place with an increase in disorder, or **entropy**, of the system. Consider, for instance, two vessels of equal volume, one filled with plain water and the other with water containing a dye. When a connecting valve is opened, the dye molecules become randomly but equally distributed between the two vessels (**Figure 3-28**). The likelihood of all the dye molecules spontaneously remaining in the first vessel,

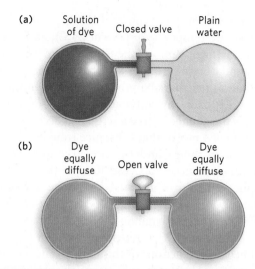

FIGURE 3-28 The spontaneous increase in disorder of a closed system. (a) Two vessels of equal volume are connected by a closed valve, one vessel filled with water containing a purple dye and the other with plain water. (b) When the connecting valve is opened, the dye molecules become randomly but equally distributed between the two vessels over time, maximizing the disorder, or entropy, of the system.

or all the dye molecules moving into the other vessel, is essentially zero, even though the energy of either of these arrangements is no different from that of the evenly distributed molecules. The spontaneity of a process, such as a biochemical reaction, cannot be predicted from knowledge of the system's entropy change alone.

Every closed system, where no reactants or products can escape, tends toward equilibrium, a state in which the forward and reverse reaction rates are exactly balanced. The approach to equilibrium is accompanied by a transfer of energy from one form to another, such as from potential (stored) energy to kinetic energy. The concept of **free energy (G)** provides a useful way to express this energy change. Free energy, denoted by the symbol G (for nineteenth-century physicist Josiah Willard Gibbs), is energy that is available to do work. The second law of thermodynamics states that free energy always decreases—that is, the **free-energy change (ΔG)** is negative—in spontaneous reactions that occur without a change in temperature or pressure; at equilibrium, ΔG is zero. Free energy lost during the approach to equilibrium is either converted into heat or used to increase the amount of entropy.

The tendency of a chemical reaction to proceed to completion can be expressed as an equilibrium constant, which is related to the **standard Gibbs free-energy change (ΔG°)** of the reaction by the expression:

$$\Delta G^{\circ} = -RT \ln K_{eq} \qquad (3\text{-}9)$$

where R is the universal gas constant (1.987 cal/mol • K), T is the absolute temperature (298 at 25°C), and ln K_{eq} is the natural logarithm of the equilibrium constant K_{eq}. The standard state (under which K_{eq} is measured) requires that all reactants be present at a concentration of 1 mol/L (1 M). For any compound, the ΔG° of formation (ΔG_f°) is the change in free energy that accompanies the formation of 1 mol of that substance from its component elements in their standard states (the most stable form of each element at 25°C and 100 kPa (kilopascals) of atmospheric pressure). Because K_{eq} can be measured experimentally, this relationship gives us a way of calculating ΔG°—the thermodynamic constant characteristic of each reaction. When K_{eq} is much greater than 1, ΔG° is large and negative; reactions of this nature tend to go to completion. In contrast, when K_{eq} is much smaller than 1, ΔG° is large and positive; reactions with this property are not spontaneous and require energy to drive them to completion.

Catalysts Increase the Rates of Biological Reactions

The ΔG of a chemical reaction reflects its equilibrium. However, ΔG, whether positive or negative, bears no relationship to the *rate* of the reaction. For example, the series of reactions that convert glucose to H_2O and CO_2 in most cells is thermodynamically favorable, with a net ΔG that is highly negative. However, a large, negative ΔG does not correlate with a fast reaction. A sterile solution of glucose will sit on a laboratory shelf for months with no detectable change. Glucose remains stable because of the substantial activation energy barrier that exists for its reaction. Virtually every chemical reaction in a cell occurs at a significant rate only because of the action of enzymes, which, like all **catalysts**, are molecules that dramatically increase the rate of specific chemical reactions without being consumed in the process.

Catalysts function by lowering the activation energy for a particular reaction without affecting the reaction equilibrium. Given that a catalyst changes the reaction's rate but not its equilibrium, it must change the rate of the reverse reaction to the same extent as the rate of the forward reaction. Catalysts can do this because they enable the reaction to proceed by a different mechanism than that of the uncatalyzed reaction. For example, enzymes bind to the transition state of the reactants by providing a molecular surface complementary to its shape and charge. Because of this favorable interaction, binding of an enzyme stabilizes the transition state, reducing the activation energy for the reaction and thus greatly enhancing the reaction rate. Additional contributions to catalysis occur when reacting molecules—substrates—bind

to an enzyme in an orientation that favors the reaction and when chemical groups in the enzyme bind metal ions or protons that participate in the reaction. As a consequence of these effects, enzymes often increase reaction rates 10^{12}-fold or more above the rate of the uncatalyzed reaction.

Most cellular enzymes are proteins, though some RNA molecules also have catalytic activity. In general, each enzyme catalyzes a specific reaction, and each reaction in a cell is catalyzed by a different enzyme. Thus, many thousands of enzymes are required in each cell. Because enzymes are exquisitely capable of discriminating between substrates, and because they are subject to various regulatory mechanisms, cells can selectively adjust reaction rates. Such selectivity is critical for the effective control of cellular processes. By enabling specific reactions to occur at particular times and locations within a cell or organism, enzymes determine how chemicals and energy are channeled into biological activities. Enzyme function is described in detail in Chapter 5.

Energy Is Stored and Released by Making and Breaking Phosphodiester Bonds

The formation and breakdown of adenosine triphosphate (ATP) (and in some cases guanosine triphosphate, GTP) links the molecule-making and molecule-degrading pathways of cellular metabolism. The formation of this critical energy-storing molecule from inorganic phosphate and adenosine diphosphate (ADP), by the creation of a phosphodiester linkage, is coupled to some of the steps of degradative metabolism and, in plants, to photosynthesis.

Because the hydrolysis of a phosphodiester bond is exothermic under physiological conditions, energy is released when ATP is hydrolyzed to ADP. In turn, the free energy stored in the phosphodiester bonds of ATP is used to drive biosynthetic reactions of metabolism. Although energy is required for bond breaking in ATP, the products of the reaction (ADP and phosphate) form highly favorable interactions with water. Thus, hydration of the breakdown products of ATP more than makes up for the input energy necessary to break the bond in the first place, resulting in an overall energetically favorable process. Almost as soon as it is formed in concert with a coupled degradative reaction (or by photosynthesis), ATP is consumed by enzymes to provide the energy necessary to propel another reaction to completion. In this way, ATP functions as a transient vehicle of intracellular energy transfer (**Highlight 3-2**).

ATP consists of an adenosine nucleoside (base + ribose) and three phosphate groups (see Figure 3-26). The phosphate groups, starting with the one directly

ATP: The Critical Molecule of Energy Exchange in All Cells

The universal role of ATP in cellular reactions has led many molecular biologists and chemists to question its origin and the reasons for its emergence as the universal mediator of energy exchange in cells. Experiments performed by Juan Oro, Stanley Miller, and Harold Urey in the 1950s and 1960s showed that adenine bases can be produced by heating concentrated hydrogen cyanide and ammonia, leading to speculation that adenine came into wide biological use in part because it arose very early in the evolution of life. The yields of adenine in the laboratory experiments ranged from ~1% to upward of 20%, depending on the reaction conditions. However, the likelihood of concentrated cyanide and ammonia

Juan Oro, 1923–2004
[Source: Special Collections, University of Houston Libraries. University of Houston Digital Library. Web. May 12, 2014. http://digital.lib.uh.edu/collection/p15195coll38/item/230.]

existing in the environment of early Earth is uncertain. Adenine has been found in meteorites, though, providing evidence that it is produced naturally in space. Some researchers have speculated that various inorganic clays could have helped sequester adenine and foster its reaction with ribose to form adenosine nucleosides (see the How We Know section in Chapter 1). The uncatalyzed phosphorylation of nucleosides to nucleotides (i.e., of adenosine nucleoside to AMP, ADP, and ATP) has been observed under hot, dry experimental conditions. To date, however, such proposed prebiotic reactions have not been found to proceed efficiently. More research is required to determine whether other synthetic conditions could suggest more plausible pathways for prebiotic accumulation of nucleotides.

---

bonded to ribose, are referred to as the alpha (α), beta (β), and gamma (γ) phosphates, respectively. Like other such high-energy molecules, ATP contains bonds—in this case, the phosphodiester bonds between the phosphate groups—that undergo breakdown by water (hydrolysis) to release significant free energy. The second and third (β and γ) phosphate groups of ATP are unusually rich in chemical energy: the net change in energy (ΔG) on hydrolysis of ATP to produce ADP and inorganic phosphate (P_i) inside a living cell is -12 kcal/mol. This large negative change in free energy makes the breakdown of ATP thermodynamically favorable and hence ATP is valuable for chemically storing energy that can be used to do work. The stored energy is captured when the cleaved phosphate group is transferred to another small molecule or protein as part of a metabolic pathway. Many enzymes harness ATP hydrolysis in this way to perform the work of the cell.

Note that a single hydrolytic reaction produces one P_i and ADP, and the ADP can be broken down further to yield a second P_i and adenosine monophosphate (AMP). But ATP can also be hydrolyzed to AMP directly, with the release of inorganic diphosphate, or pyrophosphate (PP_i). This latter reaction is effectively irreversible in the cellular environment, because enzymes called pyrophosphatases rapidly hydrolyze pyrophosphate into two phosphates. As a result, it would be very difficult to accumulate sufficient concentrations of pyrophosphate in the cell to drive the reaction in the reverse direction. Thus, the hydrolysis of ATP to AMP can be used to render a coupled process effectively irreversible.

For example, DNA and RNA are synthesized from nucleoside triphosphate precursors through a reaction that forms a phosphodiester bond and releases PP_i. The required free energy of bond formation comes in part from the concomitant splitting of the high-energy pyrophosphate group (by pyrophosphatase) as it is released. Interestingly, when no pyrophosphatase is present, such as in a laboratory test tube, robust DNA and RNA synthesis can still occur in this reaction. This is because the thermodynamics of phosphate bond formation and cleavage is only part of the story; base stacking and base-pair formation in polynucleotides are energetically favorable and therefore contribute some of the push toward polynucleotide synthesis. During protein synthesis in cells, amino acids are activated for peptide bond formation by linkage to AMP with the release of PP_i. Again, the splitting of the pyrophosphate group helps drive the reaction irreversibly toward formation of the activated amino acid–AMP (aminoacyl adenylate), the substrate for protein synthesis.

In addition to phosphorus–oxygen bonds, phosphorus–nitrogen and sulfur–carbon bonds also release significant free energy on hydrolysis, and these bonds are found in other important classes of energy-storing compounds that are used to drive reactions in biology. For example, the sulfur–carbon (thioester) bond in acetyl-coenzyme A is the primary source of chemical energy for fatty acid biosynthesis. The free energy released by hydrolysis of high-energy bonds ranges in value, and its utility comes from the coupling of the released energy to another

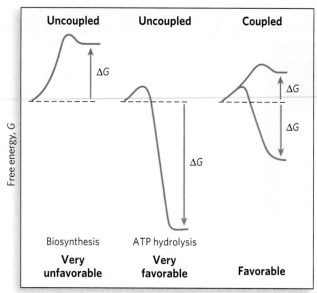

FIGURE 3-29 Energy coupling in biological processes. The reaction coordinate diagram on the left shows an energetically unfavorable biosynthetic reaction. The middle diagram shows a favorable reaction such as the hydrolysis of ATP. By coupling biosynthetic reactions having a positive (unfavorable) ΔG value with the breakage of high-energy bonds having a negative ΔG of greater absolute value, the overall reaction has a negative ΔG (shown in the diagram on the right). At equilibrium, the coupled reaction favors synthesis of the biomolecule over its breakdown.

reaction to drive it forward. Thus, the coupling of biosynthetic reactions having a positive (unfavorable) ΔG value with the breakage of high-energy bonds that has a negative ΔG of greater absolute value ensures that the equilibrium favors synthesis of the biomolecule over its breakdown (**Figure 3-29**).

The overall free-energy change associated with a series of linked reactions determines whether a particular reaction in the series will occur. Reactions with small positive ΔG values, which in isolation would be unfavorable and would not take place, are often embedded in metabolic pathways in which they precede reactions having large negative ΔG values. It is critical to bear in mind that no single biochemical reaction, and no single pathway, takes place in isolation. Instead, the overall equilibria of the huge network of pathways within a cell are constantly adjusting to changing concentrations of substrates as the cell grows and responds to its environment.

SECTION 3.6 SUMMARY

- Chemical reactions involve the breakage of covalent bonds and formation of new bonds. They are written with reactants on the left and products on the right. Half-headed arrows ($\rightleftharpoons$) indicate that the reaction can proceed in either direction.

- The starting and ending points of a chemical reaction are bridged by an activation energy, a barrier that separates reactants from products. As reactants come together and bonds are breaking and forming, the reacting species progress through a high-energy transition state, at which point decay to substrates or decay to products are equally probable, because it is downhill either way.

- Reaction rates depend on the individual steps in the reaction. Usually, one step is slower than the others, and this rate-limiting step governs the overall reaction rate.

- Chemical reactions obey the laws of thermodynamics. The first law states that energy can never be created or destroyed. The second law states that all spontaneous processes take place with an increase in disorder, or entropy, of the system.

- Free energy, G, is energy that can do work. According to the second law of thermodynamics, free energy always decreases (ΔG is negative) in spontaneous reactions that occur without a temperature or pressure change. The tendency of a chemical reaction to proceed to completion is expressed by an equilibrium constant (K_{eq}), which is related to the standard free-energy change (ΔG°) of the reaction by the expression:

$$\Delta G^\circ = -RT \ln K_{eq}$$

- Virtually every chemical reaction in a cell occurs at a significant rate only because enzymes dramatically increase the rate of chemical reactions without being consumed in the process.

- Breakdown of the phosphodiester bonds between the phosphates of adenosine triphosphate (ATP) by reaction with water (hydrolysis) produces significant free energy, which can be used to change the structure or binding properties of enzymes and thus assist in catalyzing other cellular reactions. Phosphate groups are transferred from ATP to other metabolites or proteins in a coupled reaction, yielding new high-energy phosphate bonds that can be hydrolyzed to provide the free energy for further reactions.

HOW WE KNOW

Single Hydrogen Atoms Are Speed Bumps in Enzyme-Catalyzed Reactions

Cha Y., C.J. Murray, and J.P. Klinman. 1989. Hydrogen tunneling in enzyme reactions. *Science* 243:1325–1330.

Klinman J.P., and A. Kohen. 2014. Hydrogen tunneling links protein dynamics to enzyme catalysis. *Annu. Rev. Biochem.* 82:471–496.

Judith Klinman [*Source: Courtesy of Judith Klinman.*]

Understanding what limits the rates of biochemical reactions, and how enzymes speed them up, has long fascinated scientists. Thanks to a phenomenon called the kinetic isotope effect, researchers can deduce how single atoms affect reaction rates. Kinetic isotope effects are observed when different isotopes of an atom (such as of hydrogen or carbon), incorporated into a reactant, alter the rate of a chemical reaction. Substituting one isotope for another in a chemical bond that is broken or formed in the rate-limiting step will significantly change the observed reaction rate. This is exactly what happened when Judith Klinman and her colleagues initially studied the conversion of benzyl alcohol to benzaldehyde, a reaction catalyzed by yeast alcohol dehydrogenase (Figure 1). In these experiments, the reaction rate constant measured for the substrate containing hydrogen (^{1}H) was faster than the rate constants measured for the substrates containing deuterium (^{2}H) or tritium (^{3}H). These observed rate differences result from the effect of the mass of an atom on the vibrational frequency of the chemical bond that it forms. Heavier atoms lead to lower vibrational frequencies, and more energy must be supplied to break the bond. This results in a higher activation energy for bond cleavage, which in turn slows the measured reaction rate. The magnitude of these effects in Klinman's experiments indicated that the transfer of a hydrogen atom is the rate-limiting, or slowest, step of the alcohol dehydrogenase reaction, and hydrogen transfer is the part of the reaction influenced by the enzyme. If transfer of the hydrogen were *not* the rate-limiting step, the rate wouldn't change enough with different isotopes to be noticeable.

In experiments of this type, the isotope-dependent rate changes are largest when the difference between isotope masses is maximized, because of the vibrational frequency effects noted above. A deuterium atom (D) has twice the mass of a hydrogen atom, and a C—D bond reacts 6 to 10 times more slowly than the corresponding C—H bond, which provides an easily measurable difference.

These initial findings for alcohol dehydrogenase, and later for additional enzymes, led Klinman and other researchers to conclude that many enzymes enhance chemical reaction rates by speeding up the movement of hydrogen atoms in a quantum-mechanical process known as tunneling.

FIGURE 1 Alcohol dehydrogenase, an enzyme essential for metabolizing ethanol and other alcohols, catalyzes the conversion of an alcohol to an aldehyde. The reaction uses a molecule called a cofactor (in this case, nicotinamide adenine dinucleotide, or NAD$^+$) as a hydrogen atom (proton) acceptor.

HOW WE KNOW

Peptide Bonds Are (Mostly) Flat

Edison, A.S. 2001. Linus Pauling and the planar peptide bond. *Nat. Struct. Mol. Biol.* 8:201–202.

MacArthur, M.W., and J.M. Thornton. 1996. Deviations from planarity of the peptide bond in peptides and proteins. *J. Mol. Biol.* 264:1180–1195.

Pauling, L., R.B. Corey, and H.R. Branson. 1951. The structure of proteins: Two hydrogen-bonded helical configurations of the polypeptide chain. *Proc. Natl. Acad. Sci. USA* 37:205–211.

Janet Thornton
[Source: Courtesy of Janet Thornton.]

More than 50 years ago, Linus Pauling realized that the planar nature (flatness) of peptide bonds was an important constraint on polypeptides, leading him to the prediction of key elements of protein structure: the α helix and the β-pleated sheet (described in Chapter 4). But are peptide bonds truly flat? Two dominant resonance structures of the N—C bond, as measured in small molecules by spectroscopic methods, result in ~40% double-bond character, supporting the idea that peptide bonds and their covalently attached atoms lie in a plane (see Figure 3-13). But Pauling did this work before any protein structures were known at high resolution, so the planarity of peptide bonds in proteins couldn't be tested.

Today, the availability of thousands of protein and peptide structures makes it possible to conduct statistical surveys of peptide bonds in natural proteins. Janet Thornton and her colleagues showed that many such structures contain deviations from planar peptide bonds. Using a subset of available high-resolution protein structures, they estimated the energies of peptide bond rotation (Figure 2). This work revealed a small but statistically significant trend away from absolute planarity. Furthermore, previous experimental studies of small peptides had shown that nonplanar peptide bonds do occur in both cyclic and linear peptides. Pauling realized this, of course. As a brilliant chemist, he wrote about the calculated low energetic barrier to small rotations about the peptide bond, which provided proteins with some flexibility—the extent depending on the structural environment of a particular segment of the polypeptide chain. Thus, theory, calculation, and experimentation all led to the same conclusion: the peptide bond is (mostly) planar.

(a)

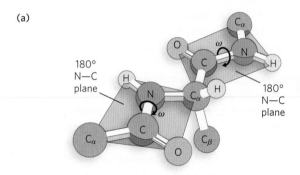

(b)

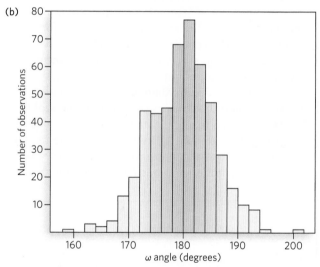

FIGURE 2 Peptide bond rotational energies and the distribution of angular frequencies. (a) The angle ω represents the rotation of a bonded atom about the peptide bond; 180° is planar, because the bonded atoms point to opposite corners of a (planar) rectangle. (b) This histogram represents the angular frequency distribution of ~200,000 ω values obtained from coiled regions of ~4,000 high-resolution protein structures in the January 2001 release of the Protein Data Bank (see Highlight 4-2). There is clearly some deviation from 180°—some ω values are as much as 20° lower or higher. *[Source: Data from M. W. MacArthur and J. M. Thornton, J. Mol. Biol. 264:1180–1195, 1996.]*

KEY TERMS

deoxyribonucleic acid (DNA), 62

ribonucleic acid (RNA), 62

nucleotide, 62

phosphodiester bond, 62

nucleoside, 62

deoxyribonucleotide, 62

thymine (T), 62

adenine (A), 62

guanine (G), 62

cytosine (C), 62

ribonucleotide, 62

uracil (U), 64

amino acid, 64

peptide bond, 64

alpha carbon atom (α carbon, C_α), 64

chemical bond, 68

mole, 72

van der Waals interaction, 74

hydrophobic effect, 74

pi stacking, 75

hydrogen bond, 75

achiral, 78

chiral, 78

pH, 81

buffer solution, 82

pK_a, 82

activation energy, 84

transition state, 84

reaction mechanism, 84

reaction intermediate, 84

reaction kinetics, 84

rate constant, 85

first law of thermodynamics, 85

second law of thermodynamics, 85

entropy, 85

free energy (G), 85

standard Gibbs free-energy change (ΔG°), 85

catalyst, 86

PROBLEMS

1. Consider the O—O and O=O bonds. Is the O—O bond stronger or weaker? Are the oxygen atoms in the O=O bond closer together or farther apart than those in the O—O bond?

2. Do two enantiomers of a chemical have the same density? The same melting point? If the chemical is an acid, do they have the same pK_a?

3. Which of the following statements about a catalyst is correct?

   (a) It can change the equilibrium constant of a chemical reaction.
   (b) It speeds up the rate of the forward but not the reverse reaction.
   (c) It is used up in the course of a reaction.
   (d) It lowers the activation energy for a reaction.

4. A solution with pH 7 is 100 times more basic than a solution with a pH of what value?

5. Amino acids are joined by peptide bonds, the formation of which is accompanied by the loss of water. Is the dipeptide alanylglycine the same as the dipeptide glycylalanine?

Why or why not? (Note that peptides are always written with the amino-terminal residue on the left.)

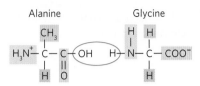
Alanine — Glycine

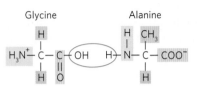
Glycine — Alanine

6. For the reaction coordinate diagram shown below, the activation energy is larger when the reaction proceeds in which direction?

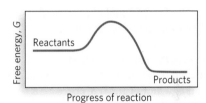

7. A flask contains 10 mL of salt water. If 10 mL of distilled water is added, does the number of moles of sodium chloride increase by 50%, decrease by 50%, or remain unchanged?

8. Which law of thermodynamics explains why living things require the input of energy to maintain their ordered structure?

9. One of the two resonance structures for a formate ion is shown below. Which carbon–oxygen bond, A or B, is longer?

10. The activation energy for a chemical reaction can be determined in which of the following ways?

   (a) Measuring product amounts
   (b) Measuring rates
   (c) Calculating energy of bond hydrolysis
   (d) Calculating change-in-entropy values

11. What is the pH of the solutions with the following hydrogen ion concentrations?

   (a) 1.75×10^{-5} M
   (b) 6.50×10^{-10} M

(c) 1.0×10^{-4} M
(d) 1.50×10^{-5} M

12. What is the hydrogen ion concentration of the solutions with the following pH values?

(a) 3.82
(b) 6.53
(c) 11.11

13. Calculate the pH of dilute solutions that contain the following molar ratios of acetate to acetic acid ($pK_a = 4.70$).

(a) 2:1
(b) 1:3
(c) 5:1
(d) 1:1
(e) 1:10

14. A buffer contains 0.01 mol of lactic acid ($pK_a = 3.60$) and 0.05 mol of sodium lactate per liter.

(a) Calculate the pH of the buffer.
(b) Calculate the change in pH after 5 mL of 0.5 M HCl is added to a liter of the buffer.
(c) Calculate the change in pH after the same quantity of this acid is added to a liter of pure water.

15. An unknown compound is thought to have a carboxyl group with a $pK_a = 2.0$ and a second ionizable group with a pK_a between 5 and 8. When 75 mL of 0.1 M NaOH was added to 100 mL of a 0.1 M solution of this compound at pH 2.0, the pH increased to 6.72. Calculate the pK_a of the second group.

16. The base thymine (see Figure 3-1) contains a six-membered ring. From your understanding of bond structures, is this ring flat/planar or bent? Explain your reasoning.

17. Free rotation is possible around single bonds, but not around double bonds. The repeating structure of the backbone of a polypeptide is shown below, as it is usually drawn. The bond torsion angles, describing rotation about these bonds, are labeled phi (φ), psi (ψ), and omega (ω). In reality, significant rotation can occur about just two of these bonds. For which bond is free rotation most restricted, and why?

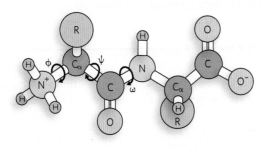

18. The two sets of reactants shown below represent the starting points for (a) amide formation and (b) phosphoryl group transfer. In each panel, draw the curved arrows needed to indicate the first step in each reaction. Do not draw any additional intermediates or steps.

(a)   **(b)**   H⁺

19. The structure of one of the two stereoisomers of alanine is shown below. A pure solution of this stereoisomer is dextrorotatory. Is this molecule D-alanine or L-alanine?

20. A biochemist wishes to use histidine to buffer a solution at pH 6.6. Histidine has three ionizable groups, with pK_a values of 1.82, 6.00, and 9.17. The biochemist has 100 mL of a 0.1 M histidine solution at pH 1.82. What volume of 1.0 M NaOH must she add to bring the pH to 6.6?

ADDITIONAL READING

Chemical Building Blocks of Nucleic Acids and Proteins
Adams, R.L., J.T. Knowler, and D.P. Leader. 2009. *The Biochemistry of the Nucleic Acids*, 11th ed. New York: Academic Press.
Saenger, W. 1984. *Principles of Nucleic Acid Structure.* New York: Springer-Verlag.

Chemical Bonds and Weak Chemical Interactions
Pauling, L. 1960. *The Nature of the Chemical Bond.* Ithaca, NY: Cornell University Press. A classic text covering the details of chemical bonding and the properties of molecules.
Pauling, L. 1988. *General Chemistry.* New York: Springer-Verlag. This text, also a classic, provides a great general introduction to the principles of chemistry.

Chemical Reactions in Biology
Fersht, A. 2005. *Structure and Mechanism in Protein Science: A Guide to Enzyme Catalysis and Protein Folding.* New York: Macmillan. A discussion of enzyme catalytic mechanisms from the standpoint of protein structure and folding.
Jencks, W.P. 1987. *Catalysis in Chemistry and Enzymology.* Mineola, NY: Courier Dover Publications. A clear and cogent discussion of catalytic mechanisms and experimental approaches to understanding how enzymes work.

4 Protein Structure

MOMENT OF DISCOVERY

I'll never forget one of our early breakthroughs in computational protein design. Our idea was to write a mathematical description of the protein structure and then optimize its thermodynamic stability by adjusting the amino acid sequence. At the time, several high-profile theoreticians said this would be impossible because protein-folding rates—kinetics—would also need to be considered.

Undaunted, we started by showing that regions of proteins could be designed using our methods. In 1996, we attempted to design a 20 amino acid polypeptide that would form a zinc finger structure, a characteristic polypeptide fold that is held together by zinc ions. After many attempts, student Bassil Dahiyat finally generated a sequence called FSD1 that was predicted to form a zinc finger fold without requiring any zinc. He synthesized this peptide in the laboratory and late that evening analyzed it by circular dichroism, a method that measures the amount of secondary structure in a protein. We had made many unsuccessful attempts at protein design by this time, so we were very familiar with the CD [circular dichroism] spectra of unfolded proteins! At about midnight, Bassil called me at home and said, "Steve, you've got to see this spectrum!" On my home computer over an incredibly slow Internet connection, I watched as a gorgeous spectrum with exactly the shape expected for a folded protein came up on my screen. We realized at that moment that we had achieved something many had considered impossible. When we later solved the molecular structure of the peptide using NMR spectroscopy, the peptide had exactly the structure we had predicted.

—Steve Mayo, on his discovery of the first successful method for computational protein design

Steve Mayo [*Source: Courtesy of the Archives, California Institute of Technology.*]

The beauty of the DNA double helix is indisputable, but to a trained eye, protein structures are even more compelling. Proteins have wonderfully complex architectures, sculpted over time to perform their tasks to near perfection. The fact that a protein adopts a unique conformation is amazing: despite the astronomical number of ways in which even a small protein could possibly fold, it usually folds into a single shape. The instructions for the unique shape of a protein are contained entirely within the linear amino acid sequence. Exactly how the folding instructions are encoded is still not understood; it remains the holy grail of the protein-folding field, given that the conformation of proteins is essential to their proper function.

Part of the explanation of how proteins fold lies in their reaction to an aqueous environment. Most proteins reside in the cell's aqueous cytoplasm, yet many amino acids are hydrophobic, or water-fearing. Hydrophobic residues scattered throughout the length of a protein tend to gather together, thus helping to fold the protein. In this way, proteins form highly compact molecules with hydrophobic interiors. The polar amino acids are oriented toward the outer surface, where they may interact with water. The final, overall protein structure is held together by weak noncovalent forces, which include hydrophobic effects, hydrogen bonds, ionic interactions, and van der Waals forces (these chemical interactions are discussed in Chapter 3). As a consequence, proteins are only marginally stable and tend to unfold quite easily.

One might wonder why protein structures did not evolve to be more stable. In fact, thermophilic organisms—those that live at near-boiling temperatures—have very stable proteins. Why didn't evolution select for high stability in proteins for organisms living at lower temperatures? Interestingly, studies of proteins from thermophilic organisms provide an explanation: many proteins isolated from thermophiles are simply not active at 20°C to 40°C and require high temperatures for optimal activity. Thus, conformational flexibility must be important to the function of many proteins, and too much stability may compromise that flexibility.

Protein structure is commonly defined in terms of four hierarchical levels (**Figure 4-1**). Primary structure is essentially the sequence of amino acid residues. Secondary structure includes particularly stable hydrogen-bonded arrangements of amino acid residues that give rise to regular, repeating patterns. Tertiary structure includes all aspects of the three-dimensional folding pattern of the protein. And, in proteins that have two or more polypeptides, quaternary structure describes how the various polypeptides come together to form the final protein.

In this chapter, we explore how proteins are constructed, starting with the features of the peptide bond, which links amino acids together. Then we look at how weak forces mold protein chains into shape, and discover that, despite the bewildering array of different structures that proteins can form, all proteins contain only a few types of secondary structural elements. We will also

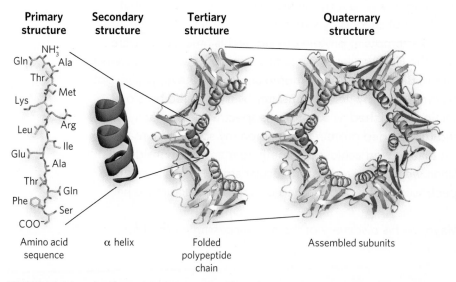

FIGURE 4-1 Levels of structure in proteins. The primary structure consists of a sequence of amino acids linked together by peptide bonds. The resulting linear polypeptide can be coiled into units of secondary structure, such as an α helix. The helix and other secondary structural elements fold together and define the polypeptide's tertiary structure. The folded polypeptide shown here is one of the subunits that make up the quaternary structure of a multisubunit protein (composed of more than one polypeptide), the dimeric *Escherichia coli* β processivity factor, which is involved in DNA replication. *[Source: PDB ID 2POL.]*

see that there are some common ways in which these elements are stitched together to generate a diversity of folded proteins. A discussion of the two methods currently used to solve—that is, to determine—the atomic structure of proteins completes the chapter.

4.1 PRIMARY STRUCTURE

The **primary structure** of a protein is the sequence of amino acids that make up the polypeptide chain. Most proteins range in size from 100 to 1,000 amino acid residues, although there are many examples of proteins that fall outside this range. In this section, we first examine the properties of the amino acids and take a close look at how amino acids are linked together, then examine how protein sequences hold information about their evolutionary heritage.

KEY CONVENTION

The terms *peptide*, *polypeptide*, and *protein* are often used interchangeably. As generally defined, however, a peptide usually consists of a very short segment of 10 or fewer amino acids. A polypeptide usually consists of fewer than 100 amino acids, and "polypeptide chain" can refer to a polypeptide of any size. A protein is a large macromolecule that can be composed of one or more polypeptide chains.

Amino Acids Are Categorized by Chemical Properties

All amino acids have a central carbon atom, designated C_α (the α carbon), which is bonded to a hydrogen, an amino group, a carboxyl group, and a side chain called an **R group (Figure 4-2)**. The R group distinguishes one amino acid from another and ranges from a simple hydrogen atom (in glycine) to relatively complex arrangements of carbon, hydrogen, nitrogen, oxygen, and sulfur. Side chains can be assorted into four groups according to their polarity and charge. The 20 most common amino acids found in proteins are shown in **Figure 4-3** and **Table 4-1**.

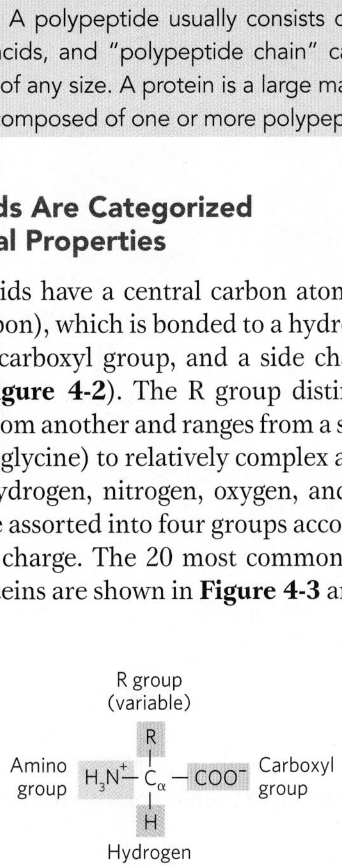

FIGURE 4-2 The general structure of an amino acid. The R group, or side chain, attached to the α carbon (the central carbon as shown here) is different in each amino acid. The name "amino acid" derives from the presence of both an amino group and a carboxylic acid group on the α carbon.

Sometimes, other, much less common amino acids are also found in protein sequences. For example, selenocysteine contains a selenium atom in place of the sulfur atom of cysteine and is inserted into proteins by an unusual mechanism. Selenocysteine is an essential amino acid for humans, although it is known to occur in only 25 of the thousands of proteins encoded by the human genome.

Amino acids are often abbreviated using a three-letter name or a one-letter symbol. Some amino acids have an R group that is ionizable, which may give it a positive charge when protonated and a neutral charge when unprotonated, or a neutral charge when protonated and a negative charge when unprotonated. When the pK_a of the side-chain R group is lower than the pH of its surroundings, the group will be unprotonated (see Section 3.5 for an explanation of pK_a). The overall charge of a protein is mostly due to side-chain groups, because the amino and carboxyl groups are involved in peptide bonds—except for those of the N-terminal and C-terminal amino acid residues, respectively, of the polypeptide chain.

Nonpolar, Aliphatic R Groups Aliphatic side chains are those composed only of hydrocarbon chains ($-CH_2-$), which are nonpolar and quite hydrophobic. Methionine, with a nonpolar thioether group ($R-S-CH_3$), is also included here. These residues tend to cluster inside proteins and stabilize the structure through hydrophobic effects. Glycine is also nonpolar, but having only a single hydrogen atom as its side chain, it contributes little to hydrophobic effects. Proline has an aliphatic side chain, too, and its rigid cyclic structure constrains and limits its possible conformations.

Polar, Uncharged R Groups Polar, uncharged R groups can interact extensively with water, or with atoms in other side chains, through hydrogen bonds. Recall from Chapter 3 that hydrogen bonds are interactions between a donor hydrogen atom that is covalently bonded to an electronegative atom and an acceptor atom that usually has a lone pair of electrons. Examples of donor groups are the hydroxyl groups of serine and threonine and the sulfhydryl group ($R-S-H$) of cysteine. Asparagine and glutamine contain an amide group that can act as donor or acceptor. Two Cys residues brought in close proximity may be oxidized to form a **disulfide bond** (see the How We Know section at the end of this chapter).

Polar, Charged R Groups Three amino acids carry a positive charge at pH 7.0 (i.e., they are basic). Lysine contains a side-chain amino group, arginine has a guanidinium group, and histidine contains an imidazole group (see Figure 4-3). The side chains of two amino acids, aspartate and glutamate, contain a carboxyl group and therefore carry a negative charge at pH 7.0 (i.e., they are acidic).

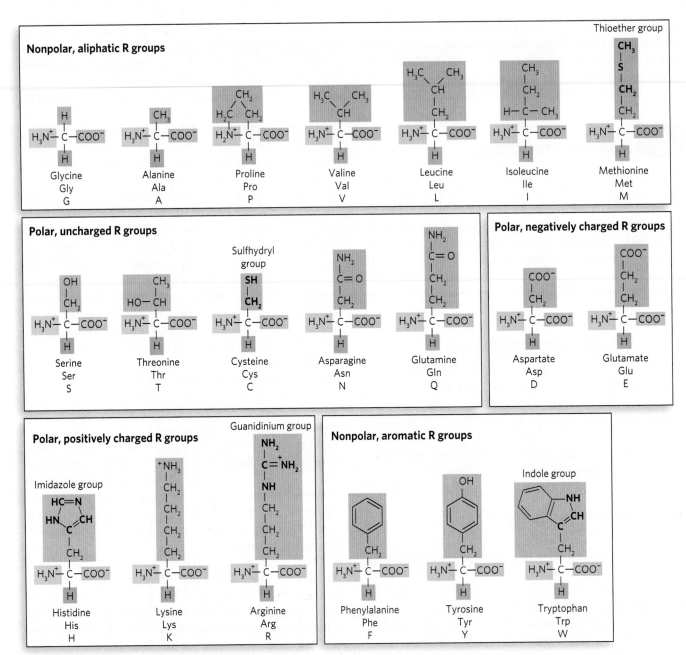

FIGURE 4-3 The 20 common amino acids. The structural formula for each amino acid shows its ionization state at pH 7.0. The groups attached to the α-carbon atom that are shaded pink, blue, or gray—the carboxyl group, amino group, and single proton, respectively—are common to all the amino acids. The side chains (R groups), which are unique to each amino acid, are shaded purple. The R group of histidine is drawn as uncharged, but its pK_a is such that a significant fraction of the His side chains will be protonated and thus positively charged at pH 7.0. Several important functional groups in some amino acid side chains are highlighted.

Charged side chains can form hydrogen bonds and can form ionic interactions with amino acids of opposite charge.

Nonpolar, Aromatic R Groups Phenylalanine, tyrosine, and tryptophan contain aromatic side chains and therefore are hydrophobic. Phenylalanine is the most hydrophobic among them, whereas the tyrosine hydroxyl group and the tryptophan nitrogen can form hydrogen bonds and thus impart some polarity to these residues.

Amino Acids Are Connected in a Polypeptide Chain

The covalent link between two adjacent amino acids is called a **peptide bond,** and the result of many such

TABLE 4-1

The 20 Common Amino Acids

Name	Abbreviation	Symbol	M_r*	pK_a values pK_COOH	pK_NH3+	pK_R
Alanine	Ala	A	89	2.34	9.69	—
Arginine	Arg	R	174	2.17	9.04	12.48
Asparagine	Asn	N	132	2.02	8.80	—
Aspartate	Asp	D	133	1.88	9.60	3.65
Cysteine	Cys	C	121	1.96	10.28	8.18
Glutamine	Gln	Q	146	2.17	9.13	—
Glutamate	Glu	E	147	2.19	9.67	4.25
Glycine	Gly	G	75	2.34	9.60	—
Histidine	His	H	155	1.82	9.17	6.00
Isoleucine	Ile	I	131	2.36	9.68	—
Leucine	Leu	L	131	2.36	9.60	—
Lysine	Lys	K	146	2.18	8.95	10.53
Methionine	Met	M	149	2.28	9.21	—
Phenylalanine	Phe	F	165	1.83	9.13	—
Proline	Pro	P	115	1.99	10.96	—
Serine	Ser	S	105	2.21	9.15	—
Threonine	Thr	T	119	2.11	9.62	—
Tryptophan	Trp	W	204	2.38	9.39	—
Tyrosine	Tyr	Y	181	2.20	9.11	10.07
Valine	Val	V	117	2.32	9.62	—

*M_r values reflect the structures shown in Figure 4-3. The elements of water (M_r 18) are deleted during peptide bond formation, when the amino acid is incorporated into a polypeptide.

linkages is known as a **polypeptide chain**. The peptide bond is formed by condensation of the α-carbon carboxyl group of one amino acid with the α-carbon amino group of another. Therefore, the linear sequence of a polypeptide chain has an **amino terminus**, or **N-terminus**, and a **carboxyl terminus**, or **C-terminus**.

> ### KEY CONVENTION
>
> When an amino acid sequence is given, it is written and read from the N-terminus to the C-terminus, left to right.

The C_α atoms of two adjacent amino acids in a polypeptide chain are separated by three covalent bonds: C_α—C—N—C_α. The C—N connection is the peptide bond that joins two amino acids, but these four atoms constitute the covalent bonds that connect all the residues of a polypeptide chain and thus make up the polypeptide "backbone." Single bonds between atoms typically allow free rotation, but not so for the peptide bond. Linus Pauling and Robert Corey's analysis of dipeptides and tripeptides by x-ray crystallography revealed that the four atoms of the peptide backbone lie in the same plane. Another key observation was that the peptide C—N bond length (1.32 Å; 1 Å (angstrom) is 1×10^{-10} m) is significantly shorter than a single C—N bond (1.49 Å) and approaches the length of a C=N double bond (1.27 Å). These observations are explained by resonance, the sharing of electrons between the carboxyl oxygen and amide nitrogen, creating partial double bonds (**Figure 4-4a**; also see the How We Know section in Chapter 3).

Atoms are not free to rotate about a double bond. A partial double bond gives rise to two possible configurations, referred to as the cis and trans isomers. In peptide bonds, the trans isomer is favored about

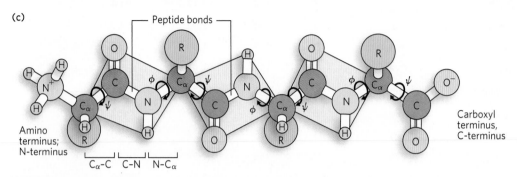

FIGURE 4-4 Peptide backbone atoms. (a) Resonance of the peptide bond gives it a partial double-bond character. (b) The cis and trans isomers of a peptide bond. The bonds in most proteins are trans. The peptide backbone is shaded in orange and the peptide bond is in red. (c) The three bonds that separate sequential α carbons in a polypeptide chain lie in a plane. The N—C$_\alpha$ and C$_\alpha$—C bonds can rotate, with torsion angles designated φ and ψ.

1,000:1 over the cis isomer. The trans isomer of a peptide bond is one in which the two C$_\alpha$ atoms of adjacent amino acids lie on opposite sides of the peptide bond, as do the carbonyl oxygen and the amide hydrogen (**Figure 4-4b**). The double-bond character of the peptide bond explains why the atoms in a peptide bond lie in the same plane. Therefore, a chain of amino acid residues can be envisioned as a series of connected planes (**Figure 4-4c**). The C$_\alpha$—C and N–C$_\alpha$ bonds are free to rotate. However, in a polypeptide, the angles between these bonds are constrained. These angles are referred to as torsion angles (or dihedral angles): φ (phi) for the N–C$_\alpha$ bond and ψ (psi) for the C$_\alpha$—C bond.

KEY CONVENTION

Rotation around a double-bonded pair of atoms is restricted, placing the other atoms that adjoin them in one plane. Two atoms or groups adjoining the double-bonded atoms can lie either in cis (Latin for "same side") or in trans ("other side"). The two forms are isomers because there is no difference between them other than their configuration. The amide hydrogen and carbonyl oxygen can be used to specify the cis and trans isomers of the peptide bond, as can the C$_\alpha$ atoms of adjacent amino acid residues. For example, in the trans isomer, the C$_\alpha$ atoms of adjacent amino acids lie on opposite sides of the peptide bond that joins them.

In reality, however, rotational movements are restricted, because the size of a bulky side chain may preclude a close approach to nearby atoms in the polypeptide backbone. This "steric clash" between an amino acid side chain and neighboring atoms limits φ and ψ and thus the permissible orientations of one peptide-bond plane relative to another. G. N. Ramachandran developed a way to represent graphically the allowed values of φ and ψ for each amino acid. The **Ramachandran plot** for alanine is shown in **Figure 4-5**. The plots for most other amino acids look quite similar, with two exceptions. Glycine has a broader range of allowed angles, because its side chain is a single hydrogen atom and therefore very small. In contrast, proline, with its side chain in a cyclic structure that is covalently bonded to the α-amino group, is greatly restricted in its allowed range of conformations. Conformations deemed possible are those that involve little or no interference between atoms, based on known van der Waals radii and bond angles.

Evolutionary Relationships Can Be Determined from Primary Sequence Comparisons

As organisms evolve and diverge to form different species, their genetic material remains almost the same at first, but differs increasingly as time passes. For this reason, the amino acid sequences of proteins can be used to explore evolution. The premise is simple. If two

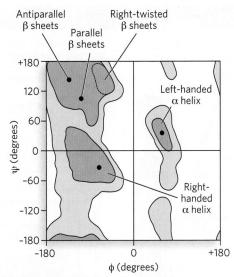

FIGURE 4-5 A Ramachandran plot: torsion angles between amino acids. The conformations of peptides are defined by the values of ψ and φ for each amino acid residue. Allowable conformations are those that involve little or no steric hindrance between atoms of the amino acid side chain and nearby atoms of the peptide backbone. Shown here is the Ramachandran plot for Ala residues. Easily allowed conformations are in dark blue; medium blue signifies bond conformations that approach unfavorable values; light blue, conformations that are allowed if some flexibility is permitted in the torsion angles. Unshaded regions indicate conformations that are not allowed. With the exception of Gly and Pro residues, the plots for all other amino acid residues are very similar to this plot for alanine. The range of allowed φ and ψ values is characteristic for each type of secondary structure, as shown. Secondary structural elements are discussed in Section 4.2. *[Source: Adapted from T. E. Creighton, Proteins, p. 166. © 1984 by W. H. Freeman and Company.]*

organisms are closely related, the primary sequence of the same protein in two different organisms should be similar, but the sequences will diverge as the evolutionary distance between the organisms—that is, the time since they arose from a common ancestor—increases. The wealth of whole-genome sequences now available, from bacteria to humans, can be used to trace evolutionary lineages.

Amino acid substitutions occurring through mutations do not appear at random, and this opens up any analysis to interpretation. Some proteins have more amino acid variation among species than others, indicating that proteins evolve at different rates. At some positions in the primary structure, the need to maintain protein function limits amino acid substitutions to a few that can be tolerated. In other words, amino acid residues essential for the protein's activities are conserved over evolutionary time. Residues that are less important to function vary more over time and among species, and these residues provide the information needed to trace evolution.

Protein sequences are superior to DNA sequences for exploring evolutionary relationships. DNA has only four different nucleotide building blocks, and a purely random alignment of unrelated sequences would produce matches at about 25% of the nucleotides in the alignment. In contrast, the 20 common amino acids used in proteins greatly lower the probability of such chance, uninformative alignments. An example of how protein sequences can be used to trace evolutionary origins is presented in the How We Know section at the end of this chapter. Genomics, proteomics, and the use of sequences to study the molecular evolution of cells are discussed in detail in Chapter 8.

Before moving on, we should note that many of the protein analyses used to determine the information presented in this chapter require that the protein be purified—completely separated from all the other proteins in the cell. Protein purification typically takes several fractionation steps. Particularly powerful techniques used to purify and analyze proteins include column chromatography and polyacrylamide gel electrophoresis, as summarized in **Highlight 4-1**.

SECTION 4.1 SUMMARY

- The primary structure of a protein is its sequence of amino acids, along with any disulfide linkages between cysteine residues.
- An amino acid consists of an amino group and a carboxyl group with a central carbon atom (C_α) between them. Also connected to C_α are a side chain (R group) and a hydrogen atom.
- There are 20 common amino acids, with characteristic side chains that differ in their chemical properties. Side chains can be charged or uncharged, polar or nonpolar, aliphatic or aromatic.
- Amino acid residues in a protein are linked by peptide bonds. The atoms of a peptide bond and the α carbons connected to them lie in one plane, due to the partial double bond between the carbonyl and amide groups, giving rise to cis and trans isomers. The trans isomer of the peptide bond is the most common in proteins.
- The planar configuration of the peptide bond limits how close the R groups of adjoining amino acids can approach one another. This leads to preferred, or allowed, torsion angles of the single bonds that connect the C_α atom to the carbonyl carbon (C_α–C) and the amide nitrogen (N–C_α): angles ψ (psi) and φ (phi), respectively.
- Protein sequences reveal evolutionary relationships among species. The more similar the primary sequence of the same protein between two species, the more recently the species diverged from a common ancestor.

Purification of Proteins by Column Chromatography and SDS-PAGE

To study the structure of a protein, the researcher must first purify it from all other proteins in the cell. First, cells are lysed and particulate matter, such as cell wall debris and insoluble protein, is removed by centrifugation to yield a "crude extract." The crude extract, containing soluble proteins, is then fractionated to separate the proteins and isolate the one that is of particular interest.

There are many ways to separate proteins. One method is **chromatography**, and one of the most powerful chromatographic techniques is **column chromatography.** In this technique, a protein mixture is applied to a column containing a resin, or matrix, that interacts differently with the various proteins (Figure 1). After the protein solution is applied to the top of the column, a buffer is passed through the column to thoroughly wash away any proteins that do not bind to the matrix. Then another buffer is applied that causes bound proteins to dissociate from the matrix; the proteins are carried out in the buffer flow, a process referred to as "elution" of proteins from the column. The proteins come off the column at different times, depending on how they interact with the resin. The column matrix and "elution buffer" are carefully chosen so that different proteins dissociate from the matrix at different times. The eluted proteins are collected in a fraction collector, which gradually moves test tubes under the column, thus keeping the proteins that elute at different times separate from one another.

Several types of resin can be used in column chromatography, which separate proteins based on different properties. In **ion-exchange chromatography**, proteins are separated by charge. The resin contains either cation groups (in a process called anion exchange) or anion groups (in cation exchange). Proteins are usually eluted from the column with an increasing concentration of salt solution, and their release depends on the nature of charged amino acid residues on their surface. In **gel-exclusion chromatography**, proteins are separated by size. The resin is composed of hollow beads with pores of a particular size; large proteins move around the beads and so elute earlier than smaller proteins that can enter the resin pores and thus take a longer path through the column. In **affinity chromatography**, proteins are sorted by the type of ligand they bind. A selected ligand is covalently coupled to the column resin, and the protein mixture is applied. Elution can be performed with a salt solution but is often done with a solution of the ligand itself, which binds to the active site of the protein, releasing it from the resin-bound ligand. Because ligand binding can be very specific to a protein, this technique is often highly selective for the protein of interest.

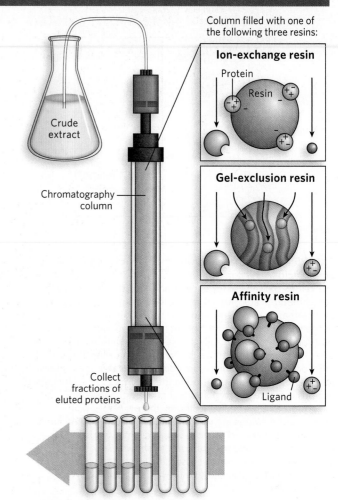

FIGURE 1 Column chromatography is performed in a glass or plastic tube containing one type of fractionating resin (matrix). The protein mixture is applied to the top of the column, and as buffer flows through, different proteins bind to the matrix according to the properties selected by the particular resin. These properties are typically the size or charge of the protein or the specific ligand to which the protein binds. Proteins are then dissociated from the matrix by eluting with a buffer that releases them at different times, and the proteins are collected in separate fractions.

After column chromatography, the fractions are analyzed for the protein of interest; for example, if the protein is catalytic, its presence in fractions can be analyzed by an assay that measures that particular protein activity. Next, proteins in the fractions obtained at various stages of the purification process can be visualized by **sodium dodecyl sulfate–polyacrylamide gel electrophoresis (SDS-PAGE)** (Figure 2a). The process begins with the preparation of a polyacrylamide gel. A

(a)

(b)

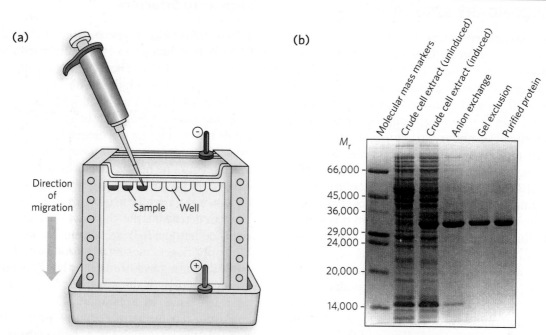

FIGURE 2 (a) In SDS-PAGE, the cross-linked gel is contained in a device to which an electric current can be applied, such that the proteins migrate through the gel matrix. (b) A Coomassie Blue–stained SDS-PAGE gel, tracking the gradual purification of glycine *N*-methyltransferase. The "induced" cell extract sample is from cells that were induced to produce the protein, while the "uninduced" sample is from cells prior to inducing the protein. In practice, the protein bands are stained blue, but in publications the image is almost always shown in black and white.
[*Source: (b) Courtesy of Hirofumi Ogawa.*]

solution of acrylamide and *bis*-acrylamide is prepared, then ammonium persulfate is added, which supplies free radicals that initiate polymerization of the acrylamide. The *bis*-acrylamide cross-links the polyacrylamide to form a matrix. The cross-linked gel acts like a sieve to sort proteins by size. A typical polyacrylamide gel contains around 6% to 10% acrylamide, but different percentages can be used; "high percent" gels resolve small proteins better than "low percent" gels. Before the acrylamide mixture polymerizes, SDS is added and the solution is poured between two glass plates, and wells (which will later hold the protein samples) are created by inserting a comb, or mold. The protein samples are treated with SDS, a negatively charged detergent that binds proteins and denatures them, giving all proteins in the sample a similar shape. Because the amount of SDS binding to a protein is usually related to protein size, SDS also gives all proteins a similar charge-to-mass ratio. The treated samples are loaded into the wells at the top of the gel (which also contains SDS), and an electric field is applied to the gel, which pulls the charged proteins through the matrix. The proteins migrate through the gel at different rates according to their relative molecular mass, with smaller proteins migrating faster than larger ones. Larger proteins migrate more slowly because they cannot weave through the matrix as quickly as smaller proteins.

After the proteins have been separated within the gel, the gel is removed from the glass "sandwich" and soaked in an acidic buffer to precipitate (or fix) the proteins, which prevents their diffusing out of the gel. The gel is then treated with a dye that selectively binds to proteins. A common dye for this purpose is Coomassie Blue. Figure 2b shows a Coomassie Blue–stained SDS-PAGE gel containing protein samples taken at different stages of a protein purification. The rightmost lane in the gel shows only the subunits of the pure protein, the enzyme glycine *N*-methyltransferase; samples taken earlier in the purification procedure show additional proteins. Proteins of known molecular mass are typically applied to one lane of the gel to serve as "molecular mass markers" (as in the leftmost lane in Figure 2b), which allows the researcher to estimate the mass of other proteins in the gel. Electrophoresis is an invaluable technique for molecular biologists, and we'll encounter many variations of it throughout this book.

4.2 SECONDARY STRUCTURE

Secondary structure refers to regularly repeating elements within a protein, in which hydrogen bonds form between polar atoms in the backbone chain. These hydrogen-bonded structures shield, or neutralize, many charges and allow the intrinsically polar polypeptide chain to traverse the nonpolar interior of a protein. The main secondary structures are the α helix, typically 10 to 15 residues long, and the β conformation, composed of individual segments (called β strands) of 3 to 10 residues. A typical protein contains about one-third α helix and one-third β conformation, although there are plenty of exceptions to this general rule, including proteins that have only one of these types of secondary structure. The portion of a protein that has neither α helices nor β conformation is composed of loops and turns that allow secondary structural elements to reverse direction back and forth to form a folded, globular protein. Here we describe the structure and properties of α helices and the β conformation and briefly discuss the structure of reverse turns, which allow secondary structures to fold.

The α Helix Is a Common Form of Protein Secondary Structure

The **α helix** was originally predicted by Pauling and Corey in 1951, based on x-ray studies of keratin by William Astbury in the 1930s. The α helix contains 3.6 amino acid residues per turn (**Figure 4-6a**). One full turn of the α helix is 5.4 Å (1.5 Å per residue) long, and the R groups protrude outward from the helix. The hydrogen on the amide nitrogen forms a hydrogen bond with the carbonyl oxygen of the fourth residue toward the N-terminus, which makes about one helical turn. The α helix forms a right-handed spiral, which, moving away from an observer looking down the spiral, corresponds to a clockwise rotation. You can determine the chirality of a spiral (i.e., whether right- or left-handed) using your hands (**Figure 4-6b**). With your fingers making a fist and your thumbs sticking out and pointing away from you, a left-handed spiral would appear to curve in the same direction as the fingers on your left hand, in a counterclockwise rotation, as the spiral projects in the direction of your thumb. A right-handed spiral, such as the α helix, curves in the same direction as the fingers of your right hand, as the spiral projects in the direction of your right thumb.

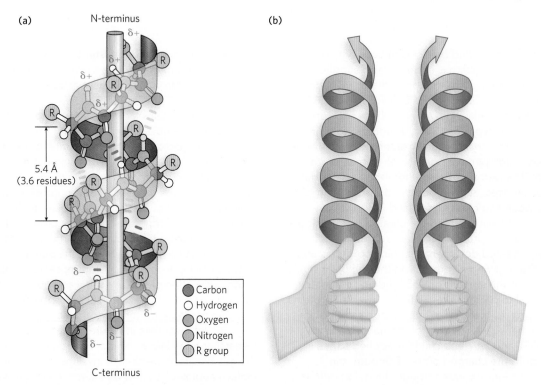

FIGURE 4-6 The structure of the α helix. (a) Peptide bonds form a right-handed spiral; intrachain hydrogen bonds are shown. The electric dipole of the helix, established through intrachain hydrogen bonds, propagates to the amino and carbonyl constituents of each peptide bond. The partial charges of the electric dipole are indicated by δ+ and δ−. (b) An easy way to distinguish left- and right-handed helices (see text).

All the hydrogen bonds of an α helix point in the same direction, and this sets up an electric dipole that gives a partial positive charge to the N-terminus and a partial negative charge to the C-terminus of the helix. Because the last four residues at either end of an α helix are not fully hydrogen-bonded, the dipole charges are spread out on these residues. For this reason, the conformations at the ends of an α helix are often irregular or form a more strained version of the α helix with less favorable torsion angles.

Some general guidelines enable us to predict from a protein sequence the sections where an α helix will form. Consecutive stretches of amino acid residues with long or bulky R groups cannot approach one another closely enough to form the tightly packed α helix. Also, polar side chains can hydrogen-bond to the peptide backbone, thereby destabilizing the helix. For this reason, serine, asparagine, aspartate, and threonine are found less frequently in α helices than most other amino acids. In addition, consecutive like-charged R groups repel one another in the close confines of the α helix. Glycine, due to its conformational flexibility, is also infrequently found in α helices. Finally, proline is infrequent in α helices because its cyclic structure lacks an amide hydrogen-bond donor and restricts N–C$_\alpha$ bond rotation. Proline is often referred to as a helix-breaking residue. The relative frequency of the 20 amino acids in different types of secondary structure is shown in **Figure 4-7**.

Some arrangements of amino acid residues can stabilize the helix. For example, side chains spaced four residues apart are stacked upon one another in the helix. Oppositely charged side chains that are close together can form an ion pair, which stabilizes the helix. Likewise, aromatic side chains with this four-residue spacing can form hydrophobic effects that stabilize the helix. Amino acids with a charge opposite to the partial charge of the helix dipole are sometimes located at the ends of a helix, which adds stability.

The β Conformation Forms Sheetlike Structures

A segment of peptide in the **β conformation** rarely occurs alone. Instead, multiple segments of β conformation are arranged side by side to form a **β sheet** consisting of at least two, and frequently many more, β strands (**Figure 4-8**). The β sheet, like the α helix, is formed by hydrogen bonds between backbone amide and carbonyl groups, but unlike the α helix, the β sheet structure cannot form from one β strand. Instead, all hydrogen bonds are formed between the backbones of two different β strands. The peptide bonds in a β sheet are arranged in a remarkably extended form, with a distance of 3.5 Å per residue. The R groups of adjacent amino acid residues in a β strand lie on opposite sides of the sheet, and this alternating geometry prevents the

(a) **Antiparallel**

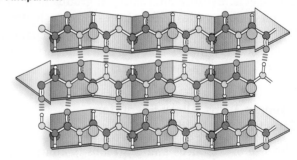

(b) **Parallel**

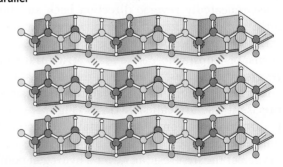

FIGURE 4-8 The structure of the β sheet. R groups extend out from the β sheet, emphasizing the pleated shape. Hydrogen bonds between adjacent β strands are also shown. (a) In an antiparallel β sheet, the N- to C-terminal orientation of the β strands alternates (shown by the arrowheads). (b) In a parallel β sheet, the β strands align in the same direction.

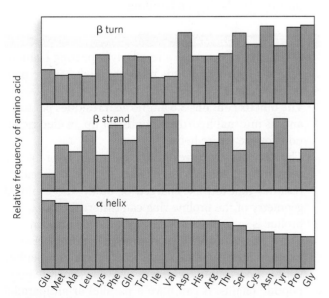

FIGURE 4-7 The relative frequency of amino acids in secondary structural elements. The plot shows the observed relative frequencies of the 20 common amino acids in three types of secondary structure. [*Source: Adapted from G. Zubay,* Biochemistry, *Macmillan, 1988, p. 34.*]

interaction of R groups of adjacent residues. This sets up a zigzag pattern and, along with the side-to-side arrangement of the strand segments, resembles a series of pleats. Thus, the β sheet is often referred to as a "β-pleated sheet."

Often, the many β segments or strands that compose a β sheet are covalently connected in a single polypeptide. The strands of a β sheet may be close together in the polypeptide sequence, but they can also be far apart, separated by other secondary structures in the same polypeptide chain. The formation of β sheets between two different polypeptides is also common. In either case, when the β strands are oriented in the opposite N- to C-terminal directions, the structure is known as an **antiparallel β sheet**; when they run in the same direction, it is called a **parallel β sheet** (see Figure 4-8). The sheets can also be composed of a mixture of parallel and antiparallel strands. Sheets made with antiparallel strands form nearly straight hydrogen bonds and are thought to be slightly more stable than parallel sheet structures. The β sheet structure can readily accommodate large aromatic residues, such as Tyr, Trp, and Phe residues. In addition, proline, which is unfavored in the α helix, is often found in β sheets, especially in the "edge" strands, perhaps to prevent association between proteins that are not meant to bind one another. The most common sheet structures are antiparallel, followed by mixed sheets, then purely parallel sheets. Because alternating R groups in β sheets are on opposite sides of the sheet structure, hydrophobic residues that alternate with polar side chains can yield a sheet structure that acts as a boundary between greasy and watery environments.

Reverse Turns Allow Secondary Structures to Fold

The size of α helices and β strands is limited by the diameter of a globular protein, and these structures must repeatedly reverse direction back and forth to form a properly folded protein. Approximately one-third of a polypeptide chain is composed of **reverse turns**, or loops, where secondary structural elements reverse themselves. Sometimes these turns are large and irregular, but many reverse turns are small and precise; these are called **β turns** (**Figure 4-9a**). The β turn makes a complete reversal of direction using only four residues, in which the backbone carbonyl oxygen of the first residue forms a hydrogen bond with the amide hydrogen of the fourth residue. The second and third residues form no inter-residue hydrogen bonds, and this typically places β turns on the surface of the protein, where the backbone can hydrogen-bond to water.

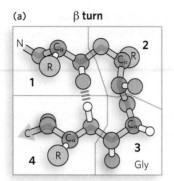 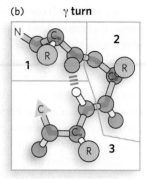

FIGURE 4-9 The linking of secondary structural elements by reverse turns. (a) An example of the β turn, involving four amino acid residues. (b) The much less common γ turn, accomplished by three residues.

There are eight types of β turns, classified according to the torsion angles between the inner two amino acid residues. Often, β turns contain a Pro residue at position 2 and Gly residue at position 3. Glycine is prevalent in β turns because its R group (hydrogen) allows it to accommodate many conformations that are not possible for other amino acids (see Figure 4-3). Although the conformation of a Pro residue is highly restricted compared with other amino acids, the imino nitrogen of proline can readily assume a cis configuration (see Figure 4-4b), a form that is especially suitable for tight turns.

A much less common reverse turn is the γ turn, consisting of only three amino acids (**Figure 4-9b**). The backbone carbonyl and amide groups of the first and third amino acid residues form a hydrogen bond, and the second (middle) amino acid of the γ turn is not involved in inter-residue hydrogen bonding.

SECTION 4.2 SUMMARY

- Secondary structure is a regularly repeating element that has hydrogen bonds between atoms of the peptide backbone.
- The α helix contains 3.6 amino acid residues per turn, and its internal hydrogen bonds set up an electric dipole for the entire helix.
- The compact structure of the α helix is unfavorable for certain combinations of residues, such as consecutive residues that are like-charged or bulky; the confined geometry of the proline ring can distort the helix, and proline is sometimes referred to as a helix breaker.
- The β sheet is formed by hydrogen bonding between two peptide segments in the β conformation, either within a polypeptide chain or between separate polypeptide chains. It can have a parallel or antiparallel configuration.
- The extended structure of a β sheet can accommodate most amino acid residues, including bulky aromatic amino acids and proline.

- Secondary structural elements in a polypeptide are connected by reverse turns. These may consist of long, unstructured stretches of amino acid residues or tighter, precise structures consisting of four amino acids (β turns) or, more rarely, three amino acids (γ turns).

4.3 TERTIARY AND QUATERNARY STRUCTURES

Despite the use of only a few types of regular secondary structure, proteins exhibit a diverse spectrum of three-dimensional shapes, honed by evolution to perform their particular roles in the cell. And despite the great diversity of protein structure and function, general principles of protein shape apply. In overall shape, proteins are of two main types: the globular proteins, found in the aqueous environment of cells and in membranes, and the fibrous structural proteins, such as keratin and collagen. We concern ourselves here mainly with globular proteins. Globular proteins are roughly spherical, but contain sufficient irregularities to yield a surface area that is about twice that of a perfect sphere of equivalent volume. These nooks and crannies often form active sites or protein-protein interaction surfaces and are essential to protein function.

Knowledge of the tertiary and quaternary structures of a protein at atomic resolution offers a wealth of information and greatly helps our understanding of how the protein functions. Further, with a structure in hand, the biochemist can direct amino acid substitutions to defined architectural positions to test hypotheses about function. Atomic resolution of proteins that have medical relevance also enables the design of drugs aimed specifically at an active site.

Tertiary and Quaternary Structures Can Be Represented in Different Ways

The **tertiary structure** of a protein is defined by the three-dimensional orientation of all the different secondary structures and the turns and loops that connect them. The **quaternary structure** of a protein is defined by the connections between two or more polypeptide chains. A common quaternary structure is the association of two identical subunits. A protein consisting of two polypeptide subunits is called a **dimer**.

The tertiary and quaternary structures of the *E. coli* DNA polymerase β subunit, a dimer of two identical polypeptides that encircles the bacterial DNA, are shown in **Figure 4-10**. The β subunit is part of the replication apparatus that duplicates the genome, and it holds the apparatus to the DNA (we discuss DNA replication and the role of the β subunit in Chapter 11). The figure shows six different representations of the β dimer. Each representation emphasizes one or more of the many structural features of a complete protein, and no single diagram can represent them all. In most of the representations, two colors differentiate the two identical polypeptide chains that comprise the complete protein.

Figure 4-10a shows the van der Waals radius for each atom, but because there are thousands of atoms in the structure, only the surface of the protein is visible. Figure 4-10b shows the atoms as sticks, allowing us to view the inside, but it is still somewhat bewildering to view the protein all at once. Typically, this representation is used to study the structure of just a small section of a protein. Figure 4-10c shows a thick-line trace of only the α carbons in the backbone, which simplifies the structure considerably and gives a view of the overall architecture. Figure 4-10d is a ribbon diagram, in which β sheets are shown as broad arrows pointing in the N- to C-terminal direction, α helices as coils, and loops and turns as narrow tubes, with the two polypeptides in different colors. The ribbon representation summarizes the secondary structural elements and overall architecture of the protein. Figure 4-10e is also a ribbon diagram, with the α helices and β sheets colored differently. Finally, Figure 4-10f is an electrostatic surface representation, showing the surface charges (red for acidic, blue for basic). The location of DNA-binding sites in proteins is often revealed by a basic patch in the electrostatic surface representation (note the blue area inside the β dimer).

Domains Are Independent Folding Units within the Protein

For a protein with 150 to 200 amino acid residues (M_r ~20,000), the polypeptide chain usually folds into two folding units known as **domains**. The larger the protein, the more domains it usually contains. The secondary structures that comprise a domain are typically adjacent to one another in the primary sequence, although this is not always the case. A protein with two or more domains may perform a single function in the cell, but sometimes, different domains within one protein have different functions. Although domains are independent folding units that can form a folded structure when separate from the full-length protein, various domains in the same protein often interact. The boundaries between domains are usually assessed by examining the protein's structure, but they can also be defined by limited proteolysis (a technique that cleaves the polypeptide backbone), which sometimes separates a protein into its domains.

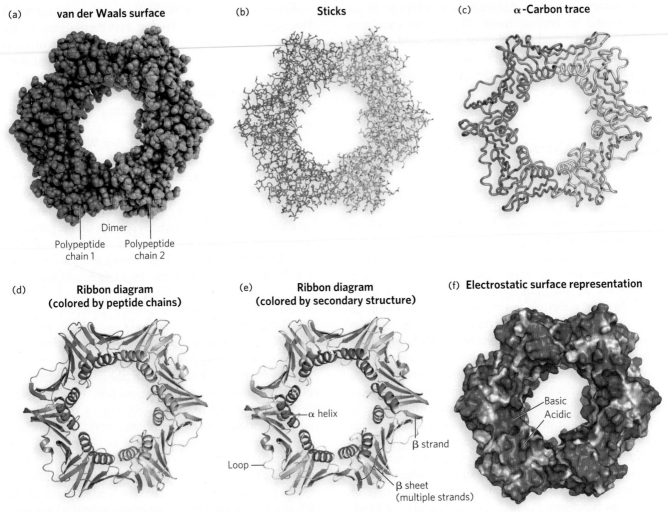

(a) **van der Waals surface**

(b) **Sticks**

(c) **α-Carbon trace**

Dimer

Polypeptide chain 1

Polypeptide chain 2

(d) **Ribbon diagram (colored by peptide chains)**

(e) **Ribbon diagram (colored by secondary structure)**

α helix

Loop

β strand

β sheet (multiple strands)

(f) **Electrostatic surface representation**

Basic

Acidic

FIGURE 4-10 Different representations of tertiary and quaternary structure of the _E. coli_ DNA polymerase β subunit. This protein dimer surrounds the _E. coli_ DNA (not shown here). [Source: PDB ID 2POL.]

Domains, then, can have independent functions. For example, the zinc finger domain is often used to bind DNA. (See this chapter's Moment of Discovery and Figure 4-20a; also see Chapter 19 for more on zinc-binding domains.) An individual domain can also catalyze a reaction, such as a nuclease activity. There are also proteins that require several domains that come together to perform a single function. Structures of proteins containing one, two, three, and four domains are shown in **Figure 4-11**. Myoglobin (M_r 16,700) is a small, oxygen-binding protein that folds into a single domain. An example of a two-domain protein is γ crystallin (M_r 21,500), a component in the lens of the eye. DnaA (M_r 52,000), a bacterial protein involved in the initiation of DNA synthesis, contains three domains. A proteolytic fragment of _E. coli_ DNA polymerase I (M_r 68,000), an enzyme that synthesizes new DNA strands, consists of four domains with two

separate enzymatic activities. One domain comprises an exonuclease, which proofreads the product of the DNA polymerase and removes any mistakes. The other three domains cooperate to form the DNA polymerase activity. (DNA polymerase structure and function are described in Chapter 11.)

Supersecondary Structural Elements Are Building Blocks of Domains

Particularly stable and common arrangements of multiple secondary structural elements are called **supersecondary structures**, also referred to as **structural motifs** or **folds**. Supersecondary structures are linked together to form sections of domains, or even whole domains. Some supersecondary structures are formed only of β sheets, some only of α helices, and others have a combination of α helices and β sheets.

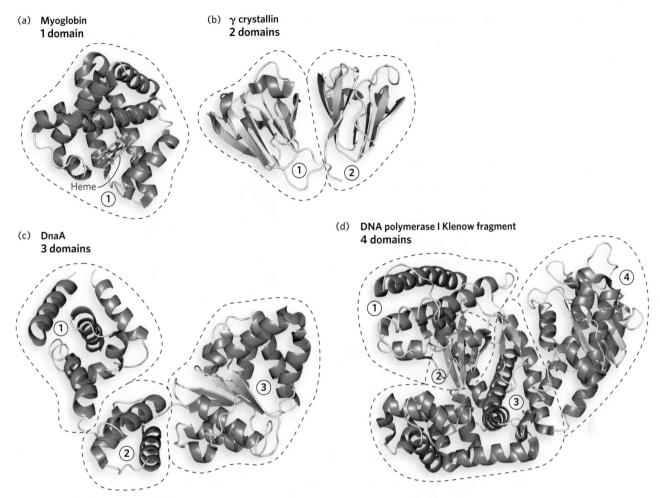

(a) **Myoglobin**
1 domain

(b) **γ crystallin**
2 domains

(c) **DnaA**
3 domains

(d) **DNA polymerase I Klenow fragment**
4 domains

FIGURE 4-11 Domains. Proteins fold into one or more domains, depending on their size. (a) One domain: sperm whale myoglobin. (b) Two domains: human γ crystallin. (c) Three domains: *Aquifex aeolicus* DnaA. (d) Four domains: *E. coli* Pol I Klenow fragment; the DNA polymerase activity is contained in the domains numbered 1, 2, and 3, and the exonuclease activity is in the fourth domain. *[Sources: (a) PDB ID 2BLH. (b) PDB ID 1H4A. (c) PDB ID 2HCB. (d) PDB ID 1D8Y.]*

Structural Motifs Containing a β Sheet The smallest motif containing a β sheet is the **β hairpin**, in which two antiparallel β strands are connected, often by a β or γ turn (**Figure 4-12a**). In proteins that contain this motif, the β hairpin can be found alone or as a repeated structure that forms a larger, antiparallel β sheet. Whether they are parallel or antiparallel, β sheets tend to follow a right-handed twist (**Figure 4-12b**). When the strands of a β sheet contain hydrophobic R groups at every second residue, the groups lie on the same side of the sheet, thus facilitating layer formation in the folded state. For example, a β sheet with a hydrophobic surface may pack against the hydrophobic side of another sheet. A β sheet of eight or more strands, and with one surface that is hydrophobic, can form a cylinder in which the first β strand hydrogen-bonds with the last β strand. This structure, referred to as a **β barrel**, sequesters the hydrophobic side chains inside the cylinder

(**Figure 4-12c**). In proteins with a β barrel, the barrel is a single supersecondary structure that also forms a complete domain. The simplified diagram below each supersecondary structure in Figure 4-12 shows the chain-folding pattern of the structure in two dimensions and is known as a **chain topology diagram**.

The **Greek key motif**, named for a design on Greek pottery, is another common β motif, consisting of four antiparallel β strands (**Figure 4-13** on p. 109). One example is found in the OB-fold domain, which mediates DNA binding in many different types of proteins. We will encounter the OB fold again in Chapter 5, where the functions of many DNA-binding proteins are discussed.

Motifs with α Helix and α Helix/β Sheet Structures In parallel β sheets, the connection between adjacent strands requires a much longer linker than in antiparallel β sheets. A basic unit of parallel β sheets is the **β-α-β**

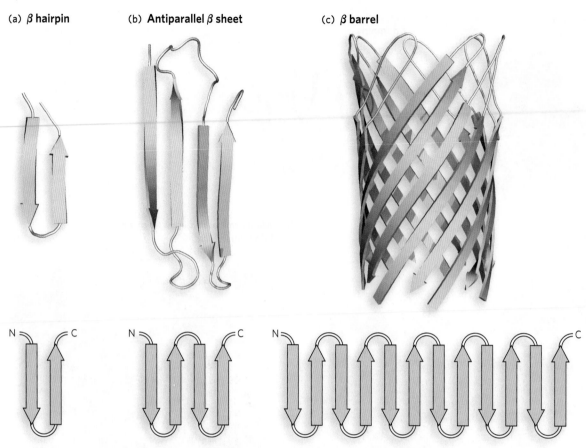

(a) **β hairpin**   (b) **Antiparallel β sheet**   (c) **β barrel**

FIGURE 4-12 Supersecondary structures of antiparallel β sheets. (a) The β hairpin motif is an antiparallel β sheet composed of two β strands adjacent in the primary structure; the strands are often connected by β or γ turns. (b) The β strands of antiparallel (shown here) and parallel β sheets tend to have a right-handed twist. (c) The β barrel motif is formed by the twist in a β sheet with a connection between the first and last strands. This example is a single domain of hemolysin (a pore-forming toxin that kills a cell by creating a hole in its membrane) from the bacterium *Staphylococcus aureus*. The diagram below each structure shows the folding topology of the polypeptide chain. The β strands in a β barrel can have several different chain topologies; shown in (c) is an "up-and-down barrel," reflecting the chain topology. In a simple β barrel, the strands would be connected as shown in the topology diagram (bottom), but the bottom strands of the particular β barrel shown here (top) are not connected, because the barrel is associated with additional domains (not shown). [Sources: (b, top) PDB ID 1LSH. (c, top) PDB ID 7AHL.]

motif, which consists of two parallel β strands connected by an α helix. The β-α-β motifs can be stitched together by the α helix linker in two different ways. In one, the α helix linker connects two β strands that are adjacent and hydrogen-bonded to each other (**Figure 4-14a**). This linkage forms a very common domain architecture called an **α/β barrel,** consisting of eight β strands surrounded on the outside by eight α helices (**Figure 4-14b**). The active site of an α/β barrel is almost always found on the loops at one end of the barrel.

The second way that the β strands in a β-α-β motif are joined by an α helix is shown in **Figure 4-15** on p. 110. This arrangement does not allow β strands in adjacent β-α-β motifs to hydrogen-bond and therefore prevents circularization of the β-α-β motifs. The result is a domain with a central parallel β sheet that contains α helices on both sides of the sheet. This architecture is commonly observed in ATP- or GTP-binding proteins and is sometimes referred to as a **Rossmann fold**. The active site in these proteins is usually located on the loops at the junction formed by β strands that are not directly connected by an α helix (see the How We Know section at the end of this chapter).

Several motifs have only α helices. One supersecondary structure that uses two α helices is the **helix-turn-helix motif,** sometimes found in proteins that bind specific DNA sequences (**Figure 4-16a** on p. 110). It is commonly found in bacterial transcription factors (proteins that regulate gene expression), as well as in some eukaryotic transcription factors. Another common α helix supersecondary structure is the **four-helix bundle** (**Figure 4-16b**). This structure consists of four α helices; the way they interact depends on the protein. Sometimes the bundle is formed by antiparallel

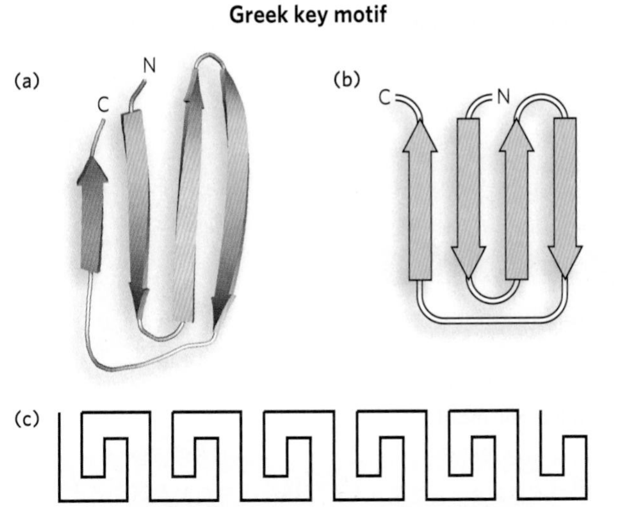

Greek key motif

(a)

(b)

(c)

FIGURE 4-13 The Greek key motif. (a) A portion of the subunit of *E. coli* DNA polymerase III adopts the Greek key conformation. (b) A chain topology diagram of the Greek key motif. (c) An example of the design found on Greek pottery from which this motif derives its name. *[Source: (a) PDB ID 2POL.]*

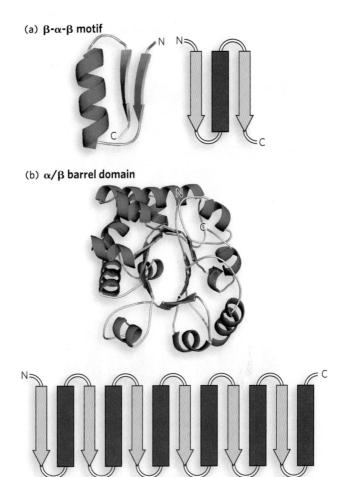

(a) β-α-β motif

(b) α/β barrel domain

FIGURE 4-14 The β-α-β motif. (a) The β-α-β motif contains two parallel β strands connected by an α helix. (b) Four β-α-β motifs, interconnected by α helices, underlie the α/β barrel structure. This α/β barrel is from triosephosphate isomerase of *Trypanosoma brucei*. The topology diagram in (b) shows the general arrangement of β strands and α helices; the actual structure of this α/β barrel domain contains eight β strands connected by α helices. *[Source: (b, top) PDB ID 4TIM.]*

helices, and other times by parallel helices or a mixture of the two. Some four-helix bundles are even formed by dimerization of two different subunits, each of which contributes two α helices to the motif. In the **coiled-coil motif**, two α helices pack against each other at an angle of 18° and gently twist around one another in a left-handed supercoil (**Figure 4-16c**). The two α helices interact through hydrophobic contacts along the sides of each helix that form the coiled-coil interface.

The interaction between α helices in a protein occurs through hydrophobic surfaces that face one another. A useful way to visualize the alignment of residues along the edges of an α helix is the helical wheel, a two-dimensional representation of the residues in the α helix (**Figure 4-17a**). The α helix has a nonintegral number of residues (3.6) per turn, but a nearly integral number of residues (7.2) in two turns. For this reason, the helical wheel representation consists of seven positions along two turns of α helix, designated by letters *a* through *g*.

Residues that are spaced every two turns, or every seven residues, lie on approximately the same side of the helix. When two α helices interact with each other by forming a hydrophobic interface, they often contain a repeating pattern of a hydrophobic residue every seventh amino acid. This seven-residue spacing pattern of hydrophobic residues in an α helix is sometimes referred to as a heptad repeat, or a leucine repeat, because the residue occupying the *d* position in a heptad repeat such as this is often leucine. Coiled-coils usually contain two heptad repeats, one set within the other,

with *a*-to-*a* and *d*-to-*d* spacing that allows contact with another helix with a similar arrangement of hydrophobic residues, as illustrated by the α-helical wheel diagrams in **Figure 4-17b**. This four-and-three hydrophobic repeat pattern is a hallmark of coiled-coils and forms a helix that is hydrophobic on one side and hydrophilic on the rest of the surface, which is referred to as an **amphipathic helix** (**Figure 4-17c**).

Some proteins, such as keratin, are long, extended, fibrous proteins in which the main structural element is a very long coiled-coil. However, there are many examples of globular proteins with short coiled-coils that mediate dimer formation. For example, the common **leucine zipper motif** in some eukaryotic transcription factors consists of four to five heptad repeats of leucine. These globular proteins, each containing a leucine zipper motif, can dimerize.

Rossmann fold

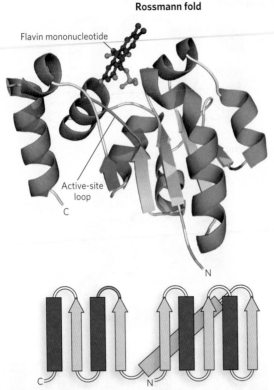

FIGURE 4-15 The nucleotide-binding Rossmann fold.
The Rossmann fold, a common nucleotide-binding motif, is
minimally composed of a β-α-β-α-β motif. This example is
from a section of EpiD, a decarboxylase enzyme from the
bacterium *Staphylococcus epidermis*. The β-α-β motifs are
connected as shown in the chain topology diagram (bottom).
In EpiD, the active-site loop binds a flavin mononucleotide.
[*Source: PDB ID 1G5Q.*]

Quaternary Structures Range from Simple to Complex

A protein composed of multiple polypeptide chains is
referred to as an **oligomer**, or multimer, and the individual polypeptide chains are referred to as subunits,
or **protomers**. An oligomer with identical subunits
is referred to as a **homooligomer**, while an oligomer
with nonidentical subunits is a **heterooligomer**. The
quaternary structures of three proteins are shown in
Figure 4-18. An example of a homodimer (i.e., a protein
with two identical subunits) is *E. coli* cAMP receptor
protein (CRP), also called catabolite gene activation
protein (CAP) (M_r 22,000). Each CRP subunit has two
domains; one domain binds DNA and the other binds
the cyclic nucleotide cAMP. In the homodimer, the two
CRP protomers interact through a coiled-coil, and the
two DNA-binding domains are oriented adjacent to
each other (Figure 4-18a). An example of a homotrimer is eukaryotic PCNA, in which the three subunits
(each subunit M_r 29,000) are joined by an intermolecular β sheet and by helix-helix packing (Figure 4-18b).

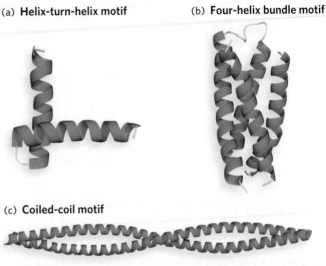

(a) Helix-turn-helix motif

(b) Four-helix bundle motif

(c) Coiled-coil motif

FIGURE 4-16 Supersecondary structures of α helices. (a) The
helix-turn-helix motif is common in DNA-binding proteins. (b) The
four-helix bundle is formed by the interaction of four α helices.
(c) The coiled-coil motif consists of two α helices twisted in a left-
handed supercoil. [*Sources: (b) PDB ID 1UM3. (c) PDB ID 1D7M.*]

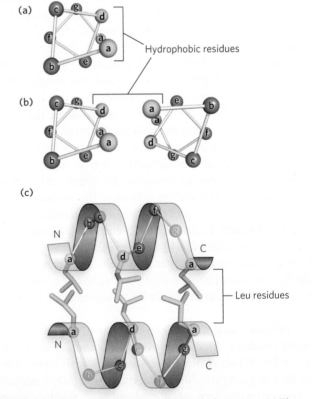

FIGURE 4-17 The helical wheel and heptad repeats. (a) The
seven residues that comprise two turns of the α helix can be
represented by a helical wheel and are labeled *a* through *g*.
(b) Residues in positions *a* and *d* lie on the same side of the helix;
hydrophobic residues in these positions create an amphipathic
helix and form the interface with a second helix to form a coiled-
coil. (c) The packing of *a* and *d* residues of two α helices in a
coiled-coil. Here, the hydrophobic residues are leucines.

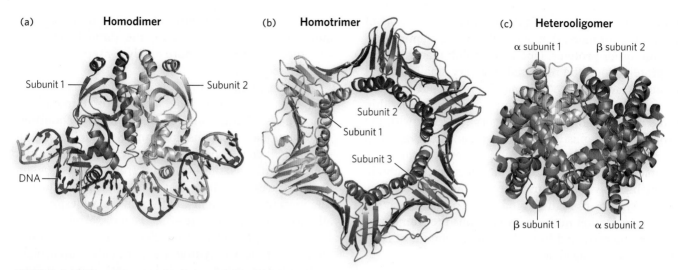

FIGURE 4-18 The quaternary structure of three proteins. Protomers (subunits) are represented in different colors. (a) Homodimeric *E. coli* CRP protein, as a complex with DNA. (b) Human PCNA, a homotrimer. (c) Human hemoglobin, a heterooligomer of two α chains and two β chains. [*Sources: (a) PDB ID 1CGP. (b) PDB ID 1AXC. (c) PDB ID 1GZX.*]

PCNA functions like the *E. coli* β subunit, anchoring the replication apparatus to DNA. Hemoglobin is a well-studied example of a heterooligomer (tetramer M_r 64,500). Hemoglobin contains two α protomers and two β protomers (Figure 4-18c).

Some oligomers are made up of numerous subunits. One example is the ribosome, a large, multiprotein oligomer that contains many nonidentical subunits, each often present as a single copy, as well as functional RNAs—all of which, together, form a molecular machine that translates a nucleic acid sequence into a protein sequence. The GroEL chaperonin is another example of a large, machinelike oligomer, which we discuss in Section 4.4.

Why do cells assemble such large oligomers from multiple subunits, rather than producing a large protein as a single, multidomain polypeptide chain? There are several reasons. For one, the folding of the many different domains in a single, very large polypeptide chain may be problematic. In addition, if one domain did not fold properly, the entire protein would lack function, so the investment of energy in making it would be wasted. A multisubunit composition avoids these problems. If a protein subunit misfolds, it will not be included in the oligomer, but at least only the investment of cellular resources to make this one domain will be wasted. In fact, a similar argument can be made even if the entire machinery is folded correctly. If one subunit subsequently denatures (becomes unfolded) or becomes inactive for any reason, it can be replaced by another subunit. The accuracy of translation is also an issue for very large proteins. During protein synthesis, approximately one mistake occurs in every 100,000 peptide-bond joining events. Therefore, a large protein of M_r exceeding 10^6 may accumulate mistakes, perhaps becoming inactive; smaller, individual subunits that do not fold properly can simply be discarded. Finally, the architecture of many large oligomers includes multiple copies of some subunits. More DNA would be required to encode a single large protein than is needed to encode multiple copies of individual subunits.

Intrinsically Unstructured Proteins Have Versatile Binding Properties

In the past 10 years, a very unconventional and new class of protein has been discovered that functions without having a defined structure. These **intrinsically unstructured proteins** appear to carry out some of the most important duties of the cell. Intrinsically unstructured proteins lack any detectable tertiary structure at all and normally would be regarded as inactive or denatured protein. Yet, because of their unstructured state, they are capable of binding to several different types of sites on different partners, and thus the same protein can be reused in multiple pathways. This diversity in binding is believed to play important roles in cell signaling, placing this class of protein at the center of protein interaction networks, where it can act as a hub for information transmission. Disease datasets indicate that intrinsically unstructured proteins may underlie pathological states with altered regulation of signaling, transcriptional control, cell division, and translation. Although an unstructured protein may acquire some structure on binding another protein, this is not always so. It is also thought that many normally structured proteins have a section of unstructured polypeptide, of 50 residues or more, that may serve the same purpose—to bind other proteins. One might think that this class of protein would be relegated to a very small niche of proteins encoded by a genome. But this is not the case. In fact, the prevalence of proteins, or sections within proteins, that fall into this category increases with the evolutionary sophistication of the organism. It has been proposed that up to one-third of proteins in the more complex eukaryotes either are unstructured or contain an intrinsically unstructured region. Indeed, the ability of this type of protein to bind many partners and act as an information network hub extends the complexity that can be extracted from the limited amount of DNA that can be packaged into a cell.

Protein Structures Help Explain Protein Evolution

Proteins with a similar primary sequence and similar function usually share a common evolutionary heritage and are said to be in the same **protein family**. For example, the globin family consists of many different proteins with structural and sequence similarity to myoglobin (e.g., myoglobin and the α and β subunits of hemoglobin have the same folding pattern). In many cases, however, even though the primary sequence does not show an evolutionary relationship, the protein structures are similar, indicating that they are related through a common ancestor. This is because the three-dimensional structure of a protein is more highly conserved than the primary sequence.

Protein families with little sequence similarity but with the same supersecondary structural motif and functional similarities are referred to as **superfamilies**. An evolutionary relationship among the families of a superfamily is considered probable, even though time and functional distinctions—resulting from different adaptive pressures—may have erased many telltale sequence relationships. A protein family may be widespread in all three domains of life—the Bacteria, Archaea, and Eukarya—suggesting a very ancient origin. For example, members of the α/β hydrolase superfamily contain eight β strands connected by α helices with three catalytic residues occurring in a particular order. This supersecondary structure is found in various proteases, esterases, lipases, and other types of proteins. Other families may be present in only a small group of organisms, indicating that the protein structure arose more recently. An example of this is the eukaryotic Ras superfamily of small GTPases, members of which contain a conserved "G domain" and are involved in cell proliferation, vesicle transport, and cell morphology. Tracing the natural history of structural motifs, using systematic structural classification databases, provides a powerful complement to sequence analysis (**Highlight 4-2**).

HIGHLIGHT 4-2 | **A CLOSER LOOK**

Protein Structure Databases

Each protein structure contains thousands of atoms arranged in three-dimensional space. When a scientist determines a protein structure, the x, y, and z coordinates for each atom are stored in a database called the **protein Data Bank (PDB)** and each is assigned a PDB identification number (ID). To view the atomic structure of a protein on a computer screen, the PDB ID file of atomic coordinates is imported into a computer program that displays the atomic model. There are many types of computer programs written just for this purpose, including PyMol, which is easily used on desktop and laptop computers.

Proteins can be classified according to their secondary structural elements, supersecondary structural motifs, and sequence homologies, and there are several databases that classify protein structures. One of the oldest is the **Structural Classification of Proteins (SCOP) database**. The highest level of classification in the SCOP database places all proteins into one of four classes: all α helix, all β sheet, α/β (where α and β segments are interspersed), and α + β (where α and β segments are somewhat segregated). For instance, myoglobin has α helices but no β sheet and therefore is placed in the α class. Likewise, proteins consisting only of a β barrel are in the β class. DnaA and Pol I have α helices mixed in with β sheets and thus are in the α/β class. In the circular β and PCNA clamps, the β sheet and α helices are somewhat segregated, placing these proteins in the α + β class.

Each class contains tens to hundreds of distinct substructure folding arrangements. Some substructures are very common, and others are found in only one protein. The number of unique folding motifs in proteins is far lower than the number of proteins—perhaps fewer than 1,000 different folds. As new protein structures are elucidated, the proportion of those containing a new structural motif has been declining. Below the levels of class and fold (or motif), which are purely structural, categorization in the SCOP database is based on evolutionary relationships.

The SCOP database is curated manually, with the objective of placing proteins in the correct evolutionary framework on the basis of conserved structural features. Two similar enterprises, the CATH (Class, Architecture, Topology, and Homologous superfamily) and FSSP (Fold Classification Based on Structure-Structure alignment of Proteins) databases, make use of more automated methods and can provide additional information.

- The tertiary structure of a protein is its three-dimensional structure, consisting of all of its secondary structural elements and loops.
- A protein domain is an independent folding unit within a protein and typically consists of up to 150 residues.
- Supersecondary structural elements, also called structural motifs, are arrangements of multiple secondary structural elements commonly found in proteins. Supersecondary structures are the building blocks associated with particular functions.
- The quaternary structure of a protein includes all the connections between two or more polypeptides and can range from a simple homodimer to a large, multiprotein assembly such as a ribosome.
- Some proteins have no apparent tertiary structure, yet are known to exist in cells and function to organize other proteins.
- The three-dimensional structure of proteins that evolved from a common ancestor is often more conserved than the primary sequence, making protein structures very useful in determining evolutionary heritage.

4.4 PROTEIN FOLDING

The structures of more than 50,000 proteins have been solved, and in all cases the polypeptide backbone folds to adopt a particular conformation, a process known as **protein folding**. Decades have passed since the classic studies by Christian Anfinsen showing that the amino acid sequence determines the folding pattern of a protein. He showed that ribonuclease that had been completely unfolded in a denaturing solution could rapidly fold into a biologically active protein after removal of the denaturant. Ribonuclease contains eight Cys residues that form four intrachain disulfide bonds. Amazingly, on refolding, all four disulfide links formed in the correct places, even though there are 105 possible combinations. The results suggested that other, weak interactions direct protein folding and precisely position the Cys residues prior to disulfide bond formation.

Not all proteins renature as easily as ribonuclease, and the exact process by which most proteins fold is still unknown. Furthermore, some proteins require the assistance of other proteins for proper folding. Even when correctly folded, a protein structure is constantly in flux, and all of its atoms and structural elements vibrate rapidly. This inherent flexibility is essential to protein function—to achieve, for instance, a different thermodynamic state when substrate is converted to product in an enzyme reaction. As a further example, some proteins are triggered by the binding of a substance known as an allosteric effector to adopt another conformational state that has substantially different activity (we explore protein function in Chapter 5). No matter how the folding is achieved, a particular amino acid sequence nearly always folds into the same conformation, and this conformation is required for proper protein function. Misfolding of proteins, whether due to mutations within the sequence or other factors, can affect protein function and could have serious consequences for the cell or organism.

Predicting Protein Folding Is a Goal of Computational Biology

The primary sequence holds the instructions for protein folding, so, theoretically, researchers should be able to predict a protein's tertiary structure from its amino acid sequence alone. This is not yet possible, however—but not for want of trying. The protein folding "code" is complex, and the problem is complicated by the small difference in free energy between the folded and unfolded states of a protein. Although hydrogen bonds are strong, hydrogen bonding is not very different in the folded and unfolded states, because water is plentiful in the cell and can also hydrogen-bond with the protein. The main force driving protein folding is therefore derived from van der Waals contacts and the hydrophobic effect, but the free-energy difference between folded and unfolded protein states is still only about 5 to 15 kcal/mol. This small energy difference makes it difficult to understand and quantify the protein folding "code" that determines three-dimensional structure from primary sequence.

Inspection of many protein structures has provided some guidelines for protein folding. Hydrophobic residues avoid water by becoming buried, and this drives the polar polypeptide backbone into the interior, where it must form secondary structures with internal hydrogen bonds. The interior of a protein is amazingly well packed: about 75% of atoms in the interior of a protein are packed together as close as their van der Waals radii, the theoretical limit. Not even crystals of free amino acids have closer packing. Polar residues of a protein are usually at the surface, where the side chains can interact with water. These guidelines are used when predicting the extent to which different residues are buried in protein structures (**Figure 4-19**).

The exceptions to this general rule are easily explained. The hydrophobic side chain of proline is sometimes found on the surface, but Pro residues are useful for making sharp turns, which are usually on the surface. Cysteine is very moderately polar, but some Cys residues form disulfide bridges that are hydrophobic and therefore easily buried. Also, some residues that do not at first seem hydrophobic contribute greatly to hydrophobic forces. For example, lysine, a polar, charged amino

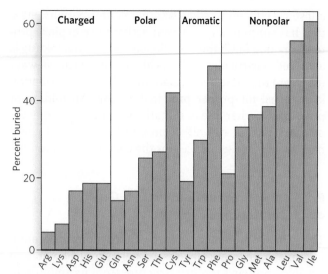

FIGURE 4-19 Buried residues. This plot shows the relative extent to which the 20 different amino acid residues are found in buried positions in globular proteins. *[Source: Adapted from P. Y. Chou and G. D. Fasman, Biochemistry 13:222–245, 1974.]*

acid, has a long hydrocarbon side chain that is often buried and contributes to hydrophobic interactions, while only its charged amino group protrudes to the surface.

Computational biologists have developed powerful algorithms to predict protein structure from amino acid sequences. The computations make use of the allowed values of torsion angles for different amino acids, their polar, ionic, and hydrophobic features, and the large body of structural information contained in the protein structure database. In fact, there is an international competition each year, called CASP (Critical Assessment of Protein Structure Prediction), which evaluates different protein structure prediction algorithms. To make this competition possible, research groups that have determined a protein structure but have not yet published it provide only the sequence of the protein to the CASP competition. Then computational biologists apply their algorithms to these sequences and arrive at structural models. When the experimentally determined structures are published, the degree to which the computed models agree with the experimentally determined structures is evaluated, and the scientist with the most accurate algorithm wins.

Besides CASP, there is a compelling protein-folding computational program set up as an online video game that enlists nonscientists in solving important questions in protein structure. The Foldit program for online gamers has even resulted in publication of original research papers in scientific journals. The Foldit puzzle game takes the beginner through a series of introductory puzzles that teach certain basic principles, then the user can choose from among several puzzles that enable nonscientists to help solve difficult computational science questions. This puzzle game is an experimental research project of

the University of Washington Center for Game Science, in collaboration with the Biochemistry Department at the University of Washington. The premise is to add the human "insight" factor to computer-based simulation. At some point just prior to solution of a protein structure by x-ray crystallography, the protein sequence is presented to gamers. The object is to fold it correctly, and the proof is in the actual structure solved by a research lab. The algorithm has rules of folding from known structures, and the player can juggle, shake, and move sections of the protein around, scoring points as the protein backbone approaches a better fit to known parameters. Gamers typically form groups that chat and work together to achieve a high score within a set clock countdown. At the end of the competition, the highest score wins. But more exciting is when the actual structure is published, and gamers can see whether the high-scoring result actually arrived at the correct structure. In fact, winners of this game have also won first place in several categories of the CASP competition—all without the formal training of a biochemical or protein structural background.

The Foldit game also includes puzzles with experimental data and asks the gamer to fit the protein sequence into it. In 2011, Foldit players helped determine the crystal structure of an important drug target of a monkey AIDS virus, the retroviral protease of the Mason-Pfizer monkey virus. Technical hurdles prevented a solution of the crystal structure by conventional means, but the help of online gamers provided a solution that fit all known data in less than two weeks after the game was posted. This work led to a publication in a scientific journal.

Structure-based protein design is another goal of computational biology. The scientist starts with a "target" structure and then computationally designs a sequence that adopts the target folding pattern. Most of the algorithms developed for protein design focus on the redesign of a preexisting protein or "protein core," the section that contains the central hydrophobic residues and defines how the protein folds. The complete de novo design of a protein is a difficult computational problem, given the vast number of possible conformations each amino acid residue can adopt.

A striking advance in de novo protein design was made by Steve Mayo's group. An algorithm they had developed was applied to compute a sequence that folds into the structure of a zinc finger domain (**Figure 4-20a**). As we noted earlier in the chapter, zinc finger domains consist of a β-α-β motif, composed of about 30 amino acids. Despite its small size, the zinc finger domain contains sheet, helix, and turn structures, and the zinc atom is not required for proper folding. For a target structure, Mayo and colleagues used a 28-residue zinc finger domain in a protein called Zif268. There are 1.9×10^{27} possible combinations for 28-residue proteins, given the

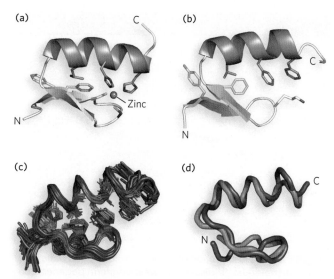

FIGURE 4-20 A protein structure designed by computation. (a) The structure of a zinc-binding domain from a DNA-binding protein (Zif268). (b) The predicted structure of the FSD1 sequence, derived entirely by computation. (c) The NMR structure of FSD1. (d) Superimposition of the Zif268 zinc-binding domain (red) and FSD1 (blue). Only residues 3 through 26 are shown. [Source: B. I. Dahiyat and S. L. Mayo, Science 278:82–87, 1997.]

allowed torsion angles of the 20 common amino acids. The extent of the problem can be appreciated by the fact that just one molecule of each of these possible peptides, lumped together, would amount to 11.6 metric tons! The sequence to emerge from 90 hours of computing time gave a novel protein sequence that was unrelated to any known sequence in the database, yet was predicted to have a folding pattern similar to that of the target protein (compare the two patterns in **Figure 4-20a, b**). The protein having the computed sequence, named FSD1 (full sequence design 1), was then synthesized and the structure was experimentally determined by nuclear magnetic resonance (NMR) (**Figure 4-20c**); the actual FSD1 structure corresponded amazingly well with the

target protein (**Figure 4-20d**). The overall deviation in backbone atoms of the target and FSD1 structures was only 0.98 Å over residues 8 through 26. The fact that completely different sequences can serve the same function supports the idea that once evolution arrives at a particular solution, it often uses this solution repeatedly in other proteins, rather than coming up with an entirely new one.

Polypeptides Fold through a Molten Globule Intermediate

In 1968, Cyrus Levinthal reasoned that a protein should not be capable of randomly folding into a unique conformation in our lifetime. For example, starting from a 100-residue polypeptide in a random conformation, and assuming that each amino acid residue can have 10 different conformations, 10^{100} different conformations for the polypeptide are possible. If the protein folds randomly, by trying out all possible backbone conformations, and if each conformation is sampled in the shortest time possible ($\sim 10^{-13}$ seconds, the time scale of a single molecular vibration), it would still take about 10^{77} years to sample all possible conformations! Yet, E. coli makes a biologically active protein of 100 amino acids in about 5 seconds at 37°C. This apparent contradiction is referred to as Levinthal's paradox, and the astronomical disparity between calculation and observation reveals that protein folding is far from random: it must follow an ordered path that side-steps most of the possible intermediate conformations.

Intensive studies indicate the different ways that intermediate conformations could be side-stepped. We discuss here two pathways that are supported by experimental data. A **hierarchical model** of folding proposes that local regions of secondary structure form first, followed by longer-range interactions (e.g., between two α helices), and that this process continues until complete domains form and the entire polypeptide is folded (**Figure 4-21**).

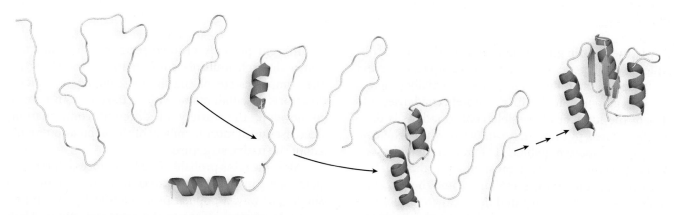

FIGURE 4-21 A hierarchical model of protein folding. In this model, which has been demonstrated for a few small proteins, some secondary structural elements form locally, and they nucleate the folding of the rest of the protein.

Prion-Based Misfolding Diseases

Stanley Prusiner [*Source: Russ Fischella Photography.*]

A misfolded protein seems to be the cause of several rare, degenerative brain diseases in mammals. Perhaps the best known is bovine spongiform encephalopathy (BSE), also known as mad cow disease. An outbreak of BSE made international headlines in the spring of 1996. Related diseases are kuru and Creutzfeldt-Jakob disease in humans, scrapie in sheep, and chronic wasting disease in deer and elk. These diseases are referred to as **spongiform encephalopathies**, because the diseased brain frequently becomes riddled with holes and appears spongelike (Figure 1). Symptoms include dementia and loss of coordination, and the diseases are usually fatal.

In the 1960s, investigators found that preparations of the disease-causing agents seemed to lack nucleic acids, suggesting that the agent was a protein. Initially, the idea seemed heretical. All disease-causing agents known up to that time—viruses, bacteria, fungi, and so on—contained nucleic acids, and their virulence was related to genetic reproduction and propagation. However, four decades of investigation, pursued most notably by Stanley Prusiner, provided evidence that spongiform encephalopathies are different.

The infectious agent has been traced to a single protein (M_r 28,000), dubbed **prion** (from *proteinaceous infectious only*; analogous to "virion"). The infected tissue contains abnormal spots with densely packed protein fibers called plaques. Fibers in these plaques are resistant to protease, and thus the protein is referred to as PrP (*protease-resistant protein*). The normal prion protein, called PrP^C (*C* for cellular), is found throughout the body in healthy animals, and it does not form plaques. Its role

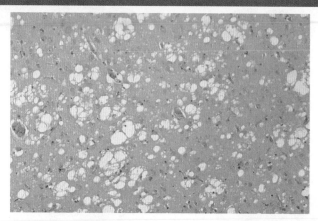

FIGURE 1 A stained section of cerebral cortex from a patient with Creutzfeldt-Jakob disease shows spongiform degeneration (holes in the tissue), the most characteristic neurohistological feature. [*Source: Stephen J. DeArmond.*]

in the mammalian brain (or any tissue) is not known, but it may have a molecular signaling function. Strains of mice lacking the *Prnp* gene that encodes PrP (and thus lacking the protein itself) suffer no obvious ill effects. Illness occurs only when the animal produces a PrP protein that misfolds, or converts, to an altered conformation, called PrP^{Sc} (*Sc* for scrapie). The misfolded PrP^{Sc} aggregates to form fibers that presumably lead to the plaques associated with spongiform encephalopathy. Although researchers do not completely understand how the misfolded state of PrP is propagated during infection, it is commonly thought that the interaction of PrP^{Sc} with PrP^C converts the latter to PrP^{Sc}, initiating a domino effect in which more and more of the brain protein converts to the disease-causing form.

The structure of the C-terminal region of normal PrP^C is known (Figure 2); it contains three α helices and two β strands. In contrast, fibers of PrP^{Sc}, as shown by circular

---

In the **molten globule model**, the hydrophobic residues of a polypeptide chain rapidly group together and collapse the chain into a condensed, partially ordered, "molten globule state." With the protein condensed into a ball, the number of possible conformations is drastically limited to those that can occur within the confines of the molten globule. The molten globule state is not as compact as the final state; only about half of the hydrophobic residues are buried, and those that remain on the surface must find their way to the core. As hydrophobic residues are buried, the polar backbone atoms must form hydrogen-bonded secondary structures. The molten globule is therefore an ensemble of different partially ordered segments that shift and churn through multiple conformations, searching for the compact state of the native—completely folded—structure. As subdomains with tertiary structure begin to develop, the variety of different conformations decreases until most members of the population finally adopt the native structure.

Most proteins probably fold by a process that incorporates features of both models. Instead of following a single pathway, a population of identical polypeptides may take several routes to the same end point. Thermodynamically, the folding process can be viewed as a kind of free-energy funnel (**Figure 4-22** on p. 118).

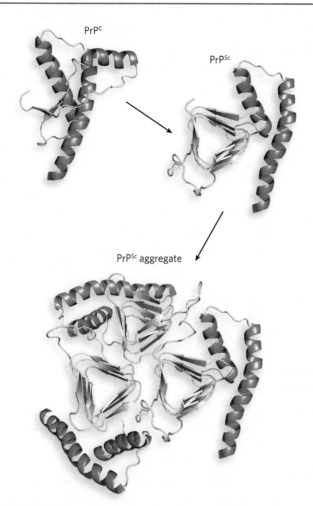

dichroism, contain a core of tightly packed β sheets. The exact structure of PrP^{Sc} is unknown because it is an aggregate and difficult to work with, but it is thought to contain several β sheets—accounting for the prevalence of β sheet structure in the PrP^{Sc} aggregate inferred from CD measurements. Future insights about the structure and function of PrP^{Sc} will provide treatment strategies for this disease.

Spongiform encephalopathy can be inherited, can occur spontaneously, or can be transmitted through infection. Most cases are spontaneous. Infectious transmission accounts for fewer than 1% of cases and occurs through direct contact with diseased tissue. Inherited forms of prion diseases, which account for 10% to 15% of cases, are due to a variety of point mutations in the *Prnp* gene, each of which is believed to make the spontaneous conversion of PrP^{C} to PrP^{Sc} more likely. A detailed understanding of prion diseases awaits new information on how prions affect brain function. One recent advance is the development of mouse models of the disease. With the alteration of just one amino acid residue, the mutant prion protein leads to disease that faithfully mimics the full disease spectrum, from formation of plaques in the brain to transmission of disease to healthy animals by direct contact. Mouse models hold promise for developing and testing therapies to prevent these fatal neurodegenerative diseases.

FIGURE 2 The globular domain of human PrP^{C} monomer (top left) contains three α helices and two short β strands. To its right is a proposed structure of the corresponding region of PrP^{Sc}, containing many more β strands, which may promote aggregation (bottom). *[Sources: (top left) PDB 1QLX; (top right and bottom) adapted from S. B. Prusiner, Sci. Am. 291(1):86–93, 2004.]*

Unfolded forms rapidly collapse to more compact forms, after which the number of conformational possibilities decreases. These rapid first stages quickly narrow the funnel. Small depressions along the sides of the free-energy funnel represent semistable intermediates that briefly slow the folding process. At the bottom of the funnel, the ensemble of folding intermediates has been reduced to the single conformation of the final, native state of the protein.

Defects in protein folding may be the molecular basis for a wide range of human genetic disorders. For example, cystic fibrosis is caused by the misfolding of a chloride channel protein called cystic fibrosis transmembrane conductance regulator (CFTR). Many disease-related mutations in collagen are also caused by defective folding. In addition, defective protein folding causes the prion-related diseases of the brain (**Highlight 4-3**).

Thermodynamic stability is not evenly distributed over a protein. For example, a protein may have two stable domains joined by a segment with lower structural stability. The regions of low stability may allow the protein to alter its conformation between two (or more) states. As we shall see in Chapter 5, variations in the stability of regions within a given protein are often essential to protein function.

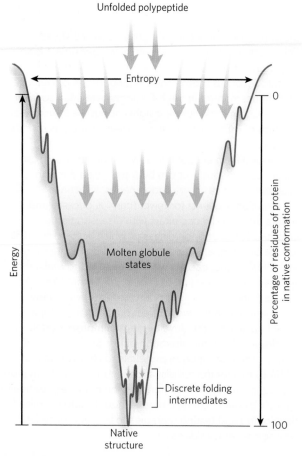

Unfolded polypeptide

Figure 4-22 Protein folding as a free-energy funnel. The top of the funnel represents the unfolded state of a protein, the highest entropy state. As it folds, the protein progresses to lower free energy. The more compact the protein gets, the more rapidly it arrives at the properly folded and lowest free-energy state, at the bottom of the funnel. [*Source: Adapted from P. G. Wolynes, J. N. Onuchic, and D. Thirumalai, Science 267:1619, 1995.*]

Chaperones and Chaperonins Can Facilitate Protein Folding

Not all proteins fold spontaneously. The folding of many proteins is facilitated by **chaperones**, specialized proteins that bind improperly folded polypeptides and facilitate correct folding pathways or provide microenvironments in which folding can occur. Two classes of molecular chaperones have been well studied, and both are conserved across species, from bacteria to humans. Proteins in the first class, a family of proteins called **Hsp70** (heat shock proteins of M_r 70,000), become more abundant in cells stressed by elevated temperature. Hsp70 proteins bind to unfolded regions that are rich in hydrophobic residues, preventing aggregation and thus protecting denatured proteins and newly forming proteins that are not yet folded. Some chaperones also facilitate the assembly of subunits in oligomeric proteins.

Hsp70 proteins bind and release polypeptides in a cycle that involves several other proteins (including the class Hsp40) and ATP hydrolysis. **Figure 4-23** diagrams chaperone-assisted folding of a protein mediated by the DnaK and DnaJ chaperones of *E. coli*, functional equivalents (homologs) of the eukaryotic Hsp70 and Hsp40 proteins.

Chaperonins, the second class of chaperones, are elaborate protein complexes required for the folding of some cellular proteins. In *E. coli*, an estimated 10% to 15% of cellular proteins require the resident chaperonin system—GroEL/GroES—for folding under normal conditions. Up to 30% of proteins require assistance when the cell is heat-stressed. Unfolded proteins are bound in pockets in the GroEL complex, and the pockets are capped transiently by GroES (**Figure 4-24**). GroEL undergoes substantial conformational changes, coupled with ATP hydrolysis and the binding and release of GroES, which together promote folding of the bound polypeptide. Although the structure of the GroEL/GroES chaperonin is known, many details of its mechanism of action remain unresolved.

Protein Isomerases Assist in the Folding of Some Proteins

The folding of certain proteins requires enzymes that catalyze isomerization reactions. **Protein disulfide isomerase** is a widely distributed enzyme that catalyzes the interchange, or shuffling, of disulfide bonds until the bonds of a protein's final conformation have formed. Among its functions, protein disulfide isomerase catalyzes the elimination of folding intermediates with inappropriate disulfide cross-links.

The cis and trans isomers of peptide bonds are rapidly interchangeable (see Section 4.1). However, proline's cyclic structure makes interconversion between cis and trans isomers a slow process. **Peptide prolyl cis-trans isomerase** catalyzes the interconversion of the cis and trans isomers of proline peptide bonds (**Figure 4-25**). Most peptide bonds in proteins are in the trans isomeric form, but the cis isomer of proline is often found in tight turns between secondary structural elements.

SECTION 4.4 SUMMARY

- The polypeptide chain of a protein folds into a unique conformation, and the instructions for this folding are inherent in its primary sequence.
- The folding pattern of a protein is hard to predict from the primary sequence, because the forces that stabilize the folded state are weak and cannot be recognized from the amino acid sequence.
- In a protein's folded state, hydrophobic residues are usually found in the interior of the protein, and polar residues often localize to the surface.

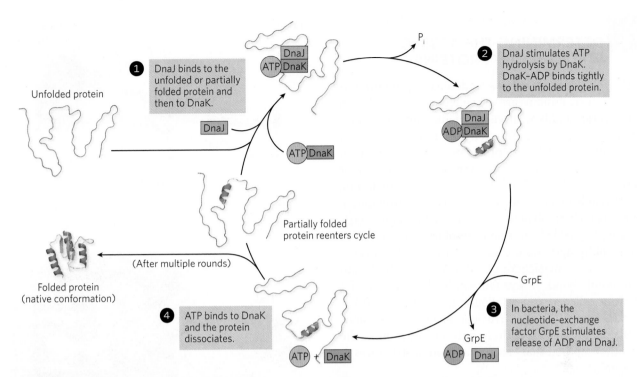

FIGURE 4-23 Chaperone-assisted protein folding. Chaperones bind and release polypeptides by a cyclic pathway, shown here for *E. coli* chaperone proteins DnaK and DnaJ, homologs of the eukaryotic chaperones Hsp70 and Hsp40. The chaperones mainly prevent the aggregation of unfolded polypeptides. Some polypeptides released at the end of a cycle are in their native conformation; the rest are rebound by DnaJ and DnaK to repeat the cycle for further attempts at refolding. In bacteria, GrpE interacts transiently with DnaK late in the cycle (step 3), promoting dissociation of ADP and possibly DnaJ. No eukaryotic analog of GrpE is known.

- The protein-folding pathway is not random. The folding of some proteins is thought to proceed through a condensed molten globule state that limits folding to conformations that are compatible with a compact volume.
- Protein folding is sometimes assisted by chaperones and chaperonins. Chaperones are proteins of the Hsp70 class that bind unfolded proteins and use cycles of ATP-binding and hydrolysis to help the proteins refold. Chaperonins are complex multisub-unit structures that engulf the protein in a chamber during the refolding process.
- Protein folding is also assisted by isomerases. Protein disulfide isomerase catalyzes the breakage and re-formation of disulfide bonds, and peptide prolyl cis-trans isomerase facilitates interchange between the cis and trans isomers of Pro residues.

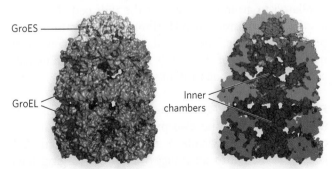

FIGURE 4-24 Chaperonin-assisted protein folding. The *E. coli* chaperonins GroEL and GroES. Each GroEL complex consists of two chambers formed by two heptameric rings (each subunit M_r 57,000). GroES, also a heptamer (each subunit M_r 10,000), blocks one of the GroEL pockets. Surface (left) and cut-away (right) images of the GroEL/GroES complex are shown. *[Source: PDB ID 1AON.]*

FIGURE 4-25 The trans and cis isomers of proline in a peptide bond. Most peptide bonds (>99.95%) are in the trans configuration, but about 6% of bonds involving Pro residues are in the cis configuration; many of these occur in β turns.

4.5   DETERMINING THE ATOMIC STRUCTURE OF PROTEINS

There are very few methods to deduce a protein's tertiary structure. Proteins are too small to allow resolution of structural details with visible light. The lower limit of visible light has a wavelength of about 400 nm (400×10^{-9} m) and therefore cannot resolve objects of a size less than about half this wavelength (200 nm, or 2,000 Å). Even huge ribosomes, with a radius of 18 nm, are not visible in a light microscope. The electron microscope has high resolving power, but at the high-energy wavelengths needed for atomic resolution, the electron beam rapidly destroys the sample. True atomic resolution requires a wavelength of ~1.5 Å, about the length of an atomic bond. X rays fall within this range, and a technique called x-ray crystallography can provide atomic resolution of proteins. Nuclear magnetic resonance (NMR) operates in an entirely different way and is the only other method that can reveal protein structures at the atomic level.

Most Protein Structures Are Solved by X-Ray Crystallography

The use of **x-ray crystallography** to determine the structure of proteins was pioneered by Max Perutz and John Kendrew, who solved the structures of hemoglobin and myoglobin. This was an enormously difficult task at the time. The equipment and techniques continue to be improved, and more than 50,000 protein structures are now available in the Protein Data Bank, a repository of information on protein structures. Of all known protein structures, over 90% were determined by x-ray crystallography (see Highlight 4-2). There is no theoretical limit to the size of protein that can be analyzed by this method.

Max Perutz, 1914–2002 (left); John Kendrew, 1917–1997 (right) [Source: © Hulton Deutsch Collection / Corbis.]

Amplifying Diffracted X Rays X-ray crystallography illuminates a protein crystal with an x-ray beam, and the diffracted x rays are collected for analysis (**Figure 4-26a**). Diffracted x rays travel in every direction and thus normally create a blur on a detector. But in a crystal, trillions of protein molecules are aligned in a regular lattice, and therefore some of the diffracted x rays combine and add up in a process called **constructive interference**, forming a **reflection spot** on a film or detector. Each reflection spot in a **diffraction pattern** is produced

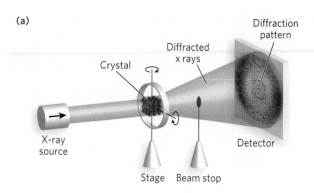

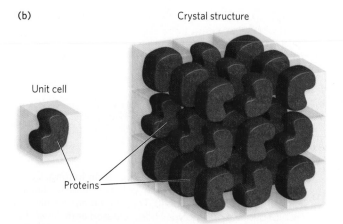

FIGURE 4-26 Protein crystals and diffraction patterns. (a) Protein crystals produce a diffraction pattern when x rays are passed through them. In x-ray crystallography, the crystal is rotated in all directions, and diffracted x rays are collected by a detector. X rays that pass through the crystal without diffraction are blocked from hitting the detector by a beam stop. (b) The unit cell, or repeating unit in the lattice of a protein crystal, may contain one or more protein molecules.

by the summation of diffracted x rays in the **unit cell**, the smallest regularly repeating unit in the crystal (**Figure 4-26b**). The unit cell can be as small as a single protein molecule, but it often consists of two or more identical protein molecules. The spacing of reflections is related to atomic distance in the unit cell. To obtain a sufficient number of reflections for determining the protein structure, the crystal is rotated during x-ray irradiation. Tens of thousands of reflections are usually collected to solve one protein structure.

Reconstructing the Protein Image An object illuminated in a light microscope also produces a diffraction pattern, but it is not visible, because the diffracted light is recombined into an image with a converging lens. Electron microscopy works in a similar fashion, using magnets to refocus the diffracted electrons into an image. However, no lens can recombine diffracted x rays. Instead, the diffraction pattern is recombined into an image by a mathematical converging series, called a

Fourier series, that acts, mathematically, like a converging lens. X rays are photons and therefore behave as sine waves, each of which has an amplitude, a wavelength, and a phase. The amplitude is the height of the wave and the wavelength is the distance between waves. The phase describes how wavelengths align. For example, two waves of the same wavelength could be completely out of phase with one another, with the crest of one occurring in the trough of the other; or they could be in phase, in which case their wave crests are additive. These three parameters—amplitude, wavelength, and phase—are required for the Fourier series. The wavelength, λ (lambda), is the same as that of the x-ray beam used to illuminate the crystal, and the amplitude, A, is calculated from the spot intensity, I ($I = A^2$). But when a diffracted x-ray wave hits the detector, the wave collapses and the phase is lost. The experimenter must determine the phase of the x-ray waves before the crystal structure can be solved. There are several methods for determining the phase of each reflection, which we will not go into here.

The reconstructed image is displayed on a molecular graphics console as a volume encased in a meshwork referred to as an **electron density map (Figure 4-27)**: the higher the resolution, the greater the detail contained in the electron density map. Paradoxically, the reflections in the diffraction pattern that carry the highest resolution are those that are farthest from the center.

Examples of electron density maps of a tryptophan side chain in a protein, obtained by using reflections at increasing distances from the center of a diffraction

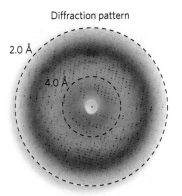

Diffraction pattern

2.0 Å

4.0 Å

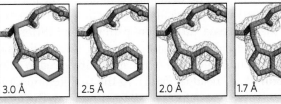

3.0 Å   2.5 Å   2.0 Å   1.7 Å

FIGURE 4-28 The relationship between resolution and diffraction pattern. The dotted circles in the diffraction pattern (top) represent the locations of reflections responsible for two different resolutions of an electron density map. The maps below show four different resolutions. The area of electron density corresponds to Trp[123] of the *E. coli* DNA polymerase β subunit. Positions of the atoms of the Trp residue in the final model are shown in the four panels. *[Source: Based on data and images courtesy of Roxana Georgescu, laboratory of Mike O'Donnell, Rockefeller University.]*

pattern, are shown in **Figure 4-28**. Using reflections in the diffraction pattern that correspond to a resolution of 4 Å between atoms in an electron density map, the resolution is insufficient to trace the path of the peptide backbone or to place most of the side chains. Using reflections that correspond to a resolution of 3 Å, the peptide backbone is discernible as a continuous ribbon. Secondary structural elements are visible, and the general shape of side chains is often apparent, but there are usually some disordered regions in loops that cause breaks in the density and prevent a continuous chain trace. A range of 2.2 to 3.0 Å resolution is required to get the most complete information on a protein's structure.

The Initial Model The three-dimensional protein structure inferred from the electron density map is known as the **initial model**. In the early stages of analysis, the initial model is hypothetical. To build the model, the known amino acid sequence of the protein must be fitted into the electron density mesh. Model building is aided by graphics on a computer screen, but it is mostly performed manually and requires the skill and patience of the experimenter. Because the peptide bond is planar, a peptide bond "ruler" helps identify the C_α atoms. To position the primary sequence in the electron density

Electron density map

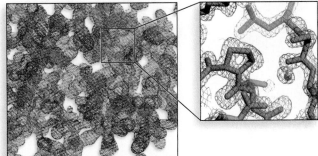

FIGURE 4-27 An electron density map. An electron density map (left) contains too much information to analyze when viewed all at once. The experimenter focuses instead on one small region at a time (right) and fits the polypeptide backbone into the density. Shown here are molecules that lie in the outlined portion of the electron density map. The small red sphere is an ordered water molecule—a water molecule that is held in one place and thus can be visualized. *[Source: Based on images courtesy of Roxana Georgescu, laboratory of Mike O'Donnell, Rockefeller University.]*

map, the experimenter looks for unusual arrangements of large, characteristic side chains. The remaining side chains are then filled in, and each is adjusted into the electron density. The resulting initial model is far from perfect, but errors are minimized in the next stage, the refinement process.

Refinement Improvements in the electron density map are generated by **refinement**. Refinement is an iterative process (**Figure 4-29**). It starts by taking the model and building a model crystal from it computationally (in silico). Then the Fourier series is used to compute a diffraction pattern for the model crystal, and the position and intensity of each calculated reflection are compared with the observed diffraction pattern. The difference between the calculated and observed values yields a measurement of the error in the model, referred to as an **R factor** (R for residual error). At the first iteration, the R factor value is usually 0.4 to 0.5. Although refinement theoretically has no ending, in practice, structures are refined to an R factor value of 0.15 to 0.25.

The physical environment within a crystal is not identical to that in a solution or in a living cell, so the conformation of a protein in a crystal could, in principle, be affected by nonphysiological factors such as incidental protein-protein contacts. However, when structures derived from crystal analysis are compared with structural information obtained by NMR (described below), the crystal-derived structure almost always represents a functional conformation of the protein.

Smaller Protein Structures Can Be Determined by NMR

An important complementary method for determining protein structure is **nuclear magnetic resonance (NMR)**. NMR is performed on proteins in solution, which is an advantage over x-ray crystallography because protein crystals can be difficult to obtain. However, only relatively small protein structures can be solved by NMR ($M_r < 25,000$, or ~200 amino acids).

Obtaining Primary Data The primary data from NMR are radio-frequency lightwave emissions from atomic nuclei. Only certain atoms, including ^1H, ^{13}C, ^{15}N, ^{19}F, and ^{31}P, possess the kind of nuclear spin that gives rise to an NMR signal. The technique involves placing the protein sample in a strong magnetic field, which aligns the spins of all the nuclei of the particular atom under study. Then the sample is pulsed with radio-frequency radiation to excite the nuclei. As the nuclei relax, they emit radiowaves, which are detected and recorded. After many repetitions in rapid succession, the data are averaged. Repeated pulses of irradiation and collected emitted radiowaves are summed, thus increasing the signal-to-noise ratio to produce an NMR spectrum. The emissions are plotted as a spectrum of **chemical shifts**, expressed as parts per million (ppm). The chemical shift of a nucleus is sensitive to its environment and therefore carries environmental signatures that can be used to obtain structural information.

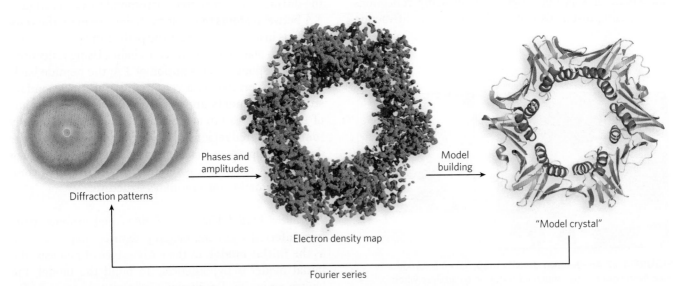

Diffraction patterns → Phases and amplitudes → Electron density map → Model building → "Model crystal"

Fourier series

FIGURE 4-29 Refinement. During refinement in x-ray diffraction analysis, the initial model of the protein structure is used to calculate the theoretical diffraction pattern it would produce, using a Fourier series. The phases are then adjusted to obtain a pattern close to the observed diffraction pattern. The adjusted phases generate a more detailed electron density map, thus allowing more precise positioning of amino acid residues in the model. The process is repeated several times until the residual error (R factor) between observed and calculated diffraction patterns is reduced to an acceptable value. [Source: Based on images courtesy of Roxana Georgescu, laboratory of Mike O'Donnell, Rockefeller University.]

¹H is particularly important in NMR experiments because of its high sensitivity and natural abundance. However, even a small protein has hundreds of ¹H atoms, typically resulting in a one-dimensional NMR spectrum too complex for analysis (**Figure 4-30a**). Structural analysis of proteins became possible with the advent of **two-dimensional NMR** techniques.

Many variations of two-dimensional NMR are performed by using different combinations of radio-frequency pulses and delays to separate the signals. In two-dimensional NMR, the data derived from the different pulses and delays are plotted along x and y axes, yielding a two-dimensional spectrum (**Figure 4-30b**). Instead of the plotting of peak height along the y axis, each spot carries a unique intensity that correlates with the peak height in the one-dimensional spectrum. The signals along the diagonal line through the two-dimensional spectrum are the same signals as in the one-dimensional

spectrum, and the variation in intensity along the diagonal correlates with their peak heights. The signals that lie off the diagonal, called nonsequential signals, are derived by magnetization transfer between two protons that are close in space. In one type of two-dimensional NMR, called **correlation spectroscopy (COSY)**, the signals allow the identification of protons connected by covalent bonds (**Figure 4-30c**). In two-dimensional **nuclear Overhauser effect spectroscopy (NOESY)**, these nonsequential signals allow the measurement of distances through space between nearby atoms (**Figure 4-30d**). All we need to know for now is that the COSY signals allow the researcher to trace the polypeptide backbone and thus to assign signals to particular amino acids. The NOESY signals occur through space and do not travel through the bonds between atoms. Therefore, NOE signals can arise from residues that are far apart in the primary structure but close together in the tertiary

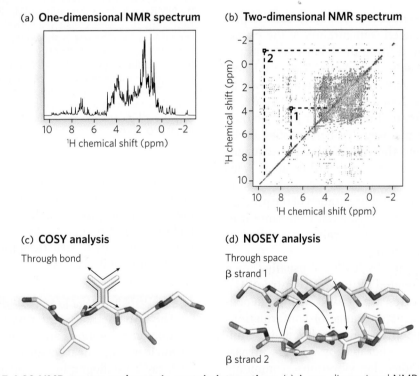

FIGURE 4-30 NMR spectra and protein-protein interactions. (a) A one-dimensional NMR spectrum of a globin from a marine bloodworm. The spectrum represents the amount of chemical shift for each proton in a peptide segment. For a protein, the proton signals do not resolve in a one-dimensional spectrum, as indicated by the many overlapping peaks. (b) A two-dimensional NMR spectrum of the same globin molecule. The spots and their intensities that lie along the diagonal line are equivalent to the data contained in the peaks of the one-dimensional spectrum. The off-diagonal peaks (e.g., peaks 1 and 2) are nuclear Overhauser effect (NOE) signals generated by close-range interactions of ¹H atoms that generate signals quite distant in the one-dimensional spectrum. (c) The two-dimensional COSY analysis identifies proton-proton signals through one or two covalent bonds ("through-bond" signals) and thus is limited to individual amino acid units. (d) The NOESY analysis yields NOE signals resulting from proton-proton interactions occurring through empty space ("through-space" signals) and thus identifies protons close in space but not necessarily close in the primary sequence. *[Source: (c), (d) Based on images courtesy of Roxana Georgescu, laboratory of Mike O'Donnell, Rockefeller University.]*

structure (see Figure 4-30d). These through-space signals carry the most important information about the three-dimensional structure of the protein and are the main data used to solve protein structures by NMR.

Tertiary Structure Determination Once the chemical shifts that derive from the primary sequence have been assigned by COSY, the through-space NOE signals provide information that restrains the possible tertiary structure solutions—information that is referred to as a "restraint." Restraints are absolutely essential to the prediction of tertiary structure. More than 1,000 restraints are required to predict a structure containing 100 residues or more. Although most of these restraints are NOE signals that represent protons close in space but distant in the primary sequence, another type of restraint is the torsion angles between residues, as obtained from the COSY spectrum. A third type of restraint is the known geometric restraints of all amino acids, such as chirality, van der Waals radii, and bond lengths.

With sufficient restraints, a structure can be predicted. First, a randomized configuration of the primary sequence is produced using the known geometry of the peptide bond and side-chain atoms. This still leaves a huge number of possible configurations, however, because the backbone $N–C\alpha$ and $C\alpha–C$ bonds are, to some degree, free to rotate. The computer program then tries to fold the chain in a way that best satisfies all the restraints, starting from those nearby in the sequence and proceeding to those that are farther apart. This procedure is repeated several times, each time starting with a different randomized configuration of the primary sequence. If the structure is substantially the same after each trial, then the number of restraints was sufficient to arrive at a unique solution.

Structures determined by NMR are usually shown as a group of closely related structures (**Figure 4-31**). The individual structures are arrived at by independent trials and represent the range of conformations consistent with the list of restraints. Although the uncertainty in structures generated by NMR is in part a reflection of the molecular vibrations (commonly called breathing) within a protein structure in solution, the observed variation is also due to errors or insufficiencies in the list of restraints. For example, the areas of greatest variation between different structures of a group usually signify areas with fewer restraints. For this reason, in NMR analyses, the total number of restraints is far more important than the accuracy of individual restraints.

Whenever a protein structure is determined by both x-ray crystallography and NMR, the structures generally agree well. In some cases, the precise locations of

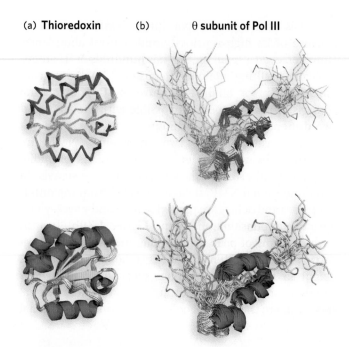

(a) Thioredoxin **(b)** **θ subunit of Pol III**

FIGURE 4-31 The structure of two proteins as determined by NMR. (a) Human thioredoxin (M_r 12,000). Multiple lines represent structures consistent with the restraints from the NMR data. One line is shown thicker than the rest to show the secondary elements within the structure. (b) The θ (theta) subunit (M_r 8,600) of DNA polymerase III (Pol III). The divergent models reflect the lack of restraints in disordered areas. The protein contains a region that lacked sufficient restraints to arrive at a unique solution and probably signifies a region of disordered residues. [Sources: (a) PDB ID 4TRX. (b) PDB ID 1DU2.]

particular amino acid side chains on the protein exterior are different, often because of effects related to the packing of adjacent protein molecules in a crystal. The two techniques together are at the heart of the rapid increase in the availability of structural information on the macromolecules of living cells.

SECTION 4.5 SUMMARY

- The two methods that reveal protein structure at atomic resolution are x-ray crystallography and nuclear magnetic resonance.
- X-ray crystallography can be applied to a protein of any size, but it requires a protein crystal.
- The diffraction pattern of x rays that have passed through a protein crystal must be recombined into an image mathematically, using the Fourier series.
- NMR is performed on proteins in solution and can be applied only to small proteins ($M_r < 25,000$).
- In NMR, the atomic nuclei are excited in a magnetic field, and emitted radiation is collected; some of the signals are sensitive to environment and contain structural information.

• In NOESY and COSY, two types of two-dimensional NMR, atoms that are covalently bonded or otherwise in close proximity to one another are identified, and the distances between them are used to create a list of restraints from which a structure can be generated.

? UNANSWERED QUESTIONS

Numerous protein structures have been determined, and one might think that, by now, researchers would have deciphered most of the rules about how proteins fold into their unique shapes. Yet, the information that directs how proteins fold and how they associate with their proper partners in a cell remains largely unknown and continues to be a highly active area of research. Here are some of the many questions being actively pursued.

1. What is the "code" in the primary sequence that determines how a protein folds? We know that the instructions for folding lie in the primary sequence. However, despite the large database of protein structures, we still do not know how these instructions are read. The problem lies in the relatively small difference in energy between the folded and unfolded states. Researchers remain hopeful that the "rules" of protein folding will someday be understood. Perhaps the accurate prediction of the structure adopted by a given sequence will be obtained by computations that draw on the vast empirical knowledge of structural folding patterns in proteins, combined with theoretical energy computations.

2. How do proteins "know" they are to form multiprotein complexes? Many of the important functions in a cell are performed by multiprotein complexes that act as machines to carry out complicated tasks. These tasks include central jobs such as transcription, replication, and translation. Given the thousands of different proteins in a cell, it is perplexing that particular subunits "know" how to join up, to the exclusion of others, to form these large complexes.

3. How do chaperones and chaperonins "know" when to bind a protein? Proteins that denature, or newly synthesized proteins that require assistance with folding, are targeted by chaperones and chaperonin complexes. However, most proteins contain disordered regions even when they are properly folded. We know little about how chaperones and chaperonins specifically target unfolded proteins, and even less about how these protein-folding assistants recognize when their job is done, or when to keep working.

HOW WE KNOW

Sequence Comparisons Yield an Evolutionary Roadmap from Bird Influenza to a Deadly Human Pandemic

Taubenberger, J.K., A.H. Reid, R.M. Lourens, R. Wang, G. Jin, and T.G. Fanning. 2005. Characterization of the 1918 influenza virus polymerase genes. Nature 437:889–893.

Worldwide pandemic outbreaks of flu can lead to millions of deaths. A virus has little to gain by killing its host, and, usually, the more deadly a virus is, the more recently it evolved. The evolution of a deadly virus has been intensively studied for influenza strains that cause pandemics.

A given influenza virus is typically confined to a certain host species, such as birds, horses, pigs, or humans—partly because cell surface receptors that allow entry of a virus are different in each species. However, some influenza strains evolve and jump the species barrier. Trouble for humans starts when the viruses also acquire efficient human-to-human transmissibility. This rare event can result in a worldwide influenza pandemic. About one to three such pandemics occur every century. How do these deadly viruses evolve? This can be determined from their genome sequences and comparisons with other influenza viruses.

The influenza virus genome consists of eight segments of RNA that encode 10 different proteins. Evolution is facilitated in a couple of ways: through errors introduced by the viral replicase, an RNA-dependent RNA polymerase that copies the RNA genome, and through genetic reassortment of RNA segments between two different viruses to form a novel virus. Genetic reassortment of RNA segments occurs when one host animal becomes infected by two different viruses at the same time. The pig, for example, has cell surface receptors that allow infection by both avian and human influenza viruses and thus may act as a "mixing vessel" to produce recombinant influenza viruses.

Comparative sequence analysis of the avian and human viruses responsible for the influenza pandemics of 1957 and 1968 reveals that the viruses evolved by genetic reassortment of two or three genes between an avian and a human virus. Both pandemic viral strains contained an avian *PB1* gene, which encodes part of the viral replicase, a 1:1:1 protein complex composed of products of the *PB1*, *PB2*, and *PA* genes. However, a comparison of the PB2 protein sequences of influenza viruses from

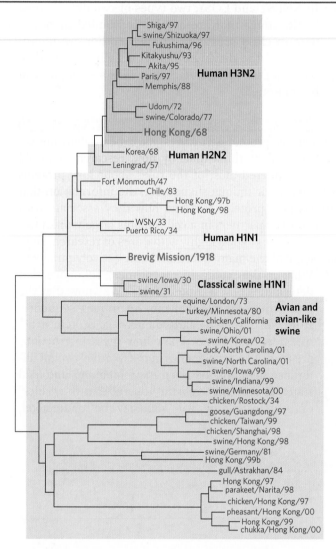

FIGURE 1 This phylogenetic tree of avian and human influenza viruses is based on sequence comparisons of the *PB2* gene. Branch points indicate the place where two *PB2* sequences diverged from a common ancestor. The 1918, 1957, and 1968 pandemics mentioned in the text are highlighted in red. [*Source: J. K. Taubenberger et al., Nature 437:889–893, 2005.*]

multiple sources reveals a human origin of the *PB2* gene for the Asian flu pandemic of 1957 and the Hong Kong flu pandemic of 1968 (Figure 1). This result supports the hypothesis of a mixing vessel in the evolution of these viruses. Interestingly, the 1918 Brevig Mission strain, which resulted in the worst pandemic so far, follows a different evolutionary history. Hundreds of millions of people were infected during the 1918–1919 "Spanish flu" pandemic, resulting in the deaths of about 50 million people worldwide. Comparative analysis of all the genes of the 1918 Brevig Mission virus indicates that it did not evolve by genetic reassortment with a second virus but, instead, adapted to humans directly from an avian source.

We Can Tell That a Protein Binds ATP by Looking at Its Sequence

Koonin, E.V. 1993. A superfamily of ATPases with diverse functions containing either classical or deviant ATP-binding motif. *J. Mol. Biol.* 229:1165–1174.

Saraste, M., P.R. Sibbald, and A. Wittinghofer. 1990. The P-loop: A common motif in ATP- and GTP-binding proteins. *Trends Biochem. Sci.* 15:430–434.

Wouldn't it be handy to be able to figure out what role a protein plays without having to perform complicated experiments? Sequence comparisons can do just that. A classic example is the identification of an ATP/GTP-binding site, based on some characteristic features of ATP/GTP-binding proteins.

Many proteins that bind ATP or GTP have several structural features in common. The P-loop is a conserved, glycine-rich sequence that forms a loop and connects a β strand to an α helix (Figure 2). The P-loop (phosphate-binding loop) interacts with the phosphates of ATP or GTP (Figure 3), and the presence of a P-loop sequence usually indicates that the protein's function involves the use of ATP or GTP. The P-loop is found in association with the nucleotide-binding Rossmann fold motif.

The P-loop is also referred to as a Walker A sequence. It is often followed in the protein's primary sequence (after a variable number of residues) by a Walker B sequence, sometimes called a DEAD box

$$...(G/A)XXGXGK(T/S)...$$

FIGURE 2 The P-loop consensus sequence (i.e., the sequence found in many ATP-binding proteins). X means any amino acid; a pair of residues in parentheses means that one can substitute for the other—for example, G/A means either G or A (Gly or Ala) in that position.

(for the amino acid sequence Asp–Glu–Ala–Asp, in one-letter symbols), which contains acidic residues that bind magnesium ions and assist in ATP or GTP hydrolysis. These two sequences are widespread in proteins that function with ATP or GTP. A few examples are eukaryotic Ras family GTP-binding proteins and eukaryotic and bacterial recombinases and mismatch repair proteins. Despite the widespread use of the P-loop motif, however, some proteins bind ATP using sequences that are unrelated to the Walker A motif.

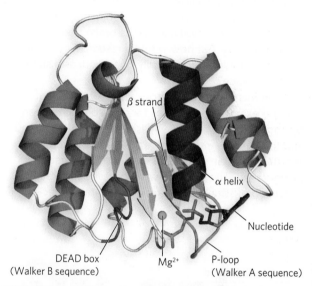

FIGURE 3 This ribbon representation of the ATP-binding RFC2 subunit of the yeast RFC clamp loader, which loads sliding clamps onto DNA for DNA polymerase (discussed in Chapter 11), highlights the features of the Rossmann fold motif that participate in nucleotide binding. The P-loop interacts with the nucleotide, and the DEAD box binds ions that assist in nucleotide hydrolysis. [*Source: PDB ID 1SXJ.*]

HOW WE KNOW

Disulfide Bonds Act as Molecular Cross-Braces to Stabilize a Protein

Matsumura, M., G. Signor, and B.W. Matthews. 1989. Substantial increase of protein stability by multiple disulfide bonds. *Nature* 342:291–293.

In building construction, cross-braces make bridges and walls stronger and sturdier. Proteins, too, utilize cross-braces. A disulfide bond connects two regions of one or more polypeptide chains within a protein and probably acts as a molecular cross-brace to enhance protein stability. But how would a researcher determine whether a disulfide bond really does work as a stabilizing cross-brace?

Brian Matthews's laboratory at the University of Oregon examined disulfides for cross-brace function by engineering pairs of Cys residues into T4 lysozyme and then measuring their effect on protein stability (a mutant protein with three disulfide bonds is shown in Figure 4a). The proteins were crystallized to allow detection of structural alterations by x-ray diffraction,

and protein stability was measured by circular dichroism (CD) spectroscopy at different temperatures. CD spectroscopy measures the amount of secondary structure in a protein and allows the researcher to follow the loss of α-helical content that accompanies protein denaturation.

Mutant proteins with a single disulfide bond had the same structure as wild-type lysozyme, with only small distortions at the replacement sites. Reduction of the disulfide bond (forming two unlinked Cys residues) resulted in a less-stable protein compared with wild-type lysozyme, indicating that Cys substitution had slightly destabilized each mutant protein. But in the oxidized form, the disulfide cross-link greatly increased the stability of several mutant proteins over the wild-type lysozyme. Addition of multiple disulfide cross-links to the wild-type lysozyme gave an additive effect in the stability of the mutant proteins (Figure 4b).

These elegant structural and biochemical studies demonstrated that disulfide bonds really do act as molecular cross-braces to enhance the stability of a protein. Antibodies, for example, contain many disulfide bonds, and we may assume that these bonds stabilize the proteins as they circulate in blood, outside the protective confines of the cell membrane.

(a)

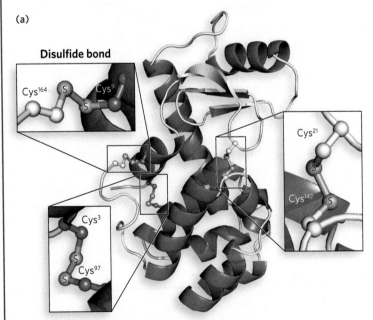

(b)

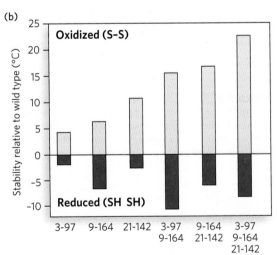

FIGURE 4 (a) Pairs of Cys residues (numbered for their position in the primary sequence) were engineered into T4 lysozyme. This mutant has three disulfide bonds. (b) Additional disulfide bonds stabilize protein structure relative to wild-type lysozyme. The numbers below the plot indicate the positions of the Cys residues involved in the disulfide bonds. The rightmost column shows the results for the three-disulfide mutant protein in (a). The stability of these proteins is indicated by the temperature at which the protein loses its activity, compared with wild type. [*Source: M. Matsumura, G. Signor, and B. W. Matthews, Nature 342:291–293, 1989.*]

KEY TERMS

primary structure, 95	amphipathic helix, 109
disulfide bond, 95	oligomer, 110
Ramachandran plot, 98	protomer, 110
secondary structure, 102	molten globule model,
α helix, 102	116
β sheet, 103	chaperone, 118
reverse turn, 104	x-ray crystallography,
tertiary structure, 105	120
quaternary structure, 105	diffraction pattern, 120
domain, 105	unit cell, 120
supersecondary	electron density map,
structure, 106	121
structural motif, 106	nuclear magnetic
Rossmann fold, 108	resonance (NMR), 122
helix-turn-helix motif,	nuclear Overhauser
108	effect spectroscopy
coiled-coil motif, 109	(NOESY), 123

PROBLEMS

1. Each ionizable group of an amino acid can exist in one of two states, charged or neutral. The electric charge on the functional group is determined by the relationship between the group's pK_a and the pH of the solution. This relationship is described by the Henderson-Hasselbalch equation (discussed in Chapter 3).

   (a) Histidine has three ionizable groups. Write the equilibrium equations for its three ionizations and assign the proper pK_a for each. Draw the structure of histidine in each ionization state. What is the net charge on the histidine molecule in each ionization state?

   (b) Draw the structures of the predominant ionization states of histidine at pH 1, 4, 8, and 12. Note that the ionization state can be approximated by treating each ionizable group independently.

   (c) What is the net charge of histidine at pH 1, 4, 8, and 12? For each pH, will histidine migrate toward the anode (+) or cathode (−) when placed in an electric field?

2. A quantitative amino acid analysis reveals that bovine serum albumin (BSA) contains 0.58% tryptophan (M_r 204) by weight.

   (a) Calculate the *minimum* molecular weight of BSA (i.e., assume there is only one Trp residue per protein molecule).

   (b) The BSA protein is purified and its molecular weight is estimated to be 70,000. How many Trp residues are present in a molecule of serum albumin?

3. A peptide has the following sequence:

   E-H-W-S-G-G-L-R-P-G

   (a) What is the net charge of the molecule at pH 3, 8, and 11? (Use pK_a values for side chains and terminal amino and carboxyl groups as given in Table 4-1.)

   (b) Purification of a peptide or protein is often easier if you understand its ionization properties. At a pH called the isoelectric point (pI), the net charge of the peptide or protein is zero. At lower or higher pH, it has a net positive or net negative charge, respectively. Estimate the pI for the above peptide.

4. A biochemist isolates a peptide hormone with the following sequence:

   ADSERNCQLVILLAWLPGVKVQCALLDRET

   (a) Which of these residues could contribute a positive charge? (Assume the residues are numbered 1 through 30, left to right.)

   (b) Which residues contribute a negative charge?

   (c) Which residues can be connected by a disulfide bond?

5. The sequence shown below, with 86 amino acid residues, folds into a β sheet substructure within a protein. The residues that form the β strands are noted (by residue number) above the sequence, and those outside the sheet are shown below. What type of β sheet structure is likely to form, parallel or antiparallel? Explain. What types of secondary structure are possible in the sequences between the β strands?

6. Given that the β strands in the β sheet structure of Problem 5 contain hydrophobic residues in most of the even-numbered positions and that most of the odd-numbered residues have polar R groups, predict how the β sheet structure will fold in three dimensions.

7. Ramachandran plots can help increase the accuracy of models derived from x-ray crystallography data. The plot below was created by measuring ψ and φ for each amino acid residue in a 2.2 Å resolution crystal structure. In the plot, which excludes Gly and Pro residues, selected

residues (dots) are numbered 1 through 5. What types of secondary structure are most consistent with the location of each of the numbered residues? Which residue(s) would you suspect of being incorrectly modeled into the electron density map? How would your suspicions change if the plot had included Gly residues?

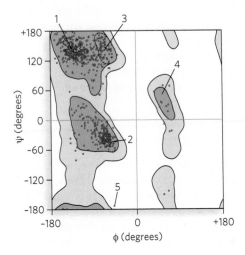

8. Inspect the 20-residue sequence below and predict the most favorable region for an α helix that is 10 residues long. Explain your reasoning. Point out any stabilizing interactions that might occur.

AIPRKKREFICRFGAIRPNT

9. Predict which of the following sequences would bind ATP (or GTP), and explain your answer. (See the How We Know section in this chapter.)

(a) YLFGGTRGVGKTSIA
(b) LLIQALPGMGDDARL
(c) LLIFGPPGLPKTTKL
(d) FINAGSQGIGKTACL

10. A polypeptide chain has 140 amino acid residues. How long will the polypeptide chain be if it is entirely α-helical? How long will it be if it is one continuous β strand?

11. Compare and contrast four aspects of the use of NMR and x-ray crystallography in protein structure determination.

12. Five proteins are listed below, each a monomer containing the number of amino acid residues indicated. How many domains would you expect each protein to have? Explain your reasoning.

(a) 70
(b) 110
(c) 150

(d) 200
(e) 250

13. For the following 20-residue sequence in a protein, list five amino acid residues that are likely to be buried in the protein, inaccessible to water. Pick five that are good candidates for surface residues.

DLKFTISVGAPVLTREQLLE

14. Consider an α helix in isolation, apart from the rest of the protein:

NRGAAEGAFCRAN

How would the following amino acid substitutions (indicated by residue number) affect the stability of the helix?

(a) Change N1 to K.
(b) Change N1 to E.
(c) Change R2 to K and E6 to R.
(d) Change R2 to K and E6 to D
(e) Change both G3 and G7 to F.
(f) Change G7 to P.

15. A simple protein structure is shown below, from two different angles. In (a), label the N-terminus and C-terminus and the two β turns. In (b), indicate which side of the β sheet, left or right, is likely to be more hydrophobic.

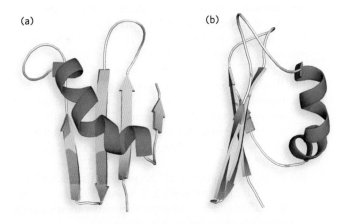

16. Draw a topology diagram for a 10-stranded, up-and-down β barrel.

17. Three Ramachandran plots are shown below. One of these is a guide; the other two, (a) and (b), are plots for two proteins, also shown below. Match each Ramachandran plot with its protein: which plot is most likely to be derived from bovine serum albumin, and which from green fluorescent protein? Both plots were downloaded from the Protein Data Bank. They are drawn a bit differently, but the *axes are the same as in the guide.*

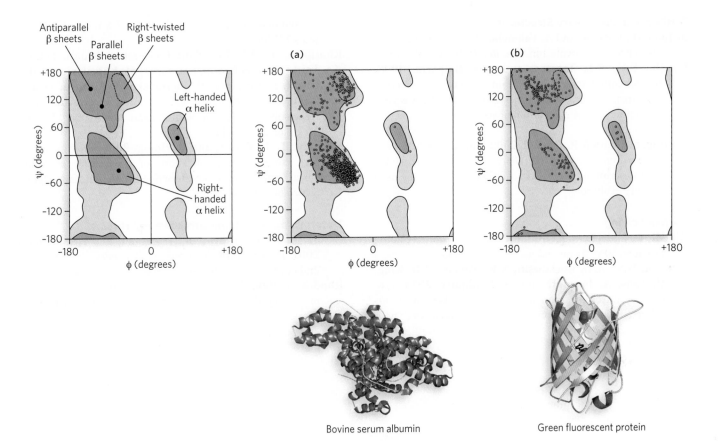

Bovine serum albumin

Green fluorescent protein

18. When proteins are synthesized in cells, all of the peptide bonds are synthesized in the trans configuration. Some proteins, however, have peptide bonds—especially adjacent to Pro residues—that are in the cis configuration. How do the peptide bonds in the cis configuration arise?

19. Food for thought: In this chapter, we introduce quaternary structure as the assembly of multiple protein subunits. This assembly is generally enforced through weak, noncovalent interactions—mainly the hydrophobic effect. Later in this book, we will deal with many very large multimeric complexes. For example, the ribosomes that are responsible for protein synthesis in all cells are assemblies of dozens of proteins and multiple large RNA molecules, with aggregate molecular weights in the millions. Are weak, noncovalent interactions sufficient to keep such complexes together?

ADDITIONAL READING

General

Branden, C., and J. Tooze. 1999. *Introduction to Protein Structure*, 2nd ed. New York: Garland Publishing, Inc. This is a classic, and the illustrations are very helpful.

Primary Structure

Doolittle, J., J. Abelson, and M. Simon, eds. 2009. *Molecular Evolution: Computer Analysis and Nucleic Acid Sequences.* Methods in Enzymology, vol. 183 (Amsterdam: Elsevier). A collection of articles on computational analysis of DNA and protein sequences and the construction of phylogenetic trees.

Wolf, M.Y., Y.I. Wolf, and E.V. Koonin. 2008. Comparable contributions of structural-functional constraints and expression level to the rate of protein sequence evolution. *Biol. Direct* 3:40–55. A comprehensive resource for comparison of protein sequences and how they relate to evolution.

Zuckerkandl, E., and L. Pauling. 1965. Molecules as documents of evolutionary history. *J. Theor. Biol.* 8:357–366. This report is widely considered to be the founding paper in the field of molecular evolution.

Secondary Structure

Ramachandran, G.N., C. Ramakrishnan, and V. Sasisekharan. 1963. Stereochemistry of polypeptide chain configurations. *J. Mol. Biol.* 7:95–99.

Rost, B. 2001. Review: Protein secondary structure prediction continues to rise. *J. Struct. Biol.* 134:204–218.

Tertiary and Quaternary Structures

de Juan, D., F. Pazos, and A. Valencia. 2013. Emerging methods in protein co-evolution. *Nat. Rev. Genet.* 14:249–261.

Koonin, E.V., R.L. Tatusov, and M.Y. Galperin. 1998. Beyond complete genomes: From sequence to structure and function. *Curr. Opin. Struct. Biol.* 8:212–217. A review on accuracy in the correlation of sequences with function in genomics.

Ponting, C.P., and R.R. Russell. 2002. The natural history of protein domains. *Annu. Rev. Biophys. Biomol. Struct.* 31:45–71. A description of how structural databases can be used to study evolution.

Protein Folding

Dill, K.A., and J.L. MacCallum. 2012. The protein-folding problem, 50 years on. *Science* 338:1042–1046.

Jackson, W.S., A.W. Borkowski, N.E. Watson, O.D. King, H. Faas, A. Jasanoff, and S. Lindquist. 2013. Profoundly different prion diseases in knock-in mice carrying single PrP codon substitutions associated with human diseases. *Proc. Natl. Acad. Sci. USA,* doi: 10.1073/pnas.1312006110.

Khatib, F., F. DiMaio, Foldit Contenders Group, Foldit Void Crushers Group, S. Cooper, M. Kazmierczyk, M. Gilski, S. Krzywda, H. Zábranská, I. Pichová, et al. 2011. Crystal structure of a monomeric retroviral protease solved by protein folding game players. *Nat. Struct. Mol. Biol.* 18:1175–1177.

Koloday, R., D. Petrev, and B. Honig. 2006. Protein structure comparison: Implications of the nature of "fold space," and structure and function prediction. *Curr. Opin. Struct. Biol.* 16:393–398.

Determining the Atomic Structure of Proteins

Cavanagh, J., W. Fairbrother, A. Palmer, and A. Skelton. 2007. *Protein NMR Spectroscopy: Principles and Practice,* 2nd ed. San Diego: Academic Press.

Rhodes, G. 2006. *Crystallography Made Crystal Clear: A Guide to Users of Molecular Models,* 3rd ed. San Diego: Academic Press.

5 Protein Function

Smita Patel [*Source: Courtesy Smita Patel.*]

MOMENT OF DISCOVERY

My story shows the insights one can achieve by using the right tools. During my postdoctoral years with Ken Johnson, I studied the DNA polymerase of bacteriophage T7. Charles Richardson demonstrated that replication of the T7 chromosome requires a DNA helicase, encoded by the T7 genome as gene 4. Collaborating with Bill Studier (who had recently constructed the first of a set of excellent cloning vehicles called pET vectors), I cloned and expressed gene 4. With this clone, I was able to purify the gene 4 protein product. I took this with me when I started my first faculty position at Ohio State University in 1992.

In the early days of getting my lab set up, I ran the lab with a small army of undergraduates. These students carried out a range of assays on the T7 DNA helicase. The enzyme bound stoichiometrically to DNA oligonucleotides that were 10 to 20 nucleotides in length, and hydrolyzed TTP and ATP. However, we were confused that the binding stoichiometry was consistently six to seven monomers of T7 helicase to each DNA molecule. We initially thought we had quite a lot of inactive enzymes.

Helicases were fairly new enzymes in those days. The DNA helicases getting the most attention were monomers or dimers in their active form. The RNA helicase Rho was known to be a hexamer, as was the DnaB helicase, but this did not seem to be a pattern. Then, my first graduate student, Manju Hingorani, decided to look at the protein directly by electron microscopy. This finally answered our questions. I remember my excitement as we looked at the beautiful ring structures in the electron micrographs. We could count up to five subunits. Later, collaborating with Ed Egelman, higher-resolution micrographs helped to establish that T7 helicase was a hexamer and the central channel was binding the DNA. We realized that rings are general in nature and widely used as replicative helicases that travel through thousands of base pairs of DNA without falling off. We published the results in a series of papers from 1993 to 1995. Since then, I have been endlessly captivated by these fascinating ring-shaped motor proteins.

—*Smita Patel, on her early work with the T7 gene 4–encoded DNA helicase*

133

Biological information—in the form of the genome of every organism and virus—is the focus of molecular biology, and of this textbook. As is true for all cellular functions, the packaging, function, and metabolism of this genomic information involve a wide range of proteins and RNA molecules. These macromolecules can be broadly divided into three classes, depending on function: reversible binding, catalysis, or motor activities. In the first three sections of this chapter, we review these macromolecular functions in succession, focusing on their roles in DNA and RNA metabolism. First, some proteins and RNAs simply bind reversibly to nucleic acids; this binding often has a structural or regulatory function. Second, another large class of proteins (and some RNAs) act as biological catalysts, accelerating the reactions needed to sustain and propagate living systems. These are the **enzymes**, as critical to life as are the information-containing DNA and RNA genomes. And third, motor proteins do the work of moving cellular molecules from one location to another, of separating molecules, and of bringing molecules together.

The great majority of macromolecules that carry out these three functions are proteins, although several RNA enzymes are known and are increasingly well understood. The functions of proteins are particularly important to the topics of every chapter in this book, and an introduction to protein function now becomes our focus. The various functions of RNAs are described in Chapters 15 and 16, although the general principles described here apply to RNA molecules as well as to proteins. In this chapter, after exploring each of the three major functions of proteins, we conclude with a discussion of protein regulation.

5.1 PROTEIN-LIGAND INTERACTIONS

Sometimes, a simple reversible interaction of two macromolecules is all that is needed to elicit major changes in a cell or cellular process. A protein bound to another macromolecule can alter structure and/or function in many different ways. A few examples should suffice to illustrate the principle. A protein bound to a specific DNA sequence can regulate the expression of an adjacent or nearby gene. Proteins bound without sequence specificity can condense DNA in a chromosome or package a DNA molecule into a virus head. A protein subunit bound to an enzyme can increase or decrease that enzyme's activity. Polymeric structures built up of many noncovalently linked protein subunits help guide cell division. A protein bound reversibly to a small molecule can act as a transporter, facilitating the movement of that molecule within or between

cells. Whether the binding association is prolonged or fleeting, it is often the basis of complex physiological processes, such as gene regulation, immune function, or cellular signaling. Molecular biology deals with countless such interactions.

Reversible Binding of Proteins to Other Molecules Follows Defined Principles

The interactions between proteins and other molecules are not random. We can summarize several principles that guide reversible binding processes:

1. The molecule bound by a given protein is referred to as its **ligand**. A ligand can be any kind of molecule, including nucleic acid or another protein. In some cases, the ligand may even be larger than the protein itself, although the "ligand" term is used most often for molecules that are smaller. The transient nature of protein-ligand interactions is critical to life, allowing an organism to respond rapidly and reversibly to changing environmental and metabolic circumstances.

2. A ligand binds at a site on the protein called the **binding site**, which is complementary to the ligand in size, shape, charge, and hydrophobic or hydrophilic character. The interaction is specific; the protein discriminates among the thousands of different molecules in its environment and selectively binds only one or a few. A given protein may have separate binding sites for several different ligands. These specific molecular interactions are crucial in maintaining the high degree of order in a living system. Our discussion here excludes the binding of water, which may interact weakly and nonspecifically with many parts of a protein.

3. Proteins exhibit conformational flexibility. Changes in conformation may be subtle, reflecting molecular vibrations and small movements of amino acid residues throughout the protein. A protein flexing in this way is sometimes said to "breathe." Conformational changes may also be dramatic, with major segments of the protein structure moving as much as several nanometers. Specific conformational flexibility is frequently essential to a protein's function.

4. Many protein-ligand interactions require a conformational change known as **induced fit**, in which a conformational change in the protein alters a binding site so that it becomes more complementary to the ligand, permitting tighter binding. The induced fit is the adaptation that occurs between the protein and the ligand.

5. The subunits in a multisubunit protein often exhibit **cooperativity**. A conformational change in one subunit can affect the conformation of other subunits. Thus, a conformational change triggered by the binding of a ligand to one subunit can increase or decrease the affinity of a neighboring subunit for the same ligand, giving rise to cooperative binding.

6. The activities of many proteins are subject to regulation. Interactions of ligands and proteins may be regulated, usually through specific interactions with one or more additional ligands. These other ligands may cause conformational changes in the protein that affect the binding of the first ligand.

Protein-Ligand Interactions Can Be Quantified

The function of many proteins depends on their ability not only to bind to a ligand but also to release the ligand when and where it is needed. Function in molecular biology often revolves around a reversible protein-ligand interaction of this type. A quantitative description of this interaction is therefore a central part of many investigations.

In general, the reversible binding of a protein (P) to a ligand (L) can be described by a simple **equilibrium expression (Figure 5-1a)**:

$$P + L \rightleftharpoons PL \tag{5-1}$$

The reaction is characterized by an equilibrium constant, K_a, such that:

$$K_a = \frac{[PL]}{[P][L]} = \frac{k_a}{k_d} \tag{5-2}$$

where k_a and k_d are rate constants that describe, respectively, the rate of association and dissociation of the ligand and the protein. K_a is an **association constant** (not to be confused with the K_a that denotes an acid dissociation constant, described in Chapter 3). It describes the equilibrium between the complex and the separate, unbound components of the complex. The association constant provides a measure of the affinity of the ligand L for the protein P. K_a has units of M^{-1}; a higher value of K_a corresponds to a higher affinity of the ligand for the protein.

It is more common (and intuitively simpler), however, to consider the **dissociation constant, K_d**, which is the reciprocal of K_a ($K_d = 1/K_a$) and is given in units of molar concentration (M). K_d is the equilibrium constant for the release of ligand. Note that a lower value of K_d

(a)

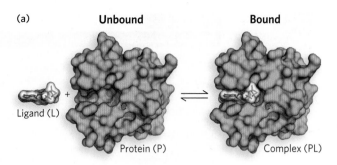

(b)

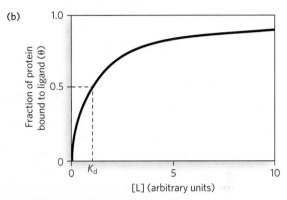

FIGURE 5-1 Ligand binding. (a) Reversible binding of a protein (P) to a ligand (L). The protein shown here is nucleoside diphosphate kinase and the ligand is ATP. (b) The fraction of ligand-binding sites occupied, θ, is plotted against the concentration of free ligand, [L]. A hypothetical binding curve is shown. The [L] at which half the available ligand-binding sites are occupied is equivalent to $1/K_a$, or K_d. The curve has a horizontal asymptote at θ = 1 and a vertical asymptote (not shown) at [L] = $-K_d$. [Source: (a) PDB ID 2BEF (ATP-binding domain only).]

corresponds to a higher affinity of ligand for the protein. The relevant expression changes to:

$$K_d = \frac{[P][L]}{[PL]} = \frac{k_d}{k_a} \tag{5-3}$$

We can now consider the binding equilibrium from the standpoint of the fraction, θ (theta), of ligand-binding sites on the protein that are occupied by ligand:

$$\theta = \frac{\text{binding sites occupied}}{\text{total binding sites}} = \frac{[PL]}{[PL] + [P]} \tag{5-4}$$

Substituting $K_a[P][L]$ for [PL] (see Equation 5-2) and rearranging terms gives:

$$\theta = \frac{[L]}{[L] + K_d} \tag{5-5}$$

Any equation of the form $x = y/(y + z)$ describes a hyperbola, and θ is thus found to be a hyperbolic function of [L] (**Figure 5-1b**). The fraction of ligand-binding sites occupied approaches saturation asymptotically as [L] increases. The [L] at which half of the available

TABLE 5-1

Protein Dissociation Constants

Protein	Ligand	K_d (M)*
Avidin (egg white)	Biotin	1×10^{-15}
Replication protein A (RPA; eukaryotes)	ssDNA	$1 \times 10^{-9} - 1 \times 10^{-11}$
SSB (as the SSB$_{65}$ binding mode)	ssDNA	2×10^{-7}
Lactose repressor	dsDNA (nonspecific)	5×10^{-4}
	dsDNA (specific)	2×10^{-11}

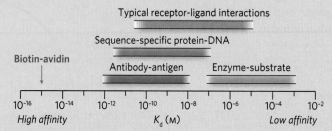

The range of dissociation constants for typical interactions in biological systems, denoted by color for each class. A few interactions, such as that between the protein avidin and the enzyme cofactor biotin, fall outside the typical range. The avidin-biotin interaction is so tight that it may be considered irreversible.

Note: SSB, single-stranded DNA–binding protein; ssDNA, single-stranded DNA; dsDNA, double-stranded DNA.
*A reported dissociation constant is valid only for the particular solution conditions under which it was measured. K_d values for a protein-ligand interaction can be altered, sometimes by several orders of magnitude, by changes in the solution's salt concentration, pH, interactions with additional proteins, or many other variables.

ligand-binding sites are occupied (i.e., $\theta = 0.5$) corresponds to the K_d: when $[L] = K_d$, half of the ligand-binding sites are occupied. As $[L]$ falls below K_d, progressively less of the protein has ligand bound to it. For 90% of the available ligand-binding sites to be occupied, $[L]$ must be nine times K_d.

The mathematics can be reduced to simple statements: K_d equals the molar concentration of ligand at which half the available ligand-binding sites are occupied. At this point, the protein is said to have reached half-saturation with respect to ligand binding. The more tightly a protein binds a ligand, the lower is the concentration of ligand required for half the binding sites to be occupied, and thus the lower the value of K_d. Some representative dissociation constants are given in **Table 5-1**, along with a range of such constants found in typical biological systems.

DNA-Binding Proteins Guide Genome Structure and Function

DNA-binding proteins are a key example of proteins that simply bind to a ligand (in this case DNA)

reversibly without altering its covalent structure. DNA-binding proteins protect DNA, organize DNA, regulate genes or groups of genes, alter the conformation of DNA, facilitate all aspects of the metabolism of DNA, and ensure the proper segregation of chromosomes during cell division. DNA-binding proteins fall into two principal categories. Some bind to DNA nonspecifically, independent of DNA sequence; others recognize particular DNA sequences and bind tightly at the genomic locations where those sequences occur. The distinction is not absolute. "Nonspecific" DNA-binding proteins often display a measurable bias for binding of DNA sequences with particular features. "Specific" DNA-binding proteins generally exhibit some measurable (albeit much weaker) binding to nonspecific sequences (an example is the lactose repressor; see Table 5-1).

The binding of proteins to DNA, and thus the measured K_d of these interactions, is almost always sensitive to parameters such as pH and salt concentration. DNA is a polyelectrolyte (a polymer with multiple ionizable groups). In a cell, the negative charges of the phosphates in the DNA backbone are neutralized by interaction with counterions, and there is generally a high concentration of ions such as Mg^{2+} and K^+ surrounding the DNA. As a protein binds to DNA some of these ions are released, and some bound water molecules are released from both the protein and the DNA as well. The release of ions and water has both positive and negative effects on the association of a protein with DNA. The positive effects come from a general gain in entropy ($\Delta S >> 0$) as the water and ions are released. The negative effects reflect the energy of interactions between the water and ions and the macromolecules, interactions that must be eliminated to make the protein-DNA complex. Protein-DNA interactions are thus rarely as simple as the coming together of two complementary macromolecules. The interactions are affected in important ways by additional interactions of each macromolecule with water and with ions.

Two additional parameters are notable and are characteristic of interactions between proteins and nucleic acids. First, the number of nucleotides (in single-stranded DNA) or base pairs (in double-stranded DNA) that are occluded by the bound protein defines the binding site size, n. This parameter helps determine the number of binding sites on the DNA that might be available to a protein, and a knowledge of n for a particular protein is necessary for any complete description of its binding equilibrium. Second, some DNA-binding proteins, particularly certain proteins that bind to DNA nonspecifically, exhibit cooperativity in binding; that is, when one protein molecule binds to a nucleic acid, it facilitates the binding of another protein molecule. Cooperativity can also have important effects on binding equilibria.

The examples that follow focus on proteins with physical and structural properties that are particularly well studied.

Nonspecific DNA-Binding Proteins As we will discuss in Chapter 9, chromosomes are the largest macromolecules found in cells. If chromosomal DNA molecules were laid out linearly, they would typically be hundreds or even thousands of times longer than the cells in which they are housed. The protection and compaction of chromosomal DNA is largely the job of myriad nonspecific DNA-binding proteins found in every cell. These proteins also organize some key chromosomal functions, such as facilitating DNA replication and repair or guiding chromosomal segregation at cell division.

In most cases, nonspecific DNA-binding proteins exhibit only limited hydrogen-bonding interactions with bases in the DNA. Instead, electrostatic interactions with the negatively charged phosphate groups, hydrogen bonds to the backbone deoxyribose, and the

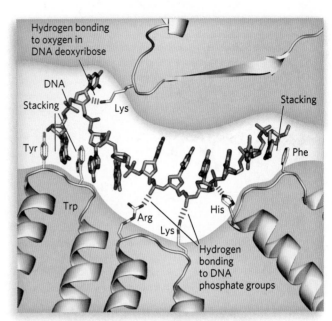

FIGURE 5-2 Nonspecific interactions of proteins with nucleic acids. Electrostatic interactions occur between proteins and the DNA backbone. The charged phosphate groups are exposed at the exterior surface of a single-stranded DNA molecule, where positively charged amino acid side chains (Arg, Lys, His) can interact. Hydrogen bonds occur between the protein and the deoxyribose groups in the DNA backbone. Hydrophobic interactions involving the intercalation of Tyr, Trp, or Phe side chains between two stacked bases are also prominent in many cases of nonspecific DNA-protein binding. Similar interactions occur with RNA.

nonspecific hydrophobic effect with the bases predominate, to varying degrees (**Figure 5-2**). The hydrophobic interactions often take the form of an aromatic amino acid side chain (Tyr, Trp, or Phe) intercalating between two adjacent bases.

An example of a nonspecific DNA-binding protein is the bacterial single-stranded DNA–binding protein (SSB). SSB binds and protects single-stranded DNA as it is transiently created during DNA replication and repair. The importance of this protein is illustrated by a simple observation: if SSB function is lost, the cell dies. The structure of the SSB from *Escherichia coli* is typical for bacteria, consisting of four identical subunits around which the single-stranded DNA wraps (**Figure 5-3a**). Each subunit contains one oligonucleotide/oligosaccharide-binding fold, or OB fold, a structural unit that often binds single-stranded DNA. The OB fold is common to all proteins in the SSB class (**Figure 5-3b**), as well as many others that associate with single-stranded DNA as part of their function.

In addition to binding to single-stranded DNA, bacterial SSBs interact directly and reversibly with a range

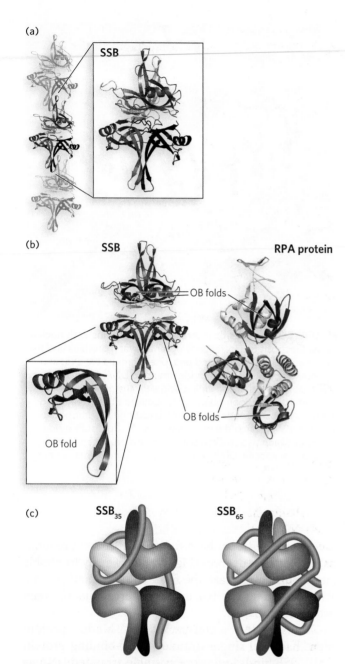

(a)

(b)

SSB

RPA protein

OB folds

OB fold

OB folds

(c)

SSB$_{35}$

SSB$_{65}$

FIGURE 5-3 Binding of single-stranded DNA to single-stranded DNA–binding proteins. (a) Single-stranded DNA–binding protein (SSB) can bind in a filamentous form on single-stranded DNA. A single SSB tetramer is highlighted. (b) The key structural element that functions in single-stranded DNA binding is an OB fold. The basic fold consists of two three-stranded antiparallel β-sheets that share strand 1, with the strands ordered 1-2-3-5-4-1 (with the numbering coinciding to their order in the linear protein sequence). The sheets are tightly curved to form a small β-barrel, and an α-helix often connects strands 3 and 4. The structure can vary in appearance due to differences in lengths of the β-strands. The four OB folds in an SSB tetramer are highlighted in blue (left). The eukaryotic replication protein A (RPA) binds to single-stranded DNA in a similar fashion, using the OB folds in each subunit of the heterotrimer (right). (c) SSB can bind to single-stranded DNA in multiple binding modes, with the two most prominent modes shown in this schematic. The blue tube represents bound single-stranded DNA. *[Sources: (a) and (b) (left) PDB ID 1EYG. (b) (right) PDB ID 1L1O.]*

of other proteins in DNA metabolism. The *E. coli* SSB interacts with at least 15 different proteins (which we will encounter in Chapters 11–14), thereby helping to organize the functions of DNA replication and repair (**Table 5-2**) All of these interactions occur through a conserved C-terminus of SSB, a segment that features multiple negatively charged amino acid residues among the final eight to nine residues of the polypeptide.

SSBs are found in every class of organism, and they always play essential roles in DNA metabolism. The eukaryotic SSB is called **replication protein A (RPA)**. It consists of three different subunits (i.e., it is a heterotrimer) containing a total of six OB folds. Its function is quite similar to that of the bacterial SSBs, and it also interacts with a range of other proteins as part of its function in eukaryotic DNA metabolism.

In the test tube, the *E. coli* SSB binds to single-stranded DNA according to several different binding modes, depending on the concentrations of salt and protein. Two of these binding modes are notable. At relatively low concentrations of salt, SSB binds with a binding site size (*n*) of ~35 nucleotides and with a very high degree of cooperativity between tetramers. In this mode, called SSB$_{35}$, the single-stranded DNA is bound to two of the four subunits in each SSB tetramer (**Figure 5-3c**), and the tetramers are arranged on the DNA as a fairly regular filamentlike structure (see Figure 5-3a). When the salt concentration is higher, the SSB$_{65}$ (*n* = 65 nucleotides) binding mode predominates. Here, the single-stranded DNA is wrapped around all four SSB subunits (see Figure 5-3c); the cooperativity between SSB tetramers is reduced, and filaments form less readily. The SSB binding modes affect SSB function and interactions with other proteins in vitro and presumably in vivo. In both cases, SSB binds to single-stranded DNA with a combination of electrostatic interactions with the phosphoribose backbone and intercalation of particular Trp and Phe side chains between adjacent DNA bases. The *E. coli* SSB binds tightly to single-stranded DNA, with measured K_d values generally in the range of 8 to 700 nM, depending on the solution conditions.

Many other proteins that bind to single-stranded or double-stranded DNA with little specificity also bind such that the DNA is wrapped or bent around the protein. For example, duplex DNA wraps tightly around the nucleosomes of eukaryotic chromosomes, as we will see in Chapter 10. Although the histone proteins that make up a nucleosome are considered nonspecific DNA-binding proteins, the positioning of nucleosomes on double-stranded DNA is not entirely random. In particular, DNA sequence elements that facilitate the bending or wrapping of double-stranded DNA around a protein, such as regions with several contiguous A=T base pairs,

TABLE 5-2

Proteins That Interact with Bacterial Single-Stranded DNA–Binding Protein

Protein	Function
χ subunit of DNA polymerase III	DNA replication
DnaG primase	DNA replication
RecQ helicase	Recombinational DNA repair
RecJ nuclease	Recombinational DNA repair
RecG helicase	Recombinational DNA repair
RecO recombination mediator	Recombinational DNA repair
PriA replication restart protein	Replication restart after repair
PriB replication restart protein	Replication restart after repair
Exonuclease I	DNA replication and repair
Uracil DNA glycosylase	DNA repair
DNA polymerase II	Mutagenic replication under stress
DNA polymerase V	Mutagenic replication under stress
Exonuclease IX	DNA repair
Bacteriophage N4 virion RNA polymerase	Viral nucleic acid metabolism

can have strong effects on the locations of bound nucleosomes along the DNA.

Specific DNA-Binding Proteins Proteins that bind with an enhanced affinity to particular DNA sequences are critical to the regulation of many processes in DNA metabolism. Many of these proteins regulate the expression of genes. Their affinity for specific target sequences is roughly 10^4 to 10^6 times their affinity for any other DNA sequence. Most regulatory proteins have discrete DNA-binding domains containing substructures that interact closely and specifically with the DNA. These binding domains usually include one or more of a relatively small group of recognizable and characteristic structural motifs (described in Chapter 4), often called binding motifs to reflect their function.

To bind to specific DNA sequences, regulatory proteins must recognize and distinguish surface features on the DNA. Most of the chemical groups that differ among the four bases and thus permit discrimination between base pairs are hydrogen-bond donor and acceptor groups exposed in the major groove of DNA. These interactions are illustrated by the binding of a eukaryotic regulatory protein with its DNA binding site (**Figure 5-4**). Most of the protein-DNA contacts that impart specificity are hydrogen bonds. A notable exception is the nonpolar surface near C-5 of thymine and cytosine, where the two are readily distinguishable by the protruding methyl group of thymine. Protein-DNA contacts are also possible in

the minor groove of DNA, but the hydrogen-bonding patterns there generally do not allow ready discrimination between base pairs.

In specific DNA-binding proteins, the amino acid side chains that most often hydrogen-bond to bases in the DNA are those of Asn, Gln, Glu, Lys, and Arg residues. Is there a simple recognition code in which a particular amino acid always pairs with a particular base? Two hydrogen bonds can form between Gln or Asn and the N^6 and N-7 positions of adenine but not any other base. An Arg residue can form two hydrogen bonds with N-7 and O^6 of guanine (**Figure 5-5**). However, examination of the structures of many DNA-binding proteins has revealed that the proteins can recognize each base pair in more than one way, leading to the conclusion that there is no simple amino acid–base code. For some proteins, the Gln-adenine interaction can specify A=T base pairs; for others, a van der Waals pocket for the methyl group of thymine can recognize A=T base pairs. As yet, researchers cannot examine the structure of a DNA-binding protein and predict the DNA sequence to which it binds.

An example of a specific DNA-binding protein is the well-studied lactose (Lac) repressor of *E. coli* (see the How We Know section at the end of this chapter). This protein is part of a regulatory network that controls the expression of three consecutive genes in the *E. coli* chromosome, all of them involved in some aspect of lactose metabolism. The three genes are transcribed

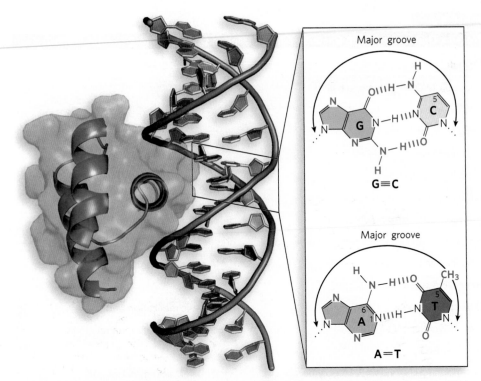

FIGURE 5-4 **Groups in DNA that can guide specific protein binding.** The eukaryotic DNA-binding protein called Engrailed interacts with the major groove of double-stranded DNA. Shown in the inset are functional groups on base pairs displayed in the major and minor grooves. Red indicates groups that can be used for base-pair recognition by proteins. Most specific binding is through interactions with the major groove. [Source: PDB ID 2HDD.]

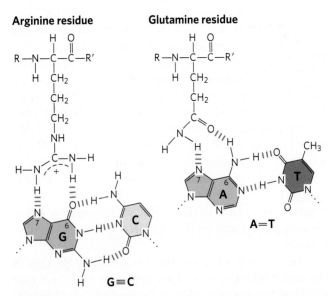

FIGURE 5-5 **DNA-protein binding.** Two examples of amino acid–base pair interactions that have been observed in DNA-protein binding. An asparagine residue can participate in the same type of interaction as the glutamine residue.

together in a unit described as an operon (**Figure 5-6**), and regulation of transcription occurs in and around a specific sequence, called the Lac operator, where the Lac repressor binds to the DNA. When the Lac repressor is bound to the Lac operator, transcription of the operon genes is blocked. The lactose operon is described in detail in Chapter 20, and we use the Lac repressor here simply to illustrate a common property of sequence-specific DNA-binding proteins. The DNA binding sites for regulatory proteins are often inverted repeats of a short DNA sequence, where multiple (usually at least two) subunits of a regulatory protein bind cooperatively. The Lac repressor functions as a tetramer, with two dimers tethered together at the end distant from the DNA-binding sites.

An *E. coli* cell usually contains about 20 tetramers of the Lac repressor protein. Each of the tethered dimers can independently bind to an operator sequence, in contact with 17 base pairs of a 22 base pair region in the *lac* operon (see Figure 5-6). The tetrameric Lac repressor binds to two proximal operator sequences in vivo with an estimated K_d of about 10^{-10} M. The repressor discriminates between the operators and other sequences by a

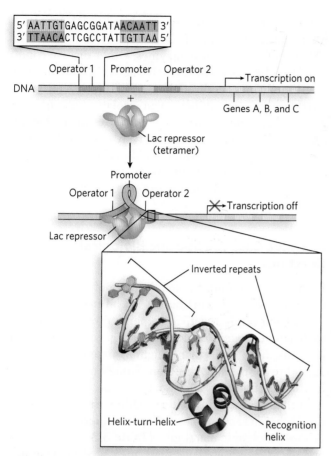

FIGURE 5-6 Simplified representation of the *lac* operon. When the Lac repressor is not bound to the operon's promoter region, RNA polymerase binds to the promoter and transcribes several linked genes, here labeled *A*, *B*, and *C*. The Lac repressor binds to two operators on either side of the promoter to shut down transcription, apparently forming a loop in the DNA that prevents RNA polymerase from binding to the promoter. Each operator consists of an inverted repeat (top inset). The helix-turn-helix motif of the repressor protein binds specifically in the major groove of the operator recognition sequences (bottom inset). See Chapter 20 for a more in-depth discussion of the *lac* operon.

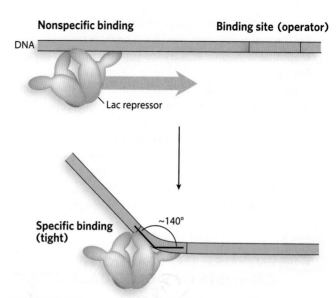

FIGURE 5-7 Nonspecific versus specific DNA binding. The Lac repressor interacts transiently and nonspecifically with the DNA phosphoribose backbone during the search for its specific DNA binding site. When the sequence of its normal binding site is located, the repressor interacts specifically with the nucleotide bases in that site. Binding results in bending of the DNA.

factor of about 10^6, so binding to these few dozen base pairs among the 4.6 million or so of the *E. coli* chromosome is highly specific.

The Lac repressor interacts with DNA through a **helix-turn-helix motif**, a DNA-binding motif that is crucial to the interaction of many bacterial regulatory proteins with DNA; similar motifs occur in some eukaryotic regulatory proteins. The helix-turn-helix motif comprises about 20 amino acid residues in two short α-helical segments, each 7 to 9 residues long, separated by a β turn (Figure 5-6, bottom inset). This structure generally is not stable by itself; it is simply the functional portion of a somewhat larger DNA-binding domain. One of the two α-helical segments is called the recognition helix, because it usually contains many of the amino

acid residues that interact with the DNA in a sequence-specific way. This α helix is stacked on other segments of the protein structure so that the helix protrudes from the protein surface.

One set of amino acid residues in the recognition helix of the Lac repressor's helix-turn-helix domain participates in both nonspecific and specific interactions with DNA. The nonspecific interactions, even though they are weaker, usually occur first and play an important role in accelerating the search for the specific DNA binding site. In the nonspecific binding mode, these residues interact electrostatically with the DNA's phosphoribose backbone (**Figure 5-7**). When bound to the specific recognition sequences in the Lac operator, the recognition helix is positioned in, or nearly in, the major groove. A network of specific hydrogen bonds and hydrophobic interactions governs the stability of this complex.

There are many proteins that bind to specific sequences in a nucleic acid, and we will encounter numerous examples of these in later chapters.

SECTION 5.1 SUMMARY

- Many proteins bind reversibly to other molecules, known as ligands. Ligand binding often involves protein conformational changes, in the process of induced fit.
- Ligand binding can be quantified, and the key parameter is the dissociation constant, K_d, which is the

concentration of ligand at which half of the protein's binding sites are occupied by ligand. A lower K_d corresponds to higher affinity for (tighter binding to) the ligand.

- A DNA- or RNA-binding protein may bind to a nucleic acid either nonspecifically or in a sequence-dependent manner. Nonspecific binding usually involves interactions with the phosphoribose backbone of the nucleic acid. Specific DNA or RNA binding requires an interaction of amino acid side chains in the protein with functional groups in the nucleic acid bases, especially those exposed in the major groove of DNA.

5.2 ENZYMES: THE REACTION CATALYSTS OF BIOLOGICAL SYSTEMS

Rare indeed is the organic chemical reaction that proceeds unaided at a rate sufficient to support living systems. Enzymes have extraordinary catalytic power, often far greater than that of synthetic or inorganic catalysts. They have a high degree of specificity for their substrates, they substantially accelerate chemical reactions, and they function in aqueous solutions under very mild conditions of temperature and pH. Few nonbiological catalysts have all these properties.

Enzymes are central to every cellular process. Acting in organized sequences, they catalyze the hundreds of stepwise reactions that degrade nutrient molecules, conserve and transform chemical energy, make biological macromolecules from simple precursors, and carry out the various processes of DNA and RNA metabolism.

In molecular biology, the study of enzymes has immense practical importance. In some diseases, especially hereditary genetic disorders related to DNA or RNA metabolism, there may be a deficiency or even a total absence of one or more enzymes. Other disease conditions may be caused by excessive activity of an enzyme. Many medicines act through interactions with enzymes. Furthermore, researchers can isolate and harness enzyme functions to suit their purposes in the laboratory. The set of methods collectively described as "biotechnology" (many of which are described in Chapter 7 and elsewhere throughout this book) is made possible by our understanding of the enzymes of DNA and RNA metabolism.

Enzymes Catalyze Specific Biological Reactions

With the exception of a small group of catalytic RNA molecules (described in Chapter 16), all enzymes are proteins. Their catalytic activity depends on the integrity of their native protein conformation. Enzymes, like other proteins, have molecular weights ranging from about 12,000 to more than 1 million. Some enzymes require for their activity no additional chemical groups other than their amino acid residues. Others require an additional chemical component called a **cofactor**, either one or more inorganic metal ions (**Table 5-3**) or a complex organic or metallo-organic molecule called a **coenzyme**, which acts as a transient carrier of specific functional groups (**Table 5-4**). Most coenzymes are derived from vitamins, organic nutrients required in small amounts in the human diet. Some enzymes require *both* a coenzyme and one or more metal ion cofactors for activity. A coenzyme or inorganic cofactor that is very tightly or even covalently bound to the enzyme protein is known

TABLE 5-3

Inorganic Elements as Cofactors for Enzymes and Regulatory Proteins		
Cofactor	*Enzyme Example*	*Function*
Cu^{2+}	Superoxide dismutase	Cellular protection from reactive oxygen species
Fe^{2+} or Fe^{3+}	AlkB	DNA repair
Mg^{2+}	RecA protein	Recombinational DNA repair
	ATPases (all)	Many functions
	Nucleases	DNA cleavage
	DNA and RNA polymerases	Nucleic acid synthesis
Mn^{2+}	Ribonucleotide reductase	Biosynthesis of deoxynucleotides
Mo^{6+}	Molybdate sensor protein	Gene regulation
	Certain bacterial riboswitches	Gene regulation
Zn^{2+}	Many DNA-binding proteins	Gene regulation

TABLE 5-4

Coenzymes: Transient Carriers of Specific Atoms or Functional Groups

Coenzyme*	Examples of Chemical Group(s) Transferred	Dietary Precursor in Mammals
Biocytin	CO_2	Biotin
Coenzyme A	Acyl groups	Pantothenic acid and other compounds
5'-Deoxyadenosylcobalamin (coenzyme B_{12})	H atoms and alkyl groups	Vitamin B_{12}
Flavin adenine dinucleotide (FAD)	Electrons	Riboflavin (vitamin B_2)
Lipoate	Electrons and acyl groups	Not required in diet
Nicotinamide adenine dinucleotide (NAD)	Hydride ion (:H^-)	Nicotinic acid (niacin)
Pyridoxal phosphate	Amino groups	Pyridoxine (vitamin B_6)
Tetrahydrofolate	One-carbon groups	Folate
Thiamine pyrophosphate	Aldehydes	Thiamine (vitamin B_1)
S-Adenosylmethionine (adoMet)	Methyl groups	Not required in diet

*The structures and modes of action of these coenzymes are described in most biochemistry textbooks.

as a **prosthetic group**. A complete, catalytically active enzyme, with its bound coenzyme and/or inorganic cofactor, is referred to as a **holoenzyme**. The protein part of such an enzyme is called the **apoenzyme** or **apoprotein**.

An enzyme catalyzes a reaction by providing a specific environment in which the reaction can occur more rapidly. Here are some key principles of enzyme-catalyzed reactions:

1. A molecule that undergoes an enzyme-catalyzed reaction is referred to as a **substrate**. A substrate differs from a ligand in that it can undergo a chemical transformation while bound to the enzyme, whereas a ligand does not.

2. The substrate interacts with the enzyme in a pocket known as the **active site** (**Figure 5-8**). The active site is typically lined with multiple chemical groups—amino acid side chains, metal ion cofactors, and/or coenzymes—all oriented to facilitate the reaction.

3. An enzyme-catalyzed reaction is highly specific for that particular reaction. An active site that is set up to catalyze one reaction with one substrate will not interact well with other substrates. Specificity is an important property of every enzyme. The catalysis of a different reaction requires a different enzyme.

4. Catalysis often requires conformational flexibility. As we have seen for proteins that reversibly bind ligands, conformational changes have an essential role in enzyme function. Induced fit and cooperativity also play roles in enzyme catalysis.

5. Many enzymes are regulated. The panoply of enzymes available to a given cell confers the opportunity not just to accelerate reactions but to control them. In this way, cellular metabolism can be modulated as resources and circumstances demand.

The enormous and highly selective rate enhancements achieved by enzymes can be explained by the many types of covalent and noncovalent interactions between enzyme and substrate. Chemical reactions of many types may take place between substrates and the functional groups (specific amino acid side chains, metal ions, and coenzymes) on enzymes. The particular

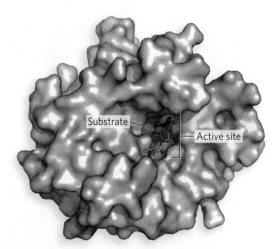

FIGURE 5-8 Binding of a substrate to an enzyme at the active site. This is the enzyme T4 RNA ligase, with a bound substrate, ATP (stick representation). Active sites are typically pockets in the surface of enzymes. [*Source: PDB ID 2C5U.*]

reactions that occur depend on the requirements of the overall reaction to be catalyzed. An enzyme's catalytic functional groups may form a transient covalent bond with the substrate and activate it for reaction. Or a group may be transiently transferred from the substrate to the enzyme. The most common type of group transfer involves the transfer of protons between ionizable amino acid side chains in the active site and groups on the substrate molecule, a process called general acid and base catalysis (**Figure 5-9**). In the enzymes important to molecular biology, phosphoryl group transfers are also common. In many cases, these group transfer reactions occur only in the enzyme active site. The capacity to facilitate multiple interactions and transfers of this type, sometimes all at once, is one of the factors contributing to the rate enhancements provided by enzymes.

Covalent interactions are only part of the story, however. Much of the energy required to increase the reaction rate is derived from weak, noncovalent interactions between substrate and enzyme, including hydrogen bonds and hydrophobic and ionic interactions. The formation of each weak interaction is accompanied by the release of a small amount of free energy that stabilizes the interaction. The energy derived from enzyme-substrate interaction is called **binding energy, ΔG_B**. Its significance extends beyond a simple stabilization of the enzyme-substrate interaction. *Binding energy is a major source of the free energy used by enzymes to increase the rates of reactions.*

In the context of nucleic acid metabolism, one type of weak interaction merits special mention. Ionic interactions can include interactions between bound metals (such as Mg^{2+}, Mn^{2+}, and Fe^{2+} or Fe^{3+} ions) and substrates. About one-third of all enzymes utilize metals in their catalytic mechanisms, and that proportion is much higher for the enzymes that act on DNA and RNA. For example, the active sites of DNA and RNA polymerases universally feature two metal ions, usually two Mg^{2+} ions, that help orient substrates and facilitate the overall reaction in multiple ways (see Figure 5-9). Mg^{2+} ions play key roles at the active sites of a wide range of enzymes discussed in later chapters.

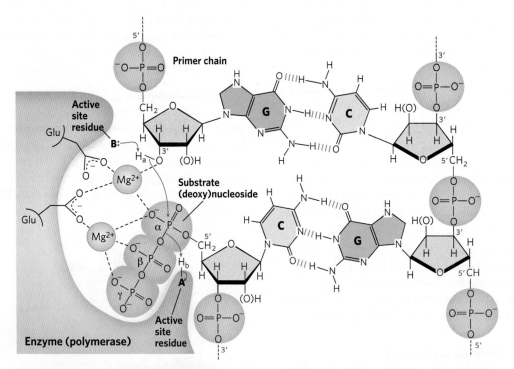

FIGURE 5-9 An enzyme-catalyzed reaction. Shown here is the key step in the formation of a new phosphodiester bond in the active site of a DNA or RNA polymerase. The end of a growing chain of nucleic acid (the primer chain) is at the top left, and an incoming (deoxy)nucleoside triphosphate is at the lower left. The reaction begins with general base catalysis by an active-site residue (B) that abstracts a proton (H_a) from the attacking 3′ hydroxyl at the end of the primer chain. The oxygen of the hydroxyl group concurrently attacks the phosphorus of the α-phosphoryl group of the nucleoside triphosphate, displacing pyrophosphate (PP_i). The pyrophosphate is protonated by another active-site residue (usually a Lys, shown here as A), an example of general acid catalysis, which facilitates ejection of the PP_i. Two metal ions, usually two Mg^{2+}, are in the active site. One metal ion lowers the pK_a of the primer 3′ hydroxyl to facilitate the general base catalysis. The other metal ion coordinates with and orients oxygens of the triphosphate and also aids catalysis by stabilizing the transition state of the reaction.

Enzymes Increase the Rate of a Reaction by Lowering the Activation Energy

A simple enzyme reaction might be written like this:

$$E + S \rightleftharpoons ES \rightleftharpoons EP \rightleftharpoons E + P \qquad (5\text{-}6)$$

where E, S, and P represent the enzyme, substrate, and product, and ES and EP are transient complexes of the enzyme with the substrate and with the product.

To understand catalysis, we must recall the important distinction between reaction equilibria and reaction rates. The equilibrium between S and P reflects the difference in the free energies of their ground states. Any reaction, such as $S \rightleftharpoons P$, can be described by a reaction coordinate diagram, a picture of the energy changes during the reaction (**Figure 5-10a**). Energy in biological systems is described in terms of free energy, G (see Chapter 3). In the diagram, the free energy of the system is plotted against the progress of the reaction (the reaction coordinate). In this example, the free energy of the ground state of P is lower than that of S, so the **biochemical standard free-energy change**, or $\mathbf{\Delta G'^{\circ}}$, for the reaction is negative and the equilibrium favors P.

KEY CONVENTION

To describe the free-energy changes for reactions, chemists define a standard set of conditions (temperature 298 K; partial pressure of each gas 1 atm, or 101.3 kPa; concentration of each solute 1 M) and express the free-energy change for a reacting system under these conditions as ΔG°, the standard free-energy change. Because living systems commonly involve H^+ concentrations far below 1 M, biochemists and molecular biologists define a biochemical standard free-energy change, $\Delta G'^{\circ}$, the standard free-energy change at *pH 7.0*; we use this definition throughout the book.

(a) Reaction coordinate diagram

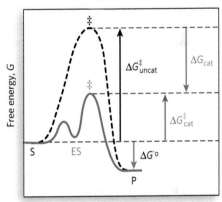

FIGURE 5-10 The use of noncovalent binding energy to accelerate an enzyme-catalyzed reaction. (a) A reaction coordinate diagram. The free energy of a system is plotted against the progress of the reaction $S \rightarrow P$. This kind of diagram describes the energy changes during the reaction; the horizontal axis (reaction coordinate) reflects the progressive chemical changes (e.g., bond breakage or formation) as S is converted to P. The activation energies, $\Delta G^{\ddagger}$, for the $S \rightarrow P$ and $P \rightarrow S$ reactions are indicated. $\Delta G'^{\circ}$ is the overall biochemical standard free-energy change in the direction $S \rightarrow P$. The ES intermediate occupies a minimum in the energy progress curve of the enzyme-catalyzed reaction. The terms $\Delta G^{\ddagger}_{uncat}$ and $\Delta G^{\ddagger}_{cat}$ correspond to the activation energy for the uncatalyzed reaction (black, dashed curve) and the overall activation energy for the catalyzed reaction (blue, solid curve), respectively. The activation energy is lowered by the amount ΔG_{cat} when the enzyme catalyzes the reaction. (b) An imaginary enzyme (stickase) designed to catalyze the breaking of a metal stick. Before the stick is broken, it must be bent (transition state). Magnetic interactions replace weak enzyme-substrate bonding interactions. A stickase with a magnet-lined pocket that is structurally complementary to the stick (substrate) stabilizes the substrate (middle). Bending is impeded by the magnetic attraction between stick and stickase. An enzyme with a pocket complementary to the reaction transition state helps destabilize the stick (bottom), contributing to catalysis. The binding energy of the magnetic interactions compensates for the increase in free energy needed to bend the stick. In enzyme active sites, weak interactions that occur only in the transition state aid in catalysis.

(b)

No enzyme

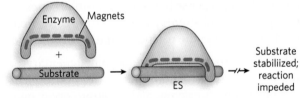

Enzyme complementary to substrate

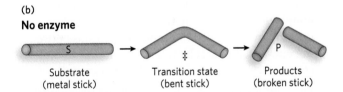

Enzyme complementary to transition state

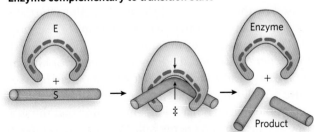

A favorable equilibrium does not mean that the S → P conversion will occur at a fast or even detectable rate. An unfavorable equilibrium does not mean that the reaction will be slow. Instead, the *rate* of a reaction depends on the height of the energy hill that separates the product from the substrate. At the top of this hill lies the **transition state** (denoted by ‡ in Figure 5-10a). The transition state is not a stable species but a transient moment when the alteration in the substrate has reached a point corresponding to the highest energy in the reaction coordinate diagram. The difference between the energy levels of the ground state and the transition state is the **activation energy, $\Delta G^{\ddagger}$**. A higher activation energy corresponds to a slower reaction.

The function of a catalyst is to increase the *rate* of a reaction. Catalysts do not affect reaction equilibria, and enzymes are no exception. The bidirectional arrows in Equation 5-6 make this point: any enzyme that catalyzes the reaction S → P also catalyzes the reaction P → S. The role of enzymes is to *accelerate* the interconversion of S and P. The enzyme is not used up in the process, and the equilibrium point is unaffected. However, the reaction reaches equilibrium much faster when the appropriate enzyme is present, because the rate of the reaction is increased. Enzymes increase reaction rates by lowering the activation energy of the reaction. To achieve this, enzymes utilize noncovalent and covalent interactions in somewhat different ways.

Two fundamental and interrelated principles provide a general explanation for how enzymes use noncovalent binding energy to accelerate a reaction:

1. Much of the catalytic power of an enzyme is ultimately derived from the free energy released in forming many weak bonds and interactions between the enzyme and its substrate. This binding energy contributes to specificity as well as to catalysis.

2. Weak interactions are optimized in the reaction transition state; enzyme active sites are complementary not to substrates per se but to the transition states through which substrates pass as they are converted to products during the reaction (**Figure 5-10b**).

When the enzyme active site is complementary to the reaction transition state, some of the noncovalent interactions between enzyme and substrate occur only in the transition state. The free energy (binding energy) released by the formation of these interactions partially offsets the energy required to reach the top of the energy hill. The summation of the unfavorable (positive) activation energy $\Delta G^{\ddagger}$ and the favorable (negative) binding energy ΔG_{B} results in a lower net activation energy (see Figure 5-10a). Even on the enzyme, the transition state is not a stable species but a brief point in time that the substrate spends atop an energy hill. The enzyme-catalyzed reaction is much faster than the uncatalyzed process, however, because the hill is much smaller. The groups on the substrate that are involved in the weak interactions between the enzyme and transition state can be at some distance from the substrate bonds that are broken or changed. The weak interactions formed only in the transition state are those that make the primary contribution to catalysis (see Figure 5-10b).

Covalent interactions can accelerate some enzyme-catalyzed reactions by creating a different, lower-energy reaction pathway. When the reaction occurs in solution in the absence of the enzyme, the reaction takes a particular (and usually very slow) path. In the enzyme-catalyzed reaction, if a group on the enzyme is transferred to or from the substrate during the reaction, the reaction path is altered. The new pathway results in acceleration of the reaction only if its overall activation energy is lower than that of the uncatalyzed reaction.

The Rates of Enzyme-Catalyzed Reactions Can Be Quantified

The oldest approach to understanding enzyme mechanisms, and the one that remains most important, is to determine the rate of a reaction and how it changes in response to changes in experimental parameters, a discipline known as **enzyme kinetics**. We provide here a brief review of key concepts related to the kinetics of enzyme-catalyzed reactions (for more advanced treatments, see the Additional Reading list at the end of the chapter.)

Substrate concentration affects the rate of enzyme-catalyzed reactions. Studying the effects of substrate concentration [S] in vitro is complicated by the fact that it changes during the course of the reaction, as substrate is converted to product. One simplifying approach in kinetics experiments is to measure the **initial velocity**, designated V_0 (**Figure 5-11a**). In a typical reaction, the enzyme may be present in nanomolar quantities, whereas [S] may be five or six orders of magnitude higher. If just the beginning of the reaction is monitored (often no more than the first few seconds), changes in [S] can be limited to a small percentage, and [S] can be regarded as constant. V_0 can then be explored as a function of [S], which is adjusted by the investigator. The effect on V_0 of varying [S] when the enzyme concentration is held constant is shown in **Figure 5-11b**. At relatively low concentrations of substrate, V_0 increases almost linearly with an increase in [S]. At higher substrate concentrations, V_0 increases by smaller and smaller amounts in response to increases in [S]. Finally, a point is reached beyond which increases in V_0 are vanishingly small as [S] increases. In this plateau-like V_0 region, the reaction approaches its **maximum velocity, V_{max}**.

(a)

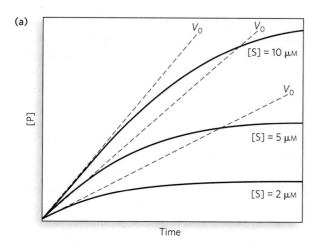

(b)

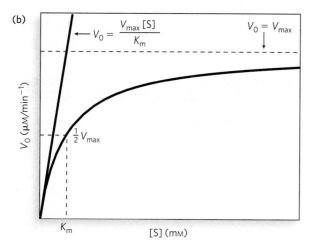

FIGURE 5-11 The initial velocity of an enzyme-catalyzed reaction. (a) A theoretical enzyme catalyzes the reaction $S \rightleftharpoons P$. Progress curves for the reaction (product concentration, [P], vs. time) measured at three different initial substrate concentrations ([S]) show that the rate of the reaction declines as substrate is converted to product. A tangent to each curve taken at time zero (dashed lines) defines the initial velocity, V_0, of the reaction. (b) The maximum velocity, V_{max}, is indicated as a horizontal dashed line. The straight solid line describes the linear dependence of the initial velocity, V_0, on [S] at low substrate concentrations. However, as [S] increases, the line in reality becomes nonlinear, as depicted by the curved line. V_0 approaches but never quite reaches V_{max}. The substrate concentration at which V_0 is half maximal is K_m, the Michaelis constant. The concentration of enzyme in an experiment such as this is generally so low that [S] >> [E] even when [S] is described as low or relatively low. At low [S], the slope of the line is defined by $V_0 = V_{max}[S]/K_m$, and this is where V_0 exhibits a linear dependence on [S]. The units shown here are typical for enzyme-catalyzed reactions and help illustrate the meaning of V_0 and [S]. (Note that the curved line describes part of a rectangular hyperbola, with one asymptote at V_{max}. If the curve were continued below [S] = 0, it would approach a vertical asymptote at [S] = $-K_m$.)

evident in Figure 5-11a. The reaction quickly achieves a **steady state** in which [ES] (and the concentration of any other intermediates) remains approximately constant over time. The concept of a steady state was introduced by G. E. Briggs and J. B. S. Haldane in 1925. The measured V_0 generally reflects the steady state, even though V_0 is limited to the early part of the reaction, and analysis of these initial rates is referred to as **steady-state kinetics**.

J. B. S. Haldane, 1892–1964 [Source: Hans Wild/Time Life Pictures/ Getty Images.]

The curve expressing the relationship between [S] and V_0 (see Figure 5-11b) has the same general shape for most enzymes (approaching a rectangular hyperbola). It can be expressed algebraically by an equation developed by Leonor Michaelis and Maud Menten, called the **Michaelis-Menten equation**:

$$V_0 = \frac{V_{max}[S]}{K_m + [S]} \tag{5-7}$$

The important terms are [S], V_0, V_{max}, and a constant called the **Michaelis constant, K_m**. All these terms are readily measured experimentally.

The Michaelis-Menten equation is the rate equation for a one-substrate enzyme-catalyzed reaction. It states the quantitative relationship between the initial velocity V_0, the maximum velocity V_{max}, and the initial substrate concentration [S], all related through the Michaelis constant K_m. Note that K_m has units of concentration. Does the equation fit experimental observations? Yes; we can confirm this by considering the limiting situations where [S] is very high or very low, as shown in Figure 5-11b. The K_m is functionally equivalent to the [S] at which V_0 is one-half V_{max}. At low [S], $K_m \gg$ [S], and the [S] term in the denominator of the Michaelis-Menten equation (Equation 5-7) becomes

Leonor Michaelis, 1875–1949 [Source: Rockefeller University Archive Center.]

Maud Menten, 1879–1960 [Source: University of Toronto Archives.]

When the enzyme is first mixed with a large excess of substrate, there is an initial period called the **pre-steady state**, when the concentration of ES (enzyme-substrate) builds up. This period is usually too short to be observed easily, lasting just microseconds, and is not

insignificant. The equation simplifies to $V_0 = V_{max}[S]/K_m$, and V_0 exhibits a linear dependence on [S]. At high [S], where $[S] \gg K_m$, the K_m term in the denominator of the Michaelis-Menten equation becomes insignificant and the equation simplifies to $V_0 = V_{max}$; this is consistent with the plateau observed at high [S]. The Michaelis-Menten equation is therefore consistent with the observed dependence of V_0 on [S], and the shape of the curve is defined by the terms V_{max}/K_m at low [S] and V_{max} at high [S].

Kinetic parameters are used to compare enzyme activities. Many enzymes that follow Michaelis-Menten kinetics have different reaction mechanisms, and enzymes that catalyze reactions with six or eight identifiable intermediate steps within the enzyme active site often exhibit the same steady-state kinetic behavior. Even though Equation 5-7 holds true for many enzymes, both the magnitude and the real meaning of V_{max} and K_m can differ from one enzyme to the next. This is an important limitation of the steady-state approach to enzyme kinetics. The parameters V_{max} and K_m can be obtained experimentally for any given enzyme, but by themselves they provide little information about the number, rates, or chemical nature of discrete steps in the reaction. Nevertheless, steady-state kinetics is the standard language with which biochemists compare and characterize the catalytic efficiencies of enzymes.

Figure 5-11b shows a simple graphical method for obtaining an approximate value for K_m. The K_m, as noted, can vary greatly from enzyme to enzyme, and even for different substrates of the same enzyme. The magnitude of K_m is sometimes interpreted (often inappropriately) as an indicator of the affinity of an enzyme for its substrate. However, K_m is not equivalent to the substrate K_d (the simple binding equilibrium for the enzyme substrate) for many enzymes.

HIGHLIGHT 5-1 ━━━ **A CLOSER LOOK**

Reversible and Irreversible Inhibition

Enzyme inhibitors are molecules that interfere with catalysis by slowing or halting enzyme reactions. The two general categories of enzyme inhibition are reversible and irreversible. There are three types of **reversible inhibition**: competitive, uncompetitive, and mixed.

A **competitive inhibitor** competes with the substrate for the active site of an enzyme (Figure 1a). While the inhibitor (I) occupies the active site, it prevents binding of the substrate to the enzyme. Many competitive inhibitors are structurally similar to the substrate and combine with the enzyme to form an EI complex, but without leading to catalysis. Even fleeting combinations of this type will reduce the efficiency of the enzyme. The two other types of reversible inhibition, though often defined in terms of one-substrate enzymes, are in practice observed only with enzymes having two or more substrates. An **uncompetitive inhibitor** binds at a site distinct from the substrate active site and, unlike a competitive inhibitor, binds only to the ES complex (Figure 1b). A **mixed inhibitor** also binds at a site distinct from the substrate active site, but it binds to either E or ES (Figure 1c).

All of these inhibition patterns can be analyzed with the aid of a single equation derived from the Michaelis-Menten equation:

$$V_0 = \frac{V_{max}[S]}{\alpha K_m + \alpha'[S]}$$

where α and α' reflect the interaction of an inhibitor with the free enzyme (through K_I) and with the ES complex (through K'_I), respectively. These terms are defined as:

$$\alpha = 1 + \frac{[I]}{K_I}, \text{ and } K_I = \frac{[E][I]}{[EI]}$$

(a) Competitive inhibition

$$E + S \rightleftharpoons ES \longrightarrow E + P$$
$$+$$
$$I$$
$$\Big\updownarrow K_I$$
$$EI$$

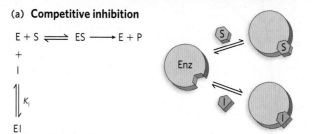

(b) Uncompetitive inhibition

$$E + S \rightleftharpoons ES \longrightarrow E + P$$
$$+$$
$$I$$
$$\Big\updownarrow K'_I$$
$$ESI$$

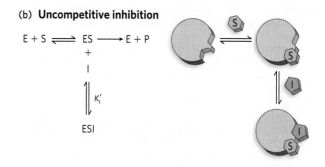

(c) Mixed inhibition

$$E + S \rightleftharpoons ES \longrightarrow E + P$$
$$+ \qquad +$$
$$I \qquad I$$
$$\Big\updownarrow K_I \qquad \Big\updownarrow K'_I$$
$$EI + S \rightleftharpoons ESI$$

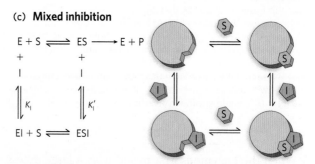

FIGURE 1 The three types of reversible inhibition. (a) Competitive inhibitors bind to the enzyme's active site. (b) Uncompetitive inhibitors bind at a separate site, but bind only to the ES complex. (c) Mixed inhibitors bind at a separate site, but may bind to either E or ES.

The term V_{max} depends on both the concentration of enzyme and the rate of the rate-limiting step in the reaction pathway. Because the number of steps in a reaction and the identity of the rate-limiting step can vary, it is useful to define a **general rate constant, k_{cat}**, to describe the limiting rate of any enzyme-catalyzed reaction at saturation. If the reaction has several steps and one is clearly rate-limiting, k_{cat} is equivalent to the rate constant for that limiting step. When several steps are partially rate-limiting, k_{cat} can become a complex function of several of the rate constants that define each individual reaction step. In the Michaelis-Menten equation, $V_{max} = k_{cat}[E_t]$, where $[E_t]$ is the total concentration of enzyme, and Equation 5-7 becomes:

$$V_0 = \frac{k_{cat}[E_t][S]}{K_m + [S]} \qquad (5\text{-}8)$$

The constant k_{cat} is a first-order rate constant and hence has units of reciprocal time. It is equivalent to the number of substrate molecules converted to product in a given unit of time on a single enzyme molecule when the enzyme is saturated with substrate. Hence, this rate constant is also called the **turnover number**. A k_{cat} may be 0.01 s^{-1} for an enzyme with an intrinsically slow function (some enzymes with regulatory functions act very slowly) or as high as 10,000 s^{-1} for an enzyme catalyzing some aspect of intermediary metabolism.

In cells, there are many situations in which enzyme activity is inhibited by specific molecules, including other proteins. From a practical standpoint, the development of pharmaceutical and agricultural agents almost always involves the development of inhibitors for particular enzymes. Some aspects of enzyme inhibition are reviewed in **Highlight 5-1**.

TABLE 1		
Effects of Reversible Inhibitors on Apparent V_{max} and Apparent K_m		
Inhibitor Type	*Apparent V_{max}*	*Apparent K_m*
None	V_{max}	K_m
Competitive	V_{max}	αK_m
Uncompetitive	V_{max}/α'	K_m/α'
Mixed	V_{max}/α'	$\alpha K_m/\alpha'$

and

$$\alpha' = 1 + \frac{[I]}{K'_I}, \text{ and } K'_I = \frac{[E][I]}{[ESI]}$$

For a competitive inhibitor, there is no binding to the ES complex, and $\alpha' = 1$. For an uncompetitive inhibitor, there is no binding to the free enzyme (E), and $\alpha = 1$. For a mixed inhibitor, both α and α' are greater than 1. Each class of inhibitor has characteristic effects on the key kinetic parameters in the Michaelis-Menten equation, as summarized in Table 1. The altered K_m or V_{max} measured in the presence of an inhibitor is often referred to as an *apparent K_m or V_{max}*.

Many reversible enzyme inhibitors are used as pharmaceutical drugs; two examples are shown in Figure 2. The human immunodeficiency virus (HIV) encodes a DNA polymerase that can use either RNA or DNA as template; this enzyme is a reverse transcriptase (discussed in Chapter 14). It uses deoxynucleoside triphosphates as substrates, and it is competitively inhibited by the drug AZT—the first drug to be used in treating HIV infections. Similarly, quinolone antibiotics widely used to treat bacterial infections are uncompetitive inhibitors of enzymes called topoisomerases (described in Chapter 9).

An **irreversible inhibitor** can bind covalently with or destroy a functional group on an enzyme that is essential for the enzyme's activity, or it can form a particularly stable noncovalent association. The formation of a covalent link between an irreversible inhibitor and an enzyme is common. Because the enzyme is effectively inactivated, irreversible inhibitors affect both V_{max} and K_m. An inhibitor that does not form a covalent link but binds so tightly to the enzyme active site that it does not dissociate within hours or days is also effectively an irreversible inhibitor. Given that enzyme active sites bind most tightly to the transition state of the reactions they catalyze, a molecule that mimics the transition state can be a tight-binding inhibitor. Inhibitors designed in this way are called transition state analogs. Many drugs used to treat people with HIV/AIDS are designed in part as transition state analogs that bind tightly to the HIV protease.

Note that uncompetitive and mixed inhibitors should not be confused with allosteric modulators (see Section 5.4). Although the inhibitors bind at a second site on the enzyme, they do not necessarily mediate conformational changes between active and inactive forms, and the kinetic effects are distinct.

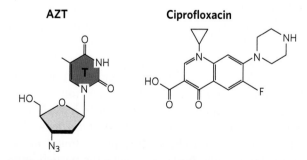

FIGURE 2 Examples of inhibitors with medical applications. AZT (3'-azido-3'-deoxythymidine) is used in the treatment of HIV/AIDS; ciprofloxacin is a quinolone antibiotic.

DNA Ligase Activity Illustrates Some Principles of Catalysis

An understanding of the complete mechanism of action of a purified enzyme requires the identification of all substrates, cofactors, products, and regulators. Moreover, it requires a knowledge of (1) the temporal sequence in which enzyme-bound reaction intermediates form, (2) the structure of each intermediate and each transition state, (3) the rates of interconversion between intermediates, (4) the structural relationship of the enzyme to each intermediate, and (5) the energy contributed by all reacting and interacting groups to intermediate complexes and transition states. As yet, there is probably no enzyme for which we have an understanding that meets all these requirements.

It is impractical, of course, to cover all possible classes of enzyme chemistry, and we focus here on an enzyme reaction important to molecular biology: the reaction catalyzed by DNA ligases. The discussion concentrates on selected principles, along with some key discoveries that have helped bring these principles into focus. We also use the DNA ligase example to review some of the conventions used to depict enzyme mechanisms. Many mechanistic details and pieces of experimental evidence are necessarily omitted; an entire book would be needed to document the rich experimental history of enzyme research.

DNA ligases were discovered in 1967, and reports were published from four different research groups in that year. These enzymes catalyze the joining of DNA ends at strand breaks (also called nicks) that have a phosphorylated 5′ terminus and a 3′ terminus with a free hydroxyl group. In cells, these enzymes provide the critical links between discontinuous segments of replicated DNA (called Okazaki fragments; see Chapter 11) and carry out the final step in most DNA repair reactions (see Chapter 12). Since their discovery, many details of the reaction mechanism of these enzymes have been elucidated. DNA ligases have become essential tools of biotechnology, used by laboratories around the world to covalently join DNA segments to create recombinant DNA (see Chapter 7). RNA ligases have also been characterized; they use a similar reaction mechanism.

DNA ligases make use of two cofactors: Mg^{2+} ions and either ATP or nicotinamide adenine dinucleotide (NAD^+). ATP-dependent DNA ligases are found in eukaryotes, viruses, and some bacteria and archaea. NAD^+-dependent ligases are found in most bacteria, as well as in some viruses and archaea, but are not found in eukaryotes. Commonly, NAD^+ is a cofactor participating in oxidation-reduction reactions. Its role in DNA ligase reactions is quite different, however, and it parallels the role of ATP in the ATP-dependent ligases.

All DNA ligases promote a reaction that involves three chemical steps (**Figure 5-12** on p. 152). In step 1, an adenylate group, adenosine 5′-monophosphate (AMP), is transferred from either ATP or NAD^+ to a Lys residue in the enzyme active site. This process occurs readily in the absence of DNA. In step 2, the enzyme binds to DNA at the site of a strand break and transfers the AMP to the 5′ phosphate of the DNA strand. This activates the 5′ phosphate for nucleophilic attack by the 3′-hydroxyl group of the DNA, in step 3, leading to displacement of the AMP and formation of a new phosphodiester bond in the DNA that seals the nick. Each of the three steps has a highly favorable reaction equilibrium that renders it effectively irreversible. In the absence of DNA, the adenylated enzyme formed in step 1 is quite stable, and it is likely that most DNA ligases in a cell are already adenylated and ready to react with DNA. In addition to illustrating the reaction pathway, Figure 5-12 introduces the conventions commonly used to describe enzyme-catalyzed reactions.

The overall picture of the DNA ligase reaction mechanism is a composite derived from kinetic and structural studies of many closely related enzymes, including those isolated from bacteriophages T4 and T7, bacteria, eukaryotic viruses (that is, viruses with eukaryotic host cells), and mammals. The ligases typically consist of a DNA-binding domain, a nucleotidyltransferase (NTase) domain, and an OB-fold domain. In DNA ligases, the OB fold, normally associated with binding to single-stranded DNA (see Section 5.1), interacts with the minor groove of double-stranded DNA. These domains provide a flexible structure that closes to completely encircle a nicked DNA molecule, with all three domains in contact with the DNA (see Figure 5-12). During step 1, some residues in the OB fold become part of the active site for transfer of AMP to the active-site Lys residue. As the covalent link between the enzyme and AMP is formed, the adenine base of AMP is fixed in a binding site in the NTase domain, where it stays throughout the remaining steps. In steps 2 and 3, a conformational change rearranges this part of the OB fold so that the same residues now face the solvent, while other parts of the OB fold bind to the DNA and interact primarily with the strand adjacent to the 5′-phosphate end of the strand break. At the same time, the conformational change closes the enzyme around the nicked DNA, and the *N*-glycosyl bond between the adenine base and the ribose moiety of AMP is rotated, thereby realigning the phosphate of AMP for reaction, in step 2, with the 5′ phosphate of DNA. Step 3 follows closely behind step 2 within the same complex.

This one example cannot provide a complete overview of the broad range of strategies that enzymes use, but it does serve as a good introduction to the complexity of enzyme reaction mechanisms. In addition, it illustrates the transfer of phosphoryl groups, a reaction catalyzed by protein enzymes and RNA enzymes (ribozymes) involved in almost every aspect of molecular biology—from DNA polymerases and RNA polymerases to nucleases, topoisomerases, spliceosomes, and ligases.

SECTION 5.2 SUMMARY

- Most enzymes are proteins. They facilitate the reactions of substrate molecules. The catalyzed reaction occurs in an active site, a pocket on the enzyme that is lined with amino acid side chains and, in many cases, bound cofactors that participate in the reaction.
- Enzymes are catalysts. Catalysts do not affect reaction equilibria; they enhance reaction rates by lowering activation energies.
- Enzyme catalysis involves both covalent enzyme-substrate interactions and noncovalent interactions. Enzyme active sites bind most tightly to the transition states of the reactions they catalyze.
- The rates of most enzyme-catalyzed reactions are described by the Michaelis-Menten equation, which relates the key kinetic parameters V_{max}, K_m, and k_{cat}.
- A DNA ligase catalyzes a series of phosphoryl transfer reactions to seal nicks in the DNA backbone; its reaction mechanism illustrates several general principles of enzyme-catalyzed reactions.

5.3 MOTOR PROTEINS

Organisms move. Cells move. Organelles and macromolecules within cells move. Most of these movements arise from the activity of **motor proteins**, a fascinating class of protein-based molecular motors. Fueled by chemical energy, usually derived from ATP, organized groups of motor proteins undergo cyclic conformational changes that create a unified, directional force—the tiny force that pulls apart chromosomes in a dividing cell and the immense force that levers a quarter-ton jungle cat into the air.

As for all proteins, interactions among different motor proteins, or between motor proteins and other types of proteins, include complementary arrangements of ionic bonds, hydrogen bonds, hydrophobic interactions, and van der Waals interactions at protein-binding sites. In motor proteins, however, these interactions achieve exceptionally high levels of spatial and temporal organization.

Motor proteins promote the contraction of muscles (actin and myosin), the migration of organelles along microtubules (kinesin and dynein), the rotation of bacterial flagella, and the movement of some proteins along DNA. In molecular biology, the motor proteins of most interest are those that function in the transactions involving nucleic acids. These include the helicases, proteins that unwind double stranded DNA or RNA; DNA and RNA polymerases that synthesize polynucleotide chains; DNA topoisomerases that relieve the tension created by overwound or underwound DNA; and other proteins that move along DNA as they carry out their functions in DNA metabolism. These motor proteins are described in detail in Chapters 9–16. Here, we focus on helicases, motor proteins that are highly relevant to the information pathways explored in this text. A great many human genetic diseases have been traced to defects in motor proteins of this class, attesting to their general importance.

Motor proteins combine the functions of ligand binding and catalysis. Each motor protein must interact transiently with another macromolecular ligand, binding and releasing it in a reversible process. To bring about productive motion, the sequence of binding and release cannot be random; it must have a unidirectional component. This requires energy, usually derived from ATP hydrolysis, which is coupled to a directed bind-and-release process between protein and ligand. For the motor proteins we are concerned with, the ligand is generally DNA or RNA.

Helicases Abound in DNA and RNA Metabolism

A **helicase** is a protein that separates the paired strands of a nucleic acid, converting a duplex into two single strands. Helicases are part of a larger family of motor proteins that promote reactions by translocating along the DNA (or RNA) substrate, resulting in the displacement of proteins from nucleic acids, the separation of DNA strands that is required for replication and recombination, conformational changes in nucleic acids, and the remodeling of chromatin. All of these processes are coupled to the hydrolysis of ATP; enzymes, including helicases, that hydrolyze ATP are often referred to as ATPases.

Two structural classes of ATPase domain are found in helicases. The first is structurally related to the core domain of the bacterial RecA recombinase (see Chapter 13); ATP is often bound in a site near the intersection

How to Read Reaction Mechanisms—A Refresher

Chemical reaction mechanisms, which trace the formation and breakage of covalent bonds, are communicated with dots and curved arrows, a convention known informally as "electron pushing." A covalent bond consists of a shared pair of electrons. Nonbonded electrons important to the reaction mechanism are designated by a pair of dots (:). Curved arrows ($\frown$) represent the movement of electron pairs. For movement of a single electron (as in a free radical reaction), a single-headed fishhook-type arrow is used ($\frown$). Most reaction steps involve an unshared electron pair (as in the ligase mechanism).

Some atoms are more electronegative than others; that is, they more strongly attract electrons. The relative electronegativities of atoms encountered in this text are $F > O > N > C \approx S > P \approx H$. For example, the two electron pairs making up a C=O (carbonyl) bond are not shared equally; the carbon is relatively electron-deficient as the oxygen draws away the electrons. Many reactions involve an electron-rich atom (a nucleophile) reacting with an electron-deficient atom (an electrophile). Some common nucleophiles and electrophiles in biochemistry are shown in the box in the center of the diagram.

In general, a reaction mechanism is initiated at an unshared electron pair of a nucleophile. In mechanism diagrams, the base of the electron-pushing arrow originates near the electron-pair dots, and the head of the arrow points directly at the electrophilic center being attacked. Where the unshared electron pair confers a formal negative charge on the nucleophile, the negative charge symbol itself (−) can represent the unshared electron pair and serves as the base of the arrow. In some cases, the electron pair is the pair that makes up a covalent bond, and the base of the arrow is then shown at the middle of the bond. In the ligase mechanism, the nucleophilic electron pair in step 1 is provided by the nitrogen of the ε-amino group of the Lys residue. This electron pair provides the base of the curved arrow. The electrophilic center under attack is the phosphorus atom of the α-phosphoryl group of ATP. The C, O, P, and N atoms have a maximum of 8 valence electrons, and H has a maximum of 2. These atoms are occasionally found in unstable states with less than their maximum allotment of electrons, but C, O, P, and N cannot have more than 8. Thus, when the electron pair from the ligase's Lys residue N attacks the substrate's phosphorus, an electron pair is displaced from the phosphorus valence shell. These electrons move toward the electronegative oxygen atoms. The oxygen of the P=O bond has 8 valence electrons both before and after this chemical process, but the number shared with the phosphorus is reduced from 4 to 2, and the oxygen acquires a negative charge. To complete the process (not shown), the electron pair conferring the negative charge on the oxygen moves back to re-form a double bond with phosphorus and reestablish the P=O linkage. Again, an electron pair must be displaced from the phosphorus, and this time it is the electron pair shared with the oxygen that bridges the α- and β-phosphoryl groups so that pyrophosphate is released. The remaining steps follow a similar pattern.

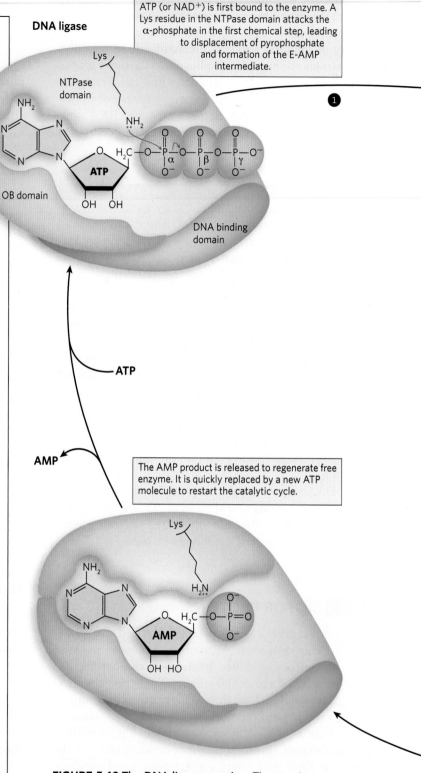

ATP (or NAD$^+$) is first bound to the enzyme. A Lys residue in the NTPase domain attacks the α-phosphate in the first chemical step, leading to displacement of pyrophosphate and formation of the E-AMP intermediate.

DNA ligase

1

ATP

AMP

The AMP product is released to regenerate free enzyme. It is quickly replaced by a new ATP molecule to restart the catalytic cycle.

FIGURE 5-12 The DNA ligase reaction. The reaction creates a new phosphodiester bond at the site of a break, or nick, in the DNA. The same series of three chemical steps is used by every RNA or DNA ligase. In each of the three steps, one phosphodiester bond is formed at the expense of another. Steps 1 and 2 lead to activation of the 5′ phosphate in the nick. In the *E. coli* DNA ligase reaction, AMP is derived from NAD$^+$ rather than ATP, and the reaction releases nicotinamide mononucleotide (NMN) rather than pyrophosphate.

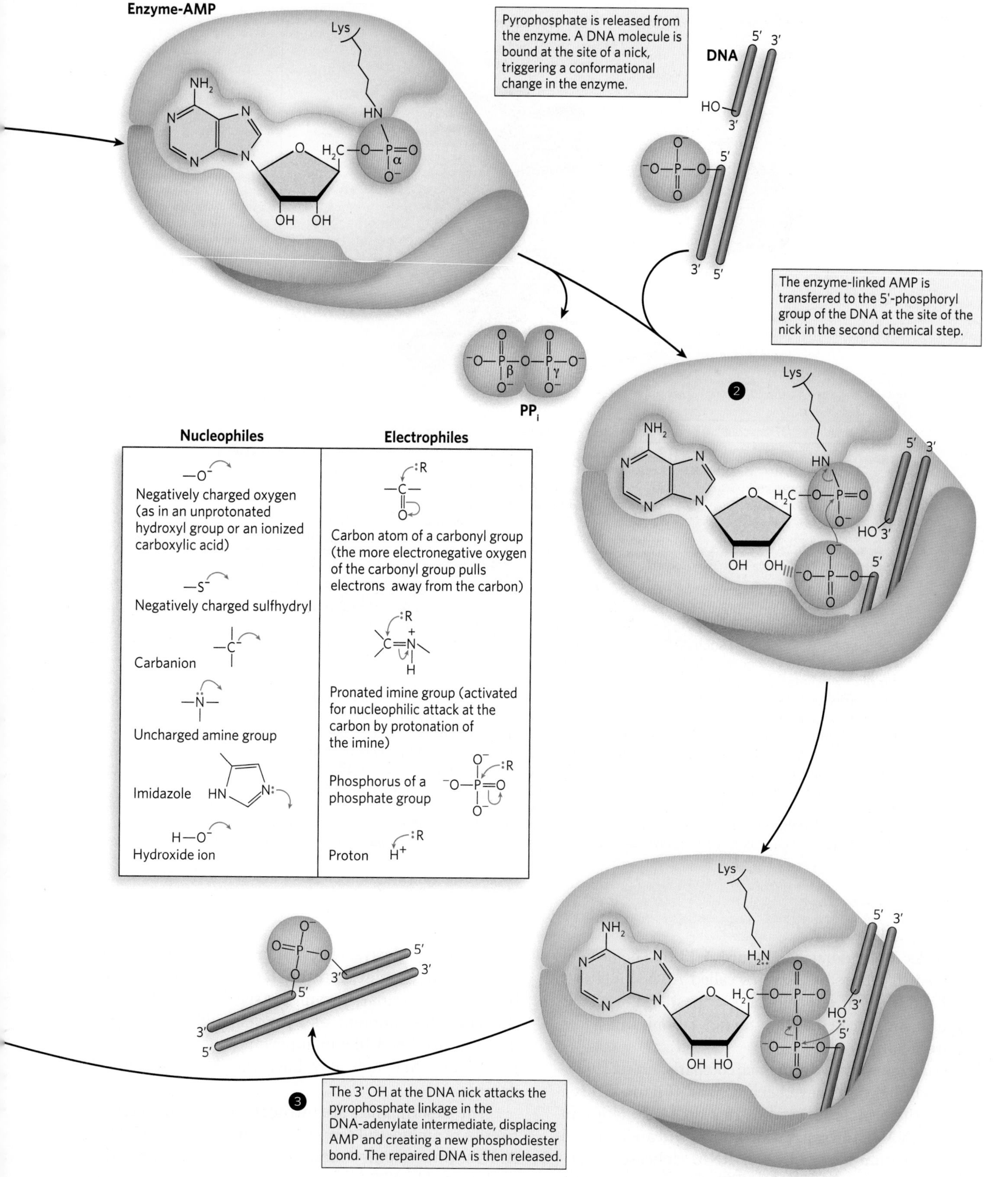

Enzyme-AMP

Lys

NH₂

H₂C—O—P—O
α

OH OH

Pyrophosphate is released from the enzyme. A DNA molecule is bound at the site of a nick, triggering a conformational change in the enzyme.

DNA

5′ 3′

HO
3′

O—P—O
5′

3′ 5′

O—P—O—P—O
β γ

PPᵢ

The enzyme-linked AMP is transferred to the 5′-phosphoryl group of the DNA at the site of the nick in the second chemical step.

❷ Lys

NH₂

HN
H₂C—O—P—O

HO 3′

OH OH

5′ 3′

O—P—O
5′

Nucleophiles	Electrophiles
—O⁻	:R, —C=O
Negatively charged oxygen (as in an unprotonated hydroxyl group or an ionized carboxylic acid)	Carbon atom of a carbonyl group (the more electronegative oxygen of the carbonyl group pulls electrons away from the carbon)
—S⁻	:R, C=N⁺—H
Negatively charged sulfhydryl	
—C⁻	Pronated imine group (activated for nucleophilic attack at the carbon by protonation of the imine)
Carbanion	
—N:	O⁻, ⁻O—P=O, :R
Uncharged amine group	Phosphorus of a phosphate group
Imidazole HN—N:	:R, H⁺
H—O⁻	
Hydroxide ion	Proton

Lys

NH₂

H₂N

H₂C—O—P—O

HO 3′

OH HO

5′ 3′

O—P—O—O
5′

O—P—O
O⁻
3′ 5′
5′ 3′
3′

❸ The 3′ OH at the DNA nick attacks the pyrophosphate linkage in the DNA-adenylate intermediate, displacing AMP and creating a new phosphodiester bond. The repaired DNA is then released.

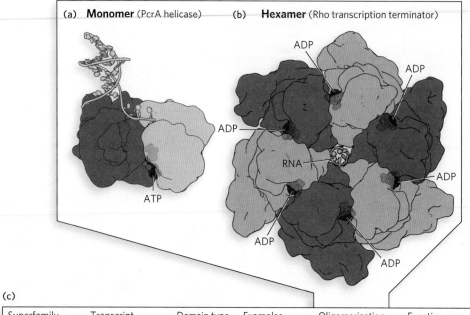

(a) Monomer (PcrA helicase)   **(b) Hexamer** (Rho transcription terminator)

(c)

Superfamily	Transcript	Domain type	Examples	Oligomerization	Function
1*		RecA-like	PcrA, Rep, UvrD	Monomer or dimer	Recombination, repair
2*		RecA-like	WRN, NS3	Monomer or dimer	Replication, repair
3		AAA+	T-ag, BPV E1	Hexamer	Replication
4		RecA-like	T7gp4, DnaB	Hexamer	Replication
5		?	Rho	Hexamer	Transcription
6		AAA+	MCM	Hexamer	Replication

*Superfamilies 1 and 2 contain a tandem repeat of a RecA-like fold, with the consecutive repeats shown in blue and beige.

FIGURE 5-13 DNA helicases. (a) Enzymes of helicase superfamilies 1 and 2 (SF1 and SF2) have a conserved core structure, the RecA-like fold. Here, an ATP analog (black and red) that mimics ATP structure but is not hydrolyzed is bound at the interface of the two domains in a monomeric helicase PcrA. The use of such ATP analogs is often necessary to obtain a structure that illustrates the complex. (b) The core of helicases of SF3 through SF6 consists of six subunits with nucleotide-binding pockets at the subunit-subunit interfaces. Shown here is the Rho transcription termination factor (discussed in Chapter 15). In the structure shown, ADP was present in the crystallization mixture, and its bound location in the resulting structure indicates the ATPase active site. (c) Helicase superfamilies. Core domains (blue and beige) and positions of the conserved sequence motifs that characterize a protein as a helicase (red) are shown for each family ("Transcript"). The sequence motifs are universal structural elements in all helicases and include motifs involved in the binding and hydrolysis of a nucleoside triphosphate. There are two different ATPase domain types in helicases, as shown in the "Domain type" column. Well-studied examples of each family, typical oligomeric structure, and functions are listed in the remaining three columns. [Sources: (a) PDB ID 3PJR (dimer). (b) PDB ID 3ICE (hexamer). (c) Data from M. R. Singleton et al., Annu. Rev. Biochem. 76:23–50, 2007.]

of two RecA-like domains (**Figure 5-13a**). The second is related to a class of enzymes called AAA⁺ (ATPases associated with various cellular activities); again, ATP is bound in sites located at the subunit-subunit interfaces (**Figure 5-13b**).

Based on extensive sequence comparisons, six superfamilies of helicases have been defined (**Figure 5-13c**). Most of them include proteins that do not function in DNA or RNA strand separation but instead are involved in translocation along DNA or RNA. Helicases of superfamilies 1 and 2 (SF1 and SF2), the most common, usually have two RecA-like domains. These proteins

often, but not always, function best as oligomers. Helicases of superfamilies 3 through 6 (SF3–SF6) generally function as circular hexamers (see Figure 5-13b and this chapter's Moment of Discovery). The individual domains are RecA-like in superfamily 4. The subunits in SF3 and SF6 helicases have the AAA⁺ structural domain.

The superfamilies are further defined on the basis of a series of amino acid sequence motifs (see Figure 5-13c). At least three motifs found in all helicase superfamilies are involved in ATP binding and hydrolysis. Other motifs are involved in DNA or RNA binding or oligomerization.

Helicase Mechanisms Have Characteristic Molecular Parameters

Any discussion of helicase mechanisms must take into account several key biochemical properties of these motor proteins: oligomeric state, rate of nucleic acid unwinding or translocation, directionality, processivity, step size, and ATP-coupling stoichiometry.

For some helicases, the **oligomeric state** in which they are active has been quite controversial. A few helicases exhibit some activity as monomers but are greatly stimulated by the addition of more subunits or auxiliary proteins. The observed rates of DNA unwinding or translocation can thus be a complicated function of reaction conditions and proteins added.

Helicases are associated with DNA **unwinding**, the separation of the two paired strands of a nucleic acid. Closely related proteins are often involved in **translocation**, which is movement along duplex DNA or RNA without separating the paired strands. These latter proteins are sometimes called **translocases**, but many other names are used that are more closely attuned to a protein's specific function. In some cases, these proteins are bound to a structure such as a membrane or a viral coat and function to pump DNA or RNA through it. Others function to eject bound proteins from a nucleic acid.

Helicases move unidirectionally on a strand of nucleic acid, and **directionality**—the direction in which the enzyme moves along the strand—is an important distinguishing characteristic of a helicase mechanism. Some helicases move only in the 3′→5′ direction, and some only in the 5′→3′ direction (**Figure 5-14**). With double-stranded DNA, the direction of movement is defined by only one strand. The strand that guides the direction of movement is established during the process of loading the helicase onto the DNA. Helicases that translocate in the 3′→5′ direction seem to be more common than those moving in the 5′→3′ direction.

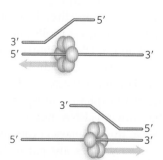

FIGURE 5-14 Direction of movement by helicases. Helicases can be classified based on translocation directionality: 3′→5′ (top) or 5′→3′ (bottom).

Processivity refers to the number of base pairs unwound or the number of nucleotides translocated, on average, each time an enzyme of this type binds to a DNA molecule. Some helicases may unwind only a few base pairs before dissociating, whereas a much greater processivity is the norm for hexameric helicases that encircle DNA and function in such processes as chromosomal DNA replication.

The **step size** is the average number of base pairs or nucleotides over which the helicase moves for each ATP molecule hydrolyzed. The **ATP-coupling stoichiometry** is the average number of ATP molecules consumed per base pair or nucleotide traversed. This may seem like another way of describing step size, and the two can be closely related. However, if coupling is not perfect and some ATP is hydrolyzed unproductively, the step size measured experimentally can be lower than the true step size that reflects the coupled ATP hydrolysis–movement cycle.

Most helicases seem to use a translocation mechanism that can be described as "stepping." It requires the helicase to have at least two DNA-binding sites. Using a series of conformational changes facilitated by ATP binding and hydrolysis, the enzyme moves along the DNA. Different sites on the same or different helicase subunits bind alternately to the DNA strand defining directionality, in a closely orchestrated reaction sequence (**Figure 5-15**). When multiple subunits are employed (dimers or hexamers), DNA molecules may get passed between the subunits and step sizes can be larger.

Helicase translocation is necessary but not sufficient for the unwinding of a nucleic acid. Unwinding mechanisms are classified as active or passive. In an active mechanism, the enzyme interacts directly with double-stranded DNA or RNA to destabilize the duplex structure. In a passive mechanism, the enzyme interacts only with single strands, moving onto the strands and stabilizing them as they form through normal thermal base-pair opening and closing. Although active mechanisms are not yet well understood, they seem to be used by many helicases. In many cases, the destabilization of a double-stranded DNA or RNA is facilitated by a structure often described as a pin, a kind of wedge that pries the two strands apart. The ATPase motor pushes or pulls the paired strands past the pin (as seen in Figure 5-15a). Some dimeric helicases may employ an unwinding mechanism in which one subunit interacts with and destabilizes the duplex DNA, while the other subunit acts as a translocase on adjacent single-stranded DNA (see Figure 5-15b).

A great variety of motor proteins, many of them in the helicase families, carry out a range of functions other than nucleic acid unwinding. We will encounter many

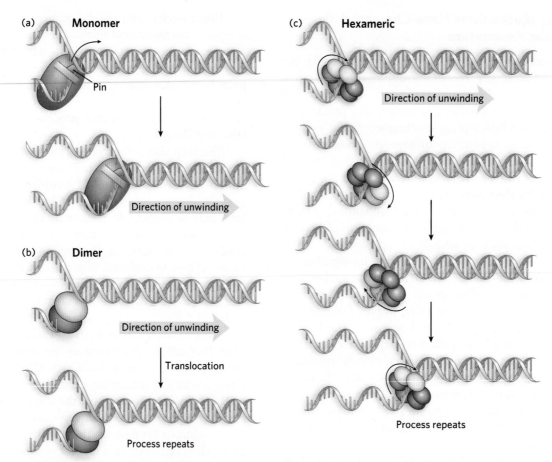

FIGURE 5-15 Unwinding reaction mechanisms for DNA helicases. (a) A monomeric helicase hydrolyzes ATP, causing conformational changes that couple unidirectional DNA translocation with destabilization of the duplex. Some of these enzymes have pinlike structures that function in strand separation. (b) A functional dimer model. A dimeric helicase has two identical subunits. These may alternate in binding to single- and double-stranded DNA or may translocate linearly along one strand of DNA (as shown here). In either case, the protein unwinds the DNA as it moves. (c) Hexameric helicases move along one DNA strand of a duplex DNA, separating the strands as it moves. The bound DNA strand is usually bound alternately by three pairs of dimers within the hexameric structure. The passing of the strand from one dimer to another is coupled to conformational changes driven by ATP hydrolysis. In the diagram, the red, green, and yellow subunit pairs represent the three states: one state has bound ATP, the second has bound ADP (immediately after hydrolysis), and the third has no bound nucleotide.

examples in later chapters, and we consider just a few here.

The RuvB protein (SF6 class; its name derives from *r*esistance to *UV* irradiation) is a DNA translocase involved in the movement of four-armed DNA junctions called Holliday intermediates that appear during DNA recombination (we discuss recombination and Holliday intermediates in Chapter 13). Acting with the RuvA protein, two hexamers of RuvB are arranged symmetrically to propel the DNA, catalyzing a rapid migration of the DNA branch to which they are bound (**Figure 5-16a**).

The eukaryotic Snf proteins (SF2 class; named for sucrose *non*fermenting) are involved in the alteration and/ or disruption of a variety of protein-DNA interactions.

Each of the many Snf proteins has a particular target protein or protein complex, or set of targets, including RNA polymerase, nucleosomes, and regulatory DNA-binding proteins. Snf proteins interact directly with their targets, utilizing their motor functions to displace or reposition the target proteins on the DNA (**Figure 5-16b**). Snf proteins are also heavily involved in chromatin remodeling (see Chapter 10). The NPH-II RNA helicase (SF2 class; *N*TP (nucleoside triphosphate) *p*hospho*h*ydrolase) is a viral helicase that unwinds RNA, but it is also very effective as a translocase that displaces proteins from single-stranded RNA. SF2 enzymes are particularly common and important in RNA metabolism. Many are critical to the maturation of large ribonucleoprotein complexes, such as ribosomes and spliceosomes.

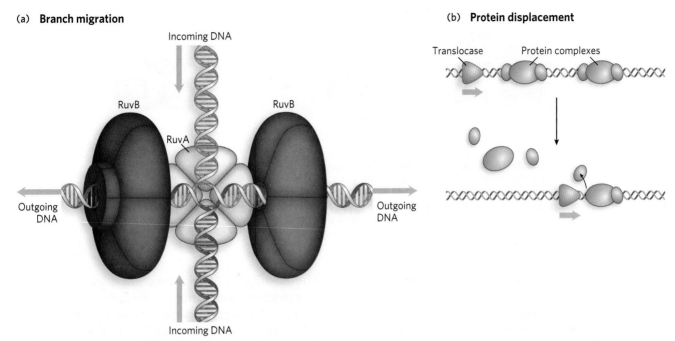

(a) Branch migration

Incoming DNA

RuvB   RuvB

RuvA

Outgoing DNA   Outgoing DNA

Incoming DNA

(b) Protein displacement

Translocase   Protein complexes

FIGURE 5-16 DNA translocases. (a) Branch migration during double-stranded DNA translocation. The bacterial RuvA and RuvB proteins form a complex that binds to a four-armed DNA junction that often forms during recombination (a Holliday intermediate; see Chapter 13), and the RuvA protein binds to the junction itself. The RuvB protein is the translocase, propelling the DNA to the left and right, as shown (arrows). (b) Protein displacement during DNA translocation. Translocase proteins in both bacteria and eukaryotes displace proteins from nucleic acids.

With respect to ATP hydrolysis, motor proteins generally exhibit the standard steady-state kinetic behavior of an enzyme. An example is the Michaelis-Menten kinetics seen for ATP hydrolysis by the helicase PcrA in the presence of single-stranded DNA (**Figure 5-17**). PcrA is an SF1 class helicase found in gram-positive bacteria.

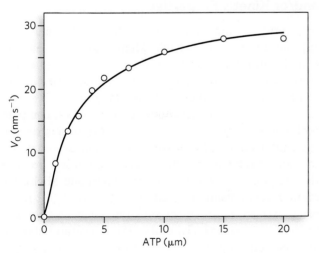

FIGURE 5-17 A plot of V_0 versus [S] for the hydrolysis of ATP by the helicase PcrA. This hydrolytic reaction follows classic Michaelis-Menten kinetics. Compare this with Figure 5-11b. [Source: Data from C. P. Toseland et al., J. Mol. Biol. 392:1020–1032, 2009.]

SECTION 5.3 SUMMARY

- Motor proteins combine the functions of ligand binding and catalysis. The ligand, often RNA or DNA, is bound reversibly. Movement involves protein conformational changes triggered by ATP hydrolysis catalyzed by the motor protein.

- Helicases are ubiquitous motor proteins involved in DNA or RNA strand separation. Helicase mechanisms are discussed in terms of oligomeric state, rate of nucleic acid unwinding or translocation, directionality, processivity, step size, and ATP-coupling stoichiometry.

- Many motor proteins closely related to helicases are involved in translocation along nucleic acids, movement of branches in double-stranded DNA during replication, protein displacement from nucleic acids, chromatin remodeling, and other functions.

5.4   THE REGULATION OF PROTEIN FUNCTION

In the flow of biological information, groups of enzymes often work together in sequential and interconnected pathways to carry out a given process, such as the replication of a chromosome or the splicing of an intron in a messenger RNA. Such processes use

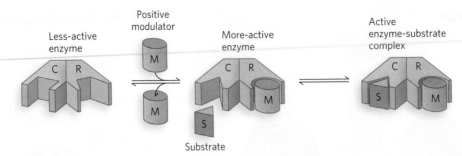

FIGURE 5-18 Allosteric enzyme interactions. In many allosteric enzymes, the substrate-binding site and the modulator-binding site(s) are on different subunits: catalytic (C) and regulatory (R) subunits, respectively. Binding of a positive (stimulatory) modulator (M) to its specific site on the R subunit is communicated to the C subunit through a conformational change, which renders the C subunit active and capable of binding the substrate (S) with higher affinity. On dissociation of the modulator from the regulatory subunit, the enzyme reverts to its inactive or less-active form.

enormous amounts of chemical energy in the form of nucleoside triphosphates (NTPs). It is critical not only that these processes occur, but that they do so only at a specific time and in a specific place, so that resources are not wasted. In addition, the many reactions that carry out these complex processes must be precisely coordinated. Faulty coordination or poor timing could damage or alter the cellular genome. Regulation is thus an important aspect of virtually every process in molecular biology.

Most enzymes follow the Michaelis-Menten kinetic patterns described in Section 5.2. However, the timing and rate of many processes are governed by **regulatory enzymes** that exhibit increased or decreased catalytic activity in response to certain signals. The resulting regulation conserves cellular resources and prevents inappropriate alterations in the genetic material.

The activities of many proteins in DNA and RNA metabolism are regulated, including binding proteins that simply bind reversibly to DNA or RNA, enzymes, and motor proteins. Their functions can be modulated in several ways: by the noncovalent binding of allosteric modulators, autoinhibition by a segment of the protein, reversible covalent modification, proteolytic cleavage, or interaction with special regulatory proteins. We touch on all of these mechanisms, while focusing on those that are most common in enzymes and proteins involved in nucleic acid metabolism.

Modulator Binding Causes Conformational Changes in Allosteric Proteins

Allosteric enzymes or **allosteric proteins** are those having "other shapes" or other conformations induced by the binding of modulators. This property is found in certain regulatory enzymes, as conformational changes induced by one or more **allosteric modulators** interconvert more-active and less-active forms of the enzyme.

The modulators for allosteric enzymes may be inhibitory or stimulatory. Often the modulator is the substrate itself, and a regulatory protein or enzyme for which substrate and modulator are identical is referred to as **homotropic**. When the modulator is a molecule other than the normal ligand or substrate, the enzyme or protein is said to be **heterotropic**.

The properties of allosteric enzymes are significantly different from those of simple, nonregulatory enzymes. Some of the differences are structural. In addition to active sites, allosteric enzymes generally have one or more regulatory, or allosteric, sites for binding the modulator (**Figure 5-18**). Just as an enzyme's active site is specific for its substrate, each regulatory site is specific for its modulator. Enzymes with several modulators generally have different specific binding sites for each.

Allosteric Enzymes Have Distinctive Binding and/or Kinetic Properties

Allosteric enzymes show relationships between V_0 and [S] that differ from Michaelis-Menten kinetics. They do exhibit saturation with the substrate, but for some allosteric enzymes, plots of V_0 versus [S] produce a sigmoid saturation curve, rather than the hyperbolic curve typical of nonregulatory enzymes (**Figure 5-19a**; compare with Figure 5-11b). On the sigmoid saturation curve, we can find a value of [S] at which V_0 is half maximal, but we cannot refer to it with the designation K_m, because the enzyme does not follow the hyperbolic Michaelis-Menten relationship. Instead, the symbol $[S]_{0.5}$ or $K_{0.5}$ is often used to represent the substrate concentration giving half-maximal velocity of the reaction catalyzed by an allosteric enzyme.

For homotropic allosteric proteins or enzymes, the substrate often acts as a positive modulator (an activator), because the subunits act cooperatively. Cooperativity occurs when the binding of a substrate to one binding site

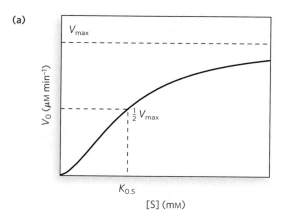

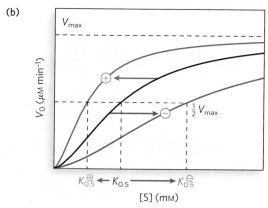

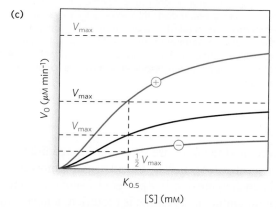

FIGURE 5-19 Substrate-activity curves for allosteric enzymes.
(a) The sigmoid curve of a homotropic enzyme, in which the substrate also serves as a positive (stimulatory) modulator, or activator. (b) The effects of a positive modulator (+) and a negative modulator (−) on an allosteric enzyme in which $K_{0.5}$ is altered without a change in V_{max}. The central curve shows the substrate-activity relationship without a modulator. (c) A less common type of modulation, in which V_{max} is altered and $K_{0.5}$ is nearly constant.

alters an enzyme's conformation and affects the binding of subsequent substrate molecules. Most commonly, the binding of one molecule enhances the binding of others, an effect called positive cooperativity. This accounts for the sigmoid rather than hyperbolic change in V_0 with increasing [S]. One characteristic of sigmoid kinetics is

that small changes in the concentration of a modulator can be associated with large changes in enzyme activity. A relatively small increase in [S] in the steep part of the curve causes a comparatively large increase in V_0 (see Figure 5-19a). Much rarer are cases of negative cooperativity, where the binding of one substrate molecule impedes the binding of subsequent molecules.

For heterotropic allosteric proteins or enzymes, those with modulators that are molecules other than the normal substrate, it is difficult to generalize about the shape of the binding or substrate-saturation curve. An activator may cause the curve to become more nearly hyperbolic, with a decrease in $K_{0.5}$ but no change in V_{max}, resulting in an increased reaction velocity at a fixed substrate concentration (V_0 is higher for any value of [S], as shown in **Figure 5-19b**, top curve). A negative modulator (an inhibitor) may produce a *more* sigmoid substrate-saturation curve, with an increase in $K_{0.5}$ (Figure 5-19b, bottom curve). Other heterotropic allosteric enzymes respond to an activator by increasing V_{max} with little change in $K_{0.5}$ (**Figure 5-19c**, top curve) and to an inhibitor by decreasing V_{max} with little change in $K_{0.5}$ (Figure 5-19c, bottom curve). Heterotropic allosteric proteins and enzymes therefore show different kinds of response in their substrate-activity curves, because some have inhibitory modulators, some have activating modulators, and some have both.

Many of the allosteric effects encountered by molecular biologists are heterotropic. Numerous examples can be found among the regulatory proteins that bind to specific DNA sequences adjacent to genes, such as the **cAMP receptor protein**, or **CRP** (also called CAP, for *catabolite gene activation protein*; see Figure 4-18a). CRP is a dimeric DNA-binding protein that participates in the regulation of genes involved in bacterial carbohydrate metabolism. Each subunit has separate domains that bind a modulator, cAMP (cyclic AMP), and a specific site in the DNA. The binding of cAMP exhibits negative cooperativity, in which the binding of cAMP to the modulator-binding site of one subunit reduces the affinity of the cAMP-binding site of the other subunit by two orders of magnitude. Bound cAMP also promotes conformational changes in the DNA-binding domain that facilitate the binding of CRP to its DNA binding site. This is just one example of the variety of allosteric effects observed in gene repressor and activator proteins that help modulate the sensitivity of these proteins to their environment.

Autoinhibition Can Affect Enzyme Activity

Many processes of DNA and RNA metabolism are precisely targeted; they are limited to particular locations and circumstances, some of which appear transiently and unpredictably. For example, if DNA is damaged, the lesion must be repaired before the next replication cycle. The enzymes that repair DNA cleave the DNA strands near the lesion, remove damaged nucleotides, and replace them. It is essential that these enzymes be available on short notice and that they act only at DNA lesions.

One way to make such enzymes generally available but not active until needed is by **autoinhibition**. A segment of the protein molecule, sometimes an entire domain, can reduce or eliminate the activity of the enzyme. The enzyme may be present in the cell at all times, but it functions weakly or not at all under normal circumstances. Activation of the enzyme requires self-assembly into a more functional oligomer, the binding of an auxiliary protein, or interaction with a particular ligand. In all cases, the interaction results in a conformational change that repositions the auto-inhibitory protein segment. The activity of the enzyme then increases.

Autoinhibition has been documented for several proteins, including certain helicases and the bacterial RecA recombinase. In bacteria, the activity of a helicase known as Rep, involved in DNA replication, is autoinhibited by a subdomain called 2B. The bacterial RecA recombinase is autoinhibited by a short segment of polypeptide at its C-terminus (**Figure 5-20**). For most RecA proteins, this segment includes a high concentration of negatively charged amino acid residues, and it may interact with other parts of the RecA protein through electrostatic interactions.

Autoinhibition may be just one aspect of a broader regulatory strategy. The autoinhibited enzyme can be maintained in the cell without its activity causing problems when it is not needed. Activation by interaction with additional proteins can occur quickly when its activity is required, and the enzyme's function can be more readily targeted to specified locations and situations.

Some Proteins Are Regulated by Reversible Covalent Modification

In another important class of regulatory mechanism, activity is modulated by **covalent modification** of one or more amino acid residues in a protein molecule. More than 300 different types of covalent modification have been found in proteins. Common modifying groups include phosphoryl, acetyl, adenylyl, uridylyl, methyl, amide, carboxyl, myristoyl, palmitoyl, prenyl, hydroxyl, sulfate, and adenosine diphosphate ribosyl groups. There are even entire proteins that are used as specialized modifying groups, such as ubiquitin. Some

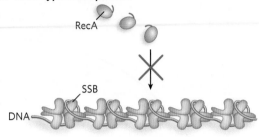

(a) **Wild-type RecA protein**

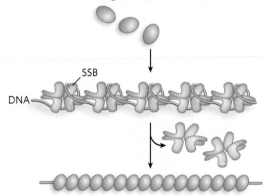

(b) **RecA mutant lacking its C-terminal residues**

FIGURE 5-20 Autoinhibition. (a) The RecA recombinase (see Chapter 13) forms filaments on DNA, characterized by distinct nucleation and filament extension phases. In the bacterial RecA protein, a C-terminal segment prevents efficient nucleation of binding to single-stranded DNA when single-stranded DNA–binding protein (SSB) is bound to the DNA strand. (b) If the C-terminal segment is removed (by recombinant DNA techniques; see Chapter 7) to create a truncated RecA protein, nucleation and subsequent displacement of SSB to form a filament on the DNA are rapid.

of these modifications are shown in **Figure 5-21**. These varied groups are generally linked to and removed from a regulated enzyme or other protein by separate enzymes. When an amino acid residue is modified, a novel amino acid with altered properties is effectively introduced into the protein. The introduction of a charge can alter an enzyme's local properties and induce a conformational change. The introduction of a hydrophobic group can trigger association with a membrane. The changes are often substantial and can be critical to the function of the altered enzyme.

The variety of protein modifications is too great to cover in detail, but we present a few examples here. In eukaryotic cells, histones are important modification targets. As described in detail in Chapter 10, many histones and histone variants are subject to precise patterns of methylation, acetylation, phosphorylation, and ubiquitination. Such modifications play an important role in altering chromatin structure in

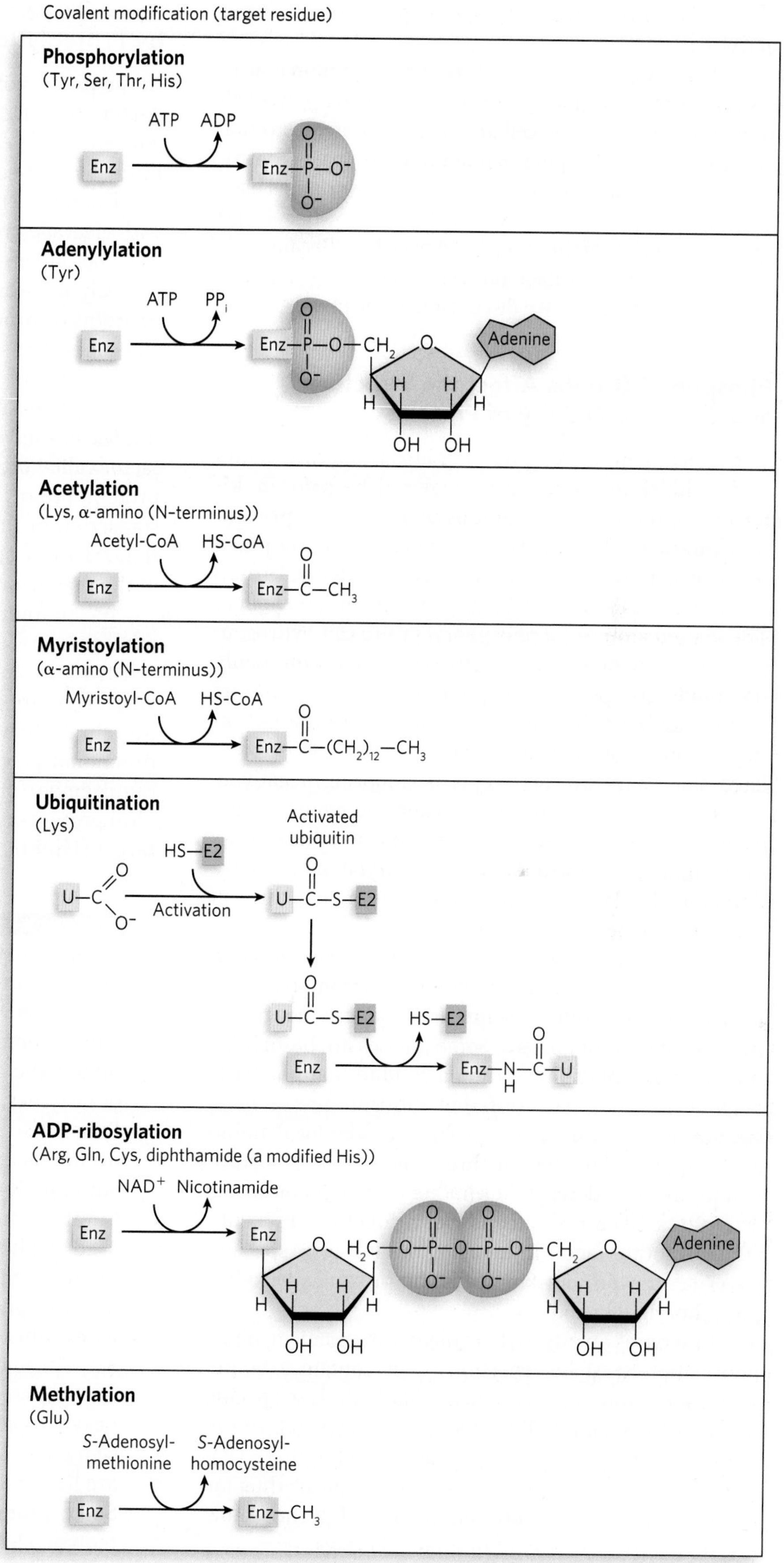
Covalent modification (target residue)
Phosphorylation
(Tyr, Ser, Thr, His)
ATP ADP
Enz
Enz—P—O⁻
Adenylylation
(Tyr)
ATP PPi
Enz
Adenine
OH OH
Acetylation
(Lys, α-amino (N-terminus))
Acetyl-CoA HS-CoA
Enz
Myristoylation
(α-amino (N-terminus))
Myristoyl-CoA HS-CoA
Enz
Ubiquitination
(Lys)
Activated
ubiquitin
HS—E2
Activation
U—C—S—E2
HS—E2
Enz
Enz—N—C—U
ADP-ribosylation
(Arg, Gln, Cys, diphthamide (a modified His))
NAD⁺ Nicotinamide
Enz
Adenine
Methylation
(Glu)
S-Adenosyl-methionine
S-Adenosyl-homocysteine
Enz
Enz—CH3

FIGURE 5-21 Some enzyme modification reactions. In the ubiquitination pathway, E2 is a carrier protein for activated ubiquitin. See text for details.

specific regions, to facilitate gene expression and other activities.

Phosphorylation is probably the most common type of regulatory modification. It is estimated that one-third of all proteins in a eukaryotic cell are phosphorylated, and one or (often) many phosphorylation events are part of virtually every regulatory process. Some of these proteins have only one phosphorylated residue, others have several, and a few have dozens of sites for phosphorylation. Because this mode of covalent modification is central to a large number of regulated processes, we discuss it in some detail.

Phosphoryl Groups Affect the Structure and Catalytic Activity of Proteins

The attachment of phosphoryl groups to specific amino acid residues of a protein is catalyzed by **protein kinases**; the removal of the groups is catalyzed by **protein phosphatases**. The addition of a phosphoryl group to a Ser, Thr, or Tyr residue introduces a bulky, charged group into a region that was previously only moderately polar. The oxygen atoms of a phosphoryl group can hydrogen-bond with one or several groups in a protein, commonly the amide groups of the peptide backbone at the start of an α helix or the charged guanidinium group of an Arg residue. The two negative charges on a phosphorylated side chain can also repel neighboring negatively charged (Asp or Glu) residues. When the modified side chain is located in a region of an enzyme critical to its three-dimensional structure, phosphorylation can have dramatic effects on enzyme conformation and thus on substrate binding and catalysis.

The Ser, Thr, or Tyr residues that are phosphorylated in regulated proteins occur within common structural motifs called **consensus sequences** that are recognized by specific protein kinases. Some kinases are basophilic, preferentially phosphorylating a residue that has basic neighbors; others have different substrate preferences, such as for a residue near a proline. Besides local amino acid sequence, the overall three-dimensional structure of a protein can determine whether a protein kinase has access to a given residue and can recognize it as a substrate. Another factor influencing the substrate specificity of certain protein kinases is the proximity of other phosphorylated residues.

To serve as an effective regulatory mechanism, phosphorylation must be reversible. Cells contain a family of phosphoprotein phosphatases that hydrolyze specific ℗–Ser, ℗ Thr, and ℗ Tyr esters (℗ is shorthand for the phosphoryl group), releasing inorganic phosphate (P_i). The phosphoprotein phosphatases we know of thus far act only on a subset of phosphoproteins, but they show less substrate specificity than protein kinases.

Some Proteins Are Regulated by Proteolytic Cleavage

In the process of **proteolytic cleavage**, an inactive precursor protein is cleaved to form the active protein. Many eukaryotic proteases (proteolytic enzymes) are regulated in this way. A subunit of the *E. coli* DNA polymerase V (see Chapter 12) called MutD is also activated in this way, with cleavage producing the active form, MutD′.

The larger, uncleaved precursor proteins, before proteolytic cleavage, are generally referred to as **proproteins** or **proenzymes**, as appropriate. For example, a class of proteins known as transcription factors facilitate the function of RNA polymerases in all organisms. The process of sporulation (spore formation) in the bacterium *Bacillus subtilis* is controlled in part by a transcription factor called σ^E, which is synthesized as an inactive proprotein, pro-σ^E. The conversion of pro-σ^E to the mature transcription factor involves the regulated proteolytic removal of 27 amino acid residues from the N-terminus of the precursor protein.

As another example, the small number of proteins that are encoded by the genomes of eukaryotic retroviruses, including the human immunodeficiency virus, HIV, are generally synthesized as one large polyprotein, which must be cleaved into the individual functional proteins by a virus-encoded protease. The requirement for the HIV protease to activate the viral proteins has made this enzyme an important drug target (**Highlight 5-2** on p. 164).

SECTION 5.4 SUMMARY

- Specific enzymes, motor proteins, and other proteins are subject to various types of regulation.
- The noncovalent binding of allosteric modulators, either homotropic or heterotropic, can facilitate or inhibit the activity of nucleic acid–binding proteins or enzymes.
- Parts of a protein's own structure can reduce the overall activity of the protein in the process of autoinhibition.
- Covalent modification is a common mechanism used to alter the function of proteins and enzymes. Common modifications involve the addition and removal of phosphoryl, methyl, acetyl, ubiquitinyl, and many other types of groups.
- Ser, Thr, and Tyr residues can be phosphorylated and dephosphorylated by protein kinases and phosphatases, respectively. Kinases are specific to consensus sequences in the target protein, but phosphorylases are less specific.
- Some proteins and enzymes are regulated by proteolytic cleavage. These proteins are synthesized as

larger, inactive proproteins or proenzymes and are activated by the proteolytic removal of one or more amino acid residues.

? UNANSWERED QUESTIONS

The study of protein function is, arguably, the oldest subdiscipline in biochemistry and molecular biology. But there is still much to learn. The relatively young science of genomics keeps pointing to genes that encode proteins about which we know little or nothing at all. Some shortcuts to functional discovery are discussed in later chapters.

1. How does protein structure relate to function? This is an old but still very relevant question for every scientist who studies proteins. Advanced methods of structural analysis are providing more information than ever before, but many of these structural pictures are static. A clear picture of a complete binding or catalytic cycle can require a detailed knowledge of the structure of multiple protein conformations. Certain structural motifs and domains (e.g., the OB fold of single-stranded DNA–binding proteins and other proteins, the AAA$^+$ ATPase domain, and simple β-barrel structures) appear in proteins that often have seemingly unrelated functions. The manner in which particular structures are adapted to different functions is an ongoing area of investigation.

2. How do proteins function in the context of large protein assemblies? Many proteins act only as a part of a much larger protein complex, involving anywhere from a few to many dozens of additional proteins. Unraveling the individual contributions of the subunits of these large complexes has become one of the major challenges of modern molecular biology.

3. Within the context of molecular biology, how many types of protein function remain to be discovered? A textbook such as this one might leave a student with the impression that the fundamental protein/enzyme activities underlying information pathways are now understood. This is not so! Although major processes such as DNA replication, RNA transcription, and protein synthesis are increasingly well understood, new types of proteins with important functions are continually being discovered. Many of the newer discoveries involve proteins that have regulatory functions, or facilitate changes in the bacterial nucleoid or eukaryotic chromatin during cell division, or carry out functions in RNA metabolism. There are no boundaries to these frontiers of protein research.

4. How do proteins in the small concentrations found in cells find their interacting partners—particularly, specific sequences on very large nucleic acids—in the complex cellular environment? This is an issue that remains of great interest to investigators in many areas of molecular biology.

HIV Protease: Rational Drug Design Using Protein Structure

Human immunodeficiency virus (HIV), the causative agent of AIDS, kills cells of the immune system. The development of a vaccine has been unsuccessful, because the surface glycoproteins targeted by antibodies change rapidly, in part due to the extremely high mutation rate of HIV (about one replication mistake in every 10,000 nucleotides of HIV genome per generation). However, there has been substantial success in the development of drugs targeting HIV-encoded enzymes that are essential for viral propagation. HIV is a retrovirus, an RNA virus that converts, or reverse transcribes, its RNA genome into DNA. Before HIV was discovered, various laboratories had already studied the life cycle of retroviruses, many of

which cause cancer in humans and other animals. With the arrival of HIV, researchers studying this new virus had a head start—a vast amount of established research that had already identified key enzymes required for retroviral propagation. Among these enzymes is a protease that digests long, precursor polypeptides into smaller, active viral proteins.

The usual route for developing a drug that inhibits enzyme activity starts with the random screening of hundreds of thousands of chemical compounds. Possible inhibitors are then chemically optimized for potency, availability in an oral form, and low toxicity. The U.S. Food and Drug Administration's approval procedure for the use of a drug in humans usually takes well over a dozen years. This type of process has led to drugs that inhibit certain HIV enzymes, including the reverse transcriptase. Rational drug

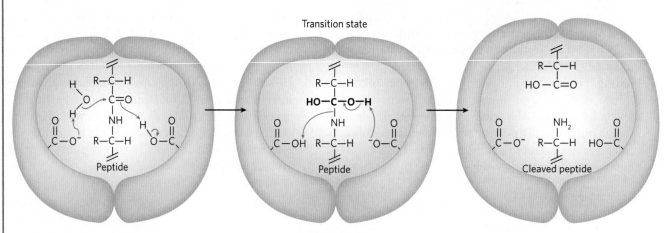

FIGURE 1 In the chemical mechanism of peptide hydrolysis by the HIV protease, one active-site Asp residue is contributed by each of the identical subunits. One Asp activates a water molecule, while the other stabilizes the leaving group.

design is another, shorter route to drug discovery. It starts with the target protein structure and designs chemicals that plug the active site and shut the enzyme down. This process short-circuits the random, brute-force approach to searching for chemical inhibitors and has the potential to cut years off the drug-discovery process.

An astounding success in rational drug design has been achieved with the HIV protease. This is partly due to the unique architecture of the protein. HIV protease is a dimer of identical subunits, but unlike typical dimers, which contain two active sites, the HIV protease dimer has a single active site located in the central hydrophobic chamber at the dimer interface. The active site has twofold symmetry; each subunit contributes a catalytic Asp residue, and the two cooperate to hydrolyze the peptide bond (Figure 1). Information obtained from the biochemical studies and crystal structures of HIV protease made possible the rational design of chemical inhibitors. By 1996, the FDA had approved three HIV protease inhibitor drugs: indinavir (Crixivan), ritonavir (Norvir), and saquinavir (Invirase). Notably, all of these drugs are effective, in part, because they mimic the transition state of the proteolytic reaction catalyzed by the enzyme and thus bind to the enzyme virtually irreversibly (Figure 2).

Protease inhibitor drugs have helped reduce the viral titer in the plasma of HIV-infected individuals. However, because of the high mutation rate of the virus, these drugs are effective only for a limited time. Viral mutability can be countered by using different drugs at different times or in a combination cocktail with drugs that inhibit other HIV enzymes. An intensive study of the structure of HIV protease mutants that can circumvent drug action is underway to identify regions of the protein that cannot endure mutation, with the hope of designing inhibitors to which the virus cannot develop resistance.

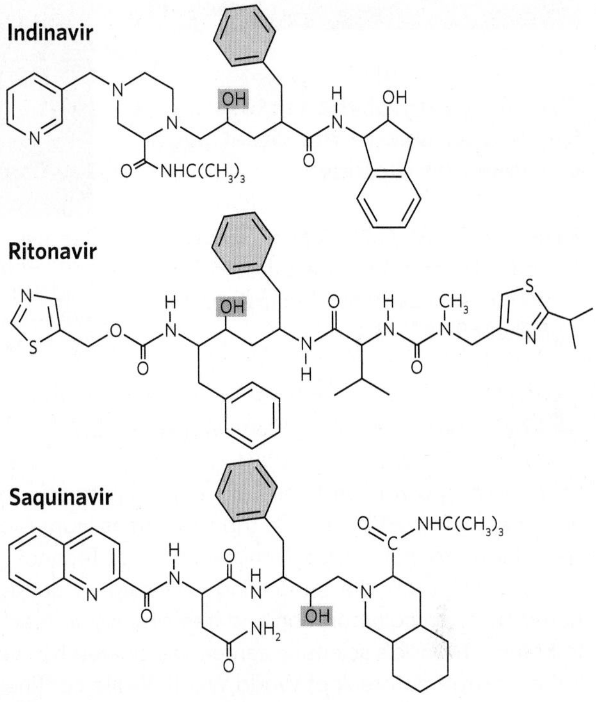

FIGURE 2 Three HIV protease inhibitors used in the treatment of HIV infection. The hydroxyl group (light red) is designed to mimic the tetrahedral transition state of the enzyme-catalyzed reaction shown in Figure 1. The aromatic groups (beige) are designed to fit into other pockets on the surface of the enzyme.

HOW WE KNOW

The Discovery of the Lactose Repressor: One of the Great Sagas of Molecular Biology

Rickenberg, H.V., G.N. Cohen, G. Buttin, and J. Monod. 1956. La galactoside-perméase d'Escherichia coli. *Ann. Inst. Pasteur* 91:829–857.

Jacques Monod, 1910–1976 (left); André Lwoff, 1902–1994 (middle); and François Jacob, 1920–2013 (right).
[Source: © Bettmann/Corbis.]

Jacques Monod began his scientific career in the 1930s, as a graduate student with André Lwoff, studying the capacity of *E. coli* to adapt its metabolism to different growth conditions. His approach to science was shaped in part by a 1936 trip to Thomas Hunt Morgan's laboratory at the California Institute of Technology, where he found a stimulating environment dominated by open collaboration and free discussion. Back in France, Monod's scientific career was slowed but not halted by the outbreak of World War II. While continuing his research, Monod was an active member of the French Resistance. His laboratory at the Sorbonne doubled as a meeting place and propaganda printing press. In the lab, he hit on the idea of an inducer, a cellular signal that would trigger the production of new enzymes needed for adapting to new metabolic circumstances. After the war, he turned back to science full-time. With the help of Lwoff, he obtained a position at the Pasteur Institute in Paris. The metabolism of lactose soon caught his attention.

The disaccharide lactose is cleaved to the monosaccharides glucose and galactose by the enzyme β-galactosidase. Monod found that when lactose was not present in the *E. coli* growth medium, β-galactosidase was barely detectable in cell extracts. When lactose was added as the sole carbon source for bacterial growth, the levels of β-galactosidase enzyme increased dramatically. Monod wondered how this might occur. In a 1940s' scientific world in which DNA sequencing, PCR (the polymerase chain reaction), and the structure of DNA were unknown, and messenger RNA still remained to be discovered, the question was not trivial.

Many bacterial enzymes of intermediary metabolism exhibit this inducible pattern. What recommended lactose metabolism as a subject of investigation? Like many other laboratories in the late 1940s, the Pasteur Institute lacked modern cold rooms and other facilities now commonly used to keep proteins active

during purification. The one inducible enzyme stable enough to survive the summer heat of an attic lab in Paris was β-galactosidase, later shown to be encoded by a gene called *lacZ*. The enzyme was also fairly easy to assay. When it was present at high levels, it would cleave an alternative substrate, 5-bromo-4-chloro-3-indolyl-β-D-galactopyranoside (denoted, more simply, X-gal), and the indole released would color the bacterial colonies blue. The Monod group initiated a study of bacterial lactose metabolism that featured a creative union of biochemistry and genetics.

Mutations were soon found that affected the induction of β-galactosidase. Colonies of these mutants turned blue on X-gal plates even in the absence of lactose. Most of the mutant cells had a mutation in a gene that came to be known as the *i* (inducer) gene, and later the *lacI* gene. In these mutations, the *lacZ* gene was expressed (produced β-galactosidase) all the time; this is known as constitutive expression (Figure 1). Moreover, the appearance of β-galactosidase activity was paralleled by the appearance of a function that transported lactose into the cell, an activity that Monod called a galactoside permease (this is encoded by the gene *lacY*). The coordinated regulation of two genes, and the loss of regulation in constitutive mutants, led to the concept that some genes regulate other genes. When Monod explained this then-revolutionary idea to his wife, a nonscientist, he was somewhat pained at her reply: "Of course, it is obvious!" Monod now had to figure out what the *lacI* gene was doing, as the next part of the story recounts.

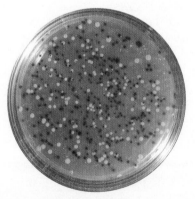

FIGURE 1 Bacterial colonies growing on an agar plate containing X-gal. Cells in the blue colonies have a mutation that results in constitutive expression of the *lac* genes.
[Source: Courtesy Edvotek.]

The *lacI* Gene Encodes a Repressor

Jacob, F., and J. Monod. 1961. Genetic regulatory mechanisms in the synthesis of proteins. *J. Mol. Biol.* 3:318–356.

Next door to Monod at the Pasteur Institute was the laboratory of François Jacob. Wounded as a member of the Free French Forces during the Normandy campaign, Jacob could not pursue his planned career in surgery and instead turned to science. He studied a phenomenon dubbed the "erotic induction" of bacteriophage λ. Bacteriophage λ is a prophage, a bacterial virus that integrates its DNA into its host chromosome and remains quiescent there.

Some years earlier, Joshua Lederberg had discovered the phenomenon of bacterial conjugation (bacterial sex), in which DNA is transferred from a donor to a recipient cell. The transfer is mediated by a genetic element, separate from the bacterial chromosome, known as the F plasmid. In some cells, this plasmid becomes integrated into the host chromosome, creating a strain that mediates high-frequency (Hfr) transfer of chromosomal genes. The transfer begins at the location where the F plasmid is integrated and proceeds linearly along the bacterial chromosome.

Jacob, working with Elie Wollman, used this technique to produce some of the first genetic maps, and they even deduced that the *E. coli* chromosome was circular. They also noticed that when an Hfr strain also contained an integrated prophage, the bacteriophage λ was transferred into the recipient along with the chromosomal genes. When the dormant bacteriophage λ was thus transferred to a strain that lacked an integrated prophage, the recipient cells were soon lysed. The prophage was somehow activated when it entered the recipient, leading to lytic reproduction—that is, it was induced. It gradually dawned on both Monod and Jacob that the induction phenomena they were studying were closely related, and one of the great scientific collaborations of the twentieth century was born. Their subsequent work defined the regulation of the lactose operon (described in detail in Chapter 20) and established many key regulatory principles.

Monod first liked the idea that mutation in the *lacI* gene (*lacI⁻*) produced some kind of inducer that made the addition of lactose unnecessary. If this was true, then the *lacI⁻* gene should be dominant over the normal (wild-type) *lacI* gene (*lacI⁺*). To get the two variations of *lacI* (*lacI⁻* and *lacI⁺*) into the same cell, Monod worked with Jacob and Arthur Pardee (on sabbatical leave from the University of California, Berkeley) to use the bacterial conjugation method. The resulting "PaJaMo" experiments at first supported Monod's idea. When an Hfr strain was used to introduce the *lacI⁺* and *lacZ⁺* genes into a recipient that had both the *lacI⁻* gene and another mutation that prevented β-galactosidase synthesis (*lacZ⁻*), β-galactosidase production occurred immediately after the *lacZ⁺* gene arrived in the recipient. This seemed to indicate that the *lacI⁻* gene of the recipient was dominant. However, after an hour or so, β-galactosidase production was halted as the product of the *lacI⁺* gene built up. Clearly, *lacI⁺* encoded a substance that shut off the *lacZ* gene. Continued experimentation refined the idea that *lacI⁺* encodes some kind of repressor and that the interaction of lactose with the repressor is needed to induce the *lacZ* gene and the additional genes that are coregulated with it. A similar repressor normally suppresses activation of an integrated λ prophage. The prophage is activated when transferred to the recipient cell during conjugation because no repressor is present in the recipient. Cell lysis results.

The work continued. If *lacI* was encoding a repressor, this repressor had to interact with something to shut down the lactose genes. Monod and Jacob now mutagenized cells that had two good copies of the lactose genes. It was unlikely that both copies of *lacI* would be inactivated, but inactivation of one repressor target would be enough to induce the function of one set of lactose genes. The researchers predicted that the resulting mutations would appear at a site on the chromosome distinct from *lacI* and would lead to constitutive synthesis of the lactose gene products. The mutants were found, and they defined a site that Monod and Jacob called the operator.

In a famous 1961 paper, Jacob and Monod laid out these ideas and others as part of their operon model. The concepts have guided our thinking about gene regulation ever since. Other experiments, carried out in parallel, showed that bacteriophage λ also encodes a repressor, and this repressor is needed to keep most other bacteriophage λ genes from being expressed.

HOW WE KNOW

Discovery of the Lactose Repressor Helped Give Rise to DNA Sequencing

Gilbert, W., and B. Müller-Hill. 1966. Isolation of Lac repressor. *Proc. Natl. Acad. Sci. USA* 56:1891–1898.

Walter Gilbert [Source: Ira Wyman/Sygma/Corbis.]

By 1961, it was clear that a repressor existed, but not at all clear what it was. It could be RNA or protein, or some other kind of molecule that was synthesized by a *lacI* enzyme. By 1966, nonsense-type mutations, those that prematurely halt protein synthesis, had been found in the *lacI* gene that produced a *lacI⁻* effect. This finding had convinced most scientists that the repressor was a protein, but it was still necessary to isolate one to prove it. The isolation work was carried out at Harvard, by physicist-turned-biologist Walter Gilbert.

To isolate a protein, you need a way to measure its presence (an assay), but that was difficult to construct. The repressor presumably bound to operator DNA, but that DNA sequence was not yet defined. The repressor also bound to the inducer (then thought to be lactose, later shown to be allolactose, a metabolic by-product of lactose metabolism). Because lactose was metabolized in the cell, it would be destroyed in crude cell extracts and thus would be of little use to the researchers. Gilbert turned to isopropyl β-D-1-thiogalactopyranoside (IPTG), a molecule known to induce the lactose operon without being metabolized in the cell.

Using a technique known as equilibrium dialysis, the researchers suspended a dialysis bag containing a bacterial cell extract in a solution containing radioactive IPTG. Pores in the dialysis bag were sufficiently small to prevent the protein molecules from escaping, but smaller molecules such as IPTG could diffuse through. If the lactose repressor was present in the extract, it would bind to IPTG, and the concentration of IPTG would increase inside the dialysis bag relative to the surrounding solution. The assay eventually worked, but only after Gilbert and his colleagues developed methods to greatly increase its sensitivity. In 1966, they reported their detection of a repressor protein that bound IPTG.

The lactose repressor was finally purified to homogeneity by the Gilbert group and several other research groups in the early 1970s. Gilbert used the repressor problem to develop several new methods to define the DNA binding sites of proteins that bound DNA specifically. For example, his group found that when dimethylsulfate (DMS) was used to modify adenine or guanine residues in the DNA, the adjacent DNA backbone became more labile to cleavage in mild alkali. DMS methylates guanine residues at N-7 and adenine residues at N-3. If the treatment is done briefly so that only one A or G per DNA strand is labeled, on average, then subsequent cleavage will break each DNA strand at just one position. If all the strands are radioactively labeled at the same end, a banding pattern is produced that acts as a map of the A and G residues in the DNA strand (Figure 2; note that G residues are methylated preferentially and generate stronger bands). If a protein is bound to the DNA prior to addition of DMS, it partially protects the adenines and guanines from methylation in the region where it is bound. The resulting disruption of the base-methylation pattern helps to define the binding site of the protein; in Figure 2, the protein is the lactose repressor.

In this technique, the control lane (lane C in Figure 2) proved to be as important as the protein that was the subject of the experiment. As Gilbert and his colleague Allan Maxam looked at one of these gels, they realized that the gel could be read to reveal the positions of all the A and G residues in the DNA strand. If they could find a technique to break the strands at C and T residues as well, they would have a new way to sequence DNA. The methods were developed and published as a new sequencing technology in 1977. The Maxam-Gilbert DNA-sequencing procedure was eventually supplanted by the Sanger sequencing method, published in the same year; that method, in turn, has given way to newer technologies (see Chapter 7). The effects of all these advances were profound, and they helped define the science of molecular biology in their time.

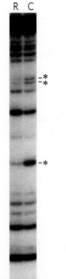

R C

FIGURE 2 In this polyacrylamide gel, the C lane contains a control to which no repressor was bound. The darker bands pinpoint G residues; the less intense bands result from cleavage at A residues. The R lane includes added repressor. Several bands (*) exhibit diminished intensity in the R lane, defining sites to which the repressor was bound. [Source: R. T. Ogata and W. Gilbert, *J. Mol. Biol.* 132:709–728, 1979. Photo courtesy of Ronald Ogata.]

KEY TERMS

enzyme, 134
ligand, 134
binding site, 134
induced fit, 134
cooperativity, 135
dissociation constant
(K_d), 135
cofactor, 142
coenzyme, 142
prosthetic group, 142
holoenzyme, 143
apoenzyme (apoprotein),
143
substrate, 143
active site, 143
biochemical standard
free-energy change
$(\Delta G'^\circ)$, 145
transition state, 146
maximum velocity
(V_{max}), 146

steady-state kinetics,
147
Michaelis-Menten
equation, 147
Michaelis constant
(K_m), 147
turnover number, 149
helicase, 151
regulatory enzyme,
158
allosteric enzyme
(allosteric protein),
158
allosteric modulator,
158
homotropic, 158
heterotropic, 158
autoinhibition, 160
protein kinase, 162
protein phosphatase,
162

PROBLEMS

1. Protein A has a binding site for ligand X with a K_d of 10^6 M. Protein B has a binding site for ligand X with a K_d of 10^9 M. Which protein has a higher affinity for ligand X? Explain your reasoning. Convert K_d to K_a for both proteins.

2. The lactose (Lac) repressor has a binding site for DNA and binds to a specific DNA site with a K_d of approximately 10^{-10} M. The Lac repressor also has a binding site for the galactoside allolactose. When the Lac repressor interacts with allolactose, it dissociates from the DNA. When allolactose is bound to the Lac repressor, does the K_d for the repressor's specific DNA binding site increase or decrease? Explain.

3. The Lac repressor interacts with DNA through its helix-turn-helix motif, primarily through the recognition helix. If amino acid residues in the recognition helix that interact with DNA are changed, is the K_d for the repressor's specific DNA binding site likely to increase or decrease? Explain.

4. The binding of any protein to DNA invariably involves the displacement of other atoms or molecules. What types of atoms or molecules are most commonly displaced?

5. When a protein binds to DNA nonspecifically, with which parts of the DNA does the protein generally interact? When a protein binds to DNA at a site defined by a particular nucleotide sequence, with which parts of the DNA does the protein generally interact?

6. Which of the following situations would produce negative cooperativity? Explain your reasoning in each case.

   (a) The protein has multiple subunits, each with a single ligand-binding site. The binding of ligand to one site decreases the binding affinity of other sites for the ligand.
   (b) The protein is a single polypeptide with two independent ligand-binding sites, each having a different affinity for the ligand.
   (c) The protein is a single polypeptide with a single ligand-binding site. As purified, the protein preparation is heterogeneous, containing some protein molecules that are partially denatured and thus have a lower binding affinity for the ligand.

7. To approximate the actual concentration of enzymes in a bacterial cell, assume that the cell contains equal concentrations of 1,000 different enzymes in solution in the cytosol and that each protein has a molecular weight of 100,000. Assume also that the bacterial cell is a cylinder (diameter 1.0 μm, height 2.0 μm), that the cytosol (specific gravity 1.20) is 20% soluble protein by weight, and that the soluble protein consists entirely of enzymes. Calculate the *average* molar concentration of each enzyme in this hypothetical cell.

8. Which of the following effects would be brought about by any enzyme catalyzing the simple reaction $S \underset{k_2}{\overset{k_1}{\rightleftharpoons}} P$, where $K_{eq} = [P]/[S]$? (a) Decreased K_{eq}; (b) increased k_1; (c) increased K_{eq}; (d) increased $\Delta G^\ddagger$; (e) decreased $\Delta G^\ddagger$; (f) more negative $\Delta G'^\circ$; (g) increased k_2.

9. In the late nineteenth century, the famous chemist Emil Fischer proposed that an enzyme should possess a pocket that is complementary in shape to the substrate it interacts with in its catalytic function. This "lock and key" hypothesis was highly influential at the time. Although Fischer made many important contributions to the development of enzymology, this particular idea was largely wrong. Explain why.

10. If an irreversible inhibitor inactivates an enzyme, what is the effect on that enzyme's k_{cat} and K_m?

11. (a) At what substrate concentration would an enzyme with a k_{cat} of 30.0 s^{-1} and a K_m of 0.0050 M operate at one-quarter of its maximum rate?
    (b) Determine the fraction of V_{max} that would be achieved at the following substrate concentrations: [S] = $\frac{1}{2}K_m$, $2K_m$, and $10K_m$.
    (c) An enzyme that catalyzes the reaction X $\rightleftharpoons$ Y is isolated from two bacterial species. The two enzymes have the same V_{max} but different K_m values for the

substrate X. Enzyme A has a K_m of 2.0 μM, and enzyme B has a K_m of 0.5 μM. The plot below shows the kinetics of reactions carried out with the same concentration of each enzyme and with [X] = 1 μM. Which curve corresponds to which enzyme?

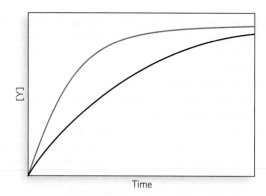

12. A research group discovers a new enzyme, which they call happyase, that catalyzes the chemical reaction HAPPY $\rightleftharpoons$ SAD. The researchers begin to characterize the enzyme.

   (a) In the first experiment, with a total concentration of enzyme, $[E_t]$, at 4 nM, they find that $V_{max} = 1.6$ μM s^{-1}. Based on this experiment, what is the k_{cat} for happyase? (Include the appropriate units.)

   (b) In another experiment, with $[E_t]$ at 1 nM and [HAPPY] at 30 μM, the researchers find that $V_0 = 300$ nM s^{-1}. What is the measured K_m of happyase for its substrate HAPPY? (Include the appropriate units.)

   (c) Further research shows that the supposedly purified happyase used in the first two experiments was actually contaminated with a reversible inhibitor called ANGER. When ANGER is carefully removed from the happyase preparation, and the two experiments are repeated, the measured V_{max} in (a) increases to 4.8 μM s^{-1}, and the measured K_m in (b) is 15 μM. For the inhibitor ANGER present in the original preparation, calculate the values of α and α′ (see Highlight 5-1).

   (d) Based on the information given above, what type of inhibitor is ANGER?

13. An enzyme is discovered that catalyzes the reaction A $\rightleftharpoons$ B. Researchers find that the K_m for substrate A is 4 μM, and the k_{cat} is 20 min^{-1}.

   (a) In an experiment, [A] = 6 mM, and the initial velocity, V_0, is measured as 480 nM min^{-1}. What is the $[E_t]$ used in the experiment?

   (b) In another experiment, $[E_t]$ = 0.5 nM, and the measured V_0 = 5 nM min^{-1}. What is the [A] used in the experiment?

   (c) The compound Z is found to be a very strong competitive inhibitor of the enzyme. In an experiment with the same $[E_t]$ as in part (a) but a different [A],

an amount of Z is added that produces an α = 10. This reduces V_0 to 240 nM min^{-1}. What is the [A] in this experiment?

14. Although graphical methods are available for accurate determination of the V_{max} and K_m of an enzyme-catalyzed reaction (see the Additional Reading list), sometimes these quantities can be quickly estimated by inspecting values of V_0 at increasing [S]. Estimate the V_{max} and K_m of the enzyme-catalyzed reaction for which the following data were obtained.

[S] (m)	V_0 (μM/min)
2.5×10^{-6}	28
4.0×10^{-6}	40
1×10^{-5}	70
2×10^{-5}	95
4×10^{-5}	112
1×10^{-4}	128
2×10^{-3}	139
1×10^{-2}	140

15. The bacterial RuvB protein, a DNA translocase, belongs to helicase superfamily 6. RuvB functions as a circular hexameric complex with a central opening. Duplex DNA is bound in the opening. RuvB is also an ATPase, and it moves along the DNA when ATP is hydrolyzed. An amino acid substitution in the ATP-binding site of RuvB generates a protein that binds to DNA but does not hydrolyze ATP and does not translocate along the DNA. When normal RuvB protein subunits are mixed with an equal amount of mutant RuvB subunits, heterohexamer complexes are formed that contain both normal and mutant subunits. These heterohexamers can hydrolyze ATP but do not move along the DNA. What conclusions can you draw from these observations?

16. When eukaryotic DNA replication is prematurely halted, due to DNA damage or other causes, two proteins—ATM (ataxia telangiectasia mutated) and ATR (ATM related)—initiate a response called a checkpoint, which involves regulated changes in the functions of many cellular proteins to facilitate DNA repair. ATM and ATR are enzymes with a regulatory function. They alter the covalent structure of the proteins they regulate, increasing their measured molecular weight. Suggest what kind of enzymatic activity they might possess.

17. A biochemist is purifying a new DNA helicase encoded by a pathogenic virus. The enzyme activity is readily detected after the first several purification steps that generate fractions *a*, *b*, and *c*, each with successively greater purity.

The researcher then puts fraction c on an ion-exchange column, and this process generates two new fractions, d and d'. The helicase activity cannot be detected in either of the new fractions. However, when fractions d and d' are mixed, the activity reappears. Explain.

ADDITIONAL READING

General

Kornberg, A. 1989. Never a dull enzyme. *Annu. Rev. Biochem.* 58:1–30. An especially illuminating essay for young scientists.

Kornberg, A. 1990. Why purify enzymes? *Methods Enzymol.* 182:1–5.

Kornberg, A. 1996. Chemistry: The lingua franca of the medical and biological sciences. *Chem. Biol.* 3:3–5. This and the two articles above provide inspiration from one of the great biochemists of the past century.

Nelson, D.L., and M.M. Cox. 2013. *Lehninger Principles of Biochemistry*, 6th ed. New York: W. H. Freeman. See Chapters 3 through 6 for more detailed background on enzyme kinetics.

von Hippel, P.H. 2007. From "simple" DNA-protein interactions to the macromolecular machines of gene expression. *Annu. Rev. Biophys. Biomol. Struct.* 36:79–105.

Protein-Ligand Interactions

Jayaram, B., and T. Jain. 2004. The role of water in protein-DNA recognition. *Annu. Rev. Biophys. Biomol. Struct.* 33:343–361.

Kalodimos, C.G., N. Biris, A.M.J.J. Bonvin, M.M. Levandoski, M. Guennuegues, R. Boelens, and R. Kaptein. 2004. Structure and flexibility adaptation in nonspecific and specific protein-DNA complexes. *Science* 305:386–389.

Lohman, T.M., and D.P. Mascotti. 1992. Thermodynamics of ligand–nucleic acid interactions. *Methods Enzymol.* 212:400–424.

Raghunathan, S., A.G. Kozlov, T.M. Lohman, and G. Waksman. 2000. Structure of the DNA binding domain of *E. coli* SSB bound to ssDNA. *Nat. Struct. Biol.* 7:648–652.

Enzymes: The Reaction Catalysts of Biological Systems

Arabshahi, A., and P.A. Frey. 1999. Standard free energy for the hydrolysis of adenylylated T4 DNA ligase and the apparent pK_a of lysine 159. *J. Biol. Chem.* 274:8586–8588.

Ellenberger, T., and A.E. Tomkinson. 2008. Eukaryotic DNA ligases: Structural and functional insights. *Annu. Rev. Biochem.* 77:313–338. A complete summary of many details gleaned from structural analysis.

Lehman, I.R. 1974. DNA ligase: Structure, mechanism, and function. *Science* 186:790–797.

Liu, P., A. Burdzy, and L.C. Sowers. 2004. DNA ligases ensure fidelity by interrogating minor groove contacts. *Nucleic Acids Res.* 32:4503–4511.

Motor Proteins

Lohman, T.M., and K.P. Bjornson. 1996. Mechanisms of helicase-catalyzed DNA unwinding. *Annu. Rev. Biochem.* 65:169–214.

Lohman, T.M., E.J. Tomko, and C.G. Wu. 2008. Nonhexameric DNA helicases and translocases: Mechanisms and regulation. *Nat. Rev. Mol. Cell Biol.* 9:391–401.

Pyle, A.M. 2008. Translocation and unwinding mechanisms of RNA and DNA helicases. *Annu. Rev. Biophys.* 37:317–336.

Singleton, M.R., M.S. Dillingham, and D.B. Wigley. 2007. Structure and mechanism of helicases and nucleic acid translocases. *Annu. Rev. Biochem.* 76:23–50.

Regulation of Protein Function

Bialik, S., and A. Kimchi. 2006. The death-associated protein kinases: Structure, function, and beyond. *Annu. Rev. Biochem.* 75:189–210.

Elphick, L.M., S.E. Lee, V. Gouverneur, and D.J. Mann. 2007. Using chemical genetics and ATP analogues to dissect protein kinase function. *ACS Chem. Biol.* 2:299–314.

Gelato, K.A., and W. Fischle. 2008. Role of histone modifications in defining chromatin structure and function. *Biol. Chem.* 389:353–363.

Martin, C., and Y. Zhang. 2005. The diverse functions of histone lysine methylation. *Nat. Rev. Mol. Cell Biol.* 6:838–849.

Millar, C.B., and M. Grunstein. 2006. Genome-wide patterns of histone modifications in yeast. *Nat. Rev. Mol. Cell Biol.* 7:657–666.

Moorhead, G.B.G., L. Trinkle-Mulcahy, and A. Ulke-Lemee. 2007. Emerging roles of nuclear protein phosphatases. *Nat. Rev. Mol. Cell Biol.* 8:234–244.

Shahbazian, M.D., and M. Grunstein. 2007. Functions of site-specific histone acetylation and deacetylation. *Annu. Rev. Biochem.* 76:75–100.

Tonks, N.K. 2006. Protein tyrosine phosphatases: From genes, to function, to disease. *Nat. Rev. Mol. Cell Biol.* 7:833–846.

6 DNA and RNA Structure

Jamie Cate [Source: Michael Barnes, UC Berkeley College of Chemistry.]

MOMENT OF DISCOVERY

When I first started my lab at the Whitehead Institute, my dream was to *crystallize the bacterial ribosome and to solve its molecular structure at high resolution.* Although a lot had been learned about RNA structure from work on catalytic RNAs and the individual subunits of the ribosome, the possibility of seeing the complete structure of the protein-synthesizing machinery was irresistible.

Working closely with graduate student Steve Santoso, we eventually got one small perfect-looking crystal to grow from a purified ribosome sample. We took the crystal to a synchrotron x-ray beamline and saw the first diffraction pattern indicating that the crystal indeed contained ribosomes. That was so exciting! But just to be sure, we recovered the crystal from the x-ray diffraction apparatus, dissolved it in water, and checked the contents on an agarose gel—and there was the ribosomal RNA, clear as could be.

We figured out how to grow more of those crystals, and eventually we solved the crystal structure of the complete ribosome. Staring in awe at the electron density map, we saw how the RNA helices wove through the molecule like great curving spiral staircases. I felt chills down my spine, realizing I was the first person to see such incredible molecular beauty. This was the culmination of six years of challenging—and at times frustrating—experiments. I felt the sweet joy of success, and also contemplated the many new discoveries that would result from this work.

—Jamie Cate, on determining the molecular structure of the bacterial ribosome

Online resources related to this chapter:

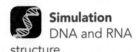

Simulation DNA and RNA structure

Nature exercise Polymerase chain reaction, DNA polymerization

Discovered in the nineteenth century, **DNA (deoxyribonucleic acid)** would be proposed, by the early twentieth century, as the molecule that stores biological information (see Chapter 2). At that time, however, the way in which the particular properties of its molecular structure could produce traits and behaviors in living organisms was unimaginable. By midcentury, hoping to determine how DNA carried genetic messages that are faithfully transmitted to the next generation when cells divide, researchers in several laboratories had made it their goal to solve the molecular structure of DNA. In 1953, James Watson and Francis Crick, at Cambridge University, used x-ray diffraction data obtained by Rosalind Franklin to deduce DNA's simple and beautiful double-helical structure (**Figure 6-1**). This landmark discovery, for which Watson and Crick (together with Maurice Wilkins, for his work on the x-ray diffraction) received the Nobel Prize in Physiology or Medicine in 1962, gave rise to all of modern molecular biology. It was immediately apparent to scientists how this unique structure of DNA could allow biological information to be easily and faithfully duplicated and transmitted from generation to generation.

Like DNA, **RNA (ribonucleic acid)** was first isolated in the nineteenth century from the nuclei of cells. Scientists later recognized that RNA is chemically distinct from DNA, because it contains a different kind of sugar in its nucleotide building blocks (see Chapter 3). As described in Chapter 2, ribosomal RNAs (rRNAs) were found to be components of ribosomes, the complexes that carry out protein synthesis. Messenger RNAs (mRNAs) were known to be intermediaries, carrying genetic information from genes to ribosomes. And transfer RNAs (tRNAs) had been identified as

James Watson
[Source: AP Photo.]

Francis Crick,
1916–2004
[Source: AP Photo]

adaptor molecules that translate the information in mRNA into a specific sequence of amino acids. We now know that RNA molecules have many other biological functions as well. For example, they comprise the genomes of certain viruses, such as the human immunodeficiency virus (HIV) and hepatitis C virus (HCV). Some RNA molecules have the ability to work as catalysts—a discovery that provided, for the first time, a plausible scenario for the evolution of early life forms based on self-replicating RNA. (The diversity of functional RNAs and their roles in evolution are discussed in Chapters 15 and 16.) In the quest to understand how RNA could perform such a range of functions, researchers have determined the structures of numerous types of RNA molecules and RNA-protein complexes, including the structure of the ribosome itself. Unlike DNA, RNA molecules are almost always single-stranded, and they consist of much shorter chains of nucleotides. They also have a propensity to fold back on themselves, creating many discrete double-helical regions that can assemble into complex three-dimensional structures.

As we will see, there is no single generic structure of DNA or RNA. Rather, there are numerous variations on a common structural theme, resulting from the chemical and physical properties of the polynucleotide chain. Indeed, the structural stability of DNA and the structural diversity of RNA explain why these molecules have evolved to function in all aspects of maintaining and transmitting biological information. In this chapter, we first explore the general properties of nucleotides, then turn to the structures of DNA and RNA. We conclude by looking at the chemical behavior of nucleic acids under biological conditions.

6.1 THE STRUCTURE AND PROPERTIES OF NUCLEOTIDES

All nucleic acids are chemically linked chains of nucleotides, the basic building blocks of DNA and RNA. To understand the structures, functions, and replication of nucleic acids, we first need to understand the structure of their nucleotide components and how they behave in the context of a DNA or RNA polymer. We therefore begin our discussion of DNA and RNA by considering the nature of the nucleotide.

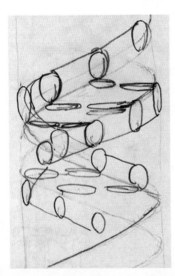

FIGURE 6-1 Francis Crick's first drawing of DNA structure. Two base-paired strands of DNA form a helical structure in which the phosphate and sugar groups are on the outside and the bases are on the inside. The helix twists in a right-handed direction. [Source: © Photo Researchers/Alamy.]

Nucleotides Comprise Phosphates and Characteristic Bases and Sugars

A **nucleotide** is a molecule consisting of three characteristic components: a heterocyclic base, a five-carbon sugar called a pentose, and a phosphate group. (A **heterocyclic compound** is a cyclic compound with one or more ring structures that contain atoms of at least two different elements.) The same molecule without the phosphate group is called a **nucleoside**. Each base is a derivative of one of two parent compounds, a **purine** or a **pyrimidine** (**Figure 6-2a**), which are nitrogenous (nitrogen-containing) bases. They are called bases because free purines and pyrimidines are weakly basic compounds. The carbon and nitrogen atoms in the parent structures are numbered according to convention to facilitate the naming and identification of the many derivative compounds. The carbon atoms in the pentose are also numbered; in nucleotides and nucleosides, these numbers are given a prime (′) designation to distinguish the sugar carbons from the numbered carbon and nitrogen atoms of the nitrogenous bases.

In nucleosides, the covalent joining of a base (at N-9 of purines and N-1 of pyrimidines) to the 1′ carbon (C-1′) of the pentose forms a **glycosidic bond** (specifically, an N-β-glycosyl bond), which involves the loss of a molecule of water. To form a nucleotide, a phosphate group is covalently joined to the 5′ carbon (C-5′) of the pentose to form an ester, also with the concomitant loss of a water molecule (**Figure 6-2b**).

Four different bases are found in DNA: two are purines, **adenine (A)** and **guanine (G)**, and two are pyrimidines, **cytosine (C)** and **thymine (T)**. RNA also contains four types of bases. The two purines are the same as those in DNA: adenine and guanine; and, as in DNA, one of the pyrimidines is cytosine. However, the second major pyrimidine in RNA is **uracil (U)** instead of thymine. Only rarely does thymine occur in RNA, or uracil in DNA. The structures of the five major bases are shown in Figure 6-2a; the nomenclature of their corresponding nucleotides and nucleosides was summarized in Chapter 3 (see Table 3-1).

Nucleic acids have two kinds of pentoses. The recurring nucleotide units of DNA contain 2′-deoxy-D-ribose, whereas the nucleotide units of RNA contain D-ribose. The D-ribose has a hydroxyl group attached to the 2′ carbon, whereas 2′-deoxy-D-ribose lacks this functional group (see Figure 6-2b). In nucleotides, both types of pentoses are in their β-furanose (closed five-membered ring) form (**Figure 6-3a**). As **Figure 6-3b** shows, the pentose ring is not planar but exists in one of a variety of conformations generally described as "puckered." The predominant type of sugar pucker that characterizes DNA differs from that found in RNA, resulting in the different shapes and geometries of the DNA and RNA double helices, as we describe later in this chapter.

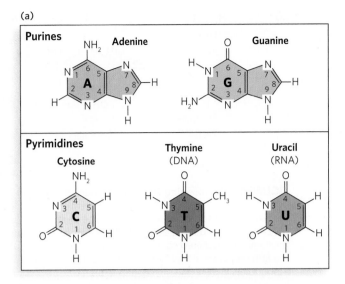

(a)

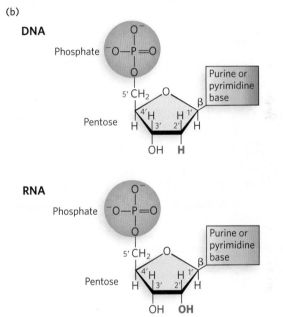

(b)

FIGURE 6-2 The chemical composition of nucleotides. (a) The bases are purines, with nine-membered rings, or pyrimidines, with six-membered rings, with numbering systems as shown. In DNA and RNA, the purines are adenine and guanine; in DNA, the pyrimidines are cytosine and thymine; in RNA, the pyrimidines are cytosine and uracil. The ring atoms in the bases that connect to the ribose (N-9 for purines and N-1 for pyrimidines) are indicated with bold red numbers. (b) Nucleotides consist of a phosphate group, a pentose sugar, and a heterocyclic base; carbons in the pentose rings are numbered as shown, with numbers followed by a prime (′) to distinguish them from the numbered atoms of the bases. In DNA, the pentose is 2′-deoxyribose, which has no hydroxyl group on the 2′ carbon (red); in RNA, the sugar is ribose, which has a 2′ hydroxyl (red). A glycosidic bond links the 1′ carbon of ribose or deoxyribose to the base; β indicates the configuration of the base relative to the pentose ring.

Because of their different pentose components, the structural units of DNAs and RNAs are known as **deoxyribonucleotides** (or deoxyribonucleoside 5′-monophosphates) and **ribonucleotides** (or ribonucleoside 5′-monophosphates), respectively (**Figure 6-4**).

(a)

D-Ribose

β-D-Furanose

(b)

C-2' endo
predominates in DNA

C-3' endo
predominates in RNA

FIGURE 6-3 Pentose ring structures in nucleic acids. (a) The linear and closed-ring forms of ribose are in equilibrium in solution. When incorporated into nucleosides, nucleotides, or polynucleotides, the pentose exists only in the ring form. The pentose ring is formed by reaction of the hydroxyl group on C-4 with the aldehyde at C-1. (b) The pentose rings in nucleosides and nucleotides can exist in four predominant puckered conformations. In each case, four of the five ring atoms are nearly coplanar, but the fifth ring atom, either C-2' or C-3', is out of the plane. The C-2' endo configuration, in which the C-2' atom points in the same direction as the C-5' atom, predominates in DNA. The C-3' endo configuration, in which the C-3' atom points in the same direction as the C-5' atom, predominates in RNA.

(a) Deoxyribonucleotides

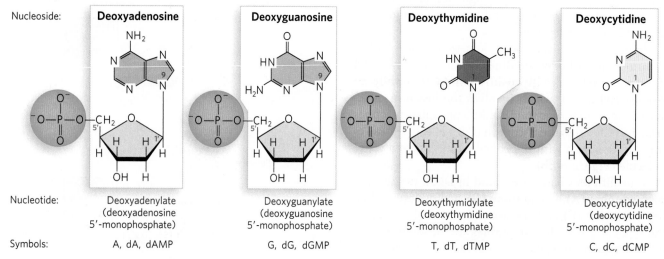

Nucleoside: **Deoxyadenosine**   **Deoxyguanosine**   **Deoxythymidine**   **Deoxycytidine**

Nucleotide: Deoxyadenylate   Deoxyguanylate   Deoxythymidylate   Deoxycytidylate
                (deoxyadenosine   (deoxyguanosine   (deoxythymidine   (deoxycytidine
                5'-monophosphate)   5'-monophosphate)   5'-monophosphate)   5'-monophosphate)

Symbols: A, dA, dAMP   G, dG, dGMP   T, dT, dTMP   C, dC, dCMP

(b) Ribonucleotides

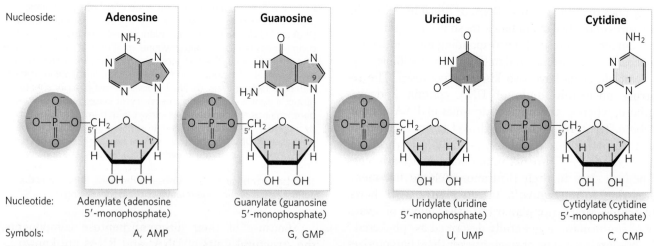

Nucleoside: **Adenosine**   **Guanosine**   **Uridine**   **Cytidine**

Nucleotide: Adenylate (adenosine   Guanylate (guanosine   Uridylate (uridine   Cytidylate (cytidine
                5'-monophosphate)   5'-monophosphate)   5'-monophosphate)   5'-monophosphate)

Symbols: A, AMP   G, GMP   U, UMP   C, CMP

FIGURE 6-4 Deoxyribonucleotides and ribonucleotides of nucleic acids. All nucleotides are shown in their predominant form at neutral pH. Note that the nucleoside is the boxed portion and the nucleotide is the form with the phosphate group included. (a) Deoxyribonucleotides of DNA. (b) Ribonucleotides of RNA.

Although the major purine and pyrimidine nucleotides are the most common, both DNA and RNA molecules also contain some minor bases. In DNA, the minor bases are usually methylated forms of the major bases. These unusual bases in DNA molecules often have roles in regulating or protecting the genetic information. Minor bases of many types are also found in RNA molecules, particularly in tRNAs, rRNAs, and other RNAs whose function requires a specific three-dimensional structure. In cells, minor bases in RNA can be formed by enzymatic modification of one of the common nucleotides to add or remove a functional group, or by complete replacement of a standard base with a less common one. Chemical modifications of DNA and RNA and their effects on nucleotide structure and function are discussed in Section 6.4.

Cells also contain nucleotides with phosphate groups in positions other than on the 5′ carbon (**Figure 6-5**). For example, **ribonucleoside 2′,3′-cyclic monophosphates** are stable intermediates, and **ribonucleoside 2′-monophosphates** or **ribonucleoside 3′-monophosphates** are the end products of the hydrolysis of RNA

by enzymes called **ribonucleases**. Other variations are adenosine 3′,5′-cyclic monophosphate (cAMP) and guanosine 3′,5′-cyclic monophosphate (cGMP), which are important chemical signals of the metabolic state of the cell (further discussed later in this section).

Phosphodiester Bonds Link the Nucleotide Units in Nucleic Acids

The successive nucleotides of DNA and RNA are covalently joined through phosphate group "connectors" in which the 5′-phosphate group of one nucleotide unit is linked to the 3′-hydroxyl group of the next nucleotide, creating a **phosphodiester bond** (**Figure 6-6**); this involves the loss of water, and the joined nucleotides are therefore referred to as "residues." As we will see, these 5′-to-3′ links give every DNA or RNA chain a directionality, or polarity. The alternating phosphate and sugar residues form the backbone of the nucleic acid, and the bases can be viewed as side groups joined to this sugar–phosphate backbone at regular intervals.

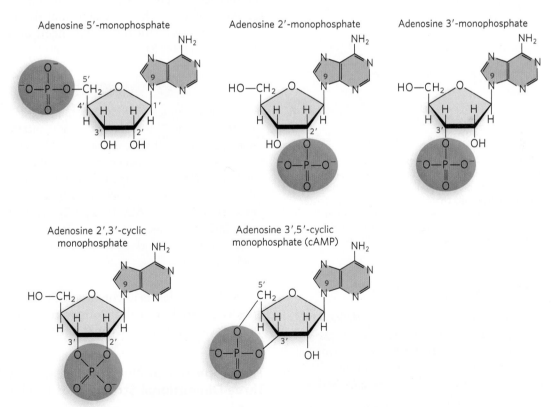

FIGURE 6-5 Examples of adenosine monophosphates. Adenosine 5′-monophosphate, with a phosphate group on C-5′, is the most common adenine-containing nucleotide and the one found in RNA. Adenosine 2′-monophosphate, adenosine 3′-monophosphate, and adenosine 2′,3′-cyclic monophosphate are formed during enzymatic or alkaline hydrolysis of RNA (see Figure 6-7). Adenosine 3′,5′-cyclic monophosphate (cAMP) is a signaling molecule that accumulates when the cell has a limited supply of nutrients.

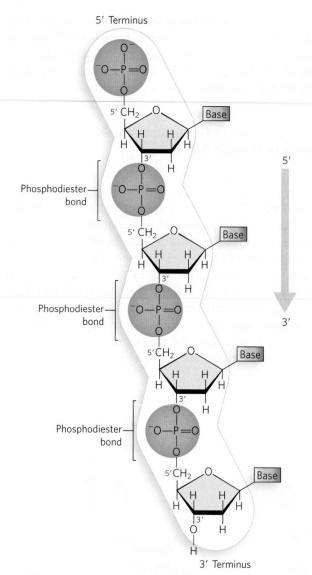

FIGURE 6-6 The phosphodiester linkages in nucleic acids. Phosphodiester bonds covalently connect the nucleotide units in DNA and RNA. The backbone of alternating sugars and phosphate groups is highly negatively charged. All the phosphodiester linkages in a polynucleotide chain have the same orientation, giving each linear nucleic acid strand a specific polarity and distinct 5′ and 3′ termini. Shown here is a strand of DNA.

KEY CONVENTION

The polarity of a single DNA or RNA chain is defined by the chemical groups—a free 5′ phosphate or 3′ hydroxyl— at the termini of the chain, not by the 5′ and 3′ oxygens of internal phosphodiester bonds. Linear DNA and RNA molecules each have a single unique 5′ terminus and 3′ terminus.

The backbone of both DNA and RNA is hydrophilic. The hydroxyl groups of the sugar residues form hydrogen bonds with water. The phosphate groups, with a pK_a

near 2, are completely ionized and negatively charged at pH 7, and the negative charges are generally neutralized by ionic interactions with positive charges on proteins, metal ions, or short, linear organic molecules called polyamines, which contain two or more amine groups.

The covalent backbone of DNA and RNA is subject to slow, nonenzymatic hydrolysis of the phosphodiester bonds. In the test tube, RNA is hydrolyzed rapidly under alkaline conditions, but DNA is not; the 2′-hydroxyl group on the sugars in RNA is directly involved in the hydrolytic process. Cyclic 2′,3′-monophosphates are the first products of the action of alkali on RNA, and these are rapidly hydrolyzed further to yield a mixture of nucleoside 2′- and 3′-monophosphates (**Figure 6-7**). The sugar component of DNA does not have a 2′-hydroxyl group and is not as easily hydrolyzed, making the DNA backbone inherently more stable than that of RNA. The inherent instability of RNA relative to DNA serves a purpose in gene expression: RNA molecules can be synthesized and degraded many times during the life of a cell, whereas the corresponding DNA is maintained during cell division and during extended periods in nonreplicating cells. It is the stability of DNA that allows the recovery and analysis of DNA from historically preserved samples, such as Neanderthal tissues, which in recent years has enabled the sequencing of Neanderthal DNA (see Chapter 8).

KEY CONVENTION

The structure of a single strand of nucleic acid is always written with the 5′ terminus at the left and the 3′ terminus at the right—that is, in the 5′→3′ direction. When a double-stranded sequence is shown, the top strand is written in the 5′→3′ direction. The various representations of a nucleotide sequence, using a pentanucleotide as example, are: 5′-ACGTA-3′, ACGTA, pA-C-G-T-A$_{OH}$, pApCpGpTpA, and pACGTA, where *p* denotes a monophosphate, and a subscript *OH* denotes a 3′-hydroxyl group.

A short nucleic acid containing 50 or fewer nucleotides is generally called an **oligonucleotide**. A longer nucleic acid is called a **polynucleotide**.

The Properties of Nucleotide Bases Affect the Three-Dimensional Structure of Nucleic Acids

Purines and pyrimidines have a variety of chemical properties that affect the structure, and ultimately the function, of nucleic acids. The purine and pyrimidine bases common in DNA and RNA are conjugated ring systems, with alternating single and double bonds

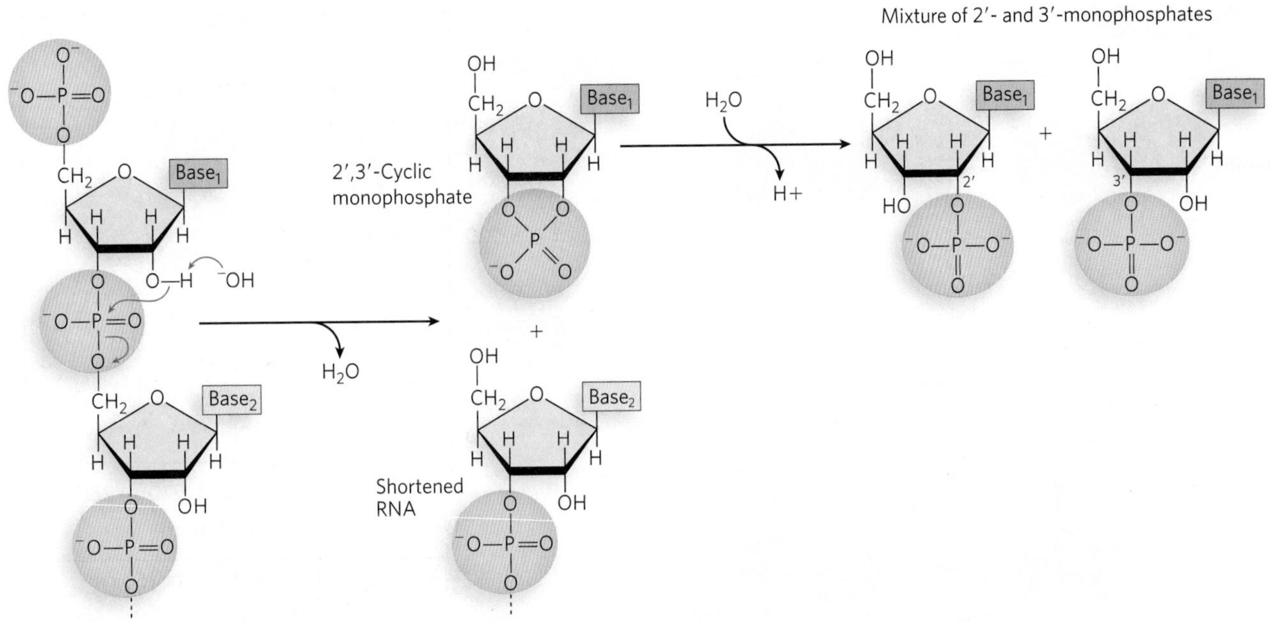

FIGURE 6-7 The hydrolysis of RNA. The 2'-hydroxyl group can be activated as a nucleophile under alkaline (pH >7) conditions or by ribonucleases. The 2',3'-cyclic monophosphate product is further hydrolyzed to a mixture of 2'- and 3'-monophosphates.

between ring atoms (see Figure 6-2). Resonance among atoms in the rings gives most of the bonds a partial double-bond character. One result is that pyrimidines are planar molecules and purines are very nearly planar, with just a slight pucker. Free pyrimidine and purine bases can exist in two or more forms, called tautomers, depending on pH (**Figure 6-8**). The structures shown on the left in Figure 6-8 are the tautomers that predominate at physiological pH (pH ~7). As a result of resonance, delocalized electrons in the conjugated rings are available to absorb ultraviolet (UV) light at wavelengths near 260 nm (**Figure 6-9**). Ultraviolet light absorbance is used as a method for detecting nucleic acids (see Section 6.4).

The chemical properties of the purines and pyrimidines also give rise to two important modes of interaction between bases in nucleic acids. The first, called **hydrophobic stacking**, arises because the bases are hydrophobic and thus relatively insoluble in water at the near-neutral pH of the cell. As a result, the bases align such that two or more are positioned with the planes of their rings in parallel, like a stack of coins (**Figure 6-10**). Base stacking helps minimize the contact of the bases with water, and base-stacking interactions are very important in stabilizing the three-dimensional structure of nucleic acids. Such stacking also involves a combination of van der Waals and electrostatic interactions among the bases.

The second important mode of base interaction in nucleic acids is **base pairing**, which results from the hydrogen-bonding capacity of the ring nitrogens, ring

carbonyl groups, and exocyclic (i.e., outside the ring structure) amino groups of the pyrimidines and purines. Hydrogen bonds between bases, involving the amino and carbonyl groups, permit a complementary association of two (and occasionally three or four) nucleic acid strands. The most important hydrogen-bonding patterns are those defined by Watson and Crick in 1953, in which A hydrogen-bonds specifically with T (or U), and G with C (**Figure 6-11** on p. 181). These two types of base pairs predominate in double-stranded DNA and RNA (and thus are considered the canonical base pairs). The purine and pyrimidine tautomers that predominate at physiological pH, shown on the left in Figure 6-8, readily adopt these hydrogen-bonding patterns. It is this specific pairing of bases in the double-stranded DNA helix that permits the duplication of genetic information.

Nucleotides Play Additional Roles in Cells

As we have mentioned, nucleotides have functions in cells beyond providing the building blocks for DNA and RNA. The phosphate group covalently linked to the 5' hydroxyl of a nucleoside may have one or two additional phosphates attached. The resulting molecules are referred to as nucleoside mono-, di-, and triphosphates (**Figure 6-12** on p. 181). Starting from the phosphate closest to the ribose, the three phosphates are generally labeled α, β, and γ. Nucleoside triphosphates are the activated precursors of DNA and RNA synthesis (see Chapters 11 and 15). Furthermore, transfer of the

Predominant forms at physiological pH

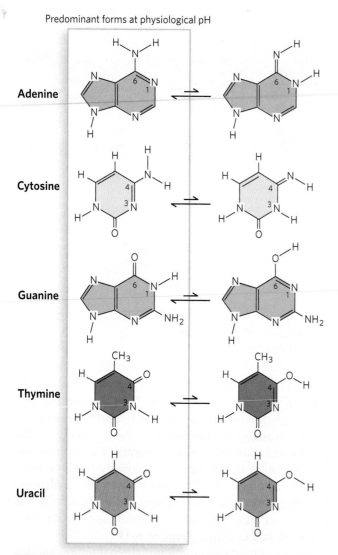

FIGURE 6-8 Tautomers of pyrimidine and purine bases. Each purine or pyrimidine can exist as one of several isomers that differ in the placement of a hydrogen atom and a double bond (tautomers). Shown here are two tautomers for each of the common bases found in nucleic acids. The predominant tautomer of each base at physiological pH is on the left. These predominant tautomeric forms are found in DNA and RNA and participate in canonical Watson-Crick base pairing (see Figure 6-11).

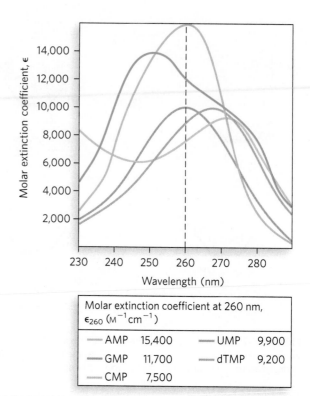

Molar extinction coefficient at 260 nm, ϵ_{260} ($M^{-1}cm^{-1}$)			
—— AMP	15,400	—— UMP	9,900
—— GMP	11,700	—— dTMP	9,200
—— CMP	7,500		

FIGURE 6-9 Absorption spectra of the common nucleotides. The plots show molar extinction coefficients at pH 7.0 as a function of wavelength for the nucleoside 5'-monophosphates. The molar extinction coefficient, ε (epsilon; units $m^{-1}cm^{-1}$), measures the amount of light absorbed by a 1 M solution with a light path length of 1 cm. The table shows the molar extinction coefficients at 260 nm for the plotted nucleotides.

form of vitamin B_2 (riboflavin), which transfers electrons in some biosynthetic reactions. Enzymatic reactions that involve the transfer of a methyl group from one molecule to another often involve the substrate *S*-adenosylmethionine (adoMet), which consists of an adenosine linked to a methionine.

γ-phosphoryl group from a nucleoside 5'-triphosphate (typically ATP) to another molecule provides the chemical energy to drive a wide variety of cellular reactions (as discussed in Chapter 3).

The nucleoside adenosine also forms part of the structure of otherwise unrelated enzyme cofactors that perform a wide range of chemical functions (**Figure 6-13**). For example, nicotinamide adenine dinucleotide (NAD⁺) plays a crucial role in cellular energy production in both animal cells and plant cells. A related cofactor, NADP⁺, contributes to the synthesis of lipids and nucleic acids, and participates in photosynthesis. Flavin adenine dinucleotide (FAD) is the active

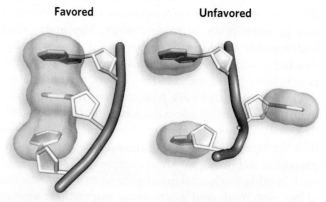

Favored **Unfavored**

FIGURE 6-10 Base stacking in nucleic acids. Hydrophobic, van der Waals, and electrostatic interactions favor the alignment of bases in an aqueous solution or within a polynucleotide chain (three nucleotides in an RNA chain are shown here); the unstacked orientation is unfavored. Van der Waals radii are shown in gray.

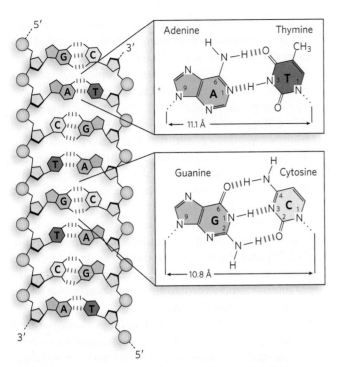

FIGURE 6-11 Hydrogen-bonding patterns in canonical Watson-Crick base pairs. Hydrogen bonds are represented by three blue lines.

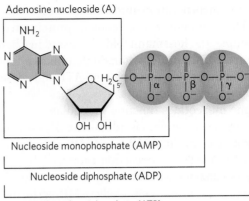

FIGURE 6-12 Nomenclature for nucleotides. The phosphate group covalently linked to the 5' carbon of a nucleoside may have one or two additional phosphates attached; the resulting nucleotide molecules are referred to as nucleoside mono-, di-, and triphosphates. Starting from the phosphate closest to the ribose, the three phosphates are designated α, β, and γ.

In these adenosine-containing compounds, the adenosine portion does not participate directly in the molecule's primary function. Instead, it seems to be a molecular "handle" that allows the cofactor or substrate to bind tightly in an enzyme active site. Adenosine may

have taken on this role partly because of the abundance of adenine in the environment of the early Earth (see the How We Know section in Chapter 1). The importance of adenosine probably lies not so much in some special chemical characteristic as in the evolutionary advantage of using one compound for multiple purposes. Once ATP became the universal source of chemical energy, biological systems developed to synthesize ATP in greater abundance than the other nucleotides; because adenosine was abundant, it became the logical

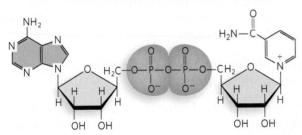

FIGURE 6-13 Some cofactors and a substrate containing adenosine. The adenine base (blue) and the ribose attached to it form the adenosine portion of each molecule. See text for details.

choice for incorporation into a wide variety of structures. This economy also extends to protein structure. For example, the Rossmann fold (discussed in Chapter 4; see Figure 4-15), a protein domain that binds adenosine, is found in many enzymes that bind ATP and adenosine-containing enzyme cofactors.

Some nucleotides function as regulatory molecules. One of the most common is **adenosine 3′,5′-cyclic monophosphate** (**cyclic AMP**, or **cAMP**) (see Figure 6-5), formed from ATP in a reaction catalyzed by adenylyl cyclase—an enzyme whose activity is closely linked to the metabolic state of the cell. Cyclic AMP performs regulatory functions in virtually every cell outside the plant kingdom. **Guanosine 3′,5′-cyclic monophosphate (cGMP)** occurs in many cells and also has regulatory functions.

Both cAMP and cGMP are called **second messengers**, because they are produced or degraded in response to the interaction of extracellular chemical signals ("first messengers") with receptors on the cell surface. The second messengers induce adaptive changes in the cell interior. In this way, cells can respond quickly to environmental changes by taking cues from hormones or other external chemical signals. For example, light entering the photoreceptors of the human eye activates an enzyme that degrades cGMP, causing sodium channels in the photoreceptor cell membrane to close and thereby triggering visual information to be sent to the brain. Another regulatory nucleotide, guanosine tetraphosphate (ppGpp), is produced in bacteria in response to a slowdown in protein synthesis during amino acid starvation. This nucleotide inhibits synthesis of the rRNA and tRNA molecules needed for protein synthesis, preventing the unnecessary production of nucleic acids.

SECTION 6.1 SUMMARY

- A nucleotide consists of a nitrogenous base (a purine or a pyrimidine), a pentose sugar, and one or more phosphate groups. Nucleic acids are polymers of nucleotide units, joined by phosphodiester linkages between the 3′-phosphate group of one unit and the 5′-hydroxyl group of the next. Polynucleotides have a directionality defined by a 5′ terminus and a 3′ terminus.
- DNA and RNA are two types of nucleic acids. The nucleotides in DNA contain 2′-deoxy-D-ribose, whereas the nucleotides in RNA contain D-ribose. The hydroxyl group at the 2′ position in D-ribose makes the RNA backbone more susceptible to hydrolysis than DNA.
- Both DNA and RNA contain four different bases, two purines and two pyrimidines. The purines in DNA and RNA are the same: adenine and guanine. DNA contains the pyrimidines cytosine and thymine, and RNA contains the pyrimidines cytosine and uracil.

- In addition to A, G, C, T, and U, numerous minor bases occur in nature, often differing from the canonical bases by the presence of a functional group at a particular position on the base; these bases can play central roles in nucleic acid structure and biochemical function.
- The chemical properties of the nitrogenous bases affect nucleotide and nucleic acid structure. As a result of resonance, the bases in a nucleotide chain are planar and tend to stack. The hydrogen-bonding capabilities of the conjugated rings allow the formation of specific base-pair interactions: A pairs with T (or U), and C pairs with G.
- Adenosine is a building block for some important enzyme cofactors, such as nicotine adenine dinucleotide (NAD^+) and flavin adenine dinucleotide (FAD). The presence of an adenosine component in a variety of cofactors enables recognition by enzymes that share common structural features.
- Cyclic AMP, formed from ATP in a reaction catalyzed by adenylyl cyclase, is a common second messenger produced in response to hormones and other chemical signals.

6.2 DNA STRUCTURE

In Chapter 2 we discussed the experiments that revealed DNA as the genetic material in cells. Recognizing that DNA is the primary carrier of biological information in cells and viruses, researchers were motivated to determine its molecular structure. What they learned about DNA's structure explained how it functions as the key molecule of inheritance, thereby paving the way for many investigations into the mechanisms of DNA replication and metabolism.

DNA Molecules Have Distinctive Base Compositions

In the 1940s, Erwin Chargaff and his colleagues made an important discovery that provided clues to the structure of DNA. Using DNA samples isolated from many different organisms, they observed that the four bases of DNA occur in different ratios that are characteristic of each species. They also observed that for specific *pairs* of bases, the amounts of each base are closely related. Their data showed that:

1. The base composition of DNA generally varies from one species to another.

2. DNA specimens isolated from different tissues of the same species have the same base composition.

3. The base composition of DNA in a given species does not change with an organism's age, nutritional state, or environment.

4. In *all* cellular DNAs, regardless of the species, the number of adenosine residues equals the number of thymidine residues (i.e., A = T), and the number of guanosine residues equals the number of cytidine residues (G = C). From these relationships it follows that the sum of the purine residues equals the sum of the pyrimidine residues: A + G = T + C.

Referred to as **Chargaff's rules**, these quantitative relationships were confirmed by many subsequent researchers. Not only were these findings a key to establishing the three-dimensional structure of DNA, but they also yielded clues about how genetic information is encoded in DNA and passed along from one generation to the next.

DNA Is Usually a Right-Handed Double Helix

Chargaff's discoveries in the 1940s imposed important constraints on possible models for the structure of DNA. At the same time, Rosalind Franklin and Maurice Wilkins were using the powerful method of x-ray diffraction to analyze DNA fibers (see the How We Know section at the end of this chapter). Franklin's data showed that DNA produces a characteristic x-ray diffraction pattern. From this pattern, Watson and Crick deduced that DNA molecules are helical, with two periodicities along their long axis: a primary one of 3.4 Å and a secondary one of 34 Å. The challenge then was to formulate a three-dimensional model of the DNA molecule that could account not only for the x-ray diffraction data but also for the specific A = T and G = C base equivalences discovered by Chargaff.

The x-ray diffraction data obtained by Franklin and Wilkins enabled Watson and Crick to formulate a model in which DNA is composed of two polynucleotide strands entwined in the form of a right-handed double helix (**Figure 6-14**). Alternating 2′-deoxy-D-ribose and phosphate units make up the backbone of each strand, from which the bases project inward, toward the center of the helix. The bases are thus positioned to form hydrogen-bonding interactions with each other according to the preferred pairings of A with T and G with C. The twist of the helix forms two unequal surfaces, which are called the **major groove** and the **minor groove**. DNA strands always have a defined directionality, or polarity, due to the asymmetric shape and chemical linkage of the component nucleotides. In the double helix, the two strands have opposite directionality, and the helix is said to be **antiparallel**. In chemical terms, this means that one strand runs in the 5′→3′ direction and the other runs in the 3′→5′ direction. The antiparallel orientation of the DNA strands is more energetically favorable than the parallel orientation, due to the

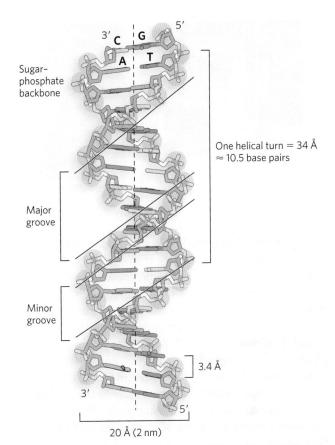

FIGURE 6-14 The double-helical structure of DNA. The original model proposed by Watson and Crick had 10 base pairs and a length of 34 Å per turn of the helix. Later measurements of DNA in solution (as opposed to in a crystal or fiber) showed 10.5 base pairs per helical turn. The major and minor grooves, where most interactions with proteins or other nucleic acids occur, are shown.

geometry of the component bases. Furthermore, the DNA double helix almost always twists in a right-handed direction (see Figure 4-6b for an explanation of helix handedness). Rarely, left-handed helices are observed in DNA, with the twist to the left. By convention, DNA helices are assumed to be right-handed unless otherwise specified.

Watson and Crick's double-helical model of DNA makes chemical sense, accounting for the known properties of the component nucleotides. The hydrophilic backbones of alternating sugar and phosphate groups are on the outside of the helix, facing the surrounding water. The pentose ring of each deoxyribose is in the C-2′ endo conformation (see Figure 6-3), and this sugar pucker defines the distance between adjacent phosphate groups in the DNA backbone. The purine and pyrimidine bases of both strands are stacked inside the double helix, with their hydrophobic and nearly planar ring structures very close together and more or less perpendicular to the long axis. Each nucleotide base of one strand is paired in the same plane with a base of the other strand. Watson

Rosalind Franklin, 1920–1958 [Source: Science Source.]

Maurice Wilkins, 1916–2004 [Source: AP Photo.]

and Crick found that the hydrogen-bonded base pairs, A with T and G with C, are those that agreed best with the x-ray diffraction data, providing a rationale for Chargaff's rule that in any DNA, A = T and G = C. It is important to note that three hydrogen bonds can form between G and C, symbolized G≡C, but only two can form between A and T, symbolized A=T. By always pairing a purine (A or G) with a pyrimidine (T or C), consistent spacing is maintained between the two antiparallel DNA backbones, giving a regular, uniform shape to the double helix. This has significant consequences for the stability of any double-stranded DNA sequence, as we will see in Section 6.4.

The double-helical structure of DNA also explains the periodicities observed in the x-ray diffraction patterns of DNA fibers. The vertically stacked bases inside the double helix are 3.4 Å apart; the secondary repeat distance of about 34 Å is accounted for by the presence of 10 base pairs in each complete turn of the double helix. In aqueous solution the structure differs slightly from that in fibers, having 10.5 base pairs per helical turn.

The stability of the DNA double helix arises primarily from the hydrophobic base-stacking interactions, which are largely nonspecific with respect to sequence. The configuration of planar purine–pyrimidine base pairs at the center of the helix allows their flat surfaces to stack on top of each other (see Figure 6-10), through dipole-dipole interactions and pi orbital stacking (discussed in Chapter 3). This energetically favorable situation stabilizes the double helix relative to single-stranded DNA by minimizing contact of the hydrophobic purines and pyrimidines with water. Furthermore, extensive networks of weak bonds in double-stranded DNA, such as van der Waals interactions and hydrogen bonds, are arranged so that for most of these bonds, they cannot break without simultaneously breaking many others. Consequently, DNA double helices that are 10 or more base pairs long are quite stable at room temperature. As mentioned in Section 6.1, DNA can persist in fossil samples over long periods of time, making possible the sequencing of DNA samples from long-extinct species—including Neanderthal hominids and woolly mammoths.

The most significant property of the double helix as an information carrier is the hydrogen bonding between the bases. Because adenine is always hydrogen-bonded to thymine, and guanine is always hydrogen-bonded to cytosine, exact copies of encoded information can be replicated. This specific base pairing gives the two helical strands a complementary relationship in which the base sequence of one strand defines the sequence of its partner. For example, the sequence 5'-GTAACGC-3' on one strand specifies the complementary sequence 5'-GCGTTAC-3' on the other strand.

Thus, the discovery of the DNA double helix immediately suggested a mechanism for the transmission of genetic information. As Watson and Crick proposed, this structure could logically be reproduced by separating the two strands and synthesizing a complementary strand for each. Because nucleotides in each new strand are joined in a sequence specified by the base-pairing rules stated earlier, each preexisting strand functions as a template to guide the synthesis of a complementary strand (**Figure 6-15**). These expectations were

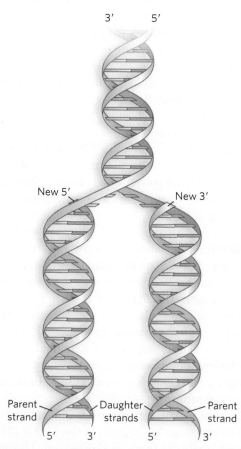

FIGURE 6-15 The mechanism for DNA replication. The newly synthesized complementary strands ("daughter" strands) are shown in red.

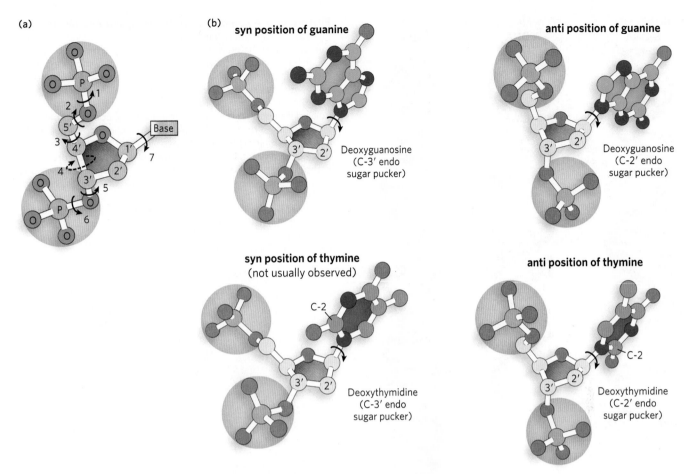

FIGURE 6-16 Factors contributing to structural variation in DNA. (a) DNA nucleotide conformation is affected by rotation about seven different bonds. Six of the bonds rotate freely; rotation about bond 4 is constrained by the sugar ring, giving rise to the sugar pucker. (b) The syn and anti positions of deoxyguanosine, and the syn and anti positions of deoxythymidine. Note that pyrimidines are restricted to the anti position; the carbonyl at C-2 causes steric clash in the syn conformation.

experimentally confirmed, inaugurating a revolution in our understanding of biological inheritance. (DNA replication is discussed in detail in Chapter 11.)

DNA Adopts Different Helical Forms

Nucleic acids are inherently flexible molecules. Numerous bonds in the sugar–phosphate backbone can rotate, and thermal fluctuation can lead to bending, stretching, and localized unpairing of the two strands. As a result, cellular DNA contains significant deviations from the Watson-Crick DNA structure, some or all of which may play important roles in DNA metabolism. Generally, such structural variations do not affect the key properties of strand complementarity: antiparallel strands and the requirement for A=T and G≡C base pairs.

Variation in the three-dimensional structure of DNA reflects three things: the different possible conformations of the deoxyribose (see Figure 6-3), rotation about

the contiguous bonds that make up the sugar–phosphate backbone (**Figure 6-16a**), and free rotation about the glycosidic bond. Because of steric constraints, purines in purine nucleotides are restricted to two stable conformations with respect to deoxyribose, called syn and anti (**Figure 6-16b**). Pyrimidines are generally restricted to the anti conformation, because of steric interference between the sugar and the carbonyl oxygen at C-2 of the pyrimidine.

The Watson-Crick structure is known as **B-form DNA,** or **B-DNA.** As the most stable structure for a random-sequence DNA molecule under physiological conditions, B-DNA is the standard point of reference in any study of the properties of DNA. Structures of short B-DNA duplexes have been studied in depth, revealing many details about the double helix (see the How We Know section at the end of this chapter). Two structural variants well characterized by x-ray crystallography are **A-form DNA (A-DNA)** and **Z-form DNA (Z-DNA).** These three DNA conformations

(a)

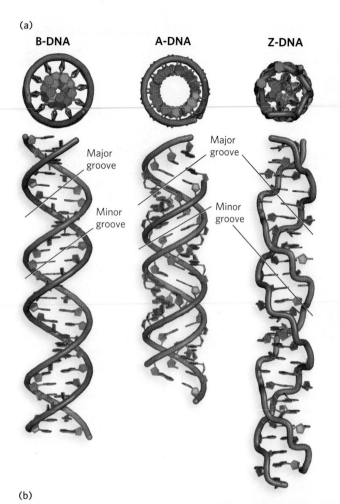

B-DNA   **A-DNA**   **Z-DNA**

Major groove

Minor groove

Major groove

Minor groove

FIGURE 6-17 A comparison of the B, A, and Z forms of DNA. (a) In each case, the sugar–phosphate backbones wind around the exterior of the helix (red and blue), with the bases pointing inward. The same 25 base pair DNA sequence is shown in all three forms. (As described later in the chapter, double-stranded RNA typically assumes the A-form geometry shown here.) Differences in helical diameter can be seen in the end-on views (top); differences in helical rise and groove shape are apparent in the side views (bottom). B-DNA, the most common form in cells, has a wide major groove and a narrow minor groove. A-form helices, common for RNA and certain DNA structures, are more compact than B-DNA. The major groove is deeper and the minor groove is shallower than in B-DNA. Z-DNA, which forms only under high salt conditions or with C≡G-rich DNA sequences, is left-handed, and its backbone has a zigzag pattern. It is less compact than B-DNA, with a very shallow major groove and a narrow and deep minor groove. (b) The table summarizes some properties of the three forms of DNA.

(b)

	B-DNA	A-DNA	Z-DNA
Helix sense	Right-handed	Right-handed	Left-handed
Diameter	~20 Å	~23 Å	~18 Å
Base pairs per helical turn	10.5	11	12
Helix rise per base pair	3.4 Å	2.6 Å	3.7 Å
Base tilt in relation to the helix axis	−6°	+20°	−7°
Sugar pucker conformation	C-2′ endo	C-3′ endo	C-2′ endo for pyrimidines; C-3′ endo for purines
Glycosyl bond conformation	Anti	Anti	Anti for pyrimidines; syn for purines

are shown in **Figure 6-17**, with a summary of their properties.

A-DNA is favored in many solutions that are relatively devoid of water. In this case, the DNA is still arranged in a right-handed double helix, but the helix is wider (23 Å in diameter, compared with 20 Å for B-DNA) and the number of base pairs per helical turn is 11, rather than 10.5 as in B-DNA. Whereas base pairs in B-DNA tilt slightly in a negative direction—that is, below the plane—with respect to a plane perpendicular to the helical axis, the base pairs in A-DNA are tilted above the plane by about +20°. In addition, the distance between adjacent phosphates in the polynucleotide chain, a direct consequence of the

sugar pucker, decreases from 7 Å in B-form helices to 5.9 Å in A-form helices (**Figure 6-18**). These structural changes deepen the major groove while making the minor groove shallower. The reagents used to promote crystallization of DNA tend to dehydrate it, and thus most short DNA molecules tend to crystallize in the A form.

Z-DNA is a more radical departure from B-DNA; the most obvious distinction is the left-handed helical rotation. There are 12 base pairs per helical turn, and the structure appears more slender and elongated. The DNA backbone takes on a zigzag appearance (hence the Z designation). Certain nucleotide sequences fold into left-handed Z helices much more readily than do others.

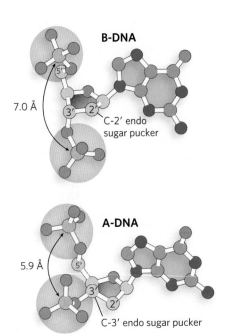

FIGURE 6-18 The effect of sugar pucker on the distance between phosphates in nucleic acids. The configuration of the pentose ring, known as the sugar pucker, is different in the backbones of B-DNA and A-DNA. As a result, the distance between the phosphate groups attached to the 5′ and 3′ positions of each nucleotide differs in the two forms, giving rise to distinct helical geometries.

Prominent examples are sequences in which pyrimidines alternate with purines, especially alternating C and G residues or 5-methyl-C and G residues (methylated bases are discussed in Section 6.4). To form the left-handed helix in Z-DNA, the purine residues flip to the syn conformation, alternating with pyrimidines in the anti conformation. The major groove is barely apparent in Z-DNA, and the minor groove is narrow and deep.

Whether A-DNA occurs in cells is uncertain, but there is evidence for some short stretches of Z-DNA in the chromosomes of both bacteria and eukaryotes. The evidence comes in part from experimentally prepared antibodies against short Z-form DNA segments, which can selectively bind to sequences in chromosomal DNA. Some evidence suggests that these potential Z-tracts correspond to actively transcribed regions of the genome, but it is also possible that antibody detection in these experiments is misleading due to capture of transient DNA structures or even spurious binding reactions. Thus, the function of these potential Z-DNA tracts remains unclear.

Certain DNA Sequences Adopt Unusual Structures

Other sequence-specific DNA structures have been detected, within larger chromosomes, that may affect the function and metabolism of the DNA segments in their immediate vicinity. For example, certain repetitive

sequences can bend the DNA helix in a distinct way. The stability and geometry of base-pair stacking influences the preferred direction of DNA bending. This was first observed in repetitive-sequence DNA isolated from trypanosomes, the protozoa that cause African sleeping sickness. Typical sequences that cause pronounced bending contain stretches of four to six A and T residues separated by C- and G-rich segments. The A- or T-tracts, each corresponding to a half-turn of the double helix, are spaced such that the geometry of the repetitive base pairs tends to curve the DNA helix in one direction (**Figure 6-19**). This DNA bending helps certain proteins—such as transcription factors, which promote the synthesis of mRNAs—bind to their target DNA binding sites.

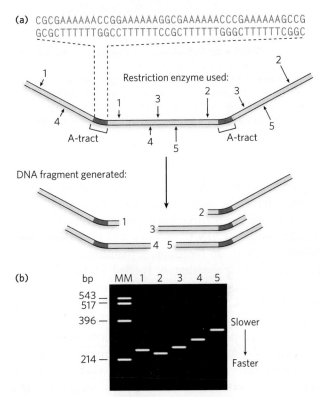

FIGURE 6-19 DNA bending at A-tracts. Bending in DNA segments containing A-tracts of six adenosines in a row can be detected by using enzymes to cut the DNA at specific sites. (a) The DNA sequence used in this experiment contains two A-tracts. Arrows point to sites that can be cleaved by various restriction enzymes (numbered 1 to 5) to generate DNA fragments of equal length that contain an A-tract at various positions in the fragment. (b) DNA fragments are analyzed by gel electrophoresis. Even though all DNA fragments are the same size (~215 base pairs), their rate of migration through the gel depends on the relative location of the A-tract. When the A-tract is located in the middle, the DNA fragment is more bent and migrates slowly; when the A-tract is near the end, the fragment is less bent and migrates faster. The lane marked MM contains molecular markers that provide a reference for DNA fragment size in base pairs (bp). [*Source: (b) Data from Asayama et al., J. Biochem. (Tokyo) 1125:460–468, 1999, by permission of Oxford University Press.*]

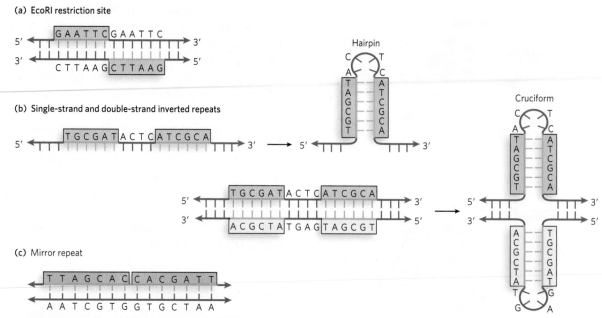

FIGURE 6-20 Palindromes, hairpins, cruciforms, and mirror repeats in DNA. (a) A palindromic sequence in DNA, such as the recognition site for the EcoRI restriction enzyme, is the same when reading from 5′ to 3′ on either strand of the DNA. (b) Inverted repeats within a single strand of DNA can be converted into a hairpin, which is double-stranded in the stem region. Inverted repeats within a double-stranded DNA sequence can form a cruciform (double-hairpin) structure. (c) DNA can also contain mirror repeats, such as TTAGCACCACGATT, which are not palindromic.

Regions of DNA where the two complementary strands have the same sequence when read in the 5′→3′ or the 3′→5′ direction occur relatively frequently in chromosomal DNA and are called **palindromes** (**Figure 6-20a**). In language, a palindrome is a word, phrase, or sentence that is spelled identically when read either forward or backward; two examples are ROTATOR and NURSES RUN. In biology, the term applies to double-stranded regions of DNA where one strand's sequence is identical to its complement; for example, 5′-GAATTC-3′ is a palindrome because its complementary sequence is also 5′-GAATTC-3′. **Inverted repeats** are complementary sequences that occur on the same strand of a DNA or RNA molecule but in inverse directions, and often with some sequence separation between the repeats (**Figure 6-20b**). Such inverted repeats have the ability to form **hairpin** or **cruciform** (cross-shaped) structures through base pairing between the adjacent repeats. A related arrangement is a **mirror repeat**, in which the inverted repeat sequence is nonpalindromic (**Figure 6-20c**).

Palindromic sequences, inverted repeats, and mirror repeats play important biological roles, such as slowing or blocking protein synthesis by the ribosome—a process called translation attenuation (see Chapter 18)—or forming recognition sites for restriction enzymes, which catalyze double-stranded DNA cleavage (see

Chapter 7). Inverted repeats, with their potential to form hairpins and cruciforms, are found in virtually every large DNA molecule and can encompass a few base pairs or thousands. The extent to which inverted repeats occur as cruciforms in cells is not known, but some cruciform structures have been demonstrated in vivo in *Escherichia coli*. Cruciforms are also known to form transiently during recombination of DNA molecules, in which genetic information on two chromosomes is exchanged during cell division. For this reason, palindromes can be involved in pathological chromosomal translocations, in which sequences from one chromosome are appended aberrantly to those of a nonhomologous chromosome, due to short segments of overlapping sequence between the two. The consequences of such events can be disastrous, leading to genetic dysregulation and cancer.

Several unusual DNA structures involve three or even four DNA strands. Although rare, these structural variants merit investigation because there is a tendency for them to form at sites where important events in DNA metabolism (replication, recombination, transcription) are initiated or regulated. Nucleotides participating in Watson-Crick base pairing have the potential to form additional hydrogen bonds, particularly with functional groups arrayed in the major groove. For example, a thymidine can pair with the adenosine of

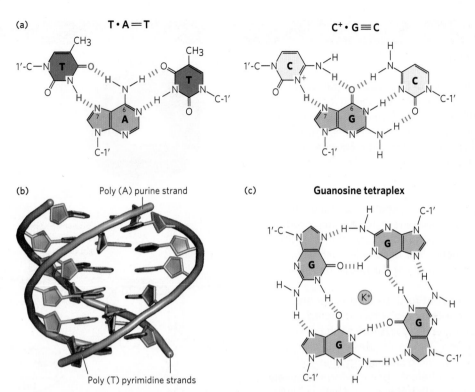

FIGURE 6-21 Three- and four-stranded DNA structures. (a) Base pairing in triplex DNA. Atoms participating in Hoogsteen pairing are in red; canonical Watson-Crick base pairs are in black. (b) Side view of a triple-helical DNA containing two poly(T) strands and one poly(A) strand. The dark blue and light blue strands in the foreground are antiparallel and engage in canonical Watson-Crick base pairing. The poly(T) strand in the background is parallel to the poly (A) strand and is paired through Hoogsteen hydrogen bonding. (c) One layer of a guanosine tetraplex (quadruplex) structure, showing hydrogen bonding of the bases. A K⁺ ion in the center of the tetraplex stabilizes the structure by coordinating the bases' functional groups. [*Source: (b) PDB ID 1BCE.*]

an A=T nucleotide pair, and a cytidine (if protonated) can pair with the guanosine of a G≡C pair, forming "base triples" (**Figure 6-21a**). The N-7, O^6, and N^6 of purines, the atoms that participate in this additional hydrogen bonding, are often referred to as **Hoogsteen positions**, and the non-Watson-Crick pairing is called **Hoogsteen pairing**. Karst Hoogsteen, in 1963, was the first to recognize the potential for these unusual pairings. Hoogsteen pairing allows the formation of **triplex DNAs** (**Figure 6-21b**).

The triplexes form most readily within long sequences containing only pyrimidines or only purines in a given strand. Some triplex DNAs contain two pyrimidine strands and one purine strand; others contain two purine strands and one pyrimidine strand. DNA triplex formation can be highly sequence-specific. For example, triplex formation between a small section of chromosomal DNA and a chemically modified single-stranded DNA enabled Peter Dervan and his colleagues to cleave a human chromosome at a single site. They did this by synthesizing short oligonucleotides that could form triplex interactions with various segments of the chromosomal

DNA. Because each oligonucleotide had a chemical modification at one end that triggered DNA strand breakage, the researchers were able to fragment the DNA at specific sites—which helped in mapping the location of the gene for the inherited neurological disorder known as Huntington disease.

Four DNA strands can also associate to form a tetraplex (or quadruplex), but this occurs readily only for DNA sequences with a very high proportion of G residues (**Figure 6-21c**). The guanosine tetraplex, or **G tetraplex**, is quite stable over a wide range of conditions. The DNA regions at the ends of linear chromosomes, called telomeres, typically consist of G-rich segments that have a propensity to form tetraplex structures when tested in the laboratory. Whether such structures contribute to the stability and recognition of telomeres in vivo is not known.

In the DNA of living cells, sites recognized by many sequence-specific DNA-binding proteins are arranged as palindromes, and polypyrimidine or polypurine sequences that can form triple helices are found within regions involved in regulating the expression of some

DNA Nanotechnology

The ability of a single strand of DNA to base-pair specifically with its complementary sequence is essential for the accurate replication of encoded information. Scientists have also recognized that base pairing is an extremely useful property for assembling various three-dimensional structures from DNA. These observations have given rise to the field of DNA nanotechnology, the design and production of nucleic acid structures to be used for a range of technological applications. Instead of carrying genetic information, the nucleic acids used in this field of research are employed as engineering materials by taking advantage of their ability to base-pair and form various kinds of complex architectures. An example is shown in Figure 1 and is further described below.

Based on concepts put forth by Nadrian Seeman in the 1980s, DNA nanotechnology has attracted increasing attention in recent years due to the ease and low cost of synthesizing DNA molecules of different lengths and sequences. The base-pairing rules for DNA, together with its inherent ability to form relatively rigid helical structures (both between and within DNA strands), enables the rational design of sequences that automatically assemble to form complex molecular architectures. For this reason, DNA nanotechnology building blocks—the DNA sequences used to build molecular devices—have been called "programmable matter."

DNA nanotechnology continues to advance toward various kinds of useful applications. For example, DNA devices might one day be the basis for "smart drugs" that allow targeted delivery of therapeutics. Scientists envision that a hollow DNA container

Gerald Joyce *[Source: Courtesy Gerald F. Joyce, MD, PhD/The Scripps Research Institute.]*

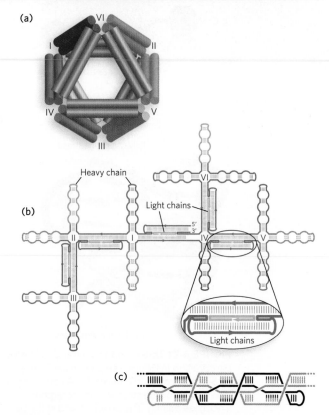

FIGURE 1 (a) A three-dimensional DNA octahedral structure as an example of DNA nanotechnology. The structure has 12 struts—the octahedron edges (double cylinders; each cylinder is a double-stranded DNA)—connected by six flexible joints (labeled I–VI). The joints are four-way junctions that connect the core-layer double helices of each strut. Colors correspond to the colored segments in (b), which shows the secondary structure of the branched-tree folding intermediate. This consists of a single 1,669-nucleotide DNA chain, or heavy chain, and five unique 40-nucleotide light chains (orange). (c) A close-up of the base-pairing scheme for each strut. Black and gray indicate two separate parts of the DNA strand that interact to form a strut. The cross-over base pairing gives the strut its structure and strength. *[Source: Data from W. M. Shih, J. D. Quispe, and G. F. Joyce, Nature 427:618–621, 2004, Fig. 1.]*

eukaryotic genes. In principle, synthetic DNA strands designed to pair with these sequences to form triplex DNA could disrupt gene expression. This approach to controlling cellular metabolism is of growing commercial interest for its potential application in medicine and agriculture. Unusual DNA structures can also be engineered, raising the possibility of using DNA as a container to deliver drugs or proteins (**Highlight 6-1**).

SECTION 6.2 SUMMARY

- Chargaff's rules state that in double-stranded DNA, the number of adenine nucleotides equals the number of thymine nucleotides (A=T), and the number of guanine nucleotides equals the number of cytosine nucleotides (G=C).

- Using Franklin's and Wilkins's x-ray diffraction data from DNA fibers, Watson and Crick proposed that native DNA consists of two antiparallel chains in a right-handed double-helical arrangement. The hydrophilic sugar–phosphate backbone of each strand is on the outside of the helix, and the planar purine and pyrimidine bases project inward, perpendicular to the backbone axis. Complementary base pairs, A=T and G≡C, are formed by hydrogen bonding within the helix, consistent with Chargaff's rules. The helical structure is further

encapsulating proteins capable of inducing cell death could be programmed to open only when in contact with cancerous cells. It might even be possible to produce these artificial structures in engineered living cells. This would then enable the targeted evolution of improved self-assembling DNA nanostructures.

The trick in bringing such concepts into being is to control the DNA base-pairing interactions to favor the formation of desired shapes. A particular challenge is to design sequences that can be cloned and copied by DNA polymerase enzymes in cells and that are also capable of self-assembling into specific three-dimensional shapes. Gerald Joyce and his colleagues at the Scripps Research Institute succeeded in designing a DNA sequence that could fold up into an octahedron in the presence of short complementary oligonucleotides. The research group produced a 1,669-nucleotide, single-stranded DNA molecule containing many such self-complementary sequences to enable specific base pairing between segments (see Figure 1). To induce three-dimensional folding, five 40-nucleotide oligonucleotides were added to form the struts of an octahedron by base pairing to distinct sites within the 1,669-nucleotide DNA. DNA mixtures were heated to disrupt base pairing and make single strands accessible, and then cooled to induce formation of base pairs. The changes in DNA structure resulting from base-pair formation were monitored by observing changes in the mobility of the DNA through a gel matrix and visualizing assembled DNA octahedra by electron microscopy (Figure 2). The microscopic analysis showed that, as expected from the sequence design, the DNA strands folded with 12 struts or edges joined at six four-way junctions to form hollow octahedra approximately 22 nm in diameter.

In theory, such DNA structures could be chemically modified to provide binding sites for small molecules or proteins within the enclosed space. If this idea can be shown to work, simple DNA base pairing will have been harnessed as a practical tool for drug delivery.

(a)
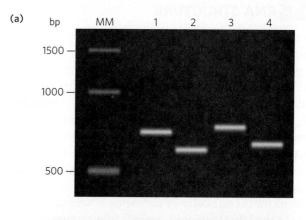

(b)

FIGURE 2 Gel electrophoresis and electron microscopy revealed the DNA octahedron assembly through base pairing. (a) Agarose gel electrophoresis of octahedron-forming DNA under different conditions. Lane MM: molecular marker lane with DNA size standards (number of base pairs (bp) is indicated on the left). Lane 1: 1,669-nucleotide strand (heavy chain) folded in the absence of Mg^{2+}. Lane 2: 1,669-nucleotide strand folded in the presence of Mg^{2+}. Lane 3: 1,669-nucleotide strand and 40-nucleotide light chains folded in the absence of Mg^{2+}. Lane 4: 1,669-nucleotide strand and 40-nucleotide light chains folded in the presence of Mg^{2+}. The Mg^{2+} shields the negative charges of the backbone phosphates and also promotes base pairing, creating a more compact structure that migrates more quickly through the gel. The resolution of the gel is not sufficient to detect differences in mobility of the DNA with or without the 40-nucleotide light chains. (b) Three views of the three-dimensional images of assembled DNA octahedra generated computationally, using electron micrographs as the starting point. [*Source: Reprinted by permission from Macmillan Publishers Ltd. William M. Shih, Joel D. Quispe and Gerald F. Joyce Nature 427, 618–621 (12 February 2004) © 2004. (a) Data from Fig. 2; (b) Fig. 3b.*]

stabilized by shared electrons between the stacked planar base pairs.

- In B-DNA, the most common form of DNA in cells, the base pairs are stacked nearly perpendicular to the long axis of the double helix, 3.4 Å apart, with 10.5 base pairs per turn (in solution).
- Two other variations on DNA structure are A-DNA and Z-DNA. Like B-DNA, A-DNA is a right-handed helix, but it is more compact, with 11 base pairs per turn. A-DNA is favored in solutions that lack water, such as reagents used to crystallize DNA. Z-DNA forms a left-handed helix that contains 12 base pairs per turn and occurs only in sequences rich in C and G residues. Evidence suggests that eukaryotic DNA contains short stretches of Z-DNA, which might function in genetic recombination or the regulation of gene expression.
- Repetitive sequences, such as tracts of A or T residues separated by G- or C-rich segments, cause bends in the DNA molecule. Bending can help facilitate DNA-protein binding.
- DNA strands with inverted repeat sequences can form hairpin or cruciform structures that play roles in recombination and regulation of gene expression. Triplex or tetraplex forms of DNA can occur, though rarely, and may function in DNA metabolism.

6.3 RNA STRUCTURE

The discovery that RNA molecules play key roles in converting the genetic information contained in DNA into the proteins that perform structural and catalytic functions in cells motivated the quest to determine the molecular structure of RNAs. In the early 1970s, Alexander Rich, Aaron Klug, and Sung-Hou Kim independently solved the structures of transfer RNAs, revealing how tRNAs carry the amino acids that are used in protein synthesis on the ribosomes. The field of RNA structural biology then languished for almost 20 years, due in part to the technical difficulty of preparing RNA samples for study in the laboratory and the lack of awareness of the wide variety of biological functions of RNA.

The situation changed in 1990, driven by the discoveries of many new kinds of RNA molecules in cells and viruses. In fact, many viral genomes, including those of HIV, HCV, and the influenza viruses, are made entirely of RNA, which can be either single-stranded or double-stranded. Questions about the stability of RNA genomes and how their structures might differ from those of genomic DNA increased the urgency of discovering how RNA molecules are structured.

RNAs Have Helical Secondary Structures

The wide-ranging functions of RNA reflect a structural diversity much richer than that observed in DNA molecules. The propensity of RNA to form compact folded shapes was first revealed in the 1970s with the determination of the molecular structures of several tRNA molecules (**Figure 6-22**). These analyses showed that a single strand of RNA folds back on itself to form short base-paired or partially base-paired segments connected

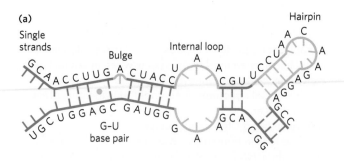

FIGURE 6-23 Double-helical characteristics of RNA. (a) Some of the diversity in the secondary structure of RNA is shown in these examples of G–U base pairs, bulges, internal loops, and hairpin loop structures. (b) An RNA helix in the form of a hairpin structure similar to the one shown in (a); notice how the unpaired bases at the hairpin loop are incorporated into the structure while maintaining the overall helical geometry of an A-form duplex.

by unpaired regions (**Figure 6-23**). This property, called **RNA secondary structure**, enables RNA molecules to fold into many different shapes that lend themselves to many different biological functions.

As in DNA, the paired strands in RNA are antiparallel and tend to assume a right-handed helical conformation dominated by base-stacking interactions. Unlike DNA, however, the base-paired segments of RNA are interspersed with a variety of other, non-Watson-Crick, base pairings such as A–A and G–U (**Figure 6-24**; also see the How We Know section at the end of this chapter). In addition, RNA secondary structures include regions of unpaired nucleotides, which can interact

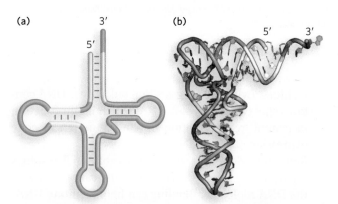

FIGURE 6-22 The three-dimensional structure of tRNA. (a) The secondary structure of tRNA forms a cloverleaf containing four helices that meet at a central junction. The structure contains several non-Watson-Crick base pairs and loops. (b) The three-dimensional structure of tRNA^{Phe} (a tRNA specific for phenylalanine in protein synthesis), determined by x-ray crystallography, is shown as a ribbon diagram. [*Source: (b) PDB ID 1EHZ.*]

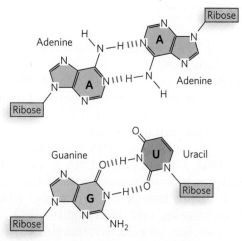

FIGURE 6-24 A–A and G–U base pairs in RNA.

with noncontiguous sequences to stabilize the three-dimensional folding. Such interactions produce compact shapes containing surfaces or crevices that bind other molecules or form sites capable of catalyzing chemical reactions, much like protein enzymes.

The greater structural variety in RNA relative to DNA reflects the three main chemical differences between the two polynucleotides: the pentose (2'-deoxyribose in DNA vs. ribose in RNA), the base composition (thymine in DNA vs. uracil in RNA), and the sugar pucker of the pentose (C-2' endo in DNA vs. C-3' endo in RNA). The presence of the 2'-hydroxyl group on the sugar of RNA nucleotides provides an extra site for hydrogen bonding, potentially stabilizing three-dimensional folding. This hydroxyl group also influences the sugar pucker, leading to more closely spaced phosphates on the 5' and 3' sides of each sugar and hence to a more compact, A-form helical structure (see Figure 6-17).

As in the DNA double helix, RNA base stacking is made energetically favorable by the resulting burial of hydrophobic surfaces away from the hydrophilic surroundings. In tRNA, as well as in the more recently discovered structures of catalytic RNA molecules and ribosomes, virtually all of the bases are stacked, even when they are not part of Watson-Crick base pairings. In each case, structural motifs, such as base-triple interactions (see Figure 6-21a, b) and helix-helix packing, allow stable three-dimensional folding.

Double-stranded RNAs, such as those that form the genomes of some viruses, do exist in nature. These RNAs exist as long helical structures analogous to the DNA double helix. In addition, some RNAs do not seem to form stable three-dimensional structures through local base-pairing interactions. For example, mRNAs seem to perform their function as transient carriers of genetic information without adopting any specific three-dimensional structure. These RNAs may fold into three-dimensional structures only in the presence of bound proteins, forming complexes called **ribonucleoproteins (RNPs)**.

RNAs Form Various Stable Three-Dimensional Structures

Most of the highly structured RNAs contain noncanonical base pairs and backbone conformations not observed in DNA. In many cases, the 2'-hydroxyl group on ribose, a chemical feature that distinguishes RNA from DNA, seems to be directly or indirectly responsible for these unique structural properties. Recall that the presence of this 2' hydroxyl gives ribose its C-3' endo geometry (its sugar pucker), as distinct from the C-2' endo characteristic of deoxyribose. This seemingly small difference in chemical conformation leads to a distinct helical geometry for RNA relative to DNA. Furthermore, it enables direct or water-mediated hydrogen bonding between the 2'-hydroxyl of one RNA nucleotide and the adjacent ribose or phosphate (**Figure 6-25**). As a result, RNA helices are more thermodynamically stable than DNA helices of the same length and sequence.

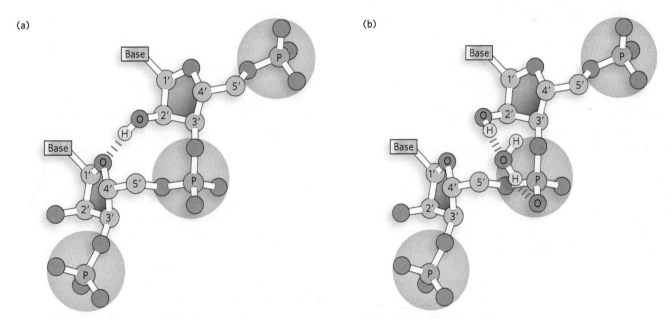

(a)

(b)

FIGURE 6-25 Stabilization of RNA secondary structure. The 2'-hydroxyl group on ribose causes RNA to favor the C-3' endo sugar pucker, due to (a) direct hydrogen bonding between a 2' hydroxyl on one ribose and the ring oxygen on an adjacent ribose, and (b) hydrogen bonding, through a water molecule, between a 2' hydroxyl and a phosphate oxygen. *[Source: Data from W. Saenger, Principles of Nucleic Acid Structure, Springer-Verlag, 1984, Fig. 4-11.]*

Where complementary sequences are present in an RNA molecule, the predominant double-stranded structure in the RNA is an A-form, right-handed double helix (see Figure 6-17). The A-RNA helix has a wider, shallower minor groove and a narrower, deeper major groove than the B-form helix observed for most DNA. The A-form geometry in RNA has a shorter distance between adjacent phosphates in the sugar–phosphate backbone than in B-DNA, a consequence of the C-3′ endo sugar pucker of ribose. A B-form RNA helix has not been observed in nature. Z-RNA helices have been induced to form in the laboratory under high-salt or high-temperature conditions but are not known to occur in cells.

As mentioned previously, mismatched or unmatched bases are common in base-paired segments of RNA, locally disrupting the regular A-form helix and resulting in bulges or internal loops (see Figure 6-23). The potential for base pairing within a single strand of RNA frequently produces thermodynamically stable secondary structures that consist of hairpinlike conformations capped by connecting loops. Such hairpins are the most common type of RNA secondary structure, often containing specific short sequences at their ends (such as UUCG or GAAA) that form particularly energetically favorable loops. The nucleotides in the loops are arranged to maximize hydrogen bonding and base stacking, thereby enhancing thermodynamic stability. Important additional structural contributions are made by hydrogen bonds that are not part of canonical Watson-Crick base pairs. These properties, evident in the structures of tRNAs and catalytic RNAs, were also evident in the ribosome structures solved by Jamie Cate and others (see this chapter's Moment of Discovery). The functions of these highly structured RNAs, like those of proteins, depend on their three-dimensional properties.

Weak interactions, especially base-stacking interactions, play a major role in stabilizing RNA structures, just as they do in DNA. The 2′ hydroxyl is often involved in hydrogen bonds and van der Waals interactions that stabilize alternative helical shapes and conformations, such as loops and kinks that require the close approach of the two phosphodiester backbones. Divalent and monovalent metal ions—such as Mg^{2+}, Ca^{2+}, K^+, and Na^+—bind to specific sites in RNA and help shield the negative charge of the backbone, allowing parts of the molecule to pack more tightly together (**Figure 6-26**).

The diversity of RNA three-dimensional structures found in nature is illustrated by riboswitches, non-protein-encoding RNA sequences with the unique ability to directly bind cellular metabolites and form alternative metabolite-free and metabolite-bound shapes. Found widely among bacteria and sporadically in other organisms, riboswitches can change their

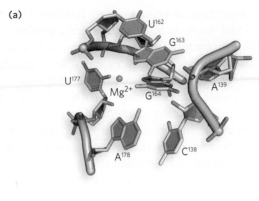

(a)

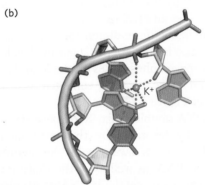

(b)

FIGURE 6-26 Mg^{2+} and K^+ binding in RNA structure.
(a) Divalent magnesium ions (Mg^{2+}) coordinate to phosphate groups and stabilize the close approach of phosphate backbones in the folded structure of the P4–P6 domain of the *Tetrahymena* group I ribozyme (a catalytic RNA). (b) Monovalent potassium ions (K^+) bind to specific sites in the P4–P6 domain, where they favor interactions between both backbone atoms and bases. [*Source:* (a) and (b) PDB ID 1GID.]

structure on binding to a particular metabolite and, as consequence, alter the efficiency of synthesis of proteins involved in pathways utilizing that metabolite. Such metabolite-sensing RNAs function through complex three-dimensional folding to create structures that can selectively recognize a small molecule within the mixture of molecules in a cell. Due to their ability to adopt a wide range of structures, riboswitches can bind to molecules ranging in size and type from tiny ions to large organic molecules. Although riboswitches employ common structural principles to form their metabolite-binding shapes, different riboswitch classes possess unique features for ligand recognition, even those that have evolved to recognize the same metabolites (**Figure 6-27**). We will encounter further examples of the wide-ranging structures of RNA when discussing ribosomes and catalytic introns in Chapters 16 and 18.

The analysis of RNA structure, along with the relationship between structure and function, is an emerging field of inquiry with many of the same complexities as the analysis of protein structure. The importance of understanding RNA structure has grown as we have

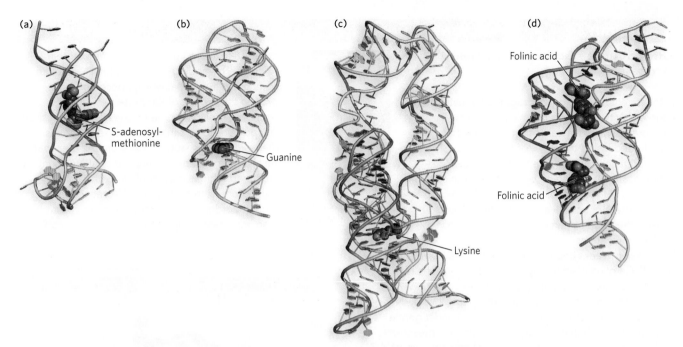

FIGURE 6-27 Examples of the diversity of RNA three-dimensional structures in riboswitches. The phosphodiester backbone in each case is shown as a solid tube; bound metabolites are shown in red. (a) The *S*-adenosylmethionine-II riboswitch; (b) a guanine-binding riboswitch; (c) a lysine-binding riboswitch; and (d) a tetrahydrofolate-binding riboswitch, bound here to two molecules of folinic acid.

become increasingly aware of the large number of functional roles for RNA molecules. For example, we now know that the extensive structural features found in the genomic RNA of HIV control viral gene expression (**Highlight 6-2**).

❇ ■ **SECTION 6.3 SUMMARY**

• RNA is chemically better suited than DNA to forming stable three-dimensional folds, due to the 2′-hydroxyl group of ribose in the RNA backbone. The presence of the 2′ hydroxyl makes RNA vulnerable to hydrolysis, but it also allows additional hydrogen bonding between segments of the molecule.

• RNA molecules can exist as long double-helical structures, typical of viral genomic RNA, or, more commonly, as single strands that fold up into short helical regions connected by loops and unpaired segments.

• Base-paired segments of RNA generally adopt the compact geometry of A-form helices. This structure arises because RNA favors a different sugar pucker from that found in DNA (C-3′ endo in RNA vs. C-2′ endo in DNA), which causes the phosphates in the backbone to become more closely spaced than in B-DNA. In some large RNAs, short helices undergo RNA-RNA and RNA–metal ion interactions to form complex three-dimensional structures.

• Base pairs other than canonical A=U and G≡C pairs are common in RNAs, including A–A and G–U. In all cases, base pairs or single bases are most stable when stacked on top of one another in a helix.

• Divalent and monovalent metal ions (Mg^{2+}, Ca^{2+}, K^+, and Na^+) bind to specific sites in RNA and help shield the negative charge of the backbone, allowing parts of the molecule to pack more tightly together.

6.4 CHEMICAL AND THERMODYNAMIC PROPERTIES OF NUCLEIC ACIDS

To understand how nucleic acids function, we must understand their chemical properties as well as their structures. The role of DNA as a repository of genetic information depends in part on its inherent stability. The chemical transformations that do happen are generally very slow in the absence of an enzyme catalyst. The long-term storage of information without alteration is so important to a cell, however, that even very slow changes in DNA structure can be physiologically significant. Other, nondestructive alterations of DNA do occur and are essential to function, such as the strand separation that must precede replication or transcription. For RNAs, chemical modifications can play significant roles in ensuring correct structure and function.

HIGHLIGHT 6-2 — MEDICINE

RNA Structure Governing HIV Gene Expression

Kevin Weeks [Source: Courtesy Kevin Weeks.]

RNA structures can be investigated by testing for reactivity to various chemical reagents. This is possible because bases that are paired or are folded inside the RNA, or those that are conformationally rigid, are protected against modification or cleavage by reactive chemicals. By analyzing which sites in a folded RNA molecule are resistant or sensitive to chemical reagents, researchers can obtain information about its structural features.

Kevin Weeks and his colleagues at the University of North Carolina used this approach to analyze the role of RNA structure in regulating gene expression in HIV-1 (a strain of the human immunodeficiency virus). Viral RNA was extracted from viral particles (virions) under gentle conditions, such that the native structure was preserved. The RNA was then treated with a chemical reagent, 1-methyl-7-nitroisatoic anhydride (1M7), which preferentially acylates conformationally flexible nucleotides at the 2′-OH of ribose. After the RNA was allowed to react with 1M7, acylated sites were detected by reverse transcription. This procedure uses the purified enzyme reverse transcriptase to make a DNA copy of the viral RNA. The researchers found that acylation blocked the progression of polymerization, generating a truncated DNA fragment extending up to the site of acylation in the RNA sequence. Following fractionation of the resulting DNAs, the sites of acylation were detected and mapped onto the HIV-1 genome sequence.

Analysis of these data revealed that in addition to encoding viral proteins in its nucleotide sequence, the viral genomic RNA is also three-dimensionally structured to optimize protein production (Figure 1). Highly structured regions in the RNA—and hence a dearth of experimentally acylated sites—occur within sequences that encode interprotein linkers, or loops, between protein domains. This finding implies that RNA structure slows down the rate at which ribosomes move through these parts of the protein-coding sequence, providing time for the newly synthesized viral proteins to fold into their active structures. The results of this study underscore the idea that the HIV genome and, perhaps, the mRNAs contain structured regions that regulate expression of the proteins they encode. RNA secondary and higher-order structure may therefore constitute an important component of the genetic code.

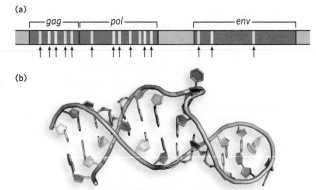

FIGURE 1 The organization and structure of the HIV-1 genomic RNA influences viral protein expression. (a) This portion of the HIV-1 genome shows three open reading frames, *gag*, *pol*, and *env*, that are translated into long polyprotein precursors; proteolytic cleavage generates multiple proteins from each reading frame. Blue bars represent sites that code for interprotein linkers, and yellow bars indicate sites coding for loops between protein domains. Black arrows mark regions of the RNA genome that are highly structured; note that these areas coincide with areas that link proteins or protein domains. (b) An example of RNA secondary structure in the HIV-1 genome. [Sources: (a) Data from J. M. Watts et al., Nature 460:711–716, 2009, Fig. 1; (b) PDB ID 2KMJ.]

In addition to providing insights about physiological processes, our understanding of nucleic acid chemistry gives us a powerful array of technologies that have applications in molecular biology, medicine, and forensic science. We examine here the chemical properties of DNA and RNA, as well as some of these technologies. Many more techniques and applications are discussed in Chapter 7.

Double-Helical DNA and RNA Can Be Denatured

Solutions of carefully isolated, double-stranded DNA are highly viscous at pH 7.0 and room temperature (25°C). When such a solution is subjected to extremes of pH or to temperatures above 80°C, its viscosity decreases sharply, indicating that the DNA has undergone a physical change. This change is due to **denaturation**, or **melting**, of the double-helical DNA and can also occur with RNA. Disruption of both the hydrogen bonding between paired bases and the base stacking causes the double helix to unwind, forming two single strands that are completely separate from each other along the entire (or partial) length of the molecule. No covalent bonds in the nucleic acid are broken during denaturation (**Figure 6-28**).

Renaturation of a DNA or RNA molecule is a rapid, one-step process, as long as a double-helical segment of at least a dozen residues still unites the two strands. When the temperature or pH is returned to the range in which most organisms live, the unwound segments of the two

Double-helical DNA

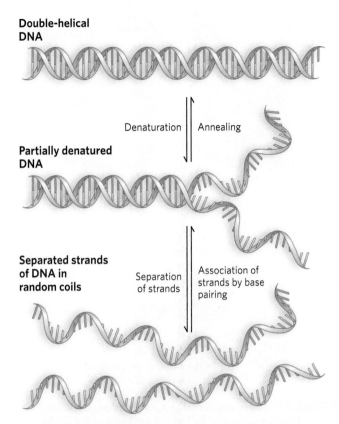

Denaturation | Annealing

Partially denatured DNA

Separation of strands | Association of strands by base pairing

Separated strands of DNA in random coils

FIGURE 6-28 Reversible denaturation of DNA. DNA is shown here, but RNA is also capable of denaturation and reannealing.

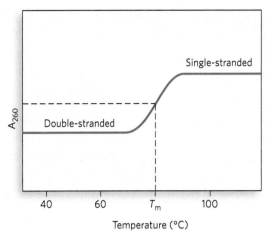

FIGURE 6-29 UV light absorption by DNA. The transition from double-stranded DNA to the single-stranded, denatured form can be detected by monitoring UV light absorption by the sample (shown here as A_{260}, absorbance at a wavelength of 260 nm). The melting point (T_m) is the temperature at which half the DNA in the sample is denatured.

strands spontaneously rewind to yield the intact duplex. This process, called **annealing**, involves re-formation of all the base pairs in the double helix. If the strands were completely separated, renaturation occurs in two steps. In the first step, which is relatively slow, complementary sequences in the two strands "find" each other by random collisions and form a short segment of double helix. The second step is much faster: the remaining unpaired bases successively come into register as base pairs, and the two strands "zipper" themselves together to form the double helix.

The close interaction of stacked bases in a nucleic acid has the effect of decreasing its absorption of UV light relative to that of a solution with the same concentration of free nucleotides, and the absorption is further decreased by the pairing of two complementary strands. Hydrogen bonding and base stacking in the double helix limit the resonance of the aromatic rings of the bases, thereby decreasing UV light absorption. This is known as the **hypochromic effect (Figure 6-29)**. When DNA is denatured, the base pairs are disrupted and the two strands separate into randomly coiled chains. The resonance of the bases in each strand is no longer constrained, as it is when the bases are part of a double helix. As a result, the UV light absorption of single-stranded DNA is approximately 40% higher than

that of double-stranded DNA at the same concentration. This increase in absorption is called the **hyperchromic effect**. The transition from double-stranded DNA to the single-stranded, denatured form can thus be detected by monitoring the absorption of UV light.

DNA molecules in solution denature when they are heated slowly. Each species of DNA has a characteristic denaturation temperature, or **melting point (T_m)**, defined as the temperature at which half the DNA is denatured. In general, the higher the content of G≡C base pairs in the DNA, the higher its melting point. This is because G≡C base pairs, with three hydrogen bonds, require more heat energy to dissociate than do A=T base pairs. Careful determination of the melting point of a DNA specimen, under fixed conditions of pH and ionic strength, can yield an estimate of its base composition. If denaturation conditions are carefully controlled, regions that are rich in A=T base pairs will specifically denature while most of the DNA remains double-stranded. Such denatured regions, or bubbles, can be visualized with electron microscopy (**Figure 6-30**). Strand separation of DNA must occur in vivo during processes such as DNA replication and transcription. As we will see in Chapter 11, the DNA sites where these processes are initiated are often rich in A=T base pairs.

RNA duplexes or RNA-DNA hybrid duplexes can also be denatured. Notably, RNA duplexes are more stable than DNA duplexes. At neutral pH, the denaturation of a double-helical RNA often requires higher temperatures, by 20°C or more, than those required to denature a DNA molecule with a comparable sequence. The stability of an RNA-DNA hybrid is generally intermediate between that of double-stranded RNA and DNA.

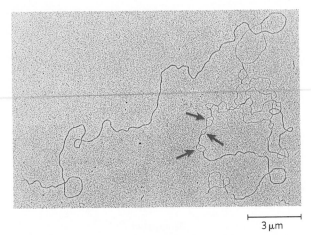

3 μm

FIGURE 6-30 Partially denatured DNA. The DNA shown in this electron micrograph was partially denatured, then fixed to block renaturation during sample preparation. The arrows point to some single-stranded bubbles where denaturation has occurred. The regions that denature are reproducible in repeated experiments and are rich in A = T base pairs. [*Source: Ross B. Inman, University of Wisconsin–Madison, Department of Molecular Biology.*]

Nucleic Acids from Different Species Can Form Hybrids

The ability of two complementary DNA strands to pair with each other can be used to detect similar DNA sequences in different species or within the genome of a single species. If double-stranded DNAs isolated from human cells and from mouse cells are completely denatured by heating, then mixed and kept at 25°C for many hours, much of the DNA will anneal. Most of the mouse DNA strands anneal with complementary mouse DNA strands to form mouse duplex DNA; similarly, most human DNA strands anneal with complementary human DNA strands. However, some strands of the mouse DNA will associate with human DNA to yield **hybrid duplexes**, in which segments of a mouse DNA strand form base-paired regions with segments of a human DNA strand (**Figure 6-31**). The ability to hybridize DNA from different species is a valuable laboratory tool for exploring evolutionary relationships. Different species have proteins and RNAs with similar functions— and often similar structures. In many cases, the DNAs encoding these proteins and RNAs have similar sequences. The closer the evolutionary relationship between two species, the more extensively their DNAs will hybridize. For example, human DNA hybridizes much more extensively with mouse DNA than with DNA from yeast.

The hybridization of DNA strands from different sources forms the basis for powerful techniques used in classical molecular genetics. Although these hybridization techniques are used less often as new, high-throughput approaches based on DNA sequencing become available (as described in Chapters 7 and 8), they are still widely used in research laboratories worldwide. Such techniques

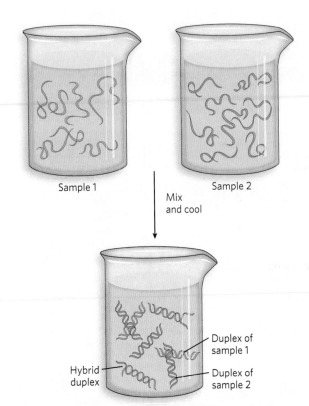

Sample 1 Sample 2

Mix and cool

Duplex of sample 1

Hybrid duplex

Duplex of sample 2

FIGURE 6-31 Cross-species DNA hybridization. Two DNA samples can be compared by heating a mixture of the DNAs to denature the strands and then cooling the mixture to allow complementary strands to form duplexes. The greater the sequence similarity between the two DNA samples, the more hybrid duplexes will form, with one strand derived from the first species and the other from the second.

have laid the foundation for understanding the study of nucleic acids.

A specific gene's DNA or RNA sequence can be detected in the presence of many other sequences by hybridization with a **probe**, a carefully chosen nucleic acid sequence complementary to the gene of interest. To be visualized in the laboratory, the probe must be labeled in some way, usually radioactively or with a fluorophore (a compound carrying a fluorescent group). The probe that is selected depends on what is known about the gene under investigation. Sometimes, a gene from another species that has sequence similarity to the gene of interest makes a suitable probe. If the protein product of a gene has been purified, probes can be designed and synthesized by working backward from the amino acid sequence, deducing the DNA sequence that would code for it. Or, researchers can often obtain the DNA sequence information necessary for creating a probe from sequence databases that detail the structure of millions of genes from a wide range of organisms. Because base-pairing stability is sensitive to pH and temperature, these parameters can be adjusted experimentally to detect nucleic acid sequences with varying degrees of complementarity to the probe. The technique is sensitive enough to reveal sequences that differ by a single base pair. This can be critically important in medical and forensic applications, given that two people can share genetic information that differs at only one or a few base pairs, called single-nucleotide polymorphisms (SNPs).

In other common hybridization methods, **gel electrophoresis** is used to separate DNA or RNA molecules by size (**Figure 6-32a**). A variation of gel electrophoresis, used to detect proteins under denaturing conditions, was discussed in Chapter 4 (see Highlight 4-1). Here, the gel matrix is not denaturing but instead is made of agarose, a kelp-derived material that does not disrupt nucleic acid base pairing. The starting nucleic acid sample, in solution in a test tube, is applied to a slot at one end of the gel, and a voltage is applied. Because DNA and RNA molecules are negatively charged, they migrate toward the positive end of the gel matrix in the electric field. Larger molecules tend to move more slowly than smaller ones, so this provides a means of separating nucleic acids by size. Following electrophoresis, the DNA fragments are transferred to a nitrocellulose membrane so that their positions in the gel relative to each other are preserved on the membrane. Once on the membrane, the nucleic acid can be hybridized with a DNA or RNA probe, labeled so that it can be detected by measuring radioactivity or fluorescence.

When used to detect DNA, this method is known as **Southern blotting**, named for Edwin Southern, who invented the technique at the University of Edinburgh. When used for detecting RNA, the technique is called **Northern blotting**, because of its similarity to the Southern method. Applications of these techniques include identifying a person on the basis of a single hair left at the scene of a crime or predicting the onset of a disease in an individual decades before symptoms appear. Northern blotting can also be used to detect the levels of a particular type of RNA in different body tissues (**Figure 6-32b**), providing fascinating insight into how cells regulate the expression of genes. Notably, these classical methods of Southern and Northern blotting are still used to answer specific experimental questions, despite the development of high-throughput strategies based on DNA sequencing technology (see Chapters 7 and 8).

Nucleotides and Nucleic Acids Undergo Uncatalyzed Chemical Transformations

Purines and pyrimidines, and the nucleotides of which they are a part, can undergo spontaneous alterations in their covalent structure. The rate of these reactions is generally very slow, but as noted earlier, they are physiologically significant because of the cell's low tolerance for changes in its genetic information. Alterations in DNA structure that produce permanent genetic changes are known as **mutations**. Extensive evidence suggests an intimate link between the accumulation of mutations in an individual organism and the processes of aging and carcinogenesis.

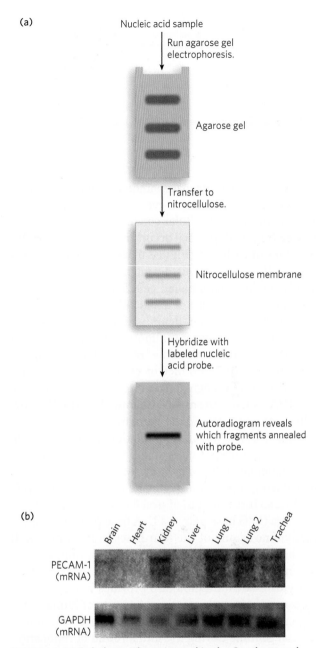

FIGURE 6-32 Gel electrophoresis used in the Southern and Northern blotting techniques. (a) Gel electrophoresis is used to size-fractionate a DNA (Southern blotting) or RNA (Northern blotting) mixture. The samples are then transferred to (i.e., are blotted onto) a nitrocellulose membrane, where they are detected using short radiolabeled oligonucleotide probes that base-pair to the samples on the membrane. (b) Northern blot analysis of RNA isolated from various human tissues. For each sample, approximately 10 µg of total RNA was separated on a 1.2% agarose-formaldehyde gel, transferred to a membrane, and hybridized to a ^{32}P-labeled probe—an mRNA for human platelet endothelial cell adhesion molecule (PECAM-1). The same blot was also probed with a cDNA (complementary DNA, a DNA copy of an mRNA sequence; see Chapter 7) for glyceraldehyde 3-phosphate dehydrogenase (GAPDH) to control for the amount of material in each lane. (GAPDH mRNA is used as a control because it is found in all tissues, in almost equal amounts.) Note the differences in PECAM-1 RNA levels detected in the different tissues; two bands are observed for PECAM-1 in each lane because there are two distinct forms of the mRNA for this gene. [Source: (b) Courtesy of Sheibani Lab.]

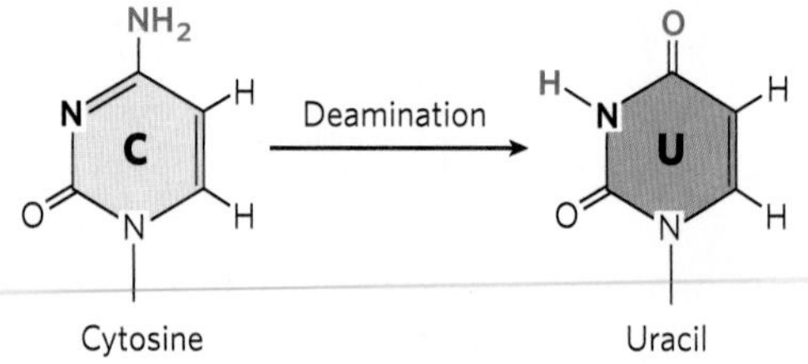

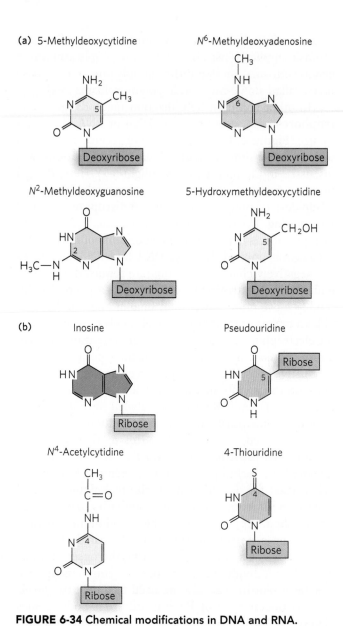

FIGURE 6-33 Cytosine deamination to uracil. Only the base is shown.

Several nucleotide bases undergo **deamination**, a spontaneous loss of their exocyclic amino groups. For example, under typical cellular conditions, deamination of cytosine (in DNA) to uracil occurs in about 1 of every 10^7 C residues in 24 hours (**Figure 6-33**). This corresponds to about 100 spontaneous deamination events per day, on average, in a mammalian cell. Deamination of adenine and guanine occurs at about 1/100th this rate.

The slow cytosine deamination reaction seems innocuous enough, but it is almost certainly the reason that DNA contains thymine rather than uracil. In DNA, uracil is the product of cytosine deamination, and it is readily recognized as foreign and is removed by a DNA repair system (see Chapter 12). If DNA normally contained uracil, the recognition of U residues resulting from cytosine deamination would be more difficult, and unrepaired uracils would lead to permanent sequence changes as they were paired with adenines during replication (cytosine normally pairs with guanine, so introduction of uracil into DNA effectively changes a C≡G base pair to a U–A base pair). Establishing thymine as one of the four bases in DNA may well have been a crucial turning point in evolution, making the long-term storage of genetic information possible.

Cytosine deamination also provides innate cellular defense against viral infection. A family of human proteins called APOBECs catalyze cytosine deamination in the viral genome during the initial round of replication by HIV. This hypermutation results in many nonviable viral particles, eventually destroying the coding capacity of the virus. In HIV and related viruses, the viral protein Vif binds to APOBECs and triggers their degradation. Vif has therefore become an important antiviral target, because viruses lacking this protein are much less capable of establishing chronic infection in human cells.

Base Methylation in DNA Plays an Important Role in Regulating Gene Expression

Certain nucleotide bases in DNA molecules are enzymatically methylated, usually after DNA synthesis is complete. Adenine and cytosine are methylated more

FIGURE 6-34 Chemical modifications in DNA and RNA. (a) Modified nucleotides in DNA. The most common postsynthetic modification to DNA is base methylation. 5-Methyldeoxycytidine occurs in the DNA of animals and higher plants; the other methylated bases shown here are produced by specific enzymes. (b) Modified nucleotides in RNA. Enzyme-catalyzed RNA base modifications are common in tRNA and rRNA, although the function of such alterations is not always clear. The presence of N^4-acetylcytidine in bacterial tRNAs may enhance protein synthesis.

frequently than guanine (**Figure 6-34a**). Methylation is generally confined to certain sequences or regions of a DNA molecule. For example, more than half of all CpG sequences in mammalian genomes are methylated on the C residue. Methylation tends to inhibit gene expression, because the methylated DNA is not efficiently copied into RNA. In many cancers, gene regulatory regions in DNA become abnormally hypermethylated. This can result in the silencing of genes that would otherwise control cell growth. DNA methylation may affect gene transcription by physically blocking the binding of proteins that facilitate transcription. Other proteins, however, can specifically bind to methylated DNA and recruit additional proteins that help form compact, inactive regions of chromosomal DNA.

All known DNA methylases (methyltransferases) use S-adenosylmethionine as a methyl group donor (see Figure 6-13). *E. coli* has two prominent methylation systems. One serves as part of a defense mechanism that helps the cell distinguish its DNA from foreign DNA by marking its own DNA with methyl groups and destroys foreign, nonmethylated DNA, a process known as restriction modification (discussed in Chapter 7). The other system methylates A residues in the sequence 5′-GATC-3′ to form N^6-methyldeoxyadenosine. Methylation in this case is mediated by the Dam (DNA adenine methylation) methylase, a component of a system that repairs the mismatched base pairs that occasionally form during DNA replication (see Chapter 12).

In eukaryotic cells, about 5% of all C residues in DNA are methylated to form 5-methyldeoxycytidine (see Figure 6-34a). As noted above, methylation is most common at CpG sequences, producing methyl-CpG symmetrically on both strands of the DNA. The observed 5% methylation frequency of C residues is lower than would be expected based on the random presence of CpG sequences in a genome, and, in fact, CpG dinucleotides occur much less often in vertebrate genomes than predicted by chance. For example, in the human genome, which has a G≡C content of ~40%, the frequency of CpG would be predicted as $0.2 \times 0.2 = 0.04$, or 4%, but the frequency is closer to just 1%. The extent of methylation of CpG sequences varies with molecular region in large eukaryotic DNA molecules. In addition to the tendency of methylation of C residues to inhibit gene expression, methylation suppresses the migration of segments of DNA called transposons (see Chapter 14).

RNA Molecules Are Often Site-Specifically Modified In Vivo

Like DNA, many functional RNAs are posttranscriptionally modified at specific nucleotides (**Figure 6-34b**). Some of the first examples were discovered in ribosomal and transfer RNAs. In some cases, modifications involve the addition of a functional group to an existing nucleotide in the sequence. For example, a methyl group can be added to the 2′ hydroxyl of ribose, thereby blocking its ability to form a hydrogen bond. In bacteria, some C residues of tRNAs are modified to N^4-acetylcytidine in a process thought to contribute to the accuracy of protein synthesis. In other cases, the base itself is changed, or its linkage to the ribose—the glycosidic bond—is altered. For instance, inosine, 4-thiouridine, and pseudouridine are relatively common in tRNAs and rRNAs.

Many of the enzymes that catalyze these chemical modifications of RNA have been identified. They are often evolutionarily conserved, indicating that RNA modification has been occurring in biological systems for a long time. More difficult to figure out is the function of these chemical changes in RNA. Molecular biologists can produce unmodified versions of RNAs in the laboratory and compare their functions with those of the chemically altered counterparts isolated from cells. This approach has only rarely discerned much of an effect of a modified base. However, genetic experiments in which an RNA-modifying enzyme is mutated or deleted from an organism suggest that these enzymes give cells a subtle but important selective advantage over organisms that do not modify their RNA. Some evidence supports the hypothesis that RNA modifications stabilize RNA structures and help RNAs interact with proteins in the cell.

The Chemical Synthesis of DNA and RNA Has Been Automated

Knowledge of DNA and RNA chemistry provided the basis for devising methods to synthesize nucleic acids in the laboratory. This technology has paved the way for many biochemical advances that depend on the ability to synthesize oligonucleotides with any chosen sequence. The chemical methods for synthesizing nucleic acids were developed primarily by H. Gobind Khorana and his colleagues in the 1970s. Refinement and automation of these methods have made possible the rapid and accurate synthesis of DNA strands.

DNA (or RNA) synthesis is carried out with the growing strand attached to a solid support (**Figure 6-35**). First, a nucleotide is attached to the support, a glass or polystyrene bead, through its 3′-hydroxyl group, and polynucleotide synthesis proceeds in the 3′→5′

KEY CONVENTION

When a chemical group attached to an atom in the purine or pyrimidine ring is altered, the ring position of the substituent is indicated by the number of that atom—for example, 5-methylcytosine, 7-methylguanine, and 5-hydroxymethylcytosine; the element to which the substituent is attached (N, C, O) is not identified. When a chemical group is altered on an exocyclic atom, the type of atom is identified and the ring position to which it is attached is denoted with a superscript. For example, the amino nitrogen attached to C-6 of adenine is N^6; the carbonyl oxygen and amino nitrogen at C-6 and C-2 of guanine are O^6 and N^2, respectively.

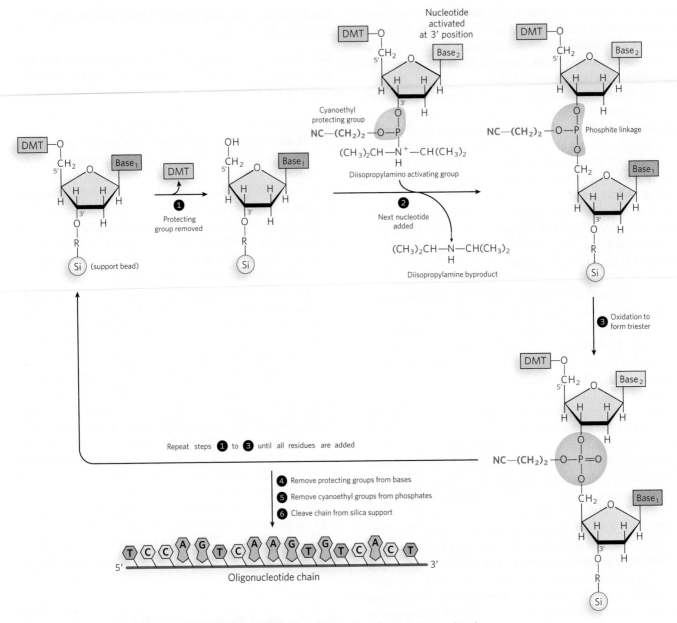

FIGURE 6-35 Solid-phase synthesis of nucleic acids. The oligonucleotide is synthesized in the 3′→5′ direction, starting with a single nucleotide that is covalently attached to a solid support, such as a glass bead (Si). In a repeated series of chemical reactions, nucleotides protected by the dimethoxytrityl (DMT) group from unwanted reactions are sequentially deprotected and reacted to produce a new phosphodiester linkage.

direction. This is the opposite of the direction of biological polynucleotide synthesis by polymerase enzymes, which is 5′→3′. Functional groups on the bases and phosphates, including hydroxyl and amine groups, are transiently protected with chemical groups that are readily removed after synthesis is complete. The 5′-hydroxyl group is temporarily protected by a dimethoxytrityl (DMT) group; the DMT group is removed from the end of the growing polymer chain at the beginning of each cycle (step 1 in Figure 6-35) to permit extension of the chain by another nucleotide (step 2).

Oxidation of the phosphite linkage between the nucleotides completes the cycle (step 3). When chain synthesis is complete, protecting groups are removed from the bases and phosphates, and the oligonucleotide chain is cleaved from its solid support (steps 4, 5, 6). The efficiency of each addition step is very high, allowing the routine laboratory synthesis of polymers containing 70 to 80 nucleotides and, in some laboratories, much longer strands.

Oligonucleotide synthesis is very useful for techniques such as Southern and Northern blotting and for the polymerase chain reaction (PCR) and DNA sequencing, which are discussed in Chapter 7. In addition, chemical synthesis makes it possible to incorporate chemical modifications in the polymer product, such as biotin groups, extra phosphates, sulfhydryl groups, and methyl groups. These functional groups are useful for such applications as specific labeling of a DNA strand or stabilization of an RNA oligonucleotide against enzymatic degradation in cells.

SECTION 6.4 SUMMARY

- Native DNA undergoes reversible unwinding and separation of the strands (melting) on heating or at extremes of pH. DNAs rich in G≡C base pairs have higher melting points than DNAs rich in A=T base pairs.
- Hybridization, or base pairing between two strands of nucleic acid from different sources, is the basis for important techniques used to study and isolate specific genes and RNAs.
- Southern blotting is a method by which a specific DNA sequence can be identified in a mixture, following size-based fractionation of the DNA sample by agarose gel electrophoresis. A probe complementary to the DNA of interest is labeled with a radioactive or fluorescent functional group. After transferring the size-fractionated DNA from the gel to a membrane, the probe is hybridized to the sample on the membrane so that the sequence of interest can be visualized.
- Northern blotting, analogous to Southern blotting, is used for detecting specific RNA sequences.
- Mutations are alterations in DNA structure that produce permanent changes in the encoded genetic information. Deamination of cytosine is a common chemical mutation in DNA that can damage the genetic code if not corrected by the cell. Deamination of viral nucleic acid can be used to defend against viral infection.
- Certain A and C residues in DNA are often enzymatically methylated after DNA synthesis. *E. coli* uses methylation to distinguish between host and foreign DNA and to facilitate the repair of mismatched base pairs that arise from replication errors. In eukaryotes, DNA methylation often inhibits gene expression.
- Some residues in RNAs are chemically modified by enzymes that introduce methyl or acetyl groups at specific sites or alter a nucleotide base in other ways. These modifications may stabilize RNA structures and can also influence RNA recognition by proteins.
- DNA and RNA polymers of any sequence can be synthesized with simple, automated procedures involving chemical and enzymatic methods. Solid-phase synthesis of DNA and RNA occurs in the $3' \rightarrow 5'$ direction, using chemically protected nucleotides that are selectively deprotected and coupled to the growing polynucleotide chain in successive cycles.

? UNANSWERED QUESTIONS

Although many details of nucleic acid structure are well understood, future challenges involve linking the chemistry of these molecules to their behavior in biological systems. Here are several interesting questions in the field.

1. What are the functions of noncanonical DNA structures in cells? We do not yet know whether non-B-DNA functions in specific cellular processes. For example, some evidence suggests that with its left-handed twist, Z-DNA relieves some of the torsional strain that would otherwise build up during DNA transcription. Perhaps for this reason, the potential to form Z-DNA structures correlates with genomic regions of active transcription. But definitive proof of these ideas has been elusive. Whether three-stranded or four-stranded structures are biologically relevant is also a topic that remains ripe for experimentation.

2. Do mRNAs have stable three-dimensional structures? Although mRNAs were once thought to be spaghetti-like molecules, increasing evidence hints that they may have stable structures that contribute to biological function. For example, many mRNAs include long sequences that extend beyond the coding region of the gene and are critical for proper gene regulation. Specific proteins bind to these regions and probably recognize structures within them.

3. How widespread is chemical modification of RNA? Modified nucleotides in tRNA and rRNA have been recognized for a long time, but we do not know whether other RNAs in cells contain such chemical changes. This is an important question, because modifications could influence the function of RNAs that play various roles in controlling gene expression and therefore might be relevant to understanding disease pathways.

HOW WE KNOW

DNA Is a Double Helix

Sayre, A. 1975. *Rosalind Franklin and DNA.* New York: W.W. Norton & Co.

Watson, J.D. 1968. *The Double Helix: A Personal Account of the Discovery of the Structure of DNA.* New York: Atheneum.

Watson, J.D., and F.H. Crick. 1953. Molecular structure of nucleic acids: A structure for deoxyribose nucleic acid. *Nature* 171:737–738.

By the early 1950s, DNA had been confirmed as the genetic material in cells, but its structure was unknown. Given that structural information would be key to understanding heredity, the race was on to solve the mystery of DNA structure. Researchers including Rosalind Franklin, Raymond Gosling, and Maurice Wilkins were measuring x-ray diffraction of DNA fibers generated by drawing them from solution using a glass rod. The diffraction patterns reflected the symmetry of the DNA molecules in the fibers, thereby providing an important clue to the molecules' overall arrangement. By examining the diffraction pattern from fibers oriented perpendicular to the x-ray beam, investigators could deduce the helical symmetry of the molecules inside. These data were interpreted in the light of Chargaff's rules, which state that in a given DNA sample, the fraction of A equals the fraction of T, and the fraction of C equals the fraction of G. Possible models of the DNA in the fiber were produced, and their calculated diffraction patterns were compared with the experimentally derived patterns.

Rosalind Franklin's famous Photograph 51 revealed a particularly well-resolved x-ray diffraction pattern of a DNA fiber that was interpreted to determine the 3.4 Å distance between base pairs and the 34 Å periodicity of the helix (characteristic of B-form DNA; see Section 6.2) (Figure 1). The darker spots are areas where the film was hit repeatedly with diffracted x-rays from repeating parts of the DNA molecule. At the top and bottom of the photograph, for example, dark patches represent the nucleotide bases of DNA—the patches are dark because the many bases in the DNA fiber are arranged in a regular pattern. The distance between bases in the DNA structure could be determined by measuring the distance between the dark patches on the film and then making a calculation based on how far the DNA sample was from the x-ray film and how it was positioned relative to the direction of the incident x-ray beam. Watson and Crick made extensive use of this image, along with related diffraction data, to develop a model of the three-dimensional structure of DNA that proved to be correct (Figure 2). This was done at a time before sophisticated computer modeling

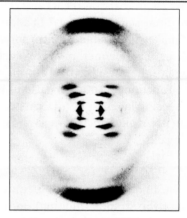

FIGURE 1 Franklin's "Photograph 51" provided the information necessary to solve the double-helical structure of DNA. *[Source: Omikron/Science Source.]*

was possible: Watson and Crick presented their work as a physical model of the double helix constructed on a wire support! Unlike other, competing models of DNA, the Watson-Crick structure had the sugar–phosphate backbone winding around the outside of the helix, with the bases pointing to the interior, where they formed base-pairing interactions between the two strands.

Watson and Crick's work was published in a letter to the British journal *Nature* in 1953. In the same issue, several other papers provided experimental support for the Watson-Crick model. The double-helical structure immediately suggested a mechanism by which DNA strands could be faithfully copied from one generation to the next. In a famously understated final sentence of their paper, Watson and Crick wrote: "It has not escaped our notice" that the specificity of base pairing could ensure accurate DNA replication.

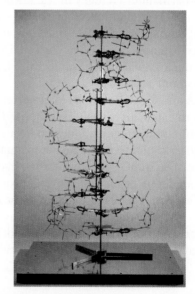

FIGURE 2 A replica of the DNA model built by Watson and Crick. The original model is on display in the Science Museum in London. *[Source: Science Museum/SSPL.]*

DNA Helices Have Unique Geometries That Depend on Their Sequence

Wing, R., H. Drew, T. Takano, C. Broka, S. Tanaka, K. Itakura, and R.E. Dickerson. 1980. Crystal structure analysis of a complete turn of B-DNA. Nature 287: 755–758.

The discovery of the DNA double helix marked the dawn of molecular biology. However, it was not until 1980 that the first single crystal of a DNA molecule was obtained. This was an important landmark in its own right, because for the first time it became possible to determine the exact helical parameters of a defined DNA sequence. Why did it take almost 30 years after the work of Watson, Crick, Franklin, Wilkins, and Gosling for specific DNA sequences to be crystallized?

The answer is technology. Until the late 1970s, it wasn't possible to synthesize DNA molecules in the laboratory, so investigators could not produce enough of a specific sequence to make growth of single crystals feasible. Once the methodology was available to synthesize DNA oligonucleotides on solid supports, short DNA molecules of specific length and sequence could be produced in milligram quantities. This material could be purified, and it crystallized readily when concentrated slowly in the presence of suitable buffers. Single crystals of DNA offered some distinct advantages over the DNA fibers analyzed by Franklin, Gosling, and Wilkins. DNA fibers can readily form from a mixture of DNAs of different lengths and sequences, and the structures obtained by analyzing the fiber diffraction patterns produce an "averaged" structure of all the molecules in the fiber. In contrast, single crystals, by definition, are formed by arrays of identical molecules.

Richard Dickerson and his colleagues recognized the wealth of information to be gained by solving a molecular structure of single DNA crystals. They used a self-complementary dodecamer sequence, CGCGAATTCGCG, to solve the first single-crystal structure of DNA. The overall double-helical structure agreed well with that determined by Watson and Crick, but many new details about the geometry of the helix were revealed (Figure 3). This structure, known as the Dickerson dodecamer, ushered in an era of high-resolution structural determinations of DNA and, eventually, the crystallographically determined structures of specific DNAs bound to protein partners. The study of individual DNA sequences also led to extensive studies of DNA–small molecule interactions and to research on the effects of DNA mutations on helical geometry. This work guided the development of certain anticancer drugs, such as cisplatin, that bind and distort DNA and thereby disrupt its replication in rapidly growing cells.

FIGURE 3 The Dickerson dodecamer structure revealed, for the first time, the details of helical geometry for a specific DNA sequence. The drawings are oriented to show the major groove (left) and minor groove (right) in the B-DNA helix. [*Source: Reprinted by permission from Macmillan Publishers Ltd. Wing R., Drew H., Takano T., Broka C., Tanaka S., Itakura K., Dickerson R.E. Nature. 1980 Oct 23; 287(5784):755-8. Figure 4. © 1980.*]

HOW WE KNOW

Ribosomal RNA Sequence Comparisons Provided the First Hints of the Structural Richness of RNA

Gutell, R.R., N. Larsen, and C.R. Woese. 1994. Lessons from an evolving rRNA: 16S and 23S rRNA structures from a comparative perspective. *Microbiol. Rev.* 58:10–26.

Ribosomes have been around for a long time, and the sequences of the RNAs they contain have been constrained over the course of evolution by the requirements of making functional ribosomes to catalyze protein synthesis. Carl Woese recognized in the 1960s that comparing ribosomal RNA sequences would provide valuable information about the evolutionary relationships among different organisms. Working over many years, he and his colleague Harry Noller assembled careful alignments of the 16S and 23S rRNAs from a large number of microbes. This work led Woese to propose the three-domain theory of life: Eubacteria (now classified simply as Bacteria), Archaebacteria (now Archaea), and Eukarya (eukaryotes).

The comparative analysis approach begun by Woese and Noller was continued by Robin Gutell, who expanded the comparison to include 16S and 23S rRNA sequences from multicellular organisms, including humans. Gutell's critical analysis provided the first hints that these RNAs form specific three-dimensional structures important to their function. One of the key insights from comparative rRNA sequence analysis was the discovery of noncanonical (i.e., non-Watson-Crick) base pairings. Although these had already been observed in tRNA structures, the much larger sizes of rRNA sequences provided vastly more data. For both 16S and 23S rRNAs, much of the sequence could be folded up into base-paired segments (Figure 4). Comparisons between species showed that the base pairings were much more conserved than were the actual nucleotide sequences. This was because a change in the identity of a nucleotide on one side of a base-paired stretch was typically matched by a mutation in its base-pair partner such that base pairing was maintained. Gutell and Woese also noticed that in many cases, such compensatory base changes occurred for base "pairs" not previously thought to form, such as G–U, A–A, and G–A (see Figure 6-24). In this way, long before high-resolution structures of large RNAs became available, it was clear that RNA molecules are much more tolerant of non-Watson-Crick base pairings than is DNA.

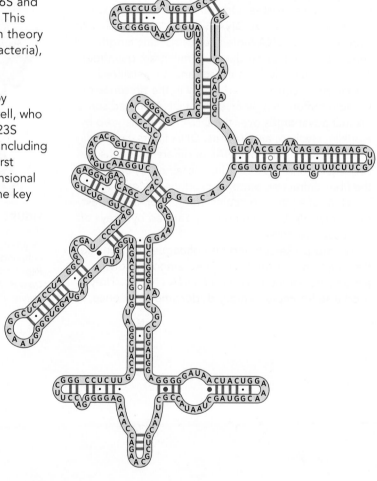

FIGURE 4 This model of a portion of the secondary structure for *E. coli* 16S rRNA shows canonical base pairs connected by red lines, G–U pairs connected by small black dots, A–G pairs connected by open circles, and other noncanonical pairings connected by solid circles. [Source: J. J. Cannone et al., The Comparative RNA Web (CRW) Site: An Online Database of Comparative Sequence, 2002.]

KEY TERMS

nucleotide, 175
heterocyclic compound, 175
adenine (A), 175
guanine (G), 175
cytosine (C), 175
thymine (T), 175
uracil (U), 175
deoxyribonucleotide, 175
ribonucleotide, 175
phosphodiester bond, 177
oligonucleotide, 178
polynucleotide, 178
base pairing, 179
Chargaff's rules, 183
antiparallel, 183
B-form DNA (B-DNA), 185

A-form DNA (A-DNA), 185
Z-form DNA (Z-DNA), 185
palindrome, 188
hairpin, 188
RNA secondary structure, 192
denaturation, 196
melting, 196
annealing, 197
hypochromic effect, 197
hyperchromic effect, 197
melting point (T_m), 197
hybrid duplex, 198
probe, 198
gel electrophoresis, 199
Southern blotting, 199
Northern blotting, 199

PROBLEMS

1. In the 1980s, Tom Cech and Sidney Altman discovered that RNA could function as an enzyme. List two properties of enzymes that must hold true for this characterization to be correct (see Chapter 5).

2. What positions in the ring of a purine nucleotide in DNA have the potential to form hydrogen bonds but are not involved in Watson-Crick base pairing?

3. During his studies of the base composition of DNAs from various species, Erwin Chargaff obtained the following data for several human samples. The data show ratios of moles of each base to moles of phosphate in samples from various tissue types. Note that the error in the molar ratios is about ±0.03.

	Sperm 1	Sperm 2	Thymus	Liver Carcinoma
Adenine	0.29	0.27	0.28	0.27
Guanine	0.18	0.17	0.19	0.18
Cytosine	0.18	0.18	0.16	0.15
Thymine	0.31	0.30	0.28	0.27
Total DNA recovered	0.96	0.92	0.91	0.87

What can you conclude from these data?

4. A part of one strand of a double-helical DNA molecule has the sequence 5′-GATTACAGCCTTAGTTAAATTCTAA GGCTGGTA-3′.

(a) Write out the sequence of the complementary strand of DNA.

(b) Does the strand shown above have the potential to form any kind of alternative DNA structure? Does the double-stranded DNA of which it is a part have the potential to form an alternative structure? If so, what structure or structures might form?

5. A double-stranded DNA oligonucleotide in which one of the strands has the sequence 5′-TAATACGACTCAC TATAGGG-3′ has a melting temperature (T_m) of 59°C. If a double-stranded RNA oligonucleotide of identical sequence (substituting U for T) is constructed, will its T_m be higher or lower?

6. If the DNA and RNA oligonucleotides of Problem 5 are both present in an aqueous solution near neutral pH, how will their structures differ (apart from the presence of U in RNA vs. T in DNA)?

7. Why does DNA predominantly contain thymine rather than uracil as one of its pyrimidine bases?

8. A trinucleotide structure is shown below.

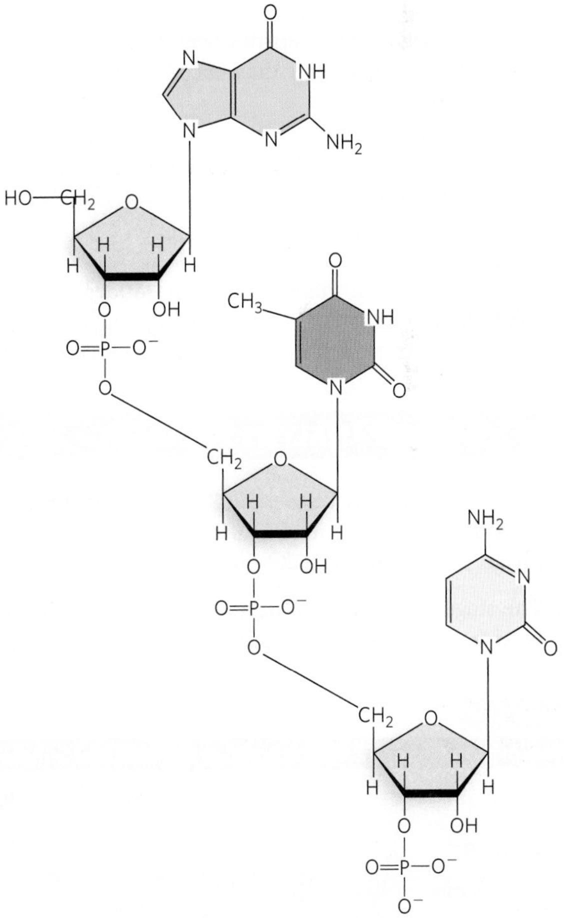

(a) As drawn, which end of the trinucleotide, 3′ or 5′, is at the top?

(b) What is the three-letter sequence of this trinucleotide?

(c) Is this trinucleotide RNA or DNA?

9. Part of a chromosome has the following sequence (on one strand): 5'-ATTGCATCCGCGCGTGCGCGCGCGATCCCGT TACTTTCCG-3'. Underline the part of this sequence that is most likely to take up the Z conformation.

10. Why does DNA form a double-helical structure?

11. Why are DNA and RNA considered acids?

12. The cells of many eukaryotic organisms have highly specialized systems that repair G–T mismatches in DNA. The mismatch is always converted to a $G \equiv C$, never to an $A = T$ base pair. This G–T mismatch repair system occurs in addition to a more general repair system that fixes virtually all types of mismatches. Suggest why cells might require a specialized G–T mismatch repair system, and why cells would specifically convert the G–T to a $G \equiv C$.

13. Food for thought: If a tRNA sequence were synthesized as DNA (to form a tDNA) instead of RNA, could it function in protein synthesis? Why or why not?

14. What dictates the strength of association between two DNA strands?

15. What sequence characteristics would you expect for regions of a chromosome that encode highly structured RNA molecules?

16. Why were single crystals critical to determining the molecular structure of DNA? (Hint: See the How We Know section in this chapter.)

17. In the structure of the modified nucleotides used in the solid-phase method for synthesizing oligonucleotides, what is the function of the dimethoxytrityl (DMT) group?

DATA ANALYSIS PROBLEM

Hershey, A.D., and M. Chase. 1952. Independent functions of viral protein and nucleic acid in growth of bacteriophage. *J. Gen. Physiol.* 36:39–56.

18. The stage occupied by Watson and Crick in 1953 was set by many other scientists, primarily Alfred Hershey and Martha Chase (see the How We Know section in Chapter 2). By 1952, a number of experiments had pointed to DNA as the genetic material, but much controversy remained, and protein was still regarded as a prime candidate. The Hershey and Chase experiments published in 1952 are credited with eliminating any remaining doubt in the scientific community that DNA alone was the genetic material.

The experiments made use of *E. coli* and one if its viruses, bacteriophage T2. When the experiments began, it was known that T2 consisted of a DNA core surrounded by a protein coat. T2 attached itself to the bacterial host and injected its core material into the host cell. New copies of T2 were made within the host, and the bacterial cell was lytically destroyed when the new T2 copies were released. Although it was clear that the T2 protein coat remained outside the bacterial cell, many workers thought that both DNA and some protein were introduced into the host when the genetic material was injected by T2, or that the attached protein shells played a role in the production of the T2 progeny. In either case, it was possible that protein was the genetic material.

Prior to their famous blender experiment (described shortly), Hershey and Chase carried out a series of experiments to better establish what parts of the T2 bacteriophage were introduced into the bacterial cells. They knew that T2 could be inactivated by osmotic shock, leaving behind T2 "ghosts" consisting of the viral coats bereft of their internal contents (which were released into the medium). Two different batches of T2 bacteriophage were grown, one labeled with ^{35}S and the other with ^{32}P. Half of each preparation was subjected to osmotic shock by incubation in 3 M NaCl followed by rapid dilution into distilled water. The authors described preparations thus treated as "plasmolyzed." Each of the four resulting preparations was further subjected to four additional treatments: (1) Addition of acid, followed by centrifugation; the supernatant was monitored for acid-soluble radioactivity. (2) Treatment with DNase (an enzyme that reduces DNA to nucleotides), followed by the addition of acid and centrifugation; acid-soluble radioactivity was again determined. (3) Addition of bacterial host cells, followed by centrifugation and determination of radioactivity in the bacterial cell pellet. (4) Treatment with T2 immune sera (antibodies), followed by centrifugation of the immune-precipitated material. The results are shown in Table 1.

TABLE 1

Composition (%) of T2 Bacteriophage Ghosts and Solution of Plasmolyzed Phage

Preparation	Whole Phage Labeled with:		Plasmolyzed Phage Labeled with:	
	^{32}P (%)	^{35}S (%)	^{32}P (%)	^{35}S (%)
1. Acid-soluble	—	—	1	—
2. Acid-soluble after treatment with DNase	1	1	80	1
3. Adsorbed to sensitive bacteria	85	90	2	90
4. Precipitated by T2 antibodies	90	99	5	97

TABLE 2

Effect of Multiplicity of Infection on Elution of Phage Membranes from Infected Bacteria

Running Time in Blender (min)	Multiplicity of Infection	^{32}P-labeled Phage		^{35}S-labeled Phage	
		Isotope Eluted (%)	Infected Bacteria Surviving (%)	Isotope Eluted (%)	Infected Bacteria Surviving (%)
0	0.6	10	120	16	101
2.5	0.6	21	82	81	78
0	6.0	13	89	46	90
2.5	6.0	24	86	82	85

(a) What macromolecules are labeled by ^{35}S? What macromolecules are labeled by ^{32}P?

(b) Why didn't the researchers use other common radioactive labels that were available at the time (^{14}C and ^{3}H)?

(c) Little or no radioactivity appears in the first acid-soluble supernatant in any of the preparations. Explain.

(d) Most of the ^{32}P label, but little of the ^{35}S, appears in the acid-soluble supernatant after DNase treatment. Explain.

(e) In the samples not treated with osmotic shock, most of both labels was found adsorbed to bacteria (in the cell pellet), but only the ^{35}S label appeared in the cell pellet in the plasmolyzed samples. Explain.

(f) In the control samples, both labels were precipitated by the T2 antibodies, but only the ^{35}S was precipitated in the plasmolyzed samples. Explain.

(g) What general conclusions can you draw from these experiments?

Several additional experiments, combined with those above, established that the bulk of the phage DNA was introduced into the cells during infection and thus might contribute to the production of progeny. About 30% of the ^{32}P label ended up in the progeny phage. But what happened to the phage protein? This consideration led to the blender experiment. Electron micrographs had previously shown that T2 attaches to the outside of bacterial cells and is attached to them by a long tail. The researchers reasoned that this "precarious attachment" could be eliminated by shearing.

Bacteria were infected with either ^{32}P- or ^{35}S-labeled phage. Two different multiplicities of infection (number of phage added per cell) were used, 0.6 and 6.0. After a few minutes, the cells were centrifuged and resuspended in fresh medium. Some were subjected to a 2.5-minute treatment in a blender. Samples were then taken. For one sample, the cells were centrifuged and the supernatant measured to determine how much isotope had been removed from the cells. Another sample was left to produce phage progeny and titrated to determine what fraction of

the infected cells was producing phage. The results are shown in Table 2 above.

(h) Why did the investigators centrifuge and then resuspend the phage after infection?

(i) How much of the ^{35}S label is stripped from the cells by the blender?

(j) What general conclusions can you draw from this experiment?

A careful subsequent examination of the progeny phage indicated that less than 1% of the ^{35}S label ended up in the progeny. Combining this with the other data, the researchers could make a clear case that DNA was the genetic material guiding the production of phage progeny.

ADDITIONAL READING

The Structure and Properties of Nucleotides

Frank-Kamenetskii, M.D., and S.M. Mirkin. 1995. Triplex DNA structures. *Annu. Rev. Biochem.* 64:65–95.

Friedberg, E.C., G.C. Walker, and W. Siede. 1995. *DNA Repair and Mutagenesis.* New York: W.H. Freeman and Company. A good source for more information on the chemistry of nucleotides and nucleic acids.

Keniry, M.A. 2000. Quadruplex structures in nucleic acids. *Biopolymers* 56:123–146. A good summary of the structural properties of quadruplexes (tetraplexes).

Rich, A., and S. Zhang. 2003. Timeline: Z-DNA—The long road to biological function. *Nat. Rev. Genet.* 4:566–572.

Shafer, R.H. 1998. Stability and structure of model DNA triplexes and quadruplexes and their interactions with small ligands. *Prog. Nucleic Acid Res. Mol. Biol.* 59:55–94.

Wells, R.D. 1988. Unusual DNA structures. *J. Biol. Chem.* 263:1095–1098. Minireview, presenting a concise summary.

DNA Structure

Collins, A.R. 1999. Oxidative DNA damage, antioxidants, and cancer. *Bioessays* 21:238–246.

Kornberg, A., and T.A. Baker. 2005. *DNA Replication*, 2nd ed. New York: W.H. Freeman and Company. The best place to start to learn more about DNA structure.

Marnett, L.J., and J.P. Plastaras. 2001. Endogenous DNA damage and mutation. *Trends Genet.* 17:214–221.

Olby, R.C. 1994. *The Path to the Double Helix: The Discovery of DNA*. New York: Dover Publications.

Sayre, A. 1978. *Rosalind Franklin and DNA*. New York: W.W. Norton & Company.

Sinden, R.R. 1994. *DNA Structure and Function*. San Diego: Academic Press. A fine discussion of many topics covered in this chapter.

Watson, J.D. 1968. *The Double Helix: A Personal Account of the Discovery of the Structure of DNA*. New York: Atheneum (paperback edition, New York: Simon & Schuster / Touchstone Books, 2001).

RNA Structure

Chang, K.Y., and G. Varani. 1997. Nucleic acids structure and recognition. *Nat. Struct. Biol.* 4(Suppl.):854–858. A description of the application of NMR (nuclear magnetic resonance) to the determination of nucleic acid structure.

Hecht, S.M., ed. 1996. *Bioorganic Chemistry: Nucleic Acids.* Oxford: Oxford University Press. A very useful text.

Moore, P.B. 1999. Structural motifs in RNA. *Annu. Rev. Biochem.* 68:287–300.

Saenger, W. 1984. *Principles of Nucleic Acid Structure.* New York: Springer-Verlag. A somewhat dated but classic text, containing one of the best in-depth compilations of nucleic acid structural data available.

7 Studying Genes

MOMENT OF DISCOVERY

Norman Arnheim [Source: Courtesy of USC.]

My lab at SUNY Stony Brook studied genetic variation in mouse ribosomal genes in the days before genomic sequencing or even the ability to readily synthesize DNA oligonucleotides. Using Ed Southern's method for hybridizing short DNA fragments to complementary sequences in genomic DNA samples, *we made restriction enzyme maps of mouse and human ribosomal DNA.*

My students borrowed some lambda phage DNA size markers from my colleague Ken Marcu to use in their Southern blotting experiments with the human and mouse DNA. Soon my students began telling me that a particular probe sequence from the nontranscribed region of a mouse ribosomal gene was lighting up a specific lambda phage DNA fragment. I can still see myself standing in the lab, looking at my students' Southern blots showing the lambda fragment—but knowing that it couldn't be lambda DNA that was complementary to the mouse sequence! So what was going on?

Ken told me he had modified the size marker for his work by adding a cloned sequence fragment from a mouse immunoglobulin gene, which immediately set us off in an interesting and unexpected new direction. Using a method called in situ hybridization, in which DNA probes are hybridized to intact chromosomes, we soon discovered that the mouse ribosomal DNA fragment was complementary to sequences on all the mouse chromosomes! So whatever this sequence was, it was interspersed throughout mouse genomic DNA. We ultimately found that this sequence was everywhere in human, bird, and amphibian DNA. We sequenced one of the complementary fragments and discovered that the sequence contained 17 tandem CA repeats. Although the investigation of CA repeat function is still ongoing, the repeats were used extensively for years in the genetic mapping of human diseases and for forensic purposes. I love the serendipity of science!

—*Norman Arnheim, on the discovery of interspersed CA repeats in genomic DNA*

Online resources related to this chapter:

***Nature* exercises**
Genes, Polymerase chain reaction, Sanger sequencing

Simulations
PCR, Sanger sequencing, CRISPR

The set of methods encompassed by biotechnology is fundamental to the advancement of modern biology, defining present and future frontiers and illustrating many important principles of the life sciences. With the ability to elucidate the laws governing enzyme catalysis, macromolecular structure, cellular metabolism, and information pathways, researchers can focus on the mechanisms of increasingly complex processes. Cell division, immunity, embryogenesis, sensory perception, oncogenesis, cognition—all are orchestrated in an elaborate symphony of molecular and macromolecular interactions that we are beginning to understand with increasing clarity. The real implications of the scientific journey begun in the nineteenth century are found in the ever-increasing power to analyze and alter living systems.

A major source of molecular insights about complex biological processes is the cell's own information archive: its DNA. The complement of genetic information in a cell—one complete copy of the information required to specify that organism—is known as the organism's **genome**. A molecular biologist is often interested in the function of one or a few genes in a genome. The sheer size of the genome, however, presents an enormous challenge: how does one find and study a particular gene among the tens of thousands of genes nested in the billions of base pairs of a mammalian genome?

In the 1970s, decades of advances by scientists working worldwide in genetics, biochemistry, cell biology, and physical chemistry came together in the laboratories of Paul Berg, Herbert Boyer, and Stanley Cohen. Solutions began to emerge, yielding techniques for locating, isolating, preparing, and studying small segments of DNA derived from much larger chromosomes. Techniques for DNA cloning paved the way to the modern fields of genomics, transcriptomics, and proteomics: the study of genes, mRNA transcripts, and proteins on the scale of whole cells and organisms (see Chapter 8). These approaches are transforming basic research, agriculture, medicine, ecology, forensics, and many other fields, while occasionally presenting society with difficult choices and ethical dilemmas.

Every student and instructor, when considering the topics we present in this text, encounters a conflict. The methods we describe were made possible by advances in molecular biology. Hence, one must understand the fundamental concepts of DNA replication, RNA transcription, protein synthesis, and gene regulation to appreciate how these methods work. But, at the same time, modern molecular biology relies on these same methods to such an extent that a current treatment of the subject becomes impossible without a proper introduction to the technology. By presenting these methods early in the book, we acknowledge that they are inextricably interwoven with

Clockwise from top left:
Paul Berg *[Source: Courtesy of the National Library of Medicine.]*
Herbert Boyer *[Source: Courtesy Genentech, Inc.]*
Stanley Cohen *[Source: Ted Streshinsky/Time Life Pictures/Getty Images.]*

both the advances that gave rise to them and the newer discoveries they now make possible. The background we necessarily provide makes the discussion here not just an introduction to technology but also a preview of many of the fundamentals of molecular biology encountered in later chapters.

This chapter is not designed to consider every method that is relevant to molecular biology. Indeed, we will be discussing many additional techniques in later chapters where they are uniquely relevant to a specific area of study. However, a particular set of methods—a set often associated with biotechnology—has become a critical engine of discovery driving the advances discussed throughout this text. Those methods now become our focus. We begin by outlining the principles of the now-classic discipline of DNA cloning, then illustrate the range of applications and the potential of many newer technologies that support and accelerate the advance of molecular biology. Finally, we take a look at some of the technologies that allow us to elucidate gene function.

7.1 ISOLATING GENES FOR STUDY (CLONING)

A **clone** is an identical copy. The term was originally applied to cells produced when a cell of a single type was isolated and allowed to reproduce to create a population of identical cells. **DNA cloning** involves separating a specific gene or DNA segment from a larger chromosome, attaching it to a small molecule of carrier DNA, introducing this modified DNA into a host cell, then replicating the DNA by increasing both the cell number and the copy number of the cloned DNA in each cell. The result is selective amplification of a particular gene or DNA segment.

The cloning of DNA from any organism entails five general steps:

1. *Obtaining the DNA segment to be cloned.* Enzymes called **restriction endonucleases** act as precise molecular scissors, recognizing specific sequences in DNA and cleaving genomic DNA into smaller fragments suitable for cloning. Alternatively, genomic DNA may be sheared randomly into fragments of a desired size. Or, since the sequence of targeted genomic regions is often known, some DNA segments to be cloned are simply synthesized.

2. *Selecting a small molecule of DNA capable of self-replication.* These small DNAs are called **cloning vectors** (a vector is a carrier or delivery agent). Most cloning vectors used in the laboratory are modified versions of naturally occurring small DNA molecules found in bacteria and lower eukaryotes such as yeast. Small viral DNAs may also play this role.

3. *Joining two DNA fragments covalently.* The enzyme **DNA ligase** links the cloning vector to the DNA fragment to be cloned. Composite DNA molecules of this type, comprising covalently linked segments from two or more sources, are called **recombinant DNAs**.

4. *Moving recombinant DNA from the test tube to a host organism.* The host organism provides the enzymatic machinery for DNA replication.

5. *Selecting or identifying host cells that contain recombinant DNA.* The cloning vector generally has features that allow the host cells to survive in an environment where cells lacking the vector would die. Cells containing the vector are thus "selectable" in that environment.

The methods used for accomplishing these and related tasks are collectively referred to as **recombinant DNA technology** or, more informally, **genetic engineering**.

Much of our initial discussion focuses on DNA cloning in the bacterium *Escherichia coli*, the first organism used for recombinant DNA work and still the most common host cell. *E. coli* has many advantages. Its DNA metabolism (like many of its other biochemical processes) is well understood, many naturally occurring cloning vectors associated with this bacterium are well characterized, and techniques are available for easily moving DNA from one bacterial cell to another. The principles discussed here are also broadly applicable to DNA cloning in other organisms, as we will see later in the chapter.

Genes Are Cloned by Insertion into Cloning Vectors

DNA can be cloned from any cellular or viral source. Although the approaches are determined partly by the DNA source and what is known about it, all cloning efforts have a few enzymes and procedures in common. Recombinant DNA technology relies on a set of enzymes made available through decades of research on nucleic acid metabolism (**Table 7-1**). Two classes of enzymes are particularly important: the restriction endonucleases (restriction enzymes) and DNA ligase (**Figure 7-1**). First, restriction endonucleases recognize DNA at specific **recognition sequences** (or restriction sites) and cleave it to generate a set of smaller fragments. Second, a DNA

TABLE 7-1

Some Enzymes Used in Recombinant DNA Technology	
Enzyme	*Function*
Type II restriction endonuclease	Cleaves DNA at specific base sequences
DNA ligase	Joins two DNA molecules or fragments
DNA polymerase I (*E. coli*)	Fills single-stand gaps in duplex DNA by stepwise addition of nucleotides to 3′ ends
Reverse transcriptase	Makes a DNA copy of an RNA molecule
Polynucleotide kinase	Adds a phosphate to the 5′-OH end of a polynucleotide, to label it or permit ligation
Terminal transferase	Adds homopolymer tails to the 3′-OH ends of a linear duplex
Exonuclease III	Removes nucleotide residues from the 3′ ends of a DNA strand
Bacteriophage λ exonuclease	Removes nucleotides from the 5′ ends of a duplex to expose single-stranded 3′ ends
Alkaline phosphatase	Removes terminal phosphates from the 5′ end, the 3′ end, or both

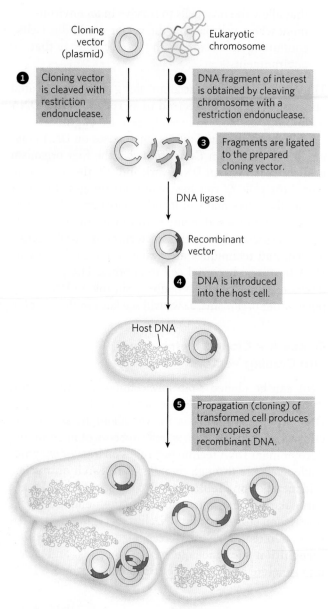

① Cloning vector is cleaved with restriction endonuclease.

② DNA fragment of interest is obtained by cleaving chromosome with a restriction endonuclease.

③ Fragments are ligated to the prepared cloning vector.

DNA ligase

Recombinant vector

④ DNA is introduced into the host cell.

Host DNA

⑤ Propagation (cloning) of transformed cell produces many copies of recombinant DNA.

FIGURE 7-1 DNA cloning. The process involves cutting two DNAs with restriction enzymes, joining (ligating) the fragments together with DNA ligase, and using the recombinant DNA products to transform a suitable host cell. (This drawing is not to scale; the size of the *E. coli* chromosome relative to that of a typical cloning vector (such as a plasmid) is much greater than depicted here.)

fragment of interest can be joined to the DNA of a suitable cloning vector by DNA ligase. The recombinant vector is then introduced into a host cell, which amplifies the DNA fragment in the course of many generations of cell division.

Restriction endonucleases are found in a wide range of bacterial species. Werner Arber discovered in the early 1960s that the biological function of these enzymes is to recognize and cleave foreign DNA (the DNA of an infecting virus, for example); such DNA is said to be *restricted*.

Acting in a system with other enzymes that protect the host DNA, restriction endonucleases participate in a kind of immune system in bacteria. There are three types of restriction endonucleases, distinguished by their complexity and the typical distance between recognition sequence and cleavage site. **Type II restriction endonucleases**, first reported by Hamilton Smith in 1970, are the simplest, require no ATP for their activity, and cleave the DNA within the recognition sequence. Daniel Nathans quickly put this group of restriction endonucleases to use, demonstrating their extraordinary utility by developing novel methods for mapping and analyzing genes and genomes.

Thousands of restriction endonucleases have been discovered in different bacterial species, and more than 100 different DNA sequences are recognized by one or more of these enzymes. The recognition sequences are usually 4 to 8 base pairs (bp) long and palindromic (the recognition sequence, read in the 5′→3′ direction, is the same on both strands of DNA). However, a few of them fall slightly outside this norm. **Table 7-2** lists the sequences recognized by a few Type II restriction endonucleases.

Some restriction endonucleases make staggered cuts across the two DNA strands, leaving 2 to 4 nucleotides of one strand unpaired at each resulting end. Depending on which restriction enzyme is used, cleavage might occur such that the extended strand has either a 5′ or a 3′ end (called a 5′ or 3′ overhang). These unpaired strands are referred to as **sticky ends**, because they can base-pair with each other or with the complementary sticky ends of any other DNA fragments (**Figure 7-2a** on p. 216). Other restriction endonucleases cleave both strands of DNA straight across at the opposing phosphodiester bonds, leaving no unpaired bases on the ends, and thus produce what are often called **blunt ends** (**Figure 7-2b**).

The average size of the DNA fragments produced by cleaving genomic DNA with a restriction endonuclease depends on the frequency with which a particular recognition sequence occurs in the DNA molecule; this in turn depends largely on the length of the recognition sequence. In a DNA molecule with a random sequence in which all four nucleotides are equally abundant, a 6 bp sequence recognized by a restriction endonuclease would occur, on average, once every 4^6 (4,096) bp. A 4 bp recognition sequence would occur much more often, about once every 4^4 (256) bp. In laboratory experiments, the fragment size can be increased by terminating the reaction before completion—that is, before the enzyme molecules have cleaved every recognition sequence in the DNA sample. The result is a partial digest. Fragment size can also be increased by using a special class of endonucleases called homing endonucleases (see Figure 13-24), which recognize and cleave much longer recognition sequences (12 to 40 bp).

TABLE 7-2

Recognition Sequences for Some Type II Restriction Endonucleases

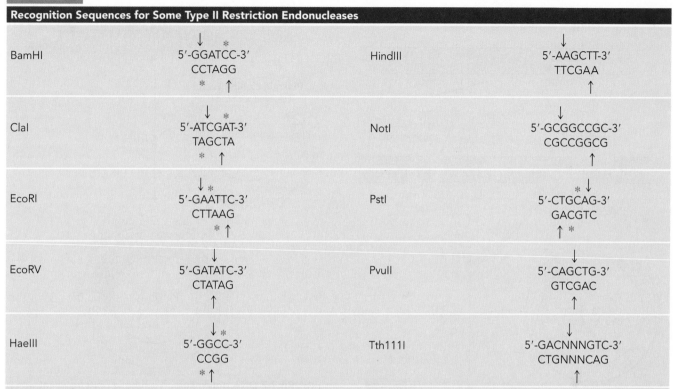

Note: Arrows denote phosphodiester bonds cleaved by each restriction endonuclease. Asterisks mark bases that are methylated by the corresponding methyltransferase (where known). N denotes any base. Each enzyme name consists of a three-letter abbreviation of the bacterial species from which it is derived, sometimes followed by a strain designation and a Roman numeral to distinguish restriction endonucleases isolated from the same bacterial species or strain. Thus, BamHI is the first (I) restriction endonuclease characterized from *Bacillus amyloliquefaciens*, strain H.

Other ways to obtain fragments of DNA for cloning are nonspecific shearing of the DNA, synthesis of the desired fragment, or use of the polymerase chain reaction (PCR). Many protocols are used to shear DNA including sonication, which uses sound energy to bring about hydrodynamic shearing, or simply forcing long DNA strands through a fine-gauge needle. Once a mixture of DNA fragments has been generated, fragments of a known size range can be separated by agarose or acrylamide gel electrophoresis (see Chapter 6). We describe later in the chapter the methods for synthesizing DNA and the use of PCR to amplify DNA fragments or whole genes in a form that makes cloning and isolation simpler.

After a target DNA fragment is obtained, DNA ligase can be used to join it to a cloning vector. The ligation reaction is greatly facilitated if the ends to be joined (ligated) have complementary sticky ends, as was apparent in the earliest recombinant DNA experiments (see the How We Know section at the end of this chapter). This is normally accomplished by cleaving the vector DNA with the same restriction enzyme used to prepare the target DNA fragments. DNA ligase catalyzes the formation of a phosphodiester bond between a 3′ hydroxyl at the end of one DNA strand and a 5′ phosphate at the end of another strand (see Figure 5-12).

Researchers can create new DNA sequences by inserting synthetic DNA fragments, called **linkers**, between the ends that are being ligated. Inserted DNA fragments with multiple recognition sequences for restriction endonucleases (often useful later in the experiment as points for inserting additional DNA by cleavage and ligation) are known as **polylinkers** (**Figure 7-3**).

Cloning Vectors Allow Amplification of Inserted DNA Segments

Genes or genomic segments are cloned for many different reasons. This is reflected in the use of a large variety of cloning vectors. The principles that govern the delivery of recombinant DNA in clonable form to a host cell, and its subsequent amplification in the host, are well illustrated by considering some popular cloning vectors used in experiments with *E. coli* and yeast: plasmids,

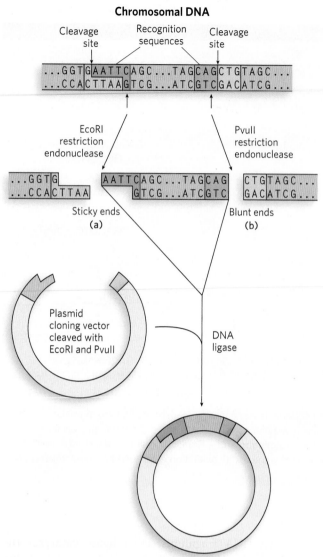

FIGURE 7-2 Cleavage of DNA molecules by restriction endonucleases. When Type II restriction endonucleases cleave DNA, they leave either (a) sticky ends (with protruding single strands) or (b) blunt ends. The restriction fragments can be ligated to other DNAs, such as the plasmid cloning vector shown here. Ligation is facilitated by the annealing of complementary sticky ends, and it is less efficient for DNA fragments with blunt ends than for those with complementary sticky ends. DNA fragments with noncomplementary sticky ends (i.e., those created by different restriction enzymes) generally are not ligated.

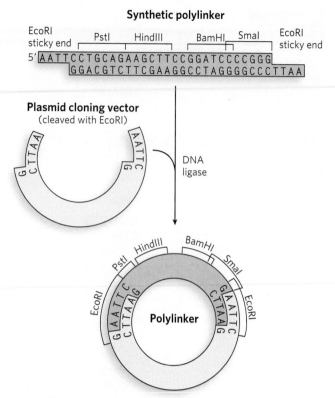

FIGURE 7-3 DNA polylinkers. A synthetic DNA fragment with recognition sequences for several restriction endonucleases—a fragment known as a polylinker—can be inserted into a plasmid that has been cleaved by a restriction endonuclease.

in size from 5,000 to 400,000 bp. Many of the plasmids found in bacterial populations are little more than molecular parasites, similar to viruses but with a more limited capacity to transfer from one cell to another. To survive in the host cell, plasmids contain or incorporate several specialized sequences that enable them to use the cell's resources for their own replication and gene expression.

Naturally occurring plasmids usually have a symbiotic role in the cell. They may provide genes that confer resistance to antibiotics or that perform new functions for the cell. For example, the Ti plasmid of *Agrobacterium tumefaciens* allows the host bacterium to colonize plant cells and make use of the plant's resources. The same properties that enable plasmids to grow and survive in a bacterial or eukaryotic host are useful to researchers who want to engineer a vector for cloning a specific DNA segment. The classic *E. coli* plasmid pBR322, constructed in 1977, is a good example of a plasmid with features useful in almost all cloning vectors (**Figure 7-4**):

1. The plasmid pBR322 has an **origin of replication**, or **ori**: a sequence where replication is initiated by cellular enzymes (see Chapter 11). This sequence is required to propagate the plasmid. An associated

bacterial artificial chromosomes, and yeast artificial chromosomes. Modern cloning vectors provide an array of options, allowing an investigator to tailor the cloning exercise to a particular goal: DNA sequencing, gene expression for protein purification, study of the effects of mutations, or creation of many kinds of gene alterations.

Plasmids A **plasmid** is a circular DNA molecule that replicates separately from the host chromosome. The wide variety of naturally occurring bacterial plasmids range

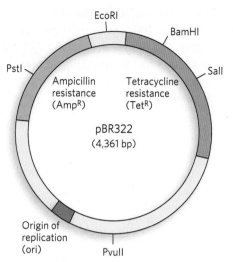

FIGURE 7-4 The constructed *E. coli* plasmid pBR322. This plasmid, one of the first to be constructed, was designed expressly for cloning in *E. coli*.

regulatory system is present that limits replication to maintain pBR322 at a level of 10 to 20 copies per cell.

2. The plasmid contains genes that confer resistance to the antibiotics tetracycline (Tet^R) and ampicillin (Amp^R), allowing the selection of cells that contain the intact plasmid or a recombinant version of the plasmid (discussed below).

3. Several unique recognition sequences in pBR322 are targets for restriction endonucleases (PstI, EcoRI, BamHI, SalI, and PvuII), providing sites where the plasmid can be cut to insert foreign DNA.

4. The small size of the plasmid (4,361 bp) facilitates both its entry into cells and the biochemical manipulation of the DNA. This small size is generated simply by trimming away many DNA segments from a larger, parent plasmid—sequences that the molecular biologist does not need.

Many variations and enhancements of these basic features of a cloning vector now exist. The replication origins inserted in common plasmid vectors were originally derived from naturally occurring plasmids. Each of these origins is regulated to maintain a particular number of plasmid copies in a cell (the plasmid copy number). Depending on the origin used, the plasmid copy number can vary from one to hundreds or thousands per cell, providing many options for investigators. Two different plasmids cannot function in the same cell if they use the same origin of replication, because the regulation of one will interfere with the replication of the other. Such plasmids are said to be incompatible. When a researcher wants to introduce two or more different

plasmids into a bacterial cell, each plasmid must have a different replication origin.

In the laboratory, small plasmids can be introduced into bacterial cells by a process called **transformation**. The cells (often *E. coli*, but other bacterial species are also used) and plasmid DNA are incubated together at 0°C in a calcium chloride solution, then subjected to heat shock by rapidly shifting the temperature to between 37°C and 43°C. For reasons not well understood, some of the cells treated in this way take up the plasmid DNA. Some species of bacteria, such as *Acinetobacter baylyi*, are naturally competent for DNA uptake and do not require the calcium chloride–heat shock treatment. In an alternative method, cells incubated with the plasmid DNA are subjected to a high-voltage pulse. This approach, called **electroporation**, transiently renders the bacterial membrane permeable to large molecules.

Regardless of the approach, relatively few cells take up the plasmid DNA, so a method is needed to identify those that do. The usual strategy is to utilize one of two types of genes in the plasmid, referred to as selectable and screenable markers. **Selectable markers** either permit the growth of a cell (positive selection) or kill the cell (negative selection) under a defined set of conditions. The plasmid pBR322 provides opportunities for both positive and negative selection (**Figure 7-5**). A **screenable marker** is a gene encoding a protein that causes a visible change in cell appearance, such as producing a color or making the cell fluoresce. Cells are not harmed whether the gene is present or not. The cells that carry the recombinant plasmid are easily identified by the colored or fluorescent colonies they produce.

Transformation of typical bacterial cells with purified DNA (never a very efficient process) becomes less successful as plasmid size increases, and it is difficult to clone DNA segments longer than about 15,000 bp when plasmids are used as the vector.

To illustrate the use of a plasmid as a cloning vector, consider a typical bacterial gene that encodes a recombinase called the RecA protein (see Chapter 13). In most bacteria, the gene encoding RecA is one of thousands of other genes on a chromosome millions of base pairs long. The *recA* gene is just over 1,000 bp long. A plasmid would be a good choice for cloning a gene of this size. As described later, the cloned gene can be altered in a variety of ways, and the gene variants can be expressed at high levels to enable purification of the encoded proteins.

Bacterial Artificial Chromosomes Large genome sequencing projects often require the cloning of much longer DNA segments than can typically be incorporated into standard plasmid cloning vectors such as pBR322. To meet this need, plasmid vectors have been developed

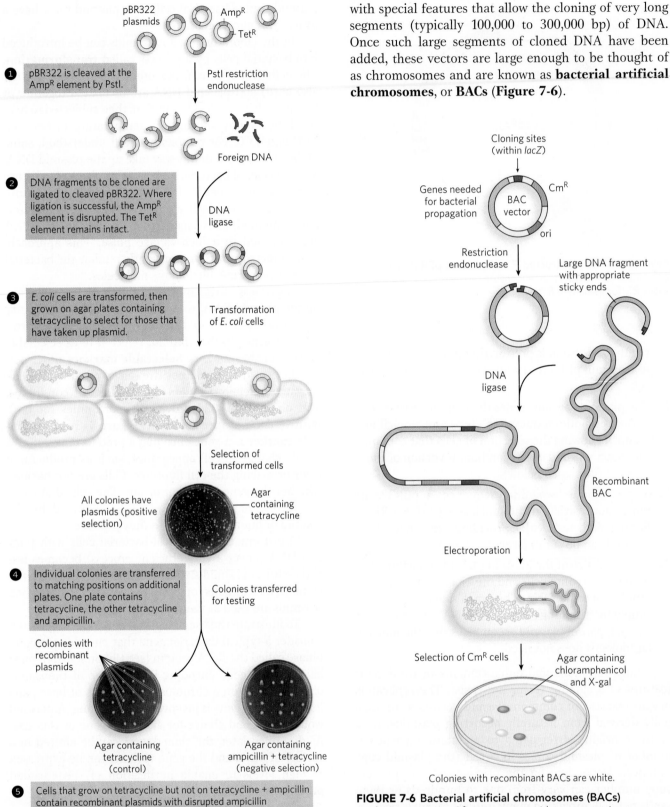

with special features that allow the cloning of very long segments (typically 100,000 to 300,000 bp) of DNA. Once such large segments of cloned DNA have been added, these vectors are large enough to be thought of as chromosomes and are known as **bacterial artificial chromosomes**, or **BACs** (**Figure 7-6**).

① pBR322 is cleaved at the Amp^R element by PstI.

Foreign DNA

② DNA fragments to be cloned are ligated to cleaved pBR322. Where ligation is successful, the Amp^R element is disrupted. The Tet^R element remains intact.

③ *E. coli* cells are transformed, then grown on agar plates containing tetracycline to select for those that have taken up plasmid.

All colonies have plasmids (positive selection)

Agar containing tetracycline

④ Individual colonies are transferred to matching positions on additional plates. One plate contains tetracycline, the other tetracycline and ampicillin.

Colonies with recombinant plasmids

Agar containing tetracycline (control)

Agar containing ampicillin + tetracycline (negative selection)

⑤ Cells that grow on tetracycline but not on tetracycline + ampicillin contain recombinant plasmids with disrupted ampicillin resistance, hence the foreign DNA. Cells with pBR322 without foreign DNA retain ampicillin resistance and grow on both plates.

FIGURE 7-5 Use of pBR322 to clone foreign DNA. The entire procedure is illustrated, including both positive and negative selection. *[Photos courtesy of Elizabeth A. Wood, Department of Biochemistry, University of Wisconsin–Madison.]*

Colonies with recombinant BACs are white.

FIGURE 7-6 Bacterial artificial chromosomes (BACs) as cloning vectors. After treatment with an appropriate restriction endonuclease, a BAC and a long fragment of DNA are ligated. The recombinant BAC is transferred into *E. coli* by electroporation, and colonies with recombinant BACs are selected by growth on media containing both the antibiotic chloramphenicol and X-gal, the substrate for β-galactosidase that produces a colored product.

A BAC vector is a relatively simple plasmid, generally not much larger than other plasmid vectors. To accommodate very long segments of cloned DNA, BAC vectors have stable origins of replication that maintain the plasmid at one or two copies per cell. The low copy number is useful in cloning large segments of DNA because it limits the opportunities for unwanted recombination reactions that can unpredictably alter large cloned DNAs over time. BACs also include *par* genes, which encode proteins that direct the reliable distribution of the recombinant chromosomes to daughter cells at cell division, thereby increasing the likelihood of each daughter cell carrying one copy, even when few copies are present. The BAC vector includes both selectable and screenable markers. The BAC vector shown in Figure 7-6 contains a gene for resistance to the antibiotic chloramphenicol (Cm^R). Positive selection for vector-containing cells occurs on agar plates containing this antibiotic. A *lacZ* gene, required for production of the enzyme β-galactosidase, is a screenable marker that can reveal which cells contain plasmids—now chromosomes—that incorporate the cloned DNA segments. The β-galactosidase catalyzes the conversion of the colorless molecule 5-bromo-4-chloro-3-indolyl-β-D-galactopyranoside (X-gal) to a blue product. If the gene is intact and expressed, the colony containing it will be blue. If gene expression is disrupted by the introduction of a cloned DNA segment, the colony will be white.

Yeast Artificial Chromosomes As with *E. coli*, yeast genetics is a well-developed discipline. The genome of *Saccharomyces cerevisiae* contains only 14×10^6 bp (less than four times the size of the *E. coli* chromosome), and its entire sequence is known. Yeast is also very easy to maintain and grow on a large scale in the laboratory. Plasmid vectors have been constructed for yeast, employing the same principles that govern the use of *E. coli* vectors. Methods are now available for moving DNA into and out of yeast cells, thus permitting the study of many aspects of eukaryotic cell biochemistry. Some recombinant plasmids incorporate multiple replication origins and other elements that allow them to be used in more than one species (e.g., in yeast and *E. coli*). Plasmids that can be propagated in cells of two or more species are called **shuttle vectors**.

Research on large genomes and the associated need for high-capacity cloning vectors led to the development of **yeast artificial chromosomes**, or **YACs (Figure 7-7)**. YAC vectors contain all the elements needed to maintain a eukaryotic chromosome in the yeast nucleus: a yeast origin of replication, two selectable markers, and specialized sequences (derived from the telomeres and centromere) that are needed for stability and proper segregation of the chromosomes at cell division

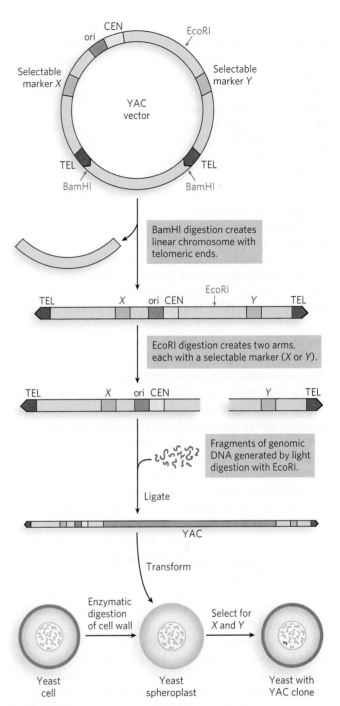

FIGURE 7-7 Construction of a yeast artificial chromosome (YAC). A YAC vector includes an origin of replication (ori), a centromere (CEN), two telomeres (TEL), and selectable markers (here designated *X* and *Y*). Two separate DNA arms are generated by digestion with BamHI and EcoRI, each arm having a telomeric end and one selectable marker. A large DNA fragment, produced by EcoRI digestion, is ligated to the two arms, creating a YAC. The YAC is transferred into yeast cells (which have been prepared by removing the cell wall to form spheroplasts). The transformed cells are selected for *X* and *Y*, and the surviving cells propagate the DNA insert.

(see Chapter 9). The YAC is a shuttle vector, initially maintained as a small circular plasmid in bacteria. It is much easier to isolate plasmids from bacteria than from yeast, and they can be maintained with very high copy numbers to facilitate vector production. Cleavage with a restriction endonuclease (BamHI in Figure 7-7) removes a length of DNA between two telomere sequences (TEL), leaving the telomeres at the ends of the linearized DNA. Cleavage at another internal site (by EcoRI in Figure 7-7) divides the vector into two DNA segments, referred to as vector arms, each with a different selectable marker.

The genomic DNA to be cloned is prepared by partial digestion with restriction endonucleases to obtain a suitable fragment size. Genomic fragments are then separated by **pulsed field gel electrophoresis**, a variation of gel electrophoresis that segregates very large DNA segments. DNA fragments of appropriate size (up to about 2×10^6 bp) are mixed with the prepared vector arms and ligated. The ligation mixture is then used to transform yeast cells (pretreated to partially degrade their cell walls) with these very large DNA molecules—which now have the structure and size to be considered yeast chromosomes. Culture on a medium that requires the presence of both selectable marker genes ensures the growth of only those yeast cells that contain an artificial chromosome with a large insert sandwiched between the two vector arms. The stability of YAC clones increases with the length of the cloned DNA segment (up to a point). Those with inserts of more than 150,000 bp are nearly as stable as normal cellular chromosomes, whereas those with inserts less than 100,000 bp long are gradually lost during mitosis (so, generally, there are no yeast cell clones carrying only the two vector ends ligated together or vectors with only short inserts). YACs that lack a telomere at either end are rapidly degraded.

As with BACs, YAC vectors can be used to clone very long segments of DNA. In addition, the DNA cloned in a YAC can be altered to study the function of specialized sequences in chromosome metabolism, mechanisms of gene regulation and expression, and many other problems in eukaryotic molecular biology.

DNA Libraries Provide Specialized Catalogs of Genetic Information

A **DNA library** is a collection of DNA clones, gathered together for purposes of genome sequencing, gene discovery, or determination of gene function. The library can take a variety of forms, depending on the source of the DNA and the ultimate purpose of the library.

One of the largest is a **genomic library**, produced when the complete genome of an organism is cleaved into thousands of fragments and *all* the fragments are cloned by insertion into a cloning vector. Building such a library has traditionally been a prelude to large sequencing projects. The first step is *partial* digestion of the DNA by restriction endonucleases, such that any given sequence will appear in fragments of a range of sizes—a range compatible with the cloning vector, ensuring that virtually all sequences are represented among the clones in the library. Fragments that are too large or too small for cloning are removed by centrifugation or electrophoresis. The cloning vector, such as a BAC or YAC, is cleaved with the same restriction endonuclease used to digest the DNA and ligated to the genomic DNA fragments. The ligated DNA mixture is then used to transform bacteria or yeast cells to produce a library of cells, each cell harboring a different recombinant DNA molecule. Ideally, all of the DNA in the genome under study is represented in the library. Each transformed bacterium or yeast cell grows into a colony, or clone, of identical cells, each cell bearing the same recombinant plasmid—one of many represented in the overall library. In some sequencing technologies, the step of introducing the library DNA into cells is skipped and the genomic DNA fragments are sequenced directly (as described later in this chapter).

With the increasing availability of genome sequences, the utility of genomic libraries is diminishing, and investigators are building more specialized libraries for studying gene function. An example is a library that includes only those sequences of DNA that are *expressed*—transcribed into RNA—in a given organism, or even just in certain cells or tissues. Such a library lacks the noncoding DNA that makes up a large portion of many eukaryotic genomes. The researcher first extracts mRNA from an organism, or from specific cells of an organism, and then prepares the **complementary DNAs (cDNAs)**. This multistep reaction, shown in **Figure 7-8**, relies on the enzyme **reverse transcriptase**, which synthesizes DNA from a template RNA. Reverse transcriptase is derived from a class of RNA viruses called retroviruses (see Chapter 14). The resulting double-stranded DNA fragments are inserted into a suitable vector and cloned, creating a population of clones called a **cDNA library**.

The search for a particular gene is made easier by focusing on a cDNA library generated from the mRNAs of a cell known to express that gene. For example, if we wished to clone globin genes, we could first generate a cDNA library from erythrocyte precursor cells, in which about half the mRNAs code for globins. A particular gene or gene segment in a library can be detected by the hybridization techniques introduced in Chapter 6. If a researcher knows something about

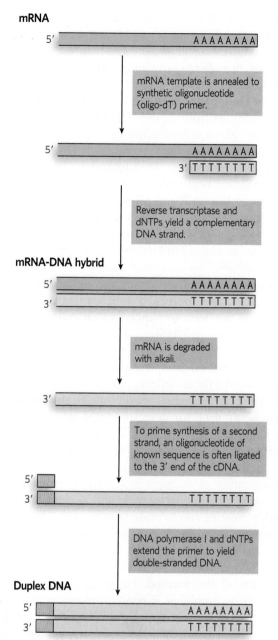

FIGURE 7-8 Building a cDNA library from mRNA. A cell's total mRNA includes transcripts from thousands of genes, and the cDNAs generated from this mRNA are correspondingly heterogeneous. Reverse transcriptase can synthesize DNA on an RNA or DNA template. Eukaryotic mRNAs end with a long sequence of A residues (poly(A); see Chapter 15), and thus a poly(dT) oligonucleotide is used to prime synthesis of the first DNA strand. To prime the synthesis of a second DNA strand, oligonucleotides of known sequence are ligated to the 3' end of the first strand, and the double-stranded cDNA produced is cloned into a plasmid.

the sequence of the DNA being sought, a short nucleic acid complementary to that sequence can be synthesized, labeled, and used to identify cells carrying a recombinant plasmid that incorporates that particular sequence.

- Genes are isolated for study by cloning them into vectors that permit their selection and amplification. A gene or genomic segment is cut out of a chromosome with a restriction enzyme and ligated into a vector. The recombinant vector is transferred into a host cell and is amplified in this transformed cell.
- Gene cloning relies on an arsenal of enzymes made available by advances in molecular biology, including restriction endonucleases, DNA ligase, DNA polymerase, and reverse transcriptase.
- Important cloning vectors include plasmids, bacterial artificial chromosomes, and yeast artificial chromosomes. BACs and YACs allow the cloning of very long DNA segments.
- DNA libraries are specialized archives used in gene sequencing, gene discovery, or the functional characterization of proteins.

7.2 WORKING WITH GENES AND THEIR PRODUCTS

The isolation of a gene, or any segment of genomic DNA, generally has one of two purposes. One is to examine the DNA itself, determine its sequence, study its structure and/or function, and compare it with other DNA segments. For example, researchers in physical biochemistry might be interested in the structure of an unusual repeated sequence. Evolutionary biologists and forensic scientists might be interested in comparing the sequence of the DNA segment with the same segment taken from other individuals in a population or with related DNA segments from other species. The other possible purpose is to work with the protein or RNA product of the isolated gene. These gene products are at the heart of every biological process. Genetic engineering provides tools not only for the isolation and study of proteins and RNA but also for their alteration for myriad purposes.

The isolation and examination of DNA segments has been greatly facilitated by PCR technology, and we discuss this first. We then explore modern DNA sequencing methods and a variety of techniques for expressing and altering gene products—primarily proteins—so as to understand their function and harness them for new purposes.

Gene Sequences Can Be Amplified with the Polymerase Chain Reaction

Genome projects continue worldwide, creating rapidly growing online databases containing the complete genome sequences of thousands of organisms. Such programs provide unprecedented access to gene sequence information. In turn, progress on this front is simplifying the process of cloning individual genes for more detailed analysis. If we know the sequence of at least the end portions of a DNA segment we are interested in, we can hugely amplify the number of copies of that DNA segment with the **polymerase chain reaction (PCR)**, a process conceived by Kary Mullis in 1983 (see the How We Know section at the end of this chapter). The amplified DNA can then be cloned by the methods described earlier or can be used in a variety of analytical procedures.

The PCR procedure has an elegant simplicity and relies on enzymes called DNA polymerases. DNA polymerases synthesize DNA strands on a pre-existing DNA template using free deoxyribonucleotides. Further, DNA polymerases do not synthesize DNA de novo, but instead must add nucleotides to preexisting strands, referred to as primers (as described in Chapter 11). Two synthetic oligonucleotides are prepared, complementary to sequences on opposite strands of the target DNA at positions defining the ends of the segment to be amplified. The oligonucleotides serve as replication primers that can be extended by a DNA polymerase. The 3′ ends of the hybridized primers are oriented toward each other and positioned to prime DNA synthesis across the targeted DNA segment (**Figure 7-9a**). Basic PCR requires four components: a DNA sample containing the segment to be amplified, the pair of synthetic oligonucleotide primers, deoxynucleoside triphosphates (dNTPs), and a DNA polymerase. The reaction mixture is heated briefly to denature the DNA, separating the two strands. The mixture is cooled so that the primers can anneal to the DNA. The high concentration of primers increases the likelihood that they will anneal to each strand of the denatured DNA before the two DNA strands (present at a much lower concentration) can reanneal to each other. The primed segment is then replicated selectively by the DNA polymerase, using the pool of dNTPs. The cycle of heating, cooling, and replication is repeated 25 to 30 times over a few hours in an automated process, amplifying the DNA segment between the primers until it can be readily analyzed or cloned. Each cycle doubles the amount of the DNA segment, so the concentration of this DNA grows exponentially. After 20 cycles, the DNA segment has been amplified up to 2^{20}, or a millionfold if reaction conditions are ideal. All other DNA in the sample remains unamplified. PCR uses a heat-stable DNA polymerase, such as the *Taq* polymerase, which remains active after every heating step and does not have to be replenished.

By careful design of the primers used for PCR, the amplified segment can be altered by the inclusion, at each end, of additional DNA not present in the chromosome that is being targeted. For example, restriction endonuclease cleavage sites can be included to facilitate the subsequent cloning of the amplified DNA (**Figure 7-9b**).

This technology is highly sensitive: PCR can detect and amplify as little as one DNA molecule in almost any type of sample—including some quite ancient ones. The double-helical structure of DNA makes it a highly stable molecule (see Chapter 6), but DNA does degrade slowly over time (through reactions described in Chapter 12). PCR has allowed the successful cloning of rare, undegraded DNA segments from samples more than 40,000 years old. Investigators have used the technique to clone DNA fragments from the mummified remains of humans and extinct animals, such as the woolly mammoth, creating the fields of molecular archaeology and molecular paleontology. DNA from burial sites has been amplified by PCR and used to trace ancient human migrations. Epidemiologists can use PCR-enhanced DNA samples from human remains to trace the evolution of human pathogenic viruses. Thus, in addition to its usefulness for cloning DNA, PCR is a potent tool in forensic medicine (**Highlight 7-1**). It is also being used for detecting viral infections before they cause symptoms and for the prenatal diagnosis of a wide array of genetic diseases.

Given the extreme sensitivity of PCR methods, contamination of samples is a serious issue. In many applications, including forensic and ancient DNA tests, controls must be run to make sure the amplified DNA is not derived from the researcher or from contaminating bacteria.

Many specialized adaptations of PCR have increased the utility of the method. For example, sequences in RNA can be amplified if reverse transcriptase is used in the first PCR cycle (see Figure 7-8). After the DNA strand is made using the RNA as a template, the remaining cycles can be carried out with a DNA polymerase by normal PCR protocols. This **reverse transcriptase PCR (RT-PCR)** can be used, for example, to detect sequences derived from living cells (which are transcribing their DNA into RNA) as opposed to dead tissues.

PCR protocols can also be made quantitative for estimating the relative copy numbers of particular sequences in a sample. The approach is called **quantitative PCR (qPCR)**. If a DNA sequence is present in higher than usual amounts in a sample—for example, certain genes may be amplified so that they are present in many copies in the cells that make up a cancerous tumor—quantitative PCR can reveal the increased representation of that sequence. In brief, the PCR is carried out in the presence of a probe that emits a fluorescent

(a)

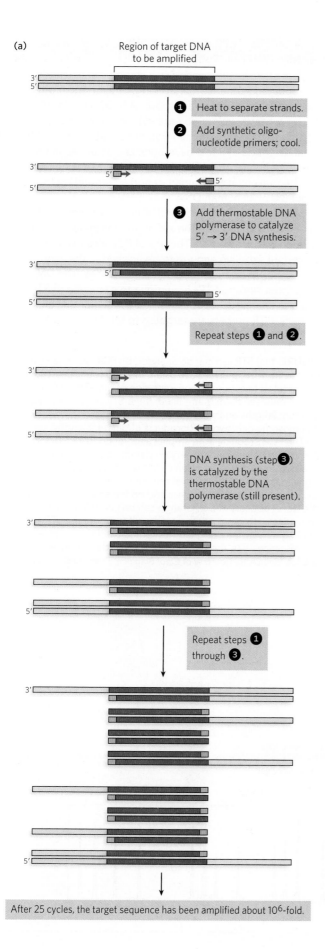

Region of target DNA to be amplified

❶ Heat to separate strands.

❷ Add synthetic oligo-nucleotide primers; cool.

❸ Add thermostable DNA polymerase to catalyze 5' → 3' DNA synthesis.

Repeat steps ❶ and ❷.

DNA synthesis (step ❸) is catalyzed by the thermostable DNA polymerase (still present).

Repeat steps ❶ through ❸.

After 25 cycles, the target sequence has been amplified about 10⁶-fold.

FIGURE 7-9 Amplification of a DNA segment by the polymerase chain reaction (PCR). PCR leads to specific amplification of DNA in a segment defined by the two designed DNA primers. If extra DNA sequences are included at the 5' end of the synthetic primers (e.g., the sequence specifying a restriction site, as shown here), those sequences are incorporated into the final product.

(b)

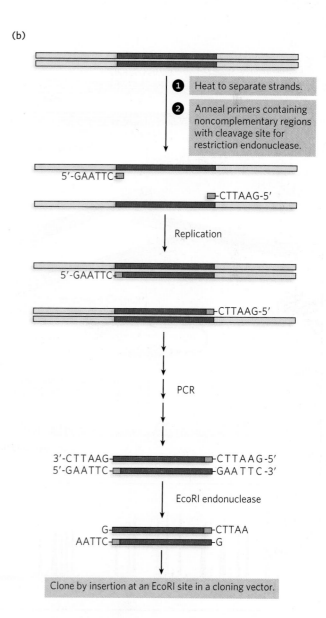

❶ Heat to separate strands.

❷ Anneal primers containing noncomplementary regions with cleavage site for restriction endonuclease.

5'-GAATTC

CTTAAG-5'

Replication

5'-GAATTC

CTTAAG-5'

PCR

3'-CTTAAG————CTTAAG-5'
5'-GAATTC————GAATTC-3'

EcoRI endonuclease

G————CTTAA
AATTC————G

Clone by insertion at an EcoRI site in a cloning vector.

A Potent Weapon in Forensic Medicine

One of the most accurate methods for placing an individual at the scene of a crime is a fingerprint. But with the advent of recombinant DNA technology, a much more powerful tool became available: **DNA genotyping** (also called DNA fingerprinting or DNA profiling). As first described by English geneticist Alec Jeffreys in 1985, the method is based on **sequence polymorphisms**, slight sequence differences among individuals—1 in every 1,000 bp, on average. Each difference from the prototype human genome sequence (the first one obtained) occurs in some fraction of the human population; every person has some differences from this prototype.

Forensic work focuses on differences in the lengths of **short tandem repeat (STR)** sequences. An STR locus is a short DNA sequence, repeated many times in tandem at a specific location in a chromosome; usually, the repeated sequence is 4 bp long. The loci most often used in STR genotyping are short—4 to 50 repeats (16 to 200 bp for tetranucleotide repeats)—and have multiple length variants in the human population. More than 20,000 tetranucleotide STR loci have been characterized in the human genome. And more than a million STRs of all types may be present in the human genome, accounting for about 3% of all human DNA.

The length of a particular STR in a given individual can be determined with the aid of the polymerase chain reaction (see Figure 7-9). The use of PCR also makes the procedure sensitive enough to be applied to the very small samples often collected at crime scenes. The DNA sequences flanking STRs are unique to each type of STR and are identical (except for very rare mutations) in all humans. PCR primers are targeted to this flanking DNA

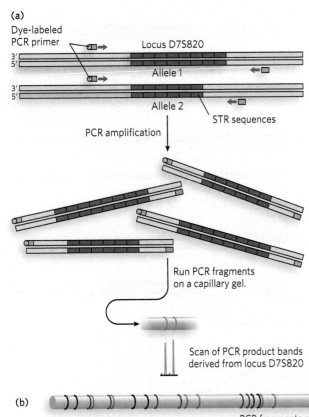

(a)

Dye-labeled PCR primer

Locus D7S820

Allele 1

Allele 2

STR sequences

PCR amplification

Run PCR fragments on a capillary gel.

Scan of PCR product bands derived from locus D7S820

FIGURE 1 (a) STR loci can be analyzed by PCR. Suitable PCR primers (with an attached dye to aid in subsequent detection) are targeted to sequences on each side of the STR, and the region between them is amplified. If the STR sequences have different lengths on the two chromosomes of an individual's chromosome pair, two PCR products of different lengths result. (b) The PCR products from amplification of up to 16 STR loci can be run on a single capillary acrylamide gel (a "16-plex" analysis). Determination of which locus corresponds to which signal depends on the color of the fluorescent dye attached to the primers used in the process and on the size range in which the signal appears (the size range can be controlled by which sequences—those closer to or more distant from the STR—are targeted by the designed PCR primers). RFU = relative fluorescence units, measured against a standard supplied with the kit. *[Source: (b) Courtesy of Carol Bingham, Promega Corporation.]*

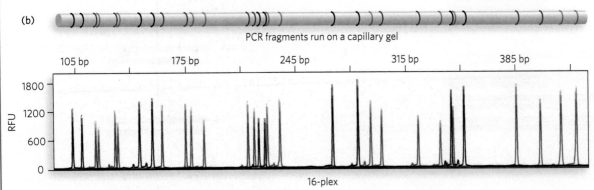

(b)

PCR fragments run on a capillary gel

105 bp   175 bp   245 bp   315 bp   385 bp

RFU
1800
1200
600
0

16-plex

and are designed to amplify the DNA across the STR (Figure 1a). The length of the PCR product then reflects the length of the STR in that sample. Because each human inherits one chromosome of each chromosome pair from each parent, the STR lengths on the two chromosomes are often different, generating two different STR lengths from one individual. The PCR products are subjected to electrophoresis on a very thin polyacrylamide gel in a capillary tube. The resulting bands are converted into a set of peaks that accurately reveal the size of each PCR fragment and thus the length of the STR in the corresponding allele. Analysis of multiple STR loci can yield a profile that is unique to an individual (Figure 1b). This is typically done with a commercially available kit that includes PCR primers unique to each locus, linked to colored dyes to help distinguish the different PCR products. PCR amplification enables investigators to obtain STR genotypes from less than 1 ng of partially degraded DNA, an amount that can be obtained from a single hair follicle, a small fraction of a drop of blood, a small semen sample, or samples that might be months or even many years old. When good STR genotypes are obtained, the chance of misidentification is less than 1 in 10^{18} (a quintillion).

The successful forensic use of STR analysis required standardization, first attempted in the United Kingdom in 1995. The U.S. standard, called the Combined DNA Index System (CODIS), established in 1998, is based on 13 well-studied STR loci, which must be present in any DNA-typing experiment carried out in the United States (Table 1). The amelogenin gene is also used as a marker in the analyses. Present on the human sex chromosomes, this gene has a slightly different length on the X and Y chromosomes. PCR amplification across this gene thus generates different-sized products that can reveal the sex of the DNA donor. By the beginning of 2014, the CODIS database contained nearly 11 million STR genotypes and had assisted more than 220,000 forensic investigations.

DNA genotyping has been used to both convict and acquit suspects, and to establish paternity with an extraordinary degree of certainty. The impact of these procedures on court cases will continue to grow as standards are refined and as international STR genotyping databases grow. Even very old mysteries can be solved. In 1996, STR genotyping helped confirm the identification of the bones of the last Russian czar and his family, who were assassinated in 1918.

TABLE 1

Properties of the Loci Used for the CODIS Database

Locus Name	Chromosome	Repeat Motif*	Repeat Length (range)[†]	Number of Alleles Seen[‡]
CSF1PO	5	TAGA	5–16	20
FGA	4	CTTT	12.2–51.2	80
TH01	11	TCAT	3–14	20
TPOX	2	GAAT	4–16	15
VWA	12	[TCTG][TCTA]	10–25	28
D3S1358	3	[TCTG][TCTA]	8–21	24
D5S818	5	AGAT	7–18	15
D7S820	7	GATA	5–16	30
D8S1179	8	[TCTA][TCTG]	7–20	17
D13S317	13	TATC	5–16	17
D16S539	16	GATA	5–16	19
D18S51	18	AGAA	7–39.2	51
D21S11	21	[TCTA][TCTG]	12–41.2	82
Amelogenin[§]	X, Y	Not applicable		

Source: Data from J. M. Butler, *Forensic DNA Typing,* 2nd ed., Elsevier, 2005, p. 96.
*Brackets indicate alternating repeats.
[†]Repeat lengths observed in the human population. Partial or imperfect repeats are seen in some alleles.
[‡]Number of different alleles observed in the human population. Careful analysis of the same locus in many individuals is a prerequisite to its use in forensic DNA typing.
[§]The amelogenin gene, of slightly different size on the X and Y chromosomes, is used to establish gender.

(a)

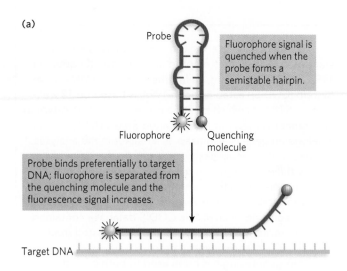

(b)

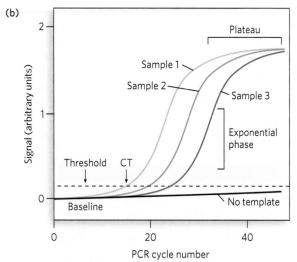

FIGURE 7-10 Quantitative PCR. PCR can be used quantitatively, by carefully monitoring the progress of a PCR amplification and determining when a DNA segment has been amplified to a specific threshold level. (a) The amount of PCR product present is determined by measuring the level (fluorescence) of a fluorescent probe attached to a reporter oligonucleotide complementary to the DNA segment that is being amplified. Probe fluorescence is not detectable initially due to a fluorescence quencher attached to the same oligonucleotide. When the reporter oligonucleotide pairs with its complement in a copy of the amplified DNA segment, the fluorophore is separated from the quenching molecule and fluorescence results. (b) As the PCR reaction proceeds, the amount of the targeted DNA segment increases exponentially, and the fluorescent signal also increases exponentially as the oligonucleotide probes anneal to the amplified segments. After many PCR cycles, the signal reaches a plateau as one or more reaction components are exhausted. When a segment is present in greater amounts in one sample than another, its amplification reaches a defined threshold level earlier. The "No template" line follows the slow increase in background signal observed in a control that does not include added sample DNA. CT is the cycle number at which the threshold is first surpassed.

signal when the PCR product is present (**Figure 7-10**). If the sequence of interest is present at higher levels than other sequences in the sample, the PCR signal will reach a predetermined threshold faster. Reverse transcriptase PCR and quantitative PCR can be combined to determine the relative transcription levels of genes in a cell under different environmental conditions or to study the regulation of transcription of one or more genes.

The Sanger Method Identifies Nucleotide Sequences in Cloned Genes

In its capacity as a repository of information, a DNA molecule's most important property is its nucleotide sequence. Until the late 1970s, determining the sequence of a nucleic acid containing even 5 or 10 nucleotides was very laborious. The development of two techniques in 1977 (one by Allan Maxam and Walter Gilbert, the other by Frederick Sanger) made possible the sequencing of larger DNA molecules. The techniques depended on the improved understanding of nucleotide chemistry and DNA metabolism and on improved electrophoretic methods for separating DNA strands that differ in size by only one nucleotide (see Figure 6-32 for a description of gel electrophoresis). In work on short DNA oligonucleotides (up to a few hundred nucleotides), polyacrylamide is often used instead of agarose as the gel matrix, because it enables researchers to detect small size differences between DNA fragments.

Although the two methods are similar in approach, the **Sanger method**, also known as the dideoxy chain-termination method, has proved to be technically easier and is in more widespread use (**Figure 7-11**). This method makes use of the mechanism of DNA synthesis by DNA polymerases (see Chapter 11). It requires the enzymatic synthesis of a DNA strand complementary to the strand under analysis, using a radioactively labeled primer and dideoxynucleotides. In the reaction catalyzed by DNA polymerase, the 3'-hydroxyl group of the primer reacts with an incoming deoxynucleoside triphosphate (dNTP) to form a new phosphodiester bond (Figure 7-11a). The identity of the added deoxynucleotide is determined by its complementarity, through base pairing, to a base in the template strand. In the Sanger sequencing reaction, nucleotide analogs called dideoxynucleoside triphosphates (ddNTPs) interrupt DNA synthesis because they lack the 3'-hydroxyl group needed for the next step (Figure 7-11b). For instance, the addition of ddCTP to an otherwise normal reaction system causes some of the synthesized strands to be prematurely terminated at the position where dC would normally be added, opposite a template dG. Given the excess of dCTP over ddCTP, the chance that the analog will be incorporated instead of dC is small. But ddCTP is present in sufficient amounts

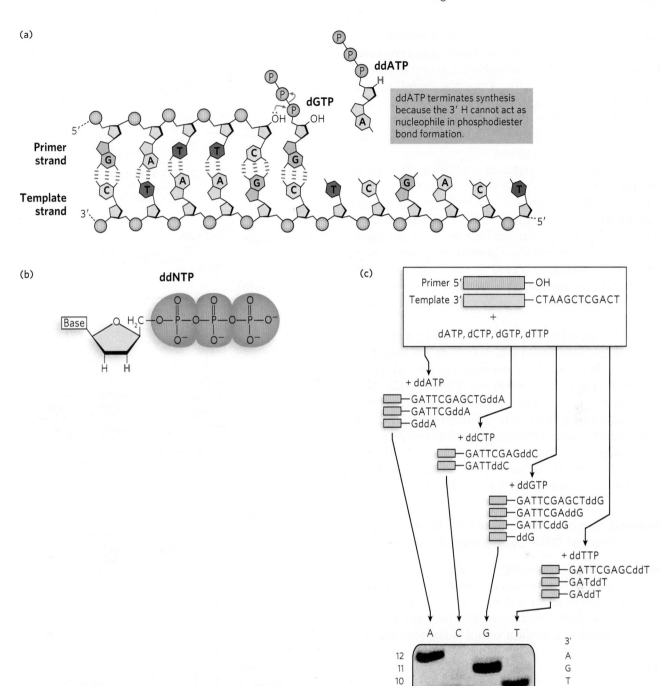

FIGURE 7-11 The Sanger method for DNA sequencing.
(a) DNA synthesis involves a reaction between the
3′-hydroxyl group of the primer dNTP and the phosphate
group of an incoming dNTP. (b) The Sanger method uses
ddNTPs, which lack the 3′-hydroxyl group, to halt DNA
synthesis at a particular nucleotide. (c) In the sequencing
reaction, DNA synthesis is carried out with a mixture of
dNTPs and a ddNTP to extend a radiolabeled primer. A
different ddNTP is used in each reaction. The products are
analyzed by autoradiography to determine the nucleotide
sequence. *[Source: (c) Photo courtesy of Lloyd Smith, Department
of Chemistry, University of Wisconsin–Madison.]*

to ensure that each new strand has a high probability of acquiring at least one ddC at some point during synthesis. The result is a solution containing a mixture of labeled fragments, each ending with a C residue. Each G residue in the template generates C-terminated fragments of a particular length. The different-sized fragments, separated by electrophoresis, reveal the location of C residues in the synthesized DNA strand.

This procedure is repeated separately for each of the four ddNTPs, and the sequence of the DNA strand can be read directly from an autoradiogram of the gel (Figure 7-11c). Because shorter DNA fragments migrate faster, the fragments near the bottom of the gel represent the nucleotide positions closest to the primer (the 5′ end), and the sequence is read (in the 5′→3′ direction) from bottom to top. Note that the sequence obtained is that of the strand *complementary* to the template strand being analyzed.

DNA sequencing was first automated by a variation of the Sanger method, in which each of the four dideoxynucleotides used for a reaction was labeled with a differently colored fluorescent tag (**Figure 7-12**). With this technology, researchers could sequence DNA molecules containing thousands of nucleotides in a few hours, and the entire genomes of hundreds of organisms were sequenced in this way. For example, in the Human Genome Project, researchers sequenced all 3.2×10^9 bp of the DNA in a human cell (see Chapter 8).

Genomic Sequencing Is Aided by New Generations of DNA Sequencing Methods

DNA sequencing technologies continue to evolve. A complete human genome can now be sequenced in a day or two, a bacterial genome in a few hours. The day when a personal genomic sequence might be a routine part of each individual's medical record is fast approaching.

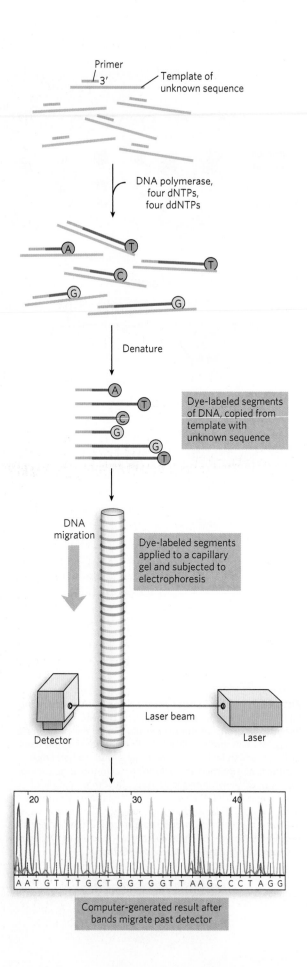

FIGURE 7-12 Automation of DNA sequencing reactions. In the Sanger method, each ddNTP can be linked to a fluorescent (dye) molecule that gives the same color to all the fragments terminating in that nucleotide, a different color for each nucleotide. All four labeled ddNTPs are added together. The resulting colored DNA fragments are separated by size in an electrophoretic gel in a capillary tube (a refinement of gel electrophoresis that allows for faster separations). All fragments of a given length migrate through the capillary gel together in a single band, and the color associated with each band is detected with a laser beam. The DNA sequence is read by identifying the color sequences in the bands as they pass the detector. This information is fed directly to a computer, and the sequence is determined. The amount of fluorescence in each band is represented as a peak in the computer output. Here, the nucleotide colors reflect the dyes actually used in the method, and thus deviate from the standard nucleotide colors used in other figures. [*Source: Data from Lloyd Smith, Department of Chemistry, University of Wisconsin–Madison.*]

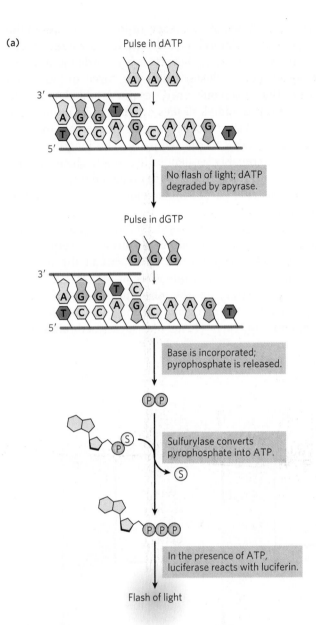

(a)

Pulse in dATP

No flash of light; dATP degraded by apyrase.

Pulse in dGTP

Base is incorporated; pyrophosphate is released.

Sulfurylase converts pyrophosphate into ATP.

In the presence of ATP, luciferase reacts with luciferin.

Flash of light

These advances have been made possible by methods sometimes referred to as next-generation, or "next-gen" sequencing. The sequencing strategy is sometimes similar to and sometimes quite different from that used in the Sanger method. Innovations have allowed a miniaturization of the procedure, a massive increase in scale, and a corresponding decrease in cost.

A genomic sequence is generated in several steps. First, the genomic DNA is broken at random locations by shearing to generate fragments that are a few hundred base pairs long. Synthetic oligonucleotides are ligated to the ends of all the fragments, providing a known point of reference on every DNA molecule. The individual fragments are then immobilized on a solid surface, and each is amplified by PCR (see Figure 7-9). The solid surface is part of a channel that allows liquid solutions to flow over the samples. The result is a solid surface just a few centimeters wide, with millions of attached DNA clusters, each cluster containing multiple copies of a single DNA sequence derived from a random genomic DNA fragment. The efficiency comes from sequencing all of these millions of clusters at the same time, with the data from each cluster captured and stored in a computer.

Two widely utilized next-generation sequencers use different strategies to accomplish the sequencing reactions. One of these, known as 454 sequencing (the numbers refer to a code used during development of the technology and have no scientific meaning), uses a strategy called **pyrosequencing** in which the addition of nucleotides is detected by flashes of light (**Figure 7-13**). The four dNTPs (unaltered) are pulsed onto the reacting surface one at a time in a repeating sequence. The nucleotide solution is retained on the surface just long enough for DNA polymerase to add that nucleotide to any cluster

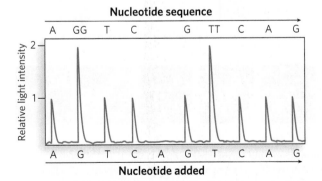

Nucleotide sequence

Relative light intensity

A GG T C G TT C A G

A G T C A G T C A G

Nucleotide added

FIGURE 7-13 Next-generation pyrosequencing. (a) Pyrosequencing detects the addition of nucleotides on the DNA to be sequenced by flashes of light. (b) An image of a very small part of one cycle of a 454 sequencing run. Each individual segment of template DNA to be sequenced is attached to a tiny DNA capture bead, then amplified on the bead by PCR. Each bead is immersed in an emulsion and placed in a tiny (~ 29 μm) well on a picotiter plate. The reaction of luciferin and ATP with luciferase produces light flashes when a nucleotide is added to a particular DNA cluster in a particular well. Circles represent the same cluster over multiple cycles. In this case, reading the top (or bottom) circle from left to right across each row gives the sequence for that cluster.

(b)

dNTP added

DNA sequence at circled clusters

A G T C A G T C A G

AGGTCGTTCAG:

_ _ _CAGTTCAG:

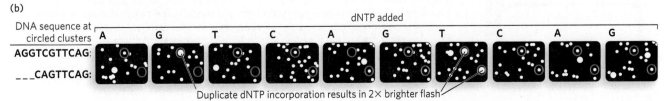

Duplicate dNTP incorporation results in 2× brighter flash

where it is complementary to the next nucleotide in the template sequence. Excess nucleotide is destroyed quickly by the enzyme apyrase before the next nucleotide pulse. When a specific nucleotide is successfully added to the strands of a cluster, pyrophosphate is released as a byproduct. Another enzyme in the solution bathing the surface is sulfurylase, which converts the pyrophosphate to ATP. The appearance of ATP ultimately provides the signal that a nucleotide has been added to the DNA. Also present in the medium is the enzyme luciferase and its substrate molecule, luciferin (luciferase is the enzyme that generates the flash of light produced by fireflies). When ATP is generated, luciferase catalyzes a reaction with luciferin that results in a tiny flash of light. When many tiny flashes occur in a cluster, the emitted light can be recorded in a captured image. For example, when dCTP is added to the solution, flashes occur only at clusters where G is the next base in the template and C is the next nucleotide to be added to the growing DNA chain.

If there is a string of two, three, or four G residues in the template, a similar number of C residues are added to the growing strand in one cycle. This is recorded as a "flash" amplitude at that cluster that is two, three, or four times greater than when only one C residue is added. Similarly, when dGTP is added, flashes occur at a different set of clusters, marking those as clusters where G is the next nucleotide added to the sequence. The length of DNA that can be reliably sequenced in a single cluster by this method—often referred to as the read length, or "read"—is typically 400 to 500 nucleotides, and is constantly increasing.

The second widely used method employs a technique known as **reversible terminator sequencing** (**Figure 7-14**), which lies at the heart of the Illumina sequencer. A special sequencing primer is added that is complementary to the oligonucleotides of known sequence that were ligated to the ends of the DNA fragments in each cluster (as described above). In

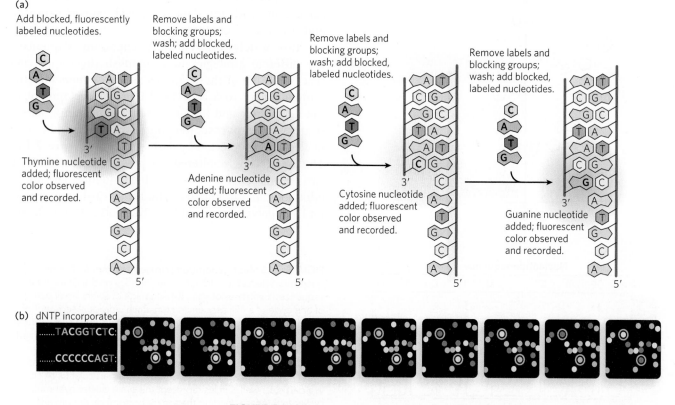

FIGURE 7-14 Next-generation reversible terminator sequencing. (a) The reversible terminator method of sequencing uses fluorescent tags to identify nucleotides. Blocking groups on each fluorescently labeled nucleotide prevent multiple nucleotides from being added per cycle. (b) Six successive cycles from one very small part of an Illumina sequencing run. Each colored spot represents the location of an immobilized DNA oligonucleotide affixed to the surface of the flow cell. The circled clusters represent the same spot on the surface over successive cycles and give the sequences indicated. Data are automatically recorded and analyzed digitally. (c) Typical flow cell used for a next-generation sequencer. Millions of DNA fragments can be sequenced simultaneously in each of the eight channels.
[Source: (c) Courtesy Michael Cox]

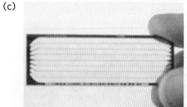

addition, fluorescently labeled terminator nucleotides and DNA polymerase are added. The polymerase adds the appropriate nucleotide to the strands in each cluster, each type of nucleotide (A, T, G, or C) carrying a different fluorescent label. These terminator nucleotides have blocking groups attached to the 3′ ends that permit addition of only one nucleotide to each strand. Next, lasers excite all the fluorescent labels, and an image of the entire surface reveals the color (and thus the identity of the base) added to each cluster. The fluorescent label and the blocking groups are then chemically or photolytically removed, in preparation for adding a new nucleotide to each cluster. The sequencing proceeds stepwise. Read lengths are shorter for this method,

typically 100 to 200 nucleotides per cluster, although refinements are ongoing.

These technologies are modern manifestations of an approach to genomic sequencing that is sometimes called **shotgun sequencing**. Many copies of the genomic DNA are sheared to generate each set of fragments. Thus, a particular short segment of the genome may be present in dozens or even hundreds of different sequenced clusters. However, there is no landmark on an individual fragment to indicate where in the genome it came from. Assembling the sequences of these millions of fragments into a genomic sequence requires the computerized alignment of overlapping fragment sequences (**Figure 7-15**). The number of times

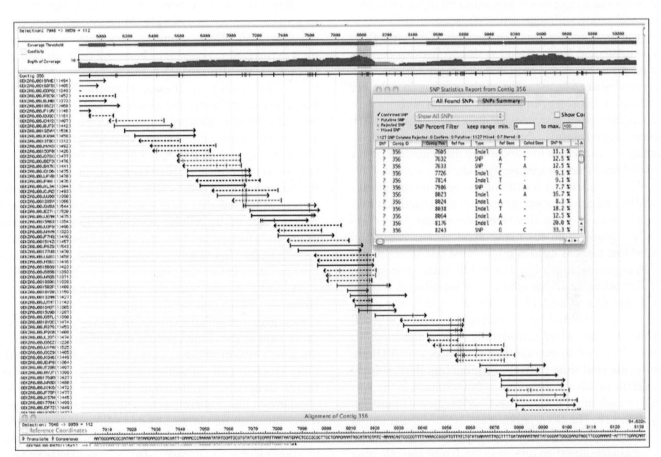

FIGURE 7-15 Sequence assembly. In a genomic sequence, each base pair of the genome is usually represented in many of the sequenced fragments, referred to as reads. Shown is a small part of the sequence of a new variant species of *E. coli*, with the reads generated by a 454 sequencer. The numbers at the top represent genomic base-pair positions, relative to an arbitrarily defined "0." The sequences all come from a particular long contig designated 356. The reads themselves are represented by horizontal arrows, with computer-assigned identifiers listed for each one at the left. DNA strand segments are sequenced at random, with sequences obtained from one strand (5′ to 3′, left to right) represented by solid arrows and sequences obtained from the other strand (5′ to 3′, right to left) represented by dashed arrows. The latter sequences are automatically reported as their complement when they are merged with the overall dataset. The "coverage threshold" at the top is a measure of sequence quality. The wider green bar indicates sequences that have been obtained enough times to generate high confidence in the results. The depth of the coverage line indicates how many times a given base pair appears in a sequenced read. The vertical blue shaded line indicates a part of the sequence that is highlighted by thin blue brackets in the sequence line at the bottom of the page. The "SNP statistics report" (inset) is a listing of positions where single nucleotide polymorphisms (SNPs; see Chapter 8) appear to be present in some of the reads. These putative SNPs are often checked by additional sequencing. They are indicated in the reads by thin, blue vertical slash marks within the horizontal lines for each read.

that a particular nucleotide in the genome is sequenced, on average, is referred to as the **sequencing depth**, or sequencing coverage. In most cases, a sufficiently large number of random fragments are sequenced so that each nucleotide in the genome is sequenced an average of 30 to 40 times (30–40 × coverage). Although the coverage of particular nucleotides may vary (some will be sequenced 100 times; perhaps a few not at all), this level of coverage ensures that most genomic nucleotides will be sequenced at least 10 times and most sequencing errors will be detected and eliminated. The overlaps allow the computer to trace the sequence through a chromosome, from one fragment to another. This permits the assembly of long contiguous sequences called **contigs**. In a successful genomic sequencing exercise, many contigs can extend over millions of base pairs. Special strategies are needed to fill in the inevitable gaps and to deal with repetitive sequences.

For some applications, the amounts of genomic DNA to be sequenced are increased so that sequencing depth is increased to 100 × or even 1,000 ×. This approach, sometimes called **deep sequencing**, can help determine whether a mutation is present in a subset of an organism's cells or whether other genomic variations are present. Deep sequencing is helpful in the characterization of genomic sequences in cancerous tumors, where the genome is highly unstable and changes frequently as the tumor grows.

DNA sequencing technologies continue to advance rapidly, and a few newer next-generation methods now complement the two described above and may eventually replace them for many applications. A method called **ion torrent** utilizes fragmented and immobilized DNA segments, much like 454 and Illumina sequencing. Deoxynucleotide triphosphates are introduced, one by one. Addition of a dNTP at a certain spot in the growing chain is detected by measuring the protons released in the reaction. More sensitive light-detection methods have given rise to yet another approach, called **single molecule real time (SMRT) sequencing**. Here, a single molecule of DNA polymerase is immobilized at the bottom of each of millions of precisely engineered pores on the flow cell. The DNA polymerase captures fragmented genomic segments as they diffuse into the pore. The labeled nucleotides then diffuse in, each new one releasing its colored fluorescent group as it is added to the chain. An innovative light-detection system records the color of the resulting light flash at the bottom of the pore, thereby revealing the identity of each added nucleotide. The method is accurate and can generate particularly long read lengths, up to nearly 10,000 base pairs.

Cloned Genes Can Be Expressed to Amplify Protein Production

Frequently, it is the product of a cloned gene, rather than the gene itself, that is of primary interest—particularly when the protein has commercial, therapeutic, or research value. Molecular biologists use purified proteins to elucidate protein function, study reaction mechanisms, generate antibodies, reconstitute complex cellular activities in the test tube with purified components, and examine protein binding partners, and for many other purposes. With an increased understanding of the fundamentals of DNA, RNA, and protein metabolism and their regulation in *E. coli*, investigators can now manipulate cells to express cloned genes in order to study their protein products. The general goal is to alter the sequences around a cloned gene to trick the host organism into producing the protein product of the gene, often at very high levels. This overexpression of a protein can make its subsequent purification a lot easier.

Here we use the expression of a eukaryotic protein in a bacterium as an example. Most cloned eukaryotic genes lack the DNA sequence elements required for their controlled expression in bacterial cells—promoters (sequences that instruct RNA polymerase where to bind to initiate mRNA synthesis), ribosome-binding sites (sequences that allow translation of the mRNA to protein), and additional regulatory sequences (see Chapter 15). Therefore, appropriate bacterial regulatory sequences for transcription and translation must be inserted at the correct positions, relative to the eukaryotic gene, in the vector DNA. In some cases, cloned genes are so efficiently expressed that their protein product represents 10% or more of the cellular protein. At these concentrations, some foreign proteins can kill the host cell (usually *E. coli*), so the cloned gene expression must be limited to the few hours before the planned harvesting of the cells.

Cloning vectors with the transcription and translation signals needed for the regulated expression of a cloned gene are called **expression vectors**. The rate of expression of the cloned gene is controlled by replacing the gene's own promoter and regulatory sequences with more efficient and convenient versions supplied by the vector. Generally, a well-characterized promoter and its regulatory elements are positioned near several unique restriction sites for cloning, so that genes inserted at the restriction sites will be expressed from the regulated promoter elements (**Figure 7-16**). Some of these vectors incorporate other features, such as a bacterial ribosome-binding site to enhance translation of the mRNA derived from the gene (see Chapter 18) or a transcription-termination sequence (Chapter 15).

Many Different Systems Are Used to Express Recombinant Proteins

Every living organism has the capacity to express genes contained in its genomic DNA; thus, in principle, any organism can serve as a host to express proteins from a different (heterologous) species. Almost every sort of organism has

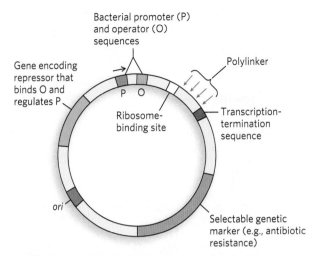

FIGURE 7-16 DNA sequences in a typical *E. coli* expression vector. The gene to be expressed is inserted into one of the restriction sites in the polylinker, near the promoter (P), with the end of the gene encoding the N-terminus of the protein positioned closest to the promoter. The promoter allows efficient transcription of the inserted gene, and the transcription-termination sequence sometimes improves the amount and stability of the mRNA produced. The operator (O) is a sequence bound by a protein called a repressor, which normally blocks gene expression from the adjacent gene. The ribosome-binding site provides sequence signals for the efficient translation of the mRNA derived from the gene. The selectable marker allows the selection of cells containing the recombinant DNA.

indeed been used for this purpose, and each host type has a particular set of advantages and disadvantages.

Bacteria Bacteria, especially *E. coli*, remain the most common hosts for protein expression. The regulatory sequences that govern gene expression in *E. coli* and many other bacteria are well understood and can be harnessed to express cloned proteins at high levels. Bacteria are easy to store and grow in the laboratory, on inexpensive growth media. Efficient methods also exist to get DNA into bacteria and extract DNA from them. Bacteria can be grown in huge amounts in commercial fermentors, providing a rich source of the cloned protein. Problems do exist, however. When expressed in bacteria, some heterologous proteins do not fold correctly, and many do not undergo the postsynthetic modifications (covalent modification, proteolytic cleavage, etc.; see Chapter 18) necessary for their activity. A variety of gene sequence features also can make a particular gene difficult to express in bacteria. For these and many other reasons, some eukaryotic proteins are inactive when purified from bacteria, or they cannot be expressed at all.

There are many specialized systems for expressing proteins in bacteria. The promoter and regulatory sequences associated with the lactose operon (see Chapters 5 and 20) are often fused to the gene of interest to direct transcription. The cloned gene will be transcribed when lactose is added to the growth medium.

However, regulation in the lactose system is "leaky": it is not turned off completely when lactose is absent—a potential problem if the product of the cloned gene is toxic to the host cells. Transcription from the Lac promoter is also not efficient enough for some applications.

An alternative system uses a promoter and RNA polymerase found in a bacterial virus called bacteriophage T7. If the cloned gene is fused to a T7 promoter, it is transcribed not by the *E. coli* RNA polymerase but by the T7 RNA polymerase. The gene encoding this polymerase is separately cloned into the same cell in a construct that affords tight regulation (allowing controlled production of the T7 RNA polymerase). The polymerase is also very efficient and directs high levels of expression of most genes fused to the T7 promoter. This system has been used to express the RecA protein in bacterial cells (**Figure 7-17**).

Yeast The yeast *S. cerevisiae* is probably the best understood eukaryotic organism and one of the easiest to grow and manipulate in the laboratory. Like bacteria, this yeast can be grown on inexpensive media. Yeast have tough cell walls that are difficult to breach in order to introduce DNA vectors, so bacteria are more convenient for doing much of the genetic engineering and vector maintenance. Several excellent shuttle vectors exist for this purpose.

The principles underlying the expression of a protein in yeast are the same as those in bacteria. Cloned genes must be linked to promoters that can direct

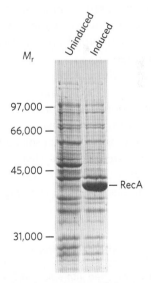

FIGURE 7-17 Regulated expression of RecA protein in a bacterial cell. The gene encoding the RecA protein, fused to a bacteriophage T7 promoter, is cloned into an expression vector. Under normal growth conditions (uninduced, left lane), no RecA protein appears when cellular proteins are separated on a polyacrylamide gel (see Chapter 6) and stained with Coomassie Blue for visualization. When the T7 RNA polymerase is induced in the cell (right lane), the *recA* gene is expressed, and the large amounts of RecA protein produced are readily observed. *[Source: Courtesy of Rachel Britt, Department of Biochemistry, University of Wisconsin–Madison.]*

high-level expression in yeast. For example, the yeast *GAL1* and *GAL10* genes are under cellular regulation such that they are expressed when yeast cells are grown in media with galactose but shut down when the cells are grown in media with glucose. Thus, if a heterologous gene is expressed using the same regulatory sequences, the expression of that gene can be controlled simply by choosing an appropriate medium for cell growth.

Some of the same problems that accompany protein expression in bacteria also occur with yeast. Heterologous proteins may not fold properly, yeast may lack the enzymes needed to modify the proteins to their active forms, or the expression of proteins may be made difficult by certain features of the gene sequence. However, because *S. cerevisiae* is a eukaryote, the expression of eukaryotic genes (especially yeast genes) is sometimes more efficient in this host than in bacteria. Folding and modification of the products may also be more accurate than for proteins expressed in bacteria.

Insects and Insect Viruses Baculoviruses are insect viruses with double-stranded DNA genomes. When they infect their insect larval hosts, they act as parasites, killing the larvae and turning them into factories for virus production. Late in the infection process, the viruses produce large amounts of two proteins (p10 and polyhedrin)—neither of which is needed for virus production in cultured insect cells, and thus both can be replaced with the gene of a heterologous protein. When the resulting recombinant virus is used to infect insect cells or larvae, the heterologous protein is often produced at very high levels—up to 25% of the total protein present at the end of the infection cycle.

Autographa californica multicapsid nucleopolyhedrovirus (AcMNPV) is the baculovirus most often used for protein expression. Its genome (134,000 bp) is too large for direct cloning. Virus purification is also cumbersome. These problems have been solved by the creation of **bacmids**, large circular DNAs that include the entire baculovirus genome along with sequences that allow replication of the bacmid in *E. coli* (**Figure 7-18**). The

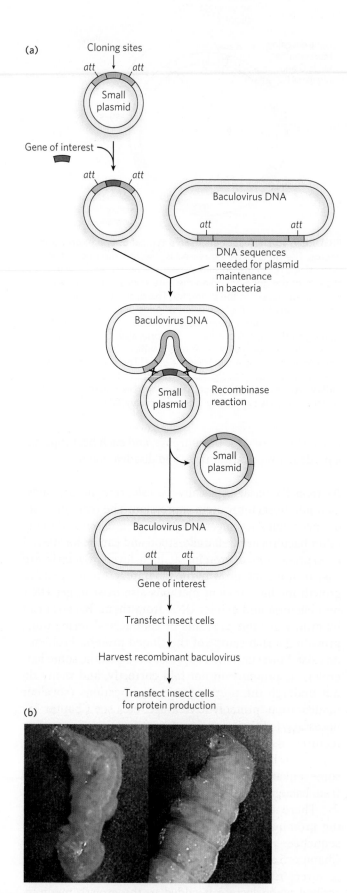

FIGURE 7-18 Cloning with baculoviruses. (a) The construction of a typical vector used for protein expression in baculoviruses. The gene of interest is cloned into a small plasmid between two sites (*att*) recognized by a site-specific recombinase, then introduced into the baculovirus vector by site-specific recombination. This generates a circular DNA product that is used to infect the cells of an insect larva. The gene of interest is expressed during the infection cycle, downstream of a promoter that normally expresses a baculovirus coat protein at very high levels. (b) Left: an insect larva infected with a recombinant baculovirus vector expressing a protein that produces a red color. Right: an uninfected larva. [Source: (b) Courtesy of Arthur McIntosh, A.H. McIntosh et al., J. Insect Sci. 4:3, 2004.]

gene of interest is cloned into a smaller plasmid and combined with the larger plasmid by site-specific recombination in vivo (described in Chapter 14). The recombinant bacmid is then isolated and transfected into insect cells (the term **transfection** is used when the DNA used for transformation includes viral sequences and the process leads to viral replication), followed by recovery of the protein once the infection cycle is finished. A wide range of bacmid systems are available commercially. Baculovirus systems are not successful with all proteins. However, with these systems, insect cells sometimes successfully replicate the protein-modification patterns of higher eukaryotes and produce active, correctly modified eukaryotic proteins.

Mammalian Cells in Culture The most convenient way to introduce cloned genes into a mammalian cell is with viruses. In this way, a molecular biologist can take advantage of the natural capacity of a virus to insert its DNA or RNA into a cell, and sometimes into the cellular chromosome. A variety of engineered mammalian viruses are available as vectors, including human adenoviruses and retroviruses. The gene of interest is cloned so that its expression is controlled by a virus promoter. The virus uses its natural infection mechanisms to introduce the recombinant genome into cells, where the cloned protein is expressed. These systems have the advantage that proteins can be expressed either transiently (if the viral DNA is maintained separately from the host cell genome and eventually degraded) or permanently (if the viral DNA is integrated into the host cell genome). With the correct choice of host cell, the proper posttranslational modification of the protein to its active form can be assured. However, the growth of mammalian cells in tissue culture is very expensive, and this technology is generally used to test the function of a protein in vivo rather than to produce a protein in large amounts.

Transgenic Animals Even large animals can be used for the commercial, large-scale production of recombinant proteins. The strategies are different from those discussed thus far and are designed to generate protein in a low-cost, renewable way, such as purification of a protein from the milk of transgenic dairy cattle (**Figure 7-19**). The gene of interest is cloned into a special vector, linked to a promoter that directs tissue-specific gene expression. For example, the gene can be placed under the control of regulatory sequences for a mammary gland–specific protein, such as casein lactoglobulin, which is normally secreted in milk in large quantities. The recombinant plasmid is injected into fertilized bovine oocytes, and some of them take up the plasmid and incorporate it into their genome. Genetic analysis or direct demonstration of heterologous protein expression then identifies

FIGURE 7-19 Cloning in transgenic animals. These cows, grazing in a field in New Zealand, were engineered to produce high levels of a recombinant protein in their milk. *[Source: Courtesy of Götz Laible.]*

animals in which the gene transfer has been successful, and these animals are bred. Heterologous proteins expressed in place of casein lactoglobulin can be secreted in the milk at levels above 50% of total milk proteins. Posttranslational protein modifications are not always carried out correctly for proteins expressed in this way, but protein production can be economical once a line of protein-expressing animals is established.

Alteration of Cloned Genes Produces Altered Proteins

Cloning techniques can be used not only to overproduce proteins but to produce protein products subtly altered from their native forms. Specific amino acids may be replaced individually by **site-directed mutagenesis**. A variety of methods, based in large measure on techniques pioneered by Michael Smith and his colleagues in the late 1970s, are now used to enhance research on proteins by allowing investigators to make specific changes in the primary structure and examine the effects of these changes on the protein's folding, three-dimensional structure, and activity. This powerful approach to studying protein structure and function changes the amino acid sequence by altering the DNA sequence of the cloned gene. If appropriate restriction sites flank the sequence to be altered, researchers can simply remove a DNA segment and replace it with a synthetic one identical to the original except for the desired change (**Figure 7-20a**).

When suitably located restriction sites are not present, **oligonucleotide-directed mutagenesis**, coupled to PCR, can create a specific DNA sequence change (**Figure 7-20b**). Two short, complementary synthetic DNA strands, each with the desired base change, are annealed to opposite strands of the cloned gene within a

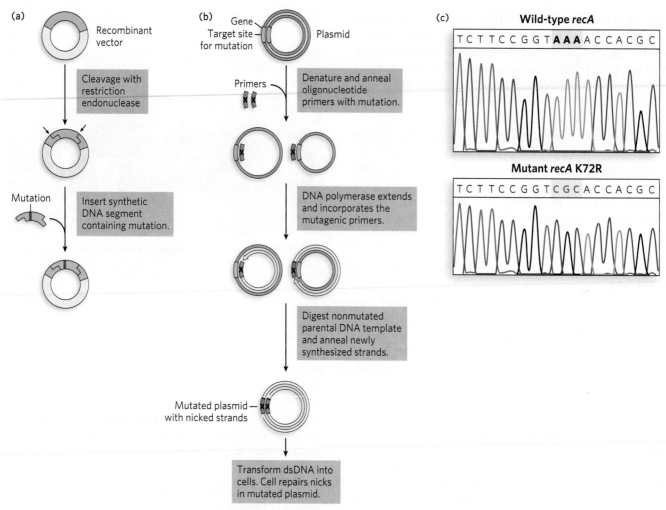

FIGURE 7-20 Two approaches to site-directed mutagenesis. (a) A synthetic DNA segment replaces a fragment removed by a restriction endonuclease. (b) A pair of synthetic and complementary oligonucleotides with a specific sequence change at one position are hybridized to a circular plasmid with a cloned copy of the gene to be altered. The oligonucleotides act as primers for the synthesis of full-length double-stranded DNA (dsDNA) copies of the plasmid that contain the specified sequence change. These plasmid copies are then used to transform cells. (c) Results from an automated sequencer (see Figure 7-12), showing sequences from the wild-type *recA* gene (top) and from an altered *recA* gene (*recA* K72R, bottom) with the triplet (codon) at position 72 changed from AAA to CGC, specifying an Arg (R) instead of a Lys (K) residue. Here, the nucleotide colors reflect the dyes actually used in the method, and thus deviate from the standard nucleotide colors used in other figures. [*Source: (c) Courtesy of Elizabeth Wood, Department of Biochemistry, University of Wisconsin–Madison.*]

suitable circular DNA vector. The mismatch of a single base pair in 30 to 40 bp does not prevent annealing. The two annealed oligonucleotides serve to prime DNA synthesis in both directions around the plasmid vector, creating two complementary strands that contain the mutation. After several cycles of PCR, the mutation-containing DNA predominates in the population and can be used to transform bacteria. Most of the transformed bacteria will have plasmids carrying the mutation. If necessary, the nonmutant template plasmid DNA can be selectively eliminated by cleavage with the restriction enzyme DpnI. The template plasmid, usually isolated from wild-type *E. coli*, has a methylated A residue in every copy of the four-nucleotide palindrome GATC.

The new DNA containing the mutation does not have methylated A residues, because the replication is done in vitro (with no methylating enzymes present). DpnI selectively cleaves DNA at the sequence GATC only if the A residue in one or both strands is methylated—that is, the enzyme breaks down only the template.

For an example, we can go back to the bacterial *recA* gene. The product of this gene, the RecA protein, has several activities (see Chapter 13). It binds to and forms a filamentous structure on DNA, aligns two DNAs of similar sequence, and hydrolyzes ATP. A particular amino acid residue in RecA (a 352-residue polypeptide), Lys[72], is involved in ATP hydrolysis. By changing this Lys residue to an Arg, a variant of RecA protein is created that will bind, but not hydrolyze, ATP (**Figure 7-20c**). The engineering and purification of this variant RecA protein has facilitated research into the roles of ATP hydrolysis in the functioning of this protein.

Changes can be introduced into a gene that involve far more than one base pair. Large parts of a gene can be deleted by cutting out a segment with restriction endonucleases and ligating the remaining portions to form a smaller gene. Parts of two different genes can be ligated to create new combinations; the product of such a fused gene is called a **fusion protein**. Researchers have ingenious methods to bring about virtually any genetic alteration in vitro. After reintroducing the altered DNA into the cell, they can investigate the consequences of the alteration.

Terminal Tags Provide Handles for Affinity Purification

Affinity chromatography is one of the most efficient methods for purifying proteins (see Highlight 4-1). Unfortunately, many proteins do not bind a ligand that can be conveniently immobilized on a column matrix. With the use of fusion proteins, almost any protein can be purified by affinity chromatography.

The gene encoding the target protein is fused to a gene encoding a peptide or protein that binds a simple, stable ligand with high affinity and specificity. The peptide or protein used for this purpose is referred to as a **tag**. Tag sequences can be added to genes such that the resulting proteins have tags at their N- or C-terminus. **Table 7-3** lists some of the peptides or proteins commonly used as tags.

The general procedure can be illustrated by focusing on a system that uses the glutathione-*S*-transferase (GST) tag (**Figure 7-21**). GST is a small enzyme (M_r 26,000) that binds tightly and specifically to glutathione. When the GST gene sequence is fused to a target gene, the fusion protein acquires the capacity to bind glutathione. The fusion protein is expressed in a host organism such as

a bacterium, and a crude extract is prepared. A column is filled with a porous matrix consisting of the ligand (glutathione) immobilized on microscopic beads of a stable polymer such as cross-linked agarose. As the crude extract percolates through this matrix, the fusion protein becomes immobilized by binding to the glutathione. The other proteins in the extract are washed through the column and discarded. The interaction between GST and glutathione is tight but noncovalent, allowing the fusion protein to be gently eluted from the column with a solution containing either a higher concentration of salts or free glutathione to compete with the immobilized ligand for GST binding. The fusion protein is often obtained with good yield and high purity. In some commercially available systems, the tag can be entirely or largely removed from the purified fusion protein by a protease that cleaves a sequence near the junction between the target protein and its tag.

A shorter tag with widespread application consists of a simple sequence of six or more histidine residues. These histidine tags, or His tags, bind tightly and specifically to nickel ions. A chromatography matrix with immobilized Ni^{2+} can be used to quickly separate a His-tagged protein from other proteins in an extract. Some of the larger tags, such as maltose-binding protein, provide added stability and solubility, allowing the purification of cloned proteins that are otherwise inactive due to improper folding or insolubility.

This technology is powerful and convenient. The tags have been successfully used in thousands of published studies; in many cases, the protein would be impossible to purify and study without the tag. However, even very small tags can affect the properties of the proteins they are attached to, thereby influencing the study results. Even if the tag is removed by a protease, one or a few extra amino acid residues can remain behind on the target protein, which may or may not affect the protein's activity. The types of experiments to be carried out, and the results obtained from them, should always be evaluated with the aid of well-designed controls to assess any effect of a tag on protein function.

TABLE 7-3

Commonly Used Protein Tags

Tag Protein	Molecular Weight	Immobilized Ligand
Protein A	59,000	Fc portion of IgG
His$_6$ (His tag)	800	Ni^{2+}
Glutathione-*S*-transferase (GST)	26,000	Glutathione
Maltose-binding protein	41,000	Maltose
β-Galactosidase	116,000	*p*-Aminophenyl-β-D-thiogalactoside (TPEG)
Chitin-binding domain	5,700	Chitin

(a)

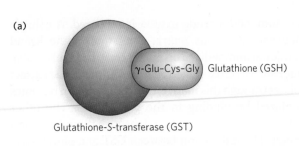

Glutathione-*S*-transferase (GST)

(b)

Gene for target protein

Gene for GST

Transcription

Gene for fusion protein

Express fusion protein in cell.

Prepare cell extract containing fusion protein as part of the cell protein mixture.

Protein mixture is added to column.

Pump

GST tag

Glutathione anchored to medium binds GST tag.

1 2 3 4 5 6 7 8 9

Other proteins flow through column.

Solution of free glutathione is added to column.

Pump

1 2 3 4 5 6 7 8 9

Fusion protein is eluted by glutathione solution.

FIGURE 7-21 Use of tagged proteins in protein purification. (a) Glutathione-*S*-transferase (GST) is a small enzyme that binds glutathione (a glutamate residue to which a Cys–Gly dipeptide is attached at the carboxyl carbon of the Glu side chain, hence the abbreviation GSH). (b) The GST tag is fused to the C-terminus of the protein by genetic engineering. The tagged protein is expressed in the cell and is present in the crude extract when the cells are lysed. The extract is subjected to chromatography through a matrix with immobilized glutathione. The GST-tagged protein binds to the glutathione, retarding its migration through the column, while the other proteins are washed through rapidly. The tagged protein is subsequently eluted with a solution containing elevated salt concentration or free glutathione.

SECTION 7.2 SUMMARY

- Genes or other DNA segments can be amplified by the polymerase chain reaction. With specialized adaptations of PCR, investigators can amplify sequences in RNA and quantify the levels of particular RNA molecules in a cell.
- Modern DNA sequencing methods enable researchers to determine the sequences of entire mammalian genomes in weeks or even days. Many thousands of genomic DNA sequences are now available in public databases.
- Cloned genes can be expressed to provide large amounts of the gene product. Systems have been developed to express genes in bacteria, yeast, insects, mammalian cells, and even entire mammalian organisms.
- Cloned genes can be altered. A gene sequence can be changed, sequences deleted, or sequences added. All changes affect the protein or RNA product of the gene.
- Added sequences can produce protein products that include fused peptide segments, called tags. With the aid of these tags, the protein can be rapidly purified.

7.3 UNDERSTANDING THE FUNCTIONS OF GENES AND THEIR PRODUCTS

One of the challenges in molecular biology is to identify the functions of the myriad genes being discovered in large genome sequencing projects, which we survey in Chapter 8. When the complete sequence of an organism's genome becomes available, we often lack functional information for half or more of the defined genes. Biotechnology provides some shortcuts to understanding the functions of gene products, and we describe some of the key technologies here (and expand on their application in Chapter 8). To begin to define the function of a new, previously unstudied protein, a molecular biologist gathers clues—defining the protein's location in cells, what it interacts with, when it is expressed, and what happens to the organism when it is absent. These clues, together with information about the activity of the purified protein in vitro, eventually paint a consistent functional picture.

Protein Fusions and Immunofluorescence Can Localize Proteins in Cells

Often, an important clue to a gene product's function comes from determining its location within the cell. For example, a protein found exclusively in the nucleus could be involved in processes that are unique to that organelle, such as transcription, replication, or chromatin condensation. Researchers often engineer fusion proteins for the purpose of locating a protein in the cell or organism. Some of the most useful fusions involve the addition of marker proteins that allow the investigator to determine the location by direct visualization or by immunofluorescence.

A particularly useful marker is the gene for **green fluorescent protein (GFP)**. A target gene (coding the protein of interest) fused to the GFP gene generates a fusion protein that is highly fluorescent—it literally lights up when exposed to blue light—and can be visualized directly in a living cell. GFP is a protein derived from the jellyfish *Aequorea victoria* (see the How We Know section at the end of this chapter). It has a β-barrel structure (**Figure 7-22a**), and the fluorophore (the fluorescent

(a)

(b)

(c)

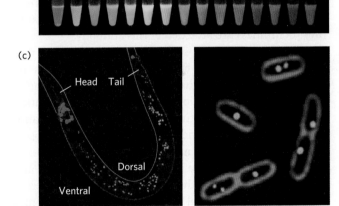

FIGURE 7-22 Green fluorescent protein (GFP). (a) The source of GFP is the jellyfish *Aequorea victoria* (left); the bioluminescent photo-organs are visible. The GFP protein has a β-barrel structure (right); the fluorophore (red) is in the center of the barrel. (b) Variants of GFP are now available in almost any color of the visible spectrum. (c) *Caenorhabditis elegans*, a nematode worm (left), containing a GLR1-GFP fusion protein. GLR1 is a glutamate receptor of nervous tissue. In the *E. coli* cells (right), the membranes are stained with a red fluorescent dye. The cells are expressing a protein that binds to a resident plasmid, fused to GFP. The green spots indicate the locations of plasmids.
[Sources: (a) (left) © Mauricio Handler/National Geographic Society/ Corbis. (b) Courtesy of Nathan Shaner, Paul Steinbach, and Roget Y. Tsien. (c) (left) Courtesy Penelope J. Brockie and Andres V. Maricq, Department of Biology, University of Utah; (right) Courtesy of Joseph A. Pogliano, Pogliano et al. (2001), Multicopy plasmids are clustered and localized in Escherichia coli PNAS 98, 4486–4491. © 2001 National Academy of Sciences. U.S.A.]

component of the protein) is in the center of the barrel. The fluorophore is derived from a rearrangement and oxidation of several amino acid residues. Because this reaction is autocatalytic and requires no other proteins or cofactors (other than molecular oxygen), GFP is readily cloned in an active form in almost any cell. Just a few molecules of this protein can be observed microscopically, allowing the study of its location and movements in a cell. Careful protein engineering, coupled with the isolation of related fluorescent proteins from other marine coelenterates, has made a wide range of these proteins available, with an array of colors (**Figure 7-22b**) and other characteristics (brightness, stability). With this technology, for example, the protein GLR1 (a glutamate receptor of nervous tissue) has been visualized as a GLR1-GFP fusion protein in the nematode *Caenorhabditis elegans* (**Figure 7-22c**).

In many cases, visualization of a GFP fusion protein in a live cell is not possible, or is not practical or desirable. The GFP fusion protein may be inactive or may not be expressed at sufficient levels to allow visualization. In this case, **immunofluorescence** is an alternative approach for visualizing the endogenous (unaltered) protein, although fusion proteins are sometimes used here, too. This approach requires the fixation (and thus death) of the cell. The protein of interest is expressed either unaltered or as a fusion protein with an **epitope tag**, a short protein sequence that is bound tightly by a well-characterized, commercially available antibody. Fluorescent molecules (fluorochromes) are attached to this antibody.

More commonly, a second antibody is added that binds specifically to the first (primary) one, and it is the secondary antibody that has the attached fluorochrome(s) (**Figure 7-23**). A variation of this indirect visualization approach is to attach biotin molecules to the primary antibody, then add streptavidin (a bacterial protein closely related to avidin, biotin's natural ligand) that is complexed with fluorochromes. The interaction between biotin and streptavidin is one of the strongest and most specific known, and the potential to add multiple fluorochromes to each target protein gives this method great sensitivity.

Highly specialized cDNA libraries (see Figure 7-8) can be made by cloning cDNAs or cDNA fragments into a vector that fuses each cDNA sequence with the sequence for a marker, also called a reporter gene. For example, libraries have been developed in which all the genes in the library are fused to the GFP gene (**Figure 7-24**). Each cell in the library expresses one of these fused genes. The cellular location of the product of any gene represented in the library can be studied—assuming that the particular fusion protein is expressed at sufficient levels and retains its normal function and location—by examining cells that express the appropriate fused gene, to detect light foci that reveal the protein's presence.

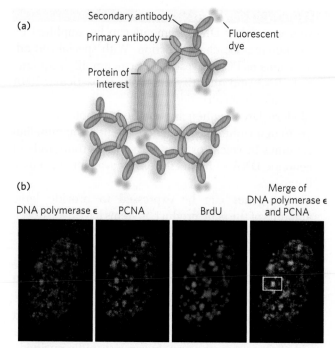

FIGURE 7-23 Indirect immunofluorescence. (a) The protein of interest is bound to a primary antibody, and a secondary antibody is added; this secondary antibody, with one or more attached fluorescent groups, binds to the primary antibody. Multiple secondary antibodies can bind the primary antibody, amplifying the signal. If the protein of interest is in the interior of the cell, the cell is fixed and permeabilized, and the two antibodies are added in succession. (b) The end result is an image in which bright spots indicate the location of the protein of interest in the cell. The images here show a nucleus from a human fibroblast, stained with antibodies and fluorescent labels for DNA polymerase ε, for PCNA, an important polymerase accessory protein, and for bromo-deoxyuridine (BrdU), a nucleotide analog. The BrdU identifies regions undergoing active DNA replication. The patterns of staining show that DNA polymerase ε and PCNA colocalize to regions of active DNA synthesis, with one example marked by a white box. [Source: (b) J. Fuss and S. Linn, J. Biol. Chem., 277: 8658–8666, 2002. © 2002 The American Society for Biochemistry and Molecular Biology.]

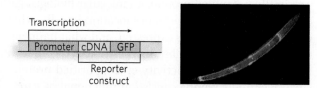

FIGURE 7-24 Specialized DNA libraries. Cloning of a cDNA next to the GFP gene creates a reporter construct. Transcription proceeds through the gene of interest (the inserted cDNA) and the reporter gene (here, GFP), and the mRNA transcript is expressed as a fusion protein. The GFP part of the protein is visible with the fluorescence microscope. Although only one example is shown, thousands of genes can be fused to GFP in similar constructs and stored in libraries in which each cell or organism in the library expresses a different protein fused to GFP. If the fusion protein is properly expressed, its location in the cell or organism can be assessed. The photograph shows a nematode worm (*C. elegans*) containing a GFP fusion protein expressed only in the four "touch" neurons that run the length of its body. [Source: Photo courtesy of Kevin Strange, PhD, and Michael Christensen, PhD, Department of Pharmacology, Vanderbilt University Medical Center.]

Proteins Can Be Detected in Cellular Extracts with the Aid of Western Blots

Western blots, also known as immunoblots, make use of antibodies to detect the presence of specific proteins in a biological sample, such as in a certain tissue or at a given time in an organism's development (**Figure 7-25**). The antibodies are obtained by purifying the protein of interest, inoculating a chicken or rabbit with the protein, and isolating the resulting antibodies from the serum. The first few steps are similar to those described for Southern blots (for nucleic acids) in Chapter 6: a protein sample is subjected to electrophoresis in a polyacrylamide gel, and the gel is then blotted on a membrane to which proteins in the gel adhere. The remaining steps of Western blotting are specific for the detection of proteins rather than nucleic acids. The membrane is first washed with a protein solution to coat the membrane, to prevent nonspecific adherence of the antibodies. Then it is washed with a solution containing the protein-specific rabbit or chicken antibodies, which bind to the immobilized target protein. A second solution is then added, containing a second antibody that specifically binds to the first (e.g., an antibody derived from goats that binds all rabbit-derived antibodies of the abundant IgG class of circulating antibodies). The second antibody has an attached radioactive or fluorescent label to allow visualization of protein-antibody complexes. In some cases, a single labeled antibody is used. The procedure is sensitive to changes in the amount of target protein present, so increases or decreases in cellular protein levels are readily monitored.

Protein-Protein Interactions Can Help Elucidate Protein Function

Another key to defining the function of a particular protein is to determine what it binds to. In the case of protein-protein interactions, the association of a protein of unknown function with a protein having a known function can provide useful and compelling "guilt by association." The techniques used in this effort are quite varied.

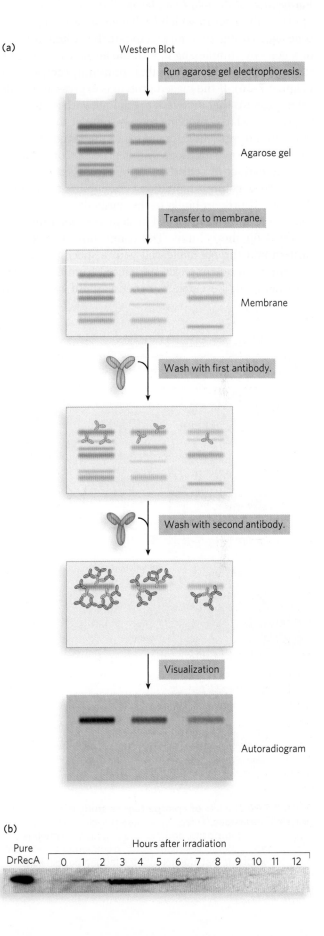

FIGURE 7-25 Western blots. (a) Proteins are subjected to electrophoresis, then transferred from the gel to a membrane. The membrane is washed successively with the first antibody and the second (labeled) antibody, allowing visualization of the protein of interest. (b) A Western blot shows levels of RecA protein in cells of the highly radiation-resistant bacterium *Deinococcus radiodurans*. The *D. radiodurans* protein DrRecA is first induced and then repressed, according to the need for DNA repair, in the hours following high-level irradiation. [Source: (b) Courtesy of Cédric Norais, Department of Biochemistry, University of Wisconsin–Madison.]

Purification of Protein Complexes With the construction of cDNA libraries in which each gene is fused to an epitope tag, investigators can precipitate the protein product of a gene by complexing it with the antibody that binds the epitope, a process called **immunoprecipitation** (**Figure 7-26**). If the tagged protein is expressed in cells, other proteins that bind to it precipitate with it. Identifying the associated proteins reveals some of the intracellular protein-protein interactions of the tagged protein. There are many variations of this process. For example, a crude extract of cells that express a tagged protein is added to a column containing immobilized antibody. The tagged protein binds to the antibody, and proteins that interact with the tagged protein are sometimes also retained on the column. The connection between the protein and the tag is cleaved with a specific protease, and the protein complexes are eluted from the column and analyzed. Researchers can use these methods to

define complex networks of interactions within a cell. In principle, the chromatographic approach to analyzing protein-protein interactions can be used with any type of protein tag (His tag, GST, etc.) that can be immobilized on a suitable chromatographic medium.

The selectivity of this approach has been enhanced with the use of **tandem affinity purification (TAP) tags**. Two consecutive tags, polypeptides known as Protein A and calmodulin-binding peptide, are fused to a target protein, and the fusion protein is expressed in a cell (**Figure 7-27**). A crude extract containing the

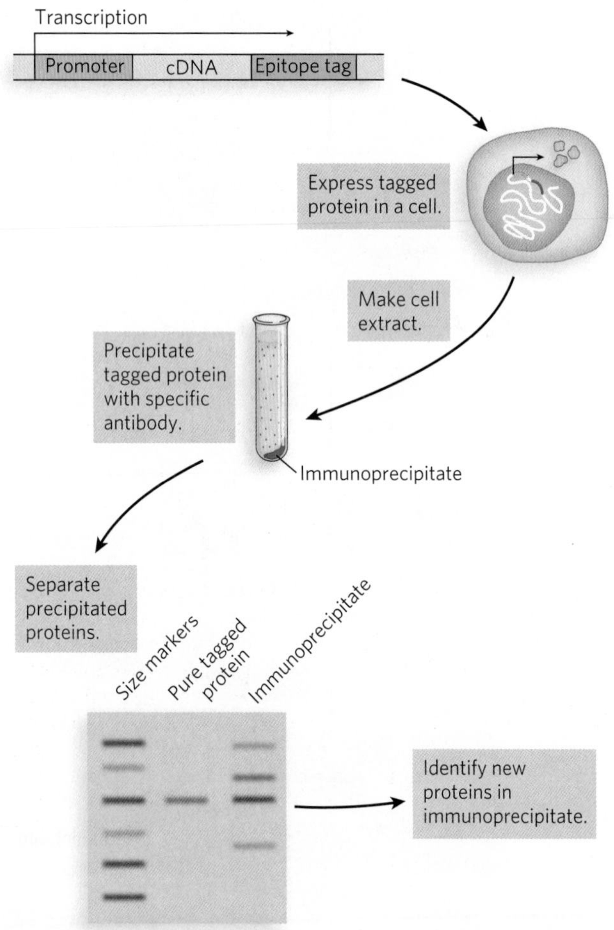

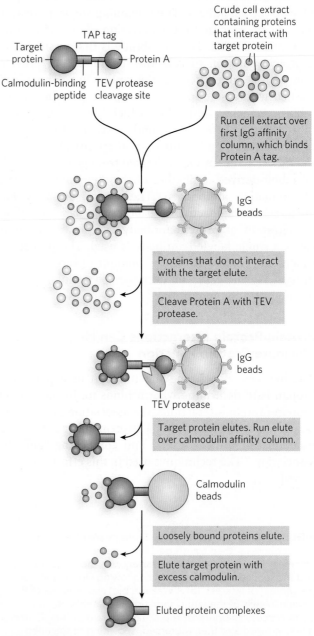

FIGURE 7-26 The use of epitope tags to study protein-protein interactions. The gene of interest is cloned next to a gene for an epitope tag, and the resulting fusion protein is precipitated by antibodies to the epitope (immunoprecipitation). Any other proteins that interact with the tagged protein also precipitate, thereby helping to elucidate protein-protein interactions.

FIGURE 7-27 Tandem affinity purification (TAP) tags. A TAP-tagged protein and associated proteins are isolated by two consecutive affinity purifications.

TAP-tagged fusion protein is passed through a column matrix that has attached IgG antibodies that bind Protein A. Most of the unbound proteins are washed through the column, but proteins associated with the target protein are retained. Protein A is then cleaved from the fusion protein with the enzyme TEV protease, and the shortened target protein and associated proteins are eluted from the column. The eluent is then passed through a second column containing a matrix of calmodulin beads. Loosely bound proteins are again washed from the column, and the target protein is eluted from the column with its associated proteins. The two consecutive purification steps eliminate any weakly bound contaminating proteins. False positives are minimized, and protein interactions that persist through both steps are likely to be functionally significant.

Yeast Two-Hybrid and Three-Hybrid Analysis A sophisticated genetic approach to defining protein-protein interactions is based on the properties of the Gal4 protein (Gal4p), which activates the transcription of *GAL* genes in yeast (genes encoding the enzymes of galactose metabolism; see Chapter 21). Gal4p has two domains: one that binds a specific DNA sequence and another that activates RNA polymerase to synthesize mRNA from an adjacent gene. The two domains of Gal4p are stable when separated, but activation of RNA polymerase requires interaction with the activation domain, which in turn requires positioning by the DNA-binding domain. Hence, the domains must be brought together to function correctly.

In **yeast two-hybrid analysis**, the protein-coding regions of the genes to be analyzed are fused to the yeast gene for either the DNA-binding domain or the activation domain of Gal4p, and the resulting genes express a series of fusion proteins (**Figure 7-28**). If a protein fused to the DNA-binding domain interacts with a protein fused to the activation domain, transcription is activated. The reporter gene transcribed by this activation is generally one that yields a protein required for growth or an enzyme that catalyzes a reaction with a colored product. Thus, when grown on the proper medium, cells that contain a pair of interacting proteins are easily distinguished from those that do not. A library can be set up with a particular yeast strain in which each cell in the library has a gene fused to the Gal4p DNA-binding domain gene, and many such genes are represented in the library. In a second yeast strain, a gene of interest is fused to the gene for the Gal4p activation domain. The yeast strains are mated, and individual diploid cells are grown into colonies. This allows large-scale screening for cellular proteins that interact with the target protein.

The function of many key regulatory proteins, especially in eukaryotic cells, involves specific interactions

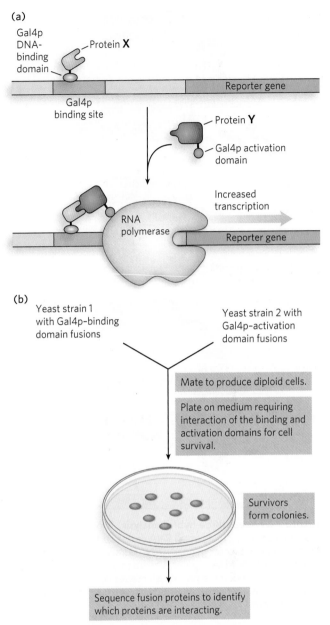

FIGURE 7-28 Yeast two-hybrid analysis. (a) The goal is to bring together the DNA-binding domain and the activation domain of the yeast Gal4 protein (Gal4p) through the interaction of two proteins, X and Y, to which one or other of the domains is fused. This interaction is accompanied by the expression of a reporter gene. (b) The two gene fusions are created in separate yeast strains, which are then mated. The mated mixture is plated on a medium on which the yeast cannot survive unless the reporter gene is expressed. Thus, all surviving colonies have interacting fusion proteins. Sequencing of the fusion proteins in the survivors reveals which proteins are interacting.

between the proteins and RNA molecules. A strategy called **yeast three-hybrid analysis** has been developed to screen for protein-RNA interactions (**Figure 7-29**). For a known RNA-binding protein, this method yields a rapid identification of all or most of the RNAs that the protein binds. The method uses three engineered

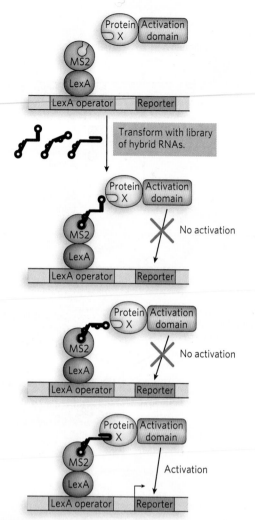

FIGURE 7-29 Yeast three-hybrid analysis. Two fusion proteins bind simultaneously to a hybrid RNA molecule to permit expression of a reporter gene. An RNA library consisting of random-sequence RNA segments fused to a hairpin recognized by MS2 is screened. If the protein of interest (X) binds the random RNA sequence expressed in a given cell, that cell will survive and produce a colony.

elements: two fusion proteins and a plasmid library. The fusion proteins are (1) the protein of interest fused to the Gal4p transcription-activation domain, and (2) the DNA-binding domain of a protein known as LexA fused to an RNA-binding protein called MS2. The LexA portion binds to a specific DNA sequence, and MS2 binds tightly to an RNA hairpin with a defined sequence. The DNA-binding site for the LexA protein is placed upstream of a reporter gene. The third element, the plasmid library, consists of the gene encoding the RNA hairpin recognized by MS2, fused to random sequences. When transcribed, each MS2 RNA is fused to another RNA segment. If a particular expressed RNA is bound by the protein of interest, the RNA serves as a tether,

linking the first fusion protein with the second and activating transcription of the reporter gene. With this method, a few dozen RNA molecules that specifically bind the target protein can be isolated from a library containing millions of cloned RNAs.

These techniques for determining cellular localization and molecular interactions provide important clues to protein function. However, they do not replace classical biochemistry and molecular biology. They simply give researchers an expedited entrée into important new biological problems. When paired with the simultaneously evolving tools of biochemistry and molecular biology, the techniques described here are speeding the discovery not only of new proteins, but of new biological processes and mechanisms.

DNA Microarrays Reveal Cellular Protein Expression Patterns and Other Information

Major refinements of the technology underlying DNA libraries, PCR, and hybridization have come together in the development of **DNA microarrays**, which allow the rapid and simultaneous screening of many thousands of genes. In the most commonly used technique, DNA segments from genes of known sequence, a few dozen to hundreds of base pairs long, are synthesized directly on a solid surface by a process called photolithography (**Figure 7-30**). Thousands of independent sequences are generated, each occupying a tiny part, or spot, of a surface measuring just a few square centimeters. The pattern of sequences is predesigned, with each of many thousands of spots containing sequences derived from a particular gene. The resulting array, or chip, may include sequences derived from every gene of a bacterial or yeast genome, or selected families of genes from a larger genome. Once constructed, the microarray can be probed with mRNAs or cDNAs from a particular cell type or cell culture to identify the genes being expressed in those cells.

A microarray can provide a snapshot of all the genes in an organism, informing the researcher about the genes that are expressed at a given stage in the organism's development or under a particular set of environmental conditions. For example, the total complement of mRNA can be isolated from cells at two different stages of development and converted to cDNA with reverse transcriptase. With the use of fluorescently labeled deoxyribonucleotides, the two cDNA samples can be made so that one fluoresces red, the other green (**Figure 7-31** on p. 246). The cDNA from the two samples is mixed and used to probe the microarray. Each cDNA anneals to only one spot, corresponding to the gene encoding the mRNA that gave rise to that cDNA.

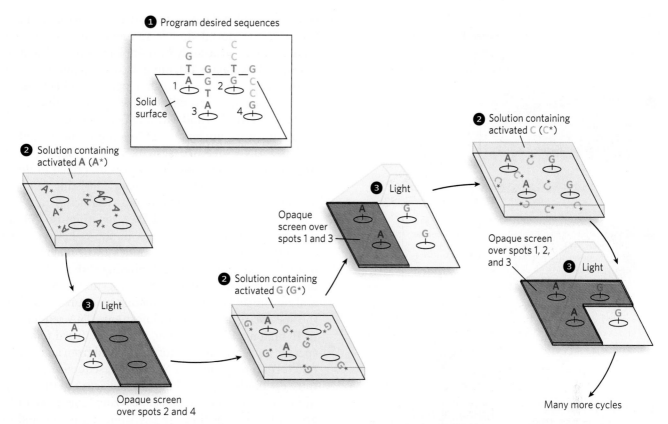

FIGURE 7-30 Photolithography to create a DNA microarray. ❶ A computer is programmed with the desired oligonucleotide sequences. ❷ The reactive groups, attached to a solid surface, are initially rendered inactive by photoactive blocking groups, which can be removed by a flash of light. An opaque screen blocks the light from certain groups, preventing their activation. Other areas or "spots" are exposed. ❸ A solution containing one activated nucleotide (e.g., A*) is washed over the spots. The 5' hydroxyl of the nucleotide is blocked to prevent unwanted reactions, and the nucleotide links to the surface groups at the appropriate spots through its 3' hydroxyl. The surface is washed successively with solutions containing each remaining activated nucleotide (G*, C*, T*). The 5'-blocking groups on each nucleotide limit the reactions to addition of one nucleotide at a time, and these groups can also be removed by light. Once each spot has one nucleotide, a second nucleotide can be added to extend the nascent oligonucleotide at each spot, using screens and light to ensure that the correct nucleotides are added at each spot in the correct sequence. This continues until the required sequences are built up on each spot on the surface.

Spots that fluoresce green represent genes that produce mRNAs at higher levels at one developmental stage; those that fluoresce red represent genes expressed at higher levels at another stage. If a gene produces mRNAs that are equally abundant at both stages of development, the corresponding spot fluoresces yellow. By using a mixture of two samples to measure relative rather than absolute sequence abundance, the method corrects for inconsistencies among spots in the microarray. The spots that fluoresce provide a snapshot of all the genes being expressed in the cells at the moment they were harvested—gene expression examined on a genome-wide scale. For a gene of unknown function, the time and circumstances of its expression can provide important clues about its role in the cell.

A Gene's Function Can Be Elucidated by Examining the Effects of Its Absence

One of the most informative paths to understanding the function of a gene is to change (mutate) it or delete it. The investigator can then examine the effects of the genomic alteration on cell growth or function. The methods available to modify genomes grow more sophisticated every year. One increasingly common strategy is to cut the gene at a site that is functionally critical, generating a double-strand break. In eukaryotes, such breaks are most commonly repaired by cellular systems that promote nonhomologous end joining (NHEJ), a process described in Chapter 13. NHEJ seals the double-strand break, but the process is imprecise. Nucleotides are often

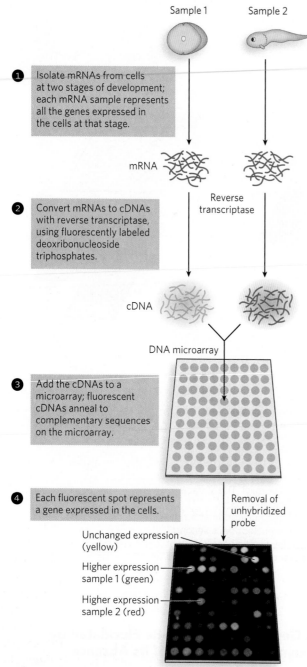

① Isolate mRNAs from cells at two stages of development; each mRNA sample represents all the genes expressed in the cells at that stage.

mRNA

② Convert mRNAs to cDNAs with reverse transcriptase, using fluorescently labeled deoxribonucleoside triphosphates.

Reverse transcriptase

cDNA

DNA microarray

③ Add the cDNAs to a microarray; fluorescent cDNAs anneal to complementary sequences on the microarray.

④ Each fluorescent spot represents a gene expressed in the cells.

Removal of unhybridized probe

Unchanged expression (yellow)

Higher expression sample 1 (green)

Higher expression sample 2 (red)

FIGURE 7-31 A DNA microarray experiment. A microarray can be prepared from any known DNA sequence, from any source. Once the DNA is attached to a solid support, the microarray can be probed with other, fluorescently labeled nucleic acids. Here, mRNA samples are collected from frog cells at two different stages of development: single-cell stage (sample 1) and a later stage (sample 2). The cDNA probes are synthesized with nucleotides that fluoresce in different colors for each sample; a mixture of the cDNAs is used to probe the microarray. The probes anneal to spots containing complementary DNA; if the spot lights up, the corresponding gene is represented in the pool of mRNA used to produce the probes. Green spots represent mRNAs more abundant at the single-cell stage; red spots, sequences more abundant later in development. The yellow spots indicate approximately equal abundance at both stages.

deleted or added during repair, inactivating the gene. In bacteria, introduced double-strand breaks are usually repaired more accurately by homologous recombination systems (see Chapter 13), but inactivating mutations can appear. Three significant technologies have been developed to cut a gene at a particular site in vivo: zinc finger nucleases, TALENs, and CRISPR/Cas systems.

Zinc finger nucleases (ZFNs) represent a kind of designer DNA cleavage system. A zinc finger (which we encountered in Chapter 4 and is described in detail in Chapter 19) is a protein domain consisting of about 30 amino acid residues. It folds into a characteristic β-α-β structure, aided by a bound Zn^{2+} ion (**Figure 7-32a**). Particular amino acid residues in the α helix contact about three consecutive nucleotides in the major groove of a DNA molecule, leading to binding with a substantial degree of selectivity. Researchers can stitch together several zinc fingers in tandem, creating structures that allow the specific recognition of almost any DNA sequence 9 to 18 bp long. When the zinc fingers are fused to a non-specific nuclease domain (often derived from an enzyme called FokI) to create a ZFN, the DNA sequence to which the ZFN binds is cleaved at a site adjacent to the recognition sequence of the associated zinc fingers. TALENs are similar, except that the DNA binding is directed by a series of TALE (transcription activator–like effector) domains (**Figure 7-32b**). These are similar in size to zinc fingers but recognize single base pairs; like zinc fingers, they can be linked together and fused to a nonspecific nuclease domain to yield a TALE nuclease, or TALEN. These enzymes can be expressed in a cell, and the resulting enzyme cleaves the target site in the genome to generate a double-strand break. TALENs can be designed to inactivate genes and even to inactivate viral DNA that is integrated into a genome. One drawback is cost: targeting a new DNA sequence necessitates the design, construction, and testing of a new TALEN enzyme, an expensive process that can take weeks or months.

A more robust and convenient approach to cutting genes in vivo is derived from a kind of bacterial immune system called CRISPR/Cas (clustered, regularly interspaced short palindromic repeats–CRISPR-associated proteins). Spacer sequences inserted between CRISPR repeats, embedded in a bacterial genome, are derived from bacteriophage pathogens. When that bacteriophage attacks a bacterium with the corresponding CRISPR/Cas system, the CRISPR sequences are transcribed to RNA, and individual spacers (plus some adjacent repeat RNA) are cleaved to form products called guide RNAs (gRNAs). A gRNA forms a complex with one or more Cas proteins and, in some cases, with another RNA called a trans-activating CRISPR RNA, or tracrRNA. The resulting complex binds specifically to the invading bacteriophage DNA, cleaving and destroying it with

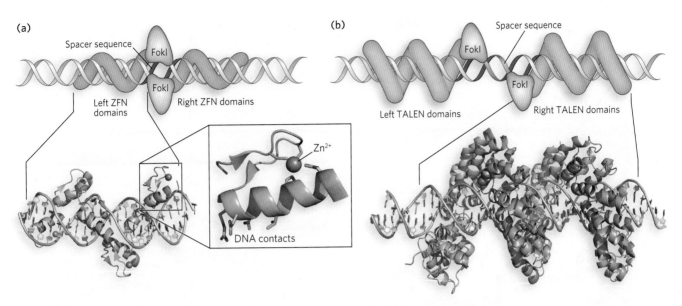

FIGURE 7-32 Designer nucleases. (a) An engineered zinc finger nuclease (ZFN) in complex with its target DNA (bottom). Surface residues that contact DNA are shown as sticks. Each zinc finger domain contacts 3 or 4 bp in the major groove of the DNA. The complexed Zn^{2+} ion is shown as a purple ball. The cartoon (top) shows a ZFN dimer bound to DNA. ZFN target sites consist of two zinc finger–binding sites separated by a 5 to 7 bp spacer sequence that is cleaved by the FokI nuclease domain. The half sites recognized by a ZFN need not be identical. (b) A TALE protein bound to its target DNA (bottom). TALEN target sites consist of two TALE-binding sites separated by a spacer sequence of 12 to 20 bp (top). [*Sources: (a) (bottom) PDB ID 2I13. (b) (bottom) PDB ID 3UGM.*]

nuclease activities associated with the Cas proteins. A relatively simple example of CRISPR/Cas, found in the bacterium *Streptococcus pyogenes*, requires only a single Cas protein called Cas9 to recognize and cleave DNA. Work in many laboratories, particularly those of Jennifer Doudna and Emmanuelle Charpentier, has produced a streamlined CRISPR/Cas9 system that is increasingly robust. The gRNA and tracrRNAs are typically fused into a what is called a single guide RNA (sgRNA). The guide sequence can be programmed (by altering its sequence) to target almost any specific genomic site (**Figure 7-33**). Cas9 has two separate nuclease domains: one domain cleaves the DNA strand paired with the sgRNA, and the other cleaves the opposite DNA strand. Inactivating one domain creates an enzyme that cleaves just one strand, creating a single-strand break, called a nick. The sgRNA is needed both to pair with the target sequence in the DNA and to activate the nuclease domains for cleavage.

Plasmids expressing the required protein and RNA components of CRISPR/Cas9 can be introduced into cells by electroporation (described earlier in this chapter). In cells from many organisms, targeted gene inactivation occurs in 10% to 50% of the treated cells. If a genomic change (mutation) rather than a simple gene inactivation is desired, it can be introduced by recombination when a DNA fragment encompassing the cleavage site and including the planned change enters the cell with the CRISPR/Cas9 plasmids. This recombination is generally more efficient if a nick rather than a double-strand break is introduced at the target site (Figure 7-33a)

Other uses for CRISPR/Cas9 are being developed. If the nuclease active sites of Cas9 are mutationally inactivated, the nuclease-deficient complex still binds tightly to its target. Such a complex can be used to block the movement of RNA polymerase, effectively silencing a gene (Figure 7-33b). In a variation of this strategy, a gene activator (see Chapter 19) can be fused to the nuclease-deficient CRISPR/Cas9. When the resulting complex is targeted to an appropriate genomic site, it can be used to activate transcription of a specific gene of interest (Figure 7-33c). Therapeutic uses of CRISPR/Cas9 are still far in the future, but the potential for alleviating the effects of genetic diseases, HIV, and many other human ailments is enormous.

SECTION 7.3 SUMMARY

- The role of a gene of unknown function can be explored by techniques that assess the location of the gene product in the cell, the interactions of the gene product with proteins or RNA, and the expression of the gene under different cellular circumstances.
- By fusing a gene of interest with the genes that encode green fluorescent protein or epitope tags, researchers can visualize the cellular location of the gene product, either directly or by immunofluorescence.

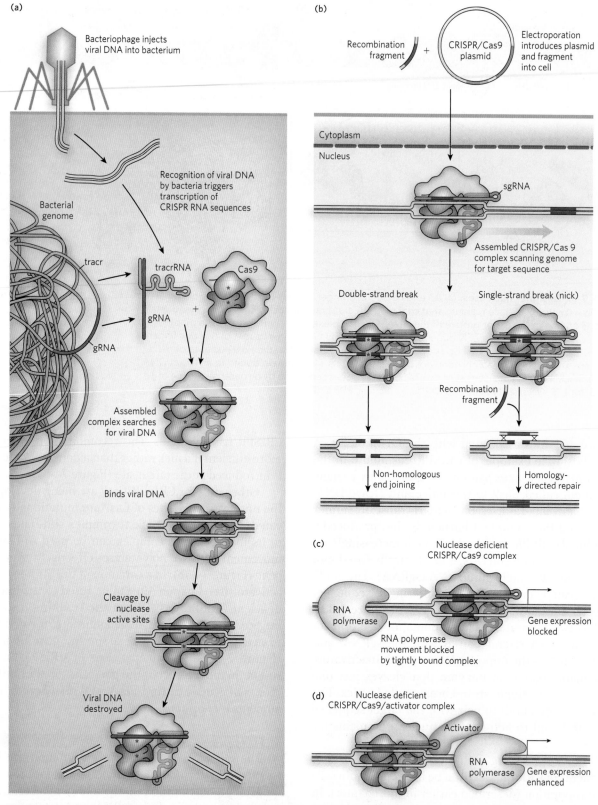

FIGURE 7-33 The CRISPR/Cas9 system for genomic engineering. (a) A complex consisting of the CRISPR sgRNA and the Cas9 protein binds to a target site in the DNA. Two nuclease active sites in the Cas9 protein are indicated by red asterisks. The cleavage produces a double-strand break or, if one active site is inactivated, a single-strand break (nick). (b) If both nuclease active sites in Cas9 are inactivated, a complex is produced that simply binds tightly to its target sequence. This bound complex can be situated so as to block transcription of a particular gene by RNA polymerase. (c) If both nuclease active sites are inactivated, the Cas9 protein can be fused to a gene activator protein (see Chapter 19). If targeted to a site near an appropriate gene, the complex can be used to trigger enhanced transcription.

- The presence of particular proteins in biological samples can be detected with the aid of Western blots.
- The interactions of a protein with other proteins or RNA can be investigated with epitope tags and immunoprecipitation or affinity chromatography. Alternatively, interactions in the cell can be probed in yeast two-hybrid and three-hybrid analyses.
- Expression patterns of genes can be probed through the use of microarrays.
- Genes can be inactivated or altered with the CRISPR/Cas9 system, as a method of elucidating or manipulating gene function.

? UNANSWERED QUESTIONS

As powerful as biotechnology has become, there are still limitations to the reach of molecular biology.

1. Can personalized genomics become commonplace? The DNA sequencing methods described in this chapter are making the personal genome a reality, at a reasonable cost. And that cost is still declining rapidly. Besides the technological developments, much thought is required about what we will do with all this DNA sequence information when we get it. Ethical considerations must play a role in these decisions.

2. How do we apply biotechnology to routinely and safely alter human genomes? The CRISPR/Cas9 system is the most recent and robust addition to the toolbox that molecular biologists can use to alter genomes. Safe and effective application to human genomes may become a reality. With more efficient methods, the treatment of human genetic disease by gene therapy may become practical.

3. Can we find out where proteins function in a cell, and what they do? There is much room for improvement in the methods based on fusion proteins, now used to track the location and function of proteins in cells. Too often, the protein tags used for this work have unintended effects on the activity of the protein to which the tag is attached. The GFP protein works well in many cases, but smaller and brighter tags would be very helpful, along with new ways to link them to proteins.

HOW WE KNOW

New Enzymes Take Molecular Biologists from Cloning to Genetically Modified Organisms

Cohen, S.N., A.C.Y. Chang, H.W. Boyer, and R.B. Helling. 1973. Construction of biologically functional bacterial plasmids in vitro. *Proc. Natl. Acad. Sci.* USA 70:3240–3244.

Jackson, D.A., R.H. Symons, and P. Berg. 1972. Biochemical method for inserting new genetic information into DNA of simian virus 40: Circular SV40 DNA molecules containing lambda phage genes and the galactose operon of *Escherichia coli. Proc. Natl. Acad. Sci.* USA 69:2904–2909.

DNA cloning is commonplace now, but the idea of combining DNA from two different species created quite a stir in the early 1970s. A convergence of research advances in many laboratories gave rise to a technological revolution.

The first experiment to join two DNA molecules together used a rather laborious strategy. Peter Lobban, a graduate student in Dale Kaiser's lab at Stanford, first used the enzyme terminal transferase to add "tails" of A residues to one DNA fragment and T residues to another. The poly(A) and poly(T) tails annealed, and the fragments could be covalently joined with DNA ligase (Figure 1). This experiment linked two segments of DNA derived from a bacterial virus, P-22. Paul Berg's lab soon used the same strategy to link two DNA segments from different species, one from the bacterial virus lambda (λ) and the other from the simian virus SV40. The Berg paper was published in 1972.

Meanwhile, Stanley Cohen, also at Stanford, began to study DNA plasmids that made certain disease-causing bacteria resistant to antibiotics (a problem then, as it remains today). His lab soon developed methods for isolating the plasmids and reintroducing them into other bacteria. To get at the genes causing the antibiotic resistance and other features of the plasmid, Cohen wanted to take the plasmids apart and reassemble them. Rather than use the laborious poly(A)–poly(T) tail approach, Cohen relied on restriction enzymes that had recently been discovered. Herbert Boyer and his colleagues at the University of California, San Francisco, had shown that one of these

enzymes, EcoRI, cleaves DNA asymmetrically at a 6 bp palindrome (see Table 7-2). The resulting sticky ends could guide the rejoining of the ends by DNA ligase.

At a November 1972 meeting in Hawaii, Cohen and Boyer began a collaboration that led to DNA cloning. They settled on a plasmid called pSC101 from the Cohen lab, which encoded a gene that conferred resistance to tetracycline. The collaborators found that Boyer's EcoRI enzyme cut the plasmid only once, and not in a region that affected the sequences needed for either replication or tetracycline resistance. By early 1973, they had demonstrated that DNA segments from any source, also derived from cleavage by EcoRI, could be linked to this plasmid. The plasmid, in turn, could be reintroduced and propagated in bacteria. These advances, reported in the *Proceedings of the National Academy of Sciences* (USA) in late 1973, gave rise to the first DNA cloning patents and a new company (Genentech). More importantly, they set the stage for the rise of biotechnology.

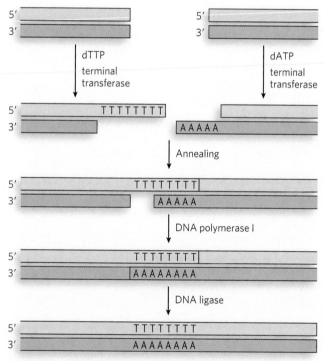

FIGURE 1 A homopolymer tail was added to each DNA segment with the aid of the enzyme terminal transferase. A poly(A) tail was added to one segment, a poly(T) tail to the other. The two complementary tails were annealed. Gaps were filled with the aid of DNA polymerase, and the DNA was joined by DNA ligase.

A Dreamy Night Ride on a California Byway Gives Rise to the Polymerase Chain Reaction

Brock, T.D., and H. Freeze. 1969. *Thermus aquaticus* gen. n. and sp. n., a nonsporulating extreme thermophile. *J. Bacteriol.* 98:289–297.

Saiki, R.K., S. Scharf, F. Faloona, K.B. Mullis, G.T. Horn, H.A. Erlich, and N. Arnheim. 1985. Enzymatic amplification of beta-globin genomic sequences and restriction site analysis for diagnosis of sickle-cell anemia. *Science* 230:1350–1354.

Sometimes, large advances in science arise through inspiration. In the spring of 1983, Kary Mullis was an employee of the Cetus Corporation in northern California. Hired in 1979 to synthesize oligonucleotides, Mullis discovered that as oligonucleotide synthesis became increasingly automated, he had more and more time to contemplate other projects. He became interested in methods to detect small sequence differences in human DNA, but initially did not make much progress. The idea for the polymerase chain reaction occurred to Mullis one night, in April 1983, as he drove with a friend up the coast. As he described it, he stopped the car and started drawing—DNA molecules hybridizing and lengthening, a chain reaction in which the products of one cycle became the templates for the next.

The first experiment was carried out a few months later, and the first report of the polymerase chain reaction came out in a 1985 paper in *Science* describing a new procedure for detecting the hemoglobin mutation that causes sickle-cell anemia. The method was spelled out in more detail in publications appearing over the next two years. In the early trials, the polymerase used was an active fragment of the *E. coli* DNA polymerase I (see Chapter 11). The heating required to denature the DNA after each PCR cycle inactivated this polymerase, so it had to be added again after each cycle.

A side story shows how basic research can contribute to major advances in surprising ways.

About two decades before the development of PCR, microbiologist Thomas Brock (at the University of Wisconsin–Madison) initiated some studies of organisms in the hot springs of Yellowstone National Park (Figure 2). In the fall of 1966, he succeeded in culturing a bacterium from a pool called Mushroom Spring, an organism that grew at higher temperatures than were thought possible for living organisms. The new thermophilic bacterium was subsequently named *Thermus aquaticus*. The heat stability of the proteins in *T. aquaticus* became highly important to the development of PCR. The DNA polymerase from this bacterium (*Taq* polymerase) is stable at very high temperatures and is not inactivated by the heating and cooling cycles that are needed to denature and reanneal the DNA during PCR. The incorporation of *Taq* polymerase into the PCR protocol in the late 1980s allowed the entire procedure to be automated.

PCR methods were quickly optimized, and new protocols gradually expanded the possible applications. By the end of the 1980s, the technology had utterly transformed the biological sciences.

FIGURE 2 Hot springs in Yellowstone National Park, one of which is shown here, were the source of the bacterium *Thermus aquaticus*. [*Source: Fox71/Dreamstime.*]

HOW WE KNOW

Coelenterates Show Biologists the Light

Chalfie, M., Y. Tu, G. Euskirchen, W.W. Ward, and D.C. Prasher. 1994. Green fluorescent protein as a marker for gene expression. *Science* 263:802–805.

Heim, R., D.C. Prasher, and R.Y. Tsien. 1994. Wavelength mutations and posttranslational autoxidation of green fluorescent protein. *Proc. Natl. Acad. Sci. USA* 91:12501–12504.

Shimomura, O., F.H. Johnson, and Y. Saiga. 1962. Extraction, purification and properties of aequorin, a bioluminescent protein from the luminous hydromedusan, *Aequorea. J. Cell. Comp. Physiol.* 59:223–239.

In 1960, shortly after joining the faculty at Princeton University, Osamu Shimomura began to study the bioluminescence produced by the jellyfish *Aequorea victoria*. Traveling regularly to the state of Washington to secure specimens, he would take the photo-organs from 20 or 30 jellyfish and squeeze them through rayon gauze. His "squeezate" was slightly luminescent, and he began to purify the molecules responsible. In 1962, his lab reported the purification of a protein associated with what was later named green fluorescent protein, which they called aequorin. The first reference to GFP appears in a footnote in that paper, in which Shimomura described isolating from squeezates a protein that formed solutions that looked greenish in sunlight and yellowish under tungsten lights, and showed very bright, greenish fluorescence in UV light. His subsequent studies gradually showed that GFP had a special property: it contained all the chemistry needed to emit fluorescence on its own. Up to that time, most other proteins known to produce bioluminescence, such as firefly luciferase, required the addition of other molecules to do so.

Douglas Prasher, at the Woods Hole Oceanographic Institution, was the first to appreciate the potential of fusing GFP to another protein and using its fluorescence as a cellular marker for that protein.

He succeeded in cloning the gene for GFP and determining its sequence, reporting this advance in 1992. Martin Chalfie, at Columbia University, had also seen the potential of GFP. Collaborating with Prasher, he expressed GFP in *E. coli*, reporting the results of the work in *Science* in 1994 (Figure 3). This work realized the potential of the system, showing that GFP expressed in a host organism produced fluorescence without the need for any other protein or factor from the jellyfish. Sergey Lukyanov, in Moscow, showed that variants of GFP could be cloned from the nonbioluminescent *Anthozoa* of coral reefs, and he managed to expand the color range of this protein class by cloning a red fluorescent protein.

Much of what we now know about the chemistry and general utility of GFP and other fluorescent proteins has resulted from the subsequent work of Roger Tsien, at the University of California, San Diego. His laboratory has constructed many mutants of GFP that produce the range of colors seen in Figure 7-22b. Many of these mutants also improve on the stability and brightness of the fluorophores in the proteins. Since the mid-1990s, GFP and its variants have become a staple of molecular and cellular biology, illuminating the mechanisms and pathways of countless other cellular proteins and processes.

FIGURE 3 The bacteria on the right side of this agar plate are glowing as a result of the expression of green fluorescent protein. [Source: From Chalfie, M. et al., Science, 263, 802–805, 1994. Reprinted with permission from AAAS. Courtesy M. Chalfie.]

KEY TERMS

PROBLEMS

1. When joining two or more DNA fragments, a researcher can adjust the sequence at the junction(s) in a variety of subtle ways, as seen in the following exercises.

   (a) Draw the structure of each end of the linear DNA fragments produced by an EcoRI restriction digest (include the sequences remaining from the EcoRI recognition sequence).

   (b) Draw the structure resulting from the reaction of these end sequences with a DNA polymerase and the four deoxynucleoside triphosphates.

   (c) Draw the sequence produced at the junction that arises when the two ends with the structure derived in (b) are ligated.

   (d) Draw the structure produced when the structure derived in (a) is treated with a nuclease that degrades only single-stranded DNA.

   (e) Draw the sequence of the junction produced when an end with structure (b) is ligated to an end with structure (d).

   (f) Draw the structure of the end of a linear DNA fragment produced by a PvuII restriction digest (include the sequences remaining from the PvuII recognition sequence).

   (g) Draw the sequence of the junction produced when an end with structure (b) is ligated to an end with structure (f).

   (h) Suppose you can synthesize a short duplex DNA fragment with any sequence you wish. With this synthetic fragment and any of the procedures described in (a) through (g), design a protocol that would remove an EcoRI restriction site from a DNA molecule and insert a new BamHI restriction site at approximately the same location. (See Figure 7-2.)

   (i) Design four different short, synthetic double-stranded DNA fragments that would permit ligation of structure (a) with a DNA fragment produced by a PstI restriction digest. In one of these fragments, design the sequence so that the final junction contains the recognition sites for both EcoRI and PstI. In the second and third fragments, design the sequences so that the junction contains only the EcoRI and only the PstI recognition site, respectively. Design the sequence of the fourth fragment so that neither the EcoRI nor the PstI recognition site appears in the junction.

2. The partial sequence of one strand of a double-stranded DNA molecule is:

   5′ − − − GACGAAGTGCTGCAGAAAGTCCGC
   GTTATAGGCATGAATTCCTGAGG − − − 3′

   The cleavage sites for the restriction enzymes EcoRI and PstI are shown below.

   EcoRI 5′-G A A T T C-3′   PstI 5′-C T G C A G-3′
          C T T A A G   G A C G T C

   Write the sequence of *both strands* of the DNA fragment created when this DNA is cleaved with both EcoRI and PstI. The top strand of your duplex DNA fragment should be derived from the strand sequence given above.

3. When cloning a foreign DNA fragment into a plasmid, it is often useful to insert the fragment at a site that interrupts a selectable marker (such as the tetracycline-resistance gene of pBR322). The loss of function of the interrupted gene can be used to identify clones containing recombinant plasmids with foreign DNA. With a YAC vector, a selectable marker is not necessary to easily distinguish cells carrying vectors that incorporate large foreign DNA fragments from those that do not. How are these recombinant vectors identified?

4. The plasmid cloning vector pBR322 (see Figure 7-4) is cleaved with the restriction endonuclease PstI. An isolated DNA fragment from a eukaryotic genome (also produced by PstI cleavage) is added to the prepared vector and ligated. The mixture of ligated DNAs is then used to transform bacteria, and plasmid-containing bacteria are selected by growth in the presence of tetracycline.

(a) In addition to the desired recombinant plasmid, what other types of plasmids might be found among the transformed bacteria that are tetracycline resistant? How can the types be distinguished?

(b) The cloned DNA fragment is 1,000 bp long and has an EcoRI site 250 bp from one end. Three different recombinant plasmids are cleaved with EcoRI and analyzed by gel electrophoresis, giving the patterns shown below. What does each pattern say about the cloned DNA? Note that in pBR322, the PstI and EcoRI restriction sites are about 750 bp apart. The entire plasmid with no cloned insert has 4,361 bp. Size markers in lane 4 have the number of nucleotides noted on the right.

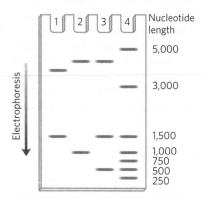

5. A new restriction endonuclease is discovered that recognizes and cleaves the palindromic sequence GGATATCC. How often does this sequence appear in a random-sequence DNA in which all four nucleotides are present in equal amounts? In a random-sequence DNA in which the G + C content is 80%, will the frequency with which this site appears increase or decrease?

6. A BAC vector is designed so that large DNA fragments can be inserted into a cleavage site for the enzyme BamHI (see Table 7-2). To prepare chromosomal DNA from a target organism for cloning into this vector, the target DNA is treated just briefly with BamHI, not long enough to cleave all of the BamHI sites present. Explain why the BamHI reaction is halted before the chromosomal DNA is completely cleaved.

7. One strand of a chromosomal DNA sequence is shown below. An investigator wants to amplify and isolate a DNA fragment defined by the segment shown in red, using the polymerase chain reaction. Design two PCR primers, each 20 nucleotides long, that can be used to amplify this DNA segment.

```
5'———AATGCCGTCAGCCGATCTGCCTCGAGTCAATCGA
TGCTGGTAACTTGGGGTATAAAGCTTACCATGGTATC
GTAGTTAGATTGATTGTTAGGTTCTTAGGTTTAGG
TTTCTGGTATTGGTTTAGGGTCTTTGATGCTATTAA
TTGTTTGGTTTTGATTTGGTCTTTATATGGTTTATGTTT
TAAGCCGGGTTTTGTCTGGGATGGTTCGTCTGATGTGC
GCGTAGCGTGCGGCG———3'
```

8. A researcher wants to amplify the same DNA segment described in Problem 7. However, to aid in cloning, she wants to add a short DNA sequence on each end of the amplified segment that includes the restriction site for the enzyme EcoRI. Design the two PCR primers that this researcher needs, incorporating 20 nucleotides complementary to the appropriate target sequences.

9. Huntington disease (HD) is an inherited neurodegenerative disorder characterized by gradual, irreversible impairment of psychological, motor, and cognitive functions. Symptoms typically appear in middle age, but onset can occur at almost any age, and the course of the disease can range from 15 to 20 years. The molecular basis of HD is becoming better understood, and the genetic mutation has been traced to a gene that encodes a protein (M_r 350,000) of unknown function. In individuals who will not develop HD, a region of the gene that encodes the N-terminus of this protein has a sequence of CAG codons (for glutamine) repeated 6 to 39 times in succession. In individuals with adult-onset HD, this codon is typically repeated 40 to 55 times; in those with childhood-onset HD, it is repeated more than 70 times. Thus, the length of this simple trinucleotide repeat indicates whether an individual will develop HD and at approximately what age the first symptoms will occur.

A small portion of the N-terminal coding sequence of the 3,143-codon HD gene is given below. The nucleotide sequence of the DNA is shown in black, the amino acid sequence corresponding to the gene is in blue, and the CAG repeat is shaded (the numbers at left indicate the starting nucleotide and amino acid residue numbers in that row). Outline a PCR-based test for HD that could be carried out on a blood sample. Assume that each PCR primer must be 25 nucleotides long. By convention, unless otherwise specified, a DNA sequence encoding a protein is written with the coding strand—the sequence identical to the mRNA transcribed from the gene (except for U in the mRNA in place of T)—on top, such that it is read 5'→3', left to right.

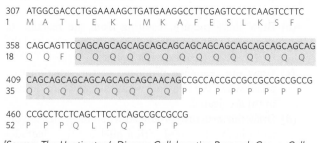

[Source: The Huntington's Disease Collaborative Research Group, Cell 72:971–983, 1993.]

10. In a species of ciliated protist, a segment of genomic DNA is sometimes deleted. The deletion is a genetically programmed reaction associated with cellular mating. A researcher proposes that the DNA is deleted in a type of recombination called site-specific recombination

(see Chapter 14), with the DNA on either side of the segment joined together and the deleted segment forming a circular DNA reaction product. Suggest how the researcher might use PCR to detect the presence of the circular, deleted DNA in an extract of the protist.

11. The short DNA shown below is to be sequenced. The asterisk represents a radioactive label. Using your knowledge of how the Sanger method works, in the gel diagram, draw in the bands that will appear when DNA polymerase is added to the reaction along with the four different nucleotide mixtures indicated (A through D; the bands for A are given). Note that some of these mixtures are *not* what would normally be used in a sequencing reaction. Dideoxynucleotides (ddNTPs) are added in relatively small amounts.

*5′———3′-OH

  3′———ACGACGCAGGACATTAGAC-5′

Nucleotide mixtures:

A. dATP, dTTP, dCTP, dGTP, ddTTP (given)

B. dATP, dTTP, dCTP, dGTP, ddATP

C. dTTP, dGTP, dCTP, ddCTP, ddATP

D. dATP, dCTP, dTTP, ddGTP

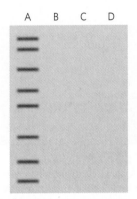

12. To express a cloned gene, the DNA encoding that gene is placed downstream of a bacterial promoter. RNA polymerase binds at the promoter and moves away from it in one direction, synthesizing RNA, using one strand of the DNA as a template. The synthesized RNA strand (mRNA) carries the information specifying a protein to the ribosome. The RNA strand is synthesized in the 5′→3′ direction. The mRNA is identical in sequence (with U replacing T) to one of the DNA strands, and complementary to the other strand. If the orientation of the cloned gene is inverted relative to the promoter, will the same protein still be expressed? Why or why not?

13. An investigator has two systems available for the cloning and expression of proteins: a bacterial plasmid designed for protein expression, and a baculovirus system. Which system would be her best choice to successfully clone and express (a) the gene encoding the *E. coli* RecA protein, and (b) the gene encoding a mammalian DNA polymerase?

14. In the protocol for Western blots, an investigator uses two antibodies. The first binds specifically to the protein of interest. The second is labeled for easy visualization and binds to the first antibody. In principle, molecular biologists could simply label the first antibody and skip one step. Why do they use two successive antibodies?

15. A group of overlapping clones, designated A through F, are isolated from one region of a chromosome. Each of the clones is separately cleaved by a restriction enzyme, and the pieces are resolved by agarose gel electrophoresis, with the results shown below. Nine different restriction fragments can be produced from this chromosomal region, with a subset appearing in each clone. Using this information, deduce the order of the restriction fragments in the chromosome.

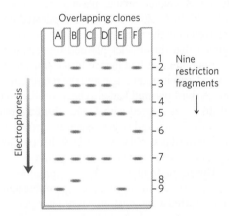

16. You are a researcher who has just discovered a new protein in a fungus. To help determine the protein's function, you want to identify the other proteins in the fungal cell with which your protein interacts. How do you design a yeast two-hybrid experiment to address this problem?

17. Figure 7-30 shows the first steps in the process of making a DNA microarray by photolithography. Describe the remaining steps for obtaining the desired sequences (a different four-nucleotide sequence on each of the four spots) shown in the first panel of the figure. After each step, give the resulting nucleotide sequence attached at each spot (numbered 1 to 4 in Figure 7-30).

18. Mammalian genomes contain some short sequences repeated over and over again. Thousands of these sequences are sometimes repeated in tandem (e.g., the sequence TTAGGG repeated thousands of times in succession). When these tandem repeat regions exceed a few hundred base pairs, it becomes difficult to define them adequately using next-generation DNA sequencing methods. Why might this be the case?

19. In the CRISPR/Cas9 nuclease system, what is the role of the sgRNA?

DATA ANALYSIS PROBLEM

Cohen, S.N., A.C.Y. Chang, H.W. Boyer, and R.B. Helling. 1973. Construction of biologically functional bacterial plasmids in vitro. *Proc. Natl. Acad. Sci. USA* 70:3240–3244.

20. The first recombinant DNA plasmid that combined two different biological activities from different sources and was readily transferred into bacteria was reported in 1973 by the research groups of Stanley Cohen and Herbert Boyer. In this work, the investigators began with the large, naturally occurring circular plasmid R6-5 (~98,500 bp), which included genes conferring resistance to multiple antibiotics, including tetracycline and kanamycin. Prior to their study, R6-5 had been sheared into fragments, the fragments used to transform *E. coli*, and tetracycline-resistant colonies selected. A smaller circular plasmid had arisen from that work, pSC101 (~9,000 bp), which replicated normally and conferred only the tetracycline resistance. The researchers assumed that the smaller plasmid represented a piece of the larger one that had ligated into a smaller circle, which included a replication origin and the tetracycline-resistance gene. A third plasmid had also been generated when plasmid R6-5 was cleaved with EcoRI, the fragments used to transform *E. coli*, and cells selected for growth on kanamycin. This third plasmid was called pSC102 (~27,000 bp), and it did not confer resistance to tetracycline.

The ultimate objective was to combine pSC101 with fragments from R6-5 or pSC102 to generate a new plasmid that conferred a demonstrable new biological activity. The investigators first cleaved all three plasmids with EcoRI. The cleavage products were subjected to electrophoresis on an agarose gel, and the DNA fragments were visualized by staining with ethidium bromide (a fluorescent molecule that binds to DNA by intercalating between adjacent base pairs). This generated the pattern in Figure 1. Electrophoresis proceeded left to right, as shown in the figure, so the bands decrease in size from left to right (lane a is pSC102; lane b, R6-5; lane c, pSC101).

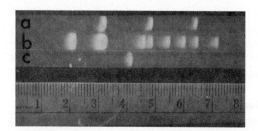

FIGURE 1
[Source: Cohen, S.N., Chang, A.C.Y., Boyer, H.W., & Helling, R.B. (1973) Construction of biologically functional bacterial plasmids in vitro. Proc. Natl. Acad. Sci. USA 70, 3240-3244. Courtesy Stanley N. Cohen.]

(a) From this pattern, how many EcoRI sites are present in each of the three plasmids? (Note that three small DNA bands from R6-5 are not visible on this gel.)

(b) Why does the brightness of the bands decrease from left to right?

(c) In the R6-5 lane (lane b), the second largest band seems to break the general pattern in that it is brighter than the band to its left. How might this be explained?

(d) The three bands in the pSC102 lane (lane a) comigrate with bands in the R6-5 lane. Explain.

The plasmids pSC101 and pSC102 were cleaved with EcoRI, and the fragments from both plasmids were mixed together and treated with DNA ligase. The ligated mixture was then used to transform *E. coli*, and the cells were grown on plates containing both kanamycin and tetracycline. Colonies appeared, and all of them contained a new plasmid, named pSC105 (~16,000 bp), that conferred resistance to kanamycin and tetracycline. Next, the plasmids pSC101, pSC102, and pSC105 were cleaved with EcoRI and subjected to gel electrophoresis. The results are shown in Figure 2. Electrophoresis proceeded, and fragment sizes decrease, from left to right (lane a is pSC105; lane b, pSC101 + pSC102; lane c, pSC102; lane d, pSC101).

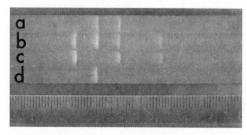

FIGURE 2
[Source: Cohen, S.N., Chang, A.C.Y., Boyer, H.W., & Helling, R.B. (1973) Construction of biologically functional bacterial plasmids in vitro. Proc. Natl. Acad. Sci. USA 70, 3240-3244. Courtesy Stanley N. Cohen.]

(e) Which of the two fragments of pSC105 (lane a) contains the gene for tetracycline resistance?

(f) Which of the two fragments of pSC105 contains the gene encoding kanamycin resistance?

(g) What is the approximate size of the second (smaller) EcoRI cleavage fragment of pSC105?

(h) How many phosphodiester bonds were created by DNA ligase to produce the circular plasmid pSC105?

(i) The largest and smallest EcoRI fragments of pSC101 (lane d) do not appear in pSC105, although they were present in the ligation mixture that gave rise to it. Why were these DNA fragments not incorporated into the recombinant plasmid?

ADDITIONAL READING

General

Cohen, S.N., A.C.Y. Chang, H.W. Boyer, and R.B. Helling. 1973. Construction of biologically functional bacterial plasmids in vitro. *Proc. Natl. Acad. Sci. USA* 70:3240–3244. The paper that gave rise to biotechnology.

Jackson, D.A., R.H. Symons, and P. Berg. 1972. Biochemical method for inserting new genetic information into DNA of simian virus 40: Circular SV40 DNA molecules containing lambda phage genes and the galactose operon of *Escherichia coli. Proc. Natl. Acad. Sci. USA* 69:2904–2909. A report of the first recombinant DNA experiment linking DNA from two species.

Lobban, P.E., and A.D. Kaiser. 1973. Enzymatic end-to-end joining of DNA molecules. *J. Mol. Biol.* 78:453–471. A report of the first recombinant DNA experiment.

Mullis, K.B. 1990. The unusual origin of the polymerase chain reaction. *Sci. Am.* 262(4):36–43. A description of that fateful night.

Sambrook, J., E.F. Fritsch, and T. Maniatis. 1989. *Molecular Cloning: A Laboratory Manual*, 2nd ed. Cold Spring Harbor, NY: Cold Spring Harbor Laboratory Press. Although supplanted by more recent manuals, this three-volume set includes much useful background information on the biological, chemical, and physical principles underlying both classical and current techniques.

Isolating Genes for Study (Cloning)

Terpe, K. 2006. Overview of bacterial expression systems for heterologous protein production: From molecular and biochemical fundamentals to commercial systems. *Appl. Microbiol. Biotechnol.* 72:211–222.

Working with Genes and Their Products

Arnheim, N., and H. Erlich. 1992. Polymerase chain reaction strategy. *Annu. Rev. Biochem.* 61:131–156.

Boyd, S.D. 2013. Diagnostic applications of high-throughput DNA sequencing. *Annu. Rev. Pathol. Mech. Dis.* 8:381–410.

Brock, T.D. 1997. The value of basic research: Discovery of *Thermus aquaticus* and other extreme thermophiles. *Genetics* 146:1207–1210. An essay on how basic research can have unexpected benefits.

Foster, E.A., M.A. Jobling, P.G. Taylor, P. Donnelly, P. de Knijff, R. Mieremet, T. Zerjal, and C. Tyler-Smith. 1999. The Thomas Jefferson paternity case. *Nature* 397:32. The last article in a series about an interesting case study in the use of biotechnology to address historical questions.

Giepmans, B.N.G., S.R. Adams, M.H. Ellisman, and R.Y. Tsien. 2006. The fluorescent toolbox for assessing protein location and function. *Science* 312:217–224.

Hofreiter, M., D. Serre, H.N. Poinar, M. Kuch, and S. Paabo. 2001. Ancient DNA. *Nat. Rev. Genet.* 2:353–359. Successes and pitfalls in the retrieval of DNA from very old samples.

Ivanov, P.L., M.J. Wadhams, R.K. Roby, M.M. Holland, V.W. Weedn, and T.J. Parsons. 1996. Mitochondrial DNA sequence heteroplasmy in the Grand Duke of Russia Georgij Romanov establishes the authenticity of the remains of Tsar Nicholas II. *Nat. Genet.* 12:417–420.

Lindahl, T. 1997. Facts and artifacts of ancient DNA. *Cell* 90:1–3. A good description of how nucleic acid chemistry affects the retrieval of DNA in archaeology.

Mardis, E.R. 2011. A decade's perspective on DNA sequencing technology. *Nature* 470:198–403.

Saiki, R.K., S. Scharf, F. Faloona, K.B. Mullis, G.T. Horn, H.A. Erlich, and N. Arnheim. 1985. Enzymatic amplification of beta-globin genomic sequences and restriction site analysis for diagnosis of sickle-cell anemia. *Science* 230:1350–1354. The first report of the polymerase chain reaction.

Understanding the Functions of Genes and Their Products

Budowle, B., M.D. Johnson, C.M. Fraser, T.J. Leighton, R.S. Murch, and R. Chakraborty. 2005. Genetic analysis and attribution of microbial forensics evidence. *Crit. Rev. Microbiol.* 31:233–254. A description of how biotechnology is used to fight bioterrorism.

Carroll, D. 2014. Genome engineering with targetable nucleases. *Annu. Rev. Biochem.* 83:409–39.

Gaj, T., C.A. Gersbach, and C.F. Barbas III. 2013. ZFN, TALEN, and CRISPR/Cas-based methods for genome engineering. *Trends Biotechnol.* 31:397–405.

Jinek, M., K. Chylinski, I. Fonfara, M. Hauer, J.A. Doudna, and E. Charpentier. 2012. A programmable dual-RNA-guided DNA endonuclease in adaptive bacterial immunity. *Science* 337:816–821.

Jinek, M., A. East, A. Cheng, S. Lin, E. Ma, and J. Doudna. 2013. RNA-programmed genome editing in human cells. *eLife* 2:e00471.

Sampson, T.R., and D.S. Weiss. 2014. Exploiting CRISPR/Cas systems for biotechnology. *Bioessays* 36:34–38.

Stoughton, R.B. 2005. Applications of DNA microarrays in biology. *Annu. Rev. Biochem.* 74:53–82.

8

Genomes, Transcriptomes, and Proteomes

MOMENT OF DISCOVERY

When the SARS virus emerged as a new infectious disease, at first no one knew what was causing the infection. There were thousands of cases around the world, and because 14% were fatal, it was clearly a public health emergency, especially because we didn't know at the time if the infection could be contained.

There were two big moments of discovery in my own lab that came about as we worked around the clock to identify the new infectious agent. The first came when we put samples from infected patients on our ViroChip, which contains fragments of conserved DNA sequences from all known types of viruses. I recall watching the spots lighting up on the scanner, and we could see that this new virus had sequences similar to coronaviruses from birds, cows, humans—it was some kind of strange new type of coronavirus that had never been seen before.

The second exciting moment came over the next couple of days as we began working nonstop to clone and sequence bits of the virus genome from the ViroChip array. Sequences were determined by a standard method in which the cloned bits were copied by a polymerase using fluorescently labeled nucleotides. I sat by the sequencer watching as each fluorescent nucleotide was detected, and I wrote down each base by hand because I couldn't wait for the run to finish to see what the sequences were! This level of intensity doesn't happen every day—we were not stopping to sleep or rest, surviving off pizza and Skittles, and it was exhausting—but it was so incredibly exciting to be the very first person to see the actual genome sequence of this new infectious agent. That was so fun. And in the end, we had to obtain enough DNA sequence to predict the encoded protein sequences that positively identified this virus as a new and highly divergent coronavirus.

—Joe DeRisi, on his discovery of the SARS virus

Joe DeRisi [Source: Courtesy of University of California, San Francisco.]

Online resources related to this chapter:

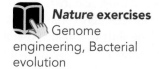

***Nature* exercises** Genome engineering, Bacterial evolution

Molecular biologists approach their subject at many levels. With the birth of biotechnology in the 1970s, the focus was on genes—their form, function, transcription, translation, and applications in medicine and agriculture. The new technologies also facilitated countless advances in our understanding of the RNA and protein products of genes. In truth, though, each gene is just a small part of a much larger genome. Understanding individual cellular components requires an examination of their function in the cellular context. At the same time, a host of new molecular questions arise that can be addressed only at the level of an organelle, cell, or organism. How does a cell or organism respond to a change in its environment, and how is the response regulated? How are the replication and transcription of genetic material coordinated with cell division? How can multiple genes be coordinately regulated? How many regulatory, replication, and repair systems for DNA exist in cells, and how does an organism's lifestyle shape its evolution? Over the past few decades, molecular biology has undergone a constant expansion of its reach. No longer limited to examining one or a few genes or gene products at a time, scientists are addressing problems dealing with increasingly complex and interconnected systems.

The shift was made possible by continuing advances in technology. The sequencing of individual genes has been supplanted by sequencing projects that encompass all of an organism's DNA—its genome. Examining the changes in the sum of a cell's RNA and proteins—its transcriptome and proteome—became a practical pursuit. The resulting efforts have spawned the new subdisciplines of genomics, transcriptomics, and proteomics, which now allow us to investigate questions on a cellular, organismal, or population scale. Huge public databases have been established, packed with information about genetic material and gene products, for species almost too numerous to count. Ongoing initiatives range from the customization of medical treatments based on an individual's genetic makeup to the detailed tracing of human evolution. The subdisciplines and databases we introduce in this chapter are an important legacy of modern molecular biology. They are also a key to its future. Biology in the twenty-first century will move forward with the aid of informational resources undreamed of just a few years ago.

8.1 GENOMES AND GENOMICS

The word "genome," coined by German botanist Hans Winkler in 1920, was derived simply by combining *gene* and the final syllable of *chromosome*. An organism's **genome** is defined today as the complete haploid genetic complement of a typical cell. In essence, a genome is one copy of the hereditary information required to specify the organism. For sexually reproducing organisms, the genome includes one set of autosomes and one of each type of sex chromosome. When cells have organelles that also contain DNA, the genetic content of the organelles is not considered part of the nuclear genome. Mitochondria are found in all eukaryotic cells, and chloroplasts occur in the light-harvesting cells of photosynthetic organisms. Each of these organelles has its own distinct genome. In viruses, which can have genetic material composed of either DNA or RNA, the genome is a complete copy of the nucleic acid required to specify the virus.

In diploid organisms, sequence variations exist between the two copies of each chromosome present in a cell. Subtle as they are (estimates range from 0.1% to 0.5% variance), these differences are used to solve crimes (see Highlight 7-1), define parentage, and help trace the path of an inherited disease through generations of an affected family. As sequencing methods become more sophisticated, sequences of the complete genetic complement of a diploid cell—diploid genomes—are beginning to appear.

The study of complete genomes was rudimentary until the advent of genome sequencing projects in the 1980s. In 1986, Thomas H. Roderick of the Jackson Laboratories in Bar Harbor, Maine, came up with *Genomics* as the name for a new journal, and the word ended up defining a new field. The modern science of **genomics** is dedicated to the study of DNA on a cellular scale. Advances in this field have been propelled by improvements in sequencing technology (described in Chapter 7), in computer technologies, and in innovative approaches to the organization and searching of stored information.

Many Genomes Have Been Sequenced in Their Entirety

The genome is the ultimate source of genetic information about an organism. Less than 10 years after the development of practical DNA sequencing methods, serious discussions began about the prospects for sequencing the entire 3×10^9 base pairs (bp) of the human genome. Although the effort was inspired by our natural curiosity about ourselves, it has become much more than a human genome project. Genome sequencing in the twenty-first century is becoming routine. The number of genomes sequenced in their entirety is now in the tens of thousands and includes organisms ranging from bacteria to mammals.

The International Human Genome Project got under way in the late 1980s. Several additional and closely linked projects focused on organisms other than humans. The first complete genome to be sequenced was that of the bacterium *Haemophilus influenzae*, in 1995 (see the

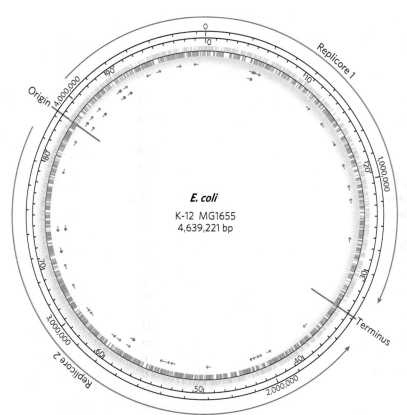

FIGURE 8-1 A snapshot of the *Escherichia coli* genome. K-12 MG1655 is the most commonly used laboratory strain of *E. coli*. The origin and terminus of replication are shown as green and red lines, respectively. Replication proceeds in two directions from the origin, dividing the genome into two regions, or replicores, which are replicated separately. The two outer black circles provide genome reference points in terms of number of base pairs (outer) and minutes (inner). The use of 100 minutes as a genomic yardstick in this chromosome is derived from the approximately 100 minutes it takes to transfer the *E. coli* chromosome from donor cell to recipient during bacterial conjugation. Yellow and orange markings on the inner circle denote protein-coding genes. Green and red arrows indicate the locations of genes for tRNA and rRNA, respectively. [*Source: Data from F. R. Blattner et al.,* Science *277:1453–1462, 1997.*]

How We Know section at the end of this chapter). The first eukaryotic genome sequenced was that of the yeast *Saccharomyces cerevisiae*, in 1996, and the genome sequence for the bacterium *Escherichia coli* became available in 1997 (**Figure 8-1**). The much larger effort directed at the human genome was also accelerating.

The Human Genome Project eventually included significant contributions from 20 sequencing centers distributed among six nations: the United States, Great Britain, Japan, France, China, and Germany. Some general coordination was provided by the Office of Genome Research at the National Institutes of Health in the United States, led first by James Watson and after 1992 by Francis Collins. However, much of the effort relied on the informal and very successful international collaborations.

Francis Collins [*Source: National Institutes of Health.*]

The published sequence of the human genome is actually a composite, derived from several anonymous donors. Although the DNA of several individuals is represented, the sequence in any given genomic region is generally from one individual. Research teams used restriction enzymes to partially digest the entire human genome, and then cloned suitably long segments into BAC and YAC vectors (bacterial and yeast artificial chromosomes are discussed in Chapter 7). Overlapping clones in the resulting libraries were identified by hybridization and other methods and organized into long contiguous stretches of chromosomal DNA called **contigs**. Each contig included at least one and usually many identifying sequences that had already been mapped to a particular region of a particular chromosome. These regions were either a unique and previously characterized sequence called a **sequence tagged site (STS)** or a gene for which expression could be monitored, known as an **expressed sequence tag (EST)**. The STS and EST landmarks in the contigs were often sequences that had been roughly mapped along a specific chromosome. These landmark-containing contigs could thus be ordered along each chromosome, gradually defining a physical map of the genome (**Figure 8-2**).

The contigs were then divided up between the international sequencing centers, and each center began sequencing the mapped BAC or YAC clones corresponding to its assigned segments of the genome. Because many of the clones were more than 100,000 bp long, and contemporary sequencing techniques resolved only 600 to 750 bp of sequence at a time, each clone had to be sequenced in pieces. The strategy was a shotgun approach, in which researchers used powerful new automated sequencers to sequence random segments of a given

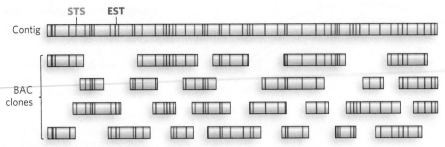

FIGURE 8-2 Mapping BACs into a contig. A contig of a chromosomal region is constructed by identifying the sequence tagged site (STS; orange) and expressed sequence tag (EST; blue) markers in each BAC and mapping them to the contig.

BAC or YAC clone, then assembled the entire clone by computerized identification of overlaps. Each clone was sequenced at least four to six times to ensure accuracy. The data were made available in the growing genome database.

A competing commercial effort to sequence the human genome had been initiated by the newly established Celera Corporation in 1997. Led by J. Craig Venter, the Celera group made use of a different strategy, called **whole-genome shotgun sequencing**, which eliminates the step of assembling a physical map of the genome. Instead, teams sequenced DNA segments from throughout the genome, at random. The sequenced segments were ordered by the computerized identification of sequence overlaps (with some reference to the public project's detailed and published physical map). Like the public genome project, the Celera effort used DNA from several human donors. About 70% of the sequence comes from one male donor—Craig Venter himself.

J. Craig Venter [Source: © Jessica Rinaldi/ Reuters/Corbis.]

At the outset of the Human Genome Project in the 1980s, shotgun sequencing on this scale had been deemed impractical because the sequence assembly was too computationally complex. However, by 1997, advances in computer software and sequencing automation had made the approach feasible. The ensuing race between private and public efforts substantially shortened the timeline for completing the project (**Figure 8-3**). After publication of the draft human genome sequence in February 2001, two years of follow-up work ensued, to eliminate nearly a thousand discontinuities and to provide high-quality sequence data, contiguous throughout the genome. The project was completed in April 2003, several years ahead of schedule.

The human genome is only part of the genome sequencing story, however, and an increasingly small part. The genomes of many other species have been sequenced in the continuing effort, gradually providing a unique look at genomic complexity throughout the three domains of living organisms: Bacteria, Archaea, and Eukarya (eukaryotes) (see Figure 8-3). Whereas many early sequencing projects focused on species commonly used in research laboratories, the programs have expanded to include a wide range of species of practical, medical, agricultural, and evolutionary interest. Completed bacterial genomes include at least one species from virtually every known bacterial family, and completed eukaryotic genomes number in the thousands. Each completed genome becomes a resource for scientists around the world who use that organism in their research, facilitating the identification of important genes. Thousands of individual human genomes have been sequenced, and the number is growing exponentially as genome-based, personalized medicine becomes a reality. The sequencing efforts have even expanded to include extinct species such as *Homo neanderthalensis*, as well as humans who died in past millennia. The many genome sequences provide a source for broad comparisons that help pinpoint both variable and highly conserved gene segments and allow the identification of genes that are unique to a species or groups of species. Efforts to map genes, identify new proteins and disease genes, elucidate genetic patterns of medical interest, and trace our evolutionary history, as well as many other initiatives, are under way.

Annotation Provides a Description of the Genome

A genome sequence is simply a very long string of A, G, T, and C residues. The value of this sequence information depends almost entirely on the manner in which the information is organized when it is stored. The critical process of **genome annotation** yields a listing of information about the location and function of genes and other critical sequences. Genome annotation converts the sequence itself to information that any researcher can use. Much of the effort focuses on genes encoding

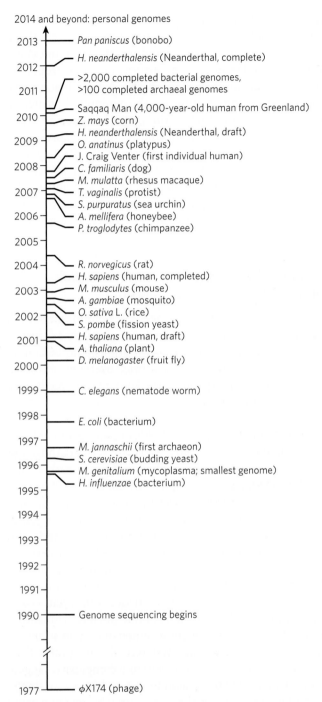

2014 and beyond: personal genomes

2013 — *Pan paniscus* (bonobo)

2012 — *H. neanderthalensis* (Neanderthal, complete)

2011 — >2,000 completed bacterial genomes, >100 completed archaeal genomes

2010 — Saqqaq Man (4,000-year-old human from Greenland)
— *Z. mays* (corn)

2009 — *H. neanderthalensis* (Neanderthal, draft)
— *O. anatinus* (platypus)
— *J. Craig Venter* (first individual human)

2008 — *C. familiaris* (dog)
— *M. mulatta* (rhesus macaque)

2007 — *T. vaginalis* (protist)
— *S. purpuratus* (sea urchin)

2006 — *A. mellifera* (honeybee)
— *P. troglodytes* (chimpanzee)

2005

2004 — *R. norvegicus* (rat)
— *H. sapiens* (human, completed)

2003 — *M. musculus* (mouse)
— *A. gambiae* (mosquito)

2002 — *O. sativa* L. (rice)
— *S. pombe* (fission yeast)

2001 — *H. sapiens* (human, draft)
— *A. thaliana* (plant)

2000 — *D. melanogaster* (fruit fly)

1999 — *C. elegans* (nematode worm)

1998 — *E. coli* (bacterium)

1997 — *M. jannaschii* (first archaeon)

1996 — *S. cerevisiae* (budding yeast)
— *M. genitalium* (mycoplasma; smallest genome)

1995 — *H. influenzae* (bacterium)

1994

1993

1992

1991

1990 — Genome sequencing begins

1977 — φX174 (phage)

FIGURE 8-3 The genome sequencing timeline. Preparatory work for the Human Genome Project, including extensive mapping to provide genome landmarks, occupied much of the 1990s. The rapid development of sequencing methods and strategies initiated a broad range of additional sequencing efforts. By the time the draft human genome was published in 2001, hundreds of other projects were already under way. Many genome projects involving species that are widely used in research have their own websites, which serve as central repositories for the most recent data. The increasing inclusion of species that are close human relatives and recent human ancestors is both clarifying human evolution and providing important resources for the discovery of genes involved in human diseases.

RNA and protein, because such genes are most often the target of scientific investigations. Every newly sequenced genome includes many genes—often 40% or more of the total—about which little or nothing is known. Here, the annotation exercise is most challenging.

Protein and RNA function can be described on three levels. **Phenotypic function** describes the effects of a gene product on the entire organism. For example, the loss or mutation of a particular protein may lead to slower growth, altered development, or even death. **Cellular function** is a description of the metabolic processes in which a gene product participates and of the interactions of that gene product with other proteins or RNAs in the cell. **Molecular function** refers to the precise biochemical activity of a protein or an RNA, such as the reactions an enzyme catalyzes, the ligands a receptor binds, or the complex formed between a specific RNA and a protein. Each of these functions can be elucidated by computational and experimental approaches, some of which are described here. Additional techniques are presented in Section 8.2.

Computational approaches involve Web-based programs that are used to define gene locations and assign tentative gene functions (where possible), based on similarity to genes previously studied in other genomes. Most of these programs are freely available on the Internet, although some are written for specific purposes by individual researchers. For investigating the function of a particular gene, resources such as the classic BLAST (Basic Local Alignment Search Tool) algorithm allow a rapid search of all genome databases for sequences related to one that a researcher has just generated. Two other prominent Internet resources are the NCBI (National Center for Biotechnology Information) site, sponsored by the National Institutes of Health, and the Ensembl site, cosponsored by the EMBL-EBI (European Molecular Biology Laboratory–European Bioinformatics Institute) and the Wellcome Trust Sanger Institute.

The availability of many genome sequences in online databases enables researchers to assign gene functions by genome comparisons, an enterprise referred to as **comparative genomics**. Sequence comparison can be done with DNA, RNA, or protein. Any two genes with a demonstrable sequence similarity, whether or not they are closely related by function, are called **homologs**. The sequence similarity (homology) implies an evolutionary relationship. Quite often, sequence similarity and a functional relationship go hand in hand. When two genes in different species possess a clear sequence and functional relationship to each other, they are known as **orthologs**—genes derived from an ancestral gene in the last common ancestor of these two species. **Paralogs** are genes that are similarly related to each other but within a single species; they arise most often from gene duplication in a single genome, followed by specialization of one or both copies of the gene over the course of evolution. If the function of any gene has been characterized for one species, this information can be used to at least tentatively assign gene function to a related gene in a second species. The entire process of annotation relies heavily on evolutionary theory. Comparing genomes from different species is productive precisely because all organisms share a common ancestor.

Gene identity is often easiest to discern when comparing genomes from closely related species, such as mice and humans, although many clearly orthologous genes have been identified in species as distant as bacteria and humans. In many cases, even the order of genes on a chromosome is conserved over large segments of the genomes of closely related species. Conserved gene order, or **synteny**, provides additional evidence for an orthologous relationship between genes at identical locations in the related segments (**Figure 8-4**). The distinction between orthologs and paralogs was introduced by Walter Fitch in 1970, and its importance was established with the advent of genome sequencing projects in the 1990s. As the number of known genome sequences increases, many genes and genomic segments can be productively annotated by using automated tools available on the NCBI and Ensembl websites.

In every newly described genome sequence, the many genomic segments and genes that have never been characterized—that unknown 40% or so of the total—represent a special challenge. Elucidation of gene function in these cases will probably take many decades. Some experimental approaches exist, and new ones are being developed. Many of the current approaches again focus on protein-coding genes; for several genomes, such as those of *S. cerevisiae* and the plant *Arabidopsis thaliana*, gene knockout (inactivation) collections have been developed by genetic engineering. Each strain

in an organism's collection has a different inactivated gene, and a high fraction of genomic genes (excluding a core of essential genes required for life at all times) are represented. If the growth patterns or other properties of the organism change when the gene is inactivated, this provides information on the phenotypic function of the protein product of the gene. In other available libraries, each gene in a specific genome is expressed as a tagged fusion protein (see Chapter 7). The tags may be designed to allow isolation of the protein, investigate interactions with other proteins, or explore subcellular localization. Some approaches for using tags to determine the function of genes are described in more detail in Section 8.2.

Genome Databases Provide Information about Every Type of Organism

The available genome sequences are helpful to research in all biological disciplines. Increasingly, they are inspiring molecular biologists to ask questions that, until now, could not be answered. Just a brief overview will illustrate the utility of expanding the Human Genome Project to essentially all species.

Viruses Viruses are not free-living organisms but obligate intracellular parasites: thus, every virus is a pathogen of some organism. The viruses that are human pathogens—such as SARS (see this chapter's Moment of Discovery)—are of special interest. However, viruses that infect farm animals, food crops, landscape plants, and many other organisms can be economically important. Bacteria also serve as hosts to viruses, which are generally termed bacteriophages. Even these are of medical interest, because bacteriophages that kill pathogenic bacteria have therapeutic potential. Viruses can be divided into seven classes, depending on whether the genomic nucleic acid is RNA or DNA, whether it is single-stranded or double-stranded, and the mechanisms employed to replicate it (**Table 8-1**). Viruses vary a great deal in genomic complexity, ranging from a mere 2,000 nucleotides (found in a few single-stranded DNA viruses that infect vertebrates) to around 1.2 million bp (in a double-stranded DNA virus that infects amoebas). The number of viral genomes that have been or are now being sequenced has grown so quickly that estimates become obsolete almost as soon as they are generated. The ongoing effort will greatly aid future progress in medicine and agriculture.

Bacteria Bacteria inhabit every environment—from polar ice to deserts, from ocean depths to kitchen counters and the soil in your backyard. Some are pathogens. Others help digest our food, convert atmospheric nitrogen to forms that all organisms can use, convert

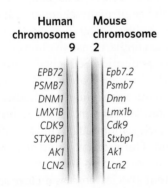

Human chromosome 9	Mouse chromosome 2
EPB72	Epb7.2
PSMB7	Psmb7
DNM1	Dnm
LMX1B	Lmx1b
CDK9	Cdk9
STXBP1	Stxbp1
AK1	Ak1
LCN2	Lcn2

FIGURE 8-4 Synteny in the human and mouse genomes. Large segments of the two genomes have closely related genes aligned in the same order on the chromosomes. In these short segments of human chromosome 9 and mouse chromosome 2, the genes exhibit a very high degree of homology, as well as the same gene order. The different lettering schemes for the gene names simply reflect the different naming conventions for the two species.

TABLE 8-1				
The Seven Classes of Viruses				
Class	Genome	Replication	Examples	Genome Size (kbp or kb)
I	Double-stranded DNA	Nuclear or cytoplasmic; host DNA polymerase	Polyomaviruses Adenoviruses Baculoviruses Papovaviruses Poxviruses I* P22* T5*	5–1,200
II	Single-stranded DNA	Nuclear in eukaryotic host; host DNA polymerases	Circoviruses Geminiviruses Parvoviruses Inoviruses* Microviruses*	2–9
III	Double-stranded DNA	Cytoplasmic; virus-encoded replicases	Birnaviruses Chrysoviruses Cystoviruses Hypoviruses Partitiviruses Reoviruses Totiviruses	3–32
IV	Single-stranded, positive-sense RNA[†]	Virus-encoded replicases	Bromoviruses Coronaviruses Picornaviruses	2–31
V	Single-stranded, negative-sense RNA	Virus-encoded replicases	Arenaviruses Bunyaviruses Bornaviruses Rhabdoviruses Paramyxoviruses	9–19
VI	Single-stranded RNA, reverse-transcribing	Virus-encoded reverse transcriptase	Retroviruses (many types)	4–12
VII	Double-stranded DNA	DNA genome generated by reverse transcription of RNA intermediates in viral particle during maturation	Caulimoviruses Hepadnaviruses	3–9

*Bacterial virus.
[†]Most abundant viral genome type. "Positive sense" means that the sequence is identical to the sequence of most mRNAs encoded by the virus. (A negative-sense RNA is one with a sequence complementary to the mRNAs encoded by the virus.)

carbon dioxide to oxygen, and carry out myriad other tasks without which all other life forms would perish. With that in mind, molecular biologists are subjecting thousands of representative bacterial species to genome sequencing.

In the past few decades, researchers have realized that a vast number of bacterial species remain uncharacterized. Many bacteria live in interdependent microbial communities and cannot be cultured in pure form in the laboratory. Examples are found in the human intestine, in the termite gut, and in the effluent of deep-sea steam vents. Many of these bacteria are important to human health, both directly and indirectly. Bacteria are so plentiful in humans—in our intestines, on our skin, in our saliva, and so on—that they represent 1% to 3% of the mass of a typical person, with symbiotic bacterial cells outnumbering their human host cells by a factor of 10 to 1. Other bacteria are of economic importance. For instance, an understanding of the microbial processes that allow termites to digest cellulose in their gut could provide new ways to convert grass and other cellulose materials to usable fuels.

The need to know more about these microbial communities has given rise to a subdiscipline of genomics: **metagenomics**. In metagenomics projects, DNA is isolated not from a single bacterial species but from an entire community of microbial species (**Highlight 8-1**). The DNA is sequenced by a shotgun technique, and the researcher uses computer programs to assemble overlapping segments derived from individual genomes.

HIGHLIGHT 8-1   **TECHNOLOGY**

Sampling Biodiversity with Metagenomics

Most of the biological diversity on our planet is found in microorganisms. However, we know surprisingly little about Earth's microbial diversity. A wealth of microbial diversity remains to be discovered in the world's swamps, deserts, and oceans, involving species that cannot yet be cultured. Assessing the diversity in communities of microorganisms is one goal of the new discipline of metagenomics.

The sampling involves shotgun DNA sequencing on a truly grand scale. Individual species are not isolated. Instead, an entire microbial population is taken from a given environment, and DNA sequences from that population are analyzed at random. Early approaches have looked at a biofilm in an acid mine, soil in Minnesota, water samples from the Sargasso Sea, whale falls (whales that have died and sunk to the ocean floor), and human feces. DNA from the bacteria and/or viruses in the sample is broken into fragments and sequenced at random. Computerized analyses identify any overlapping sequences and link them into longer contigs. These genomic snippets are assembled in a database. The diversity can be measured by focusing on specific genes. For example, the 16S rRNA gene is universal in bacteria and is often used as a benchmark for defining species.

This technology has given rise to some very ambitious metagenomics initiatives. One was carried out by Craig Venter and his coworkers at the J. Craig Venter Institute (Rockville, Maryland) in the spring of 2003. A 30-meter sailing sloop, the *Sorcerer II*, was converted into an oceanic research vessel to carry out the Global Ocean Sampling (GOS) expedition. Launched in March 2004 in Halifax, Canada, the nearly two-year research voyage circumnavigated the globe (Figure 1). Samples of ocean water were taken every 200 nautical miles. Microorganisms were strained from the water with a series of filters and sent back to the lab for extraction of DNA and sequencing. The result is the largest database of DNA sequences derived from marine organisms yet released into the public domain. More than 7.7 million sequences are included in the GOS database, encompassing more than 6.3 billion bp of DNA and representing more than 800 species—perhaps half of which were previously unknown. Millions of new proteins have been discovered, and thousands of new protein families. The voyages continue.

A second effort in metagenomics is to map the human microbiome. Major contributions come from the U.S. Human Microbiome Project of the National Institutes of Health and the European MetaHIT (*Meta*genomics of the *H*uman *I*ntestinal *T*ract) program, both launched in 2008. The goal is to define the bacteria and other microorganisms (including some archaea and yeasts) associated with humans. Perhaps even more important, the projects will determine how the microbiome changes in response to geography, diet, age, pregnancy, disease state, and many other factors. This effort is inspired in part by an increasing awareness of the role of the human microbiome in health. Clear links are becoming apparent between a disruption of homeostasis between microbiota and their human host (referred to as dysbiosis) and the development of disease states such as inflammatory bowel disease, obesity, and type 2 diabetes. Microbe replacement therapy could be in your future.

FIGURE 1 The route taken by *Sorcerer II*. [Source: Courtesy J. Craig Venter Institute.]

When the community includes only a handful of species, researchers can reconstruct multiple genomes from these mixed samples. The process has some unique complexities, many related to the close evolutionary relationship of many bacterial species. This results in similarities over large stretches of genomic nucleic acid that can complicate the computerized assembly of genomes from short sequence "reads." Assembling genomes from microbial communities with hundreds of different species will require further advances in both sequencing technologies and assembly programs.

Archaea In 1977, Carl Woese and his colleagues introduced the world to a new domain of living organisms, the Archaebacteria, now renamed Archaea. A careful study of 16S ribosomal RNA sequences led to the discovery of this previously unsuspected group. The archaea (singular, archaeon) are single-celled organisms, very similar in appearance to bacteria, and like bacteria are ubiquitous. However, many of the most interesting species are extremophiles, inhabiting hot springs or water with very high salinity or other unusual environments. Sharing some properties with both bacteria and eukaryotes, archaea have nevertheless evolved as an independent line. Their contributions to the chemistry of the biosphere make them important targets for study and genome sequencing.

Eukaryotes Many eukaryotic genomes are considerably larger than genomes in the other two domains. Nevertheless, the sequencing of even very large eukaryotic genomes is becoming routine. Databases already contain complete genomes ranging from single-celled eukaryotes such as *S. cerevisiae* to nematodes, plants, insects, and mammals. Orthologs of genes involved in important processes and disease states in humans can almost always be found in the genomes of model organisms, facilitating laboratory research into gene function. Specialized databases have been developed for the genomes of organisms that are of particular interest to science, including mouse, fruit fly, mustard weed, and yeast (see the Model Organisms Appendix). Other databases are being established that focus on plant and animal species critical to agriculture, such as corn, rice, and cattle. Some databases focus on specific types of genes. All of these databases are easily found through an Internet search or via links on the NCBI and Ensembl websites. Individual human genome sequences are also available online, including those of James Watson and Craig Venter!

The Human Genome Contains Many Types of Sequences

All of these rapidly growing databases have the potential not only to fuel advances in biology but to change the way we think about ourselves. What does our own genome, and its comparison with those of other organisms, tell us?

In some ways, we are not as complicated as we once imagined. Decades-old estimates that humans had about 100,000 genes within the approximately 3.2×10^9 bp of the human genome have been supplanted by the discovery that we have only about 20,000 protein-coding genes—less than twice the number in a fruit fly (13,601 genes), not many more than in a nematode worm (19,735 genes), and fewer than in a rice plant (~38,000 genes).

In other ways, however, we are more complex than we previously realized. The study of eukaryotic chromosome structure and, more recently, the sequencing of entire eukaryotic genomes have revealed that many, if not most, eukaryotic genes contain one or more intervening segments of DNA that do not code for the amino acid sequence of the polypeptide product. These nontranslated inserts interrupt the otherwise colinear relationship between the gene's nucleotide sequence and the amino acid sequence of the encoded polypeptide. Such nontranslated DNA segments are called **introns** (or **intervening sequences**), and the coding segments are called **exons** (**Figure 8-5**). Few bacterial genes contain introns. The process of removing introns from a primary RNA transcript to generate a transcript that can be translated contiguously into a protein product is known as splicing (described in Chapter 16). An exon often (but not always) encodes a single domain of a larger, multidomain protein. Humans share many protein domain types with plants, worms, and flies, but use these domains in more complex arrangements. Alternative modes of gene expression and RNA splicing permit the production of alternative combinations of exons, leading to the production of more than one protein from a single gene. Humans and other vertebrates engage in this process far more than do bacteria, worms, or any other form of life—thereby allowing greater complexity in the proteins generated.

In mammals and some other eukaryotes, the typical gene has a much higher proportion of intron DNA than exon DNA; in most cases, the function of introns is not clear. Less than 1.5% of human DNA is "coding" or exon DNA, carrying information for protein or RNA products (**Figure 8-6a**). However, when the much larger introns are included in the count, as much as 30% of the human genome consists of genes. Several efforts are under way to categorize the protein-coding genes by function (**Figure 8-6b**).

The relative paucity of genes in the human genome leaves a lot of DNA unaccounted for. Much of the nongene DNA is in the form of repeated sequences of several kinds. Perhaps most surprising, about half the

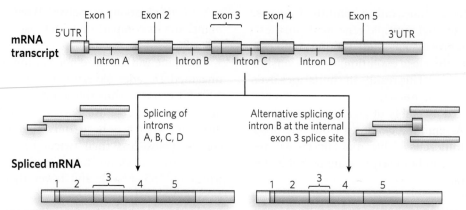

FIGURE 8-5 Introns and exons. The gene for human growth hormone 1 (*GH1*) contains five exons and four introns, along with 5′ and 3′ untranslated regions (5′UTR and 3′UTR). Two of the several alternative patterns of splicing are shown here. Alternative splicing allows cells to synthesize different variants of a protein from one gene. [*Source: Data from J. J. Kopchick et al., Nat. Clin. Pract. Endocrinol. Metab. 3:355–368, 2007.*]

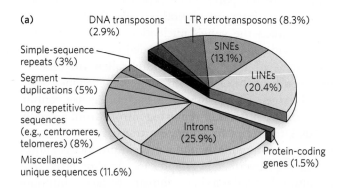

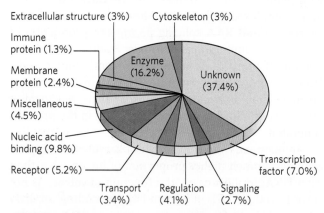

Barbara McClintock, 1902–1992 [*Source: Science Source.*]

FIGURE 8-6 A snapshot of the human genome. (a) This pie chart shows the proportions of various types of sequences in the human genome. The classes of transposons that represent nearly half of the total genomic DNA are indicated in shades of gray. LTR retrotransposons are retrotransposons with long terminal repeats. Long interspersed nuclear elements (LINEs) and short interspersed nuclear elements (SINEs) are special classes of particularly common DNA transposons (see Chapter 14). (b) The approximately 20,000 protein-coding genes in the human genome can be classified by the type of protein encoded. [*Source: (a) Data from T. R. Gregory, Nat. Rev. Genet. 6:699–708, 2005.*]

human genome is made up of moderately repeated sequences that are derived from transposable elements—segments of DNA, ranging from a few hundred to several thousand base pairs long, that can move from one location to another in the genome. Originally discovered in corn by Barbara McClintock, transposable elements, or **transposons**, are a kind of molecular parasite. They efficiently make their home in the genomes of essentially every organism. Many transposons contain genes encoding the proteins that catalyze the transposition process itself, as described in more detail in Chapter 14. There are multiple classes of transposons in the human genome. Many are strictly DNA segments, slowly increasing in number as a result of replication events coupled to the transposition process. Some, called **retrotransposons**, are closely related to retroviruses, transposing from one genomic location to another through RNA intermediates that are reconverted to DNA by reverse transcription. Some transposons in the human genome are active, moving at a low frequency, but most are inactive, evolutionary relics altered by mutations. Transposon movement can lead to the redistribution of other genomic sequences and has played a major role in human evolution.

Once the protein-coding genes (including exons and introns) and transposons are accounted for, perhaps 25% of the total DNA remains (the purple, blue, green, and yellow segments in Figure 8-6a). As a follow-up to the Human Genome Project, an initiative called ENCODE was launched by the National Human Genome Research

Institute in 2003. Involving a worldwide consortium of research groups, its purpose is to identify functional elements in the human genome. As the project proceeds, the nongene DNA in the human genome is becoming less of a mystery. ENCODE has revealed that the vast majority (>80%, including most transposons) of human genomic DNA either is transcribed into RNA in at least one type of cell or tissue or is involved in some functional aspect of chromatin structure. Much of the noncoding DNA is associated with regulatory functions that affect the expression of the 20,000 protein-coding genes and the many additional genes encoding functional RNAs. Many mutations associated with human genetic diseases lie in this noncoding DNA, where they probably affect regulation of one or more genes. As described in Chapters 16 and 19 through 22, new classes of functional RNAs are being discovered at a rapid pace. Many are encoded by genes whose existence was previously unsuspected. They are now being identified in screens using technologies such as RNA-Seq (see Section 8.2).

About 3% of the human genome consists of highly repetitive sequences referred to as **simple-sequence repeats (SSRs)**. Generally less than 10 bp long, an SSR is sometimes repeated millions of times per cell, distributed in shorter segments of tandem repeats. The most prominent examples of SSR DNA occur in centromeres and telomeres (see Chapter 9). Human telomeres, for example, consist of up to 2,000 contiguous repeats of the sequence GGTTAG. Additional repeats of simple sequences are found throughout the genome. These isolated segments of repeated sequences, often featuring up to a few dozen tandem repeats of a simple sequence, are called short tandem repeats (STRs). Such sequences are the targets of important technologies used in forensic DNA analysis (see Highlight 7-1).

What does all this information tell us about the similarities and differences among individual humans? Within the human population there are millions of single-base variations, called **single nucleotide polymorphisms**, or **SNPs** (pronounced "snips"). Each human differs from the next by, on average, 1 in every 1,000 bp. Many of these variations are in the form of SNPs, but a wide range of larger deletions, insertions, and small rearrangements also occur in the human population. From these often subtle genetic differences comes the human variety we are all aware of—such as differences in hair color, stature, eyesight, allergies to medication, foot size, and (to some unknown degree) behavior.

The process of genetic recombination during meiosis tends to mix and match these small genetic variations so that different combinations of genes are inherited (as discussed in Chapter 13). However, groups of SNPs and other genetic differences that are close together on a chromosome are rarely affected by recombination

and are usually inherited together; these groupings are known as **haplotypes**. Haplotypes provide convenient markers for certain human populations and for individuals within populations.

Defining a haplotype requires several steps. First, positions that contain SNPs in the human population are identified in genomic DNA samples from multiple individuals (**Figure 8-7a**). Each SNP may be separated from the next by many thousands of base pairs. Second, SNPs that are inherited together are compiled into haplotypes (**Figure 8-7b**). Each haplotype consists of the particular bases found at the various SNP positions in the defined haplotype. Finally, tag SNPs—a subset of the SNPs that define the entire haplotype—are chosen to uniquely identify each haplotype (**Figure 8-7c**). By sequencing just these tag positions in genomic samples from human populations, researchers can quickly identify which of the haplotypes are present in each individual. Especially stable haplotypes exist in the mitochondrial genome (which never undergoes meiotic recombination) and on the male Y chromosome (only 3% of which is homologous to the X chromosome and thus subject to recombination). As we will see in Section 8.3, haplotypes can be used as markers to trace human migrations.

Genome Sequencing Informs Us about Our Humanity

A primary purpose of most genome sequencing projects is to identify conserved genetic elements of functional significance, such as conserved exon sequences, regulatory regions, and other genomic features (centromeres, telomeres, etc.). The primary purpose of sequencing the human genome is quite distinct. Here, we are interested in the differences between our genome and those of other organisms. Relying again on the power of evolutionary theory, these differences can reveal the molecular basis of human genetic diseases. They can also help identify genes, gene alterations, and other genomic features that are unique to the human genome and thus likely to contribute to definably human characteristics.

As the genome projects have made clear, the human genome is very closely related to other mammalian genomes over large segments of every chromosome. However, for a genome measured in billions of base pairs, differences of a few percent can add up to millions of genetic distinctions. Searching among these, and utilizing comparative genomics techniques, we can begin to explore the molecular basis of our large brain, language skills, tool-making ability, or bipedalism.

The genome sequences of our closest biological relatives, the chimpanzees and bonobos, offer some important clues and can illustrate the comparative process.

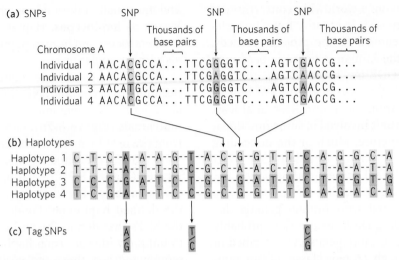

FIGURE 8-7 Haplotype identification. (a) Positions in the human genome where single nucleotide polymorphisms occur are often identified in genomic samples. The SNPs can be in any part of the genome, whether or not part of a known gene. (b) Groups of these SNPs are compiled into a haplotype. The SNPs will vary in the overall human population, such as in the four fictional individuals shown here, but the SNPs chosen to define a haplotype are often the same in most individuals of a particular population. (c) A few single nucleotide polymorphisms are chosen as haplotype-defining SNPs (tag SNPs), which are then used to simplify the process of identifying an individual's haplotype (by sequencing 3 loci instead of 20). If the positions shown here are sequenced, an A-T-C haplotype might be characteristic of a population native to a location in northern Europe, whereas a G-T-C haplotype might prevail in a population in Asia. Multiple haplotypes of this kind are used to trace prehistoric human migrations.
[*Source: Data from International HapMap Consortium,* Nature *426:789–796, 2003.*]

Humans and chimpanzees shared a common ancestor about 7 million years ago. Genomic differences between the two species are of two types: base-pair changes (SNPs) and larger genomic rearrangements of many types. SNPs in the protein-coding regions, whether or not they result in an amino acid change, can be used to construct a phylogenetic tree (**Figure 8-8a**), as described in Section 8.3. Over the course of evolution, segments of chromosomes may become inverted as a result of a segmental duplication, transposition of one copy to another arm of the same chromosome, and recombination between the two segments (**Figure 8-8b**). Such inversions have occurred in the human lineage on chromosomes 1, 12, 15, 16, and 18. Chromosome fusions can also occur. In the human lineage, two chromosomes found in other primate lineages have been fused to form human chromosome 2 (**Figure 8-8c**). The human lineage thus has 23 chromosome pairs rather than the 24 pairs typical of simians. Once this fusion appeared in the line leading to humans, it would have represented a major barrier to interbreeding with other primates that lacked it.

If we ignore transposons and large chromosomal rearrangements, the published human and chimpanzee genomes differ by only 1.23% at the level of base pairs (compared with the 0.1% variance from one human to another). Some variations are at positions where there is a known polymorphism in either the human or the chimpanzee population, and these are unlikely to reflect a species-defining evolutionary change. When we also ignore these positions, the differences amount to about 1.06%, or about 1 in 100 bp. This might seem a small number, but in large genomes it translates into more than 30 million base-pair changes, some of which affect protein function and gene regulation. Similar calculations show that we are approximately as closely related to bonobos.

The genome rearrangements that help distinguish chimpanzees and humans include 5 million short insertions or deletions involving a few base pairs each, as well as a substantial number of larger insertions, deletions, inversions, or duplications that can involve many thousands of base pairs. When transposon insertions—a major source of genomic variance—are added to the list, the differences between the human and chimpanzee genomes increase. The chimpanzee genome has two classes of retrotransposons that are not present in the human genome (see Chapter 14). Other types of rearrangements, especially segmental duplications, are also common in primate lineages. Duplications of chromosomal segments

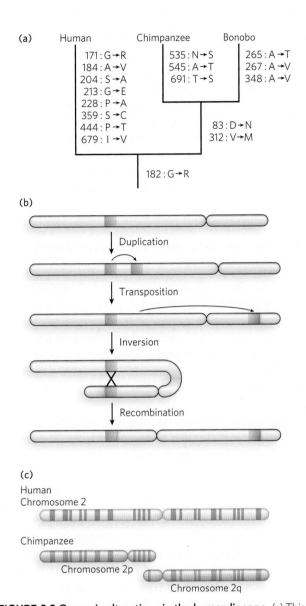

(a)

Human	Chimpanzee	Bonobo
171 : G→R	535 : N→S	265 : A→T
184 : A→V	545 : A→T	267 : A→V
204 : S→A	691 : T→S	348 : A→V
213 : G→E		
228 : P→A		
359 : S→C		83 : D→N
444 : P→T		312 : V→M
679 : I→V		

182 : G→R

(b)

Duplication

Transposition

Inversion

Recombination

(c)

Human
Chromosome 2

Chimpanzee

Chromosome 2p

Chromosome 2q

FIGURE 8-8 Genomic alterations in the human lineage. (a) This evolutionary tree is for the progesterone receptor, which helps regulate many events in reproduction. The gene encoding this protein has undergone more evolutionary alterations than most. Amino acid changes associated uniquely with humans, chimpanzees, and bonobos are listed beside each branch (with the residue number). (b) One of the multistep processes that can lead to the inversion of a chromosome segment. A gene or segment of the chromosome is duplicated, then moved to another chromosomal location by transposition. Recombination of the two segments may result in inversion of the chromosomal DNA between them. (c) The genes on chimpanzee chromosomes 2p and 2q are homologous to those on human chromosome 2, implying that two chromosomes fused at some point in the line leading to humans. Homologous regions can be visualized by bands created in metaphase chromosomes by certain dyes, as shown here. [*Source: (a) Data from C. Chen et al., Mol. Phylogenet. Evol. 47:637–649, 2008.*]

can lead to changes in the expression of genes contained in these segments. There are about 90 million bp of such differences between humans and chimpanzees, representing another 3% of the genomes. In effect, each species has segments of DNA, constituting 40 to 45 million bp,

that are entirely unique to that particular genome, with larger chromosomal insertions, duplications, and other rearrangements affecting more base pairs than do single-nucleotide changes. Thus, the total genomic difference between chimpanzees and humans amounts to about 4% of their genomes.

Sorting out which genomic distinctions are relevant to features that are uniquely human is a daunting task. If the two species share a common ancestor, then, logically, half the changes represent chimpanzee lineage changes and half represent human lineage changes (if one assumes a similar rate of evolution in both lines). When you see a difference, how do you tell which variant was the one present in the common ancestor? One way is to compare both genome sequences with those of more distantly related organisms referred to as **outgroups**. Consider a locus, X, where there is a difference between the human and chimpanzee genomes (**Figure 8-9**). The lineage of the orangutan, an outgroup, diverged from that of chimps and humans prior to the chimpanzee/human common ancestor. If the sequence at locus X is identical in orangutan and chimpanzee, this sequence was probably present in the chimpanzee/human ancestor, and the sequence seen in humans is specific to the

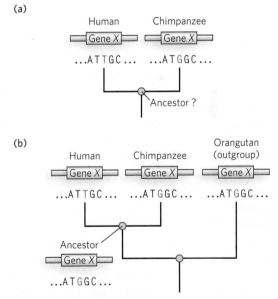

(a)

Human   Chimpanzee
Gene X   Gene X
...ATTGC...   ...ATGGC...

Ancestor ?

(b)

Human   Chimpanzee   Orangutan (outgroup)
Gene X   Gene X   Gene X
...ATTGC...   ...ATGGC...   ...ATGGC...

Ancestor
Gene X
...ATGGC...

FIGURE 8-9 Determination of sequence alterations unique to one ancestral line. (a) Sequences from the same hypothetical gene in humans and chimpanzees are compared. The sequence of this gene in their last common ancestor is unknown. (b) The orangutan genome is used as an outgroup. The sequence of the orangutan gene is found to be identical to that of the chimpanzee gene. This means that the mutation causing the difference between humans and chimpanzees almost certainly occurred in the line leading to modern humans, and the common ancestor of humans and chimpanzees (and orangutans) had the variant now found in chimpanzees.

human lineage. Sequences that are identical in humans and orangutans can be eliminated as candidates for human-specific genomic features. The importance of comparisons with closely related outgroups has given rise to new efforts to sequence the genomes of orangutans, macaques, and many other primate species. And the study of genes and alleles of special significance to humans is being refined as the bonobo genome is further analyzed.

The search for the genetic underpinnings of special human characteristics, such as our enhanced brain function, can benefit from two complementary approaches. The first searches for genomic regions where extreme changes have occurred, such as genes that have been duplicated many times or large genomic segments not present in other primates. The second approach looks at genes known to be involved in relevant human diseases. For brain function, for example, one would examine genes involved in cognition, such as those that contribute to mental disorders when mutated.

Several factors, such as the development of human-specific life history traits (e.g., a greater age of sexual maturity and thus a longer generation time), have led to an approximately 3% slower accumulation of genomic changes in the ancestral line leading to humans than in the line leading to chimpanzees. Evolution has occurred somewhat faster in other primate lines. Observed genetic changes are sometimes concentrated in a particular gene or region. In principle, human-specific traits could reflect changes in protein-coding genes, in regulatory processes, or both. A few classes of protein-coding genes exhibit evidence of accelerated divergence (more amino acid substitutions than normal). These include genes involved in chemosensory perception, immune function, and reproduction. In these cases, rapid evolution is evident in virtually all primate lines, reflecting physiological functions that are critical to all primate species. Another class of genes showing evidence of accelerated evolution is those encoding transcription factors—proteins involved in the expression of other genes (discussed in Chapter 21).

Notably, analyses of the human lineage have not detected an enrichment of genetic changes in protein-coding genes involved in brain development or size. Guided in part by the results obtained for transcription factor genes, the focus of such analyses has gradually shifted to changes in gene expression. In primates, most genes that function uniquely in the brain are even more highly conserved than genes functioning in other tissues. This may reflect some special constraints related to brain biochemistry. However, some differences in gene expression are observed. For example, the gene encoding the enzyme glutamate dehydrogenase, which plays an important role in neurotransmitter synthesis, has an increased copy number due to gene duplication. Analyses

of changes in genomic regions related to gene regulation show that genes involved in neural development and nutrition are disproportionately affected. A variety of RNA-coding genes, some with expression concentrated in the brain, also show evidence of accelerated evolution (**Figure 8-10**). Many of these genes are probably involved in regulating the expression of other genes. The many new classes of RNA that are now being discovered (see Chapter 22) are likely to radically change our perspective on how evolution alters the workings of living systems.

Genome Comparisons Help Locate Genes Involved in Disease

One of the motivations for the Human Genome Project was its potential for accelerating the discovery of genes underlying genetic diseases. That promise has been fulfilled: well over 1,600 human genetic diseases have been mapped to particular genes. Some disease-gene hunters caution that so far, the work may have uncovered mostly the relatively easy cases, with many challenges remaining.

The main approach during the past two decades has used a method called **linkage analysis**. In brief, the gene involved in a disease condition is mapped relative to well-characterized genetic polymorphisms that occur throughout the human genome, using methods firmly rooted in evolutionary biology. The search often begins with one or more large families that include several individuals affected by a particular disease. We illustrate this by describing the search for one gene involved in Alzheimer disease. About 10% of all cases of Alzheimer's in the United States result from an inherited predisposition. Several different genes have been discovered that, when mutated, can lead to early onset of the disease. One such gene (*PS1*) encodes the protein presenilin-1, and its discovery made heavy use of linkage analysis. **Figure 8-11a** on p. 274 shows two of the many family pedigrees that were used to search for this gene in the early 1990s. In studies of this type, DNA samples are collected from both affected and unaffected family members. Researchers first localize the region associated with the disease to a specific chromosome. This effort makes use of a set of genomic locations where common SNPs or other mapped genomic alterations occur in the human population, as identified by the Human Genome Project. Using a panel that includes several well-characterized SNP loci mapped to each chromosome, investigators compare the genotypes of individuals with and without the disease, focusing especially on close family members. By identifying the particular SNPs that are most often inherited with the disease-causing gene, the responsible gene can gradually be localized to a single chromosome.

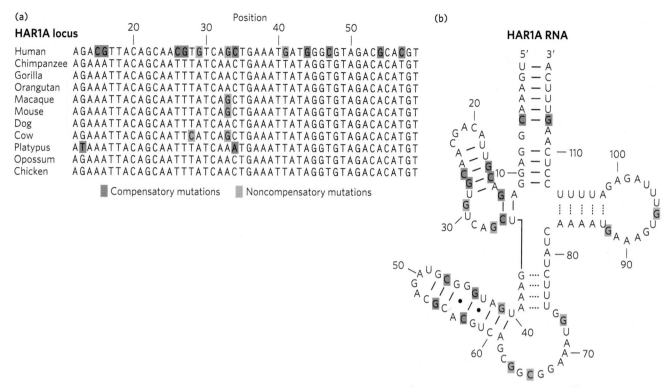

FIGURE 8-10 Accelerated evolution in some human genes. (a) The HAR1A locus specifies a noncoding RNA that is highly conserved in vertebrates. This RNA functions in the brain during neurodevelopment. In humans, the HAR1A gene exhibits an unusual number of substitutions (highlighted by color shading), providing evidence of accelerated evolution. (b) The secondary structure of the HAR1A RNA has several paired loops. Many of the sequence changes, shaded green here and in (a), are compensatory in the context of this RNA secondary structure: a change on one side of the loop is mirrored by a compensatory change on the other side of the loop, to permit proper base pairing. Noncompensatory changes are shaded red. [*Source: Data from T. Marques-Bonet et al., Annu. Rev. Genomics Hum. Genet. 10:355–386, 2009.*]

In the case of the *PS1* gene, coinheritance was strongest with markers on chromosome 14 (**Figure 8-11b**).

Chromosomes are very large DNA molecules, and localizing the gene to one chromosome is only a small part of the battle. That chromosome contains a mutation that gives rise to the disease, but in every individual human genome, each chromosome contains thousands of SNPs and other changes—representing alterations of all kinds relative to the reference sequence in the human genome database. Simply sequencing the entire chromosome would be unlikely to reveal the SNP or other change associated with the disease. The more detailed localization of a disease-causing gene on a chromosome relies instead on an even more elaborate application of linkage analysis. Statistical methods can correlate the inheritance of additional, more closely spaced polymorphisms with the occurrence of the disease, focusing on a denser panel of polymorphisms known to occur on the chromosome of interest. The more closely a marker is located to a disease gene, the more likely it is to be inherited along with that gene. This process can pinpoint a region of the chromosome that contains the gene. However, the region may still contain a long length of DNA encompassing many genes.

In our example of Alzheimer disease, linkage analysis indicated that the disease-causing gene was somewhere near a SNP locus called D14S43 (**Figure 8-11c**).

The final steps again use the human genome databases. The local region containing the gene is examined and the genes within it are identified. DNA from many individuals, some who have the disease and some who do not, is sequenced over this region. This process, with an increasing number of individuals analyzed, gradually leads to the identification of gene variants consistently present in individuals with the disease state, and not in unaffected individuals. The search can be aided by an understanding of the function of the genes in the target region, because particular metabolic pathways may be more likely than others to produce the disease state. In 1995, the chromosome 14 gene associated with Alzheimer disease was identified as gene *S182*. The product of this gene was given the name presenilin-1, and the gene itself was subsequently renamed *PS1*.

Many human genetic diseases are caused by mutations in a single gene or in sequences involved in its regulation, and the defect is inherited in Mendelian patterns (see Chapter 2). Several different mutations in a particular

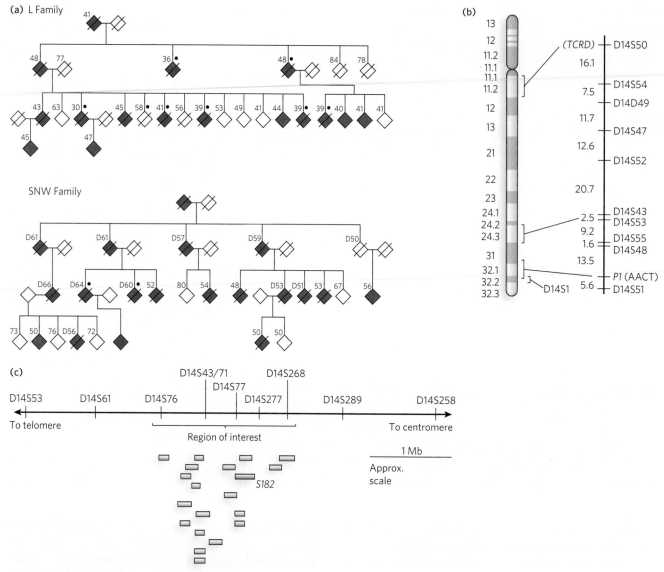

FIGURE 8-11 Linkage analysis in the discovery of disease genes. (a) These pedigrees for two families affected by early-onset Alzheimer disease are based on the data available at the time of the study. Filled symbols represent affected individuals; slashes indicate deaths. The number above each symbol is the person's age at onset of symptoms (for affected individuals), or age at the time of the study (for living unaffected individuals), or age at time of death (for deceased unaffected individuals and others marked by "D"). Black dots indicate that an autopsy was done to verify the presence of Alzheimer disease. To protect family privacy, gender is not indicated. (b) Chromosome 14 has marker positions shown at the right, with the genetic distance between them in centimorgans. *TCRD* (T-cell receptor delta) and *PI* (AACT, α1-antichymotrypsin) are genes with variations in the human population that were used as markers, along with SNPs, in chromosome mapping. (c) A region of interest containing 19 expressed genes was eventually defined near marker D14S43. The gene labeled *S182* (red) encodes presenilin-1. (Mb indicates 10^6 bases.) *[Sources: (a), (b) Data from G. Schellenberg et al., Science 258:668, 1992. (c) Data from R. Sherrington et al., Nature 375:754, 1995.]*

gene, all leading to the same or related genetic condition, may be present in the human population. There are several variants of *PS1*, for example, all giving rise to a much increased chance of early-onset Alzheimer's. Another, more extreme example is the several genes encoding different hemoglobins: more than 1,000 known mutational variants are present in the human population. Some of these variants are innocuous; some cause diseases ranging from sickle-cell anemia to thalassemias. The inheritance of particular mutant genes may be concentrated in families or in isolated populations.

More complex are cases where a disease condition is caused by the presence of mutations in two different genes (neither of which, alone, causes the disease), or where a particular condition is enhanced by an otherwise innocuous mutation in another gene. Identifying the genes and mutations responsible for such digenic diseases is exceedingly difficult, and these diseases are sometimes possible to document only within small, isolated, and highly inbred populations.

Modern genome databases are opening up alternative paths to the identification of disease genes. In many cases, we already have biochemical information about the disease. In the case of Alzheimer disease, an accumulation of the amyloid β-protein in limbic and association cortices of the brain is at least partly responsible for the symptoms. Defects in presenilin-1 (and in a related protein, presenilin-2, encoded by a gene on chromosome 1) lead to elevated cortical levels of amyloid β-protein. Focused databases are being developed that catalog such functional information on the protein products of genes, as well as on protein-interaction networks (determined by methods described in Section 8.2), SNP locations, and other data. The result is a streamlined path to the identification of candidate genes for a particular disease. If a researcher knows a little about the kinds of enzymes or other proteins likely to contribute to a disease, these databases can quickly generate a list of genes known to encode proteins with relevant functions, additional uncharacterized genes with orthologous or paralogous relationships to the genes in this list, a list of proteins known to interact with the target proteins or orthologs in other organisms, and a map of gene positions. With the aid of data from some selected family pedigrees, a short list of potentially relevant genes can often be determined rapidly.

These approaches are not limited to human diseases. The same methods can be used to identify the genes involved in diseases—or genes that produce desirable characteristics—in other animals and in plants.

SECTION 8.1 SUMMARY

- A genome is one copy of the complete genetic complement of an organism. Thousands of complete genome sequences are now available. The Human Genome Project was undertaken by two competing teams that used different strategies of shotgun sequencing.
- The sequencing of a genome is followed by genome annotation, an attempt to summarize the locations and functions of genes and other sequences.
- The human genome contains approximately 20,000 genes, fewer than expected. Only 1.1% to 1.4% of the human genome encodes proteins; the remainder is made up of transposons, functional RNA-encoding genes, introns, sequences involved in gene regulation, and tandem repeats of short sequences.
- The sequencing of multiple primate genomes is opening windows into human evolution. Genomic alterations that are specific to the human lineage occupy about 4% of our genome, with large genomic

rearrangements such as transposon insertions and segmental duplications playing a larger role than single-nucleotide polymorphisms.
- Genome sequence databases facilitate the search for genes that specifically contribute to particular traits, and for genes involved in disease.

8.2 TRANSCRIPTOMES AND PROTEOMES

A gene is not simply a DNA sequence; it is information that is converted into a useful product—a protein or functional RNA molecule—when and if needed by the cell. We now turn to methods that contribute to our understanding of the functions of these gene products. The methods can be applied to efforts to study the response of a cell or organism to particular events or changes in the environment. They can also be used to help identify the functions of the many genes in every genome for which we know very little about their roles in the cell.

The study of complex interconnected processes in biology is called **systems biology**. Genome sequencing contributes to systems biology by providing information about all the genes in an organism. The methods we now address contribute more directly—by examining the expression of genes or the interactions of many kinds of proteins under specified sets of conditions. Many of the methods are described in Chapter 7. Here, we focus on increasingly complex problems in cellular metabolism.

Special Cellular Functions Are Revealed in a Cell's Transcriptome

Only a subset of the many genes in a genome is expressed in any given cell. That subset may change in response to changes in the cellular environment or to extracellular signals of many kinds. The genes expressed in a cell under a given set of conditions constitute its **transcriptome**. Studies of the transcriptome, carried out by researchers in the subdiscipline of **transcriptomics**, can help reveal new cellular processes, as well as identify the genes and gene products involved in known processes. If the function of a gene is not known, an understanding of the circumstances that result in expression of that gene can provide an important functional clue.

Transcriptome analysis was first made practical with the advent of microarray technologies (see Figure 7-31). Microarrays can reveal, for example, the genes that are newly induced when a cell is subjected to heat shock, variations in expression patterns in different regions of a mammalian brain, or changes that occur when a pathogenic bacterium invades a host organism. The growing use of microarray-based transcriptome analysis has led to the development of online databases, some specific to a single organism, that make data available to the entire scientific community. As the quality of transcriptome data improves, the transcriptomes themselves become more than a list of expressed genes. They are also a kind of fingerprint that characterizes a class of cell under a given set of conditions. These databases are becoming very useful not only in basic research but in medicine as well.

For instance, the cells making up a tumor exhibit characteristic patterns of gene expression—a transcriptional profile—that may differ greatly from one tumor to the next. These profiles can provide a kind of tumor fingerprint, which can be used to predict a patient's prognosis and/or select the most beneficial therapies. The value of these tools to oncologists and patients will only increase as the technologies become more widespread.

Recent progress in the diagnosis and treatment of breast cancer illustrates the potential of the technology. Broad clinical studies over the past decade have used microarrays to develop transcriptional profiles of thousands of breast cancers. Treatment protocols have been tracked, and the successes and failures carefully documented. Researchers are gradually identifying specific genes and groups of genes that, when expressed at higher levels and in certain combinations, serve as prognostic indicators. The result is a growing database of correlations that allows the use of transcriptional profiles to develop both prognoses and treatments.

High-Throughput DNA Sequencing Is Used in Transcriptome Analysis

Microarrays have some disadvantages for transcriptome analysis. They can provide inaccurate information about relative levels of transcription for genes that are expressed at very low or very high levels. In addition, they can miss any RNAs that are not homologous to genes included on the microarray. A newer method, called **RNA-Seq**, has been developed to address these shortcomings, taking advantage of modern, high-throughput DNA sequencing technologies (see Section 7.2).

A typical RNA-Seq experiment is shown in **Figure 8-12**. RNA is isolated from the cell or tissue to be analyzed. In most cells, rRNA is by far the most abundant RNA, but it is usually other types of RNA that are of most interest. Thus, most protocols include a step involving subtractive hybridization of the rRNA, using complementary probes that allow removal of the hybridized material. The remaining RNA is then converted to cDNA with the enzyme reverse transcriptase (see Figure 7-8). The cDNA is fragmented (e.g., by shearing or nuclease digestion) to an appropriate average length. Short adapter DNA segments that provide target sequences for the primers needed for DNA sequencing are ligated to both ends. Each cDNA

is then "read" by DNA sequencing. Huge numbers of these short sequencing reads (typically 30 to several hundred base pairs, depending on the sequencing technology used) are produced. The gene from which each sequencing read is derived is determined by computerized alignment with the same sequence in the relevant genome database. Genes expressed at high or low levels are represented by correspondingly high or low levels of sequence reads. Gene expression levels can be mapped across genes, chromosomes, and entire genomes.

RNA-Seq provides information on gene expression levels with a much greater dynamic range and has proved highly accurate when compared with more laborious methods. The direct sequencing also provides additional information, showing the exact transcriptional boundaries of genes and revealing how exons are linked together in transcripts. In genes that have alternative splicing patterns (see Figure 8-5), the method can also reveal which exons within a single gene are being expressed at higher levels in a particular tissue. In some organisms, RNA transcripts are edited, producing new sequences not present in the DNA genes (a process described in Chapter 16). These sequence changes are directly revealed in RNA-Seq. As the costs of high-throughput DNA sequencing continue to decrease, RNA-Seq is replacing microarrays as the method of choice for transcriptome analysis.

The Proteins Generated by a Cell Constitute Its Proteome

In any effort to understand the metabolic status of a cell, the researcher must eventually look at the proteins. The complement of proteins present in that cell under a given set of conditions is called the cell's **proteome**. The word "proteome" first appeared in the research literature in 1995. The subdiscipline of **proteomics** includes efforts to define the proteome. More broadly, any effort to analyze a complex mixture of proteins, whether or not it applies to all the proteins in a cell, falls under the proteomics umbrella. For example, some studies are directed at the proteins in a specific organelle or the proteins embedded in the cytoplasmic membrane.

The problems that proteomics researchers explore can be straightforward to describe, but the solutions often are not. Each genome presents us with thousands of protein-coding genes. We wish to know which proteins are present and contributing to cellular metabolism under every possible set of circumstances. Analysis includes the structure, posttranslational modifications, cellular localization, and detailed functions of all those proteins, as well as how the various proteins interact. Many research problems focus on particular cellular

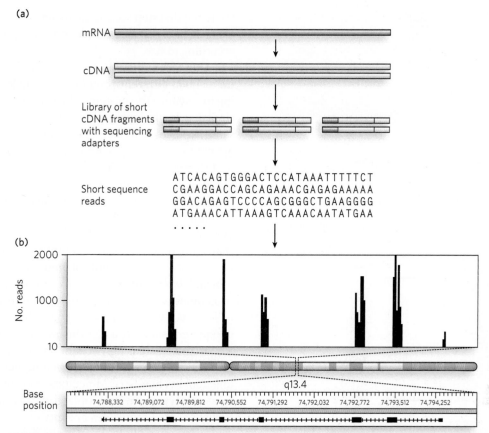

FIGURE 8-12 Use of RNA-Seq to examine transcriptomes (a) An mRNA is isolated, transcribed to cDNA, fragmented into smaller pieces, and ligated to adapter oligonucleotides that provide targets for sequencing primers. Sequencing then follows, using one of the methods described in Section 7.2. (b) The number of times a sequence from a given gene or segment of a gene appears in a sequencing read (i.e., the number of reads containing all or part of that sequence) is plotted. The number of reads from a given genomic region reflects the relative level of mRNA produced from that region. Shown here are data for a small portion of human chromosome 11, segment q13.4.

systems, encompassing a subset of proteins present at a certain time. Given that many proteins can still reveal surprises even after years of study, the investigation of an entire proteome or any complex protein mixture is daunting. Biochemists can now apply shortcuts in the form of new and updated technologies and databases that address protein function on a cellular level.

Transcriptome information tells us about RNA levels in a cell, but this does not necessarily inform us about protein levels. The expression of many genes, particularly in eukaryotes, is regulated at the level of translation. Messenger RNAs for particular genes are often stored in the cell in an inactive state until the protein product of that gene is needed. In addition, many proteins are initially synthesized in an inactive state, their function dependent on posttranslational modifications. A complete understanding of a proteome requires information about all modified states of a protein and how those states affect function.

Electrophoresis and Mass Spectrometry Support Proteomics Research

There are two principal ways to detect and identify proteins: polyacrylamide gel electrophoresis (see Highlight 4-1) and mass spectrometry. The resolving power of polyacrylamide gel electrophoresis can be amplified by carrying out two electrophoretic steps in succession, separating proteins on the basis of different properties; the technique is known as **two-dimensional gel electrophoresis** (**Figure 8-13**). The first step (or dimension) uses isoelectric focusing, a method that separates proteins on the basis of their isoelectric point, or pI (the pH at which the net charge of a protein is zero). Polyacrylamide gels containing an immobilized pH gradient are commercially available. Voltage is applied across the gel. Proteins migrate through the gel, halting where the pH of the strip equals the pI of the protein. In the second step, the gel strip is laid on top of another gel, and electrophoresis is carried out at

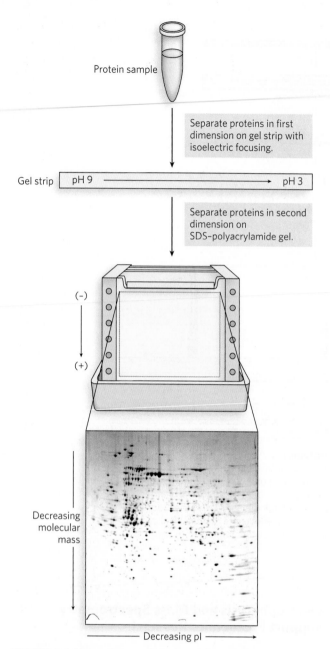

Protein sample

Separate proteins in first dimension on gel strip with isoelectric focusing.

Gel strip   pH 9 ⟶ pH 3

Separate proteins in second dimension on SDS–polyacrylamide gel.

(−)

(+)

Decreasing molecular mass

Decreasing pI

FIGURE 8-13 Two-dimensional gel electrophoresis of a complex mixture of proteins. In the first dimension, proteins are separated on a gel strip according to charge (i.e., pI) by isoelectric focusing. The strip is then laid on top of an SDS-polyacrylamide gel, and the proteins are separated according to size (i.e., molecular mass) by electrophoresis. The original protein complement is thus spread in two dimensions, aiding the separation of similar proteins into individual spots. The spots can be cut out of the gel and the proteins identified by mass spectrometry. [*Source: Courtesy Axel Mogk. Identification of thermolabile Escherichia coli proteins: prevention and reversion of aggregation by DnaK and ClpB. A. Mogk, T. Tomoyasu, et al, The EMBO Journal (1999) 18, 6934–6949, doi:10.1093/emboj/18.24.6934.]*

90° to the first step (i.e., in the second dimension), this time using an SDS-polyacrylamide protocol to separate proteins according to size.

This technique allows the separation and display of up to 1,000 different proteins on a single gel. After staining to visualize the proteins, the gel can be compared with similar gels displaying the proteins in extracts from the same types of cells but under different conditions. The appearance (or disappearance) of particular protein spots in different samples can help define the cellular function of these proteins. Individual spots on the gel can be cut out and identified, typically with the aid of mass spectrometry. Two-dimensional gel electrophoresis is a robust procedure, and modern protocols are highly reproducible. Importantly, the method resolves complete proteins. Antibodies can be used to detect multiple modified variants of a protein and determine their relative abundance. However, the proteomes of cells more complex than bacteria or single-celled eukaryotes (e.g., yeast) are often too complex for adequate resolution using this method. Proteomic analyses are thus increasingly reliant on mass spectrometry.

The mass spectrometer has long been an indispensable tool in chemistry. Molecules to be analyzed, referred to as **analytes**, are first ionized in a vacuum. When the newly charged molecules are introduced into an electric and/or magnetic field, their paths through the field are a function of their mass-to-charge ratio, *m/z*. This measured property of the ionized species can be used to deduce the mass (*M*) of the analyte with very high precision.

Mass spectrometry (MS) provides a wealth of information for proteomics research, enzymology, and protein chemistry in general. Because the techniques require only miniscule amounts of a sample, they are readily applied to the small amounts of protein that can be extracted from a two-dimensional electrophoretic gel. The accurately measured molecular mass of a protein is one of the critical parameters in its identification. Once a protein's mass is accurately known, mass spectrometry is also a convenient and accurate method for detecting changes in mass due to bound cofactors, bound metal ions, covalent modifications, and so on.

Modern mass spectrometry is by no means limited to characterizing the mass of pure proteins or identifying proteins cut out of polyacrylamide gels. Rapid evolution of this technology now allows the analysis of highly complex mixtures of proteins. The most common approach involves a refinement of mass spectrometry called tandem mass spectrometry, or **tandem MS**—effectively, two mass spectrometry procedures in succession. Typically, a protein mixture is extracted from cells or tissue and digested into peptides with a protease, and the peptides are fractionated (separated by liquid chromatography). As the peptides appear at the end of the fractionation, they are ionized and fed directly into the tandem mass spectrometer. The first of the two mass spectrometers sorts the peptides, which are then fed one at a time into a connecting chamber where each one is reduced—by collisions with helium or argon fed into the chamber—to a series of fragments, generally by breakage at peptide bonds. The fragments are next fed into the second mass spectrometer, where the m/z ratio of each is measured and its mass deduced. The collection includes a set of fragments that are related by the precise loss of mass corresponding to one, two, three, or more contiguous amino acid residues in the peptide. As the mass of each amino acid is distinctive, this allows the sequence of each peptide to be deduced. The final product is a long list of peptides, each of which has a known sequence and mass. The entire process is automated, and many thousands of peptides in a sample can be analyzed and recorded. With the wealth of genomic information available in the databases, each peptide is readily assigned to an individual protein. The relative abundance of a protein in a sample is generally proportional to the abundance of peptides derived from it in a sample. Numerous variants of the procedure have been developed that allow researchers to narrow the analysis. For example, it is possible to look only at phosphorylated peptides in a cellular protein mixture.

For the investigation of cellular systems, mass spectrometry is often paired with other techniques. For example, if a protein of interest is tagged (with a TAP tag; see Figure 7-27) and precipitated, the proteins it is associated with can be identified by mass spectrometry. A variety of methods can be used to increase or decrease the amount of a given protein in a cell, or eliminate it altogether (see Chapter 7). The researcher can then use tandem MS on the cellular proteome to determine how that protein's absence (or increased abundance) affects other proteins in the cellular proteome. The possible variations in experimental design are limitless.

Computational Approaches Help Elucidate Protein Function

With the number and size of databases increasing rapidly, the information required to answer a biological question may be right at one's fingertips. Increasingly, data-mining is complementing experimentation as a highly productive path to mechanistic and functional insights about genes, RNAs, and proteins.

Sequence Relationships A wide range of conserved amino acid sequences, many of them relatively short, have been reliably associated with particular protein functions. These sequence motifs, or structural motifs, are readily identified in sequence databases. When such a motif is found in a protein or group of proteins, an associated function is inferred and likely to be present (see Chapter 4). The motifs

often correspond to binding activities (e.g., ATP, nucleic acids, NAD⁺, metals) or catalytic activities (e.g., helicase, polymerase, ATPase). The presence of a structural motif may suggest, for example, that the protein catalyzes ATP hydrolysis, binds DNA, or forms a complex with zinc ions, thus helping define molecular function. Resources for conducting searches for such motifs are available on the NCBI and Ensembl websites.

Structural Relationships Accurate determination of the three-dimensional structure of a protein is not always successful, but efforts are so common that structural databases are replete with protein structures of all types. Determination of the structure of a newly discovered protein can help define its function, as structural relationships often remain long after clear sequence homologies have been erased by evolutionary time. When a new protein is found to have structural folds that are clearly related to motifs with known functions in the structural databases (see Highlight 4-2), this information can suggest a molecular function for the protein.

Comparisons of Genome Composition Although not evidence of direct association, the mere presence of combinations of genes in certain genomes can hint at protein function. One can simply search the genome databases for specific genes, then determine which other genes are present in the same genomes—a process known as **phylogenetic profiling** (**Figure 8-14**). (Phylogenetics is explained in more detail in Section 8.3.) The consistent appearance of two genes together in a genome suggests that the proteins they encode may be functionally related. Such correlations are most useful if the function of at least one of the proteins is known.

Phylogenetic profiling is often carried out on hundreds or thousands of genes at once, in broad studies that complement approaches such as linkage analysis. The search

	Species			
Protein	1	2	3	4
P1	+	−	+	+
P2	−	−	+	−
P3	+	+	−	+
P4	+	−	+	−
P5	+	−	−	−
P6	+	+	−	+
P7	+	+	+	−

FIGURE 8-14 Use of comparative genomics to identify functionally related genes. This example of phylogenetic profiling shows gene comparisons for four organisms. P1 through P7 indicate proteins encoded by each species. The + or − indicates presence or absence of the protein. The technique does not require homologous proteins. Because proteins P3 and P6 always appear together in a genome (red shading), they may be functionally related. In particular, they may have a function that is found in species 1, 2, and 4, but not in species 3. Further testing would be needed to confirm this inference.

for a gene called *BBS5*, involved in Bardet-Biedl syndrome (BBS), provides an example. Bardet-Biedl syndrome is a serious genetic condition characterized by retinal degeneration, obesity, a variety of physical deformities, and learning disabilities. Six *BBS* genes discovered before *BBS5* were found to be involved in the function of a cellular structure called a flagellar and basal body. *BBS5* had been localized to a region in chromosome 2. To facilitate identification of the gene in this region, the researchers did a phylogenetic profile, comparing human genes and genes of the green alga *Chlamydomonas*—species that possess flagellar and basal bodies—with genes of the plant *Arabidopsis*, which lacks this cellular structure. They generated a list of 688 genes present in humans and *Chlamydomonas* but absent in *Arabidopsis*. The region of chromosome 2 that interested the researchers had a total of 230 genes, but only 2 of them were on the list of 688 generated by the phylogenetic profile. One of these turned out to be *BBS5*.

Experimental Approaches Reveal Protein Interaction Networks

Every protein functions by interacting with other molecules, from small metabolites to nucleic acids and other proteins. One of the strongest clues to protein function is knowing which other proteins that protein interacts with. For example, if a protein of unknown function interacts with an RNA polymerase, there is a good chance that the protein is also involved in transcription. Powerful new technologies are providing information on protein interaction networks in cells.

Protein Chips For large-scale studies, proteins, like nucleic acids, can be immobilized on a solid surface, forming protein chips—a kind of protein microarray. These can be used to detect the presence or absence of other proteins in a sample. For example, an array of antibodies to a particular set of proteins is immobilized as individual spots on a solid surface. A sample of proteins is added, and any proteins that bind an antibody on the chip can be detected by a variety of methods. However, whereas DNA is consistent in its physicochemical properties and is readily immobilized on silicon chips, proteins vary a great deal in their properties, and the construction of protein chips can be challenging. The conformation of many proteins depends on solution conditions, and immobilization on a silicon chip may inactivate some proteins in a manner that is not always predictable. Nevertheless, many successful efforts have been reported.

Probing Macromolecular Interactions In Vivo The study of protein-protein interactions by the two-hybrid method (see Figure 7-28) and the study of protein-RNA interactions by the three-hybrid method (see Figure 7-29) rely on macromolecular interactions that occur in vivo. Both are important avenues for examining protein interaction networks in proteomics research. A somewhat different approach to detecting protein interactions in vivo involves immunoprecipitation of proteins from cell extracts. Antibodies are used to precipitate a given protein, and the precipitate is examined to identify (by mass spectrometry) any other proteins that were associated with the target protein in the cell and thus precipitated with it. This is a variation on the use of a TAP tag (see Figure 7-27).

Miscellaneous Approaches The proteomics literature is replete with examples of creative approaches to dissecting protein interaction networks. One approach is the search for "Rosetta stone" fusions. Sometimes, two proteins that exist as separate entities in species 1 may have orthologs in species 2 that are the product of two fused genes. This fusion in species 2 makes it highly likely that the two proteins in species 1 interact. Another approach simply mines the biochemical literature, focusing on proteins that are mentioned together in the same article. If two proteins are mentioned together in a large number of publications, the assumption is made that the two may interact.

SECTION 8.2 SUMMARY

- A transcriptome is a listing of the genes that are expressed in a given cell under a defined set of conditions. The transcriptome may change in response to environmental changes or cellular signals.
- Microarrays provide one picture of cellular transcriptomes. The RNA-Seq approach is even more effective in generating a detailed transcriptome.
- A proteome is a compilation of all the proteins present in a given cell under a defined set of conditions. Computational and experimental techniques explore the proteome, as well as the functions of the proteins it encompasses, on a cellular scale.
- The most common approaches to examining a cellular proteome under a defined set of conditions are two-dimensional gel electrophoresis and mass spectrometry.
- The generation of protein interaction networks is one of the goals of proteomics research. Techniques include protein chips, two-hybrid and three-hybrid methods, immunoprecipitation, and protein fusions.

8.3 OUR GENETIC HISTORY

Genomics research carries important implications that go to the heart of human existence. Where did we come from? How did we get to where we are now? Genomics provides an especially informative and often quantitative window on evolution, ultimately advancing the scientific answer to these and many other fundamental questions. A quest to understand how we came to this point in time is not simply an academic exercise. A better understanding of how new species evolve and how we are related to one another and to other species is highly relevant to advancing knowledge in areas ranging from ecosystems to pandemics. Answers can pay huge dividends in medicine, agriculture, resource management, and general quality of life.

All Living Things Have a Common Ancestor

One goal of modern genomics and evolutionary biology is the reconstruction of the evolutionary tree that traces the origin of every extant species. We can approach this problem from both ends—the first living organism and the current list of living organisms—and explore how genomics can help trace the path between them.

The successful living entity that gave rise to all life now on Earth is referred to as **LUCA**, the **last universal common ancestor**. Although its biological form and genome are obscured by billions of years of evolution, there are several approaches to thinking about LUCA. The first approach, which has been greatly facilitated by modern genome sequencing efforts, attempts to assemble the list of genes and other features that are currently shared by every living organism: these features were probably present in LUCA. The second approach is an effort to define the minimum set of genes necessary to support a living cell. Such a minimalist cell would help define the essence of the free-living state and would provide a more complete understanding of the basic problem of life and the threshold of complexity that had to be breached by the first viable cell.

As revealed by genomics and work in many other biological fields, current organisms share several features that permit a trace to a common ancestor. The core components of the translation and transcription machinery in all cells are demonstrably related. The use of the D isomers of sugars in cells and the L isomers of amino acids in protein synthesis is also universal. Beyond this point, the generalities start to break down. All organisms consist of cells surrounded by lipid-containing membranes, but the composition and structure of those membranes can vary greatly from one group to another. For example, bacteria have membranes consisting mostly of fatty acid esters, whereas membranes in the archaea consist of isoprene ethers. All organisms replicate their DNA, but the replication machinery varies in important ways.

Current estimates for the number of genes shared by all known species vary from 80 to about 500. The lower number focuses on genes with clearly identifiable orthologs in all organisms, based on sequence comparisons. The higher number includes genes required for processes that are found in all organisms, but for which many mechanistic and sequence similarities have been obscured by evolutionary time. A cell with 500 components would be simpler than most existing life forms, but still very complex. There must have been many intermediates of gradually increasing complexity in the process that led to LUCA.

The search for the minimal genome begins with the assumption that this cell will grow in a stress-free laboratory environment with abundant resources and constant temperature. The goal is to define the minimum group of components for supporting life, without the specialized functions required for particular world environments. Bacteria with small genomes are a useful place to start.

The bacterium *Mycoplasma genitalium*, a parasite of the genital and respiratory tracts of primates, has the smallest genome sequenced thus far for a defined organism. Its 580,000 bp DNA includes 521 genes, 482 of which encode proteins. Directed efforts to disrupt individual genes have revealed that the bacterium can dispense with only 97 of them and still retain viability in the laboratory, giving a minimal complement of 385 genes. The small genome of this *Mycoplasma* reflects its sheltered environment as a parasite.

Similar experiments have indicated that the minimal genome for an organism living autonomously includes about 1,350 genes. Attempts to define a minimal gene complement are fueling efforts to create an artificial cell from nonliving chemical components, an achievement that would mark a new level of understanding of living systems.

Genome Comparisons Provide Clues to Our Evolutionary Past

The evolutionary relationship among species, populations, or genes is known as a **phylogeny**, and the study of such relationships is called **phylogenetics**. Phylogenetics helps biologists classify organisms. It can also reveal important information about the evolution of traits in an organism or the appearance of new pathogens. It can even aid in criminal investigations (**Highlight 8-2**). Phylogenies are usually described with the aid of phylogenetic trees, which can be based on sequence information or on other attributes of a species, such as morphological characteristics. The construction of evolutionary trees was imprecise and descriptive until the 1950s, when mathematical biologists began to systematize the process. That work continues.

On one level, the branched evolutionary trees often depicted in popular or scientific literature are almost self-explanatory. However, when used scientifically they

Phylogenetics Solves a Crime

In the summer of 1994, a nurse in Lafayette, Louisiana, broke off a 10-year affair with a physician. The physician had been giving the nurse vitamin shots for fatigue. He gave her one more of these shots in August 1994, after the breakup. The nurse had donated blood to a local blood bank on several occasions; she was tested and found negative for HIV in October 1992, May 1993, and April 1994. In late 1994, the nurse became ill and tested positive for both HIV-1 and hepatitis C, although she had no history of contacts that could have led to the infections. The nurse accused the doctor of infecting her with HIV.

Investigators found records indicating that the doctor had treated and drawn blood from his only HIV-infected patient and a hepatitis C–infected patient just before giving the nurse the vitamin injection in August 1994. But how does one link the patients' blood to the nurse-victim in this case? The subsequent trial of the doctor was the first to use phylogenetics in a court case. The investigation focused on the HIV infection.

Once HIV begins to replicate in a new host, the virus mutates rapidly, an evolution that occurs within one infected individual. Samples taken from a person with HIV years after infection can be used to build a phylogenetic tree that can trace the evolution of the virus in that individual. Blood samples were collected from the doctor's HIV-infected patient and from the nurse. Control samples were collected from 30 different HIV-positive patients selected at random in the Lafayette area. The HIV in the samples was sequenced and analyzed independently by two different laboratories at Baylor University and the University of Michigan. Both analyses yielded the same result. The phylogenetic analysis of the victim's HIV strains showed that they were most closely related to, and nested within, the strains from the doctor's patient (Figure 1).

With this and other evidence, the doctor was convicted of attempted second-degree murder in 1998. The verdict was upheld by a Louisiana appeals court in 2000, and the U.S. Supreme Court refused to hear the case in 2002, ending court proceedings. The doctor is now serving a 50-year prison sentence. The same methodology has since been used in rape and child abuse cases.

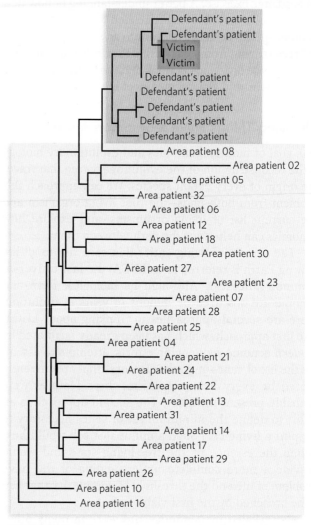

FIGURE 1 A phylogenetic tree reveals the diversity of HIV samples in the Lafayette, Louisiana, area. The part of the tree derived from the doctor's HIV-positive patient is highlighted, with the nurse-victim's DNA clearly nested within this set of sequences. [Source: Data from M. L. Metzker et al., Proc. Natl. Acad. Sci. USA 99:14,292–14,297, 2002.]

are based on a host of underlying assumptions and conventions, and the structures in the tree have specific meanings (**Figure 8-15a**). The subjects of an evolutionary tree are groupings of organisms known as **taxa**. A taxon is any such grouping, and it can refer to an individual species (*Homo sapiens*), a genus (*Homo*), a class (*Mammalia*), particular populations of a single species, and so on. The tips or ends of the branches on the tree represent the taxa being studied and often reflect species or groups of species now in existence. Each branch point, or node, represents an extinct ancestral species common to the two connected branches. The node at the base of the tree signifies the common ancestor of all the taxa represented in the tree; this is sometimes called the root of the tree. As shown later, not all trees are rooted. Generally, one species is thought to give rise to two, leading to a bifurcating tree. Uncertainty about some evolutionary relationships can lead to the generation of a tree with multiple branches at a single node; such a node is called a polytomy. Node orientation and rotation are arbitrary (**Figure 8-15b**). Many different tree depictions are in use, with different branch shapes

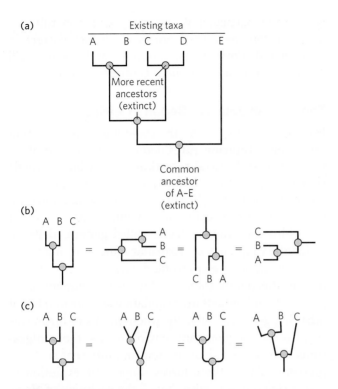

FIGURE 8-15 Phylogenetic trees. An evolutionary tree consists of branches (usually bifurcating) connected by nodes. (a) Basic conventions. The ends of external (upper) branches represent existing taxa, the nodes represent extinct ancestors, and the root end represents the common ancestor of the taxa included in the tree. (b) The orientation of the tree does not matter, and (c) there are several common (and equivalent) depiction styles.

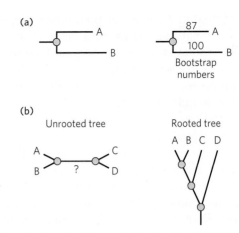

FIGURE 8-16 Time depictions and rootedness in phylogenetic trees. (a) The length of tree branches can be meaningless or, if specified, can represent some unit of evolutionary time. For example, the relative lengths of the branches correspond to differences in a time measure leading to taxa A and B. When numbers are given (right), these indicate the investigator's confidence in the information in that branch, based on statistical tests; the numbers here are typical of the common bootstrap analysis. (b) In an unrooted tree, some of the relationships between taxa may be evident, but there is uncertainty about the common ancestor of all the taxa. In a rooted tree, all taxa can be traced to a common ancestor.

(**Figure 8-15c**). The choice of representation is made simply for convenience or personal preference.

Branch length can be meaningless, but often it is used to represent some measure of evolutionary time, such as numbers of altered morphological features, numbers of mutations in one or several genes (more common in modern trees), or numbers of genomic alterations in a region of the genome (**Figure 8-16a**). For example, we can look at the differences in a gene found in both humans and chimpanzees and determine which variant existed in the common ancestor, such as by examining outgroups, as described in Section 8.1. Once the sequence of the gene in the common ancestor is determined, the common ancestor becomes a node in the tree. The lengths of the branches leading from ancestor node to humans and chimpanzees reflect the number of changes occurring between that ancestor and the living species.

Numbers next to branches on a tree usually reflect the level of confidence the investigator has in the information contained in that branch (see Figure 8-16a, right). A common method for setting confidence limits is the bootstrap analysis, which gives a range from 100 (very high confidence) to 0 (no confidence). In brief,

the bootstrap is a statistical method that starts with the set of sequences used to generate the original tree. For example, let's say a particular sequence of gene X is used to construct a tree for 100 species that all have gene X. A computer program randomly samples the original 100 sequences to create a new set of 100 sequences. In this new set, some of the original set may be missing, and other sequences may be included multiple times. A new phylogenetic tree is generated from each of the created datasets, and the number of times the same branch configuration arises for a cluster of species is tallied. The score reflects the absence (high confidence) or presence (lower confidence) of viable branching alternatives.

An unrooted tree is one for which the positioning of the common ancestor is uncertain (**Figure 8-16b**). In such a tree, the direction of evolution for parts of the tree might be unknown.

A wide range of problems arise in the construction of evolutionary trees. Mutation rates are often assumed to be constant, but this assumption is flawed. Mutation rates can be affected by environmental factors. For example, reactive oxygen species are the most common source of mutagenic DNA lesions (as described in Chapter 12). Thus, aerobic organisms are subject to more DNA damage and potential mutagenesis than anaerobic organisms. Exposure to DNA-damaging agents such as UV light can vary greatly, depending on the ecological niche occupied by a given species. Certain regions of a gene may accommodate mutations better than others, depending on the functional importance of a given segment. An occasional back-mutation to the original base or amino acid could obscure actual mutation rates. Finally, not all DNA in an organism is inherited linearly from parents to offspring. In individuals, genes can be lost, such as by genomic deletions due to DNA replication errors, and genes can be gained.

Gene gain can result from a process called **horizontal gene transfer**, which is common in bacteria and archaea (witness the very rapid spread of genes encoding antibiotic resistance in human bacterial pathogens). Early viruses may have transferred genes from one bacterium to another, and from one species to another, resulting in the sudden appearance (rather than the gradual evolution) of a gene in a particular evolutionary line. Large genome rearrangements might abruptly break up a pattern of synteny in one ancestral line, complicating the analysis. Sorting out these patterns is the job of increasingly sophisticated computer algorithms.

The complexity of the problem is evident in a current tree of life, two versions of which are shown in **Figures 8-17a and 8-17b**. These trees are based on analyses of particular genes and patterns in fully sequenced genomes. They are probably not correct in every detail. Corrections, additions, and updates will continue for decades, perhaps centuries, to come.

The Human Journey Began in Africa

Four major factors affect the evolution of any group of organisms. **Mutation rates** determine the extent of genetic diversity. **Natural selection** affects which genomic changes are inherited in a population. Many mutations are relatively neutral, however, and do not undergo positive or negative selection. Neutral mutations are subject to a third evolutionary factor called **genetic drift**, in which the frequency of particular mutations in a population changes more or less randomly over time. Genetic drift is affected by such variables as the number of reproducing individuals in a population and the number of offspring generated. Finally, when groups of organisms colonize new regions and environments, their **migrations** may subject them to new and different selective pressures. All of these forces shaped the evolution of *Escherichia coli*; they also shaped the evolution of *Homo sapiens*.

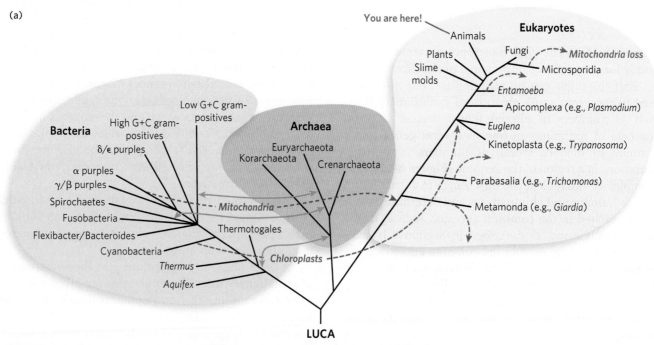

(a)

FIGURE 8-17a The tree of life. This relatively simple tree includes only a few of the many sequenced genomes, but it illustrates some of the complexities of generating a complete tree of life. One crucial complicating factor is the tendency of living systems to occasionally incorporate DNA into their genomes from other sources by horizontal gene transfer (orange arrows). Other factors are the assimilation of bacteria as organelles (mitochondria and chloroplasts; blue and green dashed arrows, respectively) and the subsequent loss of such organelles in some evolutionary lines (red dashed arrows). [*Data from J. R. Brown, "Universal tree of life," in* Encyclopedia of Life Sciences, *Wiley InterScience (online), 2005.*]

About 7 million years ago, the common ancestor of chimpanzees, bonobos, and humans lived in Africa. Groups of that ancestral species followed divergent lines of evolution, one leading to chimpanzees and bonobos, and one leading to humans (**Figure 8-18**). The path to humans first generated a series of species in a genus dubbed *Australopithecus*. The Australopithecines remained in Africa, giving rise, about 3 million years ago, to *Homo habilis*, the

first species of our own genus. The archaeological record indicates that *H. habilis* was the first species to use stone tools. About 1.7 million years ago, a successor to *H. habilis* emerged—*Homo erectus*. The hominids were a bit more adventurous than the Australopithecines. Armed with better tools and a mastery of fire, *H. erectus* spread from Africa to virtually all of Eurasia. The fossil record provides evidence of many other *Australopithecus* and *Homo* species during

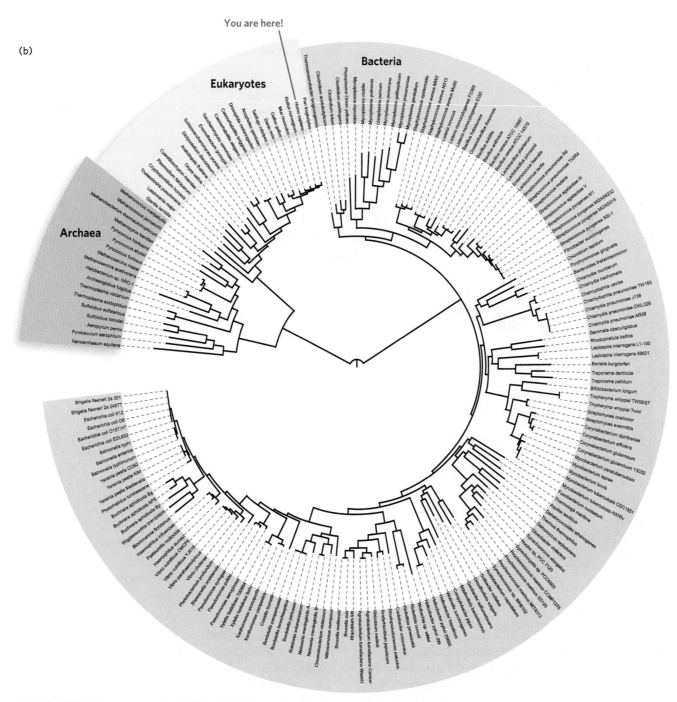

FIGURE 8-17b The tree of life (*continued*). This more complex evolutionary tree was developed using data from 191 species with sequenced genomes. [*Source: Data from F. D. Ciccarelli et al., Science 311:1283–1287, 2006.*]

(a)

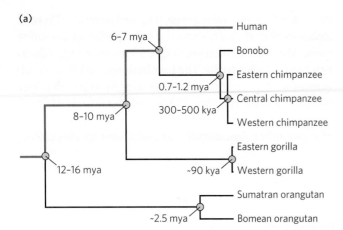

FIGURE 8-18 The evolution of humans and their close relatives.
(a) The closest living relatives of humans are chimpanzees and bonobos. The orangutan and gorilla lines branched off earlier. All of the lines shown are considered hominids. Estimated times of species divergence (in units of mya, million years ago, and kya, thousand years ago) are shown at the branch points. The evolutionary line leading to humans is in red. (b) The Hominid family diverged into Homininae and Ponginae (orangutans) subfamilies about 14 million years ago. The Homininae line later diverged into the Hominini (hominins) and Gorillini (gorillas) tribes. The hominins further split into lines leading to the *Homo* and *Pan* genuses about 6 to 7 million years ago. This figure provides some current details about the evolutionary paths in the hominin line leading to the *Homo* genus and eventually to humans. [*Source: (b) Reconstruction photos from the Smithsonian Human Origins Program, courtesy of Karen Carr Studios.*]

(b)

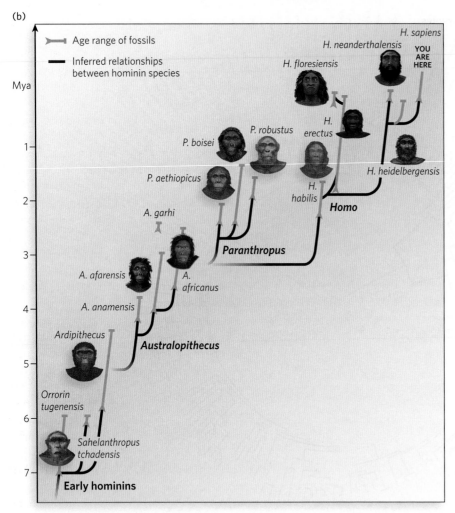

the past 3 million years. These species probably arose by **allopatric speciation**: geographic isolation of a group of individuals, followed by evolution, resulting in the formation of a distinct species that no longer can interbreed with the original one. All of these species ultimately became extinct, except for one.

Homo sapiens evolved about 500,000 years ago. For decades, scientists argued about two possible human origins. The multiregional theory proposed that humans evolved gradually in many places, with gene flow occurring constantly between the various populations. This would entail a direct evolution of *H. erectus* into

H. neanderthalensis into *H. sapiens*, occurring simultaneously in Eurasia and Africa. The alternative, "out of Africa" theory posits that the *H. erectus* and *H. neanderthalensis* expansions into Eurasia were independent of the *H. sapiens* expansion, and that the former two species represented separate evolutionary branches. Modern genomics has definitively resolved the debate in favor of the "out of Africa" theory.

A woman who lived in sub-Saharan Africa 140,000 to 200,000 years ago, sometimes called Eve by the scientists who study our ancestral tree, is the female genetic ancestor of all living humans. She was not the only human female then living, but she is the only one whose DNA has been inherited in the modern human lineage. All mitochondrial DNA is inherited maternally, deposited in the egg prior to fertilization. Mitochondrial DNA is also not subject to the scrambling effects of recombination. Thus, stable haplotypes of mitochondrial genome polymorphisms can be reliably traced back in time. The current human lineage traces back to mitochondrial Eve.

There is also a genomic Adam, but Eve never knew him. All males now living on Earth are descended from a male who lived in Africa about 60,000 years ago. Again, Adam was not the only male member of his species present. He is simply the one whose DNA survives. Our information about this individual comes from analyses of haplotypes in Y chromosome DNA, most of which is not subject to recombination.

Human Migrations Are Recorded in Haplotypes

About 50,000 years ago, a small group of humans looked out across the Red Sea to Asia. Perhaps encouraged by some innovation in small boat construction, or driven by conflict or famine, or simply curious, they crossed the water barrier. That initial colonization, involving perhaps 1,000 individuals, began a journey that did not stop until humans reached Tierra del Fuego (at the southern tip of South America) many thousands of years later. In the process, the established populations from previous hominid expansions into Eurasia, including *H. neanderthalensis* and *H. erectus*, were displaced. The Neanderthals disappeared, as did *H. erectus* (**Highlight 8-3**).

This journey can be traced by looking at our genomic polymorphisms. Efforts are under way to survey genetic diversity in human populations around the globe. One such endeavor is the International HapMap (haplotype map) project; another is the Human Genome Diversity Project. Both are international efforts to sequence thousands of human genomes taken from carefully selected populations around the world, and to accumulate information about tens of thousands of polymorphisms in the sequenced genomes. These enterprises are every bit as large and complex as the original Human Genome Project. The results have helped define mitochondrial Eve and Y chromosome Adam, and they are telling us a great deal more. Complementary analyses of mitochondrial DNA from Neanderthals and Denisovans, another hominid group recently discovered in the Denisova Cave in Siberia, have established that they were on a separate evolutionary line.

Phylogenetic analyses of species evolution generally rely on gene mutations that are fixed in a given species: all the members of species X have one gene sequence, and all the members of species Y have a different sequence. The analysis of genetic polymorphisms within a species increasingly relies on a different kind of mathematical analysis, called **coalescent theory**. Even though it is not subject to recombination, the sequence of mitochondrial DNA, like that of chromosomal DNA, changes slowly with time because of mutations. If a mutation occurred recently, it will appear in the relatively few individuals descended from that female. If the mutation appeared much earlier, it is found in many individuals across broad geographic regions. With mathematical models that take into account estimated mutation rates, selection, genetic drift, and other factors, various polymorphisms are traced back to the ancestor in which they first appeared—a coalescence.

Overall genetic diversity in the human lineage is lower than that in chimpanzees. This is one of several pieces of evidence indicating that early human populations went through evolutionary bottlenecks a few hundred thousand years ago, when only a few thousand or tens of thousands of individuals existed and genetic diversity was limited. Our mitochondrial Eve and Y chromosome Adam lived in times when there were far fewer humans than today. More than 85% of the polymorphisms in the human population appear at the same frequency in all human populations worldwide, indicating that they arose before the appearance of the first modern humans. The remaining 15% tell us about human migrations.

Genetic diversity, in terms of haplotypes that do not occur uniformly across world populations, is by far the greatest in extant African populations. When that wanderlust-possessing population of humans first colonized Asia, they took only a subset of the variable human haplotypes with them. That first colony expanded in population size, and at some point, additional migrations led to new colonies farther away. The new colonies would reflect a subset of the haplotypes present in the previous colony, but sometimes

Getting to Know the Neanderthals

Modern humans and Neanderthals coexisted in Europe and Asia as recently as 30,000 years ago. The human and Neanderthal ancestral populations permanently diverged about 370,000 years ago, but they are the closest known hominid relatives of modern humans. Neanderthals used tools, lived in small groups, and buried their dead. For hundreds of millennia, they inhabited large parts of Europe and western Asia (Figure 1). Buried in the bones and remains taken from burial sites are fragments of Neanderthal genomic DNA. Technologies developed for use in forensic science (see Highlight 7-1) and studies of ancient DNA have been combined to allow the complete sequencing of the Neanderthal genome. If the chimpanzee and bonobo genomes can tell us something about what it is to be human, the Neanderthal genome can tell us more.

This endeavor is unlike the genome projects aimed at extant species. The Neanderthal DNA is present in small amounts and is contaminated with DNA from other animals and bacteria. How does the researcher get at it, and how can we be certain that the sequences really came from Neanderthals? The answers have come from a new application of metagenomics (see Highlight 8-1). In essence, the small quantities of DNA fragments found in a Neanderthal bone or other remains are cloned into a library, and the cloned DNA segments are sequenced at random, contaminants and all. The sequencing results are compared with the existing human genome and chimpanzee genome databases. Segments derived from Neanderthal DNA are readily distinguished from segments derived from bacteria or insects by computerized analysis, because they have sequences closely related to human and chimpanzee DNA. Once a collection of Neanderthal DNA segments is sequenced, they are used as probes to identify sequence fragments in ancient samples that overlap with these known fragments. The potential problem of contamination with the closely related modern human DNA

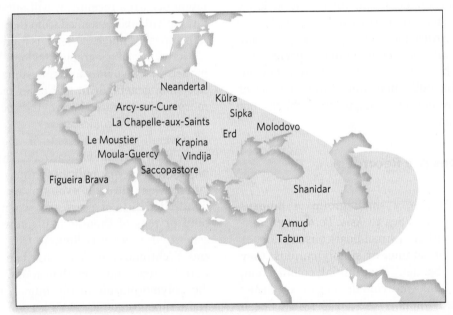

FIGURE 1 Neanderthals occupied much of Europe and western Asia until about 30,000 years ago. Major Neanderthal archaeological sites are shown here. (Note that this hominid group was named for the site at Neandertal in Germany.)

(every few thousand or tens of thousands of years) a new colony would pick up a new haplotype, due to a random new mutation, that would spread exclusively in that group (a founder event). As humans dispersed across the planet, the farthest spread (into the Americas) is characterized by the lowest overall haplotype diversity, while at the same time being marked by a few unique haplotypes picked up relatively late in the migratory process. This prevalence of key haplotypes in various populations lets us trace the paths of human migrations (**Figure 8-19** on p. 290).

Our genomic sequences also tell a story of our interaction with other hominids. During their migrations across the planet, humans encountered Neanderthals and Denisovans. The interactions were clearly complex. Some rare but identifiable interbreeding occurred between humans and these groups, and the evidence remains in our DNA.

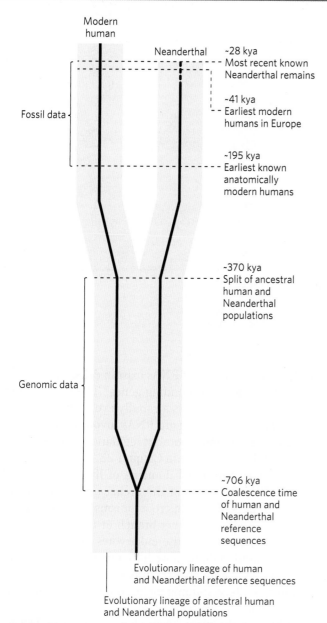

Modern human

Neanderthal

~28 kya
Most recent known
Neanderthal remains

~41 kya
Earliest modern
humans in Europe

~195 kya
Earliest known
anatomically
modern humans

Fossil data

~370 kya
Split of ancestral
human and
Neanderthal
populations

Genomic data

~706 kya
Coalescence time
of human and
Neanderthal
reference
sequences

Evolutionary lineage of human
and Neanderthal reference sequences

Evolutionary lineage of ancestral human
and Neanderthal populations

FIGURE 2 This timeline shows the divergence of human and Neanderthal genome sequences (black lines) and of ancestral human and Neanderthal populations (yellow screen). Key events in human evolution are noted (kya indicates a thousand years ago). [*Source: Data from J. P. Noonan et al., Science 314:1113–1118, 2006.*]

can be controlled for by examining mitochondrial DNA. Human populations have readily identifiable haplotypes (distinctive sets of genomic differences; see Figure 8-7) in their mitochondrial DNA, and analysis of Neanderthal samples has shown that their mitochondrial DNA has its own distinct haplotypes.

The Neanderthal genome was completed in 2013. The data provide evidence that modern humans and the Neanderthals who were the source of this DNA shared a common ancestor about 700,000 years ago (Figure 2). Analysis of mitochondrial DNA suggests that the two groups continued on the same track, with some gene flow between them, for about 300,000 more years. The lines split for good long before the appearance of anatomically modern humans, although the presence of small bits of Neanderthal DNA in human genomes indicates some relatively late interbreeding. Excluding transposons, the nucleotide sequence of the Neanderthal genome is 99.8% identical to that of the human genome. The sequence comes from a girl whose bones were left in a Siberian cave. We know that her parents were related at the level of half-siblings. Mating among close relatives was common in her immediate ancestors. We can infer that she lived in a small band of individuals, in little contact with the larger world.

Expanded libraries of Neanderthal DNA from different sets of remains should eventually allow an analysis of Neanderthal genetic diversity, and perhaps Neanderthal migrations. This look at the hominid past promises to be fascinating.

Of course, similar methods can be used to analyze the history of any species, from viruses to mammals. For example, these methods allow the tracing of viral evolution associated with human pandemics and reveal the types of mutational events that occurred in the past and are therefore possible in the future. Related methods are used to predict which influenza virus variants will be prevalent in the coming flu season, and thus guide the production of vaccines. Analysis of the genomic history of maize or rice could reveal lost genetic diversity in common production strains that might prove useful to agriculture.

The ongoing analysis of worldwide human genetic diversity—enriched by the completed genome sequencing of thousands of individuals and by new methods that incorporate information about haplotypes throughout the genome—will yield increasingly detailed information about human history. It will also aid the search for

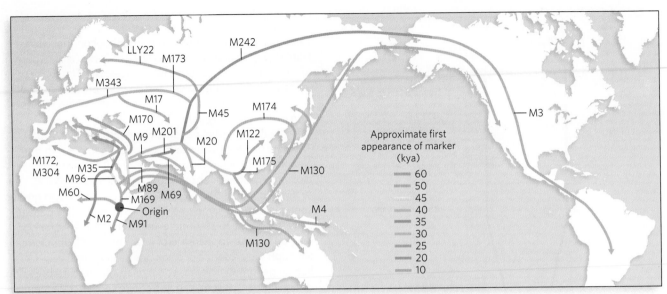

FIGURE 8-19 The paths of human migrations. This map was generated from an analysis of genetic markers (defined haplotypes with M or LLY numbers) on the Y chromosome. Genetic markers that reflect changes appearing in certain isolated populations (in "founder events") enable researchers to trace migrations from that point onward. [*Source: Data from G. Stix, Sci. Am. 299(July):56–63, 2008.*]

mutations that contribute to genetic diseases, some of which affect only certain populations. Finally, it will help pinpoint unique changes in specific populations that signal subtle adaptations to the local environment, a hallmark of ongoing evolution and a human journey that continues today.

SECTION 8.3 SUMMARY

- Genome studies have facilitated new approaches to defining the last universal common ancestor, LUCA. Genomics is used to identify genes that are common to all extant life forms and were therefore likely to exist in LUCA.
- Perhaps the ultimate challenge of genomics is to define the tree of life, a detailed description of the evolutionary history and relationships of all species now living on Earth. An evolutionary tree is referred to as a phylogeny; phylogenetics is the study of evolutionary relationships.
- The study of mitochondrial DNA, Y chromosome DNA, and genetic haplotypes in the human population enables geneticists to trace human evolution and more recent human migrations.

? UNANSWERED QUESTIONS

Genomics, transcriptomics, and proteomics are disciplines designed to obtain large amounts of information about an organism and the systems within it. The list of accomplishments is long, but the list of questions is even longer. There is a rich but largely unpredictable future in these fields.

1. How many classes of RNAs exist in cells, and how do we find the corresponding genes? The discovery of new classes of RNA is a rapidly evolving area of research. Among these are RNAs involved in all kinds of processes, most notably regulation. Some of this research is described in Chapter 22.

2. Are there additional domains of living organisms? The 1977 discovery of the archaea surprised many researchers. These microorganisms are now firmly established as a distinct branch of the evolutionary tree, separate from the eukaryotes and bacteria. Is there another domain we have missed, or even more than one? There are certainly sufficient numbers of unusual life-sustaining niches on Earth to make this plausible.

3. What are the most likely characteristics of LUCA? Ongoing endeavors to define the complete tree of life will gradually refine our understanding of how biological evolution proceeded. Aided by new sequence information, increased knowledge of mutagenesis and nonlinear events that contribute to evolution (horizontal gene transfer and transposon introduction), and complementary data from other fields (including more precise dating methods), we can expect an ever more detailed tree of life in the decades ahead. A parallel effort to better define the processes common to living systems and those that must have been present in LUCA may provide us with a look at our deepest biological past.

4. How do we investigate interdependent microbial communities? The new field of metagenomics is

starting to tackle questions about the diversity of unculturable bacterial species in environments such as the digestive tract of termites. New approaches will be needed to generate complete genome sequences of the many members of such communities and to analyze the genomes for clues about why individual species cannot survive without the others present. The human microbiome, with all of its implications for human health, is an especially important challenge.

5. What evolutionary innovations define humans as a species? Among the many subtle differences we can see between the human genome and other primate genomes are mutations of many kinds that hold the key to our capacity for higher thought, language development, and other human traits. Understanding these will advance our understanding of medicine and neurochemistry in myriad ways, some of which we cannot predict.

6. Personal genomes are in our future, but how will they be used? The science of extracting information from our genomes is still in its infancy. We don't yet understand the function of many of our own genes. The impact of personal genomes on medicine will increase as that understanding grows.

HOW WE KNOW

Haemophilus influenzae Ushers in the Era of Genome Sequences

Fleischmann, R.D., M.D. Adams, O. White, R.A. Clayton, E.F. Kirkness, A.R. Kerlavage, C.J. Bult, J.F. Tomb, B.A. Dougherty, J.M. Merrick, et al. 1995. Whole-genome random sequencing and assembly of Haemophilus influenzae Rd. Science 269:496–512.

The first genome sequencing projects, in the early 1990s, used this strategy: clone, map carefully, and then sequence. Craig Venter, who had recently established The Institute for Genome Research (TIGR), was eager to test his idea that, by using new computational methods, one could skip the time-consuming mapping steps. The result was the first complete sequence of a free-living organism—the bacterium *Haemophilus influenzae*.

H. influenzae was first described by Richard Pfeiffer in 1892 during an influenza outbreak. Until 1933, this bacterium was incorrectly thought to be the cause of the common flu. There are six types of *H. influenzae* (designated a through f) that can be immunologically distinguished by differences in their polysaccharide coat or capsule, and many other unencapsulated types. This bacterium is an opportunistic human pathogen that lives in tissues and rarely causes disease. However, *H. influenzae* type b is responsible for acute bacterial meningitis and bacteremia, primarily in children.

The organism chosen for analysis was *Haemophilus influenzae* Rd, a well-characterized type d strain often used in laboratory studies. For the purposes of Venter and his associates at TIGR, the bacterium had several advantages. As a human pathogen, it was a good target for genome sequencing. Its genome ($\sim 1.8 \times 10^6$ bp) is large, yet smaller than the genomes of other sequencing targets in use at the time. Its genomic 35% G + C content is close to that of humans, making it a good subject for developing methods for the Human Genome Project. Most important, no physical clone map existed for the *H. influenzae* genome. If the work succeeded, there would be no question that it was a victory for the overall strategy Venter had in mind.

The DNA isolated from *H. influenzae* was sheared mechanically and size-fractionated to yield random fragments of 1,600 to 2,000 bp. The fragments were cloned into a plasmid, and a library of the clones was constructed. The clones were sequenced at random. Then, 19,346 separate "forward sequence" reactions (entering the clone from the same end relative to the vector in which it was cloned) were carried out, with an 84% success rate. Just over half of the same set of clones were also sequenced in the reverse direction. The average length of DNA in each sequencing read was about 460 bp. The end result was 11,631,485 bp of DNA sequence in the random assembly.

Next, the computer algorithm tackled the immense job of assembling the genome by building a table of all 10 bp oligonucleotide subsequences and using the table to generate a list of potential fragment overlaps. With a single DNA fragment beginning the assembly of a contig, candidate overlap fragments were chosen and tested for more extended matches by strict criteria. Gradually, overlapping fragments were pieced together to generate a genome sequence. Assembling the 24,304 fragments required 30 hours of computer time. When the assembly was complete, the fragments had been ordered into 42 contigs, with 42 gaps in the genome and little information about how to order the contigs. However, many of the gaps were short. Sometimes a contig end fell within the same single gene as another contig end, both being identified by virtue of existing peptide sequences for the gene in question. Additional libraries containing long clones of *H. influenzae* DNA were probed with sequences near contig ends, to identify ends that were near each other. The gaps were closed by this method and by other targeted sequencing efforts.

By the end, the genomic sequences of the bacterium had been sequenced with more than sixfold redundancy. The final error rate was estimated at 1 in 5,000 to 10,000 bp. At $0.48 per finished base pair, the total cost was just under $900,000. Newer sequencing technologies, such as those described in Section 7.2, have lowered the cost of sequencing a typical bacterial genome by almost three orders of magnitude.

The result of Venter and colleagues' endeavor was a complete genome sequence with 1,830,137 bp, published in July 1995. The genome included 736 predicted genes, over half of which had no known function at that time. More importantly, the effort inspired a new generation of genome analysts. TIGR is now the J. Craig Venter Institute, and it remains a major force in genome sequencing. The shotgun sequencing approach successfully pioneered in the *H. influenzae* study is now routinely paired with the new sequencing technologies to provide rapid and inexpensive genome assemblies.

KEY TERMS

genome, 260
genomics, 260
contig, 261
sequence tagged site (STS), 261
expressed sequence tag (EST), 261
whole-genome shotgun sequencing, 262
genome annotation, 262
homolog, 263
ortholog, 263
paralog, 263
synteny, 264
metagenomics, 266
intron (intervening sequence), 267
exon, 267
simple-sequence repeat (SSR), 269

single nucleotide polymorphism (SNP), 269
haplotype, 269
outgroup, 271
linkage analysis, 272
systems biology, 275
transcriptome, 275
transcriptomics, 275
proteome, 276
proteomics, 276
last universal common ancestor (LUCA), 281
phylogeny, 281
phylogenetics, 281
taxa, 282
horizontal gene transfer, 284
allopatric speciation, 286

PROBLEMS

1. Three different but overlapping BAC clones (see Chapter 7) were digested with the restriction enzyme EcoRI and the fragments separated on an agarose gel, as shown in the figure below. Only the cloned DNA (not the plasmid vector) is shown. Order these three clones into a contig, and label the contig with the location of the EcoRI restriction sites and the distances between them.

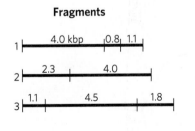
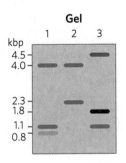

2. A researcher compares the amino acid sequences of cytochrome *c* from four vertebrates: sheep, dog, rabbit, and kangaroo. The amino acid differences among the four species are shown in the difference matrix below. From this information, build a simple evolutionary tree expressing the apparent relationships among these organisms. The branch lengths do not have to represent any measure of time.

Difference Matrix

	Sheep	Dog	Rabbit	Kangaroo
Sheep	0			
Dog	3	0		
Rabbit	4	5	0	
Kangaroo	6	7	6	0

3. In random shotgun sequencing, cloned genomic DNA from an organism is sequenced at random. Sequencing requires the use of a primer targeted to a known sequence, which can then be extended to reveal the entire sequence by the traditional Sanger method (see Chapter 7). If the researcher has no sequence information, how can any genome sequences be targeted by primers to initiate the sequencing reactions?

4. A hypothetical protein is found in orangutans, chimpanzees, and humans that has the following sequences (red indicates the amino acid residue differences; dashes indicate a deletion—the residues are missing in that sequence):

Human: ATSAAGYDEWEGGKVLIHL--KLQNRGALLELDIGAV
Orangutan: ATSAAGWDEWEGGKVLIHLDGKLQNRGALLELDIGAV
Chimpanzee: ATSAAGWDEWEGGKILIHLDGKLQNRGALLELDIGAV

What is the most likely sequence of the protein present in the last common ancestor of chimpanzees and humans?

5. For defining a cell's transcriptome, RNA-Seq provides an alternative to microarrays. In this method, cellular RNA is isolated, transcribed to complementary DNA, and sequenced.

   (a) How does the sequencing yield information about the levels of specific RNAs in a cell?

   (b) Why must rRNA be removed from most samples before conversion of the cellular RNA to cDNA?

6. A comparison between two homologous chromosomes in two closely related mammals reveals synteny over most of the length of the chromosomes. However, researchers encounter a segment of about 2,300 bp in mammal Y that is not present in mammal X. What evolutionary processes could account for this difference?

7. In large genome sequencing projects, the initial data usually reveal gaps where no sequence information has been obtained. To close the gaps, DNA primers complementary to the 5′-ending strand (i.e., identical to the sequence of the 3′-ending strand) at the end of each contig are especially useful. Explain how these primers might be used.

8. In proteomics work, two-dimensional gel electrophoresis is often used to separate the thousands of proteins in a cell on a single gel. Proteins are separated by their charge (pI) in one dimension, and then by size in the second dimension. Why do researchers use two different electrophoretic procedures, instead of simply separating proteins by size in both dimensions?

9. You have just isolated a new protein and you want to know what other proteins it might interact with. Using genetic engineering, you construct a tagged version of your protein, and then use the tag to precipitate it (sometimes called pulling it down) along with other proteins it interacts with in the cell. You subject the precipitated protein mixture to mass spectrometry and identify several hundred different peptides from over a hundred different proteins. Do all of these proteins necessarily interact functionally with your newly isolated protein? What factors might lead noninteracting proteins to precipitate with your protein?

10. You are a gene hunter, trying to find the genetic basis for a rare inherited disease. Examination of six pedigrees of families affected by the disease provides inconsistent results. For two of the families, the disease is co-inherited with markers on chromosome 7. For the other four families, the disease is co-inherited with markers on chromosome 12. Explain how this might occur.

11. In two closely related bacterial species, a cluster of five genes is found on the chromosome, with the genes arranged in the same order. However, in species B, gene X, not present in species A, is found between genes *2* and *3* of the five-gene sequence. If gene X has no sequence homology to any other gene in species B, how might it have arisen? If the nucleotide sequence of gene X is 72% identical to gene *2* in species B, how might it have arisen?

12. Native American populations in North and South America have mitochondrial DNA haplotypes that can be traced to populations in northeast Asia. The Aleut and Eskimo populations in the far northern parts of North America possess a subset of the same haplotypes that link other Native Americans to Asia, and they have several additional haplotypes that can also be traced to Asian origins but are not found in native populations in other parts of the Americas. Provide a possible explanation.

13. In the evolutionary line leading to modern humans, bottleneck periods occurred in which relatively few individuals survived. All humans living today carry mitochondrial DNA markers derived from a single female, sometimes referred to as mitochondrial Eve. All male humans living today have Y chromosome markers from an ancestor called Y chromosome Adam. Did Y chromosome Adam carry mitochondrial DNA from mitochondrial Eve?

14. DNA (haplotypes) originating from Denisovans can be found in the genomes of Australian Aborigines and Melanesian islanders. However, the same DNA markers are not found in the genomes of people native to Africa. Provide an explanation.

DATA ANALYSIS PROBLEM

Ksiazek, T.G., D.V.M. Ksiazek, D. Erdman, C.S. Goldsmith, S.R. Zaki, T. Peret, S. Emery, S. Tong, C. Urbani,

J.A. Comer, et al. 2003. A novel coronavirus associated with severe acute respiratory syndrome. *N. Engl. J. Med.* 348:1953–1966.

15. T. G. Ksiazek and other members of the SARS Working Group described their discovery of the SARS virus and its identification as a novel coronavirus. They identified the virus as a coronavirus through electron microscopy, and then confirmed it with a sequence analysis of PCR-amplified genomic segments from SARS patient samples.

    (a) The PCR method for amplifying the coronavirus genomic sample requires the use of reverse transcriptase. Why?

    (b) To amplify sequences from a new virus with an unsequenced genome, what considerations would go into the design of appropriate PCR primers?

    A sequence alignment involving a 405 bp segment of the DNA polymerase gene in four coronaviruses, BCoV, HEV, SARS CoV, and TGEV, is given on the facing page—the alignment that Ksiazek and colleagues used to generate the phylogenetic tree below (this includes coronaviruses not discussed in this problem; nt = nucleotides). In the alignment, nucleotide polymorphisms relative to the bovine coronavirus sequence (bcov) are shown in red.

    (c) How many sequence differences exist between the genome segments from HEV (hev) and BCoV (bcov)? How many between BCoV and SARS CoV (sars)? How many between TGEV (tgev) and SARS CoV? Which two coronaviruses are most closely related? Are your counts in general agreement with the phylogenetic tree?

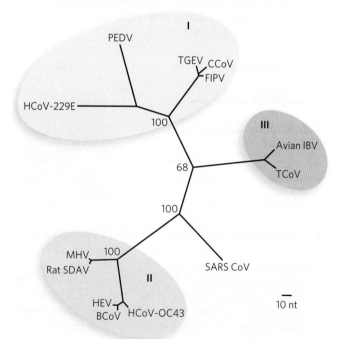

	1	60
bcov.pol.seq	TCGTGCTATGCCAAACATACTACGTATTGTTAGTAGTCTGGTTTTGGCTCGAAAACATGA	
hev.pol.seq	TCGTGCTATGCCAAACATACTACGTATTGTTAGTAGTCTGGTATTGGCCCGAAAACATGA	
sars.pol.seq	CAGAGCCATGCCTAACATGCTTAGGATAATGGCCTCTCTTGTTCTTGCTCGCAAACATAA	
tgev.pol.seq	CCGTGCTTTACCTAATATGATTAGAATGGCTTCTGCCATGATATTAGGTTCTAAGCATGT	

	61	120
bcov.pol.seq	GGCATGTTGTTCGCAAAGCGATAGGTTTTATCGACTTGCGAATGAATGCGCACAAGTTCT	
hev.pol.seq	GGCATGTTGTTCGCAAAGCGATAGGTTTTATCGACTTGCGAATGAATGCGCACAAGTTCT	
sars.pol.seq	CACTTGCTGTAACTTATCACACCGTTTCTACAGGTTAGCTAACGAGTGTGCGCAAGTATT	
tgev.pol.seq	TGGTTGTTGTACACATAATGATAGGTTCTACCGCCTCTCCAATGAGTTAGCTCAAGTACT	

	121	180
bcov.pol.seq	GAGTGAAATTGTTATGTGTGGTGGCTGTTATTATGTTAAGCCTGGTGGCACTAGTAGTGG	
hev.pol.seq	TAGTGAAATTGTTATGTGTGGTGGCTGTTATTATGTTAAGCCTGGTGGCACTAGTAGTGG	
sars.pol.seq	AAGTGAGATGGTCATGTGTGGCGGCTCACTATATGTTAAACCAGGTGGAACATCATCCGG	
tgev.pol.seq	CACAGAAGTTGTGCATTGCACAGGTGGTTTTTATTTTAAACCTGGTGGTACAACTAGCGG	

	181	240
bcov.pol.seq	TGATGCAACTACTGCTTTTGCTAATTCAGTTTTTAACATATGTCAAGCTGTTTCAGCCAA	
hev.pol.seq	TGATGCAACTACTGCTTTTGCTAATTCAGTCTTTAACATATGTCAAGCTGTTTCAGCCAA	
sars.pol.seq	TGATGCTACAACTGCTTATGCTAATAGTGTCTTTAACATTTGTCAAGCTGTTACAGCCAA	
tgev.pol.seq	TGATGGTACTACAGCATATGCTAACTCTGCTTTTAACATCTTTCAAGCTGTTTCTGCTAA	

	241	300
bcov.pol.seq	TGTATGTGCTTTAATGTCATGCAATGGTAATAAGATTGAAGATTTGAGTATACGTGCTCT	
hev.pol.seq	TGTATGTTCCTTAATGTCATGCAATGGCAATAAGATTGAAGATTTGAGTATACGTGCTCT	
sars.pol.seq	TGTAAATGCACTTCTTTCAACTGATGGTAATAAGATAGCTGACAAGTATGTCCGCAATCT	
tgev.pol.seq	TGTTAATAAGCTTTTGGGGGTTGATTCAAACGCTTGTAACAACGTTACAGTAAAATCCAT	

	301	360
bcov.pol.seq	TCAGAAGCGCTTATACTCACATGTGTATAGAAGTGATATGGTTGATTCAACCTTTGTCAC	
hev.pol.seq	TCAGAAGCGTTTATACTCACATGTGTATAGAAGTGATATGGTTGATTCAACCTTTGTCAC	
sars.pol.seq	ACAACACAGGCTCTATGAGTGTCTCTATAGAAATAGGGATGTTGATCATGAATTCGTGGA	
tgev.pol.seq	ACAACGTAAAATTTACGATAATTGTTATCGTAGTAGCAGCATTGATGAAGAATTTGTTGT	

	361	406
bcov.pol.seq	AGAATATTATGAATTTTTAAATAAGCATTTTAGTATGATGATTTTG	
hev.pol.seq	AGAATATTATGAATTTTTAAATAAGCATTTTAGTATGATGATTTTG	
sars.pol.seq	TGAGTTTTACGCTTACCTGCGTAAACATTTCTCCATGATGATTCTT	
tgev.pol.seq	TGAGTACTTTAGTTATTTGAGAAAACACTTTTCTATGATGATTTTA	

[Source: Sequence alignment courtesy of Dr. Ann Palmenberg, Department of Biochemistry and the Institute for Molecular Virology, University of Wisconsin–Madison.]

ADDITIONAL READING

Genomes and Genomics

Cordero, O.X., and M.F. Polz. 2014. Explaining microbial genomic diversity in light of evolutionary ecology. *Nat. Rev. Microbiol.* 4:263–273.

Fitch, W.M. 1970. Distinguishing homologous from analogous proteins. *Syst. Zool.* 19:99–114. The paper that introduced the concept of orthologs and paralogs.

Giallourakis, C., C. Henson, M. Reich, X. Xie, and V.K. Mootha. 2005. Disease gene discovery through integrative genomics. *Annu. Rev. Genomics Hum. Genet.* 6:381–406. A good summary of how these important searches are done.

Koonin, E.V. 2005. Orthologs, paralogs, and evolutionary genomics. *Annu. Rev. Genet.* 39:309–338.

Mardis, E.R. 2008. Next-generation DNA sequencing methods. *Annu. Rev. Genomics Hum. Genet.* 9:387–402.

Marques-Bonet, T., O.A. Ryder, and E.E. Eichler. 2009. Sequencing primate genomes: What have we learned? *Annu. Rev. Genomics Hum. Genet.* 10:355–386.

Prüfer, K., K. Munch, I. Hellmann, K. Akagi, J.R. Miller, B. Walenz, S. Koren, G. Sutton, C. Kodira, R. Winer, et al. 2012. The bonobo genome compared with the chimpanzee and human genomes. *Nature* 486:527–531.

Rasmussen, M., Y. Li, S. Lindgreen, J.S. Pedersen, A. Albrechtsen, I. Moltke, M. Metspalu, E. Metspalu, T. Kivisild, R. Gupta, et al. 2010. Ancient human genome sequence of an extinct Palaeo-Eskimo *Nature* 463:757–762.

Rusch, D.B., A.L. Halpern, G. Sutton, K.B. Heidelberg, S. Williamson, S. Yooseph, D. Wu, J.A. Eisen, J.M. Hoffman, K. Remington, et al. 2007. The Sorcerer II Global Ocean Sampling expedition: Northwest Atlantic through eastern tropical Pacific. *PLoS Biol.* 5:e77. Every molecular biologist would enjoy this kind of work.

Sikela, J.M. 2006. The jewels of our genome: The search for the genomic changes underlying the evolutionarily unique capacities of the human brain. *PLoS Genet.* 2:e80. A favorite subject for every human.

Transcriptomes and Proteomes

Altelaar, A.F.M., J. Munoz, and A.J.R. Heck. 2013. Next-generation proteomics: Towards an integrative view of protein dynamics. *Nat. Rev. Genet.* 14:35–48.

Bensimon, A., A.J.R. Heck, and R. Aebersold. 2012. Mass spectrometry-based proteomics and network biology. *Annu. Rev. Biochem.* 81:379–405.

Ly, T., Y. Ahmad, A. Shlien, D. Soroka, A. Mills, M.J. Emanuele, M.R. Stratton, and A.I. Lamond. 2014. A proteomic chronology of gene expression through the cell cycle in human myeloid leukemia cells. *eLife* 3:e01630.

Martin, J.A., and Z. Wang. 2011. Next-generation transcriptome assembly. *Nat. Rev. Genet.* 12:671–682.

Nie, L., G. Wu, D.E. Culley, J.C. Scholten, and W. Zhang. 2007. Integrative analysis of transcriptomic and proteomic data: Challenges, solutions and applications. *Crit. Rev. Biotechnol.* 27:63–75.

Wang, Z., M. Gerstein, and M. Snyder. 2009. RNA Seq: A revolutionary tool for transcriptomics. *Nat. Rev. Genet.* 10:57–63.

Our Genetic History

Acevedo-Rocha, C.G., G. Fang, M. Schmidt, D.W. Ussery, and A. Danchin. 2013. From essential to persistent genes: A functional approach to constructing synthetic life. *Trends Genet.* 29:273–279.

Cavalli-Sforza, L.L. 2007. Human evolution and its relevance for genetic epidemiology. *Annu. Rev. Genomics Hum. Genet.* 8:1–15.

Ciccarelli, F.D., T. Doerks, C. von Mering, C.J. Creevey, B. Snel, and P. Bork. 2006. Toward automatic reconstruction of a highly resolved tree of life. *Science* 311:1283–1287. Building an evolutionary tree based on sequences of 36 genes in 191 species, with methods that eliminate the effects of horizontal gene transfer.

Forster, A.C., and G.M. Church. 2006. Towards synthesis of a minimal cell. *Mol. Syst. Biol.* 2:45. An update on efforts to synthesize a living system from chemical components.

Galperin, M.Y. 2006. The minimal genome keeps growing. *Environ. Microbiol.* 8:569–573. A description of a project to define the minimum number of genes required for independent life.

Hellenthal, G., G.B.J. Busby, G. Band, J.F. Wilson, C. Capelli, D. Falush, and S. Myers. 2014. A genetic atlas of human admixture history. *Science* 343:747–751.

Li, R., Y. Li, H. Zheng, R. Luo, H. Zhu, Q. Li, W. Qian, Y. Ren, G. Tian, J. Li, et al. 2010. Building the sequence map of the human pan-genome. *Nat. Biotechnol.* 28:57–63.

Prüfer, K., F. Racimo, N. Patterson, F. Jay, S. Sankararaman, S. Sawyer, A. Heinze, G. Renaud, P.H. Sudmant, C. de Filippo, et al. 2012. The complete genome sequence of a Neanderthal from the Altai mountains. *Nature* 505:43–49.

Rannala, B., and Z. Yang. 2008. Phylogenetic inference using whole genomes. *Annu. Rev. Genomics Hum. Genet.* 9:217–231.

Wade, C.H., B.A. Tarini, and B.S. Wilfond. 2013. Growing up in the genomic era: Implications of whole-genome sequencing for children, families, and pediatric practice. *Annu. Rev. Genomics Hum. Genet.* 14:535–555.

9

Topology: Functional Deformations of DNA

MOMENT OF DISCOVERY

Carlos Bustamante *[Source: Courtesy Carlos Bustamante.]*

There was an experiment I had wanted to do for many years, but I could never convince anyone to try it. *The idea was to measure the elastic properties of DNA directly using a single molecule of DNA tethered between two opposing pipette tips, such that it could be exquisitely controlled by rotating one pipette tip relative to the other to twist the DNA to different extents.* In solution, DNA can *twist* on its long axis, and it can also *writhe* by coiling around itself—and it can be very hard to decouple twisting from writhing by measuring properties of DNA in bulk.

After many students turned me down, eventually two students, Zev Bryant and Michael Stone, were intrigued to try the DNA twisting experiment. These guys worked very hard setting up the technical aspects of the experiment. They figured out how to tether the ends of a nicked fragment of DNA to two opposing pipette tips, and they coupled a readily visualized small bead (the "rotor") to an internal position in the nicked DNA duplex. They finally got everything working late one night. Zev and Michael started using a hand crank to introduce a specific number of twists into the tethered piece of DNA. But they were so tired that they would get up to, say, 345 turns, and then they weren't sure if it was 345 or 346! They were determined to do the experiment accurately, so they had to untwist the DNA and start over. But Jan Liphardt, another student in the lab, had an idea. He offered to bring in a small motor from his Lego set at home to rotate the pipette tip, and thus twist the DNA, automatically. So Jan ran home and got the motor, hooked it up to the system, and it worked beautifully. All the data we ultimately published were measured using the Lego motor (we even listed it in the Methods section of the paper). And we learned that DNA is about 50% stiffer than had been previously estimated from bulk solution experiments!

—*Carlos Bustamante, on discovering the elasticity of DNA*

In all free-living organisms—bacteria, archaea, and eukaryotes—genomic information comes in the form of DNA; RNA genomes occur only in some classes of viruses. It takes a lot of DNA to encode all the proteins and RNAs needed in a living organism. In fact, chromosomes are *much* (orders of magnitude) longer than the biological packages—the cells, organelles, or viral particles—that contain them. Most human cells are 7 to 30 μm (micrometers, also called microns) in diameter. The nuclei that house the DNA molecules are rarely more than 10 μm in diameter. The shortest human chromosome (chromosome 21), at just under 47 million base pairs, would be about 16 mm long if stretched out in a line, or more than a thousand times longer than the nucleus. If all of the chromosomes in a diploid human cell were laid end to end, they would stretch for nearly 2 m.

Before we explore the enzymes of DNA replication, repair, recombination, and transcription and their regulation in chapters to come, we need to consider the chromosome—the stage on which these processes play out. The sheer size of chromosomes presents every cell with a fundamental problem: the DNA must be compacted, but the information within it must remain accessible. Compaction (also called condensation) of DNA is extensive, but never random. Enzymes called topoisomerases collaborate with a range of DNA-binding proteins to create layers of chromosomal structure in an orderly and highly dynamic process. Additional enzymes render parts of that structure accessible as needed. The architecture and the directed malleability of chromosomes affect nearly every other aspect of molecular biology.

As we will see, DNA is compacted primarily by coiling. The process is somewhat analogous to rolling up a length of garden hose, an old-fashioned phone cord, or a spool of electrical wire. However, the coiling of DNA occurs in the context of structural constraints peculiar to this nucleic acid, constraints that are dealt with by a unique set of proteins and enzymes.

Building on our discussion of DNA secondary structure in Chapter 6, in this chapter and the next we address the tertiary structure and function of chromosomes. First, in this chapter, we introduce the extraordinary degree of organization required for the tertiary packaging of DNA into chromosomes. We explore the principles related to the compaction of DNA, beginning with a review of the structural elements that make up viral and cellular chromosomes, and consider chromosomal size and organization in more detail. We then discuss DNA topology, for a quantitative description of the coiling and supercoiling of DNA molecules. We conclude the chapter with a discussion of the key enzymes found in all cells that are involved in creating and maintaining a very high order of compaction. As in all other areas of molecular biology, this information is not merely of

academic interest. Many of these enzymes are important targets of antibiotics and other medicines. In Chapter 10, we expand on this discussion of tertiary packaging by examining the complete structure of chromosomes in the context of the structural DNA-binding proteins unique to eukaryotes and bacteria.

9.1 CHROMOSOMES: AN OVERVIEW

Chromosomes are typically the largest macromolecules in any cell, by a considerable margin. Why must chromosomes be so large? They contain the blueprints for an organism, and thus a great deal of information is contained within them. But the genes in each chromosome constitute only part of that information. The chromosomes themselves are macromolecular entities that must be synthesized, packaged, protected, and properly distributed to daughter cells at cell division. Significant segments of every chromosome are dedicated to these functions. All aspects of chromosome function are affected by the reality of chromosome size.

Chromosome Function Relies on Specialized Genomic Sequences

The chromosomes of cells and viruses come in several forms. Bacterial chromosomes are often circular (in the sense of an endless loop rather than a perfect circle). Most eukaryotic chromosomes are linear. In viruses we find additional variations, including both single-stranded and double-stranded forms, as well as RNA genomes. Each type of chromosome structure imposes a unique set of demands on the mechanisms for replicating and transmitting the genome from one generation to the next.

Genes provide the information to specify all the RNAs and proteins produced in a given cell, but other DNA sequences in the genome are dedicated to the maintenance of the chromosome itself: initiation and termination of replication, segregation during cell division, and, where necessary, protection and maintenance of the chromosome ends. In bacteria, an origin of replication provides a start site for chromosomal replication (see Figure 8-1). Specialized replication-termination regions also exist in most known bacterial species. Within or near these regions, additional sequences serve as binding sites for proteins that ensure the faithful segregation of replicated chromosomes to daughter cells. Eukaryotic chromosomes, too, contain sequences that are critical to chromosome maintenance. Unlike bacteria, eukaryotic chromosomes often have many replication origins. (The structure and function of replication origins are discussed in Chapter 11.) Eukaryotic chromosomes also have specialized DNA sequences called centromeres and telomeres.

Metaphase chromosome

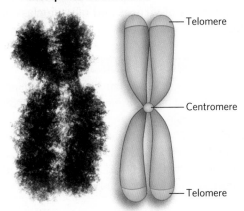

FIGURE 9-1 Linked and condensed sister chromatids of a human chromosome. The products of chromosomal replication in eukaryotes are linked sister chromatids. These are fully condensed at metaphase, during mitosis. The point where they are joined is the centromere. Telomeres are sequences at the ends of the chromatids. [Source: Don W. Fawcett/Science Source.]

TABLE 9-1

Telomere Sequences

Species	Telomere Repeat Sequence	n^*
Homo sapiens (human)	$(TTAGGG)_n$	800–2,500
Tetrahymena thermophila (ciliated protozoan)	$(TTGGGG)_n$	40
Saccharomyces cerevisiae (yeast)	$((TG)_{1-3}(TG)_{2-3})_n$	50–75
Arabidopsis thaliana (plant)	$(TTTAGGG)_n$	300–1,200

*Number of telomere repeats. Telomere length is longer and fluctuates over a wider range in multicellular eukaryotes. In vertebrates, including humans, telomere length declines with the age of the organism in most cells, but not in germ-line cells.

The **centromere** is a segment of each eukaryotic chromosome that functions during cell division as an attachment point for proteins that link the chromosome to the mitotic spindle at metaphase (**Figure 9-1**). This attachment is essential for the equal and orderly distribution of chromosome sets to daughter cells. (See Chapter 2 for a review of the events of mitosis.) The centromeres of *Saccharomyces cerevisiae* have been isolated and studied. The sequences essential to centromere function are about 130 bp long and very rich in A=T base pairs. The centromere sequences of higher eukaryotes are much longer and, unlike those of yeast, generally contain regions of simple-sequence DNA consisting of thousands of tandem copies of one or a few 5 to 10 bp sequences. This DNA serves as a binding site for centromere-binding proteins, or cen proteins. The centromere is also the site of kinetochore assembly. Built up on each centromere, the kinetochore anchors the spindle fibers as chromosomes are segregated into daughter cells during mitosis. Centromeres thus play a key role in stable chromosome segregation during cell division.

Telomeres are sequences at the ends of eukaryotic chromosomes that add stability by protecting the ends from nucleases and providing unique mechanisms for the faithful replication of linear DNA molecules. DNA polymerases cannot synthesize DNA to the very ends of a linear chromosome (see Chapter 11). Solving the end-replication problem is one key function of telomeres, which are replicated by the enzyme telomerase. Telomeres end with repeated sequences of the form

$$5'\text{-}(T_xG_y)_n$$
$$3'\text{-}(A_xC_y)_n$$

where x and y are generally between 1 and 4 (**Table 9-1**) and the number of telomere repeats, n, is in the range of 20 to 100 for most single-celled eukaryotes and generally exceeds 1,500 in mammals. As in centromeres, the telomere repeats serve as binding sites for specialized proteins that are part of telomere function. These proteins package the telomeres and help maintain them in actively dividing cells (see Chapter 11).

Artificial chromosomes provide a means of better understanding the functional significance of many structural features of eukaryotic chromosomes. A reasonably stable artificial linear chromosome requires only three components: a centromere, a telomere at each end, and an appropriate number of replication origins. Yeast artificial chromosomes (YACs) have been developed as a research tool in biotechnology (see Figure 7-7). YACs have also been useful in confirming the critical functions of centromeres and telomeres. Building on this foundation, researchers have constructed human artificial chromosomes (HACs). HACs are reasonably stable when introduced into a human tissue culture cell line, if they include human centromere and telomere sequences in addition to active replication origins.

Continued development of HACs, particularly of their efficient introduction into human cells, may eventually provide new avenues for the treatment of genetic diseases. Most genetic diseases can be traced to an alteration in one or more particular genes that changes or eliminates their function. Efforts to directly remove such genes in human cells and replace them with normal, functional versions at the correct chromosomal locus have met with limited success. Nonetheless, there are ongoing efforts to improve this process of correcting disease-causing genetic errors in somatic cells, a process termed somatic gene therapy. The CRISPR/Cas9 system, described in Section 7.3, provides a promising new approach to this effort. A simpler and more traditional path for gene therapy is to introduce the functional genes into random locations on chromosomes through recombination mechanisms (see Chapters 13 and 14). However, this technique has a number of problems. The inserted gene can run afoul of regulatory mechanisms that suppress gene expression over large segments of a chromosome, effectively silencing any new gene that is inserted there. Random integration can also result in insertion into the coding sequence of another gene, inactivating that gene. If the inactivated gene has a role in the regulation of cell division, uncontrolled cell division and tumor development can result. The introduction of functional gene copies on stable HACs may eventually circumvent these problems. Success will depend on further clarifying the mechanisms by which chromosomes are stably maintained in cells, and on the development of more efficient procedures for introducing large DNAs into the nuclei of a large number of cells in a living human being.

Chromosomes Are Longer Than the Cellular or Viral Packages Containing Them

The observation that genomic DNAs are orders of magnitude longer than the cells or viruses that contain them applies to every class of organism and viral parasite. Lengths of double-stranded nucleic acids are often described in terms of contour length, or the length measured along the axis of the double-helical DNA. For a closed-loop DNA, contour length is the circumference the chromosome would have if it were laid out in a perfect circle. Lengths are more difficult to describe for a single-stranded nucleic acid, particularly when segments of that nucleic acid fold up into secondary structures. These lengths are sometimes approximated by assuming that the single strand is arrayed in the helical path that would be described by one strand of a double helix, then measuring along the resulting axis.

Given the magnitude of the one-dimensional length of a typical chromosome, how can it be accommodated within the three-dimensional volume of a viral particle, cell, or nucleus? The compaction mechanisms required for this are highly conserved across the spectrum of living systems. Compaction entails the coiling and structural organization of the chromosome resulting from the action of enzymes; the structural organization is maintained by DNA-binding proteins, including the histones of eukaryotic chromosomes (see Chapter 10), the DNA-binding proteins of bacteria, and

the coat proteins of viral particles. We consider here the chromosomes of viruses and of each class of living organism.

> ### KEY CONVENTION
>
> Molecular biology involves structures with dimensions that are small fractions of a meter. One-thousandth of a meter is 1 millimeter (mm); 1 mm = 1,000 μm (micrometer, or micron) = 1,000,000 nm (nanometer) = 10,000,000 Å (angstrom). Nucleotides, segments of chromosomes, and cells are most often discussed in terms of angstroms, nanometers, and micrometers, respectively.

Viruses Viruses are not free-living organisms; they are infectious parasites that use the resources of a host cell to carry out many of the processes they require to propagate. Many viral particles consist of no more than a genome (usually a single RNA or DNA molecule) surrounded by a protein coat.

Almost all plant viruses and some bacterial and animal viruses have RNA genomes, and they are quite small. For example, the genomes of mammalian retroviruses, such as HIV, have about 9,000 nucleotides, and the genome of the bacteriophage Qβ has 4,220 nucleotides. However, even these small nucleic acids have total lengths of about 3 and 1.4 μm, respectively. In comparison, the protein coat of HIV is about 100 nm in diameter, and that of Qβ is about 26 nm, so the RNAs are 30 to 50 times longer than their viral protein coats. Both types of virus have linear, single-stranded RNA genomes. Some of the viral coat proteins are effectively RNA-binding proteins, and by binding to the genome they enforce a highly compacted folding arrangement of the RNA within the viral particle. An example can be seen in the tobacco mosaic virus (TMV), a pathogen of tobacco plants. The single-stranded RNA genome of TMV, 6,400 nucleotides long, is wound into a tight left-handed helix by its packaging within the rodlike helical protein coat (**Figure 9-2a**).

The genomes of DNA viruses vary greatly in size and form (see Table 8-1), but all are longer than the viral capsid heads that enclose them. Many viral DNAs are circular for at least part of their life cycle. During viral replication inside a host cell, specific types of viral DNA, called replicative forms, may appear; for example, many linear DNAs become circular, and all single-stranded DNAs become double-stranded. Bacteriophage T2 has a double-stranded linear DNA genome of 160,000 bp, a molecule more than 50 μm long that must be packaged into a virus head about 100 nm across in its longest dimension (**Figure 9-2b**). Bacteriophage φX174 is much smaller, a 5,386 nucleotide single-stranded circle (about 1.9 μm long) enclosed within a capsid 25 nm in diameter.

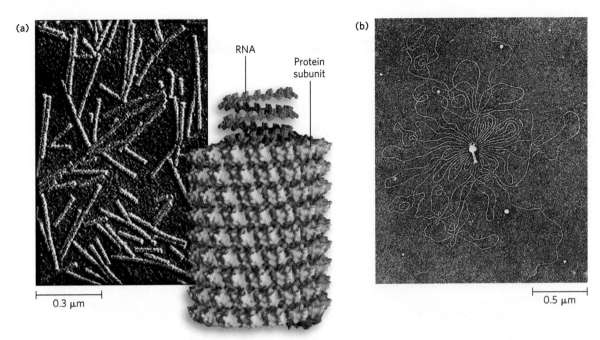

FIGURE 9-2 Genome packaging in a virus. (a) The tobacco mosaic virus has an RNA genome coiled inside a rod-shaped viral coat, packaged by RNA-binding proteins, as shown in an electron micrograph and molecular model. (b) A bacteriophage T2 particle was lysed and its DNA allowed to spread on the surface of distilled water in this electron micrograph. All the DNA shown here is normally packaged inside the phage head. *[Sources: (a) Science Source. (b) Reprinted from Kleinschmidt, A.K., Land, D., Jackerts, D., & Zahn, R.K. (1962) Biochem. Biophys. Acta 61, pp. 857–864, with permission from Elsevier.]*

Table 9-2 summarizes the genome and particle dimensions for several DNA viruses.

Bacteria A single *E. coli* cell contains almost 100 times more DNA than a bacteriophage λ particle (see Table 9-2). The chromosome of the most common *E. coli* strain studied in laboratories worldwide (K-12 MG1655) is a single, double-stranded, circular DNA molecule (**Table 9-3**). Its 4,639,221 bp have a contour length of about 1.7 mm, some 850 times the length of the *E. coli* cell, 2 μm. In

addition to the very large, circular DNA chromosome, many bacteria contain one or more **plasmids**, much smaller circular DNA molecules that are free in the cytosol (**Figure 9-3**; see also Chapter 7). Most plasmids are only a few thousand base pairs long, but some have more than 100,000 bp. Most do not encode genes essential to their host, but they are often symbiotic. They carry genetic information and undergo replication to yield daughter plasmids, which pass into the daughter cells at cell division. The spread of bacterial plasmids (and

TABLE 9-2

The Sizes of DNA and Viral Particles for Some Bacteriophages

Virus	Number of Base Pairs in Viral DNA*	Length of Viral DNA (nm)	Long Dimension of Viral Particle (nm)†	Chromosome Form
φX174	5,386	1,939	25	Circular
T7	39,936	14,377	78	Linear
λ (lambda)	48,502	17,460	190	Linear
T4	168,889	60,800	210	Linear

*Data are for the replicative form (double-stranded). The φX174 chromosome is single-stranded inside the viral particle. The bacteriophage λ chromosome is circularized after it enters a host cell. Calculation of contour length assumes 3.4 Å per base pair (see Figure 6-14).
†This measurement includes the head and the tail, where relevant.

TABLE 9-3			
DNA, Gene, and Chromosome Content in Some Genomes			
Species	*Total DNA (bp)*	*Chromosomes**	*Genes*
Escherichia coli (bacterium)	4,600,000	1	~4,300
Saccharomyces cerevisiae (yeast)	12,068,000	16†	~5,800
Caenorhabditis elegans (nematode)	97,000,000	12‡	~19,000
Drosophila melanogaster (fruit fly)	180,000,000	18	~13,600
Arabidopsis thaliana (plant)	125,000,000	10	~25,500
Oryza sativa (plant)	480,000,000	24	~57,000
Mus musculus (mouse)	2,500,000,000	40	~26,000–29,000
Homo sapiens (human)	~3,200,000,000	46	~25,000

*Diploid chromosome number for all eukaryotes except yeast.
†Haploid chromosome number; wild yeast strains generally have eight (octoploid) or more sets of chromosomes.
‡Number for females, with two X chromosomes; males have an X but no Y, for 11 total.

transposons) that confer antibiotic resistance among pathogenic bacteria has reduced the utility of standard antibiotics in medicine and agriculture (**Highlight 9-1**).

Eukaryotes A yeast cell, one of the simplest eukaryotes, has 2.6 times more DNA in its genome than an *E. coli* cell (see Table 9-3). Cells of *Drosophila melanogaster*, the fruit fly used in classical genetics studies, contain more than 35 times as much DNA as *E. coli* cells, and human cells have almost 700 times as much. The cells of many plants and amphibians contain even more.

All of this DNA must fit into a eukaryotic cell that is typically 10 to 20 μm across (although size can vary greatly, even within a single organism). The genetic material of eukaryotic cells is apportioned into multiple chromosomes, the diploid ($2n$) number depending on the species. A human somatic cell, for example, has 46 chromosomes (**Figure 9-4**). Each chromosome of a eukaryotic cell contains a single, very large, duplex DNA molecule. The DNA molecules in the 24 different types of human chromosomes (22 matching pairs plus the X and Y sex chromosomes) vary in length over a

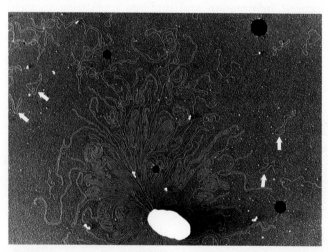

FIGURE 9-3 DNA from a lysed *E. coli* cell. In this electron micrograph, white arrows indicate several small, circular plasmid DNAs. The black spots and white specks are artifacts of the preparation. [*Source: Huntington Potter, University of Colorado, Anschutz Medical Campus and David Dressler, Balliol College, Oxford University.*]

FIGURE 9-4 Eukaryotic chromosomes. This is a complete set of chromosomes from a leukocyte of one of the authors. There are 46 chromosomes in every normal human somatic cell. [*Source: Courtesy Michael M. Cox.*]

The Dark Side of Antibiotics

Over the course of the twentieth century, the average life expectancy for people in the developed countries increased by 10 years, and the development of antibiotics for the treatment of infectious diseases was a major contributor to this improved longevity. Ironically, an overuse of antibiotics is now leading to their demise as useful therapeutics, as bacterial pathogens evolve to develop antibiotic resistance.

The most common vehicles for transmitting antibiotic-resistance elements between bacterial populations are plasmids, and large numbers are present in the environment. Some plasmids confer no obvious advantage on their host, and their sole function seems to be self-propagation. However, many plasmids carry genes that are useful to the host bacterium. These may include genes that extend the range of environments that can be exploited by the host, such as conferring resistance to naturally occurring antibiotics or conferring new metabolic properties or the ability to synthesize toxins or agents that facilitate tissue colonization—and thus make the host bacterium pathogenic to other organisms. Given that most antibiotics are natural products (e.g., penicillin is derived from the mold *Penicillium notatum*), it is not surprising that genes conferring antibiotic resistance occur in natural bacterial populations.

Plasmids contain a range of sequences involved in their own propagation. These sequences often function in several related bacterial species, and the host range can increase with the aid of small numbers of mutations. When genes conferring antibiotic resistance are integrated into a plasmid, the plasmid becomes a vehicle for transferring the resistance element from one bacterium to another, and even between species. For example, plasmids carrying the gene for the enzyme β-lactamase confer resistance to β-lactam antibiotics, such as penicillin, ampicillin, and amoxicillin. Transfer of plasmids to other bacteria can occur through horizontal gene transfer, in which plasmids pass from an antibiotic-resistant cell to an antibiotic-sensitive cell of the same or another bacterial species (see Figure 1-11). This can occur when cells of a resistant strain die and rupture, releasing their DNA into the environment. If an antibiotic-sensitive strain or species takes up the DNA, it may acquire the antibiotic resistance. In some cases, the antibiotic-resistance gene is located on a conjugational plasmid (see Chapter 13) that encodes its own machinery for transfer from one bacterium to another. The transfer of antibiotic resistance between bacteria becomes particularly efficient on such plasmids. Many antibiotic-resistance elements are also harbored in transposons, which can move from cellular chromosomes to plasmids and back again, further facilitating the dispersal of these elements.

Under the strong selective pressure brought about by widespread antibiotic treatments, bacterial pathogens can acquire antibiotic resistance rapidly. The extensive use of antibiotics in some human populations has encouraged the spread of antibiotic resistance–coding plasmids (as well as transposable elements that harbor similar genes) in disease-causing bacteria. Physicians are becoming increasingly reluctant to prescribe antibiotics unless a clear medical need is confirmed. For similar reasons, the widespread use of antibiotics in animal feeds is being curbed.

25-fold range. Each type of chromosome in eukaryotes carries a characteristic set of genes.

Eukaryotic cells also have organelles that contain DNA. Mitochondria and chloroplasts carry their own genomic DNAs **(Figure 9-5)**. The evolutionary origin of mitochondrial and chloroplast DNAs has been the subject of much speculation. A widely accepted hypothesis, proposed by Lynn Margulis, is that they are vestiges of the chromosomes of ancient bacteria that gained access to the cytoplasm of host cells and became the precursors of these organelles (see Figure 8-17a). Mitochondrial DNA (mtDNA) codes for mitochondrial tRNAs and rRNAs and a few

Lynn Margulis, 1938–2011
[Source: Nancy R. Schiff/Getty Images.]

mitochondrial proteins; more than 95% of mitochondrial proteins are encoded by nuclear DNA. Mitochondria and chloroplasts divide when the cell divides. Their DNA is replicated before and during cell division, and the daughter DNA molecules pass into the daughter organelles.

Mitochondrial DNA molecules are much smaller than nuclear chromosomes. In animal cells, mtDNA contains fewer than 20 kbp (16,569 bp in human mtDNA) and is a circular duplex. Each mitochondrion typically has 2 to 10 copies of the mtDNA, but the number can be much higher: hundreds in muscle cells, and 100,000 in a mature oocyte. In a few organisms (e.g., trypanosomes, the parasites that cause sleeping sickness), the mitochondrial DNA is particularly abundant and organized. These mitochondria contain thousands of copies of mtDNA organized into a complex interlinked matrix known as a kinetoplast. Plant cell mtDNA is much larger than that in animal cells, ranging from 200 to 2,500 kbp. Chloroplast DNA (cpDNA) exists as circular duplexes of 120 to 160 kbp. Organelle DNA,

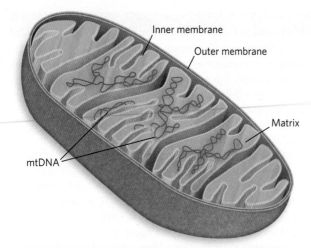

FIGURE 9-5 Mitochondrial DNA. Some mitochondrial proteins and RNAs are encoded by the multiple copies of mtDNA in the mitochondrial matrix. The mtDNA is replicated each time the organelle divides, before cell division.

like nuclear DNA, undergoes considerable compaction: DNA molecules 5 to 500 μm long must be accommodated in organelles about 1 to 5 μm in diameter.

Some eukaryotes also contain plasmids; they have been found in yeast and some other fungi.

SECTION 9.1 SUMMARY

- All cellular chromosomes contain sequences required for chromosome function, including replication origins. Bacterial chromosomes also contain termination sequences and other sequences necessary for chromosomal segregation during mitosis.
- Eukaryotic chromosomes contain centromeres, which are attachment points for the mitotic spindle, and telomeres, specialized sequences at the ends of a chromosome that protect and stabilize the entire chromosome.
- All genomic DNA and RNA molecules are longer—often orders of magnitude longer—than the viral coats, organelles, and cells in which they are packaged.
- Viral genomes vary in nucleic acid (DNA or RNA), structure (single-stranded or double-stranded), and length.
- Bacterial cells contain both genomic DNA (usually circular) and plasmids; both types are compacted in the cell and are replicated and segregated into daughter cells at cell division.
- Eukaryotic chromosomes are linear and vary in number, depending on the species. Humans have 46 chromosomes, varying in length and condensed to fit into the cell nucleus. Mitochondria and chloroplasts contain their own circular genomes, in numbers ranging from several copies to hundreds of thousands of copies per organelle, depending on cell type.

9.2 DNA SUPERCOILING

Cellular DNA, as we have seen, is a very long molecule that must somehow be made to fit inside the cell, implying a high degree of structural organization. The folding mechanism must not only pack the DNA but permit access to the information in the DNA in processes such as replication and transcription. Any consideration of how this is accomplished requires an understanding of an important property of DNA structure known as supercoiling.

Supercoiling simply means the coiling of a coil. An old-fashioned telephone cord, for example, is typically a coiled wire. If one end of the cord is manually twisted, the coiled cord forms supercoils (**Figure 9-6**). DNA is coiled in the form of a double helix, with both strands coiling around an axis. **DNA supercoiling** is the further coiling of that axis upon itself (**Figure 9-7**). As described below, **supercoiled DNA** (sometimes called superhelical DNA) is generally a manifestation of structural strain. The DNA can be overwound to generate supercoils. However, underwinding of the DNA also imparts strain and results in supercoils. When there is no net coiling of the DNA axis upon itself, the molecule is referred to as **relaxed DNA** (see the How We Know section at the end of this chapter). Supercoiling occurs in all chromosomal DNAs in all cells, as well as in viruses that have a double-stranded DNA genome or generate double-stranded DNA as a replication intermediate.

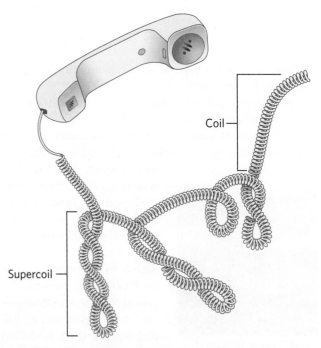

FIGURE 9-6 Supercoils. An old-fashioned phone cord is coiled like a DNA helix, and the coiled cord can supercoil.

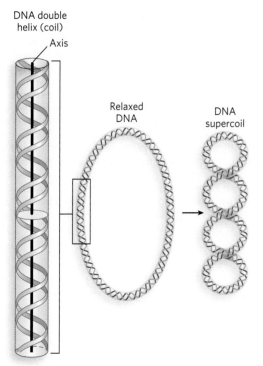

DNA double
helix (coil)

Axis

Relaxed
DNA

DNA
supercoil

FIGURE 9-7 DNA supercoiling. When the DNA double helix is coiled on itself, it forms a new helix, or superhelix. The DNA superhelix is usually referred to as a supercoil. *[Source: Data from N. R. Cozzarelli, T. C. Boles, and J. H. White, in* DNA Topology and Its Biological Effect *(N. R. Cozzarelli and J. C. Wang, eds), Cold Spring Harbor Laboratory Press, 1990, pp. 139–184.]*

It would be difficult to compact DNA without some form of supercoiling. Perhaps less evident is that replication and transcription of DNA also affect, and are affected by, supercoiling. Both processes require a separation of DNA strands, which is complicated by the helical interwinding of the strands (**Figure 9-8**). As discussed in Section 9.3, specialized enzymes found in all organisms alleviate the stress of replication and transcription by introducing or relaxing supercoils.

Inside the cell, a DNA molecule must bend, and it inevitably becomes supercoiled as its axis twists on itself. However, even when extracted and purified, many circular DNA molecules remain highly supercoiled, even in the absence of protein and other cellular components. This indicates that supercoiling is an intrinsic property of DNA tertiary structure, as opposed to an incidental result of spatial constriction. Supercoiling is the direct result of structural strain caused by the active underwinding of the DNA—that is, the removal of helical turns relative to the most stable structure of B-form DNA. DNA underwinding is catalyzed by enzymes called topoisomerases, and the degree of DNA underwinding is highly regulated in every cell.

Several measurable properties of supercoiling have been established, and the study of supercoiling has provided many insights into DNA structure and function. This work has drawn heavily on concepts derived from topology, a branch of mathematics that studies the properties of an object that do not change under continuous deformations (deformations that do not involve breaking the DNA). In the context of **DNA topology**, continuous deformations include conformational changes due to stretching, thermal motion, or interaction with proteins or other molecules. The twisting experiments in the Bustamante laboratory, described in this chapter's Moment of Discovery, provide an example of continuous deformation. Discontinuous deformations involve DNA strand breakage. Topological properties of DNA can be changed only by the breakage and rejoining (religation) of the backbone of one or both DNA strands.

Most Cellular DNA Is Underwound

The simplest examples of supercoiling are provided by small circular DNAs. These might include plasmids and small, double-stranded viral DNAs. When these DNA molecules have no breaks in either strand, they are referred to as **closed-circular DNAs**. If the double helix of a closed-circular molecule has the features of B-form structure (see Figure 6-17), with one turn of the double helix per 10.5 bp, it is relaxed rather than supercoiled (**Figure 9-9** on p. 307). Supercoiling results when DNA is subject to some form of structural strain. However, purified closed-circular DNA is rarely relaxed, regardless of its biological origin. Furthermore, all DNA derived from a given cellular source has a characteristic degree of supercoiling. DNA structure is therefore maintained with a degree of strain that is regulated by the cell and, in turn, induces the observed supercoiling.

In almost every instance in nature, the strain is a result of underwinding of the DNA double helix in the closed circle. In **DNA underwinding**, the molecule has *fewer* helical turns than would be expected for the B-form structure. Consider, for example, an 84 bp segment of a circular DNA in the relaxed state: it would contain eight double-helical turns, one for every 10.5 bp (**Figure 9-10a** on p. 307). If one of these turns were removed, there would be (84 bp)/7 = 12.0 bp per turn, rather than the 10.5 found in B-DNA (**Figure 9-10b**). This is a deviation from the most stable DNA form, and the molecule would be thermodynamically strained as a result. Generally, much of this strain would be accommodated by coiling the axis on itself, forming a supercoil (**Figure 9-10c**). Some of the strain in this 84 bp segment would simply become dispersed in the untwisted structure of the larger DNA molecule. In principle, the strain induced by this degree of net DNA underwinding could also be accommodated

(a)

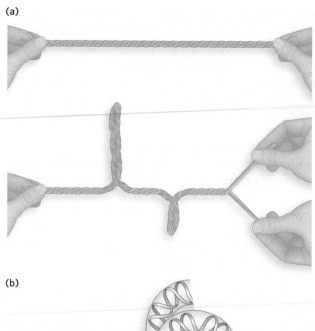

FIGURE 9-8 The effects of replication and transcription on DNA supercoiling. Because DNA is a double-helical structure, strand separation leads to added stress and supercoiling if the DNA is constrained (not free to rotate) ahead of the strand separation. (a) The general effect can be illustrated by twisting two strands of a rubber band about each other to form a double helix. If one end is constrained, separating the two strands at the other end will lead to twisting. (b) In a DNA molecule, the progress of a DNA polymerase or RNA polymerase (as shown) along the DNA involves separation of the strands. Chromosomal DNAs are typically complexed with many proteins that inhibit free rotation. As the polymerase moves on a DNA molecule that is rotationally constrained, the DNA becomes overwound ahead of the enzyme (upstream; positive supercoils) and underwound behind it (downstream; negative supercoils). Red arrows indicate the direction of winding. [Sources: (a) Data from W. Saenger, *Principles of Nucleic Acid Structure, Springer-Verlag, 1984, p. 452.]*

(b)

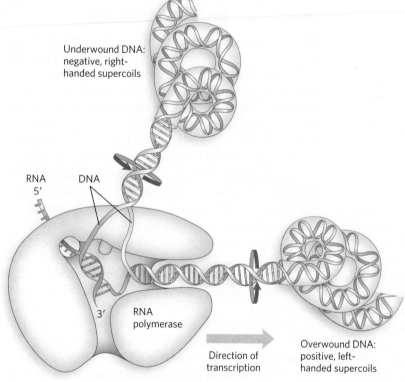

Underwound DNA: negative, right-handed supercoils

RNA 5′

DNA

3′   RNA polymerase

Direction of transcription

Overwound DNA: positive, left-handed supercoils

by separating the two DNA strands over a distance of about 10 bp (**Figure 9-10d**). In isolated closed-circular DNA, strain introduced by underwinding is generally accommodated by supercoiling rather than strand separation, because coiling the axis of the DNA usually requires less energy than breaking the hydrogen bonds that stabilize paired bases. Note, however, that the underwinding of DNA in vivo eases the task of DNA strand separation by enzymes such as DNA and RNA polymerases, and thus facilitates access to the information in the DNA.

Every cell actively underwinds its DNA with the aid of enzymes (see Section 9.3), and the resulting strained state represents a form of stored energy. Underwinding thus accomplishes two things. First, cells maintain DNA in an underwound state in part to promote its compaction by coiling. Second, underwinding facilitates strand separation and enzymatic access to the encoded information; underwinding is thus important to the enzymes of DNA replication and transcription, which must bring about strand separation as part of their function. In

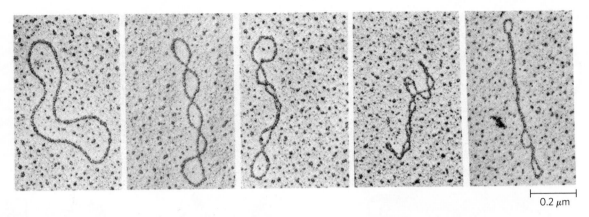

0.2 μm

FIGURE 9-9 Relaxed and supercoiled closed-circular plasmid DNAs. The scanning electron micrograph on the far left shows relaxed DNA. Increased supercoiling is shown, from left to right.
[Source: Laurien Polder from Kornberg, A. (1980) DNA Replication, *p. 29, W.H. Freeman, New York.]*

this way, two biological problems related to DNA—compaction and information access—are addressed at once by active DNA underwinding.

The underwound state can be maintained only if the DNA is a closed circle or, if linear, is bound and stabilized by proteins so that the strands are not free to rotate about each other. If there is a break in one strand of an isolated, protein-free circular DNA, free rotation at that point will cause the underwound DNA to revert spontaneously to the relaxed state. In a closed-circular DNA molecule, however, the number of helical turns

cannot be changed without at least transiently breaking one of the DNA strands. The number of helical turns in a DNA molecule therefore provides a precise description of supercoiling. In the linear chromosomes of eukaryotic cells, DNA underwinding is maintained by bound proteins that constrain the DNA in an elaborate structure called chromatin. In chromatin, large loops of DNA are constrained at their base, such that each loop is topologically fixed as if it were circular (**Figure 9-11**; we discuss this further in Chapter 10).

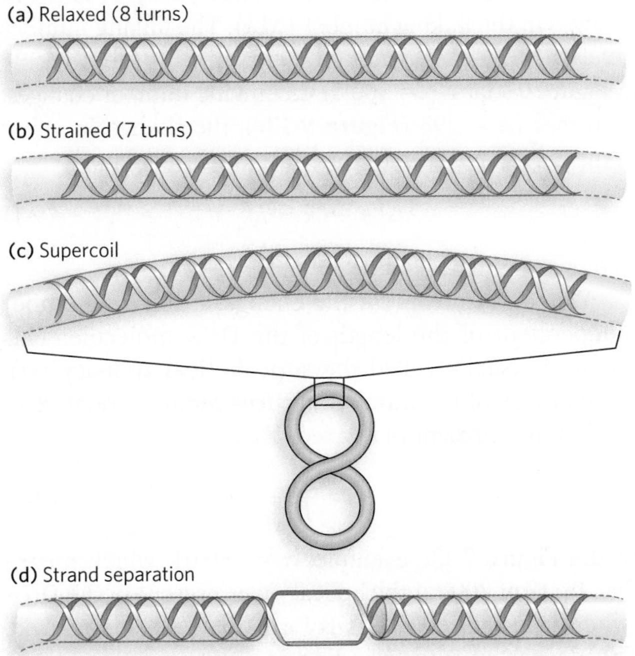

DNA Underwinding Is Defined by the Topological Linking Number

The **linking number (Lk)** is a topological property of double-stranded DNA—that is, it does not change when the DNA is bent or deformed. To define linking number, imagine the separation of the two strands of a double-stranded circular DNA. If the two circular strands are interlocked as shown in **Figure 9-12a**, they are effectively

(a) Relaxed (8 turns)

(b) Strained (7 turns)

(c) Supercoil

(d) Strand separation

FIGURE 9-10 The effects of DNA underwinding. *[Source: Data from N. R. Cozzarelli, T. C. Boles, and J. H. White, in* DNA Topology and Its Biological Effect *(N. R. Cozzarelli and J. C. Wang, eds), Cold Spring Harbor Laboratory Press, 1990, pp. 139–184.]*

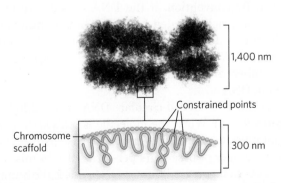

1,400 nm

Constrained points

Chromosome scaffold

300 nm

FIGURE 9-11 Loops in a eukaryotic chromosome constrained by scaffold proteins. The chromatin scaffold is attached to the chromosome at intervals, with the DNA between the attachment points defining loops that are topologically constrained.
[Source: Don W. Fawcett/Science Source.]

(a) *Lk* = 1

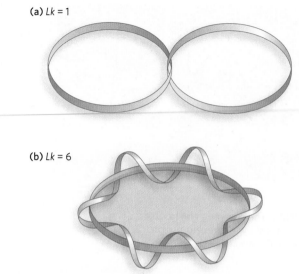

(b) *Lk* = 6

FIGURE 9-12 Linking number (*Lk*). Each blue ribbon represents one strand of a double-stranded DNA molecule. (a) The linking number of the molecule is 1. (b) The linking number is 6. One of the strands is kept untwisted for illustrative purposes, to define the border of an imaginary surface (solid blue oval). The number of times the twisting strand (its relative length exaggerated here) penetrates this surface is the linking number. [Source: Data from N. R. Cozzarelli, T. C. Boles, and J. H. White, in DNA Topology and Its Biological Effect (N. R. Cozzarelli and J. C. Wang, eds), Cold Spring Harbor Laboratory Press, 1990, pp. 139–184.]

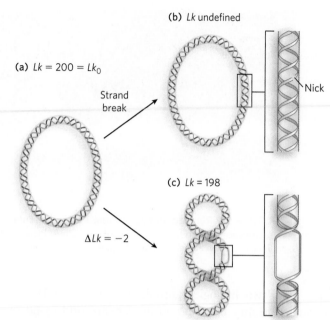

(b) *Lk* undefined

(a) *Lk* = 200 = *Lk*$_0$

Strand break

Nick

(c) *Lk* = 198

$\Delta Lk = -2$

FIGURE 9-13 Linking number of closed-circular DNAs. A 2,100 bp molecule is shown in three forms: (a) relaxed, *Lk* = 200; (b) relaxed with a nick in one strand, *Lk* undefined; (c) underwound by two turns, *Lk* = 198. The underwound molecule is generally supercoiled, but underwinding also facilitates the separation of DNA strands.

linked by what we define as a topological bond. Even if all hydrogen bonds and base-stacking interactions were removed so that the strands were not in physical contact, the two strands would still be interlocked like the links of a chain link fence. This is the topological linkage. If we think of one of the circular strands as the boundary of a surface (such as the soap film on the loop of a bubble wand before you blow a bubble), the linking number can be defined as the number of times the second strand pierces this surface (**Figure 9-12b**). The linking number for a closed-circular DNA is always an integer. By convention, if the DNA strands are interwound in a right-handed helix, the linking number is positive (+); for strands interwound in a left-handed helix, the linking number is negative (−). Negative linking numbers are, for all practical purposes, not encountered in DNA.

Consider a closed-circular DNA with 2,100 bp (**Figure 9-13a**). When the molecule is relaxed, the linking number is equal to the number of base pairs divided by the number of base pairs per turn, 10.5; in this case, *Lk* = 200. In unstrained B-form DNA, this is the number of times that one strand would pierce the imaginary surface defined by the other strand (see Figure 9-12). DNA can have a topological property such as a linking number only if both strands are intact. If there is a break in either strand, the strands can, in principle, be unraveled and

separated; in this case, no topological bond exists and *Lk* is undefined (**Figure 9-13b**).

We can now describe DNA underwinding in terms of changes in the linking number (Δ*Lk*). The linking number in relaxed DNA, *Lk*$_0$, is used as a reference. For the molecule in Figure 9-13a, *Lk*$_0$ = 200. If the linking number changes, such that *Lk* = 198 (**Figure 9-13c**), the change (number of turns removed) can be described by the equation:

$$\Delta Lk = Lk - Lk_0 \qquad (9\text{-}1)$$

In our example, Δ*Lk* = 198 − 200 = −2.

We can also express the change in linking number independent of the length of the DNA molecule. This quantity, usually called the **superhelical density (σ)**, is a measure of the number of turns removed relative to the number present in relaxed DNA:

$$\sigma = \frac{\Delta Lk}{Lk_0} \qquad (9\text{-}2)$$

In the Figure 9-13c example, σ = −0.01, which means that 1% (2 of 200) of the helical turns present in the DNA (when it is relaxed, in its B form) have been removed. The degree of underwinding in cellular DNAs generally falls in the range of 5% to 7%; that is, σ = −0.05 to −0.07. The negative sign indicates that the change in the linking number is due to underwinding of the DNA. The supercoiling induced by underwinding is therefore

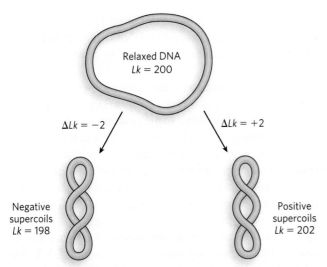

FIGURE 9-14 Negative and positive supercoils. For the relaxed DNA in Figure 9-13a, underwinding or overwinding by two helical turns ($\Delta Lk = \pm 2$) produces negative or positive supercoiling, respectively. Notice that the DNA axis in the two forms twists in opposite directions.

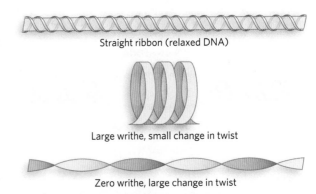

FIGURE 9-15 Writhe and twist. The beige ribbon represents the axis of a relaxed DNA molecule. Strain from underwinding of the DNA can manifest as writhe or twist. Topological changes in the linking number are usually accompanied by geometric changes in both writhe and twist.

defined as **negative supercoiling**. Conversely, under some conditions DNA can be overwound, resulting in **positive supercoiling**.

Notice that negative supercoiling results in a twisting of the axis of the DNA to form a right-handed superhelix, and positive supercoiling results in a left-handed superhelix (**Figure 9-14**). Supercoiling is not a random process; it is largely prescribed by the torsional strain imparted to the DNA by decreasing or increasing the linking number relative to that of B-DNA.

The linking number can be changed by ± 1 by breaking one DNA strand, rotating one of the ends 360° about the unbroken strand, and rejoining the broken ends. This reaction is catalyzed by topoisomerases (see Section 9.3). The change in linking number has no effect on the number of base pairs or the number of atoms in the circular DNA molecule. Two forms of a circular DNA that differ only in a topological property such as linking number are referred to as **topoisomers**.

We can break down the linking number into two structural components, writhe (Wr) and twist (Tw) (**Figure 9-15**). **Writhe (Wr)** is a measure of the coiling of the helical axis, and **twist (Tw)** describes the local twisting or spatial relationship of neighboring base pairs. When the linking number changes, some of the resulting strain is usually compensated for by writhe (supercoiling) and some by changes in twist, giving rise to the equation:

$$Lk = Tw + Wr \qquad (9\text{-}3)$$

Tw and Wr need not be integers. Twist and writhe are geometric rather than topological properties, because they may be changed by deformation of a closed-circular DNA molecule. Tw and Wr may change in a reciprocal manner without altering Lk. That is, Lk can remain unchanged when either Tw or Wr increases by a given amount and the other decreases by that same amount.

In addition to causing supercoiling and making strand separation somewhat easier, the underwinding of DNA facilitates structural changes in the molecule. Although these changes are of less physiological importance, they help illustrate the effects of underwinding. Recall that a cruciform generally contains a few unpaired bases (see Figure 6-20b); DNA underwinding helps maintain the required strand separation in regions where palindromic sequences allow cruciform formation (**Figure 9-16**). Underwinding of a right-handed DNA helix also facilitates the formation of short stretches of left-handed Z-DNA in regions where the base sequence is consistent with the Z form (see Chapter 6).

DNA Compaction Requires a Special Form of Supercoiling

All supercoiled DNA molecules are similar in several respects. In the laboratory, the supercoils are right-handed in a negatively supercoiled DNA molecule, such as a supercoiled plasmid isolated from a bacterium. The supercoiled plasmids tend to be extended and narrow rather than compacted, often with multiple branches (**Figure 9-17a**). At the superhelical densities normally encountered in cells, the length of the supercoil axis (the axis about which the supercoils turn), including branches, is about 40% of the length of the DNA. This type of supercoiling is referred to as **plectonemic supercoiling (Figure 9-17b)**. This term can be applied to any structure with strands intertwined in some simple and regular way, and it is a good description of

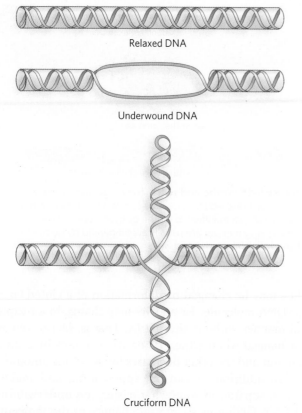

Relaxed DNA

Underwound DNA

Cruciform DNA

FIGURE 9-16 Promoting cruciform structures by DNA underwinding. Cruciforms can form at palindromic sequences, but they seldom occur in relaxed DNA because linear DNA accommodates more paired bases than the cruciform structure. DNA underwinding facilitates the partial strand separation required for promoting cruciform formation at appropriate sequences.

the general structure of supercoiled DNA in solution. In cells, it is found in smaller DNAs such as plasmids. Larger genomic DNAs are generally compacted in a different way.

Plectonemic supercoiling does not produce sufficient compaction to package genomic DNA in the cell. A second form, **solenoidal supercoiling**, can be adopted by an underwound DNA in the cell (**Figure 9-17c**). Instead of the extended right-handed supercoils characteristic of the plectonemic form, solenoidal supercoiling involves tight left-handed turns, much like a garden hose neatly wrapped on a reel. Although their structures are very different, plectonemic and solenoidal supercoiling are two forms of negative supercoiling that can be adopted by the *same* segment of underwound DNA. The two forms are readily interconvertible. The plectonemic form is more stable in solution, but the solenoidal form can be stabilized by protein binding and provides a much greater degree of compaction. Solenoidal supercoils are formed when DNA is wrapped around the nucleosomes that make up eukaryotic chromatin (see Chapter 10). Similarly, in bacteria, the tight wrapping of DNA around a variety of DNA-binding proteins gives rise to solenoidal supercoils. Solenoidal supercoiling is the primary mechanism by which underwinding contributes to genomic DNA compaction.

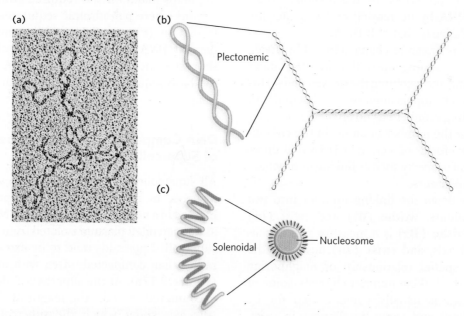

(a)

(b)

Plectonemic

(c)

Solenoidal

Nucleosome

FIGURE 9-17 Plectonemic and solenoidal supercoiling. (a) Electron micrograph of plectonemically supercoiled plasmid DNA. (b) Plectonemic supercoiling consists of extended right-handed coils. (c) Solenoidal supercoiling of the same DNA molecule depicted in (b), drawn to scale. Solenoidal negative supercoiling consists of tight left-handed turns about an imaginary tubelike structure. Solenoidal supercoiling provides a much greater degree of compaction. [*Source: (a) Reprinted from Journal of Molecular Biology, 213, James H. White, T. Christian Boles, & N. R. Cozzarelli, Structure of plectonemically supercoiled DNA, 931–951, Copyright 1990, with permission from Elsevier.*]

- Most cellular DNAs are supercoiled. Underwinding decreases the total number of helical turns in the DNA relative to the relaxed, B form. To maintain an underwound state, DNA must be either a closed circle or, if linear, bound to protein.

- Underwinding is quantified by the linking number (Lk), a topological parameter that describes the number of times two DNA strands are intertwined.

- Underwinding is measured in terms of σ, the superhelical density; $\sigma = \Delta Lk/Lk_0$. For cellular DNAs, σ is typically -0.05 to -0.07, which means that about 5% to 7% of the helical turns in the DNA have been removed. DNA underwinding facilitates strand separation by enzymes of DNA metabolism.

- Plectonemic supercoiling includes right-handed branches and is the most common type of supercoiling in isolated DNA. Solenoidal supercoiling, an alternative form that produces a greater degree of compaction, is characterized by tight left-handed turns that are stabilized by wrapping the DNA around proteins; this occurs in eukaryotic and bacterial chromosomes.

9.3 THE ENZYMES THAT PROMOTE DNA COMPACTION

DNA supercoiling—or, more specifically, DNA underwinding—is a precisely regulated process that influences many aspects of DNA metabolism. As we have seen, underwinding allows access to DNA during replication and transcription, and it contributes to DNA condensation. The underwinding and relaxation of DNA are catalyzed by DNA topoisomerases, enzymes that break one or both DNA strands to allow a topological change, and then religate them. Additional condensation of cellular DNA is facilitated by SMC proteins, a class of enzymes that reversibly form protein loops that link DNA segments, affecting both the condensation/compaction of chromosomes and the cohesion of daughter DNA molecules for periods following eplication. The maintenance of the underwound and condensed state of chromosomes by structural DNA-binding proteins such as histones is discussed in Chapter 10.

Topoisomerases Catalyze Changes in the Linking Number of DNA

All cells, from bacteria to eukaryotes, have enzymes with the sole function of underwinding and relaxing DNA. **Topoisomerases** increase or decrease the extent of DNA underwinding by changing the linking number. They play an especially important role in the complex changes in DNA topology during replication and DNA packaging.

As we show in later chapters, topoisomerases are crucial to every aspect of DNA metabolism. As a consequence, they are important drug targets for the treatment of bacterial infections and cancer (**Highlight 9-2**). Inactivating mutations in genes that encode key cellular topoisomerases, the enzymes responsible for the degree of supercoiling in cells, often result in severe growth deficiencies or cell death.

There are two classes of topoisomerases (**Table 9-4**). **Type I topoisomerases** break one of the two DNA

TABLE 9-4

Topoisomerases in Bacteria and Eukaryotes

Topoisomerase	Class	Function
Bacteria		
Topoisomerase I	Type I	Relaxes negative supercoils
Topoisomerase II (DNA gyrase)	Type II	Introduces negative supercoils
Topoisomerase III	Type I	Has specialized functions in DNA repair and replication
Topoisomerase IV	Type II	Unlinks replicated chromosomes
Eukaryotes		
Topoisomerase I	Type I	Relaxes negative supercoils, especially during DNA replication
Topoisomerase IIα	Type II	Relaxes positive or negative supercoils; functions in chromatin condensation, replication, transcription
Topoisomerase IIβ	Type II	Relaxes positive or negative supercoils; functions in chromatin condensation, replication, transcription
Topoisomerase III	Type I	Has specialized functions in DNA repair and replication

HIGHLIGHT 9-2 | MEDICINE

Curing Disease by Inhibiting Topoisomerases

The topological state of cellular DNA is intimately connected with its function. Without topoisomerases, cells cannot replicate or package their DNA, or express their genes—and they die. Inhibitors of topoisomerases have therefore become important pharmaceutical agents, targeted at infectious organisms and malignant cells.

Two classes of bacterial topoisomerase inhibitors have been developed as antibiotics. The coumarins, including novobiocin and coumermycin A1, are natural products derived from *Streptomyces* species. They inhibit the ATP binding of the bacterial type II topoisomerases, DNA gyrase and topoisomerase IV. These antibiotics are not used to treat infections in humans, but research continues to identify clinically effective variants.

The quinolone antibiotics, also inhibitors of bacterial DNA gyrase and topoisomerase IV, first appeared in 1962 with the introduction of nalidixic acid (Figure 1). This compound had limited effectiveness and is no longer used clinically in the United States, but the continued development of this class of drugs eventually led to the introduction of the fluoroquinolones, exemplified by ciprofloxacin (Cipro). The quinolones act by blocking the last step of the topoisomerase reaction in bacteria, the resealing of the DNA strand breaks. Ciprofloxacin is a broad-spectrum antibiotic that works on a wide range of disease-causing bacteria. It is one of the few antibiotics reliably effective in treating anthrax infections and is considered a valuable agent in protection against possible bioterrorism. Quinolones are selective for the bacterial topoisomerases, inhibiting the eukaryotic enzymes only at concentrations several orders of magnitude greater than the therapeutic doses.

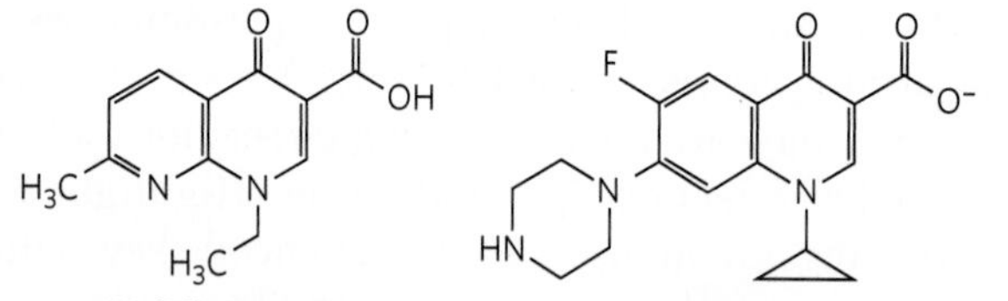

Some of the most important chemotherapeutic agents used in cancer treatment are inhibitors of human topoisomerases. Tumor cells generally contain elevated levels of topoisomerases, and agents targeted to these enzymes are much more toxic to the tumors than to most other tissue types. Inhibitors of both type I and type II topoisomerases have been developed as anticancer drugs.

Camptothecin, isolated from a Chinese ornamental tree and first tested clinically in the 1970s, is an inhibitor of eukaryotic type I topoisomerases. Clinical trials indicated limited effectiveness in treating cancer, however, despite its early promise in preclinical work on mice. Two effective derivatives were developed in the 1990s: irinotecan (Campto) and topotecan (Hycamtin), used to treat colorectal cancer and ovarian cancer, respectively (Figure 2). Additional derivatives are likely to be approved for clinical use in the coming years. All of these drugs act by trapping the topoisomerase-DNA complex in which the DNA is cleaved, inhibiting religation.

Nalidixic acid   **Ciprofloxacin**

FIGURE 1 Inhibitors of bacterial type II topoisomerases.

---

strands, pass the unbroken strand through the break, and ligate the broken ends; they change *Lk* in increments of 1. **Type II topoisomerases** break both DNA strands and change *Lk* in increments of 2. The DNA is never released from the enzyme during these topological transactions, so uncontrolled relaxation of the DNA does not occur.

The activity of these enzymes can be observed with agarose gel electrophoresis, which separates DNA species according to their topoisomeric form (**Figure 9-18**). A population of identical plasmid DNAs with the same linking number migrates as a discrete band during electrophoresis. DNA topoisomers that are more supercoiled are more compact and migrate faster in the gel. Topoisomers with *Lk* values differing by as little as 1 can be separated by this method, so the changes in linking number induced by topoisomerases are readily detected.

E. coli has at least four individual topoisomerases, I through IV. Topoisomerases I and III are of type I, and they generally relax DNA by introducing transient single-strand breaks to remove negative supercoils

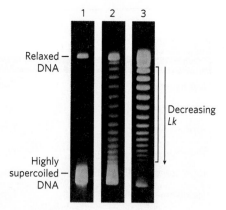

FIGURE 9-18 Visualizing topoisomers. In this experiment, DNA molecules (plasmids) have an identical number of base pairs but differ in the degree of supercoiling. In lane 1, highly supercoiled DNA migrates as a single band. Lanes 2 and 3 show the effect of treating supercoiled DNA with a type I topoisomerase; the DNA in lane 3 was treated for a longer time than the DNA in lane 2. Each individual band in the bracketed region of lane 3 contains DNA plasmids with the same linking number; *Lk* changes by 1 from one band to the next. [*Source: W. Keller,* Proc. Natl. Acad. Sci. USA *72:2553, 1975. Courtesy Michael M. Cox.*]

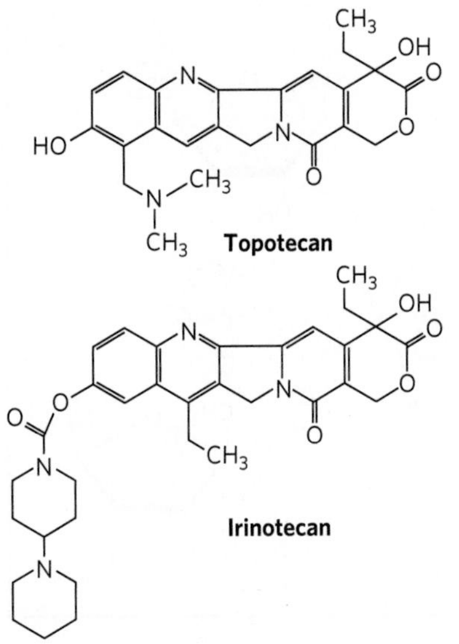

FIGURE 2 Inhibitors of eukaryotic topoisomerase I that are used in cancer chemotherapy.

The human type II topoisomerases are targeted by a variety of antitumor drugs, including doxorubicin (Adriamycin), etoposide (Etopophos), and ellipticine (Figure 3). Doxorubicin, effective against several kinds of human tumors, is in clinical use. Most of these drugs stabilize the covalent topoisomerase–cleaved DNA complex.

All of these anticancer agents generally increase the levels of DNA damage in targeted, rapidly growing tumor cells, but noncancerous tissues can also be affected, leading to a more general toxicity and unpleasant side effects that must be managed during therapy. As cancer therapies become more effective and survival statistics for cancer patients improve, the independent appearance of new tumors is becoming a greater problem. In the continuing search for new cancer therapies, topoisomerases are likely to remain prominent targets.

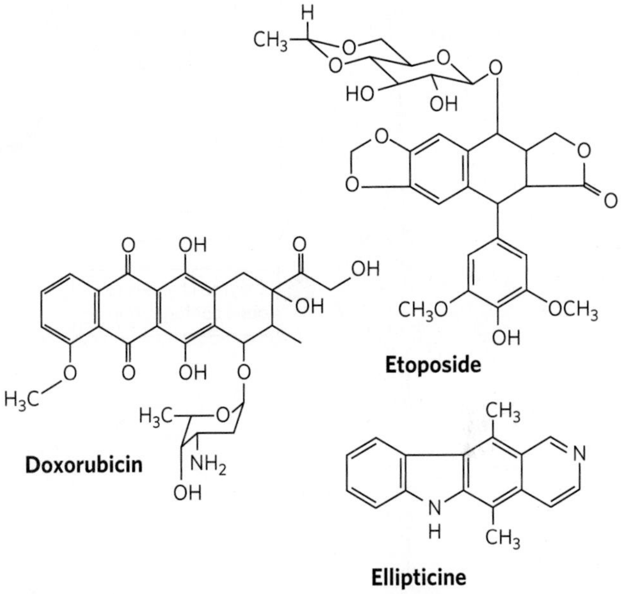

FIGURE 3 Inhibitors of human topoisomerase II that are used in cancer chemotherapy.

(increasing Lk). **Figure 9-19** shows the steps in the reaction catalyzed by bacterial type I topoisomerases (also see the How We Know section at the end of this chapter). A DNA molecule binds to the topoisomerase, and one DNA strand is cleaved (step 1). The enzyme changes conformation (step 2), and the unbroken DNA strand moves through the break in the first strand (step 3). Finally, the DNA strand is ligated and released (step 4). ATP is not used in this reaction. The enzyme promotes the formation of a less strained, more relaxed state by removing supercoils.

The topoisomerase must both cleave a DNA strand and religate it after the topological change is complete. The phosphodiester bond is not simply hydrolyzed, because this would entail loss of a high-energy bond, and an activation step would then be required to promote the subsequent ligation. Instead, a nucleophile on the enzyme (usually a Tyr residue, as in the case of *E. coli* topoisomerase I) attacks the phosphodiester bond, displacing the 3' hydroxyl of one nucleotide and forming a covalent 5'-phosphotyrosyl linkage with the next nucleotide in the DNA strand at the break. Strand passage brings about the topological change. The broken strand is then ligated by a direct attack of the free 3'-hydroxyl group on the phosphotyrosyl linkage. In this scheme, one high-energy bond is replaced by another at each chemical step. The resulting conservation of energy allows strand ligation without an activation step that would otherwise consume ATP.

The Two Bacterial Type II Topoisomerases Have Distinct Functions

Bacterial topoisomerase II, also known as DNA gyrase, can introduce negative supercoils (decrease Lk). This enzyme cleaves both strands of a DNA molecule (thus is a type II topoisomerase) and passes another duplex through the break (see the How We Know section at the end of this chapter). The introduction of negative supercoils alone would put additional strain on the DNA molecule, but gyrase has an additional activity that uses

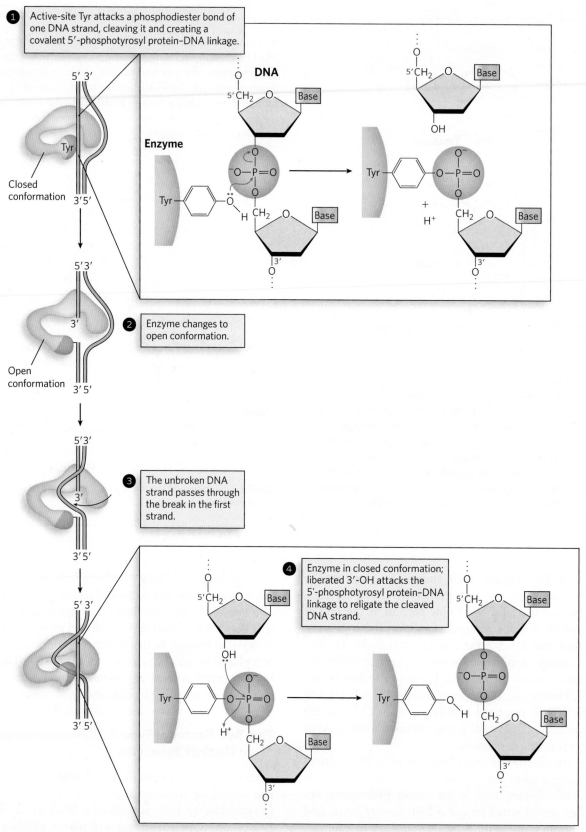

1 Active-site Tyr attacks a phosphodiester bond of one DNA strand, cleaving it and creating a covalent 5'-phosphotyrosyl protein–DNA linkage.

Closed conformation

2 Enzyme changes to open conformation.

Open conformation

3 The unbroken DNA strand passes through the break in the first strand.

4 Enzyme in closed conformation; liberated 3'-OH attacks the 5'-phosphotyrosyl protein–DNA linkage to religate the cleaved DNA strand.

FIGURE 9-19 The type I topoisomerase reaction. Bacterial topoisomerase I increases *Lk* by breaking one DNA strand, passing the unbroken strand through the break, then resealing the break. Nucleophilic attack by the active-site Tyr residue breaks one DNA strand. The ends are ligated by a second nucleophilic attack. At each step, one high-energy bond replaces another. [*Source: Data from J. J. Champoux,* Annu. Rev. Biochem. *70:369, 2001, Fig. 3.*]

the energy of ATP to drive key conformational changes that counteract the thermodynamically unfavorable introduction of negative supercoils. Bacterial DNA gyrases are the only topoisomerases known to actively introduce negative supercoils.

Gyrase is composed of two types of subunits, GyrA and GyrB, functioning as a $GyrA_2GyrB_2$ heterotetramer (**Figure 9-20a**). GyrB interacts with DNA and ATP and catalyzes ATP binding and hydrolysis. Parts of GyrB form the entry point for DNA, called the N-gate. The DNA exits through a domain in GyrA called the C-gate. A separate domain of GyrA binds DNA and promotes DNA wrapping. Reaction steps are detailed in **Figure 9-20b**. To introduce negative supercoils, a gyrase complex first binds to a DNA segment via the N-gate (step 1) and wraps the DNA around itself (step 2). The DNA is bound such that a positive node (a crossover of two DNA segments that cross in the sense of a positive supercoil) is created in the active site. Active-site Tyr residues bind ATP and cleave both strands of one of the DNA segments (step 3), forming two 5′-phosphotyrosine intermediates. ATP hydrolysis is coupled to the passage of the second segment of DNA through the cleaved DNA strands, entering at the N-gate and exiting at the C-gate. To complete the reaction (step 4), the DNA strands are ligated by attack of the free 3′-hydroxyl groups on the phosphotyrosine intermediates. The complex is then poised to initiate another reaction cycle. The degree of supercoiling of bacterial DNA is maintained by regulation of the net activity of topoisomerase I, which increases Lk, and DNA gyrase, which decreases Lk. Decreased activity of one of these enzymes (e.g., by mutation) causes growth deficiencies, which can be partially relieved by compensating mutations in the gene encoding the other enzyme.

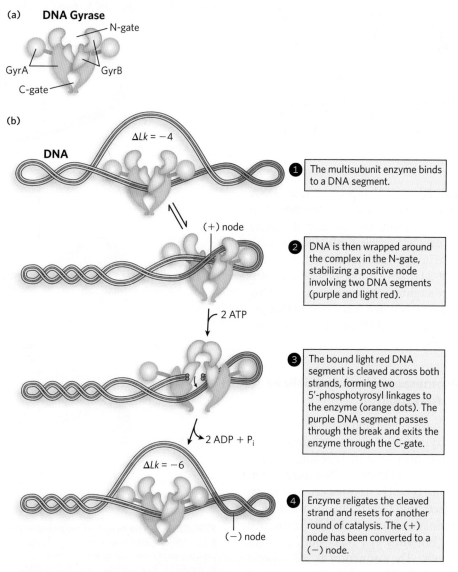

FIGURE 9-20 Introduction of negative DNA supercoils by DNA gyrase. (a) The structure of DNA gyrase. (b) The mechanism of gyrase action. [*Source: Data from N. P. Higgins, Nat. Struct. Mol. Biol. 14:264–271, 2007.*]

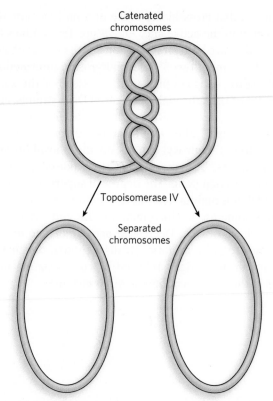

FIGURE 9-21 Solving topological problems with type II topoisomerases. Type II topoisomerases resolve knots and catenanes that arise in DNA by passing one duplex through a transient double-strand break in another duplex.

The second bacterial type II topoisomerase, DNA topoisomerase IV, has a very specialized function. Immediately following replication, the circular daughter chromosomes of bacteria are topologically intertwined. Circular DNAs that are intertwined in this way are called **catenanes** (**Figure 9-21**). DNA topoisomerase IV unlinks the catenated daughter chromosomes, allowing their proper segregation at cell division. Unlike DNA gyrase, this enzyme does not use ATP and does not introduce negative supercoils.

Eukaryotic Topoisomerases Have Specialized Functions in DNA Metabolism

Eukaryotic cells also have type I and type II topoisomerases. The type I enzymes are called topoisomerases I and III. They function primarily in relieving tension and resolving topological problems in DNA during replication and repair. The type II enzymes are topoisomerases IIα and IIβ (see Table 9-4). Eukaryotic type II topoisomerases cannot underwind DNA (introduce negative supercoils), but they can relax both positive and negative supercoils. They function in all aspects of eukaryotic DNA metabolism, resolving a range of topological problems that occur during replication, transcription, and repair. They play an especially

important role in the condensation of chromosomes into highly structured chromatin.

Although eukaryotes do not have an enzyme that can introduce negative supercoils into DNA, when a circular DNA is isolated from a eukaryotic cell (e.g., a plasmid from yeast), it is negatively supercoiled. This reflects the generally underwound state of cellular DNA in eukaryotic cells. One probable origin of negative supercoils in eukaryotic DNA is the tight wrapping of the DNA around a nucleosome in chromatin, which introduces a negative solenoidal supercoil without changing the number of turns in the molecule (see Chapter 10). Because the *Lk* remains unchanged, the negative solenoidal supercoil has to be compensated for by a positive supercoil elsewhere in the DNA (**Figure 9-22**). The type II topoisomerases relax the unbound positive supercoils that arise in this way. The bound and stabilized negative supercoils are left behind, conferring a net negative superhelicity on the DNA. Next to the histones that make

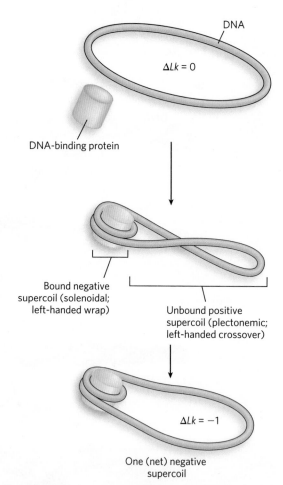

FIGURE 9-22 The origin of negative supercoiling in eukaryotic DNA. When DNA is wrapped tightly around a DNA-binding protein or protein complex, a solenoidal negative supercoil is fixed in the DNA. In a constrained DNA molecule, positive supercoils must develop elsewhere to compensate for the resulting strain. Relaxation of unbound positive supercoils by topoisomerases leads to development of a net negative superhelicity in the DNA.

up the nucleosomes, type II topoisomerases are the most abundant proteins in chromatin.

Figure 9-23 shows the reaction catalyzed by eukaryotic type II topoisomerases. The multisubunit enzyme binds a DNA molecule (step 1). The gated cavities above and below the bound DNA are the N-gate and C-gate, respectively. The second segment of the same DNA is bound at the N-gate (step 2). Both strands of the first DNA are now cleaved (step 3; the chemistry is similar to that in Figure 9-19), forming phosphotyrosine intermediates. The second DNA segment passes through the break in the first segment (step 4), and the broken DNA is ligated and the second segment released through the C-gate (step 5). Two ATPs are bound and hydrolyzed during this cycle; it is likely that one is hydrolyzed in the step leading to the complex in step 4. Additional details of the ATP hydrolysis have yet to be worked out.

SMC Proteins Facilitate the Condensation of Chromatin

Whereas topoisomerases influence supercoiling by changing the linking number of chromosomes, **SMC proteins** (structural *m*aintenance of *c*hromosomes) promote chromosome condensation by creating physical contact between segments of DNA that may otherwise be quite distant from each other in the chromosome, or even on different chromosomes. This class of protein is found in all organisms. In eukaryotes, where their function is best understood, these enzymes have integral roles in DNA condensation and chromosome segregation during mitosis, as well as in DNA repair. They perform their tasks by lining up along the DNA and binding to each other, providing a link between distant parts of the chromosome.

SMC proteins have five distinct domains (**Figure 9-24a**). The amino-terminal (N) and carboxyl-terminal (C) domains each contain part of an ATP-hydrolytic site, and they are connected by two regions of α-helical structure (see Figure 4-16c) joined to a hinge domain. When the hinge bends, the α-helical regions form a coiled-coil motif, and the N and C domains come together to form a head structure at one end with a complete ATP-binding site. SMC proteins are generally dimeric, forming a V-shaped complex linked through the hinge domains (**Figure 9-24b**). Thus the dimeric SMC complex contains two head domains and two ATP-binding sites. ATP is not hydrolyzed until the two heads come together. Although many details of SMC protein function have yet to be elucidated, the head-head association between the two subunits seems to be critical.

All bacteria have at least one SMC protein that functions as a homodimer to assist in compacting the genome, whereas eukaryotes generally have six SMC proteins that work in defined pairs as heterodimers with different functions (**Figure 9-24c**). Electron microscopy reveals the flexible V shape of these proteins (**Figure 9-24d**). The SMC1-SMC3 and SMC2-SMC4 pairs have roles in mitosis, and the SMC5-SMC6 pair is involved in DNA repair, but its molecular role is not

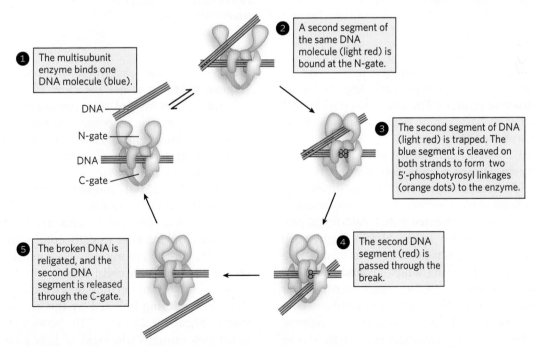

FIGURE 9-23 Alteration of the linking number by eukaryotic type II topoisomerases. The general mechanism is similar to that of the bacterial DNA gyrase (see Figure 9-20b), with one intact duplex DNA segment being passed through a transient double-strand break in another segment. The enzyme structure and use of ATP are distinct to this reaction. [Source: Data from J. J. Champoux, Annu. Rev. Biochem. *70:369, 2001, Fig. 11.*]

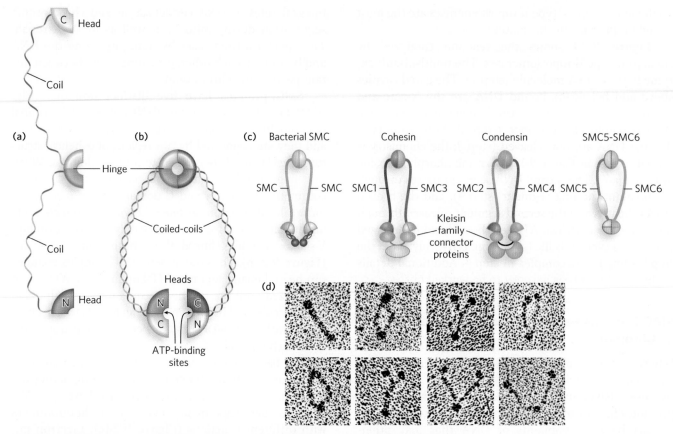

FIGURE 9-24 The structure of SMC proteins. (a) SMC proteins have five domains. (b) Each SMC polypeptide is folded so that the two coiled-coil domains wrap around each other and the N and C domains come together to form a complete ATP-binding site. Two polypeptides are linked at the hinge region to form the dimeric V-shaped SMC molecule. (c) Bacterial SMC proteins form a homodimer. The three different eukaryotic SMC proteins form heterodimers. Cohesins are made up of SMC1-SMC3 pairs, and condensins consist of SMC2-SMC4 pairs. (d) Electron micrographs of different bacterial SMC dimers show the variety of shapes these dimers can take. [*Source: (d) Courtesy Harold Erickson.*]

well understood. All these complexes are bound by regulatory and accessory proteins, including the kleisin family of connector proteins. The interactions with DNA involve patches of basic amino acid residues near the hinge regions of the SMC proteins.

The SMC1-SMC3 pair forms a functional unit called a **cohesin**. During mitosis, cohesins link two sister chromatids immediately after chromosomal replication and keep them together as the chromosomes condense to metaphase (**Figure 9-25**). Additional proteins, particularly proteins in the kleisin family such as SCC1, bridge the cohesin head units to form a ring (see Figure 9-24c). The ring wraps around the sister chromatids, tying them together until separation is required at cell division. The ring may expand and contract in response to ATP hydrolysis. As chromosome segregation begins, the cohesin tethers are removed by enzymes known as separases.

Interaction among the head groups of multiple SMC proteins has the potential to produce several different architectures, such as rings, rosettes, and filaments (**Figure 9-26**). It is not yet clear whether the ringed cohesin tethers around sister chromatids are intra- or intermolecular. The associated proteins may modulate intermolecular interactions, or, for intramolecular rings, they may perform a gatekeeping function in bringing DNA molecules into the ring.

The SMC2-SMC4 complex is called a **condensin**. The bacterial SMC proteins are most closely related to condensins. The condensins are critical to chromosome condensation as cells enter mitosis (see Figure 9-25). In the laboratory, condensins bind DNA to create positive supercoils; that is, condensin binding causes the DNA to become overwound, in contrast to the underwinding induced by the binding of nucleosomes. **Figure 9-27** on p. 320 shows a current model (with two minor variations) of how condensins may interact with DNA to promote chromosome condensation. The condensin complexes (SMC2-SMC4 plus associated proteins) first bind to the DNA in a closed

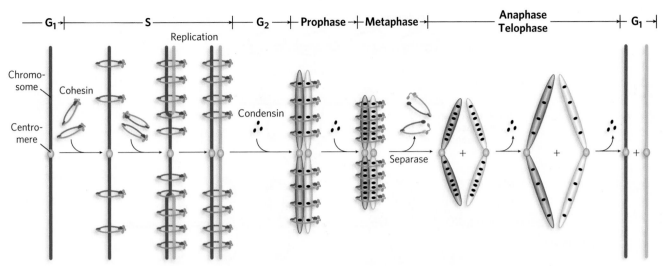

FIGURE 9-25 The roles of cohesins and condensins in the eukaryotic cell cycle. Cohesins are loaded onto the chromosomes during G1 (see Section 2.2), tying the sister chromatids together during replication. At the onset of mitosis, condensins bind and maintain the chromatids in a condensed state. During anaphase, the enzyme separase removes the cohesin links. Once the chromatids separate, condensins begin to unload and the daughter chromosomes return to the uncondensed state. [*Source: Data from D. P. Bazett-Jones, K. Kimura, and T. Hirano, Mol. Cell 9:1183, Fig. 5.*]

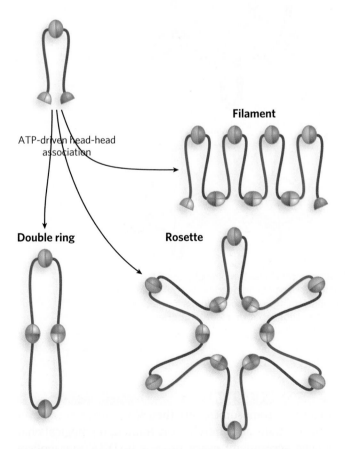

FIGURE 9-26 Proposed architectural arrangements of SMC proteins. Head-to-head association results in the formation of rings, rosettes, and filaments. [*Source: Data from T. Hirano, Nat. Rev. Mol. Cell. Biol. 7:311–322, 2006.*]

form. ATP hydrolysis then opens the intramolecular ring and brings the DNA inside. Head-to-head association creates a structure that traps DNA with a positive superhelical tension. Finally, aggregation of the condensins into rosettes forms a condensed chromatid with a defined architecture.

The topoisomerases and SMC proteins enable cells to deal with the complex topological changes occurring as DNA strands separate during replication, repair, and transcription, as well as the extraordinary degree of DNA compaction required in every cell. The compaction is maintained by additional specialized DNA-binding proteins, and we turn to these proteins, and their organization and function, in Chapter 10.

SECTION 9.3 SUMMARY

- Topoisomerases catalyze the underwinding and relaxation of DNA. On a molecular level, topoisomerases catalyze changes in the linking number.
- The two classes of topoisomerases, type I and type II, change *Lk* in increments of 1 or 2, respectively, per catalytic event.
- The reactions catalyzed by DNA topoisomerases involve the formation of transient covalent DNA-enzyme intermediates, usually in the form of a phosphotyrosyl linkage.
- Bacterial DNA gyrases introduce negative supercoils.
- Topoisomerases have functions specific to DNA metabolism, such as unlinking catenated bacterial DNA

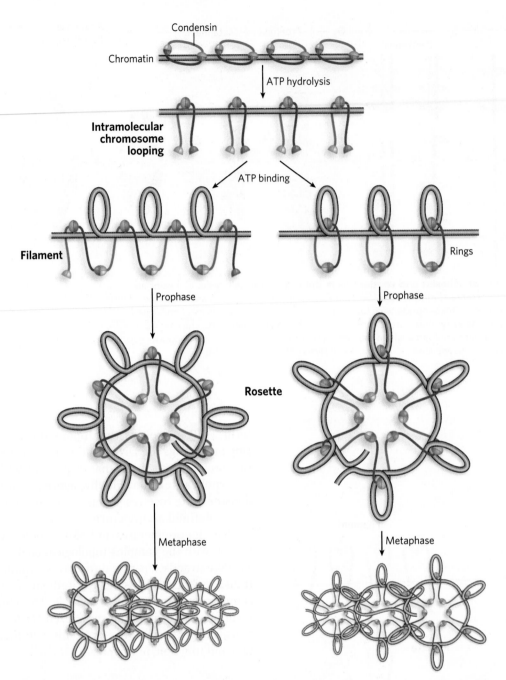

FIGURE 9-27 A proposed role of condensins in chromatin condensation. Initially, the DNA is bound at the hinge region of the SMC protein, in the interior of what can become an intramolecular SMC ring. ATP binding leads to head-to-head association, forming supercoiled loops in the bound DNA. Subsequent rearrangement of the head-to-head interactions to form rosettes condenses the DNA. [Source: Data from T. Hirano, Nat. Rev. Mol. Cell. Biol. 7:311–322, 2006.]

after replication or relaxing supercoils formed by unwinding during replication and transcription.

- Condensation of cellular chromosomes is facilitated by SMC proteins, including cohesins and condensins. Cohesins tether the sister chromatid products of DNA replication, and condensins provide a general structural scaffold for chromosome condensation.

? UNANSWERED QUESTIONS

In large chromosomes, with their DNA highly complexed with proteins and thereby constrained, topological challenges accompany every process in DNA metabolism. The simple movement of a DNA polymerase through the DNA during replication (defining a structure called a replication fork) leads to overwinding ahead of the fork and underwinding behind it. Some of the challenges are extraordinary, such as when two replication forks converge in a eukaryotic chromosome, or when intertwined chromosomal DNAs must be separated at cell division. A complete understanding of DNA metabolism in cells will require more detailed information on how every system interfaces with the proteins that solve topological problems.

1. What is the role of bacterial SMC proteins? Bacteria generally have at least one SMC protein, and sometimes several. Mutational loss of the major bacterial SMC protein usually leads to defects in the condensation and segregation of chromosomes at cell division. Recent work indicates that this SMC protein is recruited to the daughter replication origins and participates in the mechanism that ensures proper segregation. However, much remains to be learned about this process.

2. What is the function of the eukaryotic SMC5-SMC6 proteins? The most enigmatic of the eukaryotic SMC proteins is the SMC5-SMC6 pair. So far, we know that this protein pair functions primarily in processes such as DNA recombination and repair, and much evidence now implicates SMC5-SMC6 in chromosome dynamics during meiosis. However, the precise contribution of this pair of proteins remains unknown.

3. How does topoisomerase III function in DNA metabolism? In both bacteria and eukaryotes, topoisomerase III is closely tied to the function of helicases in the RecQ family. In humans, defects in the genes encoding RecQ family helicases lead to genetic diseases, including Bloom and Werner syndromes, that are characterized by genomic instability and an increased propensity to develop cancer. These enzymes are essential to many aspects of DNA metabolism. For example, the topoisomerase III–2RecQ pairing in bacteria plays a critical role in resolving topological problems that accompany the convergence of replication forks. Again, much remains to be learned about the mechanics of these complicated transactions.

HOW WE KNOW

The Discovery of Supercoiled DNA Goes through Twists and Turns

Lebowitz, J. 1990. Through the looking glass: The discovery of supercoiled DNA. *Trends Biochem. Sci.* 15:202–207.

Vinograd, J., J. Lebowitz, R. Radloff, R. Watson, and P. Laipis. 1965. The twisted circular form of polyoma viral DNA. *Proc. Natl. Acad. Sci. USA* 53:1104–1111.

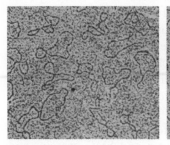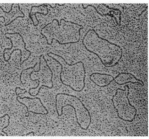

FIGURE 1 These electron micrographs of polyoma virus DNA, form I (20S) at left and form II (16S) at right, show the unexpected circular patterns. *[Source: Courtesy Vinograd J. et al., Proc. Natl. Acad. Sci. USA 53: 1104–1111, 1965. Images provided by Dr. P. Laipis.]*

By 1962, the double-helical structure of DNA was established, but little was known about the detailed structure of chromosomes. Researchers had been surprised by reports that the *E. coli* chromosome was a continuous circle. Were circular chromosomes unusual, or were they widespread? Two research groups, led by Renato Dulbecco and Jerome Vinograd, both at the California Institute of Technology, took up the problem of DNA structure, using the mammalian polyoma virus. The experiments with polyoma DNA led the Vinograd group to the concept of supercoiling.

Analytical ultracentrifugation measures the migration of molecules through a density gradient when ultracentrifugal force is applied (see Figure 16-2); molecules that migrate farther through the gradient have a larger sedimentation coefficient (S). Using this and other methods, researchers found three forms of polyoma DNA in the isolated preparations, which migrated with sedimentation coefficients of 20S, 16S, and 14S. Because the 20S and 16S forms (forms I and II, respectively) were most abundant, the researchers focused on those two. The two forms had the same molecular weight, so the different sedimentation velocities had to involve different conformations. Form I could be isolated in almost pure preparations, if care was taken while preparing the DNA.

Both the Dulbecco and the Vinograd groups published reports indicating that form I could be converted to form II by the addition of reagents that promote DNA strand cleavage. The researchers developed a model in which they assumed form I to be circular and form II to be linear. The observed kinetics suggested that the conversion occurred in a single step. This challenged the model, because two strands would have to be cleaved to generate the linear molecule, and both would have to be broken at the same position.

The Vinograd group decided to examine all three forms of polyoma DNA (20S, 16S, and 14S)

by electron microscopy. Philip Laipis carried out the first experiments. His examination of forms I and II yielded the unexpected result that both were circular (Figure 1). Laipis was a relatively inexperienced undergraduate, and some of the other researchers in the lab initially assumed he had made a mistake in preparing the samples. However, several repetitions yielded the same result. Only form III (14S) was linear. Careful controls eliminated any possibility that the result reflected a selective elimination of linear forms during preparation of the DNA for examination; when the researchers premixed measured amounts of forms III and II and then examined them, the linear and circular DNAs were always there in the expected proportions. Additional kinetic studies showed that only one strand break was needed to convert form I to form II. Cleaving form I with endonuclease I (which cleaves both strands) produced only the form III (14S) DNA, with no form II. Forms I and II also had identical buoyant densities, making it unlikely that some non-DNA mass was removed in converting form I to form II.

The important clue lay in the electron micrographs. The form I molecules appeared much more twisted on themselves than the form II molecules. Denaturation experiments showed that the strands of form II could be separated, but those of form I could not. The researchers gradually established that form II was a nicked DNA circle. Understanding form I required a little more work. A comment from colleague Robert Sinsheimer led Vinograd to focus on the twisted nature of the form I DNA. His subsequent modeling with phone cords was documented in 1990 in a delightful retrospective article by Jacob Lebowitz, one of the authors on the 1965 paper. Although the term "supercoiling" was not yet in common use, the discovery of the supercoiled nature of polyoma DNA opened an entirely new field of investigation.

The First DNA Topoisomerase Unravels Some Mysteries

Wang, J.C. 1971. Interaction between DNA and an *Escherichia coli* protein ω. *J. Mol. Biol.* 55:523–533.

Over the course of the twentieth century, life scientists were getting used to a fundamental idea: if a change occurs in any cellular structure, there is at least one enzyme that catalyzes it. The first discovery of an enzyme involved in DNA supercoiling was made by James Wang and reported in 1971.

Working at the University of California, Berkeley, Wang had initiated studies of superhelicity in small DNAs that could be isolated from bacteria. Focusing on the DNA from bacteriophage λ, he noticed that *E. coli* extracts contained a macromolecule that converted the superhelical form of the circular DNA to a relaxed form. He did what any good molecular biologist would do: he purified the macromolecule. The result was a protein that he dubbed the ω (omega) protein. Later, it became known as DNA topoisomerase I.

Initially, the purification was incomplete. However, Wang could establish that the macromolecule was a protein, because its activity was not lost after extended dialysis (using a membrane that allowed small molecules to escape but retained larger proteins); activity was lost when the preparation was heated to 50°C (a temperature that denatures most proteins). Wang demonstrated that the conversion of a DNA circle with 150 superhelical turns to a relaxed circle did not happen in one step. Using sedimentation velocity studies and electron microscopy, he showed that the change was progressive, with one or a few superhelical turns lost in each catalytic step. The activity affected only negative, not positive, superhelicity.

Wang concluded that the enzyme had two activities: a nicking activity and a strand-joining activity. He proposed that the ω protein acted by transiently introducing a break into one strand, creating a swivel in the unbroken strand that would allow the removal of superhelical turns. The reaction required no enzymatic cofactors, suggesting that the reaction pathway featured a transient covalent intermediate. His speculation, shown in Figure 2, proved to be largely correct. Overall, the study produced a remarkable set of insights that thoroughly framed the subsequent study of this and related topoisomerases.

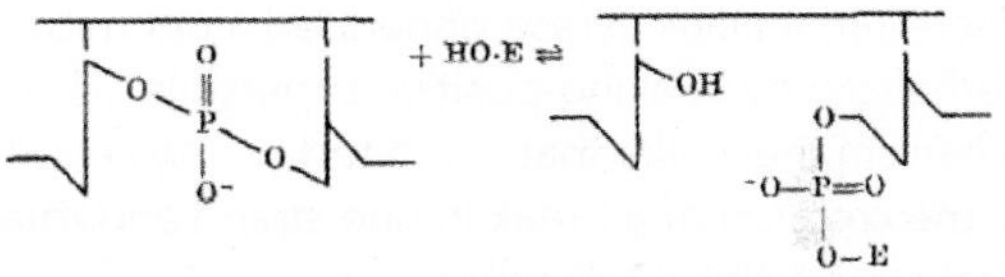

FIGURE 2 James Wang proposed this simple reaction mechanism for the chemistry of DNA strand nicking and closing by the ω protein (later renamed DNA topoisomerase I). HO·E represents a hydroxyl group on the enzyme (ω protein) [*Source: Reprinted from J.C. Wang, J. Mol. Biol. 55:523–533, 1971, with permission from Elsevier.*]

HOW WE KNOW

DNA Gyrase Passes the Strand Test

Brown, P.O., and N.R. Cozzarelli. 1979. A sign inversion mechanism for enzymatic supercoiling of DNA. *Science* 206: 1081–1083.

Nicholas Cozzarelli, 1938–2006 [*Source: Noah Berger Photography, www. noahbergerphoto.com.*]

In 1976, Martin Gellert and colleagues reported the discovery of a second topoisomerase in *E. coli*. The enzyme, DNA gyrase, had the novel property that it could introduce negative supercoils into DNA, hydrolyzing ATP in the process. DNA gyrase was quickly shown to be critical to DNA replication and other processes, and there was great interest in determining how it worked. Many researchers expected that DNA gyrase generated a net negative superhelicity by relaxing positive supercoils, using a mechanism much like that exhibited by the ω protein, with the creation of a break in one strand and rotation of that strand about the other.

Nicholas Cozzarelli and colleagues, at the University of Chicago, began to focus on experimental observations that did not fit this scheme. First, when active DNA gyrase was acting on a DNA, and the gyrase-DNA combination was treated with a protein denaturant, double-strand breaks were introduced into the DNA. Gyrase molecules were covalently linked to the 5'-phosphoryl groups in the DNA on *both* ends of the break. This implied that the normal mechanism of gyrase action involves an intermediate in which both strands, not just one, are cleaved. The research group also noticed that gyrase has the unusual capacity to catenate (interlink) two DNA circles. Such a reaction would require the formation of at least a transient double-strand break in one of the DNAs.

Pooling this and other information, Cozzarelli proposed a very different mechanism for gyrase action, one he dubbed "sign inversion" (Figure 3). He imagined that in a circular DNA, gyrase would bind to two segments of DNA that crossed over each other, thereby stabilizing a positive crossover, or node. The creation of such a node would necessarily create a compensating negative node elsewhere in the DNA molecule. Gyrase would then invert the sign of the bound node by breaking both DNA strands, passing the unbroken DNA segment through the break, and resealing the break on the other side. This would change the sign of the node to negative and effectively fix two negative supercoils in the DNA.

The sign inversion model made a novel and unique prediction. DNA gyrase would do something very different from the ω protein: it would change superhelicity in increments of 2 rather than increments

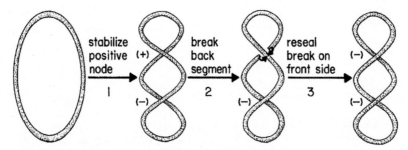

FIGURE 3 The sign inversion model for the generation of negative supercoils by DNA gyrase. [*Source: From Brown, P.O., & Cozzarelli, N.R. (1979) A sign inversion mechanism for enzymatic supercoiling of DNA. Science 206, no. 4422, 1081–1083. Reprinted with permission from AAAS.*]

of 1. This prediction was not trivial to test. Supercoiled circular DNA (such as plasmid DNA) is isolated from cells as a heterogeneous mixture of topoisomers with a roughly Gaussian distribution of linking numbers. Gyrase could shift the center of that distribution, but highlighting individual reaction steps to observe the predicted *Lk* increments of 2 would be difficult. Cozzarelli and his student, Patrick O. Brown, found a way to overcome the problem.

They initially focused on a particularly small circular DNA, a plasmid of about 2,400 bp called p15. Such a small DNA limited the total number of topoisomers in the Gaussian distribution and facilitated the separation of one topoisomer from another on an agarose gel. Using the ω protein, Brown and Cozzarelli took a naturally supercoiled preparation of p15 DNA and relaxed it completely. They then ran the DNA on an agarose gel under conditions in which the topoisomers of p15 were well separated. They cut the most abundant topoisomer out of the gel and extracted it, effectively isolating a DNA preparation with one topoisomer only.

With a topologically pure DNA in hand, the key experiment could be done. The researchers added enough DNA gyrase (about two heterotetramers per DNA molecule) to ensure that essentially every DNA circle had a bound gyrase. After incubating the DNA and gyrase for 3 minutes, they added ATP, but only enough (30 μM, or about one-tenth of the K_m) to support a slow reaction. The results, shown in Figure 4, are most striking at the 5 second time point. The major product is a species with a change in linking number (ΔLk) of 22. A small amount of DNA with a ΔLk of 24 is also evident. Markers showing topoisomers differing by a ΔLk of 1 are shown in the lanes marked MW. At later time points, the DNA becomes more supercoiled, but topoisomers with an odd-numbered ΔLk do not appear.

Brown and Cozzarelli carried out one additional test. After 5 minutes, the p15 DNA was highly supercoiled. They then added novobiocin, an antibiotic that inhibits supercoiling but not relaxation by gyrase (see Highlight 9-2). After another 30 minutes of incubation, much of the DNA had been substantially relaxed (see Figure 4). The topoisomers present included species

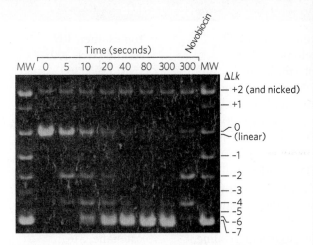

FIGURE 4 The p15 plasmid DNA was topologically pure; the minor band at the top of the *t* = 0 column is a small amount of nicked DNA, due to damage inflicted during purification. The p15 plasmid was mixed with gyrase for 5 to 300 seconds. After 300 seconds, a sample of DNA was treated with novobiocin and incubated for another 30 minutes. Marker lanes (MW) show the p15 plasmid DNA with change in linking number (ΔLk) in increments of 1. [*Source: From Brown, P.O., & Cozzarelli, N.R. (1979) A sign inversion mechanism for enzymatic supercoiling of DNA. Science 206, 1081–1083. Reprinted with permission from AAAS.*]

with superhelicity changes of 0, –2, and –4 relative to the starting material. This demonstrated that gyrase promoted both supercoiling and relaxation of DNA in increments of 2. The result fulfilled a key expectation of any enzymatic reaction—that the reaction pathway is the same in the forward and reverse directions. Overall, the experiments constituted a compelling case for the sign inversion model and provided the impetus to eventually define two separate classes of topoisomerases.

These advances helped explain the mechanism of action of a range of important antibiotics and antitumor drugs (see Highlight 9-2). They were among a string of important contributions from the Cozzarelli lab, first at Chicago and later at the University of California, Berkeley. With an ebullient personality and a creative intellect, Cozzarelli inspired a generation of scientists, both as a mentor and a colleague. Cozzarelli died due to complications of a treatment for Burkitt's lymphoma in 2006, at the age of 67. But his lab motto "Blast Ahead" lives on.

KEY TERMS

PROBLEMS

1. The diameter of a typical human cell nucleus is approximately 6 μm. Human chromosome 2 consists of approximately 243 million contiguous base pairs. Calculate the length of chromosome 2 if it were extended in a relaxed B-form structure (refer to structural parameters provided in Figure 6-17). How does this compare with the diameter of the nucleus?

2. What is the superhelical density (σ) of a closed-circular DNA with a length of 4,200 bp and a linking number (*Lk*) of 374? What is the superhelical density (σ) of the same DNA when *Lk* = 412? In each case, is the molecule negatively or positively supercoiled?

3. The T4-like bacteriophage JS98 has a DNA of molecular weight 1.11×10^8 contained in a head about 100 nm long.

   (a) Calculate the length of the DNA (assume the molecular weight of a nucleotide pair is 650) and compare it with the length of the JS98 head.
   (b) Consult the online database Genome (www.ncbi.nlm.nih.gov/genome). What is the exact number of base pairs in the JS98 genome?

4. The base composition of phage M13 DNA is A, 23%; T, 36%; G, 21%; C, 20%. What does this tell you about the DNA structure of phage M13?

5. The complete genome of the simplest bacterium known, *Mycoplasma genitalium*, is a circular DNA molecule with 580,070 bp. Calculate the molecular weight (assume the molecular weight of a nucleotide pair is 650) and contour length (when relaxed) of this molecule. What is Lk_0 for this *Mycoplasma* chromosome? If σ = −0.06, what is *Lk*?

6. A closed-circular DNA molecule in its relaxed form has an *Lk* of 500. Approximately how many base pairs are in this DNA? How does the linking number change (increases, decreases, doesn't change, becomes undefined) in each of the following situations?

   (a) A protein complex binds, wrapping the DNA around it to form a solenoidal supercoil.
   (b) One DNA strand is broken.
   (c) DNA gyrase and ATP are added to the DNA solution.
   (d) The double helix is denatured by heat.

7. In the presence of a eukaryotic condensin and a type II topoisomerase, the *Lk* of a relaxed closed-circular DNA molecule does not change. However, the DNA becomes highly knotted, as shown in the figure below. The formation of knots requires breakage of the DNA, passage of a segment of DNA through the break, and religation by the topoisomerase. Given that every reaction of the topoisomerase would be expected to result in a change in linking number, how can *Lk* remain the same?

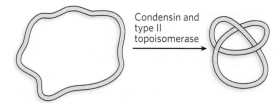

8. Bacteriophage λ infects *E. coli* by integrating its DNA into the bacterial chromosome. The success of this recombination depends on the topology of the *E. coli* DNA. When the superhelical density (σ) of the *E. coli* DNA is greater than −0.045, the probability of integration is <20%; when σ is less than −0.06, the probability is >70%. Plasmid DNA isolated from an *E. coli* culture is found to have a length of 13,800 bp and an *Lk* of 1,222. Calculate σ for the plasmid DNA (which reflects the superhelical density of all DNA in the cell, plasmid and chromosome), and predict the likelihood that bacteriophage λ will be able to infect this culture.

9. (a) What is the *Lk* of a 5,250 bp circular duplex DNA molecule with a nick in one strand?
   (b) What is the *Lk* of the molecule in (a) when the nick is sealed (forming a relaxed molecule)?
   (c) How would the *Lk* of the molecule in (b) be affected by the action of a single molecule of *E. coli* topoisomerase I?
   (d) What is the *Lk* of the molecule in (b) after eight enzymatic turnovers by a single molecule of DNA gyrase in the presence of ATP?
   (e) What is the *Lk* of the molecule in (d) after four enzymatic turnovers by a single molecule of bacterial type I topoisomerase?
   (f) What is the *Lk* of the molecule in (d) after binding of one protein that wraps DNA around it to form a solenoidal supercoil, with no other changes in the DNA?

10. Explain how the underwinding of a B-DNA helix might facilitate or stabilize the formation of Z-DNA.

11. **(a)** Describe two structural features required for a circular DNA molecule to maintain a negatively supercoiled state.

    (b) List three structural conformations that become more favorable when a DNA molecule is negatively supercoiled.

    (c) What enzyme, with the aid of ATP, can generate negative supercoiling in DNA?

    (d) Describe the physical mechanism by which this enzyme acts.

12. YACs are used to clone large pieces of DNA in yeast cells. What three types of DNA sequence are required to ensure proper replication and propagation of a YAC in a yeast cell, and what is the function of each?

13. When DNA is subjected to electrophoresis in an agarose gel, shorter molecules migrate faster than longer ones. Closed-circular DNAs of the same size but with different linking numbers can also be separated on an agarose gel; topoisomers that are more supercoiled, and thus more condensed, migrate faster through the gel. In the gel shown below, purified plasmid DNA has migrated from top to bottom. There are two bands, with the faster band much more prominent.

[Source: Courtesy Michael M. Cox.]

    (a) What are the DNA species in the two bands?

    (b) If topoisomerase I is added to a solution of this DNA, what will happen to the upper and lower bands after electrophoresis?

    (c) If DNA ligase is added to the DNA, will the appearance of the bands change?

    (d) If DNA gyrase and ATP are added to the DNA after adding DNA ligase, how will the band pattern change?

14. As in the experiment described in Problem 13, gels are run to separate closed-circular DNAs of the same size but different linking number. The new gels include two different concentrations of the dye chloroquine:

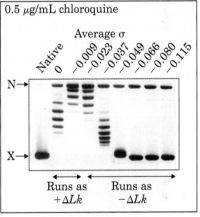

Chloroquine

Chloroquine intercalates between base pairs and stabilizes a more underwound DNA structure. When the dye binds to a relaxed, closed-circular DNA, the DNA is underwound where the dye binds, and unbound regions take on positive supercoils to compensate. In the experiment shown below, topoisomerases were used to make preparations of the same closed-circular DNA with different superhelical densities (σ). Completely relaxed DNA migrated to the position labeled N (nicked), and highly supercoiled DNA (above the limit where individual topoisomers can be distinguished) migrated to the position labeled X.

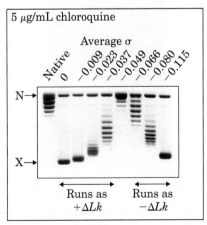

Gel A

Gel B

[Source: R. P. Bowater (2005) Supercoiled DNA: structure. In Encyclopedia of Life Sciences, doi: 10.1038/npg.els.0006002, John Wiley & Sons, Inc./ Wiley InterScience, www.els.net Copyright © 2010, John Wiley and Sons.]

(a) In gel A, why does the σ = 0 lane (i.e., DNA prepared so that σ = 0, on average) have multiple bands?

(b) In gel B, is the DNA from the σ = 0 preparation positively or negatively supercoiled in the presence of the intercalating dye?

(c) In both gels, the σ = −0.115 lane has two bands, one a highly supercoiled DNA and one relaxed. Propose a reason for the presence of relaxed DNA in these lanes (and others).

(d) The native DNA (leftmost lane in each gel) is the same closed-circular DNA isolated from bacterial cells and untreated. What is the approximate superhelical density of this native DNA?

15. In the early electron microscopy experiments of Vinograd and colleagues (see this chapter's How We Know section), the polyoma virus DNA was clearly circular in both form I and form II. However, the DNA in form II tended to be spread out on the electron microscope grid, whereas the DNA in form I tended to twist on itself, often repeatedly. Explain these observations.

16. A small plasmid DNA is isolated, and one negatively supercoiled topoisomer is purified (see this chapter's How We Know section). A small amount of bacterial topoisomerase III or topoisomerase II (DNA gyrase) plus ATP is added to the plasmid DNA in two separate experiments, allowing only limited reaction of the topoisomers. The experiments yield the DNA banding patterns shown below. Which pattern is produced by DNA topoisomerase III, and which by DNA gyrase?

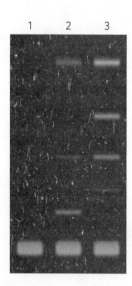

17. If bacterial topoisomerase I is inactivated by mutating the gene (*topA*) encoding it, will the superhelical density of the cell's chromosomal DNA increase or decrease? When such mutations are introduced into the *topA* gene, compensating mutations appear rapidly in another gene, which serve to partly ameliorate the deleterious cellular

effects of the *topA* mutations. Which gene is likely to be the site of the new mutations, and why?

DATA ANALYSIS PROBLEM

Boles, T.C., J.H. White, and N.R. Cozzarelli. 1990. Structure of plectonemically supercoiled DNA. *J. Mol. Biol.* 213:931–951.

18. The structural parameters of plectonemically supercoiled DNA in solution were first examined by Cozzarelli and coworkers. Using a combination of electron microscopy and gel electrophoresis, they examined the properties of plasmid DNAs with different superhelical densities (σ). DNAs were spread for electron microscopic examination. DNA twists on itself to form a superhelix. The length of the superhelical axes of the entire molecule can be determined by measuring and adding up the lengths of the axes of all segments and branches. In the example shown below, there are five segments with three branch points. The superhelix is seen as the crossing of the DNA as it winds about itself, with the crossing points on the DNA defined as nodes.

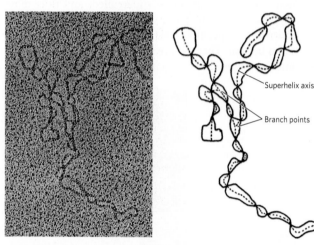

[Source: Reprinted from Boles, T.C., White, J.H., Cozzarelli, N.R. (1990), Structure of plectonemically supercoiled DNA. J. Mol. Biol. 213: 931–951, with permission from Elsevier.]

(a) How many nodes are there in this DNA? Each node is roughly equivalent to a supercoil.

(b) If one assumes that each supercoil results from an underwinding of the DNA by one turn, and the DNA has a length of 7,000 bp, calculate σ.

The investigators found a linear relationship between the number of nodes (*n*) and Δ*Lk*, such that *n* = −0.89 Δ*Lk* (implying that most, but not all, of the change in linking number results from changes in the writhe or twist of the DNA helix axis on itself).

(c) With this in mind, calculate the σ for this molecule.

(d) Is there any evidence of solenoidal supercoiling?

ADDITIONAL READING

General

Cozzarelli, N.R., and J.C. Wang, eds. 1990. *DNA Topology and Its Biological Effects*. Cold Spring Harbor, NY: Cold Spring Harbor Laboratory Press. A definitive resource.

Kornberg, A., and T.A. Baker. 1991. *DNA Replication*, 2nd ed. New York: W.H. Freeman & Company. A good place to start for further information on the structure and function of DNA.

DNA Supercoiling

Boles, T.C., J.H. White, and N.R. Cozzarelli. 1990. Structure of plectonemically supercoiled DNA. *J. Mol. Biol.* 213:931-951. A study that defines several fundamental features of supercoiled DNA.

Garcia H.G., P. Grayson, L. Han, M. Inamdar, J. Kondev, P.C. Nelson, R. Phillips, W. Widom, and P.A. Wiggins. 2007. Biological consequences of tightly bent DNA: The other life of a macromolecular celebrity. *Biopolymers* 85:115-130. A nice description of the physics of bent DNA.

Lebowitz, J. 1990. Through the looking glass: The discovery of supercoiled DNA. *Trends Biochem. Sci.* 15:202-207. A short and interesting historical note.

Marko, J.F. 2007. Torque and dynamics of linking number relaxation in stretched supercoiled DNA. *Phys. Rev.* E76(2, Pt. 1):021926. A quantitative look at DNA properties.

Proteins Involved in DNA Compaction

Berger, J.M. 1998. Type II DNA topoisomerases. *Curr. Opin. Struct. Biol.* 8:26-32.

Champoux, J.J. 2001. DNA topoisomerases: Structure, function, and mechanism. *Annu. Rev. Biochem.* 70:369-413. An excellent summary of the topoisomerase classes.

Hirano, T. 2006. At the heart of the chromosome: SMC proteins in action. *Nat. Rev. Mol. Cell Biol.* 7:311-322.

Nitiss, J.L. 2009. DNA topoisomerase II and its growing repertoire of biological functions. *Nat. Rev. Cancer* 9:327-337.

Pommier, Y. 2009. DNA topoisomerase I inhibitors: Chemistry, biology, and interfacial inhibition. *Chem. Rev.* 109:2894-2902.

Thanbichler, M. 2009. Closing the ring: A new twist to bacterial chromosome condensation. *Cell* 137:598-600.

Wang, J.C. 2002. Cellular roles of DNA topoisomerases: A molecular perspective. *Nat. Rev. Mol. Cell Biol.* 3:430-440.

10 Nucleosomes, Chromatin, and Chromosome Structure

C. David Allis [Source: Courtesy C. David Allis.]

MOMENT OF DISCOVERY

In the early 1990s, there was a huge attempt by many top-notch laboratories to *purify an enzyme that would modify histones*. In particular, histones were known to be acetylated, but the presumptive "histone acetylase" enzyme that carries out this important task eluded the grasp of biochemists who had tried for years to identify it. Did it even really exist? Was there another way nature devised to modify histones? I didn't think that would be the case, and my laboratory was one of those dedicated to solving this mystery. Then in 1996 it finally happened, and it was a really neat succession of events that I'll always remember. Working with my graduate student, Jim Brownell, *we isolated a 55,000 dalton protein (p55) that displayed histone acetyltransferase (HAT) activity from enriched extracts of* Tetrahymena *in an in-gel HAT assay*. This was it! Finding this evidence provided the stimulus needed for a really Herculean effort. Brownell grew about 200 L of *Tetrahymena* culture, and after purification of p55 there was just a couple of gel lanes worth of purified protein. But it was enough for microsequencing of p55 peptides by R. Kobayashi (working at Cold Spring Harbor Laboratory), which provided the information to clone the p55 gene.

I vividly recall the weekend day that Brownell biked to the lab to run all of his precious, most highly purified p55 on a gel, then transferred it to a sequencing-compatible membrane and sent it to Kobayashi. This gel was so precious to us that we both kept watching it every minute to make sure all was going well. As his advisor, I tried to appear calm and collected on the outside; on the inside, I was going crazy. What an important gel this turned out to be! Totally unexpected to us, p55 from *Tetrahymena* was homologous to Gcn5, a known transcriptional coactivator in yeast. This discovery essentially established that Gcn5/p55 performed its transcriptional function by being a HAT, an enzyme that adds an acetyl group to regulatory lysines on histones. This seminal finding was published in *Cell* on March 22, 1996, which, remarkably, happened to be my birthday—I couldn't imagine a better birthday present. With help from Sharon Dent, Brownell expressed and purified the yeast Gcn5, and it, too, was active in the in-gel HAT assay! I remember when Brownell developed the x-ray film of the gel, and upon seeing the positive result, he was barely able to talk: he could only run around wildly, jumping up and down the hallway. I felt the same way. The image of Jim, and the gel result, too, make for a day I will never forget.

—C. David Allis, *on establishing that p55 from* Tetrahymena *is a histone acetylase, as is transcription factor Gcn5*

Eukaryotes contain thousands of times more DNA than do bacteria, and as a result, the DNA condensation problems of eukaryotes—compacting the DNA so that it fits in the cell nucleus—are more complex than those of bacteria. In Chapter 9 we introduced DNA topoisomerases, enzymes that can untwist DNA and keep the long DNA molecules within cellular chromosomes from becoming intertwined. We also discussed the ring-shaped condensins and cohesins that encircle DNA segments to hold them tightly together in loops, thus increasing compaction. This chapter focuses on a specific DNA condensation particle of eukaryotes—the nucleosome—around which DNA is wrapped. Bacteria do not contain nucleosomes, although they have small, basic (positively charged) proteins that are involved in condensing their DNA.

DNA compaction must be dynamic: changes in the degree of condensation must occur quickly and when needed, as the cell passes through the stages of the cell cycle (see Figure 2-10). Furthermore, when in its most highly compacted form, DNA is not accessible to transcription or replication enzymes, so it must be able to rapidly expose regions containing genes that are required at any given moment, and then condense again. Changes in DNA compaction in a cell can occur on a global level (such as during mitosis or replication) or on a local level (such as giving access to specific genes for transcription regulation). To accommodate these essential activities, modification enzymes have evolved that alter the state of DNA condensation by various means, and these enzymes can target their activity to specific regions of the chromosome that must be transcribed or replicated.

In this chapter we explore how nucleosome units are arranged in higher levels of chromosome structure and how the cell manipulates nucleosomes in various ways to change the state of DNA condensation, which affects the regulation of gene expression. We then look at how nucleosomes are modified by enzymes that attach various small chemicals to the nucleosome proteins. These chemical alterations regulate genes and, in fact, are inherited, passing down this information from one cell generation to the next. This is especially important during an organism's development, to maintain new transcriptional programs of differentiated cell types. Genetic information not coded by the DNA sequence is referred to as *epigenetic* information, and defects in epigenetic information are associated with cancer.

10.1 NUCLEOSOMES: THE BASIC UNITS OF DNA CONDENSATION

Scientists have been fascinated with the structure and behavior of chromosomes for more than 100 years. Chromosomes were first made strikingly visible through the use of dyes that stain specific subcellular structures. The deeply stained colored bodies (thus the name "chromosomes") appeared in pairs and separated into two new daughter cells at cell division, correlating with Mendel's observations that the particles of heredity come in pairs. As we learned in Chapter 2, scientists eventually came to understand that DNA is the hereditary substance inside chromosomes.

The material of chromosomes, both protein and DNA, is often referred to as **chromatin**. The protein component is about equal in mass to the DNA component. Some chromatin proteins are SMC proteins (see Chapter 9), topoisomerases, and transcription regulatory molecules; however, the largest protein component of chromatin is the histones. **Histones** are highly conserved, basic proteins that assemble into octameric complexes containing two each of four different histone subunits. DNA wraps around the histones to form condensed **nucleosomes**. Beginning with nucleosomes, eukaryotic chromosomal DNA is packaged into a succession of higher-order structures that ultimately yield the compact chromosome seen under the light microscope (see, for example, Figure 2-8, which shows the chromosomes of plant cells).

Histone Octamers Organize DNA into Repeating Units

Evidence that DNA is packaged into regularly organized units came from several types of studies. In one approach, chromosomal DNA was treated with a nonspecific DNA nuclease, such as micrococcal nuclease, that cuts DNA wherever it is not associated with proteins. The digested DNA fragments were then analyzed for size in an agarose gel. If DNA were packaged by proteins into units of a particular size, the nuclease would cleave only the DNA between these units, and the protected DNA segments (i.e., those bound to protein) would migrate in the gel as a ladder of unit-sized bands. If there were no regular repeating unit of protein-DNA packaging and proteins were distributed on DNA in a random way, then nuclease digestion would produce a smear of DNA fragments with no regular pattern. The results of the experiments revealed a series of regularly spaced DNA bands about 200 bp apart, indicating that DNA is packaged by proteins into units that encompass approximately 200 bp (**Figure 10-1**).

When the protein-DNA units, or nucleosomes, were examined by SDS-polyacrylamide gel electrophoresis (SDS-PAGE), four histone proteins (designated H2A, H2B, H3, and H4) were found in approximately equimolar ratios (**Figure 10-2**). A fifth histone (H1) was present in about half the amount relative to the other four histones. The five histones have molecular weights (M_r) between 11,000 and 21,000 (Da). Histones are rich in the basic amino acids arginine and lysine, which together make up about 25% of the amino acid residues in any given

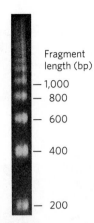

Fragment length (bp)

— 1,000

— 800

— 600

— 400

— 200

FIGURE 10-1 Evidence of DNA packaging into repeating units obtained from an experiment using a nuclease. Carefully isolated chromatin was treated with micrococcal nuclease and analyzed by agarose gel electrophoresis. The result was a DNA ladder of fragments that differed in length by 200 bp, suggesting that DNA packaging involves a repeat unit of 200 bp. [*Source: Roger Kornberg, MRC Laboratory of Molecular Biology.*]

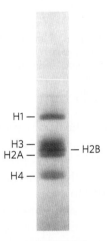

H1 —

H3 —
H2A —   — H2B

H4 —

FIGURE 10-2 Histone representation in nucleosomes. Histones were extracted from chromatin, and the histone subunits were separated by SDS–polyacrylamide gel electrophoresis. Measurement of the band intensity showed histones H2A, H2B, H3, and H4 present in equal amounts, and histone H1 at about half the level of the other histones. [*Source: Sakol Panyim, Roger Chalkley, High resolution acrylamide gel electrophoresis of histones. Archives of Biochemistry and Biophysics, Volume 130, 1969, Pages 337-346.*]

histone protein (**Table 10-1**). The amino acid sequences of histone proteins are highly conserved among eukaryotic organisms. Histones H3 and H4 are nearly identical in all eukaryotes, suggesting strict conservation of their functions. For example, only 2 of the 102 amino acid residues differ between the H4 histones of peas and cows, and only 8 residues differ between the H4 histones of humans and yeast. Histones H1, H2A, and H2B show less sequence similarity among organisms, but on the whole, they are more conserved than other types of proteins. Eukaryotes

TABLE 10-1

Types and Properties of Histones

Histone	M_r (Da)	Number of Amino Acid Residues	Content of Basic Amino Acids (% of total)	
			Lys	Arg
H1	21,130	223	29.5	1.3
H2A	13,960	129	10.9	9.3
H2B	13,774	125	16.0	6.4
H3	15,273	135	9.6	13.3
H4	11,236	102	10.8	13.7

Note: For H1, H2A, and H2B, size varies from species to species; the data here are for bovine histones, except H1, which is from rabbit.

also have several variant forms of certain histones, notably histones H2A and H3, which, as we see later in the chapter, have specialized roles in DNA metabolism.

To understand how the histones are organized within the nucleosome, the native state of the nucleosome unit must be preserved. Early studies of nucleosomes used denaturing methods of extraction that disrupted their native state. Later, by extracting chromatin with mild salt solutions (2 M sodium chloride and 50 mM sodium acetate), researchers kept nucleosomes intact and could investigate the composition and organization of the nucleosome unit.

Some of the key studies involved protein **cross-linking**. In this technique, a chemical with two reactive groups is used to react with a protein complex. Because the chemical has two reactive groups, it can covalently attach to two proteins, but because the chemical is a small molecule, it can only react with two proteins that are close together. Thus, identification of the cross-linked proteins reveals which proteins are next to each other in an oligomer. Based on findings from these procedures, Roger Kornberg proposed how histones are organized within the nucleosome (see the How We Know section at the end of this chapter).

Kornberg suggested that most of the 200 bp DNA segment in a protein-DNA unit is wrapped around a **histone octamer** composed of two copies each of histones H2A, H2B, H3, and H4. These four histones have come to be known as the **core histones** (**Figure 10-3a**). The remainder of the DNA serves as a linker between nucleosomes, to which histone H1 binds. Early studies of nucleosome-DNA structure included visualization of nucleosomes in the electron microscope by Ada Olins and Donald Olins at Oak Ridge National Laboratory. These studies revealed a structure in which the DNA is bound tightly to beads of protein, often regularly spaced like beads on a string (**Figure 10-3b**)—a result that defined the nucleosome as a basic unit of DNA compaction.

(a)

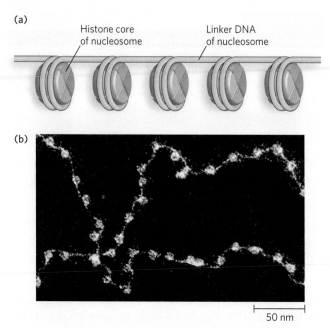

(b)

50 nm

FIGURE 10-3 Nucleosomes as beads on a string. (a) Regularly spaced nucleosomes consist of core histone proteins bound to DNA. (b) In this electron micrograph, the DNA-wrapped histone octamers are clearly visible, with linker DNA between them. *[Source: (b) Don W. Fawcett/Science Source.]*

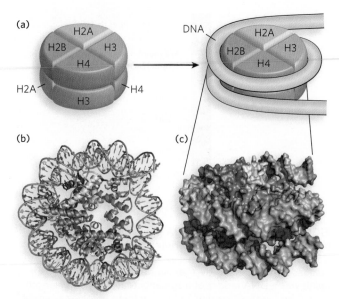

FIGURE 10-4 Structural model of the nucleosome.
(a) The simplified structure of a nucleosome octamer (left), with DNA wrapped around the histone core (right). (b) A ribbon representation of the structure of the nucleosome from the African frog *Xenopus laevis*, with different colors representing the five histones, matching the colors in (a). The view in (b) is rotated relative to the view in (a) to show one face of the nucleosome. (c) Surface representation of the nucleosome. A 146 bp segment of DNA binds in a left-handed solenoidal supercoil that circumnavigates the histone complex 1.67 times. The orientation in (c) is the same as that in (a) to show the DNA wrapping around the nucleosome. *[Source: (b) PDB ID 1AOI.]*

Under physiological conditions, formation of the histone octamer from individual histone proteins requires the presence of DNA. In the absence of DNA, the highly conserved H3 and H4 subunits form a tightly associated heterotetramer that contains two of each subunit, and the less conserved H2A and H2B subunits form a heterodimer. Without DNA, these components do not assemble into a histone octamer unless incubated under nonphysiological conditions at high salt concentrations. In the presence of DNA, however, two H2A-H2B heterodimers assemble with one H3-H4 heterotetramer and the DNA to form the nucleosome.

DNA Wraps around a Single Histone Octamer

The crystal structure of a histone octamer in a complex with 146 bp of DNA reveals the detailed architecture of a nucleosome particle (**Figure 10-4**). The most striking feature is the tight wrapping of DNA around the octamer in the form of a left-handed solenoidal supercoil (described in Chapter 9). Overall, the supercoil arrangement of DNA on the nucleosome compacts the DNA six- to sevenfold. The DNA is not uniformly bent, but instead follows a pattern of relatively straight 10 bp segments joined by bends.

Each histone contains a **histone-fold motif**, three α helices linked by two short loops (**Figure 10-5**). The elemental structural unit of the nucleosome is a head-to-tail dimer of histone-fold motifs of either the H3-H4 pair or the H2A-H2B pair. Each histone-fold dimer forms a V-shaped structure that contains three DNA-binding

sites (**Figure 10-6a**). The octamer structure shows that the connections between each dimer (the two H3-H4 pairs and two H2A-H2B pairs) of the four core histones are also mediated mainly by the highly conserved histone fold (**Figure 10-6b**).

The contacts between histones and DNA are mainly between the conserved histone fold and the phosphodiester backbone or minor groove of the DNA, in keeping with the relatively nonspecific binding of nucleosomes to DNA. Approximately half of the 142 hydrogen bonds of the histone fold occur between the DNA and peptide backbone atoms, rather than amino acid side chains. This seems counterintuitive, given the many basic side chains of histones. A possible explanation is that amino acid side chains are not as rigidly held in place as is the peptide backbone, and therefore hydrogen bonding of DNA to peptide backbone atoms more firmly secures the DNA to the protein. The basic histone side chains are needed to neutralize the negative charge of the DNA's phosphodiester backbone. Charge neutralization is important in DNA condensation, especially in the further stages of compaction.

The average DNA twist when wrapped around the histone octamer is 10.2 bp per turn, compared with the 10.5 bp per turn of unrestricted DNA; thus, the DNA structure must adapt to the octamer. This overtwisting of DNA results in a narrowing of the minor groove.

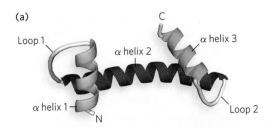

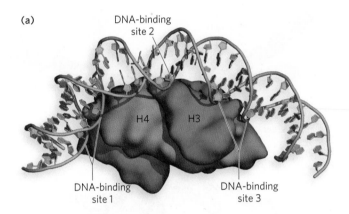

FIGURE 10-5 The histone-fold motif. (a) This internal structure of each core histone is formed from three α helices connected by two loops. (b) The amino acid sequences of the histone folds of the four core histones. Residues that are identical in the different histone subunits are shaded in red. *[Sources: (a) PDB ID 1AOI. (b) Data from V. Ramakrishnan, Annu. Rev. Biophys. Biomol. Struct. 26:83–112, 1997.]*

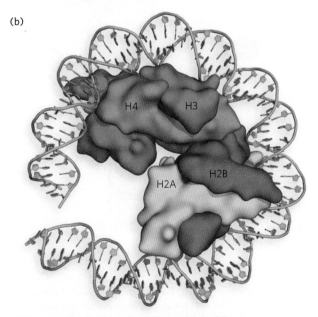

FIGURE 10-6 The histone-fold dimer. (a) A ribbon representation of the histone folds of one pair of H3-H4 subunits within the octamer, bound to DNA. DNA is in contact with the three DNA-binding sites formed by the two histone folds (i.e., the histone-fold dimer). (b) One face (one half) of the nucleosome contains two histone-fold dimers. *[Source: PDB ID 1AOI.]*

The change in shape of the DNA as it binds the histone octamer through bending and overtwisting implies that the octamer is more likely to bind DNA sequences that readily conform to such changes. For example, a local abundance of A=T base pairs in the minor groove of a DNA helix, where it is in contact with the histones, facilitates the compression of the minor groove that is needed for tight wrapping of DNA around the histone octamer (**Figure 10-7**). In fact, histone octamers assemble particularly well with sequences where two or more A=T base pairs are staggered at 10 bp intervals, because DNA is naturally bent at these sequences, and where two or more consecutive A=T base pairs are spaced along the same face of the helix, the DNA bends into a circle. Tracts of G≡C base pairs have the opposite effect, preventing compression of the minor groove, and thus are preferred at positions not facing nucleosomes.

The left-handed solenoidal supercoil of the 146 bp duplex that winds 1.67 times around the nucleosome reveals why eukaryotic DNA is underwound, even though eukaryotic cells lack topoisomerases that underwind DNA. Recall that the solenoidal wrapping of DNA is just one form of supercoiling that underwound (negatively supercoiled) DNA can assume (see Chapter 9). The tight wrapping of DNA around the histone core requires the removal of about one helical turn in the DNA. When the histone core of a nucleosome binds in vitro to relaxed, closed-circular DNA, the binding introduces a negative supercoil. Because binding in this fashion does not break DNA or change the linking number, formation of the negative supercoil around the histones must be accompanied by a compensatory positive supercoil in the unbound region of the DNA (see Figure 9-22). As described in Chapter 9, eukaryotic topoisomerases can relax positive supercoils and are required for the assembly of chromatin from purified histones and closed-circular DNA in vitro. Relaxing the unbound positive

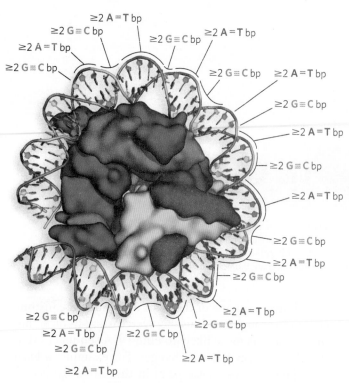

≥2 A=T bp
≥2 G≡C bp
≥2 A=T bp
≥2 G≡C bp
≥2 G≡C bp
≥2 A=T bp
≥2 G≡C bp
≥2 A=T bp
≥2 A=T bp
≥2 G≡C bp
≥2 A=T bp
≥2 G≡C bp
≥2 A=T bp
≥2 G≡C bp
≥2 A=T bp
≥2 G≡C bp
≥2 A=T bp
≥2 G≡C bp
≥2 A=T bp
≥2 G≡C bp'
≥2 A=T bp
≥2 G≡C bp
≥2 G≡C bp
≥2 A=T bp

FIGURE 10-7 The effect of DNA sequence on nucleosome binding. Runs of two or more adjacent A=T base pairs facilitate the bending of DNA, whereas runs of two or more G≡C base pairs have the opposite effect. When spaced at about 10 bp intervals, consecutive A=T base pairs help bend DNA into a circle. When consecutive G≡C base pairs are spaced 10 bp apart and offset by 5 bp from runs of A=T base pairs, the DNA binds tightly to the nucleosome. [*Source: PDB ID 1AOI.*]

supercoil leaves the negative supercoil fixed through its binding to the nucleosome histone core. Overall, this results in a decrease in linking number.

Histone Tails Mediate Internucleosome Connections That Regulate the Accessibility of DNA

Most of the mass in the histone octamer forms a tightly packed particle, but the N-termini of the histones protrude from the core particle and are less ordered (**Figure 10-8**). These N-terminal **histone tails** are flexible and therefore mostly disordered in the crystal structure. The parts of the histone tails that are visible in the crystal structure appear as irregular chains extending outward from the nucleosome disk. The tails exit the DNA superhelix through channels formed by the alignment of minor grooves of adjacent DNA helices every 20 bp. The histone tails do not contribute much strength to DNA binding, but they form intermolecular contacts with adjacent nucleosome particles and organize nucleosomes into a higher-order chromatin structure (**Figure 10-9**).

Chromatin structure is far from static and can, in fact, change in quite a dynamic fashion with changes in the environment and state of cellular differentiation. The ability

of chromatin to change structure affects gene expression and is a very active area of research. The histone tails are at the heart of the dynamic regulation of chromatin structure, because they are the target of numerous chemical modifications that control the access of regulatory proteins to DNA. These modifications affect the net electrical charge, shape, and other properties of histones, which in turn affect the structural and functional properties of chromatin. Histone tails can also be recognized by particular enzymes. As we will see, these modifications play important roles in the regulation of transcription, replication, and DNA repair.

Histone tail modifications impact the molecular interactions between adjacent nucleosomes, thereby changing the level of chromatin compaction and thus

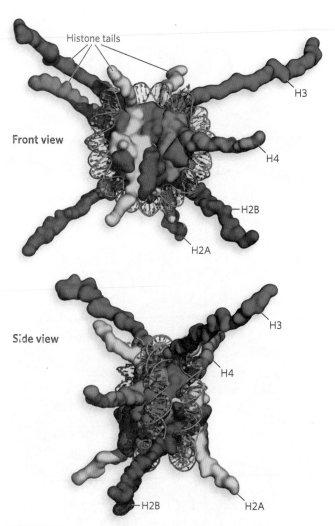

Histone tails

Front view

H3

H4

H2B

H2A

Side view

H3

H4

H2B

H2A

FIGURE 10-8 Histone tails. Two views of histone N-terminal tails protruding from between the two DNA duplexes that supercoil around the nucleosome. (a) Ribbon representation, looking at the circular face of the nucleosome. (b) Space-filling representation of the nucleosome as viewed from the side. Some tails pass between the supercoils through holes formed by alignment of the minor grooves of adjacent helices. The H3 and H2B tails emerge between the DNA turns wrapped around the histone; the H4 and H2A tails emerge between adjacent histone cores. [*Source: PDB ID 1AOI.*]

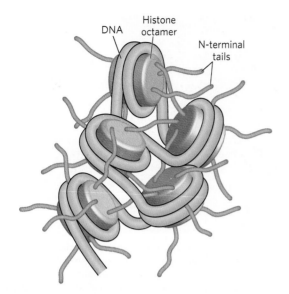

FIGURE 10-9 Internucleosomal contacts through N-terminal tails. The N-terminal tails of the histones of one nucleosome protrude from the particle and interact with adjacent nucleosomes, resulting in higher-order DNA packaging.

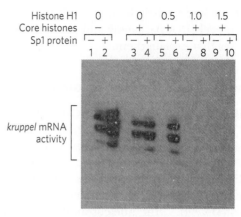

Histone H1	0		0	0.5	1.0	1.5
Core histones	−		+	+	+	+
Sp1 protein	−	+	−	+	−	+	−	+	−	+
	1	2	3	4	5	6	7	8	9	10

kruppel mRNA activity

FIGURE 10-10 The repression of transcription by histones. An in vitro transcription extract is used to monitor transcription from a promoter in the presence or absence of histone proteins and the transcription activator Sp1. The first two lanes are controls in the absence of histones: transcription is optimal, and Sp1 stimulates. Lanes 3 and 4 are also controls, showing results with core histones present but no H1. In lanes 5 to 10, histone H1 was added in concentrations of 0.5, 1.0, or 1.5 H1 molecules per nucleosome. Sp1 prevents gene repression by binding specific sequences in the promoter. [Source: P. J. Laybourne and J. T. Kadonaga, Science 254:238–245, 1991, Fig. 7B. Reprinted with permission from AAAS.]

the access of proteins to the altered chromatin structure. The tighter the internucleosome connections mediated by the histone tails, the less accessible DNA is to transcription factors and other proteins. Histone H1 also plays a role in DNA sequestration. Histone H1 is not as extensively modified by chemical groups as the core histones are, but it facilitates the general repressive effect of histones on transcription. This effect can be demonstrated experimentally in an in vitro assay that uses a cellular transcription extract, where transcription can proceed when DNA templates are added. As shown in **Figure 10-10**, plasmid DNA containing the *kruppel* gene of *Drosophila melanogaster* was transcribed in the absence (lanes 1 and 2) or presence (lanes 3 to 10) of histones. Adding the core histone proteins resulted in decreased RNA synthesis (compare lanes 1 and 3). Adding the linker histone H1 also had a profound effect: as histone H1 was titrated into the assay (as shown along the top of Figure 10-10), transcription was diminished even further (compare lanes 1, 3, and 5).

Transcription factors that bind specific sites on DNA can modulate the repressive effect of histones on transcription. The site-specific DNA-binding protein Sp1, which enhances transcription, protects against histone-mediated transcription repression (compare lanes 5 and 6 in Figure 10-10). One may infer that Sp1 binds DNA and blocks the binding of histones to DNA in its immediate vicinity. However, Sp1 is not effective in the presence of higher levels of H1 (compare lanes 6, 8, and 10).

SECTION 10.1 SUMMARY

• Chromatin refers to the protein and DNA components that comprise the material of chromosomes.

• The basic unit of DNA packaging is the nucleosome, composed of DNA wrapped around a histone octamer.

• The histone octamer contains a tetramer of histones H3-H4 and two dimers of histones H2A-H2B. The amino acid sequences of the histones are highly conserved among eukaryotes.

• Nucleosomes are linked together by DNA with one bound molecule of H1. The spacing of nucleosomes along DNA results in a beads-on-a-string appearance as seen by electron microscopy.

• DNA is wrapped around the histone octamer in a left-handed solenoidal supercoil, making 1.67 turns. Histones contain a conserved histone-fold motif, which contacts the DNA (mainly the phosphodiester backbone and minor groove).

• N-terminal histone tails are flexible and can be modified to affect DNA transcription. These modifications probably alter the molecular interactions of adjacent nucleosomes, thereby changing the access of proteins to DNA in the altered chromatin structure.

10.2 HIGHER-ORDER CHROMOSOME STRUCTURE

As we have seen, the nucleosome compacts DNA six- to sevenfold, yet condensing a 2 m thread of DNA into a 5 μm diameter nucleus ultimately requires a 10,000-fold reduction in length. This remarkable level of condensation is achieved in stages—none of which are defined at high resolution—as is DNA wrapping in the nucleosome.

Here we explore what is known about these processes of higher-level DNA packaging and some of the current hypotheses that explain them.

Histone H1 Binds the Nucleosome

Treating chromatin with a nonspecific nuclease to digest all naked DNA (DNA not protected by protein) yields a segment of about 168 bp to which all five histones are bound—the four core histones plus histone H1. More extensive nuclease digestion releases H1, leaving the core histone octamer bound to about 147 bp of DNA (i.e., the nucleosome). Indeed, the addition of H1 to a nucleosome results in protection of an additional 20 to 22 bp of linker DNA adjacent to the nucleosome, and thus H1 is often referred to as the **linker histone**. Only one H1 subunit is present per histone octamer, unlike the core histones, which are present in two copies each.

Compared with the core histones, H1 is more variable in sequence. Most organisms even have multiple H1 subtypes. For example, mammals have eight H1 variants that differ in their ability to condense chromatin. The H1 subtypes are expressed at different times of development or are present in different cell types. The avian counterpart of H1 is known as H5.

H1 consists of a short, 20 to 35 residue N-terminal region, a central globular domain of about 80 amino acid residues, and a long C-terminal region of about 100 residues. DNA binding is intrinsic to the central globular region, which contains two DNA-binding sites. It was originally thought that the two DNA-binding sites in H1 were used to bind each of the two linker DNA strands at the sites where DNA enters and exits the nucleosome. However, more recent studies indicate that H1 binds only one of the linker DNA strands, and the second DNA-binding site in histone H1 binds to the central region of the DNA supercoil in the nucleosome (**Figure 10-11**).

By binding an additional 20 bp of DNA, histone H1 alters the DNA entry and exit angles, facilitating the

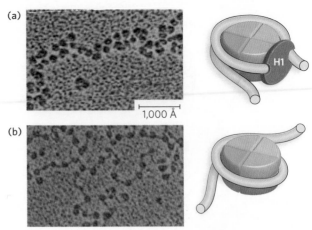

FIGURE 10-12 The zigzag appearance of nucleosomes in the presence of histone H1. Electron micrographs of nucleosomes (a) in the presence of H1 and (b) in the absence of H1. Histone H1 increases the zigzag appearance by decreasing the DNA entry/exit angles, as shown in the cartoon models to the right of each micrograph. In these electron micrographs, the samples were stained by rotary shadowing using carbon-platinum. Scale bar = 1,000 Å. [*Source: Photos © 1979 The Rockefeller University Press.The Journal of Cell Biology, 1979, 83:403-427. doi:10.1083/ jcb.83.2.403*]

packing of DNA into higher-order chromatin structures (**Figure 10-12**). When H1 is extracted from chromatin, DNA seems to enter and exit nucleosomes at different places, thus creating the beads-on-a-string appearance (see Figure 10-3b). The presence of H1 causes DNA to enter and exit the nucleosome at nearly the same place, resulting in a zigzag pattern with a DNA entry/exit angle between 40° and 100°, depending on the conditions of sample preparation. Overall, the level of condensation provided by H1 is six to seven times that of the nucleosome, for a total length reduction of 35- to 40-fold in chromatin. H1 helps nucleosomes condense into a higher level of packaging, the 30 nm filament (described below).

Nucleosomes generally repress transcription by sterically preventing the access of regulatory proteins to promoter sequences. H1 participates in this activity by stabilizing nucleosomes on the DNA and promoting higher-order chromatin structure that further compacts nucleosomal DNA. Indeed, most regions of actively transcribed DNA are known to lack histone H1.

Chromosomes Condense into a Compact Chromatin Filament

Under certain experimental conditions, such as increased ionic strength or the presence of particular divalent cations, nucleosomes condense into a compact filament with a width of about 30 nm, referred to as the **30 nm filament (Figure 10-13a)**. The 30 nm filament

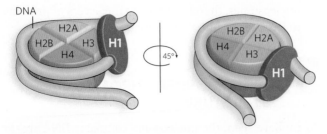

FIGURE 10-11 The binding of DNA by histone H1. Two views of the nucleosome containing histone H1. H1 has two DNA-binding sites, through which it makes contact with one arm of linker DNA and the central region of the DNA wrapped around the histone octamer.

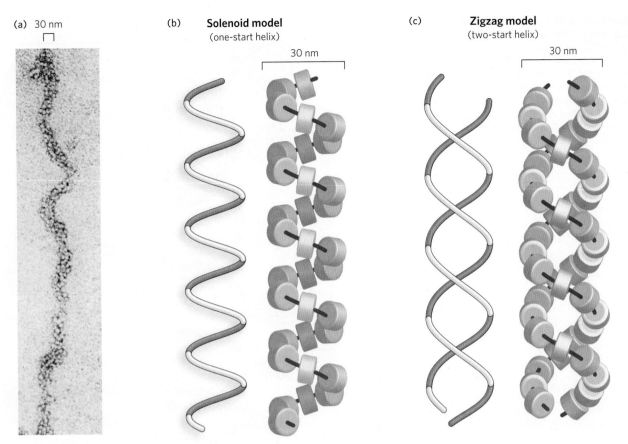

FIGURE 10-13 The 30 nm filament, a higher-order organization of nucleosomes. The compact filament is formed by the tight packing of nucleosomes. (a) The 30 nm filament as seen by electron microscopy. Two proposed models of filament structure are (b) the solenoid model and (c) the zigzag model. The path of DNA is shown as a dark brown line through the histones, for clarity of histone visualization. [*Source: (a) Barbara Hamkalo, Department of Molecular Biology and Biochemistry. University of California, Irvine.*]

is thought to exist in living cells, but this has yet to be rigorously proven. Although histone H1 promotes condensation into the 30 nm filament, it is not essential for forming it. In contrast, the N-terminal tails of the core histones are absolutely required, suggesting that the tails provide important nucleosome-nucleosome contacts needed for 30 nm filament formation. Recall that the crystal structure of the nucleosome shows that the terminal tails of H4, H3, and H2A make contact with adjacent nucleosomes, and these contacts may be involved in the condensation of nucleosomes into a filament.

The exact arrangement of nucleosomes in a 30 nm filament is still unclear, although any model should accommodate the following observations: (1) neutron diffraction studies place histone H1 in the center of the filament; (2) linker DNA should also be placed in the center of the filament, because H1 binds linker DNA; and (3) electron microscopy and x-ray diffraction studies indicate that the nucleosome units in the 30 nm filament

are arranged with a helical pitch of about 11 nm, the width of a nucleosome.

The two most widely accepted models for nucleosome arrangement in the 30 nm filament fit these criteria, and both are supported by substantial experimental evidence. The two models may in fact be alternative ways in which nucleosomes pack in different areas of the same 30 nm filament. In the **solenoid model** (also called the one-start helix model), the nucleosome array adopts a spiral shape, in which the flat sides of adjacent nucleosome disks are next to each other (**Figure 10-13b**). The linker DNA is presumed to bend inside the center of the filament to account for the observed constant thickness of the fiber with different linker lengths.

The second model for the 30 nm filament is the **zigzag model** (also called the two-start helix model), in which zigzag histone pairs stack on each other and twist about a central axis (**Figure 10-13c**). The zigzag model was inspired by the appearance of nucleosomes under

FIGURE 10-14 The crystal structure of a tetranucleosome. (a) The four nucleosomes are arranged in two zigzag pairs that stack on top of each other. Shown here are the DNA and histone octamers (top) and the DNA only (bottom), in orientations that illustrate the geometry of the two stacks. (b and c) A model of the 30 nm filament made by repeating the tetranucleosome in a continuous zigzag configuration. [*Source: Data from T. Schalch et al., Nature 436:138, 2005, Figs. 1, 3.*]

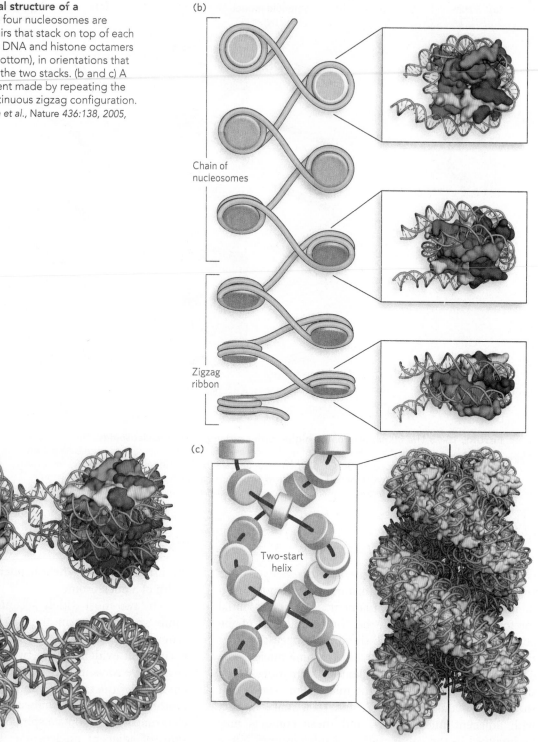

(b)

Chain of nucleosomes

Zigzag ribbon

(a)

(c)

Two-start helix

the electron microscope (see Figure 10-12). This model is supported by the crystal structure of a tetranucleosome, in which four nucleosomes are bound to one DNA molecule (**Figure 10-14a**). The structure shows two histone-pair zigzags stacked on top of each other, with the linker DNA passing through the central axis connecting two nucleosomes on opposite sides of the filament. A zigzag

model of a chromatin filament of many nucleosomes, based on the tetranucleosome structure, is shown in **Figure 10-14b** and **c**.

The two models for nucleosome stacking in a 30 nm filament are fundamentally very different, so it might seem they would be easy to distinguish experimentally. The inability of investigators to resolve this issue reflects the many irregularities in natural chromatin fibers, leading to the poor quality of experimental data. However, as noted above, both types of nucleosome packing might occur in the same chromatin filament.

Higher-Order Chromosome Structure Involves Loops and Coils

DNA is much more condensed inside chromosomes than in the 30 nm filament. Treating chromosomes with a low-salt buffer causes them to expand, and the edges of these swollen chromosomes reveal 30 nm filaments that appear to be organized in loops estimated at 40 to 100 kbp long (**Figure 10-15a**). The existence of loops of DNA as a substructure within chromosomes is also supported by electron micrographs of histone-depleted chromosomes. Histones can be selectively extracted from chromosomes by treatment with negatively charged polymeric chemicals, such as heparin and dextran sulfate, which compete with the DNA for binding histones. After histone extraction, a proteinaceous residue remains that retains the size and shape of the original chromosome. This residue is called the **chromosomal scaffold** (**Figure 10-15b**). One of the major components of the chromosomal scaffold is the SMC proteins that hold DNA strands together, keeping eukaryotic chromosomes topologically constrained (see Figure 9-11). The histone-depleted DNA can be seen to form large loops

anchored in the scaffold (**Figure 10-15c**), which is consistent with looping as a higher-order arrangement of DNA in chromosomes.

Although very little is known about further steps of DNA condensation beyond the 30 nm filament, there are probably several more layers of organization in eukaryotic chromosomes, each increasing the degree of compaction. Many speculative models have been proposed, one of which is shown in **Figure 10-16**. In this model, loops of the 30 nm filament are connected to the scaffold in a radial fashion. One radial turn forms a rosette composed of six loops, held together by SMC proteins (see Figure 9-27). A spiral of 30 rosettes per turn forms a coil, and a chromosome consists of several coils.

In reality, the higher-order structure of chromatin probably varies from chromosome to chromosome, from one region to the next in a single chromosome, and from moment to moment in the life of a cell. No single model can adequately describe these structures. Nevertheless, the principle is clear: DNA condensation in eukaryotic chromosomes probably involves coils upon coils upon coils.

Bacterial DNA, Like Eukaryotic DNA, Is Highly Organized

Bacterial DNA is compacted in a structure called the **nucleoid**, which can occupy a significant fraction of the

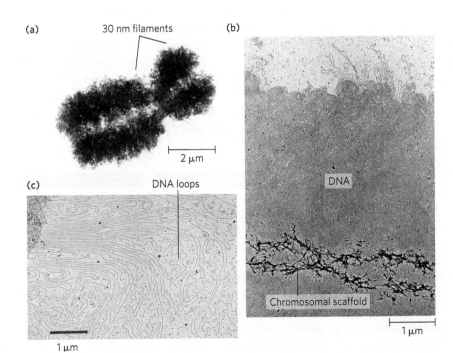

(a) 30 nm filaments
2 μm

(c) DNA loops
1 μm

(b) DNA
Chromosomal scaffold
1 μm

FIGURE 10-15 Loops of DNA attached to a chromosomal scaffold. (a) A swollen chromosome, produced in a buffer of low ionic strength, as seen in the electron microscope. Notice the appearance of 30 nm filaments (chromatin loops) at the margins. (b) Extraction of the histones leaves a proteinaceous chromosomal scaffold surrounded by naked DNA. (c) The DNA appears to be organized in loops attached at their base to the scaffold in the upper left corner. Note the different magnifications for the three images. [Sources: (a) Don W. Fawcett/Science Source. (b) & (c) Laemmli et al. 1978 Metaphase Chromosome Structure: The Role of Nonhistone Proteins, Cold Spring Harbor Symposia on Quantitative Biology, vol. 42, p. 351. Fig. 4 & 5.]

DNA	Nucleosome	30 nm filament	Extended form of chromosome	Condensed section of chromosome	Mitotic chromosome

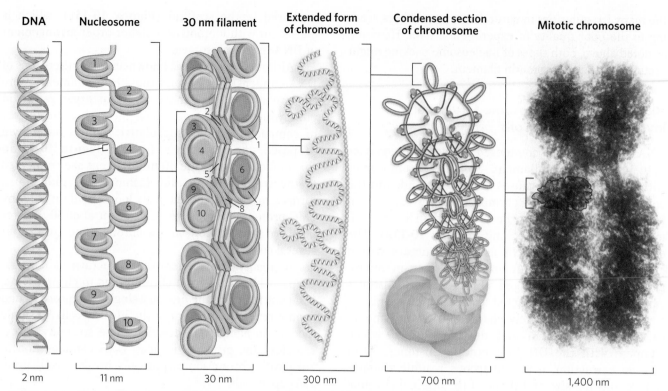

2 nm	11 nm	30 nm	300 nm	700 nm	1,400 nm

FIGURE 10-16 Higher-order DNA compaction in a eukaryotic chromosome. This model shows the levels of organization that could provide the observed degree of DNA compaction in the chromosomes of eukaryotes. First the DNA is wrapped around histone octamers, then H1 stimulates formation of the 30 nm filament (solenoid model is shown). Higher levels of organization are not well understood but seem to involve further coiling and loops in the form of rosettes, which also coil into thicker structures. Overall, progressive levels of organization take the form of coils upon coils upon coils. In cells, the higher-order structures (above the 30 nm filament) are unlikely to be as uniform as depicted here. [Source: Don W. Fawcett/Science Source.]

cell's volume (**Figure 10-17a**). The DNA seems to be attached at one or more points to the inner surface of the cytoplasmic (plasma) membrane. Much less is known about the structure of the nucleoid than of eukaryotic chromatin. Bacteria contain SMC proteins, and studies in *E. coli* reveal a scaffoldlike structure that seems to organize the circular chromosome into a series of about 500 looped domains, each encompassing 10 kbp, on average (**Figure 10-17b**). Like the looped domains in eukaryotic chromosomes, the looped DNA domains in the bacterial chromosome are topologically constrained. For example, if the DNA is cleaved in one domain, only the DNA in that domain becomes relaxed. However, bacterial DNA does not seem to have any structure comparable to the local organization provided by nucleosomes in eukaryotes.

Although bacteria lack nucleosomes, they do contain abundant histonelike proteins. A well-studied example is the two-subunit protein HU (M_r 19,000). Bacterial histonelike proteins do not seem to form stable oligomeric structures, and this may reflect the need for bacteria to respond very rapidly to their environment,

requiring more ready access to their genetic information. For example, bacterial cell division can be as short as 15 minutes, whereas a typical eukaryotic cell may not divide for hours or even months. In addition, a much greater proportion of bacterial DNA than eukaryotic DNA is used to encode protein or functional RNA. Furthermore, higher rates of cellular metabolism in bacteria mean that a much larger proportion of their DNA is being transcribed or replicated at a given time than in most eukaryotic cells.

SECTION 10.2 SUMMARY

- The histone octamer and associated DNA that form the nucleosome combine with histone H1. H1 binds additional DNA, altering the entry and exit angles of the DNA so that the nucleosomes pack in a zigzag pattern.
- Two models describe how nucleosomes might pack into a 30 nm filament. In the solenoid model, nucleosome disks are next to one another, forming a spiral shape. In the zigzag model, pairs of nucleosomes stack on top of each other and twist about a central axis.

(a)

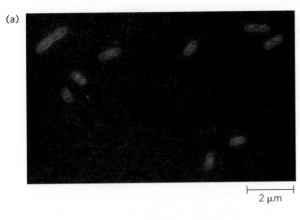

2 μm

(b)

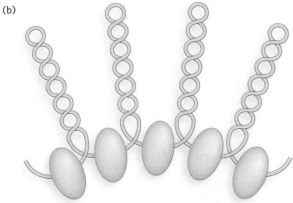

FIGURE 10-17 Highly condensed bacterial DNA. (a) The DNA is stained with a dye that fluoresces when exposed to UV light. The light areas define the nucleoids. Notice that some cells have replicated their DNA but have not yet undergone cell division and hence have multiple nucleoids. (b) Looped domains of DNA in the bacterial chromosome are separated by points of connection to the scaffold, shown as beige ovoids. [Source: (a) Courtesy Michael Cox Lab.]

- Higher-order chromatin structure is largely undetermined, but most likely involves loops that form topologically constrained domains. Looped domains form rosettes, which form coils. Chromosome shape is determined by a rigid proteinaceous scaffolding containing SMC proteins, among other components.
- Bacteria lack nucleosomes, but their chromosomal DNA is compacted into looped domains.

10.3 REGULATION OF CHROMOSOME STRUCTURE

Nucleosomes control the accessibility of DNA to decoding proteins such as RNA polymerase and to transcription activators and repressors. For example, nucleosome arrangements that lead to a more "open" chromatin state are associated with active gene transcription, whereas a more "closed" chromatin state typically represses transcription. The arrangement of nucleosomes on DNA is regulated by two main classes of enzymes, the activity of which usually depends on other functional subunits within a multiprotein complex.

One class of proteins that alter nucleosome arrangement is the **chromatin remodeling complexes**, which slide the nucleosome to a different location, eject it from the DNA, or replace it with a new nucleosome that contains a variant histone subunit (**Figure 10-18a**). Variant histone subunits impart special properties to the chromatin. The other class is the **histone modifying enzymes**, which covalently modify the N-terminal tails of histones (**Figure 10-18b**). Histone modifications that affect chromatin structure solely through the modification itself, without involving other molecules, are referred to as cis acting. Such a modification may result in opening or closing of the chromatin by tightening or loosening the arrangement of nucleosomes along the DNA. Histone modifications that are transacting (those that act through molecules other than the chromatin itself) attract proteins such as transcription factors or chromatin remodeling factors, which produce the chromatin change. Additional types of chemical modifications of histones are being discovered all the time.

Enzymes that modify histones are often contained in a multiprotein complex, along with transcription activators or repressors. Nucleosome modifying enzymes and chromatin remodeling complexes can also work together. Nucleosomes are altered by the factors in specific regions of chromatin that detect environmental signals, such as transcriptional activating histone acetylases that modify histones for increased expression of particular genes. In their simplest form, nucleosome alterations control DNA accessibility by altering the structure of nucleosome-DNA complexes. DNA accessibility is essential to all types of genetic processes, including gene expression, replication, repair, and recombination.

Histone modifications are inheritable, which is especially important during an organism's development, when a cellular program to form a specific cell type or tissue must be propagated over numerous cell divisions. Inheritance of genetic properties that are not encoded in the DNA sequence is referred to as **epigenetic inheritance** (further discussed later in this section).

Histone variants, along with the many covalent modifications that histones undergo, define and stabilize the chromatin state in localized regions throughout the genome. They mark the chromatin, facilitating or suppressing specific functions in a dynamic fashion, including transcription, replication, DNA repair, and chromosome segregation.

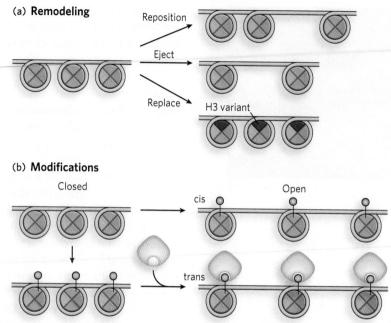

FIGURE 10-18 Modification of nucleosome arrangements. (a) Nucleosome position can be altered by chromatin remodeling complexes that move, eject, or replace nucleosomes. (b) Histone modifying enzymes attach chemical groups to specific amino acid residues of nucleosome subunits, as indicated by small circles attached to histones. If the outcome is a direct result of the modification, such as making the chromatin structure more open, the modification is acting in cis. If the modification attracts another protein (blue) that performs the histone-modifying function, it is acting in trans.

Nucleosomes Are Intrinsically Dynamic

Given the numerous contacts between nucleosomes and DNA, and the high level of DNA compaction, it seems reasonable to imagine that the condensed DNA is no longer available for other proteins to bind. However, for promoters to become accessible, the chromatin structure within the promoter region must become accessible. Molecular interactions that hold nucleosome arrays together in the 30 nm filament are likely to be relatively weak, in which case regions within nucleosome arrays will be dynamic, with hydrogen bonds between the histones and DNA breaking and re-forming quite easily.

Clever experimental designs have allowed measurements of the force on the individual molecules in chromatin. Single chromatin fibers can be isolated and chemically modified at either end, and the two ends attached to different polystyrene beads (**Figure 10-19a**). The bead at one end is held in a laser optical trap, and the bead at the other end is captured by a micropipette. The force required to stretch the chromatin fiber and thus disrupt the nucleosome array can be measured by moving the optical trap with a known force (measured in

piconewtons; 1 piconewton (pN) = 10^{-12} newtons). The stretching force on a chromatin fiber produces a structural transition at just 5 pN (**Figure 10-19b**)—substantially less than the 10 to 15 pN needed to unzip double-stranded DNA, and much less than the 1,600 pN required to break a covalent bond. Once the chromatin fiber is stretched, the force can be relieved and the fiber then contracts to its original length, thus demonstrating that condensation of the nucleosome array is spontaneous and reversible.

ATP-Driven Chromatin Remodeling Complexes Can Reposition Nucleosomes

Although nucleosomes can bind almost any DNA sequence, the position of nucleosomes in the genome is not entirely random. Some DNA sequences, as noted earlier, are preferentially bound by nucleosomes, but nucleosome positioning is also dictated by chromatin remodeling complexes, which use the energy of ATP to move nucleosomes around on the DNA. Chromatin remodeling complexes consist of 2 to 18 subunits and can be divided into three main classes: SWI/SNF (switch-sniff, the first remodeling complex discovered), ISWI (imitation switch), and Mi2/NURD (**Table 10-2** on p. 346). All chromatin

(a)

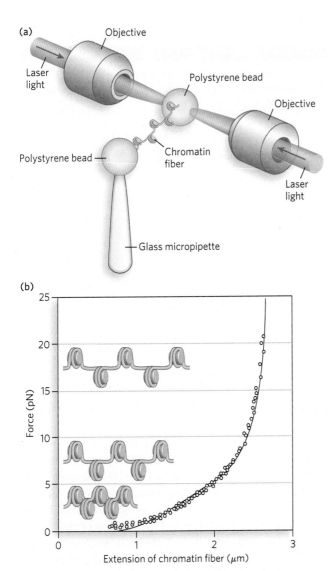

(b)

FIGURE 10-19 **Weak internucleosomal interactions in chromatin as revealed by optical trapping.** (a) The experimental setup uses DNA packaged into chromatin, with a bead coupled to each end of the DNA. One bead is held in place with a micropipette, and the other is held by an optical trap formed by a focused laser beam. The optical trap can be moved to apply a specific force on the DNA, and the length to which the DNA is stretched is measured directly. (b) Fiber length is measured (from the known magnification) at different applied forces (in piconewtons), and the process can be reversed, showing that chromatin can stretch and snap back to its original structure. [*Source: Data from Y. Cui and C. Bustamante, Proc. Natl. Acad. Sci. USA 97:127–132, 2000, Fig. 1.*]

remodeling complexes contain a conserved ATPase subunit and use the energy from ATP hydrolysis to disrupt the many contacts between nucleosomes and DNA: this allows nucleosomes to be ejected or repositioned on the DNA. Although we lack details about the mechanisms, all chromatin remodeling complexes can be thought of as mobilizing nucleosomes on DNA. The activity of chromatin remodeling complexes can result in gene

activation or repression. In general, the SWI/SNF class is associated with gene activation, and the ISWI and Mi2/NURD classes with gene repression.

The different chromatin remodeling complexes seem to have distinct mechanisms of action. Some mobilize nucleosomes by forming a DNA loop within the nucleosome (**Figure 10-20a**). The loop is propagated around the histone octamer, causing the nucleosome to slide to a new segment of the DNA. This can result in enhanced DNA accessibility—for example, by exposing a promoter that was previously blocked by the nucleosome. Alternatively, the same mechanism can reposition a nucleosome over a previously exposed promoter region, thereby repressing gene transcription. Likewise, repositioning nucleosomes into a regular array tends to condense the DNA, making it less accessible, whereas disrupting regularity in nucleosome arrays decondenses chromatin and thus facilitates transcription. Some chromatin remodeling complexes can also eject a histone octamer to generate a nucleosome-free region for transcriptional activation.

The mechanistic details of how chromatin remodeling complexes function are largely unknown, and this is an exciting area of research. The low-resolution structure of a remodeling complex of yeast, for example, reveals a complex containing a jaw sufficiently wide to accommodate an entire nucleosome (**Figure 10-20b**).

The position of nucleosomes within a genome can be determined through the use of large-scale array technologies, such as the ChIP-Seq and ChIP-Chip techniques (**Figure 10-21a** on p. 347). Briefly, cells are treated with formaldehyde to covalently cross-link nucleosomes to DNA. The cells are then disrupted, and genomic DNA is digested with micrococcal nuclease. An antibody to a histone is then used to immunoprecipitate the nucleosome-DNA complex. Any DNA not bound to the histone is digested and washed away, the protein-DNA cross-links are broken, and the released DNA is sequenced. This technique of chromatin immunoprecipitation followed by DNA sequencing is known as **ChIP-Seq**. Alternatively, the released DNA is labeled and used to probe a microarray representing the genomic sequences of that particular cell type. The pattern of hybridization on the microarray reveals the DNA sequences that associate with the nucleosomes. Because microarrays are often referred to as chips, this technique is called a **ChIP-Chip** experiment.

Application of the ChIP-Seq technique has defined the position of every nucleosome in the genomes of some species of yeast, some worms, fruit flies, and humans. A consensus among this genome-wide data shows a surprisingly well-defined pattern of nucleosome position at active promoter elements (**Figure 10-21b**). Transcriptionally active promoters are usually free of bound

TABLE 10-2

Examples of Chromatin Remodeling Complexes

Class	Chromatin Remodeling Complex	Organism	ATPase	Number of Subunits	Domain That Associates with Histones*	Effect on Transcription
I. SWI/SNF	SWI/SNF	S. cerevisiae	Swi2/Snf2	11	Bromodomain	Activation
	RSC	S. cerevisiae	Sth1	15	Bromodomain	Activation
	Brahma	D. melanogaster	Brahma	Unknown	Bromodomain	Activation
	SWI/SNF	H. sapiens	hBRM	10	Bromodomain	Activation
	NRD	H. sapiens	CHD4	18	Bromodomain	Activation
II. ISWI	ISWI	S. cerevisiae	ISWI	4	—	Repression
	ACF	D. melanogaster	ISWI	2	—	Repression
	NURF	D. melanogaster	ISWI	4	—	Repression
	CHRAC	D. melanogaster	ISWI	5	—	Activation
III. Mi2/NURD	Mi2/NURD	H. sapiens	Mi2	8–10	Chromodomain	Repression

*Bromodomains and chromodomains are discussed later in this section.

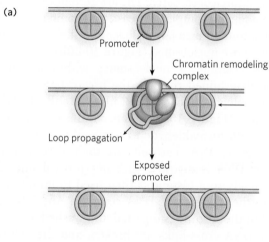

(a)

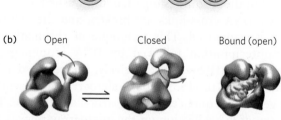

(b)

FIGURE 10-20 A mechanism for the action of chromatin remodeling complexes. (a) When a nucleosome covers a promoter, a chromatin remodeling complex can expose the promoter by pulling the DNA out of the nucleosome and into a loop. The loop can be propagated around the nucleosome, with the net effect of moving the nucleosome away from the promoter. (b) The shape of the chromatin remodeling complex RSC of yeast, based on three-dimensional image reconstruction from numerous images taken in the electron microscope with different views and tilts. By grouping the microscopically observed shapes into distinctive categories, different conformations can be deduced, as shown here. [Source: (b) Data from A. E. Leschziner et al., Proc. Natl. Acad. Sci. USA 104:4913–4918. © 2007 National Academy of Sciences, U.S.A.]

nucleosomes and are flanked on either side by a nucleosome with a well-defined position. This arrangement facilitates transcription by making the promoter region accessible to transcription factors. The upstream nucleosome is the most well defined, and the DNA sequence often conforms quite well to sequences known to bind nucleosomes (see Figure 10-7), as determined by Jonathan Widom and other researchers. The position of the downstream nucleosome is also well defined; nucleosome positions farther downstream are less defined and become irregular. This pattern is observed in the genomes of all eukaryotes.

Variant Histone Subunits Alter DNA-Binding Affinity

Eukaryotes contain several variants of H2A and H3 that differ in their N- and C-terminal sequences and replace the wild-type (canonical) subunit; these changes confer special properties on the chromatin structure (**Figure 10-22** on p. 348). The primary difference in the histone variants that replace H3 is the susceptibility of residues in the N-terminal tail to modifications such as methylation and phosphorylation. In contrast, variants of H2A differ primarily in the C-terminal tail region, which can recruit various proteins to the nucleosome. The H2A variants differ slightly in amino acid sequence, and often result in a different phosphorylation state. It is not yet clear how histone variants function, but genetic models have linked each histone variant with a specific cellular process.

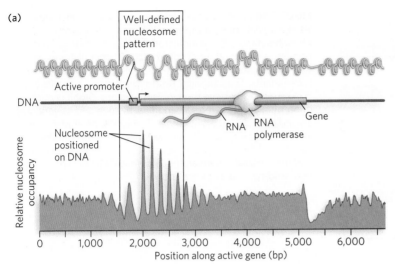

(a)

FIGURE 10-21 Determining the position of nucleosomes on genomic DNA with ChIP techniques. (a) Genome-wide analysis of the position of histones at all active promoter elements of yeast by the ChIP-Seq technique reveals a pattern of DNA accessibility at the promoter region. Peaks indicate regions of DNA bound to histones, and valleys indicate regions of DNA relatively free of histones. (b) Chip-Chip experiments define the location of nucleosomes positioned around an active gene. [Source: (b) Data from T. N. Mavrich et al., Genome Res. 18:1073–1083, 2008.]

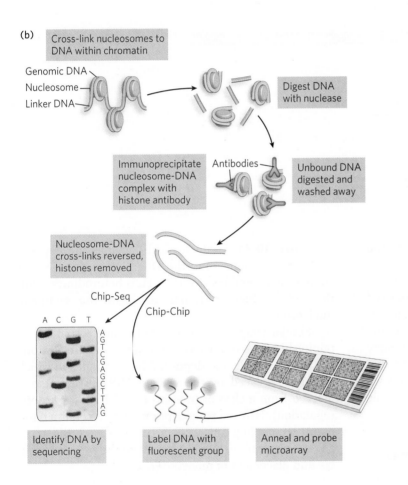

The wild-type H3 subunit is replaced by the H3.3 variant in regions where active gene expression is occurring (**Figure 10-23a** on p. 349). The H2A variant H2AZ is also associated with nucleosomes located at actively transcribed genes. Based on these findings, H3.3 and H2AZ are thought to stabilize the open state of chromatin, thereby facilitating access of the transcriptional machinery to DNA in actively transcribed regions.

The H3 histone variant CENPA is associated with the repeated DNA sequences in centromeres and contains a large extension that connects to the kinetochore (**Figure 10-23b**), the site where spindle fibers attach and pull chromosomes apart during cell division (see Chapter 2). Modification of the H2A variant H2AX is associated with DNA repair and genetic recombination (**Figure 10-23c**). Modest amounts of H2AX seem to be

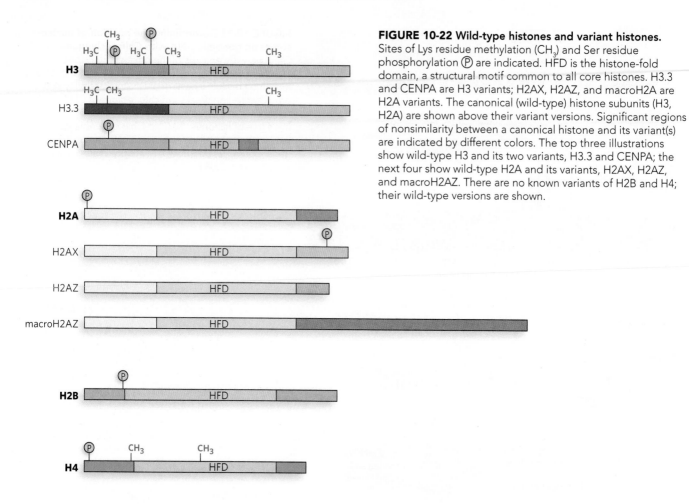

FIGURE 10-22 Wild-type histones and variant histones. Sites of Lys residue methylation (CH$_3$) and Ser residue phosphorylation (P) are indicated. HFD is the histone-fold domain, a structural motif common to all core histones. H3.3 and CENPA are H3 variants; H2AX, H2AZ, and macroH2A are H2A variants. The canonical (wild-type) histone subunits (H3, H2A) are shown above their variant versions. Significant regions of nonsimilarity between a canonical histone and its variant(s) are indicated by different colors. The top three illustrations show wild-type H3 and its two variants, H3.3 and CENPA; the next four show wild-type H2A and its variants, H2AX, H2AZ, and macroH2AZ. There are no known variants of H2B and H4; their wild-type versions are shown.

scattered throughout the genome. When a double-strand break occurs, nearby molecules of H2AX become phosphorylated at Ser[139] in the C-terminal region, attracting DNA repair proteins. If this phosphorylation is blocked experimentally, formation of the protein complex necessary for DNA repair is inhibited. Another H2A variant, macroH2A, is abnormally large and contains a unique C-terminal domain. MacroH2A is involved in X chromosome inactivation, a fascinating process that shuts down one of the two X chromosomes in the cells of female mammals (**Highlight 10-1** on p. 350).

Nucleosome Assembly Requires Chaperones

Isolated histones are difficult to assemble onto DNA in vitro; only low yields are obtained in the absence of other factors. In cells, **histone chaperones** are required to assist the assembly of histone octamers on DNA. These histone chaperones are acidic proteins that bind either the H3-H4 tetramer or the H2A-H2B dimer. Histone chaperones also function in assembling new histones on DNA during DNA replication. Chaperone-mediated nucleosome assembly occurs in two steps, as shown

in **Figure 10-24**. First, the CAF-1 chaperone deposits an H3-H4 tetramer onto the DNA; second, the NAP-1 chaperone assembles two H2A-H2B heterodimers with the H3-H4 heterotetramer to form the complete nucleosome.

Similar chaperones assemble the nucleosomes containing histone variants. Nucleosomes containing histone H3.3, for example, are deposited by a complex in which CAF-1 is replaced by the protein HIRA (the name is derived from a class of proteins called HIR, for *histone regulation*). HIRA, a transcriptional repressor, is a chromatin remodeling complex but can also be considered a histone chaperone, helping to ensure the proper assembly and placement of nucleosomes.

Modifications of Histone Tails Alter DNA Accessibility

For many years, scientists thought that histones performed only a structural role—condensing DNA. We now know that chromatin structure regulates essentially all genetic transactions and that modification of the N-terminal tails of histones plays a major role in altering

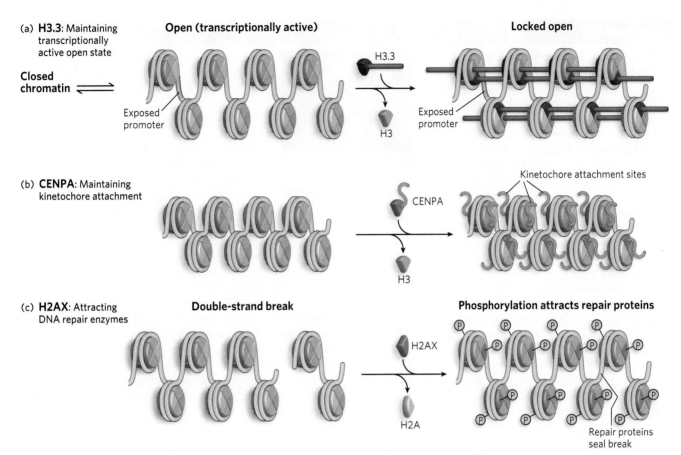

FIGURE 10-23 Proposed mechanisms for the roles of three histone variants. (a) The H3.3 variant is remodeled into octamers at sites of active transcription, where it helps stabilize chromatin in an open, transcriptionally active form. (b) The CENPA variant is localized to the centromere region; it contains a unique extension that may help attach the chromosome to the kinetochore. (c) H2AX localizes to sites of double-strand breaks, where it becomes phosphorylated on Ser[139] and attracts DNA repair complexes to the site.

chromatin structure. The amino acid residues of histone tails can be chemically modified in many ways; the most intensively studied are the acetylation of lysines, methylation of lysines and arginines, phosphorylation of serines, and ubiquitination of lysines

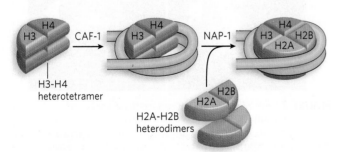

FIGURE 10-24 Chaperone-mediated nucleosome assembly. The H3-H4 heterotetramer is placed on DNA by the CAF-1 chaperone. Two H2A-H2B heterodimers then assemble in a reaction promoted by NAP-1.

(**Figure 10-25** on p. 351). **Figure 10-26** on p. 351 shows some of the most common amino acid modifications in histones.

Acetylation of Lys residues, shown in Figure 10-26a, is performed by enzymes called **histone acetyltransferases**, or **HATs** (see the How We Know section at the end of this chapter). Several different types of HATs modify histone subunits, and most often they acetylate specific residues in a histone tail. Acetylation is generally associated with enhanced accessibility to DNA and subsequent transcriptional activation. Histone deacetylases (HDACs) remove acetyl groups from histones, and deacetylation of Lys residues generally results in transcriptional repression (**Table 10-3** on p. 352). Initially, investigators thought that acetylation, which neutralizes the positive charge on lysine, simply relaxed the grip of a nucleosome on the DNA, thereby opening chromatin to regulatory proteins. More recent research indicates that acetylation has a larger effect on nucleosome-nucleosome contacts

HIGHLIGHT 10-1 ┤ **A CLOSER LOOK**

The Use of a Histone Variant in X Chromosome Inactivation

Female mammals have two X chromosomes, and males have only one. Therefore, we might expect all the genes on the X chromosome to be expressed at twice the level in females compared with males. Most fertilized cells harboring an extra autosomal (non-sex) chromosome are lethal at the embryonic stage, due to the enhanced gene dosage from the extra chromosome. So, how do female cells, with two X chromosomes, survive with double the gene dosage? Or, how do male cells survive with half the necessary gene dosage? The answer lies in a fascinating process called X chromosome inactivation, in which one of the two X chromosomes in every cell of the female is converted into highly condensed heterochromatin, a form of very tightly compacted DNA, which silences its genes. Inactivation of the X chromosome is also known as lyonization, after its discoverer, Mary Lyon.

Mary Lyon [Source: Courtesy of Medical Research Council, London.]

In the two- or four-cell stage of development of female mammals, only the paternally derived X chromosome is inactivated; gene silencing of this type, specific to either paternal or maternal genes, is known as imprinting (see Chapter 21). Both X chromosomes become active again in the early blastocyst, but one is permanently inactivated a short time later. This permanent inactivation is random; either the maternal or paternal X chromosome is targeted for inactivation. Thus, the female body is a mosaic of cells, some of which have an active maternal X chromosome and others an active paternal X chromosome. Because of the random nature of X chromosome inactivation, an animal that is heterozygous for an X-linked trait has a mosaic phenotype, as seen in the mottled coat color of a calico cat (Figure 1).

The inactive X chromosome contains an abundance of the histone variant macroH2A. This variant contains a large C-terminal region that doubles the size of the

protein. The current hypothesis is that the extra protein sequence binds RNA. Essential to X chromosome inactivation is the presence of a noncoding RNA called XIST (see Figure 21-20). XIST RNA is transcribed from only one X chromosome, whereupon it acts in cis to coat the entire chromosome, encapsulating it. The macroH2A histone variant binds XIST and may facilitate this process. Recent evidence suggests that macroH2A1 may be involved in maintenance of the inactive state. The condensed and inactivated X chromosome, called a Barr body, is easily identified in cells stained with an antibody specific for macroH2A1. Figure 2 shows two blastocyst cells (top and bottom) stained for total DNA (left) and the same two cells stained using an antibody to the unique C-terminal region of macroH2A1 (right), thus revealing the X-inactivated chromosome containing macroH2A1. Other modifications associated with X chromosome inactivation are cytosine methylation at CpG residues, low levels of histone acetylation, and methylation of Lys^9 on histone H3. These modifications are typically associated with transcriptional repression.

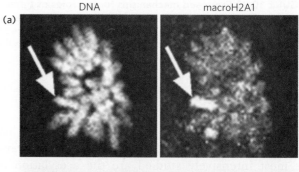
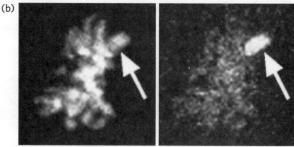

FIGURE 2 Cells of two different mouse blastocysts are stained for total DNA (left) or for macroH2A1 (right), using an antibody to macroH2A1. The arrows point to the inactive X chromosome containing macroH2A1. [Source: From C. Costanzi et al., Development 127:2283–2289, 2000. © The Company of Biologists. Courtesy of J. R. Pehrson.]

FIGURE 1 The coat color variation of a calico cat is an example of a mosaic phenotype. [Source: Linn Currie/Shutterstock.]

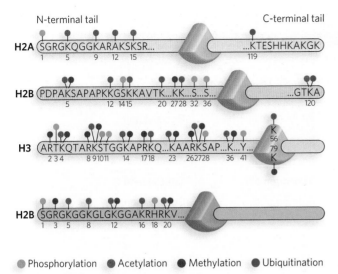

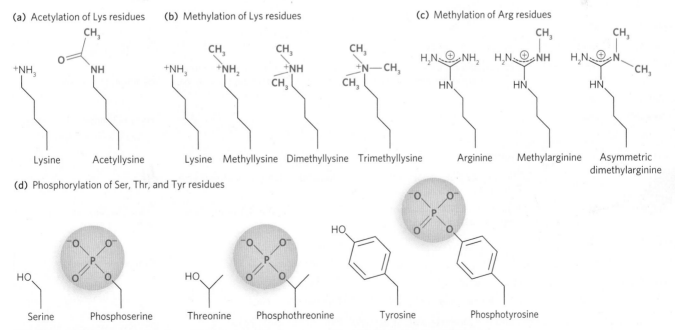

FIGURE 10-25 The modification of nucleosome N-terminal tails by small molecules. Modifications are shown on just one tail of each type of core histone. Residue numbers are given for the modified residues (here, shown by one-letter abbreviations and subscript numbers—S_1, K_5, K_9, etc.). The N-terminal tail is on the left, connecting to the globular domain (indicated as a triangle), followed by the short C-terminal tail. Modifications occur at all these positions, but the most common are those in the N-terminal tails. The major modifications are represented as follows: P, phosphorylation; A, acetylation; M, methylation; and U, ubiquitination. [*Source: Data from Yong Zhong Xu, Cynthia Kanagaratham and Danuta Radzioch (2013). Chromatin Remodelling During Host-Bacterial Pathogen Interaction, Chromatin Remodelling, Dr. Danuta Radzioch (Ed.), ISBN: 978-953-51-1087-3, InTech, DOI: 10.5772/55977. Available from: http://www.intechopen.com/books/chromatin-remodelling/chromatin-remodelling-during-host-bacterial-pathogen-interaction.*]

and alters higher-order chromatin structure that way, rather than changing nucleosome-DNA affinity.

Lysine residues can also be methylated, to methyl-, dimethyl-, or trimethyllysine (Figure 10-26b). Methylation (by methylases) was once thought to be irreversible, but recent experiments have identified enzymes (histone demethylases) that remove methyl groups from histone tails. Some amino acid residues, such as Lys[9] or Lys[14] of H3, can be either methylated or acetylated, and this seems to stabilize the open or closed state of the nucleosome. Arginine can also be methylated, to methylarginine or two forms of dimethylarginine; double methylation can result in one methyl on each nitrogen of the guanidinium group or two methyls on one of the nitrogen atoms of the guanidinium group (Figure 10-26c).

Phosphorylation is another type of modification commonly found on histone tails of H3 and H4. The phosphorylation of Ser, Thr, and Tyr residues incorporates a negative charge into a histone tail (Figure 10-26d). As noted above, histones can also be modified by ubiquitination (not shown in Figure 10-26). Recently, histone modifications by sumoylation, proline isomerization, and ADP ribosylation have also been observed (see Chapter 21). The function of many of these modifications is still unclear.

Certain individual modifications can be correlated with specific phenotypes. While many studies are performed in yeast, our main interest is the phenotype in a multicellular organism such as ourselves. **Table 10-4**

FIGURE 10-26 Chemical structures of some modified amino acid residues in histones. There are many types of chemical modifications to nucleosomal tails. Only the most common are shown here.

TABLE 10-3
Histone Modifying Complexes

Histone Acetyltransferases (HATs)

HAT	Organism	Protein Complex	Domain That Associates with Histones	Effect on Transcription
Gcn5	Yeast	SAGA, ADA, SILK	Bromodomain	Coactivation
Gcn5L	Mammal, fruit fly	STAGA, TFTC	Bromodomain	Coactivation
PCAF	Mammal	PCAF	Bromodomain	Coactivation
ESAI	Yeast	NuA4	Bromodomain	Coactivation
SAS2	Yeast	SAS-1	Bromodomain	Coactivation
Tip60	Mammal	Tip 60	Bromodomain	Coactivation

Histone Deacetylases (HDACs)

HDAC	Organism	Protein Complex	Effect on Transcription (or other effect)
HOS2	Yeast	SETSC	Repression
HDAC1	Mammal	mSin3, Nurd, N-CoR-2	Repression
HDAI	Yeast	Hdal complex	Repression
HDAC4	Mammal	Unknown	Repression
HST1	Yeast	Set3C	Repression
SIR2	Yeast	Sir4, REMT	Repression
SIRT6	Mammal	Unknown	DNA repair

TABLE 10-4
Phenotypes Associated with Modifications at Particular Residues in Histone Subunits

Residue	Modification	Chromatin Function In Vitro	Mutant Phenotype in Yeast	Correlation with Process in Mammals
H3Y41	P		Reduced viability	Transcriptional regulation
H3T45	P	Point mutant stimulates RSC remodeling	Reduced viability	Apoptosis
H3R52	Me		Reduced viability	
H3R53	Me			
H3K56	Ac	DNA breathing, DNA accessibility, RSC and SWI/SNF remodeling	DNA damage sensitivity, chromatin assembly, DNA repair, transcription	Genome stability, DNA damage, transcription
H3R63	Me			
H3K64	Me			Heterochromatin formation
H3R83	Me			
H3K115	Ac	Disassembly	Transcription	
H3T118	P	Thermal mobility, disassembly, Poll II processivity	Reduced viability, SIN phenotype	
H3K122	Ac	Thermal mobility, nucleosome disassembly, transcription	Transcription, nucleosome occupancy	Transcription, transcription enhancement
H4K31	Me		Reduced viability	
H4S47	P		SIN phenotype	Chromatin assembly
H4K77	Ac	DNA accessibility	Telomeric silencing	
H4K79	Ac	DNA accessibility	Telomeric silencing	

Source: Data from P. Tropberger and R. Schneider, *Nat. Struct. Mol. Biol.* 20:657–661, 2013.

Note: This table includes amino acid residues in histones H3 and H4 and their known modifications. Functions are shown for in vitro and in vivo phenotypes in yeast and for the processes in mammals that correlate with the yeast phenotypes.

provides some examples of the effects of particular histone modifications on the phenotypes of yeast and mammals.

Proteins with Bromodomains and Chromodomains Bind Modified Histones

A more striking way in which histone tail modifications function is by recruiting enzymes that recognize particular modified amino acid residues. Proteins with **bromodomain** motifs recognize acetylated Lys residues (**Figure 10-27**). Bromodomain proteins are usually contained within a larger, multiprotein complex, such as a chromatin remodeling complex (see Table 10-2) or a complex that acetylates histones (see Table 10-3). If a chromatin remodeling complex contains a subunit with a bromodomain and also a subunit with histone acetylase activity, when the complex binds to an acetylated

nucleosome it acetylates histones in a neighboring nucleosome. By this means, a specific pattern of acetylation can be propagated in a targeted area of the chromosome. Propagation of the acetylation pattern typically leads to higher levels of gene expression. Creation of a specific chromatin state underlies the epigenetic inheritance of gene expression patterns not encoded by DNA.

As described earlier, a common modification of histone tails is methylation of Lys and Arg residues, which can result in either gene activation or gene repression. For example, in H3, methylation of Lys^4 is associated with transcriptional activation, whereas methylation of Lys^9 typically results in repression by recruiting chromatin remodeling complexes that condense the chromatin. Proteins that bind to methylated Lys residues contain motifs called **chromodomains** (see Figure 10-27). Like bromodomain proteins,

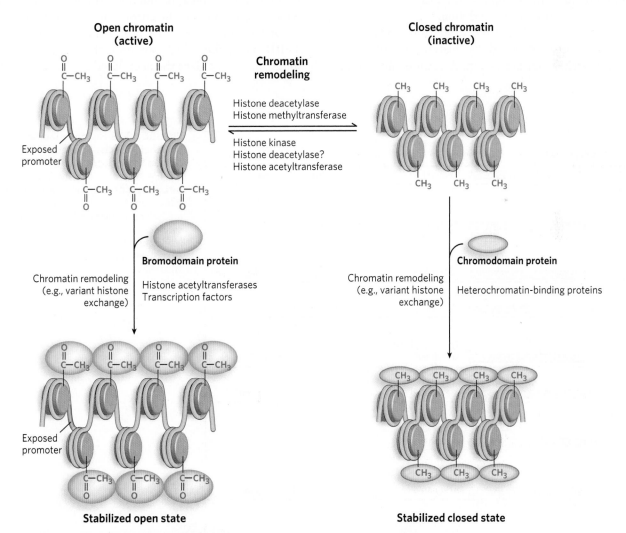

FIGURE 10-27 Bromodomains and chromodomains. Chromatin structure is made more open or closed by various modifications performed by the enzymes shown here. Acetylated nucleosomes (top left) are recognized by bromodomain proteins that may help stabilize the open chromatin state (bottom left). Methylated histones (top right) are recognized by chromodomain proteins that may help promote the closed state (bottom right).

chromodomain proteins can be recognized by this specific sequence motif. Chromodomain proteins are found in complexes with other enzymes that further modify chromatin structure.

Histone Modifications and Remodeling Complexes May Read a Histone Code

Some researchers have proposed that chemical modifications of histone tails can recruit proteins in a stepwise fashion, establishing a pathway that can activate the expression of a particular gene. For example, imagine a gene activation pathway that begins when a histone is acetylated. This modification recruits (via a bromodomain subunit) a HAT-containing complex that propagates the modification to neighboring histones and partially opens the chromatin structure. The resulting modifications then recruit a chromatin remodeling complex (also via a bromodomain) that completes the task of opening the chromatin structure for transcription.

The basis of the hypothetical **histone code** is that successive events directed by histone modifications drive transcriptional activation. A reasonable amount of research evidence supports two-step switches, in which one histone modification leads to another type of modification. Many researchers consider the histone code as still a working hypothesis, but the numerous types of histone modifications and their resulting effects provide a rich library for combinatorial regulatory signals (**Figure 10-28**).

Supporting evidence for a histone code came from studies of the expression of the human IFN-β gene (**Figure 10-29**). First, the Gcn5 coactivator binds within a complex of proteins to a specific DNA sequence upstream of the gene. The Gcn5 subunit is an acetyltransferase (HAT) that acetylates Lys^8 of H4 and Lys^9 of H3. Next, a protein kinase phosphorylates Ser^{10} of H3. The kinase may be a subunit of a protein complex that binds to the promoter region. This modification then promotes acetylation of Lys^{14} of H3 by Gcn5. These modifications, in turn, recruit a SWI/SNF chromatin remodeling complex, through a bromodomain subunit, which then remodels the chromatin and exposes a DNA sequence called the TATA box. Transcription activator TFIID binds the TATA box and, through a bromodomain, also recognizes the acetylated Lys^9 and Lys^{14}. The

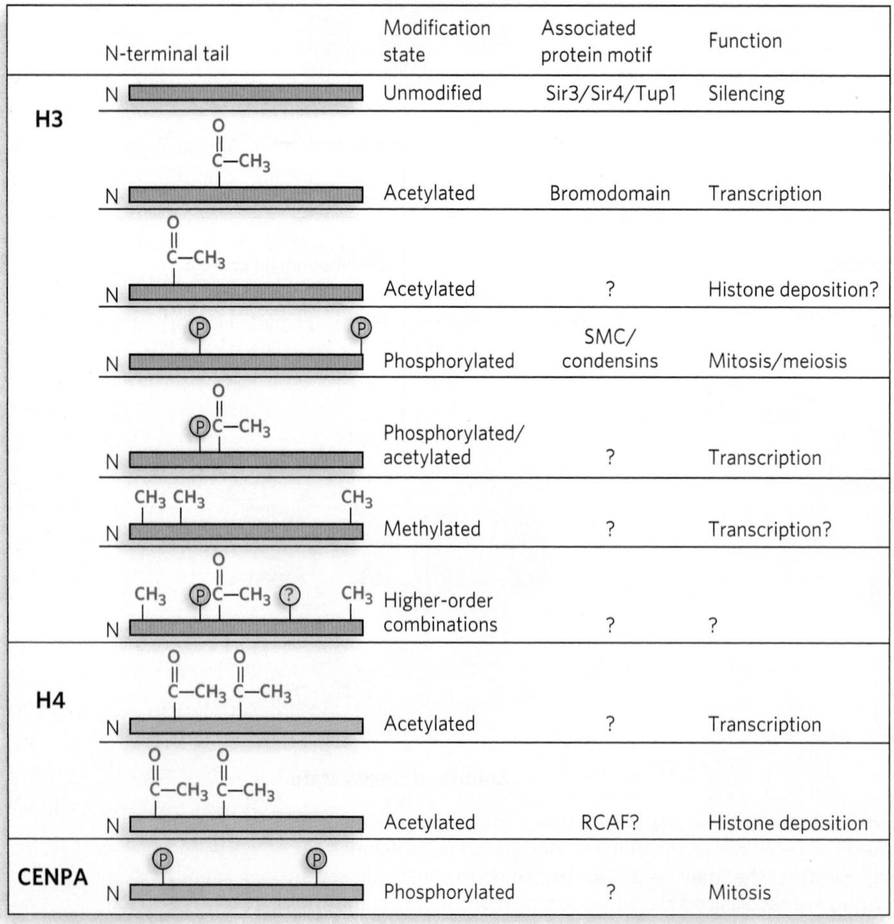

FIGURE 10-28 Histone modifications and their effects. Specific combinations of modifications of histones H3, H4, and CENPA (an H3 variant) are recognized by different proteins and have diverse effects on chromatin. Acetylated residues are lysines, phosphorylated residues are serines, and methylated residues are lysines and arginines. [Source: Data from B. D. Strahl and C. D. Allis, Nature 403:41–45, 2000, Fig. 2.]

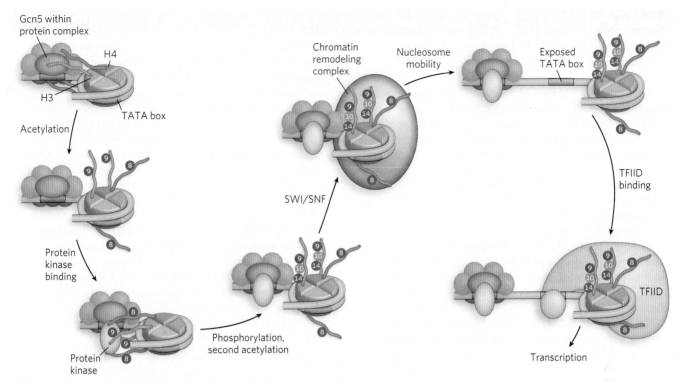

FIGURE 10-29 An example of a histone code. Activation of the IFN-β promoter requires an ordered succession of events driven by histone modifications that occur in a temporal sequence. [*Source: Data from T. Agalioti, G. Chen, and D. Thanos,* Cell *111:381–392, 2002, Fig. 5.*]

specific interactions of TFIID with the DNA and modified histones promote transcription. The proteins and DNA sequences involved in the initiation of transcription are discussed in greater detail in Chapter 15.

Histone Modifying Enzymes Maintain Epigenetic States through Cell Division

Certain modifications of nucleosomes are passed down to daughter cells during cell division, in mitosis and meiosis. These modifications can activate or repress certain genes. Inheriting the control of gene expression is essentially like inheriting any other trait: in addition to inheriting genes, offspring also inherit the nucleosome modifications that control expression of those genes. The underlying basis of this epigenetic inheritance is very different from classical Mendelian genetics, which is defined solely by the DNA sequence of a gene.

Histone modifications underlie the majority of epigenetic traits, although the direct methylation of cytosine bases in DNA also plays a role in epigenetic inheritance. The diversity of histone modifications provides a rich combination of modification patterns that can be coupled to epigenetic inheritance. Variant histone subunits also play a role in epigenetic inheritance in chromatin, facilitating or suppressing specific functions such as chromosome segregation, transcription,

and DNA repair. Histone modifications do not disappear at cell division or during meiosis, and thus they are a part of the information transmitted from one generation to the next in all eukaryotic organisms. These modifications help define which genes are expressed by daughter cells. Epigenetic modifications are important during an organism's development and in maintaining cell and tissue types. Epigenetic traits are also important in the process of imprinting in germ cells, in which one copy (allele) of a gene is silenced in the fertilized egg and remains turned off in every cell of the developing organism. Humans contain at least 80 known imprinted genes that are controlled by epigenetic marks, without which the fertilized egg cannot proceed past the blastula stage (see Chapter 21). The control of epigenetic marks can also be influenced by environmental factors, and when disturbed this can lead to disease, including cancer (**Highlight 10-2**).

Epigenetic inheritance of a chromatin state during cell division requires the preservation of histone modification patterns during DNA replication. If histone octamers were completely displaced during replication, all the epigenetic information encoded in the histone modifications would be lost in the daughter cells. Studies on the fate of individual histone subunits during replication demonstrate that histone octamers split apart during replication, but the H3-H4 tetramers remain bound to

Defects in Epigenetic Maintenance Proteins Associated with Cancer

The development of cells into tissues, organs, and whole organisms requires individualized programs of gene expression for different cell types. Many of these programs are controlled by inheritable epigenetic mechanisms. Epigenetic controls often rely on chromatin structures based on specific histone tail modification patterns that provide or restrict the accessibility of DNA to transcription factors. Epigenetic controls are also maintained by chromatin remodeling complexes, histone chaperones, and DNA methyltransferases.

Molecular studies of cancer cells reveal that many types of cancer are associated with mutations in genes encoding proteins that act in epigenetic pathways. Presumably, the disruption of epigenetic pathways can activate genes that promote cell growth or repress genes that limit cell growth, leading to tumor formation. For example, aberrant histone modification patterns are observed in mammalian cancers that include mutations in certain subunits of chromatin remodeling complexes. Mutations that inactivate the SNF5 subunit, common to all SWI/SNF complexes, lead to a predisposition to familial cancer and a condition known as rhabdoid predisposition syndrome. Aberrations in certain HAT genes, as well as in particular HDACs, are associated with various hematological malignancies. Abnormal histone methyltransferases have also been linked to a predisposition to cancer. For example, loss of function of the enzyme that targets Lys^9 in the N-terminal tail of H3 disrupts heterochromatin formation and could affect chromosome stability.

Epigenetic control as a basis for cancer has a positive aspect: altered epigenetic states are reversible. Cancers based in epigenetic states offer the promise of strategic therapies to block the progression of cancer cells by reestablishing the correct epigenetic state. Regaining normal gene expression by resetting epigenetic states in cell lines has already been accomplished with certain inhibitors of HATs and HDACs. These findings hold promise for future diagnostics and treatments for cancer.

---

the DNA (**Figure 10-30**). The parental (marked) H3-H4 heterotetramers are distributed randomly on the two new daughter DNA duplexes that are made during replication, and coat only half of the total DNA after replication. New H3-H4 heterotetramers that lack the particular modification pattern of those they replace are assembled onto the replicated DNA by the CAF-1 chaperone. Current research suggests that the parental H2A-H2B dimers remain in the vicinity after being displaced by the replication fork and quickly reassemble with H3-H4 heterotetramers on the newly replicated DNA, chaperoned by NAP-1. New, unmodified H2A-H2B dimers must also be assembled on the newly replicated DNA. Hence, four types of nucleosomes form on the daughter DNA strands: nucleosomes containing (1) old (parental) H3-H4 and new H2A-H2B, (2) new H3-H4 and old H2A-H2B, (3) entirely parental H2A-H2B-H3-H4 histones, or (4) entirely new H2A-H2B-H3-H4 histones (see Figure 10-30). Overall, the newly replicated DNA has only half of the parental epigenetic information in its histones, but the daughter DNA duplexes are "salted" with the parental histone modification pattern.

If half of the epigenetic histone modifications are lost during replication, how is the epigenetic state preserved in daughter cells? The answer lies in the ability of the cell to propagate the histone modification pattern of the parental nucleosomes to new histone subunits. Histone modifications are transmitted to nearby nucleosomes through the action of histone modifying complexes that recognize and bind modified residues on the parental histone subunits. For example, acetylated or methylated residues are recognized by protein complexes with specific bromodomains or chromodomains. The multiprotein histone modifying complex also contains an enzyme (e.g., HAT or methyltransferase) that spreads the modification to adjacent, unmodified nucleosomes (**Figure 10-31** on p. 358). The different epigenetic marks are presumed to recruit specific histone modifying complexes that promote the modification of other, unmodified nucleosomes. These actions transmit and propagate the histone modification pattern to new histone subunits in the same vicinity, thereby preserving the epigenetic state of the parental cell in the two new daughter cells.

SECTION 10.3 SUMMARY

- Nucleosomes influence gene transcription by controlling the open or closed state of chromatin. They can be positioned at specific locations on the DNA and can slide along the duplex to allow or block access to specific regions of the DNA.

- ATP-driven chromatin remodeling complexes can reposition, eject, or replace a nucleosome. The three classes of chromatin remodeling complexes are SWI/SNF, ISWI, and Mi2/NURD.

- Variant histones can replace individual histone subunits in the nucleosome, thus altering the chromatin's open or closed state.

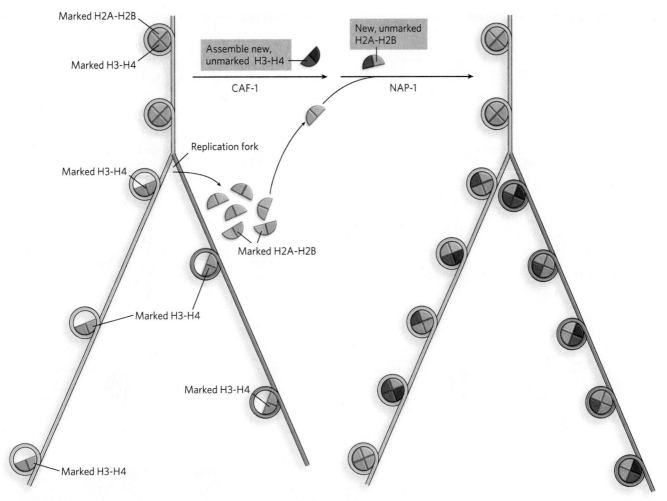

FIGURE 10-30 Preserving histone modification patterns during DNA replication. After DNA replication, the two new daughter duplex DNAs that are produced during replication lack histones H2A-H2B, and the H3-H4 heterotetramers distribute among the two new strands (left). The parental (marked) H3-H4 heterotetramers bind quickly to both daughter strands, along with new (unmarked) H3-H4 heterotetramers (right). The old (marked) and new (unmarked) H2A-H2B heterodimers reassemble randomly on daughter strands with old and new H3-H4 heterotetramers. The epigenetic marks are subsequently spread to adjacent nucleosomes on the two daughter strands, preserving the histone modification pattern of the parental DNA.

- The position of every nucleosome in a genome can be determined by ChIP-Seq and ChIP-Chip techniques.
- Chemical modifications to histone tails alter the open or closed state of chromatin. Generally, acetylation is associated with enhanced accessibility of the DNA.
- Proteins with bromodomains bind to acetylated Lys residues on the histone tail, recruiting other histone modifying complexes and spreading the pattern of acetylation to nearby histones. Methylated Lys residues can be bound by proteins with chromodomains, which likewise recruit histone modifying complexes that spread the methylation pattern.

- The set of histone modifications in a certain region of chromatin may be read as a code that directs transcription.
- Chaperone proteins are necessary for assembling histones into nucleosomes.
- Epigenetic inheritance of information not encoded in the DNA relies on the transmission of histone variants and modifications through successive cell generations. Maintenance of the epigenetic state of chromatin during DNA replication requires equal distribution of H3-H4 heterotetramers on each daughter DNA. The remaining half of the parental epigenetic information is then spread to nearby newly formed histones on the two daughter strands.

(a) Maintaining active (open) chromatin state

(b) Maintain repressed (closed) state

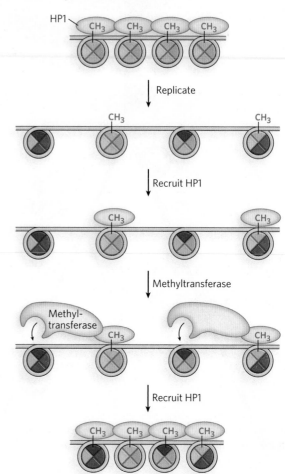

FIGURE 10-31 The spread of epigenetic states after replication. (a) The open state may be propagated through the binding of a bromodomain-containing complex that also contains a HAT subunit, which acetylates the same residues in neighboring histone octamers. (b) The closed state may be spread through binding of HP1, which reassociates with parental histones after replication and is associated with the repressed state of tightly packed heterochromatin. HP1 contains a chromodomain that binds methylated Lys residues in histones and the chromodomain also recruits a methyltransferase to propagate the parental methylation pattern in neighboring new histones.

? UNANSWERED QUESTIONS

The organization of DNA in chromosomes is still largely a mystery. At the level of individual DNA segments, the details of how chromatin remodeling complexes and histone modifying enzymes control access to the DNA have yet to be explored. The way in which epigenetic marks are interpreted is only now coming into focus; how they are read and how they relate to normal functions and disease are poorly understood. We can expect a rapid pace of new discoveries in this exciting field of inquiry.

1. How is DNA packaged in a chromosome? We have an exceptional understanding of DNA condensation at the level of the nucleosome. But the arrangement of nucleosomes and their packing into the 30 nm filament account for only a 35- or 40-fold compaction of the DNA, far from the 10,000-fold compaction needed to pack the genome into the cell.

2. How do nucleosomes control individual gene expression? The expression of genes is controlled by many factors, including modification of the histone tails. Do individual modifications produce a series of distinct changes in chromatin, as suggested by the histone code hypothesis? If so, what are the steps that distinguish the expression of individual genes?

3. How are different epigenetic states specified? The exact type and combination of histone modifications that encode various epigenetic states are poorly understood, yet they are of utmost importance to human health and disease. Are epigenetic states that are inherited through generations altered by environmental factors? Can diseases that derive from an epigenetic source be reversed by therapeutic intervention? Understanding how epigenetic states are achieved and maintained is a most exciting avenue of future research.

HOW WE KNOW

Kornberg Wrapped His Mind around the Histone Octamer

Kornberg, R.D. 1974. Chromatin structure: A repeat in unit of histones in DNA. *Science* 184:868–871.

Kornberg, R.D., and J.O. Thomas. 1974. Chromatin structure: Oligomers of the histones. *Science* 184:865–868.

Roger Kornberg [*Source: Courtesy of Roger Kornberg.*]

Roger Kornberg is best known for his discovery of the structure of eukaryotic RNA polymerase II (see Chapter 15). But well before that Nobel Prize–winning work, he performed seminal studies showing that the nucleosome consists of a histone octamer with DNA wrapped around it. By the time of this work in the 1970s, scientists had already determined that the nucleosome contains four "core" histones in a 1:1:1:1 stoichiometry, and they thought that histones were arranged in some type of spiral. Kornberg's goal was to determine the crystal structure of the histone-DNA complex. But when he purified the separate histone subunits according to established purification protocols and mixed them together, they did not assemble into an oligomer. Kornberg deduced that the pure histones failed to assemble because they were denatured by the acid and guanidine hydrochloride used in the purification.

Around this time, a new histone purification protocol was developed that did not use denaturing solvents, but the procedure went largely unnoticed because it did not separate the four histones; rather, it split them into two peaks, one containing H2A and H2B, and the other containing H3 and H4. Kornberg realized that this mild purification method probably kept the histones intact and that the two peaks might be heterodimer complexes of histone subunits that lie adjacent to each other in the nucleosome.

To test this idea, Kornberg treated the fraction containing H3 and H4 with a chemical cross-linking agent—a molecule that has two reactive groups and can covalently link two proteins, provided they are close together. In SDS–polyacrylamide gel electrophoresis

(SDS-PAGE), a cross-linked H3-H4 heterodimer should produce a band that migrates at a position representing the sum of the masses of H3 and H4. Cross-linking is not 100% efficient, so the expected experimental result would be three bands, corresponding to H3, H4, and cross-linked H3-H4. Instead, Kornberg observed eight bands! The eight bands corresponded to the masses expected for an $H3_2$-$H4_2$ heterotetramer: H3, H4, H3-H4, H3-H3, H4-H4, H3-H3-H4, H3-H4-H4, and H3-H3-H4-H4 (Figure 1). Supporting this conclusion, sedimentation analysis yielded a molecular mass of 53.9 kDa for the largest H3-H4 complex, very close to the 53.2 kDa expected for a $H3_2$-$H4_2$ heterotetramer.

Combining the known 1:1:1:1 ratio of core histone subunits with his finding that H3-H4 is a heterotetramer, Kornberg hypothesized that the histone subunits assemble into an octamer containing two of each subunit. He proposed how they are arranged and surmised that about 200 bp of DNA wraps around the histone octamer, because micrococcal nuclease treatment of chromatin was known to give a repeat fragment length of 200 bp. We now know that Kornberg's insightful conclusions were right on target!

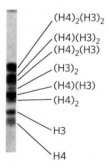

$(H4)_2(H3)_2$
$(H4)(H3)_2$
$(H4)_2(H3)$
$(H3)_2$
$(H4)(H3)$
$(H4)_2$
H3
H4

FIGURE 1 In Kornberg's experiments, chemical cross-linking revealed that H3-H4 forms a heterotetramer. An H3-H4 complex was treated with a chemical cross-linker, followed by analysis by SDS-PAGE to separate the cross-linked proteins according to mass, which decreases from top to bottom of the gel shown here. [*Source: Data from R. D. Kornberg and J. O. Thomas, Science 184:865–868, 1974, Fig. 1. Reprinted with permission from AAAS.*]

HOW WE KNOW

A Transcription Factor Can Acetylate Histones

Brownell, J.E., and C.D. Allis. 1995. An activity gel assay detects a single, catalytically active histone acetyltransferase subunit in *Tetrahymena* macronuclei. *Proc. Natl. Acad. Sci. USA* 92:6364–6368.

Brownell, J.E., J. Zhou, T. Ranalli, R. Kobayashi, D.G. Edmondson, S.Y. Roth, and C.D. Allis. 1996. *Tetrahymena* histone acetyltransferase A: A homolog to yeast Gcn5p linking histone acetylation to gene activation. *Cell* 84:843–851.

For many years, histones were thought to have only a structural function—packing DNA into the confines of a nucleus. We now know that histones regulate gene expression and are involved in epigenetic inheritance. Early genetic and biochemical studies of transcription factors had identified the basal RNA polymerase machinery and several transcription coactivators, but little did anyone know that many of these factors are enzymes that covalently modify histones.

With hindsight, we can see that as far back as the 1960s, researchers had correlated histone acetylation with gene regulation, suggesting that histones do not just play a structural role. The idea that histone modification could regulate transcription heated up significantly with the discovery of a nuclear histone acetyltransferase (HAT) by David Allis's laboratory in 1995 (see this chapter's Moment of Discovery). The Allis group chose macronuclear extracts of *Tetrahymena thermophila* (each cell of this organism has two nuclei: a macronucleus and a micronucleus) as an enzyme source, because this ciliate contains large amounts of acetylated histones. HAT activity was followed during purification by an activity gel assay. The assay used SDS-PAGE, starting with a gel in which histones were uniformly distributed by adding them to the polyacrylamide solution before pouring the gel and polymerizing it into a solid matrix. Column fractions of the *Tetrahymena* macronuclear extract were analyzed on this gel, then the gel was soaked in buffer lacking SDS, allowing proteins to renature. The buffer also contained [3H]-labeled acetyl-CoA, the substrate for HAT activity. Covalent attachment of [3H]acetyl groups to the uniformly distributed histones in the gel occurred only where a HAT enzyme renatured in the gel. Unused [3H]acetyl-CoA was removed by soaking the gel, and this was followed by fluorography to visualize the position of the HAT activity, where the 3H-labeled histones overlapped with the HAT (Figure 2). The Allis

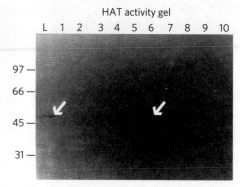

FIGURE 2 Fractions of *Tetrahymena* macronuclear proteins eluted from an ion-exchange chromatography column were analyzed by SDS-PAGE. Lane numbers (1 to 10) represent fraction numbers; L is the starting extract. Numbers on the left indicate molecular mass in kilodaltons. The p55 protein contains HAT activity and can be visualized in the activity gel both in the initial extract (lane L) and in the column fraction in lane 6 (arrows). [Source: Data from J. E. Brownell and C. D. Allis, Proc. Natl. Acad. Sci. USA 92:6364–6368, 1995, Fig. 5.]

lab referred to the *Tetrahymena* HAT as p55 because it migrates as a 55 kDa protein.

Next, the researchers identified the gene encoding p55. The sequence finding was astonishing: p55 was homologous to a very well-studied transcription activator, the yeast Gcn5 protein. This result implied that Gcn5 has HAT activity and that this enzymatic activity may account for the activator's regulatory function. In fact, Allis's group used their gel assay to demonstrate that Gcn5 has HAT activity (Figure 3). Since this discovery, several transcription factors have been shown to contain a subunit with HAT activity.

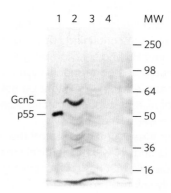

FIGURE 3 The activity gel result showing that Gcn5 has HAT activity. Macronuclear extracts of *Tetrahymena* show the expected HAT activity of p55 (lane 1). The gene for Gcn5 was cloned into an expression plasmid and then induced. The induced Gcn5 (60 kDa) protein has HAT activity (lane 2). Lane 3 shows the result for uninduced cells; lane 4, induced cells lacking the expression plasmid containing the Gcn5 gene. The numbers on the right indicate molecular mass in kilodaltons. [Source: Data from J. E. Brownell et al., Cell 84:843–851, 1996, Fig. 4.]

PROBLEMS

1. When proteins are incubated with the detergent sodium dodecyl sulfate (SDS), they are partially denatured, losing much of their structure, and tend to take on a consistent mass-to-charge ratio (see Highlight 4-1). In SDS-polyacrylamide gel electrophoresis, proteins are separated almost entirely as a function of their mass. Histones are an exception. On these same gels, many histones migrate more slowly than they should, as though they were much larger than they actually are. Suggest an explanation for this behavior.

2. In eukaryotic chromatin, approximately what length of DNA is typically associated with a single histone octamer?

3. Which of the following protein modifications—acetylation, phosphorylation, and methylation—could change the net charge on the surface of a modified histone?

4. Within the histone structure, do protein modifications occur primarily near the N-terminus or near the C-terminus? What features distinguish the structure where modifications occur?

5. In bacteria, the transcription of a subset of genes is affected by DNA topology, with expression increasing or (more often) decreasing when the DNA is relaxed. When a bacterial chromosome is cleaved at a specific site by a restriction enzyme (one that cuts at a long, and thus rare, sequence), only nearby genes (within 10,000 bp) exhibit either an increase or a decrease in expression. The transcription of genes elsewhere in the chromosome is unaffected. Explain.

6. In different regions of chromatin, the ratio of histone H1 to histone H2A may vary, but the ratio of histone H2A to histone H2B is generally the same. If the amount of H1 increases in a region of chromatin, will transcription of genes in that region increase or decrease? Explain your answer.

7. In chromatin, nucleosomes are organized in higher-order structures: 30 nm filaments. Although the detailed structure is not known, which features of the 30 nm filament have been experimentally determined?

8. In eukaryotes, chromosomes are packaged into successively higher-order structures, such as the 30 nm filaments. In bacteria, DNA is not packaged into such stable proteinaceous structures, and histonelike proteins bind to the DNA less tightly. Suggest an explanation for this difference.

9. Describe at least three differences between chromatin regions that are transcriptionally active and those in which genes are transcriptionally silent.

10. How does epigenetic inheritance differ from Mendelian inheritance?

11. During replication, nucleosomes are partially displaced and distributed on the daughter DNA strands. New histone subunits are added to bring the entire complement of nucleosomes up to the required level. Nucleosomes on the DNA to be replicated may have modified histone subunits, but the new histones that appear after replication lack the modifications (at least transiently). Which of the following statements describes how the modified and unmodified histone subunits are distributed in nucleosomes after replication?

    (a) The modified and unmodified histones are assembled randomly into nucleosomes.
    (b) The modified histone subunits stay together in nucleosomes, separate from unmodified nucleosomes.
    (c) The H3-H4 modified pairs stay together, the H2A-H2B modified pairs stay together, and nucleosomes assemble with modified and unmodified H3-H4 and H2A-H2B pairs. The various combinations occur at random on each daughter DNA molecule.
    (d) Modified nucleosomes are segregated to one daughter chromosome, and completely unmodified nucleosomes are segregated to the other daughter chromosome.

12. The human genome contains about 3.1×10^9 bp of DNA. Assuming that the DNA is covered with nucleosomes spaced as described in this chapter, how many molecules of histone H2A are present in one somatic human cell? (Do not consider any reductions in H2A due to its replacement by H2A variants.) How would the number change after DNA replication but before cell division?

13. Roger Kornberg's histone cross-linking experiments defined an H3-H4 heterotetramer as a nucleosome substructure (see the How We Know section for this chapter). Suppose nucleosomes actually contained two H3 subunits but only one H4 subunit, forming a stable H3-H3-H4 heterotrimer. How would the cross-linking results have been different?

DATA ANALYSIS PROBLEM

Lowary, P.T., and J. Widom. 1998. New DNA sequence rules for high affinity binding to histone octamer and sequence-directed nucleosome positioning. *J. Mol. Biol.* 276:19–42.

14. In their efforts to define the rules for nucleosome positioning on DNA, Lowary and Widom obtained a library of 5×10^{12} DNA molecules, each with a different randomized sequence. All of the molecules were assembled from three synthesized segments of 72, 76, and 72 bp each, with two different 6 bp linkers that contained restriction sites. Each molecule was flanked by short segments of DNA that served as targets for PCR amplification. The total length of all the random DNA segments was 220 bp. The entire library was PCR-amplified to increase the total amount of DNA. Then, histone octamers were mixed with the DNA, in a ratio of 10 DNA molecules per histone octamer. The nucleosome-DNA complexes were isolated after dialysis into a relatively high concentration of salt, the bound DNA was amplified by PCR, and the procedure was repeated. The same cycle was repeated 12 to 15 times, and eventually the bound DNAs that came through the selection were cloned and sequenced to determine which DNA sequences bound most tightly to nucleosomes. This led to a series of nucleosome positioning rules. For example, the dinucleotide TA tended to appear at 10- or 20-nucleotide intervals in sequences that were strongly bound by nucleosomes.

   (a) Why was the random DNA length set at 220 bp?
   (b) Why were nucleosomes added at a level that could bind only 10% of the DNA molecules present?
   (c) What was the function of the high salt concentration?
   (d) Why was the DNA amplified by PCR in each cycle?

ADDITIONAL READING

General

Maze, I., K.M. Noh, A.A. Soshnev, and C.D. Allis. 2014. Every amino acid matters: Essential contributions of histone variants to mammalian development and disease. *Nat. Rev. Genet.* 15:259–271.

Poirier, M.G., E. Oh, H.S. Tims, and J. Widom. 2009. Dynamics and function of compact nucleosome arrays. *Nat. Struct. Mol. Biol.* 16:938–944.

Nucleosomes: The Basic Units of DNA Condensation

Kornberg, R.D., and Y. Lorch. 1999. Twenty-five years of the nucleosome, fundamental particle of the eukaryote chromosome. *Cell* 98:285–294.

Richmond, T.J., and C.A. Davey. 2003. The structure of DNA in the nucleosome core. *Nature* 423:145–150.

Schalch, T., S. Duda, D.F. Sargent, and T.J. Richmond. 2005. X-ray structure of a tetranucleosome and its implications for the chromatin fibre. *Nature* 436:138–141.

Higher-Order Chromosome Structure

Cui, Y., and C. Bustamante. 2000. Pulling a single chromatin fiber reveals the forces that maintain its higher-order structure. *Proc. Natl. Acad. Sci. USA* 97:127–132.

Li, G., and D. Reinberg. 2011. Chromatin higher-order structures and gene regulation. *Curr. Opin. Genet. Dev.* 21:175–186.

Pepenella, S., K.J. Murphy, and J.J. Hayes. 2014. Intra- and inter-nucleosome interactions of the core histone tail domains in higher-order chromatin structure. *Chromosoma* 123:3–13.

Robinson, P.J.J., and D. Rhodes. 2006. Structure of the "30 nm" chromatin fibre: A key role for the linker histone. *Curr. Opin. Struct. Biol.* 16:336–343.

Schalch, T., S. Duda, D.F. Sargent, and T.J. Richmond. 2005. X-ray structure of a tetranucleosome and its implications for the chromatin fibre. *Nature* 436:138–141.

The Regulation of Chromosome Structure

Couture, J.F., and R.C. Trievel. 2006. Histone-modifying enzymes: Encrypting and enigmatic epigenetic code. *Curr. Opin. Struct. Biol.* 16:753–760.

Ducasse, M., and M.A. Brown. 2006. Epigenetic aberrations and cancer. *Mol. Cancer* 5:60.

Fischle, W., Y. Wang, and C.D. Allis. 2003. Histone and chromatin cross-talk. *Curr. Opin. Cell Biol.* 15:172–183.

Formosa, T. 2013. The role of FACT in making and breaking nucleosomes. *Biochim. Biophys. Acta* 1819:247–255.

Leschziner, A.E., A. Saha, J. Wittmeyer, Y. Zhang, C. Bustamante, B.R. Cairns, and E. Nogales. 2007. Conformational flexibility in the chromatin remodeler RSC observed by electron microscopy and the orthogonal tilt reconstruction method. *Proc. Natl. Acad. Sci. USA* 104:4913–4918.

Lorch, Y., B. Maier-Davis, and R.D. Kornberg. 2010. Mechanism of chromatin remodeling. *Proc. Natl. Acad. Sci. USA* 107:3458–3462.

Sarma, K., and D. Reinberg. 2005. Histone variants meet their match. *Nat. Rev. Mol. Cell Biol.* 6:139–149.

11

DNA Replication

Robert Lehman [Source: Courtesy of American Society for Biochemistry and Molecular Biology.]

MOMENT OF DISCOVERY

I started my laboratory at Stanford to study DNA recombination using the biochemical methods established by Arthur Kornberg for working with DNA polymerase. Matt Meselson's lab at Harvard had evidence that DNA recombination involved breaking and joining of DNA strands, so *we set out to find an enzyme that sealed the break by forming a phosphodiester bond.*

Using *E. coli* cell extracts and a substrate DNA duplex with single-strand breaks (nicks) labeled with radioactive ^{32}P at their 5′ termini, we measured conversion of the 5′-^{32}P to a form that was protected from removal by an enzyme, presumably due to incorporation into a new phosphodiester bond. This assay worked, and the DNA strands were joined after incubation in the extract! But I wanted to get definitive proof that the strands were sealed by a phosphodiester bond. Although it sounds obvious now, at the time we thought it possible that a protein linker might be the cause of the DNA strand joining.

We used two different enzyme nucleases to degrade the DNA. One enzyme digested DNA to mononucleotides leaving the ^{32}P at the 5′ end; the other enzyme digested DNA in the opposite direction to give a mononucleotide with ^{32}P at the 3′ end. If the linkage in the DNA were due to a phosphodiester bond, both of these products would be observed, depending on which nuclease was used. But if the linkage was due to a protein, neither of these would show up in the analysis, and we would have seen ^{32}P attached to a protein instead. I'll never forget the electric moment when my students and I looked at the two chromatograms used to analyze the DNA degradation products, and there they were staring right at us—the 5′-^{32}P-labeled mononucleotides and 3′-^{32}P-labeled mononucleotides, the exact products expected for a true phosphodiester bond. It was wonderfully exciting to discover a new enzyme, and we now know that DNA ligase not only is involved in recombination, but has also turned out to be one of the central players involved in DNA replication and repair in all cells.

—Robert Lehman, on discovering DNA ligase

Online resources related to this chapter:

***Nature* exercise** Single molecule studies, DNA polymerization

Simulations DNA replication, DNA polymerase

Having covered the properties of DNA, RNA, and protein—their chemical structures and functions and the methods for studying them—we now make a transition and begin to delve deeply into the heart of molecular biology. In this chapter and Chapters 12-18, we explore the mechanisms of biological information transfer. Proteins and nucleic acids enable the flow of biological information through a network of processes that collectively allow the cell to grow, reproduce, maintain its structures and organization, acquire energy, sense and respond to its environment, and, in multicellular organisms, diversify into tissues and organs. Although we know a lot about the individual molecules and reactions that conduct this information flow in the cell, we have only recently begun to understand how the pieces fit together.

This chapter focuses on the process of **DNA replication**—the duplication of the cellular genome, in which the stored genomic information is handed down to the next generation. DNA replication is central to life and to evolution; without it, there could be no transfer of information across generations. Any modifications to the genome that help or hurt the organism would be lost instead of inherited by its offspring. DNA replication is also highly regulated in response to the environment. Replication should occur only when the cell has sufficient resources to divide and form two new cells. In multicellular organisms, loss of control over this process leads to cancer—uncontrolled cell division that eventually kills the entire organism.

The structure of DNA is so elegant and simple that one might think the process of duplicating DNA would reflect this simplicity. But the replication of DNA is far from simple. Imagine the evolutionary pressure to develop robust enzymatic machinery that duplicates this large set of instructions with the high fidelity imperative to maintaining the species. The accuracy of replication is particularly crucial because even a seemingly low error rate of one incorrect base pair in a thousand would produce three million mutations after a single replication cycle of the DNA in a human cell! As we will see, this evolutionary pressure has resulted in novel enzymatic architectures working in ways that are beautiful to behold. The interplay of numerous replication proteins follows a complex choreography to produce two identical DNA molecules from one.

The principles of DNA replication are surprisingly similar in bacteria, archaea, and eukaryotes alike. We begin the chapter with some classic studies that provide an overview of the replication process, and then take a look at the chemistry of the reaction, the structure of DNA polymerase (the enzyme that joins nucleotides into a DNA chain), and the many other proteins needed to replicate double-stranded DNA. We next explore how replication starts at specific origin sequences and how replication forks—the sites of

DNA polymerization—are established. Initiation is a key regulatory step in replication because, once replication is initiated, it does not stop until the entire DNA molecule is successfully duplicated. Finally, we discuss how the ends of chromosomes are replicated, a process that contributes to the control of the lifespan of eukaryotic cells.

11.1 DNA TRANSACTIONS DURING REPLICATION

Replication begins at specific locations on the chromosome where DNA polymerase is recruited, then travels bidirectionally along the DNA until the whole chromosome is copied. During this process, each of the two complementary strands of duplex DNA is used as a template for a newly synthesized strand. We review here a few of the exciting, now classic, discoveries about the replication process.

DNA Replication Is Semiconservative

Watson and Crick hypothesized that during DNA replication, the double helix is unwound and each parental DNA strand is used as a template to generate a new daughter strand, and thus two daughter duplexes, following the A=T and G≡C base-pairing rules. This mode of replication is called **semiconservative**, because each daughter duplex conserves one strand of parental DNA; the other strand is completely new. However, two other hypotheses were also proposed to explain how DNA could be replicated (**Figure 11-1a**). In one alternative hypothesis, conservative replication, the two parental strands would remain together while acting as template for production of a new DNA duplex. Another possible mechanism was dispersive replication, in which the duplex would be replicated in a random patchwork.

An elegant experiment to distinguish among these three replication mechanisms was devised by Matthew Meselson and Franklin Stahl. First, they grew *Escherichia coli* cells for multiple generations in a medium containing $^{15}NH_4$ as the sole nitrogen source to uniformly label the cellular DNA with the heavy nitrogen isotope, ^{15}N. Although DNA contains only about 17% nitrogen by mass, and the mass of ^{15}N is only 7% greater than that of ^{14}N (the nitrogen isotope normally found in DNA), the resulting 1.2% increase in the mass of DNA labeled with ^{15}N is enough to separate [^{15}N]DNA from [^{14}N]DNA in a density gradient. **Figure 11-1b** shows one of their control experiments in which an equal mixture of [^{14}N]DNA and [^{15}N]DNA was added to a solution of CsCl (cesium chloride) in a tube and subjected to ultracentrifugation, forming a density gradient of CsCl from the top to the bottom of the tube. As the density gradient

(a) Three proposed models:

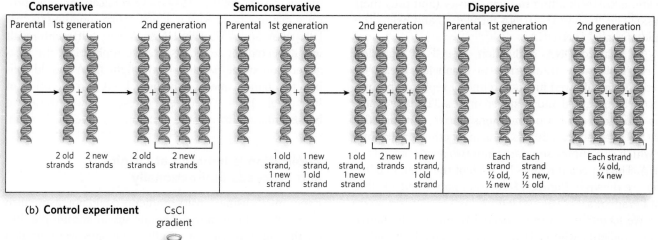

(b) Control experiment

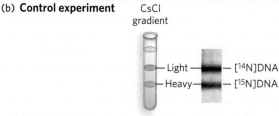

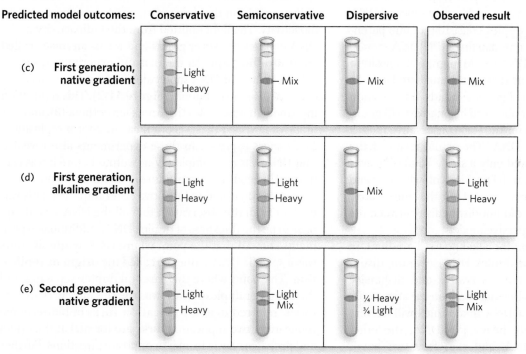

FIGURE 11-1 Semiconservative DNA replication. (a) Three possible mechanisms of replication, and the distribution of parental DNA (blue) and daughter DNA (red) expected for each, after one and two generations of growth. (b) In one of Meselson and Stahl's control experiments, equal amounts of double-stranded [^{14}N]DNA (both strands light) and [^{15}N]DNA (both strands heavy) were mixed and subjected to ultracentrifugation in a native CsCl density gradient. At right is a picture of the centrifuge cell, taken while the centrifuge is running, that visualizes DNA bands using UV absorption (DNA absorbs UV light). (c) Expected and observed results in a native CsCl density gradient (DNA is double-stranded) for conservative, semiconservative, and dispersive replication following one round of replication (first generation). (d) Expected and observed results in an alkaline CsCl density gradient (DNA strands separate) after one round of replication. (e) Expected and observed results in a native CsCl density gradient following two rounds of replication (second generation). [Source: Data from M. Meselson and F. W. Stahl, Proc. Natl. Acad. Sci. USA 44:671–682, 1958, Fig. 2.]

formed, DNA molecules coalesced into sharp bands at positions corresponding to their density (and thus their mass). The "heavy" [^{15}N]DNA formed a band below the band of "light" [^{14}N]DNA.

To analyze DNA replication in cells, ^{15}N-labeled *E. coli* cells were transferred to unlabeled (i.e., ^{14}N) medium and allowed just one round of replication, then DNA was extracted and analyzed in a CsCl density gradient under either of two conditions: native, in which the two DNA strands of a duplex remain together, or alkaline, in which the duplex separates into two single strands of DNA. The expected results for each of the three hypotheses are shown in **Figure 11-1c–e**, for native and alkaline CsCl gradients.

We focus first on the expected results for the native CsCl gradient, where DNA remains double-stranded, illustrated in Figure 11-1c. If replication were conservative, two bands of duplex DNA would be observed: one of lower density, containing two new strands of [^{14}N]DNA (light/light), and one of higher density, consisting of the original parental [^{15}N]DNA duplex (heavy/heavy). If replication were semiconservative, only one band of duplex DNA would be observed, each duplex consisting of one parental [^{15}N]DNA strand and one daughter [^{14}N]DNA strand. This heavy/light band of DNA would migrate at a position in the native CsCl gradient that is intermediate between the heavy/heavy and light/light DNA bands. If replication were dispersive, the result observed in a native CsCl gradient would be the same as for the semiconservative model: one band of heavy/light DNA. The actual result of the native CsCl gradient showed only a single band of heavy/light DNA, located between the positions of light/light and heavy/heavy DNA. This result ruled out the conservative model of replication, but did not distinguish between the semiconservative and dispersive models.

The experiment using the alkaline CsCl gradient, in which the two strands of duplex DNA separate, distinguishes between the semiconservative and dispersive models of replication. Following semiconservative replication, the two strands of DNA in a duplex will be very different—one completely heavy [^{15}N]DNA, the other completely light [^{14}N]DNA—and therefore two bands should be observed. In contrast, the two strands of a DNA duplex produced by dispersive replication would both be composed of 50% heavy and 50% light DNA, and thus only one band would be observed and would migrate at the heavy/light position in the alkaline CsCl gradient. The observed results supported the semiconservative mechanism and ruled out the dispersive mechanism.

To be sure of their conclusion, Meselson and Stahl performed one more experiment to distinguish between semiconservative and dispersive replication. Cells were allowed to proceed through a second round of replication in unlabeled medium, and the DNA was again analyzed

in a native CsCl gradient. The expected results for each model were as follows. Dispersive replication would produce just one band with a density corresponding to 25% [^{15}N]DNA and 75% [^{14}N]DNA. A very different result was expected for semiconservative replication: the first-generation single band of heavy/light DNA should split into two bands, one of light/light DNA and the other of heavy/light DNA. The observed result showed these two bands, thus confirming the semiconservative mechanism.

Replication Is Initiated at Origins and Proceeds Bidirectionally

Experiments using density gradients could not distinguish whether synthesis initiated at one site or at multiple sites on the chromosome, or whether the sites were defined or random. These questions were answered by observing replication of a circular bacterial plasmid DNA in an electron microscope, using autoradiography. In this procedure, radioactively labeled DNA was extracted and spread on an electron microscope grid, then overlaid with a silver emulsion that formed an image on exposure to radioactivity. The light emitted from the radioactivity in the DNA exposes the silver grains and forms an image called an **autoradiograph**. In the plasmid experiments, the resulting images of DNA molecules undergoing replication contained a loop or bubble (**Figure 11-2**). This replication intermediate resembles the Greek letter theta (θ), and this mode of DNA synthesis is often called θ-form replication. The images captured in these experiments also revealed that DNA was not completely unwound before it was replicated. Instead, as the strands separated, they were simultaneously replicated to form two sister duplexes. This was evident from the observation that all the DNA strands appear to be the thickness of duplex DNA. Additional experiments showed that the point where DNA synthesis starts was a specific DNA sequence called the **origin of replication**. The point where the parental duplex separates and the daughter duplexes form was the site of new DNA synthesis, referred to as a **replication fork**. In bidirectional replication, two replication forks are formed at the origin and propagate away from it in opposite directions. Further studies showed that the circular chromosomes of bacteria are replicated in the same way as plasmids, and bidirectional replication occurs from the origin in the bacterial chromosome as well as in most plasmid DNAs.

A single origin of replication is sufficient for the comparatively small bacterial chromosome and plasmids, but eukaryotic cells contain much more DNA. Also, eukaryotic replication forks travel at about one-tenth the speed of bacterial replication forks, possibly due to the complex compaction of eukaryotic chromosomes (see Chapters 9 and 10). It would take more than 100 hours to duplicate the DNA in a human cell if each chromosome

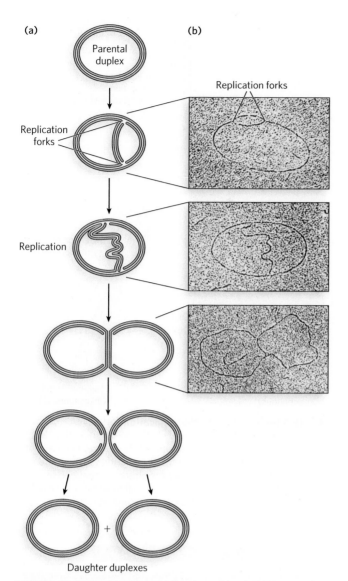

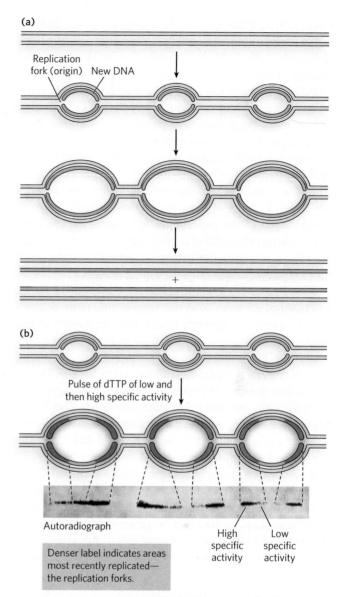

FIGURE 11-2 Simultaneous replication of both DNA strands.
(a) Replication stages of a circular, double-stranded DNA plasmid, based on autoradiography and electron microscopic observations. Blue strands are parental DNA; red strands are new DNA. The two replication forks proceed in opposite directions (bidirectionally) around the circular plasmid until they meet at the opposite side, forming two duplex circles. (b) Autoradiographs viewed in the electron microscope, showing ³H-dTTP-labeled plasmid DNA undergoing replication from a single origin. The origin is located in the center of the small DNA bubble in the first micrograph. [*Source: (b) Bernard Hirt, Institut Suisse de Recherches Experimentales sur le Cancer.*]

FIGURE 11-3 Multiple eukaryotic replication forks. (a) In eukaryotes, bidirectional replication forks from multiple origins form replication bubbles that expand and meet head-on to produce two new linear duplex chromosomes. Parental DNA is blue; newly replicated DNA is red. (b) Chinese hamster ovary cells were labeled with two pulses of [³H]dTTP, first of low radioactivity, then high radioactivity. DNA was then extracted and imaged by autoradiography. The image shows the newly replicated, ³H-labeled DNA initiated at multiple origins. Increased density of radioactivity at both ends of a bubble, at the replication forks, reveals that each new DNA strand is growing away from the midpoint and thus that replication is bidirectional. Replication forks had formed prior to addition of [³H]dTTP, so the new DNA is not completely radioactive along its entire length (the DNA closest to the origin is not labeled). On addition of [³H]dTTP, the new DNA becomes labeled. The label intensity increases at each forked junction because of the increase in specific activity of the label during the experiment. Therefore, the taper of the signal in the autoradiograph reveals that the direction of fork movement is bidirectional, away from the midpoint (origin) between two sets of replication forks. [*Source: (b) Reprinted from J. Mol. Biol. 75, Huberman, J. and Tsai, A., Direction of DNA replication in mammalian cells, pages 5-12, copyright 1973, with permission from Elsevier. Courtesy Joel Hu.*]

contained only one replication fork that started from a single origin. To duplicate the larger eukaryotic genomes within a biologically relevant timeframe, numerous replication forks from numerous origins of replication are generated on each chromosome, as shown in experiments using autoradiography (**Figure 11-3**).

As in bacteria, replication bubbles grow bidirectionally in eukaryotes, with replication forks moving in opposite directions on either side of an origin, as illustrated in Figure 11-3a. An example of how bidirectional replication was tested in mammalian cells is shown in Figure 11-3b. Cells were supplied with ³H-labeled thymidine ([³H]dTTP) in two separate pulses—first with [³H]dTTP having a low level of radioactivity (low specific activity), followed by a short pulse of [³H]dTTP with high radioactivity (high specific activity). The resulting autoradiograph shows a gradient of radioactive density that is heaviest at forked junctions moving in opposite directions, indicating that replication forks travel bidirectionally from each origin. Only the replication bubbles, not the entire DNA, are visible in the autoradiograph, because the duration of the experiment was short—far less than needed for a full round of replication. Experiments of this type have also been performed in bacteria, archaea, and other eukaryotic cells, all with similar results.

We know that specific origin sequences are present in bacteria, archaea, and the simple eukaryote *Saccharomyces cerevisiae* (baker's yeast), but the nature of the origins of replication in complex eukaryotes remains uncertain. More details on replication origins, including how they initiate replication, are given in Section 11.4.

Bidirectional replication is the most common form of replication, but not completely universal. The *E. coli* Col E1 plasmid is an example of a circular plasmid with a defined origin that replicates in a single direction. The genomes of some organisms, such as bacteriophage Φ29 and adenovirus, are single, linear, double-stranded DNAs with no internal origin. These genomes are replicated starting from their ends. Another form of replication, used by certain phages with a circular genome, involves a single replication fork that proceeds multiple times around the circular DNA to generate numerous copies of the genome.

Replication Is Semidiscontinuous

Now that we have seen that replication occurs bidirectionally with one replication fork at each end of the bubble, we take a closer look at how two daughter strands are produced at a single replication fork. The Y-shaped replication fork consists of a parental duplex DNA stem and two prongs, the new daughter duplexes. The parental DNA strands are antiparallel, so the links between nucleotides in the two daughter strands also run in opposite directions (**Figure 11-4**). However, all known **DNA polymerases**, the enzymes that synthesize DNA, extend DNA in only one direction: 5′→3′. Hence, both daughter strands cannot be replicated in the same direction in which the replication fork moves. Furthermore, as we discuss later in the chapter, DNA polymerase cannot initiate DNA chains; these must be initiated by short sections of RNA or DNA, referred to as primers. In the cell, the primers at a replication fork are RNA, synthesized by an enzyme called primase.

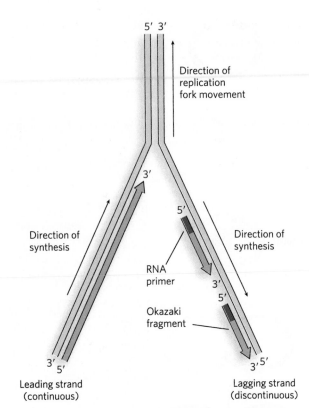

FIGURE 11-4 Structure of the replication fork. DNA polymerases can extend DNA only in the 5′→3′ direction, but the two parental strands are antiparallel. Therefore, one daughter strand (the leading strand) is synthesized continuously, in the direction of fork movement, while the strand synthesized in the opposite direction (the lagging strand) must be replicated discontinuously as a series of Okazaki fragments.

KEY CONVENTION

The direction of synthesis by DNA polymerases refers to the direction in which each new nucleotide is chemically linked to the growing daughter strand. All DNA polymerases function in the 5′→3′ direction, linking the α-5′-phosphate of a new dNTP to the 3′ position of the nucleotide residue at the end (i.e., the 3′ end) of the chain.

Reiji Okazaki and his coworkers investigated the proposal that one of the daughter strands is continuously extended in the same direction in which the replication fork moves, while the other strand is synthesized in the opposite direction. The strand that is made in the direction opposite to fork movement is still synthesized in the 5′→3′ direction, and thus the strand must be reinitiated at intervals and synthesized as a series of fragments (see Figure 11-4). This mode of replication is described as **semidiscontinuous**, because only one daughter strand is synthesized continuously; the other is made as a series of discontinuous fragments. The continuously synthesized daughter strand is called the

leading strand, and the discontinuous daughter strand is the **lagging strand**. The "lagging" designation is based on the fact that some unreplicated single-stranded DNA is generated on the lagging strand by the moving fork, so the conversion to duplex DNA on this strand lags behind that of the leading strand.

The model of semidiscontinuous replication predicts that during chromosome replication, short DNA fragments are produced on one of the strands. These fragments are expected to exist only transiently before being connected together. To look for these lagging-strand fragments, Okazaki used *E. coli* cells infected with T4 phage. Because T4 makes many copies of itself simultaneously, detection of lagging-strand fragments is made easier by their abundance. T4-infected *E. coli* cells were labeled with radioactive nucleotide precursors for brief time intervals, then the DNA was analyzed in alkaline CsCl gradients to separate the new, radioactive DNA strands from the unlabeled parental strands. Small fragments in the 1,000 to 2,000 nucleotide (1 to 2 kb) range were observed, as predicted, and have come to be known as **Okazaki fragments**. Lagging-strand Okazaki fragments are 1 to 2 kb long in bacteria, but are shorter—only 100 to 200 nucleotides—in eukaryotes. Each Okazaki fragment is primed at the 5′ end by a short RNA of 10 to 13 nucleotides. As we discuss in detail later in the chapter, the RNA primer is removed by nuclease action, and the Okazaki fragments are joined by ligase soon after the replication fork has passed.

SECTION 11.1 SUMMARY

- DNA replication is semiconservative; each daughter chromosome contains one parental strand and one newly synthesized strand.

- Replication is initiated at specific sites on the DNA, the origins of replication. Circular bacterial chromosomes usually have only one origin, whereas long, linear eukaryotic chromosomes have numerous origins.

- Replication usually proceeds bidirectionally from the origin, at growing points known as replication forks.

- At a replication fork, the parental DNA strands are separated and used as templates to simultaneously form two new daughter duplexes.

- The two strands of DNA in a duplex are antiparallel, yet DNA polymerases extend DNA only in the 5′→3′ direction. Therefore, only one strand, the leading strand, is extended "continuously" in the direction of replication fork movement; the other strand, the lagging strand, is synthesized "discontinuously" in the opposite direction as a series of Okazaki fragments, which subsequently are joined by ligase. This is the process of semidiscontinuous replication.

11.2 THE CHEMISTRY OF DNA POLYMERASES

The first DNA polymerase was identified in the 1950s by Arthur Kornberg and his postdoctoral fellow, Robert Lehman (see the How We Know section at the end of this chapter). In so doing, they initiated what would become an entire field of study on DNA replication enzymology. Initially, *E. coli* DNA polymerase I (Pol I) was simply called "DNA polymerase," as it was presumed to be the only DNA polymerase in the cell. We now know, after decades of study, that *E. coli* contains five different DNA polymerases, involved in a variety of cellular processes. In fact, Pol I mainly functions in the repair of damaged DNA, although it also carries out an important function in connecting Okazaki fragments during replication, as we will see later in this chapter. We focus here on Pol I because the study of this enzyme revealed features of DNA synthesis that are common to all DNA polymerases, and it remains the most intensively studied and well-characterized DNA polymerase.

DNA Polymerases Elongate DNA in the 5′→3′ Direction

Early work on Pol I led to the definition of two central requirements for DNA polymerization. First, all DNA polymerases require a **template strand** that guides the polymerization reaction according to the Watson-Crick base-pairing rules: where dC is present in the template, dG is added to the new strand, and so on (**Figure 11-5a**). Second, DNA polymerases require a **primer strand** that is complementary to the template and contains a free 3′-OH group to which a nucleotide can be added. In other words, DNA polymerases can only add nucleotides to a preexisting strand; they cannot synthesize DNA starting from only a template strand. Most primers are oligonucleotides of RNA rather than DNA. The free 3′ end of the primer, where nucleotides will be added, is called the **primer terminus**. The double-stranded RNA-DNA formed by the primer and template is the **primed template**. Specialized enzymes (primases) synthesize RNA primers when and where they are required (as discussed in Section 11.3).

Studies of Pol I confirmed that the nucleotide precursors to DNA are the four deoxyribonucleoside 5′-triphosphates (dNTPs). The studies also showed that the different dNTPs bind the same active site on Pol I. Pol I differentiates among dNTPs only after it undergoes a conformational change that checks for the proper geometry of the base pair formed between the bound dNTP and the matching base on the template strand. Only the correct geometry of an A=T or G≡C base pair fits into

(a)

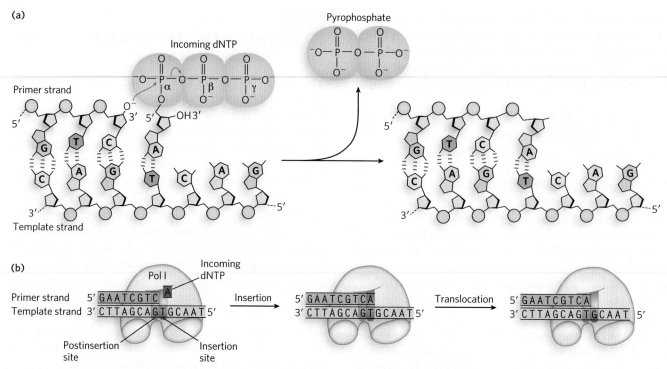

(b)

FIGURE 11-5 **The DNA polymerase reaction.** (a) DNA polymerases require a primer strand and a template strand (i.e., a primed template). As dNTPs are added to the 3'-OH group of the primer strand, this strand grows in the 5'→3' direction. Incoming dNTPs are complementary to the template strand. (b) The insertion and postinsertion sites in DNA polymerase properly align the primed template for sequential addition of incoming nucleotides.

the active site. An incorrect fit results in dissociation of the dNTP and binding of a new one. Normally, the polymerase is able to distinguish the correct nucleotide with this method, but one in every 10^4 to 10^5 nucleotides is added incorrectly. In the event of a misincorporated nucleotide, the polymerase has a way to remove it, which we discuss shortly.

The 3'-hydroxyl group of the primer strand is activated to attack the α phosphorus of the incoming dNTP, resulting in attachment of a dNMP to the primer 3' terminus and release of pyrophosphate (PP_i) (see Figure 11-5a). The overall reaction—where the primer strand is denoted as $(dNMP)_n$—is:

$$(dNMP)_n + dNTP \rightarrow (dNMP)_{n+1} + PP_i \quad (11\text{-}1)$$

Following incorporation of a dNMP, Pol I must slide forward on the new 3' terminus to incorporate another dNTP. Therefore, the DNA polymerase active site must be divided into at least two distinct sites (**Figure 11-5b**). The template nucleotide that will pair with the incoming dNTP is positioned in the **insertion site**. The primer strand 3'-terminal base pair is positioned in the **postinsertion site**. The 3' OH on the terminal ribose of the primer strand then attacks the phosphodiester bond that connects the α and β phosphates of the incoming dNTP. This results in addition of one dNMP to the

primer strand and the release of pyrophosphate. After addition of a dNMP to the primer terminus, the new terminal base pair occupies the insertion site and must be translocated to the postinsertion site, allowing the next template nucleotide and a new dNTP to occupy the insertion site. Translocation of DNA can occur by sliding of the enzyme or dissociation of the enzyme from the DNA, followed by rebinding with the terminal base pair in the postinsertion site.

DNA synthesis proceeds with only a minimal change in free energy, given that one phosphodiester bond is formed (with the addition of one dNMP to the 3' primer terminus) at the expense of another, similar bond (between the α and β phosphates of the dNTP). However, noncovalent base-pairing and base-stacking interactions of the newly added nucleotide residue provide additional stabilization that favors incorporation of the correct dNMP into the growing DNA molecule. Pol I can also catalyze the reverse reaction, called pyrophosphorolysis, in which PP_i and primed DNA drive Pol I to remove dNMPs from the primer strand and release them as dNTPs. In the cell, pyrophosphorolysis is largely prevented by the action of the enzyme pyrophosphatase, which splits pyrophosphate into two molecules of P_i, thereby removing PP_i so that the reverse reaction cannot occur. Pyrophosphate hydrolysis is energetically favorable and goes to near completion.

At first glance, the use of a dNTP to form one dNMP link to DNA might seem to be a waste of energy. Why not use dNDP as the nucleotide precursor, which would produce the same DNA product and one inorganic phosphate (P_i) instead of PP_i, which is then split into two P_i? The drawback would be that the reverse reaction could easily be initiated at any time, because P_i, the molecule that would initiate the reverse reaction, is abundant in the cell. Using triphosphate precursors ensures that the reverse reaction will not occur, because, as described above, PP_i is eliminated by pyrophosphatase—making the DNA polymerase reaction irreversible under cellular conditions. This is probably why all known DNA polymerases use triphosphate precursors. The same strategy also applies to RNA polymerases, which use NTPs and release PP_i during RNA synthesis.

The initial studies of Pol I, performed decades ago, demonstrated that it requires a primed template and dNTP precursors as substrates. Many different DNA polymerases have been studied over the years, from sources as diverse as phages and other viruses, various types of bacteria, archaea, and eukaryotes, and mitochondria from many different species. All known DNA polymerases use the mechanism shown in Figure 11-5. Indeed, RNA polymerases also use this same basic mechanism (described in Chapter 15).

Most DNA Polymerases Have DNA Exonuclease Activity

DNA nucleases are a class of enzymes that degrade DNA. Nucleases that shorten DNA from the ends are called **exonucleases**; **endonucleases** cut DNA at internal positions. All cells have several nucleases of both types that are used in various tasks. A biochemist who wants to study any other type of enzyme that uses DNA as a substrate must carefully purify the enzyme and remove all nucleases, otherwise the DNA will be destroyed by the nuclease activity, and the researcher will be unable to observe other enzymatic reactions that require DNA. This careful purification is what Kornberg thought he was doing as he purified Pol I from a cell extract, but he was troubled by his inability to separate what he thought was a contaminating exonuclease activity from Pol I. He finally had no choice but to come to the paradoxical conclusion that the same enzyme that makes DNA also degrades it. In fact, Kornberg found that Pol I has two different exonuclease activities: one starts at the 3′ end and degrades DNA in the 3′→5′ direction (opposite to the direction of DNA synthesis), and the other starts at the 5′ end and degrades DNA in the 5′→3′ direction. The active sites of the two DNA exonucleases of Pol I are distinct from each other and from the DNA polymerase active site.

Exonucleases that digest a DNA strand from the 3′ terminus are called 3′→5′ exonucleases, because the strand shortens at the 3′ end while the 5′ end remains intact. In contrast, 5′→3′ exonucleases digest DNA from the 5′ terminus, while the 3′ end remains intact.

The 3′→5′ Exonuclease As noted previously, DNA polymerases are typically very accurate and produce only about one error every 10^4 to 10^5 nucleotides incorporated, by incorrect base selection. This error rate is improved 10^2- to 10^3-fold by a polymerase's 3′→5′ exonuclease activity. When an incorrect dNMP is incorporated, the 3′→5′ exonuclease removes the mismatched nucleotide, giving the polymerase a second chance at incorporating the correct one (**Figure 11-6**). This activity, known

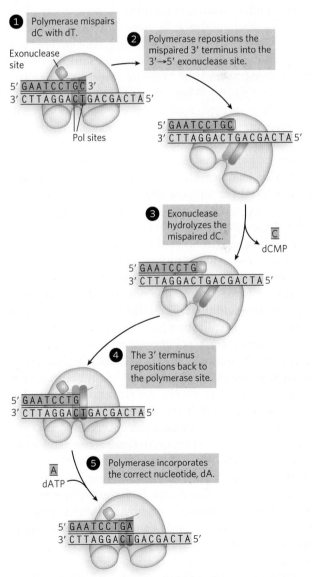

1 Polymerase mispairs dC with dT.

Exonuclease site

5′ GAATCCTGC 3′
3′ CTTAGGACTGACGACTA 5′

Pol sites

2 Polymerase repositions the mispaired 3′ terminus into the 3′→5′ exonuclease site.

5′ GAATCCTGC
3′ CTTAGGACTGACGACTA 5′

3 Exonuclease hydrolyzes the mispaired dC.

C
dCMP

5′ GAATCCTG
3′ CTTAGGACTGACGACTA 5′

4 The 3′ terminus repositions back to the polymerase site.

5′ GAATCCTG
3′ CTTAGGACTGACGACTA 5′

A
dATP

5 Polymerase incorporates the correct nucleotide, dA.

5′ GAATCCTGA
3′ CTTAGGACTGACGACTA 5′

FIGURE 11-6 The 3′→5′ proofreading exonuclease of DNA polymerases. The 3′→5′ exonuclease active site is distinct from the polymerization site and it proofreads the DNA polymerase product.

as **proofreading**, is not the same thing as pyrophosphorolysis, the reverse of the polymerization reaction—pyrophosphate is not involved, and the mismatched nucleotide is released as a dNMP, not a dNTP. Also, proofreading by the 3′→5′ exonuclease occurs at a separate active site from polymerization. The mismatched primer strand terminus repositions from the polymerase site to the exonuclease site, where water is activated to hydrolyze the 3′ nucleotide from the primer strand. Most DNA polymerases have this proofreading exonuclease activity, although some do not and thus become more capable of extending a DNA mismatch (leaving one mispaired base in the DNA product) or bypassing a damaged nucleotide in DNA, as discussed in Chapter 12.

One way in which DNA polymerase makes errors is by incorporating dNTP tautomers (see Chapter 6) at the polymerase active site. Purine and pyrimidine tautomers can form non–Watson-Crick base pairs with nucleotides in the template strand. Sometimes these incorrect base pairs are indistinguishable in shape and size from correct A=T or G≡C base pairs, thus "fooling" the enzyme into incorporating the incorrect nucleotide. Tautomers exist only transiently, and they rapidly revert to their usual bonding structure. When a tautomer of an incorrect nucleotide is incorporated into the growing DNA chain and then rapidly reverts to its normal structure, the primer strand terminus becomes unpaired and no longer fits into the polymerase active site for addition of the next dNTP. This significantly slows further chain extension and gives time for the mispaired 3′ terminus to relocate from the polymerase site to the 3′→5′ exonuclease site, where the mispaired nucleotide is quickly removed. The DNA can then move back to the polymerase site, allowing the polymerase to have another try.

When base selection and proofreading are combined, Pol I leaves behind one net error for every 10^6 to 10^8 nucleotide additions. The DNA polymerase involved in chromosome replication has a similar error rate. How accurate must a DNA polymerase be to replicate the *E. coli* genome without making a mistake? Replication of the 4.6×10^6 bp (4.6 Mbp) chromosome requires polymerization of 9.2×10^6 dNTPs. An error rate of about 1 in 10^7 would result in only one incorrect nucleotide insertion per cell division. In fact, the observed accuracy of the overall replication process in *E. coli* is one error in 10^9 to 10^{10} polymerization events. The additional accuracy derives from a repair system that recognizes and removes mismatches that escape both the polymerase and the proofreading exonuclease activities (see Chapter 12). At this level of accuracy, only a single error is acquired in every 100 to 1,000 new cells.

The 5′→3′ Exonuclease Pol I also has a second exonuclease that degrades DNA in the 5′→3′ direction, the same direction as DNA synthesis. The 5′→3′ exonuclease is

unique to Pol I and reflects the enzyme's role in DNA repair. Pol I performs a host of clean-up functions during replication, recombination, and repair, which require the trimming of single-stranded DNA ends and the removal of RNA primers or DNA lesions. Both exonucleases of Pol I are applied to these tasks. However, only the 5′→3′ exonuclease is capable of functioning at the same time as the polymerase, because the two activities act in the same direction. As the 5′→3′ exonuclease degrades a DNA or RNA strand in the duplex, the polymerase simultaneously adds dNTPs behind it. This concerted action of 5′→3′ excision and DNA polymerization is called **nick translation** (shown in **Figure 11-7** for removal of an RNA primer). "Nick translation" should not be confused with the term "translation," the process of protein synthesis. "Nick translation" is simply a descriptive term that refers to the fact that a nick in the DNA gets translated (moved) along the length of the strand by repeated cycles of excision and polymerization. The nick in DNA during nick translation is a discontinuity in the phosphodiester backbone between the 3′ hydroxyl

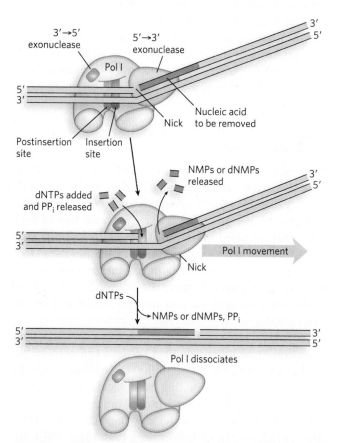

FIGURE 11-7 Nick translation by Pol I. DNA polymerase I (Pol I) is organized into three major domains: DNA polymerase, 3′→5′ proofreading exonuclease, and 5′→3′ exonuclease. At a nick—here, the gap between lagging-strand fragments—Pol I degrades the RNA primer in the 5′→3′ direction, releasing rNMPs, and simultaneously extends the 3′ terminus with dNTPs in the same direction. The net result is movement of the nick in the 5′→3′ direction along the DNA until all RNA is removed. DNA ligase can then seal the fragments (not shown here).

TABLE 11-1

The Five DNA Polymerases of *E. coli*

DNA Polymerase	Number of Subunits	Mass (kDa)	Gene(s)	Function	3'→5' Exonuclease?	5'→3' Exonuclease?
Pol I	1	103	*polA*	Okazaki fragment processing and DNA repair	Yes	Yes
Pol II	1	88	*polB*	Translesion synthesis	Yes	No
Pol III	3	167	*dnaE, holE, dnaQ*	Chromosome replication	Yes	No
Pol IV	1	40	*dinB*	Translesion synthesis	No	No
Pol V	2	69	*umuC, umuD*	Translesion synthesis	No	No

of one nucleotide and the 5' phosphate of the adjacent nucleotide. After Pol I dissociates from DNA, the nick is sealed by DNA ligase (see Section 11.3).

Five *E. coli* DNA Polymerases Function in DNA Replication and Repair

Escherichia coli contains five different DNA polymerases (**Table 11-1**). The large excess of intracellular Pol I delayed the discovery of the other DNA polymerases. Then, in the 1970s, DNA polymerase II (Pol II) and DNA polymerase III (Pol III) were discovered in studies using a mutant strain of *E. coli* called *polA*, which is depleted of most Pol I. These studies showed that Pol III is the DNA polymerase that replicates the chromosome; it is sometimes referred to as a replicase, or chromosomal replicase. Pol II seems to be involved in DNA repair.

Pol IV and Pol V were not discovered until 1999. They are different from the other DNA polymerases in that they lack a 3'→5' proofreading exonuclease and thus often incorporate the wrong nucleotide. These low-fidelity polymerases are produced in cells when the DNA sustains damage that stalls the replication fork (see Chapter 12). The low accuracy of Pol IV and Pol V enables them to insert an incorrect nucleotide opposite a damaged template base. Although this results in an error, it gets the replication fork moving again. The ability of the replication fork to move past a damaged site is a matter of life and death, and all cells—bacterial, archaeal, and eukaryotic alike—contain these error-prone "translesion" DNA polymerases.

DNA Polymerase Structure Reveals the Basis for Its Accuracy

The crystal structure of *E. coli* Pol I resembles a right hand, with domains referred to as the palm, thumb, and fingers. All DNA polymerases have these same structural features. The bound DNA lies on the palm domain, which contains the polymerase active site and is the most conserved feature among all DNA polymerases. The fingers domain contains the dNTP-binding site, and the thumb domain partially curves around the duplex portion of the primed template, tightening the grip on DNA (**Figure 11-8**).

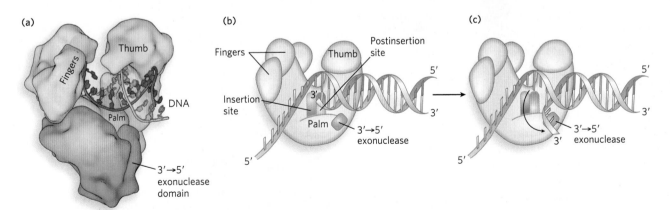

FIGURE 11-8 The structure of Pol I. (a) Crystal structure of *Thermus aquaticus* Pol I bound to DNA. All DNA polymerases are shaped like a right hand and have domains referred to as fingers, thumb, and palm. The 3'→5' exonuclease is in a separate domain from the polymerase active site. *E. coli* Pol I (not shown) also has a 5'→3' exonuclease. (b) The 3' terminus of the DNA binds to the palm, but neither the duplex DNA nor the template 5' single-stranded DNA enters the cleft between the fingers and thumb. The dNTP-binding site is located in the fingers. (c) A mispaired 3' terminus frays by about four nucleotides to insert into the 3'→5' exonuclease site. [*Source: (a) PDB ID 4KTQ.*]

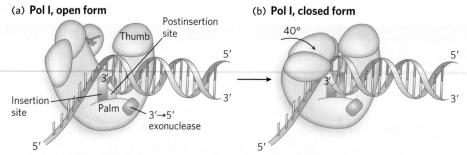

(a) **Pol I, open form**

Thumb

Postinsertion site

Insertion site

Palm

3′→5′ exonuclease

5′

5′

3′

3′

(b) **Pol I, closed form**

40°

5′

5′

3′

3′

FIGURE 11-9 Open and closed forms of Pol I. (a) In the open form, a dNTP (red) binds to the fingers domain. (b) In the closed form, the fingers domain undergoes a 40° rotation that moves the dNTP into base-pairing position with the template and forms an active-site cavity that fits the shape of a correct Watson-Crick base pair.

A dramatic conformational change occurs on the binding of a correct dNTP to the fingers domain. The domain rotates inward about 40°, carrying the dNTP down to the DNA template. In so doing, the enzyme forms an active-site enclosure with a shape that corresponds to a correct base pair. This conformation of Pol I is often referred to as the **closed form**, to distinguish it from the **open form** prior to dNTP binding (**Figure 11-9**). The A=T and G≡C base pairs have similar shapes, and either pair fits into the active-site cavity of the closed form. However, incorrect base pairs cannot be accommodated, and this prevents Pol I from completely closing around the DNA (**Figure 11-10**).

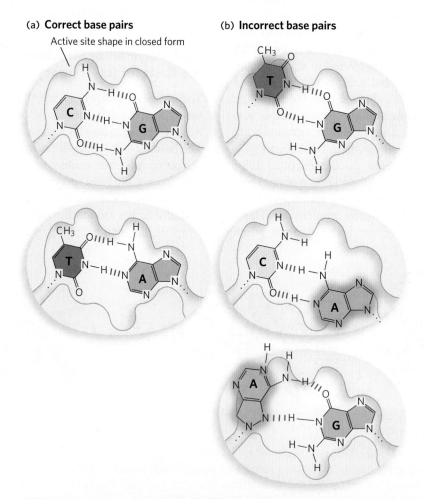

(a) **Correct base pairs**

Active site shape in closed form

(b) **Incorrect base pairs**

FIGURE 11-10 Base pairing in the Pol I active-site cavity. The shape of the active site in the closed conformation. (a) Correct G≡C and A=T base pairs fit into the active site. (b) Incorrect base pairs do not fit.

Only when Pol I is fully closed do the catalytic moieties at the active site properly align for rapid catalysis. Thus, the accuracy of the polymerase functions at the level of shape recognition and provides a classic example of an induced-fit catalytic mechanism (see Chapter 5).

Incorporation of an incorrect base pair is 10- to 1,000-fold slower than incorporation of a correct base pair, which causes the polymerase to pause prior to incorporation of an incorrect nucleotide. This kinetic pause gives time for the incorrect dNTP to dissociate from Pol I and for Pol I to bind another dNTP. Even when a mismatched dNTP is incorporated, the proofreading $3' \rightarrow 5'$ exonuclease usually excises it. The $3' \rightarrow 5'$ exonuclease is a separate domain located 20 to 30 Å from the polymerase active site. Stalling of catalysis by an incorrectly incorporated dNTP buys time for the mismatched primer strand terminus to relocate to the $3' \rightarrow 5'$ exonuclease domain for proofreading (see Figure 11-8). The slow incorporation yet rapid removal of a mispaired dNTP underlies the inherent accuracy of DNA polymerases. Accuracy is further enhanced by a vastly diminished rate of dNTP incorporation at a mismatched 3' terminus (**Figure 11-11**).

The polymerase active site contains two magnesium ions that are held in place by conserved Asp residues (**Figure 11-12a**). One Mg^{2+} deprotonates the primer 3'-OH group to form the $3'$-O^- nucleophile. The other binds the incoming dNTP and facilitates departure of the pyrophosphate leaving group. This two-metal-ion-catalyzed reaction is remarkable in that no amino acid side chain plays a direct role in catalysis; the two metal ions do it all. The $3' \rightarrow 5'$ exonuclease uses a similar two-metal-ion mechanism, in which one Mg^{2+} deprotonates H_2O to form the nucleophile (HO^-) for hydrolysis of the 3' dNMP, and the other promotes departure of the leaving group by stabilizing the charge of the dNMP product (**Figure 11-12b**). The exclusive use of metal ions to catalyze DNA synthesis suggests that the first polymerase may have been an RNA molecule, as RNA is highly effective in the coordination of metal ions (as discussed in Chapter 16).

Surprisingly, certain mutations result in DNA polymerases that are even more accurate than the wild-type polymerases. Many of these "antimutator" DNA polymerases have a hyperactive $3' \rightarrow 5'$ exonuclease; they even excise perfectly good bases. Cells with antimutator DNA polymerases display lower levels of spontaneous mutation. Why haven't cells with the more accurate DNA polymerases been selected by evolution? The energy cost of using only a highly accurate polymerase must outweigh the benefit; spontaneous mutation also provides variation within a population, which is a necessary component of evolution.

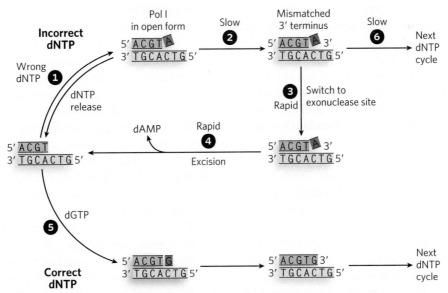

FIGURE 11-11 Favored incorporation of a correct dNTP over an incorrect dNTP.
❶ When an incorrect dNTP enters the insertion site on Pol I, binding is readily reversed (back arrow). ❷ In the rare instance of incorporation of the incorrect dNTP, the process is slow due to the imperfect active-site fit in the closed form. ❸ In the favored (rapid) route, the mispaired 3' terminus is shifted to the $3' \rightarrow 5'$ exonuclease, and in ❹ the mismatched nucleotide is excised to re-form the original primed site. ❺ This allows Pol I to insert the correct nucleotide on the second try. Binding and incorporation of a correct dNTP is rapid and paves the way for the next round of incorporation. ❻ If the incorrect nucleotide remains, the mismatched DNA is slow to act as substrate for the next round of dNTP incorporation.

(a) **DNA synthesis**

(b) **3′→5′ exonuclease**

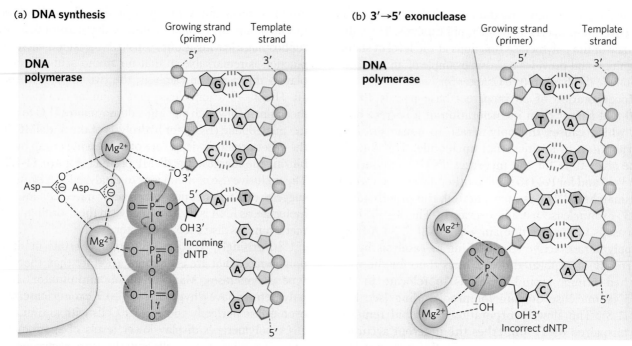

FIGURE 11-12 The role of metal ions in DNA polymerase catalysis. (a) The DNA polymerase active site contains two divalent metal ions (Mg²⁺) held in place by residues in the palm, including conserved Asp residues. The two ions play an essential role in catalysis, as described in the text. (b) The 3′→5′ exonuclease also uses a two-metal-ion-catalyzed reaction.

Processivity Increases the Efficiency of DNA Polymerase Activity

During DNA synthesis, the DNA product must be moved to the postinsertion site so that the new 3′ terminus lies in the insertion site for addition of the next dNTP. To accomplish this repositioning, a polymerase can take either of two paths. It can fully dissociate from the DNA and rebind the primer strand terminus in the postinsertion site; this dissociation followed by rebinding at each nucleotide addition is referred to as **distributive synthesis**. Alternatively, the polymerase may simply slide forward one base pair along the DNA to reposition the 3′ terminus, without dissociating from the DNA. When a polymerase remains attached to DNA during multiple catalytic cycles, the process is known as **processive synthesis**.

The "processivity number" is the average number of nucleotides incorporated before the enzyme dissociates from the DNA. Processivity can result in exceedingly efficient polymerization, because much time is wasted by a dissociated polymerase in locating and rebinding a 3′ primer strand terminus. Pol I has a processivity number of 10 to 100 nucleotides, depending on conditions. In contrast, Pol III, like most DNA polymerases that replicate chromosomes (i.e., replicases), has a processivity number in the thousands, which, as we will see, results from a protein ring that encircles the DNA.

SECTION 11.2 SUMMARY

- DNA polymerases require a primed template and extend the 3′ terminus of the primer strand by reaction with dNTPs.
- A dNTP that correctly base-pairs to the template strand is incorporated into the primer strand with the release of pyrophosphate. Pyrophosphate is hydrolyzed by pyrophosphatase, which reduces the concentration of pyrophosphate in the cell and makes the reverse reaction extremely unlikely.
- DNA polymerases are inherently very accurate and are made even more accurate by a proofreading 3′→5′ exonuclease.
- Pol I also has a 5′→3′ exonuclease that degrades DNA while the polymerase synthesizes DNA, in the process of nick translation.
- *E. coli* has five DNA polymerases. Pol III is responsible for chromosome replication, and Pol I is used to remove RNA primers and fill in the resulting gaps with DNA. The other three (II, IV, and V) are involved in DNA repair and in moving replication forks past sites of DNA damage.
- Binding of a dNTP to a DNA polymerase results in a large conformational change, yielding an active site in which only correct base pairs fit.
- DNA polymerases often have high processivity, in which many nucleotides are added to a DNA chain in one polymerase-binding event.

11.3 MECHANICS OF THE DNA REPLICATION FORK

The advance of a replication fork requires the coordinated action of several different types of proteins, in addition to DNA polymerase. Many of these proteins work together as a highly dynamic "replisome" machine. Some proteins at a replication fork dissociate from the DNA and are replaced by new ones every few seconds, whereas others are left behind at intervals to direct clean-up processes after the replication fork has passed. In addition, unwinding of the parental DNA requires it to be broken and resealed hundreds of thousands of times, to relieve torsional stress.

Replication fork mechanics are best understood in the *E. coli* system, where the process is streamlined and illustrates the fundamental mechanism of a moving replication fork. Even so, more than a dozen different proteins are involved (**Table 11-2**). We describe here the individual replication proteins and how they act together at a replication fork in *E. coli*, then describe replication mechanisms in eukaryotes. Cellular processes in eukaryotes are generally more complex, and thus it is not surprising that several more proteins are involved at the replication fork.

DNA Polymerase III Is the Replicative Polymerase in *E. coli*

Pol III is responsible for replicating the *E. coli* chromosome. The **Pol III core** is a heterotrimer composed of three subunits called α, ε, and θ. The DNA polymerase activity is in the α subunit; the ε subunit has the proofreading $3' \rightarrow 5'$ exonuclease activity. The role of the θ subunit is currently unknown. The crystal structure of the Pol III α subunit, shown in **Figure 11-13**, reveals the hand shape common to all DNA polymerases and the presence of a polymerase domain and a histidinol phosphate (PHP) domain. The PHP domain, unique to bacterial Pol III, has a chain folding pattern similar to that of certain phosphodiesterases, suggesting that this domain may be a vestigial $3' \rightarrow 5'$ exonuclease. In *E. coli*, as in many bacteria, the PHP domain is not

TABLE 11-2

Proteins Involved in Replication of the *E. coli* Chromosome

Protein	Number of Subunits	Mass, Total (kDa)	Gene(s)	Function(s)
DnaA	1	52	*dnaA*	Initiator, binds *oriC*
HU	2	19	*hupA, hupB*	Stimulates open complex at *oriC*
DnaC	1	29	*dnaC*	Helicase loader
DnaB	6 (homohexamer)	300	*dnaB*	Helicase
Gyrase	4	400	*gyrA, gyrB*	Type II topoisomerase (A$_2$B$_2$)
SSB	4 (homotetramer)	76	*ssb*	Stabilizes and protects single-stranded DNA
Primase	1	60	*dnaG*	Synthesizes lagging-strand RNA primers
Pol III holoenzyme	17	796	See Table 11-3	Chromosomal replicase, consists of Pol III core, β clamp, and clamp loader
Pol I	1	103	*polA*	Okazaki fragment processing: removes RNA and replaces it with DNA
RNaseH	1	18	*rnhA*	Okazaki fragment processing: removes RNA
Ligase	1	74	*lig*	Okazaki fragment processing: joins segments
Dam methylase	1	32	*dam*	Methylates adenines in GATC sequences
RNA polymerase	5	454	*rpoA, B, C, D*	Stimulates open complex at *oriC*
SeqA	1	20	*seqA*	Binds hemimethylated GATC, inhibits *oriC*
Hda	1	28	*hda*	Induces DnaA to hydrolyze ATP
Tus	1	36	*tus*	Participates in replication termination
Topoisomerase IV	4	308	*parC, parE*	Removes supercoil tension and decatenates completed chromosomes (C$_2$E$_2$)

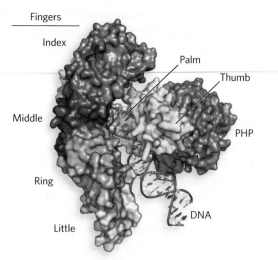

FIGURE 11-13 The *E. coli* Pol III α subunit. A space-filling model of the Pol III α subunit, with the palm, fingers, thumb, and PHP domains labeled. *[Source: PDB ID 3E0D.]*

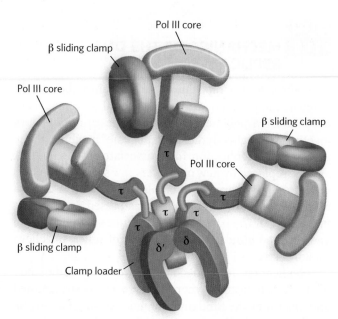

FIGURE 11-14 The architecture of *E. coli* Pol III holoenzyme. The C-termini of the τ subunits protrude from the clamp loader and bind the Pol III cores. Each Pol III core also attaches to a β sliding clamp. *[Source: Data from N. Yao and M. O'Donnell, Mol. Biosyst. 4:1075–1084, 2008.]*

enzymatically active, and the α subunit recruits the ε subunit for 3′→5′ proofreading activity. In some bacteria, however, the PHP domain in the α subunit is, in fact, the 3′→5′exonuclease proofreading activity, and these α subunits might not recruit an ε subunit.

The Pol III core is just one part of a much larger protein assembly called the **Pol III holoenzyme**, which replicates both leading and lagging strands (**Table 11-3**). The Pol III holoenzyme includes three Pol III cores, two ring-shaped **β sliding clamps** that improve processivity, and one **clamp loader** (also called **τ complex**) that

assembles β clamps onto the DNA. The clamp loader includes three τ (tau) subunits with C-terminal domains that protrude from the clamp loader and bind to the Pol III cores (**Figure 11-14**).

The Pol III core itself is capable of DNA synthesis at a slow rate, but DNA synthesis by the Pol III holoenzyme

TABLE 11-3

Subunit Composition of the Three Components of the *E. coli* Pol III Holoenzyme

Holoenzyme Component(s)	Subunit	Number of Subunits	Mass per Subunit (kDa)	Gene	Function(s) of Subunit
3 Pol III cores	α	1/Pol III core	130	*dnaE*	DNA polymerase
	ε	1/Pol III core	28	*dnaQ*	3′→5′ exonuclease
	θ	1/Pol III core	9	*holE*	Binds ε
3 β clamps	β	2 subunits/clamp	41	*dnaN*	Sliding clamp
1 clamp loader*	τ	3	71	*dnaX*	ATPase; binds Pol III and DnaB
	δ	1	39	*holA*	Opens clamp
	δ′	1	37	*holB*	Binds δ
	χ	1	17	*holC*	Binds SSB
	ψ	1	15	*holD*	Binds χ and γ
Totals		22 subunits	1,068 kDa		

*The clamp loader in the holoenzyme contains three τ subunits and no γ subunits. The γ subunit may be used to form a clamp loader that loads clamps onto DNA for other processes outside the replication fork.

is exceedingly rapid, nearly 1 kb/s. The β clamps and clamp loader help maintain contact between the Pol III core and DNA, making the Pol III core highly processive (~100 kb per binding event). Having two DNA polymerases in one holoenzyme assembly facilitates the coordinated synthesis of the leading and lagging strands at the replication fork. The β sliding clamps are assembled onto both DNA strands by the single clamp loader.

A DNA Sliding Clamp Increases the Speed and Processivity of the Chromosomal Replicase

The processivity (1 to 10 nucleotides) and rate of synthesis (10 nucleotides per second) of the Pol III core are dramatically enhanced by the β sliding clamp. The β clamp is a homodimer, shaped like a ring that encircles the duplex DNA (**Figure 11-15a**). The β clamp has sixfold symmetry; it is constructed from a single domain repeated three times in each monomer. The topological binding of the β clamp to DNA allows the clamp to slide along the duplex while staying tightly attached (see the How We Know section at the end of this chapter). The β clamp binds to the Pol III core (**Figure 11-15b**) and thereby holds Pol III to the DNA while sliding along the duplex. This converts Pol III core from a distributive enzyme, which moves on and off DNA as it works, to a processive enzyme that stays attached to DNA during repetitive cycles of dNMP incorporation (**Figure 11-16**).

An example of the remarkable speed and processivity conferred on the Pol III core by the β clamp is shown in **Figure 11-17**. The DNA substrate used in this experiment was the large (5.4 kb), single-stranded, circular DNA

genome of phage φX174. With the Pol III core (within the holoenzyme) bound to the β clamp at a primed site, [^{32}P] dNTPs labeled on the α phosphate were added to start the polymerase reaction. At different times, reaction samples were analyzed by agarose gel electrophoresis and autoradiography to visualize the newly synthesized radioactive DNA. The results show that Pol III goes full circle within a few seconds. When the β clamp was omitted in a control reaction, no DNA synthesis was observed, because without the clamp, Pol III runs into secondary structures that block synthesis (data not shown). However, even in the absence of these secondary structures, Pol III would require about 9 minutes to complete replication of the 5.4 kb DNA without the help of the β clamp.

The β sliding clamp does not assemble onto DNA by itself; it requires the multiprotein clamp loader to open and close the ring around the DNA. The clamp loader within the Pol III holoenzyme is the τ complex, which consists of several subunits, $τ_3$ (τ trimer), δ, and δ′, arranged in a circular pentamer (see Figure 11-14), and two small subunits, χ (chi) and ψ (psi), that connect to other proteins at the replication fork. The clamp loader uses the energy of ATP binding to open the β sliding clamp. In this operation, the clamp loader binds to one of the flat surfaces of the clamp and forces it to open (**Figure 11-18a** on p. 381). The gene that encodes the τ subunit (*dnaX*) also produces a shorter version, called γ, through a translation frameshift event. During translation frameshifting, the mRNA slips and places the mRNA in a different reading frame. In the case of *dnaX*, the new reading frame results in a stop codon after one amino acid, thus yielding a δ subunit that is about 2/3

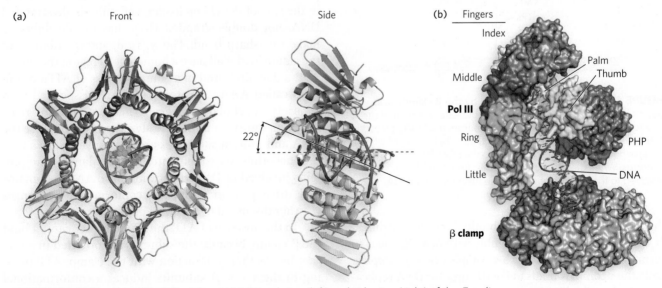

FIGURE 11-15 The *E. coli* β sliding clamp. (a) Frontal view (left) and side view (right) of the *E. coli* β clamp bound to DNA. The DNA is tilted 22° from the perpendicular. Subunits of β are shades of purple; the DNA is blue. (b) Model of *E. coli* Pol III α bound to the β clamp. [*Sources: (a) PDB ID 3BEP. (b) PDB ID 3BEP and 3E0D.*]

(a) Distributive synthesis

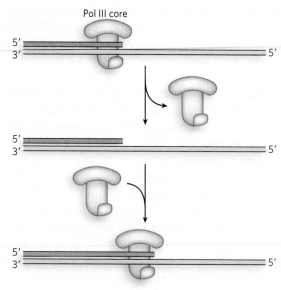

(b) Processive synthesis

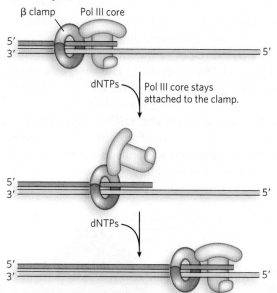

FIGURE 11-16 Processivity conferred by the β clamp. (a) In distributive synthesis, the Pol III core would extend a primed site by a few nucleotides and then dissociate from the DNA. It must rebind the primed site to continue synthesis. (b) Binding of a Pol III core to the β clamp tethers the core to the DNA for processive synthesis. When the Pol III core detaches from the DNA, it stays attached to the β clamp and rapidly reattaches to the DNA.

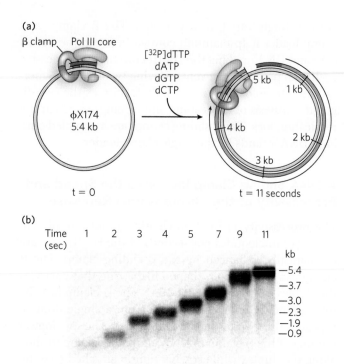

FIGURE 11-17 Rapid DNA synthesis by *E. coli* Pol III. The β sliding clamp is loaded onto a primed site by a clamp loader in the Pol III holoenzyme. Synthesis of the 5.4 kb circular DNA of bacteriophage φX174 is completed within 11 seconds, as shown in the autoradiograph from the experiment described in the text. [*Source: Data from O'Donnell and Kornberg (1985), J. Biol. Chem. 260, 12875. Fig. 2A. © 1985 The American Society for Biochemistry and Molecular Biology.*]

the size of τ. The τ and γ subunits form homotrimers ($τ_3$ and $γ_3$), and each can assemble with δ, δ′, χ, and ψ to form a clamp loader, either a τ complex or a γ complex. Only the τ complex binds to Pol III cores for DNA replication. As described in Chapter 12, sliding clamps are used by many different proteins for a variety of purposes, not just for replication. The γ complex is used to assemble β clamps onto DNA for use by enzymes other than Pol III.

The clamp loader has an inner chamber that binds primed DNA, and this positions the DNA through the sliding clamp. The clamp-loading activity is specific to a primed site, because the DNA must bend out of a gap in the side of the clamp loader, and only single-stranded DNA, not double-stranded DNA, has the flexibility to make this sharp bend. The $γ_3/τ_3$, δ, and δ′ subunits of the clamp loader all have a homologous region that binds ATP, a domain found in a common class of ATPase proteins called **AAA+ proteins** (ATPases associated with a variety of cellular activities). The clamp-loading reaction is an unusual enzyme-catalyzed process because the DNA and protein substrates are not converted to a new product—they are simply intertwined. Many of the proteins involved in DNA replication have AAA+ domains and drive protein and DNA conformational changes (further discussed in Section 11.4).

In the absence of ATP, the clamp loader cannot bind the β clamp, because the subunits are oriented in a way that blocks their interaction with the clamp. ATP binding to the τ (or γ) subunits induces a conformational change that enables the clamp loader to bind and open the clamp (as shown for the γ complex in **Figure 11-18b**). ATP is also needed for the clamp loader to bind DNA. ATP hydrolysis causes the clamp loader to revert to the

(a)

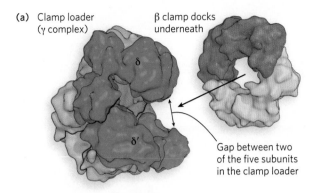

Clamp loader
(γ complex)

β clamp docks
underneath

δ

δ'

Gap between two
of the five subunits
in the clamp loader

(b)

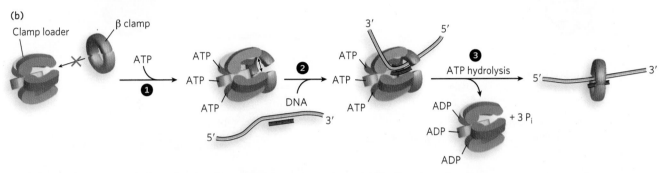

FIGURE 11-18 The *E. coli* clamp loader. The γ complex is shown here. (a) The five clamp-loading subunits are arranged in a circle, with a gap between parts of two subunits, δ' and δ. The β clamp (purple) docks onto the clamp loader. (b) The clamp loading mechanism. ❶ ATP binding to the γ subunits powers a conformational change that enables the binding and opening of the β clamp. ❷ The combined γ complex–ATP–β clamp binds primed DNA in a central chamber, and the single-stranded template DNA passes through the gap in the side of the clamp loader. ❸ ATP hydrolysis ejects the clamp loader, allowing β to close again around the DNA. *[Sources: (a) PDB ID 1JR3. (b) Data from N. Yao and M. O'Donnell, in* Encyclopedia of Biological Chemistry, *2nd ed. (W. J. Lennarz et al., eds.), Elsevier, 2013.]*

form that cannot bind the β clamp or DNA, thereby ejecting the clamp loader and allowing the clamp to close around DNA. It is important that the clamp loader be ejected at the end of the reaction because it binds to the same spot on β to which the Pol III core must attach.

Many Different Proteins Advance a Replication Fork

The simultaneous replication of both DNA strands at a replication fork requires the interaction of many proteins in addition to the Pol III core and β clamp (**Figure 11-19a**). Here we give particular attention to the *E. coli* system, but all cells contain these basic protein components for chromosome replication.

DNA Helicase The two strands of the parental DNA duplex are separated by a class of enzymes known as **DNA helicases**, which harness the energy of NTP hydrolysis (usually ATP) to drive strand separation (**Figure 11-19b**). Helicases are used in a wide variety of DNA and RNA transactions. DNA helicases usually load onto DNA at a single-strand gap in the duplex and move along the

DNA strand in one direction (fueled by NTP hydrolysis), unwinding the duplex as they move. The direction of translocation along the DNA is characteristic of the particular helicase (see Chapter 5).

KEY CONVENTION

A helicase translocates along a single strand of DNA in one direction, parting the duplex as it moves. The direction of movement is specified, by convention, as the direction along the strand to which the enzyme is bound. If the helicase binds to a DNA strand and progresses from the 5' end toward the 3' end, it is said to be a 5'→3' helicase.

Helicases that function at a replication fork are typically ring-shaped hexamers that encircle one DNA strand. The ring shape of replicative helicases is thought to enhance their grip on DNA for processive unwinding. The *E. coli* replicative helicase is a hexamer of DnaB protein; it encircles the lagging strand and translocates in the 5'→3' direction. An

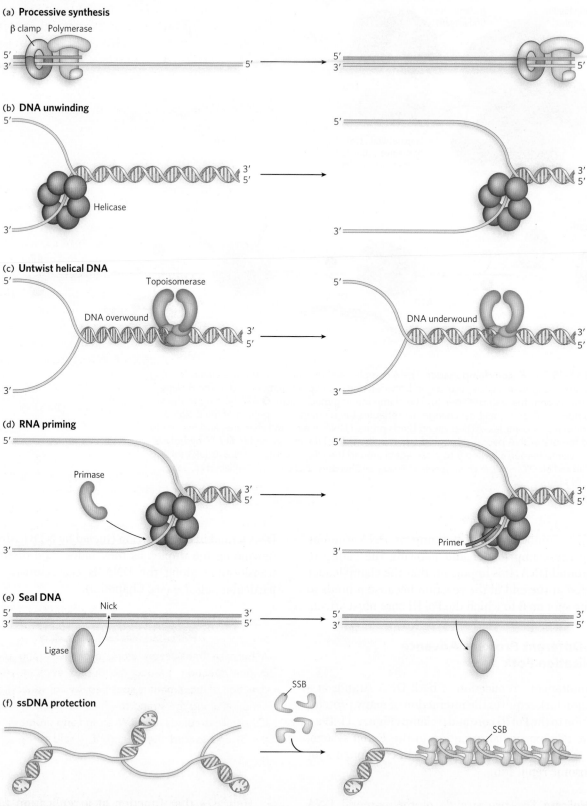

FIGURE 11-19 Activities required at a DNA replication fork. (a) DNA polymerase and the β sliding clamp are required for processive DNA synthesis. (b) Helicases that function at replication forks are hexamers that encircle single-stranded DNA and translocate on the DNA strand to separate the strands of the parental duplex. (c) A topoisomerase removes twists in the DNA. In *E. coli*, the topoisomerase is DNA gyrase. (d) Primase (an RNA polymerase) makes short RNA primers to initiate DNA synthesis. Primases typically bind the helicase, thus localizing primers to the replication fork. (e) Ligase seals DNA nicks, joining Okazaki fragments together (after removal of the RNA primers). (f) Single-stranded DNA–binding protein (SSB) binds cooperatively to single-stranded DNA (ssDNA), removing secondary structure in the DNA strand and protecting it from the action of nucleases.

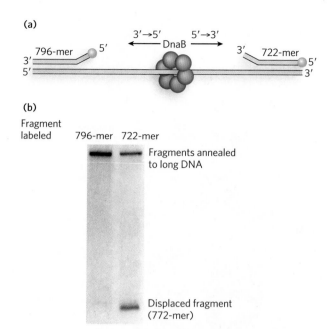

(a)

796-mer   5′   3′→5′   DnaB   5′→3′   3′   722-mer
3′   5′
5′   3′

(b)

Fragment
labeled

796-mer   722-mer

Fragments annealed
to long DNA

Displaced fragment
(772-mer)

FIGURE 11-20 An assay for determining the direction of DNA helicase activity. (a) DnaB helicase is added to a long, single-stranded DNA that has short ^{32}P-labeled DNA strands of different sizes annealed to its ends (shown here are a 796-mer and 722-mer). Each annealed DNA strand has a short single-stranded tail to mimic a replication fork. DnaB initially binds to the single-stranded DNA region, then translocates in one direction to unwind one of the two annealed-DNA duplexes. (b) The DNA-unwinding products are analyzed in a polyacrylamide gel. The result here shows that DnaB displaced only the 722-mer, revealing that DnaB translocates in the 5′→3′ direction along single-stranded DNA. *[Source: (b) LeBowitz, J.H. and McMacken, R. (1986) The Escherichia coli DnaB replication protein is a DNA helicase. J. Biol. Chem. 261, 4738–4748, © 1986, The American Society for Biochemistry and Molecular Biology.]*

assay that demonstrates the action of DnaB is shown in **Figure 11-20**. In this experiment, the DNA substrate included a central single-stranded DNA region for helicase assembly, flanked on each side by duplex DNA of different lengths; one strand of each duplex was radioactively labeled. The direction of helicase translocation was determined by observing which DNA fragment was unwound from the substrate, as analyzed on a polyacrylamide gel. The resulting autoradiograph shows that DnaB translocates in the 5′→3′ direction along a single strand of DNA and unwinds the duplex at only one of the two ends. Helicases require NTPs for translocation, and a control reaction conducted in the absence of ATP showed no unwinding.

Topoisomerase As helicase separates the two strands of DNA, the duplex DNA ahead of the helicase becomes overwound, and this creates superhelical tension ahead of the fork. This occurs because the helicase cannot untwist the DNA strands. The action of untwisting DNA requires a special type of enzyme that cuts one

or both strands of DNA to unwind it, then reseals it. This untwisting action is performed by **topoisomerases** that act on duplex DNA ahead of the replication fork (**Figure 11-19c**). There are many different kinds of topoisomerase. In *E. coli*, gyrase is the primary replicative topoisomerase, although topoisomerase IV also participates (see Chapter 9).

Primase As discussed earlier, DNA polymerase requires a preformed primer from which to elongate. Cells contain specialized RNA polymerases called **primases** that synthesize short RNA primers specifically for initiating DNA polymerase action. In *E. coli*, an RNA primer of 11 to 13 nucleotides is synthesized by the DnaG primase. RNA primers are needed to initiate each of the thousands of Okazaki fragments on the lagging strand. The leading strand is initiated by primase at a replication origin. *E. coli* DnaG primase must bind the DNA helicase for activity, and this localizes primase action to the replication fork (**Figure 11-19d**). RNA synthesis is less accurate than DNA synthesis (see Chapter 15), and the use of RNA to prime DNA synthesis provides a way for DNA polymerase I to recognize and remove the less accurate primer before Okazaki fragments are joined together. Note that although DNA polymerases will extend DNA primers as well as an RNA primer, in cells, only RNA is used to initiate DNA synthesis. It is not clear why this is the case. One possibility is that the task of forming the first phosphodiester bond, to create a dinucleotide, is more difficult than forming subsequent bonds, because two NTPs must be held by the enzyme at the same time, and the much higher concentration of rNTPs than dNTPs in the cell may be required for this first step.

Pol I and Ligase RNA primers must be removed at the end of each Okazaki fragment and replaced with DNA. This is achieved through the nick translation activity of Pol I (see Figure 11-7), which removes the ribonucleotides of the primer while simultaneously replacing them with deoxyribonucleotides. A ribonuclease called **RNaseH** can also remove RNA that is base-paired to DNA, but it cannot remove the last rNMP attached to the DNA. Hence, in *E. coli*, RNaseH may help remove RNA, but another enzyme (e.g., Pol I) is needed to complete the task.

Once all RNA is replaced with DNA, the nick in the phosphodiester backbone is sealed by **DNA ligase** in a reaction that requires ATP (or NAD⁺ in *E. coli*) (**Figure 11-19e**; see also Figure 5-12). Ligase acts only on a 5′ terminus of DNA, not on RNA. This specificity ensures that all the RNA at the end of an Okazaki fragment is removed before the nick is sealed. Both ligase and Pol I interact with the β sliding clamp.

SSB Single-stranded DNA produced by helicase-catalyzed unwinding is quickly bound by **single-stranded DNA–binding protein (SSB)**, protecting the DNA from endonucleases (see Figure 5-3 for a more detailed look at how SSB binds DNA). SSB stimulates DNA polymerase activity by melting small DNA hairpin structures (i.e., separating base pairs) in the single-stranded template (**Figure 11-19f**). SSB is found in all cell types and binds DNA in a sequence-independent fashion. *E. coli* SSB is a homotetramer, but other SSBs range from monomers (e.g., gene 32 protein in T4 phage) to heterotrimers (e.g., RPA in eukaryotes, discussed shortly).

Helicase Activity Is Stimulated by Its Connection to the DNA Polymerase

The *E. coli* DnaB helicase connects to the Pol III holoenzyme through the τ subunits of the clamp loader within the Pol III holoenzyme. Without this connection to polymerase, DnaB helicase is slow, unwinding about 35 bp/s. On connection of DnaB to the Pol III holoenzyme, unwinding proceeds at a rate of approximately 700 bp/s. The complex of Pol III holoenzyme, DnaB helicase, and primase forms a **replisome**. The three τ subunits of the τ complex clamp loader bind three Pol III cores, and these same τ subunits also bind the DnaB helicase. **Figure 11-21** shows one Pol III core associated with the leading strand and one with the lagging strand; as we will see later, the third Pol III core also participates in replication of the lagging strand. The leading-strand Pol III–β clamp complex moves continuously with DnaB helicase, while the lagging-strand Pol III–β clamp complex repeatedly moves on and off the DNA to extend the multiple RNA primers made by primase.

DNA Loops Repeatedly Grow and Collapse on the Lagging Strand

The lagging-strand polymerase must repeatedly extend RNA primers into full-length, 1 to 2 kb Okazaki fragments. As we have seen, however, the direction of chain growth on the lagging strand is opposite to that on the leading strand. How can the lagging-strand Pol III synthesize DNA in the opposite direction to replication fork movement, yet remain tethered to the replisome? To accommodate these opposed directions, the lagging-strand template is pulled up through the polymerase during chain extension to form a loop (**Figure 11-22**). As an Okazaki fragment is extended, the double-stranded portion of the loop grows longer. As the replication fork generates more single-stranded DNA, it also adds to the growing DNA loop. When the Okazaki fragment is complete, the polymerase bumps into the fragment

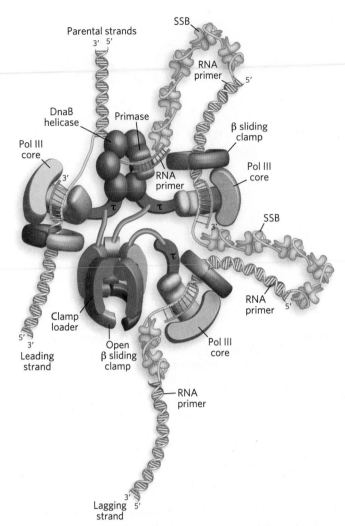

FIGURE 11-21 The architecture of the *E. coli* replisome at a replication fork. DnaB helicase encircles the lagging strand, and the Pol III holoenzyme connects to DnaB via the τ subunits of the clamp loader. The holoenzyme contains three Pol III cores. In the illustration, two Pol III cores connect to β clamps on the leading and lagging strands. The function of the third Pol III is discussed in the text. Primase (DnaG) transiently associates with DnaB for the synthesis of an RNA primer on the lagging strand. The lagging-strand Pol III core–β clamp travels with the replisome, yet extends DNA in the 5'→3' direction, resulting in a DNA loop. The lagging strand is bound by SSB. The clamp loader is shown bound to a β clamp that it has opened in preparation for loading the clamp onto the RNA primer. [*Source: Data from N. Y. Yao and M. O'Donnell, Cell 141:1088–1088e1, 2010.*]

it made previously and lets go of the DNA loop so that it can extend a new RNA primer for the next Okazaki fragment. The process of repeated loop growth and disassembly is often referred to as the **trombone model** of replication, because it resembles movement of the slide of a trombone. The trombone model was first proposed in 1980 by Bruce Alberts. At a replication fork speed of about 700 nucleotides per second, and with Okazaki fragments of 1 to 2 kb, a new loop is formed every 2 to

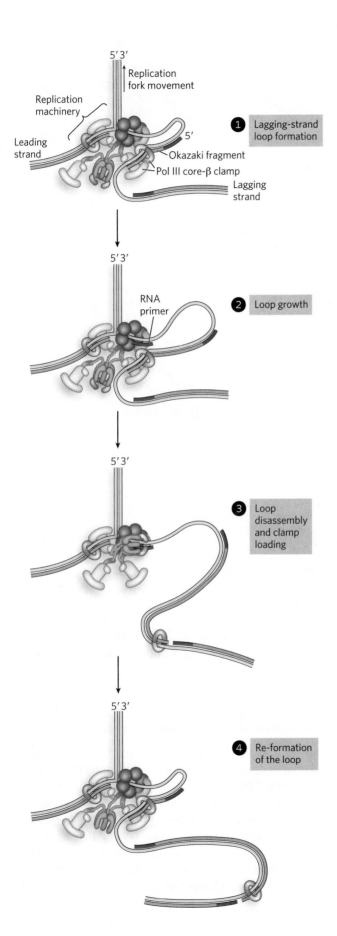

① Lagging-strand loop formation

② Loop growth

③ Loop disassembly and clamp loading

④ Re-formation of the loop

3 seconds. Two of the three Pol III cores of the Pol III holoenzyme can act on the lagging strand, so there are sometimes two loops on the lagging strand.

Bruce Alberts [Source: Painting by Jon Friedman, NAS. Courtesy Bruce Alberts.]

When the lagging-strand polymerase finishes an Okazaki fragment, it must dissociate from the DNA in order to transfer to a new RNA primer and extend the next fragment. It does so by detaching from the β clamp, leaving it behind on the DNA. Once released, the lagging-strand Pol III core can associate with a new RNA-primed site on which a new β clamp has been assembled (see Figure 11-22). This process results in a buildup of β clamps on replicated DNA. These leftover clamps perform additional functions, as we will see shortly, but ultimately they, too, must be removed and recycled.

Recent studies have shown that two of the three Pol III cores of the Pol III holoenzyme can function to replicate the lagging strand while the third continuously replicates the leading strand. These actions are illustrated in **Figure 11-23**. First, two polymerases are shown, one on each strand (step 1). As the Okazaki fragment is being extended by the first lagging-strand Pol III core, the primase synthesizes an RNA primer for a second Okazaki fragment (step 2), and the clamp loader assembles a clamp onto the new primer (step 3). Then the second lagging-strand polymerase can associate with the new clamp and extend a second Okazaki fragment while the first lagging-strand polymerase is still active, forming two lagging-strand DNA loops (step 4). On completing the first Okazaki fragment, the first lagging-strand Pol III core dissociates from its clamp and is now ready for another cycle of Okazaki fragment synthesis, while the second lagging-strand Pol III core continues to extend the second Okazaki fragment (step 5).

Experiments using a simple model system first suggested that Pol III hops from one β clamp to another to cycle among Okazaki fragments on the lagging strand (**Figure 11-24** on p. 387). The Pol III holoenzyme was

FIGURE 11-22 The trombone model of replication fork function. The lagging-strand polymerase extends the 3′ terminus of an Okazaki fragment in the opposite direction to fork movement, yet this polymerase is part of the replisome and thus moves with the fork. The opposed directions result in formation of a DNA loop for each Okazaki fragment. As multiple Okazaki fragments are synthesized, loops repeatedly grow and are released, similar to the movement of a trombone slide as the instrument is played. For simplicity, the participation of the third Pol III core in lagging-strand synthesis is not shown here.

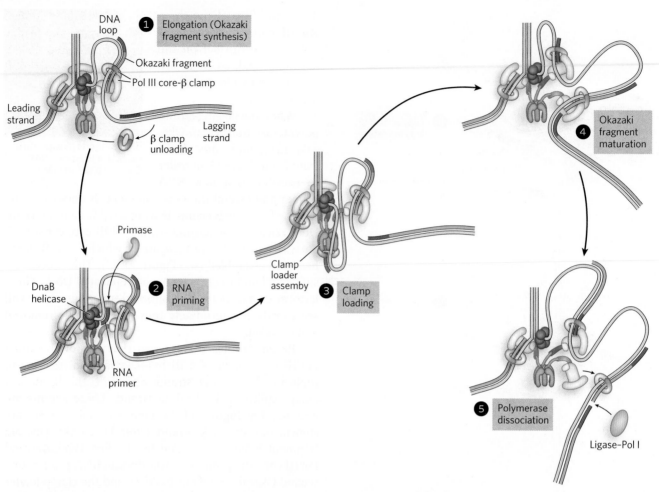

FIGURE 11-23 Function of three polymerases during the Okazaki fragment cycle. ❶ Two Pol III cores are attached to β clamps on the leading and lagging strands, creating a loop. ❷ Primase binds to DnaB helicase and synthesizes an RNA primer (purple). ❸ The clamp loader assembles a β clamp onto the new RNA primer. ❹ The second lagging-strand Pol III core (third polymerase) assembles with the new clamp while the first lagging-strand Pol III core is extending an Okazaki fragment, creating two DNA loops. ❺ The first lagging-strand Pol III core ejects from the β clamp on the first Okazaki fragment, leaving the clamp on the DNA.

assembled on a 5.4 kb DNA substrate (M13mp18), then mixed with two competing DNA substrates of different sizes (M13Gori and φX174), each with one site primed for DNA synthesis. In each of two experiments, only one of the competing DNAs included a β clamp at the primed site. Replication was initiated using [³²P]dNTPs, and timed aliquots were analyzed by agarose gel electrophoresis and autoradiography. As the result in **Figure 11-24b** shows, Pol III replicated the initial 5.4 kb DNA, then transferred to the DNA substrate that included the preassembled clamp. Because the competing DNA with a preassembled β clamp was preferentially replicated compared with the DNA lacking a clamp, the result suggests that Pol III leaves the β clamp behind on the first template, then hops to another DNA that has a new β clamp. Further

studies confirmed that Pol III indeed hops from one clamp to another, leaving clamps behind on the DNA as it does so.

Okazaki Fragments Require Removal of RNA and Ligase-Mediated Joining of DNA

The RNA at the 5′ terminus of each Okazaki fragment must be removed and the gap filled in with DNA. This job is performed by the nick translation activity of Pol I (see Figure 11-7). As noted earlier, the cell also has a backup enzyme, RNaseH, which can remove the RNA, in which case the single-strand gap must be filled in by a DNA polymerase. Processed fragments are then joined together by ligase to form a continuous duplex. Both Pol I and ligase interact with the β clamp, and the β clamps left behind by

(a)

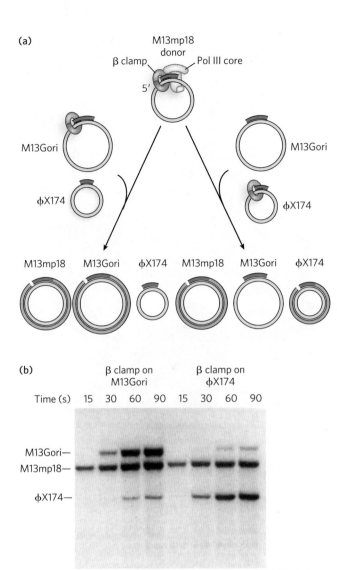

(b)

FIGURE 11-24 Transfer of Pol III from an old β clamp to a new β clamp. In these experiments, the Pol III holoenzyme (just the β clamp and Pol III core are shown for simplicity) is assembled onto a primed donor DNA circle (M13mp18), then the Pol III–DNA is mixed with two competing primed acceptor DNA circles of different sizes, only one of which includes a β clamp. In the experiment shown on the left side, the M13Gori DNA has a β clamp and the other (φX174) DNA does not. In the experiment on the right, φX174 DNA has a β clamp and the other (M13Gori) does not. Replication is initiated and timed. Aliquots of the reaction mixtures are analyzed on an agarose gel. The acceptor DNA with a β clamp is replicated in preference to the acceptor DNA without a β clamp. *[Source: (b) Reprinted from Cell, 78(5), Stukenberg, P. T., Turner, J., and O'Donnell, M., An explanation for lagging strand replication: polymerase hopping among DNA sliding clamps. 877–887, copyright 1994, with permission of Elsevier. Courtesy P. T. Stukenberg.]*

the replisome are thought to attract Pol I and ligase for the removal of RNA and sealing of the fragments.

Okazaki fragments outnumber β clamps in the cell by about 10 to 1, so clamps must be recycled during chromosome replication. Clamps are removed from DNA by the δ subunit of the clamp loader, which is produced in excess in the cell (relative to the other clamp loader subunits). By itself, the δ subunit can open and unload a β clamp from DNA, but it cannot assemble a clamp onto DNA.

The β clamp binds many proteins, including all five *E. coli* DNA polymerases, ligase, the clamp loader complex, and several proteins not described in this chapter. These proteins all bind to the same spot on β. Therefore, when the β clamp is being used by a DNA polymerase or another protein, the recycling of β is blocked. Only when the clamp is no longer bound by other proteins is it available to be recycled (**Figure 11-25**).

The Replication Fork Is More Complex in Eukaryotes Than in Bacteria

Many eukaryotic replication proteins have counterparts in bacteria (e.g., the clamps and clamp loader), but the replication fork machinery of eukaryotes includes more proteins beyond those used in the comparatively simple bacterial machinery (**Figure 11-26**). New eukaryotic replication factors are still being identified, and details of the replication fork in eukaryotic cells are only now coming into focus.

The **MCM complex** is thought to function at the replication fork. Like *E. coli* DnaB, the MCM subunits form a ring-shaped hexamer, but each subunit is different (named Mcm2 through Mcm7, or Mcm2–7). The six subunits are homologous AAA+ proteins (each subunit $M_r \sim 100,000$). The Mcm2–7 complex associates with a heterotetramer called GINS and with the Cdc45 protein, forming a complex referred to as the **CMG complex** (for Cdc45-*MCM*-*GINS*). The CMG complex is an active helicase, while the MCM complex alone has only feeble helicase activity. The mechanism by which GINS and Cdc45 activate the CMG helicase is not yet understood, and these proteins have no sequence-related homologs in bacteria. The direction of activity of the CMG helicase is 3′→5′, opposite to that of *E. coli* DnaB, and therefore the CMG helicase must act on the leading strand to unwind the parental DNA. Some of the eukaryotic replication proteins, including the CMG helicase, are targets of cell cycle kinases (cyclin-dependent protein kinases), enzymes that phosphorylate specific proteins and are active at certain phases of the cell cycle. For example, phosphorylation of a replication protein may activate it, and this modification may occur only on entering S phase of the cell cycle. Even though most eukaryotic chromosomes are linear, they are still topologically constrained, and helicase unwinding creates torsional stress that is relieved by topoisomerases.

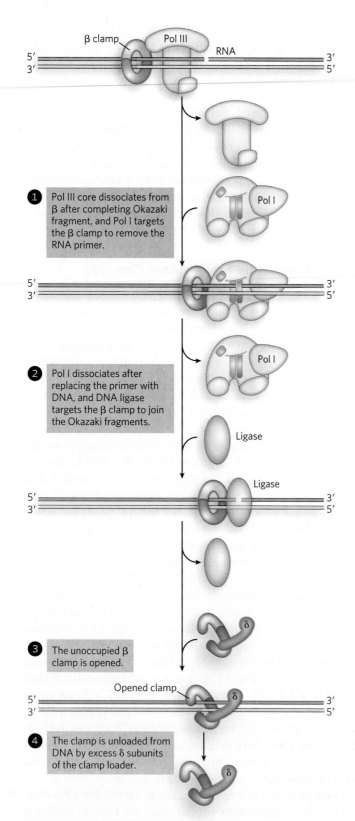

1. Pol III core dissociates from β after completing Okazaki fragment, and Pol I targets the β clamp to remove the RNA primer.

2. Pol I dissociates after replacing the primer with DNA, and DNA ligase targets the β clamp to join the Okazaki fragments.

3. The unoccupied β clamp is opened.

4. The clamp is unloaded from DNA by excess δ subunits of the clamp loader.

FIGURE 11-25 β clamp recycling in *E. coli*.

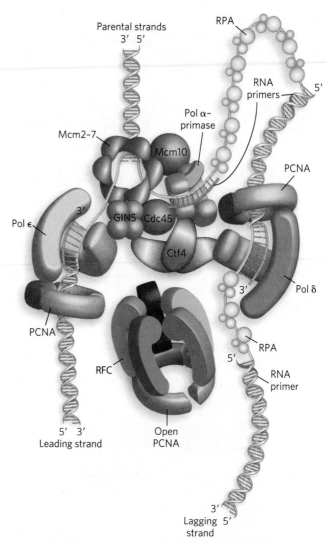

FIGURE 11-26 A hypothetical model of the eukaryotic replication fork. Eukaryotes contain all the proteins that function in a bacterial replisome, but most components have more subunits than the bacterial proteins, and several additional proteins function at the eukaryotic replication fork. The components and their functions are described in the text. [*Source: Data from N. Y. Yao and M. O'Donnell, Cell 141:1088–1088e1, 2010.*]

The eukaryotic primase is a four-subunit complex called **DNA polymerase α (Pol α)**. Priming activity is located in the smallest subunit, and it makes an RNA primer of about a dozen nucleotides. The largest subunit of Pol α is a DNA polymerase that extends the RNA primer with DNA, to a total length of 25 to 40 nucleotides. As in bacteria, the RNA is excised before Okazaki fragments are joined. The DNA made by Pol α may contain errors, because the enzyme has no 3′→5′ proofreading exonuclease, and it is thought that the DNA made by Pol α is also replaced or is corrected by repair proteins.

Eukaryotes have two different chromosomal replicases: **DNA polymerase δ (Pol δ)** and **DNA polymerase**

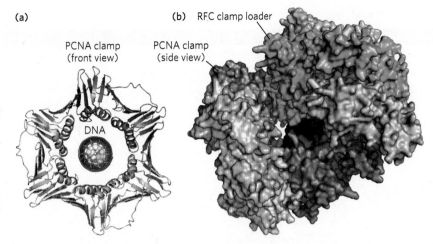

FIGURE 11-27 The eukaryotic PCNA clamp and RFC clamp loader. (a) PCNA is a homotrimer; the monomer units are shown in different colors. (b) The RFC clamp loader is homologous to the bacterial clamp loader. Compare this with Figure 11-18a. [*Sources: (a) PDB ID 1AXC. (b) PDB ID 1SXJ.*]

ε (**Pol ε**). Both Pol δ and Pol ε are four-subunit enzymes in higher eukaryotes, and the largest subunit of each has both a DNA polymerase and a 3′→5′ exonuclease activity (Pol δ in yeast has only three subunits). Current research suggests that Pol δ and Pol ε operate on different strands at the replication fork: Pol ε on the leading strand and Pol δ on the lagging strand.

Both Pol δ and Pol ε interact with a DNA sliding clamp called **PCNA** (proliferating cell nuclear antigen). PCNA looks remarkably like the *E. coli* β clamp (**Figure 11-27a**). The two proteins share no sequence homology, but the three-dimensional structures are nearly superposable. Both proteins are constructed from a domain that is repeated six times around the ring. The three monomer units of PCNA have only two domains and trimerize to form a ring, while the two monomer units of the β clamp consist of three domains and dimerize to form the ring. The eukaryotic clamp loader, **replication factor C (RFC)**, has five subunits similar in shape and function to those of the *E. coli* clamp loader (**Figure 11-27b**). In a fascinating twist, eukaryotes contain alternative forms of RFC in which one of the subunits is replaced by another protein. These alternative clamp loaders usually function with PCNA, and their intracellular role is not entirely clear. In one case, the alternative clamp loader loads an entirely different clamp onto the DNA.

The eukaryotic replication machinery also includes several other proteins associated with the CMG complex. The complex was isolated from cells using highly selective antibodies directed against CMG subunits. The large protein assemblage is referred to as the **replisome progression complex (RPC)**, and its subunit composition has been identified by mass spectrometry. Among the proteins identified in this fashion are Ctf4 and Mcm10, proteins that move with replication forks in vivo and are essential for cell viability (except in yeast, where Ctf4 is not essential). Ctf4 binds to both GINS and Pol α, acting as a bridge to affix Pol α in the replisome. The role of Mcm10 is presently unknown. RPCs also contain several nonessential proteins that are thought to control the rate of replication during times of cellular stress.

Eukaryotic replication forks proceed at a rate of about 30 to 50 nucleotides per second, far slower than bacterial forks. Also, eukaryotic Okazaki fragments are considerably shorter than bacterial Okazaki fragments, only 100 to 200 nucleotides. The heterotrimeric **replication protein A (RPA)** is the functional equivalent of *E. coli* SSB.

Additional proteins of the eukaryotic replication machinery have functions in finishing and sealing Okazaki fragments. On completing an Okazaki fragment, Pol δ performs limited strand displacement, lifting the RNA primer synthesized by Pol α. The RNA is then excised by a 5′→3′ nuclease called Fen1, and DNA ligase I then joins the fragments. This process often removes some of the DNA made by Pol α. An alternative pathway for RNA primer removal comes into play when strand displacement has proceeded farther than normal; this is accomplished by the Dna2 nuclease, which can excise longer tracts of displaced RNA and DNA than Fen1.

The identity and arrangement of proteins that function at the replication fork in eukaryotes are still the subject of intense investigation. Several proteins with functions that, as yet, lack clear definition seem to be involved in the architecture of the eukaryotic replication fork. The numbers and types of proteins currently thought to participate in eukaryotic chromosome replication are summarized in **Table 11-4**.

TABLE 11-4

Proteins That Function in Eukaryotic Replication

Protein	Number of Subunits	Function(s)	Complex
ORC	6	Initiator	Prereplication complex (preRC)
Cdc	1	Helicase loader	Prereplication complex (preRC)
Cdt1	1	Helicase loader	Prereplication complex (preRC)
Mcm2–7	6	Presumed helicase	Prereplication complex (preRC)
CDK	1	S-phase cyclin kinase	
DDK	1	S-phase cyclin kinase	
Mcm2–7	6	Presumed helicase	Replication complex (RC)
Pol α	4	Primase	Replication complex (RC)
Pol δ	3–4	Replicase	Replication complex (RC)
Pol ϵ	4	Replicase	Replication complex (RC)
PCNA	3	Sliding clamp	Replication complex (RC)
RFC	5	Clamp loader	Replication complex (RC)
Ctf4	3	Scaffold	Replication complex (RC)
RPA	3	Single-stranded DNA–binding protein	Replication complex (RC)
Cdc45	1	Loading of DNA polymerase onto origin; helix destabilization	Replication complex (RC)
GINS	4	Loading of DNA polymerase onto origin; helix destabilization	Replication complex (RC)
Sld2	1	Loading of DNA polymerase onto origin; helix destabilization	Replication complex (RC)
Sld3	1	Loading of DNA polymerase onto origin; helix destabilization	Replication complex (RC)
Dpb11	1	Loading of DNA polymerase onto origin; helix destabilization	Replication complex (RC)
Mcm10	1	Loading of DNA polymerase onto origin; helix destabilization	Replication complex (RC)
DNA ligase I	1	Seals Okazaki fragments	
FenI	1	Removes RNA primers	
Dna2	1	Processes Okazaki fragment	
Topoisomerase I	1	Removes supercoil stress	
Topoisomerase II	2	Removes supercoil stress	

SECTION 11.3 SUMMARY

- The *E. coli* chromosomal replicase, the Pol III core, connects to the ring-shaped β sliding clamp that encircles DNA for processive DNA synthesis. The β clamp is assembled onto DNA by a multiprotein clamp loader. Three Pol III cores, three β clamps, and one clamp loader complex form the Pol III holoenzyme assembly.
- The Pol III holoenzyme, DnaB helicase, and DnaG primase form the replisome complex. The hexameric DnaB helicase encircles the lagging strand and uses ATP to unwind DNA at the replication fork. DnaG primase forms RNA primers to initiate DNA synthesis.
- Topoisomerases act ahead of the replication fork to remove superhelical tension generated by DNA unwinding. SSB binds the single-stranded DNA created by the unwinding action of helicase, preventing the formation of secondary structures in the DNA and protecting it from endonucleases.
- RNA primers are removed from finished Okazaki fragments by the nick translation action of Pol I, and the processed fragments are joined by DNA ligase.
- Simultaneous replication of the two antiparallel strands of duplex DNA by two Pol III cores in the replisome requires loops to form on the lagging strand that repeatedly grow and reset for each Okazaki fragment.

- Eukaryotes have two different multiprotein DNA polymerases (Pol ε and Pol δ) that function on the leading and lagging strands. These DNA polymerases connect to PCNA sliding clamps that are loaded onto the DNA by the RFC clamp loader.

- Eukaryotes have functional counterparts for each *E. coli* replication fork protein, but the eukaryotic replisome is more complex. The eukaryotic primase is a four-subunit enzyme (Pol α) that has both DNA polymerase and primase activities. The CMG helicase has 11 subunits: an Mcm2–7 heterohexamer, GINS heterotetramer, and Cdc45. The eukaryotic SSB homolog, RPA, has three different subunits. Several other eukaryotic proteins that travel with the fork have no known homologs in bacteria.

11.4 INITIATION OF DNA REPLICATION

The site (or multiple sites) on a chromosome where replication is initiated is called the origin. In bacteria, it is the primary point at which regulatory mechanisms control DNA replication. Control of initiation is more complex in eukaryotic cells than in bacteria, because eukaryotes have numerous origins on each chromosome. The total length of DNA replicated from one origin is called a **replicon**. Many bacteria have only one origin, and the replicon is the entire chromosome. In eukaryotic chromosomes, each replicon is the section of DNA replicated from one of its many origins.

Early genetic studies by François Jacob and his coworkers showed that replication starts at a particular place on the DNA, which they termed a replicator (now known as an origin). Genetic studies have since revealed numerous genes encoding the proteins needed for replication. These proteins fall into two classes: those that affect initiation and those that affect replication.

The two classes of proteins were identified by the speed at which their depletion, in *E. coli* mutants, affected DNA synthesis (**Figure 11-28**). Temperature-sensitive mutants for these proteins allowed incorporation of [³H]thymidine during DNA synthesis at a permissive temperature, but no DNA synthesis at a nonpermissive temperature. Temperature-sensitive genes encoding proteins that play a direct role in the replisome caused an abrupt end, or "fast stop," of DNA synthesis when the mutant cells were transferred to nonpermissive growth conditions. For other genes, however, a "slow stop" of replication was observed, which suggested that these genes encode factors needed for initiation, not for fork progression (replication). These "slow-stop" mutants allowed already-initiated DNA synthesis to continue at the nonpermissive temperature until replication of the chromosome was finished.

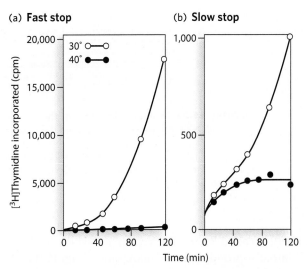

FIGURE 11-28 Two types of replication genes revealed through genetic studies. Temperature-sensitive mutants of *E. coli* were analyzed for the time needed for replication to stop after shifting cells to a nonpermissive temperature. DNA replication was observed by the uptake of [³H]thymidine into cellular DNA (measured as counts per minute, cpm) at a permissive temperature (30° C, open circles) and nonpermissive temperature (40° C, solid circles). (a) A gene giving a fast-stop phenotype encodes a protein involved in progression of the replication fork. (b) A gene giving a slow-stop phenotype encodes a protein involved in the initiation of replication. [*Source: Data from Y. Hirota, A. Ryter, and F. Jacob, Cold Spring Harb. Symp. Quant. Biol. 33:678, 1968, Cold Spring Harbor Laboratory Press, Cold Spring Harbor, NY.*]

The **initiator protein**, which binds specific sites at the origin, is an example of a protein encoded by a slow-stop gene. Binding of the initiator protein to an origin provides a foothold for other proteins to bind, and often results in strand separation in a small region of DNA at the origin. Helicases are assembled at the unwound region, paving the way for more extensive DNA unwinding and the assembly of bidirectional replication forks.

Replication from an origin is a carefully orchestrated and controlled process that involves many different proteins. We first examine initiation at the *E. coli* origin, which is understood in great detail and serves to outline the basic events involved in initiation of replication in all cells. Then we describe our current understanding of how replication from the multiple origins of eukaryotic chromosomes is initiated and controlled.

Assembly of the Replication Fork Follows an Ordered Sequence of Events

The single *E. coli* origin, called *oriC*, was identified by a genetic technique using a recombinant DNA plasmid (see the How We Know section at the end of this chapter). The minimal *E. coli* origin is 245 bp long and contains four copies of a nine-nucleotide (9-mer) consensus

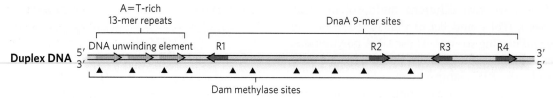

FIGURE 11-29 Structural elements of the *E. coli* origin. The *E. coli* origin, *oriC*, contains four 9-mer DNA sites (blue arrows) that bind the DnaA initiator protein. A possible fifth site deviates from the consensus sequence (not shown). The arrowheads indicate the relative direction of the 9-mers. The three 13-mer direct repeats are A=T-rich and are the locus of initial DNA strand separation (orange arrows). The *oriC* sequence also contains 11 GATC sites that are methylated by the Dam methylase (methylation sites indicated by arrowheads).

sequence (called R sites, for "repeat") to which the bacterial initiator protein, called DnaA, binds (**Figure 11-29**). To one side of the DnaA 9-mer sites are three A=T-rich direct repeats (sequences repeated with the same directionality) of 13 bp each. These A=T-rich repeats, referred to as the DNA unwinding element, are the first area in *oriC* to unwind after the initiator binds. Many origins of replication contain A=T-rich repeats that probably function in a similar way.

The *E. coli* initiator protein, DnaA, is a member of the AAA+ family. Like most AAA+ proteins, it binds and hydrolyzes ATP, although turnover is very slow. After one DnaA binds the origin, it oligomerizes and wraps the origin DNA around the oligomer (**Figure 11-30**, step 1). In the presence of ATP, DnaA destabilizes the A=T-rich 13-mer repeats, forming a single-stranded DNA bubble. Formation of this bubble is stimulated by HU, a small, basic, histonelike protein. Because the

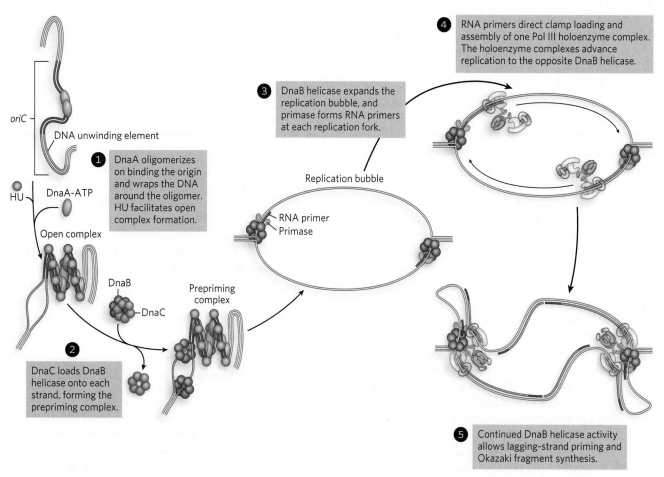

FIGURE 11-30 Activation of *oriC* and assembly of bacterial replication forks.

DnaA-ATP-*oriC*-HU complex forms a bubble at the origin, it is referred to as the **open complex**.

The DNA bubble in the open complex is the site where two hexamers of DnaB helicase are assembled (see Figure 11-30, step 2). Interaction of DnaB helicase with the DnaA initiator protein helps target DnaB to the origin, but it also requires a helicase-loading protein to assemble onto single-stranded DNA in the open complex. The helicase-loading protein is DnaC, another AAA+ protein. The helicase-loading activity of DnaC is thought to function by prying open the hexameric DnaB ring and slipping it onto the single-stranded DNA at the bubble. The ATP-bound form of DnaC binds DnaB helicase tightly and represses its helicase activity. Hydrolysis of ATP ejects DnaC from DnaB helicase, releasing DnaB for its DNA unwinding activity. Two hexamers of DnaB helicase are assembled on the origin, one on each strand of the single-stranded bubble. This assembled group of proteins on the DNA at *oriC* is referred to as the **prepriming complex**.

With the addition of ATP, the DnaB helicases translocate and unwind the DNA, dislodging the DnaA protein (see Figure 11-30, step 3). Unwinding generates positive supercoil stress in the DNA ahead of the replication fork, and this stress must be removed by topoisomerase action (e.g., gyrase). The newly unwound DNA is coated with SSB. Before DNA synthesis can begin, RNA primers must be synthesized by primase. The first RNA primer can be formed when the bubble grows to 100 to 200 bp. An RNA primer on each strand directs β clamp loading and the assembly of a Pol III core–β clamp complex within the holoenzyme (step 4). Each Pol III holoenzyme extends its RNA primer until each of them connects with the DnaB helicases traveling in the same direction. The coupled helicase-polymerase now moves rapidly, producing single-stranded DNA for primase to act upon, followed by clamp loading and engagement of another Pol III core–β clamp complex within each holoenzyme (step 5). This completes the assembly of two bidirectional replication forks at *oriC*.

Replication Initiation in *E. coli* Is Controlled at Multiple Steps

Cell division requires sufficient nutrients and cell mass to support two new cells, so replication must be coordinated with the cell's nutritional status and growth. Regulation occurs at the initiation step, because once replication has begun, the cell is committed to division. It is also of paramount importance that the origin, once replicated, can be inactivated to prevent a second round of replication during the first round, which would commit the cell to splitting twice (resulting in four cells).

Binding of the DnaA initiator protein at *oriC* is a central point at which initiation is controlled. One mechanism for controlling initiation at *oriC* is through DNA methylation. Both strands of the palindromic sequence GATC are recognized by the enzyme **Dam methylase (DNA adenine methyltransferase)**, which methylates the N^6 position of A residues on both strands of the GATC site. Occurring at random, the average frequency of a GATC sequence would be once every 256 bp, yet the 245 bp *oriC* contains 11 GATC sites (see Figure 11-29). Immediately after a GATC site is replicated, the new strand is not yet methylated and the GATC site is thus hemimethylated. This hemimethylated state of newly replicated DNA is only temporary, until Dam methylase acts, but the high density of GATC sites in *oriC* delays complete methylation. The SeqA protein (Seq for *seq*uestration) binds specifically to hemimethylated GATC sites and thereby sequesters the newly replicated *oriC*, preventing DnaA from rebinding the replicated origin and initiating another replication event (**Figure 11-31a**). Dam methylase, working between SeqA dissociation-reassociation cycles, eventually methylates the GATC sites in *oriC*, and this blocks SeqA binding and opens up the origin to DnaA binding once more.

Initiation depends on the nucleotide-bound state of DnaA, which uses the energy of ATP binding to form the open complex at *oriC* (**Figure 11-31b**). When replication forks dislodge DnaA from the origin, it can rebind. But DnaA hydrolyzes the bound ATP after initiation, and even though ADP-DnaA can rebind the origin, it is unable to form the open complex, thus preventing reinitiation. Important to this regulatory step is that the exchange of free ATP for bound ADP on DnaA is slow, requiring up to half an hour—time for the cell division cycle to finish. ATP hydrolysis by DnaA is ensured by the Hda protein (**Figure 11-31c**). After replication forks start moving, Hda binds the β sliding clamp and stimulates ATP hydrolysis by DnaA, thus inactivating DnaA.

The number of DnaA-binding sites in the cell may also play a role in the control of reinitiation. The chromosome contains numerous DnaA-binding sites in various promoters, because DnaA is also a transcription regulator. In aggregate, these other DnaA sites far outnumber the few at the origin. Therefore, as chromosome duplication proceeds, the total number of DnaA-binding sites doubles, and these sites may act as a sink to lower the free DnaA available for binding to *oriC*.

In laboratory experiments, the RNA polymerase inhibitor rifampicin blocks replication in cells, suggesting that RNA polymerase, too, plays a role in chromosome replication. As it unwinds DNA ahead of the sequence it is transcribing, RNA polymerase creates supercoil strain in the DNA template. When RNA polymerase is

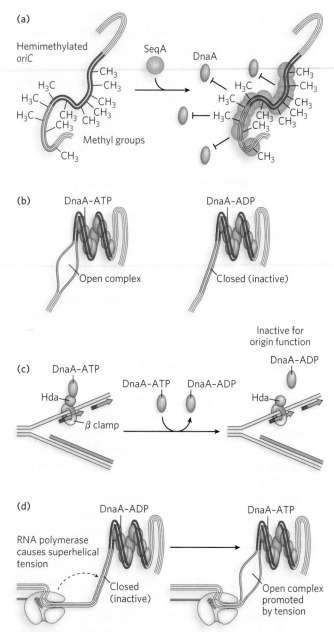

FIGURE 11-31 Regulation of the *E. coli* origin. Initiation at the *E. coli* origin, *oriC*, is regulated in several ways. (a) SeqA protein binds hemimethylated DNA and sequesters the newly replicated origin, preventing DnaA binding. (b) DnaA-ADP, formed when DnaA hydrolyzes its ATP, cannot destabilize the A=T-rich region to maintain the open complex containing a single-stranded DNA bubble, thus forming a closed complex in which the bubble has collapsed. (c) The Hda protein binds the β clamp on the DNA, causing DnaA to hydrolyze its ATP and become inactive (DnaA-ADP). (d) RNA polymerase produces superhelical tension that promotes DnaA-induced melting of the A=T-rich region.

near the origin, it stimulates the initiation of replication through this supercoil strain, probably by helping DnaA destabilize the A=T-rich 13-mer repeats involved in forming the open complex bubble (**Figure 11-31d**). For this reason, rifampicin is an effective antiviral for some

RNA viruses that code for RNA polymerase as their sole polymerase.

Despite these several layers of control, there can be instances in which a bacterial cell is able to initiate replication again before the first cycle is complete. This occurs in times of abundant nutrients. However, as we will see shortly, eukaryotes have elaborate control schemes to prevent re-replication at origins until the first cycle of genome duplication is complete.

Eukaryotic Origins "Fire" Only Once per Cell Cycle

The much greater DNA content of eukaryotes, coupled with their slower replication forks, necessitates multiple origins on each chromosome to allow complete replication in the 24-hour division time of, for example, a human cell. Origins are spaced 10 to 40 kbp apart along each chromosome, and multiple replication forks eventually meet to yield the two daughter chromosomes. "Firing" (activation) of an origin is under tight control, as is reinitiation at an origin that has already been duplicated.

Eukaryotes have defined cell cycle phases, with chromosome replication occurring in S phase and separation of the duplicated chromosomes in M phase. (For an overview of the cell cycle, see Chapter 2.) A protein complex essential to replication is assembled on chromosome origins even before S phase. The assembly process occurs late in G_1 phase and marks origins that will be used for replication during S phase. This separation of events in the cell cycle is critical to the exquisite coordination that eukaryotes require to duplicate their long, linear chromosomes.

The simple eukaryote *S. cerevisiae* has well-defined replication origins referred to as ARS (autonomously replicating sequences), which are 100 to 200 bp long and contain four common components: a highly conserved A sequence and the B1, B2, and B3 elements (**Figure 11-32**, top). Two-dimensional gel electrophoresis can be used to identify DNA segments that contain an origin (**Highlight 11-1** on p. 396). The identification of discrete origins in yeast has made it a convenient model organism for studying origin function in eukaryotes. Furthermore, homologs of the yeast replication proteins exist in all eukaryotes, indicating that the lessons we learn from yeast will probably generalize to more complex eukaryotic organisms. However, unlike yeast, other eukaryotes have origins that lack easily identifiable sequence motifs.

The eukaryotic initiator is a heterohexamer called the **origin recognition complex (ORC)**. Several subunits of ORC are, like the *E. coli* DnaA initiator protein, AAA+ proteins. ATP is required for ORC binding

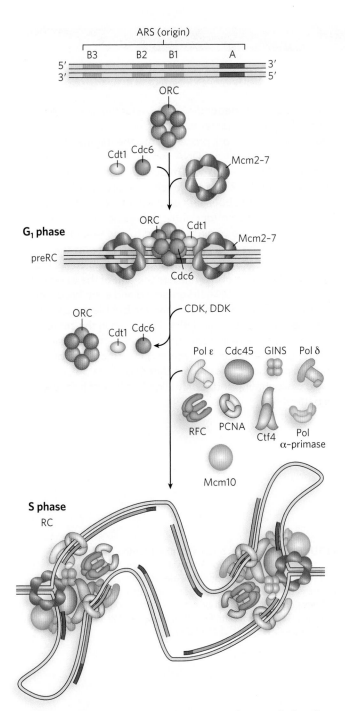

FIGURE 11-32 Assembly of eukaryotic replication forks. The generalized structure of an origin in *S. cerevisiae* is shown at the top. The prereplication complex (preRC) assembles in G$_1$ phase (middle). The initiator, ORC, binds to the conserved A element and the B1 element. MCM helicases are loaded onto the DNA by Cdc6 and Cdt1. After the cell progresses to S phase (bottom), the origin forms replication forks, as cyclin-dependent protein kinases (CDK and DDK) facilitate assembly of other proteins to form the replication complex (RC) from which replication fork movement commences.

to the origin (see Figure 11-32, middle). After ORC binds to the DNA, the Cdc6 protein (also an AAA+ protein) binds to ORC. The ORC-Cdc6 complex then loads the Mcm2–7 complex onto DNA. The Mcm2–7 complex is a circular heterohexamer that binds one molecule of Cdt1, which is required before the ORC-Cdc6 complex can load the Mcm2–7 complex onto the DNA. Two Mcm2–7 complexes are loaded such that they encircle duplex DNA adjacent to the ORC-Cdc6 complex. These events occur only in G$_1$ phase, and the resulting complex of ORC, Cdc6, Cdt1, and Mcm2–7 is referred to as the **prereplication complex (preRC)**. The Cdc, Cdt, and Mcm names derive from genetic experiments that defined DNA metabolic pathways; the functions of these proteins were determined later, and in many cases the exact functions are still being worked out.

Cyclin-dependent protein kinases (also called cyclin kinases or cell cycle kinases) that phosphorylate certain target proteins are central to the separation of cell cycle phases. In G$_1$ phase there is very low kinase activity, and proteins are generally not phosphorylated. On entering S phase, *S. cerevisiae* S-phase cyclin kinases phosphorylate some of the preRC proteins. Mcm4 and Mcm6 are targets of the cyclin kinase DDK. Phosphorylation by DDK enables Mcm2–7 complexes to dissociate from one another and move in opposite directions. The cyclin kinase CDK inactivates ORC by phosphorylating one of its subunits. In addition to inactivating ORC, these events lead to degradation of Cdc6 and Cdt1, and all these events prevent further preRC assembly until the cell has divided and reentered G$_1$.

Several other replication factors associate with Mcm2–7 at origins early in S phase. Among these are Sld3 and Cdc45, which bind to the double hexameric Mcm2–7 complexes. Also, a preloading complex (pre-LC) forms, consisting of Pol ε, Sld2, Dpb2, and possibly other proteins. CDK phosphorylates Sld2 and Sld3, and the complex of Mcm2–7 and Cdc45 combines with GINS and Pol ε to form the CMG helicase at a replication fork, ejecting Dpb11, Sld2, and Sld3; full assembly of the replication forks can then proceed with association of Mcm10, Ctf4, Pol α, and Pol δ (see Figure 11-32, bottom). Phosphorylation by S-phase cyclin kinases is necessary for replication fork assembly and confines the initiation of replication to S phase. Only after the S, G$_2$, and M phases are complete, and cell division is accomplished, is S-phase cyclin kinase activity reduced and Cdc6 and Cdt1 made available to direct the assembly of the preRC on the chromosomes of new cells in G$_1$ phase.

SECTION 11.4 SUMMARY
- The assembly of bacterial replication forks at the origin occurs in steps, starting with the binding of

HIGHLIGHT 11-1   TECHNOLOGY

Two-Dimensional Gel Analysis of Replication Origins

Origins in the process of replication generate DNA molecules that contain bubbles and replication forks. These unusually shaped DNAs produce characteristic patterns in two-dimensional agarose gels (see Chapter 8). In this technique, the section of DNA to be examined for a replication origin is cut on either side of the origin with a restriction enzyme. In the first dimension of the gel, molecules are sorted mainly by size by low-voltage electrophoresis through a low-percentage agarose gel. An unreplicated 1 kbp fragment of DNA will travel farther through the gel than a replicated (2 kbp) fragment (Figure 1a). The second dimension is run at higher voltage to sort molecules mainly by shape. DNA fragments containing replication forks are less streamlined and will travel more slowly through the gel than unreplicated or completely replicated fragments. This two-dimensional assortment by size and shape of the same piece of DNA undergoing replication generates arc patterns. The DNA is analyzed by Southern blotting, in which DNA in the gel is transferred to nitrocellulose and probed with a radioactive DNA fragment that hybridizes to the region of interest (see Chapter 6).

Figure 1b shows DNA structures that result from replication initiating at two different origins on one section of DNA. Vertical dashed lines represent restriction sites. After digestion by restriction enzymes, three different restriction fragments are produced, RF1, RF2, and RF3. The three panels below the DNA structures represent the results of two-dimensional gel analysis of the cut DNA using a radioactive DNA probe—probe 1, 2, or 3. These probes hybridize specifically to RF1, RF2, or RF3. The lower left panel shows the Y arc pattern, using probe 1 and RF1. This pattern is produced by DNA that contains no origin of its own. Replication forks that enter the DNA produce Y-shaped DNAs that form differently sized and shaped fragments, depending on how far the fork travels into the fragment (fragments d, e, f, and g are produced in succession as the fork proceeds into the fragment). The top of the arc results from DNA containing three arms of equal length (fragment f). The middle panel shows the results when probe 2 is used, which hybridizes to RF2; this produces the bubble-to-Y arc pattern. This pattern is most indicative of an origin within the restriction fragment and occurs when the origin is located to one side of the center point of the fragment. Restriction fragments that contain bubbles (fragments a, b, and c) produce an arc that "breaks" when the bubble reaches the end of the fragment, to produce a Y-form DNA (fragments d and e are produced at this point). The right panel shows the bubble arc pattern generated with probe 3 and RF3. The origin in the center produces bubbles of increasing size. Due to the central location of the origin in this restriction fragment, the bubble does not produce a Y form on either end; therefore, the arc has no discontinuity and is only slightly different from the Y arc pattern.

FIGURE 1 Replication origins can be identified based on their mobility in a two-dimensional agarose gel. (a) The steps involved in two-dimensional gel electrophoresis, showing how an arc is generated. (b) The expected patterns from analysis of linear chromosomal DNA with more than one origin—the typical case. [Source: Data from B. J. Brewer and W. L. Fangman, Cell 51:463–471, p. 464, © 1987 Cell.]

DnaA initiator protein, which melts an A=T-rich region. A DnaB helicase is then loaded onto each of the single strands of DNA by the DnaC helicase loader. As DNA is unwound by DnaB, DnaG primase synthesizes RNA primers; this is followed by entry of two Pol III holoenzymes to form a bidirectional replication fork.

• Origin activation in bacteria is regulated at the initiation step by various means, including DNA methylation that results in SeqA sequestering the origin, ATP

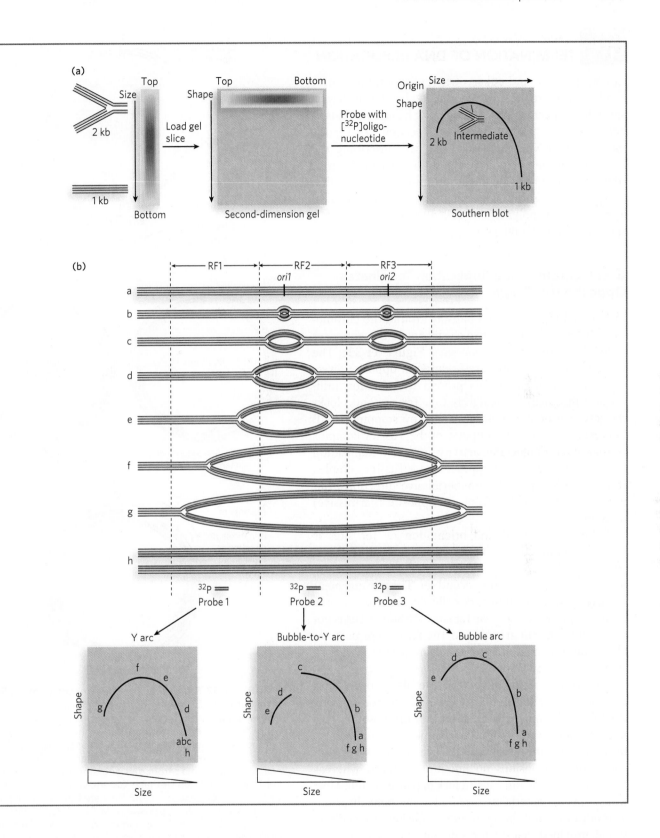

turnover by DnaA, and the activity of Hda protein, which signals DnaA to hydrolyze ATP after forks are formed.

• Eukaryotes have many replication origins, and tight initiation control is achieved by dividing the activation of origins into different cell cycle phases. Some proteins can bind the origin only in G_1 phase, when cyclin kinase activity is low (preRC). Further assembly of proteins to form replication forks occurs only in S phase and is associated with phosphorylation by S-phase cyclin kinases.

11.5 TERMINATION OF DNA REPLICATION

In both bacteria and eukaryotes, replication forks meet head-on when replication is complete. *E. coli* cells have a specialized mechanism for preventing head-on collisions between DNA polymerases, and between DNA polymerase and RNA polymerase that is transcribing the DNA. Eukaryotic cells have the additional problem of replicating the ends of linear chromosomes. Evolution has provided a solution to this problem, in the form of a novel DNA polymerase—called telomerase—that is specialized for this purpose.

E. coli Chromosome Replication Terminates Opposite the Origin

In *E. coli*, a region located halfway around the chromosome from *oriC* contains two clusters of 23 bp sequences called **Ter sites** (*ter*mination sites; **Figure 11-33**). The two clusters of Ter sites are oriented in opposite directions. The monomeric Tus protein (*ter*mination utilization substance) binds tightly to a Ter site and blocks the advance of the replication fork by stopping DnaB helicase. A fascinating property of the Tus-Ter complex is that its fork-blocking activity is polar. Replication forks are blocked when approaching a Tus-Ter complex from one direction (the nonpermissive direction), but not when approaching from the opposite (permissive) direction.

The arrangement and orientation of Ter sites is such that bidirectional replication forks from *oriC* can pass through the first set of Ter sites that they encounter, but are blocked by the second set. This arrangement localizes the replication fork collision zone to the area between the two clusters of Ter sites. Although Tus is not essential for *E. coli* growth, the Tus-Ter system presumably evolved to confer a growth advantage in the natural (nonlaboratory) setting.

Actively replicating bacteria are also growing and metabolizing, and therefore are actively transcribing RNA from promoters throughout the chromosome. This means that collisions between RNA polymerase and replication forks are inevitable. In vivo studies show that codirectional collisions do not impede forks, whereas head-on collisions can cause a fork to pause or stall. Most transcripts in bacteria are oriented in the same direction as replication, and therefore most collisions are codirectional, provided that the forks do not proceed more than halfway around the chromosome (see Figure 11-33c). Perhaps the Tus-Ter system evolved to prevent replication forks from going too far around the circular chromosome, where the direction of transcription would result in head-on collisions.

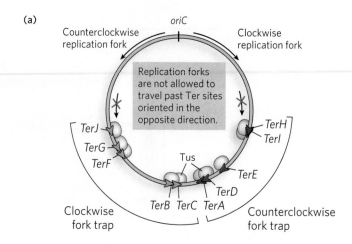

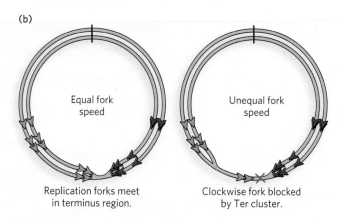

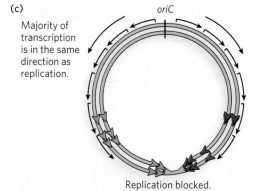

FIGURE 11-33 The role of Ter sites in control of replication termination in *E. coli*. (a) Ter sites are located in two clusters, halfway around the chromosome from *oriC*. Each cluster contains multiple Ter sites oriented in the same direction, but the two clusters have opposite polarity, indicated by the arrows. Tus protein binds to a Ter site, and Tus-Ter blocks helicase approaching from one direction and not the other. Replication forks can displace Tus and pass through the first Ter cluster they encounter, but they are blocked at the second cluster, which has the opposite polarity. (b) Replication forks of equal speed meet in the terminus region of the chromosome (left). The Tus-Ter system ensures that even replication forks moving at unequal speed will meet in the terminus region (right). (c) The Tus-Ter system prevents a replication fork from extending much beyond the halfway point around the chromosome, and thus ensures that the fork always moves in the same direction as transcription.

Replicating the last bit of DNA between converging replication forks presents certain topological problems that must be solved to disentangle the two daughter chromosomes. Specialized type II topoisomerases unlink the catenated daughter chromosomes (see Chapter 9).

Telomeres and Telomerase Solve the End Replication Problem in Eukaryotes

The replication of linear chromosome ends poses a unique problem. At the end of a chromosome, after the leading strand has been completely extended to the last nucleotide, the lagging strand has a single-strand DNA gap that must be primed and filled in. The problem arises when the RNA primer at the extreme end is removed for replacement with DNA (**Figure 11-34**). There is no 3' terminus for DNA polymerase to extend from, so this single-strand gap cannot be converted to duplex DNA. The genetic information in the gap will be lost in the next round of replication, and repeated rounds will

cause the ends to progressively shorten until the genes near the ends are entirely lost. The resulting loss of gene function can be detrimental to a cell, disrupting cellular functions or contributing to the formation of cancerous tumors. This **end replication problem** does not occur in circular DNA, which has no ends. Indeed, avoidance of the end replication problem may underlie the widespread occurrence of circular DNAs in bacteria and their plasmids and phages.

The ends of the linear eukaryotic chromosomes are called **telomeres** (see Chapter 12). How were telomeres discovered, and what is so special about them? This story starts in the 1930s with Barbara McClintock's research on maize. McClintock noticed that when chromosomes become damaged and break in two, the broken chromosomes can rapidly "heal" by joining the two chromosome ends. When parts of two different chromosomes become joined, it sometimes creates an unusual chromosome with two centromeres. On entering cell division, chromosomes with two centromeres are pulled to opposite poles by the mitotic spindles, breaking the chromosomes, and this is followed by another round of healing. The process of healing and breaking occurs over and over as cells divide, leading to chromosome rearrangements, loss of genetic information, and cell death. It would, of course, be disastrous if the ends (telomeres) of normal chromosomes underwent this "healing" process, because all the chromosomes would break on entering cell division. Therefore, McClintock surmised, there must be something special

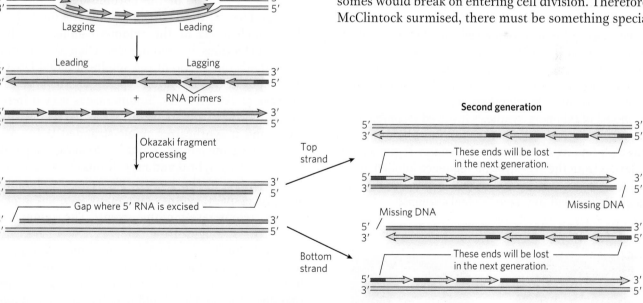

FIGURE 11-34 The end replication problem in linear chromosomes. Just two rounds of chromosome replication are shown here. In the first generation of replication (left; red indicates new DNA), lagging-strand synthesis results in an RNA primer at or near one 5' end of each new chromosome. After removal of the RNA, the 5' single-strand gap in the DNA cannot be filled, thus a 3' single-stranded DNA overhang remains. In the second generation (right; yellow indicates new DNA), each first-generation chromosome produces two new chromosomes, for a total of four new duplexes. Two of the new chromosomes have lost DNA at one end. All four chromosomes terminate with a 5' single-strand gap in the DNA after the RNA is removed. Further losses will be sustained with each new generation.

about the ends of normal chromosomes, preventing them from being joined together by the healing process that mends broken chromosomes. Independently, Hermann Mueller, working with irradiated flies, made similar observations on the fate of broken chromosomes (in this case, broken by x rays) and hypothesized that the ends of chromosomes have a protective quality about them. He called the ends "telomeres," derived from the Greek *telos* ("end") and *meros* ("part").

In 1978, Elizabeth Blackburn identified the first telomere sequence in the single-celled ciliate *Tetrahymena*, which contains thousands of linear chromosomes and thus became the model organism to study telomere biology. In fact, further study has shown that the telomerase enzyme functions in broadly similar ways in all eukaryotes. *Tetrahymena* telomeres are composed of numerous direct repeats of TTGGGG. We now know that the telomeres of different organisms consist of tracts of similar short repeats that vary from a few hundred base pairs in single-celled eukaryotes to over 10,000 bp (10 kbp) in higher eukaryotes (see Table 9-1). The sequence TTAGGG is repeated for a length of 5 to 15 kbp in human telomeres. The significance of the telomere sequence was demonstrated in a key experiment conducted jointly by Blackburn and Jack Szostak. Szostak had found that linearized plasmids, when transformed into yeast, were either degraded or integrated into the genome. But after ligation of the *Tetrahymena* telomere sequence to a linearized plasmid, the linear DNA was maintained in the cell like a very small chromosome. Interestingly, the telomeres grew longer as the cells divided in log-phase growth, and sequencing of the plasmids revealed a different (i.e., not *Tetrahymena*) repeat. This repeat turned out to be the telomere sequence of yeast! This had enormous implications for how telomeres are formed. The popular model of telomere synthesis involved recombination (a process discussed in Chapter 13). But Blackburn and Szostak's finding that the added telomere sequence was different from the *Tetrahymena* sequence implied an enzymatic mechanism, since recombination would have yielded repeats of the same original sequence.

The finding that yeast telomere sequences were added to the *Tetrahymena* telomere sequence suggested that telomeres may be synthesized by a polymerase that can synthesize DNA from an exogenous template, presumably an RNA molecule. DNA polymerases that use RNA as a template were first found in certain viruses and are called reverse transcriptases, because they transcribe RNA into DNA—the opposite of the usual process (see Chapter 14). Carol Greider, a graduate student in Blackburn's laboratory, set out to identify the putative "telomere reverse transcriptase." Using a small synthetic oligonucleotide and radioactive dNTPs, Greider looked for and found an end-extending activity in extracts from *Tetrahymena* cells. The enzyme added increments of six residues onto the oligonucleotide, as illustrated in the sequencing gel shown in **Figure 11-35**.

After purification and characterization, the **telomerase reverse transcriptase (TERT)** was found to carry a tightly bound, noncoding **telomerase RNA (TR)**. The TR in humans is 451 nucleotides long, but can range from 150 to 1,300 nucleotides, depending on the organism. The TR contains about 1.5 telomere repeat units that it uses as a template to extend the 3′ terminus of the telomere (**Figure 11-36** on p. 402). The TERT-TR holoenzyme is referred to as **telomerase**. The reaction cycle of *Tetrahymena* telomerase is shown in Figure 11-36. The reaction occurs in S phase, as part of the chromosome duplication process. First, at the 3′-terminal end of a linear DNA, three nucleotides of the telomere anneal to three RNA nucleotides in telomerase. Then the telomerase extends the 3′ end of the single-stranded DNA by the length of one telomere repeat—six nucleotides in the case of *Tetrahymena*. Telomerase is very different from other DNA polymerases; it carries its own template, and it synthesizes single-stranded DNA. After adding a six-nucleotide repeat, telomerase separates the RNA-DNA hybrid and repositions on the telomere for extension of the next 6-mer repeat. Telomerase acts processively, synthesizing many telomere repeats in one telomerase-binding event. The telomerase-extended, 3′ single-stranded DNA terminus is then converted to duplex DNA by the same priming and polymerization machinery used in chromosome replication. Because the telomere DNA is a simple repeat that does not encode a biomolecule, the cell can tolerate a certain amount of variability in final telomere length. It is important to note, however, that the DNA product is not completely duplex. The 3′ terminus of a new telomere still has single-stranded DNA, due to the same RNA primer–removal problem discussed earlier.

Elizabeth Blackburn
[Source: Elisabeth Fall/Fallfoto.com.]

Carol Greider
[Source: Courtesy of Johns Hopkins University.]

(a)

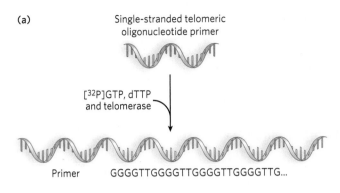

Single-stranded telomeric
oligonucleotide primer

[³²P]GTP, dTTP
and telomerase

Primer   GGGGTTGGGGTTGGGGTTGGGGTTG...

(b)

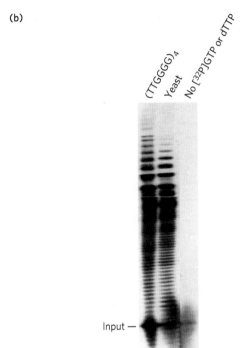

(TTGGGG)₄

Yeast

No [³²P]GTP or dTTP

Input —

FIGURE 11-35 Assay for *Tetrahymena* telomerase activity.
(a) In the assay, an oligonucleotide of the *Tetrahymena* telomere
sequence is added to the extract, along with [³²P]dGTP and
unlabeled dTTP. Telomerase activity adds telomere repeats to
the oligonucleotide. (b) Autoradiograph of a polyacrylamide
gel analysis of reaction assays using different oligonucleotide
sequences as substrate. The result shows that oligonucleotides
of *Tetrahymena* and yeast telomere sequences are extended in
six-nucleotide increments. [*Source: Reprinted from Cell, 43, Greider,
C.W., Blackburn, E.H., Identification of a specific telomere terminal
transferase activity in tetrahymena extracts, 405-413, Copyright 1985,
with permission from Elsevier.*]

Telomere Length Is Associated with Immortality and Cancer

Tetrahymena, like other single-celled organisms, is immortal and can divide innumerable times. On mutagenesis of the telomerase gene, *Tetrahymena* telomeres shorten with each cell division, until, after 20 to 25 generations, the telomeres have shrunk below a

critical level and the cells die. Similar observations apply to mammalian cells. Most somatic cells have little or no telomerase, and cells in culture divide about 40 to 60 times before losing their telomeres and dying. This suggests that telomeres act as a clock that counts down the number of cell divisions. Indeed, the cells of individuals with premature aging syndromes have shorter-than-normal telomeres (**Highlight 11-2**). But what if telomerase were expressed in all cells? Does the activation of telomerase hold promise as our ticket to immortality, the proverbial "fountain of youth"? Probably not. Studies in mice have shown that activation of telomerase in somatic cells leads to an increased incidence of tumors, and lifespan is shortened by cancer. Several additional mutations are needed to form a cancer cell, not just telomerase activation. These other mutations involve suppression of the programmed cell death pathway and activation of the mitotic pathway (i.e., mutations in tumor suppressors or in oncogenes; see Chapter 12). But activation of telomerase would decrease the number of mutations needed for a cell to become cancerous. Perhaps more informative is the finding that mice without telomerase have a normal lifespan. Despite these mixed findings, research on telomerase and antiaging therapy will probably continue past the lifespan of all those who read this book.

The fact stands that telomerase is associated with cancer. HeLa cells are an immortal human cell line derived from a woman named Henrietta Lacks, who died of ovarian cancer in 1951. HeLa cells express telomerase and have been grown in cell culture for decades in laboratories around the world. If telomerase could be inhibited by a drug, the telomeres in cancer cells should shorten with each cell division until the telomeres collapse, causing cell death. Therefore, drugs that inhibit telomerase hold promise as an anticancer therapy, an active area of current research.

Telomeres are Protected and Regulated by Proteins

The linear ends of eukaryotic chromosomes present another problem: they could be mistaken for sites of chromosome breakage and thereby induce the cell's DNA repair systems that would join the ends of chromosomes together. The cell has two main repair systems that use recombination to join broken DNA ends (see Chapter 13). This would have disastrous consequences because, as described earlier, the joined chromosomes would be torn apart during cell division. In the cell, telomeric ends are protected by specialized telomere DNA-binding proteins that prevent chromosomes from joining and

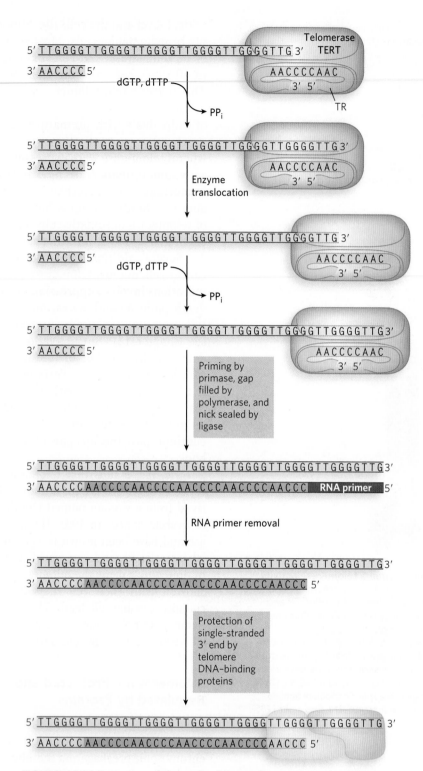

FIGURE 11-36 Extension of the ends of linear chromosomes by telomerase.
Telomeres at the ends of eukaryotic chromosomes are composed of a short repeating DNA sequence. Shown here is the repeating 5'-TTGGGG-3' sequence of *Tetrahymena*. Telomerase extends the 3' single-stranded DNA end with dNTPs, using its internal RNA molecule as template. The extended 3' single strand of DNA is filled in by RNA priming and DNA synthesis. Removal of the RNA primer for this fill-in reaction still leaves a short 3' single-stranded DNA overhang; this end is sequestered by telomere DNA–binding proteins. These proteins protect the chromosome ends from becoming substrates for the cell's double-strand break repair machinery.

Short Telomeres Portend Aging Diseases

Loss of chromosome ends during successive cell divisions leads to cell senescence or apoptosis, whereas expression of telomerase imparts immortality to cells. Hence, we may think of telomeres as a type of molecular clock that counts down the age of our cells. Can immortality be achieved by activating telomerase activity? Unfortunately, activation of telomerase in every cell of the body is not an option. Activation of telomerase is associated with cancer, and, in fact, mice that are engineered to express telomerase in somatic cells develop tumors and die early. We are caught in a delicate balance between requiring telomeres for life but also requiring that they have a finite lifetime. Given the connection between telomeres and cancer, one possible explanation for the telomere clock is that telomere shortening may have evolved as a way to suppress tumor growth in multicellular organisms. An equally feasible explanation is that natural selection favors a finite lifespan, because it ensures diversity in the genetic pool for evolution.

Does telomere length correlate with longevity? Several observations indicate a role of telomeres in cellular aging. A large body of epidemiological studies on human telomeres in blood cells reveals that short telomeres correlate with a number of diseases related to aging. Abnormally short telomeres are associated with diabetes, osteoporosis, obesity, cancers, impaired function of the immune system, a variety of cardiovascular diseases, and stroke. Thus, short telomeres correlate with a broad general syndrome of diseases that reflect, or perhaps even cause, aging in a fundamental way. In sum, long telomeres may impart not necessarily a longer lifespan but rather a healthier life for people in their seventies. Telomere-associated diseases seem to be influenced by both genetic and nongenetic factors; among the nongenetic factors are psychological stress, behavior, and possibly even diet. Indeed, telomere length may more clearly reflect the biological age of cells rather than chronological age. Someday, monitoring of individuals' telomeres could become one of the "yardsticks of general health," much like keeping track of weight and blood pressure.

Some hereditary human diseases have their basis in telomerase or in proteins that bind telomeres. Such mutations are present in individuals with dysfunctional telomeres and particular types of degenerative diseases. Importantly, these degenerative diseases are age-related because the short-telomere defect that they share is acquired with age (Figure 1). One of the most common manifestations of defective telomere biology is mutations that result in the lung disease pulmonary fibrosis. Additional degenerative diseases that correlate with dysfunctional telomeres are dyskeratosis congenita, Hoyeraal-Hreidarsson syndrome, and aplastic anaemia.

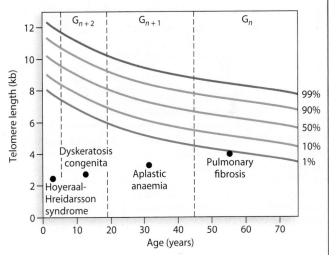

FIGURE 1 The association between dysfunctional telomeres and age-related diseases. The lines represent changes in typical percentage telomere length with age in lymphocytes. With age, telomere length decreases. The percentage normal distribution of telomere length among humans is given on the right. The red dots indicate the telomere length at the given age for four different telomere-related genetic diseases. The vertical dashed lines, G_n, G_{n+1} and G_{n+2}, designate three successive generations, revealing that the onset of the disease occurs earlier in life with each generation. [Source: Data from M. Armanios and E. H. Blackburn, Nat. Rev. Genet. 13:693–704, 2012, Fig. 2.]

---

also regulate telomerase activity to prevent telomeres from growing abnormally long.

Telomere-binding proteins bind the duplex and cover the 3′ single-stranded DNA overhang. There are differences among the telomere-binding proteins of various organisms, but all serve the same function: to preserve the telomere ends. In mammals, the telomere is bound by a set of proteins called **shelterin (Figure 11-37)**. TRF1 and TRF2 (*telomere repeat factors 1 and 2*) bind the direct repeat sequences in the double-stranded DNA, and POT1 (*protection of telomeres 1*) binds the single-stranded DNA. These proteins are held together by TIN2 (*TRF1 interacting nuclear protein-2*) and TPP1 (*TIN2 interacting protein 1*). RAP1 (*repressor/activator protein 1*) is another component of shelterin.

Shelterin not only protects telomeres against chromosome joining but also regulates telomere length. POT1 inhibits telomerase and becomes a stronger inhibitor with each additional POT1 protein bound to the single-stranded DNA. Thus, short telomeres are not good inhibitors of telomerase and they become longer, but long telomeres are good telomerase inhibitors and

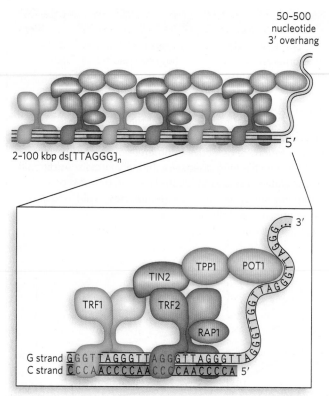

50–500 nucleotide 3' overhang

2–100 kbp ds[TTAGGG]ₙ

G strand GGGTTAGGGTTAGGGTTAGGGTTA
C strand CCCAACCCCAACCCCAACCCCA 5'

FIGURE 11-37 Mammalian telomeres are bound by shelterin complex. Three copies of the shelterin complex are shown at the upper left, and one shelterin at a human telomere at the lower left. TRF1 and TRF2 bind the duplex portion, and Pot1 binds the single-stranded portion. TIN2 and TPP1 form a bridge to POT1, and RAP1 binds TRF2. [*Source: Data from T. de Lange,* Science *326:948–952, 2009.]*

telomere extension stops. Thus, telomeres grow to a length regulated by proteins that bind them. Another facet of telomere regulation lies in a special DNA loop structure. Titia de Lange's group at Rockefeller University, along with Jack Griffith's laboratory at the University of North Carolina, discovered that, in mammals, the 3'-terminal single-stranded DNA is folded back and hybridized to the duplex to form a loop called a **t-loop** (telomere loop) **(Figure 11-38)**. The t-loop buries the 3' terminus of the telomere, which further regulates telomere length. The t-loop is thought to be a dynamic structure. When telomeres are long, the t-loop is more stable; when short, the t-loop is destabilized and telomerase can gain access for telomere elongation. At this point, there are also fewer shelterins, and telomere elongation commences.

Telomere DNA–binding proteins have yet another

Titia de Lange
[*Source: Courtesy Titia de Lange/de Lange & Konarska Labs.]*

important function. Broken DNA ends induce two important signal responses that, when activated, lead to cell cycle arrest. These signal pathways are referred to as the ATM and ATR pathways, both of which detect damaged DNA and stop the cell cycle. The signal for these pathways is either a broken DNA end (ATR) or a section of single-stranded DNA (ATM). Telomeres have both of these structures, yet must be prevented from inducing the ATM and ATR signal pathways—otherwise the cell could not divide. The proteins that bind to telomeres hide the ends and the single-stranded DNA and thus prevent these signal responses. Studies indicate that TRF2 represses ATM and POT1 represses ATR. In mammalian cells, the t-loop provides yet further protection against activation of these signaling pathways.

In contrast to the broad conservation of telomerase among different species, the proteins that bind telomeres often differ from one species to another. The different telomere-binding proteins in different organisms may be based in the diversity of organisms' repair and signal pathways. So far, we have focused on how shelterin functions on mammalian chromosomes to protect and regulate telomeres. Yeast serves as an example of an organism that uses a different set of telomere-binding proteins. Yeast telomeres are bound by a three-subunit complex composed of Rap1, Rif1, and Rif2. Rap1 has homology to mammalian RAP1, but differs in that yeast Rap1 binds the telomeric duplex DNA directly. Neither Rif1 nor Rif2 shares homology with mammalian telomere-binding proteins. Yeast Cdc13 binds the single-stranded portion of the telomere. Unlike mammalian POT1, yeast Cdc13 positively regulates telomerase,

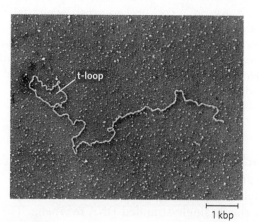

1 kbp

FIGURE 11-38 Mammalian t-loops. The t-loop at the end of a linear chromosome of mammalian cells is visible in this electron micrograph. The telomere has been separated from the rest of the chromosomal DNA by a restriction enzyme. [*Source: Reprinted from* Cell, 97, *Griffith, J. D., Corneau, L., Rosenfield, S., Stansel, R. M., Bianchi, A., Moss, H., and de Lange, T. Mammalian Telomeres End in a Large Duplex Loop, 503–519, Fig. 3B, copyright 1999, with permission from Elsevier.]*

recruiting it for telomere extension. Rap1 negatively regulates telomerase activity.

SECTION 11.5 SUMMARY

- Termination of replication in *E. coli* occurs halfway around the circular chromosome from *oriC*. Bidirectional replication forks meet head-on within a terminus region bordered on both sides by multiple Ter sites. Tus binds to Ter and blocks replication forks in one direction but not the other, thus localizing termination to the terminus region.
- Replication of eukaryotic linear chromosomes cannot be completed at the extreme ends with the replication fork machinery. To solve this end replication problem, telomeres are synthesized at the chromosome ends by telomerase, which carries its own RNA template strand and adds multiple 6-mer repeats to the 3′ terminus, extending the 3′ single-stranded DNA. The single strand is then converted to duplex DNA by priming and chain extension.
- After telomere synthesis, removal of the last RNA primer still leaves a small single-strand gap in the DNA that cannot be filled. Because the telomeric repeats are noncoding and can be replaced by further telomerase action, their loss is of no consequence.
- Somatic cells lack telomerase, and they die when their telomeres become too short. Telomerase is activated in cancer cells, which become immortal and form tumors. Harnessing the activity of telomerase to kill cancer cells or rejuvenate normal but aging somatic cells is an important subject of medical research.
- Telomere DNA–binding proteins protect the ends of chromosomes from nucleases and recombination.

? UNANSWERED QUESTIONS

Although we know the major actors that replicate bacterial chromosomes, the mechanics of advancing a replication fork in the highly condensed DNA of a chromosome in the cell still raises several questions. Regulation of the various steps in replication affects cell division and thus is central to preventing uncontrolled cell growth in diseases such as cancer. Control of replication will undoubtedly be an important subject of future studies.

1. How do replication forks respond to DNA-bound proteins? Chromosomal DNA contains many DNA-bound proteins, including repressors, transcription activators, and nucleosomes. We know that the replisome can displace and bypass RNA polymerase in *E. coli*, but only if the direction of replication is the same as the direction of transcription. What happens when the replication and transcription machineries collide head-on? In eukaryotes, do nucleosomes stay bound to DNA during replication, and how is the epigenetic information contained within them sustained in the daughter chromosomes?

2. What protein modifications control replication? The impact of protein modifications on replication control in eukaryotic cells is extremely important and probably involves the phosphorylation of replication proteins, because their phosphorylation state can be seen to change with phases of the cell cycle. The identity of the kinases, which proteins and amino acid residues they modify, and the change in activity these modifications bring about are nearly unexplored territory.

3. What is the relationship between telomerase, aging, and cancer? The loss of telomeres leads to chromosome instability and cell death. Most normal somatic cells lack telomerase and die when their telomeres become too short. Immortal cancer cells express telomerase, and their telomeres are maintained. These observations imply that telomerase, aging, and immortality are related. The clinical ramifications of controlling telomere length in cells is a highly active area of current research.

HOW WE KNOW

DNA Polymerase Reads the Sequence of the DNA Template to Copy the DNA

Bessman, M.J., I.R. Lehman, E.S. Simms, and A. Kornberg. 1958. Enzymatic synthesis of deoxyribonucleic acid: II. General properties of the reaction. *J. Biol. Chem.* 233:171–177.

Lehman, I.R., M.J. Bessman, E.S. Simms, and A. Kornberg. 1958. Enzymatic synthesis of deoxyribonucleic acid: I. Preparation of substrates and partial purification of an enzyme from *Escherichia coli. J. Biol. Chem.* 233:163–170.

Arthur Kornberg, 1918–2007 [*Source: © Bettmann/Corbis.*]

Arthur Kornberg did not intend to discover how DNA was made, or even to become a scientist. He was a physician on a naval ship, but soon after setting out to sea, his single publication as a medical student led to an offer to transfer to the National Institutes of Health. After jumping ship, he began an incredible scientific odyssey that founded the field of replication enzymology.

Kornberg and his group wanted to understand how the DNA polymer was made. They developed an assay for DNA synthesis using bacterial cell extracts to which they added [14C]thymidine to ensure that any radioactive polymer recovered would be DNA and not RNA. Though radioactive incorporation was feeble, it was reproducible. During fractionation of the extract, Kornberg's group discovered that several heat-stable factors (i.e., not proteins) were needed for the DNA synthesis reaction. They identified these as nucleoside triphosphates. Kornberg also found that excess unlabeled DNA had to be added to the cell extracts in order to observe DNA synthesis. These insights allowed the purification of what we now know as DNA polymerase I (Pol I).

On characterizing Pol I, the researchers were initially puzzled that it required all four dNTPs for robust DNA synthesis. If DNA were serving only as a primer, why couldn't a DNA polymer be made from just one, two, or three types of nucleotides? The finding implied that the enzyme received instructions from existing DNA acting as a template, as suggested by Watson and Crick, but at that time, the idea of an enzyme receiving direction from its substrate was preposterous. Kornberg's group conducted experiments to test whether this was in fact the case. They tested DNAs that varied in A=T versus G≡C content, and the result was astounding. Regardless of the mix of dNTPs, the ratio of A=T and G≡C in the product matched that in the template DNA. That settled it! The DNA was serving not only as primer but also as a template. To support this conclusion, they used Pol I to convert the 5.4 kb single-stranded φX174 bacteriophage genome into the duplex viral form. The double-stranded DNA product was infectious! This finding set off a flurry of newscasts: "Life created in a test tube!" More importantly, it marked the beginnings of biotechnology.

Then came a discovery by John Cairns that *polA* mutant *E. coli* cells, with less than 1% residual Pol I activity, had no growth defects. This result, combined with identification of numerous genes required for replication, revealed a process far more complex than anyone had imagined. Unsettling to Kornberg and his colleagues was the questioning of their work on DNA polymerase I in pointed editorials in *Nature New Biology.* Did the assays used to purify Pol I result in a red herring? Do "real" polymerases need a primed template? Are dNTPs the true precursors of DNA? Is a 3'→5' exonuclease proofreader needed by the "real" DNA polymerase?

Kornberg's son Tom identified both Pol II and Pol III from extracts of *polA* mutant cells. These polymerases were just like Pol I in the use of a primed template and dNTPs and the presence of a 3'→5' proofreading exonuclease. Fortunately, the controversial issues raised in *Nature New Biology* soon vanished. Coincidentally, so did the journal itself.

Polymerase Processivity Depends on a Circular Protein That Slides along DNA

Kong X.P., R. Onrust, M. O'Donnell, and J. Kuriyan. 1992. Three-dimensional structure of the β subunit of *E. coli* DNA polymerase III holoenzyme: A sliding DNA clamp. *Cell* 69:425–437.

Stukenberg, P.T., P.S.-V. Studwell, and M. O'Donnell. 1991. Mechanism of the sliding β clamp of DNA polymerase III holoenzyme. *J. Biol. Chem.* 266:11,328–11,334.

DNA polymerases that replicate chromosomes were long known to require "accessory proteins" that somehow confer rapid and processive polymerase activity. However, it seemed a contradiction that proteins that increase the affinity of polymerase for DNA also enable rapid motion along DNA. Specifically, how can a polymerase bind DNA tightly and also rapidly slide along it?

Surprisingly, experiments showed that the β subunit of Pol III holoenzyme, by itself, binds to DNA. This required the γ complex and ATP. However, the β subunit bound only circular DNA and could not bind linear primed DNA. This suggested that the β subunit binds DNA by encircling it, and thus slides off linear DNA. No protein was known to encircle DNA at the time, so this idea was not taken seriously. However, the test was rather simple. A [³H]β subunit was loaded onto circular primed DNA, and the reaction mixture was divided. In one tube, the DNA was linearized using BamHI, and in the other tube, the DNA was untreated and remained circular. The two reaction mixtures were then analyzed on gel filtration columns. The [³H]β bound to the large DNA molecule eluted much earlier (fractions 7 to 16) than [³H]β not bound to DNA (fractions 20 to 40).

If the [³H]β subunit encircles DNA like a doughnut, it should slide off linear DNA but remain on circular DNA. This was exactly the result observed. The solid circles in the upper plot in Figure 1 show the sample treated with BamHI. Most of the [³H]β in this sample eluted late, as [³H]β not associated with DNA. In the untreated sample (open circles), the early fractions show [³H]β bound to DNA. The result is clear: β remains on circular DNA but

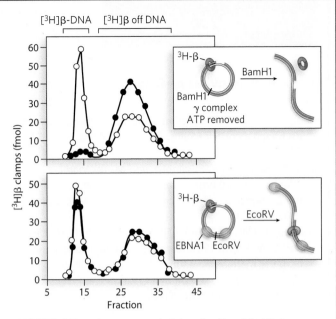

FIGURE 1 Experiments revealed that the *E. coli* Pol III β subunit binds to DNA by encircling it and slides along the duplex. *[Source: Data from P. T. Stukenberg, P. S.-V. Studwell, and M. O'Donnell, J. Biol. Chem. 266:11328–11334, 1991, Fig. 3.]*

falls off linear DNA. This behavior suggests that β is shaped like a ring.

This hypothesis was tested using DNA with two sites for a DNA-binding protein known as EBNA1 (lower plot in Figure 1). The [³H]β was loaded onto DNA (with γ complex and ATP), then EBNA1 was added. One half of the reaction mixture was treated with EcoRV (open circles), which cuts the DNA between the two EBNA1 proteins. The result from gel filtration analysis (also known as gel-exclusion chromatography; see Highlight 4-1) showed that [³H]β was retained on the linear DNA by EBNA1 bound to each end, supporting the idea that β encircles the DNA and slides along it. Another experiment, not shown here, demonstrated that Pol III core binds to β, implying that β acts as a clamp that encircles DNA and tethers the polymerase to the template for high processivity during synthesis. The hypothesis that β is circular was confirmed by its crystal structure (see Figure 11-15).

HOW WE KNOW

Replication Requires an Origin

Hiraga, S. 1976. Novel F prime factors able to replicate in *Escherichia coli* Hfr strains. *Proc. Natl. Acad. Sci. USA* 73:198–202.

Oka, A., H. Sasaki, K. Sugimoto, and M. Takanami. 1984. Sequence organization of replication origin of the *Escherichia coli* K-12 chromosome. *J. Mol. Biol.* 176:443–458.

Zyskind, J.H.W., J.M. Cleary, W.S. Brusilow, N.E. Harding, and D.W. Smith. 1983. Chromosomal replication origin from the marine bacterium *Vibrio harveyi* functions in *Escherichia coli*: oriC consensus sequence. *Proc. Natl. Acad. Sci. USA* 80:1164–1168.

To identify the replication origin of a host cell, plasmid, or virus, a plasmid with a known origin and a selectable marker (e.g., the gene conferring resistance to the antibiotic ampicillin) is first treated with restriction enzymes to excise the origin (Figure 2). DNA extracted from host cells is cut with a restriction enzyme to produce many fragments. Individual fragments are inserted into the cut plasmid, and DNA ligase is used to recreate plasmid DNA circles. These recombinant plasmids are transferred into *E. coli*, and the transformed cells are plated on selective media (e.g., plates containing ampicillin). To survive the antibiotic in the medium and form a colony, cells must contain the plasmid with the ampicillin-resistance gene. In turn, the plasmid must contain a functional origin of replication in order to continue duplicating itself over multiple cell generations. Surviving plasmids are isolated from the bacteria and sequenced to identify the origin required for replication.

The recombinant plasmid approach has identified numerous origins of bacterial chromosomes, plasmids, and bacteriophage. Yeast (a eukaryote) has defined origins that can be isolated in a similar way. However, eukaryotic plasmids cannot be selected using antibiotics. Instead, genes needed for the metabolism of a particular amino acid are used, and cells are plated on media lacking that amino acid. This experimental approach has not been successful in identifying replication origins in eukaryotes more complex than yeast. It is possible that higher eukaryotes do not have defined origins and that chromatin structure defines replication start sites. Alternatively, the origins of higher eukaryotes are too large to be determined by this method.

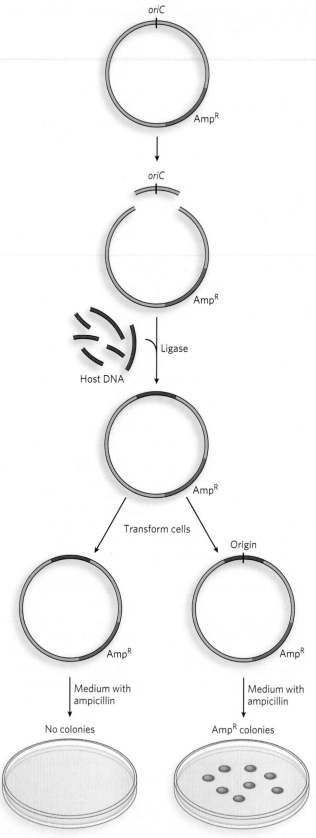

FIGURE 2 The recombinant plasmid method for identifying replication origins. Amp^R indicates a gene for ampicillin resistance.

PROBLEMS

1. Shown below are cartoon drawings of four molecules observed in the electron microscope after cutting a circular plasmid once with a restriction enzyme. Does this plasmid replicate bidirectionally or unidirectionally from the origin? Explain. Order the molecules by time of replication, from the earliest to the latest.

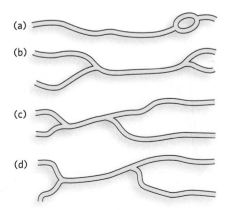

2. A DNA polymerase cannot synthesize a DNA strand de novo from a mixture of deoxynucleotide triphosphates. Explain the role of the primer and template in the reaction promoted by this enzyme.

3. Arthur Kornberg and his colleagues incubated an extract of *E. coli* with a mixture of dATP, dTTP, dGTP, and dCTP. Only the dTTP was labeled with ^{32}P in the α-phosphate group. After incubation, the mixture was treated with trichloroacetic acid to precipitate DNA, but not the dNTPs. The precipitate was collected, and the extent of [^{32}P]dTMP incorporated into DNA was determined.

 (a) If any one of the dNTPs were omitted from the incubation mixture, would radioactivity be found in the precipitate? Explain.
 (b) Would radioactivity be found in the precipitate if ^{32}P labeled the β- or γ-phosphate position? Explain.

4. Provide at least two observations suggesting that Pol I is not the chromosomal replicase.

5. You are characterizing a new DNA polymerase. When the enzyme is incubated with [^{32}P]DNA and no dNTPs, you observe the release of [^{32}P]dNMPs. This release is prevented by adding unlabeled dNTPs. Explain the reactions that most likely underlie these observations. What would you expect to observe if you added pyrophosphate instead of dNTPs?

6. Explain how the enzymes that join together Okazaki fragments ensure that all the RNA is removed from the ends of fragments before they are sealed.

7. A key step in the initiation of replication in bacteria is to load the DnaB helicase onto the DNA at the replication origin. What proteins function at the replication origin during initiation prior to the loading of DnaB?

8. A DNA fragment containing an origin has been identified, using the recombinant plasmid approach (see this chapter's How We Know section). The origin contains six short palindromes: four A sites of similar sequence, and two B sites unrelated to the A sites.

 (a) Assuming that three of the A sites are essential, explain an approach to identify which ones are necessary for origin function.
 (b) Plasmids with a mutation in either of the B sites still replicate. What additional experiment would you do to determine whether B-site function is necessary at all?

9. Five DNA polymerases have been identified in *E. coli*. Two of them—DNA polymerases IV and V—lack a 3′→5′ exonuclease activity. What is the function of the 3′→5′

exonuclease activity present in many DNA polymerases? How does the absence of such an activity in DNA polymerases IV and V make sense in the context of the cellular function of these enzymes?

10. In eukaryotes, DNA replication occurs in the S phase of the cell cycle. However, important preparatory events occur in the G_1 phase, prior to S phase. Describe these events and the status of eukaryotic replication origins as a cell enters S phase.

11. Eukaryotic origins are tightly regulated so as not to fire more than once during S phase. Explain the key points of regulation of this process in eukaryotic cells.

12. Explain the role of the τ subunit of Pol III.

   (a) What is the minimum number of τ subunits that must be present in Pol III to coordinate leading- and lagging-strand DNA synthesis at a replication fork? Explain your answer.

   (b) How many τ subunits must be present to allow Pol III to extend an oligonucleotide primer processively on a single-stranded template in vitro? Explain your answer.

   (c) How many τ subunits are actually present in an active Pol III replisome acting at a replication fork in vivo? Explain the function of these τ subunits.

13. The AAA+ ATPases are the wrenches and crowbars of DNA metabolism. Briefly indicate what is accomplished when ATP is hydrolyzed by each of the following AAA+ ATPases: DnaA, DnaC, and the γ or τ subunits of Pol III.

14. The replication termini (Ter sites) on the *E. coli* chromosome are oriented as shown below. Briefly describe what would happen in one round of replication if the orientations of all the Ter sites in the chromosome were reversed.

15. The enzyme telomerase is considered a reverse transcriptase, synthesizing DNA on an RNA template. What is the source of the RNA used in the synthesis of telomeres, where is it located, and how many nucleotides of the RNA are actually used as template for DNA synthesis?

DATA ANALYSIS PROBLEM

Olivera, B.M., and I.R. Lehman. 1967. Linkage of polynucleotides through phosphodiester bonds by an enzyme from *Escherichia coli*. *Proc. Natl. Acad. Sci. USA* 57:1426–1433.

16. The discovery of DNA ligase provides some classic examples of the development of enzyme assays (see this chapter's Moment of Discovery). In the paper first reporting their detection of a DNA-joining activity, Lehman and colleagues described the use of partially purified enzyme extracts from *E. coli* that had this joining activity. For a DNA substrate, they used polydeoxyadenosine, poly(dA), along with polydeoxythymidine, poly(dT), the latter digested with micrococcal nuclease (which degrades DNA strands at random locations) until the average length of the poly(dT) strands was about 250 nucleotides. These DNA strands were then labeled at their 5′ ends with a ^{32}P-labeled phosphate group. When the labeled DNA was precipitated with HCl and filtered, the label remained with the precipitated DNA on the filter. The labeled DNA was then treated with an enzyme called alkaline phosphatase, which removes terminal phosphates from DNA; with this treatment, the label is freed from the DNA and washes through the filter.

   In their experiments, Olivera and Lehman found that the poly(dT) strands were linked together, because the label became resistant to the alkaline phosphatase treatment. Some of their results are shown below, for a reaction containing 1 µmol (micromole) of labeled DNA ends.

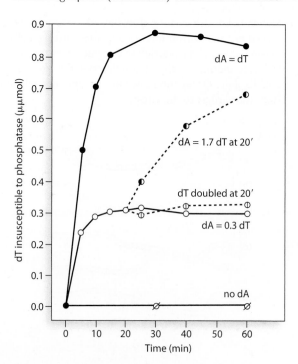

(a) Why does the linking reaction make the label resistant to alkaline phosphatase treatment?

(b) The reaction required the addition of poly(dA) in addition to the labeled poly(dT) strands. Suggest a reason.

(c) In one experiment, the investigators added only enough poly(dA) to allow linking of 30% of the poly(dT). After 20 minutes, they added either more poly(dA) or more unlabeled poly(dT). The reaction recommenced only in response to the poly(dA). What does this reveal about the enzyme activity?

(d) A cofactor in the crude extract was needed for the reaction, but it could not be replaced by added ATP. Suggest a reason.

ADDITIONAL READING

General

Kornberg, A. 1989. *For the Love of Enzymes: The Odyssey of a Biochemist.* Cambridge, MA: Harvard University Press.

Kornberg, A., and T. Baker. 1991. *DNA Replication*, 2nd ed. New York: W.H. Freeman and Company.

Lehman, I.R. 2006. Wanderings of a DNA enzymologist: From DNA polymerase to viral latency. *Annu. Rev. Biochem.* 75:1–17.

DNA Transactions during Replication

Cairns, J. 1963. The chromosome of *Escherichia coli. Cold Spring Harb. Symp. Quant. Biol.* 28:43–46.

Meselson, M., and F.W. Stahl. 1958. The replication of DNA in *Escherichia coli. Proc. Natl. Acad. Sci. USA* 44:671–682. A classic study, very well written, with elegant experimental rationale and results.

Okazaki R., T. Okazaki, K. Sakabe, K. Sugimoto, R. Kainuma, A. Sugino, and N. Iwatsuki. 1968. *In vivo* mechanism of DNA chain growth. *Cold Spring Harb. Symp. Quant. Biol.* 3:129–143.

The Chemistry of DNA Polymerases

Doublie, S., S. Tabor, A.M. Long, C.C. Richardson, and T. Ellenberger. 1998. Crystal structure of a bacteriophage T7 DNA replication complex at 2.2 Å resolution. *Nature* 391:251–258. Description of the first time that a DNA polymerase was crystallized with a primed template and bound dNTP.

Joyce, C.M., and S.J. Benkovic. 2004. DNA polymerase fidelity: Kinetics, structure, and checkpoints. *Biochemistry* 43:14,317–14,324.

Steitz, T.A. 1999. DNA polymerases: Structural diversity and common mechanisms. *J. Biol. Chem.* 274:17,395–17,398.

Overview that compares and contrasts DNA polymerase structures in the different polymerase families.

Mechanics of the DNA Replication Fork

Benkovic, S.J., A.M. Valentine, and F. Salinas. 2001. Replisome-mediated DNA replication. *Annu. Rev. Biochem.* 70:181–208.

McHenry, C.S. 2003. Chromosomal replicases as asymmetric dimers: Studies of subunit arrangement and functional consequences. *Mol. Microbiol.* 49:1157–1165.

Mok, M., and K.J. Marians. 1987. Formation of rolling-circle molecules during phi X174 complementary strand DNA replication. *J. Biol. Chem.* 262:2304–2309.

Stillman, B. 2008. DNA polymerases at the replication fork in eukaryotes. *Mol. Cell* 30:259–260.

Yao, N.Y., and M. O'Donnell. 2008. Replisome dynamics and use of DNA trombone loops to bypass replication blocks. *Mol. Biosyst.* 4:1075–1084.

Initiation of DNA Replication

Boos, D., J. Frigola, and J.F. Diffley. 2012. Activation of the replicative DNA helicase: Breaking up is hard to do. *Curr. Opin. Cell. Biol.* 24:423–430.

Li, H., and B. Stillman. 2012. The origin recognition complex: A biochemical and structural view. *Subcell. Biochem.* 62:37–58

O'Donnell, M., L. Langston, and B. Stillman. 2013. Principles and concepts of DNA replication in bacteria, archaea, and eukarya. *Cold Spring Harb. Perspect. Biol.*, doi:10.1101/cshperspect.a010108. http://cshperspectives.cshlp.org/content/5/7/a010108.full.pdf+html.

Tsakraklides, V., and S.P. Bell. 2010. Dynamics of pre-replicative complex assembly. *J. Biol. Chem.* 285:9437–9443.

Termination of DNA Replication

Armanios, M., and E.H. Blackburn. 2012. The telomere syndromes. *Nat. Rev. Genet.* 13:693–704.

Greider, C. 2009. Telomerase Discovery: The Excitement of Putting Together Pieces of the Puzzle. Nobel lecture. The Nobel Prize in Physiology or Medicine 2009: Elizabeth H. Blackburn, Carol W. Greider, Jack W. Szostak. Video. www.nobelprize.org/nobel_prizes/medicine/laureates/2009/greider-lecture.html.

Nandakumar, J., and T.R. Cech. 2013. Finding the end: Recruitment of telomerase to the telomere. *Nat. Rev.* 14:69–82.

Palm, W., and T. de Lange. 2008. How shelterin protects mammalian telomeres. *Annu. Rev. Genet.* 42:301–334.

Skloot, R. 2010. *The Immortal Life of Henrietta Lacks.* New York: Crown Publishers.

12 DNA Mutation and Repair

Rose Byrne [*Source: Courtesy of Rose Byrne.*]

MOMENT OF DISCOVERY

One of the defining moments in my career as a scientist happened at no less magical a place than Cold Spring Harbor. I had been invited to give a talk at the Phage meeting. My results showed that *E. coli, normally a radiation-sensitive organism, could become a radiation-resistant extremophile by directed evolution.* Only three simple mutations were needed. I will always love telling the scientific story that framed my graduate career. However, to tell it at a special place like Cold Spring Harbor, and at a meeting where so much of microbial physiology had been unveiled, I was both nervous and exhilarated. These feelings were compounded when I realized that Richard Lenski, one of the major players in microbial evolution and my science idol, was going to be in the audience. I had been following his work for years and had taught myself almost everything I knew about microbial evolution from his publications. As I finished my talk and the lights came up, I watched Dr. Lenski's hand shoot into the air and gesture for the microphone. My heart stopped. His first words were: "Fascinating stuff . . ." The next day, when he was chairing the session on microbial evolution, he referenced my work as incredible—twice. When the work came out in *eLife* a few months later, the story went viral. One columnist even suggested that we had found a way to genetically modify astronauts so they are less susceptible to radiation while exploring the galaxy. While that may be a bit of a stretch, the entire project was exciting. But that moment at Cold Spring Harbor planted a seed, and it is still growing as I explore a new field as a postdoctoral associate.

—*Rose Byrne, on her discovery that* E. coli *could become a radiation-resistant extremophile*

Online resources related to this chapter:

 Simulation Mutation and repair

413

ADNA genome encodes the instructions for the production of every molecule in the cell and, as such, is also essential for reproduction of the cell for future generations. Because each cell contains only one or two copies of its DNA, the DNA sequence is highly protected from harm. DNA is a relatively stable molecule, but damage to it is inevitable: our dependence on oxygen and water makes us vulnerable to a continuous barrage of oxidative and hydrolytic reactions, many of which strike intracellular molecules, including DNA. DNA is also subject to attack by many other reactive chemicals in the environment, both natural and synthetic, as well as by various types of irradiation from the sun and the radioactive decay of terrestrial elements. DNA can also be altered by mistakes made during its own replication or recombination. Damage and sequence alterations to DNA are often quickly repaired, but when they are not, the DNA becomes permanently altered and harbors a mutation.

Mutations are changes in DNA sequence, and when mutations occur in germ-line cells, or in single-celled organisms, these changes are inheritable. Indeed, some frequency of mutation is necessary to produce the variability on which natural selection acts to drive evolution. However, a mutation that confers an advantage on a cell or organism is rare. In multicellular organisms, a cell that accumulates many mutations usually dies. A cancer cell has mutations that prevent cell death, resulting in loss of cell cycle control and unregulated cell division, which leads to malignant tumors that can end the life of the entire organism.

Mutations occur through many different mechanisms, but all originate as an alteration in DNA. Only after the alteration is converted through replication into an incorrect base pair (such as an A=T pair where a G≡C pair should be) does it become a stable, inheritable mutation. Therefore the cell has a limited amount of time to fix the initial alteration and restore the DNA to its normal sequence, before replication converts the alteration into a mutation that will be passed on to the next generation. In all organisms, an army of repair enzymes has evolved that holds a constant vigil over the DNA. Indeed, the vast majority of damaged nucleotides that occur in a mammalian cell every day are repaired; fewer than one in a million become a mutation. DNA repair often takes advantage of the double-stranded DNA structure to restore a damaged nucleotide on one strand to the original residue, using the complementary strand as a template.

The enormous selective pressure favoring enzyme systems that repair DNA damage has led to some of the most fascinating enzyme reactions in biology. DNA repair reactions are also among the most costly in terms of energy—testimony to the importance of their job in ensuring survival of the species. In some cases, an entire enzyme is used only once per repair event. In other words, the information contained in a cell's DNA sequence is to be preserved at all costs; the energy expended in the process is irrelevant.

We begin this chapter by defining mutations and describing how replication of a damaged nucleotide base can become an inheritable mutation. We explore different types of DNA damage that arise from agents inside the cell, as well as from those in the external environment, and then consider many of the DNA repair processes that restore the original sequence. As we will see, certain types of DNA damage can be repaired by more than one enzyme system, whereas others are repaired by specific processes dedicated to a particular type of DNA damage. Remarkably, some repair mechanisms can recover the original, correct DNA sequence even when it cannot be obtained from a complementary DNA strand. Other types of DNA damage, however, prevent recovery of the original sequence, and repair gives rise to mutations. Although it seems counterintuitive that a repair process would produce a mutation, the alternative could spell death to a cell. For example, some DNA lesions halt the replication machinery. This is the worst thing that can happen to a cell and must be avoided at all costs; if replication cannot be completed, daughter cells will not receive a full complement of DNA, and they will die. We explain how cells resolve these conflicts, enabling them to complete replication even at the expense of incurring a mutation.

Most DNA repair processes evolved before the emergence of different cell types from the last common ancestor, and thus the major DNA repair pathways are similar in bacterial, archaeal, and eukaryotic cells. Because bacterial repair systems have been studied most intensively, they provide us with the highest level of mechanistic detail. However, there are several aspects of eukaryotic repair that differ from bacterial repair. We discuss these differences, as well as the human diseases that arise from defective DNA repair or from DNA alterations that do not lend themselves to repair.

12.1 TYPES OF DNA MUTATIONS

A **mutation** is a change in a DNA sequence that is propagated through cellular generations. Mutations can be as small as a single base pair or can range from a few base pairs to thousands. Mutations of one or a few base pairs usually result from errors in replication or damaged nucleotides. Those that span large sections of DNA are typically due to chromosomal rearrangements that arise from errant recombination.

Mutations can have different effects on gene structure and function. A mutation in a gene product can result in a loss of function or a gain of function. For example, loss of function can be the result of mutations that destroy the active site of an enzyme, produce a truncated protein, or disrupt the regulation of gene expression. Gain-of-function mutations might increase the affinity of an enzyme for its substrate, remove the regulatory portion of a protein, or increase gene expression and thus produce more protein. Loss-of-function mutations are generally recessive in a diploid organism, whereas gain-of-function mutations are often dominant over the wild-type allele. Examples of every type of mutation can be observed in human disease. But mutations are also important to life as we know it. Mutations that duplicate entire genes can eventually lead to entirely new proteins with different functions. Mutations have given us new, higher yields of grain. And mutations also lead to new species, providing the diversity that drives evolution. Without mutations you would not be here.

A Point Mutation Can Alter One Amino Acid

A change in a single base pair is often referred to as a **point mutation**. Point mutations fall into two categories depending on the types of base substitution in the DNA sequence (**Figure 12-1**). A **transition mutation** is the exchange of a purine-pyrimidine base pair for the other purine-pyrimidine base pair: C≡G becomes T=A, or T=A becomes C≡G. A **transversion mutation** is the replacement of a purine-pyrimidine base pair with a pyrimidine-purine base pair, or vice versa. For example, C≡G becomes either G≡C or A=T. Transition mutations exchange bases that have a similar size, which is easier for a polymerase to accommodate, and thus transitions are nearly 10 times more frequent than transversions. A point mutation in the protein-coding region of a gene can result in an altered protein with partial or complete loss of function. If the protein is central to cell viability, the cell could die.

Point mutations in a protein-coding region can be classified by their effect on the protein sequence. The DNA sequence encoding a protein is read in triplets, or codons. Each codon corresponds to an amino acid (see Chapter 17). A **silent mutation** is a nucleotide change that produces a codon for the same amino acid. For example, GAA and GAG both code for glutamate. A **missense mutation** is a nucleotide change that results in a different amino acid, such as a change from glutamate (GAA) to glutamine (CAA). A **nonsense mutation** changes the nucleotide sequence so that instead of encoding an amino acid, the triplet functions as a stop codon, terminating the translation process and generating a truncated protein without a complete amino acid sequence.

KEY CONVENTION

When a point mutation results in an altered protein sequence, the change in that protein's amino acid sequence is denoted by letters and a number—for example, E214A. The first letter is the single-letter abbreviation for the amino acid residue in the wild-type protein (E); the number is the position of the residue, numbering from the N-terminus of the protein sequence (214); and the second letter is the amino acid residue in the mutant protein (A). A nonsense mutation is identified by an X as the second letter—for example, E214X.

Point mutations are produced in a two-step process. In the first step, an incorrect nucleotide is incorporated by a DNA polymerase (**Figure 12-2**). This can happen when DNA polymerase encounters a damaged base in the template strand that no longer forms a normal base pair. It can also occur without DNA damage, because DNA polymerases, despite their high accuracy, sometimes make a mistake. For example, as discussed in Chapter 11, a DNA polymerase can incorporate a tautomeric form of a dNTP, resulting in a mismatch. Regardless of how the mismatch is formed, it is not yet a mutation, because it can be detected and repaired by processes described later in this chapter. However, if the mismatch is not repaired, it becomes a mutation in a second step during replication, which incorporates the mismatch into a fully base-paired duplex DNA. This results in a new, correctly paired base pair that can no longer be detected by repair enzymes but alters the original sequence and is therefore an inheritable mutation.

Evolution depends on mutations that can confer a selective advantage, but it is relatively rare for a mutation to have a positive outcome for the organism. Studies in *Drosophila melanogaster* suggest that mutations that alter the protein sequence are most likely to be harmful; about 70% have a negative effect, and the rest either are neutral or have a weak beneficial effect. As shown by studies in yeast, however, for mutations outside the protein-coding region, fewer than 7% are harmful.

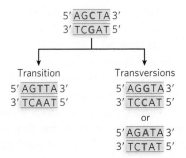

FIGURE 12-1 Transition and transversion point mutations. The parental DNA (top) contains a C≡G base pair. There are two possible point mutations: a transition (left), in which a purine (in this case, G) is replaced with a different purine (A), producing a T=A base pair on replication; or a transversion (right), in which a pyrimidine (in this case, C) is replaced with a purine (G or A) to produce either a G≡C or an A=T base pair. (To review hydrogen bonding between base pairs, see Figure 1-3.)

(a) Replication error

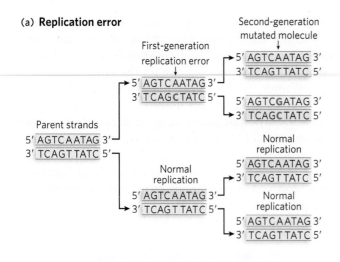

(b) One possible effect of mutagen

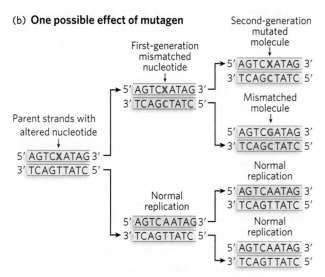

FIGURE 12-2 The two-step process of a point mutation. (a) DNA polymerase incorporates an incorrect nucleotide (C instead of T), leading to a mismatch in one of the first-generation cells. If the mismatch is not corrected before this cell replicates, one of the second-generation cells will incur a mutation. (b) The parental DNA contains a damaged adenosine base (X). The damaged A more readily pairs with a C than a T, so replication over the damaged base results in a mismatched nucleotide in one of the first-generation cells. Further replication can result in a permanent mutation in a cell of the second generation.

Point mutations are known to cause a wide variety of human diseases. One example is sickle-cell anemia (discussed in Highlight 2-1), resulting from a transversion that produces an amino acid change in hemoglobin: a GAG, encoding glutamate (E), at residue 6 changes to GTG, encoding valine (V) (E6V). Another example of a point mutation in human disease is Werner syndrome, which causes premature aging due to genetic instability. The *WRN* gene encodes a helicase. Several different point mutations in the *WRN* gene lead to Werner syndrome. Some are of the missense type (e.g., K32R),

although most are nonsense mutations that lead to a shorter protein product (e.g., E41X).

The most harmful mutations are those occurring in the genes involved in DNA repair, because these often result in cancer. As we discuss later in this chapter, the repair of mismatch errors requires many different proteins. A mutation in a gene encoding a mismatch repair protein can result in the production of many more mutations in the cell, as subsequent mutations are no longer corrected. The occurrence of many mutations in a cell can result in cancer, because, eventually, a mutation will occur in a gene (or genes) that encodes a protein needed to control cell division. In normal cells, **oncogenes** encode proteins that drive the cell division cycle forward, and **tumor suppressor genes** encode proteins that suppress cell division. Many tumor suppressors are transcription factors that regulate the expression of genes that drive the cell cycle. The transcription factor p53 and the retinoblastoma protein are examples of tumor suppressors that are mutated in many types of cancer.

Small Insertion and Deletion Mutations Change Protein Length

Another type of mutation is the gain or loss of one or more base pairs. **Insertion mutations** occur when one or more base pairs are added to the wild-type sequence; conversely, **deletion mutations** are due to the loss of one or more base pairs. Insertion and deletion mutations are collectively referred to as **indels**. Indels are caused by aberrant recombination or by template slippage by the DNA polymerase during replication. As we discuss in more detail in Chapter 17, proteins are encoded (via a messenger RNA intermediary) in a series of codons that starts with ATG and ends with one of three different codons that signal translation to stop (**Figure 12-3a**). The protein is synthesized from its N-terminus to its C-terminus. The DNA sequence from the start codon to the stop codon is referred to as a **reading frame**. Because nucleotides are decoded in triplets, an indel mutation of only one or two base pairs in the coding sequence of a protein throws off the reading frame after the mutation, resulting in a **frameshift mutation** (**Figure 12-3b**). Given the frequency of stop codons in a random sequence (1 in every 20 codons), these mutations usually produce a truncated protein, especially if the gene is large and the mutation occurs a few hundred base pairs prior to the wild-type stop codon. Frameshift mutations often destroy the protein's function, especially if the truncation is close to the N-terminus.

Indel mutations that occur in multiples of three base pairs preserve the reading frame of the gene. The most common such mutation is an insertion of three nucleotides. This is thought to be caused by template slippage

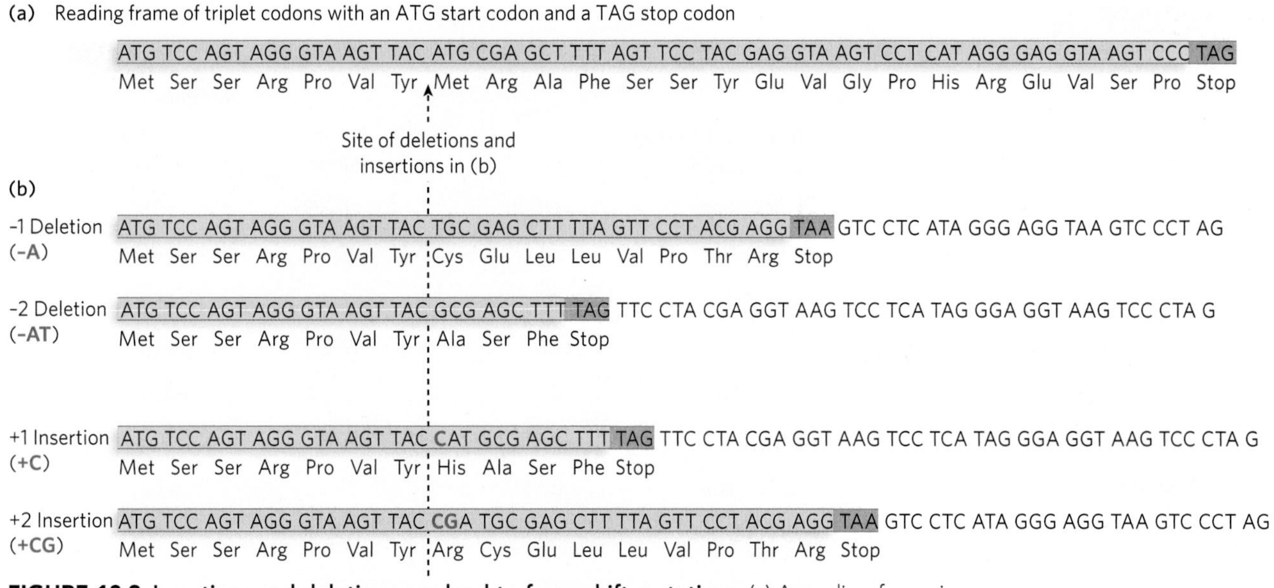

(a) Reading frame of triplet codons with an ATG start codon and a TAG stop codon
ATG TCC AGT AGG GTA AGT TAC ATG CGA GCT TTT AGT TCC TAC GAG GTA AGT CCT CAT AGG GAG GTA AGT CCC TAG
Met Ser Ser Arg Pro Val Tyr Met Arg Ala Phe Ser Ser Tyr Glu Val Gly Pro His Arg Glu Val Ser Pro Stop
Site of deletions and insertions in (b)
(b)
-1 Deletion (-A)
ATG TCC AGT AGG GTA AGT TAC TGC GAG CTT TTA GTT CCT ACG AGG TAA GTC CTC ATA GGG AGG TAA GTC CCT AG
Met Ser Ser Arg Pro Val Tyr Cys Glu Leu Leu Val Pro Thr Arg Stop
-2 Deletion (-AT)
ATG TCC AGT AGG GTA AGT TAC GCG AGC TTT TAG TTC CTA CGA GGT AAG TCC TCA TAG GGA GGT AAG TCC CTA G
Met Ser Ser Arg Pro Val Tyr Ala Ser Phe Stop
+1 Insertion (+C)
ATG TCC AGT AGG GTA AGT TAC CAT GCG AGC TTT TAG TTC CTA CGA GGT AAG TCC TCA TAG GGA GGT AAG TCC CTA G
Met Ser Ser Arg Pro Val Tyr His Ala Ser Phe Stop
+2 Insertion (+CG)
ATG TCC AGT AGG GTA AGT TAC CGA TGC GAG CTT TTA GTT CCT ACG AGG TAA GTC CTC ATA GGG AGG TAA GTC CCT AG
Met Ser Ser Arg Pro Val Tyr Arg Cys Glu Leu Leu Val Pro Thr Arg Stop

FIGURE 12-3 Insertions and deletions can lead to frameshift mutations. (a) A reading frame in the DNA begins with the start codon, ATG, and stops with a stop codon (TAG in this case). When the coding strand of the DNA is transcribed into mRNA, these sequences are AUG (start) and UAG (stop). (b) Insertions or deletions of one or two base pairs in a protein-coding sequence result in frameshift mutations, in which the reading frame becomes out of register relative to the wild-type sequence and typically leads to a premature stop codon.

during replication, with the inserted triplet embedded in a region of triplet repeats (**Figure 12-4**). Thus, the wild-type gene has a repeating array of codons, and the protein product contains a region of repeats of the same amino acid. Although the mutant protein, with an added amino acid residue, is slightly larger than the wild-type protein, it usually retains some degree of function.

There are many examples of human disease caused by the insertion of triplet sequences, often referred to as **triplet expansion diseases** (**Table 12-1**). More than half of the triplet expansion diseases involve expansion of the CAG codon for glutamine (Q) and are known as **polyglutamine (polyQ) diseases**. An example is Huntington disease. When the number of CAG repeats increases to 36 copies or more, degeneration of the cerebral cortex may occur in midlife (see Figure 2-28). The number of triplet repeats correlates with the severity and time of onset of this neurological disorder.

The first triplet expansion disease to be identified was fragile X syndrome, which causes a form of mental impairment. The defect in fragile X syndrome maps to the X chromosome and involves a CGG repeat. Humans normally have 5 to 54 repeats, but some individuals have more than 230 repeats (up to 2,000), and this leads to the fragile X mental impairment. Individuals with 54 to 230 repeats show no symptoms but are carriers of the disease, because the genetic instability leading to the triplet expansion occurs in germ-line cells and therefore becomes worse

Repeated sequence of CAG (polyglutamine tract)
5′ ACAGTTCAGCAGCAGCAGTTACCT 3′
3′ TGTCAAGTCGTCGTCGTCAATGGA 5′
Normal replication
3′ TGTCAAGTCGTCGTCGTCAATGGA 5′
5′ ACAGTTCAGCAGCAGCAGTTACCT 3′
Normal
Thr Val Gln Gln Gln Gln Leu Pro
Template slippage introduces insertion of extra CAG
A
C G
5′ ACAGTTCAGCAGCAGCAGTTACCT 3′
3′ TGTCAAGTCGTCGTCGTCAATGGA 5′
Abnormal protein with extra Gln residue
Thr Val Gln Gln Gln Gln Gln Leu Pro

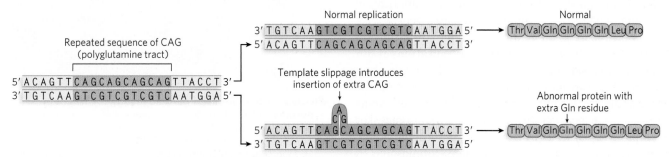

FIGURE 12-4 A three-nucleotide insertion mutation. Template slippage on repeated sequences during replication can lead to small insertions and deletions. Shown here is a three-nucleotide (CAG) insertion in a repeated CAG region of the coding strand of a gene sequence, resulting in an additional Gln residue in a polyglutamine tract. The protein sequence shown is the sequence encoded by the coding strand of DNA, the strand that is transcribed into mRNA.

TABLE 12-1
Triplet Expansion Diseases

Disease	Repeated Codon (amino acid)	Gene	Codon Repeats	
			Normal Range	Threshold for Disease
Huntington disease	CAG (Gln)	*HTT*	11–35	>35
Spinocerebellar ataxia type 1	CAG (Gln)	*ATXN1*	6–35	>49
Machado-Joseph disease	CAG (Gln)	*ATXN3*	12–40	>55
Kennedy disease	CAG (Gln)	*AR*	9–36	>38
Fragile X syndrome	CGG (Arg)	*FMR1*	5–54	>230
Fragile X-E syndrome	CCG (Pro)	*AFF2*	6–35	>200
Myotonic dystrophy	CTG (Leu)	*DMPK*	5–37	>50
Friedreich's ataxia	GAA (Glu)	*FXN*	7–34	>100

with each generation—until the threshold of 230 is passed and the disease manifests. Diagnostic genetic screening for triplet expansion diseases can be performed with PCR primers that anneal to the unique sequences known to lie on either side of the repeated region (**Figure 12-5**).

Some Mutations Are Very Large and Form Abnormal Chromosomes

Some types of mutations involve extensive changes in the DNA sequence, most commonly caused by aberrant recombination events (**Figure 12-6a**). These types of mutations generally cannot be repaired, but are mentioned here because they form an important class of mutations that cause disease, and they also drive the process of evolution. One type of mutation that changes an extensive amount of DNA is the deletion of a large tract of DNA that can lead to a complete loss of genes and also bring genes into proximity that were once far apart. The opposite of a deletion is a **duplication mutation**, the amplification of a large tract of DNA, leading to increased gene dosage effects (i.e., increased amounts of product from the amplified gene). Chromosome **inversion mutations** result from the inversion of a large section of DNA in a chromosome and can have varied effects, especially on the genes at the break points. Aberrant recombination events can also occur between two different chromosomes. For example, a region of DNA from one chromosome can be transferred as an insertion to another chromosome. A chromosome **translocation mutation** occurs when two nonhomologous chromosomes exchange large regions of DNA. In particularly rare instances, entire chromosomes may be fused to each other. Such large rearrangements

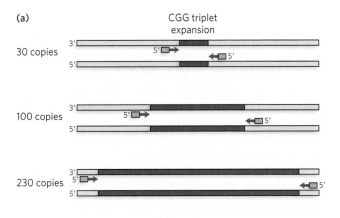

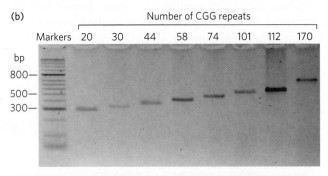

FIGURE 12-5 PCR-based genetic testing for fragile X syndrome, a triplet expansion disease. (a) The number of repeats in the fragile X locus can be determined by length analysis of PCR products. This uses PCR primers that hybridize to unique sequences bordering the region of interest, then the PCR products are analyzed by gel electrophoresis. The size of a PCR product correlates with the size of the repeated region in the gene. (b) Size of the repeated unit in the gene for eight different patients. *[Source: (b) Data from M. S. Khaniani et al., Mol. Cytogenet. 1(1):5, 2008. Courtesy of Paul Kalitsis.]*

(a)

Single chromosome mutations

Deletion   Duplication   Inversion

(b)

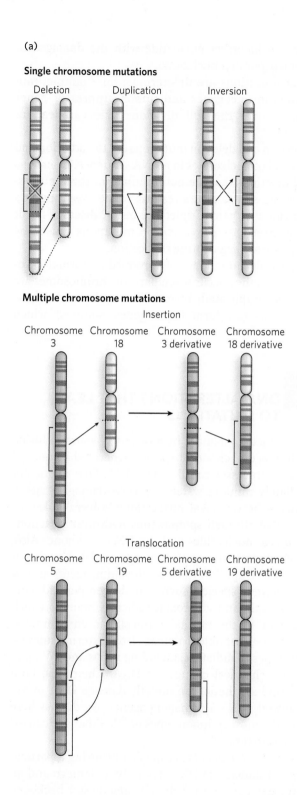

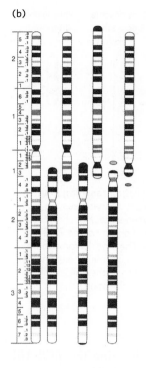

Multiple chromosome mutations

Insertion

| Chromosome 3 | Chromosome 18 | Chromosome 3 derivative | Chromosome 18 derivative |

Translocation

| Chromosome 5 | Chromosome 19 | Chromosome 5 derivative | Chromosome 19 derivative |

FIGURE 12-6 Large-scale mutations. (a) Mutations that lead to alterations in a chromosome can occur internally through deletion, duplication, or inversion events (top), or they can be due to an insertion (middle) or translocation (bottom), exchanging DNA with another chromosome. (b) Simians such as chimpanzees have a genome consisting of 25 different chromosomes, rather than the 24 different chromosomes (22 autosomes plus the sex chromosomes X and Y) found in humans. The distinction is not as great as it sounds. Two simian chromosomes were fused during the evolution of hominids, producing human chromosome 2—shown next to the related chromosomes from (left to right) chimpanzee, gorilla, and orangutan. [*Source: (b) From Yunis, J. J. and Prakash, O., The Origins of Man: A Chromosomal Pictorial Legacy. Science, 215: 1525–1530, 1982, Fig. 2-2. Reprinted with permission from AAAS.*]

are almost always deleterious and almost never inherited. However, when the product is viable and is inherited, it can have major evolutionary consequences. A fusion of two simian chromosomes was an important event in the evolution of hominids (**Figure 12-6b**).

Chromosomal abnormalities can result in the formation of a **fusion gene**, a hybrid of two different genes. Several types of fusion genes are associated with various forms of cancer, including lymphoma, sarcoma, and prostate cancer. An example of a chromosomal translocation that forms a carcinogenic fusion gene is the formation of *BCR-ABL*, in which one end of chromosome 9 breaks and rejoins (i.e., recombines) with part of chromosome 22, forming a very small derivative chromosome 22 (**Figure 12-7**). The break point in chromosome 9 occurs within the *ABL* gene, which becomes fused with the

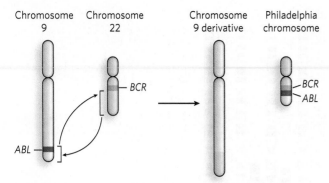

FIGURE 12-7 A chromosomal translocation resulting in a fusion gene. The chromosomal translocation that fuses the *ABL* gene with the *BCR* gene occurs when a piece of chromosome 9 (with *ABL*) is exchanged with a piece of chromosome 22. The *BCR-ABL* fusion gene causes leukemia. The small chromosome resulting from the translocation (the chromosome 22 derivative), which carries the fusion gene, is referred to as the Philadelphia chromosome, after the city where the translocation was first identified and studied.

BCR gene in chromosome 22, forming the *BCR-ABL* fusion gene. The ABL protein is a cell cycle protein kinase, a tyrosine kinase that helps control the cell cycle by phosphorylating certain proteins (on specific Tyr residues) in response to cellular signals. When *ABL* becomes fused with *BCR*, the sections of the ABL protein that regulate the kinase activity are lost; the resulting unregulated tyrosine kinase activity leads to the uncontrolled cell division associated with lymphoblastic leukemia.

Sometimes a fusion gene retains production of the normal protein but comes under the control of a strong, unregulated promoter that produces too much of the protein. An example involves the *C-MYC* gene, an oncogene, on chromosome 8. A common translocation involving this gene, found in certain types of cancer, occurs between chromosomes 8 and 14.

SECTION 12.1 SUMMARY

- A mutation is a change in DNA sequence. It can be a point mutation, affecting a single base pair, an insertion or deletion affecting more than one base pair, or a chromosomal rearrangement that affects many genes on the chromosome.
- Point mutations are classified as transitions and transversions. A transition converts a purine-pyrimidine base pair to the other purine-pyrimidine pair, or pyrimidine-purine to pyrimidine-purine. A transversion converts a purine-pyrimidine base pair to a pyrimidine-purine pair, or vice versa.
- Most mutations are produced in a two-step process. First, a nucleotide is either damaged or misincorporated during replication, then subsequent replication pairs an incorrect nucleotide with the damaged or misincorporated nucleotide.
- Most mutations are deleterious; for example, mutations in oncogenes and tumor suppressor genes that control the cell division cycle can lead to cancer.
- Insertion and deletion mutations are the addition and removal of nucleotides in a DNA sequence. Insertions and deletions that are not multiples of three nucleotides can shift a gene's reading frame, resulting in a truncated protein. Triplet expansion diseases result from three-base-pair insertion mutations caused by template slippage during replication.
- Large-scale mutations can produce abnormal chromosomes; they occur when parts of chromosomes are deleted, duplicated, inverted, or exchanged. These mutations can form fused genes, some of which cause cancer.

12.2 DNA ALTERATIONS THAT LEAD TO MUTATIONS

DNA is subject to damage by a variety of sources. Inside the cell, reactive chemical species generated in normal metabolic processes can damage DNA. One might not immediately think of water as a DNA-damaging agent, yet water—the universal intracellular solvent—takes its toll on DNA through spontaneous hydrolysis reactions that damage nucleotides and the DNA backbone. Also, the high degree of negative charge on DNA makes it prone to electrophilic attack by alkylating agents and by reactive oxygen species (such as hydrogen peroxide, hydroxyl radicals, and superoxide radicals), many of which are present in the normal intracellular environment. DNA is also susceptible to damage from external sources. Various types of radiation, including x rays and UV light, can cause chemical changes in DNA. Chemicals, both natural and synthetic, can directly damage DNA or are metabolized to DNA-damaging agents. We discuss here some of the most common types of DNA damage caused by these sources.

Each of these agents is capable of causing a particular type of damage to DNA. Hydrolytic damage and alkylating agents can affect the phosphodiester backbone or a nucleotide base. Irradiation can cause slightly larger lesions, such as cross-links between bases or breaks in the DNA strand. And the DNA sequence can even be altered by the very enzymes responsible for its preservation. In fact, some of the errors occurring during replication and recombination can be among the most difficult for the cell to detect and sometimes escape cellular mechanisms of repair.

Spontaneous DNA Damage by Water Can Cause Point Mutations

Hydrolysis is the cleavage of a molecule by the addition of water. The reactive hydroxide ion (HO⁻) is formed when a proton (H⁺) dissociates from water (see Chapter 3). Hydrolysis reactions are often initiated by nucleophilic attack by a hydroxide ion; the leaving group acquires the proton, giving a net addition of water to the substrate (**Figure 12-8a**). At physiological pH, the concentration of hydroxide ion is low, so hydrolysis is slow—otherwise, life as we know it could not exist. However, given the large number of nucleotides in a genome, damage to nucleotide bases by even a slow rate of hydrolysis quickly adds up and becomes significant to the cell.

One result of hydrolysis is the deamination of nucleotide bases (see Figure 12-8a). **Deamination** is the removal of an amino group from a compound, and all molecules that contain an amino group are possible targets of hydrolytic attack. Three of the nucleotide bases contain an amino group—cytosine, adenine, and guanine—and thus can be deaminated by hydrolysis (**Figure 12-8b**). Deamination of cytosine is the most common, and it results in uracil (see Chapter 6). Adenine and guanine also undergo deamination, but at only one-hundredth the rate of cytosine deamination. Deamination of adenine and guanine, unlike that of cytosine, produces bases not normally found in nucleotides: adenine yields hypoxanthine, which pairs with cytosine if left uncorrected; guanine forms xanthine, which is less deleterious because it still pairs with cytosine, through two hydrogen bonds instead of three.

Left unrepaired, a C → U change would be highly mutagenic, because uracil pairs with adenine more readily than with guanine. Hence, if the change is not repaired before replication, the cell will replicate the U to form a U=A base pair without pausing, replacing the original C≡G base pair. Repair of the uracil after replication would then insert thymine in place of uracil, completing the C≡G to T=A transition mutation. However, because DNA contains thymine rather than uracil, uracil in DNA is readily recognized as foreign. A biological need to avoid mutations during long-term maintenance of genomic DNA may have driven the evolution of the use of T in place of U in DNA so that the spontaneous appearance of U might be readily repaired.

In higher eukaryotic cells, in about 5% of C residues in DNA the cytosine is methylated to 5-methylcytosine (forming 5-methylcytidine, 5-meC), a modification that is linked to gene expression. Methylation is most common on C residues that are followed by a G, in **CpG sequences**. Methylation at CpG sequences produces 5-meCpG symmetrically on both strands of the DNA. Deamination of 5-methylcytosine produces thymine

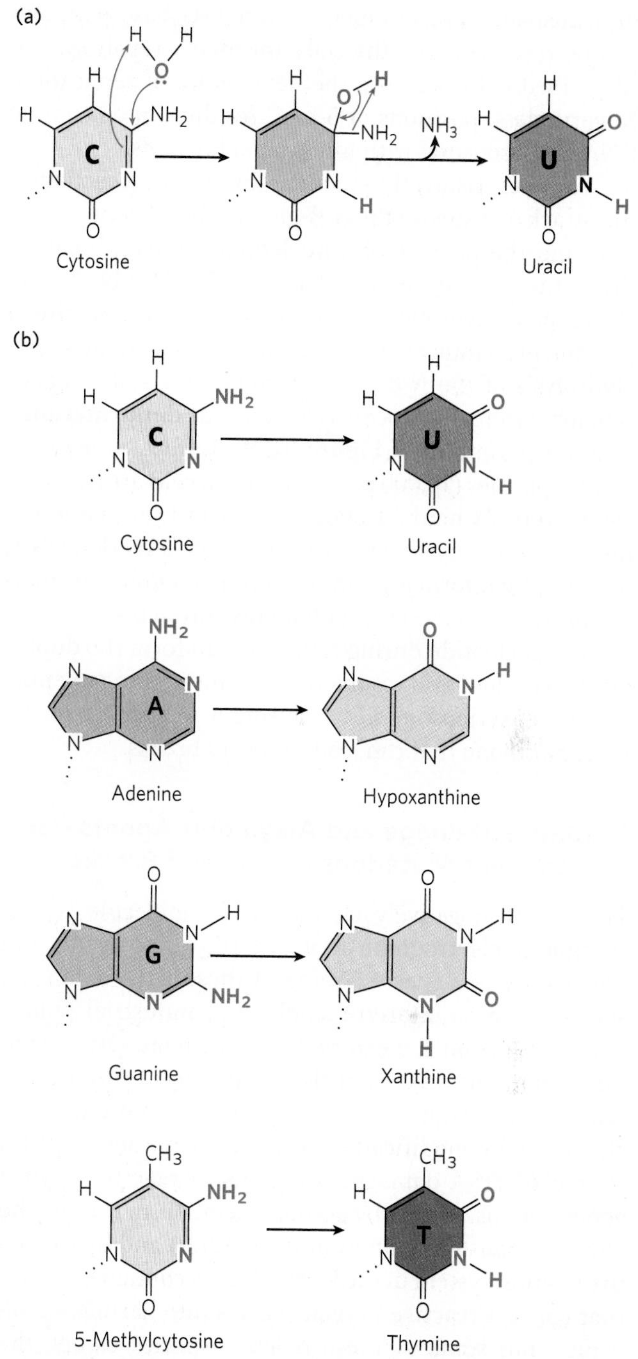

FIGURE 12-8 Deamination of nucleotide bases by spontaneous hydrolysis. In these deamination reactions, only the base is shown for each nucleotide residue. (a) A hydrolysis reaction in which water is added to cytosine, resulting in deamination to uracil and ammonia. A similar mechanism occurs in other deaminations. (b) Common deamination reactions resulting from hydrolysis of nucleotides in DNA.

instead of uracil (see Figure 12-8b). Because thymine is a naturally occurring base in DNA, the modification is not recognized as damage and has the potential to become a transition mutation in which a G≡C base pair is changed to an A=T base pair. However, the damage does result in a mispair, from a 5-meC–G to a T–G base pair, and can be recognized by the cell's mismatch repair system (described below) before the next round of replication. Nevertheless, positions of 5-meC residues in eukaryotic DNA are associated with mutational hotspots.

Another relatively frequent hydrolytic reaction is the attack of water on the N-β-glycosyl bond between the base and the pentose of a nucleotide residue, breaking the connection between the base and the DNA backbone. This type of hydrolytic reaction leaves an **abasic site**, a position in an intact DNA backbone that is without a base. Hydrolysis of the N-β-glycosyl bond occurs at a higher rate for purines—a process referred to as **depurination**—than for pyrimidines (**Figure 12-9**). In fact, as many as 1 in 10^5 purines (10,000 per mammalian cell) are lost from DNA every 24 hours. Depurination can lead to a mutation when the abasic site forms in single-stranded DNA, because the information in the complementary strand is not present and the DNA polymerase often inserts an incorrect nucleotide during replication to form the duplex DNA. Base loss also results in spontaneous ring opening in the deoxyribose moiety, causing a weakening of the DNA backbone that can lead to strand breaks.

Oxidative Damage and Alkylating Agents Can Create Point Mutations and Strand Breaks

The highly negatively charged DNA molecule is susceptible to electrophilic attack by alkylating agents and reactive oxygen species. Some of these DNA-damaging agents, such as cigarette smoke and industrial pollutants, come from the external environment. Often, synthetic and natural chemicals that damage DNA are not reactive per se but are transformed into DNA-damaging chemicals by modification reactions. Another plentiful source of DNA-damaging reagents is reactive oxygen generated inside cells by aerobic metabolism (i.e., by the electron transfer chain in mitochondria) and by the detoxification system in the liver. All cells contain enzymes that convert reactive oxygen species into harmless molecules, but some of these reactive species escape the cellular cleansing systems and can damage DNA and other biomolecules.

Nitrous Acid–Induced Deamination Environmental pollutants that are metabolized to DNA-reactive forms include sodium nitrate ($NaNO_3$), a common food preservative, which is converted in the stomach to nitrous acid (HNO_2). Nitrosamines and nitrate salts are also converted to nitrous acid. Nitrous acid acts as a mutagen by reacting with A and C residues, altering their ability to form correct base pairs (**Figure 12-10**). If the modified bases are not repaired, a transition mutation will result. Bisulfite (HSO_3^-) has similar effects and is also used as a food preservative. Sodium nitrate and bisulfite do not seem to increase the risk of cancer in humans when used in this way, perhaps because, in such small amounts, they make only a minor contribution to the overall level of DNA damage.

Oxidative Damage Possibly the most important source of mutagenic alterations in DNA is oxidative damage. The DNA of each cell in the body is subjected to thousands of damaging oxidative reactions every day. Reactive oxygen species, such as hydrogen peroxide (H_2O_2), hydroxyl radicals ($^{\cdot}OH^-$), and superoxide radicals ($^{\cdot}O_2^-$), arise during irradiation or as byproducts of aerobic metabolism. Oxidative DNA damage ranges from oxidation of the base to oxidation of the deoxyribose sugar to the removal of bases (forming abasic sites), and it can also cause strand breaks.

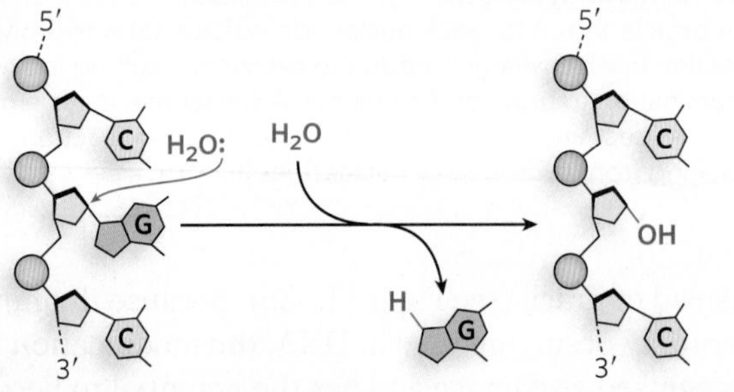

FIGURE 12-9 Depurination resulting from hydrolysis. In depurination, a purine (in this case, guanine) is lost from DNA by hydrolysis of the N-β-glycosyl bond.

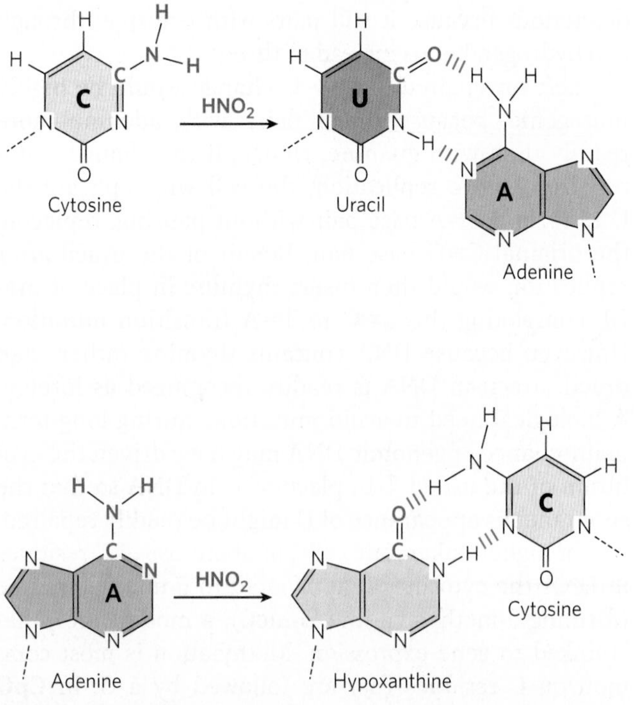

FIGURE 12-10 Deamination by nitrous acid. Nitrous acid can deaminate C and A residues, causing them to base-pair with the wrong nucleotide during replication. The incorrect base pairing after deamination is shown on the right.

The hydroxyl radical reacts with both purine and pyrimidine bases. In pyrimidines, the double bond between C-5 and C-6 is highly susceptible to attack, resulting in a variety of oxidized bases, including 5-hydroxyuracil, 5-hydroxycytosine, uracil glycol, and thymine glycol (**Figure 12-11a**). In purines, too, oxidation generates a spectrum of products. Among these reactions is the oxidation of guanine to 8-oxoguanine, which is extremely mutagenic. In the syn conformation (see Figure 6-16), an 8-oxoG residue base-pairs with A, and replication that occurs before 8-oxoG is repaired results in a G≡C to T=A transversion mutation (**Figure 12-11b**). Such G≡C to T=A transversions are among the most common mutations in human cancers.

Damage by Alkylation The addition of an alkyl group to atoms in nucleotide bases or to the phosphodiester backbone is known as **alkylation**. The N-3 of adenine and O^6 of guanine are common sites of alkylation, although many other positions in all four nucleotide bases, as well as the phosphodiester backbone of DNA, can also be modified, depending on the alkylating agent (**Figure 12-12a**). Alkylation of a nucleotide base can have a range of effects on base pairing, from no effect to completely preventing base pairing of the alkylated base with another base.

Any substance directly involved in promoting cancer is called a **carcinogen**. A known carcinogen in the smoke of burning cigarettes, wood, and coal tar that reacts with DNA is benzo[*a*]pyrene. Benzo[*a*]pyrene is an intercalating agent and is in a class of chemicals referred to as polycyclic aromatic hydrocarbons, which are hydroxylated in the liver as part of a detoxification process. However, hydroxylation of some polycyclic aromatic hydrocarbons results in a highly reactive epoxide. In the case of benzo[*a*]pyrene, hydroxylation forms an epoxide that reacts with purines: with N^6 of adenine or N^2 of guanine (**Figure 12-12b**). Although N^2 of guanine is not a base-pairing atom, these bulky alkylated bases no longer form correct base pairs and can lead to mutations if not corrected.

Nitrogen mustard gas is an alkylating agent that was used as a weapon in World War I. Nitrogen mustards are cross-linking agents, and they react with adjacent G residues to form interbase cross-links (**Figure 12-12c**).

The Ames Test Identifies DNA-Damaging Chemicals

DNA-reactive compounds are referred to as **genotoxic**, because they cause chemical changes in genomic DNA. Many sources of cancer in humans can be traced to DNA-damaging agents. In a seeming contradiction, not only can DNA-damaging agents cause cancer, but some are used in chemotherapy to treat some types of cancer. Small

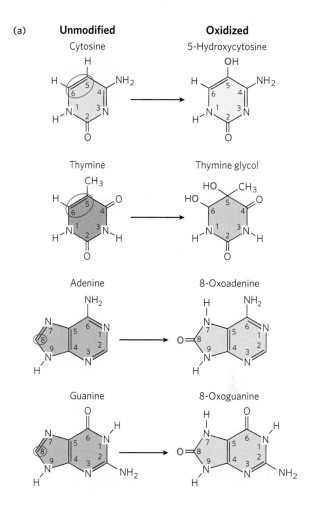

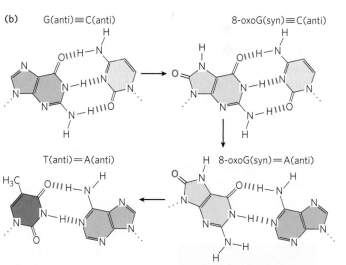

FIGURE 12-11 Oxidative damage to purine and pyrimidine bases. (a) Common positions (circled in red) of oxidative damage to nucleotide bases (left) and the products of oxidation at these positions (right). (b) Oxidation of the guanine of a G residue to 8-oxoguanine is extremely mutagenic. The 8-oxoG residue base-pairs with either A or C and thus can result in a G≡C to T=A transversion.

(a)

Cytosine

Thymine

(b)

Benzo[*a*]pyrene

Benzo[*a*]pyrene–guanine adduct

Adenine

Guanine

(c)

Nitrogen mustard
(mechlorethamine)

$Cl{-}CH_2{-}CH_2$

$N{-}CH_3$

$Cl{-}CH_2{-}CH_2$

FIGURE 12-12 Alkylation of DNA. Only the bases of the nucleotide residues are shown. (a) Atoms that are most susceptible to alkylation (circled in red). (b) Benzo[a]pyrene is hydroxylated in the liver to a reactive form (an epoxide) that alkylates nucleotide bases; the product of its reaction with guanine is shown. (c) A nitrogen mustard forms interstrand cross-links between two G residues.

amounts of a mutagen can do just enough damage to cause cancer. But large amounts can do more than cause mutations: they can kill the cell—they are **cytotoxic**—and can sometimes be used to kill cancer cells.

It clearly is in our interest to identify carcinogenic substances so that we can choose to avoid them and/or put them to use in treating cancer. Many mutagens are also carcinogens. A rapid and inexpensive test for a chemical mutagen is provided by a bacterial screen invented by Bruce Ames. The **Ames test** makes use of a strain of *Salmonella typhimurium* that contains a mutation in the biosynthetic pathway for histidine and therefore requires histidine in the growth medium. These mutants are called **auxotrophs**, which are cells that have lost the capacity to synthesize various organic compounds such as amino acids. Histidine-auxotrophic *S. typhimurium* cells plated on a medium

Bruce Ames [*Source: Courtesy Bruce Ames.*]

lacking histidine do not survive. But a small number of the cells may acquire mutations that reverse the original mutation, and cells with a **reversion mutation** (or back mutation) of this type can synthesize histidine. These mutant cells can survive on a histidine-free medium and form a small number of colonies that grow in random positions when the cell culture is grown on a petri plate (**Figure 12-13a**).

To test a potential mutagen (and thus a potential carcinogen), the compound is absorbed on a filter-paper disk placed in the center of a plate that contains *S. typhimurium* in the growth medium (**Figure 12-13b**). The chemical diffuses from the disk into the medium, creating a concentration gradient. The clear zone immediately adjacent to the disk contains the highest concentration of compound, too high to support life (**Figure 12-13c**). Beyond this lethal zone, the mutagen produces reversion mutations in some cells that enable *S. typhimurium* to form colonies. Because some compounds are mutagens only after they have been hydroxylated in the liver, the Ames test includes a step in which the compound being tested is first incubated with a liver extract. Most known

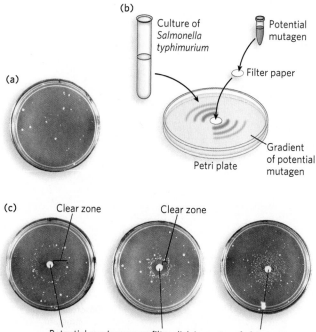

FIGURE 12-13 The Ames test for carcinogens, based on mutagenicity. The Ames test uses a strain of *Salmonella typhimurium* lacking an enzyme needed to synthesize histidine; the bacterium is plated on a histidine-free medium. (a) In the absence of the chemical to be tested, a few cells spontaneously develop a reversion mutation and form colonies. (b) The chemical to be tested is soaked into a filter-paper disk placed in the center of the plate, and the chemical diffuses to create a concentration gradient in the plate. (c) Identical nutrient plates are inoculated with an equal number of cells, but have progressively lower concentrations of the putative mutagen on the filter-paper disk (from left to right). At lower concentrations, the clear zone around the disk is smaller because fewer cells are killed by the chemical. [*Source: Bruce N. Ames, University of California, Berkeley, Department of Biochemistry & Molecular Biology.*]

human carcinogens result in increased mutation in the Ames test. In fact, numerous common industrial chemicals and natural products produce increased mutation in this test. The Ames test is therefore only the start of a process that identifies a compound as a human carcinogen. Compounds identified as mutagens in an Ames test require further testing in animals to determine whether they are likely to be human carcinogens.

DNA-Damaging Agents Are Used in Cancer Chemotherapy

DNA-reactive agents used in chemotherapy for cancer kill cells by creating broken chromosomes or stalled replication forks, either of which leads to cell death during cell division. The cytotoxic effect of DNA-damaging agents therefore requires the cell to be actively dividing. Chemotherapeutic agents are toxic to cancer cells because these cells must divide to form a tumor, but not toxic to most somatic cells that are not dividing. Nondividing or slowly dividing cells do sustain DNA damage during chemotherapy, but the damage is often repaired before replication occurs. The adverse side effects of chemotherapeutic DNA-damaging agents (hair loss, anemia, and nausea) are due largely to their effects on the few cell types in the body that rapidly divide, such as hair follicle cells, blood cells, and the cells that line the digestive tract.

Some types of DNA-damaging agents are particularly efficient at blocking replication forks or breaking chromosomes. A common and potent chemotherapeutic agent is the cross-linking drug cisplatin. Cisplatin is an alkylating agent that forms covalent adducts with the N-7 position of the two purine residues (**Figure 12-14a**). The intrastrand and interstrand cross-links resulting from reaction with cisplatin can be hard to repair, and they persist until they are encountered by a replication fork, leading to arrest of replication and cell death. Cisplatin is used in the treatment of many cancers, including bone cancer, lung cancer, certain lymphomas, and ovarian cancer.

Bleomycin, a complex biomolecule isolated from a bacterium, binds an atom of iron and activates molecular oxygen to form hydroxyl radicals that damage DNA. Bleomycin also binds to DNA, and its proximity to DNA directs the hydroxyl radical–mediated damage (**Figure 12-14b**). Bleomycin is used to treat Hodgkin lymphoma and testicular cancer.

Anthracyclines are chemotherapeutic agents that function by intercalating into DNA, reversibly inserting between base pairs. Doxorubicin is one such agent, and intercalation leads to double-strand breaks by inhibiting the resealing step of topoisomerase II (**Figure 12-14c**). Doxorubicin is used to treat leukemias, Hodgkin lymphoma, and several other types of cancer, including ovarian and breast cancer and tumors of the stomach, bladder, and thyroid gland.

Chemotherapeutic DNA-damaging agents are frequently used in combination and kill cells by blocking replication or transcription, often by triggering programmed cell death. Unfortunately, the damage incurred by chemotherapy also leads to mutations in some normal cells, and people who survive the primary cancer are at increased risk of developing a secondary tumor later in life. Therapies that more precisely target cancer cells alone are under development.

Solar Radiation Causes Interbase Cross-Links and Strand Breaks

Virtually all forms of life are exposed to energy-rich radiation that can cause chemical changes in DNA. We are subject to a constant field of ionizing radiation in the form of UV light (in sunlight) and cosmic rays, which can penetrate deep into the Earth, as well as radiation

(a)

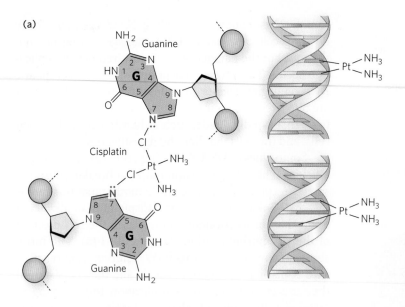

(b)

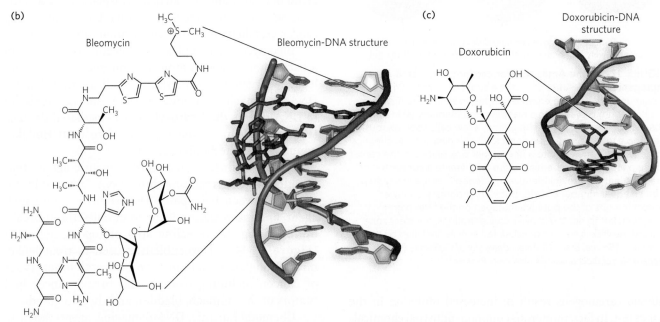

FIGURE 12-14 Chemotherapeutic DNA-damaging agents. Chemotherapeutic agents act by preferentially damaging the DNA of fast-dividing tumor cells. (a) Cisplatin is a cross-linking reagent that reacts with N-7 of two G residues to form intrastrand or interstrand cross-links. (b) Bleomycin binds to DNA and forms reactive oxygen species (not shown) that cause strand breaks. Iron normally binds bleomycin, but cobalt was used to prevent DNA modification in the crystal structure shown here. (c) Doxorubicin is a DNA intercalator, inserting between adjacent nucleotide residues. Intercalators can cause frameshift mutations during replication, or they can block replication forks, making the forks susceptible to nucleases that cause strand breaks. *[Sources: (b) PDB ID 1XMK. (c) PDB ID 1D12.]*

emitted from radioactive elements such as radium, plutonium, uranium, radon, carbon-14, and tritium (^{3}H). X rays used in medical and dental examinations, and in radiation therapy to treat cancer and other diseases, are other sources of exposure. UV and other ionizing radiation is estimated to be responsible for about 10% of all DNA damage caused by environmental agents.

Nucleotide bases interact very strongly with UV light (wavelengths of 200 to 400 nm), which can promote reactions that chemically change the DNA. **Pyrimidine dimers** are formed by UV-induced covalent cross-links between neighboring pyrimidines on the same DNA strand. Pyrimidine dimers form through the condensation of two ethylene groups on adjacent pyrimidines to form a **cyclobutane ring** (**Figure 12-15a**, left). Adjacent T residues or adjacent C residues can react to form a **6-4 photoproduct** (see Figure 12-15a, right). Pyrimidine dimers cause a significant distortion in the DNA and

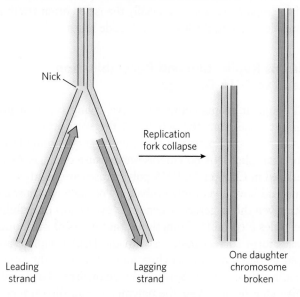

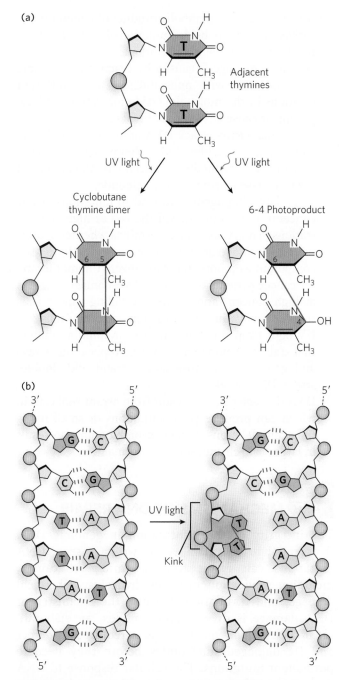

(a)

Adjacent thymines

UV light   UV light

Cyclobutane thymine dimer   6-4 Photoproduct

(b)

UV light

Kink

FIGURE 12-15 Pyrimidine dimers and their effect on the DNA duplex. (a) One type of reaction caused by UV light results in a cyclobutane ring, which involves C-5 and C-6 of adjacent pyrimidine (in this case, thymine) bases. An alternative reaction results in a 6-4 photoproduct that links C-6 and C-4 of adjacent pyrimidines. (b) Formation of a pyrimidine dimer introduces a bend or kink in the DNA.

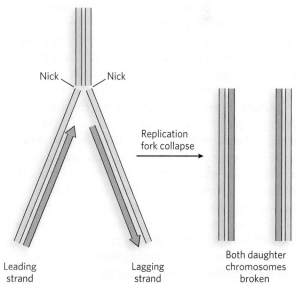

(a)   A nick in one strand fragments one chromosome.

(b)   Nicks in two strands fragment both daughter chromosomes.

FIGURE 12-16 DNA damage caused by strand breaks during replication. (a) When a replication fork encounters a single-strand break (nick), one daughter chromosome is broken, while the other daughter chromosome remains intact. The intact chromosome can become a template for repair of the broken one through homologous recombination (described in Chapter 13). (b) When the break occurs in both strands, both daughter chromosomes are broken, and neither daughter chromosome is completed.

thus can no longer base-pair with another nucleotide (**Figure 12-15b**). On encountering a pyrimidine dimer during replication, the DNA polymerase stalls; repair is therefore a matter of urgency.

Gamma rays and x rays are much higher-energy radiation than UV light and generate reactive oxygen

species that can break one or both strands of DNA (**Figure 12-16**). A break in a single strand of duplex DNA is easily repaired because it can be rejoined by ligase. But when both strands are broken, the repair job is much more difficult. Rejoining of a broken chromosome is performed by either homologous recombination or nonhomologous end-joining reactions, as we describe in detail in Chapter 13. These repair reactions, especially the latter, often result in mutations due to the loss of nucleotide bases.

Errant Replication and Recombination Lead to DNA Damage

Although it seems counterintuitive, many types of mutations arise from the activity of the very proteins that have evolved to maintain the integrity of genomic DNA. For example, DNA polymerases can cause damage. As we learned in Chapter 11, DNA polymerases are highly accurate and have a very low probability of making a mistake. But given the large size of genomes, even a low probability becomes a certainty. Sometimes, tautomers of nucleotide bases form correct matches with the DNA template and lead to base-pair mismatches, and sometimes the proofreading $3' \rightarrow 5'$ exonuclease misses an incorrect base, and a mismatch results. Also, the polymerase can slip on repetitive sequences (template slippage), giving rise to insertion or deletion mutations. Recall, too, that not all DNA polymerases are as accurate as the primary replicase. There are rare instances in which the replication fork encounters a damaged base and low-fidelity translesion DNA polymerases are recruited that can insert an incorrect nucleotide in order to move the replication fork past the damage—allowing replication to continue, but leading to a mutation.

During cell division, homologous chromosomes are lined up in pairs so that they can be partitioned into daughter cells. The proximity of similar DNA sequences can lead to the exchange of DNA segments through homologous recombination during meiosis or mitosis. Errors in this process sometimes occur. For example, a repeated nucleotide sequence on a chromosome can lead to recombination between two regions on homologous chromosomes or between two nonhomologous chromosomes. This can lead to chromosomal aberrations such as deletions, duplications, inversions, insertions, and translocations.

When chromosomal abnormalities arise during development, the anomaly is present in every cell in the body that is descended from the mutant cell. Chromosomal abnormalities can also arise in a somatic cell of an adult, and when this happens, the abnormalities are present only in a subset of cells, such as those in a tumor formed from the mutated cell.

SECTION 12.2 SUMMARY

- Hydrolysis can deaminate nucleotide bases, altering their ability to base-pair and leading to a mismatch during replication. Deamination of cytosine to uracil is the most common type and, if not repaired, can lead to a $C \equiv G$ to $T = A$ transition mutation. Hydrolysis can also sever the glycosyl bond between the pentose and the base, leaving an abasic site.

- Nitrous acid, the metabolic product of a food preservative, can induce the deamination of A or C residues, resulting in a transition mutation.
- Oxidative damage is caused by reactive oxygen species that react with nucleotides at many different positions in the molecule. Oxidation can affect base pairing or cause DNA strand breaks.
- Alkylating agents attack DNA at any of several electron-rich atoms, adding bulky chemical groups to the base or the phosphodiester backbone. Alkylation can alter the base pairing of the nucleotide.
- DNA-damaging agents can lead to mutations at low concentrations and can kill the cell at high concentrations. The Ames test determines whether a compound is mutagenic in bacteria, thus identifying the compound as a potential carcinogen. Chemotherapy for cancer patients often uses DNA-damaging agents at high concentrations, thereby killing cancer cells that are replicating faster than most normal cells.
- UV light from the sun can form pyrimidine dimers that stall DNA polymerase during replication. X rays and gamma rays cause single-strand and double-strand DNA breaks.
- DNA damage can also result from errant replication, which can produce point mutations or small insertions and deletions. Errant recombination can result in large-scale chromosomal abnormalities.

12.3 MECHANISMS OF DNA REPAIR

The integrity of the information in genomic DNA is essential to cell viability. As we have seen, DNA mutations can result from a variety of causes, including replication mistakes, hydrolysis, chemical damage, and irradiation. Because the chemistry of DNA damage is diverse and complex, an elaborate set of DNA repair mechanisms is required to detect and fix damaged nucleotides before they become permanent mutations. The cellular response to DNA damage includes a wide range of enzyme systems that catalyze some of the most interesting chemical transformations in DNA metabolism. Mistakes made during replication and recombination do not involve damaged bases, yet there are enzyme systems that detect and repair these errors before they become mutations. We consider here the major repair systems in bacteria and eukaryotic cells.

Mismatch Repair Fixes Misplaced-Nucleotide Replication Errors

DNA polymerase III of *E. coli* contains a proofreading activity that confers a very low mutation rate of one error in 10^6 to 10^8 nucleotides. However, the observed accuracy of replication in *E. coli* is even higher: one error in 10^9 to 10^{10} polymerization events. The additional accuracy comes from an efficient repair process that recognizes and corrects mismatches that escape Pol III.

Mismatched nucleotides incorporated by the replication apparatus are corrected by the **mismatch repair (MMR)** system, which is conserved in all cell types from bacteria to humans. The mismatches are nearly always corrected to reflect the information in the parent strand. Given that neither strand contains a *damaged* base, the cell must discriminate between the parental template and the newly synthesized strand, and replace only the nucleotide base in the new strand. Besides mismatches,

the *E. coli* MMR system can also recognize small loops of up to 4 bp of unpaired nucleotides, formed by template slippage during replication or by recombination. Left unrepaired, these small loops of extra DNA result in deletions or insertions. Loops of more than 4 bp are not recognized by the MMR system, and there is no other mechanism to recognize these mistakes. Thus, larger indels are simply not corrected. The MMR system of *E. coli* includes at least 12 protein components that function in either the strand discrimination reaction or the repair process itself (**Table 12-2**).

The mechanism by which the newly synthesized strand is identified and targeted for correction has not been worked out for most bacteria or eukaryotes, but it

TABLE 12-2

Proteins of Mismatch Repair in *E. coli* and Eukaryotes

E. coli

Protein	Function
MutS (as dimer MutS$_2$)	Recognizes single base mismatches
MutL (as dimer MuL$_2$)	Binds MutS and coordinates repair
MutH	Cleaves hemimethylated DNA
Helicase II	Unwinds DNA
SSB	Binds unwound ssDNA
Exonuclease I	$3' \to 5'$ exonuclease
Exonuclease X	$3' \to 5'$ exonuclease
RecJ	$5' \to 3'$ exonuclease
Exonuclease VII	$5' \to 3'$ exonuclease
Pol III holoenzyme	Fills in gap
DNA ligase	Seals DNA

Eukaryotes

Protein		Function
Yeast	Human	
MSH2/MSH6	MSH2/MSH6	Repairs single base mismatches, small loops
MSH2/MSH3	MSH2/MSH3	Repairs larger loops; functions with MSH2/MSH6 and MSH2/MSH3
MLH1/PMS1	MLH1/PMS2	Functions with MSH2/MSH6 and MSH2/MSH3
MLH1/MLH2	MLH1/MLH2	Unknown
MLH1/MLH3	MLH1/MLH3	Unknown
RPA		ssDNA-binding protein
Exonuclease I		$5' \to 3'$ exonuclease
RFC, PCNA, polymerase		Fill in gap
DNA ligase		Seals DNA

is well understood for *E. coli* and some closely related bacterial species. In these bacteria, strand discrimination is based on the action of **Dam methylase**, the enzyme that methylates DNA at the N^6 position of adenine within a 5′-GATC-3′ sequence (see Chapter 11). The GATC sequence on both strands of the parental DNA is methylated, but during replication, the newly synthesized strand is unmethylated for a short period (a few seconds or minutes) immediately after passage of the replication fork. During this interval of hemimethylation, proteins in the MMR complex can identify and distinguish the unmethylated new strand from the methylated parent strand. Replication mismatches in the vicinity of a hemimethylated GATC sequence are then repaired according to the information in the methylated (parental) template strand (**Figure 12-17**).

Studies of the MMR proteins have outlined the process by which mismatch repair occurs (see the How We Know section at the end of this chapter). The mismatched base pair creates a distortion in DNA that is recognized by the MutS protein. This enables MutS to bind the MutL protein, and the MutS-MutL complex, using ATP, scans bidirectionally along the DNA, forming a DNA loop. The crystal structure of MutS bound to a mismatch shows that MutS forms a homodimer (MutS$_2$) that binds DNA at the dimer interface (**Figure 12-18** on p. 432). The MutS dimer also contains a hole large enough to surround the DNA, but whether MutS does encircle DNA during scanning is unknown. On arriving at a hemimethylated GATC site, the complex recruits and activates MutH, a site-specific endonuclease that cleaves unmethylated GATC sites. After strand cleavage, MutS-MutL recruits helicase II (also called UvrD), which unwinds DNA in the direction of the mismatch. During the unwinding, an exonuclease degrades the displaced DNA strand. Different exonucleases are used, depending on whether the enzyme needs to travel in the 5′→3′ or 3′→5′ direction along the DNA (**Figure 12-19** on p. 432). Unwinding and DNA degradation stop shortly after the mismatch is excised, leaving a single-strand gap that extends from the mismatch to the original incision at the GATC site. The single-strand gap is coated with single-stranded DNA–binding protein (SSB), filled in by Pol III holoenzyme, and sealed by ligase.

Mismatch repair is a particularly costly process for *E. coli* in terms of energy expended. The distance between the mismatch and the GATC cleavage site can be more than 1,000 bp. The degradation and replacement of a strand segment of this length requires an enormous investment in dNTPs to repair a single mismatched base. But this energy consumption is affordable relative to the cost of incurring a mutation. The conservation of such a high-cost repair system in all cells is a good illustration of the importance to cells of maintaining the sequence of their genomic DNA.

Eukaryotic cells have several proteins that are structurally and functionally analogous to bacterial MutS and MutL (see Table 12-2). The MutS homologs work in heterodimers, and each has a specialized function. For instance, in yeast, heterodimers of MSH2 and MSH6 generally bind to single base-pair mismatches and bind less well to slightly longer mispaired loops. The eukaryotic homolog of bacterial MutL is also a heterodimer that binds to the MutS homologs.

Eukaryotic cells lack homologs to bacterial MutH and Dam methylase, and they do not use methylation to distinguish between new and old strands. It is thought that in eukaryotes, strand discrimination relies on the fact that only the newly replicated strand has nicks, and we know that these nicks are frequent on the lagging strand, given the 100 to 200 bp size of Okazaki fragments in eukaryotic cells (see Chapter 11). The nick on a newly synthesized DNA strand can be used as a starting point to excise the DNA strand past the mismatch, and the single-strand gap is then filled in by a polymerase, much as in bacterial mismatch repair; unlike in bacteria, however, the exonucleolytic reaction in eukaryotes does not appear to require the assistance of a helicase. Although this may explain how eukaryotes use nicks to distinguish new from old DNA on the lagging strand, what about the new DNA of the leading strand? The origin of strand nicks in the newly synthesized leading strand is not yet clear, but recent studies show that unlike bacterial MutL, the human MutL homolog has an endonuclease activity that depends on the PCNA clamp, suggesting that strand nicking is coordinated with DNA replication. Because PCNA has distinct "front" and "back" sides, the orientation of a PCNA clamp used during replication may direct the endonuclease to the newly synthesized strand (see Chapter 11).

Mutations in the genes that encode mismatch repair proteins result in the accumulation of mutations throughout the human genome, because misinsertions and short indels can no longer be repaired by the MMR system. Indeed, mutations in MMR genes result in some of the most common inherited cancer-susceptibility syndromes, such as hereditary nonpolyposis colorectal cancer (**Highlight 12-1** on p. 433). Approximately 15% of all colon cancers are of this type.

Direct Repair Corrects a Damaged Nucleotide Base in One Step

Some types of DNA damage that would normally lead to a base substitution or a one-nucleotide deletion are repaired directly, without removing a base or a nucleotide. The best-characterized example of direct repair is the **photoreactivation** of cyclobutane pyrimidine dimers, first recognized in the late 1940s, before the discovery of

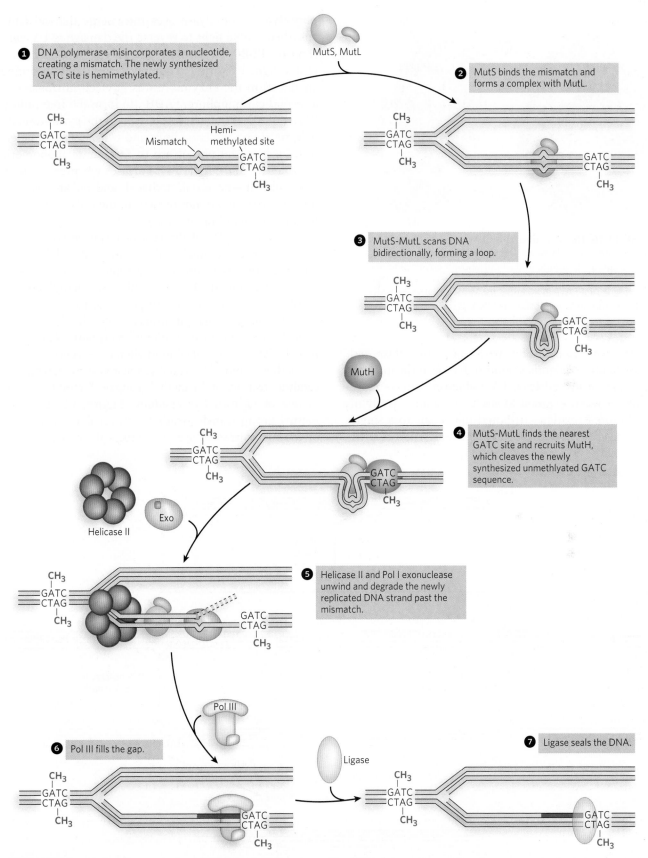

FIGURE 12-17 Mismatch repair of a nucleotide misincorporated by DNA polymerase. In *E. coli*, MutS₂MutL₂ binds a mismatch and scans the DNA for a GATC site. MutH nicks the DNA at the nearest unmethylated GATC site, facilitating repair of the mismatch on the newly synthesized strand by excision, filling in of the gap, and ligation.

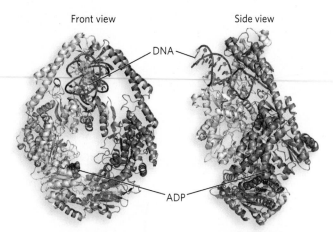

Front view   Side view

DNA

ADP

FIGURE 12-18 The structure of MutS bound to a mismatch in DNA. Front and side views of the MutS homodimer, MutS$_2$ (identical subunits, orange and white), show it enveloping and kinking DNA that contains a mismatched base pair (red). The nucleotide is ADP; ATP is essential for MutS function, but ADP keeps the enzyme from moving on DNA. *[Source: PDB ID 1E3M.]*

DNA structure (see the How We Know section at the end of this chapter). Scientists noticed that bacteria and bacteriophage recovered from UV radiation damage more efficiently when exposed to sunlight. Genetic study of this photoreactivation attributed the repair to a single gene. The gene product is an enzyme referred to as **DNA photolyase**. Photolyase uses the energy derived from absorbed visible light to reverse the damage of UV light (**Figure 12-20** on p. 434). The energy absorbed from visible light by a first chromophore (a light-absorbing chemical) in the enzyme results in electron transfer to a second chromophore, FADH$^-$, to form the free radical FADH$^·$. FADH$^·$ donates its electron to the pyrimidine dimer, reversing the cross-links and transferring the electron back to the photolyase to regenerate monomeric pyrimidines and FADH$^-$. Photolyases are present in almost all cells—bacterial, archaeal, and eukaryotic—yet, for some reason, are not present in the cells of placental mammals (including humans).

More examples of direct repair can be seen in the repair of oxidized nucleotides. The modified base O^6-methylguanine is a common and highly mutagenic lesion that results from alkylation (in this case, methylation) of O^6 of a G residue. It tends to pair with thymine rather than cytosine during replication, resulting in G≡C to A=T (via O^6-meG–T) transition mutations (**Figure 12-21a** on p. 434). Direct repair of O^6-methylguanine is performed by O^6-methylguanine–DNA methyltransferase, an enzyme that catalyzes transfer of the methyl group of O^6-methylguanine to one of its own Cys residues (**Figure 12-21b**). The methyl group transfer leads to irreversible inactivation of the methyltransferase and targets it for degradation

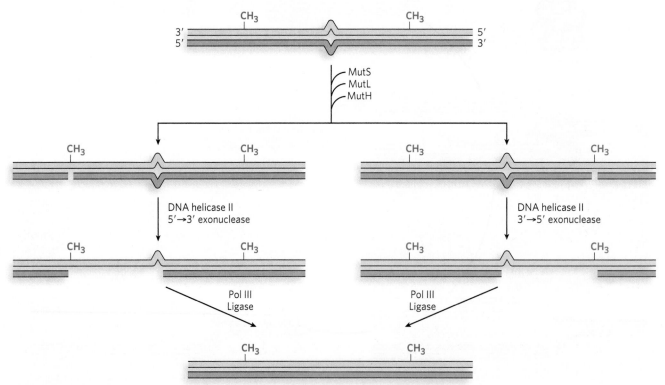

FIGURE 12-19 Multiple exonucleases involved in methyl-directed MMR. When cleavage at a hemimethylated GATC is on the 3′ side of the mismatch, a 5′→3′ exonuclease is recruited, either RecJ or exonuclease VII. When cleavage is on the 5′ side of the mismatch, a 3′→5′ exonuclease is required, either exonuclease I or exonuclease X.

Mismatch Repair and Colon Cancer

Most cancer cells have mutations in genes that regulate cell division (oncogenes and tumor suppressor genes). However, no single mutation is responsible for the progression from a normal cell to a malignant tumor. This progression requires an accumulation of mutations, sometimes over several decades, and is fairly well understood in the case of colon cancer (Figure 1).

Discovery of the link between MMR and hereditary nonpolyposis colorectal cancer (HNPCC) was made by the laboratories of Richard Kolodner and Bert Vogelstein, where MMR gene mutations were identified in HNPCC cells. The inherited entity is a loss of function in one allele, usually of the gene encoding MLH1 or MSH2. These genes are essential to mismatch repair (see

Tom Kunkel *[Source: Courtesy Thomas Kunkel.]*

Richard Kolodner *[Source: Courtesy Richard Kolodner.]*

Table 12-2). Mutation of the second allele leads to the rapid accumulation of multiple new mutations that produce a malignant cell. HNPCC mutant cells have an increased frequency of small insertions and deletions in microsatellite repeats—1 to 6 bp sequences that are repeated 10 to 100 times. This is referred to as microsatellite instability.

The exact number of microsatellite repeats varies from one person to the next, but in one individual, all cells normally contain the identical number of repeats. However, in a person with HNPCC, cells contain different numbers of microsatellite repeats. Independent studies in the laboratories of Tom Kunkel and Paul Modrich showed that extracts of cells displaying microsatellite instability were defective in mismatch repair.

A test of microsatellite length is a simple indication of whether an individual has a mutation in the MMR genes of a tumor (Figure 2). PCR primers are used to amplify specific regions containing microsatellite sequences in the genome. The tumor cell contains some microsatellite DNAs that are longer or shorter than those of a normal cell from the same individual, thus indicating a defective MMR gene. The person was born with two good alleles, but both copies of MLH1 or MSH2 became inactive during his or her lifetime. Microsatellite instability and defects in mismatch repair have now been correlated with several types of cancer other than colon cancer, including ovarian, stomach, cervical, breast, skin, lung, prostate, and bladder cancers.

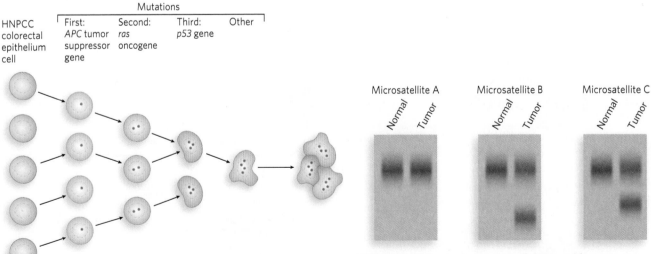

FIGURE 1 The development of colorectal cancer has several recognizable stages, each associated with a mutation. If mismatch repair becomes nonfunctional (through mutation), new mutations accrue quickly.

FIGURE 2 PCR primers are designed to amplify genomic DNA of three different microsatellite repeats in normal cells and tumor cells from the same individual. In this illustration of a possible result, two of the three microsatellite repeats that were tested have different sizes in the tumor cells—evidence that the tumor originated from a mutation in an MMR gene.

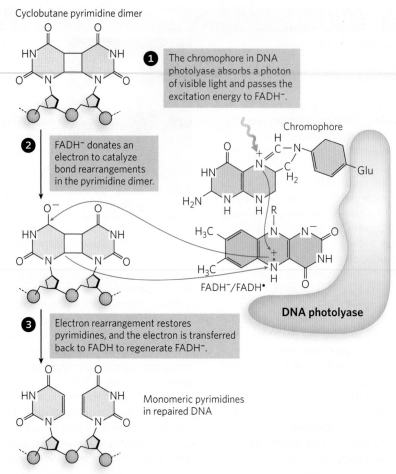

FIGURE 12-20 Photorepair of a pyrimidine dimer. *E. coli* photolyase has two chromophores (light-absorbing groups) that work in sequence to use the energy of light to repair a pyrimidine dimer.

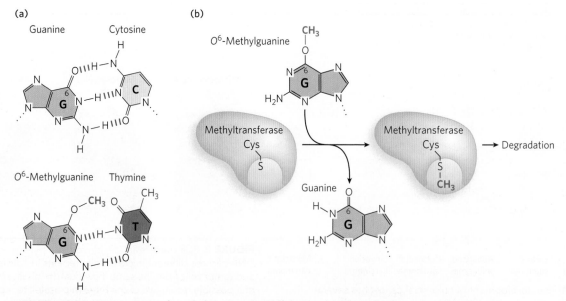

FIGURE 12-21 Direct repair of methylated nucleotide bases. (a) When the G residue of a normal G≡C base pair (top) is methylated on O^6, the resulting O^6-methylguanine base-pairs with thymine (bottom) and thus is highly mutagenic. (b) O^6-Methylguanine–DNA methyltransferase transfers the methyl group from the O^6-methylguanine onto one of its own Cys residues, and the enzyme is thereby targeted for degradation.

(an unusual property for an enzyme). Consumption of an entire protein molecule to correct a single damaged base is another vivid illustration of the priority given to maintaining the integrity of cellular DNA. Direct repair is also used to dealkylate other alkylated nucleotides.

Base Excision Repairs Subtle Alterations in Nucleotide Bases

Excision repair is the most prevalent method that cells use to repair damaged DNA. There are two types: base excision and nucleotide excision repair. **Base excision repair (BER)** functions at the level of a single damaged nucleotide that distorts DNA very little. It is also the main pathway for the repair of single-strand DNA breaks that lack a ligatable junction and therefore require "cleaning" of the 3′ or 5′ terminus for ligation.

In bacterial BER, recognition of the damaged base is performed by a **DNA glycosylase**, which cleaves the nucleotide base from the pentose by hydrolyzing the *N*-β-glycosyl bond, leaving an apurinic or apyrimidinic site (AP site). Insertion of the correct nucleotide base does not occur by re-forming the glycosyl bond with a new, correct base. Instead, the single-stranded DNA is cleaved at the abasic site by **AP endonuclease**, creating a nick with a 3′ hydroxyl and a 5′ deoxyribose phosphate. In *E. coli*, a segment of DNA is removed by the nick translation activity of Pol I, and DNA ligase seals the remaining nick (**Figure 12-22a**).

Eukaryotic BER proceeds by either of two paths; in each case, the first two steps are the same as in bacteria (**Figure 12-22b**). One eukaryotic BER mechanism, similar to bacterial BER, is often referred to as long patch repair because up to 10 nucleotides are replaced. Eukaryotic DNA polymerases lack 5′→3′ exonuclease activity, and therefore a special "flap endonuclease" is recruited to remove the displaced 5′ terminus. The second eukaryotic BER mechanism is used the most; it replaces only the damaged nucleotide base and thus is sometimes referred to as short patch repair. The one-nucleotide fill-in reaction is performed by Pol β, which also removes the 5′ deoxyribose phosphate, leaving a 5′ phosphate for ligation.

Most damaged bases repaired by the BER system remain base-paired in the helix and stack with adjacent bases. This brings up the question of how a damaged base that is buried in the DNA helix can be identified by an enzyme for repair. The crystal structure of a uracil DNA glycosylase (UDG; discussed below) reveals a fascinating recognition process in which the enzyme scans the minor groove of the helix, and damaged-base recognition is performed by kinking the DNA and "flipping" the damaged base completely out of the helix and into the enzyme's active site (**Figure 12-23a**). An experiment demonstrating UDG activity is shown in **Figure 12-23b**.

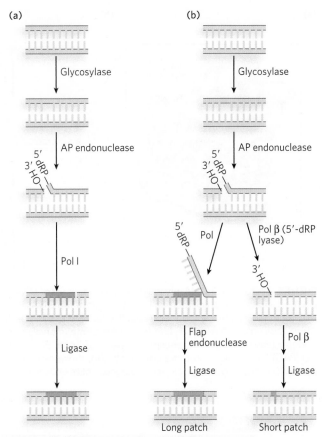

FIGURE 12-22 Base excision repair. (a) In bacteria, a glycosylase excises a damaged nucleotide base, then an AP endonuclease nicks the backbone at the abasic site. Nick translation by Pol I excises the 5′ deoxyribose phosphate (5′-dRP) and some dNMPs, and synthesizes a new strand. Ligase seals the gap. (b) Eukaryotic BER, after the first two steps (similar to those in bacteria), can take either of two paths. In long patch repair (left), a DNA polymerase extends the DNA strand from the 3′ terminus, displacing the 5′ single-stranded DNA; this is followed by cleavage by a flap endonuclease and ligation. In short patch repair (right), only one nucleotide is inserted (by Pol β) prior to ligation.

The substrate for this reaction is a synthetic 23-mer duplex in which one strand has an internal dU residue and a 5′-terminal ^{32}P label. The UDG removes the uracil, forming an abasic site in the [^{32}P]DNA strand. Treatment with an AP endonuclease then results in cleavage of the ^{32}P-labeled strand. Analysis in a DNA sequencing gel, which separates the strands, reveals the smaller, cleaved [^{32}P]DNA strand.

All cells contain several different glycosylases that recognize different types of damaged bases (**Table 12-3**). There are two main types of DNA glycosylases. One type is highly specific to a particular damaged base; the other type recognizes oxidative damage, and the substrate spectrum is more diverse. As discussed earlier, spontaneous deamination of C residues to U residues in DNA is fairly frequent, and chromosomal DNA is in need of constant repair of uracil bases. A uracil DNA glycosylase

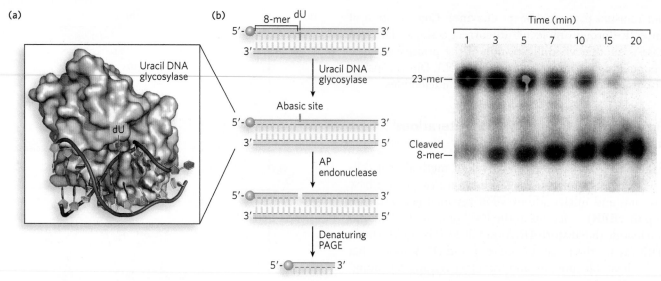

FIGURE 12-23 Uracil DNA glycosylase. (a) Human uracil DNA glycosylase (brown) bound to its DNA substrate. The uracil (purple) is flipped out of the DNA duplex and fits into the enzyme active site. (b) In this experiment, the DNA duplex, labeled at the 5′ end with ^{32}P (orange balls), is treated with *E. coli* uracil DNA glycosylase for different time intervals to produce the abasic site, then with AP endonuclease to cleave the phosphodiester backbone. Electrophoresis in a denaturing polyacrylamide gel (PAGE), followed by autoradiography (right), shows results for the various time periods. [*Sources: (a) PDB ID 4SKN. (b) Courtesy of Roxana Georgescu, Mike O'Donnell Laboratory, Rockefeller University.*]

TABLE 12-3		
DNA Glycosylases in Bacteria and Mammals		
		Glycosylase
Lesion	Bacteria	Mammals
Uracil	Uracil DNA glycosylase	UNG1, UNG2, MBD4, SMUG1, TDG
3-Methyladenine	AlkA	MYH
A base-paired to 8-oxoG	8-OxoguanineDNA glycosylase	8-OxoguanineDNA glycosylase
Oxidized bases	Endoglycosylase III	NTH1
	Endoglycosylase VIII	OGG1
	MutM	NEIL1
	Tag	NEIL2

is found in most cells, and it specifically removes uracil bases from DNA. This glycosylase acts only on DNA; it does not remove uracil from RNA. As may be expected, *E. coli* strains with mutations in this enzyme have a high rate of G≡C to A=T mutations.

Most bacteria have just one UDG, whereas humans and other mammals have several types (see Table 12-3), with different specificities for removing U residues. Specific UDGs remove uracils incorporated during replication, or those formed by deamination of cytosine in double-stranded DNA or single-stranded DNA, or those formed in DNA during transcription. There is also a human DNA glycosylase that removes T residues generated by the deamination of 5-meC. Mismatch repair can also recognize T–G and U–G mismatches, and it corrects them with different levels of efficiency, depending on the sequence context.

A wide variety of damaged bases can be removed by other DNA glycosylases that have evolved to recognize lesions such as formamidopyrimi line and 8-oxoguanine (both arising from purine oxidation), hypoxanthine (arising from adenine deamination), alkylated bases including 3-methyladenine and 7-methylguanine, and even some pyrimidine dimers. The BER pathway can also repair the thousands of abasic sites that arise from spontaneous hydrolysis, as well as breaks in single-stranded DNA that require processing at the 3′ or 5′ terminus before ligation.

Nucleotide Excision Repair Removes Bulky Damaged Bases

Nucleotide excision repair (NER) targets large, bulky lesions and removes DNA on either side of them. In contrast to base excision repair, NER does not require specific recognition of a damaged nucleotide and therefore can remove DNA lesions, even those caused by chemicals that did not exist in the environment until recently. It is the predominant repair pathway for removing pyrimidine dimers, 6-4 photoproducts, and several other bulky base adducts, including benzo[a]pyrene-guanine, which is formed on exposure to cigarette smoke (see Figure 12-12b). The nucleolytic activity of the NER system is novel in the sense that two incisions are made in one strand of DNA, excising the lesion and several nucleotides on either side of it; this unique enzymatic activity is called an **excinuclease**.

In *E. coli*, the NER pathway makes use of four *uvr* gene products—UvrA through UvrD—as well as several other factors (**Table 12-4**; **Figure 12-24**). First, a UvrA₂UvrB complex scans the DNA for damage. When it encounters a bulky damaged base, the strands become separated to form a single-stranded DNA bubble containing the lesion, and UvrA dissociates, leaving UvrB tightly

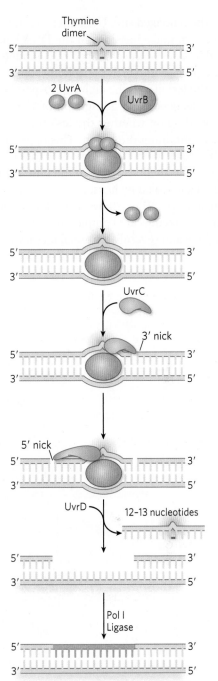

FIGURE 12-24 Nucleotide excision repair in *E. coli*. The NER pathway uses several proteins, including UvrA (red), UvrB (purple), and UvrC (green), that recognize the lesion and make incisions on either side, allowing UvrD (helicase II) to displace a section of lesion-containing DNA. The single-strand gap is filled in by Pol I, and the DNA is sealed by ligase. A transcription-coupled repair (TCR) path can also be taken, in which RNA polymerase stalls at the lesion on the coding strand (not shown). After the RNA polymerase is displaced, the reaction proceeds as shown here, using UvrA–D, Pol I, and ligase.

TABLE 12-4

Proteins Involved in Nucleotide Excision Repair	
Protein	*Function*
Bacteria	
UvrA	Recognizes lesion
UvrB	Unwinds DNA
UvrC	Excinuclease
UvrD	Helicase
Pol I	Fills in gap
DNA ligase	Seals DNA
Eukaryotes	
XPC	Recognizes lesion
RNA polymerase	Recognizes lesion: TCR
XPA	Verifies lesion
XPB	Unwinds DNA (TFIID subunit)
XPD	Unwinds DNA (TFIID subunit)
XPF	5′ excinuclease
XPG	3′ excinuclease
RPA	Stabilizes bubble
Pol δ or ε, RFC, PCNA	Fill in gap
Ligase I or IV	Seals DNA

bound to the damaged site. UvrB then recruits the UvrC excinuclease to make incisions in the DNA backbone on the 5′ and 3′ sides of the damaged nucleotide(s). The incisions are precise: the fifth phosphodiester bond on the 3′ side of the lesion and the eighth phosphodiester bond on the 5′ side, generating a fragment of 12 to 13 nucleotides (depending on whether the lesion involves one or two bases) containing the lesion. This oligonucleotide is released through the helicase action of UvrD (also called helicase II). The small gap is then filled by Pol I, and the resulting nick is sealed by ligase.

In eukaryotes, NER follows a similar chemical path, although the enzymes are completely different in amino acid sequence and several additional factors are involved

(**Figure 12-25**; see Table 12-4). The main factors were discovered through research on the human genetic disease xeroderma pigmentosum (XP) (**Highlight 12-2**). Individuals with XP are thousands of times more likely to develop skin cancer from exposure to sunlight. Studies of such individuals have identified at least seven different genes that, when any one of them is defective, can contribute to XP. All of these genes are involved in nucleotide excision repair.

Genetic studies of XP implicate the genes *XPA* through *XPG* in eukaryotic NER. Studies of the proteins encoded by these genes have revealed their roles. The XPC protein initiates the repair process by recognizing the lesion, acting like bacterial UvrA. Then XPB and

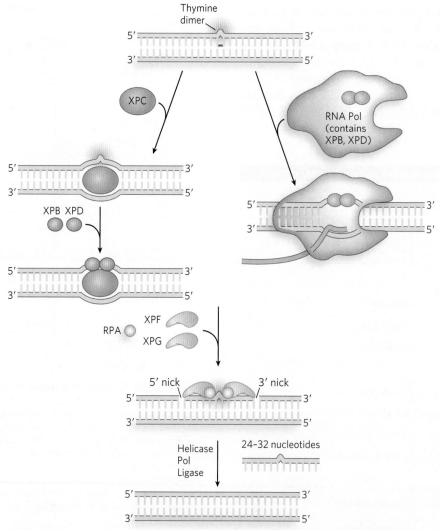

FIGURE 12-25 Nucleotide excision repair in eukaryotes. NER can be initiated by two slightly different methods in eukaryotes. One pathway (left) is similar to that occurring in bacteria, except that a larger section of lesion-containing DNA is removed. The other pathway (right) is referred to as transcription-coupled repair (TCR), because the lesion is first encountered by RNA polymerase, which then stalls. Some of the NER factors are bound to the RNA polymerase itself.

HIGHLIGHT 12-2 — MEDICINE

Nucleotide Excision Repair and Xeroderma Pigmentosum

Early studies of UV-damaged *E. coli* cells showed that their survival was enhanced if the cells were incubated in growth

Robert Painter
[Source: Courtesy James Cleaver.]

James Cleaver
[Source: Courtesy James Cleaver.]

medium before being plated. In genetic studies of this effect, researchers isolated strains with mutations in three different genes, *uvrA*, *uvrB*, and *uvrC*. Further study demonstrated that the repair of UV damage produced small sections of DNA synthesis, which later studies showed were at sites of DNA damage. In Robert Painter's laboratory at the University of California, San Francisco, a similar process was shown to occur in mammalian cells. James Cleaver, working in Painter's lab, recognized that the mammalian repair might be related to the repair of UV lesions in bacteria, but he needed similar mutants in mammalian cells to cement this connection. After reading an account of the genetics of human xeroderma pigmentosum (XP), Cleaver realized that the disease might be the sought-after connection to the UV repair pathway in bacteria. He obtained skin biopsies from patients with XP and developed cell lines that could be grown in culture.

Using the XP cell lines and methods for studying nucleotide excision repair in bacteria, Cleaver was able to identify the major protein components of NER in humans. Cell extracts from each patient were used to measure repair. No extract by itself could repair a UV lesion, but when two extracts were mixed together, repair was observed. These extracts, each missing a different protein of the repair pathway, complemented each other. The researchers were eventually able to group types of XP by complementation and identify which protein was missing in each complementation group. Defects in genes encoding any of seven different proteins of NER can result in XP; the proteins are denoted XPA to XPG. Some of these—XPB, XPD, and XPG—also play roles in transcription-coupled repair. Because NER is the sole repair pathway for pyrimidine dimers in humans, people with XP are extremely sensitive to light and readily develop sunlight-induced skin cancers with intense freckling (Figure 1). Most people with XP also have neurological abnormalities, possibly because of an inability to repair lesions caused by the high rate of oxidative metabolism in neurons.

Various facets of NER in humans await further study. For example, the function of XPE is not yet known. Also, it seems incongruent that bacteria have a second pathway for repair of pyrimidine dimers, making use of DNA photolyase, but humans and other placental mammals do not. However, mammals do have a pathway that bypasses pyrimidine dimers, involving the translesion polymerase, Pol η. This enzyme preferentially inserts two A residues opposite a T–T dimer and does not result in a mutation. Indeed, it is tempting to speculate that the appearance of Pol η replaced the need for photolyase in humans, allowing the photolyase gene to be discarded during evolution.

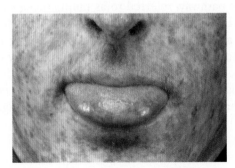

FIGURE 1 This patient exhibits the characteristics of xeroderma pigmentosum. [Source: Wellcome Image Library/Custom Medical Stock.]

XPD, which normally act as helicases in RNA transcription, are recruited to the lesion, where they separate the DNA strands to form a single-stranded DNA bubble, acting much like *E. coli* UvrB. RPA, the eukaryotic equivalent of SSB, then binds to the bubble and positions two nucleases, XPF and XPG, on either side of the lesion. XPG cleaves on the 3′ side and XPF on the 5′ side. The 24- to 32-nucleotide fragment containing the lesion is displaced, and the PCNA clamp recruits a DNA polymerase to fill the gap, which is then sealed by ligase.

In eukaryotes, a process has evolved that targets NER to a damaged template nucleotide that has stalled RNA polymerase. This process, referred to as **transcription-coupled repair (TCR)**, differs from NER only in the way the damaged site is recognized (see Figure 12-25). In TCR, the damage is recognized by the RNA polymerase that is stalled at the lesion. TCR is particularly efficient because it specifically targets repair to actively transcribed DNA that is currently yielding information needed for cell survival, rather than correcting lesions that may lie in vast untranscribed regions of the genome. Bacteria also have a type of transcription-coupled repair. When bacterial RNA polymerase stalls at a lesion, it is displaced by the Mfd helicase, which then recruits the UvrABC proteins for lesion repair.

Recombination Repairs Lesions That Break DNA

Lesions that block the replication fork can lead to cell death if not repaired before the next round of replication. One replisome-blocking lesion in DNA is a double-strand break. The typical route by which polymerase-blocking lesions are repaired is through a high-fidelity homologous recombination pathway (see Chapter 13). Repair by homologous recombination makes use of the sister chromosome to recover the original sequence. The chromosomes are paired, and the lesion can be repaired by using the homologous strand for the correct information. The role of homologous recombination in DNA repair may even have been the main selective force that drove the evolution of recombination enzymes.

Double-strand breaks can also be repaired by nonhomologous recombination, which uses a different set of proteins, conserved from bacteria to humans. Sealing the ends of broken DNA by this process usually incurs deletions or insertions and therefore produces mutations. This pathway, referred to as nonhomologous end joining (NHEJ; see Chapter 13), may be particularly useful when a sister chromosome is unavailable for high-fidelity homologous recombination. Typically, only small deletions or insertions are observed as a result of NHEJ, although large deletions of more than 1 kbp can occur. The sequence at the site of the DNA religation suggests that NHEJ occurs through short, 1 to 6 bp regions of homology.

Specialized Translesion DNA Polymerases Extend DNA Past a Lesion

Most DNA repair occurs on double-stranded DNA, where the original sequence can be restored using information in the complementary strand. However, sometimes a lesion occurs at a replication fork after the DNA strands have been unwound. In this case, the replicating polymerase stalls at the lesion. One mechanism that has evolved to resolve this potentially lethal situation is a pathway known as **translesion synthesis (TLS)**. Translesion

Wei Yang *[Source: Courtesy Wei Yang.]*

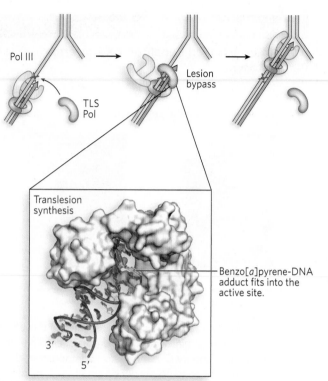

FIGURE 12-26 Translesion synthesis by TLS DNA polymerases. When the high-fidelity *E. coli* replicase, Pol III (yellow), encounters a leading-strand lesion (red X), it stalls, and a TLS polymerase (TLS Pol, gray) takes over the β sliding clamp (purple) to extend the leading strand across the lesion. After lesion bypass, Pol III resumes its function with the β clamp. The structure of a TLS DNA polymerase is shown (bottom), the archaeal Dbo4 (white) binding to DNA that contains a benzo[a]pyrene adduct (yellow). *[Source: (bottom) PDB ID 1S0M.]*

synthesis uses a bypass DNA polymerase, or TLS DNA polymerase, which usually lacks a proofreading 3′→5′ exonuclease and is capable of extending the DNA strand across a bulky template lesion. Structural studies of a TLS polymerase by Wei Yang reveal a wider-than-normal active-site architecture, which may explain how this class of enzymes can misincorporate nucleotides opposite noncoding DNA lesions. **Figure 12-26** shows the structure of the complex of an archaeal Pol IV (a TLS polymerase) with DNA containing a benzo[a]pyrene attached to an A residue. The TLS polymerase takes over from the high-fidelity DNA polymerase stalled at the template lesion and extends the DNA strand over the lesion. Because the lesion may be noncoding, lesion bypass by a TLS DNA polymerase often results in a mutation. The first TLS DNA polymerase was discovered in Myron Goodman's laboratory (see the How We Know section at the end of this chapter).

TABLE 12-5

Specialized DNA Polymerases Involved in Translesion Synthesis

Polymerase	Family
Bacteria (E. coli)	
Pol IV	Y
Pol V	Y
Pol II	B
Yeast (S. cerevisiae)	
Rev1	Y
Pol ξ	B
Pol η	Y
Human (H. sapiens)	
Rev1	Y
Pol ξ	B
Pol η	Y
Pol κ	Y
Pol ι	Y
Pol λ	X
Pol μ	X
Pol β	X
Pol θ	A
Pol ν	A

Given the emphasis throughout this chapter on the importance of genomic integrity, the existence of a cellular system that increases the rate of mutation may seem incongruous. However, we can think of this as a desperation strategy. The mutations resulting from translesion synthesis are the biological price a species must pay to overcome an otherwise insurmountable barrier to replication, as it permits a few mutant cells to survive. In *E. coli*, the main

TLS polymerase is Pol V, with components encoded by the *umuC* and *umuD* genes. It can extend DNA over the most common lesions, including pyrimidine dimers, 6-4 photoproducts, and abasic sites. All cells contain multiple TLS DNA polymerases (**Table 12-5**), each suited to bypassing particular types of lesions.

Humans have at least 10 different TLS polymerases. Pol ξ (xi) is an error-prone TLS polymerase that places nucleotides at random across noncoding lesions. In contrast, Pol η (eta) bypasses pyrimidine dimers in an error-free event, incorporating two A residues opposite the thymine dimer. Pol η goes no farther than the pyrimidine dimer; another DNA polymerase is required to extend the chain to a length that can be used by Pol δ or Pol ε. Pol η has very low fidelity on undamaged DNA, and this may be why another polymerase is needed to extend DNA after Pol η has incorporated A residues opposite a thymine dimer. In other words, Pol η has evolved to be specific and accurate at a thymine dimer, not on normal, unmodified DNA, and therefore any other polymerase will do a better job than Pol η of extending DNA over an undamaged template strand.

Several other low-fidelity DNA polymerases, including Pol β, Pol ι (iota), and Pol λ, have specialized roles in eukaryotic base excision repair. Each of these enzymes has a 5′ deoxyribose phosphate lyase activity in addition to its polymerase activity. After base removal by the glycosylase and backbone cleavage by the AP endonuclease, these low-fidelity polymerases remove the abasic site (a 5′ deoxyribose phosphate) and fill in the gap. The frequency of mutation is minimized by the very short lengths (often just one nucleotide) of DNA synthesized.

This chapter has outlined several of the major types of DNA damage and the DNA repair pathways known to correct, or traverse, the damage. The variety of repair pathways, types of damage they repair, and notable enzymes involved are summarized in **Table 12-6**.

TABLE 12-6

Overview of DNA Repair Processes

Repair Process	Examples of Damage Repaired	Notable Enzyme(s)
Mismatch repair	Mismatched base pairs incorporated by polymerase	MutS/L (Table 12-2)
Base excision repair	Oxidative damage; alkylated bases	DNA glycosylases (Table 12-3)
Photoreactivation	Pyrimidine dimers caused by UV light	DNA photolyase
Nucleotide excision repair	Bulky alkylated bases; pyrimidine dimers	Excinuclease (Table 12-4)
Double-strand break repair	DNA double-strand breaks	RecA recombinase
Translesion synthesis	Abasic sites; alkylated bases; pyrimidine dimers	TLS Pols (Table 12-5)

SECTION 12.3 SUMMARY

- Cells have diverse and robust systems that repair DNA and usually restore it to its original sequence.
- The mismatch repair system corrects nucleotide residues misincorporated during replication.
- Some types of lesions are repaired directly, such as photoreversal of pyrimidine dimers by photolyase.
- The base excision repair pathway corrects relatively small, single-base lesions and uses different DNA glycosylases to recognize particular lesions.
- The nucleotide excision repair system repairs bulky lesions by using an excinuclease that makes strand incisions on either side of the lesion. Transcription-coupled repair adapts the nucleotide excision repair system to lesions identified by a stalled RNA polymerase.
- Double-strand DNA breaks result in fragmented chromosomes and are usually repaired by homologous recombination, a high-fidelity process. Double-strand breaks can also be processed by error-prone nonhomologous end joining.
- Cells contain multiple specialized DNA polymerases that extend DNA across lesions, but translesion synthesis is usually an error-prone process that results in a mutation.

? UNANSWERED QUESTIONS

Many mysteries about DNA repair remain to be solved. Basic research in this area currently focuses on the detailed mechanisms of the various pathways. Future studies may well uncover entirely new pathways of DNA repair. We also need better ways of confirming whether particular chemicals are true human carcinogens. Only then can legal controls be put in place as safeguards against these potentially lethal agents in the environment.

1. Could anticancer drugs overcome the checkpoint controls of the cell cycle and cause the cell to divide before DNA damage has been repaired? Little is known about the detailed workings of how cell cycle checkpoint control is initiated and regulated by DNA damage.

2. How does DNA repair interweave with other DNA metabolic processes? Repair must be coordinated with other processes of DNA metabolism. For example, enzymes of the recombination, replication, and transcription pathways may encounter lesions, and they need to couple their actions to repair enzymes. Our current understanding of these multiprotein DNA metabolic pathways—transcription, replication, and repair—suggests that they occur in a linear order of independent events. However, it seems likely that the several protein participants in any given pathway are highly coordinated and may even function within some type of superstructure.

3. How do translesion DNA polymerases coordinate their low-fidelity actions with the high-fidelity polymerase at a replication fork? The primary replicase must have high fidelity, by its very nature. When encountering a template lesion, the replisome must somehow yield to low-fidelity translesion DNA polymerases to carry the DNA across the lesion, otherwise the fork might collapse and the cell could die. How do low-fidelity translesion polymerases take over at a replication fork, yet prevent fork collapse? How does the primary replicase regain access to the replication fork after the translesion polymerase has finished its work?

4. How does the eukaryotic cell control the timing and activity of all of its many DNA polymerases? The eukaryotic cell contains numerous TLS polymerases. How does the cell know when to use each of these different polymerases? Is the activity of the many DNA polymerases in the cell regulated by posttranslational modification, and if so, how does this affect their function?

HOW WE KNOW

Mismatch Repair in *E. coli* Requires DNA Methylation

Au, K.G., K. Welsh, and P. Modrich. 1992. Initiation of methyl-directed mismatch repair. *J. Biol. Chem.* 267:12,142–12,148.

Lahue, R.S., and P. Modrich. 1989. DNA mismatch correction in a defined system. *Science* 245:160–164.

Paul Modrich *[Source: Courtesy of Paul Modrich.]*

A landmark study by Paul Modrich and his coworkers reconstituted mismatch repair from purified proteins: MutS, MutL, MutH, SSB, helicase II (UvrD), exonuclease I, Pol III holoenzyme, and ligase. A key to success was the insightful design of a circular duplex DNA (with a viral (V) strand and complementary (C) strand) containing a single mismatch and a GATC site about 1 kbp away from the mismatch. The researchers made two different DNAs: one methylated on the V strand and the other methylated on the C strand (Figure 1a).

In two sets of reactions, the two different DNAs were treated with combinations of MutS, MutL, and MutH, followed by agarose gel electrophoresis. One can make the prediction that if the mismatch repair system can distinguish new (unmethylated) from old (parental, methylated) DNA, only the unmethylated strand will be nicked. To distinguish which strand was cleaved, the agarose gel was analyzed with a ^{32}P-labeled DNA probe specific for either the C or V strand (Figure 1b). As a control, the restriction enzyme MboI was used to produce a marker (leftmost lane in Figure 1b), because it cleaves both strands at the GATC site, the same sequence that MutH should nick (on one strand). The result shows that MutH nicks DNA only when MutS and MutL are also present: cleavage is specific to the strand that is unmethylated.

These studies demonstrate that MutS-MutL activates MutH to cleave the unmethylated strand of a hemimethylated GATC site. Further study using this system showed that the mismatched nucleotide on the nicked strand is corrected in the presence of helicase II, exonuclease I, Pol III holoenzyme, SSB, and ligase. The detailed mechanism of the mismatch repair reaction is described in Section 12-3.

(a)

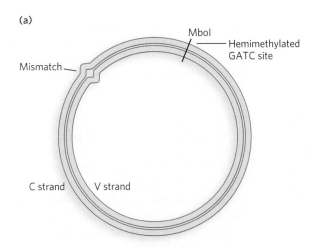

(b)

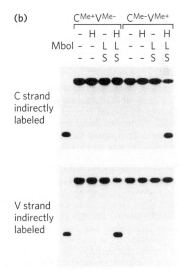

FIGURE 1 MutS and MutL are required to activate MutH, which cleaves the unmethylated strand at a GATC site. (a) The DNA substrate contained a mismatch and a GATC sequence methylated on either the viral (V) strand or the complementary (C) strand. The MboI restriction site (in GATC) is shown. (b) Gel analysis of the reaction products. Proteins added to the reactions in each experiment are indicated at the top of the gel (H, L, and S represent MutH, MutL, and MutS), with MboI fragments as control. DNAs were detected by transfer to a membrane and use of a [^{32}P]DNA oligonucleotide complementary to either the V or the C strand, to produce the autoradiograph. *[Source: Lahue, R.S., and Modrich, P. (1989) Science, 245(4914): 160–164, 1989. Reprinted with permission from AAAS. Courtesy Paul Modrich.]*

HOW WE KNOW

UV Lights Up the Pathway to DNA Damage Repair

Delbecco, R. 1949. Reactivation of ultraviolet-inactivated bacteriophage by visible light. *Nature* 163:949–950.

Kelner, A. 1949. Effect of visible light on the recovery of *Streptomyces griseus* conidia from ultraviolet-irradiation injury. *Proc. Natl. Acad. Sci. USA* 35:73–79.

The first discovery of a DNA repair reaction was that sunlight can repair UV damage. The discovery involved a serendipitous observation by Albert Kelner, an astute scientist who was studying the effects of UV light on the conidia (spores) of the mold *Streptomyces griseus*, with the hope of finding mutants that produced new varieties of antibiotics. In the course of his studies, Kelner noticed that irradiated cells sometimes had a higher rate of survival (i.e., produced more colonies) than expected. This indicated the presence of a repair pathway.

The variability in survival rate suggested that the repair pathway was inducible. During repeated attempts to identify the conditions that resulted in "induction" of the UV-damage repair pathway, Kelner obtained frustratingly irreproducible results. For example, thinking that temperature influenced the reaction, he incubated cells that received the same dose of UV light at different temperatures. At first there seemed to be an effect, but two preparations treated in the same way gave results that differed more than 100-fold. In a keen observation, Kelner noticed that repair (i.e., increased survival) correlated with culture flasks that were closer to a window, and thus the elusive variable might be sunlight. Controlled experiments confirmed that sunlight was indeed the agent underlying the repair of UV damage. Exposure to visible light yielded more than 10^5 the number of colonies as the absence of light (Figure 2).

Kelner suggested that reversal of UV damage by visible light, or photoreversal, may generalize to other organisms. Having heard of Kelner's studies, Renato Delbecco, at Indiana University, looked for photoreversal in *E. coli* using a T phage. UV irradiation of T phage reduced its ability to grow in *E. coli*, as

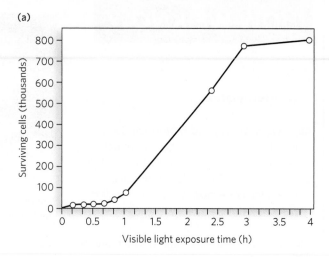

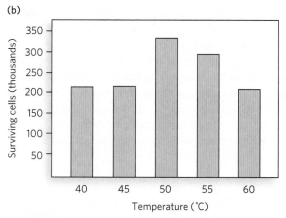

FIGURE 2 Results showing the effect on cell survival of (a) different periods of exposure to visible light and (b) different temperatures of incubation. *[Source: Data from A. Kelner, Proc. Natl. Acad. Sci. USA 35:73–79, 1949.]*

expected, but its subsequent exposure to visible light had no photoreversal effect on its viability. However, when the UV-irradiated phage were inside *E. coli* cells and the infected cells were exposed to visible light, Delbecco observed repair (increased phage viability). Hence, he concluded that visible light did not act directly to reverse UV damage, but a cellular factor was required for photorepair.

These studies were performed before scientists knew about DNA structure. We now know that UV light results in pyrimidine dimers and that photolyases directly reverse the pyrimidine cross-links.

Translesion DNA Polymerases Produce DNA Mutations

Rajagopalan, M., C. Lu, R. Woodgate, M. O'Donnell, M.F. Goodman, and H. Echols. 1992. Activity of the purified mutagenesis proteins UmuC, UmuD', and RecA in replicative bypass of an abasic DNA lesion by DNA polymerase III. *Proc. Natl. Acad. Sci. USA* 89:10,777–10,781.

Harrison (Hatch) Echols
[Source: Courtesy Carol Gross]

Bacterial cells contain proteins that produce mutations in response to DNA damage. Four genes are required for this "mutagenic response" in *E. coli*: *recA*, *lexA*, *umuC*, and *umuD*. The gene product of *umuD* is first cleaved by the enzyme RecA coprotease to produce the functional form of UmuD, called UmuD', which forms a complex with UmuC (UmuC$_1$UmuD'$_2$). Harrison (Hatch) Echols and Myron Goodman intuited that UmuC$_1$UmuD'$_2$ acts during replication, and they developed an in vitro reaction that demonstrates lesion bypass (Figure 3a). They constructed a 5.4 kb, linear, single-stranded DNA substrate with an abasic site near one end, and placed a

5'-^{32}P-labeled primer (P1) just upstream of the abasic site so that they could observe polymerization by PAGE analysis in a denaturing gel. They added RecA, or UmuC, or UmuD', along with Pol III holoenzyme and SSB, and performed each reaction in duplicate. Bypass of the abasic site is observed only in the presence of UmuC and UmuD' (Figure 3b; lanes 5 and 6 are duplicate reactions). Later studies showed that UmuC and UmuD' combine to form a distributive polymerase—one that must dissociate from and rebind to the DNA repeatedly, rather than stick to the DNA constantly as do processive polymerases (such as Pol III). The distributive nature accounts for the many bands in lanes 5 and 6 of the gel analysis. Further study also showed that UmuC is an entirely new class of DNA polymerase, with a sequence unrelated to other known DNA polymerases. The UmuC$_1$UmuD'$_2$ translesion polymerase was renamed Pol V. Soon after these studies, many other translesion DNA polymerases that are homologous to UmuC, designated Y-family polymerases, were identified from cells of all types.

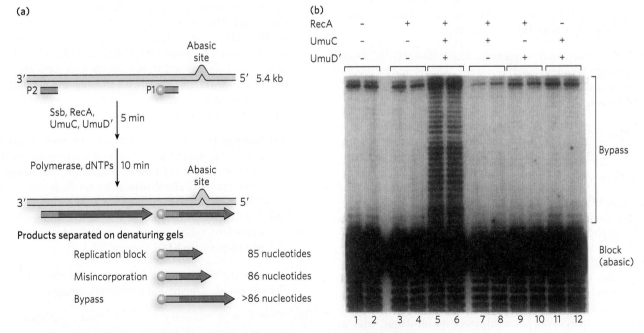

FIGURE 3 Pol V can bypass an abasic site in a DNA substrate. (a) Primer 2 (P2) is added to convert the single-stranded DNA upstream of the lesion to duplex DNA. Primer 1 (P1), labeled with ^{32}P (orange ball), is added to allow observation of extension over the template lesion. (b) PAGE analysis of the reaction products. Only in the presence of RecA, UmuC, and UmuD' is extension of primer 1 observed (lanes 5 and 6). [Source: (b) Data from M. Rajagopalan et al., Proc Natl. Acad. Sci. USA 89:10, 777–10,781, 1992, Fig. 2.]

KEY TERMS

PROBLEMS

1. The base-pairing atoms of thymine are not directly involved in the cyclobutane ring of a pyrimidine dimer, formed by UV irradiation. Why does a pyrimidine dimer stall the replicative DNA polymerase?

2. For the nucleotide sequence AAC(O^6-meG)TGCAC, with a damaged (methylated) G residue, what would be the sequence of each strand of the double-stranded DNA in the following situations?

   (a) Replication occurs before repair.
   (b) The DNA is acted upon by a glycosylase and then repaired, but only after replication has occurred.
   (c) Two rounds of replication occur, followed by repair.

3. Name three of the common ways in which DNA lesions are incurred. What is required for these DNA lesions to result in a mutation?

4. Benzo[*a*]pyrene, the cancer-causing agent in cigarette smoke, is a powerful mutagen. Benzo[*a*]pyrene itself is relatively harmless, but it is metabolized in the liver to produce active molecules that react covalently with DNA. In an experiment, benzo[*a*]pyrene is incubated with a mixture of liver enzymes to form its genotoxic metabolites. These metabolites are added to *E. coli* cells that have a mutation in a gene encoding an enzyme of the serine-synthesizing pathway (i.e., the cells are serine auxotrophs, requiring serine for growth). When the treated cells are grown on serine-containing medium, the results show that the benzo[*a*]pyrene metabolites kill cells in a dose-dependent manner. When treated and untreated serine-auxotrophic cells are plated separately on serine-free media, the cells treated with benzo[*a*]pyrene metabolites show a 10- to 100-fold increase in survivors compared with untreated cells. Explain these results.

5. In the experiment described in Problem 4, some of the untreated serine auxotrophs were able to grow on a medium lacking serine. Why?

6. In an experiment using *S. typhimurium* histidine auxotrophs, the cells are grown on a thin layer of agar with nutrient medium that lacks histidine. The culture (~10^9 cells) produces ~13 colonies over a two-day incubation period at 37°C.

   (a) How did these colonies arise in the absence of histidine?
   (b) When the experiment is repeated in the presence of 0.4 μg of 2-aminoanthracene, the number of colonies produced over two days exceeds 10,000. What does this indicate about 2-aminoanthracene?
   (c) What can you surmise about its carcinogenicity?

7. What type of mutation is most likely to result from the following lesions (if left unrepaired)?

   (a) Deamination of cytosine: G≡C → _____.
   (a) Formation of 8-oxoguanine: G≡C → _____.
   (b) Deamination of adenine: A=T → _____.

8. The human disease known as xeroderma pigmentosum (XP) arises from mutations in at least seven different genes. The resulting deficiencies are generally in enzymes involved in some part of the pathway for nucleotide excision repair. The various types of XP are denoted A through G (XPA, XPB, etc.), with a few additional variants lumped together under the label XP-V. Cultures of fibroblasts from healthy individuals and from patients with XPG are irradiated with UV light. The DNA is isolated and denatured, and the resulting single-stranded DNA is examined by analytical ultracentrifugation.

   (a) Samples from the normal fibroblasts show a significant reduction in the average molecular weight of the single-stranded DNA after irradiation, but samples from the XPG fibroblasts show no such reduction. Why might this be?
   (b) If you assume that an NER system is operative in fibroblasts, which step might be defective in the cells of patients with XPG? Explain.

9. Describe the most critical difference between global nucleotide excision repair and transcription-coupled repair.

10. What do base excision repair and repair of an abasic site have in common? How do they differ?

11. Many eukaryotes have a DNA glycosylase that specifically removes T residues from DNA, but only when they are paired with G. There is no comparable enzyme that removes the G residues from G—T mismatches. Why is it useful for a cell to always repair a G—T mismatch to G≡C rather than to A=T?

12. A gene is found that has a sequence of 11 contiguous A residues in one strand. Mutations occur at an elevated frequency in this gene, mostly in the region with the

repeated A residues. Most of these mutations result in inactivation of the encoded protein, with many amino acids either missing or altered. What type of mutations would account for these observations, and how might they occur?

13. Many bacteria, including *E. coli*, are capable of growing under both anaerobic and aerobic conditions. Some mutations are introduced into an *E. coli* strain that inactivate several enzymes involved in DNA repair. The mutant strain grows normally when kept in an incubator with a 100% nitrogen gas atmosphere. However, the strain dies when exposed to a normal laboratory atmosphere. Why?

14. In humans, a fetus lacking at least one good copy of a gene encoding any one of dozens of key DNA repair enzymes is usually nonviable. If the fetus has one good copy of the gene and one mutant (inactive) copy, the individual will have a fully functional DNA repair system, but a higher than normal probability of acquiring cancer in middle age. Explain.

15. In an *E. coli* cell, DNA polymerase III makes a rare error and inserts a G opposite an A residue at a position 850 bp away from the nearest GATC sequence. The mismatch is accurately repaired by the mismatch repair system. How many phosphodiester bonds derived from deoxynucleotides (dNTPs) are expended in this repair process? ATPs are also used in this process. Which enzymes consume the ATP?

16. If an oxidative lesion occurs spontaneously in a single-stranded DNA fragment generated on the lagging strand during replication, it is not readily repaired by nucleotide excision repair or base excision repair. Explain why.

17. O^6-Methylguanine lesions are repaired directly by transfer of the methyl group to O^6-methylguanine methyltransferase. A very high level of metabolic energy is invested in this simple methyl transfer reaction. Describe this energy investment.

DATA ANALYSIS PROBLEM

Tang, M., X. Shen, E.G. Frank, M. O'Donnell, R. Woodgate, and M.F. Goodman. 1999. UmuD$'_2$C is an error-prone DNA polymerase, *Escherichia coli* pol V. *Proc. Natl. Acad. Sci. USA* 96:8919–8924.

18. Myron Goodman and colleagues discovered the *E. coli* DNA polymerase V, the first example of a translesion DNA polymerase, using a reconstituted system including Pol III, the UmuD' and UmuC proteins, and RecA protein. A control experiment was carried out in which Pol III was left out of the reaction mixture, and translesion DNA synthesis still occurred. This was the first suggestion that UmuC and UmuD' might have DNA polymerase activity, but more work was needed to prove it.

(a) Suggest why the control experiment might not be definitive.

Goodman and colleagues cultured *E. coli* cells that expressed Pol V, using a strain with mutations giving rise to both an inactivated Pol II and a temperature-sensitive Pol III.

(b) Suggest why the investigators did not use a strain with a completely inactivated Pol III.

The researchers partially purified Pol V and then carried out a gel filtration experiment to separate Pol III from Pol V. Their results are shown in Figure 1, a Western blot analysis of the fractions from the gel filtration column, using antibodies to the α subunit of Pol III and to UmuC.

Three fractions—50, 56, and 64—were chosen for further analysis, as shown in Figure 2. DNA polymerization activity was examined on a primer template (labeled P in the gel) in which the second position in the single-stranded template had an abasic site (X). The work was done at two temperatures, 37°C and 47°C.

(c) Suggest why DNA polymerization is greater for fraction 56 than for fraction 50.
(d) Why does the DNA polymerization activity decline in fractions 50 and 56 at 47°C relative to 37°C?
(e) The activity of fraction 64 does not decline with increased temperature. Suggest why.

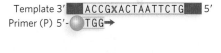

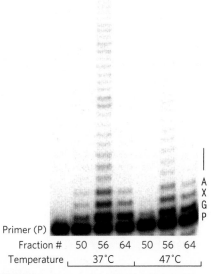

FIGURE 2 [*Source: Tang, M., Shen, X., Frank, E.G., O'Donnell, M., Woodgate, R., and Goodman, M.F. (1999) UmuD' 2C is an error-prone DNA polymerase, Escherichia coli pol V. Proc. Natl. Acad. Sci USA 96, 8919–8924, Fig. 2.]*

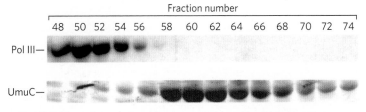

FIGURE 1 [*Source: Tang, M., Shen, X., Frank, E.G., O'Donnell, M., Woodgate, R., and Goodman, M.F. (1999) UmuD' 2C is an error-prone DNA polymerase, Escherichia coli pol V. Proc. Natl. Acad. Sci USA 96, 8919–8924, Fig. 1.]*

ADDITIONAL READING

General

Friedberg, E.C. 2007. A brief history of the DNA repair field. *Cell Res.* 18:3–7. An excellent overview of DNA repair over the past 50 or more years.

Hanawalt, P.C. 2007. Paradigms for the three Rs: DNA replication, recombination, and repair. *Mol. Cell* 28:702–707.

Types of DNA Mutations

Abeysinghe, S.S., N. Chuzhanova, and D.N. Cooper. 2006. Gross deletions and translocations in human genetic disease. *Genome Dyn.* 1:17–34.

Dion, V., and J.H. Wilson. 2009. Instability and chromatin structure of expanded trinucleotide repeats. *Trends Genet.* 25:288–297.

Maki, H. 2002. Origins of spontaneous mutations: Specificity and directionality of base-substitution, frameshift, and sequence-substitution mutageneses. *Annu. Rev. Genet.* 36:279–303.

Orr, H.T., and H.Y. Zoghbi. 2007. Trinucleotide repeat disorders. *Annu. Rev. Neurosci.* 30:575–621.

Seo, K.Y., S.A. Jelinsky, and E.L. Loechler. 2000. Factors that influence the mutagenic patterns of DNA adducts from chemical carcinogens. *Mutat. Res.* 463:215–246.

DNA Alterations That Lead to Mutations

Batista, L.F., B. Kaina, R. Meneghini, and C.F. Menck. 2009. How DNA lesions are turned into powerful killing structures: Insights from UV-induced apoptosis. *Mutat. Res.* 681:197–208.

Deweese, J.E., and N. Osheroff. 2009. The DNA cleavage reaction of topoisomerase II: Wolf in sheep's clothing. *Nucleic Acids Res.* 37:738–748.

Friedberg, E.C., G.C. Walker, W. Siede, R.D. Wood, R.A. Schultz, and T. Ellenberger. 2006. *DNA Repair and Mutagenesis,* 2nd ed. Washington, DC: American Society for Microbiology.

Ohnishi, T., E. Mori, and A. Takahashi. 2009. DNA double-strand breaks: Their production, recognition, and repair in eukaryotes. *Mutat. Res.* 669:8–12.

Mechanisms of DNA Repair

Cleaver, J.E. 2003. Classics in DNA repair: Photoreaction. *DNA Repair (Amst.)* 2:629–638. The dispute arising over priority in the discovery of photoreactivation is revisited and placed in perspective.

Cox, M.M. 2007. Regulation of bacterial RecA protein function. *Crit. Rev. Biochem. Mol. Biol.* 42:41–63.

Goodman, M.F., and B. Tippin. 2000. Sloppier copier DNA polymerases involved in genome repair. *Curr. Opin. Genet. Dev.* 10:62–168.

Heller, R.C., and K.J. Marians. 2006. Replisome assembly and the direct restart of stalled replication forks. *Nat. Rev. Mol. Cell Biol.* 7:932–943.

McCulloch, S.D., and T.A. Kunkel. 2008. The fidelity of DNA synthesis by eukaryotic replicative and translesion synthesis polymerases. *Cell Res.* 18:148–161. An excellent summary of translesion polymerases.

Sancar, A., L.A. Lindsey-Boltz, K. Unsal-Kaçmaz, and S. Linn. 2004. Molecular mechanisms of mammalian DNA repair and the DNA damage checkpoints. *Annu. Rev. Biochem.* 73:39–85.

13

Recombinational DNA Repair and Homologous Recombination

Lorraine Symington [Source: Courtesy Lorraine Symington.]

MOMENT OF DISCOVERY

We recently figured out how double-strand DNA breaks (DSBs) are processed in eukaryotic cells, as a first step in generating meiotic crossovers. This story started some years ago, when Jack Szostak showed that DSBs stimulate homologous recombination of DNA and function as natural initiators for recombination during meiosis. Researchers also found that DNA ends produced by DSBs undergo degradation of the 5′-ending strand to produce 3′ single-stranded DNA tails, which are required to initiate homologous recombination. Around the same time, Jim Haber found the same thing to be true for the very similar recombination-based process of mating-type switching in yeast. *But how are the 3′ single-stranded tails made?*

We had been using genetic and biochemical approaches over a number of years, with the expectation that a single enzyme, probably a specialized nuclease, was responsible. We anticipated that a single mutant in yeast would completely block 3′-tail formation, but we were unsuccessful in our efforts to find it. Then a new student arrived in my lab, Eleni Mimitou, and she hypothesized that helicases might also play an important role. The first helicase-encoding gene she deleted from yeast, *SGS1*, produced a profound defect in double-strand break processing! The role of Sgs1 at such an early stage of recombination was quite surprising because prior studies had shown a role for it in a late stage of recombination.

Eleni went on to show that the function of Sgs1 is partially redundant with that of a nuclease called Exo1. When she deleted both *SGS1* and *EXO1* from yeast, the resulting strain produced only partially processed DNA ends during recombination, a result that was obvious from the very first experiment that Eleni looked at. She also found that another protein, Sae2, was required to complete the initial processing of DSBs. These results and other data led us to propose a two-step mechanism for the production of 3′ single-stranded tails. After 15 years of working on recombination, all of these results came together in about six months! These "moments" of discovery in science make all of the struggles in between worthwhile.

—*Lorraine Symington, on discovering how DNA ends are processed to initiate DNA recombination*

Genetic recombination is the exchange of genetic information between chromosomes or between different chromosomal segments in a single chromosome. Such exchanges occur by several mechanisms. Homologous genetic recombination, often simply termed **homologous recombination**, encompasses genetic exchanges at sequences that are identical, or nearly so, in both DNA segments involved in the recombination. Any sequence will do, as long as it is shared by the DNAs undergoing an exchange. Chapter 14 explores other forms of recombination, including exchanges that require a specific sequence (site-specific recombination and some forms of transposition) and some that can occur almost randomly (other forms of transposition). The idea of processes with the potential to scramble genetic information may seem incompatible with the DNA replication and repair processes that so thoroughly maintain genomic integrity. But for the most part, the disconnect is illusory. First and foremost, homologous recombination is a DNA repair process, directed at the site of double-strand breaks. At their core, homologous genetic recombination and the most common and accurate form of **double-strand break repair**, called **recombinational DNA repair**, are exactly the same process and are carried out by the same enzymes. We begin our discussion by focusing on recombination as a repair process.

A **double-strand break (DSB)** is the most dangerous of all DNA lesions, and the most lethal if left unrepaired. Double-strand breaks usually arise during DNA replication, when replication forks encounter a single-strand break in a template strand. In DNA metabolism, this is a true show-stopper. Broken DNA ends make it impossible for DNA replication to continue. Double-strand breaks can also arise during exposure to UV light or γ radiation. In mammals, partial deficiencies in DSB repair systems have been linked with a genetic predisposition to many forms of cancer. Two genes most closely associated with an inherited predisposition to human breast cancer, *BRCA1* and *BRCA2* (encoding *br*east and ovarian *ca*ncer type *1* or type *2* susceptibility proteins), are intimately involved in this repair process. A wide range of human genetic diseases that are characterized by genomic instability, developmental abnormalities, light sensitivity, as well as cancer predisposition, have been traced to defects in additional genes involved in homologous recombination and recombinational DNA repair. Like the human DNA repair enzymes described in Chapter 12, the proteins involved in recombinational DNA repair are genomic guardians. If a mammalian embryo completely lacks the capacity for repairing DSBs, that embryo is never born. Its cells divide a few times and then die, the genome reduced to fragments arising from countless failed attempts to repair stalled replication forks. The capacity for the enzymatic repair of DSBs is inherent to every free-living organism.

The need to repair replication forks probably fueled the evolution of recombination systems. DNA damage is common. Oxygen first appeared in the atmosphere 2.3 billion years ago as photosynthesis evolved. However, the advantages of aerobic metabolism could not be fully realized until cells also developed the means to deal with oxidative DNA damage. A bacterial cell grown in an oxidative environment suffers more than 1,000 DNA lesions per cell per generation, and a typical mammalian cell, more than 100,000 DNA lesions every 24 hours. This omnipresent spontaneous DNA damage may have limited the size of a genome that could be replicated successfully in an aerobic organism, until the advent of systems for reconstituting and restarting collapsed replication forks.

Homologous recombination and recombinational DNA repair systems now have a broader range of functions in diploid organisms. In eukaryotes, the recombinational DNA repair machinery facilitates the accurate transmission of large chromosomes from one generation to the next, in addition to repairing DSBs. DSBs are introduced in every chromosome during meiosis. The resulting recombination provides a link between replicated sister chromosomes (chromatids) and ensures their accurate segregation at cell division. This same recombinational DNA repair process also produces chromosomal crossovers as a byproduct, exchanging large segments of genetic material between homologous chromosomes—a process that makes a significant contribution to the genetic diversity that fuels evolution. The study of homologous recombination was originally inspired by its effect on inheritance. To a great extent, the genetic recombination and recombinational DNA repair systems in every organism made the development of the entire science of genetics possible.

Cellular recombinational DNA repair systems are co-opted in additional processes, such as those that trigger changes in fungal mating types, allow pathogenic bacteria to evade host immune systems, and sometimes consummate horizontal gene transfer through genetic exchanges between cellular chromosomes and foreign DNA (see Figure 1-11). In other words, although homologous recombination began as a process of DNA repair, it has evolved into a broader mechanism that allows populations of organisms to genetically adapt more quickly to their environment.

The recombinational DNA repair of damaged replication forks is the centerpiece of our discussion. The resurrection of collapsed replication forks represents a fascinating intersection of every aspect of DNA metabolism—replication, repair, and recombination. Our examination of recombination thus begins with replication. We then expand the discussion to include the recombination processes of bacteria and eukaryotes in a variety of contexts, as well as some alternative paths to the repair of double-strand breaks.

13.1 RECOMBINATION AS A DNA REPAIR PROCESS

As discussed in Chapter 12, DNA damage is common and highly deleterious. The most important consequences of the damage do not become apparent until the DNA is replicated. When a lesion exists in the DNA template, one of several things can happen: replication continues over the lesion, leaving it in place; lesion repair is initiated but not completed, such that a break in the template strand is present when the replication fork arrives; replication stalls, requiring repair before it can continue; or replication stalls but picks up again at a location downstream of the lesion.

In the first possible outcome, DNA synthesis continues over the lesion by translesion synthesis (TLS), a process described in Chapter 12 (**Figure 13-1a**; see also

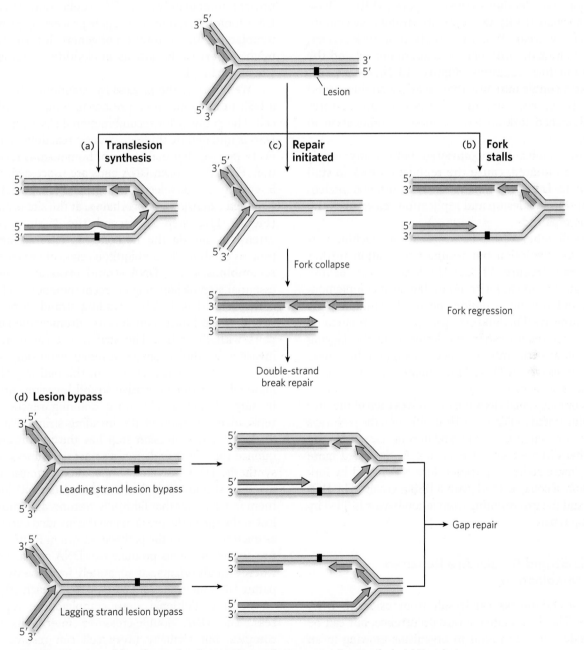

FIGURE 13-1 Possible effects of a damaged template on a replication fork. (a) Translesion synthesis occurs when the replisome encounters a lesion and keeps going. (b) If the replisome encounters a lesion that is undergoing repair, a single-strand break can cause the replication fork to collapse, creating a double-strand break. (c) Some lesions prevent progress of the replisome, resulting in a stalled replication fork. (d) In some cases, the lesion is bypassed, leaving a single-strand gap, and replication continues downstream.

Figure 12-26). Normally, this occurs only with the aid of specialized DNA polymerases capable of TLS. Translesion DNA polymerases are found in all cells and are often adapted to particular repair scenarios. More rarely, the usual DNA-replicating polymerase may replicate over lesions that do not cause significant DNA distortion. For example, an O^6-meG in the template strand pairs with T rather than C, and replication of the lesion will result in a C→T transition mutation.

If the replisome encounters a lesion that is already undergoing nucleotide excision repair (NER) or base excision repair (BER), the template strand may contain a temporary break. When the replication fork arrives, one branch of the fork becomes disconnected and the replication fork collapses (**Figure 13-1b**)—a particularly catastrophic outcome that creates a double-strand break. Here, recombinational DNA repair restores an undamaged fork structure, allowing replication to restart.

Most lesions, if encountered before any repair process is initiated, cause the replication fork to stall (**Figure 13-1c**). The replisome cannot insert a nucleotide opposite the lesion and replication ceases until the lesion is repaired.

In still other cases, the replication machinery is blocked by the lesion but resumes replication further downstream (**Figure 13-1d**). The lesion is left behind in a single-strand gap, with no undamaged complementary strand present to guide the most common DNA repair pathways. This kind of bypass-plus-restart outcome occurs most readily when the lesion is in the lagging strand, because the inherent mechanism used to initiate the synthesis of new Okazaki fragments can simply continue downstream.

Recombinational DNA repair resolves each of the situations illustrated in Figure 13-1b–d, although the pathways differ in ways large and small, and in some cases even use somewhat different sets of enzymes. As described more fully below, a stalled replication fork is repaired by fork regression, a collapsed fork with a DSB is resolved by DSB repair, and the gap resulting from lesion bypass is filled by DNA gap repair.

Double-Strand Breaks Are Repaired by Recombination

Double-strand breaks can result from oxidative DNA damage. This is a relatively rare occurrence, but can be a byproduct of respiration in organisms growing in an oxygen-rich environment or a consequence of exposure to ionizing radiation. These lesions destroy the continuity of both template strands, and they are generally lethal if not repaired. More commonly, DSBs arise when the replication fork encounters a break in one template

strand that is undergoing repair (see Figure 13-1b). We first consider a generalized pathway for the repair of DSBs—a pathway that will reappear in slightly different forms as we discuss specific repair pathways. The enzymes we encounter are described in more detail in Section 13.2.

The repair of double-strand breaks by recombinational DNA repair requires the presence of another, undamaged, homologous double-stranded DNA. In a diploid cell, that double-stranded DNA is either the second copy of each chromosome or the sister chromatid present immediately after DNA replication. This second DNA molecule guides the repair process by providing a template for the restoration of genetic information that might otherwise be lost as nucleotides missing at the site of the break.

We illustrate the process by examining the repair of a DSB in a chromosome present in a diploid eukaryotic cell. The pathway for recombinational DNA repair of the DSB is initiated by three enzymatic reactions that define every process that makes use of homologous recombination. First, the broken DNA ends are processed, with the 5′-ending strands selectively degraded to create 3′ single-stranded extensions, or overhangs, at the site of the break (**Figure 13-2**, step 1). Second, the 3′ single-stranded extensions invade the homologous chromosome, in a process catalyzed by a ubiquitous class of enzymes called **recombinases**. This **DNA strand invasion** is the quintessential step of homologous recombination and all processes related to it. The invading strand displaces one strand of the intact homologous chromosome and base-pairs with the other. The structure created by strand invasion by the 3′ single-stranded extension is sometimes referred to as a D-loop. In the pathway shown in Figure 13-2, two consecutive strand invasions are shown in steps 2 and 3. The third defining reaction is the replicative extension of the invading strand. The use of 3′ ends for the invasion step has the important consequence that these ends can also act as primers for DNA synthesis. DNA polymerase–mediated extension of the invading strands after they are paired (step 4) lengthens them in a manner that faithfully restores any information lost at the site of the break, using the invaded chromosome as the template. For the pathway shown in Figure 13-2, the two strand invasions produce two DNA crossover points, where strands originating separately from the two participating DNA molecules are base-paired to each other.

Several additional steps complete the repair process. The DNA double-crossover intermediate seems complex, but virtually every cell can resolve it in at least two ways. First, the now-lengthened invading strands can simply be displaced by the action of helicases and then anneal to each other (**Figure 13-3a** on p. 454). Any remaining gaps can be filled by DNA polymerases. DNA ligases complete the repair by

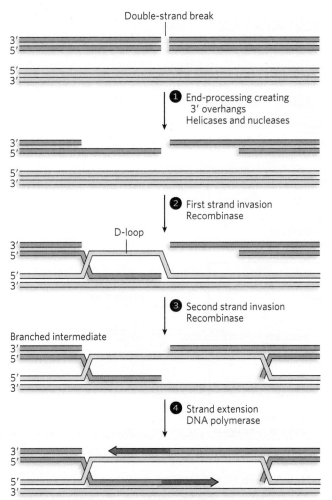

FIGURE 13-2 The repair of chromosomal double-strand breaks. Accurate repair of a DSB requires an undamaged source of duplicate genetic information—that is, homologous double-stranded DNA. The broken ends are first processed to generate 3′ single-stranded extensions, and both ends are used for strand invasion of the homologous double-stranded DNA to form D-loops. The invading 3′ ends are extended by DNA polymerases.

ligating the ends. This quite common pathway is known as **synthesis-dependent strand annealing (SDSA).**

In the second possible pathway (**Figure 13-3b**), double-strand break repair (DSBR), replication is completed by ligating the strands while they are still linked. A four-branched crossover junction, with all DNA strands intact such that each branch is a segment of duplex DNA, is called a **Holliday intermediate** (or Holliday junction)—named for the geneticist Robin Holliday, who proposed the first recombination model that included these structures. Two Holliday junctions are generated in the pathway shown in **Figure 13-3b**. Specialized

Robin Holliday, 1932–2014
[Source: Courtesy Robin Holliday.]

endonucleases present in all cells, called **Holliday junction resolvases**, recognize and cleave the Holliday intermediates in one of two ways. In both, cleavage by the resolvases creates products that are viable chromosomes with a complete set of genes. The breaks left behind by the cleavage are sealed by DNA ligase. The cleavage and rejoining may result either in the retention of the chromosomal DNA segments originally linked on either side of the repair site or in an exchange of that information, resulting in a **genetic crossover**. In Figure 3-13b, the former result occurs if both Holliday intermediates are cleaved at the sites labeled X or both are cleaved at the sites labeled Y. A genetic crossover occurs if one Holliday junction is cleaved at the X sites and the other at the Y sites, such that the genetic material extending from the site of repair to the telomere is transferred between chromosomes.

In principle, recombination can occur between any two DNA segments that share sequence similarity. Thus, deleterious events may occur. For example, recombination of two repeated sequences on the same chromosome could lead to complete deletion of all the genetic information between these sequences. Such deleterious events are rare, given the very tight regulation imposed on homologous recombination systems in all cells.

With this overview in mind, we now look at how the various steps of SDSA and DSBR are applied to the repair of replication forks. When replication forks stall or collapse, recombinational DNA repair has a major advantage over translesion DNA synthesis: it does not cause mutations.

Collapsed Replication Forks Are Reconstructed by Double-Strand Break Repair

When a replication fork encounters a break in a template strand, one arm becomes detached (see Figure 13-1b). The resulting double-strand break and collapse of the fork triggers a recombinational repair process. Here, and in some other pathways we will soon encounter, the steps shown in Figure 13-2 will become quite familiar. In brief, repair requires the reattachment of the broken arm to recreate the fork. The reattachment involves processing of the broken DNA end to create a 3′ single-stranded extension. A DNA strand invasion reaction is then mediated by a recombinase.

The reconstruction of a replication fork by recombinational DSBR is shown in **Figure 13-4a** on p. 455. The broken DNA end is processed by nucleases to remove a segment of the strand with the free 5′ end at the break (step 1). This creates the 3′ single-stranded DNA extension. A recombinase binds to the single-stranded DNA and promotes a strand invasion of the intact portion of the chromosome, so that the single-stranded extension is paired with its complementary strand (step 2). At the

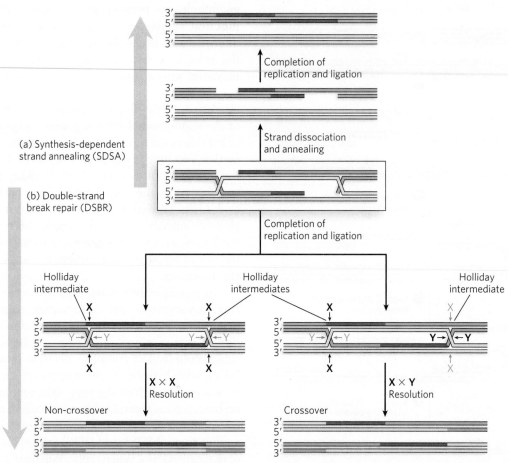

FIGURE 13-3 Two paths for completing double-strand break repair. (a) In the SDSA pathway, the invading strands dissociate and anneal to each other. Further replication and ligation complete the process. (b) In the DSBR pathway, the two Holliday intermediates are cleaved by Holliday intermediate resolvases (see Section 13.2). Resolution of the Holliday intermediates can yield two different results with respect to the DNA flanking the repair site, as described in the text.

point of insertion, the other strand of the invaded duplex DNA is displaced. Once the strand invasion is complete, the repair of a replication fork often diverges from the path shown in Figure 13-2 in that the invading strand is not extended. Instead, reconstruction of the replication fork requires the action of enzymes that promote a process called **branch migration** (step 3; **Figure 13-4b** shows the details of this process). A DNA branch is formed when separate segments of at least one DNA strand are paired with two different partner strands. In branch migration, DNA branches are moved along the DNA, with some base pairs forming and others being disrupted, but with no net increase or decrease in number of paired nucleotides. Branch migration can occur spontaneously in branched DNA molecules and takes place through a kind of random walk mechanism, with movement occurring in either direction. In a cell, branch migration is usually catalyzed such that it moves in one direction only. Branch migration may create a Holliday intermediate, which is resolved by the specialized Holliday intermediate

resolvases, along with DNA ligase (see Figure 13-3a, step 4). The reconstructed fork is then ready for renewed replication, and no mutations have been added in the process. Replication is restarted with the aid of a dedicated replication restart complex. In bacteria, the replication restart enzymes reload the replicative DnaB helicase onto the DNA. The other replisome components load spontaneously onto DnaB, and replication begins anew.

A Stalled Replication Fork Requires Fork Regression

When the replication fork encounters a template strand lesion that cannot be surmounted, the fork stalls at the lesion. In some cases, replication restarts downstream; in others, replication simply halts until the lesion is repaired. In either case, the lesion is left without a complementary strand. Most of the DNA repair processes described in Chapter 12—nucleotide excision repair, base excision repair, and mismatch repair—require that the

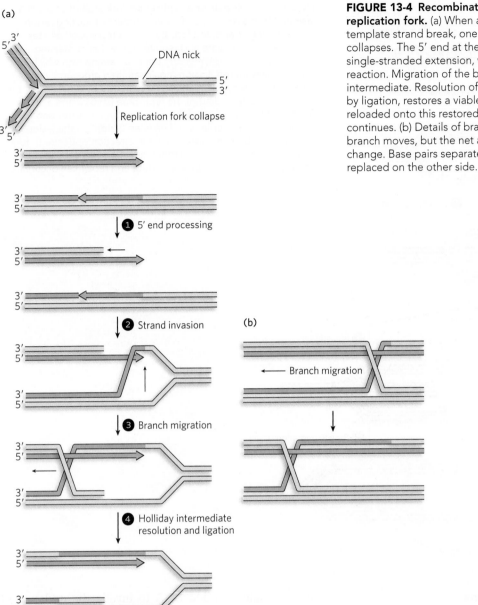

(a)

DNA nick

Replication fork collapse

① 5' end processing

② Strand invasion

③ Branch migration

④ Holliday intermediate resolution and ligation

(b)

← Branch migration

FIGURE 13-4 Recombinational DNA repair at a collapsed replication fork. (a) When a replication fork encounters a template strand break, one arm of the fork is lost and the fork collapses. The 5' end at the break is processed to create a 3' single-stranded extension, which is used in a strand invasion reaction. Migration of the branch can create a Holliday intermediate. Resolution of the Holliday intermediate, followed by ligation, restores a viable replication fork. The replisome is reloaded onto this restored fork (not shown), and replication continues. (b) Details of branch migration. In this process, the branch moves, but the net amount of duplex DNA does not change. Base pairs separated on one side of the branch are replaced on the other side.

lesion occur in only one strand of a duplex DNA. In these repair mechanisms the lesion is simply removed from the damaged strand, and the undamaged complementary strand guides the replication and ligation needed to fill in the missing nucleotide(s). For a lesion left in a single-strand gap, the reactions shown in **Figure 13-5** are directed at providing a complementary strand to allow normal repair of the lesion and to complete the process without creating a mutation.

In many cases, branch migration forces the fork backward, a process known as **fork regression** (see Figure 13-5, step 1). As the fork migrates in reverse, the lesion-containing strand is reunited with its original complementary strand, and the newly synthesized DNA strands are paired with each other. This results in a

four-branched Holliday intermediate at the point where the four strands intersect. The regression leaves the lesion paired with its undamaged complementary strand, allowing its repair by processes such as NER (see Chapter 12). If the lesion is repaired, the Holliday intermediate can be resolved by exonucleolytic degradation of the short DNA arm, as shown in Figure 13-5 (steps 2 and 3, left). A second method of resolving the Holliday intermediate can occur whether or not the lesion has been repaired. This method entails replication of the short DNA arm, followed by branch migration in the opposite direction (steps 2 and 3, right). Even if the lesion has not been repaired at this point, it is now paired with a complementary DNA strand that can be used as a template for later repair. In either case, replication must be restarted.

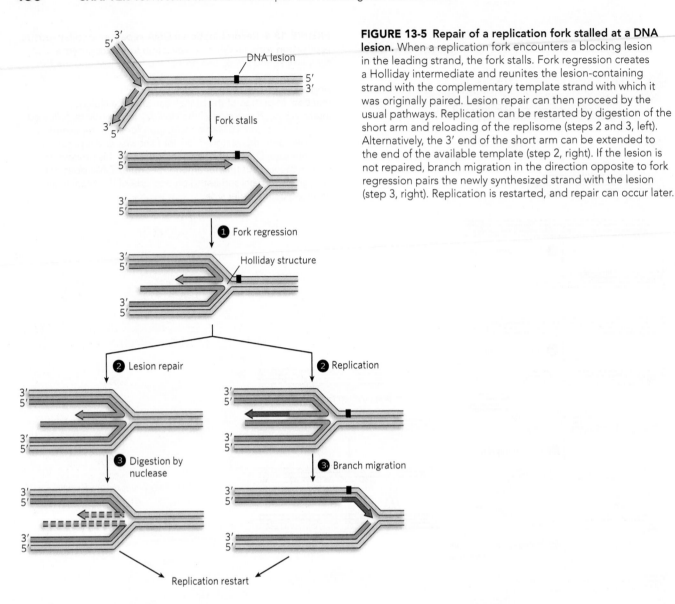

FIGURE 13-5 Repair of a replication fork stalled at a DNA lesion. When a replication fork encounters a blocking lesion in the leading strand, the fork stalls. Fork regression creates a Holliday intermediate and reunites the lesion-containing strand with the complementary template strand with which it was originally paired. Lesion repair can then proceed by the usual pathways. Replication can be restarted by digestion of the short arm and reloading of the replisome (steps 2 and 3, left). Alternatively, the 3' end of the short arm can be extended to the end of the available template (step 2, right). If the lesion is not repaired, branch migration in the direction opposite to fork regression pairs the newly synthesized strand with the lesion (step 3, right). Replication is restarted, and repair can occur later.

Single-Stranded DNA Regions Are Filled In by Gap Repair

In some cases, DNA polymerase bypasses a DNA lesion by halting and then continuing DNA synthesis upstream. This leaves the lesion within a single-strand gap, where it is not readily repaired. The repair pathway must then generate an undamaged complementary strand. The gap can be filled in by translesion synthesis (see Figure 12-26), but TLS can be mutagenic and is usually not the pathway of first choice. Alternatively, recombinational DNA repair can again be used, in a variant called **gap repair** (**Figure 13-6**). In this mechanism, the complementary strand needed for repair of the lesion is found in the undamaged arm of the replication fork.

The steps shown in Figure 13-6 follow a pattern that should now be familiar, with some subtle variations. A recombinase binds to the DNA in the single-strand gap. The recombinase promotes a strand invasion, using the bound single-stranded DNA to invade the undamaged double-stranded DNA on the other side of the replication fork (step 1). In this instance, the invading DNA has no free end. Branch migration (step 2) generates two Holliday intermediates. The displaced strand serves as an undamaged template for replication (step 3). Both Holliday intermediates are then cleaved by Holliday intermediate resolvases, and the nicks are sealed by DNA ligase (step 4). The resolution can result in a crossover; however, many cells have enzymatic systems that direct most resolution events into a non-crossover pathway. In the end, the strand that had the DNA lesion now has an undamaged complement, and nonmutagenic repair by NER or BER can begin to remove the lesion.

The pathways described in Figures 13-4, 13-5, and 13-6 provide responses to almost any deleterious event that can befall a replication fork. Recombinases, acting in concert with DNA polymerases, ligases, Holliday

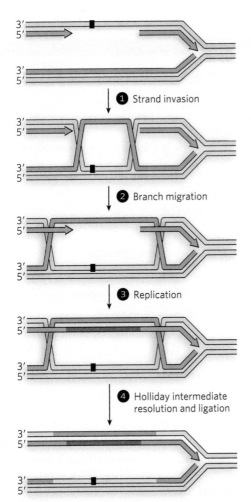

FIGURE 13-6 Repair of a DNA gap after the replication fork bypasses a lesion. If a single-strand gap is left behind by the replication fork, the information for its repair must come from the other side of the fork. A recombinase-mediated strand switch provides an undamaged template strand, which is subsequently used to extend the discontinuous strand in the gap. The process is completed by resolution of the Holliday intermediates and ligation.

intermediate resolvases, and proteins that promote branch migration, provide a repair system that readily adapts to the variety of stalled or collapsed fork structures that may present themselves. In the next section, we consider these enzymes in more detail.

SECTION 13.1 SUMMARY

- Recombination is a pathway for the repair of single-strand and double-strand breaks, both of which can arise during replication.
- Double-strand break repair entails the generation of 3′ single-stranded extensions at the broken DNA ends, which are then paired with a homologous duplex region in a strand invasion reaction catalyzed by recombinases.
- Many recombination reactions produce a four-branched DNA intermediate called a Holliday intermediate.

- Recombinational DNA repair restores DNA segments with double-strand breaks to their original genomic condition, causing no mutations.
- When a replication fork encounters a single-strand break in the template DNA, the break is converted to a double-strand break, and the replication fork collapses. The fork is then repaired by double-strand break repair, and replication is restarted.
- When a replication fork encounters a nucleotide lesion, it may bypass the lesion by translesion synthesis or it may stall. Stalled forks may be repaired by a pathway involving fork regression.
- When lesions are left in single-strand gaps, recombination provides a nonmutagenic pathway for gap repair.

13.2 ENZYMATIC MACHINES IN BACTERIAL RECOMBINATIONAL DNA REPAIR

The complex pathways shown in Figures 13-2 through 13-6 have been elucidated by genetic studies carried out in bacteria and eukaryotes. Many of the enzymes that promote the steps in these pathways have been characterized in vitro, and multiple steps of some pathways have been reconstituted with purified enzymes (**Table 13-1**). Complete reconstitution of a DSB repair or a replication fork repair remains a key goal of the field. In this section we describe the major enzyme systems found in bacteria, before moving on to eukaryotic recombination systems.

RecBCD and RecFOR Initiate Recombinational Repair

In all cells, the process of recombinational DNA repair revolves around a recombinase enzyme. In bacteria, that recombinase is the RecA protein. Before RecA can act, the stage must be set by other enzymes that degrade one strand of DNA, where necessary, and load RecA onto the DNA. The RecBCD and RecFOR complexes, with three subunits each, are the initiators of recombinational DNA repair. RecBCD loads the RecA protein onto DNA to repair double-strand breaks, and RecFOR does so to repair DNA gaps.

In *E. coli*, the *recB*, *recC*, and *recD* genes encode the heterotrimeric **RecBCD** enzyme, which has both helicase and nuclease activities. This enzyme has two jobs: (1) it prepares the 3′-ending single strand by degrading the 5′-ending strand at the site of a DSB, and (2) it directly loads the RecA protein onto the prepared single-stranded DNA tail. The RecBCD enzyme binds to linear DNA at a free (broken) end and moves inward along the double helix, unwinding and degrading the DNA

TABLE 13-1

Enzymes/Proteins Involved in Bacterial Recombinational DNA Repair

Enzyme/Protein	Size of Monomer (kDa)	Functional Form	Function
RecA	38	Filament on DNA	Bacterial recombinase
RecB	134	Part of RecBCD heterotrimer	3′→5′ helicase, forms 3′ strand extensions
RecC	129	Part of RecBCD heterotrimer	Binds chi, forms 3′ strand extensions
RecD	67	Part of RecBCD heterotrimer	5′→3′ helicase, forms 3′ strand extensions
RecF	41	Part of RecFOR mediator complex	Bind DNA, load RecA onto ssDNA gap
RecO	27	Part of RecFOR mediator complex	Bind DNA, load RecA onto ssDNA gap
RecR	22	Part of RecFOR mediator complex	Bind DNA, load RecA onto ssDNA gap
RuvA	22	Tetramer	Binds Holliday junctions, functions with RuvB
RuvB	37	Hexamer	DNA translocase, promotes branch migration
RuvC	19	Dimer	Holliday junction resolvase
RecG	76	Monomer	Helicase, promotes fork regression
SSB	19	Tetramer	Binds ssDNA
Pol I	109	Monomer	Fills in gaps
DNA ligase	75	Monomer	Seals nicks

in a reaction coupled to ATP hydrolysis. Initially, both DNA strands are degraded. The structure of RecBCD reveals a great deal about how the complex works (**Figure 13-7a**). The RecB and RecD subunits are helicase motors, with RecB moving in the 3′→5′ direction along one strand and RecD moving in the 5′→3′ direction along the other strand. One domain of the RecB subunit is also a nuclease that degrades both DNA strands as they are unwound. The activity of the enzyme is altered when it interacts with a sequence referred to as **chi**, 5′-GCTGGTGG-3′, on the 3′-ending strand, which binds tightly to a site on the RecC subunit. From that point, degradation of the strand with a 3′ terminus is greatly reduced, but degradation of the 5′-ending strand is increased (**Figure 13-7b**). This process creates the single-stranded 3′ overhang that initiates virtually all recombination events, as outlined in Figure 13-2. There are 1,009 chi sequences scattered throughout the *E. coli* genome, and these sites enhance the frequency of recombination about fivefold to tenfold in their immediate vicinity (within 1,000 bp). The chi octamer sequence is greatly overrepresented in the *E. coli* chromosome (relative to a random occurrence of a random eight-nucleotide sequence), and the chi sites are not randomly oriented in the DNA. Almost all have the orientation needed to facilitate the repair of DSBs that occur during replication.

As the 3′-ending strand is turned into a single strand, it is rapidly bound by the single-stranded DNA–binding protein (SSB) (see Figure 5-3). This protein directs traffic on segments of single-stranded DNA, blocking the access of some proteins and enzymes and facilitating the access of

others. RecA is one of the proteins that is blocked, and SSB helps prevent the binding of RecA to single-strand gaps where recombination is not required (such as the transient gaps on the lagging strand during DNA synthesis). When recombinational DNA repair becomes necessary, RecA loading is mainly a process of overcoming this SSB barrier.

As we will see, the RecA protein forms a nucleoprotein filament on the DNA. The filament is formed in two steps: first the binding of a few RecA subunits (called nucleation), then extension of the filament from that initial complex. The nucleation step limits the overall process and is the step that is blocked by SSB. As the RecBCD enzyme unwinds the DNA and degrades the 5′-ending strand, a domain within the RecB subunit recruits RecA to the 3′-ending strand. This loading function of RecB serves to nucleate the RecA binding that is needed to promote strand invasion.

The RecA protein must also be loaded onto single-strand DNA gaps, such as the one shown in Figure 13-6 (top). These gaps are also bound by SSB, blocking access by RecA. In this case, the key loading factors are the RecF, RecO, and RecR proteins, collectively called **RecFOR**. The RecF protein is recruited by an unknown mechanism to the double-stranded DNA immediately adjacent to the gap, on the free 5′ end in the discontinuous strand. With the aid of RecO and RecR, the RecA protein is activated to displace SSB and nucleate the formation of the RecA filament needed for the subsequent phases of repair (**Figure 13-8**). Proteins that have the unique function of loading other proteins onto DNA are called mediator proteins. The RecFOR proteins are recombination mediators in most bacteria.

(a)

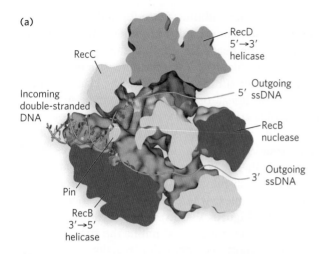

RecC

Incoming
double-stranded
DNA

RecD
5'→3'
helicase

Outgoing
5' ssDNA

RecB
nuclease

Outgoing
3' ssDNA

Pin

RecB
3'→5'
helicase

FIGURE 13-7 The RecBCD helicase/nuclease. (a) The structure of the RecBCD enzyme. (b) Activities of the RecBCD enzyme at a DNA end. The RecB and RecD subunits travel along the DNA, a process that requires ATP; RecB degrades both strands as the complex travels, cleaving the 3'-ending strand more often than the 5'-ending strand. The RecC subunit has a protrusion called a pin that facilitates separation of the DNA strands. When a chi site is encountered on the 3'-ending strand, the RecC subunit binds to it, halting the advance of this strand through the complex. Degradation of the 5'-ending strand continues as the 3'-ending strand is looped out, eventually creating a 3' single-stranded extension. RecA (not shown) is then loaded onto the processed DNA by the RecBCD enzyme. [Sources: (a) PDB ID 1W36. (b) Data from M. R. Singleton et al., Nature 432:187–193, 2004, Fig. 5.]

(b)

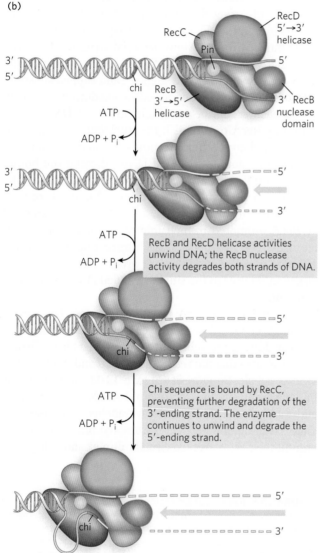

chapter). The RecA-class recombinases promote the central steps in recombination reactions. They align a single-stranded DNA with a homologous double-stranded DNA and promote a strand switch that pairs the single strand with its complement in the duplex, displacing the other duplex strand. This is the reaction that we have been referring to as strand invasion.

RecA-class recombinases are unusual among the proteins of DNA metabolism in that their active forms are ordered, helical filaments of up to several thousand subunits that assemble cooperatively (**Figure 13-9**). RecA binds most readily to single-stranded DNA. As noted above, the filament forms in two steps, first nucleating on the single-stranded DNA (the step that is blocked by SSB), then growing by the addition of subunits, primarily in the 5'→3' direction. A polar filament is created, with

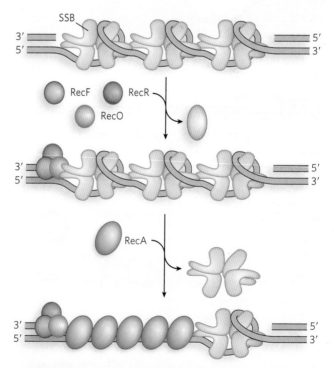

FIGURE 13-8 The role of RecFOR in loading RecA protein onto SSB-coated single-stranded DNA. SSB is displaced, and RecA is loaded.

RecA Protein Is the Bacterial Recombinase

The bacterial **RecA protein** is the prototype of a class of proteins found in all organisms, from bacteria to humans (see the How We Know section at the end of this

growth occurring at one end and most dissociation occurring at the other, trailing 5′-proximal end. RecA is a DNA-dependent ATPase. ATP is hydrolyzed by subunits throughout the RecA filament; when hydrolysis occurs in the subunit at the trailing end, it often results in

dissociation of that subunit. Once formed, the RecA nucleoprotein filament is ready to promote the DNA strand exchanges at the heart of recombination.

The RecA filament can facilitate strand exchange with a variety of substrates in vitro. The single-stranded DNA may be linear or circular, or a single-strand gap within double-stranded DNA. As RecA promotes strand invasion of these single-stranded substrates into a double-stranded DNA, a variety of structures result (**Figure 13-10a–c**). When the exchange is initiated in a DNA gap, branch migration may move the process into the adjacent duplex, where the exchange then involves four strands (**Figure 13-10d**). The reactions promoted with linear single strands best mimic the strand invasion that initiates most recombination reactions as outlined in Figure 13-2, as well as the repair of DSBs created during replication (see Figure 13-4). A circular single strand, easily purified from certain bacterial viruses, has been a particularly convenient DNA substrate for monitoring RecA activity in vitro. The single-strand-gapped substrate provides a model for reactions occurring during gap repair.

In each scenario, the single strand of DNA is first bound by RecA to establish the nucleoprotein filament. The RecA filament then promotes strand invasion into a homologous double-stranded DNA, aligning the bound single strand with the complementary strand in the duplex over a region that can involve hundreds of base pairs. The exchanged region can be further extended by RecA-promoted branch migration, in a process that requires ATP hydrolysis. The branch migration occurs at a rate of 6 bp/s at 37°C and progresses in the 5′→3′ direction relative to the single-stranded DNA in the RecA filament.

When purified RecA protein promotes DNA strand exchange in vitro between a circular single strand and a linear duplex, the substrates and products have distinctive structures that are readily separated and visualized by agarose gel electrophoresis (**Figure 13-11** on p. 462). The initial pairing of the RecA-bound single-stranded circle and the linear duplex creates a branched DNA intermediate. After a period of RecA-promoted branch migration around the circle, a nicked circular duplex and a displaced

(a)

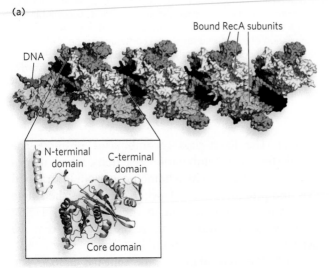

(b)

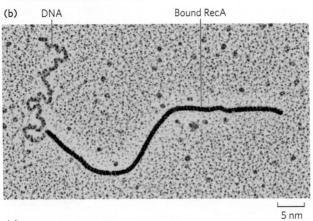

(c)

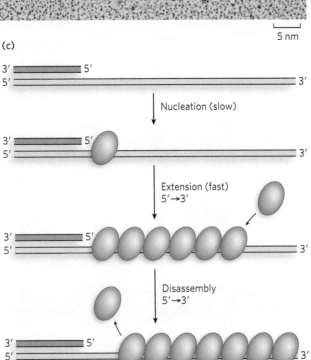

FIGURE 13-9 RecA protein filaments. RecA and other recombinases in this class function as filaments of nucleoprotein. (a) Segment of a RecA filament with four helical turns (24 RecA subunits). Notice the bound double-stranded DNA in the center. The core domain of RecA is structurally related to the domains in helicases. (b) Electron micrograph of a RecA filament bound to DNA. (c) Filament formation proceeds in discrete nucleation and extension steps. Extension occurs by adding RecA subunits so that the filament grows in the 5′→3′ direction. When disassembly occurs, subunits are subtracted from the trailing end. *[Sources: (a) Data from PDB ID 3CMX. (b) By permission of the Estate of Ross Inman. Special thanks to Kim Voss.]*

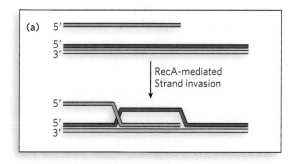

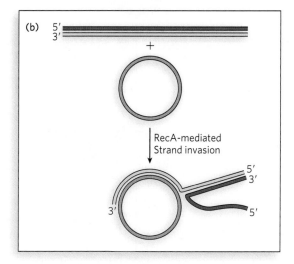

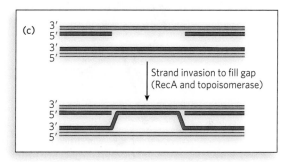

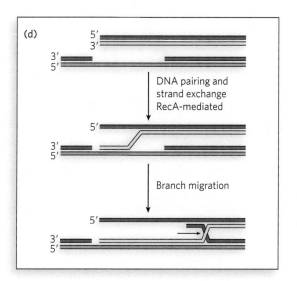

FIGURE 13-10 Recombination-related reactions promoted by the RecA protein of *E. coli*. (a) Strand invasion involving a 3′ single-stranded extension to form a D-loop. Here, RecA is promoting the strand invasion reaction outlined in Figure 13-2. (b) A DNA strand exchange involving three DNA strands. The substrates shown here are often used in research because the products are readily separated and visualized by agarose gel electrophoresis. The initial step is a strand invasion of the single-stranded DNA into one end of the duplex. Once strand invasion occurs, RecA also promotes a unidirectional (5′→3′ relative to the single-stranded circle) branch migration that completes an exchange of strands between the two DNA substrates. (c) DNA strand exchange involving a single-strand gap. Reactions of this sort also begin with strand invasion of the single strand into the duplex and are often facilitated by topoisomerases. (d) Strand exchange involving a single-strand gap and four DNA strands. The reaction begins with a strand invasion of the DNA in the single-strand gap into another duplex DNA. RecA then promotes a directional branch migration that carries the reaction into the adjacent duplex.

linear single strand are formed as reaction products. The initial DNA alignment requires ATP, but not its hydrolysis. The facilitated branch migration necessary to complete exchange is coupled to ATP hydrolysis.

If the circular substrate is a gapped double-stranded DNA (i.e., one strand is discontinuous), the DNA strand exchange is initiated in the single-strand gap, and RecA-promoted branch migration extends it into the double-stranded region (**Figure 13-12**). Because DNA is a helical structure, continued strand exchange requires an ordered rotation of the two aligned DNAs. The mechanism by which ATP hydrolysis is coupled to facilitated branch migration is not yet understood, but it probably entails the dissociation of RecA subunits at the trailing end of the RecA filament. When bound at a stalled replication fork, RecA can promote fork regression, using its capacity to promote branch migration.

RecA Protein Is Subject to Regulation

In principle, RecA-mediated recombination can occur between any two homologous DNA sequences. In every bacterial chromosome, some sequences, such as those that encode ribosomal RNAs, are repeated. Recombination between these sequences could have catastrophic consequences, leading to the deletion or rearrangement of large segments of the chromosome. For this reason, recombination, in general, and RecA protein activity, in particular, are highly regulated.

Regulation occurs at three levels: transcription of the *recA* gene, autoregulation, and regulation by other proteins. Transcriptional regulation occurs within the context of the bacterial SOS response (described in Chapter 20). Most regulation at other levels is directed at the formation, disassembly, and function of RecA protein filaments.

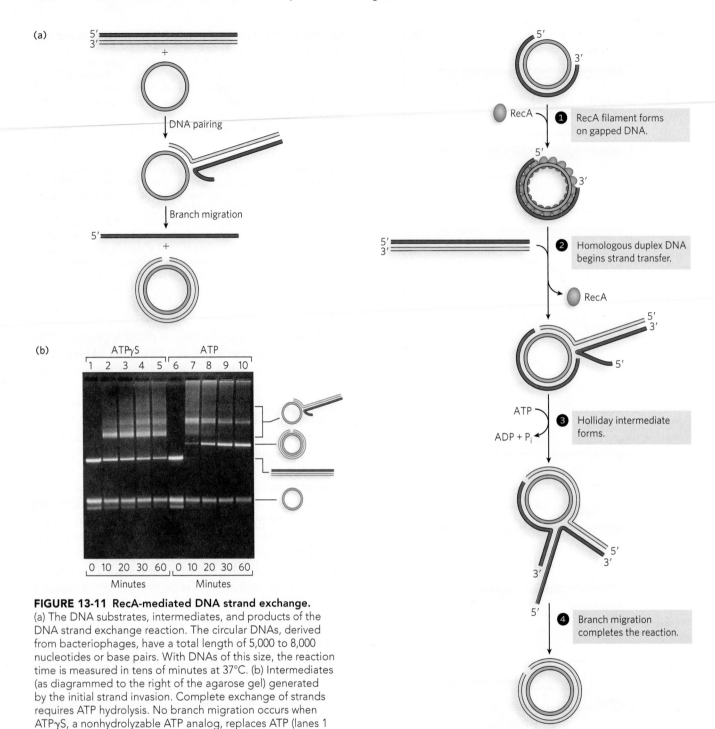

(a)

(b)

FIGURE 13-11 RecA-mediated DNA strand exchange.
(a) The DNA substrates, intermediates, and products of the DNA strand exchange reaction. The circular DNAs, derived from bacteriophages, have a total length of 5,000 to 8,000 nucleotides or base pairs. With DNAs of this size, the reaction time is measured in tens of minutes at 37°C. (b) Intermediates (as diagrammed to the right of the agarose gel) generated by the initial strand invasion. Complete exchange of strands requires ATP hydrolysis. No branch migration occurs when ATPγS, a nonhydrolyzable ATP analog, replaces ATP (lanes 1 to 5). [Source: (b) Courtesy Mike Cox.]

Autoregulation is "self" regulation. The RecA protein suppresses its own activities by means of a highly charged C-terminal peptide flap (see Figure 5-20). Removal of just 17 amino acid residues from the RecA C-terminus creates a RecA species for which almost all activities are enhanced. For example, whereas filament nucleation of native RecA protein is blocked by SSB, a C-terminal deletion mutant of RecA readily displaces SSB without the aid of RecBCD or RecFOR. In the mutant cell, this can result in elevated levels of recombination.

FIGURE 13-12 Steps in a RecA-mediated DNA strand exchange reaction involving four DNA strands. Much RecA protein remains on the DNA throughout the reaction, although filament disassembly may be coupled to the movement of the branch point. The reaction begins with only three strands, initiated in the single-strand gap of one DNA substrate. As branch migration carries the branch into the duplex adjoining the gap, a Holliday intermediate forms.

Many other proteins play a role in regulating the RecA protein. As we have seen, the RecBCD and RecFOR complexes facilitate the RecA filament nucleation process. Reliance on these loading functions helps direct RecA filament formation to DNA regions where it is needed. Another regulatory protein, RecX, binds to the growing RecA filament end and halts filament extension. A protein called DinI binds along the RecA filament and stabilizes it, while at the same time limiting the DNA strand exchange process. The helicase UvrD actively removes RecA filaments from the DNA when they are no longer needed. These and other proteins, working as an integrated system, help limit RecA function and direct it toward particular repair requirements.

Multiple Enzymes Process DNA Intermediates Created by RecA

RecA is not the only protein in a bacterial cell that can promote branch migration; other enzyme systems are specialized for that task. As one example, the processing of Holliday intermediates is facilitated by a complex called RuvAB (repair of *UV* damage). Up to two RuvA protein tetramers bind to a Holliday intermediate and form a complex with two RuvB hexamers (**Figure 13-13**). The donut-shaped RuvB hexamers surround two of the four arms of the Holliday intermediate. RuvB is a DNA translocase, related in structure and function to hexameric DNA helicases. The DNA is propelled outward

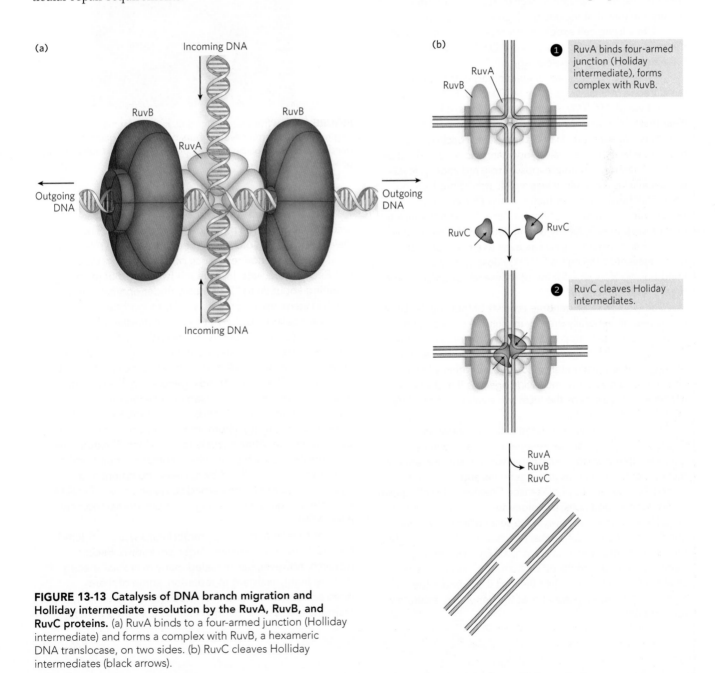

FIGURE 13-13 Catalysis of DNA branch migration and Holliday intermediate resolution by the RuvA, RuvB, and RuvC proteins. (a) RuvA binds to a four-armed junction (Holliday intermediate) and forms a complex with RuvB, a hexameric DNA translocase, on two sides. (b) RuvC cleaves Holliday intermediates (black arrows).

through the hole in the donut, away from the junction, in a reaction coupled to ATP hydrolysis. The result is very rapid branch migration that can move the position of the Holliday intermediate by thousands of base pairs in a few seconds. The RuvAB complex moves the Holliday intermediate away from the region of damaged DNA and recruits RuvC, a Holliday intermediate resolvase. RuvC replaces one of the RuvA tetramers at the junction and cleaves strands in opposing arms of the Holliday intermediate to resolve it into viable chromosomal products. The nicks in the DNA products are sealed by DNA ligase.

HIGHLIGHT 13-1   EVOLUTION

A Tough Organism in a Tough Environment: *Deinococcus radiodurans*

Some radioactive isotopes, such as ^{60}Co and ^{137}Cs, emit a type of ionizing radiation called γ rays. Gamma rays are photons; they transmit energy to atoms in solution, generating ions that include the highly reactive hydroxyl radical. In a living cell exposed to γ rays, any molecule can be damaged, including proteins and DNA. Double-strand breaks are included in the carnage.

The energy deposited by electromagnetic radiation is measured in rads or grays (1 Gy = 100 rads). For a human cell, a dose of 2 to 5 Gy is lethal. In the 1950s, it became clear that some organisms are surprisingly resistant to radiation. For example, in efforts to use radioactive sources to sterilize food, some sealed food samples were spoiled by bacterial action even after exposure to γ radiation at levels up to 4,000 Gy. The culprit was a pink, non-spore-forming, nonmotile bacterium eventually named *Deinococcus radiodurans*. *D. radiodurans* can absorb the damage inflicted by γ irradiation at 5,000 Gy with no lethality. A dose of this kind causes substantial damage even to a Pyrex beaker. More relevant to the cell, a 5,000 Gy dose produces many hundreds of DSBs, in addition to thousands of single-strand breaks and other lesions.

The *Deinococcus* genome consists of four circular DNA molecules, all generally present in multiple copies. After γ irradiation, the cells stop growing and DNA repair begins. Overlapping DNA fragments are spliced together, and the entire genome is accurately reconstituted within a few hours. The cells begin to grow and divide again as if nothing had happened. It is perhaps the most remarkable feat of DNA repair we know of so far.

This process is demonstrated in the gel shown in Figure 1. Following various treatments, *D. radiodurans* genomic DNA was isolated, treated with a restriction enzyme that cuts only a few times in the genome, and subjected to pulsed field gel electrophoresis (see Chapter 7). In cells grown under normal conditions, this procedure yields the series of large DNA fragments shown in the second lane of the gel (the first lane consists of markers of known molecular weight). Immediately after γ irradiation at 5,000 Gy, this banding pattern disappears, replaced by a smear of randomly sized, smaller DNA fragments. Over the next 3 to 4 hours, the normal band pattern reappears as the genome is accurately reconstituted.

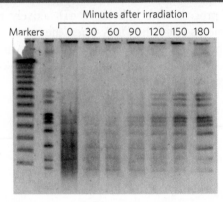

FIGURE 1 DNA from γ-irradiated *D. radiodurans* is initially fragmented, but after several hours of repair it regains its normal banding pattern. [*Source: Courtesy of John R. Battista, Louisiana State University, Baton Rouge.*]

Genome reconstitution in *D. radiodurans* is recombinational DNA repair on a massive scale. The *Deinococcus* RecA protein (DrRecA) plays a key role, and most of the radiation resistance of this organism disappears if DrRecA is inactivated. There are two stages of repair, each requiring about 90 to 120 minutes under optimal conditions. The first stage uses a process similar to synthesis-dependent strand annealing (SDSA), but with an extended phase of replication (extended SDSA, or ESDSA). Once again, the process starts with the steps outlined in Figure 13-2. The ends of the broken DNA fragments are processed so as to degrade the 5'-ending strands, generating 3' extensions (Figure 2a). These are then used in RecA-promoted strand invasion reactions. The 3' ends act as primers for extended DNA synthesis, using a homologous chromosome strand as the template. After dissociation, the long 3' extensions are annealed to each other where complementarity exists. The process is completed by nuclease treatment and ligation, as needed. The second stage again uses RecA to facilitate a larger-scale splicing of chromosomal segments (Figure 2b).

For several decades, *D. radiodurans* was considered the most radiation-resistant organism known. Recent research, however, has revealed many microbial species that are highly resistant to radiation, some of them more so than *D. radiodurans*. In addition, these highly resistant species are found in various unrelated genera,

The bacterial recombination systems that repair replication forks can also repair DSBs created by ionizing radiation, sometimes with startling proficiency. The bacterium *Deinococcus radiodurans* can survive and prosper after absorbing doses of ionizing radiation sufficient to generate thousands of DSBs (**Highlight 13-1**).

The repair of stalled or collapsed replication forks is generally followed by a restart of replication. A five-protein complex called the restart primosome loads the replicative helicase, DnaB, onto the DNA at the reconstituted replication fork. The rest of the replisome assembles around DnaB, and replication starts anew.

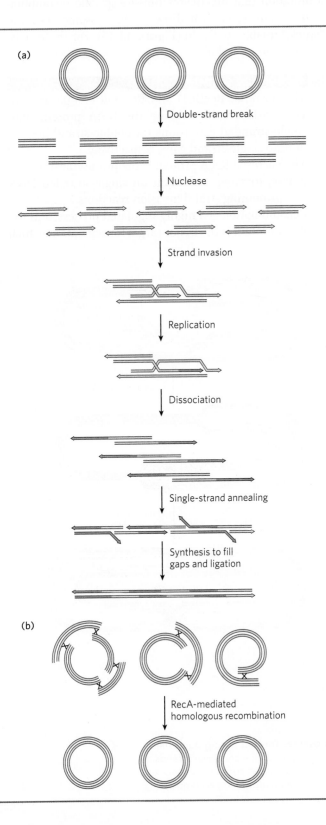

indicating that this phenotype has evolved independently many times. There are no environments on Earth that are subject to ionizing radiation at levels of thousands of grays, but these bacteria are not aliens from outer space. A few species have been found in hot springs with high radon backgrounds, where exposure to chronic low levels of radiation has forced adaptations to permit more efficient DNA repair. However, the most reliable source of bacteria with extreme radiation resistance is a desert environment, where the major selective pressure is not radiation but desiccation. When water disappears for an extended period of time, most metabolic processes, including generation of the ATP needed for DNA repair, cease. However, spontaneous DNA damage continues, and DSBs are among the accumulating lesions. The extreme radiation resistance of *D. radiodurans* and many other bacteria reflects their extraordinary capacity for rapid genomic reconstitution when desiccation gives way to conditions favorable for growth.

FIGURE 2 (a) The first stage of double-strand break repair in *D. radiodurans* closely resembles synthesis-dependent strand annealing (SDSA). (b) In the second stage of genomic reconstitution, large chromosomal segments are spliced together by DrRecA protein. [*Source: Data from M. Radman et al., Nature 443:569–573, 2006.*]

Repair of the Replication Fork in Bacteria Can Lead to Dimeric Chromosomes

Some pathways of replication fork repair lead to the creation of a Holliday intermediate behind the reconstituted fork instead of in front of it (**Figure 13-14**; see also Figure 13-5). This Holliday intermediate can be resolved by RuvC in one of two ways: by cleaving the crossover strands (shown as resolution path X in Figure 13-14) or by cleaving the template strands (resolution path Y). The first path, cleaving the crossover strands, simply leads to the completion of replication and the segregation of two monomeric chromosomes into daughter cells. The second, cleaving the template strands, has a special consequence when the genome is circular, as it is for most bacterial chromosomes: it ultimately creates a single dimeric chromosome that cannot be segregated at cell division. Under normal growth conditions, this outcome is observed in about 15% of cells in an *E. coli* culture.

Cells harboring dimeric chromosomes do not die. Instead, the stalled chromosomal segregation is detected, triggering the activity of a specialized site-specific recombination system, called XerCD, that converts the dimer back into monomeric circles (see Figure 13-14). Site-specific recombination is a class of reaction discussed in Chapter 14. Once the monomeric chromosomes are generated, cell division completes normally. If the cells have a mutation that inactivates the site-specific recombination system, cells with dimeric chromosomes become "stuck," unable to divide (**Figure 13-15**). For these cells, the mutation is lethal.

SECTION 13.2 SUMMARY

- The bacterial RecBCD and RecFOR complexes provide pathways for loading the RecA protein onto single-stranded extensions (at double-strand breaks) or onto single-strand gaps, respectively.
- The bacterial RecA protein is the prototypical recombinase, forming a filament on single-stranded DNA and promoting strand invasion reactions.
- Recombination is a highly regulated process, and the RecA protein is the major target for regulation, which

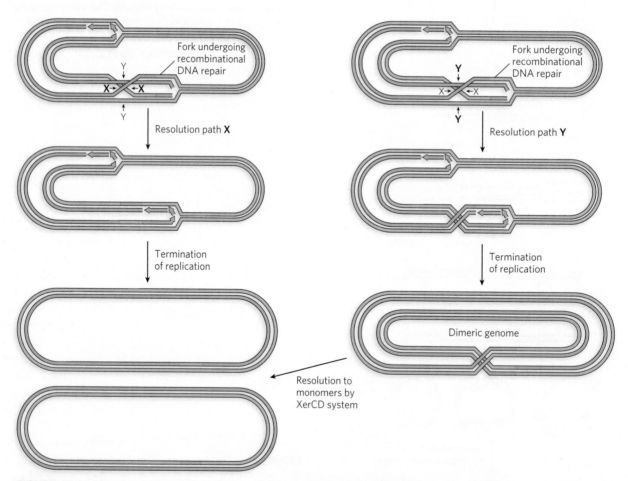

FIGURE 13-14 The generation and resolution of dimeric chromosomes formed during bacterial replication. There are two ways to resolve any Holliday intermediate. In path X, the crossover strands are cut and ligated to form two separate chromosomes. In path Y, the crossover strands are cut and ligated to form a contiguous, dimeric form of the circular chromosome, when replication is complete.

(a)   (b)

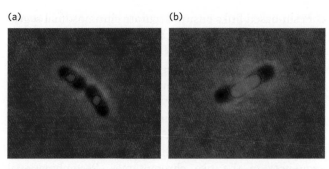

FIGURE 13-15 Cell division is hindered in bacterial cells lacking the capacity to resolve dimeric chromosomes.
(a) Wild-type *E. coli* cells immediately after cell division.
(b) *E. coli* cells with a mutation that inactivates the dimeric chromosome resolution system, after chromosomal division. In both photos, the chromosomes were condensed by treatment with chloramphenicol and stained with a blue fluorescent dye. *[Source: Courtesy of David Sherratt, University of Oxford.]*

occurs through transcriptional regulation, autoregulation, and regulation by other proteins.

- Recombination intermediates generated by RecA are processed by enzymes such as the RuvA, RuvB, and RuvC proteins.

- In the circular bacterial chromosome, resolution of Holliday intermediates associated with replication fork repair can lead to the formation of dimeric chromosomes.

13.3 HOMOLOGOUS RECOMBINATION IN EUKARYOTES

The repair of double-strand breaks that occur during replication is a key function of recombination systems in all cells. However, with the evolution of more complex cells and organisms, recombination systems underwent adaptation to carry out additional functions related to maintenance of the genome and its transmission from one generation to the next. One well-studied example is the recombination that accompanies the process of meiosis. The diploid germ-line precursor cell has two complete sets of chromosomes, or two genome equivalents. The two copies of each eukaryotic chromosome—the homologous chromosomes, or **homologs**—generally have the same genes distributed along their lengths, in the same order. However, the genes of the two homologs are not identical. The alleles inherited from each parent are often slightly different, distinguished by mutations and even by small insertions and/or deletions.

During meiosis, each pair of homologs is replicated to create four chromosome copies—two copies, or chromatids, per homolog (see Chapter 2 for a review of meiosis and the stages of the cell cycle). Homologous sister chromatid pairs are aligned during the first meiotic prophase, prophase I. The four sets of chromosomes are

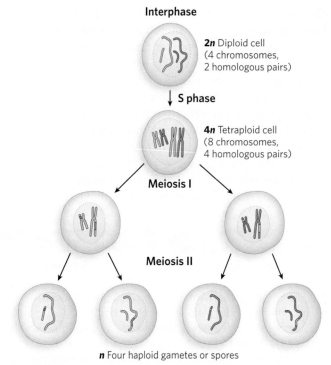

Interphase

2*n* Diploid cell (4 chromosomes, 2 homologous pairs)

↓ **S phase**

4*n* Tetraploid cell (8 chromosomes, 4 homologous pairs)

Meiosis I

Meiosis II

n Four haploid gametes or spores

FIGURE 13-16 An outline of eukaryotic meiosis. DNA is replicated in S phase. For the two pairs of chromosomes shown here, this generates four sets of sister chromatids. The homologs are aligned in meiosis I, joined by recombination (crossovers not shown), then segregated. In meiosis II, the sister chromatids segregate into the haploid cell products. (See also Figure 2-12, right.)

then segregated through two cell divisions, meiosis I and meiosis II, to create four haploid gametes (or spores, in lower eukaryotes such as fungi), each with a complete (haploid) chromosomal complement (**Figure 13-16**). Homologs segregate in the first cell division, and sister chromatids segregate in the second.

The proper segregation of chromosomes when eukaryotic cells divide requires the attachment of spindle fibers to the centromeres. Chromosomes are then drawn toward opposite poles into what will eventually become two new cells. If replicated chromosomes were scattered randomly about, and spindle fibers were attached randomly to the chromosomes, segregation of the chromosomes to daughter cells would be equally random. A given cell might receive two copies of some chromosomes and no copy of others. To segregate a complete set of chromosomes to each cell, some organizational accounting is required. Chromosome pairs are lined up and linked together in meiosis I. When spindles at opposite poles attach to the centromeres of a linked pair of chromosomes and start to pull, tension is created. This tension, sensed by a mechanism not yet understood, indicates that this pair of chromosomes is properly aligned for segregation. Once the tension is sensed, the links between the chromosomes are gradually removed so that segregation can proceed. If an improper attachment of spindle fibers occurs (e.g., if the two centromeres in the chromosome pair are linked to spindle fibers emanating from the same pole), a cellular kinase senses the lack of tension and activates a system that removes the spindle attachments, allowing the cell to try again.

To generate the needed tension, a physical link between chromosomes destined to be segregated is a requirement for any orderly cell division. During mitosis, and in meiosis II, the sister chromatids produced by replication are segregated. The physical link is provided by cohesins (see Figure 9-25). Cohesins are deposited along a chromosome as it is replicated, ensuring that only homologous sister chromatids are linked together. These cohesin-based links ensure accurate chromosomal segregation by the spindle fibers. However, the segregation of homologs during meiosis I is an event unique to meiosis. The homologs to be segregated are not related by a recent replication event in which cohesins were deposited, and thus some other molecular device is needed to ensure that only homologous chromosomes are linked. The eukaryotic answer to this problem is homologous recombination, a process that relies on closely related sequences in the two chromosomes. Recombinational crossovers provide for the accurate alignment of homologs at the metaphase plate during meiosis I (**Highlight 13-2**).

HIGHLIGHT 13-2 | **MEDICINE**

Why Proper Chromosomal Segregation Matters

When chromosomal alignment and recombination are not correct and complete in meiosis I, segregation of chromosomes can go awry. Aneuploidy, a condition in which a cell has the wrong number of chromosomes, can result. The haploid products of meiosis (gametes or spores) may have no copies or two copies of a chromosome. When a gamete with two copies of a chromosome undergoes fertilization, cells in the resulting embryo are trisomic for (have three copies of) that chromosome.

In *S. cerevisiae*, aneuploidy resulting from errors in meiosis occurs at a rate of about 1 in 10,000 meiotic events. In fruit flies, the rate is about one in a few thousand (see Figure 2-15). Rates of aneuploidy in mammals are considerably higher. In mice, the rate is 1 in 100, and it is even higher in other mammals. The rate of aneuploidy in fertilized human eggs has been estimated as 10% to 30%, mostly with monosomies or trisomies. This is almost certainly an underestimate. Most trisomies are lethal, and many result in abortive miscarriage long before the pregnancy is detected. This is the leading cause of pregnancy loss. The few trisomic fetuses that survive to birth generally have three copies of chromosomes 13, 18, or 21 (trisomy 21 is Down syndrome). Abnormal complements of the sex chromosomes are also found in the human population. Almost all monosomies are fatal in the early stages of fetal development. The societal consequences of aneuploidy in humans are considerable. Aneuploidy is the leading genetic cause of developmental and mental disabilities. At the heart of these high rates is a feature of meiosis in female mammals that has special significance for the human species.

In a human male, germ-line cells begin to undergo meiosis at puberty, and each meiotic event requires a relatively short period of time. In contrast, meiosis in the germ-line cells of human females is a highly protracted process. The production of an egg begins with the onset of meiosis in the fetus, at 12 to 13 weeks of gestation. This initiation of meiosis occurs in all the developing fetal germ-line cells over a period of a few weeks. The cells proceed through much of meiosis I. Chromosomes line up and generate crossovers, continuing just beyond the pachytene phase (see Figure 13-17)—and then the process *stops*. The chromosomes enter an arrested phase called the dictyate stage, with the crossovers in place, a kind of suspended animation where they remain as the female matures—for anywhere between 13 and 50 years. It is not until sexual maturity that individual germ-line cells continue through the two meiotic cell divisions to produce egg cells.

Between the onset of the dictyate stage and the final completion of meiosis, something can happen that disrupts or damages the crossovers linking homologous chromosomes in the germ-line cells. As a woman ages, the rate of trisomy in the egg cells she produces increases, dramatically so as she approaches menopause (Figure 1). There are many hypotheses on why this occurs, and several different factors may play a role. However, most of the hypotheses are centered on recombination crossovers in meiosis I and their stability over the protracted dictyate stage.

It is not yet clear what medical steps could be taken to reduce the incidence of aneuploidy in females of child-bearing age. What is revealed is the inherent importance of recombination and crossover generation in human meiosis.

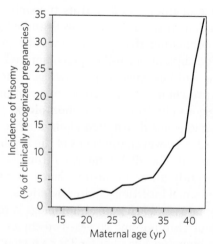

FIGURE 1 The increasing incidence of trisomy with increasing age of the mother. [*Source: Data from T. Hassold and P. Hunt, Nat. Rev. Genet. 2:280–291, 2001.*]

Many of the same enzymes are involved in both recombinational DNA repair during replication and meiotic recombination. The process again begins with DSBs, but in meiosis I, these breaks do not happen by chance. Instead, they are programmed events.

As we discussed in Section 13.1, eukaryotic recombination is not restricted to meiosis. Less frequent recombination events occur during mitosis, at least some of which are associated with the repair of stalled or collapsed replication forks. In addition, recombination systems have been appropriated during evolution to carry out some highly directed exchanges of genetic information between different segments of chromosomes, as we discuss later in this section.

Meiotic Recombination Is Initiated at Double-Strand Breaks

Following the premeiotic S-phase replication cycle, homologous chromosomes are brought together. As the cell enters meiosis, in early prophase I, DSBs are introduced at multiple locations along one chromatid of each chromatid pair (**Figure 13-17**). The breaks are not random, and yet are not entirely predictable. Certain chromosomal sites, often referred to as hot spots, are much more likely to undergo a break than others. The hot spots are defined by features of chromosome structure not yet

fully characterized. An open chromatin configuration, active transcription, and G≡C-rich sequences can all affect the process. The DSBs lead first to processing of the broken ends to generate 3′ extensions and then to strand invasion, as outlined in Figure 13-2. Some of these strand invasions are eliminated, but some undergo additional reactions to become stable crossovers, linking the two pairs of chromatids, as described below. The strand invasions are generally completed early in prophase I, during the leptotene subphase. They are processed into double Holliday intermediates and, finally, to stable crossovers during the succeeding zygotene and pachytene subphases (see Figure 13-17).

The enzymes that carry out this process are best characterized in the yeast *Saccharomyces cerevisiae*, although similar enzymes exist in all eukaryotic cells. A protein called **Spo11** (so named because inactivation of this protein causes defects in yeast sporulation), closely related to eukaryotic type II topoisomerases, catalyzes formation of DSBs (**Figure 13-18a**; see the How We Know section at the end of this chapter). Spo11 is found in all eukaryotes. Acting as a dimer, it uses an active-site Tyr residue as a nucleophile in a transesterification reaction (**Figure 13-18b**). Each subunit cleaves one DNA strand, with the phosphodiester bond replaced by a 5′-phosphotyrosyl linkage. The reaction halts at this point, and Spo11 does not carry out the additional steps

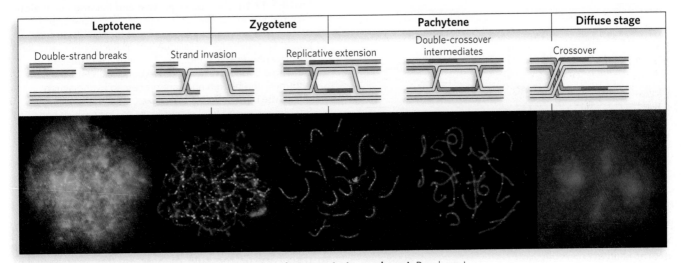

FIGURE 13-17 Homologous genetic recombination during meiotic prophase I. Prophase I includes a directed recombination process. The double-strand breaks are created and processed early. Strand invasions and replicative extension produce double-crossover intermediates, and some of these mature into stable crossovers. Shown here are the early stages of meiosis in mouse spermatocytes. (Leptotene, zygotene, and pachytene are terms used to describe the subphases of meiosis prophase I.) The progress of the recombination reactions is shown in the upper panel, matched to the subphases of prophase in the microscopic images in the lower panel. The chromosomes are viewed by staining particular proteins through immunofluorescence. Chromosomes are stained red, and a few recombination proteins, fused to green fluorescent protein, appear as green foci. As prophase progresses, the homologous chromosome pairs, at first diffuse, become tightly aligned, producing rodlike structures. The chromosomes again become diffuse as prophase ends and the cells begin the first cell division. [*Source: From A. McDougall, D. J. Elliott, and N. Hunter, EMBO meeting report, Jan. 14, 2005, Fig. 2. © 2005 European Molecular Biology Organization.*]

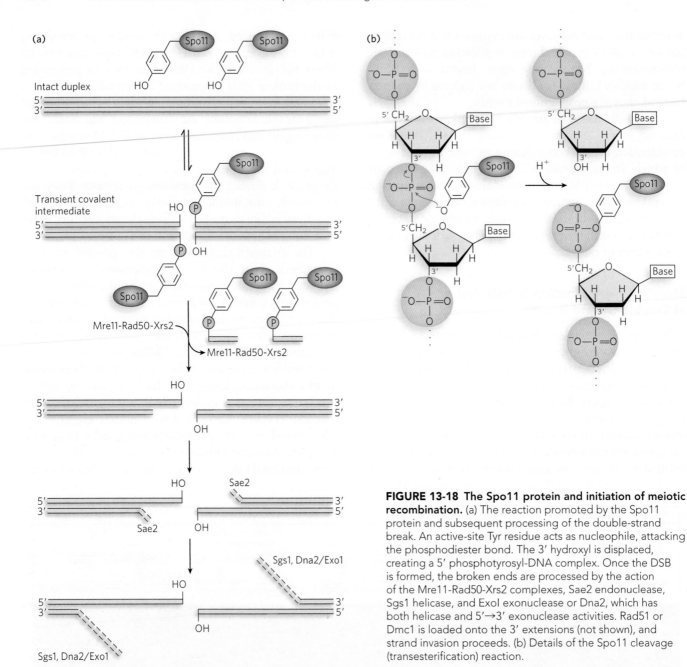

FIGURE 13-18 The Spo11 protein and initiation of meiotic recombination. (a) The reaction promoted by the Spo11 protein and subsequent processing of the double-strand break. An active-site Tyr residue acts as nucleophile, attacking the phosphodiester bond. The 3′ hydroxyl is displaced, creating a 5′ phosphotyrosyl-DNA complex. Once the DSB is formed, the broken ends are processed by the action of the Mre11-Rad50-Xrs2 complexes, Sae2 endonuclease, Sgs1 helicase, and ExoI exonuclease or Dna2, which has both helicase and 5′→3′ exonuclease activities. Rad51 or Dmc1 is loaded onto the 3′ extensions (not shown), and strand invasion proceeds. (b) Details of the Spo11 cleavage (transesterification) reaction.

of a topoisomerase reaction. A dozen or more additional proteins may cooperate in the formation of an active Spo11 complex on the DNA and in processing the DNA after it is cleaved.

To remove Spo11 from the DNA and initiate nucleolytic degradation of the 5′-ending strand, a complex of proteins consisting of Mre11 (*meiotic recombination*), Rad50 (*radiation sensitive*), and Xrs2 (*x-ray sensitive*) binds to each Spo11 complex and cleaves the DNA on the 3′ side of Spo11, liberating the linked protein along with a short segment of the attached 5′-ending DNA strand. The nuclease Sae2 degrades the same DNA strand a bit more. Three other enzymes, the helicase Sgs1 and the nucleases Dna2 or Exo1, have been implicated in

additional degradation of the 5′ ends to create long 3′ single-stranded extensions (see this chapter's Moment of Discovery for a description of experiments leading to discovery of the role of Sgs1 in this process). The single-stranded regions are bound by RPA, the eukaryotic single-stranded DNA–binding protein. Aided by mediator proteins, two RecA-class recombinases, called **Dmc1** (*disrupted meiotic cDNA*) and **Rad51**, are loaded, perhaps asymmetrically, onto the 3′ extensions on either side of the double-strand break. Dmc1 and Rad51 are the eukaryotic counterparts to RecA. They exhibit sequence and structural homology to the RecA protein and, like RecA, they form extended nucleoprotein filaments on the DNA. The site is now set up for recombination.

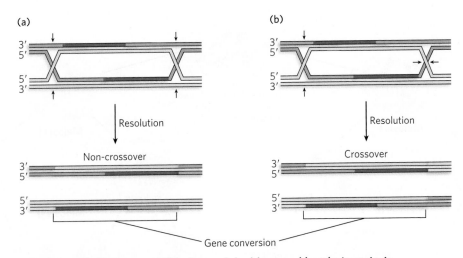

FIGURE 13-19 The two possible fates of double-strand breaks in meiosis.
(a) Recombination with no crossover but some potential for gene conversion.
(b) Recombination leading to a genetic crossover. Gene conversion can occur in some regions between the two Holliday intermediates.

Meiotic Recombination Is Completed by a Classic DSBR Pathway

Meiotic recombination is a directed process, which is slowly yielding its secrets to intensive research. The primary goal of recombination is to create physical links, or crossovers, between chromosomes. The formation of crossovers is very tightly regulated so that at least one crossover is created in every pair of homologs. At the same time, a process of interference, not yet understood, ensures that the total number of crossovers for each pair is limited. Where several crossovers occur between the same homologs, they are spaced far apart on the chromosomes. Once the DSBs have formed, subsequent recombination is regulated to occur almost exclusively between homologs rather than between sister chromatids.

The DSBs and Holliday intermediates have two possible fates, only one of which leads to a stable crossover. In one pathway, the invading strand dissociates and pairs with its complement on the other side of the break (**Figure 13-19a**). Further extension and ligation complete the process. This pathway is equivalent to SDSA. No link between the two chromosomes is created and no crossover occurs. However, even this process is not genetically neutral. Homologous regions may have different alleles of the same gene, with small base-pair differences. If the region where the two single strands are paired contains one of these base-pair differences, the final reannealed duplex will have a mismatch. This is typically repaired by the cellular mismatch repair system to create a normal base pair. The base on one strand or the other must be changed, and genetic information is lost on the changed strand. This type of outcome is referred to as **gene conversion**. It represents one byproduct of recombination, not only during meiosis but also during the recombinational repair of DSBs and single-strand gaps in other contexts. However, even though a small bit of genetic information may be transferred, the physical crossover has been eliminated.

In the alternative pathway, strand extension leads to displacement of one strand of the invaded duplex, and this strand eventually pairs with the other side of the DSB (**Figure 13-19b**). DNA ligation creates a double Holliday intermediate, and thus a physical link between the homologs. As meiosis proceeds, the double Holliday intermediate is cleaved by resolvases (similar to the bacterial RuvC), with the potential to create a stable crossover. The final crossover is embedded in a proteinaceous structure called a chiasma (see Figure 2-18). The overall pathway is closely related to that in a breakthrough model that first described meiotic recombination as a process of DSBR, proposed by Jack Szostak, Terry Orr-Weaver, Rodney Rothstein, and Frank Stahl in 1983.

Crossovers have two roles in meiosis. The first is to create the physical link essential for proper chromosomal segregation. The second role is a genetic one. Following segregation, the sister chromatids (now daughter chromosomes) are no longer identical. One end of at least one of each set of paired chromatids has been exchanged with the homolog—a genetic crossover, generating genetic diversity.

Meiotic Recombination Contributes to Genetic Diversity

Meiosis, then, is not simply a mechanism for reducing a diploid genome to a haploid genome in the eukaryotic

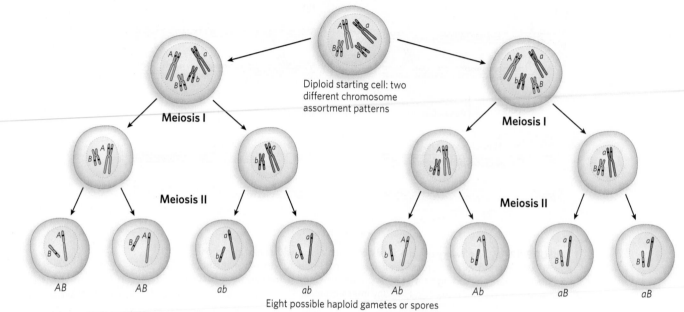

FIGURE 13-20 The contribution of independent assortment to genetic diversity. Much genetic diversity comes from the independent assortment of chromosomes during meiosis. Blue and red distinguish the chromosomes inherited from each parent. One gene on each chromosome is highlighted, with different alleles (*A* or *a*; *B* or *b*) in the homologs. Independent assortment can lead to gametes with any combination of alleles on each chromosome.

gamete (or spore). Meiosis increases the genetic diversity in a population by shuffling the genome in two ways. The first is independent assortment of unlinked genes (**Figure 13-20**). As a cell goes through meiosis I, homologous chromosomes are segregated to the daughter cells. The chromosomes assort independently. If a particular gamete obtains chromosome 1 as a result of the meiotic cycle, that has no bearing on which version of chromosome 2 also ends up in that gamete. Genes on a given chromosome are linked and are likely to be inherited together, but genes on different chromosomes are unlinked.

Recombination during meiosis I makes an additional contribution to the genetic diversity of gametes. Crossovers between homologs in prophase I shuffle the alleles on individual chromosomes. Each gamete ends up with one complete genomic complement, with all genes present. However, given the variation of crossover locations and the unpredictability of independent assortment, the resulting gamete population includes individual cells that may contain virtually any combination of alleles derived from each parent.

Recombination during Mitosis Is Also Initiated at Double-Strand Breaks

Double-strand breaks occur more rarely during mitosis, and recombination is correspondingly less frequent. The breaks can result from endonucleolytic action or exposure

to ionizing radiation. However, the most common source of DSBs in mitosis is the encounter of a replication fork with a template strand break (see Figure 13-1). When this occurs, a cellular checkpoint is activated (a signaling pathway) that halts the progression of the cell cycle, and recombinational DNA repair processes are initiated. Mitotic recombination is generally limited to the S and G_2 phases of the cell cycle, when sister chromatids are present. Recombination between homologous chromosomes in diploid cells is much less frequent—probably a simple matter of the physical distance between homologs.

The pathways in mitotic recombination are similar to those in meiotic recombination (**Figure 13-21**). When a DSB is generated by ionizing radiation, it can be repaired by two main pathways: SDSA or DSBR. In both pathways, the break is processed by nucleases to generate 3' single-stranded tails, which are coated with RPA. With the aid of recombination mediator proteins, the recombinase is loaded onto the single-stranded DNA to promote strand invasion of a double-stranded homologous chromosome, usually the sister chromatid (or, less frequently, the homolog). The Dmc1 recombinase is specific to meiosis, so mitotic recombination relies entirely on Rad51. Recombination mediators include the Rad52 protein in all eukaryotes and, in vertebrates, the **BRCA2** protein. These proteins load Rad51 onto RPA-coated single-stranded DNA and may have additional functions. Humans with a defect in the *BRCA2* gene or certain defects in the gene for Rad52 have a predisposition to breast

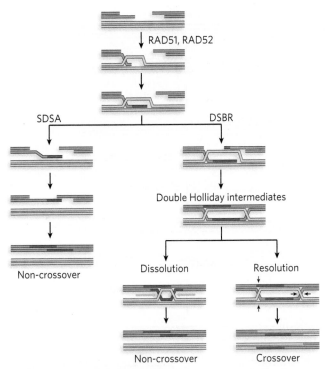

FIGURE 13-21 Mitotic recombination. The most common path of mitotic recombination is SDSA (left). After strand invasion and strand extension, the invading strand is displaced. It can then anneal with its complement on the other side of the original double-strand break. Replication, nucleolytic trimming as needed, and ligation complete the process. Crossovers occur in recombination events in the DSBR pathway (right). Rad52 protein may be involved in initiation of the second strand invasion that leads to the double Holliday intermediate. [*Source: Data from Y. Liu and S. C. West,* Nat. Rev. Mol. Cell Biol. *5:937–944, 2004, Fig. 7.*]

cancer and several other cancers. The **BRCA1** protein also plays an important, but incompletely defined, role in mediating the overall process. A cellular deficiency in either BRCA1 or BRCA2 results in the accumulation of chromosomes with aberrant structures and aneuploidy. The involvement of these key cancer suppressor genes provides a vivid illustration of the importance of mitotic recombination in maintaining the genome.

Programmed Gene Conversion Events Can Affect Gene Function and Regulation

Saccharomyces cerevisiae is a single-celled eukaryote, familiar to bakers and brewers and the thousands of scientists who have adopted this yeast as a model organism (see the Model Organisms Appendix). It can live with a haploid genome complement or as a diploid. Both forms are stable and both reproduce by mitosis, with daughter cells budding off mother cells. However, the haploid and diploid forms can also be interconverted under the right conditions. Haploid cells exist in two forms called **mating types**, designated **a** and α. A haploid cell can mate with another haploid cell of the opposite mating

type (**a** with α, or α with **a**), creating a stable diploid. When conditions are unfavorable for growth, diploid cells can undergo meiosis and its accompanying recombination, resulting in four haploid spores (two **a** spores and two α spores). The spores represent a dormant state that can survive stressful environmental conditions. When conditions improve, the spores can begin growing as haploid cells.

Cells can be converted from one mating type to another, in a process called mating type switching. This provides yet another manifestation of the recombination process. The mating type of a haploid cell is determined by two different versions of a gene expressed at a single locus called *MAT*. If the *MAT* locus is *MAT***a**, the cell has the **a** mating type. If the *MAT* locus is *MAT*α, the cell has the α mating type. The genetic information for *MAT***a** and *MAT*α is stored in nearby silent copies of each of these genes, called *HML*α and *HMR***a**. That genetic information is transferred to the *MAT* locus by an unusual mechanism, with recombination at its core (**Figure 13-22**).

Haploid cells can switch mating types, and they do so as often as every cell generation. The switch requires a directed recombination reaction initiated by a double-strand break. The overall process represents one more

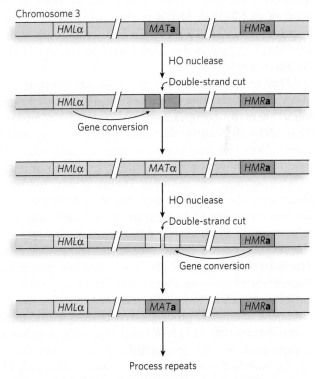

FIGURE 13-22 The mating-type loci in *S. cerevisiae.* The genetic information expressed at the *MAT* locus on chromosome 3 determines mating type (**a** or α) in yeast. That information can be changed by recombination-mediated gene conversion, moving information (normally silent) that is stored at the *HML*α and *HMR***a** loci.

example of the steps outlined in Figure 13-2. The *MAT* locus is cleaved by a nuclease called HO to generate a double-strand break at a specific location near *MAT*, and recombinational repair of the DSB is directed by recombination with one of two mating-type donor sites (see Figure 13-22). The *HMLα* locus has 700 bp of α-specific genetic information, and the *HMRa* locus has 650 bp of **a**-specific genetic information. (The locus names are derived from *homothallic locus left* and *homothallic locus right*.) The genes at *HMLα* and *HMRa* are not expressed; these loci serve only as a silent reservoir of genetic information used to change the information at the *MAT* locus by gene conversion.

The mating-type switch is a classic example of recombinational DNA repair of a double-strand break by the SDSA pathway (**Figure 13-23**; see also Figure 13-3). If the cell is initially the **a** mating type, *MATa* is cleaved by HO, and the free DNA ends are processed to generate 3′ single-stranded overhangs. The Rad51 protein binds to the overhangs and directs DNA strand invasion at a homologous part of the *HMLα* locus. As the invading 3′ end is extended by a DNA polymerase, the α mating-type information in *HMLα* is copied. The **a** mating-type information at the *MAT* locus is removed by nuclease digestion. Once the extending strand is long enough, it dissociates, the end is paired with a homologous gene segment in the original *MAT* locus, and the DNA gap is filled in and ligated. *MATa* is thus switched to *MATα*. Switching from *MATα* to *MATa* is essentially the same reaction, except that the strand invasion occurs at the *HMRa* locus.

Similar gene conversion processes arising from the SDSA form of recombination are surprisingly common, found in many bacteria and single-celled eukaryotes. They allow the cell to alternate between two or more states. Examples important to medicine can often be found in pathogens that are able to evade the human immune system. For example, the bacterium *Neisseria gonorrhoeae* is the agent that causes gonorrhea. The immune response to the pathogen is largely directed at antigens in the bacterial pili, cellular projections involved in adhesion to host cells and other bacteria. The bacterium can evade the host's immune system by alternately expressing different pilin genes, through antigenic variation. Switching from the expression of one pilin gene to another proceeds by a process very similar to the mating-type switch in yeast. One pilin gene locus is expressed, but the genetic information at that locus can be switched out by genetic recombination with alternate silent gene loci.

The use of recombination to switch expression between different sets of genes has a major advantage over the more common types of gene regulation described in Chapters 19–22. It is absolute. The mating-type locus of yeast is either **a** or α, never partially one or the other.

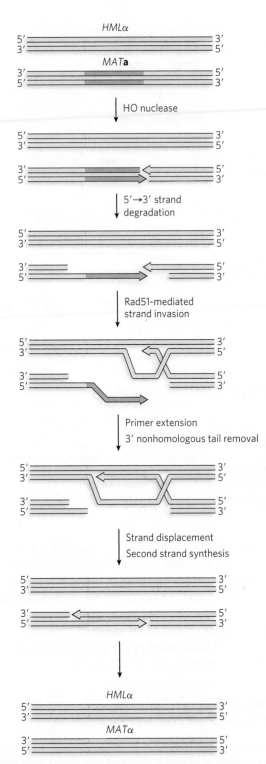

FIGURE 13-23 Pathway of the mating-type switch in yeast. The double-strand break is introduced by the nuclease HO. The subsequent strand invasion and other steps closely resemble the SDSA pathway. The *MATa* information is converted to *MATα* information by the resulting gene conversion.

The template donor genes are not expressed at all, because they are not present in the gene expression locus. For a pathogen, even low levels of leaked expression of the wrong genes could undermine its strategy for circumventing the host's immune system.

Some Introns Move via Homologous Recombination

Some of the introns that are found in many eukaryotic (and a few bacterial) genes have the interesting property that they can move—from one gene to another copy of the same gene on a homologous chromosome that lacks that particular intron. In this way, an intron that becomes associated with a specific gene (say, gene *X*) can rapidly spread through a population so that all copies of gene *X* in the population contain that intron. The movements occur whenever genetic transfers bring chromosomes together from two different sources, such as during the fusion of gametes at fertilization.

The movement can occur in several ways, but mobile introns of the group I class (these self-splicing introns are discussed in Chapter 16) use recombinational DNA repair of a targeted double-strand break (**Figure 13-24**). In brief, the introns encode a **homing endonuclease**, an enzyme that cleaves a specific sequence in any copy of the host gene that lacks the intron. The resulting DSB is repaired by recombination with the gene that *does* have a copy of the intron.

Homing endonucleases have been widely used in biotechnology. They function much like restriction enzymes (see Chapter 7), but they recognize and cleave sequences that are much larger, 12 to 40 bp, and are asymmetric. These sites occur very rarely in genomes. If a site recognized by one of these enzymes is engineered into a chromosome or viral DNA, it can be reproducibly cleaved without affecting other genomic sequences.

- During meiosis, recombination generates crossovers that create a physical link between homologous chromosomes just before the first meiotic cell division.
- Meiotic recombination events are initiated at programmed double-strand breaks, and most proceed by synthesis-dependent strand annealing.
- Meiotic recombination in eukaryotes makes an important contribution to the generation of genetic diversity in a population.
- Some meiotic recombination events do not generate crossovers, but instead result in a more subtle exchange of genetic information known as gene conversion.
- Mitotic recombination is rarer and is also initiated at DSBs.
- Directed homologous recombination promotes a mating-type switch in yeast and can promote antigenic variation in some pathogens.
- Some group I introns migrate by means of recombinational DSBR.

13.4 NONHOMOLOGOUS END JOINING

Recombination allows an accurate restoration of broken chromosomes, a considerable virtue given the importance of maintaining genomic integrity. However, recombination is complicated and requires the action of dozens of proteins. Sometimes DSBs occur when recombinational DNA repair is not feasible, such as during phases of the cell cycle when no sister chromatids are present. At these times, another path is needed to avoid the cell death that would result from a broken chromosome. That alternative is provided by **nonhomologous end joining (NHEJ)**. The broken chromosome ends are simply processed and ligated back together.

Nonhomologous End Joining Repairs Double-Strand Breaks

Nonhomologous end joining is an important pathway for DSBR in all eukaryotes, and it has also been detected in some bacteria. In general, the importance of NHEJ increases with genomic complexity. Only a few bacteria seem to have NHEJ systems. In yeast, most DSBs are repaired by recombination, and only a few by NHEJ.

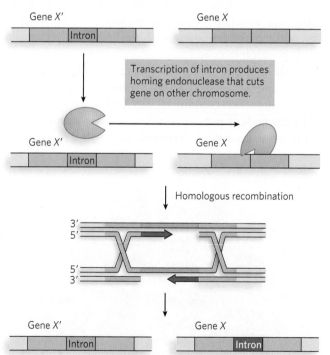

FIGURE 13-24 The homing pathway of some group I introns. In this pathway, the gene with the intron (*X'*) and the same gene without the intron (*X*) are present in the cell, and an intron-encoded endonuclease creates a double-strand break in the uninterrupted gene (*X*). The break triggers recombinational DSBR similar to that outlined in Figure 13-2, using the intron-containing gene (*X'*) as a template for repair. When the repair is complete, both genes have a copy of the intron. [Source: Data from B. S. Chevalier and B. L. Stoddard, *Nucleic Acids Res.* 29:3757–3774, 2001, Fig. 1.]

In mammals, many DSBs occurring outside meiosis are repaired by NHEJ. These patterns reflect differences in cellular lifestyles. In all eukaryotic cells, recombinational DNA repair is the preferred DSBR pathway during the S and G_2 phases of the cell cycle, when chromosomes are being replicated and paired prior to cell division. Finding a homolog to direct the repair process by recombination is readily accomplished at these times. NHEJ is critical to the repair of DSBs that arise during the G_1 and the static G_0 phases of the cell cycle, when homologous chromosomes are not readily aligned. Differentiated mammalian cells may divide rarely, if at all, and typically spend much more time in the G_1 and G_0 phases than do yeast cells, which may divide every few hours. When DSBs occur during these phases, the enzymes that promote NHEJ are rapidly activated.

Unlike homologous recombinational repair, NHEJ does not conserve the original DNA sequence. When a DSB occurs during G_1 or G_0, a protein complex forms at each broken end of the chromosome, and the two DNA-protein complexes associate to form a DNA synapse. Synapsis activates a protein kinase and helicase activity within the protein complex. The subsequent DNA unwinding may produce a short, 1 to 6 bp region of complementary sequences on each side of the break, creating what is called microhomology, that is presumably needed for end joining. Any flaps of single-stranded DNA can be trimmed by nucleases, gaps filled by DNA polymerase, and nicks sealed by DNA ligase.

NHEJ is a mutagenic process, and a smaller genome, such as that of yeast, has relatively little tolerance for the loss of information. The small genomic alterations may be tolerable in mammalian somatic cells, however, because they are not in the germ line and will not be inherited, and they are balanced by the undamaged information on the homolog in each diploid cell. Indeed, the propensity of NHEJ to create mutations has led to its being recruited in somatic cells as a source of variation in the production of genes encoding antibodies (see Chapter 14).

Nonhomologous End Joining Is Promoted by a Set of Conserved Enzymes

In eukaryotes, at least nine proteins are used in the multiple steps of NHEJ (**Table 13-2**). The reaction is initiated at a DSB by the binding of a heterodimer consisting of the proteins Ku70 and Ku80 ("KU" are the initials of the individual with scleroderma whose serum autoantibodies were used to identify this protein complex; the numbers refer to the approximate molecular weights of the subunits). The Ku proteins are conserved in almost all eukaryotes. Both subunits of the Ku70-Ku80 complex have three domains, with the central domain forming a

TABLE 13-2

Enzymes Involved in Nonhomologous End Joining

Enzyme	Function
Ku70	Binds to DNA ends
Ku80	Binds to DNA ends
DNA-PKcs	Protein kinase catalytic subunit
Artemis	Nuclease
Pol μ	Fills in gaps
Pol λ	Fills in gaps
XRCC4	Seals nicks
XLF	Seals nicks
DNA ligase IV	Seals nicks

double ring (**Figure 13-25**). The complex binds readily to double-stranded DNA blunt ends or ends with 3′ or 5′ extensions. Multiple copies of the complex may bind, sliding inward on the DNA. The Ku70-Ku80 complex binds all of the other complexes that play key roles in the subsequent steps: nuclease, polymerases, and ligase. Ku70-Ku80 thus acts as a kind of molecular scaffold. In the eukaryotic nucleus, this complex has additional roles in DNA replication, telomere maintenance, and transcriptional regulation to complement its role in NHEJ. A loss of the genes encoding NHEJ function can produce a predisposition to cancer.

NHEJ proceeds in three major stages (**Figure 13-26**). In the first stage, Ku70-Ku80 interacts with another protein complex containing DNA-PKcs (the 470 kDa DNA-dependent protein kinase catalytic subunit) and a nuclease known as Artemis. Once the complex is

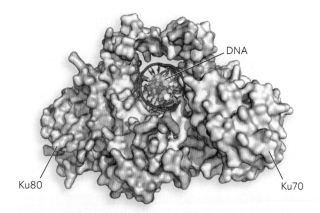

FIGURE 13-25 The Ku70-Ku80 complex. The central domains of Ku70 (yellow) and Ku80 (orange) provide an opening through which DNA can pass. The proteins slide over the DNA at a broken end. Additional domains in Ku70-Ku80 (not shown) provide interaction targets for some of the other NHEJ proteins. [Source: PDB ID 1JEY.]

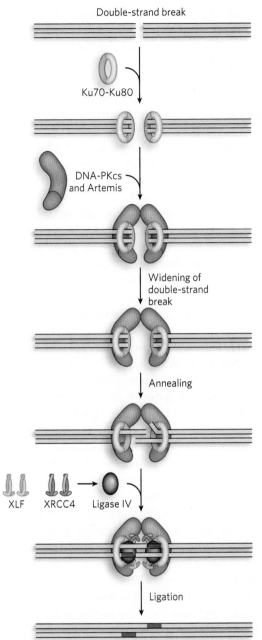

FIGURE 13-26 Nonhomologous end joining in eukaryotes. The Ku70-Ku80 complex is the first to bind the DNA ends, followed by a complex including DNA-PKcs and the nuclease Artemis. These proteins then recruit a complex of XRCC4, XLF, and DNA ligase IV. Either of two DNA polymerases, Pol μ or Pol λ (not shown), subsequently extends the annealed DNA strands, as needed, before ligation. *[Source: Data from J. M. Sekiguchi and D. O. Ferguson, Cell 124(2):260–262, 2006, Fig. 1.]*

assembled, the two broken DNA ends are synapsed (held together) and the protein kinase activity of DNA-PK is activated. DNA-PK autophosphorylates in several locations, and also phosphorylates Artemis. Artemis is generally active as a 5′→3′ exonuclease, but when phosphorylated it acquires an endonuclease function. This endonuclease can remove 5′ or 3′ single-stranded extensions, as well as

hairpins, resecting the excess DNA in overhangs at the ends. In the second stage, DNA ends are separated with the aid of a helicase, and strands from the two different ends are annealed. Artemis cleaves any unpaired DNA segments that are created. Small DNA gaps are filled in by the eukaryotic Pol μ or Pol λ. Finally, the nicks are sealed by a protein complex consisting of XRCC4 (*x-*ray *c*ross *c*omplementation group), XLF (XRCC4-*l*ike *f*actor), and eukaryotic DNA ligase IV.

DNA ends are usually not joined randomly by NHEJ. Instead, when a DSB occurs, the ends are generally constrained by the structure of chromatin and thus remain close together; they are rarely linked to the ends of other chromosomes, because all eukaryotic chromosome ends are protected by telomeres. Very rare events linking end sequences that are normally far apart in the chromosome, or on different chromosomes, may be responsible for occasional dramatic and usually deleterious genomic rearrangements.

Recombination Systems Are Being Harnessed for Genome Editing

As should be clear by now, recombination events are initiated with double-strand breaks. Once a DSB is created, many things can happen, and many of them happen efficiently. Increasingly, researchers are exploiting these processes to engineer genomic changes in cells, from bacteria to mammals. Advanced and programmable reagents to create DSBs are undergoing rapid development. These currently include zinc finger nucleases (ZFNs), TALE (transcription activator–like effector) domain nucleases (TALENS), and the CRISPR/Cas systems, all described in Section 7.3. In each case, these nucleases are tools that allow a researcher to create a DSB at any desired location. Combined with the cellular recombination systems described in this chapter, the opportunities for genome alterations are abundant and have given rise to the subdiscipline of **genome editing**. In brief, a targetable nuclease is first designed to cleave a chromosomal target site to generate a DSB. If no homologous DNA is available, a eukaryotic cell will usually repair the break with nonhomologous end joining. If the target is a gene, the errors inherent to NHEJ will often inactivate that gene. Precise changes to a gene can also be made. If a fragment of duplex DNA is made available (e.g., by injection or electroporation into cells) that contains ends homologous to either side of the break and includes a desired sequence change, recombination systems will use the DNA for DSB repair so that the change is incorporated. As this technology advances, the capacity to inactivate or modify genes in the human genome may one day alleviate the effects of many serious genetic diseases.

These goals will be met slowly, as problems and obvious ethical issues abound. The programmable nucleases have significant tendencies to cleave DNA at sites that are off-target, with the potential to create a damaged cell or a progenitor of a cancerous tumor. If these targeting issues are eventually controlled and the procedures rendered safe and reliable, the medical community and society will have to determine where the alleviation of serious disease conditions ends and a potentially unsavory practice of human eugenics begins. As in so many areas of science, promise and ethical quandaries will collide, but the potential for alleviating suffering and addressing conditions that are currently untreatable is real.

SECTION 13.4 SUMMARY

- Nonhomologous end joining is critical to the repair of double-strand breaks that arise during the G_1 and static G_0 phases of the eukaryotic cell cycle. NHEJ is more important in cells that spend a greater amount of time in G_1 and G_0, such as the somatic cells of more complex eukaryotes.
- NHEJ is promoted by the well-conserved Ku70-Ku80 protein complex, along with additional complexes containing nuclease, polymerase, and ligation activities.
- Double-strand break repair makes a critical contribution to genome editing methods that utilize programmable nucleases to generate targeted double-strand breaks.

? UNANSWERED QUESTIONS

In every area of molecular biology, there is a need to reconcile biochemistry with the observations made in living cells, using genetics and functional genomics. Recombination systems and processes are complex, and accurately reconstituting them in vitro from purified components remains a major challenge. Here are some of the significant questions in the field.

1. How often do replication forks collapse? It is not yet clear how often DNA template lesions halt replication forks in a manner requiring replication restart, and how often lesions are simply bypassed by the replication machinery. Some in vitro studies show a potential for bypass even for lesions in the leading strand. However, in vivo studies show that replication ceases for a period when sufficient DNA damage is introduced into the genome. Although we understand the outlines of the major recovery pathways, there are doubtless many variants yet to be elucidated—along with some undiscovered enzymes—that respond to different classes of lesions and the many different DNA structures found at stalled forks.

2. In a eukaryotic chromosome, what parameters determine where crossovers occur during meiotic recombination? Much remains to be discovered about the structural features of chromosomes that define crossover hot spots and how numbers of crossovers are controlled.

3. What factors remain to be discovered in double-strand break repair, and how do they work? In studies of recombinational DNA repair, many new proteins are still being discovered. Some of the newer discoveries involve protein factors that regulate recombination, or link it to replication checkpoints or other aspects of chromosome structure or cell division. The complexities are illustrated by proteins such as Spo11. The double-strand cleavage reaction of Spo11 has not yet been replicated in vitro, perhaps due to a requirement for other protein factors not yet purified (or not yet discovered). After Spo11 has cleaved a chromosome, additional enzymes must degrade the 5′-ending strand to create the 3′ single-stranded extensions needed for strand invasion. The identity of the nuclease that processes these DNA ends is not yet clear. A complete DSBR reaction has yet to be reconstituted in vitro.

4. How is recombination coordinated with other aspects of DNA metabolism? Regulation is an increasingly visible theme in this field of research. Recombination must be directed at locations where it is needed, and prevented elsewhere. When a replication fork stalls, recombinational repair systems must arrive quickly and address the situation at hand. The intricate coordination required to keep these processes on track is critical to genomic integrity, and even survival, and understanding it will keep many molecular biology laboratories engaged for decades to come.

5. What is the tension-sensing mechanism that facilitates proper chromosomal segregation in eukaryotic cell division? Some of the participating proteins have been discovered, but much remains to be done in this area of molecular biology.

HOW WE KNOW

A Motivated Graduate Student Inspires the Discovery of Recombination Genes in Bacteria

Clark, A.J. 1996. RecA mutants of *E. coli* K12: A personal turning point. *Bioessays* 18:767–772.

Clark, A.J., and A.D. Margulies. 1965. Isolation and characterization of recombination deficient mutants of *Escherichia coli* K12. *Proc. Natl. Acad. Sci. USA* 53:451–459.

Sometimes it is the student who challenges the professor. This is what happened in 1962 at the University of California, Berkeley, when first-year graduate student Ann Dee Margulies came to the office of a new assistant professor, A. John Clark. Clark later related the encounter as a career-changing moment. At a time when many molecular biologists considered the problem of recombination too complicated to address in any productive way, Margulies and Clark embarked on a project to find the genes that control recombination in bacteria.

Ann Dee Margulies, 1940/41–1980 [Source: Courtesy of Werner Maas and Renata Maas. Thanks to Alvin J. Clark.]

The two researchers decided to use bacterial conjugation as a way to measure recombination events. As Joshua Lederberg and E. L. Tatum had demonstrated in 1946, some bacteria harbor plasmids that can be transferred between cells. These F plasmids sometimes integrate themselves into the bacterial chromosome, creating strains (Hfr strains) that can convey parts of their chromosome to other cells at high frequency.

A. John Clark [Source: Courtesy Alvin J. Clark.]

When DNA is transferred, alleles from the donor DNA can be transmitted to the recipient's chromosome by recombination. Margulies and Clark used replica plating, a technique devised by Esther Lederberg and Joshua Lederberg in 1952, to search for mutants.

They used two strains of *E. coli*. The chosen Hfr donor strain could not grow unless leucine was included in the growth medium (this strain was denoted leu$^-$). The recipient strain, lacking an F plasmid, had a mutation leading to a requirement for adenine (ade$^-$). Conjugational crosses between the two strains produced recombinants that could grow in the absence of both leucine and adenine (leu$^+$ade$^+$).

The recipient strain was treated with the mutagen 1-methyl-3-nitro-l-nitrosoguanidine (MNNG) to introduce mutations at random locations in the chromosome. The researchers then had to search for those very rare mutations that affected recombination genes. The mutagenized cells were spread on agar plates containing leucine, where cells not killed by the mutagen grew into colonies. Strains were transferred one by one onto a second master plate that also contained leucine, creating a pattern of 50 to 100 colonies. On a third plate that lacked both leucine and adenine, a culture of the Hfr donor strain was spread uniformly, creating a thin "lawn" of bacteria that was alive but unable to grow, given the lack of adenine.

Using a piece of sterile velvet, Margulies replicated the pattern of colonies on the master plate onto the third plate. The transferred cells underwent conjugational mating with cells in the lawn of donor Hfr bacteria. Successful conjugation and recombination produced high-frequency ade$^+$leu$^+$ recombinants that could grow into colonies on the plates lacking leucine and adenine. Occasionally, no recombinant cells would arise where a colony was expected. If the mutagenized recipient strain continued to yield no recombinants on repeated trials, it was set aside as a candidate for a strain containing a mutation in a gene required for recombination.

The procedure was laborious, but Margulies, working under Clark's guidance, persevered. After months of careful controls and screening more than 2,000 mutagenized recipient strains, Margulies found two strains that had a recombination defect. Later work established that these strains had mutations in what became known as the *recA* gene. Clark and Margulies published their results in the *Proceedings of the National Academy of Sciences* in 1965; their paper has been cited countless times. The work launched John Clark into a productive career in elucidating recombination mechanisms. Sadly, Ann Dee Margulies, the intrepid graduate student, died of cancer in 1980, at the age of 40.

HOW WE KNOW

A Biochemical Masterpiece Catches a Recombination Protein in the Act

Keeney, S., C.N. Giroux, and N. Kleckner. 1997. Meiosis-specific DNA double-strand breaks are catalyzed by Spo11, a member of a widely conserved protein family. Cell 88:375–384.

Following the proposal of the double-strand break repair model for meiotic genetic recombination in 1983, evidence for the accuracy of major parts of the model accumulated quickly. In particular, it became clear that the process was initiated by double-strand breaks. The DSBs could be detected early in meiosis, especially in regions with recombination hot spots. But what protein created this break? For Nancy Kleckner, a Harvard biochemist who had become intrigued with genetic material as a high school student in the 1960s, this was an obvious challenge to take up. By 1995, Kleckner's postdoctoral associate, Scott Keeney, had discovered that a protein was linked to the 5′ termini at the break sites. Now, the two researchers had to identify that protein.

The answer was delivered in a biochemical exercise marked by both determination and elegance. The trick was to isolate the protein bound to the cleaved 5′ ends of the DSB, but this was no simple task. Every meiotic cell has scores of such cleavage events, and they are spaced along chromosomes containing millions of base pairs, bound by hundreds of different proteins.

The researchers' first step was one that biochemists often use: amplification of the signal. Keeney, Kleckner, and others had found that when steps subsequent to formation of the DSB, such as the rapid degradation of the 5′-ending strands, were blocked, covalent protein-DNA intermediates accumulated. A mutation in the gene encoding Rad50 (*rad50S*) served this blocking purpose well.

Using *rad50S* cells as an enriched source of the protein-DNA complexes, Keeney and Kleckner, working with collaborator Craig Giroux of Wayne State University, developed a two-step purification procedure. The first step was to eliminate bulk proteins. The researchers isolated the nuclei from the cells to remove cytoplasmic proteins and extracted the nuclear DNA with guanidinium chloride and detergent at 65°C, a treatment harsh enough to strip all but covalently linked protein from the DNA. Bulk protein was separated from the DNA in a CsCl gradient.

In the second step, the researchers separated protein-DNA complexes from bulk DNA by passing the CsCl-purified material through a glass-fiber filter, to which proteins adhere. The adhered complexes were eluted from the filter with a detergent, then treated with nucleases to remove most of the DNA. The remaining proteins were separated on a polyacrylamide gel.

The procedure was carried out in parallel on *rad50S* cells and on cells with a mutation that prevents DSB formation (as control). Doing so on a large, preparative scale yielded the results shown in Figure 1. Two bands, with apparent molecular weights of 34,000 and 45,000, were seen in the *rad50S* samples but not in the control samples. The two proteins were excised from the gel and identified by tandem mass spectrometry as a contaminant and as Spo11, respectively. More controls were carried out to solidify the case that Spo11 is the protein bound to the break sites. In one particularly compelling experiment, Spo11 was immunoprecipitated from *rad50S* cells with covalently linked DNA fragments from a known recombination hot spot.

The Spo11-mediated cleavage of DNA is the first step in the elaborate process of meiotic recombination, and its mechanism still presents a biochemical challenge. Although Spo11 is clearly the protein linked to the break sites, the actual cleavage reaction has not been observed in vitro with purified DNA and protein. The cleavage events must be regulated, and Spo11 may act only in concert with other—perhaps many other—as yet unknown proteins.

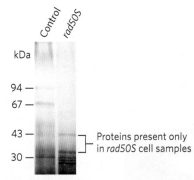

FIGURE 1 Proteins detected in the two-step purification procedure to isolate a recombination protein. [*Source: Scott Keeney, Craig N. Giroux, and Nancy Kleckner, 1997. Meiosis-specific DNA double-strand breaks are catalyzed by Spo11, a member of a widely conserved protein family. Cell 88:375–84, reprinted with permission from Elsevier.*]

KEY TERMS

homologous
 recombination, 450
double-strand break
 repair (DSBR), 450
recombinational DNA
 repair, 450
double-strand break
 (DSB), 450
recombinase, 452
DNA strand invasion,
 452
synthesis-dependent
 strand annealing
 (SDSA), 453
Holliday intermediate,
 453
genetic crossover, 453
branch migration, 454

fork regression, 455
gap repair, 456
RecBCD, 457
chi, 458
RecFOR, 458
RecA protein, 459
homolog, 467
Spo11, 469
Dmc1, 470
Rad51, 470
gene conversion, 471
BRCA2, 472
mating type, 473
homing endonuclease,
 475
nonhomologous end
 joining (NHEJ), 475
genome editing, 477

PROBLEMS

1. What are the four possible fates of a replication fork that encounters a template strand with a break or some other type of unrepaired DNA lesion?

2. A branched, circular DNA substrate is constructed to mimic one possible structure of a stalled replication fork, as shown below. An enzyme is added that promotes regression of the fork structure.

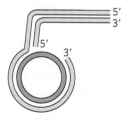

   (a) Draw the structure of the product obtained if regression proceeds halfway around the circle.
   (b) Draw the structure of the product if regression proceeds all the way around the circle. Assume the arm is the same length as the circle and has the same sequence.

3. Draw a Holliday intermediate and label the ends of each DNA strand so that the strand polarity is evident.

4. The RecBCD enzyme acts as a nuclease and a helicase in preparing DNA ends for RecA binding and strand invasion. RecBCD has several functions built into its three subunits. Indicate the subunit (RecB, RecC, or RecD) responsible for each of the following functions.

   (a) $3' \rightarrow 5'$ helicase motor
   (b) Nuclease
   (c) $5' \rightarrow 3'$ helicase
   (d) Having a "pin" structure that helps separate DNA strands
   (e) Binding to chi sites

5. Describe the three steps that initiate almost all processes related to homologous genetic recombination (the recombinational DNA repair of double-strand breaks).

6. Replication forks of a bacterial species are found to stall at double-strand cross-links, yielding a stalled fork with the structure shown below. The pathway for repair of these stalled forks involves the formation of a Holliday intermediate. Draw one step that will convert the fork into a structure with a Holliday intermediate. Place an arrowhead on all $3'$ ends (one end is so represented below). Note that the Holliday intermediate is formed without cleaving any covalent bonds in the DNA.

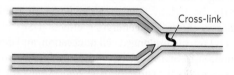

Cross-link

7. In *E. coli* cells with mutations that eliminate the RecBCD enzyme, about 20% of the cells have linearized chromosomes when grown under normal aerobic conditions. Under similar growth conditions, fewer than 3% of the chromosomes are linearized in wild-type cells. Suggest, in two or three sentences, why this difference is observed.

8. Eukaryotes have two RecA-class recombinases, called Dmc1 and Rad51. How are these enzymes utilized by cells?

9. During meiosis in yeast, if the diploid cell has alleles *a* and *A* of a particular gene, it normally forms two spores with *A* and two spores with *a*. Rarely, meiosis yields one spore with *A* and three with *a*, or three with *A* and one with *a*. How could this happen?

10. Unlike recombination, the repair of double-strand breaks by nonhomologous end joining creates mutations. Explain why.

11. At the yeast mating-type locus, the mating-type switch is initiated by introducing a double-strand break at the *MAT* locus. What would happen if the DSB were introduced at the *HMLα* locus instead?

12. In the study by Keeney, Giroux, and Kleckner (see this chapter's How We Know section), Spo11 was identified as the protein that introduces double-strand breaks to initiate meiotic recombination. To identify candidate proteins, the DNA was first extracted to remove most noncovalently bound proteins, then filtered to isolate remaining protein-DNA complexes. The samples were then extensively treated with nucleases before the samples were loaded onto a

polyacrylamide gel. Why was the nuclease treatment necessary?

13. A Holliday intermediate is formed between two chromosomes at a point between two genes, *A* and *B*, as shown below. The two chromosomes have different alleles of the two genes (*A* and *a*; *B* and *b*). Where would the Holliday intermediate have to be cleaved (points X and/or Y) to generate a chromosome with (a) an *Ab* genotype or (b) an *ab* genotype?

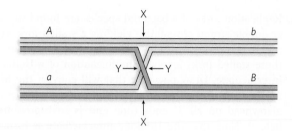

14. Ionizing radiation causes double-strand breaks in DNA, but the bacterium *Deinococcus radiodurans* is highly resistant to these DSB-generating effects. DSBs also occur (slowly) during prolonged cell desiccation, and desiccation is thought to be the selective pressure in the evolution of the extraordinary capacity of *D. radiodurans* for DNA repair. After heavy irradiation, the bacterium produces several novel proteins at high levels. One of these, called DdrA (*D*NA *d*amage *r*epair protein *A*), binds tightly to the 3′ ends of broken DNA strands and prevents their degradation by nucleases. A mutation that eliminates DdrA function has little effect on survival after irradiation, but a large effect on survival after desiccation. Suggest an explanation for the role of DdrA during desiccation. In your answer, consider the requirements of DNA repair versus DNA degradation.

DATA ANALYSIS PROBLEM

Cox, M.M., and I.R. Lehman. 1981. Directionality and polarity in RecA protein–promoted branch migration. *Proc. Natl. Acad. Sci. USA* 78:6018–6022.

15. The RecA protein promotes DNA strand exchange, as shown in Figure 13-11, with a unidirectional branch migration proceeding around the DNA circle. The direction of the branch migration was established by Cox and Lehman. A circular chromosome from bacteriophage φX174 can be isolated as either a single- or a double-stranded circle. The circular map of the φX174 genome is shown in Figure 1 (data from Cox and Lehman's paper); the single-stranded circle proceeds 5′→3′ in the clockwise direction. In the experiment, the single-stranded φX174 circle was radioactively labeled uniformly along its length. The double-stranded circular DNA was not labeled, but was cleaved at a unique site (labeled A) by a restriction enzyme to generate the substrates shown in Figure 13-11. Sites for cleavage by a second restriction enzyme (labeled B) are also noted.

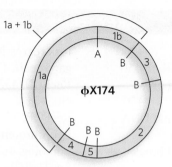

FIGURE 1

16. A DNA strand exchange reaction was initiated with the RecA protein and ATP. At various times, aliquots were removed and treated with restriction enzyme B and with a nuclease that selectively digested all single-stranded DNA, and the DNA was subjected to agarose gel electrophoresis to separate the restriction fragments. The incorporation of radioactivity (as percentage of maximum) into the various restriction fragments was measured at various time intervals; the results are shown in Figure 2.

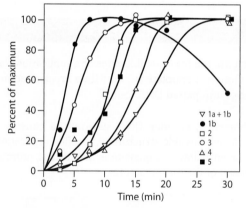

FIGURE 2

(a) Radioactivity is present throughout the single-stranded DNA circle, so why is there no radioactivity in any of the restriction fragments at the beginning of the experiment?

(b) In what direction does RecA-mediated branch migration proceed?

(c) At the end of the experiment, the amount of fragment 1b begins to decline. Suggest an explanation.

ADDITIONAL READING

General

Clark, A.J., and A.D. Margulies. 1965. Isolation and characterization of recombination-deficient mutants of *Escherichia coli* K12. *Proc. Natl. Acad. Sci. USA* 53:451–459.

Haber, J.E. 2014. *Genome Stability: DNA Repair and Recombination.* New York: Garland Science, Taylor & Francis Group.

Szostak, J.W., T.L. Orr-Weaver, R.J. Rothstein, and F.W. Stahl. **1983.** The double-strand-break repair model for recombination. *Cell* 33:25–35.

Recombination as a DNA Repair Process

Cox, M.M., M.F. Goodman, K.N. Kreuzer, D.J. Sherratt, S.J. Sandler, and K.J. Marians. **2000.** The importance of repairing stalled replication forks. *Nature* 404:37–41.

Kuzminov, A. **2001.** DNA replication meets genetic exchange: Chromosomal damage and its repair by homologous recombination. *Proc. Natl. Acad. Sci. USA* 98:8461–8468.

Michel, B., H. Boubakri, Z. Baharoglu, M. LeMasson, and R. Lestini. **2007.** Recombination proteins and rescue of arrested replication forks. *DNA Repair* 6:967–980.

Enzymatic Machines in Bacterial Recombinational DNA Repair

Cox, M.M. **2007.** Regulation of bacterial RecA function. *Crit. Rev. Biochem. Mol. Biol.* 42:41–63.

Duderstadt, K.E., R. Reyes-Lamothe, A. van Oijen, and D.J. Sherratt. **2013.** Replication fork dynamics. *Cold Spring Harb. Perspect. Biol.* 6(1):a010157.

Lusetti, S.L., and M.M. Cox. **2002.** The bacterial RecA protein and the recombinational DNA repair of stalled replication forks. *Annu. Rev. Biochem.* 71:71–100.

Homologous Recombination in Eukaryotes

deMassy, B. **2013.** Initiation of meiotic recombination: How and where? Conservation and specificities among eukaryotes. *Annu. Rev. Genet.* 47:563–599.

Filippo, J.S., P. Sung, and H. Klein. **2008.** Mechanism of eukaryotic homologous recombination. *Annu. Rev. Biochem.* 77:229–257.

Sugawara, N., and J.E. Haber. **2006.** Repair of DNA double strand breaks: In vivo biochemistry. *Methods Enzymol.* 408:416–429.

Sung, P. **2005.** Mediating repair. *Nat. Struct. Mol. Biol.* 12:213–214.

Symington, L.S., and J. Gautier. **2011.** Double-strand break end resection and repair pathway choice. *Annu. Rev. Genet.* 45:247–271.

Zickler, D., and N. Kleckner. **1999.** Meiotic chromosomes: Integrating structure and function. *Annu. Rev. Genet.* 33:603–754.

Nonhomologous End Joining

Aniukwu, J., M.S. Glickman, and S. Shuman. **2008.** The pathways and outcomes of mycobacterial NHEJ depend on the structure of the broken DNA ends. *Genes Dev.* 22:512–527.

Gu, J.F., and M.R. Lieber. **2008.** Mechanistic flexibility as a conserved theme across 3 billion years of nonhomologous DNA end-joining. *Genes Dev.* 22:411–415.

Lieber, M.R. **2008.** The mechanism of human nonhomologous DNA end joining. *J. Biol. Chem.* 283:1–5.

Pitcher, R.S., N.C. Brissett, and A. Doherty. **2007.** Nonhomologous end joining in bacteria: A microbial perspective. *Annu. Rev. Microbiol.* 61:259–282.

14 Site-Specific Recombination and Transposition

Wei Yang [Source: Courtesy Wei Yang.]

MOMENT OF DISCOVERY

I arrived at Yale University as a postdoctoral fellow in the laboratory of Dr. Tom Steitz 10 years after he and Nigel Grindley solved the crystal structure of the catalytic domain of γδ resolvase, a site-specific recombinase. *My goal was to crystallize a protein-DNA complex to aid in understanding the recombination mechanism.* The DNA recombination site *res* is 114 base pairs long. However, two *res* sites must be in supercoiled DNA for recombination to take place, making the complex very difficult to crystallize.

In the early 1990s, we had to use a "divide and conquer" strategy. We started by co-crystallizing DNA containing only one *res* cleavage site with a dimeric γδ resolvase. It was terribly exciting for me to find the initial crystals while peering into a microscope, but the experienced postdocs and students in Steitz's lab just smiled and politely wished me luck. I understood their reaction only after finding that many protein-DNA co-crystals don't diffract x rays well or at all.

After overcoming the diffraction problem, the phase problem, and the asymmetry of the γδ resolvase dimer, the moment of visualizing the electron density of the γδ resolvase–DNA complex on a computer screen for the first time was sheer joy. At that moment I was the only one in the whole world who knew how γδ resolvase bound (and bent!) its recognition site. However, obtaining the crystal structure of a true recombination intermediate took another 10 years.

Today, more than 15 years after I solved the γδ resolvase–DNA complex structure, the mechanism of how two DNA duplexes exchange partners remains a hypothesis. To fully understand a biological process, it often takes generations of scientists, with each generation making additional steps forward. I was pleased to contribute my step and eagerly look forward to following the next chapters in this story.

—Wei Yang, on researching the structure and molecular mechanisms of γδ resolvase

Life might have made its first appearance on this planet in the form of a self-replicating polymer made of RNA, or something similar. The existence of this successfully self-replicating RNA would immediately have given rise to a biosphere with two possible survival strategies. First, the polymer itself could continue to use the strategy of self-replication, drawing on resources in the environment to increase its own population. Second, in a strategy that introduced what were probably the first parasites, a smaller RNA—a primitive RNA transposable element—could insert itself into a vulnerable site in a self-replicating RNA. With this strategy, the transposable element would reproduce passively as long as the function of the self-replicator was unaffected by the addition to its mass. Thus, transposable elements may be among the most ancient of genetic entities. Always present and often deceptively simple in structure, transposons long ago perfected the art of mostly benign coexistence with their hosts. Some bear a close evolutionary relationship to viruses. The additional RNA contributed by the parasitic transposable elements was not always inconsequential; transposable elements have had a profound effect on living systems throughout evolutionary time. For example, they played a key role in the development of the vertebrate immune system, as we will see.

Perhaps most important, the fundamental chemistry employed by these elements is seen today in processes ranging from the hydrolytic cleavage of nucleic acids (Chapter 6) to DNA polymerization (Chapter 11), topoisomerization (Chapter 9), RNA transcription (Chapter 15), and intron splicing (Chapter 16). The basic reaction underlying all of these processes is a phosphoryl transfer involving a phosphodiester bond (**Figure 14-1**). Many of these processes represent a specialized form of transesterification. It is the phosphoryl transfer reactions that ultimately make every nucleic acid a dynamic purveyor of the genetic information on which all life depends. We find that basic chemistry not only in transposition but in every process described in this chapter, including site-specific recombination and the integration of retroviruses (including HIV) into genomic DNA.

Site-specific recombination and transposition are specialized types of recombination that share two key properties. First, both generally involve bringing together DNA sites without extensive homology. In that sense they complement the homologous genetic recombination discussed in Chapter 13, expanding the repertoire of DNA and RNA transactions available to every cell. Second, the key enzymes in both processes bear a notable evolutionary relationship to DNA topoisomerases. Site-specific recombinases and transposases inhabit an enzymatic world of thermodynamic equanimity. Like topoisomerases, they promote phosphoryl group transfers that are typically **isoenergetic** (i.e., $\Delta G'^{\circ} = 0$), or nearly so.

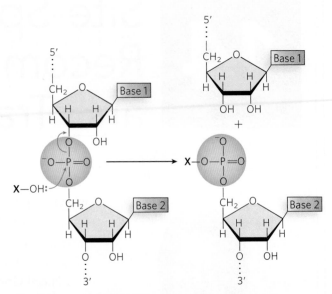

FIGURE 14-1 Phosphoryl transfer reactions involving phosphodiester bonds. In three common variants of this reaction described in this chapter, the nucleophilic hydroxyl group (X–OH) is (1) water, leading to hydrophilic cleavage; (2) the hydroxyl group of a Ser or Tyr residue in the active site of an enzyme, leading to the formation of a covalent phosphodiester linkage to that enzyme at the expense of the existing phosphodiester bond between nucleotides; or (3) the 3′-hydroxyl group of another nucleotide, leading to formation of a new phosphodiester bond at the expense of the existing one. The second and third processes are generally isoenergetic, or nearly so. The closely related phosphoryl transfer reactions leading to formation of phosphodiester bonds in the active sites of DNA and RNA polymerases are illustrated in Chapters 11 and 15, respectively.

Site-specific recombination is a precise and predictable process in which DNA is rearranged between two specific sequences. This can result in the insertion, deletion, or inversion—depending on the arrangement of the recombination sites—of a particular DNA segment. In a sense, recombinases, the enzymes that carry out the site-specific recombination reactions, are restriction endonucleases and DNA ligases combined into a single efficient package. However, unlike DNA ligation, the reactions do not require ATP or a similar cofactor; the phosphoryl transfer reactions underlying site-specific recombination are nearly isoenergetic. The reactions are often tied to genomic replication, but they have been recruited for other purposes as well. They typically provide an elegant biochemical solution to awkward topological problems that occur in plasmids and chromosomes. Site-specific recombinases are increasingly well understood, and biotechnologists love them for the precise DNA rearrangements they catalyze.

Transposable elements, also called **transposons**, are much more than the hypothetical denizens of an early RNA world. They exist today, tucked away in the genomes of essentially all organisms. They are genomic freeloaders, nucleic acid sequences that insert themselves into the genome of another organism and are replicated passively every time a cell divides. They are almost stunningly common: as noted in Chapter 8, the genomes of several million transposons or transposon remnants make up nearly half of the DNA in the human genome. Through the activity of enzymes called transposases, these genetic elements can move from one location on a chromosome to another, often with no apparent selectivity with respect to the surrounding DNA sequence. This process is known as **transposition**. The ability of transposons to move within a genome, defying Mendelian genetics, startled and then intrigued geneticists when Barbara McClintock

discovered these elements in maize plants in the 1940s. Careful work by McClintock, and other researchers' eventual discovery of closely related genetic elements in other organisms, gradually overcame the initial skepticism of colleagues. Since then, studies of transposons have placed them among the most important tools of biotechnology, yielding invaluable genetic techniques for constructing transgenic

Barbara McClintock, 1902–1992 [Source: Science Source.]

model organisms and for exploring gene function. As noted above, transposons insert themselves into genomic DNA through nearly isoenergetic phosphoryl transfer reactions like that shown in Figure 14-1.

In this chapter, we discuss site-specific recombination and transposition in succession. We begin with site-specific recombination because it is more predictable and allows us to layer biochemistry and biology onto the chemistry of Figure 14-1 in a more intuitive progression.

14.1 MECHANISMS OF SITE-SPECIFIC RECOMBINATION

Recombination between specific sequences can result in insertion, deletion, or inversion of the DNA sequence between those sites. Recombination reactions of this type occur in virtually every cell, filling specialized roles that vary greatly from one species to another but sharing a common mechanism. Each site-specific recombination system consists of a short, unique DNA sequence (20 to 200 bp) and a recombinase, an enzyme that acts specifically at that sequence. In some systems, additional proteins are required to facilitate or regulate the process. The result of a site-specific recombination reaction can be similar to that of the crossovers that sometimes accompany homologous recombination, but the process does not require extensive homology at recombination sites.

The DNA rearrangements promoted by site-specific recombinases appear in numerous and sometimes surprising roles. Examples range from prescribed roles in the replication cycles of viral, plasmid, and bacterial DNAs, to key events in the life cycle of some viruses, to regulation of the expression of certain genes.

Precise DNA Rearrangements Are Promoted by Site-Specific Recombinases

The recombination sites recognized by site-specific recombinases often consist of two inverted repeats, separated by a short asymmetric (nonpalindromic) core sequence (**Figure 14-2a**). During site-specific recombination, the asymmetric cores of two recombination sites are aligned so that their sequences proceed in the same direction. The recombinase recognizes and binds specifically to the symmetric repeats on either side of each aligned core. Since it is not directly bound by the recombinase in most of these systems, the sequence in the core itself can often be varied without affecting its recognition by the recombinase. However, recombination occurs only if the sequences of the two cores are identical.

The asymmetric core sequence in a recombination site gives each site an orientation within the surrounding DNA. The overall outcome of a site-specific recombination reaction depends on the location and relative orientation of the recombination sites within the genomic DNA in which they reside (**Figure 14-2b**). Recombination between two oppositely oriented sites on the same DNA molecule produces an inversion. Recombination between two sites with the same orientation on the same DNA molecule results in a deletion. If the sites are on different DNAs, the result is the insertion of one of the DNAs into the other. Some recombinase systems are highly specific for one of these reaction types and act only on sites with particular orientations.

Site-specific recombination systems use either a Tyr or a Ser residue as the key nucleophile in the active site. (The Cre and Flp recombinases discussed here and illustrated in Figures 14-2 and 14-3 both utilize Tyr residues.) In vitro studies of many such systems have clarified the fundamental reaction pathway. A pair of recombinases recognizes and binds to each of two recombination sites. The two recombination sites are brought together by their bound recombinases to form a synaptic complex that incorporates a total of four recombinase subunits. Within this complex, the core sequences of the two recombination sites are aligned. If the two recombination sites are on the same DNA molecule, the intervening core DNA is bent into a loop as the sites are brought together (see Figure 14-2b, left).

The site-specific recombination reaction is best understood for a family of tyrosine-class recombinases

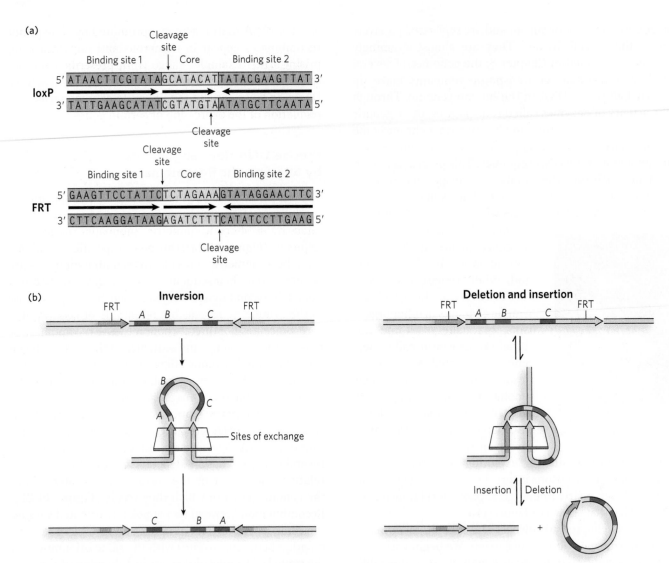

FIGURE 14-2 The structure and activity of site-specific recombination sites. (a) Shown here are both strands of the recombination sites from two well-studied recombination systems widely used in biotechnology, lox (loxP) and FRT; the inverted 13 bp repeats are binding sites for the recombinases, named Cre and Flp, respectively. (These two recombination sites and their recombinases are discussed later in the chapter.) The inverted repeats are separated by an asymmetric core sequence. The cleavage and exchange events described in Figure 14-3 occur at or near the ends of the core sequence. (b) The colored ribbons here represent double-stranded DNA. Two recombination sites flank a length of DNA to be recombined. *A*, *B*, and *C* are imaginary genes or genetic markers in the DNA separating the two FRT sites. Each orange arrow represents a complete FRT site, as illustrated in (a). Orientation (shown by the arrowheads) refers to the asymmetric nucleotide sequence in these recombination sites, *not* the 5′→3′ direction. Recombination can lead to inversion (left) or deletion and, in the reverse process, insertion (right).

that includes the enzymes Cre and Flp (**Figure 14-3a**). The active-site tyrosines of two of the four recombinases in the complex each attack a specific phosphodiester bond, each located in one DNA strand near the end of the core sequence of each recombination site. The reactions proceed as in Figure 14-1, with a new phosphodiester bond formed between the DNA and the tyrosyl oxygen (forming a phosphotyrosine bond) at each active site as the DNA itself is cleaved. One recombinase subunit thus becomes covalently linked to the DNA at each cleavage site (see Figure 14-3a, step 1). These phosphoryl transfers

do not occur at random within the complex. Instead, the reaction is structurally choreographed so that two opposing recombinase subunits (shown in light gray in Figure 14-3) are active while the other two are not. The transient protein-DNA linkage ensures that the overall reaction proceeds with a minimal free-energy change, so high-energy cofactors such as ATP are unnecessary. Each of the free 3′ hydroxyls of the cleaved DNA strands now becomes the nucleophile in step 2, attacking the phosphotyrosine linkage to free the Tyr residue and form a new phosphodiester bond. However, the reaction involves new DNA partners and leads to the formation of a Holliday intermediate closely related to the Holliday intermediates described in Chapter 13 (albeit formed via a different pathway). An isomerization then occurs in the protein complex (step 3). This step includes a branch migration through the core sequence (a step that is blocked if the cores are not identical), coupled to a conformational change such that the active sites of the

two recombinase subunits that did not participate in the first two steps become properly positioned relative to the phosphodiester bonds they must act on to complete the reaction. The sequence of two phosphoryl transfer reactions is then repeated (steps 4 and 5), one to create a new set of covalent phosphotyrosine bonds linking the DNA to the protein, and the other to resolve this protein-DNA complex and create new phosphodiester bonds. Each of these new reactions occurs on the opposite strand and at the opposite end of the two recombination sites relative to the reactions in the first two steps.

In systems that use an active-site Ser residue (see this chapter's Moment of Discovery), both strands of each recombination site are cut concurrently and rejoined to new partners, without the Holliday intermediate. All four recombinase subunits participate, each forming a phosphoserine covalent intermediate at the cleavage sites. In both types of system (serine and tyrosine), the exchange is always reciprocal and precise, regenerating a new pair of fully functional recombination sites when the reaction is complete.

Many mechanistic details of these reactions have become clear with the structural elucidation of recombinases caught at different steps of the process. The four recombinase subunits and the four DNA arms in the synaptic complex take up a square planar arrangement (**Figure 14-3b**). As shown by the crystal structure, the tyrosine-class recombinases are not in perfect fourfold symmetry. Instead, alternating subunits are in slightly different conformations—two active (with the active-site Tyr residues positioned near the phosphodiester bonds to be cleaved) and two inactive. The isomerization step described above (see Figure 14-3a, step 3), coupled with subtle conformational changes, converts the active recombinase subunits to the inactive state, and inactive subunits to active.

The overall process is closely related to the reaction mechanism promoted by topoisomerases (see Figure 9-19). For both topoisomerases and site-specific recombinases, the reaction begins with the formation of

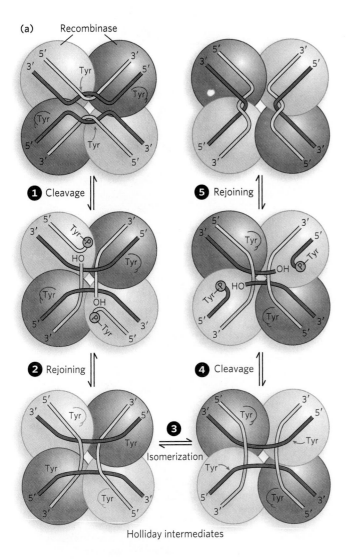

(a)

Holliday intermediates

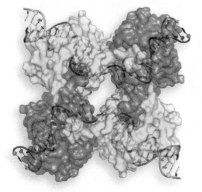

(b) **Holliday intermediate**

FIGURE 14-3 A site-specific recombination reaction. (a) The reaction proceeds within a tetramer of identical recombinase subunits. The subunits bind to the recombination site and catalyze the recombination in several steps, as described in the text. The light gray subunits are the active subunits, with the active-site Tyr residue either poised to react or covalently linked to the DNA. The darker gray subunits are in an inactive conformation in which the active-site Tyr residues are too distant from their DNA substrates to function. Isomerization (step 3) switches the conformations of both sets of subunits so that the inactive subunits (dark gray) become active (light gray) and the active-site Tyr residues are now properly positioned to promote the reaction. (b) A surface contour model of Flp recombinase, showing the four subunits bound to a Holliday intermediate; this is equivalent to the product of step 2 in (a). The protein is made transparent so that the bound DNA is visible. [Source: (b) PDB ID 3CRX.]

a protein-DNA phosphotyrosine or phosphoserine linkage at the expense of a phosphodiester bond in the DNA. In the case of topoisomerases, the same phosphodiester bond is re-created after the DNA topology has been changed. In the case of site-specific recombinases, each end of the cleaved phosphodiester bond is joined to a new partner. The different outcomes are brought about by the different architectures of the proteins promoting the two reactions, resulting in different and very precise movements of DNA segments between the phosphodiester cleavage and re-formation steps.

Site-Specific Recombination Complements Replication

Replication of the circular chromosomes of viruses, plasmids, and many bacteria poses a unique set of challenges. As noted in Chapter 13, recombinational DNA repair of stalled replication forks can give rise to contiguous dimeric chromosomes (see Figure 13-14). A specialized site-specific recombination system in *E. coli* converts the dimeric chromosomes to monomeric chromosomes so that cell division can proceed. The reaction is a site-specific deletion reaction catalyzed by a tyrosine-class recombinase, XerCD. As described here, site-specific recombination can also be used as an elegant mechanism to generate more than two copies of a chromosome during one replication cycle.

A common plasmid in *Saccharomyces cerevisiae*, the 2μ (2 micron) plasmid, has a site-specific recombination system. The recombinase, known as Flp (a shortened version of flippase, an early and somewhat whimsical name for the enzyme), is encoded by the plasmid. In this system, site-specific recombination is used to amplify the number of plasmids in the cell (the plasmid copy number) whenever necessary. When the copy number falls too low, Flp is activated to promote the recombination reaction. The key to copy-number amplification is the timing of the recombination. The replication origin of the plasmid is situated such that one Flp recombination target (FRT) site is replicated well before the other (**Figure 14-4**). If recombination occurs when only one FRT has been replicated, the result is the inversion not

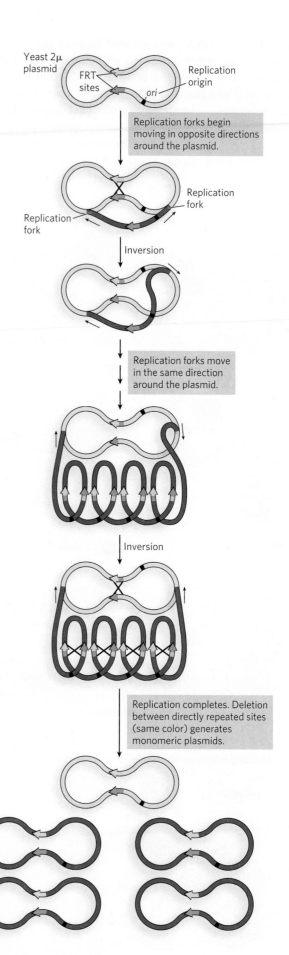

FIGURE 14-4 Coupling site-specific recombination to extensive replication in a yeast plasmid. The yeast 2μ plasmid, a circular DNA, has two FRT sites (orange) on opposite sides of the circular DNA molecule and inverted relative to each other. FRT sites are targeted by the plasmid-encoded Flp recombinase. The recombination reaction inverts the DNA in the sequences of about one half of the plasmid relative to those of the other half. The inversion also changes the direction of one replication fork relative to the other. Inversion thus leads to a double rolling-circle replication that can produce multiple copies of the plasmid in one replication cycle, increasing the plasmid copy number in the cell.

just of one segment of DNA but of one replication fork relative to the other. Instead of meeting at the opposite side of the circle, the two forks begin to follow each other around the circle, promoting an extended rolling-circle replication. This generates multiple tandem copies of the plasmid, instead of just two, from one replication initiation. The multimeric plasmid is then broken down into plasmid monomers by subsequent Flp-mediated recombination events carried out between FRT sites in the same orientation.

Site-Specific Recombination Can Be a Stage in a Viral Infection Cycle

Bacteriophages such as P1 and λ have played important roles in the development of molecular biology and biotechnology (see the How We Know section at the end of this chapter). When it enters a cell, the DNA of these phages has two potential fates (**Figure 14-5**). A **lysogenic pathway** involves incorporation of the phage genome as part of the host genome, either by integration into the host chromosome or as an autonomously replicating plasmid. In either case, phage genes are largely repressed, and the phage DNA is replicated passively by host enzymes. Lysogenized bacteriophage genomes are referred to as **prophages**, and the parasitic infection is benign as long as the phage remains in this state. In the alternative, **lytic pathway**, the bacteriophage DNA is replicated and packaged into new phage heads, and the host cell is destroyed by lysis to disperse the progeny. The specific mechanisms used in the P1 life cycle feature site-specific recombination in some key steps.

    P1 enters a bacterial cell as a linear DNA molecule containing multiple contiguous copies of the 90 kbp genome. In the host, the DNA is rapidly circularized to produce multiple genome-length circular DNAs (**Figure 14-6a**). The circularization can occur by homologous recombination, or it can be promoted by a phage-encoded site-specific recombination system. The latter system, known as **Cre-lox**, involves recombination sites called loxP (*lo*cus of crossover (*x*), *p*hage) sites—more often known simply as lox sites—and the recombinase Cre (*c*yclization *re*combination). The circularized DNA can enter a lysogenic state, maintaining the same copy number in the host cell. In addition to circularization, the Cre-lox recombination system aids in the orderly dispersal of the P1 genomes to daughter cells at cell division by resolving any circular P1 dimers to monomers. In the lytic pathway, P1 replicates in a rolling-circle mode in which the replication fork travels unidirectionally around the circularized chromosome (**Figure 14-6b**). This generates long, linear DNAs with many contiguous

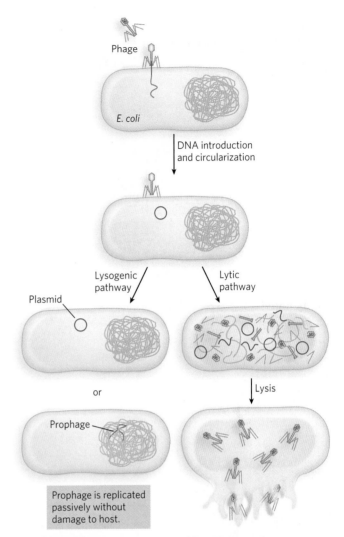

FIGURE 14-5 Two possible fates for a phage-infected host cell. Several types of bacteriophage introduce their DNA into cells in a linear form, which is circularized inside the cell. The lysogenic pathway involves either integration of the DNA (now referred to as a prophage) into the host chromosome or its passive replication as a plasmid. The alternative, lytic pathway eventually destroys the host cell and releases phage progeny.

copies of the P1 genome. The DNAs are cut and the genomes incorporated into phage heads before cell lysis. Occasionally, large pieces of host DNA are also incorporated into phage heads. This low-frequency event allows P1 to be used as an experimental vehicle to move bacterial genes from one cell to another in a process known as **bacterial transduction**.

Site-Specific Recombination Systems Are Used in Biotechnology

As noted earlier, the Flp recombination system of the yeast 2μ plasmid and the Cre-lox system of bacteriophage P1 are tyrosine-class recombinases. These are relatively simple systems in which the recombination sites are

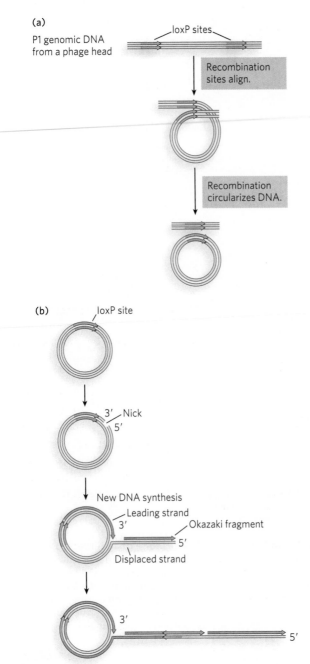

(a)

P1 genomic DNA from a phage head

loxP sites

Recombination sites align.

Recombination circularizes DNA.

(b)

loxP site

3' Nick
5'

New DNA synthesis

Leading strand
3'
Okazaki fragment
5'
Displaced strand

3'
5'

FIGURE 14-6 Circularization of P1 DNA. (a) P1 phage contains about 1.4 copies of its genome, packaged in its head. In a host cell, circularization generates monomeric, circular genomes. The orange arrows are loxP sites and serve to define one monomeric genome equivalent. Circularization may involve Cre-mediated recombination or homologous recombination at other locations. (b) Following circularization, the P1 genome can undergo rolling-circle replication, producing many copies from one template.

short (about 34 bp long) and the recombinases (Flp and Cre) are the only enzymes required. The recombination can be adapted to produce inversion, deletion, or insertion, depending on the placement and orientation of the recombination sites. The overall reactions promoted by Flp and Cre are isoenergetic, occurring without the input of ATP. The Flp and Cre reactions thus tend to approach an equilibrium in which substrates and products are in equal concentrations. In this case, simplicity gives rise to practical application.

The Cre and Flp systems will function when engineered into the cells of any organism and thus are highly useful in a wide range of biotechnological applications. A few such applications are shown in **Figure 14-7**. If the requisite lox or FRT sites are engineered into plasmids or chromosomes in the proper locations, these systems can be (and have been) used to activate a particular gene, insert a new gene into a cell at a chosen location, replace one gene with another gene or an altered version of the same gene, delete a gene, or alter the linear structure of an entire chromosome. The sequence specificity of the recombinases allows all of these transactions to be promoted with extraordinary precision. Even more elaborate manipulations are possible. For example, if you tied expression of the recombinase to a promoter expressed only in a particular tissue, you could limit the recombination event to that tissue. Deleting a gene in a certain tissue at a specified time can be a powerful tool for exploring the function of that gene. A vivid example of the application of site-specific recombination in biotechnology is described in **Highlight 14-1** on p. 494.

Gene Expression Can Be Regulated by Site-Specific Recombination

The biological uses of site-specific recombination are varied and extend even to the regulation of genes. *Salmonella typhimurium*, which inhabits the mammalian intestine, moves by rotating the flagella on its cell surface. The many copies of the protein flagellin (M_r 53,000) that make up the flagella are prominent targets of mammalian immune systems. But *Salmonella* cells have a mechanism that evades the immune response: they switch between two distinct flagellin proteins (FljB and FliC) roughly once every 1,000 generations, using the process of **phase variation**.

The switch is accomplished by periodic inversion of a segment of DNA containing the promoter for a flagellin gene. The inversion is a site-specific recombination reaction mediated by the Hin (*H* DNA *i*nvertase) recombinase at specific 14 bp sequences (*hix* sequences) at either end of the DNA segment. When the DNA segment is in one orientation, the gene for FljB flagellin and the gene for a repressor protein (FljA) are expressed; the repressor shuts down expression of the gene for FliC flagellin (**Figure 14-8a**). When the DNA segment is inverted, the *fljA* and *fljB* genes are no longer transcribed, and the *fliC* gene is induced as the repressor becomes depleted (**Figure 14-8b**). The Hin recombinase, encoded by the

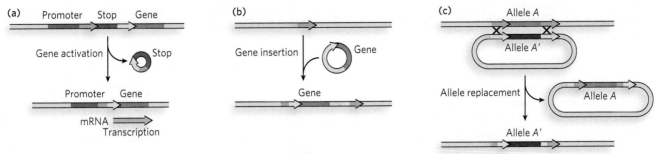

FIGURE 14-7 Some biotechnology applications of site-specific recombination. (a) Gene activation (by deletion of a transcription stop signal). (b) Gene insertion. (c) Allele replacement, using two variants of the same recombination target site at each end of the gene (or other sequence) to be replaced. In each case, the site-specific recombination target site is engineered into the chromosome or into the introduced DNA, or both, with the orientation indicated by the arrows.

hin gene in the DNA segment that undergoes inversion, is expressed when the DNA segment is in either orientation, so the cell can always switch from one state to the other.

The *Salmonella* system is by no means unique. Similar regulatory systems occur in other bacteria and in some bacteriophages. Gene regulation by DNA rearrangements that move genes and/or promoters is particularly common in pathogens that benefit by changing their surface proteins, thereby staying ahead of host immune systems.

Hin belongs to a family of recombinases that are distinct from the Flp and Cre enzymes: the serine-class recombinases. Unlike Flp and Cre, Hin promotes reactions that are highly constrained (inversions only). The reactions make use of auxiliary proteins called Fis that participate in the site-specific recombination reaction promoted by Hin.

The Hin recombinase catalyzes recombination only when its *hix* recombination sites are on the same supercoiled DNA molecule and are inverted relative to each other. This specificity is accomplished through the formation of an elaborate structure involving both Hin and the Fis proteins, which acts as a topological filter (**Figure 14-9**). The structure cannot form unless the

FIGURE 14-8 Phase variation in *Salmonella* flagellin genes.
(a) In one orientation, *fljB* is expressed along with a repressor protein (product of the *fljA* gene) that turns off transcription of the *fliC* gene; the result is production of the flagellin FljB. (b) In the opposite orientation, the *fljA* and *fljB* genes cannot be transcribed, and only the *fliC* gene is expressed, producing flagellin FliC. *Salmonella* can flip between these two flagellin-producing systems.

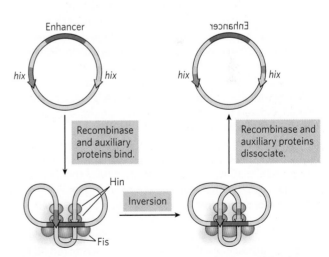

FIGURE 14-9 A topological filter in Hin-*hix* recombination. The ribbons represent double-stranded DNA. The Hin recombination complex consists of four Hin subunits bound in pairs to two *hix* sites, along with an enhancer sequence (green) bound by a host-encoded protein called Fis. Recombination cannot occur unless the entire complex comes together with the topology shown here. This particular alignment of *hix* sites occurs only if the sites are present in the same supercoiled DNA molecule and have the opposite orientation. [Source: Data from R. Harshey and M. Jayaram, Crit. Rev. Biochem. Mol. Biol. 41:387, 2006.]

Using Site-Specific Recombination to Trace Neurons

Site-specific recombination offers an opportunity to modify genomic DNA with exquisite precision. A vivid example, and one that illustrates its biotechnological potential, is the brainbow technology introduced by Jeff Lichtman, Jean Livet, and Joshua Sanes. Combining transgenic technology and site-specific recombination, they created a system to map neurons in the mouse brain.

The use of site-specific recombination to make genomic alterations always requires some genetic engineering. Recombination target sites must be installed where a biotechnologist wants them to be. The corresponding recombinase must also be present in the same cell and at the right time, so the systems have to be adapted to a particular purpose or experiment. The most widely used recombination systems are the Cre-lox system of bacteriophage P1 and the Flp-FRT system of the yeast 2μ plasmid. Both have relatively short recombination target sites (34 to 36 bp; see Figure 14-2a), and both rely on a single recombinase enzyme (Cre or Flp) that requires no auxiliary proteins.

Mario Capecchi, Oliver Smithies, and Martin Evans developed powerful procedures to generate transgenic mice with designed genomic alterations (Figure 1; see Chapter 7 for discussion of such procedures). DNA containing a gene alteration is introduced into mouse embryonic stem cells. The DNA is introduced in the form of an engineered plasmid DNA that sandwiches the desired alteration between a set of selectable markers that will allow identification of the rare cells with the desired alteration inserted at the correct location. The altered stem cells (from brown mice) are introduced into a very early embryonic stage (blastocyst) of black mice, and the embryos are implanted in a surrogate mother. The altered brown-mouse cells become part of the developing embryo and can appear in almost any tissue. Progeny with the altered brown-mouse cells in their germ line are crossed to produce all-brown progeny, signifying the presence of the desired genetic change in the germ line.

The brainbow method makes use of green fluorescent protein (GFP), the procedures for generating transgenic mice, and site-specific recombination to trace the path of the countless neurons that make up the brain (Figure 2). Inserted into the mouse genome is a gene cassette (a structured set of genes arranged for a particular biotechnological purpose) with several copies of GFP variants that encode proteins fluorescing with different colors, such as red (RFP), orange (OFP), yellow (YFP), and cyan (CFP). Variants of the loxP target site for the Cre recombinase are engineered between the genes. The loxP sites are arranged so that only one of the three possible recombination events can occur in a particular cassette, and each event will result in gene

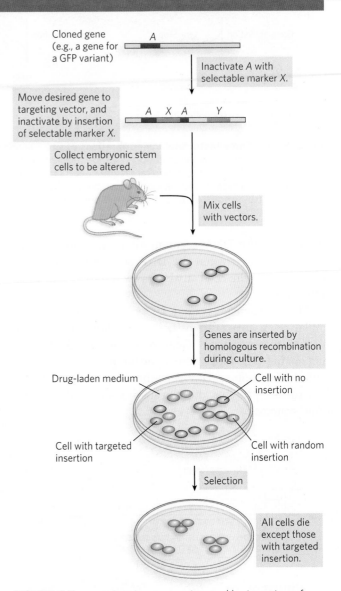

FIGURE 1 Transgenic mice are engineered by insertion of a targeting vector into embryonic stem cells. The vector contains a selectable marker, *X*, and the desired chromosomal alteration (such as a lox site to be introduced) sandwiched between two DNA segments complementary to the chromosomal site where the alteration is to be integrated. These chromosomal sequences can direct homologous recombination. A second selectable marker, *Y*, is included in the cassette, outside these homologous sequences, and is generally introduced to the chromosome only if the DNA is integrated at an incorrect (nonhomologous) chromosomal site. The targeted cells are subjected to selection in vitro, using drugs to select "for" cells that have selectable marker *X* and "against" cells that also have selectable marker *Y*. The surviving cells are then introduced into an early-stage embryo and become part of the tissue of the developing mouse. Genetic crosses allow the selection of mice expressing the desired alterations in their germ line.

(a)

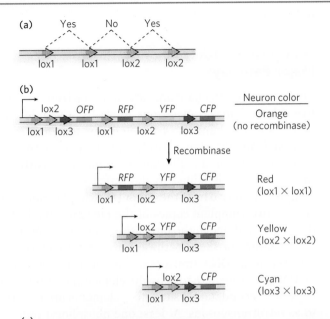

(b)

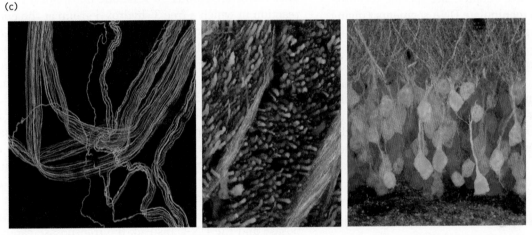

FIGURE 2 Neuronal networks can be traced by turning the network into a brainbow. Transgenic mice with genes encoding GFP variants and different (modified) intervening lox sites are mated to transgenic mice with the gene encoding the Cre recombinase, leading to heterozygous progeny that carry out Cre-mediated recombination in developing neurons. (a) The different lox sites differ in their core sequences, so lox1 reacts only with lox1, and lox2 only with lox2. (b) In this case, the cassette utilizes three or more different lox sites. Cre-mediated recombination results in different patterns of GFP variant expression, and thus a different color, in each neuron. (c) The resulting brainbow of neurons, visualized by a specialized form of light microscopy (epifluorescence). *[Source: By permission from Macmillan Publishers Ltd: Adapted from Lichtman, J.W., Livet, J., & Sanes, J.R., Nature Reviews Neuroscience 9:417–422, June 2008. © 2008.]*

(c)

expression of one of the four GFP variants. The cassette also includes a promoter that directs gene expression only in neurons. Transgenic mice have been engineered that have several of these cassettes, with the potential for expressing one GFP variant from each cassette in a given neuron.

The engineered mice with GFP cassettes are homozygous for these cassettes, and they pass them on to all their progeny. Separately, a second population of homozygous transgenic mice is engineered to express the Cre recombinase, again from a promoter directing gene expression transiently and only in developing neurons. When a mouse with the GFP cassettes is mated to a mouse expressing the Cre recombinase, all progeny are heterozygous for both the cassette and the recombinase genes. As these mouse embryos develop, the Cre recombinase is expressed early in the development of each

neuron. Recombination events occur in some or all of the cassettes in a given neuron, leading to the expression of a particular set of GFP variants. Mixing several different GFP variants in a cell increases the number of potential colors. The outcome is unpredictable for each cell. However, only one set of recombination reactions occurs in each cell, and the end result imparts a distinctive color that is expressed for the life of that neuron. Neighboring developing neurons go through the same recombination processes, but the number of possible outcomes is large and neighboring cells rarely acquire the same color. The result is a rainbowlike array of fluorescent colors in the neural network—a brainbow! Researchers use the brainbow to trace the paths of the axons through the brain. Site-specific recombinases have been used broadly in similar schemes to trace cell lineages in many organisms.

DNA is supercoiled and the sites are properly oriented. The basic site-specific recombination reaction can thus be exquisitely adapted to the particular needs of a virus or cell.

- Site-specific recombination entails the precise cleavage and rejoining of DNA ends at specific and reproducible sites in the DNA.
- There are two classes of site-specific recombinases, defined by the key nucleophilic amino acid residue at their active sites: tyrosine or serine.
- Site-specific recombination can be coupled to replication, or resolution of chromosomal dimers to monomers before cell division, or amplification of plasmid copy number.
- The mechanism of site-specific recombination can be used to facilitate DNA transactions critical to a viral life cycle.
- Biotechnologists have adapted site-specific recombination systems to manipulate DNA segments ranging from plasmids to genomes.
- Some organisms use site-specific recombination to regulate gene expression, as in phase variation in *Salmonella*.
- Site-specific recombination can be rendered specific for one particular reaction outcome (integration, deletion, or inversion) by coupling the reaction to the formation of a larger, structured complex in which only one reaction outcome is possible.

14.2 MECHANISMS OF TRANSPOSITION

We now consider another type of recombination system: recombination that allows the movement of transposable elements, or transposons—the process of transposition. Found in virtually all cells, transposons move, or "jump," from one place on a chromosome, the **donor site**, to another on the same or a different chromosome, the **target site**. DNA sequence homology is not usually required for this movement. In some cases the new location is determined more or less randomly, although most transposons exhibit a certain degree of target specificity, and some even make use of particular target sequences.

Insertion of a transposon in an essential gene could kill the cell (an outcome deleterious to the cell *and* the transposon), so transposition is tightly regulated and usually very infrequent. Transposons are perhaps the simplest of molecular parasites, adapted to replicate passively within the chromosomes of host cells. Sometimes they carry genes that are useful to the host cell and thus exist in a kind of symbiosis with the host. In almost all cases, they contribute in important ways to the evolution of the host.

Transposition Takes Place by Three Major Pathways

Transposition mechanisms are summarized in **Figure 14-10**. Some transposons move by a simple cut-and-paste mechanism. The element is completely excised from the donor DNA and then inserted into a target site. In an alternative, replicative pathway, the transposon is joined to the target site before it is completely excised from the donor DNA. Replication then creates two complete copies of the element, one in the donor site and one in the target site. Elements that make use of either of these pathways are sometimes simply referred to as DNA transposons. A third pathway uses an RNA intermediate to copy the element and insert the copy at a second, target site; these elements are referred to as retrotransposons. At least one phosphoryl transfer reaction in which a free 3′-hydroxyl group at the end of a transposon attacks a phosphodiester in genomic DNA is a feature of all three pathways. Hydrolytic cleavage of phosphodiester bonds (phosphoryl transfer to water; see Figure 14-1) is also a key reaction, often occurring to generate the free 3′ end.

The basic transposition systems consist of the transposable DNA element and an enzyme, a **transposase**, usually encoded by a gene within the transposable element. The transposable element is present at one location in a genome, part of the contiguous genomic DNA, and moves to a new location. The reactions do not involve the formation of covalent intermediates between the transposase and the DNA. Instead, the transposase catalyzes the two key reactions mentioned above, in succession: hydrolysis of a phosphodiester bond at the end of the transposable element, then attack of the liberated 3′ hydroxyl on another phosphodiester bond in a transesterification reaction (**Figure 14-11** on p. 498). The reactions generally leave behind breaks or gaps in the genomic DNA strands at the original site of the transposable element, which must be repaired by the cell. The two-reaction sequence during transposition occurs at both ends of the transposable element. Thus, two liberated 3′-hydroxyl groups attack phosphodiester bonds at the new genomic location (the target site), inserting into each strand of the target. The insertion sites on each strand of the target are offset, or staggered. The immediate product of the transesterifications is the transposable element linked to genomic DNA at both ends at the new location, but with short single-strand gaps at each end. The gaps are filled in by DNA polymerase and ligase to create short repeated sequences at each end of the inserted transposon (the dark red DNA segments in Figure 14-10).

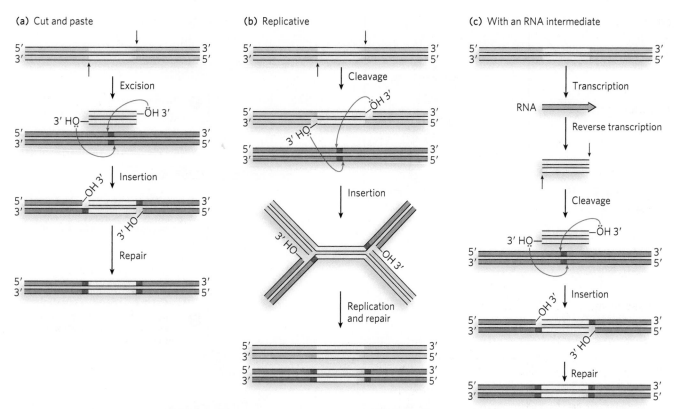

FIGURE 14-10 Three general transposition pathways. (a) Cut-and-paste transposition leaves behind a broken chromosome (double-strand break) that must be repaired. The excised transposon inserts itself into another chromosome at a new location. (b) Replicative transposition leaves a copy of the transposon at the original site, as does (c) transposition that uses an RNA intermediate.

In cut-and-paste transposition, the transposon is completely excised by double-strand cuts on each side, and the transposon moves to a new location (**Figure 14-12a**). This leaves a double-strand break in the donor DNA that must be repaired. The mechanism used by the well-characterized bacterial transposon Tn5 illustrates some additional details. Prior to the cleavage reaction, transposases bound to each end of the transposon come together to form a complex. Each transposase subunit catalyzes phosphodiester bond cleavage at one end of the transposon, generating a free 3′ end. Each of the 3′ hydroxyls then makes a nucleophilic attack on a phosphodiester bond in the complementary strand immediately across from it. Both strands are thus cleaved, and the transposon is excised along with the DNA at both ends in the form of a closed hairpin loop. Next, the transposase catalyzes the hydrolysis of a phosphodiester bond at each end, again creating free 3′-hydroxyl groups, and then the attack of these free 3′ hydroxyls on phosphodiester bonds in each strand at a new genomic location, thus inserting the transposon at the target site. The insertion points on the two strands of the target DNA are offset, as mentioned earlier, leaving single-strand gaps in the target DNA at each end of the transposon. Repair of these gaps results in the duplication of a short target-derived

DNA sequence at either end of the inserted element. The cleaved donor DNA from which the transposon was originally excised is repaired by ligation or recombinational DNA repair (see Chapter 13), or in some cases it is simply degraded.

The use of three successive phosphoryl transfer reactions to effect a double-strand break at each end of the transposon might seem overly elaborate, until we consider the economy involved. One transposase active site at each end of the transposon catalyzes all three phosphoryl group transfers, effectively cleaving two DNA strands of opposite polarity to excise the element from the donor DNA. The hairpin ends are transient intermediates, which researchers have detected in several well-studied transposition reactions. The hairpin intermediate is unique to transposons and similar systems requiring the cleavage of both strands at an element end. The steps following excision of the transposon are more broadly observed in virtually all transposon systems.

At the end of the transposon excision process, the transposase active sites at each end are poised to catalyze yet a fourth phosphoryl transfer, to insert the transposon into a new genomic site. The attack of the 3′ hydroxyl at the end of one strand on a phosphodiester

(a)

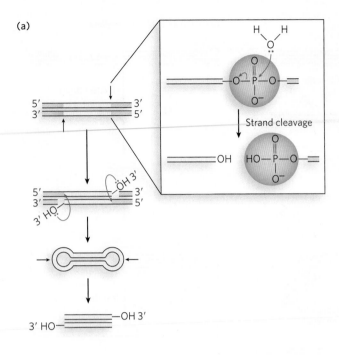

(b)

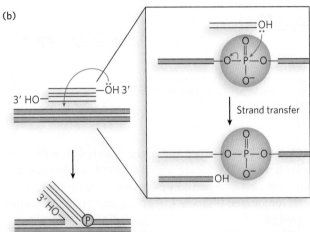

FIGURE 14-11 Hydrolytic cleavage and a transesterification reaction involved in transpositions. These are the reactions promoted by integrases and transposases. In both (a) and (b), the left side shows the fate of the DNA, and the right side shows reaction details. (a) The first step in transposition is usually the attack of a water molecule on a phosphodiester bond at each end of the transposon to create free 3' ends. This creates a nick in one DNA strand at each transposon end, but the other DNA strand remains intact. The additional steps illustrate the reactions that complete the excision of the transposon in cut-and-paste transposition. The transposase catalyzes attack of each liberated 3' hydroxyl on a phosphodiester bond in the complementary strand, immediately opposite. The transposon is fully excised, with its ends structured as closed hairpin loops, and a double-strand break is left in the DNA where the transposon originated. The closed hairpin loops are opened by yet another hydrolytic cleavage of a phosphodiester bond, again catalyzed by the transposase. (b) The 3'-OH thus liberated acts as a nucleophile, attacking another phosphodiester bond in the target DNA. The attacked bond is cleaved, and a new bond is created.

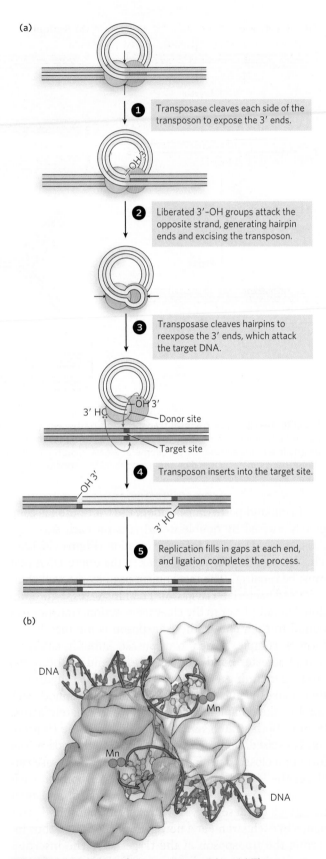

FIGURE 14-12 Cut-and-paste transposition. (a) The transposition steps. (b) Structure of the Tn5 transposase, bound to cleaved transposon ends. The manganese (Mn) ions (present during crystallization) are bound where the Mg^{2+} ions would bind in the normal reaction. [Source: (b) PDB ID 1MM8.]

bond in a contiguous DNA molecule is sometimes called strand transfer, as the phosphoryl group becomes joined to a new strand. **Figure 14-12b** shows the structure of the Tn5 transposase bound to transposon ends, ready to promote two staggered strand transfer reactions in the complementary DNA strands at the target site (see Figure 14-12a, step 4). The type of complex produced at the end of the cut-and-paste process illustrated in Figure 14-12a is broadly representative of that used in the strand transfer reactions discussed in the rest of this chapter.

In replicative transposition (**Figure 14-13**), the transposon ends are again brought together in a complex with the transposase, but only one strand is cleaved at each end of the transposon DNA to create nicks and initiate the process. The cleavage exposes the 3′ end of each transposon DNA strand, and these 3′ ends are then used as nucleophiles in two strand transfers to insert the transposon at a DNA target site in the same or a different DNA molecule. The two strand transfers occur on different DNA strands in the target, staggered a few base pairs apart. This creates a rather complex intermediate, with DNA branches at either end of the transposon. The strand transfers create free 3′ ends in the target DNA, and these are used to prime replication of both strands of the entire transposon. The replication gives rise to an intermediate called a **cointegrate**. Since the strand transfers were staggered, the replication fills in a short repeated sequence derived from the target DNA at each end of the transposon. In a cointegrate, the DNA from the donor region where the transposon originated is covalently linked to DNA at the target site to which the transposon is moving. Most important, two complete copies of the transposon are present in the cointegrate, both having the same relative orientation in the DNA. If the target site is in a different DNA molecule (plasmid or chromosome), the DNAs harboring the transposons in the donor and target sites are linked together in the cointegrate. In some well-characterized transposons, such as the bacterial Tn3, this cointegrate is resolved in a process through which the two DNAs are separated by site-specific recombination in which specialized recombinases promote the required deletion reaction. Completion of the reaction installs the transposon at the target site while leaving a copy behind at the donor location.

Transposons that produce an RNA intermediate in the process of transposition are called **retrotransposable elements**, or **retrotransposons**. These transposons can be classified according to the presence or absence of long terminal repeat (LTR) sequences at their termini. The terminal repeats range from about 100 bp to more than 5,000 bp and generally include binding sites for

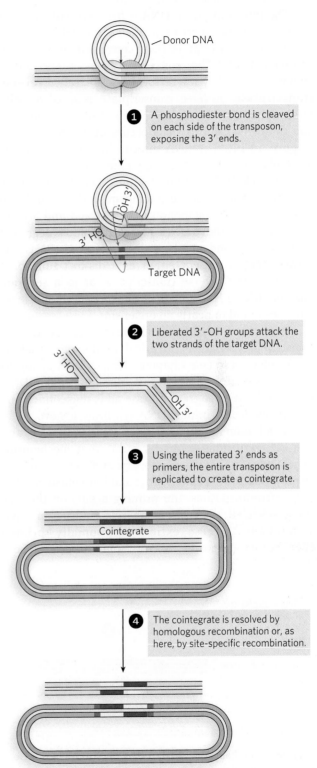

FIGURE 14-13 Replicative transposition. Only one strand at each end of the transposon is cleaved and transferred to the target site in the early steps. The liberated 3′ ends in the target site are used to prime DNA replication of both DNA strands in the transposon, creating two copies of it.

enzymes that function in the transposition process. Both LTR and non-LTR retrotransposons are widely dispersed, particularly in eukaryotic organisms.

The retrotransposon DNA is first transcribed by an RNA polymerase to create a single-stranded RNA transcript. The transcript is then converted into double-stranded DNA. The conversion uses a specialized enzyme encoded by the retrotransposon, called **reverse transcriptase**. Reverse transcriptase is a DNA polymerase that can use either RNA or DNA as a template, and the conversion to DNA occurs in two major steps (**Figure 14-14**). The RNA intermediate is first used as a template to generate a complementary DNA strand, forming an RNA-DNA hybrid duplex. The DNA strand of the duplex is then used as a template in the second step to generate a complementary DNA strand and a complete double-stranded DNA. The RNA strand is degraded during this second step by a ribonuclease H (RNase H) activity that is present either as a distinct functional domain of the reverse transcriptase or as a separate enzyme. RNase H is a class of nonspecific ribonuclease enzymes found in nearly all organisms.

The polymerization mechanism used by reverse transcriptase is largely identical to the one used by DNA polymerase (see Chapter 11), and DNA synthesis proceeds in the same $5' \rightarrow 3'$ direction. Like DNA polymerase, reverse transcriptase adds nucleotides to a primer but cannot initiate DNA synthesis de novo.

In addition to their distinct structures, LTR and non-LTR retrotransposons have different mechanisms of propagation that use primers from different sources, as proposed by Marlene Belfort and her colleagues. For LTR retrotransposons, the priming occurs on the free single-stranded RNA after the retrotransposon DNA is transcribed. These **extrachromosomally primed (EP) retrotransposons** borrow some aspects of both

Marlene Belfort [Source: Courtesy Marlene Belfort.]

cut-and-paste and replicative transposition systems (**Figure 14-15a**). First, a sequence at one end of the LTR retrotransposon (a promoter) directs transcription of the transposon DNA into RNA. (The process of transcription, promoted by RNA polymerases, is described in Chapter 15.) The resulting single strand of RNA is identical in sequence to one of the two strands of the transposon (with U replacing T). Reverse transcriptase, encoded by the retrotransposon, uses the new RNA strand as a template to synthesize a complementary strand of DNA in two steps, as shown in Figure 14-14. The primer for DNA synthesis is often a cellular tRNA with a 3' end that is complementary to, and anneals to, the retrotransposon RNA. Once the DNA strand is completed, reverse transcriptase uses this strand to generate a DNA complement (the source of primer for this synthesis also varies), resulting in a double-stranded DNA fragment that represents a complete copy of the transposon. Another enzyme, **integrase**, catalyzes insertion of this free DNA transposon copy into a DNA target site. The integrase is related to certain transposases, and it uses the 3' ends of the transposon to attack phosphodiester bonds in the target in the same kind of strand transfer process we have seen in cut-and-paste and replicative transposition systems. DNA gaps left at the ends of the inserted transposon are again repaired by host replication and ligation enzymes, leading to the generation of short sequences repeated at both ends.

In the case of non-LTR retrotransposons, the RNA intermediate is brought to and sometimes linked to the DNA target site by an enzyme encoded by the retrotransposon itself, before the reverse transcriptase reaction (**Figure 14-15b**). The enzyme makes a cut in one strand of the target DNA and uses the liberated 3' end as the primer for DNA synthesis. These non-LTR elements are sometimes called **target-primed (TP) retrotransposons**. A second cut liberates a primer for synthesis of the second DNA strand, a process coupled to elimination of the RNA strand. The product is a double-stranded DNA transposon joined to the target site.

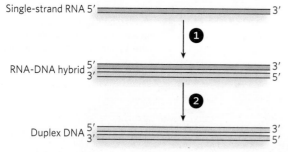

FIGURE 14-14 Reactions catalyzed by reverse transcriptase. A single strand of RNA is first converted to an RNA-DNA hybrid, using the RNA as a template to generate a complementary DNA strand. The RNA strand is then displaced (and eventually degraded) as the DNA strand is used as a template to generate a complementary DNA strand. Duplex DNA is the final product. The sources of the primers used to initiate DNA synthesis in each step are not shown.

Bacteria Have Three Common Classes of Transposons

Most of the transposons found in bacteria make use of transposition pathways that do not have RNA intermediates. Bacterial transposons are broadly classified as

(a) **Extrachromosomally primed retrotransposition**   (b) **Target-primed retrotransposition**

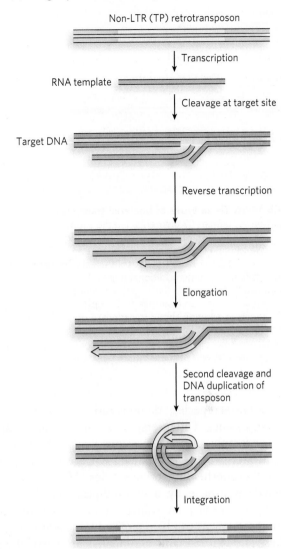

FIGURE 14-15 Retrotransposition. (a) In extrachromosomally primed (EP) retrotransposition, a tRNA or other RNA that can anneal to the transposon RNA is used as a primer for reverse transcription of the RNA to create a double-stranded DNA fragment, in two steps. This DNA fragment inserts itself into a DNA target site, as in cut-and-paste transposition. (b) In target-primed (TP) retrotransposition, the RNA migrates to the target site, where a phosphodiester bond is cleaved hydrolytically. The exposed 3′ end is used to prime reverse transcription of the RNA, which is again converted to double-stranded DNA in two steps. The insertion is completed by ligation. [*Source: Data from A. Beauregard et al., Annu. Rev. Genet. 42:587–617, 2008.*]

insertion sequences, composite transposons, or complex transposons (**Figure 14-16**).

Insertion sequences (also called IS elements) are simple transposons that contain only the sequences required for transposition and the genes for the proteins (transposases) that carry out the transposition. **Composite transposons** contain one or more genes in addition to those needed for transposition. These extra genes might, for example, confer antibiotic resistance, thereby enhancing the survival chances of the host cell. Indeed, transposition contributes substantially to the spread of antibiotic-resistance elements among pathogenic bacterial populations, which is rendering some antibiotics ineffective (see Highlight 9-1). **Complex transposons** often have larger genomes and include genes for auxiliary proteins that activate or otherwise assist the transposase, and even some enzymes that promote processes other than transposition. Some complex transposons bridge the distinction between viruses and transposons, exhibiting a capacity not only to transpose within a cell but to migrate as viruses between different cells. A bacteriophage called Mu is one example.

Many hundreds of distinct insertion sequences have been identified in bacteria. Some of these are found in

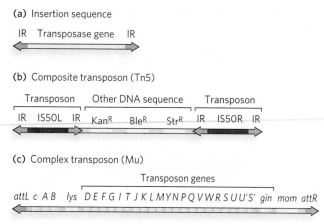

(a) Insertion sequence

IR   Transposase gene   IR

(b) Composite transposon (Tn5)

| Transposon | Other DNA sequence | Transposon |

IR   IS50L   IR   Kan^R Ble^R Str^R IR   IS50R   IR

(c) Complex transposon (Mu)

Transposon genes

attL c A B   *lys*   D E F G I T J K L M Y N P Q V W R S U U'S'   *gin* *mom* *attR*

FIGURE 14-16 Three types of bacterial transposons.
(a) Insertion sequences (IS elements) are the simplest
transposons, consisting of a transposase-encoding gene and a
transposase-binding site within the inverted repeat (IR) sequence
at either end. (b) Composite transposons usually consist of two
IS elements flanking some additional genes. The transposon
Tn5, with two IS50 elements flanking three genes specifying
antibiotic resistance, is an example. (c) Complex transposons
contain genes, in addition to the transposase gene, that are
required for general maintenance. An example is bacteriophage
Mu, which can function as either a very efficient transposon or a
bacteriophage; *attL* and *attR* are the transposase-binding sites.

many different species, demonstrating a capacity to
cross species lines. The interspecies transfer might happen, for example, on rare occasions when a bacterium
dies and pieces of its degraded DNA are taken up at
random by another bacterial species. Most insertion
sequences transpose by a cut-and-paste mechanism,
but a few use replicative transposition. Closely related
insertion sequences can be organized into subgroups or
families. The IS3, IS4, and IS5 families are particularly
widespread.

Composite transposons typically consist of two
insertion sequences flanking several other genes. An example is the transposon Tn5, which has two IS elements
called IS50 (found elsewhere in bacteria as simple
insertion sequences) flanking a group of genes conferring resistance to the antibiotics kanamycin, bleomycin,
and streptomycin (**Figure 14-16b**). Only the outer end of
each IS50 element functions in transposition, such that
the entire composite transposon is moved by cut and
paste.

Complex transposons exhibit the most variety among
the bacterial elements. Several, including bacteriophage
Mu (**Figure 14-16c**), have been studied in some detail
(see the How We Know section at the end of this chapter). Mu has a 37 kbp genome. Like bacteriophage P1,
its life cycle features both lysogenic and lytic pathways.
When Mu infects a bacterial cell, a copy of its genome
is generally inserted into a random site in the chromosome, probably by cut and paste, and can be replicated

passively there. To promote lysis of the cell, the bacteriophage DNA not only is replicated to produce new
phage particles but also undergoes rapid transposition
to additional random sites in the host chromosome by
replicative transposition. The insertion of Mu DNA into
random sites in the chromosome can create mutations,
often inactivating host genes (Mu is for *mutator*.)

Retrotransposons Are Especially Common in Eukaryotes

Eukaryotic DNA transposons are structurally similar to
bacterial transposons, and they migrate by a cut-and-
paste mechanism. Retrotransposons are much more
richly represented in eukaryotes than are other types of
transposons. In the human genome, more than 46% of
the DNA in each cell consists of transposon sequences.
More than 90% of that transposon DNA comes from retrotransposons. Of all the retrotransposons in the human
genome, just over 20% are LTR retrotransposons (about
8% of the human genome) and the remainder are non-
LTR. We consider first the DNA transposons that use the
simple cut-and-paste mechanism, then the elements that
utilize an RNA intermediate.

Eukaryotic Cut-and-Paste Transposons Transposons of
the Tc1/*mariner* family are possibly the most phylo-
genetically widespread transposons in nature, found
in eukaryotes ranging from fungi to plants to humans.
First discovered in the early 1980s in the nematode
worm *Caenorhabditis elegans* (Tc is derived from *transposon Caenorhabditis*), the family acquired the *mariner*
moniker as the ubiquitous presence of these transposons
became evident—*mariner* was the name first attached
to one of these transposons discovered in *Drosophila*.
The Tc1/*mariner* elements transpose by a cut-and-paste
mechanism and are closely related to the bacterial transposon family IS630.

The transposable elements in this family are 1,300
to 2,400 bp long, and each contains a gene encoding a
transposase. Active Tc1/*mariner* transposons can move
about the cellular genomes of just about any species.
However, few of them are active (capable of transposition), due to mutations in the transposase genes. When
active, transposons of this family jump into other DNA
sites more or less randomly, so the potential for disruption and inactivation of essential host genes is considerable. Transposons that have successfully made the
jump into a particular genome have either undergone
selection to prevent further transposition or are subject
to cellular silencing mechanisms that prevent transposition. Genetic engineering has been used to reactivate
transposons of this family, for use in genetic research
(**Highlight 14-2**).

Awakening Sleeping Beauty

After a transposon successfully integrates into a genome, evolution tends to favor inactivation of the transposase, preventing further transposition and its possible detrimental effects on the host. Thus, virtually all cut-and-paste transposons found in vertebrates are inactive. For a geneticist, this is not always a good thing. Many transposons, with their capacity to integrate into chromosomal sites more or less at random, have the potential to disrupt and thereby inactivate genes at random. Researchers can use this property to create libraries in which each individual organism has one gene inactivated, allowing broad explorations of gene function. To realize this potential, however, the transposons must be made to hop once again. Can these sleeping transposons be awakened?

The answer is clearly *yes*. Ronald Plasterk, Zsuzanna Izsvak, and their colleagues creatively used genomics to peer back into evolutionary time and deduce the structure of an active transposase. Beginning with a set of related Tc1/*mariner* elements from fish, the researchers carefully compared the sequences of the embedded and inactive transposase genes (Figure 1). Reasoning that the mutations inactivating the transposon would be different in different species of fish, they used a majority-rule kind of analysis: if a particular sequence was present in most of the transposable elements, it was likely to be the original functional sequence in the original active transposon. Sequences unique to one or a few elements were likely to represent the inactivating changes. Once the research group had established a consensus sequence for a putative active element, they began reconstructing it. Alterations were introduced into one of the inactive genes to make it identical to the consensus.

The scheme was completely successful. The engineered transposon was highly active, not just in fish but in a wide range of eukaryotic cells, including human cells. The new element was dubbed *sleeping beauty*, an apt name for an element awakened from a transposon that had probably been dormant for millennia. Researchers continue to study *sleeping beauty* to learn more about Tc1/*mariner* transposition mechanisms, and it is increasingly used as a tool to mutagenize genes in many different organisms. For example, the introduction of *sleeping beauty* into the mouse genome leads to pups that rapidly develop tumors because of inactivation of genes controlling cell division. Particular types of tumors can be screened to identify the genes in which the engineered transposon is inserted, and a list generated of all the genes that, when inactivated, produce particular tumor types.

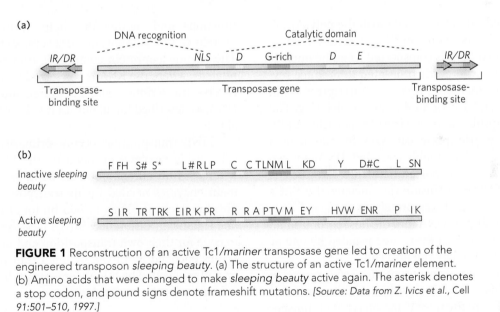

FIGURE 1 Reconstruction of an active Tc1/*mariner* transposase gene led to creation of the engineered transposon *sleeping beauty*. (a) The structure of an active Tc1/*mariner* element. (b) Amino acids that were changed to make *sleeping beauty* active again. The asterisk denotes a stop codon, and pound signs denote frameshift mutations. [Source: Data from Z. Ivics et al., Cell 91:501–510, 1997.]

Eukaryotic Retrotransposons Both the LTR (extra-chromosomally primed) and non-LTR (target-primed) retrotransposon classes are abundant in eukaryotes. Among the LTR retrotransposons are several different Ty elements of *S. cerevisiae* and the element *gypsy* of *Drosophila*. Non-LTR retrotransposons include elements classified as LINEs (long interspersed nuclear elements) and SINEs (short interspersed nuclear elements), along with certain types of introns.

LTR retrotransposons are very closely related to the retroviruses, discussed shortly. Transposition of these elements is illustrated by the Ty transposition cycle (**Figure 14-17**). A Ty element, integrated in the DNA chromosome of the host cell, is transcribed to produce

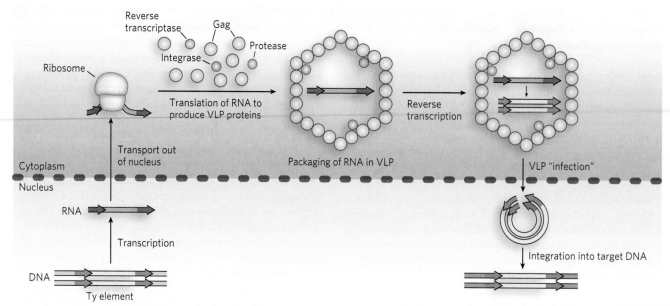

FIGURE 14-17 Retrotransposition by a Ty element in yeast. The Ty element, within the host DNA, is transcribed to produce an mRNA, which is transported to the cell cytoplasm and translated to produce a polyprotein (pol) that is cleaved to generate a protease, reverse transcriptase, integrase, and Gag (a structural protein). Within the viruslike particle (VLP), reverse transcriptase converts the RNA to double-stranded DNA. The VLP is transported back into the nucleus, the Gag protein coat is shed, and the DNA transposon is integrated into a new target site in the host chromosome. [Source: Data from A. Beauregard et al., Annu. Rev. Genet. 42:587–617, 2008.]

a single-stranded mRNA molecule with the poly(A) tail characteristic of eukaryotic mRNAs (as described in Chapter 16). The mRNA is transported from the nucleus to the cytoplasm, where it is translated to produce several Ty proteins, including reverse transcriptase (Pol), integrase (Int), and a structural protein called Gag. The mRNA is encapsulated in a viruslike particle (VLP), where reverse transcriptase catalyzes formation of a complementary linear, double-stranded DNA, in two steps. The DNA is transported back into the nucleus with the aid of the VLP. Inside the nucleus, the VLP's outer shell, consisting of Gag protein, is shed, and integrase inserts the duplex DNA into the host chromosome at a new location. Almost every step is aided by one or more proteins encoded by the host cell.

The non-LTR retrotransposons also rely on reverse transcriptase, but the transposition pathways vary. LINEs and SINEs are the most common types of transposons in the human genome. Prominent in all mammalian genomes, they also occur in other classes of eukaryotes and have played important roles in genomic evolution. The most common human LINE is L1, approximately 6,000 bp long. Present in about 520,000 copies, L1 elements account for about 17% of the entire human genome. A LINE element is first transcribed to mRNA in the nucleus. The mRNA is transported to the cytoplasm, where it is translated to produce two proteins: one that has both reverse transcriptase and endonuclease

functions, and another that helps form a ribonucleoprotein complex important to several steps in the cycle. The ribonucleoprotein complex is transported into the nucleus, where the RNA is converted to duplex DNA by reverse transcriptase and integrated into the genomic DNA, as described for all non-LTR (TP) retrotransposons above.

LINE transposition occurs primarily in germ-line cells, using enzymes encoded by the LINE. A LINE is defined as an autonomous element, as it encodes the main enzymes needed for its transposition. The transposition cycle of SINEs (<500 bp long) is similar, with one significant difference: SINEs lack the genes needed to code for their own transposition and must rely on enzymes encoded by LINEs. When a transposon relies on enzymes encoded by other elements in this way, it is said to be a nonautonomous element.

Retrotransposons and Retroviruses Are Closely Related

Retroviruses are RNA viruses that contain reverse transcriptase. Most retroviruses infect animal cells. On infection, the virus's single-stranded RNA genome (~10,000 nucleotides), along with molecules of reverse transcriptase carried in the viral particle, enters the host cell. The reverse transcriptase converts the viral RNA to double-stranded DNA (**Figure 14-18**). The resulting viral DNA

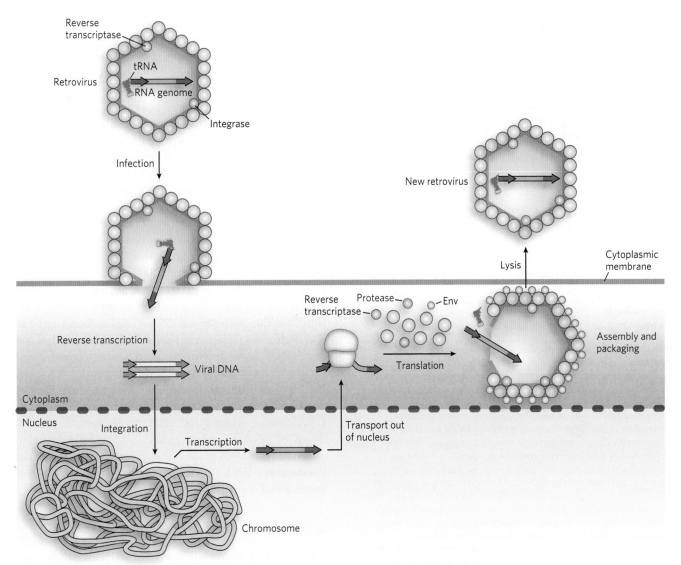

FIGURE 14-18 Retroviral infection of a mammalian cell. Viral particles entering the host cell (left) carry reverse transcriptase and a cellular tRNA (from a previous host) already base-paired to the retroviral RNA. The tRNA facilitates immediate conversion of the RNA to double-stranded DNA by the action of reverse transcriptase. The DNA enters the nucleus and is integrated into the host genome, a process catalyzed by a virally encoded integrase. On transcription and translation of the viral DNA, new viruses are formed and released by cell lysis (right).

often becomes incorporated into the genome of the eukaryotic host cell. These integrated (and dormant) viral genes can be activated and transcribed, and the gene products—viral proteins and the viral RNA genome—are packaged as new viruses.

The idea that biological information could flow from RNA to DNA in the retroviral life cycle was predicted by Howard Temin in 1962. The enzyme that promotes this reaction, reverse transcriptase, was ultimately detected by Temin and, independently, by David Baltimore in 1970. This discovery aroused much attention as dogma-shaking proof that genetic information can flow "backward," from RNA to DNA.

Retroviruses typically have three genes: *gag*, *pol*, and *env* (**Figure 14-19**). (The name *gag* is from the historical designation group-specific *a*ntigen.) In a sense, the retrovirus is simply a retrotransposon, with one

Howard Temin,
1934–1994 [Source: Hulton
Archive/Getty Images.]

David Baltimore [Source:
Scott Wintrow/Getty
Images.]

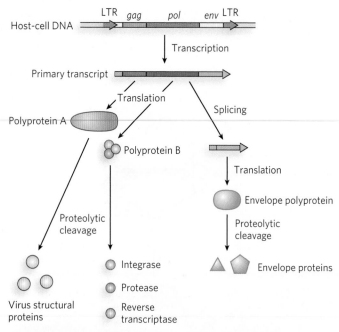

FIGURE 14-19 The structure and gene products of an integrated retroviral genome. The retroviral RNA genome is flanked by long terminal repeats (LTRs), which contain sequences needed for the regulation and initiation of transcription. The primary retroviral transcript encompasses the *gag*, *pol*, and *env* genes as well as sequences on either side of those genes. Translation of the primary transcript yields two different polyproteins. One is derived only from the *gag* gene, and this polyprotein is cleaved into three different proteins that make up the viral core, matrix, and capsid. The second polyprotein, generated at lower levels, is derived from the *gag* and *pol* genes. This polyprotein is cleaved into six distinct proteins, including those derived from the *gag* gene as well as the reverse transcriptase, integrase, and protease derived from the *pol* gene. The primary transcript is spliced to generate a shorter transcript used for translation of the *env* gene. The polyprotein derived from the *env* gene is similarly cleaved to generate viral envelope proteins.

additional gene, *env*, which encodes the proteins of the viral envelope. The additional gene is critical because it gives the element the capacity to move from cell to cell instead of just within a genome. The transcript (mRNA) that contains *gag* and *pol* is translated into a long polyprotein. This single large polypeptide is cleaved into six proteins with distinct functions. The proteins derived from the *gag* gene make up the interior core of the viral particle. The *pol* gene encodes a protease that cleaves the polyprotein, an integrase that inserts the viral DNA into the host chromosome, and reverse transcriptase. Many reverse transcriptases have two subunits, *a* and *b*. The *pol* gene specifies the *b* subunit (M_r 90,000), and the *a* subunit (M_r 65,000) is simply a proteolytic fragment of the *b* subunit. At each end of the retrovirus's linear RNA genome are LTR sequences of a few hundred nucleotides, analogous to the LTR sequences found at the ends of LTR retrotransposons. Transcribed into double-stranded DNA, these sequences facilitate integration of the viral

chromosome into the host DNA and contain promoters for viral gene expression.

Viral reverse transcriptases catalyze three different reactions: RNA-dependent DNA synthesis, RNA degradation (by a separate RNase H domain), and DNA-dependent DNA synthesis. For DNA synthesis to begin, the reverse transcriptase requires a primer. As seen for many LTR retrotransposons, the primer is a cellular tRNA obtained during an earlier infection and carried in the viral particle. This tRNA is base-paired at its 3′ end with a complementary sequence in the viral RNA. The new DNA strand is synthesized in the 5′→3′ direction, as in all RNA and DNA polymerase reactions. Reverse transcriptases, unlike DNA polymerases, do not have 3′→5′ proofreading exonucleases. They generally have error rates of about 1 per 20,000 nucleotides added. An error rate this high is extremely unusual in DNA replication and seems to be a feature of most enzymes that replicate the genomes of RNA viruses. A consequence is a higher mutation rate and a faster rate of viral evolution, which are factors in the frequent appearance of new strains of disease-causing retroviruses.

Reverse transcriptases have become important reagents in the study of DNA-RNA relationships and in DNA cloning techniques. They make possible the synthesis of DNA complementary to an mRNA template, and synthetic DNA prepared in this manner, called **complementary DNA (cDNA)**, can be used to clone cellular genes (see Figure 7-8).

A Retrovirus Causes AIDS

The human immunodeficiency virus (HIV), which causes acquired immune deficiency syndrome (AIDS), is a retrovirus. Identified in 1983, HIV has an RNA genome with the standard retroviral genes and several other, unusual genes. The virus targets mainly the T lymphocytes (T cells) of the immune system, binding specifically to a receptor protein, called CD4, on their surface. Unlike many other retroviruses, HIV kills many of the cells it infects, while quietly integrating its genome into the chromosomes of other cells, where it is passively replicated. This gradually suppresses the host's immune system, eventually leading to death. The reverse transcriptase of HIV is 10 times more error-prone than other known reverse transcriptases, resulting in high mutation rates. One or more errors are generally made every time the viral genome is replicated, so any two viral RNA molecules are rarely identical.

Many modern antiviral vaccines contain coat proteins of the virus that is targeted by the vaccine. These proteins are not infectious on their own but stimulate the immune system to recognize and resist subsequent viral invasions. Because of the high error rate of the HIV reverse transcriptase, however, the *env* gene (along

Fighting AIDS with HIV Reverse Transcriptase Inhibitors

Research into the chemistry of template-dependent nucleic acid biosynthesis, combined with modern techniques of molecular biology, elucidated the life cycle and structure of HIV, the retrovirus that causes AIDS. A few years after the isolation of HIV, this research resulted in the development of drugs that can prolong the lives of people with HIV/AIDS.

The first drug to be approved for clinical use in treating HIV infection was AZT, 3′-azido-3′deoxythymidine (Figure 1), a structural analog of deoxythymidine. AZT was first synthesized in 1964 by Jerome P. Horwitz. It failed as an anticancer drug (the purpose for which it was made), but in 1985 it was found to be a useful treatment for AIDS. AZT is taken up by T lymphocytes (T cells), immune system cells that are particularly vulnerable to HIV infection, and

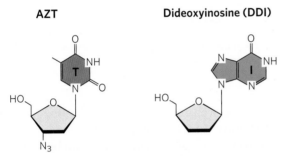

AZT   **Dideoxyinosine (DDI)**

FIGURE 1 Two HIV reverse transcriptase inhibitors used in HIV/AIDS therapy.

is converted to AZT triphosphate. (AZT triphosphate taken directly would be ineffective because it cannot cross the plasma membrane.) HIV's reverse transcriptase has a higher affinity for AZT triphosphate than for dTTP, and binding of AZT triphosphate to this enzyme competitively inhibits dTTP binding. When AZT is added to the 3′ end of the growing DNA strand, lack of a 3′ hydroxyl means that the DNA strand is terminated prematurely, and viral DNA synthesis grinds to a halt.

AZT triphosphate is not as toxic to T cells as it is to the virus, because *cellular* DNA polymerases have a lower affinity for this compound than for dTTP. At concentrations as high as 1 to 5 mm, AZT affects HIV reverse transcription but not most cellular DNA replication. Unfortunately, AZT does seem to be toxic to the bone marrow cells that give rise to erythrocytes, and many individuals taking AZT develop anemia. AZT can increase the survival time of people with advanced AIDS by about a year, and it delays the onset of AIDS in people still in the early stages of HIV infection. Some other AIDS drugs, such as dideoxyinosine (DDI; see Figure 1), have a similar mechanism of action. However, due to the rapid rate at which mutations are introduced into the HIV genome, the virus develops immunity to AZT rapidly, and the drug becomes less useful over time.

Newer drugs target and inactivate the HIV protease. Because of the high error rate of HIV reverse transcriptase and thus the rapid evolution of HIV, the most effective treatments of HIV infection use a combination of drugs directed at both the protease and the reverse transcriptase. Several different drugs targeting the protease can be used successfully, avoiding the deleterious effects of the development of viral immunity to any one of them.

---

with the rest of the HIV genome) undergoes very rapid mutation, complicating the development of an effective vaccine. In addition, integrated copies of the virus in some T cells provide a source of recurring infection. The most effective therapy currently available is directed toward the inhibition of viral enzymes, hindering the repeated cycles of cell invasion and replication needed to propagate an HIV infection. The HIV protease is targeted by a class of drugs known as protease inhibitors. Reverse transcriptase is the target of drugs widely used to treat HIV-infected individuals (**Highlight 14-3**).

SECTION 14.2 SUMMARY

- Transposons (transposable elements) are segments of DNA that can move, or "jump," from one genomic location to another.
- The three general transposition mechanisms are cut and paste, replicative transposition, and transposition with an RNA intermediate. Retrotransposons use an RNA intermediate.

- Transposition makes use of transposases, specialized enzymes that cleave phosphodiester bonds and catalyze transesterification reactions in which the 3′ end of a cleaved DNA strand attacks a phosphodiester bond.
- Bacterial transposons fall into three classes. Insertion sequences (IS elements) have only the gene encoding the transposase. Composite transposons contain several additional genes sandwiched between two IS elements. Complex transposons encode additional proteins that facilitate the function of the transposase, and some encode proteins with other functions.
- Eukaryotic transposons include elements that migrate by a cut-and-paste mechanism. Even more common are eukaryotic retrotransposons, some of which are closely related to retroviruses.
- Retrotransposons make use of the enzyme reverse transcriptase, a DNA polymerase that can use both RNA and DNA as a template for DNA synthesis.
- A retrovirus (HIV) is the causative agent of AIDS.

14.3 THE EVOLUTIONARY INTERPLAY OF TRANSPOSONS AND THEIR HOSTS

As we noted in the introduction to this chapter, transposition provides an alternative survival strategy that almost certainly has ancient roots. Evolution has given rise to many types of transposons, which have colonized the genomes of all extant organisms, and it has also given rise to other, more complex pathogens, such as viruses. Although the need to adapt to their hosts has affected transposon evolution, transposons have not always remained separate entities. Some transposable elements, and the enzymes they encode, have been appropriated by host cells and adapted to new biological tasks. The movements of transposons have driven genomic changes that have contributed in important ways to evolution.

Viruses, Transposons, and Introns Have an Interwoven Evolutionary History

Many well-characterized eukaryotic retrotransposons, from sources as diverse as yeast and fruit flies, have a structure very similar to that of retroviruses. Based on studies of reverse transcriptase genes, researchers hypothesize that retrotransposons probably gave rise to retroviruses.

The evolution of virtually all retrotransposons and retroviruses can be traced through their reverse transcriptase genes. Similarly, the evolution of an even broader range of transposable elements can be linked to the evolution of retrotransposons and retroviruses through their integrase and transposase genes. Integrases and transposases catalyze very similar reactions (**Figure 14-20**). The most widespread class of both types of enzymes uses a set of three active-site amino acids, two aspartates and one glutamate, to promote the hydrolytic cleavage and transesterification of phosphodiester bonds. These three residues (D, D, and E) are not adjacent in the primary sequence of these enzymes, but they come together in the active site and

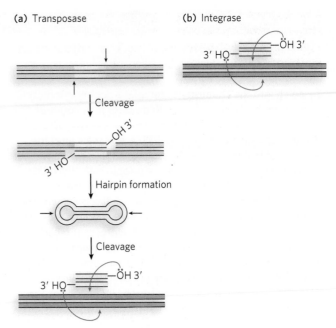

FIGURE 14-20 Transposases and integrases. (a) Transposases promote the nucleophilic attack of the 3' end of a DNA strand on a phosphodiester bond—a strand transfer reaction. (b) An integrase carries out the same reaction. (Transposases can also use a water molecule as the nucleophile.) The two types of enzyme have similar active sites.

constitute a well-recognized motif called the **DDE motif** (**Figure 14-21**). The close relationship between these enzymes, regardless of source, can have practical benefits. For example, the transposase of the bacterial transposon Tn5 can be used as a model to study the function of the HIV integrase, and even as a rapid test for drugs that might be used to inhibit the HIV enzyme.

Transposons have a long and complex history. Their dispersal may include rare events in which DNA was transferred by some means (such as bacterial conjugation, cellular fusion, viral infection, or accidental DNA uptake) among the cells of different species. When a transposon is introduced into a new species' genome, there is often a period of many host generations during which the element transposes more or less freely. The

                                     I   V   P
                      52-61 T N   F   A   L   T   C   V   T   R
Tn3 transposases   ...W--D-TY//HSDQGSQYTS--Y---L---GI--S-S--G-S-DNA--ESF-G-LK-E...
Retroviral integrases ...WQ-D-TH//-SDQGPAF-S---Q------GI-H-TGIPYNPQSQG-VER-N-TLK--...
                      46-54 T N   V   V
                                                              L

FIGURE 14-21 The DDE motif. Consensus sequences are shown for the catalytic domain of two families of transposases: the Tn3 family and retroviral integrases. The DDE motif consists of three residues (Asp, Asp, and Glu, shown in red) that are generally not adjacent in the primary sequence but come together at the active site when the protein is folded. The motif is found in most transposases and integrases. Alternative residues at any position are shown above or below the line; those shown in blue (up to two at a given position) are residues found at that position in more than 75% of the enzymes of this family. Dashes indicate any amino acid residue. Two peptide segments, with the residue numbers indicated, are omitted here. [Source: P. Polard and M. Chandler, Mol. Microbiol. 15:13–23, 1995.]

number of inserted transposons may increase, with the resulting genomic changes being passed on whenever they do not have a deleterious effect on the host. As time passes, the transposons become subject to silencing processes, including the introduction of mutations in their transposase or integrase genes that inactivate the gene products. Alternatively, the host may find a way to shut down transposition. One common silencing mechanism involves RNA interference (RNAi, a process described in Chapter 22). In brief, the cell produces short RNA molecules that are complementary to the transcripts of the transposase-encoding gene. The RNA hybridizes to the gene transcripts, preventing their translation and effectively blocking the synthesis of an enzyme required for transposition of an entire class of transposons.

Linear transmission of transposons from one host generation to the next is predominant, with transfer between species occurring rarely. Thus, many transposon families are found only in certain classes of organisms. In eukaryotes, ongoing genomic sequencing efforts have revealed 12 superfamilies of DNA transposons, including Tc1/*mariner* (**Table 14-1**). Many of these superfamilies are found in more than one eukaryotic type. Seven are closely related to transposons found in bacteria, suggesting that they appeared before the divergence of bacteria and eukaryotes.

Sometimes transposons benefit their hosts. As we have seen, the antibiotic-resistance genes encoded by the transposon Tn5 have contributed greatly to the development of bacterial pathogens that are resistant to those antibiotics. In human cells, there are more than 1 million copies of the Alu transposon (a 300 bp SINE element) in the DNA, accounting for nearly 10% of the genome. These elements are so widespread that a typical human gene includes several copies in the introns of its primary transcript. Host cells use these elements as target sites for RNA editing (see Chapter 16). Other transposon genes are appropriated by the host for other purposes. Efforts to trace the evolution of mammalian genes have identified several dozen that are derived from transposons. A dramatic case of transposon domestication—occurring in immunoglobulin formation—is described shortly.

Perhaps more important is the overall impact of transposons on the evolution of the host. Genomic changes promoted by transposons come in many forms. Transposons are set up to bring their ends together in a complex prior to any cleavage event, but this control mechanism can go awry. If transposase subunits form a complex involving two ends derived from different copies of the same transposon, on the same or different chromosomes, large genomic rearrangements can result. Genes may be captured between two transposable elements and moved to different genomic locations. If the genes are duplicated in the process, the new gene copies may evolve and acquire new functions. Transposition is not always precise; the insertion of a transposon into a gene, followed by its later excision, can add or subtract base pairs in the gene and create new alleles. Also, the insertion or excision of transposons at particular genomic sites can alter the expression of genes or sets of genes.

TABLE 14-1

DNA Transposons in Eukaryotes

Superfamily	Bacterial Relative	Catalytic Motif	Length (kbp)	Occurrence (groups of organisms)
Tc1/*mariner*	IS630	DDE (or DDD)	1.2–5.0	All except diatoms and green algae
hAT	Not determined	DDE	2.5–5.0	Vertebrates, invertebrates, plants, fungi, green algae, *Entamoeba*, *Phytophthora*
P element	Not determined	DDED	3–11	Invertebrates, green algae
MuDR/Foldback	IS256	DDE	1.3–7.4	Vertebrates, invertebrates, plants, fungi, diatoms, *Entamoeba*
CACTA	Not determined	Not determined	4.5–15	Invertebrates
PiggyBac	IS1380	DDE	2.3–6.3	*Phytophthora*
PIF/Harbinge	IS5	DDE	2.3–5.5	Vertebrates, invertebrates, plants, fungi, diatoms
Merlin	IS1016	DDE	1.4–3.5	Vertebrates, invertebrates, *Phytophthora*
Transib	Not determined	DDE	3–4	Invertebrates, fungi
Banshee	IS481	DDE	3–5	*Trichomonas*
Helitron	IS91	HHYY	5.5–17	All except diatoms, green algae
Maverick	None	DDE	15–25	All except plants, diatoms, green algae

Source: Data from C. Feschotte and E. J. Pritham, *Annu. Rev. Genet.* 41:331–368, 2007.

The transposons that seem to clutter mammalian genomes have been referred to as "selfish" or "junk" DNA, but these labels are being shed as our understanding broadens. Transposon DNA may play a key role in chromosomal structure and packaging. And far from being dormant, transposon DNA is actively transcribed in at least some cells. As new classes of functional RNAs are being discovered at a rapid pace, the RNAs produced by transposons may prove to have unexpected cellular roles.

A Hybrid Recombination Process Assembles Immunoglobulin Genes

Humans have a complex immune system capable of generating millions of different immunoglobulins (antibody proteins) with distinct binding specificities. But the human genome contains only about 25,000 genes, and just a few hundred of these are immune system genes. Somehow, the millions of different immunoglobulins are generated from these several hundred genes. As B lymphocytes (B cells) differentiate, their immunoglobulin genes recombine so that each cell will express an antibody with a unique binding specificity. Studies of the recombination mechanism reveal a close relationship to DNA transposition and suggest that this system for generating antibody diversity evolved from an ancient cellular invasion by transposons.

Immunoglobulins consist of two heavy and two light polypeptide chains (**Figure 14-22** shows the general structure of the IgG class of immunoglobulins). Each chain has two regions: a variable region, with a sequence that differs greatly from one immunoglobulin to another, and a constant region, which is virtually unchanging within a class of immunoglobulins. There are two distinct families of light chains, kappa and lambda, which differ somewhat in the sequence of their constant regions. For all three types of polypeptide chain (heavy chain, and kappa and lambda light chains), diversity in the variable regions is generated by a similar mechanism. The genes for these polypeptides are divided into segments, and the genome contains clusters with multiple versions of each segment. The joining of one version of each gene segment creates a complete gene.

Figure 14-23 depicts the organization of the DNA encoding the kappa light chain and shows how a mature kappa light chain is generated. In undifferentiated cells,

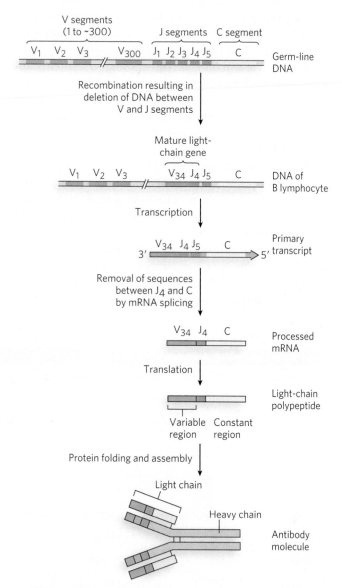

FIGURE 14-23 Recombination of the V and J gene segments of the human IgG kappa light chain. This process results in considerable antibody diversity. Shown at the top is the arrangement of IgG-coding sequences in a bone marrow stem cell. Recombination deletes the DNA between specific V and J segments. The RNA transcript is processed by RNA splicing; translation produces the light-chain polypeptide. The light chain can combine with any of several thousand possible heavy chains to produce an antibody molecule.

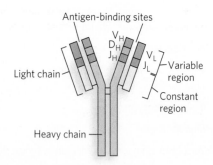

FIGURE 14-22 Immunoglobulin G (IgG). Pairs of heavy and light chains combine to form a Y-shaped molecule. Two antigen-binding sites are formed by the combination of variable domains from one light and one heavy chain. The light chains include V (variable) and J (joining) segments. The heavy chains have V (variable), D (diverse), and J (joining) segments, brought together by mechanisms similar to those for light chains, as described in the text. For heavy chains in the human genome, there are 44 V segments, 27 D segments, and 6 J segments, with one of each brought together at random in a given immunoglobulin.

the coding information for this polypeptide is separated into three segments. The V (variable) segment encodes the first 95 amino acid residues of the variable region, the J (joining) segment encodes the remaining 12 residues of the variable region, and the C (constant) segment encodes the constant region. For kappa light chains, the genome contains ~300 different V segments, 5 different J segments, and 1 type of C segment.

As a stem cell in the bone marrow differentiates to form a mature B cell, one V segment and one J segment are brought together by a specialized recombination system (**Figure 14-24**). During this programmed DNA deletion, the intervening DNA is discarded. There are about $300 \times 5 = 1,500$ possible V-J combinations. Additional variation in the sequence at the V-J junction is introduced by imprecision in the recombination reaction, as an enzyme called terminal deoxynucleotide transferase adds a few nucleotides at random to 3′ ends exposed during the recombination process. This increases the overall variation considerably. The final joining of the V-J combination to the C region is accomplished by an RNA-splicing reaction after transcription (see Chapter 16). The assembly of light chains with similarly randomized heavy chains increases diversity still further.

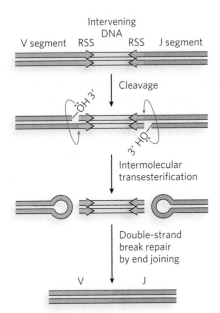

FIGURE 14-24 Immunoglobulin gene rearrangement. Proteins RAG1 and RAG2 bind to RSS (recombination signal sequences) and cleave one DNA strand between the RSS and the V (or J) segments that are to be joined. The liberated 3′ hydroxyl acts as a nucleophile, attacking a phosphodiester bond in the other strand to create a double-strand break. The resulting hairpins on the V and J segments are cleaved, and the ends are covalently linked by a complex of proteins specialized for end-joining repair of double-strand breaks, as described in Chapter 13. The steps in the generation of the double-strand break, catalyzed by RAG1 and RAG2, are chemically related to steps in transposition reactions.

The recombination mechanism for joining the V and J segments is facilitated by recombination signal sequences (RSS) that lie just downstream of each V segment and just before each J segment. These sequences are bound by proteins called RAG1 and RAG2 (products of the *recombination activating gene*). The RAG proteins catalyze the formation of a double-strand break between the RSS and the V (or J) segments to be joined. The V and J segments are then joined with the aid of a second complex of proteins.

The genes for the heavy chains and the lambda light chains form by similar processes. Heavy chains have more gene segments than light chains, with thousands of possible combinations. Because any heavy chain can combine with any light chain to generate an immunoglobulin, each human can produce at least 10^7 possible immunoglobulins. And additional diversity is generated by high mutation rates (of unknown mechanism) in the V segments during B-cell differentiation. Each individual mature B cell produces only one type of antibody, but the range of antibodies produced by the many B cells of an individual organism is clearly enormous.

The mechanism of the DNA recombination events required to generate an expressed immunoglobulin gene suggests that the immune system evolved, in part, from ancient transposons. The mechanism for generation of the double-strand breaks by RAG1 and RAG2 closely resembles several reaction steps in transposition (see Figure 14-24). The double-strand breaks that initiate the process are generated by single protein active sites and feature transient hairpin intermediates at each of the ends to be joined, as in the reaction promoted by the Tn5 transposase (see Figure 14-11). In addition, the deleted DNA, with its terminal RSS, has a sequence structure found in most transposons. In the test tube, RAG1 and RAG2 can associate with this deleted DNA and insert it, transposon like, into other DNA molecules (probably a rare reaction in vivo). In fact, a subtle rearrangement of the RSS, coupled with placement of the genes encoding the RAG proteins between the RSS ends, creates a DNA element that functions exactly like a transposon. The RAG1 protein is closely related in sequence to the transposases encoded by the *Transib* superfamily of eukaryotic transposons (see Table 14-1). The properties of the immunoglobulin gene rearrangement system point to an intriguing origin, in which the distinction between host and parasite has become blurred by evolution.

SECTION 14.3 SUMMARY

- Transposons, retrotransposons, and retroviruses have a shared evolutionary history, evident in the phylogenies of the key enzymes—reverse transcriptases, transposases, and integrases—that promote these processes.

- Important elements of the vertebrate immune system, the enzymes that promote immunoglobulin gene rearrangements and thus immunoglobulin diversity, evolved from the transposase/integrase family of enzymes. RAG1 is related to the transposases of the *Transib* transposons.

? UNANSWERED QUESTIONS

For any organism, the information required for creating a new generation is passed on through its DNA. Stable transmission is needed, yet genomes are surprisingly dynamic. Recombination processes contribute to repair and facilitate key steps of replication and cell division. A hidden world of transposons makes a home in each genome, replicating passively yet contributing to evolution in important ways. This dynamic genome still holds some secrets to unlock.

1. What is the origin of reverse transcriptase? How does it relate to the origin of retroviruses? What is its impact on genome development and diversity? For researchers interested in the origin of life, reverse transcriptase is potentially a very old enzyme that played a key role in the transition from RNA- to DNA-based life.

2. Why do the types of transposons vary so much from one class of organism to another? Each class of transposon present in a given genome represents an invasion that occurred sometime in the lineage of that organism. The study of genomic transposons may provide a rich harvest of information about the evolutionary past of all organisms.

3. How many proteins and other factors are involved in controlling transposition? Exploration of the elaborate interface between transposons and their hosts is only just beginning. The processes that silence a transposon often involve genes found in both the transposon and the host. The extent of host gene involvement has not been fully explored in most cases; functional RNA molecules may do part of the work. Similarly, elaborate processes that prevent integration of transposons into other transposons are only partially understood. AIDS is, as yet, an almost intractable disease, in part because of the capacity of HIV to integrate into a host genome and remain there, replicating passively. A permanent cure for HIV cannot occur as long as these silent HIV genomes provide a potential source of new infection. A better understanding of how this integration is regulated may eventually lead to genomic clean-up therapies to eliminate or permanently inactivate these pathogens. That understanding must come from work on a wide range of viruses and transposons to fully sample the variety of mechanisms they use, as well as to unearth host functions that play subtle roles.

4. What do retroviruses and transposons contribute to their hosts? The evolutionary history of these pathogens is clearly not entirely shaped by their own requirements. Obvious contributions to host survival have already been described, but the sheer bulk of transposon DNA in the human genome inspires new questions about function. How do all these repeated transposon sequences affect the structure and function of chromosomes? New reports suggest that much of the genomic DNA previously labeled as junk is in fact transcribed. What are all these RNA molecules doing? For example, a newly discovered class of RNAs called piwi RNAs (piRNAs) are abundant in germ-line cells (especially during spermatogenesis). They play a role in the silencing of transposon genes, but their origin and detailed function are still a mystery.

HOW WE KNOW

Bacteriophage λ Provided the First Example of Site-Specific Recombination

Echols, H. 2001. *Operators and Promoters: The Story of Molecular Biology and Its Creators.* Berkeley: University of California Press.

Gottesman, M.E., and R.A. Weisberg. 2004. Little lambda, who made thee? *Microbiol. Mol. Biol. Rev.* 68:796–813.

Nash, H.A. 1975. Integrative recombination of bacteriophage lambda DNA in vitro. *Proc. Natl. Acad. Sci. USA* 72:1072–1076.

Howard Nash, 1937–2011
[Source: Courtesy NIMH Laboratory of Molecular Biology.]

Since the 1950s, scientists have known that the DNA of bacteriophage λ (λ phage) is linked to its bacterial host chromosome at a specific chromosomal location. The correct explanation for how the λ DNA enters the chromosome appeared in 1962, before anyone knew that the linear λ DNA is circularized on entering a bacterial cell. Allan Campbell, then at the University of Rochester, had the novel insight that circularization, followed by recombination into the host chromosome, could explain many observations associated with λ lysogeny. Clearly, a uniquely precise recombination process was at work, one that used defined DNA sequences.

A molecular understanding of this, as yet unprecedented, reaction mechanism required an in vitro system—which came in a breakthrough reported separately by Howard Nash (integration) and by Max Gottesman and Susan Gottesman (excision) in 1975. The researchers were working in competing laboratories just a few buildings apart at the National Institutes of Health in Bethesda, Maryland. The Nash system was the more successful of the two, rapidly leading to a detailed biochemical definition of the λ integration reaction and its components.

In his in vitro system, Nash constructed an altered bacteriophage λ chromosome that included both recombination sites, by then defined and named *attB* and *attP* (*B* for bacterium and *P* for phage), separated by about 15% of the chromosome's length (Figure 1). As a source of the required enzymes, Nash used a concentrated extract derived from cells in which λ proteins were being produced. He then showed that integrative recombination between the two recombination sites would occur in cells to produce phage chromosomes 15% smaller than normal.

Phage with the shortened chromosome had the useful property that they were infectious in the presence of metal-chelating agents (molecules that bind to and sequester metals, effectively making them unavailable),

whereas phage with the larger chromosome were not. After an in vitro reaction, phage introduced into bacterial cells and plated on agar containing a metal chelator started an infection cycle only after successful recombination. The infection created plaques (clear spots of killed cells in the bacterial lawn) that could be counted. After years of optimizing his system, Nash reported the first in vitro site-specific recombination system in 1975 (the Gottesmans' report appeared three months later).

A successful in vitro system is a powerful thing in molecular biology. Nash, soon joined at the NIH by Kiyoshi Mizuuchi, used his system to purify the λ Int protein, discover a required host protein (IHF, for integration host factor), and define the reaction requirements. Following further work in other labs, notably that of Art Landy at Brown University, the λ integration system gradually yielded its secrets and stimulated the search for other site-specific recombination systems, such as those described in this chapter.

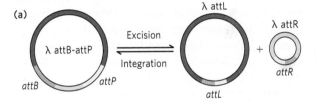

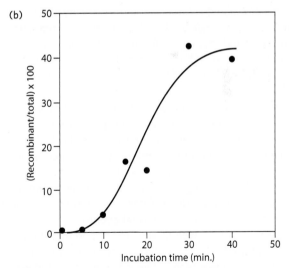

FIGURE 1 (a) The structure of the altered bacteriophage λ substrate developed by Nash (λ attB-attP), and the reaction promoted by the recombination system. A recombination reaction between the *attP* and *attB* sites produces two slightly different sites called *attL* and *attR* (products λ attL and λ attR). (b) The number of recombinant λ phages produced (as percentage of total) as a function of incubation time with the enzyme extract. [Source: Data from H. A. Nash, Proc. Natl. Acad. Sci. USA 72:1072–1076, 1975.]

HOW WE KNOW

If You Leave out the Polyvinyl Alcohol, Transposition Gets Stuck

Craigie, R., and K. Mizuuchi. 1985. Mechanism of transposition of bacteriophage Mu: Structure of a transposition intermediate. *Cell* 41:867–876.

Kiyoshi Mizuuchi *[Source: Courtesy Kiyoshi Mizuuchi.]*

Understanding how a process works often starts with focusing on the reactions that are most efficient and easiest to detect and study— which is why bacteriophage Mu was chosen as a model system for studying transposition. Although Mu is a complicated transposon, its status as a preferred research subject was based on its capacity to transpose often. When the transposon moves, it replicates itself, leaving behind a copy at the original chromosomal site and depositing a new one in the target. A few base pairs of chromosomal DNA in the target are also replicated, creating a short repeated sequence at both ends of the insertion site. The entire process seemed a little bit like magic.

In 1979, James Shapiro, at the University of Chicago, proposed a mechanism for Mu transposition that laid out the main features of the process that we now know to occur (Figure 2). It involved nicking DNA strands to expose both 3′ ends of the transposon, then making a staggered break in the target DNA, leaving 5′ overhangs on the resulting ends. The transposon 3′ ends were then joined to the target 5′ ends. The remaining 3′ ends of the target would prime replication, creating two copies of the transposon and a cointegrate intermediate. This intermediate could be resolved by homologous or site-specific recombination to yield the final products. Other researchers proposed alternatives, but Shapiro's model eventually proved to be largely correct, with the key exception that the direct attack of transposon 3′ ends on target phosphodiester bonds (the transposase-catalyzed strand transfer) was not yet known, or predicted. The key was to find the postulated reaction intermediates.

Kiyoshi Mizuuchi, working with his associate Bob Craigie, found the intermediates. In the early 1980s, they developed an in vitro system that supported Mu transposition, using a plasmid that included a

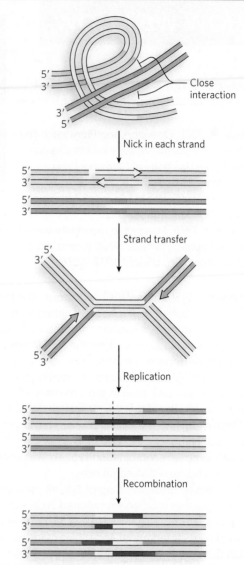

FIGURE 2 The Shapiro model for transposition. *[Source: J. A. Shapiro, Proc. Natl. Acad. Sci. USA 76:1933–1937, 1979.]*

much-abbreviated copy of Mu with both ends of the transposon (the binding sites for the Mu transposase). The Mu proteins required (MuA and MuB), as well as the *E. coli* host proteins, were made available as extracts from cells in which they were being expressed. The target DNA was another circular DNA, derived from bacteriophage φX174 (a phage that has no transposition properties). With this system, the researchers could detect both the cointegrate and the simple insertion products where a new copy of Mu was integrated in the target DNA. Although the results supported Shapiro's model, the researchers did not detect the predicted strand-transfer intermediates before replication. Either these intermediates did not exist or they were converted to the final products of transposition too quickly for detection.

There are two major steps in the process postulated by Shapiro: cleaving DNA and rejoining the ends to new partners, followed by replication (see Figure 2). We now know that the initial DNA cleavage and strand-transfer events need MuA and MuB proteins. The replication steps are more complex, requiring multiple proteins from the host cell. Completing the entire reaction required a concentrated cell extract and addition of the polymer polyvinyl alcohol. The polymer acted as a solvent-exclusion agent, further concentrating the reaction components. In controls, Craigie and Mizuuchi left out different reaction components to demonstrate that each was needed. When they left out polyvinyl alcohol, the reaction did not generate products, but a prominent new DNA band appeared on the agarose gels that they were using to analyze the reaction. Craigie and Mizuuchi knew how to capitalize on this bit of serendipity. Analysis of this new DNA species soon revealed that it was a strand-transfer intermediate, essentially equivalent to the intermediate created by end-joining as proposed by Shapiro.

What had happened in the polyvinyl alcohol control? The absence of the polymer led to a partial reaction in which strand transfer became a dead end, with the resulting transposition intermediate building up to concentrations that made it much easier to detect and study. To confirm that this species was indeed a normal reaction intermediate (and not simply produced by unusual reaction conditions), Craigie and Mizuuchi isolated the putative intermediate DNA species, added back cell extract without MuA and MuB but with replication enzymes now aided by polyvinyl alcohol, and showed that the predicted transposition products were generated. The result was a definitive study establishing key facts about the pathway of Mu transposition. More broadly, the study played a major role in developing our current understanding of replicational transposition mechanisms.

KEY TERMS

isoenergetic, 486

site-specific recombination, 486

transposable element, 486

transposon, 486

transposition, 487

lysogenic pathway, 491

prophage, 491

lytic pathway, 491

Cre-lox, 491

bacterial transduction, 491

phase variation, 492

donor site, 496

target site, 496

transposase, 496

cointegrate, 499

retrotransposable element, 499

retrotransposon, 499

reverse transcriptase, 500

extrachromosomally primed (EP) retrotransposon, 500

integrase, 500

target-primed (TP) retrotransposon, 500

insertion sequence, 501

composite transposon, 501

complex transposon, 501

retrovirus, 504

complementary DNA (cDNA), 506

DDE motif, 508

PROBLEMS

1. Holliday intermediates are associated with both homologous genetic recombination and some site-specific recombination systems. How does their formation differ in the two types of reactions?

2. Compare and contrast the DNA sequence requirements of homologous genetic recombination, site-specific recombination, and transposition.

3. If two FRT sites are present in a single DNA molecule in the same orientation, the Flp recombinase will promote a deletion of the DNA between the two sites. If the FRT sites are altered so that their core regions have a perfectly symmetric (palindromic) sequence instead of the usual asymmetric sequence, how might the fate of the intervening DNA in the reaction change? (Ignore any intermolecular reactions between the DNA molecules.)

4. When a site-specific recombinase brings two recombination sites together to form an initial complex for reaction, how many separate recombinase active sites are present in the complex? How many of those active sites are conformationally set up to catalyze a reaction?

5. Draw the products of Cre-lox–mediated site-specific recombination reactions for the recombination target sites and orientations (indicated by arrows) in the illustrations below.

6. What are the possible products of Flp recombinase-mediated site-specific recombination for the DNA molecule shown below? Arrowheads indicate the location and orientation of FRT sites. X and Y denote altered FRT sites that are functional but will not cross-react with each other. Draw all the products possible if recombination occurs at

(a) the X sites and (b) the Y sites, and if two recombination events occur at (c) the X sites and then the Y sites, and (d) the Y sites and then the X sites.

7. Which of the following DNA molecules are appropriate substrates for the Hin recombinase? Arrowheads indicate the location and orientation of *hix* sites.

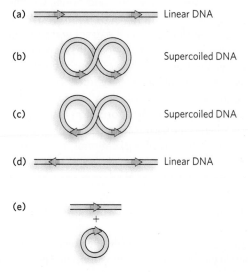

(a) ——————————————> Linear DNA

(b) [supercoiled figure-eight] Supercoiled DNA

(c) [supercoiled figure-eight] Supercoiled DNA

(d) ——————————————> Linear DNA

(e) [linear DNA + circular DNA]

8. The bacterial transposon Tn3 uses a replicative transposition pathway to move from one DNA molecule to another. Tn3 encodes not only a transposase but a site-specific recombinase. Suggest a role in transposition for the site-specific recombinase in this system.

9. The Flp and Cre-lox site-specific recombination systems are widely used in biotechnology to promote genomic rearrangements in eukaryotic cells. The bacteriophage λ site-specific recombination system (see the How We Know section in this chapter) is rarely used for this purpose, even though it was the first such system discovered. Suggest why the λ system is not used.

10. Cut-and-paste transposition results in a short repeated sequence flanking the newly inserted transposon. The DNA sequence shown below is the target site for the transposition of a cut-and-paste transposon. Indicate with arrows the phosphodiester bonds targeted for transesterification by the transposon 3′ ends to generate a 5 bp repeat, 5′-AGGCT-3′, at both ends of the newly inserted transposon.

```
5′ ...ATGCAGGCTAATGGCTACCTGA...
3′ ...TACGTCCGATTACCGATGGACT...
```

11. Many retroviruses have sequences complementary to the 3′ end of one of the tRNA molecules prominent in host cells infected by the virus. What is the purpose of this complementarity?

12. The transposase from the bacterial transposon Tn5 initiates transposition by creating a double-strand break at each end of the transposable element. How many transposase active sites are needed to generate one double-strand break, and how many phosphoryl transfers are catalyzed in that process?

13. Many TP (non-LTR) retrotransposons encode a reverse transcriptase that also has endonuclease activity. What is the primary function of the endonuclease activity?

14. The human genome has more than a million copies of the SINE element Alu. These transposons are found in the DNA between genes, and often in the introns of genes, but very rarely in gene exons. Explain why this is so.

DATA ANALYSIS PROBLEM

Livet, J., T.A. Weissman, H.N. Kang, R.W. Draft, J. Lu, R.A. Bennis, J.R. Sanes, and J.W. Lichtman. 2007. Transgenic strategies for combinatorial expression of fluorescent proteins in the nervous system. *Nature* 450:56–62.

15. Site-specific recombination and transposition are regularly used in biotechnology. An elegant use of site-specific recombination is found in a 2007 report by Livet and colleagues. One of the challenges in brain research is the sheer complexity of the neuronal network. Tracing one neuron to elucidate its connections was nearly impossible until the advent of the "brainbow" technology (see Highlight 14-1). Researchers placed genes for different colored variants of green fluorescent protein (GFP; see Figure 7-22) in cassettes, with the various GFP genes separated by lox sites recognized by the site-specific recombinase Cre (Figure 1a), and inserted the cassettes into the mouse genome. The variants in this example are red fluorescent protein (RFP), yellow fluorescent protein (YFP), and cyan fluorescent protein (CFP). Each cassette was inserted at FRT sites, using a separate site-specific recombinase, Flp. Multiple cassettes were inserted into the genomes of some mice (producing the brainbow in Figure 1b).

The Cre recombinase was inserted into the genome of a different group of mice, the cloned enzyme structured so that it was expressed uniquely in brain tissue, for a brief time, during neuron development. When mice containing the cassettes were mated to mice containing the cloned Cre recombinase, Cre-mediated recombination created a unique pattern of expression of GFP variants in each mature neuron of the heterozygous offspring, effectively coloring the neurons. The vivid results are shown in Highlight 14-1.

(a) Suggest why the Flp recombination system rather than Cre-lox was used to insert the cassettes into the mouse genome.

(a)

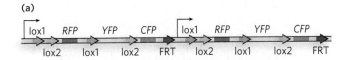

(b)

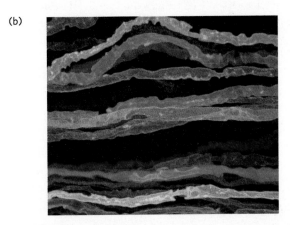

FIGURE 1 *[Source: Reprinted by permission from Macmillan Publishers Ltd: Transgenic strategies for combinatorial expression of fluorescent proteins in the nervous system. Jean Livet, Tamily A. Weissman, Hyuno Kang, Ryan W. Draft, Ju Lu, Robyn A. Bennis, Joshua R. Sanes & Jeff W. Lichtman, Nature 450:56–62, Fig. 4 (1 November 2007). Courtesy of Jeff Lichtman.]*

(b) If one GFP cassette was already present in the mouse genome, how could more cassettes be added?

(c) In the cassette shown in Figure 1a, two different lox sites are used. What sequences in the lox sites must be different to prevent the two sites from recombining with each other?

(d) When two or more different GFP variants are expressed in a neuron, the final color is a blend that reflects the amount of each GFP present. In mice with three cassettes, one GFP variant is expressed from each cassette (assuming that the lox sites preclude recombination between cassettes). Ten different colors are possible. Describe the combinations leading to the different colors.

ADDITIONAL READING

General

Craig, N.L., R. Craigie, M. Gellert, and A.M. Lambowitz. 2002. *Mobile DNA II.* Washington, DC: American Society for Microbiology.

Haber, J.E. 2014. *Genome Stability: DNA Repair and Recombination.* New York: Garland Science. See Chapter 11 for a complete discussion of site-specific recombination.

Mechanisms of Site-Specific Recombination

Biswas, T., H. Aihara, M. Radman-Livaja, D. Filman, A. Landy, and T. Ellenberger. 2005. A structural basis for allosteric control of DNA recombination by λ integrase. *Nature* 435:1059-1066.

Campbell, A. 2007. Phage integration and chromosome structure: A personal history. *Annu. Rev. Genet.* 41:1–11.

Capecchi, M. 1994. Targeted gene replacement. *Sci. Am.* 270(3):52–59.

Chen, Y., and P.A. Rice. 2003. New insight into site-specific recombination from Flp recombinase-DNA structures. *Annu. Rev. Biophys. Biomol. Struct.* 32:135-159.

Lichtman, J.W., J. Livet, and J.R. Sanes. 2008. A technicolour approach to the connectome. *Nat. Rev. Neurosci.* 9:417–422.

Sauer, B. 2002. Cre/lox: One more step in the taming of the genome. *Endocrine* 19:221-227.

Mechanisms of Transposition

Beauregard, A., M.J. Curcio, and M. Belfort. 2008. The take and give among retrotransposable elements and their hosts. *Annu. Rev. Genet.* 42:587–617.

Bordenstein, S.R., and W.S. Reznikoff. 2005. Mobile DNA in obligate intracellular bacteria. *Nat. Rev. Microbiol.* 3:688–699.

Ivics, Z., P.B. Hackett, R.H.A. Plasterk, and Z. Izsvák. 1997. Molecular reconstruction of *Sleeping Beauty*, a Tc1-like transposon from fish, and its transposition in human cells. *Cell* 91:501–510.

Peters, J.E., and N.L. Craig. 2001. Tn7: Smarter than we thought. *Nat. Rev. Mol. Cell Biol.* 5:1161–1170.

Plasterk, R.H.A., Z. Izsvák, and Z. Ivics. 1999. Resident aliens: The Tc1/*mariner* superfamily of transposable elements. *Trends Genet.* 15:326–332.

Reznikoff, W.S. 2008. Transposon Tn5. *Annu. Rev. Genet.* 42:269–286.

Shapiro, J.A. 1979. Molecular model for the transposition and replication of bacteriophage Mu and other transposable elements. *Proc. Natl. Acad. Sci. USA* 76:1933–1937.

van Opijnen, T., and A. Camilli. 2013. Transposon insertion sequencing: A new tool for systems-level analysis of microorganisms. *Nat. Rev. Microbiol.* 11:435–442.

The Evolutionary Interplay of Transposons and Their Hosts

Burns, K.H., and J.D. Boeke. 2012. Human transposon tectonics. *Cell* 149:740-752.

Feschotte, C., and E.J. Pritham. 2007. DNA transposons and the evolution of eukaryotic genomes. *Annu. Rev. Genet.* 41:331–368.

Gray, Y.H.M. 2000. It takes two transposons to tango. *Trends Genet.* 16:461–468.

Kapitonov, V.V., and J. Jurka. 2005. RAG1 core and V(D)J recombination signal sequences were derived from Transib transposons. *PLoS Biol.* 3:998–1011.

Koonin, E.V., and V.V. Dolja. 2014. Virus world as an evolutionary network of viruses and capsidless selfish elements. *Microbiol. Mol. Biol. Rev.* 78:278–303.

Lisch, D. 2012. How important are transposons for plant evolution? *Nat. Rev. Genet.* 14:49–61.

Zlotorynski, E. 2014. RNA interference: MicroRNAs suppress transposons. *Nat. Rev. Mol. Cell. Biol.* 15:298–299.

15 Transcription: DNA-Dependent Synthesis of RNA

Robert Tjian [*Source: Bonnie Azab Powell.*]

MOMENT OF DISCOVERY

In the early 1980s, it was clear that specialized proteins must exist for accurate and regulated mRNA synthesis from particular genes in mammalian cells. *However, nobody had been able to identify such "transcription factors" or determine how this process of transcriptional activation works.* The breakthrough in my laboratory came when we found out that human cell extracts contained a factor that can discriminate between two templates and somehow program the enzyme that reads DNA to choose the right promoter DNA and ignore all others. But how?

We decided to use a short piece of the active promoter DNA sequence as "bait" to fish out proteins that selectively bind this site. The challenge was to purify this activity away from the other 3,000 DNA-binding proteins present in human cell extracts! After months of struggling with this problem, I vividly recall walking into the lab and my coworkers, Jim Kadonaga and Kathy Jones, saying, "We think we know which protein it is!" They had cleverly treated the human cell extract with a huge excess of sheared calf-thymus DNA to remove most nonspecific DNA-binding proteins, enriching the treated extract for the protein we wanted. Sequence-specific DNA affinity resin was then used to bind the transcription factor in the treated extracts, leading to the purification of a single protein. We called this protein specificity protein 1 (Sp1), the first of many sequence-specific transcription factors that were to prove critical for human gene regulation.

I'll never forget the feeling of profound excitement at having shared the discovery of such a fundamental protein in biology and, at the same time, having devised with my lab members a new means to isolate hundreds more of these key gene-regulatory proteins.

—*Robert Tjian, on discovering the first specific eukaryotic transcription factor*

Online resources related to this chapter:

Simulation Transcription

Nature **exercise** mRNA processing

Information encoded in the DNA of cells and viruses provides the instructions for making the RNA and protein molecules that carry out the activities essential for life. The first step in expression of this information is **transcription**, the enzymatic production of an exact complementary strand of RNA from a DNA template. Transcription thus involves the transfer of genetic information from DNA to RNA. For protein-coding regions of DNA, transcription begins the gene expression pathway leading to the production of protein through translation of a messenger RNA (translation of mRNA into protein is discussed in Chapter 18). For non-protein-coding regions of DNA, transcription produces RNA molecules that, in many cases, are components of RNA-protein complexes, or ribonucleoproteins. Some of these are enzymes, but the majority play nonenzymatic roles in controlling gene expression on many levels. Increasing evidence shows that a much greater proportion of an organism's transcribed DNA is non-protein-coding than protein-coding. The functions of many such transcripts are just beginning to be defined.

All RNA molecules, except for the RNA genomes of certain viruses, are derived from information stored in DNA. Transcription produces three major kinds of RNA, and many other types of RNA are generated in smaller amounts. As described in Chapter 6, **messenger RNAs (mRNAs)** encode the amino acid sequence of one or more polypeptides specified by a gene or set of genes. **Transfer RNAs (tRNAs)** read the information encoded in the mRNA and provide the appropriate amino acid to a growing polypeptide chain during protein synthesis. **Ribosomal RNAs (rRNAs)** are constituents of ribosomes, the intricate cellular machines that synthesize proteins. Other specialized RNAs have regulatory or catalytic functions or are precursor forms of the three main classes of RNA (**Highlight 15-1**).

Unlike DNA replication, which involves copying the entire chromosome, transcription is selective. Only particular genes or groups of genes are transcribed at any given time, and some portions of the DNA genome may be transcribed rarely or not at all. The cell directs the transcription machinery to express genetic information as it is needed. Specific regulatory sequences mark the beginning and end of the DNA segments to be transcribed and designate which strand of the double-stranded DNA is to be used as the template for RNA synthesis. The regulation of transcription is an important and exciting aspect of gene expression. We discuss regulation of gene expression in this chapter, and in further detail in Chapters 19–22.

As one of the central cellular processes studied by molecular biologists, transcription—enzymatic RNA synthesis directed by a DNA template—has been worked out in some detail, yet some fascinating puzzles remain. We begin by examining the enzymes responsible for transcription, and then address the mechanics of transcription in bacteria and in eukaryotic cells.

15.1 RNA POLYMERASES AND TRANSCRIPTION BASICS

Transcription in cells and viruses is catalyzed by specialized enzymes called **RNA polymerases**. The transcription reaction resembles DNA replication in its fundamental chemical mechanism, in the direction of synthesis ($5' \rightarrow 3'$), and in the requirement for a template strand. And, like replication, transcription has initiation, elongation, and termination phases. In contrast to replication, however, transcription does not require a primer to begin RNA synthesis, and it involves defined sections of DNA rather than the entire molecule. Also, just one of the two strands of a DNA segment serves as the template for a given transcription reaction (**Figure 15-1**). In this section we discuss the different types of polymerases, steps in the transcription process, and how transcription can be blocked by inhibitors.

RNA Polymerases Differ in Details but Share Many Features

RNA polymerases were first discovered by testing cell extracts for activity that could form an RNA polymer from ribonucleoside 5′-triphosphates (rNTPs). Experiments demonstrated that the RNA product of this polymerase activity was complementary to the sequence of DNA supplied in the reaction mix. A particularly telling experiment involved a DNA strand with an alternating $(AT)_n$ sequence. The use of different radiolabeled rNTPs revealed that only ATP and UTP were needed for complete

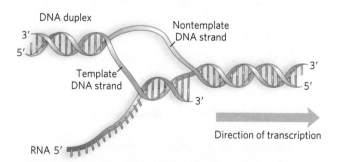

FIGURE 15-1 Transcription of DNA into RNA. The DNA duplex opens to allow a complementary RNA copy to be made from one DNA strand (the template). Synthesis proceeds in the $5' \rightarrow 3'$ direction in the growing RNA strand.

The ABCs of RNA: Complexity of the Transcriptome

When complete eukaryotic genome sequences became available, molecular biologists were excited to discover the extent of the **transcriptome**, the entire set of RNA transcripts produced in a cell. Initially, researchers focused on characterizing the transcription products of known genes. These included mRNAs and known stable noncoding RNAs (ncRNAs) such as rRNAs, tRNAs, small nuclear RNAs (snRNAs) involved in pre-mRNA splicing, and small nucleolar RNAs (snoRNAs), which guide chemical modifications in the ribosome.

Unexpected levels of complexity began to emerge, however, beginning with the discovery of naturally occurring interfering RNAs, such as small interfering RNAs (siRNAs) and microRNAs (miRNAs), which have roles in the regulation of translation (as we describe in Chapter 22). Using a combination of microarrays and RNA-Seq (see Chapter 8), researchers could detect RNA transcripts without being biased by prior expectations. Use of these techniques showed that a lot of transcription was occurring that had previously been ignored. These new technologies revealed that the transcription landscape in higher eukaryotes is much more complex than expected. Surprisingly, a large fraction of transcripts originate from intergenic regions—regions between the coding sequences of genes—that had been thought to be silent, or from sequences that run in the opposite direction (antisense) to genes. Transcription that does not map to protein-coding genes or to known ncRNA genes also occurs in yeast.

In a parallel set of experiments, arrays of synthetic DNA oligonucleotides representing all nonrepetitive sequences in human chromosomes 21 and 22 were used to map the binding sites for three human transcription factors—Sp1, cMyc, and p53—that activate the transcription of many protein-coding genes involved in cell growth and differentiation. The experiments revealed far more transcription factor–binding sites than would be predicted from the number of protein-coding genes in these chromosomes. Of these binding sites, more than one-third lie within or immediately 3′ to well-characterized genes and seem to correlate with the transcription of ncRNAs. These findings have changed our thinking about transcription: much more of it goes on than previously suspected. Just what is all that RNA doing?

Some possible roles of previously undetected transcripts have emerged. For example, long noncoding RNAs (lncRNAs) are produced from regions that are either intergenic or antisense to genes. The functional significance of lncRNAs is not known, although several studies suggest roles for these transcripts in gene regulation. Shorter transcripts, particularly those that originate near gene promoters, fall into two somewhat arbitrary categories: molecules 20 to 200 nucleotides long are called small RNAs (sRNAs), and molecules of 200 to 1,000 nucleotides are called long RNAs (lRNAs). These categories include numerous subfamilies of transcripts defined by their abundance, longevity, and genomic origin. Although the sources of these transcripts are not yet fully worked out, at least some sRNAs may result from aborted or prematurely terminated transcription.

Whether pervasive transcription is simply a consequence of low-level background transcription or has functional significance is a subject of active research. In animals, it is unclear whether transcripts that are initiated near a particular promoter are functional. One possibility is that promoter-associated transcripts may help maintain chromatin in an open state that is more accessible to the transcription and regulatory machinery. In addition, it could keep a pool of RNA polymerase available for rapid deployment to make mRNAs. This will clearly be an expanding area of research, and more surprises are certainly in store.

synthesis of RNA, and not GTP or CTP (**Figure 15-2**). Subsequent experiments using the purified *E. coli* RNA polymerase, and, later, using bacteriophage RNA polymerases, helped define the fundamental properties of transcription.

In addition to a DNA template, DNA-dependent RNA polymerases require Mg^{2+} and all four rNTPs (ATP, GTP, UTP, and CTP) as substrates for the polymerization reaction. The chemistry and mechanism of RNA synthesis closely resemble those of DNA synthesis (**Figure 15-3**). RNA polymerase extends an RNA strand by adding ribonucleotide units to the 3′-hydroxyl end, building RNA in the 5′→3′ direction. The 3′-hydroxyl group makes a nucleophilic attack on the α phosphate of the incoming rNTP, with the concomitant release of pyrophosphate. As noted above, only one of the two DNA strands serves as template. The template DNA is copied in the 3′→5′ direction (antiparallel to the new RNA strand), just as in DNA replication. Each nucleotide in the newly formed RNA is selected by Watson-Crick base pairing: U residues—not T residues, as in DNA—are inserted in the RNA to pair with A residues in the DNA template, G residues are inserted to pair with C residues, and so on (see Figure 6-11). Base-pair geometry may also play a role in nucleotide selection and the resulting fidelity of the polymerase reaction.

RNA polymerases are fascinating enzymes that continue to be actively studied. The simplest examples consist of one polypeptide chain, such as the phage T7

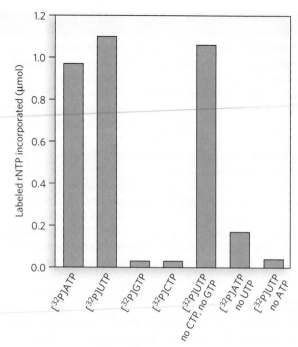

FIGURE 15-2 Early experiment demonstrating DNA-dependent RNA synthesis by an RNA polymerase. RNA polymerizing activity was partially purified from *E. coli* and analyzed for RNA synthesis, using a template DNA of sequence (AT)$_n$ and various rNTP substrates. Reactions took place in the presence of nonradioactive (unlabeled) rNTPs and one radiolabeled rNTP, as noted on the graph. Incorporation of radioactively labeled rNTP substrates was measured. [*Source: Data from J. Hurwitz et al.,* Cold Spring Harb. Symp. Quant. Biol. *26:91–100, 1961.*]

and Sp6 RNA polymerases. In contrast, all cellular RNA polymerases, from bacteria to humans, are composed of multiple polypeptides that fold together to create the functional enzyme. In *E. coli*, for example, the **RNA polymerase core** is a large, complex enzyme with five polypeptide subunits: two copies of the α subunit and one copy each of the β, β′, and ω subunits: $\alpha_2\beta\beta'\omega$ (M_r 390,000), as shown in **Figure 15-4a**. A sixth subunit, designated σ and known as **sigma factor**, binds transiently to the core and directs the enzyme to specific binding sites on the DNA. These six subunits constitute the **RNA polymerase holoenzyme.** Bacteria have multiple sigma factors, named according to their molecular weight; the most common is σ^{70} (M_r 70,000). Thus, the RNA polymerase holoenzyme of *E. coli* exists in several forms, depending on the type of σ subunit it contains. Sigma factors play an important role in the recognition of different types of bacterial genes (see Section 15.2).

In eukaryotic cells, three distinct RNA polymerases are responsible for transcribing RNAs with different functions. **RNA polymerase I (Pol I)** transcribes genes encoding large rRNA precursors. **RNA polymerase II (Pol II)** transcribes nearly all protein-coding genes to make mRNA. **RNA polymerase III (Pol III)** transcribes genes

(a) Bacterial RNA polymerase core

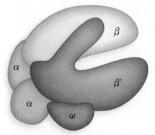

(b) Bacterial RNA polymerase   **(c)** Eukaryotic RNA polymerase II

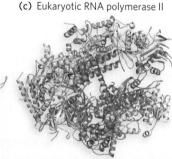

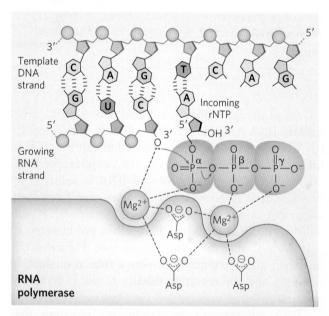

FIGURE 15-3 The chemical mechanism of RNA synthesis. The addition of an rNTP to a growing transcript is a Mg^{2+}-dependent reaction that produces a 5′→3′ phosphodiester linkage.

FIGURE 15-4 The architecture of RNA polymerases. (a) The bacterial RNA polymerase core is composed of several subunits that give the enzyme the overall shape of a crab claw. The pincers are formed from the large β and β′ subunits. At the start of transcription, the sigma factor (σ) associates with the core polymerase to form the holoenzyme (not shown). The crystal structures of (b) a bacterial RNA polymerase (*Thermus aquaticus*) and (c) a eukaryotic RNA polymerase II (*Saccharomyces cerevisiae*) are shown. [*Sources:* (b) Data from G. Zhang, Cell *98:811–824, 1999.* (c) PDB ID 1I50.]

TABLE 15-1

Some RNA Polymerase Subunits in Bacteria and Eukaryotes

Bacterial Core	Eukaryotes		
	Pol I	Pol II	Pol III
α (2)	RPC5/RPC9	RPB3/RPB11	RPC5/RPC9
β	RPA1	RPB1	RPC1
β′	RPA2	RPB2	RPC2
ω	RPB6	RPB6	RPB6

encoding smaller functional RNAs, including tRNAs, some small nuclear RNAs (snRNAs), and 5S ribosomal RNA (the naming of rRNAs is explained in Chapter 18). These enzymes are related to bacterial RNA polymerase at the level of both sequence and structure, indicating that RNA polymerase is an ancient enzyme. However, the eukaryotic RNA polymerases are larger and contain additional proteins not found in bacteria (**Table 15-1**).

The molecular structures of bacterial and yeast RNA polymerases have been determined by x-ray crystallography (**Figure 15-4b, c**). The cleft between the two pincers of the claw contains the enzyme active site and binds two Mg^{2+} ions that facilitate RNA polymerization. The more conserved parts of the polymerase complex are in the interior, whereas regions that have varied more over the course of evolution are at the exterior of the complex.

Unlike DNA polymerase, RNA polymerase does not require a primer to initiate synthesis. RNA polymerase catalyzes RNA synthesis in three distinct phases, similar to those of the DNA polymerase reaction. **Initiation** occurs as RNA polymerase binds to specific DNA sequences called **promoters**. **Elongation** is the process of adding nucleotides to the growing RNA strand. **Termination** is the release of the product RNA when the polymerase reaches the end of a gene or other transcription unit.

The two strands of a DNA duplex have different roles in transcription. The **template strand** serves as a template for RNA synthesis, and its complement, the **nontemplate strand**, is identical in base sequence to the RNA transcribed from the gene, with U in the RNA in place of T in the DNA (**Figure 15-5**). The nontemplate strand is more often called the **coding strand**. The

coding strand for a particular gene may be located in either strand of a given chromosome.

KEY CONVENTION

The template strand of the DNA is copied during transcription, and its sequence is the complement of the RNA transcript. The coding strand of the DNA has the same sequence as the RNA transcript (except for T in the DNA and U in the RNA). Hence, for example, the start codon for transcription—the beginning of the open reading frame of the gene—is 5′-ATG in the coding strand of the DNA and 5′-AUG in the mRNA. By convention, gene, promoter, and regulatory sequences in DNA are written as they appear in the coding strand.

To enable RNA polymerase to synthesize an RNA strand complementary to the template DNA strand, the DNA duplex must unwind over a short distance, forming what is known as a transcription "bubble" (**Figure 15-6**). During transcription, the *E. coli* RNA polymerase generally keeps about 17 bp of DNA unwound. In the elongation phase, the growing end of the new RNA strand base-pairs temporarily with the DNA template in the unwound region to form a short hybrid RNA-DNA double helix, 8 bp in length. The RNA in this hybrid duplex is displaced shortly after its formation as the DNA double helix re-forms. Elongation of a transcript by *E. coli* RNA polymerase proceeds at a rate of 50 to 90 nucleotides per second. Because DNA is a helix, the movement of a transcription bubble requires considerable strand rotation. Consequently, a moving RNA polymerase generates waves of positive supercoils ahead of the transcription bubble and negative supercoils behind. This has been observed both in the laboratory with purified polymerase enzymes and in live bacterial cells. In the cell, the topological problems caused by transcription are relieved through the action of topoisomerases (see Chapter 9).

RNA polymerases lack a separate proofreading $3′ \rightarrow 5′$ exonuclease active site, which exists in many DNA polymerases. Consequently, the error rate for transcription is higher than that for chromosomal DNA replication—approximately one error for every 10^4 to 10^5 ribonucleotides incorporated into RNA. Because many

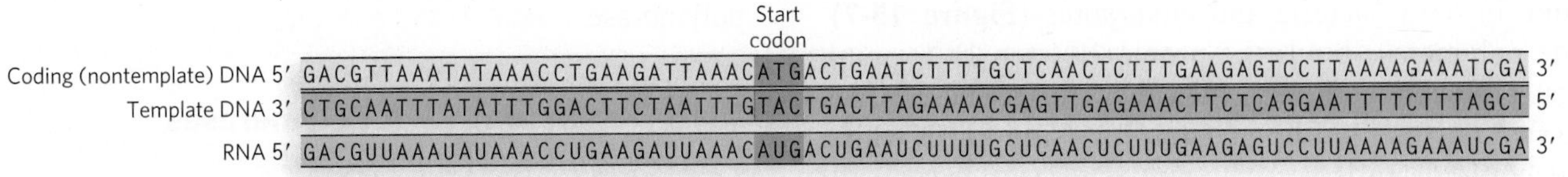

Start codon

Coding (nontemplate) DNA 5′ GACGTTAAATATAAACCTGAAGATTAAACATGACTGAATCTTTTGCTCAACTCTTTGAAGAGTCCTTAAAAGAAATCGA 3′
Template DNA 3′ CTGCAATTTATATTTGGACTTCTAATTTGTACTGACTTAGAAAACGAGTTGAGAAACTTCTCAGGAATTTTCTTTAGCT 5′
RNA 5′ GACGUUAAAUAUAAACCUGAAGAUUAAACAUGACUGAAUCUUUUGCUCAACUCUUUGAAGAGUCCUUAAAAGAAAUCGA 3′

FIGURE 15-5 The DNA template for RNA synthesis. The coding (nontemplate) strand of the DNA is identical in base sequence to the RNA transcribed from the gene, with U in the RNA in place of T in the DNA. The template strand is used to direct RNA synthesis by RNA polymerase. ATG in the coding strand (blue) is the initiation (start) codon.

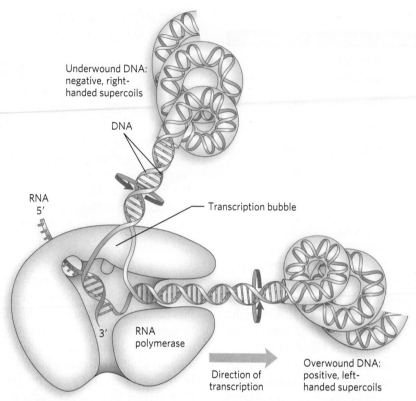

Underwound DNA:
negative, right-
handed supercoils

DNA

RNA
5′

Transcription bubble

3′ RNA
polymerase

Direction of
transcription

Overwound DNA:
positive, left-
handed supercoils

FIGURE 15-6 The transcription "bubble." The DNA duplex is unwound for about
17 bp, forming a bubble, which enables RNA polymerase to access the template
strand. DNA supercoiling occurs both in front of and behind the transcription bubble.

copies of a transcript are generally produced from a single gene, and all of these are eventually degraded and replaced, a mistake in an RNA molecule is less consequential to the cell than a mistake in the permanent information stored in DNA. Many RNA polymerases, including bacterial RNA polymerase and the eukaryotic Pol II, pause when a mispaired base is added during transcription. In addition, they can remove mismatched nucleotides from the 3′ end of a transcript by direct reversal of the polymerization reaction. However, it is still unclear whether this activity is a true proofreading function and to what extent it may contribute to the fidelity of transcription.

Transcription Initiation, Elongation, and Termination Occur in Discrete Steps

The general steps of the transcription pathway are the same in both bacteria and eukaryotes (**Figure 15-7**). The polymerase binds the promoter (step 1), forming first a **closed complex,** in which the bound DNA is intact, and then an **open complex** (step 2), in which the bound DNA is partially unwound near a region 10 bp ahead of (upstream of) the transcription start site. Transcription is initiated within the complex (step 3), leading to a conformational change that converts the

complex to the form required for elongation. **Promoter clearance**, involving movement of the transcription complex down the DNA template and away from the promoter, leads to the formation of a tightly bound **elongation complex** (step 4). Once elongation begins, RNA polymerase becomes a highly efficient enzyme, completing synthesis of the transcript before dissociating from the DNA template (step 5), then recycling for a new round of transcription. Although the steps in this pathway are conserved across species, the details of the process are somewhat more complex in eukaryotic cells.

RNA synthesis is processive, which means that once RNA polymerase begins elongating a transcript, the kinetics of the polymerization reaction greatly favor the addition of the next nucleotide over premature release of the transcript. As we will see, elongation is not a uniform process but instead occurs in fits and starts, and specific sequences trigger termination of RNA synthesis by RNA polymerase.

DNA-Dependent RNA Polymerases Can Be Specifically Inhibited

Small molecules and peptides that inhibit transcription are useful both as antibiotics and as tools for research. **Actinomycin D**, one of a class of peptide antibiotics

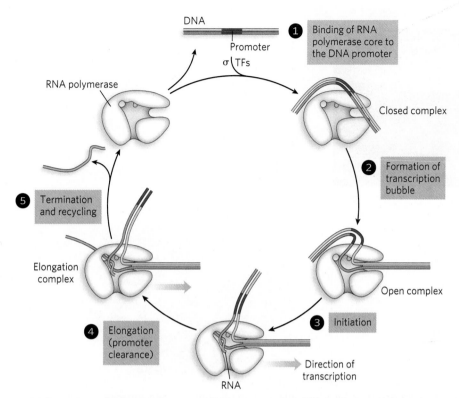

FIGURE 15-7 An overview of transcription. RNA polymerase binding at the promoter—requiring sigma (σ) factor and, in eukaryotes, transcription factors (TFs)—leads to initiation of transcription by the polymerase holoenzyme, followed by elongation and termination. The steps are further described in the text.

isolated from *Streptomyces* soil bacteria, inhibits transcription elongation by RNA polymerase in bacteria and eukaryotes (**Figure 15-8**). The planar portion of this molecule intercalates into the double-helical DNA between successive G≡C base pairs, deforming the DNA and preventing movement of the polymerase along the template. Because actinomycin D inhibits RNA elongation both in intact cells and in cell extracts, it is used in the laboratory to identify cell processes that depend on RNA synthesis. **Acridine** inhibits RNA synthesis in a similar fashion. **Rifampicin**, a small molecule isolated from *Streptomyces mediterranei*, inhibits bacterial RNA synthesis by binding to the β subunit of bacterial RNA polymerase, preventing promoter clearance. Because it does not affect the function of eukaryotic polymerases, rifampicin is sometimes used as an antibiotic to treat such bacterial diseases as tuberculosis and leprosy.

Some species rely on transcription inhibitors for natural biodefense. For example, the mushroom *Amanita phalloides* produces **α-amanitin (Figure 15-9a, b)**, a cyclic peptide that disrupts eukaryotic mRNA synthesis by blocking Pol II and, at higher concentrations, Pol III. The binding position of α-amanitin in Pol II prevents the flexibility required for translocation of the polymerase along the DNA substrate (**Figure 15-9c**). α-Amanitin is

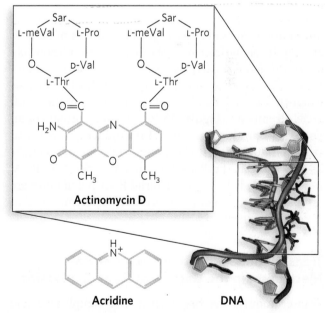

FIGURE 15-8 Inhibitors of transcription. The DNA structure can be deformed by actinomycin D (stick structure), which contains a heterocyclic group (orange) that intercalates into the DNA, inhibiting transcript elongation. (Sar is sarcosine; meVal is methylvaline.) Acridine, which also has a heterocyclic group, has a similar inhibitory effect. [*Source: PDB ID 1DSC.*]

(a)

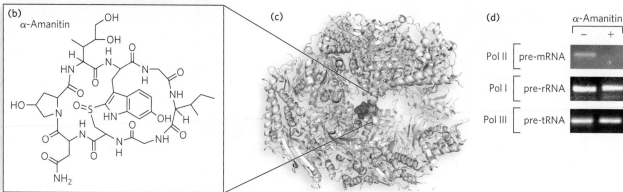

FIGURE 15-9 **Inhibition of transcription by α-amanitin.** (a) A mushroom to avoid! *Amanita phalloides* is poisonous to eukaryotes because it produces α-amanitin. (b) The chemical structure of α-amanitin. (c) Binding of α-amanitin (red) to yeast Pol II. (d) α-Amanitin inhibits synthesis of mRNA by Pol II, but not synthesis of rRNA by Pol I or tRNA by Pol III. The image shows a gel matrix in which RNA of a particular size is detected; in the presence of α-amanitin, RNA is made by Pol I and Pol III, but not by Pol II. (Note that pre-RNAs (precursor RNAs), as shown here, are the first products of transcription and are processed to form the mature RNAs.) [Sources: (a) © Wolstenholme Images/Alamy. (c) PDB ID 1K83. (d) Lee Y, Kim M, Han J, Yeom KH, Lee S, Baek SH, Kim VN. MicroRNA genes are transcribed by RNA polymerase II. EMBO Journal, Vol. 23, p. 4051, Fig. 3. © 2004 European Molecular Biology Organization.]

useful in the laboratory as a specific inhibitor of eukaryotic Pol II, or to determine the polymerase responsible for transcribing a particular gene. This can be demonstrated in an experiment monitoring RNA synthesis when an RNA polymerase and rNTPs are combined with a DNA template (**Figure 15-9d**). Under normal conditions, Pol I produces rRNA, Pol II produces mRNA, and Pol III produces tRNA. The addition of α-amanitin inhibits the synthesis of mRNA, but not that of rRNA or tRNA. Indeed, Pol I, Pol III and bacterial RNA polymerase are insensitive to α-amanitin—as is the RNA polymerase II of *A. phalloides* itself!

Transcriptional Regulation Is a Central Mechanism in the Control of Gene Expression

Transcription is the first step in the complicated and energy-intensive pathway of protein synthesis, so much of the regulation of protein levels in both bacterial and eukaryotic cells occurs during transcription, particularly its early stages. Because requirements for any gene product vary according to cellular conditions or developmental stage, cells and viruses control transcription so that gene products are made only when they are needed, and in the required proportions. Regulation can occur at any step in transcription, but much of it is directed at the promoter-binding and initiation steps.

The DNA sequence in the promoter region affects the efficiency of RNA polymerase binding and the initiation of transcription. However, differences in promoter sequences are just one of several levels of control during initiation. The binding of additional proteins to sequences both near to and distant from the promoter can also affect transcription levels. Protein binding can *activate* transcription by facilitating RNA polymerase binding or later steps in the initiation process, or it can *repress* transcription by blocking polymerase activity (activation and repression of transcription by specific proteins are discussed in detail in Chapters 20 and 21).

SECTION 15.1 SUMMARY

- Transcription is catalyzed by DNA-dependent RNA polymerases, which use ribonucleoside 5′-triphosphates to synthesize RNA complementary to the template strand of duplex DNA. The steps of transcription consist of binding of RNA polymerase to a promoter on DNA to form a closed complex, opening of the complex by local DNA unwinding near the promoter, initiation, elongation, and termination.

- The simplest RNA polymerases, with one polypeptide chain, are found in some bacteriophages. All cellular RNA polymerases are composed of multiple polypeptides that fold together to create the functional enzyme.
- Bacterial RNA polymerase uses a sigma (σ) factor to recognize and bind the promoter during initiation.
- Eukaryotic cells have three types of RNA polymerases. Pol I and Pol III transcribe genes encoding rRNAs and small functional RNAs such as tRNA, respectively. Pol II transcribes protein-coding genes to make mRNA.
- Once an elongation complex forms on a DNA template, RNA polymerase completes the synthesis of the transcript before dissociating from the DNA.
- Various naturally occurring small molecules inhibit polymerase enzymes and can be used to detect which polymerase produces specific types of RNA.
- Much of the regulation of protein levels in both bacterial and eukaryotic cells occurs during the early stages of transcription.

15.2 TRANSCRIPTION IN BACTERIA

Transcription shares many fundamental properties in all organisms. Due to their relative ease of study, RNA polymerases from bacteria and bacteriophages were the focus of the first experiments that revealed the principles of how these enzymes recognize DNA and synthesize an RNA transcript. Bacterial transcription continues to be an active area of research, in part because many experimental tools already exist for analyzing polymerase function both in vitro and in cells. One obvious difference between bacteria and eukaryotes is that bacteria have a single RNA polymerase enzyme for synthesizing all the RNA molecules in the cell, instead of the three RNA polymerases found in eukaryotes.

Promoter Sequences Alter the Strength and Frequency of Transcription

In *E. coli*, RNA polymerase binds to DNA within a 100 bp region stretching from about 70 bp before the transcription start site to about 30 bp beyond it. By convention, the DNA base pairs corresponding to the beginning of an RNA molecule are given positive numbers (+1 is the transcription start site), and those preceding the RNA start site are given negative numbers. The promoter region in *E. coli* thus extends between positions −70 and +30.

As mentioned previously, the most common sigma factor in *E. coli* is σ^{70}. Analyses and comparisons of the bacterial promoters recognized by a σ^{70}-containing RNA polymerase holoenzyme have revealed similarities in two short sequences centered about positions −10 and −35 (**Figure 15-10**). These sequences are important interaction sites for σ^{70}. Although the sequences are not identical for all bacterial promoters in this class, certain nucleotides that are particularly common at each position

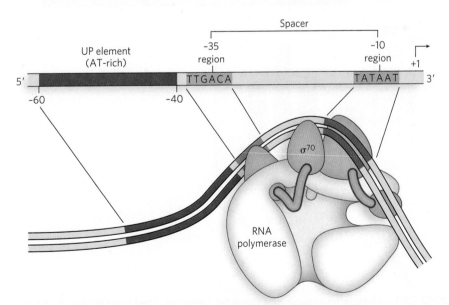

FIGURE 15-10 Features of bacterial promoters recognized by σ^{70}. Different sigma factors recognize distinct promoter elements. The σ^{70} factor recognizes the −10 and −35 sequences, shown here. Promoters recognized by other sigma factors have other consensus sequences in the promoter region. The right-angle arrow at the +1 position indicates the transcription start site.

form a **consensus sequence**. The consensus sequence centering at the −10 region is 5′-TATAAT-3′; the consensus sequence at the −35 region is 5′-TTGACA-3′. A third AT-rich recognition element, the **upstream promoter (UP) element**, occurs between positions −40 and −60 in the promoters of certain highly expressed genes. The UP element is bound by one of the α subunits of RNA polymerase. The efficiency with which an RNA polymerase binds to a promoter and initiates transcription is determined in large measure by the −10, −35, and UP sequences, the spacing between −10/−35 and the UP element, and the distance of the UP element from the transcription start site.

Many independent lines of evidence attest to the functional importance of the sequences in the −10 and −35 regions. Mutations that affect the function of a given promoter often involve a single base-pair change in these regions. Variations in the consensus sequence also affect the efficiency of RNA polymerase binding and transcription initiation. A change in just one base pair can decrease the rate of binding by several orders

of magnitude. The promoter sequence thus establishes a basal level of transcription that can vary greatly from one *E. coli* gene to the next.

Experiments with the Lac promoter in *E. coli* demonstrated the importance of promoter sequence to gene expression (**Figure 15-11**). The Lac promoter drives the expression of genes in the *lac* operon, which encodes proteins that metabolize the sugar lactose. In fact, the classic experiments that first revealed transcription to be a regulated event were performed on the *lac* operon (explored in Chapter 20). The Lac promoter sequence is close to, but not exactly the same as, the bacterial consensus promoter sequence for σ^{70}-containing RNA polymerase holoenzymes. In the experiment shown in Figure 15-11, mutations were introduced into the Lac promoter to make it conform to the consensus, and the mutated promoters were used to drive the expression of the enzyme β-galactosidase (β-gal, one of the *lac* operon proteins). Promoter activity was evaluated by a reaction in which β-gal converts a colorless substrate into a chemical with blue color: the

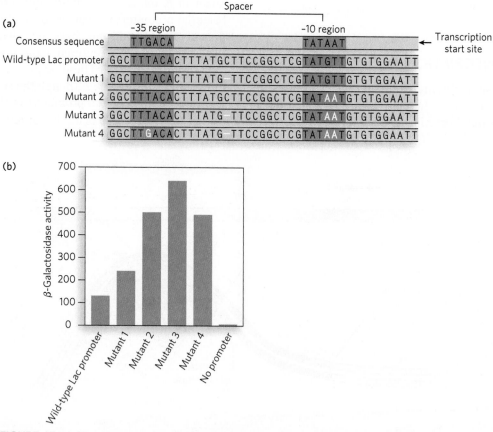

FIGURE 15-11 Mutational analysis of a bacterial promoter. The Lac promoter, driving the *lac* operon, has a sequence close to the σ^{70} consensus sequence. (a) Mutations are created that make the promoter conform to the consensus; red dashes indicate lack of a nucleotide at that position. The spacer length matters because it governs the helical orientation of the two consensus sites. (b) Mutation results in increased expression and activity of β-galactosidase. [*Source: Data from M. H. Caruthers, J. Bacteriol. 164:1353–1355, 1985.*]

TABLE 15-2					
Some *E. coli* Sigma Factors					
				Consensus Sequence	
Subunit Class	*Function of Genes Activated*		−35	Spacer (bp)	−10
σ^{70} (RpoD)	"Housekeeping" genes expressed in all growing cells		TTGACA	16–18	TATAAT
σ^{38} (RpoS)	Starvation/stationary growth phase		[Variable]		CTATACT
σ^{28} (RpoF)	Flagellar structure and movement		TAAA	15	GCCGATAA
σ^{32} (RpoH)	Heat shock		CTTAA	12–14	GGGTAT
σ^{24} (RpoE)	Extracytoplasmic stress		GGAACTT	16	TCAAA
			−24	Spacer (bp)	−12
σ^{54} (RpoN)	Nitrogen uptake and metabolism		TGGCAC	5	TTGC

Sources: For σ^{70} family (RpoD, RpoS, RpoF, RpoH) and σ^{54} (RpoN) consensus sequences: M. M. Wöstena, *FEMS Microbiol. Rev.* 22:127–150, online, January 17, 2006. For σ^{24} (RpoE): K. M. Thompson, V. A. Rhodius, and S. Gottesman, *J. Bacteriol.* 189:4243–4256, 2007, online, April 6, 2007, doi: 10.1128/JB.00020-07.

more β-gal expressed, the deeper the blue color produced. The results showed that the mutations bringing the promoter into consensus tended to confer greater levels of protein expression.

Sigma Factors Specify Polymerase Binding to Particular Promoters

Escherichia coli has at least seven different kinds of sigma factors (**Table 15-2**); the number varies for other bacteria. Each RNA polymerase molecule contains only one σ subunit, which directs the polymerase enzyme to bind a specific type of promoter sequence. The σ^{70} factor binds reversibly to RNA polymerase and is essential for general transcription in exponentially growing cells. However, it can be replaced by alternative sigma factors that trigger the transcription of genes involved in diverse functions, including stress responses, changes in cell shape, and nitrogen uptake. A sigma factor associates with the RNA polymerase core only transiently, separating from it after transcription initiation.

Through their interactions with specific promoter classes, sigma factors direct RNA polymerase holoenzymes to the transcription start sites of genes associated with particular promoters, depending on the needs of the cell. All sigma factors recognize specific variants of the −10 and −35 regions of the promoter—except for σ^{54}, which binds to sequences in the −12 and −24 regions. For example, when cells experience a sudden temperature increase or other environmental stress, RNA polymerase containing a σ^{32} (M_r 32,000) subunit binds to so-called heat shock promoters and enhances the transcription of heat shock genes. By using different σ subunits, the cell can coordinate the expression of sets of genes, permitting major changes in cell physiology.

Because different sigma factor proteins are similarly organized, researchers can exchange bits of sequence between them to examine the effect on promoter recognition and transcription activation. Carol Gross and her colleagues at the University of California, San Francisco, used this approach to dissect the mechanisms that distinguish between "housekeeping" sigma factors (σ^{70} class) and specialized sigma factors such as σ^{32}. Gross's work showed that σ^{32}, and other specialized sigma factors, have an altered 17 amino acid segment that reduces binding affinity for the promoter. As a result, these sigma factors require the exact consensus sequences at the −10 and −35 regions. In other words, σ^{70} is rather permissive in the promoters it will bind to—more deviations from the σ^{70} consensus are allowed— but σ^{32} is much more selective, binding only to promoters that have the exact σ^{32} consensus sequence. Since many promoters deviate from the consensus, specialized sigma factors bind only to the much smaller subset of promoters that contain the optimal consensus upstream sequences. Gross and colleagues found that converting the σ^{32} amino acid sequence to be the same as that found in housekeeping sigma factors such as σ^{70} decreased the requirement for −10 and −35 promoter conservation and increased transcription initiation at nonoptimal promoters.

Carol Gross [*Source: Photo by Rebecca Bartlett, Courtesy of Carol Gross.*]

Some sigma factors, such as σ^{38}, direct RNA polymerase to genes that respond to cellular stresses, including osmotic shock, temperature changes, and starvation; an example is the gene *osmY*. The process can be monitored using a combination of DNA "footprinting" and an assay for mRNA levels (**Figure 15-12a**). DNA footprint analysis can identify a region of DNA bound by a protein, such as a sigma factor or transcription factor (see this chapter's Moment of Discovery). In this

kind of experiment, DNA thought to contain sequences recognized by a DNA-binding protein is isolated and radiolabeled at one end (see Figure 15-12a, step 1). A DNA-binding protein is added to a sample of the DNA (step 2; in this case, σ^{38} and σ^{70} are the test DNA-binding proteins), and chemical or enzymatic reagents are used to cleave the DNA randomly in the samples with and without the DNA-binding protein (step 3), averaging one cut per molecule. When the sets of cleavage

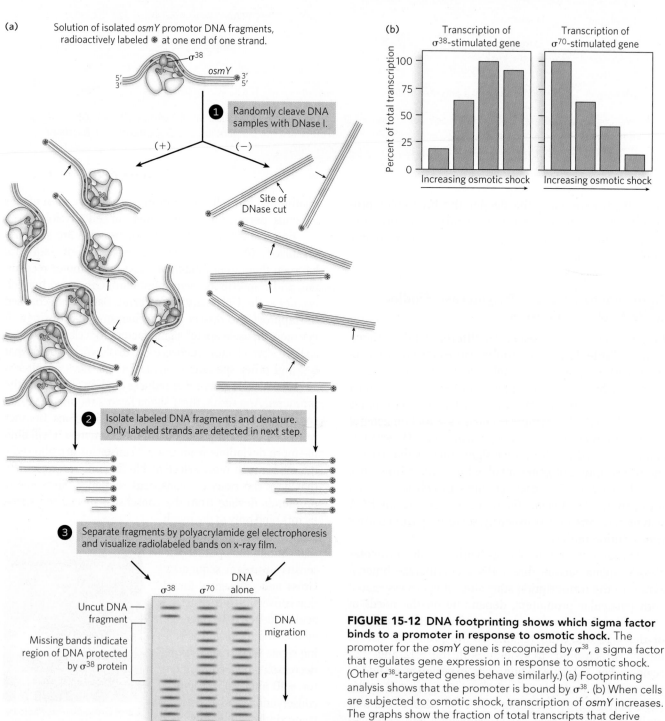

FIGURE 15-12 **DNA footprinting shows which sigma factor binds to a promoter in response to osmotic shock.** The promoter for the *osmY* gene is recognized by σ^{38}, a sigma factor that regulates gene expression in response to osmotic shock. (Other σ^{38}-targeted genes behave similarly.) (a) Footprinting analysis shows that the promoter is bound by σ^{38}. (b) When cells are subjected to osmotic shock, transcription of *osmY* increases. The graphs show the fraction of total transcripts that derive from either σ^{38} or σ^{70} promoters as a function of increasing osmotic shock. [Source: (b) Data from S. J. Lee and J. D. Gralla, Mol. Cell 14:153–162, 2004.]

products are compared side by side after separation by gel electrophoresis (step 4), a relatively uniform ladder of fragments appears for the sample not treated with the DNA-binding protein. A gap, or "footprint," in the ladder of DNA fragments in the protein-containing sample identifies the region of DNA that is bound by the protein and thus protected from cleavage. In the case of *osmY*, the promoter DNA can be bound by RNA polymerase with σ^{38}, but not σ^{70}. Monitoring of the transcription of the *osmY* gene reveals that as the osmotic strength of the culture medium is increased, the promoter for the *osmY* gene is activated to allow transcription, whereas a control promoter that uses σ^{70} (the *lacUV5* promoter) is not activated (**Figure 15-12b**). DNA footprinting is a valuable tool for mapping the binding sites of transcription factors and other DNA-binding proteins (see the How We Know section at the end of this chapter).

Structural Changes Lead to Formation of the Transcription-Competent Open Complex

The major type of bacterial RNA polymerase holoenzyme, as we have seen, contains a sigma factor of the σ^{70} class. The most common variant of the holoenzyme contains an unrelated sigma factor, σ^{54}, which is the sole representative of the σ^{54} class. The process of transcription initiation by σ^{70}-containing versus σ^{54}-containing polymerase holoenzymes is mechanistically distinct. In both cases, the holoenzyme binds to its promoter to form what is initially a closed complex, with the DNA maintaining its double-stranded structure; formation of this closed complex is readily reversible. In complexes with the σ^{70}-class factor, the closed complex can spontaneously convert to a transcription-competent open complex, in a process of isomerization (**Figure 15-13a**). In contrast, σ^{54}-containing holoenzymes require specialized activator proteins of the AAA+ family (see Chapters 5 and 11) to catalyze conversion to the open complex, with concomitant ATP hydrolysis (**Figure 15-13b**). Upstream activator sequences (UASs) are brought in contact with promoter-bound σ^{54}-RNA polymerase through DNA looping, a step that is often facilitated by auxiliary DNA-binding and bending proteins such as integration host factor (IHF). In both cases, energetically favorable conformational changes in the RNA polymerase accompany an opening of the DNA duplex, exposing the template and coding strands in the -11 to $+3$ region. In contrast to the σ^{70} open complex, formation of the σ^{54} open complex is irreversible and ensures that transcription will initiate.

To understand the structural basis for closed-to-open complex isomerization, it is helpful to examine the molecular structure of the RNA polymerase holoenzyme. In the open complex, several channels provide access to the core of the enzyme (**Figure 15-14** on p. 533). One channel enables rNTPs to enter the catalytic site,

and another allows the growing RNA polynucleotide to exit the enzyme during the elongation phase of transcription. A third channel provides space for DNA to enter the catalytic center in double-stranded form, between the two pincers of the claw-shaped complex. The two strands separate and are held apart by a cleft, or pin, in the polymerase structure that helps keep the transcription bubble open within the enzyme. But by the time the DNA exits the RNA polymerase, it is duplex DNA again.

Two significant structural changes result from the conversion of the closed to the open complex. First, the pincers close around the DNA downstream from the transcription start site. Second, the negatively charged N-terminus of the sigma factor moves from the active-site cleft of the polymerase (where the chemical reaction occurs) to an exterior position 50 Å away, allowing the DNA template strand to take its place. In the open position, the RNA polymerase is ready to begin RNA synthesis.

Initiation Is Primer-Independent and Produces Short, Abortive Transcripts

In contrast to DNA polymerases, which require an oligonucleotide base-paired with the template strand in order to prime DNA synthesis, RNA polymerases can begin transcription with a single nucleotide—they do not depend on a preexisting strand from which to extend new RNA. RNA polymerases must therefore bind and hold two nucleotides in place on the DNA template for long enough, and in the correct orientation, to catalyze phosphodiester bond formation between them. Once this occurs, and for the first 8 to 10 phosphodiester bonds formed, there is a high probability that the polymerase will release the transcript from the template without extending it further. If this happens, the assembled polymerase holoenzyme begins RNA synthesis again on the same template. This process is called **abortive initiation** (**Figure 15-15** on p. 533). Occasionally, the polymerase holds on to the transcript long enough to extend it beyond 10 nucleotides, at which point the RNA becomes stable. After a successful initiation, transcription enters the elongation phase and continues along the DNA template until it reaches a termination signal.

Interestingly, these two properties—beginning without a primer and initially producing abortive transcripts—are universal characteristics of DNA-dependent RNA polymerases. Even the single-subunit bacteriophage polymerases exhibit these properties. The molecular structures of the phage T7 RNA polymerase and the bacterial RNA polymerase suggest explanations for both of these characteristics. Abortive initiation seems to occur because, early in initiation, the RNA exit channel (see Figure 15-14) is blocked—either by part of the polymerase itself, in the case of T7 polymerase, or by part of the sigma factor. For a transcript to extend beyond about

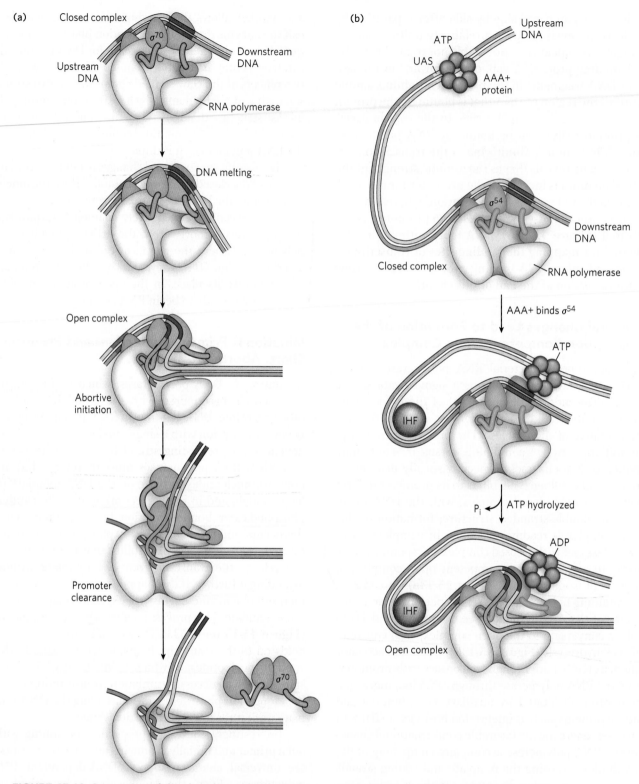

FIGURE 15-13 Conversion of the closed complex to the open complex by sigma factors. (a) σ^{70} induces strand opening and formation of the open complex, without the need for ATP or other factors. All other sigma factors except σ^{54} function in the same way. (b) σ^{54} requires an AAA+ protein to form the open complex at a promoter. The AAA+ protein is a hexamer that binds an upstream activator sequence (UAS), then also binds σ^{54} in the closed complex, creating a DNA loop. Such DNA looping is often facilitated by auxiliary DNA-binding and bending proteins such as integration host factor (IHF). AAA+ hydrolyzes a bound ATP, using the favorable energetics of this reaction to drive opening of the complex.

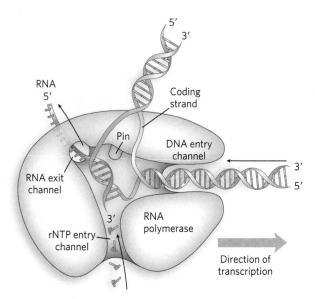

FIGURE 15-14 RNA polymerase channels. Distinct channels in RNA polymerase allow the DNA to enter as double-stranded DNA and to peel apart within the polymerase so that 8 bp form between the template strand and the growing RNA transcript. Two other channels provide entry for rNTPs and an exit for the transcript.

10 nucleotides, a structural transition must take place to unblock the RNA exit channel, a process that occurs only occasionally during the initial stage of transcription. In bacteria, this process may weaken the affinity of the sigma factor within the polymerase complex, explaining why the σ subunit often falls off the complex during elongation of the transcript.

Transcription Elongation Is Continuous until Termination

Once RNA polymerase enters the elongation phase, the enzyme does not release the DNA template until it encounters a termination sequence. In this mode the polymerase is said to be processive, moving smoothly along the template, synthesizing the complementary RNA strand and dissociating only when the transcript is complete. During transcript elongation, the DNA moves through the polymerase active site, as observed in the polymerase open complex (see Figure 15-14). The strands of the DNA double helix separate just before the site of catalysis, held apart by a structural protrusion within the polymerase, then re-form as a double helix on exiting the polymerase interior. At any given time during elongation, 8 to 9 nucleotides of the RNA transcript remain base-paired with the DNA template, while the rest of the transcript is stripped off and directed out through the RNA exit channel.

During elongation, the polymerase attempts to ensure the accuracy of transcription by **pyrophosphorolysis**, in which the catalytic reaction runs in reverse whenever the polymerase stalls along the DNA. This process, known as **kinetic proofreading**, works because the polymerase tends to stall after incorporating a mismatched base into the growing RNA chain, thus enabling pyrophosphorolysis to remove the incorrect base (**Figure 15-16a**). Pyrophosphorolysis is also used in proofreading during DNA synthesis (see Chapter 11).

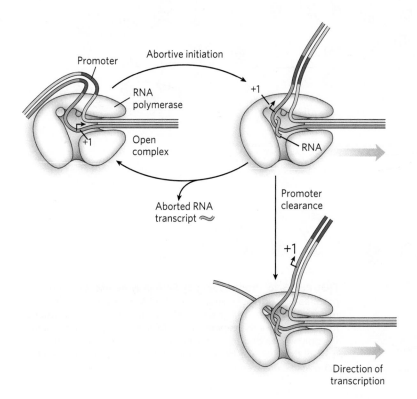

FIGURE 15-15 Abortive initiation. RNA polymerase undergoes a cycle of abortive initiation in which short RNA transcripts are synthesized and released, until the promoter site is cleared. The nascent transcript is then in position in the RNA exit channel, and initiation is successfully completed.

(a) Kinetic proofreading

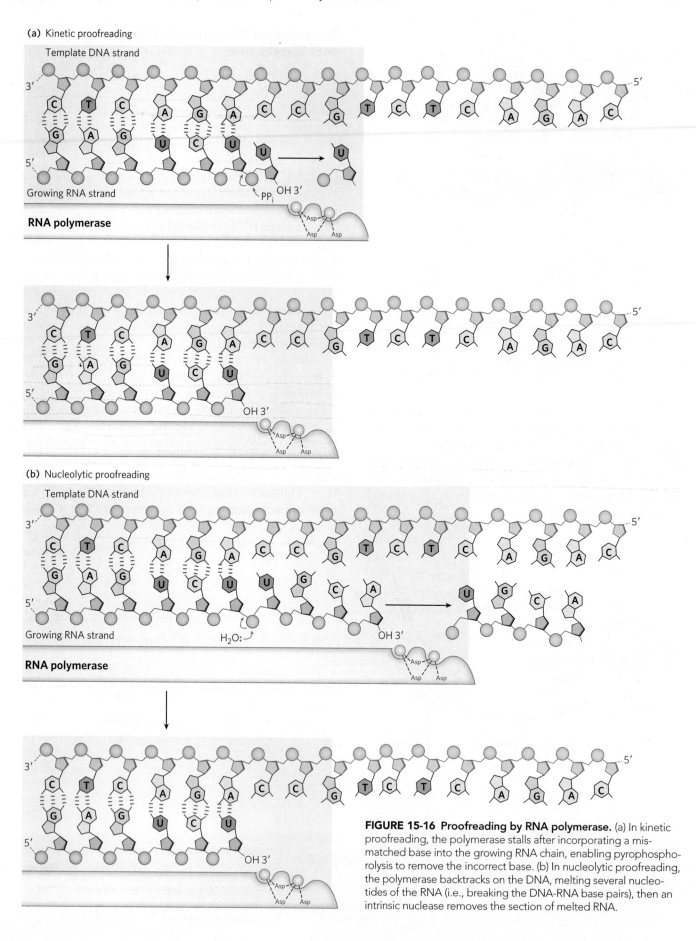

FIGURE 15-16 Proofreading by RNA polymerase. (a) In kinetic proofreading, the polymerase stalls after incorporating a mismatched base into the growing RNA chain, enabling pyrophosphorolysis to remove the incorrect base. (b) In nucleolytic proofreading, the polymerase backtracks on the DNA, melting several nucleotides of the RNA (i.e., breaking the DNA-RNA base pairs), then an intrinsic nuclease removes the section of melted RNA.

RNA polymerase contains an inherent nuclease activity, providing another means of correcting newly synthesized transcripts by hydrolysis. In this process, known as **nucleolytic proofreading**, the polymerase reverses direction by one or a few nucleotides and breaks the RNA phosphodiester bond upstream from a mismatched base, removing the error-containing strand (**Figure 15-16b**).

Despite these proofreading mechanisms, RNA polymerases are generally less accurate than DNA polymerases. Whereas, on average, DNA polymerase makes an error only once per 10^6 nucleotides incorporated, RNA polymerase makes an error about once per 10^4 to 10^5 nucleotides added. This can be tolerated because most transcripts are made in many copies, and one or two error-containing transcripts will be vastly outnumbered by correct transcripts in the cell. In RNA viruses, error-prone transcription can be a survival advantage, enabling the emergence of mutants that can escape detection by host cell immunity.

Bacterial RNA polymerase can synthesize RNA at a rate of 50 to 90 nucleotides per second. Rather than transcribing DNA at a constant pace, however, the enzyme pauses at times throughout the process. Pausing can be detected using single-molecule methods (see the How We Know section at the end of this chapter). Backtracking is a mechanism of pausing in which the RNA polymerase moves backward and peels the 3′ end of the RNA off the DNA template by inserting the end into the rNTP entry channel. This temporarily blocks further movement of the polymerase. Transcription restarts when the enzyme cleaves the peeled-back 3′ end of the transcript, using its intrinsic nuclease activity. This process differs from nucleolytic processing in that there is a time lag, or pause, between the peeling-back process and the cleavage of the RNA strand to restart transcription. Pausing has several possible functions, including providing time for the RNA transcript to fold properly and to be translated synchronously with transcription.

Specific Sequences in the Template Strand Stop Transcription

Transcription stops when the RNA polymerase transcribes through certain sequences in the DNA template. At this point, the RNA polymerase releases the finished transcript and dissociates from the DNA template. *E. coli* DNA has at least two classes of such **termination sequences,** one class that relies primarily on structures that form in the RNA transcript and another that requires an accessory protein factor called rho (ρ).

Most ρ-independent termination sequences have two distinguishing features. The first is a region that produces an RNA transcript with self-complementary sequences, permitting the formation of a hairpin structure centered 15 to 20 nucleotides before the projected end of the RNA strand, as shown in **Figure 15-17a**. (Note that these regions occur only once, at the end of the transcript, whereas the pause sites described above occur at multiple places within the transcript.) The second feature is a highly conserved segment of three A residues in the template strand that are transcribed into U residues near the 3′ end of the hairpin. When a polymerase arrives at such a termination sequence, it stalls. Formation of the hairpin in the newly transcribed RNA disrupts several A=U base pairs in the RNA-DNA hybrid segment and may disturb important interactions between RNA and RNA polymerase, leading to dissociation of the transcript.

The ρ-dependent terminators lack the sequence of repeated A residues in the template strand but typically include a CA-rich sequence called a *rut* (rho *utilization*) site. The ρ factor, a hexameric helicase, binds to RNA polymerase very early in the transcription process. Its ATP-hydrolyzing activity is used to feed the mRNA through the pore in the center of the rho complex during transcription until a rut site is encountered (**Figure 15-17b**). Here, it contributes to release of the RNA transcript from both the DNA template and the polymerase.

SECTION 15.2 SUMMARY

- Transcription begins at specific promoter sequences upstream from the coding sequence in the DNA template. A sigma factor, of which there are several classes in bacteria, binds to the polymerase holoenzyme and recruits it to a particular type of promoter, enabling transcription at subsets of genes in response to environmental stimuli and the needs of the cell.

- RNA polymerase first forms a closed complex on promoter DNA, a readily reversible state that is not yet capable of transcription.

- On conversion of the closed complex to a transcription-competent open complex, either through spontaneous isomerization or by ATP-dependent conformational change, RNA polymerase begins RNA synthesis without requiring a primer.

- Transcription initiation requires promoter clearance, in which the RNA polymerase moves beyond the promoter region of the DNA to begin rapid elongation of the transcript.

- During elongation, the RNA polymerase is highly processive, synthesizing transcripts without dissociating from the DNA template.

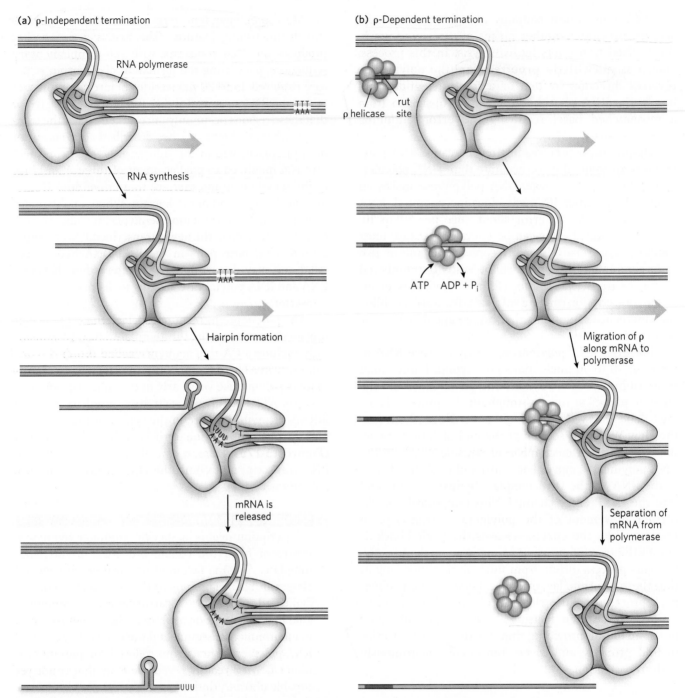

(a) ρ-Independent termination

(b) ρ-Dependent termination

FIGURE 15-17 Termination of transcription. (a) In ρ-independent termination, an mRNA sequence forms a hairpin, which is followed by three U residues, stalling the polymerase and separating it from the mRNA. (b) RNAs that include a rut site (purple) recruit the ρ helicase, which migrates in the 5'→3' direction along the mRNA and separates it from the polymerase.

- RNA polymerase corrects errors in newly synthesized transcripts in one of two ways. In kinetic proofreading, stalling of polymerase after a mismatched base is incorporated into the growing RNA chain enables pyrophosphorolysis to remove the incorrect base. In nucleolytic proofreading, the polymerase reverses direction by one or a few nucleotides and hydrolyzes the RNA phosphodiester bond upstream of a mismatched base, removing the error-containing strand.

- Termination occurs when the polymerase transcribes through certain DNA sequences, in a process that sometimes requires an accessory factor called rho (ρ).

15.3 TRANSCRIPTION IN EUKARYOTES

In eukaryotic cells, three distinct RNA polymerases—Pol I, II, and III—carry out DNA-dependent synthesis of RNA. Although the properties of these polymerases resemble those of bacterial RNA polymerase in many ways, the eukaryotic polymerases require many additional proteins, called **transcription factors**, to begin efficient transcription at promoter sequences. These factors help assemble transcription complexes on chromatin, the compacted form of DNA that makes up eukaryotic genomes. Like bacterial sigma factors, each eukaryotic transcription factor binds to a specific promoter sequence and to a particular RNA polymerase, bridging the two to initiate transcription. Using a variety of general transcription factors, eukaryotic cells promote the transcription of many sets of genes under varying conditions (**Figure 15-18**). Specific transcription factors bind DNA at a long distance upstream from the promoter, at sequences known as enhancers, and can stimulate or repress transcription in various ways. Transcription factors, both general and specific, have important roles in gene regulation and cell development. Indeed, recent studies show that differentiated cells, once believed to be committed to a particular cell type, can be converted to another cell type simply by manipulating the expression of transcription factors (**Highlight 15-2**). Transcription regulation proteins, including transcription factors, and the DNA sites to which they bind, including enhancers, are discussed in detail in Chapters 19–22.

Because transcription is a fundamental process in all cells, it is not surprising that some eukaryotic RNA polymerase subunits are homologous to those of bacterial polymerase. Furthermore, some subunits are common to all three of the eukaryotic polymerases (see Table 15-1). Relative to bacteria, eukaryotes require additional factors to help RNA polymerases find and access promoters in the cell nucleus. This is because eukaryotic DNA is packaged into chromatin through the formation of nucleosomes (see Chapter 10). In addition, the sheer size of eukaryotic genomes, and the large number of promoters to be sorted through, probably requires additional transcription machinery.

We begin with a brief discussion of all three eukaryotic polymerases and their promoters, and then focus on Pol II transcription. As the polymerase responsible for transcribing the genes that encode proteins, Pol II is the most extensively studied of the three eukaryotic polymerases.

Eukaryotic Polymerases Recognize Characteristic Promoters

Each of the three types of RNA polymerase that make up the eukaryotic transcription machinery transcribes only certain classes of genes, and thus each type binds to specific and distinct promoter sequences. Pol I binds to a single type of promoter that controls the expression of the pre-ribosomal RNA (pre-rRNA) transcript, from which rRNAs are derived. Pol II, which synthesizes mRNAs, microRNAs, and some other noncoding RNAs, can recognize thousands of promoters that vary greatly in sequence. Pol III recognizes well-characterized promoter sequences for tRNAs, the 5S rRNA, and some other small regulatory RNAs, sequences that in many cases are located *within* the transcribed region itself rather than in more conventional locations upstream from the RNA start site.

Although each polymerase works with its own unique set of transcription factors, all three types use a factor called the **TATA-binding protein (TBP)**. This protein, so-named because of its binding to a 5′-TATAAA sequence (known as the TATA box) near position −30, plays a major role in transcription initiation. Genomic sequencing studies have shown that only about a quarter of human genes include a TATA box in the core promoter, the region responsible for recruiting the essential transcription machinery. Nonetheless, TBP is used for transcription initiation of all genes, and in most of those that lack a TATA box, TBP is recruited to the gene

Cell-extrinsic (*i.e.*, endocrine, metabolic and osmotic systems)

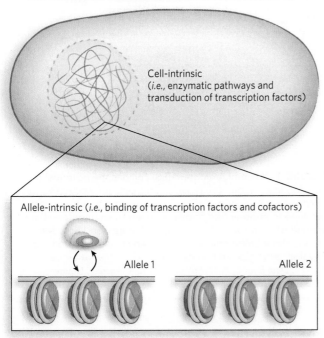

FIGURE 15-18 Multiple levels of control of gene expression. Cells respond to external signals (cell-extrinsic), then to intracellular signaling pathways (cell-intrinsic), leading to effects on the transcription of specific genes (allele-intrinsic). [*Source: Data from T. C. Voss and G. L. Hager,* Nat. Rev. Genet. *15:69-81, February 2014.*]

Using Transcription Factors to Reprogram Cells

Patterns of gene transcription largely control how cells develop into specific cell types. This process is of great importance in medicine, because the possibility of reprogramming cells to carry out specific functions could revolutionize the treatment of patients with degenerative diseases. Experimental attempts to reprogram cells began several decades ago with the discovery that engineering of an oocyte (egg cell) to contain the nucleus of an adult cell can cause the nucleus to revert to an undifferentiated state. This process, called somatic cell nuclear transfer (SCNT), can produce an embryo and embryonic stem cells with the genetic makeup of an adult cell. Presumably, these results come about through the reprogramming of transcription in the composite cells.

This idea was tested and validated in 2006, when researchers found that fibroblasts can be induced to undergo a dramatic cell-fate reversal to an undifferentiated state, becoming what are known as induced pluripotent stem cells (iPS cells), by transiently expressing four master-regulatory transcription factors in the fibroblasts. The next step was to see whether fibroblasts might be more generally susceptible to reprogramming into different kinds of cells—if the right set of transcription factors could be identified.

To test this possibility, Marius Wernig and his colleagues at Stanford University set out to convert mouse fibroblasts into neurons. Reasoning that multiple transcription factors were probably necessary to reprogram fibroblasts to a neuronal fate, the researchers cloned 19 genes that encode transcription factors expressed specifically in neural tissues or function during neural development. The genes were cloned into lentiviruses, viral vectors that could be used to introduce the genes into mouse fibroblasts by infection. To detect changes in cell fate, the researchers used fibroblasts derived from mouse embryos and tail tips of newborn or adult mice that had been genetically altered to express a green fluorescent protein marker when the gene for the protein Tau was turned on. Because the Tau gene is specifically expressed in neurons, cells that had acquired at least this property of neurons—transcription of the Tau gene—could be easily identified.

When all 19 of the candidate transcription factors were introduced into the fibroblasts, some of the cells turned green. By a process of elimination, the researchers eventually found that a combination of only three transcription factors was sufficient to convert fibroblasts

into neurons (Figure 1). These factors—Ascl1, Brn2, and Myt1l (or Zic)—caused cells to express a variety of neuronal markers and become capable of firing action potentials, a basic function of neurons. Furthermore, when cultured together with bona fide mouse neuronal cells, the reprogrammed cells received both excitatory and inhibitory synaptic connections from the mouse neurons, and could form synapses with each other.

Beyond its implications for understanding transcriptional activation and regulation, this discovery offers the intriguing possibility of creating cell types at will. If such transcription-based reprogramming proves feasible in human cells, it could be used to generate neurons that mimic particular disease states for use in drug development. Researchers are now eager to find out how many different cell types can be produced by activating distinct combinations of lineage-specific transcription factors.

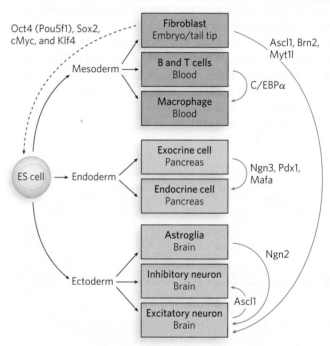

FIGURE 1 An embryonic stem (ES) cell has the potential to develop into various cell types (solid black arrows). Activation of specific transcription factors can convert one differentiated cell type to another (red arrows) or even to an undifferentiated state (dashed arrow). Mesoderm, endoderm, and ectoderm are the three embryonic layers from which all cells and tissues develop; astroglia are non-neuronal cells of nerve tissue. [Source: Data from C. R. Nicholas and A. R. Kriegstein, Nature 463:1031–1032, 2010, Fig. 1.]

through proteins called TBP-associated factors (TAFs) that recognize other promoter sequences. A summary of the eukaryotic polymerases and the types of RNA produced by each, along with their promoter elements, is given in **Table 15-3**.

RNA Polymerase I Promoters Synthesis of pre-rRNA accounts for 80% of all the transcription in eukaryotic cells (as measured for the yeast *S. cerevisiae*). The precursor rRNA transcript produced by mammalian Pol I is processed into the mature 5.8S, 18S,

TABLE 15-3

Eukaryotic RNA Polymerases and Promoter Elements

Polymerase	RNA Products	Promoter Elements
Pol I	18S, 25S, and 5.8S rRNAs	UCE, core sequence
Pol II	mRNA, microRNAs, some noncoding RNAs	BRE, TATA box, Inr, DPE
Pol III	tRNA	Box A, Box B
	5S rRNA	Box A, Box C
	7SL RNA	TATA box

and 28S rRNAs, which, together with the 5S rRNA transcribed by Pol III, are the major catalytic and architectural components of the ribosome (discussed in Chapter 18). Transcription of rRNA genes, which occurs in the nucleolus, begins with the recruitment and assembly of Pol I and transcription factors into a multiprotein complex at the rRNA gene promoter. The promoter includes a core sequence, essential for accurate transcription initiation, and an upstream control element (UCE), located 100 to 150 bp upstream from the transcription start site (**Figure 15-19**).

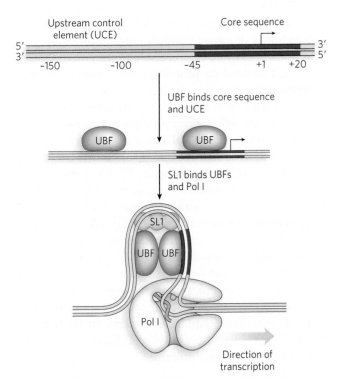

FIGURE 15-19 The Pol I promoter. The upstream binding factor (UBF) binds to the core sequence and the upstream control element (UCE). SL1, a protein complex that includes the TATA-binding protein, binds to UBF and Pol I, promoting transcription initiation.

The number of rRNA genes varies among organisms. Pol I promoter sequences also vary, but within a species, all Pol I promoters are the same. Low levels of transcription can be observed in the presence of a preinitiation complex comprising Pol I and selectivity factor 1 (SL1), which is a complex of TBP and three TAFs. Higher levels of transcription require, in addition to Pol I and SL1, an upstream binding factor (UBF). UBF binds to both the UCE and the core promoter and to SL1, stabilizing the complex with Pol I and helping recruit the polymerase to the promoter.

RNA Polymerase II Promoters Many Pol II promoters share certain sequence features, including a TATA box near −30 and an initiator sequence (Inr) near the RNA start site at +1 (**Figure 15-20**). Pol II promoters also sometimes include a sequence upstream from the TATA box, called a TFIIB recognition element (BRE), and a sequence downstream from the initiator, the downstream promoter element (DPE). These sequences comprise the **core promoter**. Other sequences are also needed for efficient Pol II recognition and transcription in the cell, such as upstream promoter elements and enhancers. These **regulatory sequences**, which can be located many thousands of base pairs away from the promoter they influence, bind a variety of specific transcription factors that either activate or repress transcription, depending on various stimuli. The proteins that bind to the elements in the promoter are discussed shortly.

KEY CONVENTION

The nomenclature for transcription factors indicates which RNA polymerase is involved. TFII is a transcription factor for RNA polymerase II, and TFIII is a transcription factor for RNA polymerase III. Individual factors are distinguished by an appended A, B, C, and so on (e.g., TFIIA, TFIIIB).

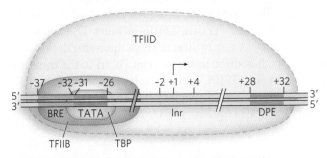

FIGURE 15-20 The Pol II core promoter. The TATA box, initiator sequence (Inr), or other sequence elements recognized by proteins that bind to the polymerase are required for transcription by Pol II. The TFIIB recognition element (BRE) and downstream promoter element (DPE) may also be involved in initiation.

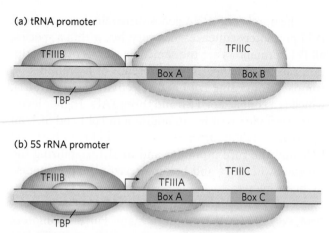

(a) tRNA promoter

(b) 5S rRNA promoter

FIGURE 15-21 Pol III promoters. Pol III promoters are found within genes. (a) The Pol III tRNA promoter uses the Box A and Box B sequence elements and is bound by transcription factors TFIIIB and TFIIIC. (b) The Pol III 5S rRNA promoter uses Box A and Box C, as well as TFIIIB, TFIIIC, and TFIIIA. Together, these factors recruit Pol III to the transcription start site.

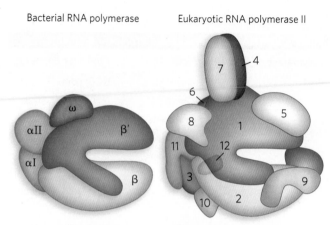

Bacterial RNA polymerase   Eukaryotic RNA polymerase II

FIGURE 15-22 Bacterial RNA polymerase and eukaryotic Pol II structural elements. Although Pol II has more subunits with additional components, it has obvious structural similarities to bacterial RNA polymerase. The numbers on the Pol II subunits indicate RPB1, RPB2, and so forth.

RNA Polymerase III Promoters Pol III is the largest RNA polymerase with the greatest number of subunits. All of its transcription products are short, untranslated RNAs, most less than 300 nucleotides long. In addition to 5S rRNA, they include tRNAs; 7SL RNA, which is required for introducing proteins into membranes as part of the signal recognition particle (see Chapter 18); and several RNAs involved in mRNA, tRNA, and rRNA processing. Perhaps reflecting their varied gene products, Pol III promoters differ in sequence and in components. The promoters of tRNA genes include two segments, Box A and Box B, located a short distance apart *within* the tRNA-coding sequence (**Figure 15-21a**). The 5S rRNA gene promoter includes Box A and Box C (**Figure 15-21b**). Other promoters contain the TATA box to which TBP can bind directly, just as for Pol II promoters.

Like the other eukaryotic polymerases, Pol III requires transcription factors. The tRNA genes require TFIIIB and TFIIIC, whereas the 5S rRNA gene requires TFIIIB, TFIIIC, and TFIIIA. Transcription of tRNA genes begins when TFIIIC binds to the promoter boxes within the gene, and then recruits TFIIIB. TFIIIB includes TBP and recognizes the DNA just upstream from the transcription start site. Together, these factors recruit Pol III to the transcription start site; TFIIIC is transiently displaced as the polymerase transcribes through its binding site in the DNA. In 5S rRNA transcription, TFIIIA binds to the DNA within the transcribed region and helps recruit TFIIIC.

Pol II Transcription Parallels Bacterial RNA Transcription

Pol II–catalyzed transcription is responsible for producing all mRNAs in the eukaryotic cell, as well as transcripts, such as microRNAs, that can base-pair with mRNAs

and help regulate their expression (see Chapter 22). Consisting of 12 subunits, Pol II is strikingly more complex than its bacterial counterpart, yet it has remarkable similarities in structure, function, and mechanism (**Figure 15-22**). The largest subunit (RPB1) exhibits a high degree of homology to the β′ subunit of bacterial RNA polymerase. Another subunit (RPB2) is structurally similar to the bacterial β subunit, and two others (RPB3 and RPB11) show some structural homology to the two bacterial α subunits (see Table 15-1). Pol II must function with genomes that have multiple chromosomes and with DNA molecules more elaborately packaged than those in bacteria. The need for protein-protein interactions with the numerous other protein factors required to navigate this labyrinth largely accounts for the added complexity of Pol II and the other eukaryotic polymerases.

An overview of the Pol II transcription complex is shown in Figure 15-23. Playing a role much like that of sigma factors in helping bacterial RNA polymerase recognize and bind promoter sequences, general transcription factors associate with promoter DNA and recruit Pol II to form a **preinitiation complex (Figure 15-23a, step 1)**. The preinitiation complex is converted to an initiation complex by unwinding the DNA (step 2). During initiation (step 3), the C-terminal domain (CTD) of Pol II is phosphorylated and some transcription factors are released (**Figure 15-23b**). Elongation (step 4) proceeds as in bacteria. Transcription is terminated (step 5) and the Pol II CTD is dephosphorylated. Each step is associated with characteristic proteins.

Transcription Factors Play Specific Roles in the Transcription Process

The transcription initiation mechanism has been most extensively studied for Pol II. Recruitment begins with

(a) Transcription at Pol II promotors

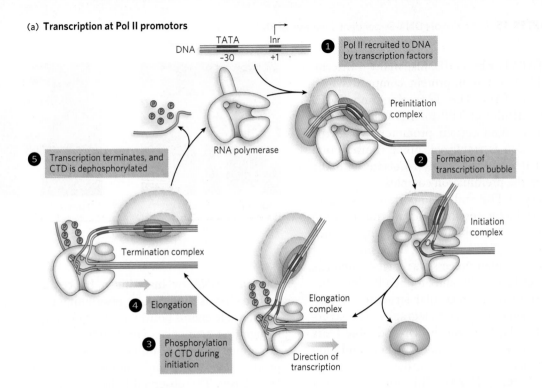

(b) Transcription at Pol IIA promotors

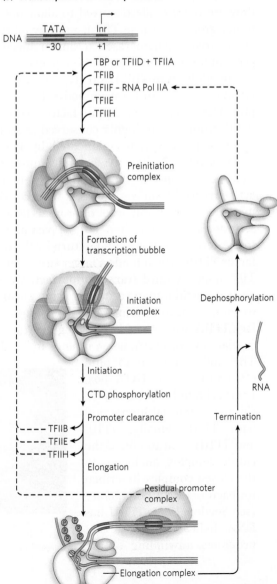

FIGURE 15-23 Transcription at Pol II promoters. (a) The phases of transcription by Pol II—assembly, initiation, elongation, and termination—are associated with characteristic proteins, as described in the text. The ordered assembly and dissociation of these factors drives the process forward. (b) After association of transcription factors with Pol II on DNA to form an open initiation complex, some factors dissociate to enable transcription elongation, termination, and recycling.

binding of the TATA box by TFIID. Like many transcription factors, TFIID is a multiprotein complex, which includes TBP and TAFs. The TAFs fine-tune TFIID by changing the affinity of TBP for DNA, helping the transcription factor bind certain promoters. Once the TBP-DNA interaction is stabilized, other transcription factors, and Pol II itself, can stably associate with the promoter to form the preinitiation complex.

The discovery of TBP and its importance as a general transcription factor required for all transcription by Pol II raised questions about how and why it binds so specifically to the TATA element. This mystery of the transcription initiation process was solved when the research groups of Paul Sigler and Stephen Burley independently determined the molecular structure of TBP bound to DNA, using x-ray crystallography. The structure revealed that TBP sits on the DNA double helix much like a saddle, with an extended β sheet and loop "stirrups" in contact with the minor groove of the TATA box sequence (**Figure 15-24**). This unconventional mode of DNA recognition—most DNA-binding proteins

Paul Sigler, 1934–2000
[Source: Courtesy Michael Marsland/Yale University.]

Stephen Burley [Source: Courtesy Rockefeller University.]

recognize DNA by inserting α helices into the major groove—bends the DNA by positioning two pairs of Phe residue side chains between base pairs at each end of the recognition sequence. The bending opens and widens the minor groove, enabling hydrogen bonding between protein side chains and the minor-groove edges of the DNA bases. The observed helical bending explains why A=T base pairs are favored in the recognition sequence: they are more easily distorted to allow opening of the minor groove. Because TBP is used by all three classes of eukaryotic polymerases, a similar mechanism may account for promoter recognition in all cases.

In addition to TBP, Pol II requires an array of transcription factors to form an active transcription complex. The general transcription factors required at every Pol II promoter are highly conserved in all eukaryotes. Using cell-free systems pioneered by Robert Roeder at Rockefeller University, in which purified proteins were added back to the reaction mix to reconstitute active transcription complexes, it was possible to determine the identity and order of proteins needed for transcription initiation. When TBP, as part of TFIID, binds to the TATA box, it is bound in turn by the transcription factor TFIIB, which binds a larger site on the DNA than TBP alone. A third transcription factor, TFIIA, is not always essential in experiments using purified proteins to monitor transcription. However, mutant cells that lack TFIIA are not viable, showing that this factor is essential in vivo. TFIIA, when it binds to TFIID, unmasks TBP and enables it to bind efficiently to the TATA box. Pol II is bound to the complex through a mutual interaction with TFIIF. Finally, TFIIE and TFIIH bind to create the closed complex, analogous to the closed complex described for bacterial RNA polymerase (see Section 15.2). TFIIH has DNA helicase activity that promotes unwinding of the

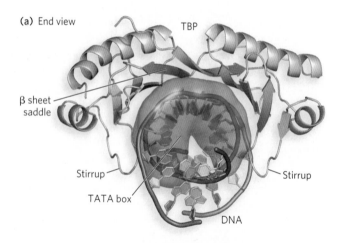

(a) End view

TBP

β sheet saddle

Stirrup

Stirrup

TATA box

DNA

(b) Side view, TBP removed

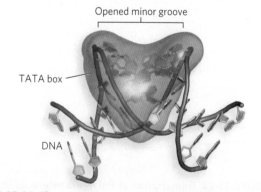

Opened minor groove

TATA box

DNA

FIGURE 15-24 The crystal structure of a TBP-DNA complex. TATA-binding protein (TBP) bends the TATA box sequence, opening the minor groove to allow sequence-specific hydrogen bonding. [Source: PDB ID 1CDW.]

Robert Roeder [Source: Zach Veilleux/Rockefeller University.]

DNA near the transcription start site. This unwinding creates an open complex that is competent to begin transcription. Counting all the subunits of the various essential factors (excluding TFIIA), this minimal active assemblage has more than 30 polypeptides!

Once assembled on a promoter, Pol II typically produces a few abortive transcripts before entering the elongation phase of transcription—behavior similar to that of bacterial RNA polymerase. In contrast to bacterial polymerase, however, Pol II must be chemically modified by the addition of phosphate groups to its CTD to disengage from the promoter and begin elongating a transcript.

The CTD, part of the largest polymerase subunit, consists of multiple repeats of the seven amino acid sequence Tyr-Ser-Pro-Thr-Ser-Pro-Ser. In yeast and animals, the CTD mainly functions as a docking platform to recruit transcription and processing factors during the transcription cycle. CTD-associated factors have a variety of functions, including to catalyze mRNA 5′ capping and 3′-end processing, pre-mRNA splicing, and histone modification. To recruit all these different proteins, the CTD uses distinct chemical codes. Reversible phosphorylations of the serines in the second and fifth positions (Ser^2 and Ser^5) of the CTD repeat sequence are the primary CTD codes and are crucial for regulating transcription and the binding of mRNA processing factors. The protein enzymes responsible for these phosphorylations, the CTD kinases, are conserved from yeast to metazoans. To enable recruitment of other kinds of proteins, the CTD adopts additional modifications, including phosphorylation of Tyr^1, Ser^7, and Thr^4, as well as cis-trans isomerization of Pro^3 and Pro^6, in the CTD repeat sequence.

TFIIE and TFIIH are released during synthesis of the initial 60 to 70 nucleotides of RNA, as Pol II enters the elongation phase of transcription. Notably, phosphorylation of Pol II CTD also influences downstream processing of the RNA transcript, providing a mechanism for coupling transcription to RNA splicing and intracellular transport (as discussed in Chapter 16).

In the elongation phase, polymerase activity is greatly enhanced by elongation factors. They suppress pausing during transcription, enhance polymerase editing of misincorporated bases by hydrolysis (as for bacterial RNA polymerase), and recruit protein complexes involved in posttranscriptional processing of the mRNA. Once the RNA transcript is completed, transcription is terminated. In eukaryotes, termination is often triggered by endonucleases that recognize and cleave specific sequences in the newly synthesized RNA, leading to disassembly and dissociation of the transcription complex. Pol II is then dephosphorylated and recycled, readying it to initiate another transcript (see Figure 15-23).

Transcription Initiation In Vivo Requires the Mediator Complex

As we have seen, the initiation and control of eukaryotic mRNA synthesis requires a large set of evolutionarily conserved general transcription factors that function at most, if not all, genes. These include initiation factors TFIIB, TFIID (which includes the TATA-binding protein, TBP), TFIIE, TFIIF, and TFIIH—which comprise the minimal set of helper proteins necessary and sufficient for in vitro selective binding and accurate transcription initiation by Pol II from core promoters. In vivo, in yeast cells, the multiprotein **Mediator complex** is also required for the regulated transcription of nearly all Pol II–dependent genes. Its presence also in humans implies a similar central role in Pol II–catalyzed transcription.

Mediator functions as an intermediary between specific transcription factors bound at upstream promoter elements or enhancers and the Pol II complex and general initiation factors bound at the core promoter (**Figure 15-25a**). First discovered and purified from yeast by Roger Kornberg and his colleagues, Mediator was found to be required for transcriptional activation by specific activators in vitro, using a reconstituted enzyme system containing purified Pol II and general initiation factors. Yeast Mediator has 20 subunits in three distinct subdomains, referred to as the head, middle, and tail modules (**Figure 15-25b**). An additional module, which includes a kinase enzyme complex, is associated with a subset of yeast Mediator complexes. The presence of the kinase corresponds to repression of a subset of genes, suggesting a role for Mediator in transcriptional down-regulation as well as in activation.

Human Mediator contains a set of consensus subunits similar to those in yeast. As in yeast, multiple forms of Mediator seem to function differently in the transcriptional control of different sets of genes. In particular, the kinase module can exert a repressive effect when associated with the mammalian Mediator, whereas other auxiliary proteins are associated with an activating form of Mediator.

The mechanisms by which Mediator complexes control mRNA synthesis involve direct interactions with DNA-binding transcription activators bound at upstream promoter elements and enhancers, interactions with Pol II, and interactions with one or more of the general initiation factors bound at the core promoter. Mediator supports transcriptional activation, at least in part, by increasing the rate and/or efficiency of assembly of the Pol II preinitiation complex. Mammalian Mediator complexes influence several steps during this assembly, including the recruitment of TFIID (or TBP), Pol II, and the other general initiation factors to the core promoter.

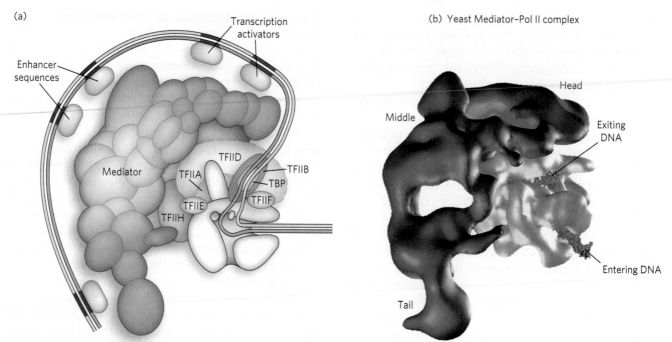

(a)

(b) Yeast Mediator–Pol II complex

FIGURE 15-25 The Mediator complex. (a) Mediator helps to bridge distant proteins bound to enhancer sequences and Pol II and its general transcription factors, bound near the transcription start site. (b) Mediator bound to Pol II. The Mediator complex consists of 20 proteins and has three subdomains. [*Source: (b) Data from J. A. Davis et al., Mol. Cell 10:409–415, 2002.*]

Termination Mechanisms Vary among RNA Polymerases

The three eukaryotic RNA polymerases use different strategies for terminating transcription, although these mechanisms have some aspects in common. The Pol III and Pol I termination pathways seem to be simpler than the Pol II pathway. Pol III terminates transcription at T-rich sequences in the DNA template located a short distance from the 3′ end of the mature RNA, assisted by just a few protein factors. Pol I terminates at a terminator site located downstream from the pre-rRNA sequence and requires terminator recognition by specific protein factors.

In contrast, Pol II termination does not occur at a conserved site or at a constant distance from the 3′ end of mature RNAs. In mammals, it takes place anywhere from a few base pairs to several kilobase pairs downstream from the 3′ end of the mature transcript. The 3′ end of the mature mRNA includes a stretch of A nucleotides, called a poly(A) tail, that is essential for translation into protein (see Chapter 18). A polyadenylation signal sequence (typically AAUAAA) is present in the primary transcript (and directly encoded by the DNA). Factors responsible for cleavage of the primary transcript bind to the AAUAAA sequence, resulting in cleavage somewhat downstream from that position. Only after this cleavage is the poly(A) tail added. Pol II

termination is coupled to 3′-end processing of precursor mRNA transcripts, and the intact polyadenylation signal is necessary for termination of transcription of protein-coding genes in human and yeast cells.

Two different models have been proposed to explain how 3′-end processing contributes to Pol II transcription termination. The first, known as the allosteric or antiterminator model, proposes that transcription through the poly(A) site triggers conformational changes in the Pol II elongation complex caused by the dissociation of elongation factors and/or association of termination factors. This is analogous to the hairpin model of termination in bacteria (see Section 15.2). According to the allosteric model, these conformational changes in Pol II cause it to fall off the DNA template. The second model, the torpedo model, suggests that after mRNA synthesis is complete, Pol II remains associated with the DNA template and continues the transcription reaction to extend the 3′ end of the mRNA. Protein complexes cleave the mRNA at the polyadenylation site, producing a new 3′ end that can be recognized and extended by the enzyme poly(A) polymerase. The new 5′ end of the downstream, or residual, mRNA strand becomes a substrate for an enzyme called Xrn2, a 5′→3′ exonuclease (an exoribonuclease) that attaches to the CTD of Pol II (**Figure 15-26**). Xrn2 proceeds to degrade the uncapped residual RNA in the 5′→3′ direction until it reaches Pol II. Similar to the ρ factor

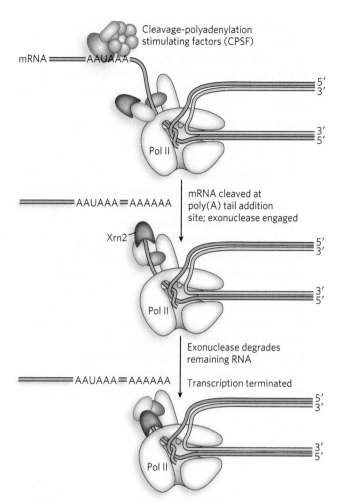

FIGURE 15-26 Torpedo model for transcription termination by Pol II. The torpedo model hypothesizes that the mRNA transcript is cleaved downstream from the poly(A) addition site by the U7 snRNP. An exonuclease (Xrn2) binds the RNA remaining on the polymerase and degrades the RNA in the 5′→3′ direction, moving closer to the polymerase and eventually causing it to release the mRNA.

in ρ-dependent termination in bacteria, Xrn2 triggers dissociation of Pol II by either pushing the polymerase off the DNA template or pulling the template out of the RNA polymerase.

Transcription Is Coupled to DNA Repair, RNA Processing, and mRNA Transport

In eukaryotes, transcription is coupled to other activities, including the repair of damaged DNA and various kinds of RNA processing and transport events. Researchers noticed that DNA damage repair and mRNA processing and transport are more efficient for genes that are actively being transcribed. Furthermore, DNA lesions in the template strand are repaired somewhat more efficiently than lesions in the coding (nontemplate)

strand. For DNA repair, these remarkable observations are explained by the alternative functions of the TFIIH subunits. Not only does TFIIH participate in forming the closed complex during assembly of a transcription complex, but some of its subunits are also essential components of the separate nucleotide excision repair complex (see Chapter 12). When Pol II transcription stalls at the site of a DNA lesion, TFIIH reassociates with the DNA and transcription machinery. TFIIH can then interact with the lesion and recruit the entire nucleotide excision repair complex. Mutations causing deletion of certain TFIIH subunits produce human diseases. Two examples are xeroderma pigmentosum, with its associated photosensitivity and tumor susceptibility, and Cockayne syndrome, which is characterized by arrested growth, photosensitivity, and neurological disorders.

Eukaryotic mRNA is processed in a variety of ways before it is shipped across the nuclear membrane to the cytoplasm for translation. We discuss the mechanisms of these processing events in Chapter 16, but it is important to note here that like DNA repair, mRNA processing is naturally linked to transcription. This is possible because some of the same proteins required for elongating RNA transcripts are also required for 5′-end processing (5′ capping) of the RNA. Because these activities are coupled, transcripts can be processed as they are synthesized.

SECTION 15.3 SUMMARY

- The RNA polymerases of eukaryotes (Pol I, II, and III) share some structural and functional features with bacterial RNA polymerase, but they are much larger and require additional proteins—transcription factors—to begin efficient transcription at promoter sequences.
- Pol I, II, and III recognize distinct promoter sequences and require unique sets of transcription factors, with the exception of TATA-binding protein (TBP), which is used by all three polymerases.
- As in bacteria, transcription initiation in eukaryotes is highly regulated and includes multiple steps that lead to assembly of an active polymerase complex at a promoter. Pol II transcription (the most studied) proceeds through distinct phases of assembly, initiation, elongation, and termination.
- In eukaryotes, the Pol II C-terminal domain must be phosphorylated before transcription can proceed from initiation to elongation.
- Transcriptional regulation in eukaryotes is enhanced by Mediator, a large protein complex that binds simultaneously with general transcription factors associated with Pol II and specific transcription factors associated with upstream promoter elements.

- Two hypotheses for transcription termination suggest a role for mRNA sequence elements and for an exonuclease, respectively.
- TFIIH, a eukaryotic transcription initiation factor, can start nucleotide excision repair of DNA when Pol II encounters a lesion in the template strand. Transcription and processing of mRNA are coupled, because some Pol II transcription factors are also required for pre-mRNA processing events.

? UNANSWERED QUESTIONS

Many details of transcription mechanisms are known, but future challenges include discovering how, where, and when transcripts are made and how they are used in cells for functions beyond encoding and synthesizing proteins.

1. How does RNA polymerase coordinate with other enzymes and regulators during gene expression? The pausing of RNA polymerase during transcription is thought to help the enzyme coordinate with other steps in the protein-producing pathway. How does this work, and do proteins such as RNA-modifying enzymes recognize paused transcripts as substrates? Understanding these mechanisms and how they differ in bacteria and humans will provide basic information about transcription and help define steps that could be disrupted to block bacterial growth, thus serving as good antibacterial drug targets.

2. What is the mechanism of promoter sequence recognition? Are there other promoter sequences that haven't yet been identified? These questions are especially relevant given the explosion in numbers of non-protein-coding RNA transcripts produced by Pol II that are now being discovered. Pol II seems to transcribe much of the human genome at low levels. How is the polymerase recruited to the DNA for this purpose? Perhaps Pol II uses its weak, nonspecific DNA-binding affinity, or perhaps it can transcribe past termination signals at some frequency. How is such transcription controlled?

3. How exactly is transcription terminated? This is still not well understood, especially in eukaryotes. If this process were better understood, it might be possible to exploit it for therapeutic purposes, such as inducing early termination of viral transcripts.

HOW WE KNOW

RNA Polymerase Is Recruited to Promoter Sequences

William Dynan [Source: Georgia Research Alliance.]

Dynan, W.S., and R. Tjian. 1983. Isolation of transcription factors that discriminate between different promoters recognized by RNA polymerase II. *Cell* 32:669–680.

Dynan, W.S., and R. Tjian. 1983. The promoter-specific transcription factor Sp1 binds to upstream sequences in the SV40 early promoter. *Cell* 35:79–87.

One of the first experiments to demonstrate how promoters are recognized by RNA polymerase II involved separating the contents of cultured human HeLa cells into the components required for accurate gene transcription in vitro. Bill Dynan and Bob Tjian used this system to find that Sp1 is a promoter-specific transcription factor required to recruit Pol II to only certain kinds of genes. Using genes from two different mammalian viruses, the monkey virus SV40 and the human adenovirus, Dynan and Tjian found that Sp1 recruited Pol II to SV40 genes, but not to adenovirus genes (Figure 1). When SV40 and adenovirus DNA templates were present together in an in vitro transcription reaction, addition of Spl stimulated early promoter transcription of the SV40 DNA 40-fold, whereas promoter transcription of adenovirus DNA was inhibited 40%. This finding suggested that Spl is involved in promoter selection and is not merely a stimulatory general transcription factor.

Further experiments using deletion mutants of the SV40 promoter showed that transcriptional activation by Sp1 required sequences within tandem 21 bp repeats located 70 to 110 bp upstream from the transcription initiation site. DNA footprinting revealed that DNA sequences within the 21 bp repeat region were bound by Sp1 (Figure 2). In this experiment, SV40 promoter-containing DNA was incubated with increasing amounts of a cell extract enriched with the Sp1 protein. DNase I, a nuclease, was then added to digest any DNA not protected by bound protein. As the protein concentration increased, a pronounced region of the DNA around the 21 bp repeat became resistant to DNase I digestion, revealing the "footprint" left by the binding of Sp1 to the DNA.

This was an exciting result, because it indicated the presence of a specific site for Sp1 binding. Furthermore, there was a correlation between this promoter-binding activity and consequent transcription stimulation. The results suggested that Sp1 activated transcription by Pol II at the SV40 early promoter by direct binding of the Sp1 to sequences in the upstream activator sequence.

(a)

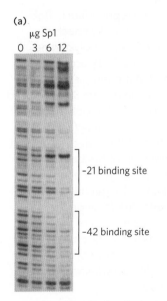

μg Sp1
0 3 6 12

−21 binding site

−42 binding site

(b)

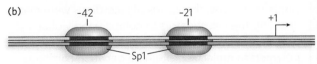

−42   −21   +1

Sp1

FIGURE 2 Sp1 leaves its footprint on a promoter. (a) The Sp1 footprint, seen as bands in the gel that decrease in intensity as the Sp1 concentration increases, is visible at sites flanking positions −21 and −42 with increasing concentrations of Sp1. The band that increases in intensity indicates a base pair in the DNA that becomes more susceptible to cleavage by DNase I on Sp1 binding—hinting at a change in the DNA structure that is induced by protein binding. (b) Sp1 binds SV40 DNA near positions −21 and −42. [Source: Dynan WS, Tijan R, Cell. 1983, Nov. 35 (1), 79–87, with permission from Elsevier.]

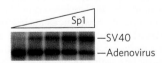

Sp1

—SV40
—Adenovirus

FIGURE 1 Sp1 activates transcription of SV40 DNA, but not human adenovirus DNA. Purified RNA polymerase and increasing amounts of Sp1 were added to a mixture of DNA containing an SV40 promoter and an adenovirus promoter; transcripts initiated from each promoter were separated and analyzed by gel electrophoresis. [Source: Data from W. S. Dynan and R. Tjian, Cell 32:669, 1983, Fig. 5, with permission from Elsevier.]

HOW WE KNOW

RNA Polymerases Are Both Fast and Slow

Neuman, K., E. Abbondanzieri, R. Landick, J. Gelles, and S.M. Block. 2003. Ubiquitous transcriptional pausing is independent of RNA polymerase backtracking. *Cell* 115:437–447.

Shaevitz, J.W., E.A. Abbondanzieri, R. Landick, and S.M. Block. 2003. Backtracking by single RNA polymerase molecules observed at near-base-pair resolution. *Nature* 426:684–687.

Molecular biologists have noticed that rather than transcribing DNA at a constant pace, an RNA polymerase hesitates at certain sites as it moves along the template. However, because the individual polymerases in a solution are not synchronized, the kinetics of pausing are difficult to study.

To circumvent this problem, Stephen Block, Bob Landick, and their coworkers chemically attached transcription elongation complexes to polystyrene beads, one polymerase to a bead. They used antibodies to attach one end of the template DNA to the surface of a microscope stage (Figure 3a), and used a laser trap to keep the bead (and RNA polymerase) in a fixed position while moving the stage (and the DNA) away, pulling the DNA taut through this constant force. They monitored the motion of the bead (and polymerase) with respect to the stage surface as the DNA was threaded through the elongation complex. This system was used to assess the force on the bead required to counteract the motion of the RNA polymerase. From these measurements, pause and arrest sites on the DNA could be mapped, and the maximal speed reached by RNA polymerase between two pause sites was measured.

Although the experiments were conducted on single molecules, thousands of recordings were made, enabling the investigators to compare individual polymerase complexes. The results showed that RNA polymerase molecules alternate between constant-velocity transcription and pausing. The velocities of individual polymerase molecules typically displayed a bimodal distribution, with one peak corresponding to the rate of transcription between pauses and a second peak, near 0 bp/s, corresponding to the pauses themselves (Figure 3b). This study of individual elongation complexes provided direct evidence that

RNA polymerases have different intrinsic speeds. The coexistence of slower and faster polymerases might explain how regulatory proteins modulate the behavior of elongation complexes during transcription, increasing or decreasing overall transcription rates in response to the needs of the cell.

(a)

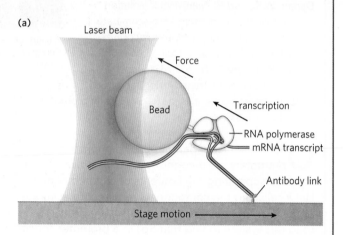

(b)

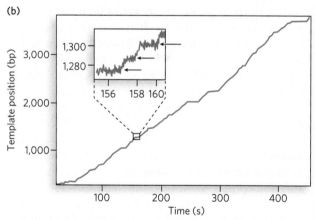

FIGURE 3 Single-molecule analysis determines the velocity of an RNA polymerase along a DNA and monitors pausing. (a) Representation of the experimental setup (not to scale). Transcribing RNA polymerase with nascent RNA is chemically attached to a polystyrene bead, and the upstream end of the duplex DNA is attached through an antibody linkage to a moveable microscope stage. The bead is held by a laser trap at a predetermined position, which results in a restoring force exerted on the bead. (b) Representative record of position and velocity for a single polymerase molecule transcribing a 3,500 bp DNA template with <18 piconewtons of hindering force. Pausing occurs on multiple timescales, with distinct pauses of seconds-long duration and shorter pauses of about 1 second, as seen in the expanded portion of the trace (arrows point to times when the polymerase transitions from paused to elongating state). [*Source: Data from K. Neuman et al., Cell 115:437–447, 2003.*]

PROBLEMS

1. The sequences of promoters tend to be rich in A and T residues. Suggest why this is so.

2. The −10 and −35 sequences in bacterial promoters are separated by about two turns of the DNA double helix. How would transcription be affected if a deletion were introduced in the promoter region that moved the −35 sequence to the −29 position?

3. The gene encoding the *E. coli* enzyme β-galactosidase begins with the sequence ATGACCATGATTACG. What is the sequence of the mRNA transcript specified by this part of the gene?

4. The gene for β-galactosidase has 3,075 bp. How long would it take for the *E. coli* RNA polymerase to transcribe this gene, assuming initiation has occurred upstream prior to its encounter with the gene?

5. The sequence of the consensus −10 region is TATAAT. If two genes, *tesA* and *tesB*, have identical promoter sequences except in the −10 region, where the *tesA* sequence is TAATAT and the *tesB* sequence is TGTCGA, which gene do you expect to be more efficiently transcribed, and why?

6. If a Pol II promoter were replaced with a promoter specific for Pol III in a human cell, what do you expect would happen to the number of transcripts produced?

7. Name the three major steps in the transcription of a typical bacterial gene, and indicate their relative rates.

8. Working in a research lab, you wish to examine the kinetics of the initiation phase of bacterial RNA polymerase, as a function of promoter sequence. You want to prevent the reaction from entering the elongation phase. How many nucleotides can be added to an RNA polymer in the initiation phase? Suggest a simple reaction strategy to limit the reaction to initiation, based on the sequence of the template and the components added to the reaction mix. With your experimental design, what would the reaction products be? Assume you have an assay to measure the production of short RNA oligonucleotides (you do not need to describe the assay).

9. People who ingest *A. phalloides* (the source of α-amanitin) initially experience gastrointestinal distress caused by other toxins also produced by this mushroom. α-Amanitin shuts down the action of RNA polymerase II, but death does not occur until about 48 hours after ingestion and usually involves liver dysfunction. Suggest a reason for the delay in lethality.

10. A drug company has discovered a natural product, cupramycin, that efficiently intercalates into DNA. How might this compound affect transcription?

11. How might an investigator search for Pol II promoters in the DNA sequence of an entire organism? Is it possible to find all such promoters computationally?

12. Gene *A* encodes protein A. A genetic engineer excises a promoter sequence for gene *A* from the DNA and reinserts it at the other end of gene *A*, oriented so that an RNA polymerase binding at the promoter will transcribe across gene *A*. Will the mRNA synthesized by the RNA polymerase still possess a sequence that produces a functional protein A? Why or why not?

13. In most organisms, specialized DNA repair systems are closely linked to transcription. Suggest a biological rationale for this close relationship.

14. In bacteria, there are many examples of two (or even more) genes being transcribed from one promoter—for example, the promoter is followed by gene *A* and then gene *B*, with both genes transcribed into a single mRNA. In some cases, the first gene in the linear sequence is transcribed at much higher levels than the second gene (i.e., many but not all of the mRNAs do not include gene *B*). What kind of DNA sequences might be present between the first and second genes to account for the lower level of transcription of gene *B*?

Grossman, A.D., J.W. Erickson, and C.A. Gross. 1984. The *htpR* gene product of *Escherichia coli* is a sigma factor for heat shock promoters. *Cell* 38:383–390.

15. In *E. coli*, σ^{70} is the major but not the only sigma factor. Several other sigma factors direct RNA polymerase to bind to different sets of promoters. The second sigma factor to be discovered was σ^{32}, which participates in the expression of genes involved in the heat shock response. When the environmental temperature suddenly rises, cellular production of about 20 proteins increases. These proteins help protect the cell from any ill effects of the higher temperature. Some of the proteins are chaperones that facilitate protein folding (see Chapter 4). In the 1970s, a gene required for normal heat shock response was discovered and named *htpR*. Carol Gross and her colleagues later identified its protein product, HtpR, as the first alternative sigma factor in *E. coli* (several had previously been discovered in *B. subtilis*); they renamed the gene *rpoH* (*rpo* genes encode RNA polymerase subunits) and named the protein σ^{32}. Was this renaming justified?

Gross and coworkers purified the 32 kDa HtpR protein. At a late stage in the purification, they ran the protein preparation over a cation-exchange column. For each fraction they collected, they ran a sample on a polyacrylamide gel (Figure 1a) and assayed the fraction for RNA synthesis directed by a heat shock promoter called P_{HS}; the RNA product of this assay was run on a separate polyacrylamide gel (Figure 1b). Fraction numbers are shown above the gel lanes. (The M lane has molecular weight markers; the H lane has purified RNA polymerase; nt indicates number of nucleotides.)

(a) Which fractions contain visible amounts of σ^{70}?

(b) Which fractions contain visible amounts of the 32 kDa protein (the putative σ^{32})?

(c) Which fractions contain other RNA polymerase subunits?

(d) Which fractions produced the most transcription from the P_{HS} promoter?

(e) What conclusion can you draw from this experiment?

The investigators then reconstituted the RNA polymerase, using purified RNA polymerase subunits but replacing σ^{70} with the 32 kDa protein that was now fairly pure. They used σ^{70}-containing RNA polymerase as a control. To find out whether the 32 kDa protein could direct RNA synthesis from more than one heat shock promoter, they constructed a plasmid with two such promoters (P1 and P2), then cut it with restriction enzymes. Transcription from the promoters would generate RNA transcripts with lengths of 215 and 140 nucleotides, respectively. The results are shown in Figure 2. Lanes 1 and 3 show RNA synthesis by the polymerase with the 32 kDa protein; lanes 2 and 4 show synthesis by the polymerase with σ^{70}.

(a)

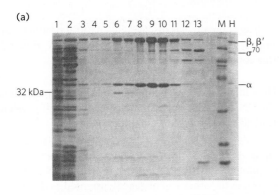

(b)

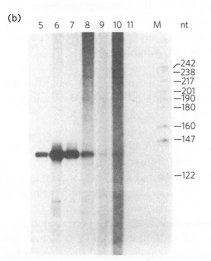

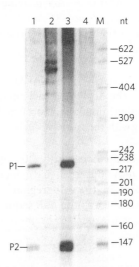

FIGURE 1 *[Source: Grossman, A.D. Erickson, J.W. and Gross, C.A. The htpR Gene Product of E. coli Is a Sigma Factor for Heat-Shock Promoters. Cell, Vol. 38, 383–390, September 1984, with permission from Elsevier. Courtesy Carol Gross.]*

FIGURE 2 *[Source: Grossman, A.D. Erickson, J.W. and Gross, C.A. The htpR Gene Product of E. coli Is a Sigma Factor for Heat-Shock Promoters. Cell, Vol. 38, 383–390, September 1984, with permission from Elsevier. Courtesy Carol Gross.]*

(f) What conclusion can you draw from these data?

(g) What is the advantage to the cell of having specialized sigma factors that recognize unique promoter sequences?

ADDITIONAL READING

RNA Polymerases and Transcription Basics

Abbondanzieri, E.A., W.J. Greenleaf, J.W. Shaevitz, R. Landick, and S.M. Block. 2005. Direct observation of base-pair stepping by RNA polymerase. *Nature* 438:460–465. A description of how individual polymerase molecules can be analyzed.

Carninci, P., T. Kasukawa, S. Katayama, J. Gough, M.C. Frith, N. Maeda, R. Oyama, T. Ravasi, B. Lenhard, C. Wells, et al. 2005. The transcriptional landscape of the mammalian genome. *Science* 309:1559–1563. This paper exemplifies a common research approach to analyzing when and where transcription occurs within an entire genome. The bottom line: many transcripts do not correspond to protein-coding genes, implying that these transcripts correspond to functional RNAs.

Darnell, J.E. Jr. 2002. Transcription factors as targets for cancer therapy. *Nat. Rev. Cancer* 2:740–749.

Gong, X.Q., C. Zhang, M. Feig, and Z.F. Burton. 2005. Dynamic error correction and regulation of downstream bubble opening by human RNA polymerase II. *Mol. Cell* 18:461–470.

Hurwitz, J. 2005. The discovery of RNA polymerase. *J. Biol. Chem.* 280:42,477–42,485. A summary of work in several different laboratories using complementary approaches.

Malik, S., and R.G. Roeder. 2006. Dynamic regulation of Pol II transcription by the mammalian Mediator complex. *Trends Biochem. Sci.* 30:256–263. This review is a good source for more information on the complexities of transcription in vivo.

Studitsky, V.M., W. Walter, M. Kireeva, M. Kashlev, and G. Felsenfeld. 2004. Chromatin remodeling by RNA polymerases. *Trends Biochem. Sci.* 29:127–135. A review of data and models for transcription in vivo.

Transcription in Bacteria

Darst, S.A. 2004. New inhibitors targeting bacterial RNA polymerase. *Trends Biochem. Sci.* 29:159–160.

Holmes, S.F., T.J. Santangelo, C.K. Cunningham, J.W. Roberts, and D.A. Erie. 2006. Kinetic investigation of *Escherichia coli* RNA polymerase mutants that influence nucleotide discrimination and transcription fidelity. *J. Biol. Chem.* 281:18,677–18,683.

Young, B.A., T.M. Gruber, and C.A. Gross. 2004. Minimal machinery of RNA polymerase holoenzyme sufficient for promoter melting. *Science* 303:1382–1384.

Transcription in Eukaryotes

Björklund, S., and C.M. Gustafsson. 2005. The yeast Mediator complex and its regulation. *Trends Biochem. Sci.* 30:240–244.

Deato, M.D., and R. Tjian. 2007. Switching of the core transcription machinery during myogenesis. *Genes Dev.* 21:2137–2149.

Halasz, G., M.F. van Batenburg, J. Perusse, S. Hua, X.J. Lu, K.P. White, and H.J. Bussemaker. 2006. Detecting transcriptionally active regions using genomic tiling arrays. *Genome Biol.* 7:R59. Description of a comprehensive method for detecting transcripts in cell extracts.

Jacquier, A. 2009. The complex eukaryotic transcriptome: Unexpected pervasive transcription and novel small RNAs. *Nat. Rev. Genet.* 10:833–844.

Vierbuchen, T., A. Ostermeier, Z.P. Pang, Y. Kokubu, T.C. Südhof, and M. Wernig. 2010. Direct conversion of fibroblasts to functional neurons by defined factors. *Nature* 463:1035–1041.

Voss, T.C., and G.L. Hager. 2014. Dynamic regulation of transcriptional states by chromatin and transcription factors. *Nat. Rev. Genet.* 15:69–81.

Willingham, A.T., and T.R. Gingeras. 2006. TUF love for "junk" DNA. *Cell* 125:1215–1220. A short, informative review of noncoding transcripts and their possible biological functions.

16 RNA Processing

Melissa Jurica [Source: Courtesy Melissa Jurica.]

MOMENT OF DISCOVERY

One of the most exciting moments I recall was the first time I "saw" spliceosomes, using electron microscopy. I had been working for a long time to purify spliceosomes, the RNA-protein complexes that produce functional messenger RNAs by removing intervening sequences, or introns. No one knew what spliceosomes look like, and *I hoped to use these samples to obtain structural information that would provide clues to the mRNA splicing mechanism.* Unfortunately, my purified samples tended to clump together when spotted onto the copper grids used for sample analysis by electron microscopy, making the images I saw uninterpretable.

Then late one night, while working alone, I developed some micrographs of a new spliceosome sample that showed uniform individual particles that finally looked like macromolecules—the spliceosome! I was so excited and moved that I felt ready to cry—but there was nobody around to show my result to! I posted the wet negative of the micrograph on my advisor Nico Grigorieff's door, and when he saw it the next morning we shared the joy of this breakthrough together.

The key to this result was a somewhat serendipitous discovery that the addition of EDTA, a magnesium ion–chelating agent, could suppress the tendency of the spliceosomes to aggregate. I think this is because SR proteins, which are associated with the spliceosome and tend to be sticky, require magnesium ions to bind to each other. Because we used EDTA to remove magnesium ions, the SR proteins lost the ability to associate tightly with the spliceosomes or each other. I still recall that overwhelming feeling of excitement mixed with happiness, and I use it as a well of motivation to this day, as my lab continues to push forward studies of the spliceosome.

—Melissa Jurica, on determining the first electron microscopic structures of spliceosomes

Online resources related to this chapter:

 Simulation RNA processing

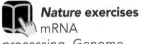 ***Nature* exercises** mRNA processing, Genome engineering

Our genome is not as large as we once imagined. Humans have only about 20,000 protein-coding genes, more than a fruit fly (14,000) and fewer than a rice plant (38,000). It turns out that much of the complexity in multicellular organisms is derived not from the number of genes but from the number of different proteins produced and the cell's ability to regulate gene expression at the level of RNA. Some bacterial RNA molecules, and virtually all eukaryotic RNAs, are synthesized as biologically inactive precursors that must be chemically modified during or immediately following transcription. These precursors, called **primary transcripts**, are processed into mature, biologically functional molecules by specialized enzymes. RNA processing events are some of the most interesting molecular activities in RNA metabolism, because they govern how, when, and whether the RNA will be used in the cell.

Although many different kinds of RNA are processed, all processing events consist of the same sort of chemical reactions: cleavage and/or joining of RNA strands and, sometimes, modification of the nucleotides themselves. RNA processing is an active and engaging area of research, because these seemingly simple reactions lie at the heart of gene expression and gene regulation, and when they go awry, diseases can result. For example, errors in splicing can lead to cystic fibrosis and spinal muscular atrophy.

Studies of RNA processing mechanisms have also shed new light on evolution and the origin of life. Some of these processing reactions are catalyzed by RNAs, not by proteins, as was once believed. The discovery of catalytic RNAs and the ability of RNA to catalyze cleavage and ligation reactions in other RNAs has provided support for the theory that life began with an RNA molecule.

RNA metabolism is much more complex in eukaryotes than in bacteria (**Figure 16-1**). Most bacterial mRNAs are not processed before being translated into protein. Indeed, they are sometimes translated even while transcription is occurring. In contrast, eukaryotic mRNAs are chemically modified in the nucleus before being transported into the cytoplasm for translation on the ribosomes (located in the cytosol or on the endoplasmic reticulum; see Chapter 18). A protective cap, made from a modified guanosine, is added to the 5′ end of the primary transcript. And the 3′ end is modified by cleavage and the addition of a string of adenylate residues to create a 3′ poly(A) "tail."

Each primary transcript for a eukaryotic mRNA typically contains sequences encompassing just one gene, although the segments encoding the polypeptide may not be contiguous. Noncoding regions that interrupt the coding region of the transcript are known as **introns**, and the coding segments, as well as 5′ and 3′ untranslated regions (UTRs), are known as **exons**. In the process of splicing, the introns are removed from the primary transcript and the exons are covalently connected (ligated) to form a continuous sequence that specifies a functional polypeptide.

(a) **Bacteria**

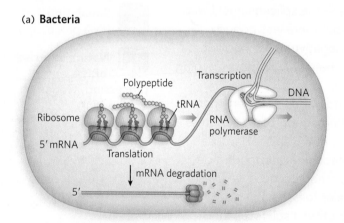

(b) **Eukaryotes**

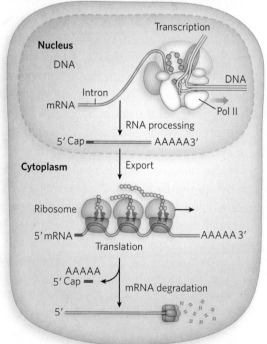

FIGURE 16-1 Messenger RNA processing in bacteria and eukaryotes. Eukaryotic mRNA undergoes several more steps than bacterial mRNA before translation. (a) Because bacterial cells lack a nucleus, transcription and translation are coupled. (b) In eukaryotic cells, transcription and pre-mRNA (precursor mRNA) processing occur in the nucleus, followed by export to the cytoplasm and subsequent translation. In both types of cells, the mRNA is eventually degraded.

The enzymes that catalyze each of these three mRNA processing reactions—capping, polyadenylation, and splicing—do not operate independently. Instead, they seem to interact with one another and with the phosphorylated C-terminal domain (CTD) of RNA polymerase II (see Figure 15-23), such that each enzyme affects the function of the others. Processing of the transcript is also coupled to its transport from the nucleus to the cytoplasm. In this way, the cell keeps track of mRNAs that have been properly processed and can immediately put them to use for translation into protein. Once an mRNA transcript has reached the end of its utility in the cell, it is subject to a final type of RNA processing—degradation. Specific enzymes remove the 5′ cap and the 3′ poly(A) tail, and the rest of the RNA is broken down by cleavage of the phosphodiester backbone.

In this chapter, we examine how and when RNA molecules are processed and how these processes are coupled. We begin with mRNA processing, one of the earliest-discovered and most ubiquitous kinds of precursor-to-mature RNA transformation that occurs in cells, describing the mechanisms of capping, polyadenylation, and splicing, and a fourth type of RNA modification known as RNA editing. We then turn our attention to the transport of mRNA from the nucleus to the cytoplasm and eventual RNA degradation. We discuss the processing of non-protein-coding RNAs, including tRNA, rRNA, and tiny regulatory RNAs called microRNAs, and, finally, RNA catalysis and its implications for life's origins.

16.1 MESSENGER RNA CAPPING AND POLYADENYLATION

In bacteria, mRNAs are often translated into protein as they are being transcribed, typically without any post-transcriptional modification. In eukaryotes, however, transcription occurs in the nucleus, and mRNAs must be transported to the cytoplasm before translation can occur. This uncoupling of transcription and translation provides opportunities for the cell to regulate when and where mRNAs are translated, and also requires that cells guard against premature degradation of the mRNA. One important way that eukaryotic cells achieve both regulation and protection of mRNAs is by chemically capping the 5′ end and adding a string of adenosines to the 3′ end. In addition to preventing degradation of the mRNA, these modifications provide chemical handles for subsequent steps in pre-mRNA (precursor mRNA) processing events and nuclear export.

Eukaryotic mRNAs Are Capped at the 5′ End

RNA polymerase II (Pol II) synthesizes single-stranded ribonucleic acids that are vulnerable to degradation at either end by exoribonucleases. The first modification made to a pre-mRNA transcript is the addition of a **5′ cap**, which prevents degradation from the 5′ end. The 5′ cap is a residue of 7-methylguanosine (7-meG) linked to the 5′-terminal residue of the mRNA through an unusual 5′,5′-triphosphate linkage. The 5′ cap helps protect mRNA by preventing recognition of the 5′ end by exoribonucleases, which digest RNA sequentially from a free 5′ or 3′ end.

The 5′ cap is formed by condensation of a molecule of GTP with the triphosphate at the 5′ end of the transcript. The guanine is subsequently methylated at N-7, and additional methyl groups are often added at the 2′ hydroxyls of the first and second nucleotides adjacent to the cap (**Figure 16-2a**). The methyl groups are derived from S-adenosylmethionine. All of these reactions occur very early during transcription, after the first 20 to 30 nucleotides of the transcript have been added (**Figure 16-2b**). Four different variants of the capping enzyme, guanylyltransferase, carry out the capping reaction. Only RNAs made by RNA Pol II are capped, because all four guanylyltransferase enzymes are associated with the CTD of Pol II (not with RNA polymerase I or III).

In the 1970s, experiments by Aaron Shatkin and his colleagues showed that the 5′ cap also plays a critical role in binding the mRNA to a ribosome. Using mRNAs isolated from reovirus, a human virus that can cause stomach flu, Shatkin's research group found that only mRNAs containing a 5′ cap are competent to initiate protein synthesis. Viral mRNAs lacking a 5′-methylated terminus are never part of actively translating ribosomes, because they fail to bind to the 40S ribosomal subunits during the translation initiation process (we discuss the details of protein synthesis in Chapter 18). Shatkin and colleagues used the reovirus RNA polymerase to synthesize reoviral mRNAs, using [^{32}P]GTP and [^{3}H]methyl-S-adenosylmethionine. Some of the resulting transcripts contained a 5′-7-meG cap (and thus were ^{3}H- and ^{32}P-labeled) and the rest lacked a cap (^{32}P-labeled only). These mRNAs were incubated with wheat germ cell extract to allow the mRNAs to bind to ribosomes. The samples were then deposited at the top of a sucrose density gradient and spun in a centrifuge, so that the samples migrated into the gradient according to their density (**Figure 16-3**). When the resulting gradients were fractionated, and each fraction measured for its content of ^{3}H- and ^{32}P-containing material, the capped mRNAs were found in the dense part of the

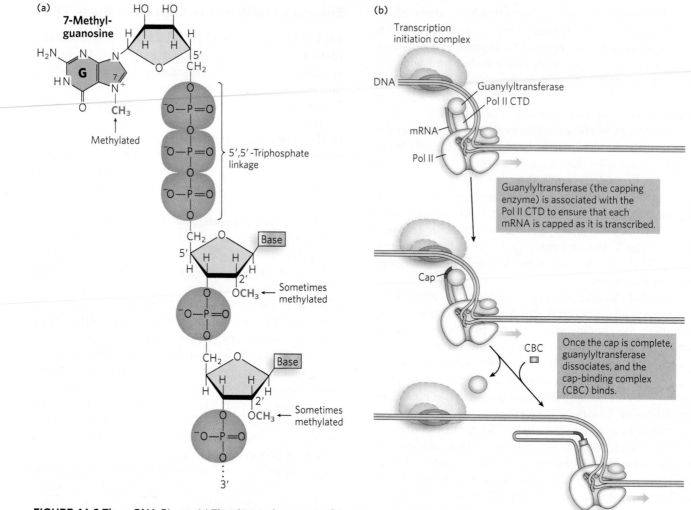

FIGURE 16-2 The mRNA 5′ cap. (a) The chemical structure of the 5′ cap of mRNA, showing the 7-methylguanosine and adjacent residues. (b) The cap is synthesized by the capping enzyme, guanylyltransferase, which is associated with the C-terminal domain (CTD) of RNA polymerase II.

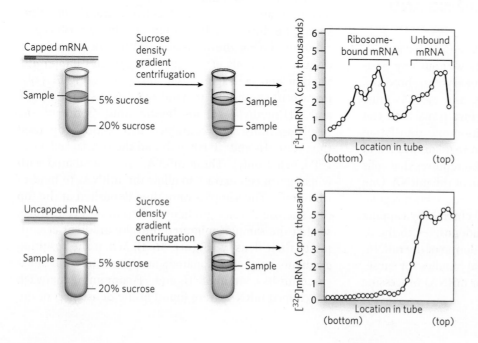

FIGURE 16-3 The 5′ cap requirement for ribosome binding and translation. Radiolabeled mRNAs with or without a 5′-methyl cap were incubated with ribosomes, and the complexes were separated from free mRNA by sucrose density gradient centrifugation. Denser particles migrate farther in the gradient, as shown on the left side of the graphs. Radioactivity levels (in counts per minute) of capped (top panel) and uncapped (bottom panel) mRNAs show that only the mRNA with a 5′ cap migrates with ribosomes. [Source: Data from G. W. Both et al., Cell 6:185–195, 1975. Copyright 1975, MIT.]

gradient where ribosomes were located, indicating that the capped mRNAs were bound to the ribosomes. In contrast, the uncapped mRNAs (containing only ^{32}P) migrated only to the top of the gradient. As we discuss in Chapter 18, the 5′ cap binds the **cap-binding complex (CBC)** (see Figure 16-2b), a specific complex of proteins that recruits capped mRNAs to the ribosome to initiate translation.

Eukaryotic mRNAs Have a Distinctive 3′-End Structure

Like the 5′ end, the 3′ end of mRNAs must be protected from nucleolytic degradation. As we discussed in Chapter 15, mRNAs contain a polyadenylation signal embedded in the sequence transcribed by RNA polymerase. An enzyme associated with the C-terminal domain of Pol II cleaves the transcript after the polyadenylation signal, and the mRNA is polyadenylated by other CTD-associated factors. The rest of the transcript dissociates from the polymerase and is degraded.

The **3′ poly(A) tail**, typically 80 to 250 A residues, serves as a binding site for one or more specific proteins that help protect mRNA from enzymatic destruction, by physically blocking the access of ribonucleases to the 3′ end. One way this was first demonstrated was by injecting mRNAs with or without a 3′ poly(A) tail into frog eggs and testing how long the mRNAs remained intact. The 3′-polyadenylated mRNAs survived for much longer, with a remarkable 40% remaining intact 48 hours after injection. Other experiments with mRNAs injected into egg cells or incubated in cell extracts showed that, like the 5′ cap, the poly(A) tail and its protein binding partners interact with the cap-binding complex during mRNA recruitment to the ribosome.

The poly(A) tail is added in multiple catalytic steps involving polyadenylation factors, polyadenylate polymerase (PAP) and poly(A) binding protein (PABP) (**Figure 16-4**). First, Pol II extends the transcript beyond the site where the poly(A) tail is to be added. The transcript is cleaved at the **poly(A) addition site** by an endonuclease component of a large enzyme complex, again associated with the CTD of Pol II. The mRNA site where cleavage occurs is marked by two sequence elements: the highly conserved sequence 5′-AAUAAA located 10 to 30 nucleotides on the 5′ side (upstream) of the cleavage site, and a less well-defined sequence rich in G and U residues located 20 to 40 nucleotides downstream from the cleavage site. Cleavage generates the free 3′-hydroxyl group that defines the end of the mRNA, to which A residues are immediately added in the PAP-catalyzed reaction:

$$\text{RNA} + n\,\text{ATP} \rightarrow \text{RNA} - (\text{AMP})_n + n\,\text{PP}_i \qquad 16\text{-}1$$

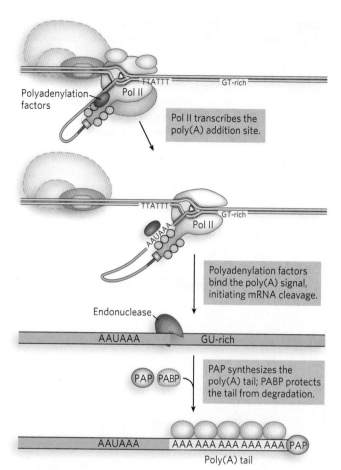

FIGURE 16-4 Addition of the 3′ poly(A) tail to the transcript. Polyadenylation factors are associated with Pol II during transcription. After transcription of the 5′-AAUAAA sequence, these proteins transfer to the transcript and help recruit cleavage factors and poly(A) binding protein (PABP) before dissociating. The poly(A) tail is synthesized by polyadenylate polymerase (PAP).

where $n = 80$ to 250. The number of A residues in the poly(A) tail varies from one species to another; it is often written simply as $(\text{A})_n$. The polyadenylation enzyme complex does not require a template, but it does require cleavage of the mRNA. A few types of eukaryotic mRNA, such as those that encode histones, are cleaved but have a different kind of 3′ tail that allows them to be regulated separately from the bulk of the cellular mRNA (**Highlight 16-1**).

mRNA Capping, Polyadenylation, and Splicing Are Coordinately Regulated during Transcription

The regulation of mRNA capping, polyadenylation, and splicing is coordinated through the association of these processes with RNA polymerase II. Capping, splicing, and polyadenylation factors are associated with the Pol II C-terminal domain, and these factors transfer to

Eukaryotic mRNAs with Unusual 3′ Tails

Unlike most eukaryotic mRNAs, those that encode histones—the small basic proteins that help package eukaryotic chromosomal DNA (see Chapter 10)—lack 3′ poly(A) tails. Instead, the 3′ ends contain a self-complementary sequence that forms a stem-loop structure, creating the binding site for a protein complex that includes the stem-loop binding protein (SLBP) (Figure 1). SLBP both stabilizes the 3′ end of histone mRNAs, preventing degradation, and provides an alternative way of recruiting the mRNA to ribosomes during translation

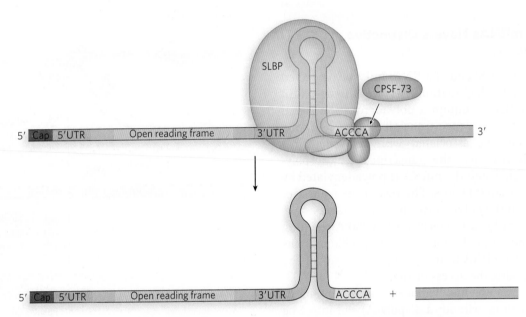

FIGURE 1 The 3′ stem-loop structure of a histone pre-mRNA binds the protein SLBP, which then assembles with CPSF-73 and other splicing factors to process the 3′ end. The open reading frame is the coding part of the gene. [*Source: Data from W. Marzluff et al., Nat. Rev. Genet. 9:843–854, 2008.*]

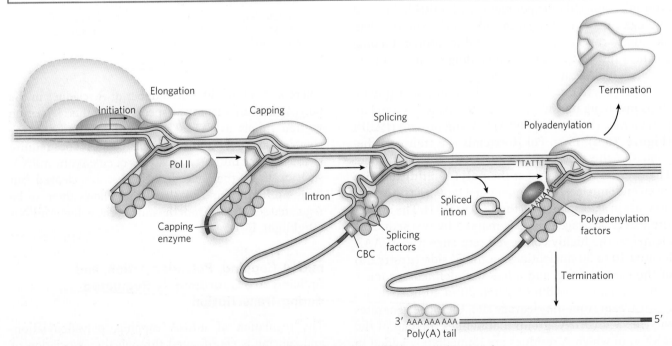

FIGURE 16-5 Coordination of transcription and pre-mRNA processing. mRNA-processing proteins associate with the C-terminal domain of Pol II. As the mRNA is synthesized, the capping enzyme (guanylyltransferase), splicing factors, and polyadenylation factors transfer from the Pol II CTD to the mRNA and process the mRNA as it is transcribed. (CBC is the cap-binding complex.) The mature transcript is used for multiple rounds of protein synthesis, then degraded.

initiation. Nascent histone-encoding transcripts have 3' extensions that are cleaved at a site five nucleotides after the conserved stem-loop, a process catalyzed by a specific endonuclease that cleaves internal bonds in the phosphodiester backbone. The upstream cleavage product corresponds to the mature histone mRNA, and the downstream product is degraded by an exonuclease, which cleaves RNA from the 5' or 3' end.

William Marzluff
[Source: Courtesy William Marzluff.]

The processing of histone pre-mRNAs is interesting, because it provides a counterpoint to the more common pre-mRNA processing pathway and therefore may illuminate evolutionary connections between the two. To identify the ribonucleases responsible for these histone pre-mRNA processing events, William Marzluff and his colleagues at the University of North Carolina conducted a cross-linking experiment (Figure 2). Histone pre-mRNA was synthesized in the laboratory to include a radioactive ^{32}P label and incubated in the presence of a total protein extract from human cells. The preparation was then exposed to ultraviolet light to induce formation of chemical bonds between the mRNA and any proteins that might have bound to it. Remarkably, the researchers found the so-called cleavage and polyadenylation specificity factor (CPSF-73), a known protein component of the machinery responsible for cleavage and polyadenylation of the vast majority of eukaryotic mRNAs. These studies suggested that CPSF-73 is

both the endonuclease and the 5'→3' exonuclease involved in histone pre-mRNA processing. The crystal structure of human CPSF-73, along with biochemical experiments on the purified protein, showed that CPSF-73 is also the enzyme responsible for cleaving the 3' ends of all pre-mRNAs prior to polyadenylation. These findings revealed an unexpected biochemical connection between 3'-end formation in histone mRNAs and in other, polyadenylated mRNAs, implying a common ancestral mechanism for processing the 3' ends of all animal mRNA transcripts.

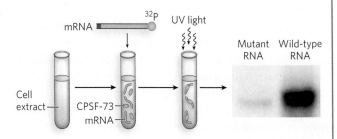

FIGURE 2 Cross-linking experiment to determine the protein bound to the radiolabeled stem-loop of histone mRNA. The cross-linked species were separated by denaturing gel electrophoresis, based on their slower migration relative to non-cross-linked molecules, revealing a prominent species that was found to contain CPSF-73. The cross-linked species corresponding to CPSF-73–pre-mRNA is present in the gel with the wild-type RNA, but is greatly reduced in a reaction that used a mutant RNA defective for 3'-end processing. [Source: Reprinted from Dominski, Z. et al, Cell 123 (1) pp. 37–48, 2005. © 2005, with permission from Elsevier. Courtesy Zbigniew Dominski.]

the nascent RNA at the appropriate sequences. This enables RNA processing to occur simultaneously with transcription, ensuring that mature mRNAs are processed rapidly as they emerge from the Pol II active site (**Figure 16-5**). Once the 5' end of the transcript has been capped, it is released from the capping enzymes and bound by the cap-binding complex, which itself associates with the Pol II CTD.

In addition, some components of the splicing apparatus appear to be tethered to the Pol II CTD, suggesting an interesting model for the splicing reaction. As the first splice junction is synthesized, it is bound by a protein splicing factor. The second splice junction is then captured by this complex as it passes by, facilitating juxtaposition of the intron ends and the subsequent splicing process. After splicing, the intron remains in the nucleus and is eventually degraded (see Figure 16-5).

SECTION 16.1 SUMMARY

- Eukaryotic mRNAs are modified by the addition of a 7-methylguanosine residue at the 5' end and by cleavage and polyadenylation at the 3' end to form a poly(A) tail.

- Modification of the ends protects the transcript from ribonucleases and is required for subsequent steps in pre-mRNA processing, as well as for export from the nucleus.

- Messenger RNA capping, polyadenylation, and splicing are coupled to transcription through the physical association of the enzymes with RNA polymerase II.

16.2 PRE-mRNA SPLICING AND EDITING

Beyond capping and polyadenylation, eukaryotic transcripts are further processed in a series of reactions referred to as RNA splicing, the splicing of pre-mRNA to produce the mature, translation-competent form of the mRNA. Then, for some eukaryotic mRNAs, the nucleotide sequence is chemically altered by a process known as RNA editing. As we will see, both RNA splicing and RNA editing can expand the coding capacity of the genome by creating mRNAs that are not directly encoded by the DNA. We discuss first the processes of RNA splicing, then RNA editing.

RNA splicing is one of the most important distinctions between gene expression in bacteria and in eukaryotes, and hence has attracted a lot of research interest. Splicing may be a relic of the way that protein-coding sequences originally evolved, providing a glimpse into how modern genes arose. Furthermore, some human diseases, including Duchenne muscular dystrophy and cystic fibrosis, can be caused by aberrant pre-mRNA splicing, suggesting possibilities for therapies if the underlying splicing mechanisms can be understood.

All pre-mRNA splicing mechanisms consist of the ordered breaking and joining of specific phosphodiester bonds to achieve the precise excision of introns. Accurate and efficient splicing relies on base pairing between the pre-mRNA and the splicing machinery to specify the bonds to be broken or formed. In all cases, splicing must be carried out quickly and correctly to produce the mRNAs required for protein production.

A complex of RNA and proteins called the **spliceosome** is responsible for most pre-mRNA splicing. Compared with ribosomes, spliceosomes are much more dynamic multipiece machines; their assembly occurs at pre-mRNA splice sites and determines whether a given set of splice sites will be used. A small number of introns found in mitochondrial, chloroplast, and bacteriophage transcripts are self-splicing, catalyzing their own excision from a primary transcript without help from proteins. The discovery of self-splicing introns, along with the discovery of the RNA-mediated catalytic activity of the enzyme ribonuclease P (discussed later in this chapter), led to a profound change in our understanding of modern biology—showing that proteins are not the only biological catalysts in cells. The realization that RNA can function both as a carrier of genetic information and as a catalyst led to the idea of an "RNA world" that might have given rise to modern cells and organisms (see Section 16.5).

Eukaryotic mRNAs Are Synthesized as Precursors Containing Introns

In bacteria, a polypeptide chain is generally encoded by a DNA sequence that is colinear with the amino acid sequence, continuing along the DNA template without interruption until the information needed to specify the polypeptide is complete. Although this might appear to be the most efficient way to encode genetic information, primary transcripts isolated from the nuclei of mammalian cells are much larger than required for simply encoding the proteins they produce. Rapid-radiolabeling experiments showed that most of the RNA in these primary transcripts is degraded before it leaves the nucleus. Independent investigations by Phillip Sharp and Richard Roberts in the late 1970s showed that ᵧ eukaryotic protein-coding genes are interrupted

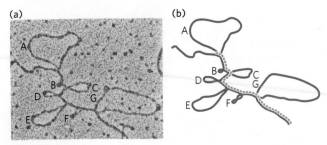

FIGURE 16-6 A DNA-mRNA hybrid revealing the presence of introns. (a) An electron micrograph and (b) drawing show the ovalbumin gene DNA (blue) hybridized to its mRNA (dashed green line). The DNA contains several loops and regions (A through G) that do not base-pair with the mRNA produced from this region, indicating regions of RNA that are processed out of the mRNA transcript. [Source: (a) Dugaiczyk, A. et al. The ovalbumin gene: Cloning and molecular organization of the entire natural gene (1979) PNAS 76, 2253–2257. Fig. 5. By permission of Bert O'Malley.]

by noncoding sequences, the introns. Electron microscopy studies of annealed chromosomal DNA and its corresponding mature mRNA revealed regions of complementarity interspersed with looped-out regions in the DNA that did not base-pair with sequences in the mRNA (**Figure 16-6**). This was the first indication that genes can include both coding regions and noncoding segments that are removed after transcription. As later became apparent, to form mature mRNA, introns must be removed from precursor transcripts and the remaining protein-coding segments, the exons, covalently connected (**Figure 16-7**).

Subsequent studies revealed that, in vertebrates, the vast majority of genes contain introns; those that encode histones are among the few exceptions. The occurrence of introns in other eukaryotes varies. Most genes in baker's yeast, *Saccharomyces cerevisiae*, lack introns, although they are more common in other yeast species. Introns are also found in a few bacterial and archaeal genes, and even occur within the genes of certain bacteriophages. Although the evolutionary

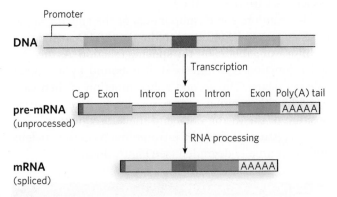

FIGURE 16-7 Interrupted genes. Introns are non-protein-coding sequences in the DNA and transcribed mRNA that are removed from the RNA during processing to form a contiguous, exon-only, protein-coding mRNA.

significance of introns is not clear, they can play roles in regulating the amount of mature mRNA produced. For example, some introns include open reading frames that encode regulatory proteins, and others are further processed, after splicing, into small RNAs that base-pair with complementary mRNAs to regulate their stability or translation (see Chapter 22). Splicing also prepares mRNAs to be recognized by proteins that export them from the nucleus and promote their translation by the ribosome.

Introns in DNA are transcribed along with the rest of the gene by RNA polymerases. Then, introns in this primary RNA transcript are excised and the exons are joined to form the mature RNA. A typical mammalian pre-mRNA includes eight introns with an aggregate length 5 to 10 times that of the flanking exons. In general, introns of animal pre-mRNAs vary in size from 50 to 20,000 nucleotides, and most exons are less than 1,000 nucleotides long, with many having just 100 to 200 nucleotides and encoding polypeptide segments of 30 to 60 amino acids. Genes of higher eukaryotes, including humans, have much more DNA devoted to introns than to exons; for example, some human genes have hundreds of introns.

Alternative RNA Splicing Can Generate Multiple Products from a Gene

The transcription of introns might seem to consume cellular resources and energy without returning any benefit to the organism, but introns may confer an advantage. One idea is that introns are vestiges of an ancient molecular parasite not unlike transposons. Another is that introns offer unique advantages to complex organisms because they provide a means of greatly increasing the number of different protein-coding sequences that can be produced, in principle, from a single gene. Comparison of an organism's genome with the complement of proteins in a given cell type—its **proteome**—shows that in many cases, the number of different proteins greatly exceeds the number of identified genes. In some cases, at least, this discrepancy is due to **alternative splicing**, a process in which exons in the primary transcript from a single gene are spliced together in various combinations to produce *different* mRNAs and thus different polypeptides (**Figure 16-8**).

Larry Zipursky
[Source: Courtesy Larry Zipursky.]

High-throughput sequencing technology such as RNA-Seq (see Figure 8-12) has revealed that more than 90% of human genes undergo alternative splicing.

Certain exons are selected for inclusion and others are not, but the order of the exons does not change relative to the primary transcript. In fruit flies, for example, the gene *Dscam* encodes an immunoglobulin (Ig) superfamily protein, a transmembrane protein required for the formation of neuronal connections, as well as participating in the immune system. Through alternative splicing, *Dscam* potentially gives rise to 19,008 different extracellular domains linked to one of two alternative transmembrane segments, resulting in 38,016 different possible forms (isoforms) of the Dscam protein! Dscam variants share the same domain structure but contain different amino acid sequences within three Ig domains in the extracellular region. Using Dscam proteins that could be recognized and distinguished by specific antibodies, Larry Zipursky and his colleagues found that each Dscam binds to molecules of the same isoform, but does not bind or binds poorly to other isoforms, contributing to the formation of complex patterns of neuronal connections. Although the *Dscam* gene is an extreme example, many mammalian genes have two or more alternatively spliced mRNAs derived from the same gene, greatly increasing the complexity of the genome and providing opportunities for regulation at the level of pre-mRNA processing.

Mechanisms of alternative splicing are not yet well understood, but we know that **splice sites**, nucleotide sequences within the intron and at the borders between introns and exons, play an important role in determining whether an exon is included in the mature mRNA. In some cases, splice sites can be masked in particular cell types, leading to intron skipping and consequent loss of protein production. For example, research on fruit flies in Don Rio's laboratory showed that in all cells except egg or sperm precursors, the protein Psi binds to a "decoy" splice site adjacent to the true splice site in a gene encoding the enzyme transposase. This prevents proper splicing of the transposase mRNA, thereby preventing transposase production in adult cells where it is not needed and could harm the integrity of the genome by disrupting gene or regulatory sequences.

Regulation by alternative splicing is implicated in some human disease pathways. For example, the inclusion of specific exons in the spliced mRNAs encoding the cell surface molecule CD44 is associated with the

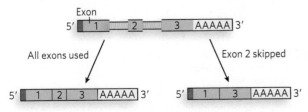

FIGURE 16-8 Different ways of assembling exons. Alternative splicing can generate multiple products from one gene.

progression of certain tumors from localized malignan-cies to invasive growths Thus, understanding what deter-mines exon choice during alternative splicing might lead to new therapeutic strategies to treat or prevent some cancers.

Complex transcripts can have more than one site for cleavage and polyadenylation, or for alternative splicing patterns, or both. If there are two or more sites for cleavage and polyadenylation, using the one closest to the 5′ end will remove more of the primary tran-script sequence. This mechanism, called **poly(A) site choice**, generates diversity in the variable domains of immunoglobulin heavy chains, which is required for an efficient immune response (see Chapter 14). In fruit flies, alternative splicing patterns produce, from a common primary transcript, three different forms of the myosin heavy chain at different stages of fly development. In rats, *both* mechanisms (alterna-tive splicing and poly(A) site choice) come into play when a single RNA transcript is processed differently to produce two different hormones: the calcium-regulating hormone calcitonin in thyroid and the calcitonin-gene-related peptide (CGRP) in the brain (**Figure 16-9**).

The Spliceosome Catalyzes Most Pre-mRNA Splicing

Each intron includes a 5′ splice site, a 3′ splice site, and an internal A residue just upstream of the 3′ splice site, at the **branch point** (**Figure 16-10**). Splicing in-volves a cascade of phosphodiester exchange reactions. First, the 2′-OH of the branch-point A residue attacks the phosphate at the 5′ splice site. Then the 3′-OH of the released exon attacks the phosphate at the 3′ splice site, thus joining the exons and separating them from the intron. These reactions are catalyzed by ribonu-cleoproteins (RNPs), complexes of non-protein-coding RNAs and proteins. The RNA components of the RNPs base-pair with the mRNA at the 5′ splice site, 3′ splice site, and branch point, positioning the associated pro-teins for catalysis.

Most introns in eukaryotic pre-mRNAs are re-moved by the spliceosome, a complex of five **small nuclear ribonucleoproteins**, or **snRNPs** (pronounced "snurps"), and hundreds of additional protein compo-nents. At the heart of each snRNP is a single **small nuclear RNA (snRNA)** belonging to a class of non-protein-coding eukaryotic RNAs 100 to 200 nucleotides

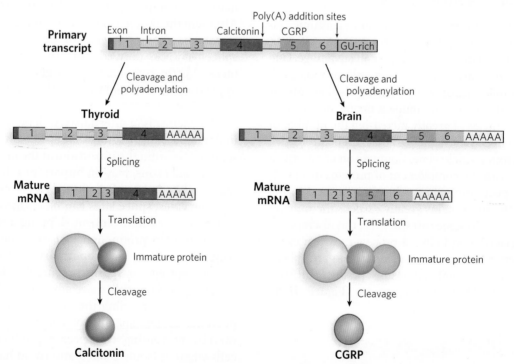

FIGURE 16-9 Alternative processing of the rat calcitonin gene transcript. Calcitonin and calcitonin-gene-related peptide (CGRP) have completely different protein sequences but are encoded by the same gene. The primary transcript contains two poly(A) addition sites: one used in the thyroid and the other in the brain. In the thyroid, exon 4 is retained; in the brain, exon 4 is eliminated and exons 5 and 6 are retained. The resulting mRNAs encode different polypeptides that are further processed to yield the final hormone products: calcitonin in the thyroid and CGRP in the brain.

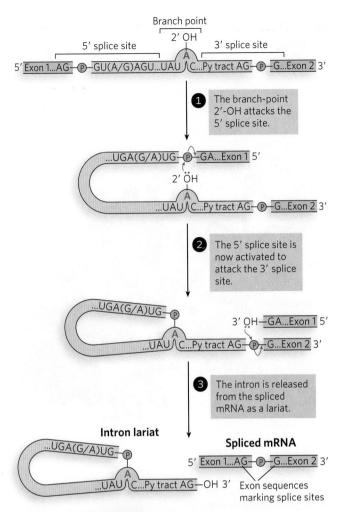

FIGURE 16-10 An overview of the splicing reaction.
Pre-mRNA splicing occurs through two site-specific transesterification reactions that result in phosphodiester bond cleavage and ligation. The 5′ and 3′ splice sites consist of the conserved sequence elements shown; the Py tract in the intron is a string of pyrimidine residues.

long (**Figure 16-11a**). Five snRNAs (U1, U2, U4, U5, and U6) involved in splicing reactions are abundant in eukaryotic nuclei. The RNAs and proteins found in snRNPs are highly conserved in eukaryotes from yeast to humans, suggesting that the splicing machinery, and hence introns themselves, were present in the earliest eukaryotes.

Each snRNA includes a binding site for a set of proteins—Sm proteins—that are common to all snRNPs (see the How We Know section at the end of this chapter). Seven Sm proteins (SmB, D1, D2, D3, E, F, and G) form a ringlike structure that binds adjacent to a hairpin fold near the 3′ end of the snRNA (see Figure 16-11a). In addition, the snRNA contains sequences that are uniquely recognized by proteins specific to that snRNP. The snRNA–Sm protein complex forms a structure

called the Sm core domain. Electron microscopy and x-ray crystallography have been used to visualize individual snRNP proteins and the Sm core (**Figure 16-11b**), as well as intact snRNPs and assembled spliceosomes (see this chapter's Moment of Discovery).

Spliceosomes assemble from snRNPs and include other proteins that are not specifically associated with any snRNA. Chief among these are the SR proteins, so named for their serine/arginine-rich region, which play central roles in selecting and regulating splice sites in pre-mRNAs. SR proteins have a common architecture consisting of an RS domain containing repeats of Arg and Ser, and an RNA-recognition motif (RRM). The RS domain is a site of protein-protein interactions that can be enhanced by the phosphorylation of Ser residues. The RRM domain binds to sequences in the pre-mRNA, often within the exons.

One of the key questions to be addressed by biochemical and structural studies is how introns are recognized and cleaved by the spliceosome. Introns have the dinucleotide sequences GU and AG at their 5′ and 3′ ends, respectively, and these sequences mark the sites where splicing occurs. However, these dinucleotide sequences at splice junctions are, by themselves, not sufficiently information-rich to specify splice sites accurately. The surrounding sequences and perhaps the structure of the pre-mRNA itself must play a role in the selection of splice sites, and the ability of snRNPs to bind to these sequences affects the likelihood of splicing. However, the details of how this occurs in the cell are as yet undetermined. Researchers must verify splice sites by comparing the genomic sequence with the corresponding sequence of the mRNA or protein it encodes.

Base pairing between the snRNAs of the spliceosome and the pre-mRNA allows cells to select correct splice sites (**Figure 16-12** on p. 565). First, the U1-containing snRNP (referred to simply as U1 snRNP) binds to the GU sequence at the 5′ splice site, along with accessory proteins, including U2AF (U2 auxiliary factor), which binds to the AG and flanking sequences at the 3′ splice site. In an ATP-dependent next step, U2AF recruits U2 snRNP to the branch point. The 2′-OH of the adenosine in the branch point becomes the nucleophile that attacks the phosphodiester bond at the 5′ splice site. Then the U4-U6-U5 trimeric snRNP complex binds, with U6 binding to U2. Next, U1 snRNP is released, U5 snRNP base-pairs with the 5′ exon, and U6 snRNP moves to the 5′ splice site. Once U4 snRNP is released, the U6 and U2 snRNPs catalyze nucleophilic attack of the 2′-OH of the branch-point adenosine on the phosphodiester bond at the 5′ splice site, cleaving the 5′ exon-intron junction and shifting the U5 snRNP to the 3′ splice site.

(a)

(b)

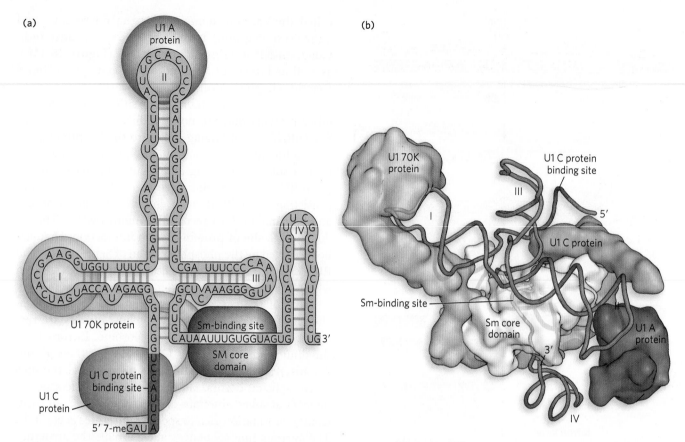

FIGURE 16-11 The structure of U1 snRNP. Each snRNP consists of an snRNA (U1 is shown here), with an Sm-binding site, as well as unique sequences and secondary structures. In addition, proteins unique to each snRNA bind to form a complex (the U1 snRNP proteins U1 A, U1 C, and U1 70K are shown here). (a) Two-dimensional and (b) three-dimensional representations of the U1 snRNP. Sm proteins bind to the Sm-binding site to form the Sm core domain; unique proteins bind elsewhere in the snRNA to form the complete snRNP. [*Source: (b) PDB ID 3CW1.*]

Formation of the simultaneous 2′ and 3′ phosphodiester linkages at the branch-point adenosine results in a lariat-shaped RNA containing the intron and the 3′ exon. The U2-U6-U5 complex remains bound to the lariat and catalyzes nucleophilic attack of the 3′-OH of the 5′ exon on the phosphodiester bond linking the intron to the 3′ exon, resulting in intron excision with concomitant joining of the 5′ and 3′ exons. ATP hydrolysis is required, presumably for unwinding the spliceosomal RNAs and proteins, as well as helping the snRNAs base-pair with each other and with the pre-mRNA.

Some pre-mRNA introns are spliced by a less common type of spliceosome, in which the U1 and U2 snRNPs are replaced by the U11 and U12 snRNPs. Whereas U1- and U2-containing spliceosomes remove introns with 5′-GU and AG-3′ terminal sequences, the U11- and U12-containing spliceosomes remove a rare class of introns that have 5′-AU and AC-3′ terminal sequences at the intron splice sites. There is no obvious pattern for the use of these so-called AT-AC introns in genes, leading

to speculation that they arose in a process parallel to the evolution of the majority of introns.

Some Introns Can Self-Splice without Protein or Spliceosome Assistance

Researchers initially assumed that all introns would be removed by protein-catalyzed reactions, but during the early 1980s, Thomas Cech and his coworkers at the University of Colorado discovered that some introns have self-splicing capability. In an in vitro system, Cech and colleagues transcribed an intron-containing piece of DNA isolated from the ciliated protozoan *Tetrahymena thermophila*, using purified bacterial RNA polymerase. Remarkably, the resulting transcript spliced itself accurately without requiring any *Tetrahymena*

Thomas Cech
[*Source: Photo by Bruce Weller.*]

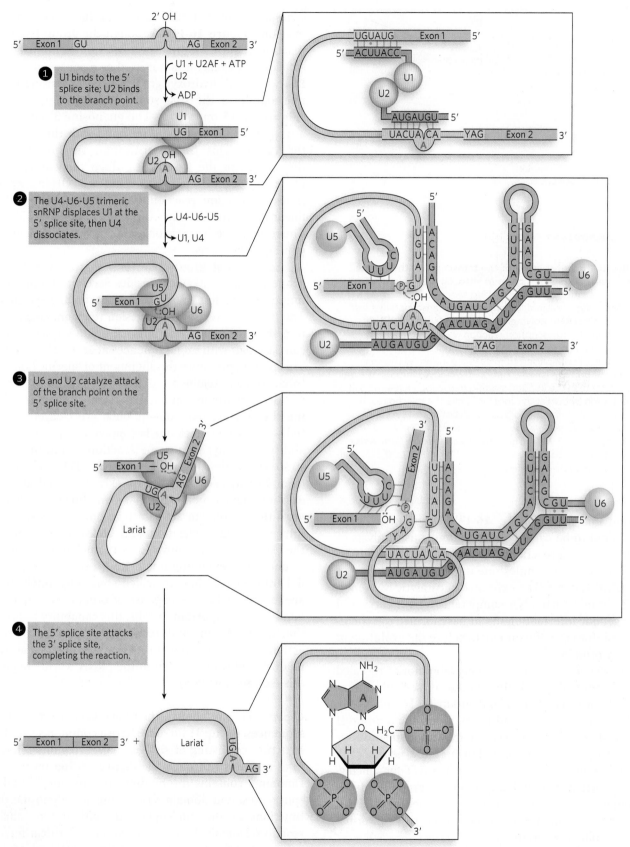

FIGURE 16-12 Spliceosome assembly on pre-mRNAs involving base pairing to snRNAs. The splicing process is described in the text. In this figure, exons are green; introns, pink; snRNAs, blue. Notice that in step 2, U6 snRNA base-pairs near the 5' exon binding site where U1 snRNA was formerly bound. As U4 dissociates, the U2-U6-U5 complex remains assembled on the pre-mRNA. U5 base-pairs to both sides of the splice junction to align the RNA for the splicing reaction, and U2 and U6 base-pair to each other. [*Source: Data from K. Nagai et al.,* Biochem. Soc. Trans. *29:15–26, 2001.*]

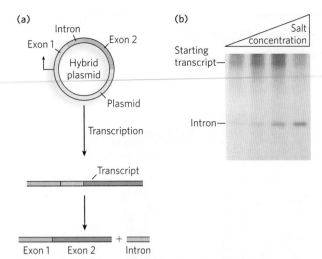

FIGURE 16-13 Self-splicing of the transcripts of the _Tetrahymena_ large rRNA gene in vitro. (a) A DNA fragment including the intron (red) and flanking sequences (blue) was cloned into a plasmid, which was then transcribed in a test tube by purified RNA polymerase. (b) Samples of pre-rRNA purified from this transcription reaction were incubated in buffer solutions containing sodium and magnesium salts. The pre-rRNA was capable of self-splicing to release the intron, as shown in this agarose gel. The (shorter) intron RNA moves through the gel matrix faster than the (longer) starting transcript. Increasing salt concentration (shown at the top) facilitates the splicing reaction because salts stabilize the folded, catalytically active form of the RNA. _[Sources: (a) Data from K. Kruger et al., Cell 31:147–157, 1982. Copyright 1982, MIT. (b) Reprinted from Thomas R. Cech, Arthur J. Zaug, Paula J. Grabowski, Cell 27(3) pp. 487–496. Fig. 1a, © 1981, with permission from Elsevier. Courtesy T. R. Cech.]_

protein enzymes (**Figure 16-13**). Other RNAs were also found to be capable of functioning as catalysts of RNA processing reactions (see Section 16.5). This exciting discovery was a milestone in our understanding of biological systems. The existence of catalytic RNA molecules implies that RNA competent to both carry and copy genetic information might have formed the chemical basis for early life on Earth, before the evolution of DNA or proteins.

Subsequent research in many laboratories revealed the existence of two distinct classes of self-splicing introns, known as group I and group II. Found mostly in bacteria, organelles, and fungi, group I and group II introns share the ability to self-splice without the involvement of any protein enzymes. Furthermore, neither class requires a high-energy cofactor such as ATP for splicing. It is important to note that group I and group II introns do require proteins for splicing in vivo—not for catalysis, but for forcing the pre-RNA into the correct conformation for splicing to occur.

Group I Introns The catalytic activities and molecular structures of several **group I introns**, such as the original example discovered in _Tetrahymena_, have been studied in detail (**Figure 16-14**). Although the group I splicing reaction requires a guanine nucleoside or nucleotide cofactor, the cofactor is not used as a source of energy. Instead, the 3′-hydroxyl group of the guanosine is the nucleophile in the first step of the splicing pathway (**Figure 16-15** on p. 568). The guanosine 3′-OH forms a normal 3′,5′-phosphodiester bond with the 5′ end of the intron. This transesterification reaction releases the 3′ end of the first exon, which then attacks the phosphate of the 3′ splice site, using its 3′-OH in a second transesterification reaction. The intron is released in linear form, with the extra G residue added to its 5′ end. The result is precise excision of the intron and concurrent joining of the exons.

Self-splicing group I introns share several other properties with enzymes, besides accelerating the reaction rate, including their kinetic behavior and specificity. The intron is precise in its excision reaction, largely due to a short sequence—the internal guide sequence—that can base-pair with exon sequences near the 5′ splice site. This pairing promotes the alignment of specific bonds to be cleaved and rejoined.

Because the intron itself is chemically altered during the splicing reaction—its ends are cleaved—it initially appeared to lack one key enzymatic property: the ability to catalyze the same reaction in multiple substrate molecules. In vivo, the intron (414 nucleotides) from _Tetrahymena_ rRNA is quickly degraded after its excision. Experiments have shown, however, that in vitro the intron can act as a true enzyme. A series of intramolecular cyclization and cleavage reactions in the excised intron remove 15 to 19 nucleotides from its 5′ end. The remaining linear RNA promotes nucleotidyl transfer reactions in which some oligonucleotides are lengthened at the expense of others. Although not known to be important in cells, this capability indicates that the RNA can catalyze RNA polymerization. Such activity hints at the possibility that RNA could catalyze its own replication—a key to the "RNA world" hypothesis (see Section 16.5).

Group II Introns Though they have very few conserved sequences, **group II introns** share a common secondary structure consisting of six base-paired regions referred to as domains I through VI (**Figure 16-16a** on p. 568). Domain I contains the binding sites for the 5′ and 3′ splice sites, and domain VI contains an adenosine that functions as the nucleophile to initiate the splicing reaction. Domain V contains sequences critical for the splicing reaction to occur efficiently.

The chemistry of splicing by group II introns is the same as that used by the spliceosome during splicing

(a)

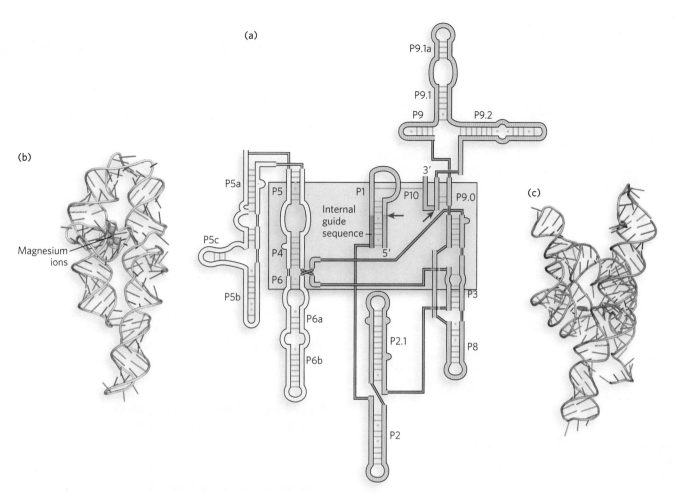

FIGURE 16-14 Secondary and tertiary structures of group I introns. (a) Group I introns share a common secondary structure in which short segments of the RNA strand fold back on themselves to form base-paired segments (exons shown in green; red arrows indicate splice sites). P segments are base-paired regions, numbered sequentially in the secondary structure; connections between P segments are shown by thinner lines indicating connectivity only, not actual sequence. The gray shaded box indicates parts of the structure that form the catalytic core, or active site, of the intron. (b) The crystal structure of the *Tetrahymena* group I intron P4-P6 domain (shown in (a) in yellow), the first large RNA structure to be solved, showing how the helices pack together to form a three-dimensional structure. (c) The complete group I intron, which forms an active site for substrate binding and transesterification. [*Sources: (b) PDB ID 1GID. (c) PDB ID 1U6B.*]

of eukaryotic pre-mRNAs (**Figure 16-17** on p. 569). Here, however, the structure of the RNA itself, rather than the assembly of multiple snRNPs, creates an active site for catalysis, in which the 2′-OH of the branch-point A residue in the intron is directed to attack the phosphodiester bond at the 5′ splice site. As in the spliceosomal reactions, a branched lariat structure forms as an intermediate. After the second step produces the joined exons, the lariat intron can be linearized and degraded or can catalyze further reactions. Thus, like group I introns, group II introns have the ability to catalyze multiple reactions.

The catalytic properties of group II introns have sparked great interest among molecular biologists. The similarity between the mechanisms of self-splicing by group II introns and spliceosome-catalyzed splicing led to the hypothesis that the two processes are evolutionarily related. Perhaps group II introns are renegade spliceosomal RNAs that found a way to survive within genomes as "selfish" genetic elements. Or perhaps group II introns are ancient precursors of the spliceosome, containing just the core RNA components required for catalysis. Either scenario is consistent with the discovery that many group II introns can readily spread to new genes and new

Primary transcript

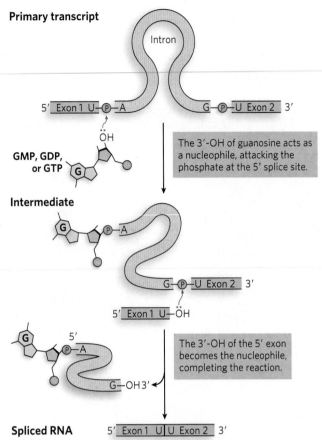

The 3'-OH of guanosine acts as a nucleophile, attacking the phosphate at the 5' splice site.

The 3'-OH of the 5' exon becomes the nucleophile, completing the reaction.

FIGURE 16-15 The self-splicing mechanism of group I introns. Group I introns use an exogenous guanosine (of a guanine nucleotide) to initiate two transesterification steps in self-splicing. The group I intron is released as a linear structure.

organisms by forming a catalytic complex with a protein encoded within the intron itself (**Highlight 16-2** on p. 570). Such a mechanism for self-propagation may have enabled a functional subset of spliceosomal RNA to escape from the spliceosome, or it may have maintained the existence of a primitive self-splicing intron long after it could have been supplanted by the spliceosome.

Although spliceosomal introns seem to be limited to eukaryotes, the self-splicing intron classes are not. Genes with group I and II introns have also been found in bacteria and bacterial viruses. Bacteriophage T4, for example, has several protein-coding genes containing group I introns. Group I introns are found in some nuclear, mitochondrial, and chloroplast genes coding for rRNAs, mRNAs, and tRNAs. Group II introns are generally found in the primary transcripts of mitochondrial or chloroplast mRNAs in fungi, algae, and plants. This prevalence is perhaps explained by the observation that both groups of introns are mobile genetic elements, capable of moving between bacterial strains and species (see Chapter 14).

Exons from Different RNA Molecules Can Be Fused by *Trans*-Splicing

Primary transcripts are sometimes covalently linked to a separate piece of RNA as introns are removed, a process referred to as ***trans*-splicing**. Although not known to occur in humans and most other eukaryotes, *trans*-splicing

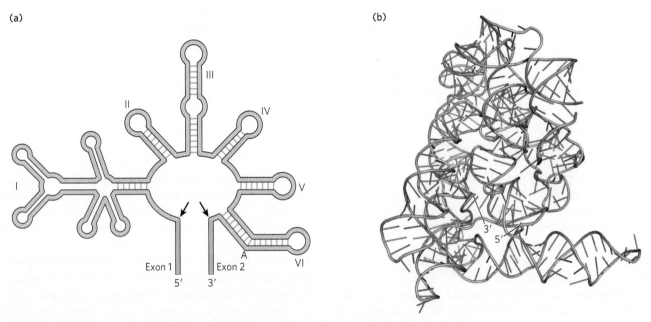

FIGURE 16-16 Secondary and tertiary structures of group II introns. (a) The overall secondary structure of group II introns is conserved, although many introns contain variable-length insertions in the loops of each base-paired segment. Black arrows indicate the splice sites marking the boundaries of the intron; green segments are the exons. The branch-point A is in domain VI. (b) The crystal structure of a spliced group II intron. [Source: (b) PDB ID 3EOH.]

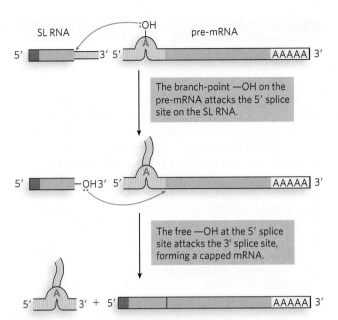

FIGURE 16-18 *Trans*-splicing. In nematodes, a short, 5′-capped leader sequence (SL RNA) is spliced onto a primary transcript (pre-mRNA) to produce the mature mRNA.

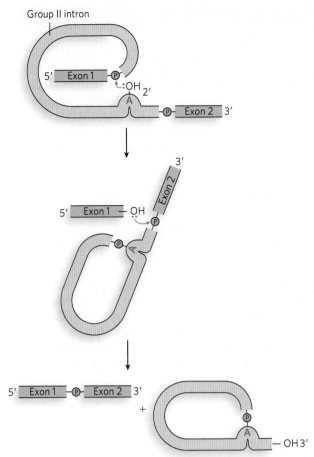

FIGURE 16-17 Self-splicing mechanism of group II introns. Group II introns self-splice by a pathway similar to that of spliceosome-catalyzed splicing. The group II intron is released as a lariat structure.

is the predominant mechanism of mRNA maturation in nematode worms. Genomic sequence data from several different organisms show that *trans*-splicing is evolutionarily conserved in the nematode family. Maturation of primary transcripts involves *trans*-splicing of a short leader sequence called SL1 or SL2 onto the 5′ end of the coding sequence for each individual gene (**Figure 16-18**).

The mechanism resembles those previously described (designated *cis*-splicing), using two transesterification steps. The SL (spliced leader) is donated by SL RNA (100 nucleotides). The 5′ splice site is on the SL RNA, and the site of SL addition, the *trans*-splice site, is the 3′ splice site on the pre-mRNA. The reaction proceeds by way of a branched intermediate similar to the lariat of *cis*-splicing. *Trans*-splicing is catalyzed by spliceosomes, including U2, U4, U5, and U6 snRNPs, but not U1. The spliced leader sequence is not translated into protein, but instead plays a regulatory role in coordinating gene expression. In *Caenorhabditis elegans*, the SL tends to be spliced adjacent to the initiation codon, AUG (often immediately next to it), so the SL is thought to play a role in initiating protein synthesis.

RNA Editing Can Involve the Insertion or Deletion of Bases

Unlike the RNA processing reactions discussed thus far, RNA editing mechanisms seem to have been acquired in recent evolutionary history and to have arisen independently in different groups of organisms. There are two types of **RNA editing**: insertion or deletion, using guide RNA as a template to add or remove bases from the mRNA, and substitution, in which one base is exchanged for another.

A particularly dramatic example of RNA editing occurs in the mitochondrial pre-mRNAs of trypanosomes, parasitic protozoa that cause human diseases such as sleeping sickness. The pre-mRNAs are edited after synthesis by an enzyme that inserts some U residues and deletes others. Like other kinds of RNA processing, editing of trypanosomal pre-mRNAs was initially discovered by comparing mtDNA sequences with corresponding mRNA sequences. RNA editing activity can be detected by isolating RNA from cells harvested at different times, converting the RNA to DNA with reverse transcriptase (see Chapter 14), then producing many copies of the DNA by the polymerase chain reaction. Comparing the sizes of DNA fragments by gel electrophoresis provides a measure of the relative amounts of edited and pre-edited targets.

RNA editing by insertion and deletion is catalyzed by the **editosome**, a complex of at least 16 proteins. A key editosome enzyme, RNA-editing terminal uridylyltransferase (TUTase), catalyzes the uridylate addition reaction. The structure of the TUTase ensures that only

HIGHLIGHT 16-2   EVOLUTION

The Origin of Introns

The origin of introns remains one of the mysteries of modern biology. Because many exons correspond to protein structural domains, it has been argued that introns are relics of early genes, in which sequences were stitched together randomly from shorter segments and only those with useful functions were maintained. According to this line of thinking, bacteria lack introns because competition for rapid growth, and consequent genome streamlining, led to intron loss from all but a few, rare bacterial and bacteriophage genes. An alternative theory is that introns arose more recently on the evolutionary timeline. Analyses of intron positions in related genes from many different species, made possible with the arrival of online whole-genome sequence databases in the 1990s, show that in many cases the introns and their positions within genes are not conserved. This might mean that introns were introduced relatively late in the evolution of modern genomes, or that introns are highly mobile.

The group II introns found in bacteria and in mitochondrial and chloroplast DNA are examples of mobile introns. Like retrotransposons, the introns encode proteins with both endonuclease and reverse transcriptase activities, allowing them to splice themselves back into DNA. In a transposition process termed **retrohoming**, the encoded protein forms a complex with the intron RNA after the intron is spliced from the primary transcript (Figure 1). Normally, the intron moves from one copy (allele) of a gene to an identical site in another copy of the same gene that lacks the intron. The initial insertion steps reprise the splicing mechanism, but in reverse. Once the RNA strand has been integrated into the DNA, the endonuclease cleaves the opposite DNA strand, and the inserted RNA is copied by the reverse transcriptase function associated with the endonuclease. The RNA is removed and replaced by DNA, converting the RNA intron to an inserted segment of DNA.

Over time, every copy of a particular gene in a population may acquire the intron. Much more rarely, the intron may insert itself into a new location in an unrelated gene. If this event does not kill the host cell, it can lead to the evolution and distribution of an intron in a new location. These mobile group II introns are thought to be the evolutionary precursors of the more widespread (and nonmobile) group II introns found in many eukaryotic genes.

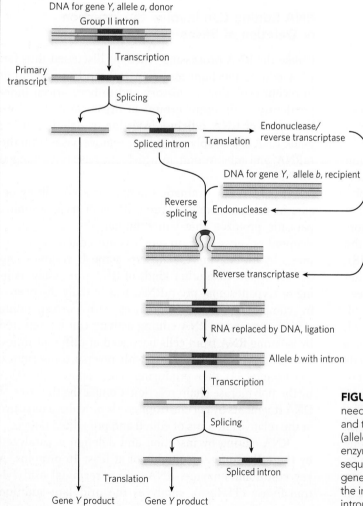

FIGURE 1 Mobile group II introns encode the enzyme activities needed to propagate the intron to new areas of the genome and to new hosts. Two alleles of a gene (*Y*) differ by the presence (allele *a*) or absence (allele *b*) of an intron. In retrohoming, an enzyme called a retrotransposase is produced from a coding sequence within the excised intron. This transposase first cleaves gene *Y*, allele *b* DNA at the site of intron insertion, then copies the intron RNA into a cDNA copy that is ligated to create a new, intron-containing allele of gene *Y*.

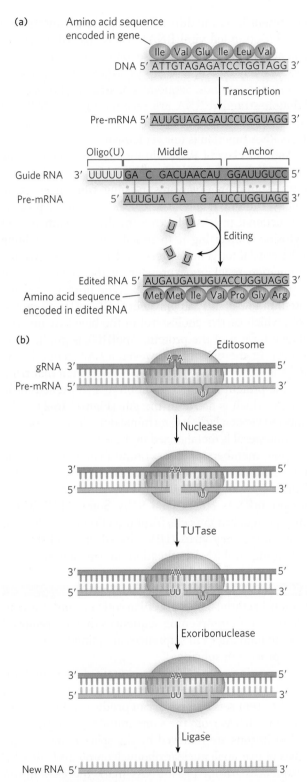

(a)

Amino acid sequence
encoded in gene

Ile Val Glu Ile Leu Val

DNA 5′ ATTGTAGAGATCCTGGTAGG 3′

Transcription

Pre-mRNA 5′ AUUGUAGAGAUCCUGGUAGG 3′

Oligo(U)   Middle   Anchor

Guide RNA 3′ UUUUU GA C GACUAACAU GGAUUGUCC 5′
Pre-mRNA 5′   AUUGUA GA G AUCCUGGUAGG 3′

Editing

Edited RNA 5′ AUGAUGAUUGUACCUGGUAGG 3′

Amino acid sequence
encoded in edited RNA

Met Met Ile Val Pro Gly Arg

(b)

Editosome

gRNA 3′ ———— AA ———— 5′
Pre-mRNA 5′ ———— U ———— 3′

Nuclease

3′ ———— AA ———— 5′
5′ ———— U ———— 3′

TUTase

3′ ———— AA ———— 5′
5′ ———— UU U ———— 3′

Exoribonuclease

3′ ———— AA ———— 5′
5′ ———— UU ———— 3′

Ligase

New RNA 5′ ———— UU ———— 3′

FIGURE 16-19 RNA editing by nucleotide insertion and deletion. (a) Before editing, a pre-mRNA is missing some U residues required in the mature RNA and contains some extra U residues. A guide RNA (gRNA) in the editosome base-pairs with the pre-mRNA, guiding insertion and deletion of U residues. The insertions and deletions change the protein sequence encoded by the mRNA. (b) The editosome has several enzymatic activities: a nuclease to cleave the pre-mRNA at an insertion site, TUTase to fill in missing U residues, exoribonuclease to delete extra U residues, and ligase to seal the nicks.

U residues are added to the mRNA. Because TUTase is essential for the survival of trypanosomes in the host bloodstream, it is a potential target for drug therapy.

The editosome uses small guide RNAs (gRNAs) that are partially complementary to the pre-mRNA regions to be changed (**Figure 16-19a**). Guide RNAs (35 to 75 nucleotides) are encoded by the trypanosomal mtDNA. Each gRNA contains three functionally important regions: an "anchor" sequence complementary to a target sequence in the pre-edited RNA, a middle segment containing the editing information, and a 3′ terminal oligo(U) (5 to 25 nucleotides) extension. The gRNAs base-pair with the target pre-mRNA, and the sequence differences are copied from the gRNA to the pre-mRNA using the gRNA as a template.

The mechanism of the editosome involves an endonucleolytic cut at the mismatch point between the gRNA and the unedited transcript (**Figure 16-19b**). The insertion step is catalyzed by TUTase, which adds U residues from UTP to the 3′ end of the mRNA. The opened ends are held in place by other proteins in the complex. Another enzyme, a U-specific exoribonuclease, removes any unpaired U residues. After editing has made the mRNA complementary to the gRNA, an RNA ligase rejoins the ends of the edited mRNA transcript. The resulting U insertions or deletions often create a frameshift in the edited mRNA, altering the sequence translated by the ribosome.

RNA Editing by Substitution Involves Deamination of A or C Residues

RNA editing also occurs in human cells and in the viruses that infect them. Many human pre-mRNAs are edited by **adenosine deaminase acting on RNA (ADAR)**, a fascinating enzyme that catalyzes the conversion of adenosine to inosine by removal of the amino group at C-6 of the adenine ring. Typically, ADAR converts just a few of the A residues to I residues within a very large transcript, but those changes are essential for creating the correct sequence to encode a functional protein. Many of the known substrates for ADAR are mRNAs coding for proteins that function in the central nervous system.

An interesting example of ADAR-mediated pre-mRNA editing occurs in the mRNAs encoding glutamate receptor channels (**Figure 16-20**). These Ca^{2+} channels, which allow the fast transmission of neural signals, are controlled by L-glutamate, the principal excitatory neurotransmitter in the brain. Two related classes of glutamate receptor channel proteins were found to differ by only a single amino acid, either an Arg or a Gln residue in a defined position of the putative channel-forming segment. However, the genomic DNA sequence encoding the channel segment of both proteins has a glutamine codon (CAG). Researchers discovered that in one

(a)

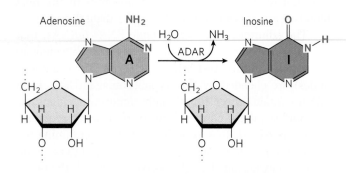

(b)

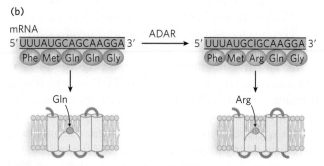

FIGURE 16-20 A-to-I editing of the mRNA for the glutamate receptor channel protein. (a) ADAR catalyzes the deamination of adenosine to inosine. (b) A-to-I editing results in replacement of a glutamine with an arginine in the protein product. The Arg residue at this position results in a protein that is much more efficient for Ca^{2+} transport.

set of mRNAs, ADAR converts the A of the CAG codon into an I. Because I can base-pair with C, the A-to-I conversion makes the CIG codon appear to be CGG when detected by standard sequencing reactions involving reverse transcription into DNA. Furthermore, the CIG codon is translated as a CGG codon during protein synthesis. Given that CGG encodes arginine, this editing event results in a glutamine-to-arginine substitution in the glutamate receptor channel protein. Such a change might seem subtle, but in fact the replacement of Gln by Arg in this segment profoundly alters the properties of ion flow across the channel.

A-to-I editing of pre-mRNAs, including the glutamate receptor channel pre-mRNA, often occurs at sites near exon-intron boundaries. This is because ADAR recognizes double-stranded RNA, such as the short duplexes formed in a pre-mRNA where an intron sequence folds back to base-pair with a complementary exon sequence. For this reason, editing necessarily occurs before the pre-mRNA is spliced. The C-terminal domain of Pol II enhances editing by preventing premature splicing, which would otherwise remove the intron recognition sites for ADAR. In some cases, editing occurs within longer regions of duplex RNA, including the large hairpin precursors of regulatory RNAs known as microRNAs

(see Section 16.4) and duplexes arising from transcription of double-stranded viral RNAs or base-paired repetitive sequences in the genome. These longer duplex RNAs can be extensively edited, with up to 50% of the adenosines converted to inosines. Sequences or structural properties of double-stranded RNA are recognized by ADAR and help specify the correct editing sites. How such editing affects RNA function is not yet known.

Another class of RNA editing enzymes, the cytidine deaminases, catalyze C-to-U conversions in mRNA substrates. Found in organisms ranging from bacteria to humans, cytidine deaminases can be critical for producing functional mRNAs. Editing involves deamination of a cytosine, converting it to uracil. Such C-to-U editing can be crucial for gene expression. For example, cytidine deamination regulates expression of the apolipoprotein B (ApoB) gene in humans. ApoB is the primary apolipoprotein in low-density lipoprotein (LDL, or "bad" cholesterol), which carries cholesterol in the bloodstream. The full-length form of the protein, ApoB100, is produced in the liver. In the intestine, however, a CAA codon in the mRNA is edited to UAA, creating a stop codon that truncates the protein during translation to a shortened form, ApoB48, which is active in the gut (**Figure 16-21**). The truncated version lacks a C-terminal domain, affecting the way cholesterol is metabolized in these tissues.

Some members of the cytidine deaminase enzyme family are found only in primates. These enzymes, called APOBECs, catalyze C-to-U conversions not only in cellular pre-mRNAs but in viral RNA. Some APOBEC enzymes protect human cells from infection by the human immunodeficiency virus (HIV), by editing and thereby inactivating viral RNA replication intermediates.

SECTION 16.2 SUMMARY

- A few bacterial and most eukaryotic pre-mRNAs contain introns, intervening sequences that are removed as the flanking exon sequences are joined together in the process of splicing.

- Eukaryotic pre-mRNAs can contain dozens or hundreds of introns. In alternative splicing, the removal of different sets of introns can produce a series of different mRNAs from the same initial transcript.

- Most introns are removed by the spliceosome, a complex of five small nuclear ribonucleoproteins (snRNPs); sequences within pre-mRNAs mark the exon-intron boundaries.

- Some introns are capable of self-splicing without assistance from any proteins; both kinds of self-splicing introns, group I and group II, also occur in bacterial and mitochondrial RNAs.

- In some organisms, *trans*-splicing produces mRNAs in which exon sequences derive from different primary transcripts.

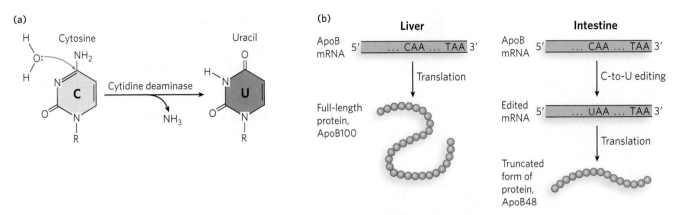

FIGURE 16-21 C-to-U editing of the mRNA for ApoB protein. (a) Cytidine deaminase catalyzes the deamination of cytidine to uridine. (b) The ApoB gene is expressed in the liver and intestine. In the intestine, the mRNA goes through an additional processing step (after splicing) in which a C residue is changed to a U residue by cytidine deaminase, resulting in a stop codon and producing a smaller protein (ApoB48) that aids in the intestinal absorption of lipids.

- The nucleotide sequence of eukaryotic mRNAs is sometimes altered by RNA editing enzymes. This can have far-reaching effects, such as altering the protein encoded by the mRNA or influencing the regulation of mRNA translation.
- One type of mRNA editing involves the insertion or deletion of U residues. Another type involves enzymatic deamination of A or C residues, converting A to I or C to U.

16.3 RNA TRANSPORT AND DEGRADATION

Once eukaryotic RNAs have been processed in the nucleus, they are ready for export to the cytoplasm. Mature mRNAs are translated into protein through the action of ribosomes, whereas other, non-protein-coding RNAs participate in various regulatory activities in the cytoplasm or undergo further modification in the cytoplasm then reenter the nucleus. In each case, the processed modifications of the nuclear RNAs identify them as being ready for export, and they are then transported to the cytoplasm.

Like other kinds of RNA processing, degradation is a highly regulated process carried out by complex enzymatic machinery. RNA degradation enzymes are essential to cell survival because they help maintain appropriate amounts of mRNAs in response to metabolic and environmental signals. These enzymes also rid the cell of defective mRNAs containing premature stop codons.

Different Kinds of RNA Use Different Nuclear Export Pathways

Nuclear export of most non-protein-coding RNAs involves members of a conserved family of transport receptors called **importins** and **exportins**, collectively known as **karyopherins**. Exportins bind to their RNA "cargo" in the nucleus and escort it through nuclear pores to the cytoplasm, where the cargo is released. Both tRNAs and some kinds of noncoding RNAs bind directly to their respective exportin. In contrast, rRNAs are exported in pre-ribosomal particles containing ribosomal proteins, several rRNA species, and nonribosomal proteins, using their own exportin. The snRNAs are transported to the cytoplasm and remain there only transiently, undergoing assembly into snRNP particles and then reentering the nucleus.

In each case, the exportin requires a small GTP-hydrolyzing protein called Ran that regulates cargo-receptor interactions (**Figure 16-22**). For nuclear export, the RNA cargo and exportin associate cooperatively with a Ran molecule bound to GTP. Once this ternary complex translocates to the cytoplasm, the Ran-bound GTP is hydrolyzed to GDP, causing release of the cargo RNA from the exportin. The Ran-GDP reenters the nucleus, where its GDP is exchanged for GTP by a guanine nucleotide exchange factor (GEF), and the cycle can begin again.

For noncoding RNAs that are processed in the cytoplasm and returned to the nucleus, the cargo RNA and its import receptor—the importin—cross through a nuclear pore into the nucleus, where the cargo is released as the importin binds to Ran-GTP. The importin-Ran-GTP then translocates back to the cytoplasm, the importin dissociates, and GTP is hydrolyzed to GDP, restarting the cycle. Thus, export and import are reverse processes for which directionality is maintained by the presence of Ran-GTP in the nucleus and Ran-GDP in the cytoplasm.

Spliced mRNA crosses through a nuclear pore via a Ran-independent pathway, involving a different set of export factors that associate with other proteins to form a much larger complex called TREX (*transcription-export*). TREX joins the machineries responsible for the

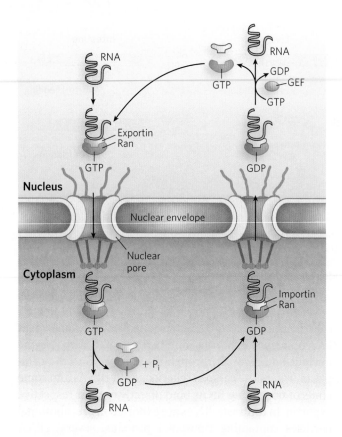

FIGURE 16-22 Nuclear export and import of RNA. The GTP-hydrolyzing protein Ran regulates RNA-receptor interactions. For nuclear export, the RNA cargo and its exportin receptor associate with Ran-GTP and the complex moves to the cytoplasm, where GTP is hydrolyzed to GDP and the cargo is released. For import, the RNA and its importin receptor cross through a nuclear pore into the nucleus, where GDP is exchanged for GTP and the RNA cargo is released.

transcription, splicing, and export of mRNA. The TREX complex is specifically recruited to actively transcribed genes through interactions with Pol II. In human cells, the TREX component protein Aly recruits the complex to the 5′ cap of nascent transcripts by interacting with the cap-binding protein CBP80. Because TREX functions as a nuclear export factor, mRNAs are transported out of the nucleus in the 5′→3′ direction through a nuclear pore.

mRNA Export from the Nucleus Is Coupled to Pre-mRNA Splicing

In eukaryotes, pre-mRNA splicing necessarily precedes export of the mature transcript from the nucleus, and researchers wondered whether pre-mRNA processing and nuclear export were closely coupled—which would provide a mechanism of mRNA quality control. To test this possibility, either pre-mRNA containing a single intron or the same mRNA lacking the intron were injected into the nuclei of frog eggs. The rate of nuclear export

of each type of mRNA was then measured by detecting the presence of the RNA in the cytoplasm over time. Results of this experiment showed that intron-containing pre-mRNAs, which were spliced in the nucleus, were exported much more rapidly and efficiently than the identical mRNAs lacking the intron. Furthermore, the spliced mRNA was found to assemble with a different set of proteins than the mRNA lacking the intron. This led to the conclusion that splicing generates a specific mRNA-protein complex that targets the mRNA for nuclear export, explaining the broader observation that an intron is required for the efficient expression of many eukaryotic protein-coding genes. In this way, only those mRNAs that have the correct end structures and spliced-exon sequence are used for protein synthesis.

Mature mRNAs produced by splicing end up in different intracellular locations and are differently translated and degraded than otherwise identical mRNAs produced from non-intron-containing genes. The explanation is that splicing influences the set of proteins that associate with the mRNA in the nucleus to form an mRNP (mRNA ribonucleoprotein particle). These proteins, in turn, ensure that the mRNA interacts with exportins for shipment out of the nucleus. Chemical cross-linking experiments with cell extracts translating mRNAs with or without introns showed that several proteins bind to exon-exon junctions only as a consequence of splicing. Spliceosomes deposit a complex of proteins called the **exon junction complex (EJC)** on mRNAs at a position 20 to 24 nucleotides upstream of exon-exon junctions (**Figure 16-23**). At its core, EJC contains four proteins: eIF4AIII, MAGOH, Y14, and MLN51. The bound complexes accompany spliced mRNA into the cytoplasm, where they are removed during the first ("pioneer") round of translation.

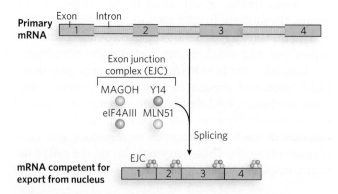

FIGURE 16-23 The exon junction complex. The EJC is a complex of four proteins responsible for mRNA quality control. An EJC is deposited on the mRNA after splicing, just upstream of each exon-exon junction. The complexes accompany the mature mRNA out of the nucleus and into the cytoplasm.

Some mRNAs Are Localized to Specific Regions of the Cytoplasm

In specialized cells, including oocytes (egg cells) and neurons, mRNAs are localized to particular sites prior to translation. The mechanism of such mRNA localization is best characterized in the fruit fly *Drosophila melanogaster*, in which maternal mRNAs are trafficked to various parts of the egg to help establish polarity during the early stages of embryo development (**Figure 16-24**). Shortly after the egg begins to mature, but before fertilization, mRNAs encoding proteins called Oskar and Bicoid bind to proteins that can move along microtubules, filamentous protein polymers that contribute to cell shape and structure. The *oskar* and *bicoid* mRNAs are shuttled to the parts of the egg where their protein products are required to form structures in the developing embryo. Analogous mechanisms of mRNA localization are thought to occur in neurons, in which mRNAs must be moved to parts of the cell very far from the nucleus for the localized protein synthesis required for proper neural function.

An obvious advantage of regulating gene expression by mRNA localization is that it allows protein production to be spatially restricted within the cytoplasm. In this way, production can be turned on (and off) as required, without waiting for transcription, mRNA export, translation, and subsequent targeting of the protein to the site where it is needed. In addition, localized mRNAs can be translated multiple times to generate many copies of a protein at the required site. Local translation can also protect the rest of the cell from proteins that would be toxic or interfere with functions in other cellular compartments.

Cellular mRNAs Are Degraded at Different Rates

As for all molecules in the cell, the amount of an mRNA transcript is determined by its relative rates of synthesis and degradation, or decay. Cellular mRNAs are degraded as part of their normal life cycle. **RNA degradation**, catalyzed by ribonucleases, is the complete hydrolysis of RNA molecules into their component nucleotides. When mRNA synthesis and decay rates are balanced, the concentration of the mRNA remains in a steady state. A change in either rate will lead to a net accumulation or depletion of the mRNA, affecting the rate of protein synthesis. Degradative pathways ensure that mRNAs do not build up in the cell and direct the synthesis of unnecessary proteins.

In eukaryotic cells, degradation rates for mRNAs produced from different genes can vary greatly, from a half-life of minutes or even seconds for a gene product that is needed only briefly, to many cell generations for a gene product in constant demand. Bacterial cells grow much faster than eukaryotic cells and must rapidly adapt to changing environmental and metabolic conditions; their mRNAs are stable for only a few minutes. Degradation rates of an RNA are affected by its primary and secondary structure. For instance, a hairpin structure in bacterial and eukaryotic mRNAs can confer stability. In eukaryotes, sequences rich in A and U residues, known as AU-rich elements (AREs), occur in the 3′UTRs of some mRNAs. These elements recruit factors such as nucleases or RNA-binding proteins that can enhance or reduce, respectively, the degradation rate of the mRNA.

For mRNAs, degradation in *E. coli* begins with one or a few cuts by an endoribonuclease, followed by 3′→5′ degradation by an exoribonuclease. In eukaryotes, the poly(A) tail is shortened, the 5′ cap is removed, then the mRNA can be degraded by ribonucleases. Eukaryotes have a complex of up to 10 conserved 3′→5′ exoribonucleases, called the **exosome**, which, besides degrading mRNAs, is involved in the processing of the 3′ end of rRNAs and tRNAs (**Figure 16-25**). The exosome is the major path of mRNA degradation in higher eukaryotes, but lower eukaryotes use primarily 5′→3′ exonucleases.

FIGURE 16-24 Transport of mRNA in *Drosophila* eggs and embryos. The building blocks of anterior-posterior axis patterning in *Drosophila* are laid out during egg formation (oogenesis), well before the egg is fertilized and deposited. The developing egg (oocyte) is polarized by differentially localized mRNA molecules. In each oocyte shown, a different mRNA is labeled with a blue fluorescent marker. Nuclear DNA is labeled with a red marker. Each mRNA contributes to pattern formation in the developing embryo. [Source: Reprinted from Eric Lecuyer, et al., Cell 131:174–187. © 2007, with permission from Elsevier. Courtesy Eric Lecuyer.]

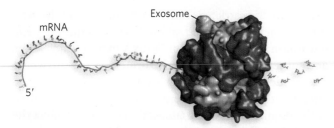

FIGURE 16-25 The human exosome. The exosome has a ringlike structure. The complex encircles the mRNA and slides along in a 3′→5′ direction, degrading the RNA as it moves. [*Source: PDB ID 2NN6.*]

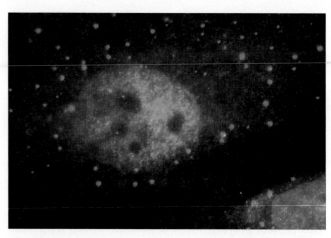

FIGURE 16-26 Processing (P) bodies in human cells. In these HeLa cells, P bodies (red spots) are visualized by staining with an antibody that binds to a protein involved in mRNA turnover. Nuclei are stained blue. [*Source: Courtesy of John Bloom and Roy Parker.*]

Decay rates and half-lives are determined by proteins that enhance or inhibit exosome binding through interactions with the 3′ UTR of the mRNA. In addition, two mRNA surveillance pathways are in place to respond to mRNAs that contain a premature stop codon or lack a stop codon. **Nonsense-mediated decay (NMD)** is triggered by exon junction complexes, which, as we have seen, are deposited during pre-mRNA splicing in the nucleus. Normally, EJCs are removed by the ribosome during the first round of translation, but if an EJC is located downstream of a stop codon, its presence indicates that splicing occurred at this position and that there should be additional coding sequence after the stop codon. The EJC thus signals that this is an mRNA with a premature stop codon, or nonsense codon, and triggers mRNA degradation. In contrast, the **non-stop decay** pathway targets mRNA molecules lacking a stop codon. Ribosomes traversing these mRNAs are released from the 3′ end of the message, and the mRNA is shunted to the exosome for degradation.

Processing Bodies Are the Sites of mRNA Storage and Degradation in Eukaryotic Cells

In eukaryotes, mRNAs that are not engaged in translation are sequestered in localized areas of the cytoplasm called **processing bodies (P bodies)**, which can be observed by light microscopy. P bodies contain proteins that catalyze removal of the mRNA 5′ cap and thus are thought to be sites of mRNA degradation. They also seem to be sites where mRNAs are temporarily stored when not being translated and may therefore play an active role in regulating which proteins are made in response to the cell's needs (**Figure 16-26**).

Recent experimental evidence supports an emerging model of cytoplasmic mRNA function in which translation and degradation rates are influenced by the relative concentrations of mRNA in polyribosomes (groups of ribosomes translating an mRNA; see Chapter 18) and in P bodies. In some cases, mRNA-specific binding factors suppress translation and promote degradation by recruiting P-body proteins to individual mRNAs.

Notably, many of the proteins necessary for microRNA (miRNA) gene silencing (discussed in Chapter 22) are localized to P bodies, including the scaffold protein GW182, Argonaute (Ago), 5′ decapping enzymes, and helicases (enzymes that unwind RNA). The current evidence suggests that P bodies are scaffolding centers of miRNA function, especially given the finding that reduction of GW182 levels disrupts P-body formation.

SECTION 16.3 SUMMARY

- Non-protein-coding RNAs are exported from the nucleus by exportin proteins in a Ran-dependent pathway that requires GTP.
- After splicing, eukaryotic mRNAs are recognized by their processing modifications and then exported to the cytoplasm through a nuclear pore.
- Spliced mRNA exits the nucleus in a Ran-independent pathway involving export factors and other proteins in a large complex called TREX, which coordinates the machineries responsible for transcription, splicing, and export of mRNA.
- Spliceosomes deposit a complex of proteins—the exon junction complex (EJC)—on spliced mRNAs upstream from exon-exon junctions, providing a physical tag indicating that splicing has occurred. The EJC promotes nuclear export and mRNA stability.
- Cellular mRNAs are degraded at different rates, depending on their interactions with ribonucleases. Nonsense-mediated decay and non-stop decay are two pathways that guard against the translation of defective mRNAs.
- Bacterial mRNA degradation begins with one or a few cuts by an endoribonuclease, followed by 3′→5′ degradation by an exoribonuclease.

- In eukaryotes, a complex of up to 10 conserved 3′→5′ exoribonucleases—the exosome—helps degrade mRNAs and processes the 3′ end of rRNAs and tRNAs.
- Processing bodies (P bodies) are cytoplasmic locations of mRNA storage and degradation.

16.4 PROCESSING OF NON-PROTEIN-CODING RNAs

We have seen how the cell produces mature mRNA transcripts through a series of processing mechanisms. Each aspect of mRNA processing, transport, and decay is carefully regulated to control how, when, and where the mRNA is made or translated. RNA processing is not unique to mRNAs. All functional RNAs in cells are produced as precursor transcripts that must be cleaved to form the mature, functional RNA. This processing could be required because RNA polymerases do not always have specific termination sites, so transcripts from the same gene can have heterogeneous ends. Processing also provides convenient opportunities for the cell to regulate RNA levels. We discuss here the processing of three different kinds of functional RNAs central to gene expression: transfer RNAs (tRNAs), ribosomal RNAs (rRNAs), and miRNAs.

Maturation of tRNAs Involves Site-Specific Cleavage and Chemical Modification

Transfer RNAs, the molecules that carry amino acids to the ribosome during protein synthesis (see Chapter 18), are coordinately expressed in response to metabolic needs. In some cases, multiple tRNAs are synthesized as a single primary transcript and are separated by enzymatic cleavage. Even tRNAs synthesized alone are derived from longer RNA precursors by the enzymatic removal of nucleotides from the 5′ and 3′ ends (**Figure 16-27**). The endonuclease ribonuclease P (RNase P), a ribonucleoprotein found in all organisms, removes nucleotides from the 5′ end of tRNAs. The RNA component of the enzyme is essential for activity. Indeed, bacterial RNase P can carry out its processing function with precision even in the absence of the protein component. The 3′ end of tRNAs is processed by one or more nucleases, including the exonuclease RNase D.

In eukaryotes, a few tRNA transcripts have introns that must be excised. These introns are spliced by an ATP-dependent mechanism distinct from that of the spliceosome: a splicing endonuclease that recognizes and cleaves the phosphodiester bonds at the intron splice sites. RNA ligase joins the two exons to complete the reaction.

Transfer RNA precursors often undergo further posttranscriptional processing. The terminal CCA-3′ to which an amino acid is attached is absent from some bacterial and all eukaryotic tRNA precursors, and is added after the initial transcript is made (see Figure 16-27). This addition is carried out by tRNA nucleotidyltransferase, an unusual enzyme that binds the three ribonucleoside triphosphate precursors in separate active sites and catalyzes the formation of the phosphodiester bonds to produce the CCA-3′ sequence. The creation of this defined sequence of nucleotides is therefore not dependent on a DNA or RNA template—the template consists of the three binding sites of the nucleotidyltransferase.

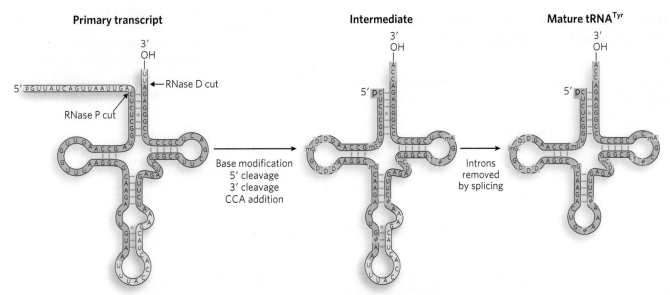

FIGURE 16-27 The processing of tRNA. Transfer RNAs are trimmed at the 5′ and 3′ ends by RNase P and RNase D, respectively. Shown here is the processing of tRNA^Tyr. The CCA trinucleotide is added to the 3′ end, and some bases are modified to provide stability and enhance the functioning of the mature tRNA. In some eukaryotic tRNAs, introns are excised during processing.

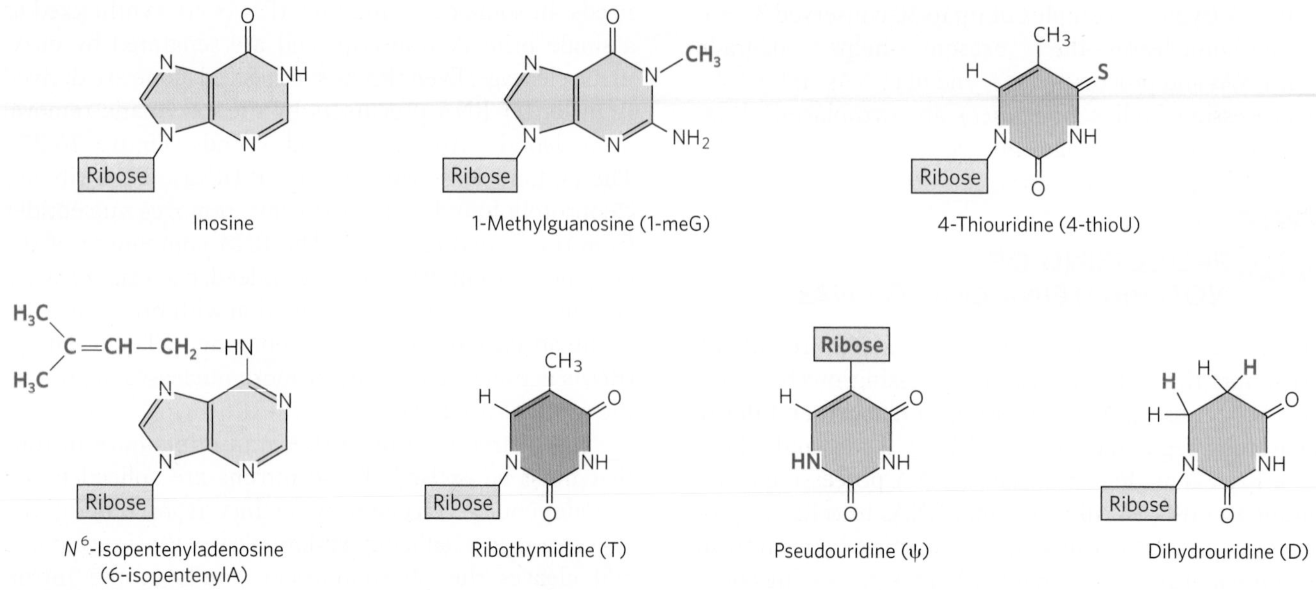

Inosine

1-Methylguanosine (1-meG)

4-Thiouridine (4-thioU)

N^6-Isopentenyladenosine
(6-isopentenylA)

Ribothymidine (T)

Pseudouridine (ψ)

Dihydrouridine (D)

FIGURE 16-28 Modified tRNA bases produced in posttranscriptional reactions. Specific nucleotide residues in tRNAs, as well as in other RNAs including rRNAs, are chemically modified by enzymes that recognize particular structures and/or sequences.

The final type of tRNA processing is the modification of some bases by methylation, deamination, or reduction. In some cases, a base is removed from a specific site in the tRNA sequence and replaced by a different, noncanonical base (**Figure 16-28**). For example, in some tRNAs, uracil is removed at a particular U residue and reattached to the ribose through C-5 to create pseudouridine (ψ). Modified bases often occur at characteristic positions in all tRNAs, suggesting their importance for structural stability or recognition of the tRNA by other enzymes (see the How We Know section at the end of this chapter).

Maturation of rRNA Involves Site-Specific Cleavage and Chemical Modification

Ribosomal RNAs of bacterial and eukaryotic cells are produced from longer precursors called **preribosomal RNAs (pre-rRNAs)**. The pre-rRNA transcripts, produced by RNA polymerase I (Pol I) in eukaryotes, are coordinately synthesized so that they are present in similar amounts for ribosome assembly. In bacteria, the three rRNAs needed to form a functional ribosome—16S, 23S, and 5S—arise from a single 30S RNA precursor of about 6,500 nucleotides. RNA at both ends of the 30S precursor and segments between the rRNAs are removed during processing (**Figure 16-29**).

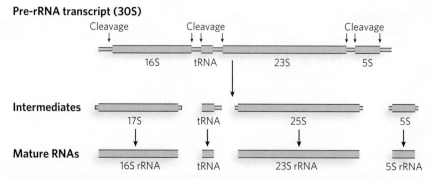

FIGURE 16-29 Processing of pre-rRNA transcripts in bacteria. The 30S pre-rRNA contains the 16S, 23S, and 5S rRNA sequences, as well as a tRNA sequence. During processing, these segments are separated by nuclease activities.

The *E. coli* genome encodes seven pre-rRNA molecules, each with essentially identical rRNA-coding regions but differing in the segments between these regions. One or two tRNA genes are found between the 16S and 23S rRNA genes, with different tRNAs arising from different pre-rRNA transcripts. Coding sequences for tRNAs are also found on the 3′ side of the 5S rRNA in some precursor transcripts.

In eukaryotes, a 45S pre-rRNA transcript is processed in the nucleolus to form the 18S, 28S, and 5.8S rRNAs characteristic of eukaryotic ribosomes. In an interesting quirk of evolution, the 5S rRNA of most eukaryotes is made as a completely separate transcript by a different polymerase (Pol III, rather than Pol I). The enzymes responsible for rRNA processing are localized in the nucleolus and are thought to begin cleaving the pre-rRNA transcript during transcription. Processing occurs in an ordered set of reactions guided in part by RNAs in **small nucleolar ribonucleoproteins (snoRNPs)**, which base-pair with the pre-rRNA transcript at specific cleavage sites. The binding of snoRNPs also specifies sites where methyl groups are added to 2′-hydroxyl groups in rRNAs. Careful experiments comparing the function of rRNAs with and without particular sites methylated have failed to reveal a dramatic difference in behavior. However, bacteria or yeast cells with all of the naturally occurring sites in rRNA methylated are found to outcompete cells lacking some of these sites.

Small Regulatory RNAs Are Derived from Larger Precursor Transcripts

Most eukaryotic cells use RNAs called **microRNAs (miRNAs)** (21 to 23 nucleotides) to regulate the expression of many genes. In addition, many eukaryotes also use short interfering RNAs (siRNAs) to prevent protein synthesis from specific mRNAs. These mechanisms of regulation by miRNAs and siRNAs, known as RNA interference, are discussed in detail in Chapter 22.

The miRNAs, like tRNAs and rRNAs, are encoded by genes that are transcribed but not translated into protein. Primary miRNA transcripts (pri-miRNAs) are typically synthesized by Pol II and then capped and polyadenylated. These transcripts have the ability to fold into extended hairpinlike structures with extended single-stranded RNA on the 5′ and 3′ sides of the hairpin. Such structures are recognized by an RNA-binding protein, which in humans is encoded by *DGCR8* (the *DiGeorge syndrome critical region gene 8*). The DGCR8 protein recruits nuclear pri-miRNAs to the enzyme **Drosha** to form a **microprocessor complex** (**Figure 16-30**). Drosha is an endonuclease that cleaves the pri-miRNA hairpin duplex to produce precursor miRNAs (pre-miRNAs), ~70 nucleotides in length and

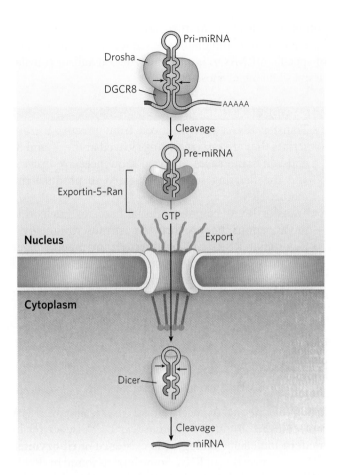

FIGURE 16-30 Processing of small regulatory RNAs. The small miRNAs and siRNAs are formed by the processing of much larger precursor transcripts by specific endonucleases. *[Source: Data from V. N. Kim, J. Han, and M. C. Siomi, Nat. Rev. Mol. Cell Biol. 10:126–139, 2009, Figs. 2–4.]*

containing characteristic dinucleotide 3′ overhangs. Pre-miRNAs then assemble with RNA export proteins, including exportin-5, for transport out of the nucleus. Most pre-miRNAs are not perfectly base-paired hairpins, but instead contain one or more unpaired bases, or non-Watson-Crick base pairs, in the stem of the hairpin—perhaps because perfectly double-stranded RNAs would activate the cell's antiviral machinery through a pathway known as the interferon response.

Once in the cytoplasm, pre-miRNAs are further cleaved by the double-stranded RNA endonuclease called **Dicer**, releasing the mature miRNA. Dicer recognizes the dinucleotide overhangs of pre-miRNAs, which help position RNA substrates such that Dicer's cleavage products have a characteristic length of 21 to 23 bp. Dicer can also cleave long double-stranded RNAs generated within the cell or introduced by viral infection or transfection, to produce siRNAs. Dicer assists in loading miRNAs and siRNAs into complexes called RNA-induced silencing complexes (RISCs). Mature miRNAs and siRNAs can base-pair with complementary sequences in the 3′UTRs of specific mRNAs to induce degradation and/or translational silencing of those mRNAs.

SECTION 16.4 SUMMARY

- Transfer RNAs are produced from precursor transcripts that are enzymatically cleaved at the 5′ and 3′ ends. In some cases, internal sequences are excised and the flanking sequences ligated to produce the mature tRNA.
- Transfer RNAs contain unusual bases or nucleotide modifications that are formed posttranscriptionally.
- Ribosomal RNAs are produced from much longer precursor transcripts by site-specific cleavage, and particular nucleotides are chemically modified.
- Regulatory RNAs, such as miRNAs, are also processed from larger transcripts.

16.5 RNA CATALYSIS AND THE RNA WORLD HYPOTHESIS

Study of the posttranscriptional processing of RNA molecules led to one of the most exciting discoveries in modern molecular biology: the existence of **ribozymes**, enzymes consisting of RNA. Some introns found in rRNA, mRNA, and tRNA genes in bacteria, mitochondria, chloroplasts and bacteriophages have intrinsic self-splicing activity, meaning that no proteins are required to catalyze their excision from precursor transcripts. In addition, in some types of cells, certain other RNA processing reactions, such as the removal of short sequences from the precursors of tRNAs, are catalyzed by RNA. RNA is the only known molecule that is capable of both encoding genetic information and functioning as an enzyme. For this reason, many scientists have proposed that RNA could have formed the basis for the early life forms that evolved into modern organisms, because, in principle, RNA could both carry and copy genetic information. We discuss here the properties of RNA that support its multiple functions and explore the implications of the RNA world hypothesis.

Ribozymes Catalyze Similar Kinds of Reactions But Have Diverse Functions

The catalytic RNAs discovered thus far include self-splicing introns, bacterial RNase P, and several additional classes of naturally occurring ribozymes. Experimentally, it has proved possible to select catalytically active RNA molecules from pools of randomized sequences prepared in the laboratory. Ribozymes selected in this way can have a variety of enzymatic functions, demonstrating the inherent catalytic capabilities of RNA. The activities of many ribozymes, natural and selected in the laboratory, consist of breaking or joining phosphodiester bonds in RNA substrates. In some cases, such as self-splicing introns, the breaking and joining are coupled, whereas other ribozymes, such as RNase P, catalyze bond cleavage only.

Because many ribozymes act on an RNA substrate, often a part of the ribozyme itself, base-pairing interactions are critical for binding and positioning the substrate for reaction. Crystallographic structures of ribozymes and their components, determined over the past two decades, show that base pairing is also central to the architecture of ribozyme active sites. The first crystal structure of a large folded RNA, the P4-P6 domain of the *Tetrahymena* group I self-splicing intron (see Figure 16-14), revealed how base-paired RNA helices can pack together through non-Watson-Crick interactions and specifically positioned Mg^{2+} ions.

The structural patterns observed in the P4-P6 RNA, including the extensive use of unpaired A residues in the stabilizing, noncovalent contacts essential to the three-dimensional structure, have been observed repeatedly in RNAs ranging from small, self-cleaving ribozymes to the RNAs of ribosomes. Thus, RNA molecules use weak interactions, including the hydrophobic interactions and hydrogen bonding inherent in base pairing and base stacking, along with site-specific metal ion coordination, to form a wide variety of structures with ligand-binding sites and catalytic centers. DNA molecules are inherently less able to form such stable three-dimensional structures, due to subtle but crucial differences in the geometry of the phosphodiester backbone and the lack of a 2′ hydroxyl in DNA nucleotides. RNA is therefore uniquely positioned in biology not only to encode genetic information but also to modify it, and possibly even to replicate it.

The known repertoire of ribozymes continues to expand. Some virusoids, small RNAs associated with plant RNA viruses, include a structure that promotes a self-cleavage reaction. (**Highlight 16-3** describes the self-cleaving ribozyme of a human virus.) The hammerhead ribozyme is in this class, catalyzing the hydrolysis of an internal phosphodiester bond important for producing unit-length virusoid RNAs. In the eukaryotic spliceosome, the splicing reaction requires a catalytic center formed, at least in part, by the U2, U5, and U6 snRNAs. And perhaps most importantly, an RNA component of ribosomes catalyzes the synthesis of proteins (see Chapter 18).

Ribozymes vary greatly in size. Ribosomal RNA is thousands of nucleotides long, the *Tetrahymena* self-splicing group I intron contains ~400 nucleotides, and the smallest active hammerhead ribozyme of virusoids

A Viral Ribozyme Derived from the Human Genome?

Ribozymes are thought to have been important in early evolution, because they provide the potential to both store and replicate genetic information within the same molecule. However, the age and origin of the known, naturally occurring ribozymes are not easy to determine.

Experiments by Jack Szostak and his colleagues at Harvard Medical School showed that, at least in one case, ribozyme evolution occurred relatively recently. The research group was looking for self-cleaving ribozymes in the human genome. Total human cellular RNA was isolated and tested for the ability to generate shorter fragments in a Mg^{2+}-dependent reaction. Cleaved fragments were selectively enriched and retested, and after many such cycles, the researchers isolated RNAs that could catalyze autocleavage at specific sites. In an experiment with one of these RNAs, the starting RNA fragment was found to cleave itself over the course of several hours, resulting in shorter fragments that could be separated by denaturing polyacrylamide gel electrophoresis (Figure 1).

Remarkably, the ribozyme discovered in this experiment is structurally and biochemically related to ribozymes first discovered in the human hepatitis delta virus (HDV). Furthermore, sequence comparisons showed that the ribozyme occurs only in mammals, implying that it may have

evolved as recently as 200 million years ago. HDV itself may have arisen sometime later, forming from fragments of human RNA.

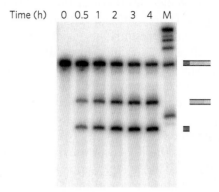

FIGURE 1 Gel revealing a viral-like ribozyme in human genomic RNA. Samples of the reaction mixture containing the transcript with the ribozyme sequence and flanking sequences were removed for analysis at different time intervals. Over time, the transcript was processed into shorter fragments corresponding to ribozyme-catalyzed strand cleavage at the junction between the 5′ flanking sequence and the boundary of the ribozyme sequence (shown in the representations to the right of the gel). M indicates a marker lane. [Source: From Salehi-Ashtiani et al. Science. 313: 1788–1792, 2006. DOI: 10.1126/science.1129308. Reprinted with permission from AAAS.]

consists of two RNA strands with only 41 nucleotides in all. Experiments have shown that in each case, a ribozyme can be inactivated by heating above its melting temperature or by the addition of denaturing chemicals or complementary oligonucleotides, which disrupt normal base-pairing patterns. Furthermore, ribozymes can be inactivated if essential nucleotides are changed, forming the basis for many experiments demonstrating the importance of particular nucleotides or base-pairing interactions in ribozyme function (see the How We Know section at the end of this chapter).

Could RNA Have Formed the Basis for Early Life on Earth?

The existence of so many kinds of ribozymes has fueled debate about why these catalysts occur in nature and what they might indicate about the origin of enzymes. Without catalysts, life would not be possible, and an understanding of how enzymes evolved is central to our understanding of life's origins.

The RNA world hypothesis proposes that organisms comprised entirely or mostly of RNA evolved on

early Earth. The base-pairing properties of RNA could have been used to store information, as occurs today in many viruses, and the capacity of RNA to form a variety of structures lent itself to the emergence of ribozymes and perhaps other functional molecules, such as those forming membrane-spanning pores. Using in vitro selection methods, Gerald Joyce and his colleagues found that even RNAs containing just two kinds of nucleotides that can base-pair with each other can become catalytically active. This observation indicates that simple nucleic acids that might have been around before life began could have given rise to activities necessary for template-dependent replication.

Eventually, DNA supplanted RNA as a data storage molecule because of its greater chemical stability. Proteins, with broader catalytic capabilities stemming from chemically diverse amino acids, became the specialized catalytic molecules. The few remaining ribozymes are remnants of the RNA world and provide clues to its former existence.

Several aspects of RNA chemistry and behavior support the RNA world hypothesis. These include the findings that RNA can function as both a genome and an enzyme, that RNA catalyzes peptide bond formation on ribosomes and thus is responsible for protein synthesis, and that components of RNA form spontaneously in "prebiotic soup" experiments designed to replicate conditions on early Earth. Furthermore, the continuing discovery of RNA molecules that function in fundamental aspects of gene expression and regulation underscores the pervasive and presumably ancient roles of RNA in virtually all aspects of life. More difficult to explain is how and why the specific sugars found in RNA and DNA were selected under prebiotic conditions, and how nucleotides could have been assembled and polymerized without the assistance of enzymes. Researchers continue to investigate these issues, using RNA and related polymers. At present, the RNA world hypothesis is considered to be the most likely explanation for the emergence and evolution of modern organisms.

SECTION 16.5 SUMMARY

- Ribozymes are important for catalyzing several RNA processing reactions, including self-cleavage of viral RNA replication intermediates and precursor tRNA processing.
- The RNA world hypothesis, based on the special properties of RNA that enable it to form stable functional structures and encode genetic information, postulates that RNA-based life predated modern DNA-based organisms.

? UNANSWERED QUESTIONS

The study of RNA processing reactions has been a long-standing and active area of research, yet much remains to be deciphered.

1. How did introns evolve and how have they achieved their current roles in biology? We don't yet know why there are introns and whether they are ancient or more recent acquisitions in genes. Some introns have been found to encode regulatory RNA molecules that function in the processing of rRNA and in the control of gene expression levels (see Chapter 22). Whether these regulatory RNAs are a cause or a result of the presence of introns is not known. Although the origin of introns may remain uncertain, further insights about their roles in the continuing evolution of genomes will be illuminating and may shed light on diseases that result from inaccurate intron removal and processing.

2. How does alternative splicing work? Experimental methods, including microarray technology and genome-wide transcript sequencing, have revealed an abundance of alternative splicing in mammalian cells. However, the mechanics of such molecular gymnastics have yet to be fully determined. Future research will focus on how particular splice sites are chosen, the frequency with which genes are alternatively spliced, and the roles of splicing regulation in disease.

3. How do ribozymes contribute to modern biology? We don't have clear information on the origin and maintenance of ribozymes in particular biological niches. Some researchers have proposed that the spliceosome is a ribozyme, but this remains unproven. Current evidence suggests that the catalytic center of the spliceosome may in fact include both RNA and protein components. If true, such close association of RNA and protein would provide the first example of a true ribonucleoprotein enzyme linking RNA-based and protein-based catalysts.

4. What governs the assembly and disassembly of P bodies? Do the proteins that localize to P bodies function there, or are they held on standby until needed elsewhere in the cell?

HOW WE KNOW

Studying Autoimmunity Led to the Discovery of snRNPs

Lerner, M.R., and J.A. Steitz. 1979. Antibodies to small nuclear RNAs complexed with proteins are produced by patients with systemic lupus erythematosus. *Proc. Natl. Acad. Sci. USA* 76:5495–5499.

Joan Steitz *[Source: Michael Marsland/Yale University, Courtesy Joan Steitz]*

The story of the discovery of snRNPs highlights the serendipity inherent in scientific research. In the early 1980s, Joan Steitz and her colleagues at Yale University were investigating the molecular basis of the puzzling autoimmune disease systemic lupus erythematosus, which causes a red facial rash, fatigue, and arthritis, among other symptoms. Working with partially purified antibodies isolated from the blood of lupus patients, Steitz discovered that the antigens—the binding targets—of these antibodies are normal cellular particles containing a single small nuclear RNA complexed with proteins: snRNPs. Specifically, the autoantibodies associated with lupus recognize a set of snRNP proteins that were eventually named Sm, for the last name of the woman (Smith) whose serum samples were tested.

The snRNPs isolated by precipitation with anti-Sm antibodies were found to bind preferentially to RNAs containing intron-exon junctions, suggesting the involvement of these snRNPs in pre-mRNA splicing. In this experiment, nuclei isolated from cultured human cells were treated with radiolabeled UTP so that any newly synthesized RNA would incorporate the radiolabel. The nuclei were then incubated with anti-Sm antibodies under conditions in which the antibodies could enter the nucleus. After incubation for an hour, RNA was purified from the nuclei and fractionated by size, using agarose gel electrophoresis. The radiolabeled pre-mRNAs and mature RNAs (i.e., unspliced and spliced RNAs, respectively) were identified with specific DNA probes, and the relative amount of each was quantified based on the amount of radioactivity in each band in the gel. Radioactivity was plotted as a function of fraction number, which

correlates with mRNA size (Figure 1). In the absence of added anti-Sm antibodies, both unspliced and spliced mRNAs were detected (top graph). In contrast, the presence of anti-Sm antibodies inhibited the production of spliced mRNAs, such that only a peak corresponding to unspliced pre-mRNAs was detected (bottom graph).

These data led to further studies that elucidated the specific roles of snRNPs in recognizing and catalyzing the removal of introns in the pre-mRNAs of all eukaryotic cells. This latter discovery was recognized by the Nobel Prize in Medicine, awarded to Phillip Sharp and Richard Roberts in 1993. We still don't know how Sm proteins can induce an autoimmune response, given that, presumably, they are sequestered within snRNPs in the nucleus.

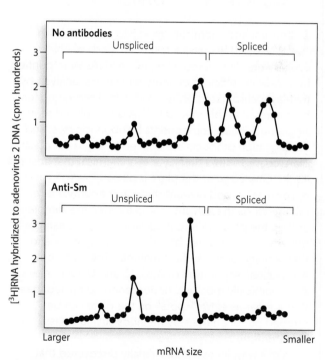

FIGURE 1 Results of labeling experiments showing that processing of pre-mRNA is blocked by anti-Sm antibodies. In the absence of antibodies (upper panel), multiple fractions corresponding to spliced mRNAs are detected; in the presence of antibodies (lower panel), only larger, unspliced pre-mRNAs are detected. *[Source: Data from V. W. Yang et al., Proc. Natl. Acad. Sci. USA 78:1371–1375, 1981.]*

RNA Molecules Are Fine-Tuned for Stability or Function

Cerutti, P., J.W. Holt, and N. Miller. 1968. Detection and determination of 5,6-dihydrouridine and 4-thiouridine in transfer ribonucleic acid from different sources. *J. Mol. Biol.* 34:505–518.

Hughes, D.G., and B.E. Maden. 1978. The pseudouridine contents of the ribosomal ribonucleic acids of three vertebrate species: Numerical correspondence between pseudouridine residues and 2'-O-methyl groups is not always conserved. *Biochem. J.* 171:781–786.

Kuchino, Y., and E. Borek. 1978. Tumour-specific phenylalanine tRNA contains two supernumerary methylated bases. *Nature* 271:126–129.

Unusual and chemically modified nucleotides in tRNAs, rRNAs, and a few other kinds of RNA molecules were discovered when these RNAs were purified from cells in sufficient quantity for classical analysis by thin-layer chromatography—a standard technique for biochemists. In this method, RNA is hydrolyzed to its single-nucleotide components by ribonucleases, and the digestion products are applied to a glass or plastic plate coated with a thin layer of a chemical that absorbs liquid. One edge of the plate is placed in a solvent that is slowly absorbed upward through the surface layer (Figure 2). As the solvent moves through the sample and up the plate, different nucleotides in the sample move at different rates based on their solubility and their affinity for the surface material. The technique allows clean separation of A, G, C, and U nucleotides. Any chemically modified or noncanonical nucleotides that appear as extra "spots" on the plate can be scraped off for analysis by methods such as mass spectrometry.

In this way, investigators initially discovered that tRNAs contain a few nucleotides other than the standard four. How are such unusual bases synthesized, and why are they maintained? So far, research shows that all cells have sophisticated molecular machinery to produce unusual bases in tRNAs and certain other RNAs. Some enzymes recognize a particular type of tRNA, excise the base from a specific nucleotide position, and replace it with another base; other enzymes chemically modify an existing base.

Modified bases seem to contribute in subtle ways to the thermodynamic stability of three-dimensionally structured RNAs such as tRNA and rRNA. For example, there are viable bacterial strains that differ only in their ability to modify rRNA at a specific site, due to the presence or absence of a particular rRNA-modifying enzyme. However, in a culture medium inoculated with equal amounts of the two strains, the strain containing the modifying enzyme eventually takes over the culture. This result implies that the modification of rRNA contributes to the efficient function of ribosomes, despite requiring extra cellular energy input.

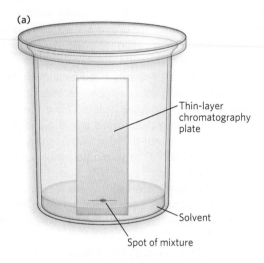

(a)

Thin-layer chromatography plate

Solvent

Spot of mixture

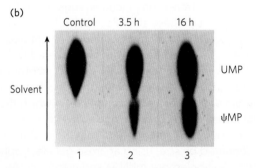

(b)

Control 3.5 h 16 h

Solvent

UMP

ψMP

1 2 3

FIGURE 2 Thin-layer chromatography to detect chemically modified nucleotides. (a) RNase-digested tRNA is spotted onto a silica-coated glass plate, and solvent and nucleotides migrate up the plate by capillary action. (b) Modified and unmodified nucleotides—in this case, pseudouridine monophosphate (ψMP) and UMP, as control—migrate at different rates, based on their different solubilities in the solvent and their different affinities for the silica gel. *[Source: (b) Yi-Tao Yu, Mei-Di Shu and Joan A. Steitz The EMBO Journal 17: 5783–5795, 1998 doi: 10.1093/emboj/17.19.5783 © 1998 European Molecular Biology Organization.]*

Ribozyme Form Explains Function

Michel, F., M. Hanna, R. Green, D.P. Bartel, and J.W. Szostak. 1989. The guanosine binding site of the *Tetrahymena* ribozyme. *Nature* 342:391–395.

Stahley, M.R., and S.A. Strobel. 2006. RNA splicing: Group I intron crystal structures reveal the basis of splice site selection and metal ion catalysis. *Curr. Opin. Struct. Biol.* 16:319–326.

The discovery of ribozymes coincided with an important technological advance for molecular biologists: the ability to transcribe RNA molecules of any sequence in vitro and thereby test the function of RNAs in the complete absence of proteins. Furthermore, they could probe the molecular structure of ribozymes with chemicals that react with RNA nucleotides only when they are not involved in base-pairing interactions or packed against other nucleotides in the folded structure of the RNA.

An early observation was that mutations in the RNA sequence that disrupted parts of the three-dimensional structure also perturbed catalytic activity. For example, researchers noticed that a specific base pair in the *Tetrahymena* group I self-splicing intron was present in the same place in the secondary structure of all related group I introns. Changing this base pair to any other base combination disrupted the self-splicing reaction, because the intron was no longer capable of binding efficiently to GTP, a cofactor in the splicing reaction (Figure 3). In this way, investigators discovered that the *Tetrahymena* group I intron (and later, other ribozymes) has a defined three-dimensional shape that is essential to catalytic activity. Using in vitro transcribed and purified RNA, it was later possible to crystallize ribozymes and their component domains, revealing how these RNAs form active sites to enhance chemical reaction rates.

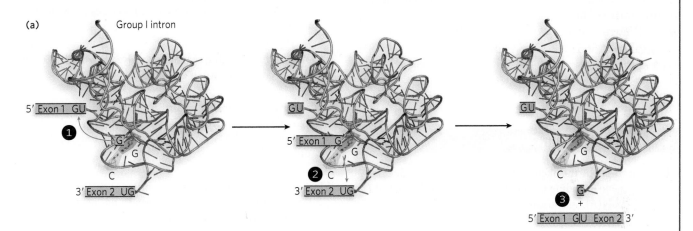

(a) Group I intron

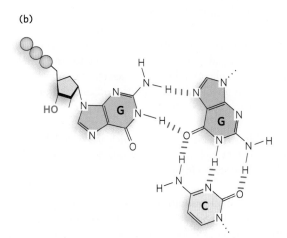

(b)

FIGURE 3 The group I intron structure includes a G≡C base pair that binds the GTP cofactor. (a) To initiate self-splicing, the group I intron structure precisely positions the G≡C base pair. (b) The G≡C binds a GTP cofactor through noncanonical base pairing. [*Sources: (a) PDB ID 1X8W. (b) Data from F. Michel et al., Nature 342:394, 1989.*]

KEY TERMS

primary transcript, 554
intron, 554
exon, 554
5′ cap, 555
3′ poly(A) tail, 557
poly(A) addition site, 557
RNA splicing, 560
spliceosome, 560
proteome, 561
alternative splicing, 561
poly(A) site choice, 562
branch point, 562
small nuclear ribonucleoprotein (snRNP), 562
small nuclear RNA (snRNA), 562
group I intron, 566
group II intron, 566
trans-splicing, 568
RNA editing, 569

editosome, 569
adenosine deaminase acting on RNA (ADAR), 571
exon junction complex (EJC), 574
RNA degradation, 575
exosome, 575
processing body (P body), 576
preribosomal RNA (pre-rRNA), 578
small nucleolar ribonucleoprotein (snoRNP), 579
microRNA (miRNA), 579
Drosha, 579
microprocessor complex, 579
Dicer, 579
ribozyme, 580

PROBLEMS

1. What would be the likely cellular effects of a large deletion in the gene encoding the polymerase responsible for adding 3′ poly(A) tails to eukaryotic mRNAs?

2. What is the minimum number of transesterification reactions required to splice an intron from a precursor transcript?

3. Compare and contrast splicing mechanisms used by spliceosomes, group I introns, and group II introns, with respect to the nucleophiles, proteins, or nucleic acids involved and how the specificity of splice sites is achieved.

4. Self-splicing introns do not require an energy source, such as ATP or GTP, to catalyze splicing. How does self-splicing proceed with a reasonable yield of products?

5. Given what you have learned about the catalytic mechanism of group I self-splicing introns, propose an experiment to identify new group I introns in the total RNA isolated from an individual organism. Include a unique feature of the group I splicing reaction mechanism in your answer.

6. Strains of bacteria lacking the enzymes required for rRNA modification have been engineered in the laboratory and seem to grow normally, despite the absence of modified rRNA. What would happen to these bacterial strains if they were forced to compete with wild-type bacteria? Explain.

7. Is it correct to call an RNA molecule that catalyzes a reaction on itself an enzyme? Explain your answer.

8. What accounts for the directionality of mRNA transport out of the nucleus?

9. What would happen to the lifetime of a human mRNA if a nonhydrolyzable phosphodiester analog were introduced near its 5′ end?

10. Some RNA editing involves the deletion and/or addition of nucleotides to an mRNA. Other RNA editing reactions involve the alteration of bases in an mRNA, without the addition or deletion of nucleotides. Which bases may be altered in RNA editing? What reactions occur to alter them, and what bases are they converted to? What changes in coding occur as a result of the base alterations?

11. What do the editing reactions catalyzed by APOBEC and ADAR enzymes have in common?

12. All living systems share key properties (see Chapter 1). Why did the discovery of RNA catalysis trigger the development of an RNA world hypothesis as a stage in the evolution of life?

13. Two different sequences in a primary mRNA transcript are critical to its cleavage in preparation for the addition of a poly(A) tail to the 3′ end. What are those sequences, and which of them, if either, is (are) retained in the mature and modified mRNA?

14. Following the first step in the splicing of a group II intron, an internal A residue in the intron is linked to other nucleotides by three phosphodiester bonds. What groups on the adenosine are involved in the phosphodiester bonds, and to what is each bond linked?

15. The double-stranded RNA endonuclease called Dicer is found in the cytoplasm of many eukaryotic cells. What type of RNA does it cleave?

16. If the tRNA nucleotidyltransferase or the enzyme that converts some U residues to pseudouridine in tRNA were inactivated in a eukaryotic cell, which inactivation would most likely be lethal?

17. RNA enzymes (ribozymes) have been discovered in cells ranging from bacteria to humans. What kinds of reactions are catalyzed by these naturally occurring ribozymes?

DATA ANALYSIS PROBLEM

Cech, T.R., A.J. Zaug, and P.J. Grabowski. 1981. In vitro splicing of the ribosomal-RNA precursor of *Tetrahymena*: Involvement of a guanosine nucleotide in the excision of the intervening sequence. *Cell* 27:487–496.

Kruger, K., P.J. Grabowski, A.J. Zaug, J. Sands, D.E. Gottschling, and T.R. Cech. 1982. Self-splicing RNA: Auto-excision and auto-cyclization of the ribosomal-RNA intervening sequence of *Tetrahymena*. *Cell* 31:147–157.

18. The discovery of RNA catalysis is sometimes cited as a classic case of serendipity, but it required a thoroughly prepared mind. In the late 1970s and early 1980s, RNA splicing was still a new concept. Thomas Cech and his colleagues set out to investigate the splicing of an intron in rRNA of the protozoan *Tetrahymena thermophila*. They selected this RNA for several reasons: rRNAs are much more abundant than most mRNAs, this rRNA has an intron, and *Tetrahymena* is a single-celled eukaryote that can be grown in large quantities.

   To carry out their early studies, Cech and colleagues devised a method to produce unspliced precursor rRNAs. As described in their 1981 paper, they isolated nuclei from *Tetrahymena* cells, lysed them, and added buffer, labeled rNTPs, and α-amanitin, then incubated the reaction mix under conditions in which the cellular RNA polymerases would synthesize RNA. They extracted the labeled RNA by a method that used phenol and chloroform, treatments that normally remove most protein.

   (a) Why did the researchers add α-amanitin to the transcription reaction?

When they incubated the extracted RNA under the right conditions, the researchers saw the production of excised intron RNA. Their result is shown in Figure 1. The bands in the first lane are size markers. The large bands at the top are the precursor rRNAs. The L and C labels indicate linear and circular forms of the excised intron RNA. Some different conditions of temperature and Mg²⁺ ion concentration were tried.

   (b) The result demonstrated that the splicing reaction was working, but it did not demonstrate RNA catalysis. Why?

The excision reaction depended entirely on the addition of a guanine nucleotide. The researchers explored this requirement further, and some of their results are shown in Figure 2. The rRNA intron excision reaction is shown in the presence of a variety of potential cofactors. The abbreviations for the nucleosides and nucleotides are the standard ones described in Chapter 3. G and dG are the guanosine nucleosides, with dG the 2′-deoxy form.

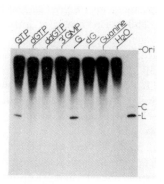

FIGURE 2 *[Source: Reprinted from Cech, TR, Zaug, AJ, Grabowski, PJ Cell 27, 487–496, Fig. 4. © 1981, with permission from Elsevier.]*

   (c) From these results, what can you conclude about the nature of the required cofactor?

Intrigued that the splicing reaction did not seem to require the addition of cell extract (proteins), the researchers set out to determine whether proteins were required for the reaction. As described in their 1982 paper, they cloned a segment of the gene for the 26S rRNA, including the intron, in a bacterial plasmid and expressed this gene segment in vitro using bacterial RNA polymerase. After deproteinizing the RNA product, they incubated the RNA with the buffer components and guanine nucleotide cofactor, as described in the earlier paper. Their results are shown in Figure 3. The C and L IVS are the circular and linear forms of the excised intron. The L-15 band is a cleaved form of the L IVS produced in a postsplicing cleavage reaction catalyzed by the IVS itself and not relevant to this problem.

   (d) Does this experiment demonstrate self-splicing (RNA catalysis) of the intron? If so, why didn't the experiment described in the first paper do so?

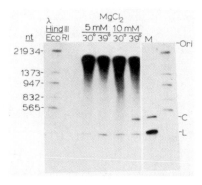

FIGURE 1 *[Source: Reprinted from Cech, TR, Zaug, AJ, Grabowski, PJ Cell 27, 487–496, Fig. 2. © 1981, with permission from Elsevier.]*

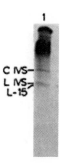

FIGURE 3 *[Source: Reprinted from Kruger, K, Grabowski, PJ, Zaug, AJ, Sands, J., Gottschling, DE, and Cech, TR, Cell 31, 147–157. Fig. 2. © 1982, with permission from Elsevier.]*

ADDITIONAL READING

General and Historical

Berget, S.M., C. Moore, and P.A. Sharp. 1977. Spliced segments at the 5′ terminus of adenovirus 2 late mRNA. *Proc. Natl. Acad. Sci. USA* 74:3171–3175.

Chow, L.T., R.E. Glinas, T.R. Broker, and R.J. Roberts. 1977. An amazing sequence arrangement at the 5′ ends of adenovirus 2 messenger RNA. *Cell* 12:1–8.

Koonin, E.V. 2006. The origin of introns and their role in eukaryogenesis: A compromise solution to the introns-early versus introns-late debate? *Biol. Direct* 1:22, doi: 10.1186/1745-6150-1-22.

Mount, S.M., I. Pettersson, M. Hinterberger, A. Karmas, and J.A. Steitz. 1983. The U1 small nuclear RNA–protein complex selectively binds a 5′ splice site in vitro. *Cell* 33:509–518.

Padgett, R.A., S.M. Mount, J.A. Steitz, and P.A. Sharp. 1983. Splicing of messenger RNA precursors is inhibited by antisera to small nuclear ribonucleoprotein. *Cell* 35:101–107.

Steitz, J.A. 1988. "Snurps." *Sci. Am.* 258(6):56–60, 63. A useful review.

Pre-mRNA Splicing and RNA Editing

Collins, C.A., and C. Guthrie. 2000. The question remains: Is the spliceosome a ribozyme? *Nat. Struct. Biol.* 7:850–854. This short review provides insight into the catalytic mechanism of the spliceosome and its possible evolutionary origins.

Le Hir, H., A. Nott, and M.J. Moore. 2003. How introns influence and enhance eukaryotic gene expression. *Trends Biochem. Sci.* 28:215–220. A summary of the ways in which splicing affects the expression of proteins in eukaryotic cells.

Nishikura, K. 2010. Functions and regulation of RNA editing by ADAR deaminases. *Annu. Rev. Biochem.* 79:321–349.

Stuart, K.D., A. Schnaufer, N.L. Ernst, and A.K. Panigrahi. 2005. Complex management: RNA editing in trypanosomes. *Trends Biochem. Sci.* 30:97–105. A short and insightful review of RNA editing mechanisms.

RNA Transport and Degradation

Belasco, J.G. 2010. All things must pass: Contrasts and commonalities in eukaryotic and bacterial mRNA decay. *Nat. Rev. Mol. Cell Biol.* 11:467–478. A discussion of the mechanistic parallels between the cellular factors and molecular events that govern mRNA degradation in eukaryotes and bacteria.

Eulalio, A., I. Behm-Ansmant, and E. Izaurralde. 2007. P bodies: At the crossroads of post-transcriptional pathways. *Nat. Rev. Mol. Cell Biol.* 8:9–22.

Kindler, S., H. Wang, D. Richter, and H. Tiedge. 2005. RNA transport and local control of translation. *Annu. Rev. Cell Dev. Biol.* 21:223–245.

Parker, R., and H. Song. 2004. The enzymes and control of eukaryotic mRNA turnover. *Nat. Struct. Mol. Biol.* 11:121–127.

Strässer, K., S. Masuda, P. Mason, J. Pfannstiel, M. Oppizzi, S. Rodriguez-Navarro, A.G. Rondón, A. Aguilera, K. Struhl, R. Reed, and E. Hurt. 2002. TREX is a conserved complex coupling transcription with messenger RNA export. *Nature* 417:304–308.

Processing of Non-Protein-Coding RNAs

Doudna, J.A., and T.R. Cech. 2002. The chemical repertoire of natural ribozymes. *Nature* 418:222–228.

Gingeras, T.R. 2009. Implications of chimaeric non-colinear transcripts. *Nature* 461:206–211.

Khalil, A.M., M. Guttman, M. Huarte, M. Garber, A. Raj, D. Rivea Morales, K. Thomas, A. Presser, B.E. Bernstein, A. van Oudenaarden, et al. 2009. Many human large intergenic noncoding RNAs associate with chromatin-modifying complexes and affect gene expression. *Proc. Natl. Acad. Sci. USA* 106:11,667–11,672.

Wilson, D.S., and J.W. Szostak. 1999. In vitro selection of functional nucleic acids. *Annu. Rev. Biochem.* 68:611–647.

RNA Catalysis and the RNA World Hypothesis

Hagiwara, Y., M.J. Field, O. Nureki, and M. Tateno. 2010. Editing mechanism of aminoacyl-tRNA synthetases operates by a hybrid ribozyme/protein catalyst. *J. Am. Chem. Soc.* 132:2751–2758.

Marvin, M.C., and D.R. Engelke. 2009. Broadening the mission of an RNA enzyme. *J. Cell. Biochem.* 108:1244–1251.

Rios, A.C., and Y. Tor. 2009. Model systems: How chemical biologists study RNA. *Curr. Opin. Chem. Biol.* 13:660–668.

17

The Genetic Code

MOMENT OF DISCOVERY

Steve Benner [Source: Courtesy Steve Benner.]

The origin of life has long been an interest of mine, particularly the evolution of nucleic acids and the reason that ribose was selected as the sugar used in RNA. In the 1950s, Stanley Miller and others showed that ribose can be produced abiotically (without enzymes), but Miller and others had noted that ribose is not very stable. This is because, in the lab, ribose and other five-carbon sugars are made under alkaline conditions from simple organic precursors, formaldehyde and glycolaldehyde; a high pH encourages reasonable reaction rates, but the ribose product tends to break down quickly into a brown tar.

Although we weren't actually studying this particular problem, a comment from a colleague about "solving the ribose problem" coincided with a trip I took to Death Valley to collect rocks. While there, I began musing about the borate-containing rock samples I was finding, and thinking about the long-known observation that borate can bind to organic molecules that contain 1,2-dihydroxyl groups—exactly the kind of chemical structure present in ribose.

When I returned to the lab, it only took about a day and a half to show experimentally that *ribose could be made stably at high pH in the presence of borate*. Because borate is abundant in nature, it seems likely that it stabilized the prebiotic production of ribose, providing a simple and logical explanation for the presence of ribose on the early Earth. It was satisfying to make this discovery, but also humbling to realize that borate-carbohydrate interactions have been known since the 1950s. So the answer to the ribose stability problem has been staring us in the face all along!

—*Steve Benner, on discovering that borate minerals stabilize ribose*

Online resources related to this chapter:

***Nature* exercises** The genetic code, Mutagenesis

The discoveries that DNA is composed of complementary strands and that it holds the instructions for all the proteins in an organism were huge advances in our understanding of the flow of biological information. However, proteins and nucleic acids are very different types of chemicals, and after the structure of DNA was solved, how the sequence in a chain of nucleotides determines the sequence of amino acids in a protein was not immediately apparent. The next 10 years brought several discoveries that revealed the fascinating processes by which DNA is decoded to produce proteins.

The linear nucleotide sequence of mRNA is translated into protein by tRNA molecules that carry amino acids and contain nucleotide sequences (called anticodons) that pair with complementary sequences (codons) in the mRNA. Different amino acid-carrying tRNAs are lined up according to the mRNA sequence, and the amino acids are stitched together by the ribosome, resulting in a polypeptide with a linear order of amino acids that corresponds to the linear order of codon sequences in the mRNA. The discovery of the translation process and the **genetic code**, the matching of each codon to the amino acid it specifies, is a fascinating story and a landmark in modern science.

Amino acids and nucleotide bases have no obvious chemical relationship, and therefore it is not at all obvious how given amino acids became matched to particular trinucleotide sequences. Yet all organisms—bacteria, yeast, amphibians, plants, archaea, and humans—use the same genetic code, with only a few minor modifications. Presumably, once the code had evolved, it resisted change. The universality of the genetic code provides amazingly strong, *molecular* evidence for evolution, much more compelling than arguments based on body shapes and the fossil record.

This chapter presents an overview of the genetic code and how it works. We first look at how the tRNA molecule functions in decoding, and how it is exquisitely designed to take advantage of the "degeneracy" of codons, enabling one tRNA to decipher more than one codon. We also examine how the genetic code can resist the harmful effects of single-nucleotide mutations. These special features indicate that the genetic code is not simply an accident of evolution, but has been fine-tuned by natural selection. The last universal common ancestor (LUCA) must have existed for sufficient time to hone the code prior to divergence of the different domains of life as we know them today. Finally, we look at exceptions to the genetic code—variations that only reinforce the idea that all life forms evolved from LUCA and its genetic code. How the genetic code came into being during evolution is still a perplexing problem. We examine this issue, too, even though there are no clear answers.

17.1 DECIPHERING THE GENETIC CODE: tRNA AS ADAPTOR

DNA and RNA each consist of only four different nucleotides, whereas proteins can have up to 20 different amino acids. For only four nucleotides to specify the 20 common amino acids, multiple nucleotides must be combined to make up a code. Combinations of two nucleotides yield only 16 (4^2) different dinucleotide code words, insufficient to encode 20 amino acids. Combinations of three nucleotides yield 64 (4^3) code words, more than enough to specify 20 amino acids. Hence, the RNA "code word," or **codon**, was hypothesized to be a combination of three nucleotides, or possibly more. Insightful experiments, described in this chapter, demonstrated that the code is indeed triplet.

In 1955, to explain how an RNA sequence codes for a sequence of amino acids, Francis Crick hypothesized the existence of an "adaptor" molecule. He proposed that adaptors can recognize specific codons in the mRNA and that each adaptor carries a specific amino acid (**Figure 17-1**). Adaptors line up on the mRNA, thus aligning the sequence of amino acids. Not long after Crick's adaptor hypothesis, Paul Zamecnik and Mahlon Hoagland discovered a small RNA that covalently attaches to amino acids in a reaction requiring ATP (see the How We Know section at the end of this chapter). These RNA–amino acid hybrids could presumably base-pair with mRNA, because they contain nucleotides and thus fit the description of the adaptor molecule needed to translate the information in an mRNA sequence into a polypeptide sequence. This small RNA, later called

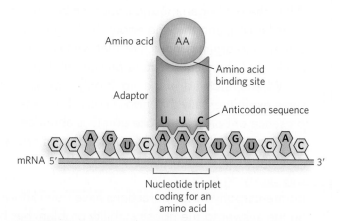

FIGURE 17-1 Crick's adaptor hypothesis. Adaptor molecules recognize codons in mRNA and carry specific amino acids. Thus, they line up amino acids in an order that depends on the sequence of codons in the mRNA. Today we know that the adaptor is a tRNA molecule. The amino acid is covalently bound at the 3′ end of the tRNA molecule, and a specific nucleotide triplet (anticodon) elsewhere in the tRNA interacts with a triplet codon in mRNA through hydrogen bonding of complementary bases.

transfer RNA (tRNA), is aminoacylated at the 3′ terminus in an ATP-dependent reaction. A tRNA with its attached amino acid is called an **aminoacyl-tRNA**, and the tRNA is said to be *charged* with that amino acid. The amino acid specificity of tRNAs is provided not by the anticodon in their nucleotide sequence but by the enzymes that attach amino acids to particular tRNAs, enzymes known as **aminoacyl-tRNA synthetases**. Thus, the association between the amino acid and the anticodon evolved, rather than being chemically predetermined.

KEY CONVENTION

In denoting tRNAs, the specificity is indicated by a superscript, and the aminoacylated-tRNA by a hyphenated name. For example, tRNA[Leu] indicates an uncharged tRNA that is specific for leucine, and leucyl-tRNA[Leu], or Leu-tRNA[Leu], indicates a leucine-specific tRNA that is charged with leucine.

All tRNAs Have a Similar Structure

The structure of the tRNA molecule reveals how it is capable of functioning as an adaptor. We briefly discuss tRNA structure here, and then explore this topic in greater detail in Chapter 18.

Transfer RNAs are relatively small, single-stranded RNA molecules. The tRNAs in bacteria and in the cytoplasm of eukaryotic cells are 73 to 93 nucleotide residues long. Mitochondria and chloroplasts contain distinctive, somewhat smaller tRNAs. All tRNAs form intramolecular base pairs and fold up into a precise three-dimensional structure. They contain the trinucleotide sequence CCA at the 3′ terminus. The 3′-terminal A residue is the nucleotide to which the amino acid attaches.

When drawn in two dimensions, the hydrogen-bonding pattern of all tRNAs forms a cloverleaf structure with four arms; the longer tRNAs also have a short fifth arm, or extra arm (**Figure 17-2a**). In three dimensions, a tRNA folds up further into the form of a twisted L (**Figure 17-2b**). Two arms of the tRNA are critical for adaptor function, and they are located at the two ends of the L shape. The 3′ terminus of the amino acid arm, in the charged tRNA, carries a specific amino acid. At the opposite end is the anticodon arm, so named because it contains the **anticodon**, the three-nucleotide sequence that base-pairs with the complementary codon in mRNA. The other major arms are the D arm, which often contains the unusual nucleotide dihydrouridine (D), and the TψC arm, containing ribothymidine (T) and pseudouridine (ψ), which has an unusual carbon–carbon bond

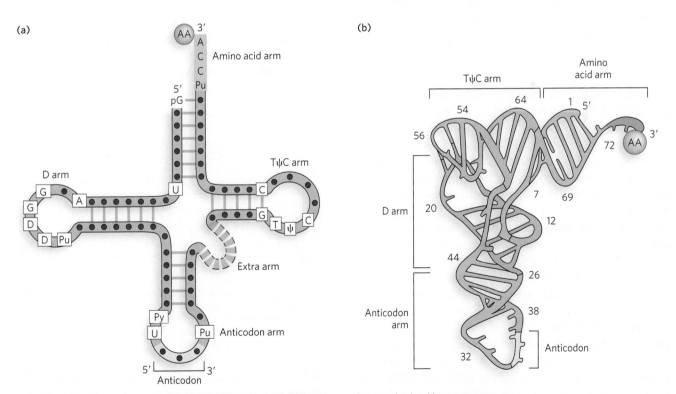

FIGURE 17-2 The structure of tRNA. (a) The cloverleaf form. Large dots on the backbone represent the nucleotide residues; blue lines indicate base pairs. Three nucleotides constitute the anticodon (at the bottom), and an amino acid is attached to the 3′-terminal amino acid arm. Unusual modified nucleotides are present in the D arm and TψC arm. Py (pyrimidine) can be either U or C; Pu (purine) can be either A or G. The D arm often contains one or more dihydrouracil residues (D). (b) The three-dimensional structure folds into a twisted L shape. The positions of the anticodon, the 3′-terminal amino acid arm, and the D and TψC arms are shown.

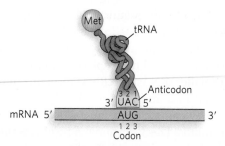

FIGURE 17-3 The pairing relationship of the codon and anticodon. The Met-tRNA^Met (methionyl-tRNA^Met) is shown. The nucleotide positions of the codon and anticodon are numbered 1, 2, and 3, in the 5'→3' direction (thus, anticodon nucleotide 3 pairs with codon nucleotide 1).

between the base and ribose. The base pairing between the anticodon in the tRNA and the codon in mRNA is antiparallel. For example, the codon for methionine is 5'-AUG, which base-pairs with the tRNA^Met anticodon 5'-CAU (i.e., 3'-UAC) (**Figure 17-3**).

> ### KEY CONVENTION
>
> The nucleotide positions of the codon (mRNA) and anticodon (tRNA) are numbered 1, 2, and 3, in the 5'→3' direction. Due to the antiparallel base pairing between the anticodon and the codon, the numbering of nucleotides in the anticodon is the reverse of that in the codon. Thus, anticodon nucleotide 3 pairs with codon nucleotide 1.

As shown in Figure 17-2, the anticodon is some distance from the 3' terminus of the amino acid arm (where the amino acid is attached), and thus the anticodon cannot directly specify the correct amino acid. Indeed, the ribosome will link any two amino acids lined up correctly on the mRNA, regardless of whether the tRNA is charged with a correct or an incorrect amino acid. It is the function of the aminoacyl-tRNA synthetases to place the correct amino acid onto the tRNA. Therefore, the specificity of the genetic code lies in the accuracy of protein-based aminoacylation of the tRNAs. Most cells contain 20 aminoacyl-tRNA synthetases, one for each amino acid. Because there are more codons than amino acids, some amino acids are specified by more than one tRNA, yet the same aminoacyl-tRNA synthetase recognizes all tRNAs that specify a given amino acid. The ribosome binds the mRNA and charged tRNAs, bringing the components into proximity for linking together the amino acids attached to adjacent aminoacyl-tRNAs as they align on the mRNA. The entire process of decoding the linear sequence of mRNA into the sequence of a protein is known as **translation**; it requires more than 100 different types of protein and RNA molecules (see Chapter 18).

The Genetic Code Is Degenerate

As we have seen, there are 64 unique ways to combine four different nucleotides in a triplet codon sequence, yet there are only 20 common amino acids. Therefore, either some codons are not found in mRNA sequences, or—as we now know—multiple codons encode the same amino acid. A **degenerate code** is one in which several codons have the same meaning. We refer to the genetic code as degenerate because a single amino acid can be encoded by more than one codon. As we will see later, the degeneracy of the genetic code is advantageous because it provides the DNA with the ability to absorb single-base mutations with minimal consequences for the protein sequences it encodes.

All 64 codons of the genetic code are used in some fashion: 61 for coding amino acids and 3 for specifying the termination of translation (**Figure 17-4**). Three amino acids—arginine, leucine, and serine—are each specified by six different codons. Five amino acids have four codons, isoleucine has three, and nine amino acids have two codons. Only two amino acids, tryptophan and methionine, are specified by a single codon (**Table 17-1**).

When several different codons specify one amino acid, the first two nucleotides of each codon are the primary determinants of specificity, and the difference between the codons usually lies at the third position. For example, alanine is specified by the triplets GCU, GCC, GCA, and GCG. When four codons specify the same amino acid, they are referred to as a **codon family**. Within a codon family, the first two nucleotides are the same, the

	U		C		A		G		
U	UUU	Phe	UCU	Ser	UAU	Tyr	UGU	Cys	U
	UUC	Phe	UCC	Ser	UAC	Tyr	UGC	Cys	C
	UUA	Leu	UCA	Ser	UAA	Stop	UGA	Stop	A
	UUG	Leu	UCG	Ser	UAG	Stop	UGG	Trp	G
C	CUU	Leu	CCU	Pro	CAU	His	CGU	Arg	U
	CUC	Leu	CCC	Pro	CAC	His	CGC	Arg	C
	CUA	Leu	CCA	Pro	CAA	Gln	CGA	Arg	A
	CUG	Leu	CCG	Pro	CAG	Gln	CGG	Arg	G
A	AUU	Ile	ACU	Thr	AAU	Asn	AGU	Ser	U
	AUC	Ile	ACC	Thr	AAC	Asn	AGC	Ser	C
	AUA	Ile	ACA	Thr	AAA	Lys	AGA	Arg	A
	AUG	Met	ACG	Thr	AAG	Lys	AGG	Arg	G
G	GUU	Val	GCU	Ala	GAU	Asp	GGU	Gly	U
	GUC	Val	GCC	Ala	GAC	Asp	GGC	Gly	C
	GUA	Val	GCA	Ala	GAA	Glu	GGA	Gly	A
	GUG	Val	GCG	Ala	GAG	Glu	GGG	Gly	G

FIGURE 17-4 The genetic code. The codon sequences are written in the 5'→3' direction. The first nucleotide of each codon is shown on the left side of the grid, the second nucleotide at the top, and the third on the right. AUG (shaded green) also serves as the start codon; UAA, UAG, and UGA (red) are stop (or nonsense) codons.

TABLE 17-1

The Degeneracy of the Genetic Code			
Amino Acid	Number of Codons	Amino Acid	Number of Codons
Arg	6	Asp	2
Leu	6	Cys	2
Ser	6	Gln	2
Ala	4	Glu	2
Gly	4	His	2
Pro	4	Lys	2
Thr	4	Phe	2
Val	4	Tyr	2
Ile	3	Met	1
Asn	2	Trp	1

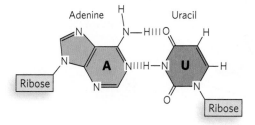

(a) Normal A═U base pair

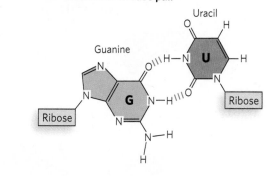

Wobble G–U base pair

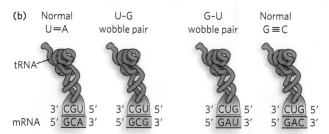

(b)

FIGURE 17-5 Wobble base pairing. (a) Wobble allows one tRNA to recognize two different codons. U normally pairs with A (top) but can form two hydrogen bonds with G to make a weak G–U wobble base pair, which occurs in the third position of the codon (bottom). (b) A tRNA pairs with two different codons through wobble pairing (shown in red) at the third nucleotide of the codon. The G of the G–U pair can be in either the anticodon or the codon.

nucleotide at the third position does not matter, and base pairing of the first two nucleotides carries the information needed to specify the amino acid. Many amino acids are specified by two codons in which the third nucleotide is either a purine in both or a pyrimidine in both.

Wobble Enables One tRNA to Recognize Two or More Codons

If all three nucleotides in an mRNA codon were needed to form Watson-Crick base pairs with their counterparts in the tRNA anticodon, 61 different tRNAs would be required in every cell. In fact, only 32 tRNAs are required to recognize all the amino acid codons, because some tRNAs recognize more than one codon. However, some cells contain considerably more than 32 different tRNAs.

As noted above, when several codons specify the same amino acid, usually the third nucleotide is the only difference. In some cases the cell uses different tRNAs for the different codons that encode the same amino acid, and in these cases a single aminoacyl-tRNA synthetase recognizes the various tRNAs and charges them all with the same amino acid. Many tRNAs can recognize more than one codon, and these tRNAs often contain either a U or a G as the 5′ nucleotide of the anticodon (i.e., in position 1, which pairs with the third nucleotide of the codon), because these nucleotides can form noncanonical (i.e., non-Watson-Crick) base pairs: U can pair with either A or G, and G can pair with either C or U (**Figure 17-5**). These noncanonical base pairs are not found in DNA because they do not fit within the tight geometric constraints of the DNA duplex, but they are accommodated in the more flexible base pairing that occurs between tRNA and mRNA. The bases that

participate in noncanonical base pairs are called **wobble bases**. The wobble bases allow a single tRNA anticodon to bind to more than one mRNA codon. The 5′ nucleotide in the anticodon is in the **wobble position**. However, it is important to note that the structure of tRNA can make the anticodon completely specific for perfect base pairing with one codon. We see this for tryptophan and methionine, each of which has only one codon. Thus, the necessary "flexibility" needed for wobble in tRNA involves more than the anticodon sequence—it also involves the way in which tRNA codon-anticodon pairs can accept particular base differences, as described shortly.

The anticodon in some tRNAs includes inosine (designated I; this nucleotide residue contains the base hypoxanthine), which can hydrogen-bond with any of three different nucleotides: U, C, or A (**Figure 17-6**). These

(a)

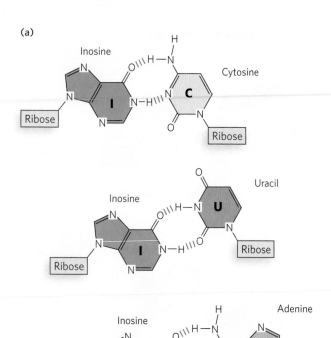

(b)

FIGURE 17-6 Inosine as a wobble nucleotide. (a) Inosine (I) can form two hydrogen bonds with either C, U, or A. (b) A tRNA containing I in the first position of the anticodon can recognize three different codons, according to the wobble rules. Wobble pairings are shown in red.

pairings are much weaker than the hydrogen bonds of Watson-Crick base pairs. When Robert Holley sequenced the yeast tRNA^Ala in 1965, he found inosine at the first position of the anticodon. This explains why the anticodon of yeast tRNA^Ala, 5′-IGC, can function with three different codons: 5′-GCA, 5′-GCU, and 5′-GCC. Inosine,

like the other modified nucleotides in tRNA, is formed posttranscriptionally—an adenosine residue is deaminated by the enzyme adenosine deaminase to produce a keto moiety in place of the amino group.

The process by which some tRNAs can recognize more than one codon was formalized by Crick, who proposed a set of four relationships known as the **wobble hypothesis**:

1. The first two bases of an mRNA codon always form Watson-Crick base pairs with the corresponding bases of the tRNA anticodon, and they confer most of the coding specificity.

2. The first base of the anticodon (reading in the 5′→3′ direction) pairs with the third base of the codon and determines the number of codons recognized by the tRNA. When the first nucleotide of the anticodon is C or A, base pairing is specific, and only one codon is recognized by that tRNA. When the first nucleotide is U or G, base pairing is less specific, and two different codons may be read by the same tRNA. When the first nucleotide of an anticodon is I, three different codons can be recognized—the maximum number for any tRNA.

3. When an amino acid is specified by several different codons, codons that differ in either of the first two bases require different tRNAs.

4. A minimum of 32 tRNAs are required to translate all 61 codons (31 tRNAs for the amino acids and 1 for initiation).

Specific Codons Start and Stop Translation

As we describe in Section 17.2 and in further detail in Chapter 18, the codons in an mRNA molecule are read by the ribosome in the 5′→3′ direction, without gaps. Because each codon has three nucleotides, an mRNA sequence has the potential to encode three different polypeptide sequences, depending on exactly where translation begins—that is, depending on which register of triplets the translation apparatus acts upon (**Figure 17-7**). Each register of triplets in mRNA is called a **reading frame**. The amino acid sequence of

AUG	GUG	CGU	AGG	GUC	GAU	UGG	CGC	AGA	AAG	UUA	GUU	AGA	GAG	UAC
Met	Val	Arg	Arg	Val	Asp	Trp	Arg	Arg	Lys	Leu	Val	Arg	Glu	Tyr

A	UGG	UGC	GUA	GGG	UCG	AUU	GGC	GCA	GAA	AGU	UAG	UUA	GAG	AGU	AC
	Trp	Cys	Val	Gly	Ser	Ile	Gly	Ala	Glu	Ser	Stop	Leu	Glu	Ser	

AU	GGU	GCG	UAG	GGU	CGA	UUG	GCG	CAG	AAA	GUU	AGU	UAG	AGA	GUA	C
	Gly	Ala	Stop	Gly	Arg	Leu	Ala	Gln	Lys	Val	Ser	Stop	Arg	Val	

FIGURE 17-7 Three possible reading frames. Shown here is a single RNA sequence translated in all three of its reading frames.

the protein encoded by the mRNA depends on which reading frame is used.

Specific sequences in mRNA signal the start of translation and thus define the reading frame. Translation almost always starts at an AUG codon, which specifies the amino acid methionine; this codon is referred to as the **initiation codon** or **start codon**. Occasionally, the codon GUG (usually encoding valine) or UUG (usually encoding leucine) is used as an initiation codon, yet the mRNA is still recognized by the initiating methionine tRNA, inserting a Met residue. The mRNA can also have internal AUG (or GUG and UUG) codons, but translation does not begin at these internal positions. In bacteria, there is a specific sequence in the mRNA next to the initiating AUG (or GUG) that binds the ribosome and directs it to start translation. In eukaryotes, the ribosome is directed to the 5′ terminus of the mRNA, after which it slides down the mRNA; translation can then be initiated at various sites, influenced by a nucleotide sequence known as the Kozak sequence (discussed in Chapter 18).

The three codons (UAA, UAG, and UGA) that signal the end of translation and do not specify any amino acid are called **termination codons** or **stop codons** (or, sometimes, nonsense codons; see Figure 17-7). Termination codons signal the ribosome to dissociate from the newly synthesized polypeptide chain. When the ribosome encounters a termination codon, a release factor associates with the ribosome and terminates protein synthesis. Release factors, even though they recognize specific codons, are proteins. In a fascinating display of molecular mimicry, the three-dimensional structure of release factor proteins is very similar to the structure of tRNA.

With 3 of the 64 codons acting as terminators, a random mRNA sequence should contain 1 stop codon for every 20 codons or so. A long sequence of nucleotide triplets with no stop codons is unlikely to occur by chance, and it generally encodes a protein. Such a sequence is known as an **open reading frame**, or **ORF** (**Figure 17-8**). For example, the average length of a gene in *E. coli* is 1,000 nucleotides, or about 333 codons that lack a termination codon.

The Genetic Code Resists Single-Base Substitution Mutations

The degeneracy of the genetic code enables it to absorb many types of point mutations without serious consequence (see Chapter 12 for a more complete discussion of types of mutation). A single-base substitution that leads to the replacement of one amino acid with another is a **missense mutation**. However, because the genetic code is degenerate, many single-base substitutions are **silent mutations** that do not result in an amino acid replacement. For example, a nucleotide change in the

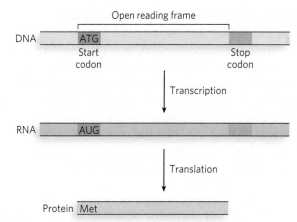

FIGURE 17-8 Start and stop signals in the open reading frame of a gene. The reading frame of a gene that encodes a protein begins at an ATG start codon in the coding strand of the DNA (AUG in the mRNA) and ends at the first stop codon in the same reading frame as the start codon.

third position of a codon results in a change in amino acid only about 25% of the time.

The ability of the code to withstand mutation is even more apparent when we consider that the most frequent mutation is a **transition mutation**, in which a purine is replaced by another purine (A·T replaced by G≡C, or G≡C by A·T). All three positions of the codon confer some type of protection from deleterious transition mutations. A transition mutation in the third position rarely cause a change at all, due to the wobble rules. Even the functioning of UAA and UAG stop codons is protected from damage by a transition mutation in the third position.

A transition mutation in the first position of most codons does result in an amino acid change, but the change is usually to an amino acid that is chemically similar to the original amino acid. This is especially evident for hydrophobic amino acids, as shown in the leftmost column of Figure 17-4. These codons contain U in the second position, and replacement of the first nucleotide results in a codon that specifies another hydrophobic residue. For example, a codon change of GUU to AUU results in an exchange of Ile for Val. Had the GUU codon been altered to CUU, the protein would contain Leu instead of Val. These amino acids have similar chemical properties and thus are much more likely to conserve the protein function than if a hydrophobic residue were replaced by a polar residue. The second position of a codon generally determines whether it encodes a polar (if nucleotide 2 is a purine) or hydrophobic (if a pyrimidine) amino acid. Therefore, transition mutations in the second position also tend to conserve the chemical nature of the protein product.

Errors produced during translation occur most frequently in the codon's first and third nucleotide positions, but the redundancy in coding due to wobble in the third position removes most errors. Eight amino acids are specified by codons that contain any of the four nucleotides in the third position. This, coupled with the fact that any purine-pyrimidine mispairing in the wobble position results in the same amino acid in all but three cases, greatly reduces the effect of reading errors at the ribosome. Just as transition mutations generally lead to a conservative change, misreading of purine-pyrimidine codon-anticodon base pairs results in conservative changes.

Computational studies that examine the theoretical ability of randomly generated genetic codes to withstand the effects of mutation show that most codes would be much less resistant to mutation than is the code actually used by cells. In fact, the probability of arriving by chance at a code that is as resistant to mutation as the genetic code of living organisms is about one in a million. These considerations suggest that the code was extensively honed by natural selection before the divergence of other life forms from LUCA, the ancestral cell.

Some Mutations Are Suppressed by Special tRNAs

Far more deleterious than missense mutations are codon changes that result in a termination codon. These **nonsense mutations** abort protein synthesis, resulting in an incomplete protein that is rarely functional. The gene can be restored to function by a second mutation that converts the nonsense codon to a missense codon or by a mutation in a tRNA that suppresses termination at the nonsense codon by inserting an amino acid at that position (**Figure 17-9**). Mutant tRNAs that function at a stop codon to allow translation to continue are called **suppressor tRNAs**. For example, a change in the anticodon of tRNA^{Tyr} from 5′-GUA to 5′-CUA results in an altered tRNA^{Tyr} that inserts tyrosine at a 5′-UAG termination codon (**Figure 17-10**). Depending on the suppressor tRNA, other amino acids could be inserted at a 5′-UAG termination codon. In theory, any tRNA with an anticodon that is one base pair different from a stop codon could become a suppressor tRNA if a single point mutation occurred in the right place in the anticodon. In fact, suppressor mutations are rare in vivo, but this phenomenon has been harnessed as a tool in the molecular biology laboratory.

Although suppressor tRNAs usually carry a single-nucleotide change in the anticodon, some mutations in suppressor tRNAs lie outside the anticodon. For example, the suppressor of UGA nonsense codons is usually a tRNA^{Trp} that usually recognizes UGG. The mutation that provides this ability to recognize UGA (and to insert tryptophan at this position) can be in the anticodon, but

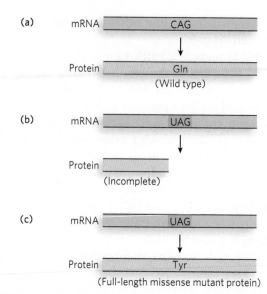

FIGURE 17-9 Suppression of a nonsense mutation. (a) The wild-type mRNA encodes a full-length protein, with CAG encoding Gln (glutamine). (b) A nonsense mutation at an internal CAG codon changes it to a UAG termination codon, resulting in an incomplete protein. (c) A tRNA^{Tyr} suppressor has a mutant anticodon that pairs with the UAG nonsense (stop) codon, resulting in a full-length protein with a Tyr residue in place of the Gln residue of the wild-type protein.

it can also be due to a change of G to A at nucleotide position 24 in the D arm of the tRNA^{Trp}. This change is presumed to lead to an altered conformation that can now recognize both the normal UGG codon and the UGA stop codon. There are other instances in which codon recognition by a tRNA is altered by mutations outside the anticodon, and these are probably mediated by effects on the larger tRNA structure, although more research is needed to clarify the mechanism.

Suppression must not be too efficient, otherwise normal termination codons would also be suppressed, leading to abnormally long protein products—an outcome that would be lethal to the cell. Suppression is limited in several ways. Many genes are terminated by multiple stop

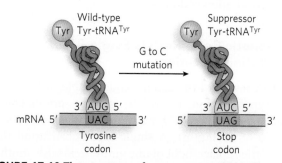

FIGURE 17-10 The structure of a suppressor tRNA. The tRNA^{Tyr} with anticodon 5′-GUA recognizes the UAC codon. The suppressor tRNA^{Tyr} contains a mutation in the anticodon, altering it to 5′-CUA, which base-pairs with the UAG nonsense (stop) codon and inserts a Tyr residue in the protein.

codons. But more importantly, there are multiple copies of each tRNA gene, even in cells that are not diploid. Some duplicate tRNA polynucleotide chains are weakly expressed and thus constitute only a small fraction of the tRNA pool for a particular amino acid. Suppressor mutants are typically found in one of these minor tRNA genes, leaving the major tRNA gene to function normally.

An example of suppression in *E. coli* is tRNA^Tyr with the anticodon 5′-GUA. *E. coli* contains three identical tRNA^Tyr genes, but one is much more highly transcribed than the others. The tRNA^Tyr suppressor mutation, which changes the anticodon to 5′-CUA and thus recognizes the 5′-UAG stop codon, is found in one of the minor, less-transcribed tRNA^Tyr genes. Therefore, the insertion of tyrosine at UAG stop codons is inefficient, but sufficient full-length protein is produced from a gene with a nonsense mutation to let the cell survive. Furthermore, UAG is used only rarely as a stop codon in *E. coli*. This allows suppression to be reasonably efficient (up to 50%) at UAG stop codons. In comparison, suppression at the more frequently used UAA and UGA stop codons must be kept below 5% to ensure cell viability. There are also examples of suppressor tRNAs for missense mutations, and of suppressor tRNAs for frameshift mutations, which place the ribosome in an incorrect reading frame by insertion or deletion of a nucleotide.

SECTION 17.1 SUMMARY

- Transfer RNAs are small RNA molecules that can covalently attach at their 3′ end to an amino acid. The triplet anticodon in tRNA pairs with a triplet codon in mRNA, and this pairing mediates translation of the nucleotide sequence in mRNA into the amino acid sequence of a protein.
- The genetic code is degenerate, because most amino acids are specified by two or more codons. One tRNA often reads two codon sequences, due to noncanonical or wobble base pairing at the third nucleotide position of the codon. When the anticodon contains inosine (I), a modified nucleotide residue, the tRNA recognizes three different codons, ending in A, C, or U.
- An AUG codon, specifying methionine, typically initiates protein synthesis. The three termination codons do not specify an amino acid, but instead instruct the ribosome to stop translation.
- Due to codon assignments and the degeneracy of the genetic code, single-base substitution mutations generally result in codons that specify the same or similar amino acids; nonsense mutations result in a stop codon that can lead to inactive protein. Mutant tRNAs that carry a single-nucleotide change in the anticodon can suppress nonsense mutations by inserting an amino acid in the polypeptide at the mutant termination codon.

17.2 THE RULES OF THE CODE

The genetic code words—the codons—must follow specific rules for protein synthesis. Even after the discovery of tRNAs, several experiments were required to determine the rules of the genetic code. These classic studies addressed whether codons are read sequentially, are overlapping, or have gaps (punctuation) between them, and they looked for confirmation of a triplet codon. Investigators also asked in what direction protein synthesis occurs. These "rules of the code" were addressed by elegant experiments performed even before the code words themselves were understood.

The Genetic Code Is Nonoverlapping

Investigators realized that the triplet codons in mRNA either overlapped with one another or were nonoverlapping. In a nonoverlapping code, each codon would be read as an independent unit; a single-nucleotide substitution in the mRNA would change only one codon, and the mutant protein would have only one amino acid change (**Figure 17-11a**). In a triplet code with maximal overlap, each codon would share two nucleotides with two other codons; a single-nucleotide change in the mRNA would alter three codons, and the resulting protein would contain three consecutive amino acid changes (**Figure 17-11b**).

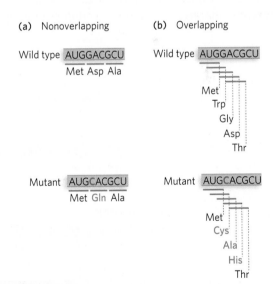

FIGURE 17-11 Mutation effects on nonoverlapping and overlapping codes. (a) In a nonoverlapping code, codons in the mRNA do not share nucleotides, so a single-nucleotide mutation alters only one codon, and the resulting protein has a single amino acid change. (b) In an overlapping code, some nucleotides are shared by several codons. In a triplet code with maximum overlap, a nucleotide can be shared by three codons, so a single-nucleotide mutation results in three codon changes and thus three amino acid alterations in the protein. The genetic code of all living systems is now known to be nonoverlapping.

A code with overlapping codons was ruled out experimentally by studies of mutant proteins. Most mutations result from single-base substitutions. Therefore, in a nonoverlapping code one amino acid would be changed, whereas in a maximally overlapping code three consecutive amino acids would be changed (see Figure 17-11). Independent studies of mutants of three different proteins—hemoglobin, tobacco mosaic virus protein, and tryptophan synthetase—demonstrated that the code must be nonoverlapping. In a combined total of nearly 100 different mutants of these proteins, almost all of the mutants had only one amino acid change. Thus, the genetic code is nonoverlapping, and any exceptions in the findings for mutant proteins are probably the result of double mutations.

There Are No Gaps in the Genetic Code

Investigators also understood that the codons in mRNA could be arranged one after the other, with no separation, or could be separated by one or more nucleotides acting as punctuation, like a comma. Commas between codons would prevent the deleterious effects of frameshift mutations that result from deletions or insertions of nucleotides. If there were no commas to set codons apart, a frameshift mutation would throw off the entire reading frame (**Figure 17-12a**). If codons were separated by a noncoding nucleotide, a frameshift mutation would result in only one amino acid change, because the next comma after the altered codon would signal the ribosome to get back on track in the correct reading frame (**Figure 17-12b**).

Francis Crick and Sydney Brenner performed a series of ingenious experiments in the early 1960s to determine whether the genetic code contains punctuation between codons. They studied the *B* gene of the T4 bacteriophage, which encodes a protein needed for the phage to grow on two different strains of *E. coli*. Severe mutations in the *B* gene restrict phage growth to one *E. coli* strain, but minor alterations produced near one end of the *B* gene are tolerated and preserve the dual-host-range phenotype. Crick and Brenner introduced mutations into the *B* gene chemically, using acridines as mutagens. These planar molecules intercalate into DNA and usually produce mutations through the insertion or deletion of a single base pair. The acridines were used at a concentration low enough to cause an average of one mutation per phage. The researchers tested the effects of these mutations by assaying for plaques (i.e., phage growth) on the two different host strains of *E. coli*.

Most mutations completely inactivated the *B* gene, even when the mutations mapped to the region of the gene that can tolerate minor alterations. These results suggested that not one but many amino acids were changed by the acridine-induced mutations: the mutations seemed to throw off the entire reading frame, indicating that the code has no commas to bring the ribosome back on track.

To study these mutations further, Crick and Brenner recombined the *B* genes from two mutant T4 phages by

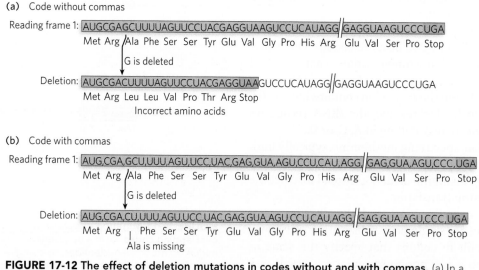

FIGURE 17-12 The effect of deletion mutations in codes without and with commas. (a) In a code without commas, a single-nucleotide deletion (here, deletion of a G residue) throws off the reading frame by one nucleotide. All amino acids added after this point are different from those in the wild-type protein. (b) In a code with commas separating the codons, the deletion should have minimal impact on the protein product because the next comma encountered after the deleted nucleotide would reset the ribosome to the correct reading frame. The genetic code of all living systems is now known to be without commas.

mixing them in a host *E. coli* culture, creating phages with a *B* gene containing two mutations instead of one. In some cases, the double-mutated gene restored the dual host range of the wild-type phage, but other pairs of mutations did not restore the dual-range phenotype. Thus, the single-mutant phages used for the recombination could be sorted into two groups, referred to as a plus or a minus group. Double mutants made by crossing two single mutants of like sign could not form a wild-type double mutant, but crossing two single mutants of opposite sign formed double mutants with the dual-host-range phenotype.

Crick and Brenner interpreted their findings as follows. The gain or loss of one base pair results in a shift in the wild-type reading frame (**Figure 17-13a**), thus changing all the amino acids after the point of mutation (**Figure 17-13b, c**). When mutants of opposite sign are combined (an insertion and a deletion), the first mutation encountered during translation causes a frameshift, but the second mutation of opposite sign restores the original reading frame (**Figure 17-13d**). Therefore, the only amino acid alterations that occur are located between the two mutations. Mutants of like sign do not complement one another, because they do not reestablish the

correct reading frame. Overall, these results implied that codons are read in a reading frame without commas. There are no gaps between words in the genetic code.

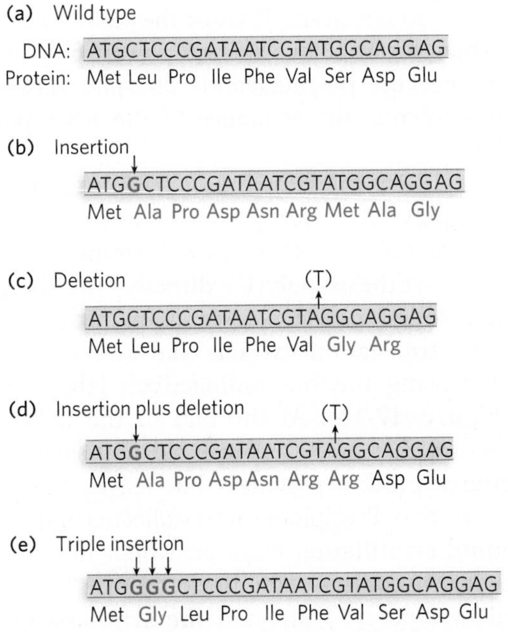

The Genetic Code Is Read in Triplets

The first indication that the code is read in groups of three nucleotides came from an extension of the acridine-induced frameshift mutant studies in T4 phage. This time, Crick and Leslie Barnett combined three different mutants. They predicted that if the codons were read in sets of three nucleotides, the crossing of three *B*-gene mutants of T4, all with the same sign, would reestablish the reading frame and produce an active *B*-gene product (**Figure 17-13e**). Indeed, crossing three *B*-gene mutants of like sign (by mixing in an *E. coli* host culture) restored wild-type, or near wild-type, activity. The results suggested that the original reading frame was restored by either the deletion or the insertion of three nucleotides.

This was the first evidence that codons consist of three nucleotides. However, this interpretation is based on the assumption that each of the mutations in the three *B*-gene mutants was indeed a single-nucleotide insertion (or deletion). Without direct sequence information, it remained possible that more than one base pair was added (or removed) in one or more of the mutants. Of course, we now know that codons really do consist of three nucleotides apiece and that the genetic code is read in triplets.

Protein Synthesis Is Linear

There are many possible ways in which a chain of amino acids could be assembled on an mRNA transcript. For example, chain growth could initiate at one end of the polypeptide, either the N-terminus or the C-terminus, or it could start in the middle and grow outward in both directions. In fact, the chain could even be synthesized in random segments that were then stitched together to form the final product. In 1961, Howard Dintzis performed elegant studies on the synthesis of hemoglobin, using extracts from rabbit reticulocytes (immature red blood cells), and demonstrated that protein synthesis proceeds linearly, from the N- to the C-terminus (see the How We Know section at the end of this chapter).

Given that the direction of protein synthesis proceeds from the N-terminus to the C-terminus, determining the direction in which codons are translated became a relatively simple matter. Chemical methods of synthesizing RNAs of defined sequence had already been developed to crack the genetic code. For example, a synthetic hexanucleotide RNA should direct

(a)   Wild type
      DNA: ATGCTCCCGATAATCGTATGGCAGGAG
   Protein: Met Leu Pro Ile Phe Val Ser Asp Glu

(b)   Insertion
      ATGGCTCCCGATAATCGTATGGCAGGAG
      Met Ala Pro Asp Asn Arg Met Ala Gly

(c)   Deletion   (T)
      ATGCTCCCGATAATCGTAGGCAGGAG
      Met Leu Pro Ile Phe Val Gly Arg

(d)   Insertion plus deletion   (T)
      ATGGCTCCCGATAATCGTAGGCAGGAG
      Met Ala Pro Asp Asn Arg Arg Asp Glu

(e)   Triple insertion
      ATGGGGCTCCCGATAATCGTATGGCAGGAG
      Met Gly Leu Pro Ile Phe Val Ser Asp Glu

FIGURE 17-13 The effects on a reading frame of combining insertion and deletion mutations. Insertion or deletion of a single nucleotide throws the ribosome into the wrong reading frame and produces a mutant protein (amino acids shown in red). (a) The wild-type protein sequence. (b) The effect of an insertion mutation. (c) The effect of a deletion mutation. (d) Combining an insertion and a deletion affects some amino acids but eventually restores the correct sequence. (e) Combining three consecutive insertion mutations (or three deletions) leaves the remaining triplets intact—evidence that a codon has three, rather than four or five, nucleotides.

the synthesis of two amino acids. Thus, an RNA of sequence 5′-AAAUUU would encode Lys-Phe if the codons were read in the 5′→3′ direction, or Phe-Lys if read in the 3′→5′ direction (recall that peptides are always written in the N- to C-terminal direction). Experiments demonstrated that codons are read in the 5′→3′ direction during translation.

SECTION 17.2 SUMMARY

- Single-nucleotide changes in a gene result in changes in a single amino acid residue in the protein product, demonstrating a nonoverlapping genetic code. Each codon is an independent unit, coding for a single amino acid.

- The genetic code has no commas. Single-nucleotide insertion or deletion mutations result in a complete loss of activity in the mutant gene; these mutations throw off the reading frame from the point of mutation onward, because the code lacks any signal (comma) to reset the reading frame. Double mutations with an insertion and a deletion restore the reading frame.

- Codons are composed of triplet sequences. Triple-insertion mutations and triple-deletion mutations result in active protein.

- Protein synthesis is linear; it proceeds from the N-terminus to the C-terminus, and mRNA is read in the 5′→3′ direction.

17.3 CRACKING THE CODE

Cracking the genetic code was one of the most significant scientific milestones of the last half of the twentieth century. Today, it would be a simple matter of comparing the sequences of nucleic acids and their corresponding proteins. But in the early 1960s, nucleic acid sequencing had not yet been invented, and although protein sequencing methods were firmly established, the process was laborious. Given the state of sequencing methodology, it is surprising that new and ingenious ways of studying and deciphering the code in its entirety were developed at all. Here we briefly review the major experimental strategies that enabled investigators to crack the code.

Random Synthetic RNA Polymers Direct Protein Synthesis in Cell Extracts

First, Marshall Nirenberg and Heinrich Matthaei made a simple but remarkable discovery that set in motion the cracking of the genetic code: they found that they could use the enzyme polynucleotide phosphorylase to synthesize RNA templates that would code for protein

polymers in cell extracts of *E. coli*. To prepare the cell extracts, endogenous mRNA had to be removed, to allow the translation machinery to work on the synthetic RNA templates. To do this, the extracts were preincubated so that the endogenous ribonuclease (RNase) activity would destroy all existing mRNA, and deoxyribonuclease (DNase) was added to the extracts to prevent further mRNA production. With these treatments, protein synthesis in the cell extracts (or "translation extracts") was dependent on the addition of exogenous synthetic RNA.

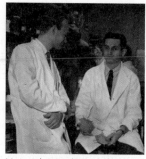

Heinrich Matthaei (left); Marshall Nirenberg, 1927–2010 (right) [Source: Courtesy of the National Institutes of Health.]

Polynucleotide phosphorylase does not require a template; it uses ribonucleoside diphosphates (NDPs) to make random polymers of RNA:

$$NDP + (NMP)_n \rightarrow (NMP)_{n+1} + P_i \qquad \text{17-1}$$

The intracellular role of polynucleotide phosphorylase is to catalyze the reverse reaction to degrade RNA, using inorganic phosphate to yield NDPs. In vitro, the enzyme can be made to synthesize RNA by the addition of excess NDPs, which at sufficient concentration are polymerized. However, because polynucleotide phosphorylase is not template-directed, the sequence of the RNA polymer is random.

Using UDP as the only ribonucleoside diphosphate substrate, polynucleotide phosphorylase catalyzes the synthesis of poly(U) RNA. To determine what type of protein synthesis poly(U) directs, Nirenberg and Matthaei added poly(U) to reaction mixtures with the *E. coli* translation extracts and amino acid mixtures, but using only one radioactively labeled amino acid (**Figure 17-14**). At the end of the incubation, the reaction mixtures were treated with acid, which precipitates protein polymers but leaves free amino acids in solution. Precipitates were collected and counted in a liquid scintillation counter, which measures radioactivity in counts per minute (cpm). The results showed that poly(U) directed the synthesis of polyphenylalanine, poly(Phe), and therefore the codon for phenylalanine must be UUU. The other homopolymers were synthesized and tested by similar methods. Poly(A) specified the synthesis of poly(Lys), identifying AAA as a codon for lysine. Poly(C) resulted in the production of poly(Pro), and thus CCC is a codon for proline. Unfortunately, poly(G) forms intramolecular hydrogen-bonded structures that prevent its use in these reactions.

(a)

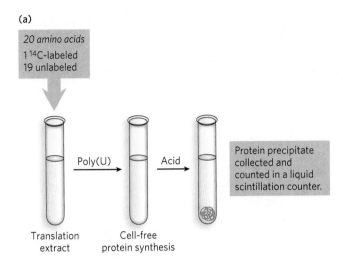

(b)

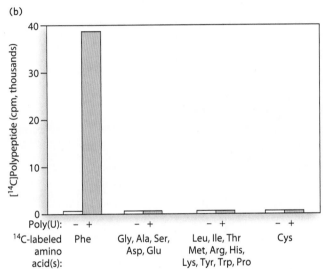

FIGURE 17-14 Poly(U)-directed synthesis of poly(Phe).
(a) Twenty different ¹⁴C-labeled amino acids were added, individually or in mixtures, with the rest of the amino acids in unlabeled form, to cell extracts that could synthesize protein in the presence of an added poly(U) template. (b) Only the extracts containing [¹⁴C]phenylalanine produced a significant amount of radioactive polypeptide in the presence of poly(U). [*Source: Data from M. Nirenberg and P. Leder, Science 145:1399, 1964.*]

Severo Ochoa, 1905–1993
[*Source: AP Photo.*]

The use of polynucleotide phosphorylase was extended to copolymers, RNAs containing more than one type of nucleotide. As an example of this type of analysis, **Table 17-2** shows the results obtained by Severo Ochoa's group, using ADP and CDP in a 5:1 molar ratio. Polynucleotide phosphorylase incorporates NDPs into RNA randomly, so these conditions produced RNA with five times more A than C. All possible codons containing A and C were generated in the random RNA copolymer, in the following distribution (setting AAA at 100): AAA (100), A_2C (60), AC_2 (12), and CCC (0.8). (Note the use of A_2C and AC_2 to indicate that the *order* of nucleotides is unknown.) Use of this heterogeneous RNA copolymer in translation extracts directed the synthesis of polypeptides containing asparagine, glutamine, histidine, lysine, proline, and threonine. From the relative amounts of these amino acids in the precipitated protein product, the investigators could deduce the respective codon compositions specifying each amino acid. No amino acid was observed at the low frequency expected of a CCC codon (0.8) relative to an AAA codon (100). But proline was observed to be incorporated at a relative frequency of 4.7, which is close to the expected frequency for two codons, CCC and AC_2.

The researchers could infer from this result that proline is specified by two codons having the compositions AC_2 and CCC, reflecting the degenerate nature of the genetic code. The result was supported by experiments using poly(C), which identified CCC as a codon specifying proline. The other codon, AC_2, could have the sequence ACC, CCA, or CAC. Because polynucleotide phosphorylase polymerizes NDPs randomly, one can

TABLE 17-2

Incorporation of Amino Acids into Polypeptides in Response to Random RNA Polymers

Amino Acid	Observed Frequency of Incorporation (Lys = 100)	Tentative Assignment for Nucleotide Composition of Corresponding Codon*	Expected Frequency of Incorporation Based on Assignment (Lys = 100)
Asparagine	24	A_2C	20
Glutamine	24	A_2C	20
Histidine	6	AC_2	4
Lysine	100	AAA	100
Proline	4.7	AC_2, CCC	4.8
Threonine	26	A_2C, AC_2	24

*These designations contain no information on nucleotide sequence (except, of course, AAA and CCC).

assign only codon compositions from these results, not codon sequences (except, of course, for codons with only one type of nucleotide).

Numerous experiments were performed using random RNA copolymers, and in this way the nucleotide composition of about 40 codons could be assigned to particular amino acids. However, to identify the nucleotide *sequence* of codons, RNA molecules of defined sequence were required.

RNA Polymers of Defined Sequence Complete the Code

Chemical synthesis of short nucleic acids of specific sequence was needed to define the complete genetic code. In 1964, Nirenberg and his coworkers at the National Institutes of Health discovered a novel method to identify an amino acid associated with a short synthetic codon. They found that during protein synthesis in *E. coli* (using translation extracts and ^{14}C-labeled amino acids), the nascent protein (the protein being synthesized) stayed attached to ribosomes bound to the mRNA and could be separated from unbound amino acids (**Figure 17-15a**). However, the ^{14}C]aminoacyl-tRNA–ribosome–RNA codon complex could not be precipitated by acid treatment, because the complex is dissociated by acid. The researchers developed a filtration method that took advantage of the fact that a nitrocellulose filter binds protein but not RNA (including tRNA, charged or uncharged). More importantly, they found that even RNA polymers as short as a trinucleotide could be used in these experiments. Maxine Singer, also working at the NIH, had developed the capability of synthesizing RNAs of defined sequence. This provided an advance that allowed assignment of several additional codons for amino acids and confirmed assignments made by the precipitation method.

Maxine Singer
[Source: The Washington Post/ Getty Images.]

Results from Nirenberg's initial study, using three different ^{14}C]aminoacyl-tRNAs and the synthetic trinucleotides UUU, CCC, or AAA, are shown in **Figure 17-15b**. Similar studies of all possible triplet RNA sequences identified approximately 50 codon assignments for specific amino acids. However, not all triplet RNA sequences induced tight binding of their corresponding ^{14}C]aminoacyl-tRNA to the ribosome, and another technique was required to completely crack the genetic code.

Another breakthrough came from H. Gobind Khorana, who developed chemical methods to synthesize

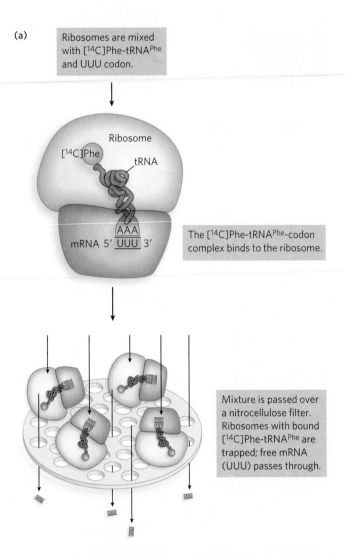

(a) Ribosomes are mixed with ^{14}C]Phe-tRNA[Phe] and UUU codon.

Ribosome

[14]C]Phe

tRNA

mRNA 5′ UUU 3′   AAA

The ^{14}C]Phe-tRNA[Phe]-codon complex binds to the ribosome.

Mixture is passed over a nitrocellulose filter. Ribosomes with bound ^{14}C]Phe-tRNA[Phe] are trapped; free mRNA (UUU) passes through.

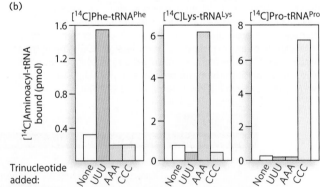

(b)

[14]C]Phe-tRNA[Phe]   [14]C]Lys-tRNA[Lys]   [14]C]Pro-tRNA[Pro]

[14]C]Aminoacyl-tRNA bound (pmol)

Trinucleotide added: None UUU AAA CCC | None UUU AAA CCC | None UUU AAA CCC

FIGURE 17-15 Use of trinucleotide sequences as mRNA. (a) Twenty ^{14}C]aminoacyl-tRNAs were formed in separate reactions, using the 20 different ^{14}C-labeled amino acids, and each was added to a preparation of ribosomes, along with a three-nucleotide RNA of defined sequence (a codon). Individual reaction mixtures were then passed through a nitrocellulose filter that binds protein, trapping the ^{14}C]aminoacyl-tRNA bound to the triplet RNA codon on the ribosome. The reaction shown here used ^{14}C]Phe-tRNA[Phe] and the UUU codon. (b) Results obtained with several codons (UUU, AAA, CCC) and ^{14}C] aminoacyl-tRNAs. *[Source: Data from M. Nirenberg and P. Leder, Science 145:1399–1407, 1964.]*

H. Gobind Khorana,
1922–2011 [Source: AP Photo.]

short polyribonucleotides with defined sequences of two, three, or four nucleotide repeats. These short RNAs were then amplified by polymerases to produce long polymers of defined repeating sequence. Results obtained with these long RNA polymers, combined with the earlier results from trinucleotide-induced ribosome binding and the use of random RNA polymers, identified the amino acids specified by all of the codons. For example, the alternating copolymer (AC)ₙ contains the alternating codons ACA and CAC. The polypeptide synthesized on this RNA copolymer contains equal amounts of threonine and histidine. Parallel studies using random RNA copolymers identified the nucleotide composition of a histidine codon as AC_2, so CAC must code for histidine and ACA for threonine. Examples of results obtained with di-, tri-, and tetranucleotide sequences are shown in **Table 17-3**.

Repeating trinucleotide sequences yield three types of homopolymeric peptides. For example, for a

repeating sequence of UUCs, the mRNA triplets can be read as repeating UUC, or UCU, or CUU codons. Poly(UUC) produced poly(Phe), poly(Lys), and poly(Leu). The ribosome-binding assay showed that UCU encodes serine and CUU encodes leucine. Therefore, the UUC codon can be assigned to phenylalanine. Poly(UAG) yielded only two homopolymers, poly(Ser) and poly(Val), because the UAG reading frame is a string of nonsense codons that yields no product. The repeating tetranucleotide poly(UAUC) produced a repeating tetrapeptide of Tyr-Leu-Ser-Ile. Knowing the codon assignments for the four amino acids in the repeating tetranucleotide also confirmed that mRNA is read in the 5′→3′ direction. If the RNA had been read in the 3′→5′ direction, the repeating tetrapeptide would have been Ile-Ser-Leu-Tyr.

Consolidation of the results from many different experiments permitted the assignment of 61 of the 64 possible codons. The other three were identified as termination codons, in part because of the results obtained with synthetic RNAs in which the codons disrupted the amino acid coding patterns. Meanings for all the triplet codons were firmly established by 1966, and they have been verified numerous times, in many different ways. In retrospect, the experiments of Nirenberg and Khorana that identified codons in translation extracts

TABLE 17-3		
Polypeptides Produced by Synthetic RNAs of Defined, Repeating Sequence		
Repeating Dinucleotide	*Codons*	*Copolymer*
(AC)ₙ	ACA-CAC-ACA-CAC-ACA	(Thr-His)ₙ
(UC)ₙ	UCU-CUC-UCU-CUC-UCU	(Ser-Leu)ₙ
Repeating Trinucleotide	*Codons*	*Homopolymer*
(UUC)ₙ	UUC-UUC-UUC-UUC-UUC	Poly(Phe)
	UCU-UCU-UCU-UCU-UCU	Poly(Ser)
	CUU-CUU-CUU-CUU-CUU	Poly(Leu)
(AAG)ₙ	AAG-AAG-AAG-AAG-AAG	Poly(Lys)
	AGA-AGA-AGA-AGA-AGA	Poly(Arg)
	GAA-GAA-GAA-GAA-GAA	Poly(Glu)
(UAG)ₙ	UAG-UAG-UAG-UAG-UAG	None
	AGU-AGU-AGU-AGU-AGU	Poly(Ser)
	GUA-GUA-GUA-GUA-GUA	Poly(Val)
Repeating Tetranucleotide	*Codons*	*Repeating copolymer*
(UAUC)ₙ	UAU-CUA-UCU-AUC	(Tyr-Leu-Ser-Ile)ₙ
	AUC-UAU-CUA-UCU	(Tyr-Leu-Ser-Ile)ₙ
	UCU-AUC-UAU-CUA	(Tyr-Leu-Ser-Ile)ₙ
	CUA-UCU-AUC-UCU	(Tyr-Leu-Ser-Ile)ₙ
(GUAA)ₙ	GUA-AGU-AAG-UAA-GUA	Val-Ser-Lys-(stop)
	UAA-GUA-AGU-AAG-UAA	Val-Ser-Lys-(stop)
	AAG-UAA-GUA-AGU-AAG	Val-Ser-Lys-(stop)
	AGU-AAG-UAA-GUA-AGU	Val-Ser-Lys-(stop)

should not have worked in the absence of initiation codons. Serendipitously, with the high magnesium concentration used in their in vitro experiments, the normal initiation requirement for protein synthesis was relaxed.

The Genetic Code Is Validated in Living Cells

We now know the sequences of entire genomes of organisms ranging from bacteria to humans, and this enormous body of information confirms that the genetic code determined in the translation extract experiments is indeed interpreted in the same way in all living organisms. Even in the 1960s, however, there was evidence that the genetic code was the same in cells as in cell extracts.

By the mid-1960s, more than 100 mutant proteins resulting from single-nucleotide substitutions had been studied, and in all cases the incorrect amino acid in the mutant, relative to the wild type, could be accounted for by a change in a single nucleotide, based on the genetic code. For example, the disease sickle-cell anemia was known to be caused by a single amino acid change, substituting glutamate for valine in human hemoglobin (see Highlight 2-1). Only one nucleotide change is needed to alter the AGU codon specifying glutamate to the UGU codon for valine: these codons specify the same amino acids in extracts of *E. coli* as they do in blood cells. An astute demonstration that the genetic code in the cell is the same as the code determined in cell extracts is described in the How We Know section at the end of this chapter.

SECTION 17.3 SUMMARY

- Researchers cracked the genetic code in experiments that used radioactively labeled amino acids and cell extracts that translate synthetic RNA templates.
- The compositions of many codons specifying amino acids were assigned on the basis of experiments using random RNA polymers synthesized by polynucleotide phosphorylase.
- Two techniques that used RNAs of defined sequence completed the codon table. An assay that induced the binding of [^{14}C]aminoacyl-tRNAs and their cognate synthetic trinucleotide RNAs to ribosomes allowed the identification of most of the codons. With the availability of long, synthetic RNA polymers of defined repeating sequence, researchers used the in vitro protein synthesis assay to assign the remaining codons.
- Studies of amino acid replacements in mutant proteins confirmed that the genetic code in living cells is the same as that determined in cell extracts.

17.4 EXCEPTIONS PROVING THE RULES

Initial studies of the genetic code suggested that it was universal and without variation. And, as we've seen, this universality implies that life evolved from a common ancestor. Presumably, once the code had developed in LUCA, it became locked in place because descendants could not tolerate changes in the genetic code. For example, a change in a codon that specifies lysine to one that specifies leucine would result in a Leu residue replacing every Lys residue in every protein in the cell. Clearly, such a global change would be fatal to the cell, so it is unlikely that such codon changes could occur, even over a long span of evolutionary time. How the translation machinery evolved in LUCA remains one of the greatest questions in evolutionary biology, and with no "missing link" cells, we may never know the answer. But even though we do not know how the process of translation evolved, it is interesting to contemplate the evolutionary hurdles that must have been overcome to arrive at this process.

We now know that there *are* some exceptions to the rules of the genetic code. It is not entirely universal after all. Does this mean that the ancestral cell did not perfect the code before modern cells diverged from it? The evidence suggests not. Most of the exceptions support the evolution of a common code in LUCA from which a few changes, in some circumstances, evolved.

Evolution of the Translation Machinery Is a Mystery

With the discovery of ribozymes, the hypothesis of the RNA world became a very plausible model for the beginning of life. In the RNA world, RNA catalyzes essentially all the chemical reactions needed for life, and there are many examples of catalytic RNAs in cells today—including the ribosome—all of which are possible vestiges of the RNA world. RNA can also catalyze its own replication. But how do we get from an RNA world to LUCA—a cell with a membrane, DNA for storage, mRNA, ribosomes, tRNA, and protein-based catalysis? Possible ways in which a cell membrane developed are discussed in Chapter 1. For the nucleic acids, we know that RNA is less stable than DNA. Despite Steve Benner's finding that borate can stabilize RNA and was probably plentiful in the prebiotic soup (see this chapter's Moment of Discovery), it is not hard to imagine the development of DNA as a more stable information storage molecule. But how can we explain the evolution of translation and the genetic code for protein synthesis?

Any hypotheses about how translation evolved must account for each part of the translation machinery. At least 20 tRNAs, 20 aminoacyl-tRNA synthetases, a coding RNA, and the entire ribosome machinery, with its numerous proteins and RNA components, are required to translate the genetic code. These components could not have evolved all at once, so a reasonable hypothesis must either reduce the complexity of the translation process or break it down into individual steps. The hypothesis must also solve the chicken-and-egg problem of how protein synthesis could evolve when the protein components (aminoacyl-tRNA synthetases and ribosomal proteins) could not be synthesized in an RNA world. Moreover, the process was probably not accurate in the beginning; accuracy most likely required evolutionary honing. If accuracy was not required initially, catalytic function is essentially ruled out as the initial role of early proteins. But there must have been an initial benefit to the cell on which natural selection could act. Plausible hypotheses must consider the forces of natural selection that would foster the evolution of proteins before their catalytic role was realized.

Finally, how did the genetic code evolve? Was it a random act of evolution, or did the amino acids somehow participate in generation of the code as we know it today? Did all 20 tRNAs appear individually, or did one appear through random mutation and then diversify into the rest? Did the first code use triplet sequences, or was it simpler, using dinucleotide sequences to code for fewer amino acids? How was a reading frame established? And how did the point of termination develop?

These are some of the challenges to hypothesizing how the translation process evolved. One hypothesis is presented in **Highlight 17-1**. All cells have the complete translation machinery; there are no cells with "missing links" to tell us the story more directly. However, certain scientific approaches can help illuminate how the process evolved. For example, as noted above, there are exceptions to the genetic code, and we can examine them for insights about its evolution. We also know that RNAs can perform many enzymatic reactions, supporting the RNA world hypothesis. Ongoing research is defining the minimal genetic and protein requirements for a living cell, which will identify the essential genetic and protein requirements for life—components that are likely to have been present in LUCA.

Mitochondrial tRNAs Deviate from the Universal Genetic Code

Phylogeny tells us that the exceptions to the genetic code are derived from a single, universal code, because the exceptions are rare and occur in different branches of the tree of life (see Chapter 8). But in what situations could changes to the genetic code be viable? Even one change would have a global impact on cellular function,

because every protein in the cell is made according to the same set of coding instructions. In other words, a single change in a codon–amino acid relationship would cause changes in all proteins encoded by a gene containing that codon.

With this in mind, we can make a few hypotheses about the types of deviations from the genetic code that might be plausible. For example, we can propose that the most easily altered codons are termination codons, which are not located in the middle of genes. If a stop codon were recruited to code for an amino acid by altering the anticodon of a tRNA, that codon could be placed in internal positions of certain genes as the organism evolved (another stop codon could be used for termination). In this way, a particular amino acid would be inserted in the middle of the gene where the stop codon is placed. We can also hypothesize that evolution of this type of exception to the code would have a higher probability in an organism of low genetic complexity (i.e., only a few genes) than in an organism of high complexity, because fewer proteins would be affected. These two hypotheses are largely confirmed.

Mitochondria are a prime example of how the genetic code can be altered. Mitochondria are thought to be the descendants of early bacterial cells that were engulfed by eukaryotic cells and proved beneficial because of their unique capacity to perform aerobic metabolism, thereby conferring this advantageous capability on early anaerobic eukaryotes. Over time, this symbiotic relationship relieved the mitochondrial genome of most of its genes, transferring them to the nucleus of the host cell. But mitochondria retained a small genome of their own, with a limited set of genes. Researchers noticed the first deviations from the genetic code when sequencing mitochondrial DNA (mtDNA). A fascinating aspect of the mitochondrial genome is that it encodes a unique set of tRNAs, just for use in decoding the mtDNA. This feature of mitochondria permits changes in their tRNAs without interfering with the information flow of the cellular genome. As predicted for alterations evolving from the standard code, the most common codon changes in mitochondria involve stop codons.

Genetic code changes in mitochondria are essentially the result of an exquisitely streamlined flow of genetic information. Vertebrate mtDNAs encode 13 proteins, 2 rRNAs, and 22 tRNAs. Instead of the minimum of 32 tRNAs needed for the standard, cellular code, the 22 mitochondrial tRNAs can decipher all possible codons by slight alterations in the rules of the code. For example, only one tRNA is used for each of four codon families. In each case, a single tRNA recognizes the four different codons, each with the same first two nucleotides. Each of these mitochondrial tRNAs has a U in the first (wobble) position of the anticodon (that base-pairs

| HIGHLIGHT 17-1 | EVOLUTION |

The Translation Machinery

How the ribosome, tRNAs, and the genetic code evolved is one of the most perplexing and fascinating areas of evolutionary history. In the RNA world, nucleic acids not only stored genetic information but also performed all the catalytic reactions necessary for life. In modern cells, most enzymes are proteins, which are more efficient catalysts than RNAs. The leap from nucleic acid to protein requires very complicated machinery, and it presumably evolved in steps. Furthermore, LUCA could not predict that proteins would be superior to RNAs as catalysts, so we can presume that the first proteins were made to serve some other purpose. What were the evolutionary steps leading to the translation apparatus, and how did natural selection produce them? As an intellectual exercise, let's consider just one of several possible explanations.

First of all, why would a cell want a protein when it is already using RNA to catalyze cellular metabolism? Harry Noller suggests that the first proteins evolved to help RNA ribozymes fold properly. For example, the high negative charge of the nucleic acid backbone hinders the close approach of nucleic acid helices, but the charge can be mitigated by a protein's basic amino acid side chains. Indeed, most ribozymes in modern cells (including ribosomes) contain a protein component, and many ribosomal proteins are at junctions of RNA helices and may stabilize their proximity. Therein lies a possible selection pressure for an RNA-based cell to develop a way to make proteins: proteins can serve a structural role that helps RNA form a more catalytically competent ribozyme.

What about the building blocks of proteins? Several amino acids were most likely present in the primordial soup, but how were they linked together in a way that is coded by an RNA without aminoacyl-tRNA synthetases (proteins)? Perhaps the anticodon of tRNA was directly involved in specifying the amino acid it carried. This is easy to envision, given what we know about cellular riboswitches. Riboswitches are small sections of RNA, usually 70 to 170 nucleotides embedded in a larger RNA molecule, that fold into complex structures and bind specific cellular metabolites (usually small molecules, including free amino acids) with high affinity (see Chapter 20). Thus, ancestral tRNAs could have selectively bound particular amino acids. However, modern tRNAs are L-shaped, with an anticodon far removed from the 3' terminus that carries the amino acid. This distance precludes participation of the anticodon in amino acid selection. But perhaps early tRNAs simply folded differently, with the 3' amino acid arm near the anticodon, enabling it to select the correct amino acid and charge itself, thus circumventing the need for aminoacyl-tRNA synthetases (Figure 1).

Now we have the rudiments of a simpler RNA world–based process of translation. The self-charged tRNAs align along mRNA through base pairing, and peptide bonds form—probably spontaneously, given the inherent reactivity of charged tRNAs. Over time, natural selection would direct the evolution of a surface (i.e., the ribosome) on which tRNAs could be more efficiently aligned on the mRNA, and this surface eventually would act as a catalyst for the peptidyl transfer reaction (see Chapter 18 for the details of protein synthesis).

The early translation process may have operated at very low fidelity, yet it served the purpose of making structural peptides that enhanced RNA catalytic function. For example, early tRNAs may not have been entirely selective for one

with position 3 of the codon). Using normal base-pairing rules, two tRNAs are needed to decode a codon family, one with a U in the wobble position (pairs with G or A) and one with a G (pairs with U or C). Therefore, the U in these mitochondrial tRNAs is not used to distinguish codons, and base pairing to the first two nucleotides of the codon specifies which amino acid is incorporated. Other mitochondrial tRNAs function with codons that contain either A or G in the third position, or either U or C, so virtually all the tRNAs recognize two or four codons.

If all mitochondrial tRNAs recognize more than one codon, yet another deviation from the rules of the standard genetic code can be inferred. Normally, tryptophan and methionine are specified by one codon each. In mitochondria, the tRNA specifying tryptophan recognizes the UGG codon, but it also recognizes UGA, which is a termination codon in the standard code. The AUG codon for methionine is used in mitochondria to initiate translation, but the standard isoleucine codon, AUA, specifies methionine at internal positions. In the mitochondria of mammals, codons AGA and AGG, which usually specify arginine, are termination codons. These same mitochondrial codons in the fruit fly specify serine. Examples of the known coding variations in mitochondria are summarized in **Table 17-4**.

The low complexity of the mitochondrial genome has allowed continued evolution, which has resulted in streamlining of the genetic code. However, there are also a few examples in which the code has been altered in free-living cells. The only bacterial variant is the use of the UGA stop codon to encode tryptophan in *Mycoplasma capricolum*. Among eukaryotes, a few species of ciliated protists use the codons UAA and UAG to specify glutamine. The most perplexing change in the genetic code is found in the yeast *Candida albicans* and several related *Candida* species, in which the CUG codon that usually encodes leucine specifies serine instead.

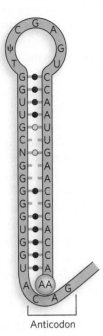

FIGURE 1 In this hypothetical primordial tRNA, the anticodon participates directly in amino acid recognition, circumventing the need for aminoacyl-tRNA synthetases. [*Source: Data from J. J. Hopfield, Proc. Natl. Acad. Sci. USA 75:4334–4338, 1978.*]

much greater chemical potential than nucleic acids for both structure and catalysis, they had developed a mechanism to code for protein. And selection could do the rest: the evolution of the protein world would proceed according to the rules of natural selection. Proteins that are better catalysts than their RNA counterparts would eventually take over much of the job of catalytic RNAs, and aminoacyl-tRNA synthetases would evolve.

The evolution of the genetic code is another area of lively debate. Proposals range from a code that simply arose at random and became frozen in time, to one in which the amino acids themselves participated directly in code development. For the latter proposal, consider the hypothetical tRNA structure in Figure 1. The anticodon would participate in selecting the correct amino acid, possibly through favorable contacts between the amino acid and a particular triplet nucleotide sequence, depending on the chemical nature of the amino acid. Indeed, experimental support for this hypothesis exists. The preferential affinity of amino acids for certain nucleic acid sequences has been studied, and it is reported that some amino acids exhibit preferential binding to the anticodons of the tRNAs that encode them. It's also possible that not all 20 amino acids were present early on, and that only a two-nucleotide code was needed for 16 (4^2) or fewer amino acids. Some modern amino acids may not have been present in the primordial soup, but instead could have arisen as side products of cellular metabolism. These "cell-invented" amino acids could have been incorporated later, expanding the code to its current triplet form. Regardless of the route taken by natural selection, the genetic code has obviously been honed over the millennia to be robust and resistant to mutations.

amino acid, but may have selected for certain amino acid properties such as charge, polarity, or hydrophobicity. This level of accuracy might be acceptable if the functions of early proteins were purely structural. But even for a structural role, the pressures of natural selection would eventually result in high fidelity in the translation process. Although cells did not "know" that amino acid side chains offered

---

TABLE 17-4

Examples of Known Variant Codon Assignments in Mitochondrial mRNA

Organisms	UGA (Stop)	AUA (Ile)	AGA, AGG (Arg)	CUN (Leu)	CGG (Arg)
Animals					
Vertebrates	Trp	Met	Stop	+	+
Drosophila	Trp	Met	Ser	+	+
Yeasts					
Saccharomyces cerevisiae	Trp	Met	+	Thr	+
Schizosaccharomyces pombe	Trp	+	+	+	+
Filamentous fungi	Trp	+	+	+	+
Trypanosomes	Trp	+	+	+	+
Higher plants	+	+	+	+	Trp

Note: Only deviations from the standard code assignment are shown; + indicates no deviation. The standard, cellular assignments for the codons are shown in parentheses. N · any nucleotide.

Initiation and Termination Rules Have Exceptions

Changes in the code need not be absolute; a codon might not always encode the same amino acid. In most organisms we find some examples of amino acids being inserted at positions that are not specified in the standard code. Two examples are the occasional use of GUG (valine) or UUG (leucine) as an initiation codon. This occurs only for those genes in which the GUG or UUG codon is properly located in the mRNA (see Chapter 18).

Another example is the insertion of selenocysteine (Sec)—sometimes referred to as the twenty-first amino acid, as it is uniquely coded for in all domains of life. When present, selenocysteine is usually found in proteins involved in oxidation-reduction reactions, such as formate dehydrogenase in bacteria and glutathione peroxidase in mammals. These enzymes require the element selenium for their activity, generally in the form of a Sec residue in the active site. Although modified amino acids are usually produced by posttranslational reactions, selenocysteine is formed at the level of an aminoacylated tRNA incorporated during translation in response to a UGA (stop) codon (**Figure 17-16**). A special type of serine-binding tRNA, tRNA^{Sec}, present at lower levels than other tRNA^{Ser} species, recognizes UGA and no other codons. The tRNA^{Sec} is charged with serine, and the serine is enzymatically converted to selenocysteine

before it is used at the ribosome. The charged tRNA does not recognize just any UGA codon; a contextual signal in the mRNA, an element referred to as SECIS (selenocysteine insertion sequence), ensures that this tRNA recognizes only the few UGA codons within certain genes that specify selenocysteine.

This process of incorporation of Sec residues has been extensively studied in *E. coli*. Of the gene products involved, SelC is a tRNA (tRNA^{Sec}) that accepts serine initially. SelA is an enzyme that uses selenophosphate to convert the serine to selenocysteine, forming Sec-tRNA^{Sec}. SelB is an elongation factor needed by the ribosome to utilize the Sec-tRNA^{Sec} and incorporate selenocysteine into the protein. For SelB to function, it must also bind a SECIS element in the mRNA, which has a secondary structure with a bulge that fits into SelB for insertion of a Sec residue. The SECIS element is adjacent to the UGA codon in *E. coli*, but it is often located in the 3′UTRs (3′ untranslated regions, discussed in Chapter 22) of eukaryotes and archaea. The SelB elongation factor is a homolog of elongation factor Tu.

The evolution of genetic code changes in small genomes such as those of mitochondria, the use of a few alternative initiation codons, and the use of a termination codon to incorporate selenocysteine—all these are relatively easy to understand as minor adjustments of a universal code. But an unusual alteration in the genetic code occurs in many fungal species of the genus *Candida*,

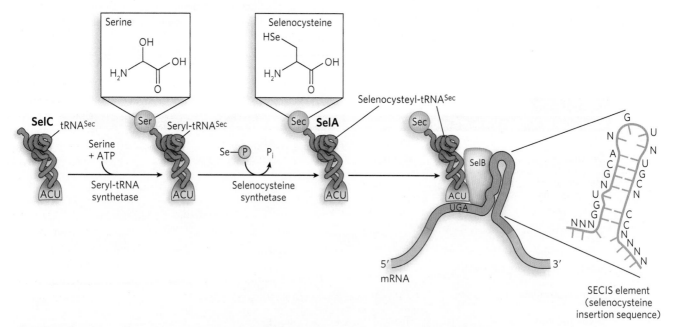

FIGURE 17-16 The incorporation of selenocysteine during translation. A serine-charged tRNA (SelC) with the anticodon 5′-UCA recognizes a UGA stop codon. The Ser-tRNA^{Sec} is enzymatically converted to selenocysteyl-tRNA^{Sec} (Sec-tRNA^{Sec}) by the protein SelA, which uses selenophosphate and releases P$_i$ on attachment of the selenium to the amino acid. SelB is an elongation factor that recognizes a specific hairpin in the mRNA and also binds Ser-tRNA^{Sec}, facilitating incorporation of selenocysteine into the protein.

as originally discovered for *Candida*. This fungus is an organism of high genomic complexity, yet its genetic code has a dramatic variation: the CUG codon that usually encodes leucine encodes serine instead. The selection pressure for this change is completely unknown. Furthermore, serine and leucine are quite different chemically. Yet, even this change can be understood based on the properties of a universal code.

When several codons encode the same amino acid and require multiple tRNAs, not all of the codons are used with equal frequency. In a phenomenon called **codon bias**, some codons for a particular amino acid are used more frequently (sometimes much more frequently) than others. The tRNAs for the frequently used codons are often present at much higher concentrations than the tRNAs for the rarely used codons. For example, there are six codons for leucine (see Figure 17-4). In bacteria, CUG is used often to encode Leu. However, in fungi closely related to *Candida*, CUG is used only rarely as a Leu codon and is often entirely absent in highly expressed proteins. A change in the coding sense of CUG would thus have a much smaller effect on fungal cell metabolism than might be expected if all Leu codons were used equally.

The coding change for the Leu codons may have occurred by a gradual loss of CUG codons in genes and of the tRNA that recognizes CUG as a Leu codon, followed by a capture event—a mutation in the anticodon of a tRNA^{Ser} that allowed it to recognize CUG. Alternatively, there may have been an intermediate stage in which CUG was recognized as both a Leu and a Ser codon, perhaps with contextual signals in the mRNAs that helped one tRNA or another recognize specific CUG codons (much like the signals used to insert selenocysteine at a particular stop codon). Phylogenetic analysis (see Chapter 8) indicates that the reassignment of CUG as a Ser codon occurred in *Candida* ancestors about 150 to 170 million years ago.

SECTION 17.4 SUMMARY

- The genetic code, tRNAs, and translation must have evolved piecemeal, without protein components, and this evolution must have had a selective advantage, even at its earliest stages. Several hypotheses exist, but evidence to support any of them is limited.
- The genetic code is largely universal. Most exceptions occur in mitochondrial DNA, a small genome genetically isolated from the nucleus and relatively free to undergo evolutionary code changes; many of these changes involve altered stop codons, yielding a streamlined genetic code that requires only 22 tRNAs.
- The few examples of genetic code alterations outside mtDNA usually involve the conversion of termination codons, in keeping with a common ancestry for the genetic code from which all variants are derived.

? UNANSWERED QUESTIONS

The genetic code has been deciphered and is of paramount importance to virtually every investigation in molecular biology. However, important fundamental questions remain. The nature of wobble pairing and the influence of tRNA structure on codon-anticodon pairing most likely will be explained through structural and mutational studies. The evolution of the genetic code remains a mystery, but creative experiments and investigations into exceptions to the code will no doubt provide further insight.

1. How do nucleotides outside the anticodon influence the structure of tRNA for wobble pairing? The use of noncanonical base pairs explains how wobble pairing can happen. But nucleotide substitutions outside the anticodon also affect wobble pairing in some way, perhaps through conformational changes when tRNA binds the ribosome.

2. Why do tRNAs have so many modified bases? The proportion of modified bases in tRNAs can approach 20%, and many genes are devoted to synthesizing these modified bases. Yet we still know very little about the functions of these many modifications.

3. Did the three classes of RNA—tRNA, mRNA, and rRNA—coevolve, as we presume? This seems like a gigantic leap. If they did not coevolve, what might the individual functions have been for each class, leading them to funnel into one central process?

4. How did the protein components that work with RNAs evolve? What functions did the earliest proteins serve? Did early proteins simply serve structural roles to bind and stabilize catalytic RNA, only to take over the catalytic roles later?

5. How did the translation machinery evolve? It is nothing short of mind-boggling to imagine how the translation machinery evolved in the first place. One challenge is the huge number of factors required for translation. The whole process could not have evolved all at once. What were the individual steps, and what forces of natural selection were at work?

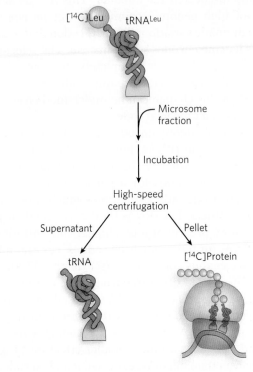

HOW WE KNOW

Transfer RNA Connects mRNA and Protein

Hoagland, M.B., M.L. Stephenson, J.F. Scott, L.I. Hecht, and P.C. Zamecnik. 1958. A soluble ribonucleic acid intermediate in protein synthesis. J. Biol. Chem. 231:241–257.

O nce Francis Crick had hypothesized the existence of an adaptor molecule to bridge the information gap between mRNA and protein, the race was on to find it. It may seem obvious now, but before the discovery of aminoacyl-tRNAs, the nature of the adaptor was mysterious—but Crick did suggest that the adaptor might be, in part, a nucleic acid that could recognize individual codons in mRNA.

The race to find the adaptor was won by Mahlon Hoagland and Paul Zamecnik. They prepared a cell extract that contained the soluble tRNA and the enzymes needed to charge tRNA with amino acids. When [^{14}C]leucine was added to the extract, the radiolabeled amino acid became attached to its tRNA. To show that this really was the adaptor molecule, Hoagland and Zamecnik demonstrated that the [^{14}C]leucyl-tRNA could incorporate the [^{14}C]leucine into a polypeptide chain. They prepared microsomes, a cell fraction containing mostly ribosomes collected as a pellet after high-speed centrifugation. On incubation of the [^{14}C]leucyl-tRNA with mRNA and microsomes, [^{14}C]leucine was transferred from the tRNA to protein—as was evident from the association of radiolabeled amino acid with ribosomes in the pellet

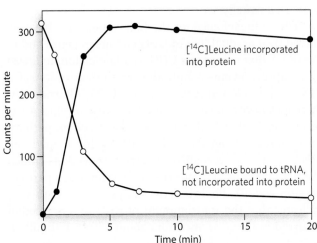

FIGURE 1 An outline of Hoagland and Zamecnik's experimental method (top) and an example of their results (bottom). During protein synthesis, the radiolabeled amino acid, [^{14}C]leucine, is transferred from the tRNA to the ribosome. [Source: M. B. Hoagland et al., J. Biol. Chem. 231:241–257, 1958.]

after centrifugation of the reaction mixture (shown in the graph in Figure 1, solid circles), not with the tRNA in the supernatant (open circles). This experiment showing that amino acids are transferred from tRNAs to polypeptides was the first step in deciphering the genetic code.

Mahlon Hoagland,
1921–2009
[Source: News Office, Massachusetts General Hospital.]

Paul Zamecnik,
1912–2009
[Source: News Office, Massachusetts General Hospital.]

Proteins Are Synthesized from the N-Terminus to the C-Terminus

Dintzis, H.M. 1961. Assembly of the peptide chains of hemoglobin. *Proc. Natl. Acad. Sci. USA* 47:247–261.

After the adaptor tRNA was discovered, the basic flow of information among the three biopolymers, DNA → RNA → protein, was understood. However, the order in which amino acids were connected to form a protein polymer was still unknown. To solve this, Howard Dintzis designed an ingenious experiment. His idea was to feed [³H]leucine to living cells and follow protein synthesis over time. Leucine was known to be one of the most abundant amino acids in proteins, and he assumed it would be located in random places from one end of a polypeptide to the other. Would incorporation of [³H] leucine start at one end of a newly synthesized protein, or would the labeled amino acid appear uniformly throughout the length of the new protein at all time points? This might sound like an easy experiment, but there were big technical hurdles. First, Dinztis had to follow only one protein, yet a cell makes many proteins all at once. Second, protein sequencing had not yet been invented, but Dintzis had to find some way to follow the sequence in which leucine was incorporated in the protein chain.

The beauty in Dintzis's experimental design lies in how he circumvented these difficulties. He knew that immature red blood cells (reticulocytes) turn off the synthesis of most proteins except hemoglobin, which is composed of two chains, α and β. Furthermore, he could use trypsin (a protease) to digest the α chain and could separate the fragments by paper electrophoresis,

although he couldn't determine the ordering of the fragments along the protein. So Dintzis added [³H] leucine to reticulocytes, lysed cells at various times, separated the α and β chains, and analyzed the α chain by tryptic digestion and electrophoresis. Full-length proteins would have [³H]leucine incorporated only in the part of the protein molecule synthesized last. To ensure that he obtained only full-length α chains, Dintzis removed incomplete chains, which remained bound to ribosomes, by centrifugation. As a control, he preincubated the cells with [¹⁴C]leucine so that he could compare new synthesis ([³H]leucine) with overall synthesis ([¹⁴C]leucine), correcting for the different leucine content in each peptide.

If protein chains are not made in a defined order, all peptides should have a similar ratio of ³H to ¹⁴C (Figure 2). But if proteins are made in a linear order, peptides will vary in their ³H:¹⁴C ratio, and the part of the protein made last should contain a higher ³H:¹⁴C ratio than the parts made earlier. The results were unambiguous and striking! The peptides differed greatly in ³H:¹⁴C ratio, ruling out a random order of synthesis. Further, the tryptic peptides could be ordered to form a gradient of radioactivity. Hence, it was apparent that hemoglobin is synthesized from one end to the other. To determine the direction of synthesis, Dintzis digested the α chain with carboxypeptidase, which specifically removes amino acids from the C-terminus. Only one tryptic peptide was affected by carboxypeptidase treatment, and it was the peptide with the highest ³H:¹⁴C ratio. This identified the C-terminus as the part of the protein that is synthesized last. Overall, the results showed that proteins are synthesized from the N-terminus to the C-terminus.

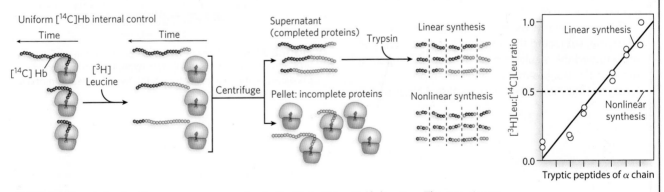

FIGURE 2 Translation is a linear process, occurring in the N- to C-terminal direction. The experiment starts with red blood cells containing hemoglobin (Hb) that is uniformly labeled with [¹⁴C]leucine. [³H]Leucine is then added, and the cells are incubated for different times before stopping the reaction. The hemoglobin is then digested with proteases, and peptide fragments are identified by electrophoresis. Fragments at the end of the protein have a higher ³H:¹⁴C ratio (as shown by the open circles in the graph), demonstrating that protein synthesis is linear. If protein synthesis were random, the ³H:¹⁴C ratio would be constant over the length of the protein (dashed line). [*Source: Data from H. M. Dintzis, Proc. Natl. Acad. Sci. USA 47:247–261, 1961. By permission of Howard Dintzis.*]

HOW WE KNOW

The Genetic Code In Vivo Matches the Genetic Code In Vitro

Terzaghi, E., Y. Okada, G. Streisinger, J. Emrich, M. Inouye, and A. Tsugita. 1966. Change of a sequence of amino acids in phage T4 lysozyme by acridine-induced mutations. Proc. Natl. Acad. Sci. USA 56:500–507.

The experiments that defined the genetic code were brilliant, and this was Nobel Prize–winning work. But the investigations that cracked the code, as beautiful as they were in experimental design, were performed outside the context of a living cell, using cell extracts and synthetic mRNA. As a result, many scientists remained skeptical about the relevance of the newly discovered code to the in vivo situation. Akira Tsugita's group designed a powerful experiment to address exactly this issue. They studied acridine-induced mutations in the lysozyme gene of phage T4, which inactivate the gene, presumably by inducing the deletion or insertion of a base and thereby throwing off the reading frame. A double mutation in the lysozyme gene creates a pseudo-wild-type T4, consistent with the hypothesis by Crick and his colleagues that an insertion combined with a deletion results in a restored reading frame. According to this hypothesis, the area between the two mutations encodes an amino acid sequence different from that of the wild-type protein. As long as the insertion and deletion are not too far apart in the linear sequence of the gene, the double-mutant protein may retain a significant level of activity.

The investigators used a protease to digest lysozyme, subjected the fragments to electrophoresis, and studied the resulting peptide maps. Comparing maps of the double-mutant and wild-type lysozyme, they identified one peptide with changed electrophoretic mobility. Sequencing of the peptide revealed a five amino acid sequence unique to the mutant:

Wild type: Lys-Ser-Pro-Ser-Leu-Asn-Ala

Mutant: Lys-Val-His-His-Leu-Met-Ala

This result supported the triplet reading frame with no internal punctuation marks, as described by Crick. It could also be used to solve whether the genetic code in living cells is the same as that determined in extracts, by comparing the wild-type and mutant nucleotide sequences. If the code is indeed the same, the codon table established by in vitro experiments should produce a nucleotide sequence for the wild type that (1) encodes the wild-type sequence of amino acids, and (2) encodes the double-mutant lysozyme sequence, given either one nucleotide insertion followed by one deletion, or the reverse. A solution within these constraints can indeed be found, providing strong evidence at the time of this study that the genetic code derived from in vitro studies is the genetic code used in living cells (Figure 3). The result also supported the conclusion that mRNA is read in the 5'→3' direction, as the solution works only if codons are read in this direction. Because of technical limitations, the solution was not validated by sequencing until a decade later.

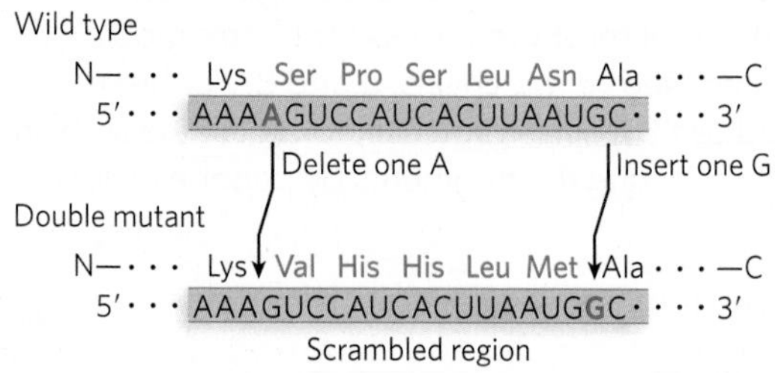

FIGURE 3 For the known difference between the wild-type and double-mutant phage T4 lysozyme, the genetic code predicts deletion of an A residue followed by insertion of a G residue in the mutant protein.

KEY TERMS

genetic code, 590
codon, 590
transfer RNA (tRNA), 591
aminoacyl-tRNA, 591
aminoacyl-tRNA synthetase, 591
anticodon, 591
translation, 592
degenerate code, 592
codon family, 592
wobble base, 593
wobble position, 593
wobble hypothesis, 594

reading frame, 594
initiation codon, 595
start codon, 595
termination codon, 595
stop codon, 595
open reading frame (ORF), 595
missense mutation, 595
silent mutation, 595
transition mutation, 595
nonsense mutation, 596
suppressor tRNA, 596
codon bias, 609

PROBLEMS

1. The following RNA polymer is added to an *E. coli* extract, where it can be translated in all three possible reading frames. Which amino acids can be polymerized into polypeptides in this system?

5′-AUAUAUAUAUAUAUAUAUAUAUAUAUAU-3′

2. Given a polynucleotide that encodes polymethionine, what other polypeptides will also be produced?

3. Translate the following mRNA into protein, starting from the first initiation codon:

5′-CCGAUGCCAUGGCAGCUCGGUGUUAC AAGGCUUGCAUCAGUACCAGUUUGAAUCC-3′

4. From the sequence of a protein, we can gain some information about the gene sequence that encodes it. However, because of the degeneracy of the genetic code, there are many possible nucleotide sequences that could encode a given protein sequence. The usefulness of genomic databases in searching for the genes for proteins of known sequence is made clear by considering the following. How many possible RNA molecules can encode the peptide Met-Asn-Trp-Tyr? How many if a Leu residue is added to the end of the peptide?

5. Shown below is the 5′ end of an mRNA molecule. What are the first three (N-terminal) amino acids of its protein product?

5′-AUGUGUUGAUGUAUCAGACCUGUC - - -

6. Translate the following mRNA, starting at the first 5′ nucleotide, assuming that translation occurs in an *E. coli*

cell. If all tRNAs make maximum use of wobble rules but do not contain inosine, how many distinct tRNAs are required to translate this RNA?

5′-AUGGGUCGUGAGUCAUCGUUAAUUGUAGCU GGAGGGGAGGAAUGA-3′

7. How does the answer to Problem 6 change if the RNA is translated in yeast mitochondria?

8. For the following RNA sequence, which positions can tolerate a mutation without resulting in a change in amino acid sequence? What changes are tolerated at each position?

5′-AUGAUAUUGCUAUCUUGGACU-3′

9. A researcher wants to explore the function of one segment of a protein, extending across six amino acid residues. Her plan is to change the amino acids at each position and determine how that affects the activity of the protein. She wants to make as many amino acid substitutions as possible, but quickly discovers that she cannot make all the changes she wants if she changes only one codon nucleotide. The sequence of the mRNA specifying that part of the protein is given below, along with the amino acid sequence it encodes. Below the amino acid sequence, list all the amino acids that could be substituted at each position if the researcher changed only one nucleotide in the codon for that particular amino acid.

5′-ACC-AUA-UUG-CUC-UCU-UCG-3′
  Thr Ile Leu Leu Ser Ser

10. What polypeptide sequence will be made from the following RNA sequence?

5′-AUGCCUCGUCAGGUGUAAAGUCAGGCUUGA-3′

What tRNA^Tyr suppressor mutation will provide read-through of the first stop codon, and what will the resulting peptide sequence be?

11. What are the sequences of the polypeptides produced from these repeating nucleotide sequences: (a) poly(AG); (b) poly(UG); (c) poly(CAA); (d) poly(AAG); (e) poly(UUAC)?

12. A researcher uses polynucleotide phosphorylase to create random RNA polymers, using a UDP:CDP ratio of 5:1. Codons should be generated in the following proportions, assuming random incorporation of the NDPs by polynucleotide phosphorylase: UUU (83.3), U_2C (16.7), UC_2 (3.3), and CCC (0.7). The following amino acids are incorporated into protein, in the proportions shown in parentheses: leucine (22.2), phenylalanine (100), proline (5.1), and serine (23.6). What are the probable codon assignments of these four amino acids? Keep in mind that

poly(U) codes for poly(Phe), and poly(C) codes for poly(Pro).

13. Polyglycine is translated from the repeating sequence 5′-(GGU-GGC-GGA)$_n$-3′. If only one tRNA is needed to make polyglycine, what can you say about the tRNA anticodon?

14. The stop codon UGA is also used to specify the amino acid selenocysteine (Sec) in a few proteins in the cells of most organisms. Why is Sec not added to all proteins at UGA codons? What protein and RNA elements are needed to permit the addition of Sec to proteins at a particular UGA codon in *E. coli*?

15. A gene with a frameshift mutation caused by the insertion of one nucleotide produces inactive protein. A second frameshift, caused by the deletion of one nucleotide at some position downstream of the original mutation, reactivates the gene. The final protein product contains four amino acid residues that differ from the wild-type protein. The two mutations occur in the following sequence:

5′- - - CATCATCATCATCATCATCATCATCAT - - -

What is the maximum number of nucleotides between the two point mutations? What is the minimum number?

16. Given the following mRNA sequence, which reading frame is most likely to encode part of a protein?

5′-ACGUCGAGUAGCAGUAUCGAUUGAGC
UCUUAGAUAAGAUCGC

17. Given the wobble rules, at least 31 tRNAs are necessary to decipher the genetic code. Only 6 tRNAs are needed to insert the four amino acids Phe, Leu, Ile, and Met. Using the table below, hypothesize the anticodon sequences of the 6 tRNAs. Multiple answers are possible.

Amino Acid	Codon
Phe	UUU
	UUC
Leu	UUA
	UUG
	CUU
	CUC
	CUA
	CUG
Ile	AUU
	AUC
	AUA
Met	AUG

DATA ANALYSIS PROBLEM

Nishimura, S., D.S. Jones, E. Ohtsuka, H. Hayatsu, T.M. Jacob, and H.G. Khorana. 1965. Studies on polynucleotides XLVII: The *in vitro* synthesis of homopeptides as directed by a ribopolynucleotide containing a repeating trinucleotide sequence—new codon sequences for lysine, glutamic acid and arginine. *J. Mol. Biol.* 13:283–301.

18. Once researchers had developed a few key strategies, the genetic code was solved within just a few years in the mid-1960s. One chapter of that story is described by Nishimura and coauthors. The work is elegant, while also demonstrating that results obtained in real-world experiments are not always as unambiguous as they may seem when presented in textbooks.

Using methods developed in the Khorana laboratory, Nishimura and colleagues examined the polypeptides generated by an oligonucleotide consisting of AAG repeats. In the first experiment (Figure 1), they examined the binding of radioactively labeled aminoacyl-tRNAs to the ribosome in response to the oligonucleotide. All of the tRNAs were tried, and only the four shown in Figure 1 gave a positive result (i.e., a labeled aminoacyl-tRNA–ribosome complex). (Note that in 1965, tRNA was often referred to as sRNA ("soluble" RNA), as shown in the figures below.)

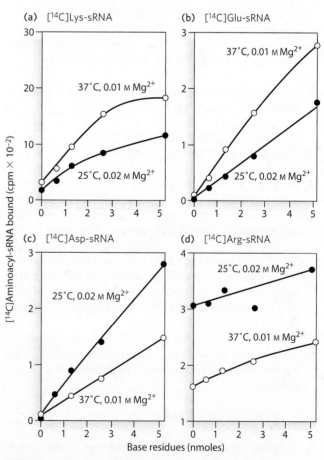

FIGURE 1 [Data from S. Nishimura et al., J. Mol. Biol. 13:283–301, 1965.]

(a) Given our present understanding of the genetic code, what is the maximum number of different labeled aminoacyl-tRNAs that could be bound to the ribosome in response to this oligonucleotide?

(b) Given the tenuous understanding of the code in 1965, can you suggest an explanation for the positive results obtained with the four different tRNAs in this experiment?

One of the possible codons present in the $(AAG)_n$ oligonucleotide was assigned to lysine, based on a series of experiments including the one shown in Figure 2. The researchers used the method advanced by Nirenberg to determine which trinucleotide would stimulate binding of [^{14}C]Lys-tRNA^{Lys} to a ribosome.

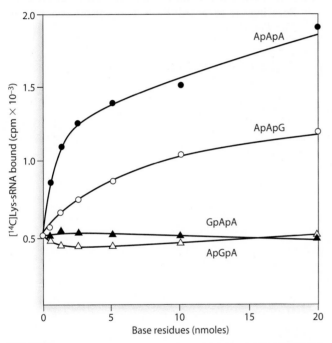

FIGURE 2 [Data from S. Nishimura et al., J. Mol. Biol. 13:283–301, 1965.]

(c) Which codon did the researchers assign to lysine?

(d) The researchers also tested the trinucleotide AAA for its response to [^{14}C]Lys-tRNA^{Lys} (see Figure 2). Suggest why they would include AAA in their experiment, even though AAA was not represented in the $(AAG)_n$ repeating oligonucleotide.

Next, the researchers turned their attention to the tRNAs for glutamine. With [^{14}C]Glu-tRNA^{Glu}, they obtained the results shown in Figure 3.

(e) From these results, which codon did the authors assign to Glu?

(f) Suggest why all three codons elicited a positive response in the experiment shown in the rightmost panel of Figure 3.

In their tests, the researchers then examined the production of homopolymers of amino acids in response to

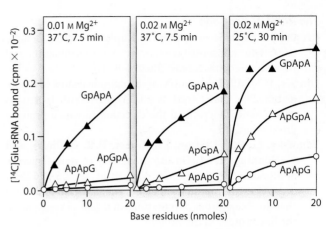

FIGURE 3 [Data from S. Nishimura et al., J. Mol. Biol. 13:283–301, 1965.]

$(AAG)_n$. They found poly(Glu), poly(Arg), and poly(Lys), but no poly(Asp).

(g) Based on all of the results shown here, what are the most likely codon assignments for the codons present in $(AAG)_n$?

(h) Go back to question (b) and suggest why Asp-tRNA^{Asp} bound to ribosomes in response to this oligonucleotide in the first experiment (see Figure 1).

ADDITIONAL READING

General

Cold Spring Harbor Symposia on Quantitative Biology. 1966. *The Genetic Code.* Vol. 31. Cold Spring Harbor, NY: Cold Spring Harbor Laboratories. The symposium was held when the genetic code was being completed; the introduction by Francis Crick gives a fascinating portrait of where things stood and where they were going.

Crick, F.H.C. 1988. *What Mad Pursuit: A Personal View of Scientific Discovery.* New York: Basic Books. A wonderful account of Crick's personal odyssey in science.

Vogel, G. 1998. Tracking the history of the genetic code. *Science* 281:329–331.

Deciphering the Genetic Code: tRNA as Adaptor

Crick, F.H.C. 1970. Central dogma of molecular biology. *Nature* 227:561–563. The classic paper in which Crick proposes the central dogma of information flow in biology.

Hoagland, M.B., M.L. Stephenson, J.F. Scott, L.I. Hecht, and P.C. Zamecnik. 1958. A soluble ribonucleic acid intermediate in protein synthesis. *J. Biol. Chem.* 231:241–257. The report that identified aminoacylated tRNA as Crick's adaptor molecule.

Holley, R.W., J.H. Apgar, G.A. Everett, J.T. Madison, M. Marquisse, S.H. Merrill, J.R. Penswick, and A. Zamir. 1965. Structure of a ribonucleic acid. *Science* 147:1462–1465. This documents the first structure of a tRNA, yeast tRNA^{Ala}; the putative anticodon contains the unusual nucleotide inosine.

The Rules of the Code

Brenner, S., A.O. Stretton, and S. Kaplan, S. 1965. Genetic code: The "nonsense" triplets for chain termination and their suppression. *Nature* 206:994–998.

Dintzis, H.M. 1961. Assembly of the peptide chains of hemoglobin. *Proc. Natl. Acad. Sci. USA* 47:247–261. Ingenious experiments determined the direction in which proteins are synthesized.

Yanofsky, C., B.C. Carlton, J.R. Guest, D.R. Helinski, and U. Henning. 1964. On the colinearity of gene structure and protein structure. *Proc. Natl. Acad. Sci. USA* 51:266–272. Different mutations in an enzyme are used to deduce that the linear order of mutations in a gene correspond to the linear order in the protein.

Cracking the Code

Khorana, H.G., H. Buchi, H. Ghosh, N. Gupta, T.M. Jacob, H. Kossel, R. Morgan, S.A. Narang, E. Ohtsuka, and R.D. Wells. 1966. Polynucleotide synthesis and the genetic code. *Cold Spring Harb. Symp. Quant. Biol.* 31:39–49.

Nirenberg, M.W., and J.H. Matthaei. 1961. The dependence of cell-free protein synthesis in *E. coli* upon naturally occurring or synthetic polyribonucleotides. *Proc. Natl. Acad. Sci. USA* 47:1588–1602. The first study that outlined an experimental approach to crack the genetic code.

Speyer, J.F., P. Lengyel, C. Basilio, A.J. Wahba, R.S. Gardner, and S. Ochoa. 1963. Synthetic polynucleotides and the amino acid code. *Cold Spring Harb. Symp. Quant. Biol.* 28:559–567.

Exceptions Proving the Rules

Fox, T.D. 1987. Natural variation in the genetic code. *Annu. Rev. Genet.* 21:67–91.

Knight, R.D., S.J. Freeland, and L.F. Landweber. 2001. Rewiring the keyboard: Evolvability of the genetic code. *Nat. Rev. Genet.* 2:49–58.

Harry Noller [Source: Courtesy Harry Noller.]

18 Protein Synthesis

MOMENT OF DISCOVERY

The defining moment in my career was the realization that *ribosomal RNA, not protein, was the functionally important component of the ribosome*—and this discovery was somewhat serendipitous. My plan when starting my laboratory, at the University of California, Santa Cruz, was to apply my training in protein biochemistry to investigating the function of ribosomal proteins. We began treating ribosomes with chemicals known to inactivate protein enzymes, hoping to recover nonfunctional ribosomes that could be analyzed to determine which proteins had been affected and thereby discover those responsible for activity. The problem was, the ribosome withstood nearly all the standard chemical treatments we tried!

An unexpected result from one experiment led me to suggest to an undergraduate researcher, Brad Chaires, to try the RNA-specific reagent kethoxal, which reacts with guanine bases to produce adducts that disrupt base pairing. To our great surprise, ribosomes were inactivated by modification of just six G residues out of the more than 4,000 nucleotides in the ribosomal RNA.

When Brad graduated, I followed up with reconstitution experiments showing that it was, in fact, the ribosomal RNA that had been functionally inactivated, and that tRNA could protect ribosomes from kethoxal inactivation. Protection resulted from the tRNA binding to the ribosome and blocking kethoxal's access to parts of the rRNA, so they couldn't be chemically modified. These results led us to propose that ribosomal RNA, rather than ribosomal protein, was responsible for the functional activity of the ribosome. Many colleagues considered this "a crackpot idea," which I found frustrating but, in the antiestablishment spirit of the 1970s, also motivating.

Our findings led us to sequence the ribosomal RNAs (which Carl Woese referred to as "Sacred Scrolls"), and we were excited to find that the sites of chemical inactivation corresponded to the most evolutionarily conserved parts of rRNA. We then embarked on a fruitful collaboration with Woese to determine the secondary structures of the ribosomal RNAs, which ultimately led us in the direction of working out the three-dimensional structure of the ribosome. After almost 40 years, we can at last see that ribosomal RNA is indeed the functional core of the ribosome.

—Harry Noller, on discovering the functional importance of ribosomal RNA

Online resources related to this chapter:

 Simulation
Translation

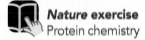 *Nature* **exercise**
Protein chemistry

The synthesis of proteins is the final step in the flow of genetic information, beginning with DNA replication and continuing with transcription into mRNA. Because of the abundance of proteins—for example, they make up roughly 44% of the dry weight of a human body—and their central catalytic, transport, structural, and regulatory roles in all organisms, substantial cellular resources are devoted to protein synthesis. Like DNA and RNA synthesis, the process of protein synthesis can be considered in terms of initiation, elongation, and termination stages. Furthermore, as in nucleic acid synthesis, substrates must be activated before polymerization, and the completed product—in this case the polypeptide chain—must be chemically modified, folded, and targeted to the correct intracellular or extracellular location to become a functional protein.

The transfer of information from the 4-letter nucleotide sequence of an mRNA to the 20-letter amino acid sequence of a protein, however, is a fundamentally more complex task than nucleotide synthesis. In contrast to the transcription of DNA into RNA, in which a direct correspondence forms by hydrogen bonding between the sequence of the template strand and that of the synthesized RNA strand, there is no obvious chemical correspondence between the three-nucleotide mRNA codons and the amino acids they represent. In the 1950s, Francis Crick suggested that amino acids are attached to "adaptor" RNA molecules that provide direct base-pairing complementarity with each codon in an mRNA sequence. And indeed, Paul Zamecnik and colleagues later showed that amino acids are covalently attached to RNA molecules, subsequently identified as tRNAs (see Chapter 17). The aminoacyl-tRNA molecules associate with the ribosomes, which were shown to be composed of both RNA and protein. Together with mRNAs, tRNAs, and aminoacyl-tRNA synthetases, ribosomes carry out the coupled tasks of recognizing each three-nucleotide codon of the genetic code and incorporating the specified amino acid into a growing polypeptide chain. Experiments in many laboratories showed that the molecular machinery of translation is essentially the same in all cells, although, as we will see, some of the mechanistic details of protein synthesis differ in bacteria and eukaryotes.

In Chapter 17 we introduced the genetic code and the mechanism of decoding by tRNAs. In this chapter, we discuss the structures and activities of ribosomes and aminoacyl-tRNA synthetases. We then explore how these remarkable molecules work together with mRNA and tRNA to carry out fast, accurate protein synthesis.

18.1 THE RIBOSOME

In 1974, George Palade received a Nobel Prize for his discovery, in the 1950s, of **ribosomes**, the macromolecular machines responsible for protein synthesis (**Figure 18-1**). Although historically considered organelles, ribosomes are not encapsulated in a lipid membrane and thus are more accurately described as very large (macro) molecules. Ribosomes are present in the cytosol of all cells, as well as in the matrix of mitochondria and the stroma of chloroplasts. Because protein synthesis is common to virtually all life forms, ribosomes are evolutionarily well-conserved. In bacteria, they can translate mRNA as it is being transcribed from DNA, but in eukaryotes, the nuclear envelope and mRNA processing steps separate

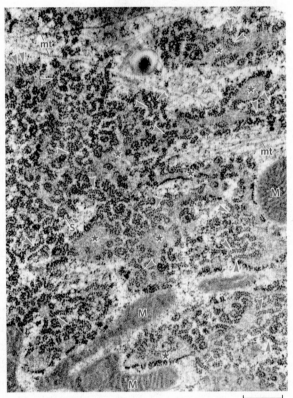

0.5 μm

FIGURE 18-1 Ribosomes. Many of the ribosomes in a cell's cytoplasm are assembled on the endoplasmic reticulum, the cell's "shipping and handling" center for membrane protein production. In this electron micrograph, arrows indicate large polyribosomes, or polysomes, which consist of many ribosomes lined up on a single messenger RNA. Mitochondria (M) and microtubules (mt) are also visible, as well as large spiral polyribosomes (S). The interior (lumen) of the rough endoplasmic reticulum often appears expanded (asterisks). [Source: From Christensen, A.K. and Bourne, C.M. Shape of Large Bound Polysomes in Cultured Fibroblasts and Thyroid Epithelial Cells, The Anatomical Record 255:116–129, 1999, Fig. 4. Copyright © 1999 Wiley-Liss, Inc.]

RNA synthesis from translation (RNA processing is discussed in Chapter 16). Ribosomes are abundant; each *E. coli* cell contains approximately 15,000 ribosomes, which make up almost 25% of the dry weight of the cell. The large number of these macromolecules is essential to the cell's ability to produce proteins when needed. The structure of the ribosome facilitates protein synthesis by bringing together the mRNA codon and the corresponding charged tRNA adaptor, and then catalyzing peptide bond formation. Although the general function of ribosomes has been well-established for some time, advances in understanding the structure and assembly of the ribosome have elucidated more of the details of protein synthesis. Here we provide an overview of ribosomal structure and function that will serve as a framework for the details in the rest of the chapter.

The Ribosome Is an RNA-Protein Complex Composed of Two Subunits

Bacterial ribosomes contain about 60% ribosomal RNA and 40% protein, organized into two unequal subunits that are named according to their sedimentation coefficients (in svedberg units, S; see Chapter 16). The 50S subunit, the larger of the two, contains the peptidyl transferase center, which catalyzes peptide bond formation between adjacent amino acids in a growing polypeptide chain. The 30S subunit contains the decoding center where aminoacylated tRNAs "read" the genetic code by base pairing with each triplet codon in the mRNA. The assembled ribosome, with a combined sedimentation coefficient of 70S, smoothly integrates the functions of each subunit to ensure rapid, accurate protein synthesis.

Each ribosomal subunit contains dozens of **ribosomal proteins (r-proteins)** and one or more large **ribosomal RNAs (rRNAs) (Table 18-1)**. Like the subunits themselves, rRNAs are named in svedberg units. The 50S subunit contains a 5S and a 23S rRNA, and the 30S subunit contains a single 16S rRNA; these RNAs together comprise more than 4,500 nucleotides. The 50S subunit in *E. coli* also contains 36 different r-proteins, named L1 through L36 (L for large subunit), and the 30S subunit contains a single copy of each of 21 different r-proteins, named S1 through S21 (S for small subunit).

Although there are many more individual r-proteins than RNAs in the ribosome, the relatively small size of most r-proteins (~15 kDa, on average, in bacteria) means that the proteins contribute only about one-third of the overall mass of the ribosome.

Decades of research on the structure and function of ribosomal proteins and RNAs have shifted the focus from the proteins to the rRNA. Despite the complexity of ribosome structure, Masayasu Nomura and colleagues demonstrated in the late 1960s that both bacterial ribosomal subunits can be broken down into their RNA and protein components, then reconstituted in vitro. Under appropriate experimental conditions, the RNAs and proteins spontaneously reassemble to form 50S or 30S subunits nearly identical in sedimentation behavior and activity to native subunits. Sequencing of rRNAs in the 1970s, and the secondary structure proposed for 16S rRNA by Harry Noller and Carl Woese in 1981, showed that rRNAs have been highly conserved by evolution, particularly in molecular regions implicated in the critical functions of the ribosome (**Figure 18-2**). Noller's biochemical

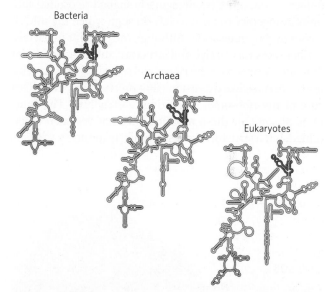

FIGURE 18-2 Conservation in the secondary structure of small-subunit rRNAs from the three domains of life. Red, yellow, and purple indicate areas where the structures of bacterial, archaeal, and eukaryotic rRNAs have diverged; conserved regions are shown in green.

TABLE 18-1				
Protein and RNA Components of the *E. coli* Ribosome				
Ribosomal Subunit	*Number of Different Proteins*	*Total Number of Protein Subunits*	*Protein Designations*	*rRNAs*
50S	33	36	L1–L36	5S rRNA and 23S rRNA
30S	21	21	S1–S21	16S rRNA

analysis of functional inactivation of ribosomes provided the first evidence for the fundamental role of rRNA in catalyzing protein synthesis (see this chapter's Moment of Discovery). Along with Woese's sequence analysis of rRNAs from a range of organisms, these data also provided an important tool for analyzing evolutionary relationships among species.

In the culmination of many years of work in multiple laboratories, cryo-electron microscopy (using frozen samples) and x-ray crystallography have now revealed the atomic-resolution structure of the bacterial ribosome and its subunits in exquisite detail (**Figure 18-3a**). Venki Ramakrishnan, Tom Steitz, and Ada Yonath received a Nobel Prize in 2009 for their crystallographic work. More recently, the crystal structure of a eukaryotic ribosome was solved, revealing a similar overall structure but with greater complexity that reflects additional levels of regulation (**Figure 18-3b**).

With a combined molecular weight of ~2.5 × 10⁶ Da, the bacterial ribosome is far more complex than the DNA or RNA polymerases, and knowledge of its structure has provided a wealth of insights about its function and evolutionary origins. The two irregularly shaped ribosomal subunits fit together to form a cleft through which the mRNA passes during translation. Although the proteins in bacterial ribosomes vary in size and structure, most have globular domains arranged on the ribosomal surface. Some proteins also have snakelike extensions that protrude into the rRNA core of the ribosome, stabilizing its structure. The functions of some of these proteins have not yet been determined, but a structural role seems likely for many of them.

Each of the three single-stranded rRNAs of *E. coli* has a specific three-dimensional conformation with extensive intrachain base pairing. This secondary structure was predicted based on the available sequences of rRNAs from many organisms (see Figure 18-2). These structures have largely been confirmed in high-resolution three-dimensional models, yet they fail to convey the extensive network of tertiary interactions evident in the complete structure.

Extensive structural, biochemical, and genetic data support the conclusion that rRNA is responsible for all ribosomal functions, including tRNA binding and peptide bond formation. The predominant location of r-proteins on the outer surface of the ribosome, away from its RNA-rich functional interface, underscores this idea (see Figure 18-3). In addition to dominating the functional centers within each ribosomal subunit, rRNA provides most of the contacts between the two subunits in the intact ribosome. The intersubunit bridges not only hold the two subunits together but also foster relative movement between the subunits that is integral to the process of polypeptide elongation (see Section 18.4).

The ribosomes of eukaryotic cells (other than the mitochondrial and chloroplast ribosomes) are larger and more complex than bacterial ribosomes. Electron microscopic and centrifugation studies show that these ribosomes have a diameter of about 23 nm and a sedimentation coefficient of about 80S. Like their bacterial counterparts, eukaryotic ribosomes have two subunits. Subunit size varies among species, but, on average, the subunits are 60S and 40S. Altogether, the cytosolic

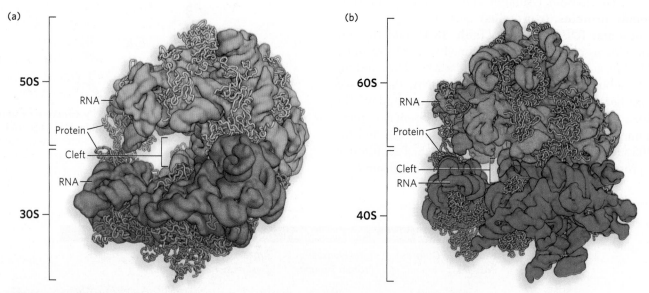

FIGURE 18-3 Crystal structures of bacterial and eukaryotic ribosomes. (a) In bacteria, the 50S (large) and 30S (small) subunits together form the 70S ribosome. The interface between the two subunits forms a cleft where the peptidyl transferase reaction (described later in this chapter) occurs. (b) The yeast ribosome has a similar structure, with increased complexity. [Sources: (a) PDB ID 1SVA and PDB ID 2OW8. (b) PDB ID 3O58 and PDB ID 3O2Z.]

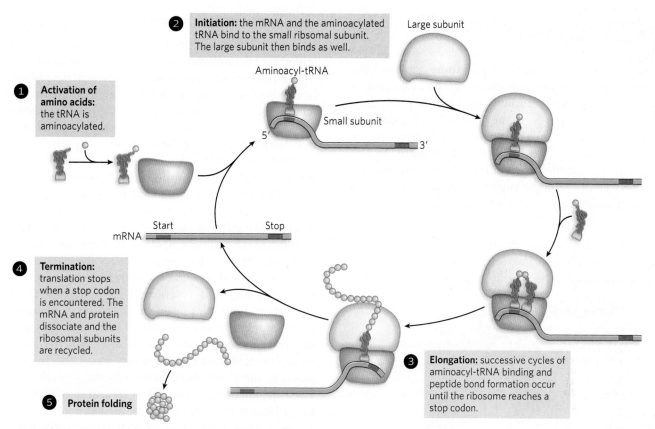

FIGURE 18-4 An overview of the main events in translation. Following the activation of amino acids by acylation to tRNAs (step 1), translation initiates with the assembly of mRNA and aminoacylated tRNA on the small ribosomal subunit, followed by joining with the large subunit to form an active ribosome (step 2). Polypeptide elongation occurs in successive cycles of aminoacyl-tRNA binding and peptide bond formation in the order directed by the genetic code in the mRNA (step 3). Translation stops when the ribosome encounters a stop codon in the mRNA, leading to release of the mRNA and dissociation of the ribosome into its two subunits (step 4); the polypeptide folds during or immediately following translation to form a functional protein (step 5).

ribosomes contain more than 80 different proteins and four types of rRNA; unlike bacterial ribosomes, they cannot be spontaneously reconstituted in vitro, suggesting a more complex assembly process. In contrast, the ribosomes of mitochondria and chloroplasts are somewhat smaller and simpler than bacterial ribosomes. Nevertheless, the rRNAs of all cell types are conserved (and, as we saw in Chapter 17, all tRNAs have similar sizes and shapes), indicating that ribosomal structure and function are strikingly similar in all organisms and organelles.

Ribosomal Subunits Associate and Dissociate in Each Cycle of Translation

The association of ribosomal subunits at the start of protein synthesis and their dissociation on release of the completed polypeptide are fundamental to the process of translation in all cells. Because the subunits are initially separated, the initiation of protein synthesis is inherently regulated by the assembly of active ribosomes on an mRNA, together with tRNAs. The recruitment of the small ribosomal subunit to an mRNA is an essential first step, and cells and viruses have a variety of mechanisms for controlling how and when this happens.

An overview of translation is shown in **Figure 18-4**. In general, translation—the production of a protein product—begins with mRNA and an initiator tRNA binding to the small ribosomal subunit. Once the small subunit is positioned at the beginning of the coding sequence of the mRNA, the large ribosomal subunit joins noncovalently to form an active ribosome that begins reading each mRNA codon in sequence in the $5' \rightarrow 3'$ direction. As each codon is encountered, a matching tRNA bearing its covalently attached amino acid enters the decoding and peptidyl transferase centers of the ribosome. After peptide bond formation between the C-terminal end of the growing polypeptide chain and the amino acid of the incoming aminoacyl-tRNA, the ribosome translocates to the next codon, and the cycle repeats. On encountering a stop codon, the ribosome is released along with the completed protein product, the ribosomal subunits dissociate, and the process is ready to begin again on another mRNA.

Because mRNAs are usually at least 300 nucleotides long and extend well beyond the ~200 Å girth of the ribosome, multiple ribosomes can occupy each mRNA during translation, forming a **polysome**, or **polyribosome** (**Figure 18-5a**). Polysomes can be visualized by electron microscopy, and because they form very high molecular weight particles, they are also readily detected in cell extracts by analyzing their sedimentation in sucrose or glycerol density gradients (**Figure 18-5b**). Polysome formation lets each mRNA molecule provide the template for multiple copies of a protein molecule at once, thus allowing the efficient use of each mRNA.

(a)

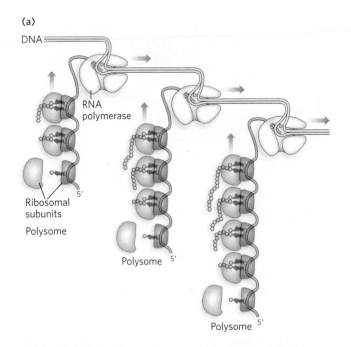

(b)

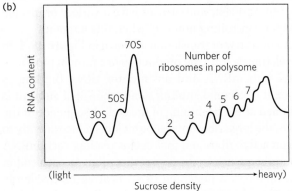

FIGURE 18-5 Polysomes. (a) Polysomes consist of multiple ribosomes associated with a single mRNA. (b) Polysomes can be separated on a sucrose density gradient. The polysome profile indicates total RNA content at increasing sucrose density. The results show that 30S and 50S ribosomal subunits and 70S ribosomes migrate near the top of the gradient, whereas polysomes, comprising two or more ribosomes associated with mRNA transcripts, migrate in the heavier (more dense) fractions of the gradient.

The Ribosome Is a Ribozyme

Noller concluded that ribosomal RNA has a fundamental importance in catalyzing protein synthesis, based on the results of two experiments. First, he inactivated the ribosome by making base changes in rRNA; second, he removed proteins from the ribosome and found that the deproteinized ribosome retained peptidyl transferase activity (see the How We Know section at the end of this chapter).

The object of the second experiment was to remove proteins without disrupting rRNA structure, which precluded the use of extraction reagents such as acid. Instead, detergents, a protease, or phenol were used, all of which are effective deproteinizing treatments that do not degrade or denature RNA. To monitor the activity of ribosomes after protein removal, Noller and his colleagues used a simplified peptidyl transferase reaction, referred to as the fragment reaction. This requires only the 50S subunit and does not need initiation or elongation factor proteins (factors discussed later in this chapter); the assay monitors the addition of [^{35}S]methionine to an amino acid mimic, puromycin. Using the fragment reaction, the researchers found that the 50S subunit of *Thermus aquaticus*, a bacterium that grows at elevated temperatures (a thermophile), retained peptidyl transferase activity even after proteins were extracted by treatments with detergent, proteinase K (a nonspecific protease that degrades virtually all proteins), or phenol (**Figure 18-6**). The thermophilic ribosomes were selected because they were expected to be inherently more stable than ribosomes from organisms such as *E. coli* that grow at moderate temperatures. However, even *E. coli* 50S subunits remained active after treatment with detergent or proteinase K, suggesting that the rRNA structure was largely intact.

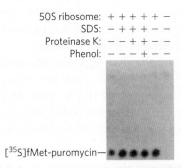

FIGURE 18-6 The effects of method of protein extraction on ribosomal function in vitro. Ribosomes from the bacterium *Thermus aquaticus* were treated with the detergent sodium dodecyl sulfate (SDS), the enzyme proteinase K, or phenol, then tested for peptidyl transferase activity using the fragment reaction. The product of the reaction, [^{35}S]fMet-puromycin, was assayed by paper electrophoresis and autoradiography. (fMet is *N*-formylmethionine, the initiating amino acid in bacterial protein synthesis, described later in this chapter.) [*Source: Noller H.F., Hoffarth, V. and Zimniak, L. Science 256:1416–9. Fig. 2b, 1992. Reprinted with permission from AAAS.*]

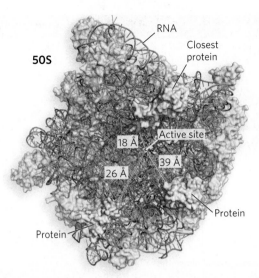

FIGURE 18-7 Ribosomal RNA in the ribosomal active site. The active site—where the peptidyl transferase reaction forms peptide bonds—is 18 Å away from the closest r-protein, evidence that the ribosome is a ribozyme. [Source: PDB ID 1Q7Y.]

These important findings foreshadowed the discovery that there is no protein within 18 Å of the peptidyl transferase active site in the crystal structure of the 50S ribosomal subunit (**Figure 18-7**). The high-resolution structure thus confirmed what had been suspected for more than a decade: the ribosome is a ribozyme. The ribosomal RNA, not protein, is responsible for catalysis of peptide bond formation.

In addition to confirming the central role of rRNA in peptide bond formation, crystal structures of the ribosome and its subunits show that most of the contacts between tRNA and the ribosome involve contacts with 16S or 23S rRNA, not with protein. Thus, with these new data, the traditional focus on the protein components of ribosomes shifted. In addition to providing the structural core, rRNAs form the functional core of the ribosome. This implies that r-proteins are secondary elements in the complex, playing a stabilizing or regulatory role rather than a catalytic role in the business of translation.

What do these findings suggest about the origin of this most fundamental cellular process? Francis Crick wondered in the 1960s about the possible existence of an all-RNA ribosome at some early point in evolution. The importance of rRNA in modern ribosomes supports this idea and is consistent with the addition of proteins to a preexisting rRNA-containing ribosome over the course of evolution. Recent studies of mitochondrial ribosomes suggest that r-proteins may restrict the variety of proteins that a ribosome can synthesize (**Highlight 18-1**).

The Ribosome Structure Facilitates Peptide Bond Formation

The ribosome must bind simultaneously to at least two tRNAs during each cycle of amino acid addition to the C-terminus of a growing polypeptide chain. In fact, experiments to measure tRNA interactions on the ribosome, as well as the parts of tRNA molecules that are protected from solvent when bound to the ribosome, showed that ribosomes contain binding sites for three tRNAs (**Figure 18-8**). The **A site** is the location of

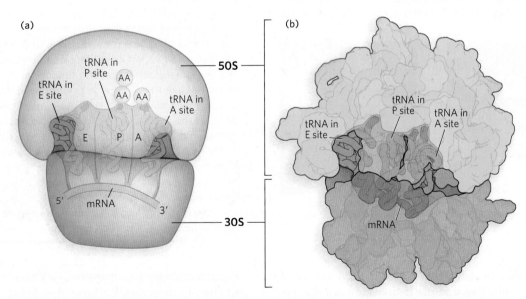

FIGURE 18-8 The A, P, and E sites of the ribosome. (a) The A, P, and E sites in relation to the bound mRNA. Both ribosomal subunits are shown. AA indicates an amino acid bound to tRNA. (b) A representation of the *E. coli* 70S ribosome crystal structure with bound tRNAs—representing aminoacyl-tRNA, peptidyl-tRNA, and free tRNA, respectively—in the A, P, and E sites, viewed from the 30S interface. [Sources: (b) PDB ID 1SVA and PDB ID 2OW8.]

HIGHLIGHT 18-1 — EVOLUTION

Mitochondrial Ribosomes: A Window into Ribosome Evolution?

Mitochondria encode their own ribosomes for synthesizing some of the mitochondrial proteins that carry out oxidative phosphorylation to produce the cell's ATP. Although mitochondria are thought to have descended from symbiotic bacteria, their ribosomes differ substantially from those of modern bacteria. In particular, the ratio of protein to rRNA in animal mitochondrial ribosomes is 2:1 by mass, rather than the 1:2 ratio observed for bacterial and archaeal ribosomes. Because the ribosomes have roughly the same mass, this means that mitochondrial ribosomes are two-thirds protein and one-third rRNA by mass, whereas bacterial ribosomes are one-third protein and two-thirds rRNA. What does this imply about the origins of the ribosome and the importance of rRNA in its function?

To begin to answer these questions, Rajendra Agrawal and Stephen Harvey proposed a structural model of the mitochondrial large ribosomal subunit, using a combination of cryo-electron microscopy and molecular modeling. Although the resolution of the electron microscopy–derived electron density, 12.1 Å, was lower than that of x-ray crystallography, it was possible to model higher-resolution bits of structure into the electron density map by using various molecular landmarks. The resulting model predicts the arrangement of individual protein and rRNA components in the large subunit of the mitochondrial ribosome (Figure 1). Although there is much less rRNA than r-protein, the rRNA forms the same overall architecture and occupies the same positions known to be essential for protein synthesis in bacterial and archaeal ribosomes. However, the r-proteins encroach on the all-RNA core of the mitochondrial ribosome, where they

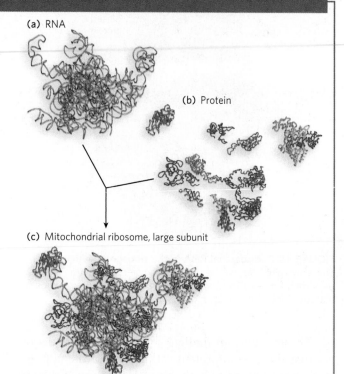

(a) RNA

(b) Protein

(c) Mitochondrial ribosome, large subunit

FIGURE 1 (a) The RNA portions and (b) the protein portions of the large ribosomal subunit of mammalian mitochondria, determined by cryo-electron microscopy and modeled using available crystal structures of the bacterial 50S ribosomal subunit. (c) The proteins and RNAs making up the complete large subunit. Notice how protein segments penetrate the interior of the subunit. [Source: PDB ID 2FTC.]

substitute for segments of rRNA that are missing relative to bacterial ribosomes.

Because mtDNA has a higher mutation rate and therefore evolves faster than the cellular genome, the mitochondrial ribosome presumably is now farther from its all-RNA origins than is the bacterial ribosome. This may reflect the greater structural and functional diversity of proteins relative to RNA. Far in the future, mitochondrial and, eventually, all ribosomes might consist primarily of protein, as their rRNA components are replaced over time. Mammalian mitochondria have evolved to synthesize just 13 proteins, so their ribosomes may be subject to different selective pressures from those at work on cytosolic ribosomes. Comparative studies of mitochondrial and cytosolic ribosome structure and activity may provide unexpected insights into the changing roles of RNA and protein in ribosome form and function.

Rajendra Agrawal [Source: Courtesy of Rajendra K Agrawal.]

Stephen Harvey [Source: Courtesy of Georgia Research Alliance.]

aminoacyl-tRNA binding, the **P site** is the location of peptidyl-tRNA binding, and the **E site** is the exit site, occupied by the tRNA molecule released after the growing polypeptide chain is transferred to the aminoacyl-tRNA. Each tRNA starts in the A site, moves to the P site after peptide bond formation, then exits through the E site.

Each of these sites spans the two ribosomal subunits and thereby functionally links the decoding center of the small subunit with the peptidyl transferase center of the large subunit. The anticodon loop at one end of each L-shaped tRNA molecule makes contact with the mRNA positioned in the small-subunit decoding site, and the

(a)

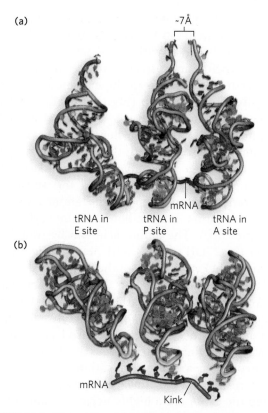

tRNA in E site   tRNA in P site   tRNA in A site

(b)

mRNA

Kink

FIGURE 18-9 Alignment of tRNAs on mRNA and the peptidyl transferase site. (a) The peptidyl-tRNA and the aminoacyl-tRNA are positioned so that the amino acids are at the optimal distance for peptide bond formation. (b) The anticodons kink the mRNA. [*Sources: PDB ID 1GIX and PDB ID 2OW8.*]

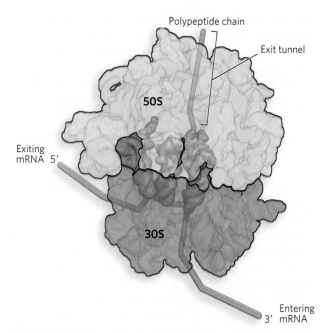

Polypeptide chain

Exit tunnel

50S

Exiting mRNA 5'

30S

Entering 3' mRNA

FIGURE 18-10 The exit tunnel for protein in the 50S subunit. The protein exit tunnel is adjacent to the P site and only wide enough to allow an α helix or an unfolded polypeptide to pass through. [*Sources: PDB ID 1VSA, PDB ID 2OW8, and PDB ID 1GIX.*]

aminoacylated 3′ end of the tRNA occupies the peptidyl transferase center in the large subunit, 70 Å away.

During elongation, the ribosome houses a growing polypeptide chain, which is covalently linked to the tRNA in the P site, as the peptidyl-tRNA. Each time the ribosome shifts from one mRNA codon to the next, it makes room for a new aminoacyl-tRNA in the A site. After the peptide bond forms (catalyzed by the peptidyl transferase), the polypeptide chain transfers from one tRNA to the other, with the aminoacyl-tRNA becoming the peptidyl-tRNA as the growing polypeptide chain is added to it.

The rate of peptide bond formation is enhanced on the ribosome largely because the 3′ ends of the two reacting tRNAs are positioned optimally for the chemical reaction to occur (**Figure 18-9a**). Note that there is no accompanying hydrolysis of a high-energy bond, such as that of a nucleoside triphosphate, at this stage of protein synthesis. This is because each amino acid has already been activated by its attachment to tRNA in an aminoacylation reaction. As we will see, the aminoacylation step requires ATP hydrolysis, so the energy cost has already been paid (see Section 18.2). Thus, each acyl group provides the high-energy bond that is hydrolyzed

to drive the formation of a new peptide bond during the peptidyl transferase reaction.

Crystal structures of the ribosome revealed the presence of channels for the movement of mRNAs and polypeptides during protein synthesis. In the small subunit, the mRNA entry and exit channels are narrow clefts with space to accommodate only a single strand of RNA. Thus, an mRNA must be unfolded and available for base pairing to tRNAs when it enters the decoding center. The A and P sites in the small subunit provide room for tRNAs to base-pair with two adjacent codons in the mRNA sequence. Here, the ribosome induces a kink in the mRNA between the two codons to help ensure accuracy of the reading frame (see **Figure 18-9b**). This kink may also contribute to correct positioning of the aminoacylated ends of the tRNA in the peptidyl transferase center of the large subunit.

In the large subunit, a tunnel (50 Å long) allows the exit of polypeptides during translation (**Figure 18-10**). The diameter of the exit tunnel seems to preclude formation of structures wider than an α helix as the polypeptide is being synthesized. Thus, nascent proteins lack tertiary structure and must assemble into their correct three-dimensional structure after exiting the ribosome.

SECTION 18.1 SUMMARY

- Protein synthesis occurs on ribosomes, large complexes of protein and rRNA. Bacteria have 70S ribosomes, with a large (50S) and a small (30S) subunit. Eukaryotic ribosomes are significantly larger, 80S (with 60S and 40S subunits), and contain more proteins.

- Protein synthesis is regulated by the ability of the small ribosomal subunit to associate independently with the mRNA before translation begins. Once the ribosome is fully assembled, it moves along the mRNA, matching a tRNA to each codon and catalyzing peptide bond formation. Multiple ribosomes can occupy a single mRNA, forming a polysome.

- In all organisms, the rRNA of the large ribosomal subunit catalyzes peptide bond formation. The genetic code is read on the small ribosomal subunit, which ensures that the correct amino acid is added to the growing polypeptide chain.

- Aminoacyl-tRNAs first bind to the A site of the ribosome. As the peptide bond is formed between the amino acid and the growing peptide chain, the newly formed peptidyl-tRNA moves to the P site and the free tRNA exits through the E site.

- The mRNA and growing polypeptide pass through separate channels in the ribosome that require them to be unfolded. The ribosome induces a kink in the mRNA between the A and P sites to allow base-pairing of the tRNAs. Polypeptides form their functional three-dimensional structure after emerging from the ribosomal exit tunnel.

18.2 ACTIVATION OF AMINO ACIDS FOR PROTEIN SYNTHESIS

For the synthesis of a polypeptide with a sequence defined by an mRNA, two chemical requirements must be met: (1) the carboxyl group of each amino acid must be activated to facilitate the formation of a peptide bond, and (2) a link must be established between each new amino acid and the information in the mRNA that encodes it. Both of these are satisfied by covalent attachment of the amino acid to a tRNA prior to protein synthesis. Attaching each amino acid to its corresponding tRNA is critical. This reaction takes place in the cytosol, not on the ribosome. Each of the 20 common amino acids is covalently linked to a specific tRNA at the expense of ATP hydrolysis, catalyzed by Mg^{2+}-dependent activating enzymes known as aminoacyl-tRNA synthetases. When attached to an amino acid, a tRNA is said to be *charged*.

Amino Acids Are Activated and Linked to Specific tRNAs

The charging of a tRNA requires the formation of an acyl linkage between the carboxyl group of an amino acid and the free 2′- or 3′-hydroxyl end of the tRNA. All tRNAs share a similar structure, including three or four arms and a 3′-terminal CCA sequence (see Figures 6-22

and 17-2). The acylation reaction results in attachment of the amino acid to the 3′-terminal adenosine.

Aminoacyl-tRNA synthetases must activate the amino acid before it is attached to the tRNA. This reaction occurs in two steps in the enzyme's active site. In the **adenylylation step** (**Figure 18-11**, step 1), an enzyme-bound intermediate, 5′-aminoacyl adenylate (5′-aminoacyl-AMP), forms when the carboxyl group of the amino acid reacts with the α-phosphoryl group of ATP to form a phosphoanhydride linkage, with displacement of pyrophosphate. In the subsequent **tRNA-charging step** (step 2), the aminoacyl group is transferred from enzyme-bound aminoacyl-AMP to its specific tRNA. The amino acid can be transferred to either the 2′-OH or the 3′-OH (left and right paths, respectively, in Figure 18-11) of the 3′-terminal adenosine of the tRNA, depending on the type of aminoacyl-tRNA synthetase. Class I synthetases attach the amino acid to the 2′-OH, and class II synthetases attach the amino acid to the 3′-OH. In the class I pathway, the aminoacyl ester linkage migrates to the 3′-OH position spontaneously by a transesterification reaction.

The resulting ester linkage between the amino acid and the tRNA has a highly negative standard free energy of hydrolysis ($\Delta G'^{\circ} = -229$ kJ/mol). Because its hydrolysis is energetically favorable, this bond between the amino acid and the tRNA provides an energetic driving force for translation. The first part of the activation reaction separates two phosphates from ATP instead of just one (as in many other ATP-driven reactions). Ultimately, the energy stored in the bond between the two phosphates is released on hydrolysis by the enzyme inorganic pyrophosphatase. Thus, *two* high-energy phosphate bonds are ultimately expended for each amino acid molecule activated, rendering the overall reaction for amino acid activation essentially irreversible.

Aminoacyl-tRNA Synthetases Attach the Correct Amino Acids to Their tRNAs

A distinct aminoacyl-tRNA synthetase is responsible for attaching each of the 20 common amino acids to its corresponding tRNA. Each enzyme is specific for one amino acid but can recognize more than one tRNA, because most amino acids are specified by more than one codon (as discussed in Chapter 17). The structures of all the aminoacyl-tRNA synthetases of *E. coli* have been determined, and the structures reflect their mode of tRNA recognition and catalytic activity (**Figure 18-12** on p. 628). Class I enzymes are typically monomeric and attach the amino acid to the 2′-OH of the 3′-terminal adenine of tRNA. As noted above, the aminoacyl group spontaneously migrates to the 3′-OH position. Class II

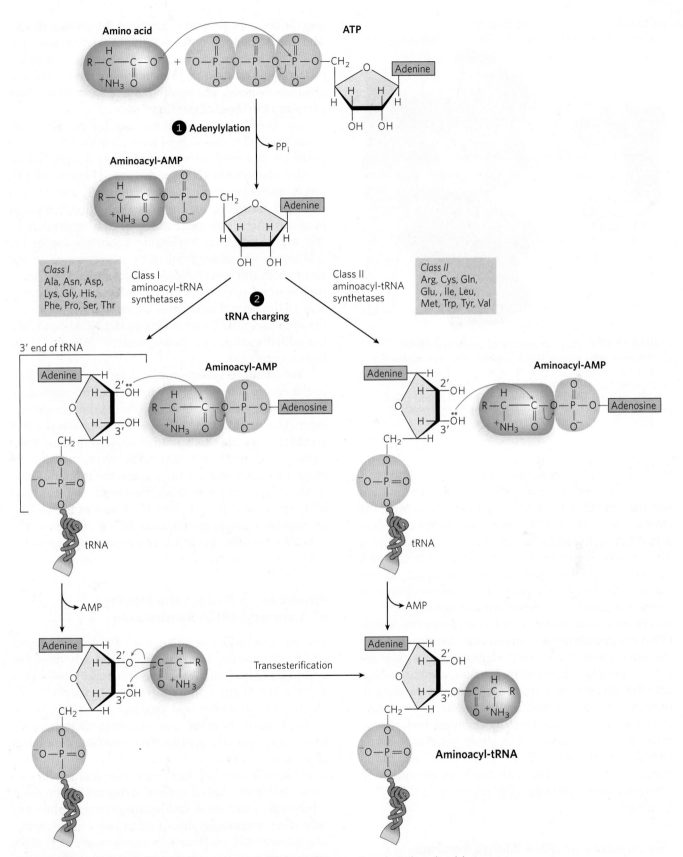

FIGURE 18-11 Charging of tRNAs by aminoacyl-tRNA synthetases. In step 1, the adenylylation step, the amino acid is linked to adenylate, forming aminoacyl-AMP. In step 2, the amino acid is transferred from the AMP to the tRNA in one of two pathways, catalyzed by a class I or class II aminoacyl-tRNA synthetase. The synthetases in each class are listed (denoted by the amino acid they transfer to tRNA).

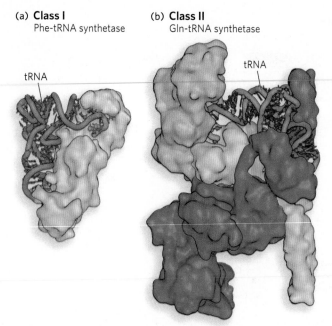

(a) Class I
Phe-tRNA synthetase

(b) Class II
Gln-tRNA synthetase

tRNA

tRNA

FIGURE 18-12 Crystal structures of aminoacyl-tRNA synthetases with bound tRNA. (a) Phe-tRNA synthetase is a class I and (b) Gln-tRNA synthetase is a class II aminoacyl-tRNA synthetase. For each structure, the bound tRNA is shown in green, and the 3′ end of the tRNA to which the amino acid attaches is shown in pink. [Sources: (a) PDB ID 1EUQ. (b) PDB ID 1EIY.]

aminoacyl-tRNA synthetases are sometimes multimeric, and they approach their tRNA substrate from a different side than the class I synthetases, typically attaching the amino acid to the 3′-OH. These two classes of aminoacyl-tRNA synthetases are the same in all organisms. There is no evidence for a common ancestor from which the two classes diverged, and the biological, chemical, or evolutionary reasons for two enzyme classes for essentially identical processes remain obscure.

There are exceptions to the rule of one aminoacyl-tRNA synthetase for one amino acid. Because the synthetase sequences are evolutionarily related, they are usually readily identified in genomic sequence data, and the absence of a gene encoding the aminoacyl-tRNA synthetase specific for glutamine in some bacteria was initially puzzling. Biochemical and genetic experiments revealed that in these bacteria, a single enzyme charges both tRNA^{Gln} and tRNA^{Glu} with glutamate. A second enzyme then catalyzes an amination reaction that converts the glutamate of Glu-tRNA^{Gln} to glutamine.

The Structure of tRNA Allows Accurate Recognition by tRNA Synthetases

The overall fidelity of protein synthesis requires that each individual aminoacyl-tRNA synthetase must be

specific for a single amino acid and for certain tRNAs. The interaction of aminoacyl-tRNA synthetases and tRNAs has been referred to as the "second genetic code," reflecting its critical role in maintaining the accuracy of protein synthesis. The "coding" rules seem to be more complex than those of the "first" code.

By observing changes in nucleotides that alter substrate specificity, researchers have identified nucleotide positions that are involved in substrate discrimination by the aminoacyl-tRNA synthetases (**Figure 18-13**). Some nucleotides are conserved in all tRNAs and therefore cannot be used for discrimination. Although in some cases the nucleotides of the anticodon itself are recognized, nucleotide positions conferring synthetase specificity tend to be concentrated in the amino acid arm and elsewhere in the anticodon arm, as well as other parts of the tRNA molecule. Determination of the crystal structures of aminoacyl-tRNA synthetases complexed with their cognate tRNAs and ATP has added a great deal to our understanding of these interactions.

Ten or more specific nucleotides may be involved in the recognition of a tRNA by its specific aminoacyl-tRNA synthetase. But in a few cases, the recognition mechanism is quite simple. For example, for alanyl-tRNA synthetase (or Ala-tRNA synthetase—a shorthand commonly used for these enzymes), across a range of organisms from bacteria to humans, the primary determinant of recognition of tRNA^{Ala} is a single G–U base pair in its amino acid arm. A short RNA with as few as 7 bp arranged in a simple hairpin mini-helix is efficiently aminoacylated by the Ala-tRNA synthetase, as long as the RNA contains the critical G–U.

Proofreading Ensures the Fidelity of Aminoacyl-tRNA Synthetases

The aminoacylation of tRNA both activates an amino acid for peptide bond formation and appends the amino acid to an adaptor tRNA that ensures appropriate placement of the amino acid in a growing polypeptide. The identity of the amino acid attached to a tRNA is not checked on the ribosome, however, so attachment of the correct amino acid to the tRNA is essential to the fidelity of protein synthesis.

Discrimination between two similar amino acid substrates has been studied in detail in the case of Ile-tRNA synthetase, which must distinguish between valine and isoleucine, amino acids that differ by just a single methylene group (—CH$_2$—). Because valine is smaller, Val-tRNA synthetase can discriminate between valine and isoleucine by having a binding pocket too small for isoleucine to bind (**Figure 18-14a**). However, the correlating strategy cannot work for Ile-tRNA synthetase, because

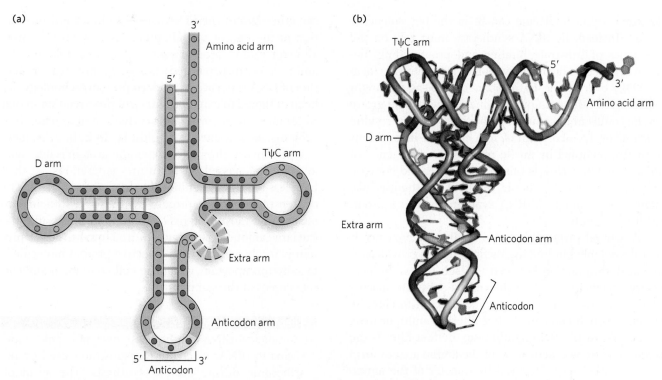

FIGURE 18-13 Sequence features of a tRNA that are recognized by aminoacyl-tRNA synthetases. (a) The two-dimensional and (b) three-dimensional structures of tRNA, showing the nucleotides recognized by only one (orange residues) or by two or more (blue residues) aminoacyl-tRNA synthetases. Nucleotides shown in purple are common to all tRNAs and therefore cannot be used to distinguish among them; nucleotides shown in black are positions in the tRNA sequence that are not specificity determinants. [Source: (b) PDB ID 1EHZ.]

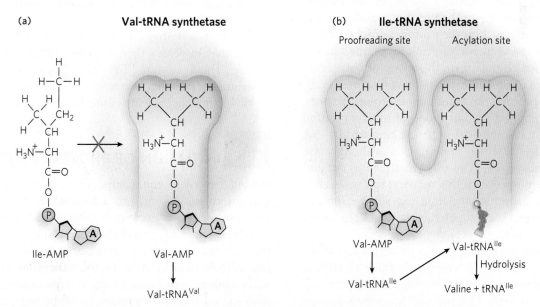

FIGURE 18-14 Proofreading by an aminoacyl-tRNA synthetase. (a) tRNA^{Val} is charged in the acylation site of the Val-tRNA synthetase. Isoleucine is larger than valine and does not fit in the acylation site. (b) tRNA^{Ile} is charged in the acylation site of the Ile-tRNA synthetase. Because valine is smaller than isoleucine, tRNA^{Ile} is sometimes charged with valine. This incorrectly charged Val-tRNA^{Ile} fits into the synthetase's proofreading site, where it is hydrolyzed to release valine from the tRNA.

the small valine molecule can fit in the big isoleucine pocket. Instead, Ile-tRNA synthetase must rely on the energetics of substrate binding and proofreading. Ile-tRNA synthetase favors activation of isoleucine (to form Ile-AMP) over valine by a factor of 200—as we might expect, given the amount by which a methylene group (in Ile) could enhance substrate binding. Yet a Val residue is erroneously incorporated into proteins in positions normally occupied by an Ile residue at a frequency of about 1 in 3,000. How is this more than tenfold increase in accuracy brought about? Ile-tRNA synthetase, like some other aminoacyl-tRNA synthetases, has a proofreading function.

A general principle of proofreading by enzymes is that if available binding interactions do not provide sufficient discrimination between two substrates, the necessary specificity can be achieved by substrate-specific binding in *two successive* steps. The effect of forcing the system through two consecutive filters is multiplicative. In the case of Ile-tRNA synthetase, the first filter is the initial binding and activation of the amino acid to form the aminoacyl-AMP, followed by transfer of the amino acid to the tRNA. The second is the binding of any *incorrect* aminoacyl-tRNA products to a separate active site on the enzyme; a substrate that binds in this second active site is hydrolyzed (**Figure 18-14b**). Because the R group of valine is slightly smaller than that of isoleucine, Val-tRNA fits the hydrolytic (proofreading) site of the Ile-tRNA synthetase but Ile-tRNA does not. Thus, only Val-tRNA is hydrolyzed in the proofreading active site.

The greatly accelerated rate of hydrolysis of incorrectly charged tRNAs provides an important mechanism for enhancing the fidelity of the overall process. This is an example of **kinetic proofreading**, in which a complex process occurs in multiple steps, the rates of which are tuned to maximize the speed of correct reactions while stalling and reversing incorrect reactions. Note that this proofreading mechanism requires energy, because it requires a second round of aminoacylation (with the correct amino acid). The few aminoacyl-tRNA synthetases that activate amino acids with no close structural relatives (e.g., Cys-tRNA synthetase) demonstrate little or no proofreading activity; in these cases, the active site for aminoacylation can sufficiently discriminate between the proper substrate and any other, incorrect amino acid.

In summary, translation relies on aminoacyl-tRNA synthetases to ensure the correct charging of tRNAs, because the ribosome does not distinguish between correctly and incorrectly charged tRNAs during protein synthesis. The decoding center of the ribosome is designed to detect and favor proper codon-anticodon base pairing, but does not link this information to the identity of the amino acid in the peptidyl transferase center at the other end of the tRNA (see the How We Know section at the end of this chapter). As a result, the overall error rate of protein synthesis ($\sim$1 mistake per 10^4 amino acids incorporated) is significantly higher than that of DNA replication. Modern protein technology has been designed to exploit this lack of proofreading so that ribosomes can incorporate synthetic amino acids into proteins, as described in **Highlight 18-2**. In cells, flaws in a protein are eliminated when the protein is degraded and are not passed on to future generations, so they have less biological significance than errors in DNA. The degree of fidelity in protein synthesis is sufficient to ensure that most proteins contain no mistakes and that the large amount of energy required to synthesize a protein is rarely wasted. One defective protein molecule is usually unimportant when the cell contains many correct copies of the same protein.

SECTION 18.2 SUMMARY

- Aminoacyl-tRNA synthetases covalently link amino acids to tRNAs to create the substrates used by the ribosome during protein synthesis. The acylation reaction results in a high-energy bond that supplies the energetic driving force for translation.

- A different aminoacyl-tRNA synthetase exists for each amino acid. Because multiple codons can specify a single amino acid, most aminoacyl-tRNA synthetases can recognize multiple tRNAs bearing anticodons complementary to the codons for a particular amino acid.

- The anticodon is responsible for the specificity of interaction between the aminoacyl-tRNA and the complementary mRNA codon, but it rarely provides the primary site for synthetase recognition.

- Two successive binding steps allow kinetic proofreading that increases the fidelity of tRNA aminoacylation. This is essential because ribosomes do not discriminate between correctly and incorrectly charged tRNAs during protein synthesis.

18.3 INITIATION OF PROTEIN SYNTHESIS

Having described the components of the translation machinery, we now turn to a detailed discussion of the initiation stage of protein synthesis—the most highly regulated step of translation. Translation **initiation** includes recruitment of the small ribosomal subunit to the mRNA; identification of the **initiation codon**, or **start codon**; association of the charged initiator tRNA with the mRNA; and recruitment of the large ribosomal subunit to form an active ribosome (**Figure 18-15** on p. 632). Each step of protein synthesis, in both *E. coli* and eukaryotes, requires several protein factors to facilitate the reaction (**Table 18-2** on p. 633). **Initiation factors**

HIGHLIGHT 18-2 — **TECHNOLOGY**

Genetic Incorporation of Unnatural Amino Acids into Proteins

Peter Schultz and his colleagues at the Scripps Research Institute wondered whether the specificity of aminoacyl-tRNA synthetases, together with the lack of discrimination of mischarged tRNAs by ribosomes, could be exploited to incorporate unnatural amino

Peter Schultz [Source: Courtesy Peter Schultz.]

acids into cellular proteins. First they introduced stop codons at sites in an mRNA where they hoped to introduce an unnatural amino acid in the corresponding polypeptide. Next they engineered suppressor tRNAs (see Chapter 17) with an anticodon sequence complementary to the stop codon. Taking advantage of the known structural determinants of the synthetases' recognition of tRNA, the researchers designed suppressor tRNAs to be charged by aminoacyl-tRNA synthetases with mutations that caused them to use unnatural amino acid substrates (Figure 1). Schultz's research group showed that bacterial, yeast, and mammalian cells containing the engineered mRNAs, tRNAs, and aminoacyl-tRNA synthetases, along with unnatural amino acids, would produce proteins with the unnatural amino acid residues at the planned positions.

This method allows the incorporation into proteins of fluorescent, glycosylated, sulfated, metal ion–binding, and redox-active (i.e., electron-transferring) amino acids, as well as amino acids with new chemical and photochemical reactivity. Why might such technology be useful? Schultz hopes to use the approach to explore protein structure and function, both in vitro and in vivo, by incorporating chemical tags or probes that can report on their local molecular/structural environment or provide molecular "beacons" in cells. Furthermore, with the ability to generate proteins with new or enhanced properties by making use of amino acids not found in nature, it may eventually be possible to enable cells to synthesize therapeutic proteins.

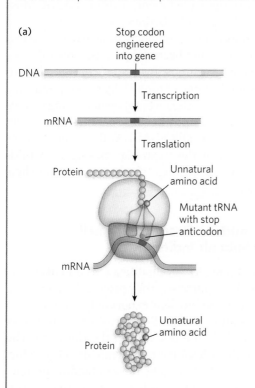

(a)

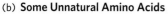

(b) **Some Unnatural Amino Acids**

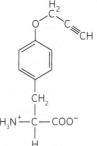

Phenylalanine   *p*-Azidophenylalanine   *p*-Propargyloxyphenylalanine

FIGURE 1 (a) An mRNA stop codon and a mutated tRNA (with a stop anticodon) are used to introduce unnatural amino acids into proteins. (b) Two unnatural amino acids, derivatives of phenylalanine (also shown, for comparison), that can be incorporated by this method.

(denoted IF in bacteria and eIF in eukaryotes) are critical to enhancing the rate and fidelity of all steps in the process, fine-tuning the underlying rRNA-based activities and the interactions among mRNA, tRNA, and rRNA. Other steps in protein synthesis are generally the same in bacteria and eukaryotes, but some aspects of initiation differ across these two groups.

Base Pairing Recruits the Small Ribosomal Subunit to Bacterial mRNAs

Translation in all organisms begins with binding of the ribosome's small subunit to an mRNA. In bacteria, the initiating 5'-AUG is guided to its correct position on the ribosome by the **Shine-Dalgarno sequence** (named

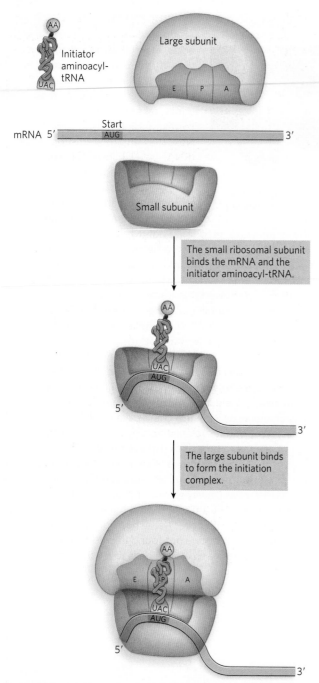

FIGURE 18-15 An overview of the events in translation initiation.

for Australian researchers John Shine and Lynn Dalgarno, who identified it), also called the **ribosome-binding site (RBS)**. This consensus sequence is an initiation signal of 4 to 9 purine residues, situated 8 to 13 nucleotides on the 5′ side of the start codon (**Figure 18-16a**). The sequence base-pairs with a complementary pyrimidine-rich sequence near the 3′ end of the 16S rRNA of the 30S ribosomal subunit. This mRNA-rRNA interaction positions the initiating 5′-AUG sequence of the mRNA in the precise location on the 30S subunit where it is required for translation initiation. Higher degrees of

base-pairing complementarity and optimal spacing between the Shine-Dalgarno sequence and the initiation codon increase the efficiency of translation of a particular mRNA. The specific 5′-AUG where the initiator aminoacyl-tRNA—a special type of methionyl-tRNA—is to be bound is distinguished from other methionine AUG codons by its proximity to the Shine-Dalgarno sequence.

The Shine-Dalgarno sequence can be used to initiate synthesis of more than one protein, if they are encoded in a single transcript called a polycistronic mRNA. Most bacterial mRNAs are polycistronic, whereas eukaryotic mRNAs are almost always monocistronic (they encode a single protein). In some bacterial polycistronic mRNAs, the open reading frames overlap (**Figure 18-16b**, bottom). Despite lacking a Shine-Dalgarno sequence for each internal start site, the internal open reading frames can be translated efficiently because of overlapping start and stop codons, typically 5′-AUGA. Ribosomes terminating translation of the upstream message can initiate the downstream message simply by shifting their reading frame.

Eukaryotic mRNAs Recruit the Small Ribosomal Subunit Indirectly

The 5′ cap and poly(A) tail modifications on eukaryotic mRNAs serve three purposes: they protect the ends from degradation, facilitate nuclear export of the mRNA, and promote translation by binding initiation factors that form a link between the mRNA and the ribosome. The 5′ terminus of the mRNA contains a modified G residue, the 5′ cap, that can be bound by the cap-binding protein, eIF4E, which in turn binds other proteins that recruit the small ribosomal subunit to the mRNA. Once associated with the 5′ end of the mRNA, the small subunit locates the 5′-AUG start codon by scanning the RNA in the 5′→3′ direction.

In addition to the 5′ cap, the presence of a purine nucleotide three residues before the start codon and a G residue immediately following the start codon is thought to enhance translation through contact with the initiator tRNA. This **Kozak sequence** was discovered by Marilyn Kozak during analysis of the sequence features of eukaryotic mRNAs that increased translation efficiency (**Table 18-3** on p. 634). At the 3′ terminus of eukaryotic mRNAs, the poly(A) tail stimulates translation efficiency by fostering reinitiation after completion of a polypeptide chain.

A Specific Amino Acid Initiates Protein Synthesis

Protein synthesis begins at the N-terminal end and proceeds by the stepwise addition of amino acids to the C-terminal end of the growing polypeptide, as found by

TABLE 18-2

Essential Components of the Main Stages of Protein Synthesis in *E. coli* and Eukaryotes

Stage	E. coli	Eukaryotes
1. Activation of amino acids	20 amino acids 20 aminoacyl-tRNA synthetases 32 or more tRNAs ATP Mg^{2+}	20 amino acids 20 aminoacyl-tRNA synthetases 32 or more tRNAs ATP Mg^{2+}
2. Initiation	mRNA *N*-Formylmethionyl-tRNA^{fmet} Start codon in mRNA (AUG) 30S ribosomal subunit 50S ribosomal subunit Initiation factors (IF-1, IF-2, IF-3) GTP Mg^{2+}	mRNA Methionyl-tRNA$_i^{met}$ Start codon in mRNA (AUG) 40S ribosomal subunit 60S ribosomal subunit Initiation factors (eIF1, eIF1A, eIF2, eIF3, eIF4B, eIF4F (complex of eIF4E, eIF4A, eIF4G), eIF5, eIF5B) GTP Mg^{2+}
3. Elongation	Functional 70S ribosome (initiation complex) Aminoacyl-tRNAs specified by codons Elongation factors (EF-Tu, ET-Ts, EF-G) GTP Mg^{2+}	Functional 80S ribosome (initiation complex) Aminoacyl-tRNAs specified by codons Elongation factors (eEF1α, eEF1βγ, eEF2) GTP Mg^{2+}
4. Termination and release	Stop codon in mRNA Release factors (RF-1, RF-2, RF-3) EF-G IF-3	Stop codon in mRNA Release factors (eRF1, eRF3)
5. Folding and posttranslational processing	Specific enzymes, cofactors, and other components for removal of initiating residues and signal sequences, additional proteolytic processing, modification of terminal residues, and attachment of phosphate, acetyl, methyl, carboxyl, carbohydrate, or prosthetic groups	Specific enzymes, cofactors, and other components for removal of initiating residues and signal sequences, additional proteolytic processing, modification of terminal residues, and attachment of phosphate, acetyl, methyl, carboxyl, carbohydrate, or prosthetic groups

(a)

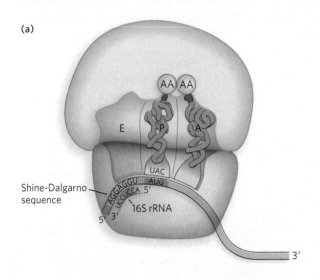

FIGURE 18-16 The Shine-Dalgarno sequence. (a) The Shine-Dalgarno sequence positions the mRNA on the bacterial ribosome by binding to the 16S rRNA. (b) In bacteria, some polycistronic genes have overlapping start and stop codons (bottom), allowing the ribosome to initiate synthesis in the absence of a Shine-Dalgarno sequence.

(b) **Non-overlapping genes**

Shine-Dalgarno sequence   Protein-coding region 1   Protein-coding region 2

UUUGAGGAGGUACGUACUACAUGGCUGA / AUCGCUAACGGGAGGAGGUUGGGAAAUGAAGCC / AGCAAUAGCUGACGUACA

Overlapping genes

Shine-Dalgarno sequence   Protein-coding region 1   Protein-coding region 2

UUUGAGGAGGUACGUACUACAUGGCUGA / GGGAAAUGAAGCCUUUUGCGCUGGGUGUUUCGUU / CUGGCUAGCUGACGUACA

TABLE 18-3
The Kozak Sequence

Organism(s)	Consensus Sequence*
Vertebrates	GCCRCCATGG
Terrestrial plants	AACAATGGC
Drosophila melanogaster (fruit fly)	CAAAATG
Saccharomyces cerevisiae (baker's yeast)	AAAAAAATGTCT
Dictyostelium discoideum (slime mold)	AAAAAAATGRNA
Plasmodium spp. (malarial protozoa)	TAAAAAATGAAN

*In this table, R = either purine; N = any base.

Howard Dintzis in 1961 (see the How We Know section at the end of Chapter 17). The 5'-AUG initiation codon specifies an N-terminal Met residue. Although methionine has only one codon, AUG, all organisms have two tRNAs for methionine. One is used exclusively when 5'-AUG is the initiation codon for protein synthesis; the other is used to code for a Met residue in an internal position in a polypeptide. An initiation factor specifically binds to the initiator tRNA and delivers it to the ribosome, thereby enabling cells to separate translation initiation from elongation.

The distinction between an initiating 5'-AUG and an internal one is straightforward. In bacteria, the two types of tRNA specific for methionine are designated tRNA^{Met}

and tRNA^{fMet}. The amino acid incorporated in response to the 5'-AUG initiation codon is N-formylmethionine (fMet). It arrives at the ribosome as **N-formylmethionyl-tRNA^{fMet}** (fMet-tRNA^{fMet}), which is formed in two successive reactions. First, methionine is attached to tRNA^{fMet} by the Met-tRNA synthetase (which in *E. coli* aminoacylates both tRNA^{fMet} and tRNA^{Met}), in a reaction of the type shown in Figure 18-11. Second, a transformylase enzyme transfers a formyl group from N^{10}-formyltetrahydrofolate to the amino group of the methionyl part of the tRNA (**Figure 18-17**). The transformylase is more selective than the synthetase: it is specific for methionyl moieties attached to tRNA^{fMet}, presumably recognizing some unique structural feature of that tRNA. Addition of the N-formyl group to methionine prevents fMet from entering interior positions in a polypeptide (only Met-tRNA^{Met} inserts methionine in interior positions) while allowing fMet-tRNA^{fMet} to bind a specific ribosomal initiation site that accepts neither Met-tRNA^{Met} nor any other aminoacyl-tRNA.

In eukaryotic cells, all polypeptides synthesized by cytosolic ribosomes begin with a Met residue (rather than fMet); however, as in bacteria, the cell uses a special initiator tRNA, tRNA$_i^{Met}$, distinct from the tRNA^{Met} used at AUG codons in interior positions in the mRNA. The Met-tRNA$_i^{Met}$ is distinct because it has a specific sequence in the anticodon arm that is recognized by an initiation factor protein. Polypeptides synthesized by mitochondrial and chloroplast ribosomes, however, begin with N-formylmethionine. This strongly supports the view that mitochondria and chloroplasts originated from bacterial ancestors symbiotically incorporated into precursor eukaryotic cells at an early stage of evolution.

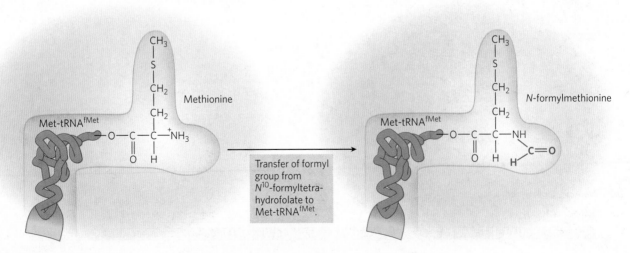

FIGURE 18-17 Formation of fMet-tRNA^{fMet}. The tRNAf^{Met} is first charged with methionine (not shown), then the Met is converted to fMet by the enzyme methionyl-tRNA formyltransferase (transformylase).

In bacteria, the enzyme deformylase typically removes the formyl group from the N-terminal Met residue during or shortly after production of the polypeptide. In both bacteria and eukaryotes, enzymes called aminopeptidases frequently remove the entire N-terminal methionine, and sometimes one or two additional amino acids, from newly synthesized polypeptides. Thus, many mature proteins do not have a Met residue at their N-terminal end.

Initiation in Bacterial Cells Requires Three Initiation Factors

As discussed above, ribosomes have three sites that bind aminoacyl-tRNAs: the A, P, and E sites. In addition to the 30S subunit, an mRNA, and an initiating fMet-tRNA^{fMet}, the initiation of polypeptide synthesis in bacteria requires a set of three initiation factor proteins known as **IF-1**, **IF-2**, and **IF-3**. Each plays a specific role in assembling the small ribosomal subunit (with the mRNA and initiator tRNA in place) and the large ribosomal subunit in a process controlled by GTP hydrolysis.

Formation of an active 70S ribosome takes place in three steps, as shown in **Figure 18-18**. In step 1, the 30S subunit binds two initiation factors, IF-1 and IF-3. IF-3 prevents premature combination of the 30S and 50S subunits. IF-1 binds at the A site and blocks tRNA binding during initiation. The mRNA then binds to the 30S subunit through base pairing of the Shine-Dalgarno sequence with 16S rRNA. This short mRNA-rRNA helix is bound in a cleft in the 30S subunit, which precisely positions the mRNA adjacent to the P site, thus accounting for the accuracy of start codon selection. The initiating 5′-AUG is now positioned at the P site. This is the only site to which fMet-tRNA^{fMet} can bind, and fMet-tRNA^{fMet} is the only aminoacyl-tRNA that can bind first to the P site. During the subsequent elongation stage, all other incoming aminoacyl-tRNAs (including the so-called elongator Met-tRNA^{Met} that binds interior AUG codons) bind first to the A site and only subsequently transfer to the P and E sites.

In step 2, the complex consisting of the 30S ribosomal subunit, IF-1, IF-3, and mRNA is joined by both GTP-bound IF-2 and the initiating fMet-tRNA^{fMet}. The anticodon of this tRNA now pairs with the mRNA's start codon in the P site. X-ray crystallography has revealed contacts between rRNA bases and three G≡C base pairs in the anticodon arm of initiator, but not elongator, Met-tRNAs. This observation suggests a mechanism by which initiator tRNA is favored in the P site during the initiation process. In step 3, a conformational change in the 30S subunit triggers the release of IF-3, enabling association with the 50S subunit; simultaneously, the GTP

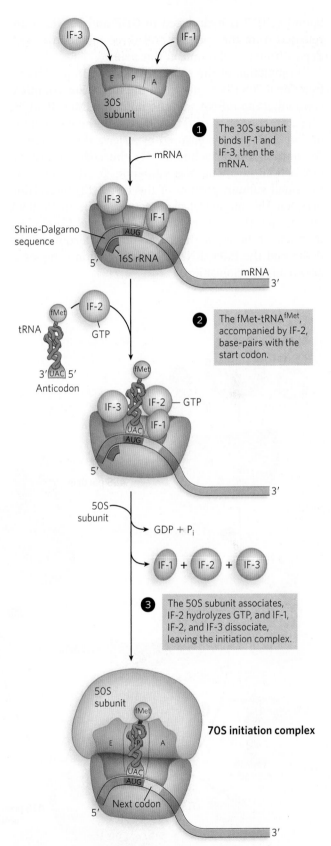

FIGURE 18-18 Translation initiation in bacteria. Initiation occurs in three steps. Initiation factors IF-1 and IF-3 prevent premature binding of the elongation aminoacyl-tRNAs (i.e., those used at the elongation stage) and the 50S subunit, respectively. IF-2 accompanies fMet-tRNA^{fMet} to the initiation site.

bound to IF-2 is hydrolyzed to GDP and P_i, which are released from the complex. All three initiation factors depart from the ribosome at this point.

Completion of the steps in Figure 18-18 produces a functional 70S ribosome called the **initiation complex**, containing the mRNA and the initiating fMet-tRNA^{fMet}. The correct location of fMet-tRNA^{fMet} in the P site in the initiation complex is ensured by at least three points of recognition and attachment: (1) the codon-anticodon interaction involving the initiation 5'-AUG fixed in the small subunit portion of the P site, (2) interaction between the Shine-Dalgarno sequence in the mRNA and the 16S rRNA of the small subunit, and (3) binding interactions between the large-subunit portion of the P site and the fMet-tRNA^{fMet}. The initiation complex is now ready for elongation.

Initiation in Eukaryotic Cells Requires Additional Initiation Factors

Translation is generally similar in eukaryotic and bacterial cells; most of the significant differences are in the mechanism of initiation. Eukaryotic initiation requires, besides separate small and large ribosomal subunits, at least 12 initiation factors and the binding and hydrolysis of ATP and GTP, as illustrated in **Figure 18-19**. In step 1, before translation begins, the ribosomal subunits are separated by initiation factors eIF3 and eIF1A; eIF1A is the functional homolog of IF-1 in bacteria. These eukaryotic factors prevent premature subunit joining and block initiator tRNA binding to the ribosomal A site, respectively. A third initiation factor, eIF1, binds to the E site. Meanwhile (step 2), the

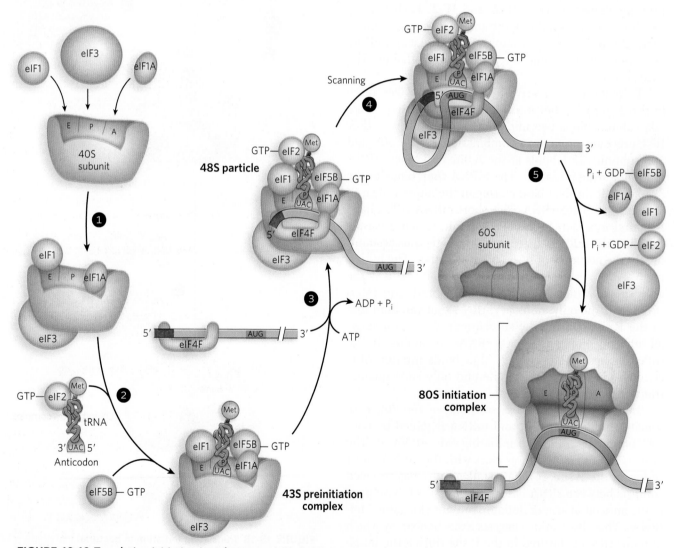

FIGURE 18-19 Translation initiation in eukaryotes. Initiation is more complex in eukaryotes. Eukaryotic initiation factors (eIFs) promote the assembly of the 43S preinitiation complex with an mRNA and subsequent scanning of the mRNA to identify the start codon. Note that circularization of the mRNA in step 3 is omitted here for clarity. It is shown in Figure 18-20.

GTP-binding factor eIF2—containing three subunits, eIF2α, eIF2β, and eIF2γ—associates with GTP and a charged initiator tRNA (Met-tRNA$_i^{Met}$) to form an eIF2–GTP–Met-tRNA$_i^{Met}$ ternary complex. An A=U base pair near the amino acid–binding site of the amino acid arm of Met-tRNA$_i^{Met}$, but not present in elongator Met-tRNA^{Met}, is critical for binding to eIF2-GTP. Two other proteins, eIF5 (not shown in Figure 18-19) and eIF5B, which are involved in later steps of ribosomal assembly, also associate at this point. Three factors, eIF1, eIF1A and eIF3, mediate interaction between the ternary complex and the 40S subunit to form a 43S preinitiation complex.

Binding of the 43S preinitiation complex to an mRNA (step 3) is mediated by a complex called eIF4F. It contains eIF4E, a factor that binds the 5′ cap; eIF4A, an ATPase and RNA helicase; and eIF4G, which binds both eIF4E and eIF3 to provide a link between the 43S complex and the mRNA. The eIF4G also binds to poly(A) binding protein (PABP), which is associated with the 3′ poly(A) tail of the mRNA, and this eIF4G-PABP association brings the 5′ and 3′ ends of the mRNA together (**Figure 18-20**). Circularization of the eukaryotic mRNA by the eIF4G-PABP interaction also facilitates the translational regulation of gene expression (as described in Chapter 21). Following association of the eIF4F complex, another initiation factor, eIF4B, binds (not shown in Figure 18-19); its function is less clear.

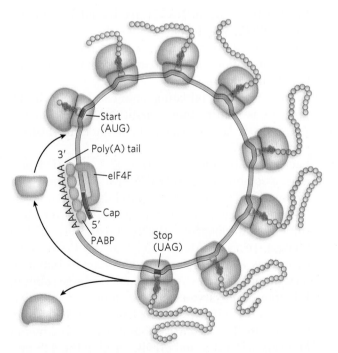

FIGURE 18-20 Circularization of eukaryotic mRNA. eIF4F (purple) consists of eIF4A, eIF4E, and eIF4G. In step 3 of the initiation process (see Figure 18-19), eIF4G binds to the 5′ cap and the poly(A) binding protein (PABP), effectively circularizing the mRNA.

This larger complex, containing the 43S components, eIF4F, and mRNA, is a stable 48S particle. Within the 48S particle, the 43S complex is capable of **scanning**, sliding along the mRNA in search of a start codon. Once assembled on the 5′ end of the mRNA (step 4 in Figure 18-19), the 43S complex scans along the mRNA in the 5′→3′ direction to the first AUG, which is almost always recognized as the start site for translation initiation. The eIF4F complex is probably involved in this scanning process, perhaps using the RNA helicase activity of eIF4A to eliminate secondary structure in the 5′ untranslated portion of the mRNA. Scanning is also facilitated by eIF4B, but the details of this process are unknown. Base pairing between the start codon and the anticodon of the initiator tRNA is primarily responsible for identification of the initiation site. Factor eIF1 is also critical to proper AUG recognition by preventing stable ribosomal association with non-AUG codons.

When the 43S complex has located the AUG start codon in a bound mRNA, the complex joins the 60S subunit to form an active 80S ribosome (step 5 in Figure 18-19). Because eIF1, eIF1A, and eIF2 occlude parts of the 40S subunit surface that must interface with the 60S subunit, they must be displaced before subunit joining. This process requires eIF5 and eIF5B. The GTPase-activating eIF5 is specific for eIF2, stimulating hydrolysis of the eIF2-bound GTP and thus reducing eIF2's affinity for the initiator tRNA. Finally, eIF5B, a ribosome-dependent GTPase homologous to bacterial IF-2, hydrolyzes GTP and triggers dissociation of eIF2-GDP and other initiation factors from the 40S subunit, with concomitant association of the 60S subunit to form the functional 80S ribosome.

The initiation complex is now complete. The efficiency of translation is affected by many properties of the mRNA and proteins in this complex, including the length of the 3′ poly(A) tail (in most cases, longer is better).

Some mRNAs Use 5′ End–Independent Mechanisms of Initiation

Some viral and eukaryotic mRNAs lack a 5′ cap but still rely on the eukaryotic translation machinery to produce their proteins. They accomplish this with an RNA segment called an **internal ribosome entry site (IRES)**, located on the 5′ side of the start codon; it recruits the 40S subunit through direct interaction with the subunit or with eIF4F (**Figure 18-21a**). Cap-independent binding can also involve other proteins, such as La protein, which binds a pyrimidine-rich segment in the 5′ region of the mRNA.

The first IRES was discovered in poliovirus, when researchers noticed that the viral mRNA is efficiently

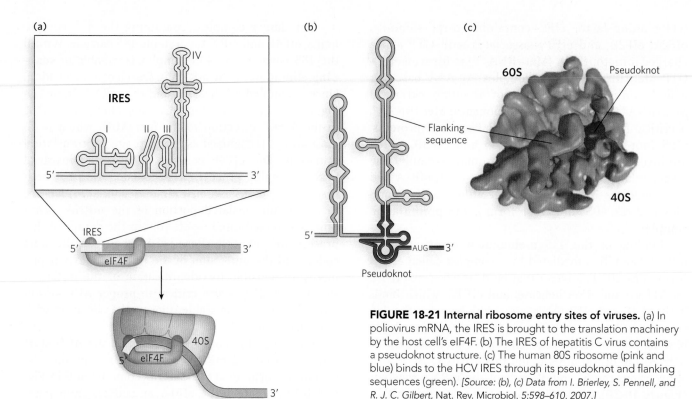

FIGURE 18-21 Internal ribosome entry sites of viruses. (a) In poliovirus mRNA, the IRES is brought to the translation machinery by the host cell's eIF4F. (b) The IRES of hepatitis C virus contains a pseudoknot structure. (c) The human 80S ribosome (pink and blue) binds to the HCV IRES through its pseudoknot and flanking sequences (green). [*Source: (b), (c) Data from I. Brierley, S. Pennell, and R. J. C. Gilbert, Nat. Rev. Microbiol. 5:598–610, 2007.*]

translated despite lacking a 5′ cap. On infection, poliovirus produces a protease that cleaves the host cell's eIF4G into two fragments. This renders eIF4G useless for the host cell's protein synthesis but does not compromise viral protein synthesis, because the poliovirus IRES requires just one fragment of eIF4G to initiate translation. In this way, the virus can simultaneously reduce levels of host protein expression and maintain its own protein synthesis in infected cells. Different viral IRES subtypes seem to form distinct three-dimensional structures that enable direct interactions with the host translation machinery. For example, hepatitis C virus (HCV) mRNA has a pseudoknot adjacent to two stem-loop structures (**Figure 18-21b**). The resulting three-dimensional structure binds to the host cell's 80S ribosome such that the start codon is positioned in the P site (**Figure 18-21c**).

Some cellular mRNAs, although they contain a 5′ cap, also contain IRES segments that enable translation under conditions that normally block initiation. Cellular mRNAs bearing an IRES can be translated even during viral infection and in other situations in which most protein synthesis is shut off, such as during starvation and programmed cell death. This is an important advantage of the IRES-mediated initiation mechanism. Internal initiation avoids the use of the 5′ cap and the initiation factors needed to recognize it. The IRES positions the mRNA start codon correctly on the 40S subunit to ensure accurate initiation during each round of translation.

SECTION 18.3 SUMMARY

- In bacteria, ribosomes are recruited to the mRNA by the Shine-Dalgarno sequence, 4 to 9 purine residues located 8 to 13 nucleotides on the 5′ side of the start codon.

- In eukaryotes, the mRNA 5′ cap is bound by initiation factor eIF4E, which then binds other proteins that recruit the ribosomal subunits. The ribosome identifies the initiation codon by scanning the mRNA in the 5′→3′ direction.

- In bacteria, the initiating aminoacyl-tRNA in all proteins is *N*-formylmethionyl-tRNA^{fMet}. In eukaryotic cells, it is a special form of methionyl-tRNA, Met-tRNA$_i^{Met}$.

- Bacterial initiation factors IF-1, IF-2, and IF-3 promote assembly of the mRNA, fMet-tRNA^{fMet}, and both ribosomal subunits to form the initiation complex. IF-1 and IF-3 prevent premature binding of the large subunit and elongator tRNAs, respectively. IF-2, a GTPase, recruits fMet-tRNA^{fMet} to the P site. GTP hydrolysis allows the release of all three initiation factors and promotes association of the large subunit.

- Initiation in eukaryotes involves a host of initiation factors: eIF1, eIF1A, and eIF3 promote association of the small ribosomal subunit with eIF2-bound Met-tRNA$_i^{Met}$, eIF5, and eIF5B, forming the 43S preinitiation complex. The mRNA 5′ cap is bound by the eIF4F complex, which also binds to the small ribosomal subunit. Once this 48S particle is formed,

activates the terminal carboxyl group for nucleophilic attack by the incoming amino acid to form a new peptide bond. As the existing ester linkage between the polypeptide and tRNA is broken during peptide bond formation, the linkage between the polypeptide and the information in the mRNA persists, because each newly added amino acid is still attached to its tRNA.

The elongation cycle in eukaryotes is quite similar to that in bacteria. Three eukaryotic elongation factors—**eEF1α**, **eEF1β**, and **eEF2**—have functions analogous to those of the bacterial elongation factors EF-Tu, EF-Ts, and EF-G, respectively.

An mRNA Stop Codon Signals Completion of a Polypeptide Chain

Elongation continues until the ribosome adds the last amino acid coded by the mRNA. Termination is signaled by the presence of a stop codon in the mRNA. Mutations in a tRNA anticodon that allow an amino acid to be inserted at a termination codon are generally deleterious to the cell.

In bacteria, once a stop codon occupies the ribosomal A site, three **termination factors**, or **release factors**—the proteins **RF-1**, **RF-2**, and **RF-3**—contribute to (1) hydrolysis of the terminal peptidyl-tRNA bond; (2) release of the free polypeptide and the last tRNA, now uncharged, from the P site; and (3) dissociation of the 70S ribosome into its 30S and 50S subunits, ready to start a new cycle (**Figure 18-27**). RF-1 and RF-2 are related factors and are referred to as class I release factors; RF-3 acts differently and is referred to as a class II release factor. RF-1 and RF-2 recognize termination codons and bind the ribosome in much the same way as tRNAs. RF-1 recognizes stop codons UAG and UAA, and RF-2 recognizes UGA and UAA. Either RF-1 or RF-2 (depending on which codon is present) binds at the stop codon and induces peptidyl transferase to transfer the growing polypeptide to a water molecule rather than to another amino acid.

RF-1 and RF-2 have domains that mimic the structure of a tRNA (**Figure 18-28**). RF-3, a GTPase, catalyzes the dissociation of RF-1 and RF-2 from the ribosome following release of the polypeptide chain. Unlike the other GTP-binding factors that regulate translation, RF-3 binds with higher affinity to GDP than to GTP.

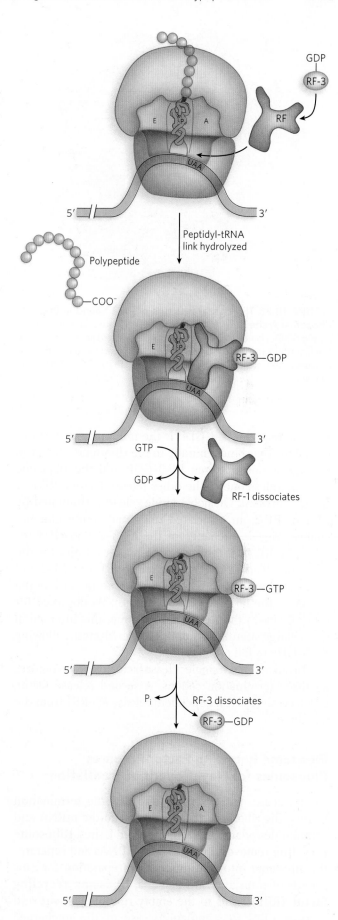

FIGURE 18-27 Termination of translation in bacteria. When the ribosome comes to a stop codon, RF-1 or RF-2 (designated here by RF) binds to the A site and induces release of the polypeptide chain. RF-3–GDP then binds to the ribosome and exchanges GTP for GDP, displacing the RF-1 or RF-2. The RF-3–GTP is tightly bound to the ribosome, but GTP hydrolysis weakens its affinity and RF-3 is released.

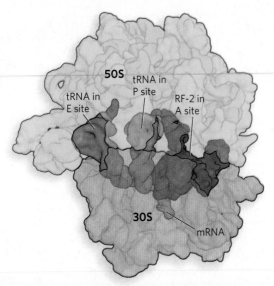

FIGURE 18-28 The crystal structure of RF-2 bound in the ribosomal A site. RF-1 (not shown) and RF-2 are proteins that mimic tRNAs, binding to the A site and mRNA, displacing the last tRNA from the A site into the P site, and releasing the polypeptide from the ribosome. [Sources: PDB ID 3FIE and PDB ID 3FIF.]

For this reason, RF-3–GDP is the predominant form of the factor that binds initially to the ribosome. However, the association between RF-3–GDP and the ribosome is weak unless one of the other release factors, RF-1 or RF-2, is present. On polypeptide release, stimulated by RF-1 or RF-2, an associated ribosomal conformational change triggers exchange of the RF-3–bound GDP for GTP. The RF-3–GTP has a much higher affinity for the ribosome, leading to displacement of RF-1 or RF-2 and creating contact between RF-3 and the factor-binding center of the large ribosomal subunit. As observed for other GTPases that bind the ribosome, this interaction on the large subunit stimulates GTP hydrolysis, allowing RF-3–GDP to fall off.

In eukaryotes, a single release factor, eRF1, recognizes all three termination codons. A second release factor, eRF3, catalyzes GTP-dependent release of eRF1 from the ribosome.

Ribosome Recycling Factor Prepares Ribosomes for New Rounds of Translation

After release of the polypeptide and the termination factors, the ribosome remains bound to the mRNA and contains deacylated tRNA in the P and E sites. **Ribosome recycling** removes the mRNA and tRNAs and separates the ribosome into its subunits in preparation for new rounds of translation. In bacteria, **ribosome recycling factor** (RRF) binds to the empty ribosomal A site and recruits EF-G to stimulate release of the uncharged

tRNAs in the P and E sites, in a process mimicking EF-G–stimulated polypeptide elongation (**Figure 18-29**).

In a continuing story of molecular mimicry, RRF, like EF-G and EF-Tu–tRNA, fits into the ribosome's tRNA-binding sites. GTP hydrolysis by EF-G is thought to result in translocation and displacement of the tRNAs, as occurs during translation. However, when RRF occupies the A site and translocates to the P site, the ribosome releases EF-G and RRF. The initiation factor IF-3 then binds the small ribosomal subunit and disaggregates the ribosome, thereby triggering release of the mRNA and preparing the small subunit for a new round of translation.

In eukaryotes, protein release factors (RFs) recognize the stop codon and stimulate hydrolysis of the ester bond linking the polypeptide chain with the peptidyl site tRNA, a reaction catalyzed by the ribosome's peptidyl transferase center. There are two classes of RFs: class 1 RFs bind the stop codon directly, whereas class 2 RFs are not codon specific but serve to enhance class 1 RF activity and make the process GTP-dependent.

Fast and Accurate Protein Synthesis Requires Energy

The synthesis of a protein true to the information specified in the cell's mRNA requires energy. Forming each aminoacyl-tRNA requires two high-energy phosphate groups. An additional ATP is consumed each time an incorrectly activated amino acid is hydrolyzed by the deacylation activity of an aminoacyl-tRNA synthetase, as part of its proofreading activity. One GTP is cleaved to GDP and P_i during the first elongation step, and another during the translocation step. Thus, on average, the energy derived from the hydrolysis of more than four NTPs to NDPs is required for the formation of each peptide bond.

This represents a large thermodynamic drive in the direction of synthesis: at least 4×30.5 kJ/mol = 122 kJ/mol of phosphodiester bond energy to generate a peptide bond, which has a standard free energy of formation of about 21 kJ/mol. The net free-energy change during peptide bond synthesis is thus −101 kJ/mol. Why would the cell need to expend so much energy on protein synthesis?

Proteins are information-containing polymers. The biochemical goal in the peptidyl transferase reaction is not simply the formation of a peptide bond but the formation of a peptide bond between two *specified* amino acids. Each of the high-energy phosphate compounds expended in this process plays a critical role in maintaining proper alignment between each new codon in the mRNA and its associated amino acid at the growing end of the polypeptide. This energy input permits very high fidelity in the biological translation of the genetic message of mRNA into the amino acid sequence of proteins.

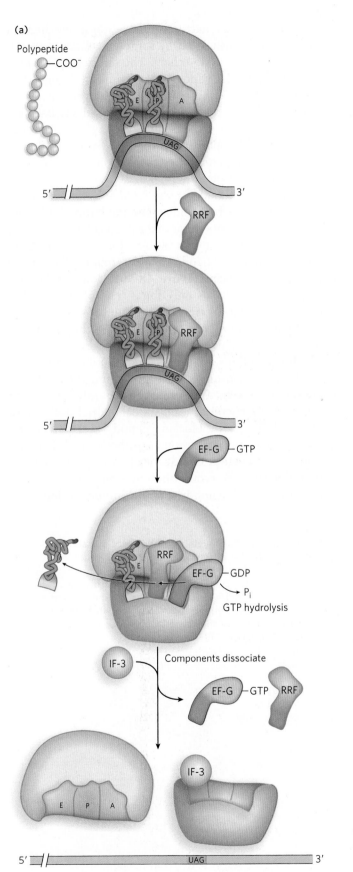

FIGURE 18-29 Ribosome recycling. (a) In bacteria, ribosome recycling factor (RRF) and EF-G separate the ribosome from the mRNA and tRNAs. (b) RRF is a protein with a three-dimensional shape very similar to that of a tRNA (shown here for comparison), allowing RRF to bind to the tRNA sites on the ribosome. *[Source: (b) PDB ID 1DD5.]*

Antibiotics and Toxins Frequently Target Protein Synthesis

Protein synthesis is the primary target of many naturally occurring antibiotics and toxins. Antibiotics are produced by bacteria or other microorganisms to inhibit protein synthesis in other bacteria—that is, these biochemical weapons are synthesized by some microorganisms and are extremely toxic to others. The differences between bacterial and eukaryotic protein synthesis, though often subtle, are sufficient that most (though by no means all) of the antibiotics discussed here are relatively harmless to eukaryotic cells. Because nearly every step in protein synthesis can be specifically inhibited by one antibiotic/toxin or another, antibiotics have become valuable tools in the study of protein synthesis.

Puromycin, made by the mold *Streptomyces alboniger*, is one of the best-understood inhibitory antibiotics. Its structure is very similar to the 3′ end of an aminoacyl-tRNA, so puromycin can bind to the ribosomal A site and participate in peptide bond formation, producing peptidyl-puromycin (**Figure 18-30**). However, because puromycin resembles only the 3′ end of the tRNA, it does not engage in translocation and dissociates from the ribosome shortly after it is linked to the C-terminus of the peptide. This prematurely stops protein synthesis.

Tetracyclines inhibit bacterial protein synthesis by blocking the ribosomal A site, preventing the binding of aminoacyl-tRNAs. **Chloramphenicol** inhibits protein synthesis by bacterial (and mitochondrial and chloroplast) ribosomes by blocking peptidyl transfer; it does not affect cytosolic protein synthesis in eukaryotes. Conversely, **cycloheximide** blocks the peptidyl transferase of 80S eukaryotic ribosomes but not that of 70S bacterial (or mitochondrial and chloroplast) ribosomes.

FIGURE 18-30 Puromycin as inhibitor of translation.
(a) Puromycin inhibits translation by binding to the A site and mimicking the 3′ end of an aminoacyl-tRNA. The puromycin participates in the peptidyl transfer reaction, but because it is not bound to a tRNA, it is not anchored to the ribosome.
(b) Crystal structure showing puromycin bound to the peptidyl transferase center of the bacterial 50S ribosomal subunit.
[Source: (b) PDB ID 1Q7Y.]

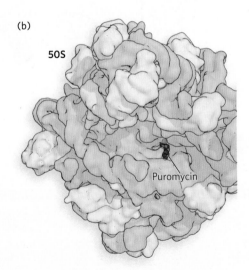

Streptomycin, a trisaccharide, causes misreading of the genetic code (in bacteria) at relatively low concentrations and inhibits initiation at higher concentrations.

Several other inhibitors of protein synthesis are notable because of their toxicity to humans and other mammals. **Diphtheria toxin** is an enzyme that catalyzes the ADP-ribosylation of a diphthamide (a modified histidine) residue of the eukaryotic elongation factor eEF2, thereby inactivating it. The toxin is secreted by the bacterium *Corynebacterium diphtheriae*. Infected individuals experience fever, chills, neck swelling, and a fast heart rate. Although common historically, diphtheria has largely been eradicated in developed countries through widespread vaccination. Among the plant-produced toxins, **ricin**, an extremely toxic protein of the castor bean, inactivates the 60S subunit of eukaryotic ribosomes by depurinating a specific adenosine in 23S rRNA (**Highlight 18-3**). **Table 18-4** lists some antibiotics and toxins that target translation, their functional consequences, and some clinical and other uses.

SECTION 18.4 SUMMARY

- In the first step of elongation in bacteria, the incoming aminoacyl-tRNA binds to the ribosome with the help of EF-Tu. This step requires GTP hydrolysis and allows time for proofreading of the codon-anticodon interaction. In the second step, a peptide bond is formed between the two amino acids bound by their tRNAs to the ribosomal A and P sites.

- Peptide bond formation involves nucleophilic attack of the α-amino group of the A-site aminoacyl-tRNA on the carbonyl carbon of the ester bond linking the growing peptide chain to the P-site tRNA. This reaction is driven by a favorable change in entropy.

- The third step of elongation is translocation of the peptidyl-tRNA from the A site to the P site. This is accomplished with the help of EF-G, a GTPase that is a structural analog of the EF-Tu–aminoacyl-tRNA complex.

- The three steps of elongation are repeated for each codon in the mRNA; each cycle consumes two GTP molecules.

- After many elongation cycles, release factors recognize the stop codon and terminate polypeptide synthesis by releasing the polypeptide, displacing the tRNAs, and separating the ribosomal subunits. In bacteria, ribosome recycling factor stimulates release of the tRNAs in the P and E sites, leading to release of the mRNA.

- At least four high-energy phosphate bonds (from ATP and GTP) must be broken to generate each peptide bond, an energy investment necessary for guaranteeing the fidelity of translation.

- Many well-studied and clinically important antibiotics and toxins inhibit some aspect of protein synthesis.

HIGHLIGHT 18-3   MEDICINE

Toxins That Target the Ribosome

The toxic effects of two small proteins called ricin and abrin, derived from castor beans and jequirity peas, respectively, have been known and studied since ancient times. After a period of extensive study of these toxins at the end of the nineteenth century, the scientific community largely lost interest—until experiments published in the late 1960s and early 1970s showed that ricin and abrin inhibited protein synthesis in rats and cultured cells. The toxins did not interfere with the structure of polysomes, so researchers concluded that they acted on a component required for elongation of the polypeptide.

Studies showed that ricin and abrin required some time to exert their effects, suggesting that they might have enzymatic activity. Both toxins consist of two distinct, disulfide-linked polypeptide chains. The finding that reducing agents enhanced the toxins' ability to inhibit protein synthesis in vitro suggested that the activity lies in one of the individual chains; in both toxins, this was found to be the smaller chain, termed the A chain. After cell fractionation to isolate ribosomes and smaller protein factors, experiments showed that the A chains act on the ribosomes. Further studies revealed that the target is the 60S ribosomal subunit.

In elegant studies in the 1980s, Ira Wool and Y. Endo found that the toxin A chain is a glycosidase (an enzyme that cleaves sugars at the glycosidic linkage) that removes the adenine from an A residue in an exposed loop of 28S rRNA (residue A4,324 in rat 28S RNA). Because this loop is involved in the binding of an elongation factor, the modified ribosomes are unable to support protein synthesis.

Ricin is legendary for its use as a murder weapon. In perhaps the most famous case, known as the Umbrella Murder, journalist Georgi Markov died as a result of an incident on a London street, reportedly killed by the Bulgarian secret police. Markov felt a pain like a bee sting on the back of his thigh. Turning, he saw a man pick up an umbrella from the ground and quickly get into a taxi. When Markov arrived at his office at the BBC World Service, the pain was increasing, and he noticed a red pimple forming at the site of the sting. Later that day he developed a fever and was admitted to a hospital, where he died a few days later. The cause of death was poisoning from a ricin-filled pellet, presumably fired from an umbrella tip. In a more recent example, ricin was used as a key plot device in the popular television series *Breaking Bad*.

TABLE 18-4

Some Antibiotics and Toxins That Inhibit Translation: Sources, Actions, and Uses

Class	Example	Source(s)	Targets	Mode of Action	Uses
Inhibitors of initiation					
Edeine	Edeine	*Bacillus brevis*	Bacteria and eukaryotic cells	Binds small-subunit P and E sites, destabilizing interaction between mRNA and initiator aminoacyl-tRNA (prevents initiator aminoacyl-tRNA binding)	Formerly an agricultural pesticide; now banned in most countries
Kasugamycin	Kasugamycin	*Streptomyces kasugiensis*	Bacteria and fungi; low toxicity to higher eukaryotes	Binds small-subunit P and E sites, destabilizing interaction between mRNA and initiator aminoacyl-tRNA (prevents initiator aminoacyl-tRNA binding)	Agricultural fungicide to prevent rice blast infection
Orthosomycins	Avilamycin	*Streptomyces viridochromogenes*	Bacteria	Bind 50S subunit, preventing association with preinitiation complex	Avilamycin used in animal feeds to promote growth
Pactamycin	Pactamycin	*Streptomyces pactum*	Bacteria and eukaryotic cells	Inhibits initiation and elongation; mechanism unknown	Potential antitumor agent
Inhibitors of elongation					
Aminoglycosides	Neomycin	*Streptomyces* spp.	Bacteria and eukaryotic cells	Promote errors in decoding by stabilizing incorrect codon-anticodon pairings	Prevent various bacterial infections; used topically (e.g., Neosporin), orally, by injection, or during surgery
Amphenicols	Chloramphenicol	*Streptomyces venezuelae* and synthetic	Bacteria	Bind A site, inhibiting peptidyl transferase reaction	Broad-spectrum antibiotics to treat serious bacterial infections
Enacyloxins and kirromycins	Aurodox	*Streptomyces* spp.	Bacteria	Stall ternary complex on ribosome by preventing conformational change in EF-Tu	Veterinary medicines and food additives to promote animal growth
Lincosamides	Clindamycin	Synthetic and actinomycetes	Bacteria	Bind A site	Treatment of bacterial infections, protozoal diseases, malaria, toxic shock syndrome
Macrolides	Azithromycin	*Streptomyces* spp.	Bacteria	Bind exit tunnel, preventing exit of polypeptide from ribosome	Treatment of bacterial infections
Oxazolidinones	Cycloserine	Synthetic	Bacteria	Bind A site; exact mechanism unknown	Treatment of bacterial infections resistant to other antibiotics

TABLE 18-4

Some Antibiotics and Toxins That Inhibit Translation: Sources, Actions, and Uses (*continued*)

Class	Example	Source(s)	Targets	Mode of Action	Uses
Pleuromutilins	Retapamulin	*Clitopilus scyphoides* and synthetic	Bacteria	Inhibit initiation and peptidyl transferase reaction by binding to A and P sites at the same time	Treatment of skin infections; veterinary medicine
Sparsomycin	Sparsomycin	*Streptomyces sparsogenes*	Bacteria and eukaryotic cells	Inhibits tRNA binding to A site and enhances tRNA binding to P site	Antitumor drug
Streptogramins	Dalfopristin	*Streptomyces* spp.	Bacteria	Bind distinct, adjacent locations on peptidyl transferase center and act synergistically to block both A and P sites in initiation and elongation	Treatment of skin and other infections
Tetracyclines	Tetracycline	*Streptomyces* spp. and synthetic	Bacteria	Bind 30S subunit, preventing ternary complex binding to ribosome	Broad-spectrum antibiotics to treat pneumonia, acne, skin infections, genital and urinary tract infections, ulcers, Lyme disease, anthrax, and other disorders
Inhibitors of translocation					
Aminoglycosides	Spectinomycin	*Streptomyces spectabilis*	Bacteria	Inhibit translocation by stabilizing an intermediate	Treatment of gonorrhea; no longer available in United States
Fusidic acid (steroid)	Fusidic acid	*Fusidium coccineum*	Bacteria	Prevents EF-G dissociation by binding to EF-G–GTP in complex with ribosome	Treatment of bacterial infections
Ricin (protein)	Ricin	*Ricinus communis*	Bacteria and eukaryotic cells	Depurinates 23S rRNA, probably disrupting GTPase-stimulating activity	Developed, but never used, by United States for biological/chemical warfare
Thiopeptides	Micrococcin	*Streptomyces azureus*	Bacteria	Bind A site, inhibiting tRNA–IF-Tu binding and EF-G binding (inhibiting initiation and translocation)	Veterinary medicine
Tuberactinomycins	Capreomycin	*Streptomyces* spp.	Bacteria	Inhibit translocation by stabilizing an intermediate	Treatment of tuberculosis

Note: "Class" includes some individual antibiotics, categorized separately because of their particular mode of action. "Example" includes just one example of antibiotics in that class. "Uses" lists treatments for humans, unless noted as veterinary or agricultural. Although not noted, many of these antibiotics/toxins are also used in the laboratory as selective markers and/or to study protein synthesis.

18.5 TRANSLATION-COUPLED REMOVAL OF DEFECTIVE mRNA

Occasionally, mRNAs with premature stop codons (nonsense mutations, as described in Chapter 17) or with no stop codon at all (called non-stop mRNAs) arise from errors in DNA replication or transcription, or from mRNA degradation. Such defective mRNAs have the potential to produce nonfunctional or even toxic proteins, and they can also prevent efficient termination of translation and recycling of the protein synthesis machinery. For these reasons, elegant mechanisms have evolved to deal with such aberrant mRNAs during translation.

Ribosomes Stalled on Truncated mRNAs Are Rescued by tmRNA

Truncated, or non-stop, mRNAs occur when DNA transcription ends prematurely or when, due to a mutation, an mRNA lacks a stop codon. Truncated proteins produced by incomplete mRNAs, if inactive, could spell disaster for the cell, because they might contain the binding determinants that allow them to take the place of the active protein in a cellular activity. The ribosome takes care of this problem in a process that recognizes and removes the defective mRNA, as well as the defective protein.

When a ribosome reaches the 3′ end of a truncated mRNA, it stalls, unable to recruit either an appropriate tRNA or the proper release factors. In bacteria and some eukaryotic organelles, a fascinating quality-control pathway solves this problem with a 457 nucleotide RNA called **tmRNA**, also known as SsrA RNA or 10Sa RNA. A versatile, evolutionarily conserved bacterial molecule, tmRNA has the combined structural and functional properties of both a tRNA and an mRNA.

The 5′ terminus of tmRNA mimics the structure of tRNA^{Ala} and is charged with alanine by the Ala-tRNA synthetase. The alanine-charged tmRNA binds like a tRNA to the A site of the stalled ribosome, together with EF-Tu–GTP (**Figure 18-31**). The Ala-tmRNA donates its alanine to the nascent polypeptide chain in the P site, in a standard peptidyl transferase reaction. The tmRNA then takes the place of mRNA in the vacated mRNA channel of the ribosome stalled at the P site. The placement of another RNA molecule instead of an mRNA in this ribosome channel could not normally occur, because an intact mRNA would occupy the channel. But a prematurely terminated mRNA presents a vacant mRNA site for a tmRNA to occupy. Thus, tmRNA can act as a surrogate mRNA, replacing the truncated mRNA, with its self-encoded peptide reading frame

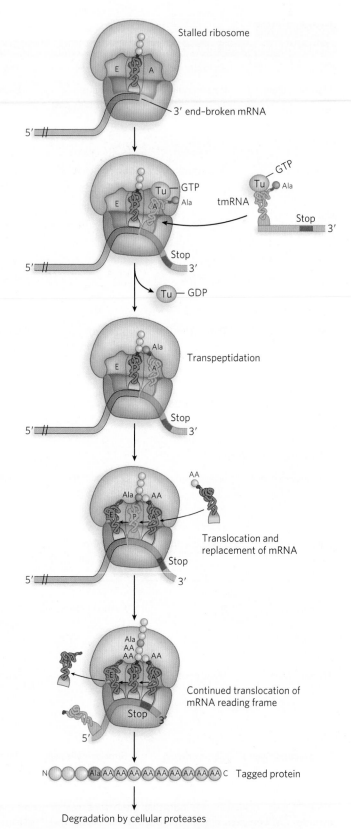

FIGURE 18-31 The rescue of stalled bacterial ribosomes by tmRNA. In bacteria, tmRNA rescues stalled ribosomes by mimicking both tRNA and mRNA, allowing release of the faulty mRNA while at the same time marking the truncated polypeptide for degradation. The encoded tag peptide (ANDENYALAA in *E. coli*) varies among different types of bacteria.

directing synthesis of a 9 amino acid sequence before encountering a stop codon. Thus, counting the alanine, the tmRNA incorporates 10 amino acids at the C-terminus of the truncated protein.

Translation terminates at the tmRNA stop codon, permitting disassembly and recycling of the ribosomal subunits. And in a remarkably strategic cellular maneuver, the 10 amino acid sequence encoded by tmRNA is a cellular signal that tags the protein for degradation—that is, the 10-residue degradation tag at the C-terminus is recognized by C-terminal–specific cellular proteases. The proteases denature and degrade tagged proteins in both the cytoplasm and the periplasmic space (between the inner plasma membrane and outer membrane of some bacteria). Furthermore, in addition to ribosome rescue and protein tagging, the tmRNA salvage system facilitates degradation of the defective mRNA by the enzyme RNase R.

Although not essential in *E. coli*, tmRNA activity is required for bacterial survival under adverse conditions and for virulence in some (perhaps all) pathogenic bacteria. Recent evidence suggests that in addition to its quality-control function, the tmRNA system might play a key role in regulating proteins for which cellular concentrations are particularly sensitive to the balance between proteolysis and the cell's translational efficiency.

Eukaryotes Have Other Mechanisms to Detect Defective mRNAs

Eukaryotes respond to non-stop mRNAs in other ways, not using the tmRNA mechanism described above. Because eukaryotic mRNAs contain a 3′ poly(A) tail, an mRNA without a stop codon is translated through the tail to produce a string of Lys residues at the C-terminus of the polypeptide (AAA encodes lysine). The ribosome stalled at the end of the non-stop mRNA binds to the protein Ski7 (related to eRF3), which initiates **non-stop mRNA decay** (**Figure 18-32**). Ski7 triggers dissociation of the ribosome and degradation of the non-stop mRNA by recruiting the exosome ribonuclease that cleaves in the 3′→5′ direction (see Figure 16-25). The defective polypeptide is rapidly degraded by a protease that recognizes the C-terminal poly(Lys) tag.

A process known as **nonsense-mediated mRNA decay** has evolved in eukaryotes to detect and destroy mRNAs that contain a premature stop, or nonsense, codon (**Figure 18-33**). This process works through the splicing machinery in the nucleus, before the mRNA is exported to the cytoplasm for translation. During pre-mRNA splicing, the site of each excised intron is marked by the presence of an exon junction complex (EJC), positioned on the 5′ side of the exon-exon boundary

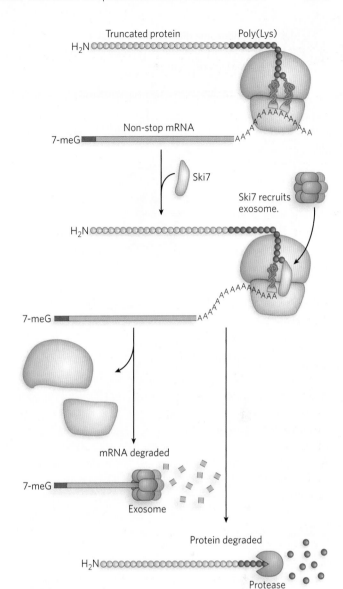

FIGURE 18-32 The rescue of stalled eukaryotic ribosomes by non-stop mRNA decay. Ribosomes continue to translate through the poly(A) tail of a non-stop mRNA, resulting in a poly(Lys) protein tail. Ski7 binds to the ribosome and recruits an exosome to degrade the non-stop mRNA and a protease to degrade the protein.

(see Chapter 16). During the first round of translation, the ribosome removes the EJCs as it moves along the mRNA from the start to the end of the coding region. If a premature stop codon is encountered, however, the ribosome is released early, before all the EJCs have been removed. In this event, Upf1 and Upf2 proteins are recruited by the leftover EJC(s), which in turn recruit an enzyme that cleaves the 5′ cap from the mRNA. Without the cap to protect it, the end of the RNA is highly susceptible to degradation by an exoribonuclease that hydrolyzes RNA from the free 5′ end.

(a) **Ribosome displacement of EJCs**

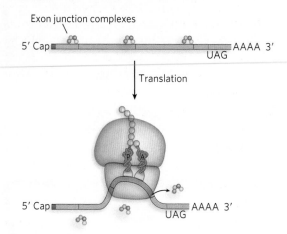

(b) **Nonsense-mediated mRNA decay at a premature stop codon**

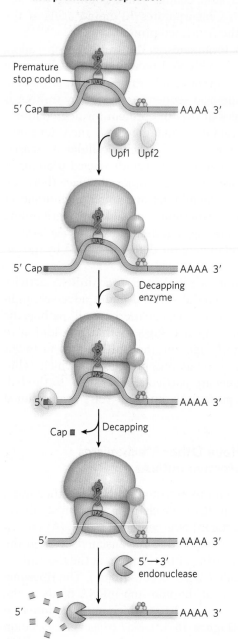

FIGURE 18-33 Eukaryotic nonsense-mediated mRNA decay.
(a) The spliceosome deposits exon junction complexes (EJCs) at the splice sites of an mRNA (see Chapter 16), and the EJCs are usually displaced by the ribosome as it travels along the mRNA. (b) If the ribosome encounters a premature stop codon upstream of an EJC, then Upf1 and Upf2 recruit a decapping enzyme and a 5′→3′ exonuclease to promote degradation of the mRNA.

Note that both of the above processes for removing damaged or truncated mRNA are performed by the ribosome, and therefore the ribosome is an effective proofreader. The recognition of defective mRNAs thus requires active translation. Recent evidence suggests that such mechanisms play significant roles in the regulation and evolution of protein function, as we will discuss in detail in Chapter 22.

SECTION 18.5 SUMMARY

• In bacteria, tmRNA rescues ribosomes stalled on non-stop mRNAs, playing the role of both tRNA and

mRNA in the A site. Translation continues, using the tmRNA as template and ending at the tmRNA stop codon.

• Eukaryotic non-stop mRNAs are translated through the poly(A) tail, encoding a string of Lys residues that tag the protein for degradation.

• In eukaryotes, pre-mRNA splicing leaves an exon junction complex on spliced mRNAs. If a ribosome encounters a stop codon before it has removed all the EJCs, the stop codon is identified as premature, and the mRNA is tagged for degradation.

PROTEIN FOLDING, COVALENT MODIFICATION, AND TARGETING

To achieve their biologically active forms, new polypeptides must fold into the correct three-dimensional conformation. Before or after folding, the new polypeptide might undergo several types of processes: enzymatic processing, including the removal of one or more amino acids (usually from the N-terminus); addition of acetyl, phosphoryl, methyl, carboxyl, or other groups to certain amino acid residues; proteolytic cleavage; and/or attachment of oligosaccharides or prosthetic groups. In this way, the linear, or one-dimensional, genetic message in the mRNA is converted into the three-dimensional structure of the protein.

Protein Folding Sometimes Requires the Assistance of Chaperones

Most proteins fold during translation (after emerging from the ribosome) or immediately after translation, typically beginning with the formation of local secondary structures, including α helices and β sheets. In a cooperative process, these secondary structural elements then interact, often through hydrophobic interactions, to produce the stable three-dimensional structure of the active protein. In many cases, proteins called **chaperones** catalyze local unfolding and refolding of polypeptide chains to enhance the rate and accuracy of overall folding. By mechanisms sometimes coupled to ATP hydrolysis, chaperones bind transiently to hydrophobic protein segments during folding to ensure that interactions occur in the proper order (see Figures 4-23 and 4-24).

After folding into their native conformations, some proteins form intrachain or interchain disulfide bonds, or bridges, between Cys residues. In eukaryotes, disulfide bonds are common in proteins to be exported from cells. The cross-links formed in this way help protect the molecule's native conformation from denaturation in the extracellular environment, which can differ greatly from the intracellular environment and is generally oxidizing (see the How We Know section at the end of Chapter 4).

Covalent Modifications Are Common in Newly Synthesized Proteins

Some newly made proteins, both bacterial and eukaryotic, do not attain their final biologically active conformation until they have been altered by one or more processing reactions known as **posttranslational modifications**. As we have seen, the first residue inserted in all polypeptides is *N*-formylmethionine (in bacteria) or methionine (in eukaryotes). The formyl group, the N-terminal Met residue, and often additional N-terminal (and, in some cases, C-terminal) residues may be removed enzymatically in forming the final functional protein. In as many as 50% of eukaryotic proteins, the amino group of the N-terminal residue is *N*-acetylated after translation. Carboxyl-terminal residues are also sometimes modified. Scientists are still learning the rules to determine which proteins will have which amino acids removed or modified.

The 15 to 30 residues at the N-terminal end of some proteins play a role in directing the protein to its ultimate destination in the cell. After this protein trafficking, these residues are often removed by specific peptidases.

Individual amino acid residues can be modified, either permanently or transiently, with significant effects on functionality, increasing or decreasing the protein's ability to bind other molecules (**Figure 18-34**). The hydroxyl groups of certain Ser, Thr, and Tyr residues of some proteins are enzymatically phosphorylated by ATP. Extra carboxyl groups may be added to Glu residues

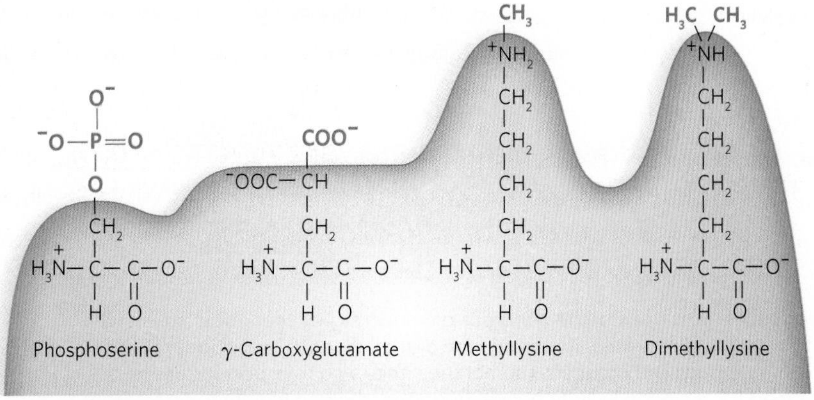

FIGURE 18-34 Modified amino acid residues. Amino acid residues can be phosphorylated, carboxylated, or methylated to alter protein function.

of some proteins, and Lys, Arg, or Glu residues can be methylated.

The carbohydrate side chains of glycoproteins are attached covalently during or after synthesis of the polypeptide. In some glycoproteins, the carbohydrate side chain is attached enzymatically to Asn residues (*N*-linked oligosaccharides), in others to Ser or Thr residues (*O*-linked oligosaccharides). Many proteins that function extracellularly, as well as the proteoglycans that coat and lubricate mucous membranes, contain oligosaccharide side chains.

Other covalent modifications to proteins include the addition of isoprenyl groups or prosthetic groups and cleavage by proteases. Many bacterial and eukaryotic proteins require covalently bound prosthetic groups for their activity. Finally, many proteins are initially synthesized as large, inactive precursor polypeptides that are proteolytically trimmed to their smaller, active forms.

Proteins Are Targeted to Correct Locations during or after Synthesis

Cells are made up of many structures and compartments, and, in the case of eukaryotic cells, they contain organelles, each with specific functions that require distinct sets of proteins and enzymes. These proteins (with the exception of those produced in mitochondria and chloroplasts) are synthesized on ribosomes in the cytosol or on the endoplasmic reticulum (ER). How are proteins directed to their final cellular destinations? Proteins destined for secretion, integration into the plasma membrane, or inclusion in lysosomes generally share the first few steps of a pathway that begins in the ER. Proteins destined for mitochondria, chloroplasts, or the nucleus use three separate mechanisms. And proteins destined for the cytosol simply remain where they are synthesized.

Günter Blobel [*Source: Courtesy Günter Blobel, Rockefeller University.*]

The most important element in many of these targeting pathways is a short sequence of amino acids called a **signal sequence**, the function of which was first postulated by Günter Blobel and his colleagues in 1970. The signal sequence directs a protein to its appropriate location in the cell and, for many proteins, is removed during transport or after arrival at its final destination. In proteins directed to mitochondria, chloroplasts, or the ER, the signal sequence is at the N-terminus of the newly synthesized polypeptide. Some signal sequences promote transport to the ER or mitochondria, others to the nucleus. The strength of a signal sequence, like that of a promoter sequence, depends on how similar it is to an unknown, hypothetical ideal sequence. In many cases, the targeting capacity of a particular signal sequence has been confirmed by fusing the sequence from one protein to a second protein and showing that the signal directs the second protein to the location where the first protein is normally found.

Some Chemical Modifications of Eukaryotic Proteins Take Place in the Endoplasmic Reticulum

The best-characterized targeting system begins in the ER. Most lysosomal, membrane, or secreted proteins are synthesized on ribosomes attached to the ER. These proteins have an N-terminal signal sequence that marks them for translocation into the ER or its lumen; hundreds of such signal sequences have been determined (**Figure 18-35**). Signal sequences vary in length from

Transmembrane proteins

Human influenza virus A hemagglutinin:
Met Lys Ala Lys | Leu Leu Val Leu Leu Tyr Ala Phe Val Ala Gly | Asp Glu

Lipoprotein:
Met Lys Ala Thr Lys | Leu Val Leu Gly Ala Val Ile Leu Gly Ser Thr Leu Leu Ala Gly | Cys Ser

Secreted proteins

Bovine growth hormone:
Met Met Ala Ala Gly Pro Arg | Thr Ser Leu Leu Leu Ala Phe Ala Leu Leu Cys Leu Pro Trp | Thr Gln Val Val Gly | Ala Phe

Bee promelittin:
Met Lys | Phe Leu Val Asn Val Ala Leu Val Phe Met Val Val Tyr Ile | Ser Tyr Ile Tyr Ala | Ala Phe

Drosophila glue protein:
Met Lys | Leu Leu Val Val Ala Val Ile Ala Cys Met Leu Ile Gly Phe Ala | Asp Pro Ala Ser Gly | Cys Lys

Human preproinsulin:
Met Ala Leu Trp Met | Arg | Leu Leu Pro | Leu Leu Ala Leu Leu Ala Leu Trp | Gly Pro Asp Pro Ala Ala | Phe Val

(Cleavage site indicated before the final two residues)

FIGURE 18-35 Signal sequences. Just a few of the hundreds of known signal sequences are shown here. Hemagglutinin is a transmembrane protein on the surface of the influenza virus. Lipoproteins are components of plasma membranes and organelle membranes; they also transport fats in the bloodstream. Proteins destined for secretion include growth hormone, the bee venom toxin melittin (as its precursor promelittin), a *Drosophila* glue protein used in forming the pupa, and insulin (as its precursor, preproinsulin). Hydrophobic residues are shown in yellow; charged residues are in blue.

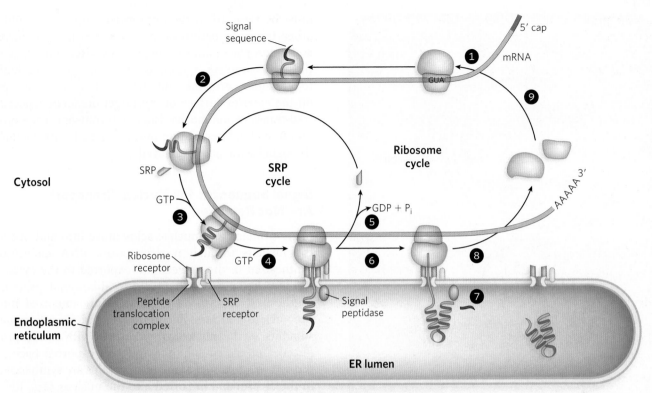

FIGURE 18-36 Trafficking of proteins from the cytosol into the ER. The signal sequence of the nascent polypeptide is bound by SRP, which targets the elongating protein to the ER lumen. After the polypeptide has been synthesized, the ribosomal subunits dissociate and are recycled.

13 to 36 amino acid residues, but all contain 10 to 15 hydrophobic residues, one or more positively charged residues near the N-terminus, and a short polar sequence near the cleavage site (for eventual removal of the signal) at the C-terminus.

The signal sequence itself helps direct the ribosome to the ER, as shown in **Figure 18-36**. The targeting pathway begins with initiation of protein synthesis on cytosolic ribosomes (step 1). The signal sequence forms early in the synthesis process: it is at the N-terminus, which is synthesized first. As the signal sequence emerges from the ribosome (step 2), the signal and the ribosome itself are bound by the large **signal recognition particle (SRP)** (step 3). The SRP, a multi-subunit complex that includes the essential 7SL RNA component, binds GTP (step 4) and halts elongation of the polypeptide when it is about 70 amino acids long and the signal sequence has completely emerged from the ribosome. The GTP-bound SRP directs the ribosome (still bound to the mRNA) and the incomplete polypeptide to GTP-bound SRP receptors in the cytosolic face of the ER. GTP is hydrolyzed, enabling delivery of the nascent polypeptide to a **peptide translocation complex** in the ER, which may interact directly with the ribosome (step 5). SRP dissociates from the ribosome, accompanied by hydrolysis of the GTP bound to the

SRP and to the SRP receptor. Elongation of the polypeptide resumes (step 6), with the ATP-driven translocation complex feeding the growing polypeptide into the ER lumen until the complete protein has been synthesized. The signal sequence is removed by a signal peptidase in the ER lumen (step 7). The ribosome dissociates (step 8) and is recycled (step 9).

Glycosylation Plays a Key Role in Eukaryotic Protein Targeting

In the ER lumen, newly synthesized proteins are further modified in several ways. Following removal of signal sequences, polypeptides are folded, disulfide bonds are formed, and many proteins are glycosylated. In many glycoproteins, Asn residues are *N*-linked to a wide variety of oligosaccharides, but the pathways by which they form share common steps. Several antibiotics, including **tunicamycin**, act by interfering with this process and have aided in elucidating the steps of protein glycosylation. A few proteins are *O*-glycosylated in the ER, but most *O*-glycosylation occurs in the Golgi complex (Golgi apparatus) or in the cytosol (for proteins that do not enter the ER).

Once a protein is suitably modified, it can move to its final intracellular destination. Proteins travel from

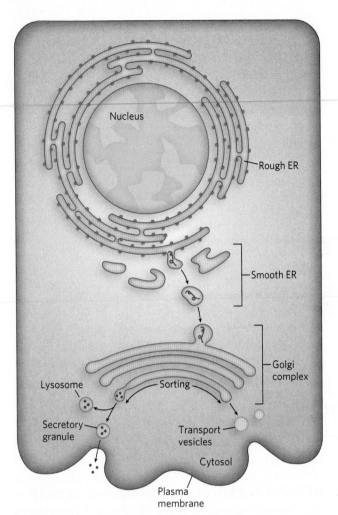

FIGURE 18-37 Movement of proteins destined for membranes or secretion. After synthesis on ribosomes of the rough ER (see Figure 18-1) and targeting to the ER lumen, proteins travel in transport vesicles through the Golgi complex. Within the Golgi complex, the proteins may be further modified before they are sorted and shipped to their final destinations in secretory granules (for export from the cell) or transport vesicles (for further transport within the cell).

the ER to the Golgi complex in transport vesicles (**Figure 18-37**). In the Golgi complex, some proteins, as noted above, are *O*-glycosylated, and some *N*-linked oligosaccharides are further modified. By mechanisms not yet fully understood, the Golgi complex also sorts proteins and sends them to their final destinations. The processes that segregate proteins targeted for secretion from those targeted for the plasma membrane or lysosomes must distinguish among these proteins on the basis of structural features other than signal sequences, which were removed in the ER lumen.

The pathways that target proteins to mitochondria and chloroplasts also rely on N-terminal signal sequences. Although mitochondria and chloroplasts contain DNA, most of their proteins are encoded by nuclear DNA and must be targeted to the appropriate organelle. Unlike other targeting pathways, however, the mitochondrial and chloroplast pathways begin only after protein synthesis is complete. Cytosolic chaperone proteins bind to precursor proteins and deliver them to receptors on the exterior surface of the target organelle. Specialized translocation mechanisms then transport each protein to its final destination in the organelle, after which the signal sequence is removed.

Signal Sequences for Nuclear Transport Are Not Removed

Many proteins and nucleic acids move into and out of the nucleus through nuclear pores. RNA molecules synthesized in the nucleus are exported to the cytosol for translation (see Chapter 16). Ribosomal proteins synthesized on cytosolic ribosomes are imported into the nucleus and assembled into 60S and 40S ribosomal subunits in the nucleolus, where the rRNAs are produced. Completed subunits are then exported back to the cytosol. A variety of nuclear proteins are synthesized in the cytosol and imported into the nucleus (e.g., RNA and DNA polymerases, histones, topoisomerases, and proteins that regulate gene expression). All of this traffic is modulated by a complex system of molecular signals and transport proteins.

In multicellular eukaryotes, cell division poses a problem for nuclear proteins. At each cell division, the nuclear envelope breaks down, and after division is completed and the nuclear envelope re-forms, the dispersed nuclear proteins must be reimported. To allow this repeated nuclear importation, the signal sequence that targets a protein to the nucleus—the **nuclear localization sequence (NLS)**—must remain on the protein after it arrives at its destination. An NLS, unlike other signal sequences, may be located almost anywhere along the primary sequence of the protein. The amino acid sequences of NLSs can vary considerably, but many consist of 4 to 8 residues and include several consecutive basic residues (Arg or Lys).

Nuclear import is mediated by proteins that cycle between the cytosol and the nucleus, including importins α and β and the Ran GTPase (**Figure 18-38**), in a mechanism like that discussed for nuclear RNA export and import in Chapter 16. A heterodimer of importins α and β functions as a carrier for cargo proteins targeted to the nucleus, with the α subunit binding cargo proteins in the cytosol. The importin-cargo complex docks at a nuclear pore and is translocated through the pore by an energy-dependent mechanism that requires the Ran GTPase. Once inside the nucleus, interaction with Ran-GTP triggers a change in the conformation of importin that leads to release of the cargo protein. The importin-Ran-

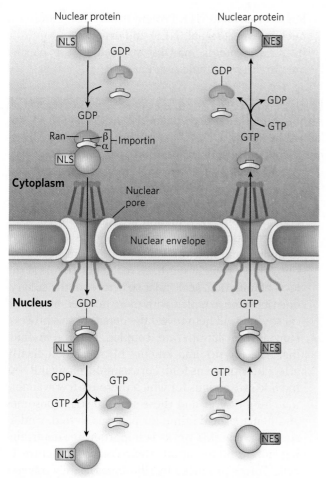

FIGURE 18-38 Targeting of nuclear proteins. Importins bind a nuclear protein through its nuclear localization sequence (NLS) and transport the protein through a nuclear pore, with the help of Ran-GDP. In the nucleus, Ran exchanges its GDP for GTP, facilitating release of the nuclear protein. The importins and Ran-GTP then shuttle back to the cytoplasm through a nuclear pore.

GTP complex then passes through the nuclear pore back into the cytosol. Here, the Ran-binding protein binds to Ran and releases importin, and a GTPase-activating protein stimulates conversion of Ran-GTP to Ran-GDP.

Some proteins contain a nuclear export signal (NES), often consisting of four hydrophobic residues that target the protein for export from the cell nucleus to the cytoplasm through the nuclear pore complex. An NES has the opposite effects of an NLS and is recognized by proteins called exportins. The most common spacing of the hydrophobic amino acids of an NES is L-X-X-X-L-X-X-L-X-L, where L is a hydrophobic amino acid (typically leucine, hence the L) and X is any other amino acid. The spacing probably reflects the structure of this signal that is required for binding to exportin complexes.

Bacteria Also Use Signal Sequences for Protein Targeting

Bacteria can target proteins to their inner or outer membranes, to the periplasmic space, or to the extracellular medium. They use signal sequences at the N-terminus of the proteins, much like those on eukaryotic proteins targeted to the ER, mitochondria, and chloroplasts.

Most proteins exported from *E. coli* make use of the pathway shown in **Figure 18-39**. Following translation, the N-terminal signal sequence may impede folding of a protein to be exported. A soluble chaperone protein, SecB, binds to the signal sequence or to other features of the protein's incompletely folded structure. The bound protein is then delivered to SecA, a protein associated with the inner surface of the plasma membrane. SecA acts as both a receptor and a translocating ATPase. Released

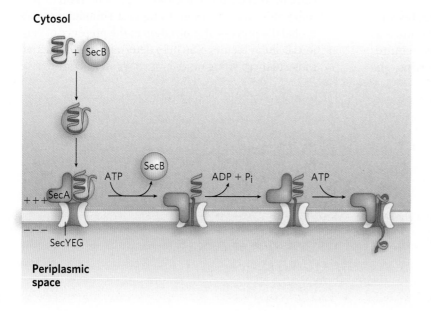

FIGURE 18-39 A model for protein export in bacteria. A partially folded protein with a signal sequence is bound by SecB and then transferred to SecA and SecYEG, the latter a component of the bacterial plasma membrane. SecYEG pushes the protein through the membrane stepwise, and the protein folds on the other side of the membrane, in the periplasmic space.

from SecB and bound to SecA, the protein is delivered to a translocation complex in the membrane, made up of SecY, SecE, and SecG, and is translocated stepwise through the membrane at the SecYEG complex, about 20 amino acid residues at a time. Each step is facilitated by the hydrolysis of ATP, catalyzed by SecA. Although most exported bacterial proteins use this pathway, some follow an alternative route that uses signal recognition and receptor proteins homologous to components of the eukaryotic SRP and SRP receptor.

SECTION 18.6 SUMMARY

- Polypeptides fold into their active, three-dimensional forms during or immediately after synthesis, often with the help of ATP-dependent chaperone proteins. Many proteins are further processed by posttranslational modification reactions that add functional groups, such as phosphates or sugars.
- During or immediately following synthesis, many proteins are directed to specific cellular locations. One targeting mechanism involves a peptide signal sequence, generally at the N-terminus of a newly synthesized protein.
- In eukaryotic cells, one class of signal sequences is recognized by the signal recognition particle, which binds the signal sequence as soon as it appears on the ribosome and transfers the entire ribosome and incomplete polypeptide to the ER. The peptides are moved into the ER lumen, where they may be modified and moved to the Golgi complex, then sorted and sent to lysosomes, the plasma membrane, or transport vesicles.
- Proteins targeted to mitochondria and chloroplasts, and those destined for export in bacterial cells, also make use of an N-terminal signal sequence. Specific enzymes remove the signal once the protein reaches its destination.
- Nuclear localization signals are not removed, because nuclear proteins must be relocalized to the nucleus each time the cell divides. Protein import requires a nuclear localization signal (NLS), importins, the Ran GTPase, and GTP. Protein export also requires a nuclear export signal (NES) and exportins.
- Bacterial proteins may be targeted to the plasma membrane by a signal sequence.

? UNANSWERED QUESTIONS

Although many details of bacterial protein synthesis are known in exquisite detail, researchers have yet to determine how eukaryotic translation is initiated and regulated, how translation assists in protein folding, and how ribosomes work together within polysomes.

1. How is the initiation of eukaryotic translation regulated? Translation initiation is much more complex in eukaryotic cells than in bacterial cells. It is critical to understand how the eukaryotic system works, because much protein synthesis regulation occurs at the level of initiation. Molecular structures of the eukaryotic ribosome, coupled with more detailed biochemical studies, will help reveal the details of this process.

2. How is translation rate coupled to protein folding? The physics and kinetics of translation clearly affect how proteins fold. For example, many mRNAs include rare codons for which there are few available matching tRNAs, and these serve to stall ribosomes and hence provide time for proteins to fold. Understanding how this works is important for determining how proteins attain their correct structure in cells. Forces produced by ribosomes as they traverse an mRNA may also be important for melting RNA structures that could otherwise impede translation.

3. How do the crowded conditions inside cells influence translation rates and accuracy? Most of the experiments that have probed translation mechanisms have been performed using purified ribosomes under relatively dilute conditions (~1 μM). In rapidly growing cells, however, ribosomes may be present at up to 100 times this concentration. Computer simulations that model the process of translation in the presence of various cellular factors may help determine the impact of molecular crowding on the rate of protein synthesis.

HOW WE KNOW

The Ribosome Is a Ribozyme

Noller, H.F., and J.B. Chaires. 1972. Functional modification of 16S ribosomal RNA by kethoxal. *Proc. Natl. Acad. Sci. USA* 69:3115–3118.

Noller H.F., V. Hoffarth, and L. Zimniak. 1992. Unusual resistance of peptidyl transferase to protein extraction procedures. *Science* 256:1416–1419.

A series of experiments conducted in the early 1970s by Harry Noller provided the first evidence that ribosomal RNA, rather than ribosomal protein, is responsible for catalyzing peptide bond formation during protein synthesis. Using chemicals that react with side chains in proteins or nucleotides, Noller discovered that the 30S ribosomal subunit of bacteria could be inactivated by a reagent called kethoxal, which primarily attacks guanosine nucleotides in RNA.

Bacterial ribosomes were purified and separated into their two subunits. After addition of kethoxal to the 30S subunits, these were mixed with unmodified 50S subunits and the ribosomes tested for the ability to stimulate in vitro protein synthesis, using tRNA and a poly(U) mRNA template. In contrast to samples with untreated 30S subunits, the kethoxal-treated sample rapidly lost activity (Figure 1). Analysis of the inactivated ribosomes showed that modification of just six G nucleotides in the rRNA was sufficient to inhibit the peptidyl transferase reaction. Inactivation was slower in the presence of bound tRNA, however, leading to the conclusion that kethoxal interferes with protein synthesis by modifying the binding site for tRNA on the 30S subunit.

In 1992, Noller and his colleagues published a study showing that virtually all the r-proteins could be removed from the ribosome with only small effects on protein-synthesizing activity, whereas damage to the rRNA destroyed the ribosome's catalytic properties. In 2000, the high-resolution crystal structure of the 50S ribosomal subunit revealed that the active site responsible for peptide bond formation is composed entirely of rRNA. Thus, the structural data confirmed what had long been suspected based on biochemical evidence: the ribosome is a ribozyme.

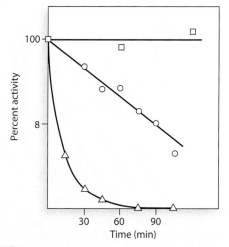

FIGURE 1 Chemical modification of rRNA inactivates the 30S ribosomal subunit. The graph shows the percentage of normal activity (synthesis of polypeptide) over time. In the absence of kethoxal (squares), the 30S subunit retained activity; in the presence of kethoxal (circles), activity was impaired over time. A control reaction lacking tRNA and mRNA had virtually no activity (triangles). The chemical structure of a kethoxal-modified G nucleoside is shown on the right. *[Source: Data from H. F. Noller and J. B. Chaires,* Proc. Natl. Acad. Sci. USA *69:3115–3118, 1972.]*

HOW WE KNOW

Ribosomes Check the Accuracy of Codon-Anticodon Pairing, but Not the Identity of the Amino Acid

Chapeville, F., F. Lipmann, G. Von Ehrenstein, B. Weisblum, W.J. Ray Jr., and S. Benzer. 1962. On the role of soluble ribonucleic acid in coding for amino acids. *Proc. Natl. Acad. Sci. USA* 48:1086–1092.

Zaher, H.S., and R. Green. 2009. Quality control by the ribosome following peptide bond formation. *Nature* 457:161–166.

Rachel Green [Source: © Paul Fetters 2007.]

Translation relies on aminoacyl-tRNA synthetases to ensure the correct charging of tRNAs, because the ribosome does not discriminate between correctly and incorrectly charged tRNAs during protein synthesis. A classic experiment by Seymour Benzer and his colleagues demonstrated that if a tRNA is aminoacylated with the wrong amino acid, this incorrect amino acid is efficiently incorporated into a protein in response to the codon normally recognized by that tRNA.

This experiment, performed in 1962, used ^{14}C labeling to track how amino acids attached to particular tRNAs were incorporated into polypeptides. For example, correctly charged [^{14}C]Cys-tRNA^{Cys} was chemically treated with Raney nickel (a nickel-aluminum alloy catalyst) to form a mischarged [^{14}C]Ala-tRNA^{Cys} (Figure 2a). Poly(UG) mRNA has UGU codons that, with wobble base pairing in the third position, match the ACG anticodon for tRNA^{Cys}, but has no codons for alanine. Using this mRNA, the researchers found that the ribosome efficiently incorporated [^{14}C]alanine into acid-insoluble polypeptide (Figure 2b). This result was an elegant and straightforward demonstration that the ribosome does not proofread tRNAs for correct aminoacylation.

For many years, protein synthesis was thought to rely on the combined accuracy of tRNA aminoacylation and aminoacyl-tRNA selection by the ribosome in cooperation with the GTPase elongation factor EF-Tu (see Section 18.4). These two processes operate before peptide bond formation to ensure that only correctly charged and correctly matched tRNAs enter the ribosomal A site.

More recently, an additional mechanism occurring *after* peptidyl transfer was found to contribute to the accuracy of protein synthesis. Using a well-defined in vitro bacterial translation system, Rachel Green and Hani Zaher showed that incorporation of an amino acid from a mismatched aminoacyl-tRNA into the elongating polypeptide leads to a general loss of specificity in the ribosomal A site. The resulting propagation of errors leads to early termination of protein synthesis, avoiding the production of a complete protein containing incorrect amino acids.

(a) **Conversion of Cys to Ala by Raney nickel**

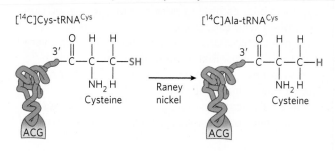

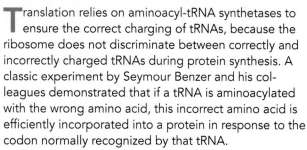

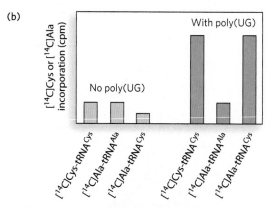

FIGURE 2 The ribosome does not proofread aminoacylated tRNAs. (a) The ^{14}C-labeled Cys residue of Cys-tRNA^{Cys} is converted to an Ala residue by Raney nickel, which removes the sulfur from cysteine. (b) Polypeptide synthesis with and without poly(UG) mRNA, using correctly charged [^{14}C]Cys-tRNA^{Cys}, or correctly charged [^{14}C]Ala-tRNA^{Ala} (as control), or mischarged [^{14}C]Ala-tRNA^{Cys}. The results in the presence of UGU codons (dark purple columns on the right) show incorporation of labeled Cys with Cys-tRNA^{Cys} (left); negligible incorporation of labeled Ala with Ala-tRNA^{Ala} (middle); and incorporation of labeled Ala with Ala-tRNA^{Cys} (right). [Source: Data from F. Chapeville et al., Proc. Natl. Acad. Sci. USA 48:1086–1092, 1962.]

representative codon in the genetic code. How might it be incorporated into proteins?

KEY TERMS

ribosome, 618
ribosomal protein
  (r-protein), 619
ribosomal RNA (rRNA),
  619
polysome, 622
polyribosome, 622
A site, 623
P site, 624
E site, 624
initiation, 630
initiation codon, 630
start codon, 630
initiation factor, 630
Shine-Dalgarno
  sequence, 631
Kozak sequence, 632
N-formylmethionyl-
  tRNA^fMet, 634
initiation complex, 636
scanning, 637

elongation, 639
termination, 639
termination codon, 639
stop codon, 639
elongation factor, 639
peptidyl transferase
  reaction, 640
translocation, 641
termination factor, 643
release factor, 643
tmRNA, 650
nonsense-mediated
  mRNA decay, 651
chaperone, 653
posttranslational
  modification, 653
signal sequence, 654
signal recognition
  particle (SRP), 655
nuclear localization
  sequence (NLS), 656

PROBLEMS

1. On average, how many phosphoanhydride bonds are hydrolyzed in the course of synthesizing a 400 amino acid protein? Assume that you begin with the mature mRNA, ribosomal subunits, tRNAs, free amino acids, and all necessary factors.

2. Name the type of chemical bonds that link (a) adjacent amino acids in a protein; (b) an amino acid to tRNA; (c) adjacent nucleotides in RNA; (d) a codon in mRNA to an anticodon in tRNA; (e) the two subunits of a ribosome.

3. A bacterial ribosome can synthesize about 20 peptide bonds per minute. If the average bacterial protein is approximately 260 amino acids in length, how many proteins can the ribosomes in an *E. coli* cell synthesize in 20 minutes with all ribosomes functioning at maximum rates?

4. Discuss the advantages to the cell of having multiple ribosomes translating a single mRNA molecule.

5. On a bacterial ribosome, only one tRNA binds directly to the P site without first interacting with the A site. Identify that tRNA and explain why the P site binding occurs.

6. The amino acid hydroxyproline, which is critical to the structure of collagen and certain other proteins, has no

7. The isoleucyl-tRNA synthetase has a proofreading function that improves the fidelity of the aminoacylation reaction, whereas the histidyl-tRNA synthetase lacks such a proofreading function. Explain why.

8. As described in Chapter 11, some DNA polymerases have proofreading activities. After a nucleotide is added to a growing nucleic acid chain, it can be removed (if incorrectly paired with the template) by hydrolysis of the phosphodiester bond that links it to the growing polymer. Ribosomes do not have similar proofreading activities; they cannot remove the last amino acid added to a growing polypeptide, regardless of whether it was correctly added or not. If ribosomes possessed such a proofreading function, would cleavage of the bond linking the last amino acid to the polymer have any effect on the rest of the polypeptide? Why or why not?

9. A researcher isolates mutant variants of the bacterial translation factors IF-2, EF-Tu, and EF-G. In each case, the mutation allows proper folding of the protein and binding of GTP, but does not allow GTP hydrolysis. At what stage would translation be blocked by each mutant protein?

10. Some aminoacyl-tRNA synthetases do not recognize and bind the anticodon of their cognate tRNAs; they use other structural features of the tRNAs to impart binding specificity. The tRNAs for alanine fall into this category.

    (a) What features of tRNA^Ala are recognized by Ala-tRNA synthetase?
    (b) Describe the consequences of a C-to-G mutation in the third position of the anticodon of tRNA^Ala.
    (c) What other kinds of mutations might have similar effects?
    (d) Mutations of these types are never found in natural populations of organisms. Why? (Hint: Consider what might happen both to individual proteins and to the organism as a whole.)

11. When a bacterial mRNA is truncated at the 3′ end so that it is missing part of its gene-encoding sequence, the ribosome stalls at the end of that truncated mRNA. What mechanism can be used to recycle the stalled ribosome to synthesize new polypeptides?

12. The gene for a eukaryotic polypeptide of 300 amino acid residues is altered so that the polypeptide has an N-terminal signal sequence recognized by SRP and an internal nuclear localization signal, beginning at residue 150. Where is the protein likely to be found in the cell?

13. Chloramphenicol binds to bacterial ribosomes and is a potent inhibitor of bacterial protein synthesis, but it does not inhibit the cytosolic ribosomes in eukaryotes. However,

because of its severe toxicity, chloramphenicol is rarely used as a human antibiotic. Suggest a reason for chloramphenicol toxicity in humans.

DATA ANALYSIS PROBLEMS

Chapeville, F., F. Lipmann, G. Von Ehrenstein, B. Weisblum, W.J. Ray Jr., and S. Benzer. 1962. On the role of soluble ribonucleic acid in coding for amino acids. *Proc. Natl. Acad. Sci. USA* 48:1086–1092.

14. The experiments demonstrating that ribosomes check the anticodon of the tRNA, but not the amino acid attached to it, were carried out well before the entire genetic code was defined. The research team, led by Seymour Benzer (see the How We Know Section in this chapter), chose tRNA^{Cys} for their demonstration, for several reasons—including the ability to reduce Cys to Ala in a simple chemical reaction using Raney nickel. For their experiment showing that Ala could be incorporated into polypeptide in place of Cys, if present in the translation reaction mix as Cys-tRNA^{Ala}, the researchers first used polynucleotide phosphorylase to synthesize an RNA polymer consisting of only U and G residues. They knew from previous reports that this polymer would yield polypeptides with the amino acid composition they required.

   (a) Which amino acids are incorporated into a polypeptide specified by a random sequence of U and G?

   (b) Why was tRNA^{Cys} a good choice for the experiment?

   Benzer and colleagues showed that incorporation of radioactively labeled amino acids into polypeptide was linearly related to the amount of labeled aminoacyl-tRNA added to the experiment. They treated one preparation of [^{14}C]Cys-tRNA^{Cys} with Raney nickel; about 60% of the charged tRNA was reduced to [^{14}C]Ala-tRNA^{Cys}. They then plotted incorporation of radioactivity, over time, in preparations with [^{14}C]Cys-tRNA^{Cys} (Figure 1) and preparations with [^{14}C]Ala-tRNA^{Cys} (Figure 2), measured in counts per minute (cpm) in the trichloroacetic acid–precipitated (TCA-ppt) preparations.

   (c) Given the amount of labeled aminoacyl-tRNA added (indicated on the vertical axis of the graphs) and the amount incorporated into polypeptide, does this result support the contention that Ala-tRNA^{Cys} is incorporated at codons specifying Cys? Why or why not?

   The researchers next treated a preparation of [^{14}C]Phe-tRNA^{Phe} with Raney nickel, then compared the incorporation of [^{14}C]Phe into polypeptide in response to a poly(U) RNA before and after Raney nickel treatment. No effect of the Raney nickel treatment was seen.

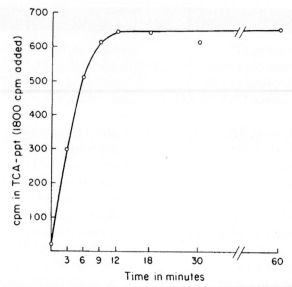

FIGURE 1 [^{14}C]Cys-tRNA^{Cys} incorporation. *[Source: F. Chapeville et al., Proc. Natl. Acad. Sci. USA 48:1086–1092.]*

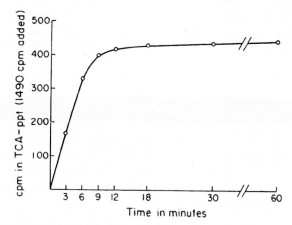

FIGURE 2 [^{14}C]Ala-tRNA^{Cys} incorporation. *[Source: F. Chapeville et al., Proc. Natl. Acad. Sci. USA 48:1086–1092.]*

   (d) Why was this experiment carried out?

   The polypeptides synthesized using poly(UG) and Ala-tRNA^{Cys} were digested to single amino acids, and the products were analyzed to directly demonstrate that about 60% of the labeled residues were Ala.

   (e) Why was this experiment needed, given the results shown in Figures 1 and 2?

Noller, H.F., V. Hoffarth, and L. Zimniak. 1992. Unusual resistance of peptidyl transferase to protein extraction procedures. *Science* 256:1416–1419.

15. Experiments strongly suggesting that the peptidyl transferase activity of ribosomes is due to rRNA, not r-proteins, were carried out by Harry Noller and his colleagues in 1992 (see this chapter's Moment of Discovery

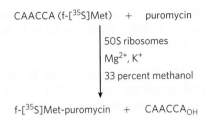

FIGURE 1 *[Source: Noller H.F., Hoffarth, V. and Zimniak, L. Science 256:1416–9. Fig. 2c. 1992. Reprinted with permission from AAAS.]*

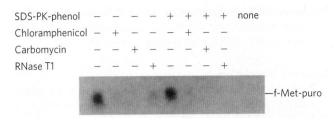

FIGURE 2 *[Source: Noller H.F., Hoffarth, V. and Zimniak, L. Science 256:1416–9. Fig. 2c. 1992. Reprinted with permission from AAAS.]*

and How We Know section). The work required the removal of most or all of the proteins in a ribosome without eliminating peptidyl transferase activity. The basic reaction used is shown in Figure 1. A ^{35}S-labeled fMet residue (shown here as f-[^{35}S]Met) was attached to the 3' end of the hexameric oligonucleotide 5'-CAACCA. This labeled fMet was transferred to puromycin in the presence of the 50S bacterial ribosome subunit, when the solvent contained about 33% methanol (with Mg^{2+} and K^+ ions).

(a) What do the substrates required for this in vitro reaction suggest about the binding properties of the 50S ribosomal subunit?

(b) What is special about the hexanucleotide sequence attached to the labeled fMet residue?

(c) In which ribosomal site must the fMet-oligonucleotide bind, and in which site must the puromycin bind?

(d) What advantages does this reaction have for this particular study, relative to protein synthesis with complete ribosomes and mRNA and charged tRNAs?

Using this assay, the researchers demonstrated that the 50S subunit or the intact 70S ribosome from *E. coli* would catalyze the reaction even after extraction with sodium dodecyl sulfate (a detergent that denatures proteins) and extensive treatment with proteinase K (which degrades virtually all proteins). Extraction with phenol (a treatment that separates protein from nucleic acid), however, led to a loss in activity. Subsequently, the investigators found that 50S subunits derived from the thermophilic bacterium *Thermus aquaticus* did *not* lose activity when extracted with phenol.

(e) Why did the researchers explore the 50S subunits from *T. aquaticus*?

Noller and colleagues went on to test the sensitivity of the reaction to chloramphenicol, carbomycin (another ribosome-binding inhibitor, with a mechanism similar to chloramphenicol), and a general ribonuclease called RNase T1, with or without (+ or −) SDS, proteinase K (PK), or phenol treatment. The results are shown in Figure 2, an autoradiograph of the product after high-voltage paper electrophoresis.

(f) What conclusions can you derive from the results of this experiment?

(g) The conclusion that ribosomes are ribozymes did not achieve general acceptance until elucidation of the three-dimensional structure of a bacterial 50S ribosomal subunit. Suggest a reason for the delay in general acceptance.

ADDITIONAL READING

Ribosome Structures

Amunts A., A. Brown, X.C. Bai, J.L. Llácer, T. Hussain, P. Emsley, F. Long, G. Murshudov, S.H. Scheres, and V. Ramakrishnan. **2014.** Structure of the yeast mitochondrial large ribosomal subunit. *Science* 343:1485-1489.

Ban, N., P. Nissen, J. Hansen, P.B. Moore, and T.A. Steitz. **2000.** The complete atomic structure of the large ribosomal subunit at 2.4 Å resolution. *Science* 289:905-920.

Greber B.J., D. Boehringer, A. Leitner, P. Bieri, F. Voigts-Hoffmann, J.P. Erzberger, M. Leibundgut, R. Aebersold, and N. Ban. **2014.** Architecture of the large subunit of the mammalian mitochondrial ribosome. *Nature* 505:515-519.

Schluenzen, F., A. Tocilj, R. Zarivach, J. Harms, M. Gluehmann, D. Janell, A. Bashan, H. Bartels, I. Agmon, F. Franceschi, and A. Yonath. **2000.** Structure of functionally activated small ribosomal subunit at 3.3 angstroms resolution. *Cell* 102:615-623.

Schuwirth, B.S., M.A. Borovinskaya, C.W. Hau, W. Zhang, A. Vila-Sanjurjo, J.M. Holton, and J.H. Cate. **2005.** Structures of the bacterial ribosome at 3.5 Å resolution. *Science* 310:827-834.

Wimberly, B.T., D.E. Brodersen, W.M. Clemons Jr., R.J. Morgan-Warren, A.P. Carter, C. Vonrhein, T. Hartsch, and V. Ramakrishnan. **2000.** Structure of the 30S ribosomal subunit. *Nature* 407:327-339.

Eukaryotic Translation

Fraser, C.S., and J.A. Doudna. **2007.** Structural and mechanistic insights into hepatitis C viral translation initiation. *Nat. Rev. Microbiol.* 5:29-38.

Isken, O., and L.E. Maquat. **2007.** Quality control of eukaryotic mRNA: Safeguarding cells from abnormal mRNA function. *Genes Dev.* 21:1833-1856.

Jackson, R.J., C.U. Hellen, and T.V. Pestova. **2010.** The mechanism of eukaryotic translation initiation and principles of its regulation. *Nat. Rev. Mol. Cell Biol.* 11:113-127.

Mathews, M.B., N. Sonenberg, and J.W.B. Hershey. 2006. *Translational Control in Biology and Medicine.* Cold Spring Harbor, NY: Cold Spring Harbor Laboratory Press.

Bacterial Translation

Noller, H.F. 2013. How does the ribosome sense a cognate tRNA? *J. Mol. Biol.* 425:3776–3777.

Pulk, A., and J.H. Cate. 2013. Control of ribosomal subunit rotation by elongation factor G. *Science* 340:1235970.

Reed, R., and H. Cheng. 2005. TREX, SR proteins and export of mRNA. *Curr. Opin. Cell Biol.* 17:269–273.

Rodnina, M.V., M. Beringer, and W. Wintermeyer. 2007. How ribosomes make peptide bonds. *Trends Biochem. Sci.* 32:20–26.

Tourigny, D.S., I.S. Fernández, A.C. Kelley, and V. Ramakrishnan. 2013. Elongation factor G bound to the ribosome in an intermediate state of translocation. *Science* 340:1235490.

Youngman, E.M., M.E. McDonald, and R. Green. 2008. Peptide release on the ribosome: Mechanism and implications for translational control. *Annu. Rev. Microbiol.* 62:353–373.

Zhou J., L. Lancaster, J.P. Donohue, and H.F. Noller. 2013. Crystal structures of EF-G–ribosome complexes trapped in intermediate states of translocation. *Science* 340:1236086.

19 Regulating the Flow of Information

Lin He [Source: Courtesy of the John D. and Catherine T. MacArthur Foundation.]

MOMENT OF DISCOVERY

One of the most exciting moments in my career happened when I was working as a postdoctoral fellow with Greg Hannon at the Cold Spring Harbor Laboratories. *We wondered whether microRNAs— short, regulatory RNAs that control the expression of some eukaryotic genes— might also be involved in promoting the development of cancer.* Indeed, one cluster of microRNAs, the *miR-17-92* polycistron, is located in a region of DNA that is amplified in human B-cell lymphomas. Furthermore, we found that B-cell lymphoma tissue samples and cultured cells contained much higher levels of primary or mature microRNAs derived from the *miR-17-92* locus compared with those found in normal, noncancerous tissues.

To test whether *miR-17-92* overexpression could actually accelerate tumor development, I used a mouse model in which hematopoietic stem cells, the precursors to B cells, were infected with a retrovirus encoding the *miR-17-92* cluster, along with the gene for green fluorescent protein (GFP), which serves as a convenient marker of cells expressing the infected virus, because the cells turn green. Our first experiments yielded only three mice that developed tumors. The next step was to determine whether the tumors came specifically from hematopoietic cells that were overexpressing the *miR-17-92* RNA. By the time I dissected the mouse tumors, made a suspension of the cells, and sent the samples to be analyzed by fluorescence-activated cell sorting (FACS), it was well past midnight—and the FACS sorting couldn't be completed until the next day.

After many anxious hours of waiting, we got the results the next morning: all the tumor cells were green! This was incredibly exciting because it indicated that these tumors came from cells that overexpressed the *miR-17-92* cluster of microRNAs, one of the first examples where functional RNA genes can promote tumorigenesis. At that moment I knew this would be a very promising direction for future research on cancer development.

—Lin He, on discovering that microRNA overexpression accelerates tumor development

Online resources related to this chapter:

Simulation
CRISPR

All cells, whether single-celled bacteria or components of a complex multicellular organism such as a human, contain every gene in that organism's genome. However, cells need the products of only some of these genes, and even those that they do need may be needed only under certain conditions. For example, a bacterium does not need the enzymes required to metabolize lactose when its current food source is glucose. Different cell types of multicellular organisms use different subsets of gene products to carry out their various functions—a liver cell does not need the same gene products as a muscle cell.

The relative amounts of each gene product required by a cell can also vary considerably. In an actively growing cell, for example, ribosomes are in high demand and can account for almost half of the cell's dry weight, whereas only a few molecules of certain DNA repair proteins are required to do the necessary repair job. From an energy standpoint, having all gene products present in the highest possible amounts at all times would overburden the cell, considering the resources required to synthesize RNA and protein. Therefore, the expression of genes must be regulated so that their products are present in the right amount and only when they are needed.

Cells have evolved to respond to environmental changes and to adapt quickly to new growth conditions. This is how organisms can colonize a wide variety of biological habitats. Changing conditions range from variation in the availability and type of food sources to complex developmental regulatory programs in multicellular organisms, for which some gene products may be needed for a surprisingly brief time and in only a few cells.

Gene regulation is also important to the prevention of diseases, including cancer. In multicellular organisms, selective cell proliferation and destruction are critical for maintaining the proper levels of each cell type. The cell division cycle and programmed cell death pathways are exquisitely controlled by genes that promote or prevent these processes in response to cellular signals. When these regulatory genes are compromised, uncontrolled cell division can lead to the development of tumors. For example, *p53* is a regulatory gene that plays a role in initiating programmed cell death. Loss-of-function mutations in *p53* are found in about 50% of all lung cancers, 70% of colon cancers, and 30% to 50% of breast cancers.

Gene expression can be regulated at many different points in the synthesis of a functional RNA or protein. Transcription initiation is the most widely used regulatory point in both bacteria and eukaryotes, as this is the least costly way to control a gene. Initiation of transcription occurs at the very beginning of the synthetic pathway, before the investment in energy needed to make either RNA or protein. Nevertheless, mechanisms

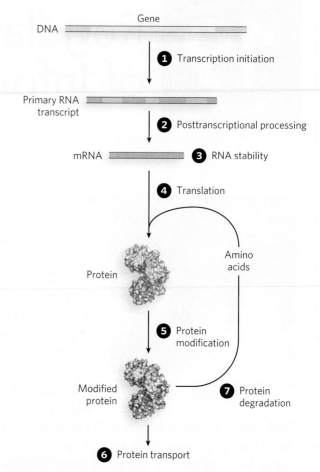

FIGURE 19-1 Seven points at which gene expression can be regulated.

to regulate gene expression are found at virtually every point along the biosynthetic pathway. The points of regulation, shown in **Figure 19-1**, include (1) transcription initiation, (2) posttranscriptional processing (RNA processing), (3) RNA stability, (4) translation (protein synthesis), (5) protein modification, (6) protein transport, and (7) protein degradation.

It is obviously important for cells to use their resources efficiently and not to waste energy synthesizing gene products that they do not need in a particular growth environment. But just as critical as efficiency is adaptability, and thus control: cells must be able to respond rapidly to changes in the need for various gene products. In other words, cells are control freaks! One could argue that control is ultimately more important than energy efficiency for the cell's adaptation and survival. Although this point is often lost in discussions of gene regulation, it is central to biology. As you can see from Figure 19-1, and as we discuss in this and the following chapters, some regulatory mechanisms are directed at mRNA or even at the protein products of mRNA translation. Why do cells "waste" their efforts in this way? Such pathways provide a means of rapidly altering the levels of active proteins in response to the cell's needs. Over the course of evolution, cells and organisms with such capability have won out over those that may have been more energy efficient but were less able to adapt to changing conditions. Thus, gene regulation involves a fine balance between efficiency and adaptability. New and surprising regulatory mechanisms continue to be discovered, and newly discovered posttranscriptional and translational regulatory processes are proving to be highly important, especially in eukaryotes.

In Chapters 15–18 we learned about the mechanics of transcription and translation, processes critical to the flow of biological information. Now we consider how these processes are regulated by the cell to conserve resources while effectively responding to changing environmental conditions and achieving optimum cell function. This chapter presents some general principles of gene regulation that are common to the mechanisms used by both bacteria and eukaryotes. We start by examining the protein-DNA interactions that hold the key to transcriptional regulation. We then discuss principles of posttranscriptional regulation, to provide a more complete overview of the rich complexity of regulatory mechanisms. Chapter 20 provides a more in-depth look into bacterial gene regulation, and Chapters 21 and 22 address the complex regulation of gene expression in eukaryotes.

19.1 REGULATION OF TRANSCRIPTION INITIATION

Regulatory processes operating at the level of transcription initiation are the best documented, and probably the most common. Elaborate mechanisms have evolved to regulate the process of transcription initiation—before large amounts of cellular energy are invested in the production of mRNAs and their protein products. But however diverse, these control mechanisms are really just variations on a common theme and boil down to simple protein-protein and protein-DNA interactions. Indeed, regulation at the step of transcription initiation can be explained simply by changes in how RNA polymerase interacts with the DNA at promoter sequences.

Regulatory proteins that bind DNA can have profound effects on the affinity of RNA polymerase for a promoter. These effects can flow in either direction, either enhancing or preventing RNA polymerase function. Given both the energy expenditure of gene expression and the need for cells to be able to respond quickly to changes in their environment, one can only imagine the enormous evolutionary pressure placed on regulatory mechanisms. We provide here an overview of the transcriptional regulatory mechanisms used by cells. The most detailed information derives from studies in bacterial systems, but eukaryotic mechanisms of gene regulation, although

more complex and using somewhat different strategies, can be explained by the same basic principles.

Activators and Repressors Control RNA Polymerase Function at a Promoter

The most basic mechanism for regulation of transcription initiation is encoded in the DNA sequence of the promoter. RNA polymerase has different intrinsic affinities for promoters of different sequence. In the absence of other controls, these differences in promoter strength correlate with the efficiency with which the genes are transcribed. Genes for products that are required at all times, such as the enzymes of central metabolic pathways, are expressed at a nearly constant level. These genes are often referred to as **housekeeping genes**, and unvarying expression is called **constitutive gene expression**. Although housekeeping genes are expressed constitutively, the expression levels of different housekeeping genes vary widely. For these genes, the RNA polymerase–promoter interaction strongly influences the rate of transcription initiation; with differences in promoter sequences, the cell can synthesize the appropriate level of each housekeeping gene product.

When the level of a gene product rises and falls with a cell's changing needs, this is known as **regulated gene expression**. **Activation** is an increase in expression and **repression** is a decrease in expression of a gene in response to a change in environmental conditions. The mechanisms of gene activation and repression, in both bacteria and eukaryotes, require the assistance of **transcription factors** (also called transcription regulators), proteins that alter the affinity of the RNA polymerase for the promoter. Transcription factors that enhance gene expression are called **activators**, and those that reduce expression are called **repressors**. As we will see later in this chapter, bacterial and eukaryotic transcription factors have many common structural and functional features. These regulators act by binding to specific DNA sequences known as **regulatory sites**.

A gene is said to be under **positive regulation** when binding of an activator protein promotes or increases expression of that gene. Conversely, a gene is under **negative regulation** when binding of a repressor protein prevents or decreases expression. Thus, positive and negative regulation refer to the type of regulatory protein involved: the bound protein either facilitates or inhibits transcription.

A repressor can lower the rate of gene transcription if its regulatory site overlaps the gene's promoter and repressor binding sterically occludes binding of the RNA polymerase to the promoter (**Figure 19-2a**). Repressors can act in other ways as well. Some prevent transcription by binding a regulatory site that is near the promoter, but the binding itself does not block RNA polymerase

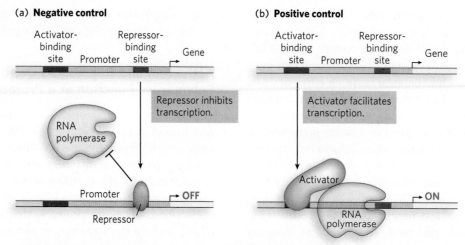

FIGURE 19-2 Negative and positive transcriptional regulation. (a) In this example of negative regulation, a repressor-binding site overlaps the promoter. When the repressor protein binds, RNA polymerase cannot initiate transcription and no mRNA is produced. (b) In this example of positive regulation, an activator protein binds near the promoter and recruits RNA polymerase to the site. Transcription is initiated and mRNA is synthesized.

binding. Recall that the RNA polymerase-promoter complex converts from the closed complex to the open complex, in which the DNA strands are melted locally, opening up the duplex for transcription (see Figures 15-13 and 15-23). Repressors can block this closed-to-open transition, thereby preventing transcription. Instead of steric occlusion, the underlying principle in these mechanisms is allostery: a conformational change in the RNA polymerase–DNA complex prevents formation of the open complex. Repressors of this type can act on the RNA polymerase or directly on the DNA to stabilize the closed complex over the open complex. Other types of repressors act by holding the RNA polymerase to the promoter site, preventing its escape from the promoter.

Activators provide a molecular counterpoint to repressors: they bind to DNA and enhance the activity of RNA polymerase at a promoter. For example, an activator may induce a conformational change in the polymerase that accelerates transition to the open complex. Alternatively, an activator may alter the torsion of DNA, making it more likely to unwind and form the open complex. Probably the most common way that activators function is through cooperativity (**Figure 19-2b**). In this case, the activator binds both the RNA polymerase and a DNA regulatory site next to the promoter, thereby increasing the affinity of the polymerase for the promoter; this activation process is referred to as recruitment.

Transcription Factors Can Function by DNA Looping

Binding sites for activators and repressors are often found at or near the promoter, particularly in bacteria. However, regulatory sites can also be found far away from the promoter. In fact, in eukaryotes, regulatory sites are sometimes thousands of base pairs upstream or downstream from a promoter. How do transcription factors exert their effects on RNA polymerase when their binding sites are so far away from the gene's promoter? Experiments directed at understanding this "action at a distance" have demonstrated that distant regulatory sites can often be placed closer to or farther from the promoter and still retain their function. Not only is distance from the promoter of little consequence (provided it is not too close and overlaps the promoter), but the regulatory sites also retain their function regardless of experimental changes in their sequence orientation relative to the promoter.

When distant regulatory sites were first discovered, scientists imagined that the regulatory proteins might bind to these sites and then slide along the DNA until they reached RNA polymerase at the promoter. However, experiments revealed that this was not the case (see the How We Know section at the end of this chapter). Instead, the DNA between the regulatory site and the RNA polymerase–promoter complex loops out to bring the regulatory protein and RNA polymerase together (**Figure 19-3**). Looping can be directly observed using an electron microscope (**Figure 19-4**). **DNA looping** is facilitated by proteins called architectural regulators that bind to DNA sequences between the regulatory site and the promoter, bending the DNA (**Figure 19-5**). The use of DNA looping is common in eukaryotic gene regulation, and it is also used by some bacterial gene regulatory proteins. In eukaryotes, these regulatory sites that bind transcription factors and exert their regulatory effect on the promoter over long distances are called **enhancers**.

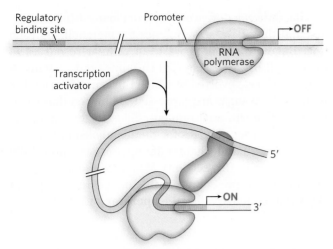

FIGURE 19-3 Action at a distance: DNA looping. After the binding of a transcription activator to a distant regulatory site, the activator also binds the promoter-bound RNA polymerase through protein-protein interactions, forming a DNA loop and activating the polymerase.

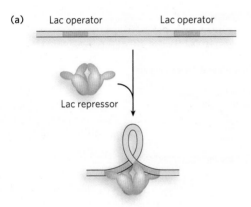

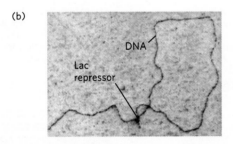

FIGURE 19-4 DNA looping mediated by a single transcription factor. (a) The bacterial Lac repressor protein, a tetramer of identical subunits, binds two distant sites on a single DNA molecule, forming a DNA loop. (b) The DNA loop is visible in this micrograph, negatively stained with uranyl acetate and imaged by dark-field electron microscopy. *[Source: (b) Reprinted from* Current Biology, *9(3), Bernard Révet et al. Four dimers of λ repressor bound to two suitably spaced pairs of λ operators form octamers and DNA loops over large distances, pp. 151–154, Fig 1e, © 1999, with permission from Elsevier.]*

Gene regulation by DNA looping can result in either activation or repression. Activation can recruit RNA polymerase to the promoter through cooperativity, in much the same way as an activator that binds near

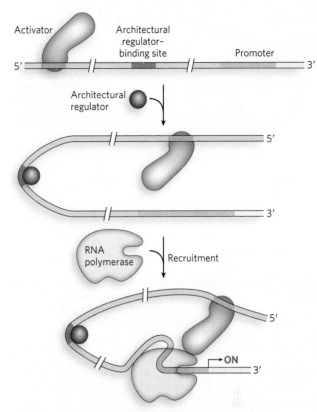

FIGURE 19-5 Transcription factors playing an architectural role. Some transcription factors, known as architectural regulators, bend the DNA when they bind their DNA site, thus promoting looping. Here, the regulator facilitates looping for recruitment of RNA polymerase by an upstream activator.

the promoter. Recruitment of RNA polymerase through DNA looping can also be mediated by a protein "bridge" between the activator and the polymerase (**Figure 19-6a**). Proteins that act by bridging activators and RNA polymerase, but do not bind DNA directly, are called **coactivators**. For example, the eukaryotic protein complex Mediator acts as a bridge between RNA polymerase II and regulatory proteins bound to distant sites and is essential for transcription activation (see Chapter 15). Repression can also occur through proteins that do not bind the DNA directly but instead bind activator proteins and prevent the recruitment of RNA polymerase (**Figure 19-6b**). Repressors that act through protein-protein interaction rather than by binding DNA directly are called **corepressors**.

There could be an unintended consequence of gene regulation that uses DNA loops over large distances. A regulator meant to target a distant promoter could act instead on a different promoter located in the opposite direction. In eukaryotes, this problem is solved by the presence of **insulators**, short sequences of DNA that prevent inappropriate cross-signaling (**Figure 19-7**). Insulators are discussed further in Chapter 21.

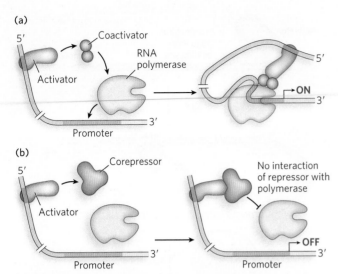

(a)

(b)

FIGURE 19-6 Transcription coactivators and corepressors acting as bridges. Coactivators and corepressors act indirectly, binding regulatory proteins without making direct contact with DNA. (a) Coactivators bind transcription activators and facilitate their function in activating RNA polymerase. (b) Corepressors bind transcription activators and inactivate their polymerase-activating function. In these examples, the activators are bound upstream from the promoter, but activator sites can also be located downstream.

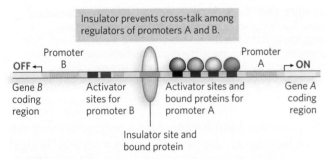

FIGURE 19-7 Insulators. Eukaryotic promoters have many regulatory elements that require DNA looping across long distances. Shown here are two genes, *A* and *B*, each with several activator-binding sites. When the regulatory sites for gene *A* are filled, the activators act on the promoter of gene *A*, but the insulator sequence blocks their action on the promoter of gene *B*. Insulators have bound proteins that enable the insulator function (see Chapter 21).

Regulators Often Work Together for Signal Integration

Activators and repressors often function at the same promoter. The use of multiple transcription factors allows the expression of a gene to be affected by more than one environmental condition. **Signal integration**, occurring in both eukaryotes and bacteria, is the control of a gene by multiple regulators in response to more than one environmental signal. A simple example in bacteria is the regulation of genes that encode products responsible for metabolizing sugar, the main energy source for bacteria (**Figure 19-8**).

Bacteria can derive energy from many different sugars, and they have sets of genes for metabolizing each one. But it would be a waste of cellular resources to express all of these genes all the time, and systems of regulation have evolved in which the genes for metabolizing a given sugar are expressed only when that sugar is present in the environment. Take, for example, the *lac* operon, a set of genes for the metabolism of lactose (see Chapter 5 and Chapter 15; operons are more fully defined later in this section). When lactose is not present, the Lac repressor protein is bound to the operon DNA at a sequence called the **operator** and ensures that

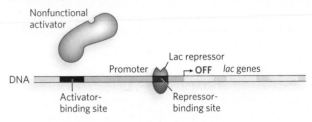

(a) Glucose present, lactose absent

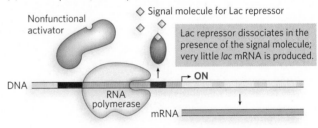

(b) Glucose present, lactose present

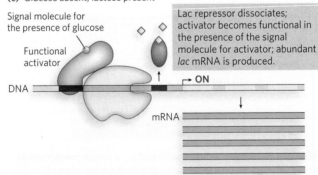

(c) Glucose absent, lactose present

FIGURE 19-8 Signal integration in gene expression. An activator and a repressor integrate two different environmental signals (the presence of glucose and of lactose) to control gene expression in the lactose-metabolic pathway of bacteria. (a) When lactose is absent, the Lac repressor binds the promoter and blocks RNA polymerase; there is no gene expression. (b) In the presence of lactose, the repressor binds a small signal molecule and separates from the DNA. The *lac* genes are now transcribed at a low, basal level. The presence of glucose keeps the activator in a nonfunctional state. (c) In the absence of glucose and presence of lactose, the activator binds a different small signal molecule, which causes it to bind DNA and recruit RNA polymerase for high-level gene expression.

the genes for lactose metabolism are not transcribed. When lactose is present, the cell sends a signal for the Lac repressor to dissociate from the operator, allowing transcription of the genes encoding lactose-metabolizing enzymes.

Though bacteria can metabolize many different sugars, their best energy source is glucose. When both glucose and lactose are present in the environment, the cell preferentially metabolizes glucose. It would be a waste of energy to continue producing the lactose-metabolizing enzymes, but the presence of lactose causes dissociation of the Lac repressor from the DNA. And yet, under these conditions, the genes encoding lactose-metabolizing enzymes are not highly transcribed. How does the cell do this? This is where signal integration comes in. The lactose-metabolizing genes are also under the control of an activator protein needed for the efficient transcription of the *lac* operon genes, even in the absence of the Lac repressor (see Figure 19-8). When glucose is present, the activator protein is kept in a nonfunctional form. But in the absence of glucose, the activator becomes functional and, provided lactose is present (and thus the Lac repressor is not bound to the operator), the genes for lactose metabolism are expressed at a high level.

This exquisite control, achieved by two different transcription factors working together, is an example of signal integration. The cell can adjust its energy resources by taking into account more than one environmental condition (the availability of glucose and of lactose).

Gene Expression Is Regulated through Feedback Loops

The regulation of gene expression usually operates as a feedback circuit. This is easier to explain in bacteria than in eukaryotes, although similar principles apply in both. Recall that genes for the metabolism of lactose are controlled by multiple transcription factors. The repressors and activators either bind DNA or do not, depending on signals received from the environment. The binding of a repressor or activator to DNA is often regulated by a molecular signal called an **effector**, usually a small molecule or another protein that binds the activator or repressor and causes a conformational change that results in an increase or decrease in transcription.

Repressors can be activated or inactivated by effectors. In one scenario, the effector binds to the repressor and induces a conformational change that results in dissociation of the repressor from its binding site on the DNA, allowing transcription to proceed (**Figure 19-9a**). Alternatively, the interaction of an inactive repressor and a signal molecule could cause the repressor to bind to DNA, shutting down transcription (**Figure 19-9b**).

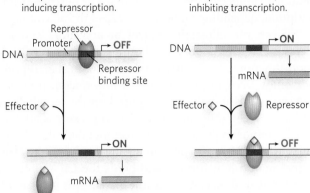

FIGURE 19-9 The role of effectors in negative regulation. The binding of signal molecules (known as effectors) to repressors can (a) relieve or (b) enhance repression. In (a), the repressor binds DNA in the absence of the effector; the external signal causes dissociation of the repressor to permit transcription. In (b), the repressor binds DNA in the presence of the signal, shutting down transcription. The repressor dissociates and transcription ensues only when the signal is removed (not shown).

The same considerations apply to activators. Some activators bind DNA and enhance transcription until dissociation of the activator is triggered by the binding of a signal molecule (**Figure 19-10a**). In other cases, the activator binds to DNA only after interaction with a

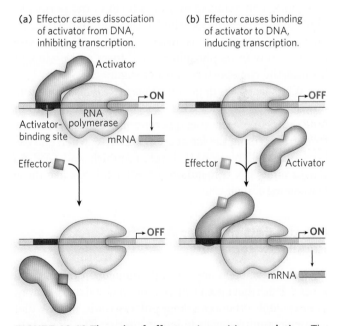

FIGURE 19-10 The role of effectors in positive regulation. The binding of effectors to activators can (a) inhibit or (b) enhance activation. In (a), the activator binds in the absence of the effector and transcription proceeds; when the signal is present, the activator dissociates and transcription is inhibited. In (b), the activator binds in the presence of the signal to stimulate transcription. The activator dissociates and transcription ceases only when the signal is removed (not shown).

signal molecule (**Figure 19-10b**). Signal molecules that bind activators can therefore increase or decrease transcription, depending on how they affect the activator.

Given the allosteric control of activators and repressors, we can understand how a regulatory feedback loop functions. In the bacterial *lac* operon, Lac repressor binds DNA in the absence of an effector, preventing the expression of genes required for the metabolism of lactose. The effector for the Lac repressor is allolactose, a minor byproduct of lactose metabolism. Therefore, when lactose is present in the environment, the signal molecule is formed and binds the Lac repressor, causing it to dissociate from the DNA. This gives RNA polymerase access to the promoter of the *lac* operon for a low, basal level of transcription. Transcription of the operon is greatly enhanced by binding of the activator cAMP receptor protein (CRP). CRP does not bind its regulatory site when glucose is available. In the absence of glucose, however, cells produce cAMP (cyclic AMP), which is an allosteric effector of CRP, producing a conformational change that enables CRP to bind its regulatory site. The bound activator then recruits RNA polymerase and boosts gene expression from the *lac* operon. A second level of control, called inducer exclusion, occurs when the lactose transporter is blocked by the glucose permease. The glucose permease is usually phosphorylated, and it becomes dephosphorylated on the transfer of phosphate to glucose during glucose transport. The dephosphorylated form of the permease binds and directly blocks the lactose and maltose transporters. When glucose is absent, the glucose permease exists mainly in its phosphorylated form and no longer inhibits the transporters of other sugars.

When lactose is depleted, allolactose is also depleted, and in the absence of this effector the Lac repressor again binds the operator site, preventing RNA polymerase from transcribing the *lac* operon. Likewise, when glucose becomes available, cAMP levels diminish and CRP no longer binds DNA. Regulatory feedback loops like these are common in all cells.

Related Sets of Genes Are Often Regulated Together

Bacterial promoters are often positioned upstream from several genes that operate in a common metabolic pathway. Transcription produces a long **polycistronic mRNA** that contains multiple genes in one transcript. The single promoter that initiates transcription of the cluster is the site of regulation for all the genes in the polycistronic message. The polycistronic DNA, its promoter, and all the additional sequences that function together in regulating its transcription are called an **operon** (**Figure 19-11**). Most operons contain 2 to 6 genes, but some have more than 20 genes.

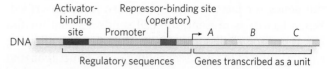

FIGURE 19-11 A bacterial operon. In this hypothetical operon, genes *A*, *B*, and *C* are transcribed as a single unit: a polycistronic mRNA. Typical regulatory sequences in the operon include binding sites for proteins that either activate or repress transcription from the promoter.

The organization of bacterial genes into operons allows small sets of genes that function together to be regulated together. But there are also instances in which multiple operons are controlled in a coordinated fashion. A group of operons with a common regulator is called a **regulon**. This arrangement allows shifts in cellular functions that can require the action of hundreds of genes—a major theme in the regulated expression of dispersed networks of genes in bacteria. Eukaryotes also exhibit global regulation of genes; genes that function together are dispersed over different chromosomes, yet are typically controlled in a coordinated way through common control elements and transcription factors. **Figure 19-12** shows a generalized view of global regulation, in which multiple genes may be turned on by the presence of the same activator or by the removal of a common repressor. Mechanisms of global transcriptional gene regulation in bacteria and eukaryotes are described in detail in Chapters 20 and 21.

Eukaryotic Promoters Use More Regulators Than Bacterial Promoters

Signal integration is important to gene regulation in both bacteria and eukaryotes. However, eukaryotic promoters for Pol II, the RNA polymerase that transcribes protein-coding genes, typically contain more regulatory binding sites than do bacterial promoters (**Figure 19-13**). The use of more transcription factors in eukaryotic gene control reflects the greater need for gene regulation in a more complex organism with a larger genome. For example, nonspecific DNA binding of regulatory proteins could become a problem in the much larger genomes of higher eukaryotes, because the chance that a specific binding sequence will occur randomly at an inappropriate site increases with genome size. Indeed, the number of transcription factor–binding sites in eukaryotic promoter regions varies with the complexity of the organism. Genes in single-celled yeasts have only a few regulator sites and are not much more complicated than bacterial genes, whereas the promoters in multicellular organisms can have 10 or more regulator-binding sites spaced over long distances, 50 kbp or more away from the transcription start site. Specificity for transcriptional regulation is

(a) **Positive regulation by activators**

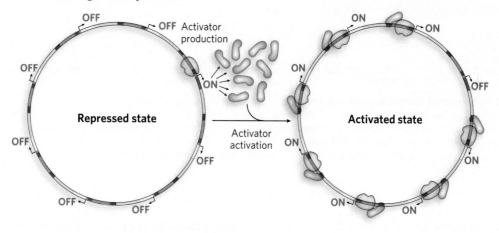

(b) **Regulation by destruction of repressors**

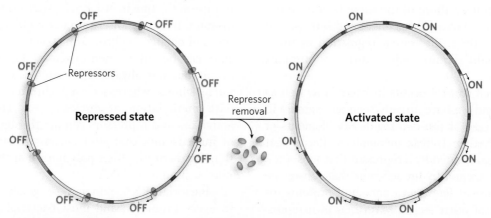

FIGURE 19-12 Global regulation of groups of genes. (a) Global regulation can occur through the binding of a common transcription activator. When needed, the activator may be produced de novo by expression of its gene, as shown, or an existing activator protein may become active for DNA binding through interaction with another protein or a small effector molecule. (b) Alternatively, global regulation can result from the removal of a common repressor bound to DNA sites, either by an allosteric change induced by binding of a small effector molecule or by proteolytic digestion of the repressor.

(a) **Bacteria**

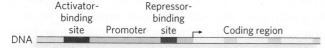

(b) **Eukaryotes**

FIGURE 19-13 Bacterial and eukaryotic regulatory regions compared. (a) Bacterial promoters are usually regulated by only one or two transcription factors, and their binding sites are typically near, or overlap, the promoter. (b) Eukaryotic genes, especially those of multicellular organisms, usually have numerous regulator-binding sites spanning a large region (sometimes more than 50 kbp) located upstream and/or downstream from the promoter, or even within the coding sequence of the gene itself (not shown).

improved by multiple regulatory proteins that must bind DNA and form a multiprotein complex to become active. This multiprotein requirement vastly reduces the probability of random gene activation or repression.

Multiple Regulators Provide Combinatorial Control

Using multiple transcription factors for every gene in a genome would be energetically costly if every gene required unique regulators, but the use of different combinations of a limited set of transcription factors to differentially regulate many genes provides an opportunity for efficiency. This is made possible by **combinatorial control**—the need for specific combinations of factors to unlock each particular gene (**Figure 19-14**). Consider the hypothetical genes *A*, *B*, and *C*, each of

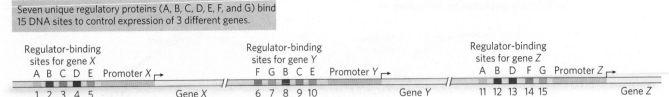

FIGURE 19-14 Combinatorial control in gene regulation. Each of these three hypothetical eukaryotic promoters requires five different regulatory proteins, to bind a total of 15 regulatory sites. Each color represents a particular transcription factor and its regulatory binding site. Each gene uses different combinations of transcription factors, and some factors are used for more than one gene. In total, there are seven unique regulatory sequences, and thus seven unique transcription factors, controlling expression of all three genes.

which requires five transcription factors. If each factor were distinct, the cell would require 15 transcription factors to control the expression of these three genes. But if genes A and B used three of the same factors, and a combination of the factors for genes A and B is used to regulate gene C, then differential regulation of these three genes would require only 7 different proteins instead of 15.

Combinatorial control occurs in bacteria as well as in eukaryotes, and we have already seen an example in bacteria in the case of the two regulatory elements of the genes involved in lactose metabolism. Recall that these genes are controlled by a repressor that senses the presence of lactose and by an activator that senses the presence of glucose. The genes encoding proteins for the metabolism of other sugars have their own repressors, but they use the same activator. For instance, the digestion of galactose requires removal of the galactose repressor from the DNA, and this occurs only when galactose is present. However, as with the lactose genes, high expression of the galactose genes is achieved only when glucose is absent from the environment. The same protein activator used at the *lac* operon also regulates the galactose genes: CRP, which becomes functional by binding its effector molecule cAMP when glucose is not present. The regulation of the genes for different sugar-metabolizing pathways by a common activator is an example of combinatorial control.

Regulation by Nucleosomes Is Specific to Eukaryotes

In eukaryotes, transcription initiation almost always depends on the action of activator proteins. One important reason for the apparent predominance of positive regulation seems obvious: packaging of DNA into chromatin renders most promoters inaccessible, and thus their associated genes are silent. Chromatin structure affects access to some promoters more than others,

but generally, repressors that prevent the access of RNA polymerase to DNA would be redundant. Therefore, eukaryotic genes are constitutively repressed and require activation in order to be transcribed. Recall from Chapter 10 that transcription is regulated by different types of change in chromatin structure. The chromatin state can be either open or closed. Open chromatin is often (but not always) associated with acetylation of nucleosomes, whereas closed chromatin is associated with methylation of nucleosomes. Thus eukaryotic activators and repressors can act through modification of nucleosomes that alter chromatin structure, rather than by recruiting RNA polymerase or preventing polymerase binding to DNA.

Bacterial RNA polymerase generally has access to every promoter, and most bacterial genes are controlled by specific repressors. In eukaryotes, however, general repression by nucleosomes, combined with the use of activators to regulate transcription, is more efficient than the use of specific repressors. If the 20,000 to 25,000 genes in the human genome were negatively regulated, each cell would have to constantly synthesize specific repressors to prevent the transcription of a great many genes. Instead, the nucleosomes that function to condense DNA also repress most genes, and the cell has to synthesize only the activators needed to promote transcription of the subset of genes required at a particular time. These arguments notwithstanding, there are examples of negative regulation in eukaryotes, from yeast to humans.

SECTION 19.1 SUMMARY

- The various mechanisms of transcription initiation are among the most well-documented regulated processes in gene expression. Transcription initiation is the step most often regulated, and regulation at this point is the most energy efficient, because it occurs before the investment of energy in mRNA and protein synthesis.

- Transcription initiation is mediated by intrinsic promoter affinity for RNA polymerase or by repressor and activator proteins that modulate promoter affinity for the polymerase.
- Repressors can hinder transcription by binding DNA at a site that prevents RNA polymerase binding or by preventing the closed-to-open transition of the polymerase-promoter complex (negative regulation).
- Activators promote RNA polymerase binding through cooperativity or promote formation of the open complex by causing a conformational change in the promoter or the polymerase (positive regulation).
- Binding sites for transcription factors need not be close to the transcription start site, and in eukaryotes they are often located thousands of base pairs from the promoter. Regulatory proteins that bind sites distant from the promoter exert their effects through DNA looping.
- Promoters may be controlled by two or more transcription factors, allowing integration of signals from more than one environmental variable.
- Small signal molecules (effectors) allosterically regulate the function of activators and repressors.
- Sets of genes that function in one pathway are often controlled simultaneously.
- Eukaryotes generally have more transcription factors than bacteria, reflecting the greater need for regulated gene expression in a complex multicellular organism. Specificity of gene expression is enhanced by the use of multiple regulators.
- In combinatorial control, the same regulatory protein is used to control different genes in combination with other regulators, forming a multiprotein regulator that is specific for individual genes.
- Chromatin structure renders most eukaryotic promoters inaccessible to RNA polymerase and plays an important role in gene expression, which typically requires proteins that modify nucleosomes and open up the chromatin structure to transcriptional activation.

19.2 THE STRUCTURAL BASIS OF TRANSCRIPTIONAL REGULATION

As we saw in Chapter 4, protein structures come in an amazingly wide assortment of shapes and sizes, but these structures can often be broken down into discrete functional domains formed by common structural motifs. Indeed, most transcription factors use a surprisingly small subset of structural motifs to interact with regulatory sites on DNA, with RNA polymerase, or with other regulatory proteins. In fact, some of the common motifs in the architecture of transcription factors can be determined from analysis of the protein's primary sequence. We explore here the most common transcription factor domain structures that are involved in DNA binding and interaction with other proteins.

Transcription Factors Interact with DNA and Proteins through Structural Motifs

The recognition of DNA by a regulatory protein almost always occurs through certain amino acid side chains of an α helix referred to as the **recognition helix**. A limited set of structural motifs function to present the recognition helix to the DNA. Amino acid side chains in this helix usually "read" the DNA sequence along the major groove, because (as you will recall from Chapter 5) more hydrogen-bond donor and acceptor atoms of nucleotide bases are found in the major groove than in the minor groove (**Figure 19-15**).

The DNA binding sites for regulatory proteins are often short inverted nucleotide repeats where multiple (usually two) subunits of the regulatory protein bind cooperatively (**Figure 19-16**). Accordingly, many bacterial and eukaryotic activators and repressors are dimers. Crystal structures of activators and repressors bound to DNA show that each monomer of a homodimer binds the same nucleotide sequence within the inverted repeat. Regulatory proteins use several structural motifs to promote dimerization.

Examples of DNA-binding and protein-dimerization motifs are described in Chapter 4. Here we focus on those that play prominent roles in the function of regulatory proteins.

Helix-Turn-Helix Motif Bacterial regulators most commonly use the **helix-turn-helix motif** to present the recognition helix to DNA, and several eukaryotic regulatory proteins also interact with DNA through this motif. The helix-turn-helix motif consists of about 20 amino acid residues that form two short α helices connected by a β turn (**Figure 19-17**). This motif lacks intrinsic stability and is generally part of a somewhat larger DNA-binding domain. Only one of the two α-helical segments serves as the recognition helix; it packs against other regions of the protein and protrudes from the protein surface for insertion into the major groove.

Homeodomain Motif Researchers first identified the **homeodomain motif** as a conserved 60 amino acid sequence in transcription activators encoded by genes that regulate body pattern development in fruit flies. We now know that the homeodomain is found in proteins from a wide variety of multicellular organisms, including

(a)

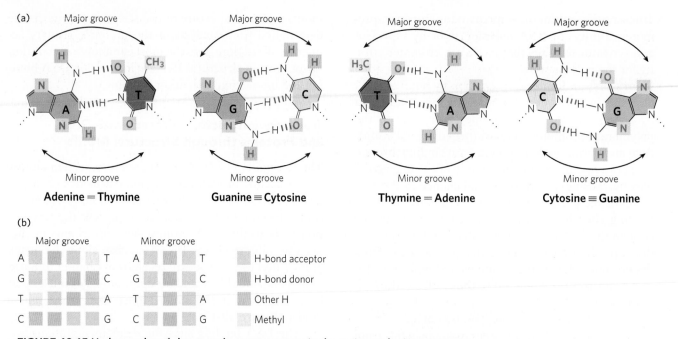

(b)

FIGURE 19-15 Hydrogen-bond donor and acceptor atoms in the major and minor grooves of DNA. (a) Shown here are the functional groups (in red) of all four base pairs as displayed in the major and minor grooves. Hydrogen-bond donor and acceptor atoms that can be used for base-pair recognition by proteins are indicated by light brown and blue squares, respectively. Other hydrogen atoms are indicated by purple squares, and methyl groups by yellow squares. (b) The atoms (color-coded as shown on the right) that can be used for protein recognition for each base pair are grouped as sets of four in each major groove (left) and as sets of three in each minor groove (middle). Notice how the sequence of the atoms in both the major and minor groove differs for the different base pairs, and how the four possible base pairs are chemically distinct in the major groove but not in the minor groove.

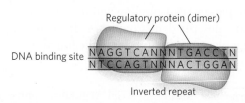

FIGURE 19-16 An inverted repeat at the site of transcription factor binding. A nucleotide sequence followed by the reverse, complementary sequence is known as an inverted repeat. It can have a variable number of base pairs that are not part of the repeat between the two repeated sequences. A palindrome is an inverted repeat with no base pairs between the two repeat sequences. Proteins that bind to inverted repeats are dimeric, and each subunit binds to one half of the repeat. Illustrated here is a homodimer binding an inverted repeat (N = any nucleotide).

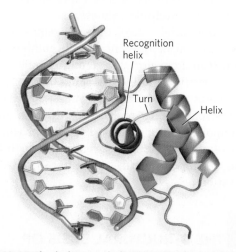

FIGURE 19-17 The helix-turn-helix DNA-binding motif. This molecular structure shows the DNA-binding domain of the bacterial Lac repressor (as a ribbon structure) interacting with the major groove of DNA. The helix-turn-helix motif is red and green; the DNA recognition helix is red. [Source: PDB ID 1LCC.]

humans. When the structure of the homeodomain was determined, it was found to contain a helix-turn-helix motif—but with some important differences. First, the homeodomain is composed of three α helices, only two of which correspond to the helix-turn-helix motif. Second, the N-terminal residues of the homeodomain reach around the DNA and interact with the minor groove **(Figure 19-18)**.

Basic Leucine Zipper and Basic Helix-Loop-Helix Motifs

The **leucine zipper motif** is an amphipathic α helix, with a series of hydrophobic amino acid residues concentrated on one side of the helix. A striking feature of this α helix is the occurrence of Leu residues at every seventh

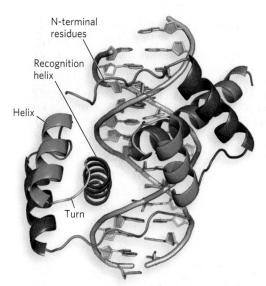

FIGURE 19-18 The homeodomain DNA-binding motif. This molecular structure shows the homeodomain motifs of the *Drosophila* transcription factor known as Paired, a dimeric protein (only a small part of the much larger Paired dimer is shown). The recognition helix in each subunit is stacked on two other α helices and can be seen protruding into the major groove. The N-terminal residues reach around the DNA and interact with the minor groove. [*Source: PDB ID 1FJL.*]

position, forming a hydrophobic surface along one side of the helix—the site where two identical subunits dimerize. Researchers initially thought that dimerization was caused by interdigitating Leu residues (hence the name "zipper"). We now know that the two subunits dimerize through packing of the residues along the inner surface of the interface, where they form a coiled-coil structure. Certain transcription factors use leucine zippers in combination with basic residues at one end of the α helices that make up the recognition helices, forming **basic leucine zipper motifs** (**Figure 19-19**). These basic leucine zipper regulators are sometimes referred to as bZIP proteins. The crystal structure of bZIPs shows that they grip and bind DNA like a set of tongs (see Figure 19-19b). However, there are many regulatory proteins that use leucine zipper helices only for dimerization and contain a separate motif for DNA binding. Leucine zippers are found in many eukaryotic transcription activators and in a few bacterial regulators.

A somewhat similar structural motif in some eukaryotic transcription factors is the **basic helix-loop-helix motif** (**Figure 19-20**). These proteins share a conserved region of about 50 residues that are important for both DNA binding and protein dimerization. The basic helix-loop-helix region contains two amphipathic α helices—one of which contains basic residues—linked by a loop of variable length. Dimer formation is mediated by one set of amphipathic α helices, and DNA binding is mediated by the amphipathic α helices that contain basic residues.

The recognition helices grip the binding sequence in DNA in much the same way as the basic leucine zipper.

Zinc Finger Motif The **zinc finger motif** comes in a few varieties, two of which are discussed here. A zinc finger domain consists of about 30 residues that form an elongated loop held together at the base by a single Zn^{2+} ion. The Zn^{2+} ion is coordinated to four amino acid side chains, usually four Cys residues or two Cys and two His residues. The zinc functions to stabilize the motif, which presents a recognition helix to DNA; the Zn^{2+} does not interact with the DNA directly. The interaction of a single zinc finger with DNA is typically weak, and this particular binding motif has the unique feature that the protein can have multiple copies of the motif that act together as a chain. Multiple zinc fingers are found in many DNA-binding proteins, and they substantially enhance binding affinity by interacting simultaneously with the DNA. In fact, one DNA-binding protein of the frog *Xenopus laevis* has 37 zinc fingers! In the mouse regulatory protein Zif268, the zinc finger domains present the recognition helix so that it winds around the DNA following the major groove (**Figure 19-21**). Zinc finger proteins in this class are among the few regulatory proteins that function as monomers. They do not require the DNA binding site to be an inverted repeat and can recognize a long sequence of DNA that contains no internal repeat sequences, because each recognition helix is unique.

Another type of regulatory zinc finger protein combines the Zn^{2+}-binding motif with the helix-turn-helix motif (**Figure 19-22** on p. 679). This type of zinc finger protein uses two Zn^{2+} ions to stabilize the DNA-binding domain, which has a helix-turn-helix motif. These proteins bind to the DNA as dimers; the example shown in Figure 19-22 uses a leucine zipper to mediate the dimer contacts. Zinc finger motifs are common in eukaryotes, and there are a few examples among bacterial regulators.

Transcription-Activation Motifs In addition to structural domains devoted to DNA binding and protein dimerization, transcription activators contain regions used for recruiting RNA polymerase or other protein factors. These recruitment regions are thought to be relatively unstructured. The first one was noted in the Gal4 protein (Gal4p), a yeast transcription activator. Researchers observed that several acidic residues were associated with the activating function, and the region could be altered by mutation without much effect on function; the region was referred to as an "acid blob."

Other activation domains of transcription factors have also been shown to contain unstructured regions characterized (like Gal4p) by acidic residues or by other types of amino acids. For example, certain activation

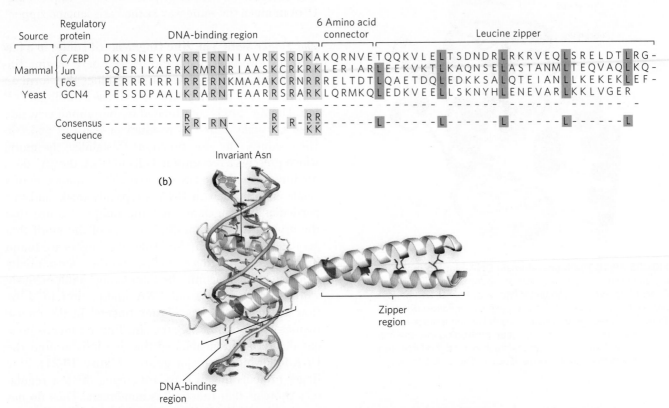

FIGURE 19-19 The basic leucine zipper motif. This motif is often used to mediate protein-protein interactions in eukaryotic transcription factors. (a) The amino acid sequences of several basic leucine zipper (bZIP) proteins. Notice the Leu (L) residues at every seventh position in the zipper region and the number of basic residues—Lys (K) and Arg (R), and one invariant Asn (N)—in the DNA-binding region. A consensus sequence is shown at the bottom. (b) A basic leucine zipper from the yeast activator protein GCN4. Only the two "zippered" α helices, each from a different subunit of the dimeric protein, are shown. The helices wrap around each other in a coiled-coil. This molecular structure shows the interacting Leu residues (red), the basic residues in the DNA-binding region (yellow), and the invariant Asn residue (blue). [Source: (b) PDB ID 1YSA.]

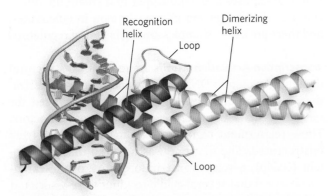

FIGURE 19-20 The basic helix-loop-helix motif. This ribbon model shows the human homodimeric transcription factor Max bound to its DNA target site. The two amphipathic α helices of each Max subunit are shown in red and white; the loop is yellow in both. The two subunits form a four-helix bundle through association of their dimerizing α helices (white), and the DNA-binding α helices (red) extend from the bundle. [Source: PDB ID 1HLO.]

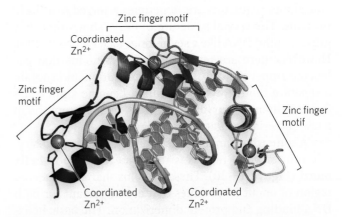

FIGURE 19-21 The zinc finger motif. This ribbon structure of a fragment from the mouse regulatory protein Zif268 shows three zinc fingers (different colors) arranged one after the other in the protein. The Zif268 recognition helices enter the major groove of the DNA, and the three fingers wind around the DNA helix. [Source: PDB ID 1ZAA.]

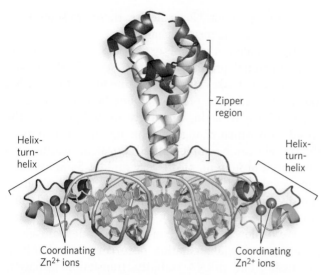

FIGURE 19-22 Zinc finger, helix-turn-helix, and leucine zipper motifs in the same regulatory protein. The Gal4 protein (Gal4p), a yeast transcription activator, is a dimer, held together by a leucine zipper. Each of the two DNA-binding domains contains two Zn²⁺ ions, which help hold the recognition helix (red) of the helix-turn-helix motif in the proper geometry for DNA recognition. [*Source: PDB ID 3COQ.*]

domains are glutamine-rich or proline-rich, and also appear to be unstructured. The current understanding is that these activation regions function as many short sections of amino acid residues that act together like a patch of Velcro: the more there are, the greater their effect. This somewhat unstructured approach to achieving activation might enable transcription factors to extend their range of protein-protein interactions for combinatorial control (see Section 19.1).

Transcription Activators Have Separate DNA-Binding and Regulatory Domains

Transcription activators typically contain a regulatory domain that is separate from and functionally independent of the DNA-binding domain. An early and now classic experiment demonstrating this property of transcription activators was performed by Mark Ptashne and his colleagues (**Figure 19-23**). They proposed that the regulatory and DNA-binding functions of a transcription activator are independent and separable. To test this idea, they spliced together DNA encoding the transcription-activation domain of a eukaryotic activator and DNA encoding the DNA-binding domain of a bacterial repressor. The prediction was that the fusion protein would activate transcription of a gene under the control of the eukaryotic activator, provided that the gene contained an upstream binding sequence recognized by the bacterial repressor. The researchers chose for their study the yeast transcription activator Gal4p, which drives the expression of genes for galactose metabolism, including the gene *GAL1*. The C-terminal activation sequences of Gal4p were fused to the N-terminal DNA-binding region of an *E. coli* repressor, LexA. The new gene encoding the LexA-Gal4 fusion protein was inserted into a plasmid and transferred to yeast, along with a second plasmid that contained the *GAL1* gene promoter fused to the bacterial β-galactosidase gene, *lacZ*. As

Mark Ptashne [*Source: Courtesy Mark Ptashne.*]

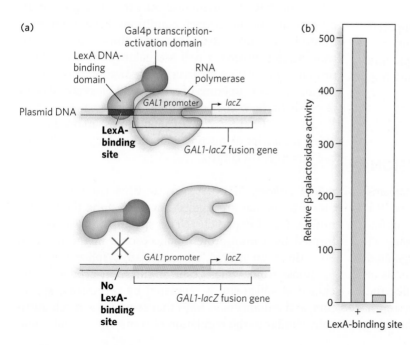

FIGURE 19-23 Experiment demonstrating the separate DNA-binding and regulatory domains of transcription activators. (a) A *GAL1-lacZ* fusion gene is inserted in a plasmid downstream from a LexA-binding site (top), or downstream from a DNA segment lacking the LexA-binding site (bottom). In both cases, cells are transformed with a plasmid encoding a fusion protein containing the bacterial LexA DNA-binding domain fused to the transcription-activation region of yeast Gal4p. (b) Expression of the *GAL1-lacZ* fusion gene is measured as described in the text. Expression of β-galactosidase is induced in cells containing plasmids with the LexA-binding site. [*Source: Data from R. Brent and M. Ptashne, Cell 43:729–736, 1985.*]

predicted, in the transformed yeast, the new LexA-Gal4 fusion protein activated the *GAL1-lacZ* gene construct containing an upstream LexA-binding sequence, but it did not activate the *GAL1-lacZ* gene on the plasmid that lacked the LexA-binding site.

As shown in Figure 19-23, the LexA-Gal4 fusion protein activated the production of β-galactosidase more than 500-fold when the LexA-binding site was present (compared with when absent) in the *GAL1-lacZ* gene construct. Production of β-galactosidase is easily measured by using a synthetic substrate that yields a blue product when cleaved by the enzyme. The LexA-Gal4 fusion protein did not activate transcription from the wild-type *GAL1* promoter, because the protein no longer recognized the Gal4p-binding site upstream from the *GAL1* gene.

This experiment not only demonstrated the modular nature of regulatory proteins but also gave molecular biologists yet another tool for studying the inner workings of the cell. The widely used yeast two-hybrid assay makes use of the fact that the DNA-binding and transcription-activation regions of the Gal4p regulatory protein are stable, separate domains (see Chapter 7).

SECTION 19.2 SUMMARY

- The DNA-binding domains of transcription factors are usually constructed from a limited set of structural motifs, including the helix-turn-helix, homeodomain, leucine zipper, helix-loop-helix, and zinc finger motifs.
- Many transcription factors form dimers and bind inverted repeat sequences, thereby increasing their affinity for DNA.
- The basic leucine zipper and basic helix-loop-helix motifs facilitate both DNA binding and protein dimerization.
- Many transcription-activation domains are composed of acidic, proline-rich, or glutamine-rich regions.
- Transcription activators typically contain separable and functionally independent DNA-binding and regulatory domains.

19.3 POSTTRANSCRIPTIONAL REGULATION OF GENE EXPRESSION

Thus far we have discussed the regulatory mechanisms involved in initiation of transcription, but as Figure 19-1 shows, regulatory mechanisms operate at many steps following transcription. Posttranscriptional processing of RNA and translation of mRNA into protein are regulated at several points. The regulation of protein synthesis at the initiation stage is common because, much like the strategy of regulating transcription at the initiation step, it saves the cell the huge energy investment of synthesizing

the protein product. Nevertheless, there are some posttranslational regulatory mechanisms.

The regulation of gene expression *after* production of a functional protein does have several advantages. For example, it takes substantial time to produce mRNA and translate it into protein, so one benefit of regulating a pathway by acting on a fully formed protein is the speed with which changes in the amount or activity of the protein can be implemented. Covalent modification can turn a protein on or off very rapidly. Some types of gene regulation can be inherited through generations of cell division through imprinting (see Chapter 21) and epigenetics (see Chapter 10).

We explore here some of the main mechanisms of posttranscriptional regulation, occurring through mRNA processing and mRNA stability, translation initiation, covalent modification, cellular localization, and protein degradation.

Some Regulatory Mechanisms Act on the Nascent RNA Transcript

After transcription initiation, there are several ways in which a gene can be regulated before the mature mRNA transcript is produced. As an overview, we briefly describe three steps at which regulation can take place: transcript elongation, mRNA splicing, and modification of mRNA termini. These, and other examples, are discussed in more detail in Chapter 20 (for bacteria) and Chapter 22 (for eukaryotes).

Transcript Elongation One bacterial example of regulation affecting transcript elongation is a process known as attenuation. Attenuation prevents movement of the transcribing RNA polymerase into the first gene of an operon unless the proper conditions have been met. Controls to stop attenuation and thus proceed with transcription involve a delicate balance of metabolites, proteins, and mRNA structure. Attenuation is relatively common for the operons of amino acid biosynthesis and is particularly well documented for the *trp* operon of *E. coli* (see Chapter 20). In eukaryotes, many factors affect transcript elongation, and these elongation factors can be targets of control.

mRNA Splicing Many eukaryotic RNA transcripts contain introns that are spliced out in forming the mature mRNA (see Chapter 16). The splicing process is performed by a multiprotein spliceosome in the nucleus. Sometimes an mRNA has alternative splice junctions to choose from, which result in different products. The choice of splice site is regulated by repressors, activators, and enhancers in ways that seem to be mechanistically similar to the regulation of transcription initiation.

Alternative splicing choice is thus another point at which gene expression can be regulated.

Modification of mRNA Termini Both the 5′ and 3′ ends of eukaryotic mRNAs are highly modified in multistep reactions (see Chapter 16). The 5′ terminus is modified by the addition of nucleotides connected by unusual phosphodiester bonds, referred to as the 5′ cap. The 3′ terminus is cleaved at a particular site prior to transcription termination, then multiple A residues are added to form a poly(A) tail. Specific proteins recognize and bind to these modifications, which are important in mRNA transport from the nucleus, mRNA stability in the cell, and efficient association of the mRNA with ribosomes and its use in translation. Exciting new discoveries are being made about control mechanisms at the level of these mRNA modification and transport steps.

Small RNAs Can Affect mRNA Stability

The amount of protein generated from a gene is dependent on the stability of the RNA message, and regulatory mechanisms have evolved to control mRNA stability. In higher eukaryotes, certain genes are "silenced" by a class of RNAs that interact with mRNAs, resulting in degradation of the mRNA or inhibition of translation. This form of gene regulation uses small RNAs. A variety of small RNAs can control developmental timing, repress the activity of transposons, or destroy invading RNA viruses, especially in plants, which lack an immune system. Small RNAs may also play a role in heterochromatin formation, which silences all the genes contained in the heterochromatin. The role of these small mRNAs in eukaryotic gene regulation is explored further in Chapter 22. Bacteria also contain a variety of small RNAs that act at several levels to regulate gene expression.

Small RNAs are sometimes called microRNAs (miRNAs). When present only temporarily, such as during development, transient small RNAs are referred to as small temporal RNAs (stRNAs). Hundreds of different miRNAs have been identified in more complex eukaryotes. They are transcribed as precursor RNAs, about 70 nucleotides long, that form hairpinlike structures (**Figure 19-24a**). An endonuclease trims the precursor RNAs to form short duplexes of 20 to 25 nucleotides, one strand of which anneals to the target mRNA. The best characterized of these endonucleases is Dicer; endonucleases in the Dicer family are widely distributed in eukaryotes.

When the expression of genes producing miRNAs goes awry, tumors can result. As described in this chapter's Moment of Discovery, overexpression of the *miR-17-92* cluster of miRNAs results in tumor formation in mice. This result was one of the first to associate

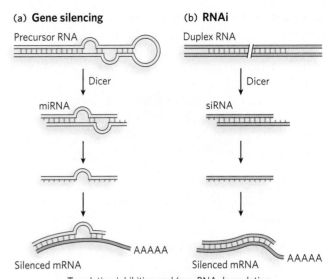

FIGURE 19-24 Gene silencing and RNA interference. (a) Dicer cleaves hairpin-shaped precursor RNAs into microRNAs (miRNAs), which bind to and silence mRNA by inhibition of translation. (b) Synthetic double-stranded RNA can also result in RNA interference (RNAi). When the double-stranded RNA is injected into a cell, Dicer cleaves it into small interfering RNAs (siRNAs), which interact with the target mRNA; the mRNA is degraded or its translation is inhibited.

functional RNAs with tumorigenesis, implicating them in the development of cancer.

Gene regulation mechanisms involving Dicer, besides their important physiological role, also have a very useful practical application. If an investigator introduces into an organism a synthetic duplex RNA corresponding to a target mRNA, Dicer cleaves the duplex into short segments called small interfering RNAs (siRNAs), which bind to and silence the mRNA (**Figure 19-24b**). This laboratory technique is called **RNA interference (RNAi)**. In plants, almost any gene can be shut down in this way. In nematodes, simply feeding the duplex RNA to the worm silences the target gene. The technique is a very important tool in studies of gene function, because any gene can be silenced without constructing a mutant organism. The study of functional RNAs (such as miRNAs) is an exciting and relatively new area of molecular biology—a field to watch for future medical advances.

Some Genes Are Regulated at the Level of Translation

Some translational regulation does occur in bacteria, but it is much more common in eukaryotes because of the long half-lives of many eukaryotic mRNAs. Most translational regulation occurs at the initiation step, for efficiency and energy conservation (**Figure 19-25**). For instance, translation of a eukaryotic mRNA requires several different initiation factors that assemble at the

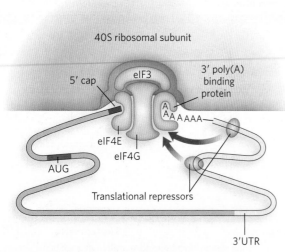

FIGURE 19-25 Regulation of translation initiation in eukaryotes. One of the most important mechanisms for translational regulation in eukaryotes involves the binding of repressors to specific sites in the 3′ untranslated region (3′UTR) of the mRNA. The repressors interact with initiation factors eIF3, eIF4E, and eIF4G or with the ribosome to prevent or slow translation. The factor eIF4G mediates an interaction between eIF4E at the 5′ cap and poly(A) binding protein at the 3′ poly(A) site (see Chapter 18). This interaction is needed for efficient translation, and factors that disrupt it repress translation.

5′ region of the mRNA on the 40S ribosomal subunit (see Chapter 18). Many initiation factors are subject to a variety of regulatory mechanisms, which in turn modulate translation at the initiation step.

Global translational control also exists in bacteria and eukaryotes. For example, the ribosomal apparatus represents a large energy investment for the cell, and synthesis of the many components of the ribosome is regulated in processes linked to the cellular demand for proteins. Translational control of dozens of genes encoding ribosomal components is regulated by protein binding to the translation start sites in the mRNAs. Furthermore, if rRNAs are not present in sufficient amounts to match ribosomal protein subunits, the excess unassembled ribosomal proteins bind and inhibit translation of their respective mRNAs, forming a feedback translational control circuit.

Some Covalent Modifications Regulate Protein Function

Protein function can be dramatically altered by many types of covalent modification. Protein modifications include phosphorylation, acetylation, methylation, glycosylation, ubiquitination, and sumoylation (further discussed later in this section). The modifications can have various effects: they may render the protein active or inactive; result in a change in oligomeric state, with functional consequences; alter the protein's affinity for

DNA or for another protein; or affect the protein's stability in the cell.

An example of proteins that are highly regulated by covalent modification is the subunits of nucleosomes. Recall from Chapter 10 that nucleosomal proteins have long N-terminal tails that are often covalently modified by phosphorylation, methylation, and acetylation. These modifications regulate transcription by changing the level of chromatin compaction, thus controlling the access of RNA polymerase and other proteins to the DNA. About 10% of the chromatin in a typical eukaryotic cell is in a more condensed form (heterochromatin) than the rest of the chromatin, and genes in these regions are strongly repressed. Most of the remaining, less-condensed chromatin (euchromatin) is transcriptionally active. Histones found in condensed and less-condensed chromatin differ in their patterns of covalent modification. These modification patterns are probably recognized by enzymes that alter the structure of chromatin (see Chapter 10). The effects of nucleosome modification on chromosome structure, and therefore on gene regulation, have no clear parallel in bacteria because bacterial chromosomes are not packaged in this way.

Modifications associated with the activation of transcription are recognized by enzymes that make the chromatin more accessible to the transcriptional machinery. When transcription of a gene is no longer required, certain modifications are enzymatically removed and others are added, marking the chromatin as transcriptionally inactive. The effect of histone modification on gene expression is discussed further in Chapter 21. Other examples of covalent modifications that direct gene expression are briefly described below and are expanded upon in Chapters 20–22.

Gene Expression Can Be Regulated by Intracellular Localization

In bacteria, transcription repressors and activators can undergo an allosteric change on binding a small effector molecule (such as allolactose or cAMP) that acts as a signal of environmental conditions, and in this way gene expression is repressed or activated in response to the signal (see Section 19.1). The compartmentation of eukaryotic cells affects the way in which gene expression can respond to environmental signals and provides opportunities for regulation at the level of transfer of proteins between intracellular locations. A prevalent pathway for communication with the extracellular environment in eukaryotes is through cell surface receptors. The receptors bind a signal molecule and relay its message through the plasma membrane by complex signal transduction pathways, eventually resulting in transcriptional regulation in the nucleus.

A relatively simple example of a signal transduction pathway is the JAK-STAT pathway (**Figure 19-26**). This

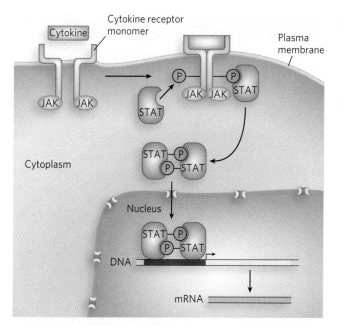

FIGURE 19-26 Signal transduction by the JAK-STAT pathway. Cytokines signal a cell to increase transcription, and they act through a membrane-bound receptor protein. Cytokine binding to the receptor causes two receptor molecules to form a dimer, resulting in a conformational change that enables the JAK kinase to phosphorylate the receptor. This attracts the STAT protein, which in turn becomes phosphorylated and dimerizes, whereupon it enters the nucleus and activates the transcription of specific STAT-regulated genes.

consists of a transmembrane receptor, a protein kinase called JAK (*J*anus *k*inase), and a transcription factor called STAT (signal *t*ransducer and *a*ctivator of *t*ranscription). The transmembrane receptor binds cytokines (e.g., interferon and interleukin), small molecules that signal cells to grow or differentiate. On cytokine binding, the receptor activates JAK, which phosphorylates the receptor. This, in turn, promotes binding and phosphorylation of STAT by the phosphorylated receptor. Once phosphorylated, STAT dimerizes and enters the nucleus, where it binds DNA and activates the expression of genes involved in cell growth and differentiation. There are at least seven different STATs in mammals, each binding a different DNA sequence. The JAK-STAT pathway is conserved in organisms ranging from worms to mammals, indicating its importance to cellular function. Genetic defects in this pathway are associated with immune diseases and cancer.

Dephosphorylation by specific protein phosphatases is also important in the regulation of transcription activator or repressor activity, and this sometimes involves cellular localization as a means of regulating gene expression—that is, access of the transcription factors to the nucleus can be regulated by their state of phosphorylation.

Phosphorylation-dephosphorylation often causes conformational changes that alter the activity of a regulatory

protein other than an activator or repressor. For example, the target could be a protein that binds an activator and masks its function. When the masking protein is phosphorylated, it dissociates from the activator and gene expression is thus enhanced. The hormone insulin regulates gene expression in this way, by phosphorylation-dephosphorylation of proteins involved in glucose metabolism. These mechanisms short-circuit the need for changes in mRNA or protein synthesis (**Highlight 19-1**). Insulin also regulates gene expression through a protein kinase signaling mechanism that ultimately activates the transcription of numerous genes involved in cellular metabolism.

Steroid hormone receptors are another example of regulation by intracellular localization. These receptors are transcription activators, held in the cytoplasm by association with a heat shock protein, Hsp70. Steroid hormones are soluble in lipids and can pass through the plasma membrane without a specific transporter. On entry of the steroid hormone into a cell that expresses the particular steroid-binding receptor, and binding to the receptor, Hsp70 dissociates and the receptor-hormone complex dimerizes and enters the nucleus (**Figure 19-27**).

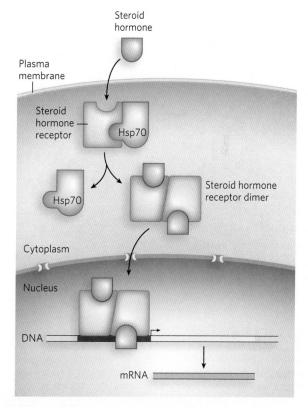

FIGURE 19-27 The regulation of a steroid hormone receptor by cellular localization. A steroid hormone enters the cell and binds its receptor, which is held in the cytoplasm by interaction with heat shock protein Hsp70. Hormone binding stimulates dissociation of the Hsp70 and dimerization of the hormone-receptor complex, which migrates into the nucleus and binds its regulatory site, activating gene transcription.

Insulin Regulation: Control by Phosphorylation

Insulin is a small peptide hormone (51 residues), produced in the pancreas, that is central to the control of energy and glucose metabolism. For example, insulin stimulates glucose uptake from the blood by muscle, fat, and liver cells, and the use of glucose (in preference to fat) as an energy source. In these duties, insulin acts as a regulator of gene transcription.

The insulin signaling pathway involves an extensive protein kinase cascade, resulting in activation of more than 100 genes. The signaling cascade is initiated when insulin binds to the membrane-bound insulin receptor, which induces the receptor to autophosphorylate specific Tyr residues within the receptor dimer. This, in turn, activates the receptor to phosphorylate other proteins. Structural analysis of the tyrosine kinase domain of the insulin receptor reveals the basis for the regulation of activity by autophosphorylation (Figure 1).

One of the target proteins in the insulin protein kinase cascade is insulin receptor substrate-1 (IRS-1). On phosphorylation by the activated insulin receptor, IRS-1 nucleates the formation of a protein complex that results in phosphorylation of a series of protein kinases. Through this protein phosphorylation cascade, the original signal of insulin binding to its membrane receptor is amplified by many orders of magnitude. The protein kinase cascade initiated by insulin is sometimes referred to as a MAPK (mitogen-activated protein kinases)

cascade. The cascade ultimately leads to phosphorylation of a transcription activator, Elk1, which initiates gene transcription (Figure 2).

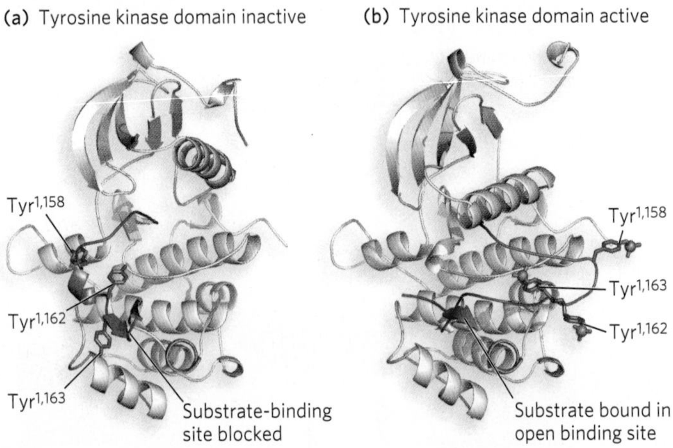

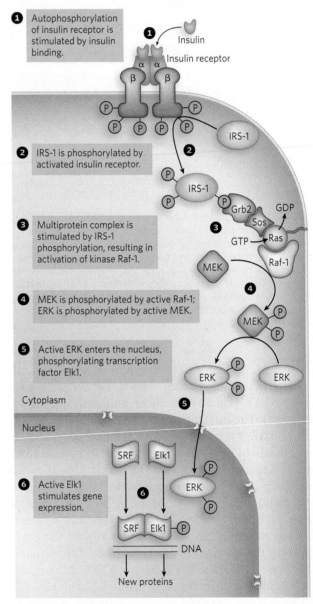

① Autophosphorylation of insulin receptor is stimulated by insulin binding.

② IRS-1 is phosphorylated by activated insulin receptor.

③ Multiprotein complex is stimulated by IRS-1 phosphorylation, resulting in activation of kinase Raf-1.

④ MEK is phosphorylated by active Raf-1; ERK is phosphorylated by active MEK.

⑤ Active ERK enters the nucleus, phosphorylating transcription factor Elk1.

⑥ Active Elk1 stimulates gene expression.

FIGURE 2 The MAPK cascade initiated by insulin regulates gene expression. Binding of insulin to the insulin receptor triggers autophosphorylation, activating the protein kinase domain in the cytoplasmic region of the receptor. The tyrosine kinase phosphorylates IRS-1, activating it to bind other proteins that then phosphorylate yet other proteins, creating a cascade of protein phosphorylation events that amplifies the original signal. Numerous "middle" factors (e.g., Grb2, Sos, Ras, Raf-1, MEK, ERK, SRF), several of which are kinases, are required to transduce and amplify the original signal. The end result is phosphorylation and activation of Elk1, a transcription factor that stimulates gene expression.

(a) Tyrosine kinase domain inactive   **(b)** Tyrosine kinase domain active

Tyr^{1,158}

Tyr^{1,162}

Tyr^{1,163}

Substrate-binding site blocked

Tyr^{1,158}

Tyr^{1,163}

Tyr^{1,162}

Substrate bound in open binding site

FIGURE 1 The tyrosine kinase domain of the insulin receptor is activated through autophosphorylation. (a) When the tyrosine kinase domain is inactive, the activation loop (blue) sits in the active site and none of the critical Tyr residues are phosphorylated. (b) When insulin binds the receptor, the tyrosine kinase activity phosphorylates Tyr^{1,158}, Tyr^{1,162}, and Tyr^{1,163}. Introducing these three phosphate groups (shown in orange) results in a 30 Å movement in the activation loop, shifting it out of the substrate-binding site, which becomes available to phosphorylate target proteins (red). [Sources: (a) PDB ID 1IRK. (b) PDB ID 1IR3.]

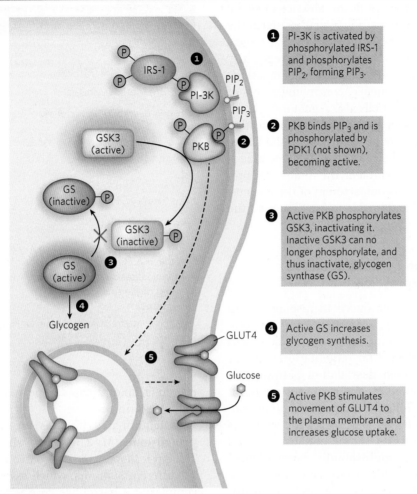

FIGURE 3 Insulin rapidly controls changes in the cell's glycogen metabolism (by increasing glycogen synthase activity) and glucose import (by moving the receptor GLUT4 to the plasma membrane), without the need for new protein synthesis. As with the MAPK cascade, many "middle" factors, as shown here, are involved in this signal transduction pathway.

Insulin and the insulin receptor also regulate glycogen (a storage form of glucose) metabolism through another phosphorylation pathway that short-circuits the need for RNA or protein synthesis in the regulation of gene expression (Figure 3). As in the pathway described above, IRS-1 is phosphorylated by the activated insulin receptor. From here, the pathways diverge. IRS-1 binds and activates the enzyme phosphoinositide 3-kinase (PI-3K), which initiates a cascade of phosphorylation events ultimately resulting in phosphorylation of glycogen synthase kinase 3 (GSK3). Active, unphosphorylated GSK3 contributes to the slowing of glycogen synthesis by inactivating the enzyme glycogen synthase. When phosphorylated, GSK3 is inactivated, and glycogen synthase remains active in liver and muscle cells. The phosphorylation cascade initiated by insulin thus results in increased glycogen synthesis. Given that glycogen is the storage form of glucose, the insulin signal effectively removes glucose from the blood by promoting cellular glycogen production.

Glucose import into cells is yet another pathway controlled by insulin in a way that short-circuits RNA and protein synthesis (see Figure 3). Cell surface receptors for glucose uptake are controlled by insulin in response to blood glucose levels. Glucose uptake is mediated by the glucose transporter protein GLUT4, which is mainly stored in intracellular vesicles. Insulin release from the pancreas in response to high blood glucose results in fusion of these cytoplasmic vesicles with the plasma membrane, introducing GLUT4 to the membrane and permitting glucose import. When blood glucose returns to normal, the GLUT4 receptors are returned to intracellular vesicles. In type 1 diabetes, the inability to release insulin (and thus to mobilize glucose transporters) results in low rates of glucose uptake into muscle and adipose tissue. One consequence is a prolonged period of high blood glucose after a carbohydrate-rich meal, which can lead to organ damage (and is also the basis for the glucose tolerance test for diagnosing diabetes).

The cell can also control the intracellular localization of a regulatory protein in the absence of signal transduction. Nuclear proteins, newly synthesized in the cytoplasm, contain a localization sequence that targets them to the nucleus. Cellular localization of a regulatory protein can thus be achieved by masking or unmasking the nuclear localization sequence, controlling access of the protein to the nucleus. Cells also regulate the localization of some proteins through covalent modification by the 101-residue polypeptide known as SUMO (small *ubiquitinlike modifier*). When the SUMO polypeptide is attached to Lys residues of a protein, the sumoylated protein is transported to a subcompartment of the nucleus, where it is sequestered and unable to function until the SUMO polypeptide is removed.

Protein Degradation by Ubiquitination Modulates Gene Expression

Once a protein has been produced in response to an environmental signal, it is important that the protein can be removed when it is no longer needed. Cells have a regulated mechanism for targeting proteins for removal through a protein degradation pathway. An efficient mechanism for proteolysis is also important for the turnover of misfolded or unfolded proteins, enabling recycling of their amino acids for the synthesis of new proteins. For protein removal, both bacteria and eukaryotes use a large, multisubunit, barrel-shaped ATP-dependent protease with a central chamber where proteins are degraded. The access of proteins to this protease machine is restricted to those specifically targeted for permanent removal.

Although we do not yet understand all the signals that trigger recognition of a protein for degradation, one simple signal has been found. For many proteins, the identity of the first amino acid residue—the one that remains after removal of the N-terminal Met residue and any other posttranslational proteolytic processing of the N-terminal end (see Chapter 16)—has a profound influence on half-life (**Table 19-1**). These N-terminal signals have been conserved over billions of years and are the same in the protein degradation systems of bacteria and eukaryotes.

In eukaryotes, but not bacteria, regulated protein degradation is directed by the attachment of the 76-residue polypeptide ubiquitin, which, as its name suggests, is ubiquitous among eukaryotes. Ubiquitin is highly conserved: it is essentially identical in organisms as different as yeast and humans. Three enzymes are involved in the covalent attachment of ubiquitin to a protein (**Figure 19-28**). Two belong to large protein families that have different specificities for target proteins and therefore regulate different cellular processes. Once a

TABLE 19-1	
The Relationship between N-Terminal Amino Acid Residue and Protein Half-Life	
N-Terminal Residue	*Half-Life*
Protein-stabilizing residues	
Ala, Gly, Met, Ser, Thr, Val	>20 h
Protein-destabilizing residues	
Gln, Ile	~30 min
Glu, Tyr	~10 min
Pro	~7 min
Asp, Leu, Lys, Phe	~3 min
Arg	~2 min

Note: Half-lives were measured in yeast for the β-galactosidase protein modified so that, in each experiment, it had a different N-terminal residue. Half-lives may vary among proteins and among organisms, but this general pattern seems to hold for all species.
Source: Data from A. Bachmair, D. Finley, and A. Varshavsky, *Science* 234:179–186, 1986.

protein is ubiquitinated, repeated cycles produce a long polyubiquitin chain.

Ubiquitinated proteins are degraded by the **26S proteasome** (M_r 2.5 $\times$ 10^6), shown in **Figure 19-29**. The proteasome consists of two copies each of at least 32 different subunits, which assort into two main subcomplexes: a barrel-like core particle and a regulatory particle at each end of the barrel. The 20S core particle consists of four rings; the outer rings are formed from seven α subunits and the inner rings from seven β subunits. Three of the seven subunits in each β ring have protease activity, each with different substrate specificity. The stacked rings of the core particle form the barrel-like structure within which target proteins are degraded. The 19S regulatory particle at each end of the core particle contains 18 subunits, including some that recognize and bind to ubiquitinated proteins. Six of the subunits are ATPases that probably function in unfolding the ubiquitinated proteins and translocating them into the core particle for degradation.

Not surprisingly, defects in the ubiquitination pathway have been implicated in a wide range of disease states. The inability to degrade certain proteins that activate cell division can lead to tumor formation, and the too rapid degradation of proteins that act as tumor suppressors can have the same effect. The ineffective or overly rapid degradation of cellular proteins also appears to play a role in a range of other conditions: renal diseases, asthma, neurodegenerative disorders (e.g., Alzheimer disease, Parkinson disease), cystic fibrosis (sometimes

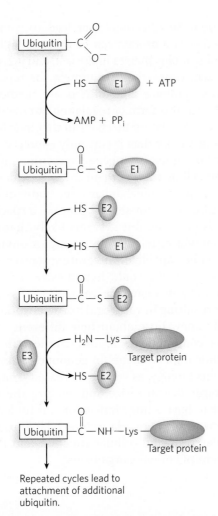

FIGURE 19-28 The protein ubiquitination pathway. In eukaryotes, three enzymes (denoted by E1, E2, and E3) carry out the polyubiquitination of proteins in a process that involves ATP and two enzyme-ubiquitin intermediates. The free carboxyl group of the ubiquitin C-terminal Gly residue is linked through an amide bond to the ε-amino group of a Lys residue of the target protein. Additional cycles produce polyubiquitin, a covalent polymer that targets the protein for destruction.

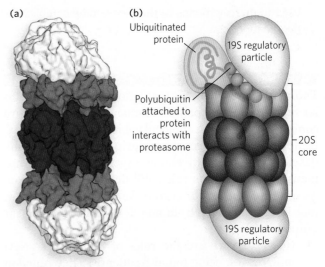

FIGURE 19-29 The regulation of proteolysis by the proteasome. (a) The three-dimensional structure of the 26S proteasome is highly conserved in all eukaryotes. The two subassemblies are the 20S core particle (light and dark brown) and the 19S regulatory particle (gray), one at each end of the core. (b) The core particle consists of four rings arranged in a barrel-like structure. Each inner ring has seven different β subunits (dark brown), three of which have protease activity; each outer ring has seven different α subunits (light brown). At each end of the core, the 19S regulatory particle forms a cap (composed of base and lid segments). The regulatory particles are thought to unfold ubiquitinated proteins and translocate them into the core particle for destruction. [Source: PDB ID 3L5Q.]

caused by overly rapid degradation of a chloride ion channel), and Liddle syndrome (in which a sodium channel in the kidney is not degraded, leading to excessive Na⁺ absorption and early-onset hypertension). Drugs designed to inhibit proteasome function are being developed as potential treatments for some of these conditions. In a changing metabolic environment, protein degradation is as important to cell survival as is protein synthesis, and much remains to be learned about these pathways.

Bacteria and eukaryotic organelles that evolved from bacteria also have proteasome-like particles; these include ClpAP, ClpXP, HslUV, Lon, and FtsH proteases. Most bacteria do not use a protein such as ubiquitin to tag proteins for degradation (although some do use a protein-tagging strategy), but their proteasomal analogs look surprisingly similar to the eukaryotic proteasome.

SECTION 19.3 SUMMARY

- Gene regulation can occur at various steps after transcription initiation. Points of regulation involving the RNA transcript include transcript elongation, splicing, modification, and stability. The stability of mRNAs can be affected by microRNAs.

- Control of gene expression can occur at the level of translation initiation or elongation. Eukaryotes are particularly adept at regulating the initiation step.

- Gene expression is also controlled at the level of protein products by several types of covalent modification, such as phosphorylation, acetylation, and methylation. Covalent modification carries the advantage of rapidly altering protein activity without waiting for changes in transcription and translation.

- Protein targeting to particular intracellular compartments is another mechanism of gene regulation. Transcription factors can be excluded from the nucleus by phosphorylation or by binding of a regulatory protein that masks a nuclear localization signal. With degradation or modification of the regulatory protein, the transcription factor can enter the nucleus.

- Gene expression can be regulated at the level of protein stability, which typically involves degradation by protease machinery. In eukaryotes, ubiquitination is used to direct proteins to the proteasome complex for degradation.

The many levels of gene regulation required in cellular function and adaptation to changing conditions are coming into focus for molecular biologists. But the extra levels and structural complexities required for the development of multicellular organisms such as humans, with 50 trillion cells, still defy the imagination. As sophisticated as our current knowledge is, when we look back some years from now, it will probably appear quite primitive.

1. How extensive are the roles of microRNAs? New miRNAs are being found frequently. They function in various ways, but the details are still scarce and the diversity of functional mechanisms is only now becoming apparent. Some miRNAs are clearly implicated in cancer, making the understanding of these small regulatory molecules extremely important to human health.

2. How often is intracellular localization used to regulate protein function? Regulatory mechanisms at steps other than transcription, such as intracellular localization, are being discovered at a rapid pace and are proving to be of great importance to cellular function. Modifications that lead to compartmentalization of a protein can be quickly implemented, enabling rapid changes in the cell, and just as rapidly reversed, conserving the protein for repeated use. Because proteins and mRNAs are neither formed nor lost in this form of regulation, it provides a fuel-efficient regulatory mechanism that may have more widespread use than is currently appreciated.

3. How do regulatory mechanisms function together in the cell or whole organism? Our understanding of regulatory mechanisms for individual genes, and sometimes for several genes in a specific pathway, is growing. But it seems likely that for a cell to function efficiently in a complex environment, it must be capable of integrating sensory inputs of many sorts. We could hypothesize that different regulatory mechanisms engage in cross-talk, possibly resulting in vast regulatory networks. We currently know little about how different regulatory paths communicate or interconnect in the cell. Further improvements in genomic techniques for systems biology, as well as increased computational power to categorize and analyze the data, are likely to have a huge impact on our understanding of how whole networks of interrelated proteins are regulated during cellular function and the development of complex organisms.

HOW WE KNOW

Plasmids Have the Answer to Enhancer Action

Dunaway, M., and P. Dröge. 1989. Transactivation of the *Xenopus* rRNA gene promoter by its enhancer. *Nature* 341:657–659.

During early studies of enhancer action, two major models were proposed for how enhancer-binding proteins might work to activate a promoter at a distance. The protein could either slide along the DNA to the promoter or it could act "through space," probably by looping out the intervening DNA so that the protein contacts the promoter and enhancer sequences simultaneously. A simple and clever test to distinguish between these two models was performed by Marietta Dunaway and Peter Dröge in a study involving plasmids in yeast.

The researchers placed an enhancer on one plasmid and a promoter on another plasmid, then topologically linked the plasmids together. They transferred these linked plasmids into *Xenopus* oocytes, along with a control plasmid containing the same promoter but no enhancer. If the enhancer-binding protein functioned through space, topological linkage would result in preferential activation of the promoter on the linked plasmid over that on the unlinked plasmid. But if the protein slid from the enhancer site to reach the promoter, it would not activate the promoter on either plasmid. The two promoter-containing plasmids had identical promoters but different gene sequences, which allowed Dunaway and Dröge to distinguish the level of transcription from each plasmid by a method called quantitative S1 mapping. In this method, cells are lysed and a ^{32}P-labeled DNA probe is hybridized to the mRNA, then the hybrid is digested with S1 nuclease, which degrades single-stranded DNA and RNA. The DNA-RNA duplex formed by the portion of the mRNA that hybridized with the probe is protected from lysis, and its length can be observed by gel electrophoresis.

To quantify transcription from the topologically linked plasmid versus the control plasmid, Dunaway and Dröge divided the lysate and performed S1 analysis using either a 40-nucleotide probe (ψ40) that specifically hybridized to the mRNA transcribed from the linked plasmid or a 52-nucleotide probe (ψ52) that specifically hybridized to the mRNA transcribed from the control plasmid (Figure 1, top left). The samples were then subjected to agarose gel electrophoresis. The results revealed that when the enhancer-containing and promoter-containing plasmids are intertwined (by topological linkage), the enhancer-binding protein preferentially stimulates transcription of the gene on the linked plasmid over the gene on the control plasmid (compare lanes 1 and 2 in Figure 1). In a control experiment (see Figure 1, top right), in which all plasmids were unlinked, transcription from the control plasmid (ψ52) was detected more than transcription from the other, also unlinked plasmid (ψ40) (compare lanes 3 and 4). In all their experiments, Dunaway and Dröge used a ψ52 probe that was more radioactive than the ψ40 probe. Further experiments (not shown) revealed that transcription from both promoters in the original control experiment (lanes 3 and 4 in Figure 1) was actually about equal. Overall, these results demonstrated that the enhancer acts through space and does not need to slide along DNA to activate the promoter.

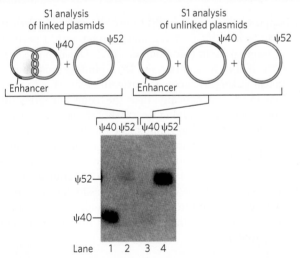

FIGURE 1 An enhancer functions through space to activate a promoter, as shown in this experiment using topologically linked plasmids. *[Source: Reprinted by permission from Macmillan Publishers Ltd. Nature, Dunaway, M. and Dröge, P., 341, 657–659, © 1989.]*

KEY TERMS

housekeeping gene, 667
constitutive gene
   expression, 667
regulated gene
   expression, 667
activation, 667
repression, 667
transcription factor, 667
activator, 667
repressor, 667
regulatory site, 667
positive regulation, 667
negative regulation, 667
DNA looping, 668
enhancer, 668
coactivator, 669
corepressor, 669
insulator, 669
signal integration, 670

operator, 670
effector, 671
polycistronic mRNA, 672
operon, 672
regulon, 672
combinatorial control,
   673
recognition helix, 675
helix-turn-helix motif,
   675
homeodomain motif, 675
basic leucine zipper
   motif, 677
basic helix-loop-helix
   motif, 677
zinc finger motif, 677
RNA interference
   (RNAi), 681
26S proteasome, 686

PROBLEMS

1. Suppose you are planning to use the yeast two-hybrid assay to identify proteins that interact with a particular target protein (see Chapter 7). The assay makes use of the ability to separate the DNA-binding domain of a typical eukaryotic activator protein from its activation domain. You genetically fuse the gene encoding the protein you are studying (the "bait") to the gene encoding the DNA-binding domain of the bacterial protein LexA, so that they are expressed as a single fusion protein. You place the binding site for LexA upstream from *lacZ* (encoding β-galactosidase) as a reporter gene—its expression can be selected for and easily detected. How might you design the rest of this genetic screen to identify the genes encoding proteins that interact with your bait protein?

2. Activator proteins A and B are required to express gene *X*. Analysis of the DNA upstream from the gene *X* promoter identified an 18 bp sequence with near twofold symmetry that is required for activation. Purification of the gene *A* and gene *B* products showed that both proteins form homodimers, but neither the A nor the B homodimer binds the 18 bp site. What are the possible functions of the A and B activators with respect to the 18 bp site? Propose a test of one of your ideas.

3. Most proteins that regulate gene expression bind at specific DNA sequences, recognizing those sequences primarily through protein-DNA interactions within in the major groove of the DNA. Why is the major groove used for

sequence recognition more often than the minor groove or the phosphoribose backbone?

4. Briefly describe the relationship between chromatin structure and transcription in eukaryotes.

5. MicroRNAs known as small temporal RNAs (stRNAs) have been discovered in higher eukaryotes. Describe their characteristics and general function.

6. An effector molecule binds to an activator protein, changing the activator's conformation so that it is no longer active. Transcription of the gene is thus shut down. Is this positive or negative regulation?

7. A transcription activator contains the following sequence:

   IARLEEKVKTLKAQNSELASTANMLTEQVAQLKQ

   The sequence includes a motif that may be used by certain transcription factors. What is this motif called? How does it function?

8. In one bacterial species, investigators find a regulon that coordinates the expression of 17 genes and identify a repressor that binds a defined site upstream from all the regulon genes. When the investigators inactivate the repressor protein, transcription of 4 of the genes increases. However, no transcription of the other 13 genes is observed, despite the presence of good promoters for RNA polymerase binding. Suggest a reason for the lack of transcription of these genes.

9. A repressor protein effectively blocks transcription from bacterial gene *X*. A mutant form of the repressor is engineered with an altered DNA-binding site in the helix-turn-helix motif. This mutant repressor does not repress transcription from gene *X*. When the mutant repressor is expressed at high levels on a plasmid that is introduced into the bacterial cell, transcription of *X* is increased even though the wild-type repressor (capable of binding its normal DNA binding site and shutting down transcription) is present in the same cell. Explain.

10. A eukaryotic transcription activator typically has separate DNA-binding and transcription-activation domains. The LexA protein is a bacterial repressor that binds to a particular LexA operator sequence. A researcher fuses random gene sequences from the *E. coli* genome to the gene encoding LexA. In yeast, the researcher replaces a binding site for a gene activator for gene *X* with the LexA operator. A small subset (about 1%) of the randomly constructed fused genes encode modified LexA proteins that activate gene *X*. Most of the "extra" gene material fused to the *lexA* gene encodes peptide segments that have high concentrations of amino acids with negatively charged side chains at neutral pH. Explain how the protein products of these gene fusions activate gene *X*. (Note: This problem reflects a classic published experiment.)

11. Zinc finger motifs have been appropriated for use in biotechnology. Several of these motifs can be strung together in an engineered protein, together with a fused nuclease domain, to create what has been dubbed a zinc finger nuclease. Such nucleases can be constructed to recognize and cleave almost any DNA sequence with high specificity. Explain why zinc finger motifs, rather than helix-turn-helix, helix-loop-helix, or homeodomain motifs, have been adapted for this purpose.

12. One of the classic ways to determine the function of a gene is to eliminate its function and determine the effect on the cell. Many approaches to eliminating gene function involve mutating or deleting the gene. In nematode worms, it is possible to shut down the function of almost any gene by synthesizing a double-stranded RNA complementary to the mRNA of the target gene and including it in the nematode's food. Explain how this works and identify the cellular system involved.

13. Steroid hormone receptors are located in the cytoplasm, where they can interact with incoming hormones. However, steroid hormones act by regulating gene function, and genes are in the nucleus. How is this regulation achieved?

14. Expression of the CRP transcription activator in *E. coli* readily leads to transcription of the lactose metabolism genes when lactose is present and glucose is not. If a particular eukaryotic activator is expressed in the appropriate eukaryotic cell, introduced on an engineered virus or plasmid, it often does not trigger transcription of its target gene. Explain.

DATA ANALYSIS PROBLEM

Brent, R., and M. Ptashne, M. 1985. A eukaryotic transcriptional activator bearing the DNA specificity of a prokaryotic repressor. *Cell* 43:729–736.

15. The concept that eukaryotic regulatory proteins have multiple functional domains was arrived at in stages. However, a few experiments stand out, such as the study by Roger Brent and Mark Ptashne published in 1985. When the study began, one known mechanism for activation of transcription by an activator protein was simply direct interaction with RNA polymerase. The investigators also considered an alternative mechanism: that the transcription activator functioned by altering the structure of the DNA to which it bound, facilitating the binding of RNA polymerase.

The study focused on two different regulatory proteins. The first was a well-characterized bacterial repressor called LexA. The LexA repressor controls a regulon in *E. coli*, the SOS response, which is activated when cellular DNA is subjected to extensive damage. The sequence of its binding site on DNA was known, and the protein had been studied by the Ptashne group and others. The second regulatory protein was a eukaryotic gene activator protein from

yeast, Gal4p, which activates transcription of the *GAL1* gene when yeast cells are grown on galactose. Ptashne and his coworkers knew that the DNA-binding element of Gal4p was located in the N-terminal 74 amino acids of the protein. They deleted these amino acids and replaced them with the first 87 amino acids of the LexA protein, which they knew contained the DNA-binding elements of that protein. They then expressed this fusion protein, LexA-Gal4, in both yeast and *E. coli*. They separately expressed the native LexA protein by itself. To monitor the effects of the fusion protein in yeast, the researchers needed to construct several variants of a second plasmid containing the β-galactosidase gene (the *lacZ* gene, encoding an enzyme activity that is easy to measure) fused to an unrelated yeast gene, *CYC1*. The different constructs contained a variety of regulatory sequences upstream from the fusion genes.

(a) Suggest why the investigators did not simply examine the effects of the LexA-Gal4 fusion protein on the *GAL1* gene already in yeast cells.

The investigators first tested the LexA-Gal4 protein in *E. coli* cells that lacked their own LexA-encoding gene, and showed that cells containing the fusion protein repressed transcription of genes normally repressed by LexA.

(b) Why was this control experiment undertaken?

Next, they carried out a series of measurements of β-galactosidase activity with the two-plasmid system in yeast cells, with the results shown in Table 1 (data from their published table). In the table, β-galactosidase activity

TABLE 1

The LexA-Gal4 Fusion Protein Activates Transcription of a *CYC1-lacZ* Fusion Gene

Growth Medium	Upstream Element	β-Galactosidase Activity of Regulatory Protein	
		LexA	LexA-Gal4
Galactose	No UAS	<1	<1
	lexA op at −178	<1	590
	lexA op at −577	<1	420
	UAS$_{C1}$ and UAS$_{C2}$	550	500
	UAS$_G$	950	950
	17mer	600	620
Glucose	No UAS	<1	<1
	lexA op at −178	<1	210
	lexA op at −577	<1	140
	UAS$_{C1}$ and UAS$_{C2}$	180	160
	UAS$_G$	<1	<1
	17mer	<1	<1

is given in units of blue color produced in the conversion of substrate to product. UAS is upstream activator sequence; UAS_{C1} and UAS_{C2} are binding sites for activator proteins that function at the *CYC1* gene; and UAS_G is the normal binding site for Gal4p. UAS_G consists of four separate binding sites for Gal4p, each 17 bp long. The 17mer is a site with just one of these sequences. The abbreviation "op" means operator, which is the LexA-binding site; −178 and −577 indicate the distance in base pairs between the operator and the transcription start site. In the yeast cells used for the study, the genes encoding the endogenous Gal4p and the *CYC1* gene activators were all present and functional.

(c) How effective is the LexA-Gal4 fusion protein (acting at the LexA operator) at activating gene expression, relative to the cellular Gal4p acting at UAS_G?

(d) Is transcription activated by the LexA protein by itself?

(e) Does the location of the LexA operator affect the activity of the LexA-Gal4 fusion protein?

(f) When UAS_G or the 17mer is upstream from the *CYC1-lacZ* reporter gene, why is expression seen only when the cells were grown in galactose?

(g) What result in Table 1 indicates that the LexA-Gal4 fusion protein is activating transcription by direct interaction with RNA polymerase, not by altering the structure of the DNA to which the polymerase binds?

ADDITIONAL READING

General

D'Alessio, J.A., K.J. Wright, and R. Tjian. 2009. Shifting players and paradigms in cell-specific transcription. *Mol. Cell* 236:924–931.

Ptashne, M. 2005. Regulation of transcription: From lambda to eukaryotes. *Trends Biochem. Sci.* 30:275–279.

Regulation of Transcription Initiation

Juven-Gershon, T., and J.T. Kadonaga. 2010. Regulation of gene expression via the core promoter and the basal transcriptional machinery. *Dev. Biol.* 339:225–229.

Pan, Y., C.J. Tsai, B. Ma, and R. Nussinov. 2010. Mechanisms of transcription factor selectivity. *Trends Genet.* 26:75–83.

Ross, W., and R.L. Gourse. 2009. Analysis of RNA polymerase-promoter complex formation. *Methods* 47:13–24.

Wade, J.T., and K. Struhl. 2008. The transition from transcriptional initiation to elongation. *Curr. Opin. Genet. Dev.* 18:130–136.

The Structural Basis of Transcriptional Regulation

Christensen, K.L., A.N. Patrick, E.L. McCoy, and H.L. Ford. 2008. The six family of homeobox genes in development and cancer. *Adv. Cancer Res.* 101:93–126.

Elhiti, M., and C. Stasolla. 2009. Structure and function of homodomain-leucine zipper (HD-Zip) proteins. *Plant Signal Behav.* 4:86–88.

He, X., L. He, and G.J. Hannon. 2007. The guardian's little helper: MicroRNAs in the p53 tumor suppressor network. *Cancer Res.* 67:11,099–11,101.

Huffman, J.L., and R.G. Brennan. 2002. Prokaryotic transcription regulators: More than just the helix-turn-helix motif. *Curr. Opin. Struct. Biol.* 12:98–106.

Posttranscriptional Regulation of Gene Expression

Breitkreutz, D., L. Braiman-Wiksman, N. Daum, M.F. Denning, and T. Tennenbaum. 2007. Protein kinase C family: On the crossroads of cell signaling in skin and tumor epithelium. *J. Cancer Res. Clin. Oncol.* 133:793–808.

Deng, S., G.A. Calin, C.M. Croce, G. Coukos, and L. Zhang. 2008. Mechanisms of microRNA deregulation in human cancer. *Cell Cycle* 7:2643–2646.

Shi, X.B., C.G. Tepper, and R.W. deVere White. 2008. Cancerous miRNAs and their regulation. *Cell Cycle* 7:1529–1538.

20 The Regulation of Gene Expression in Bacteria

Bonnie Bassler [Source: Paul Fetters Photography.]

MOMENT OF DISCOVERY

For me, science is all about those moments of clarity, when years of struggling to figure something out finally pay off with an incredible insight about how nature works. I am fascinated by how bacterial cells communicate with each other in the regulatory process known as quorum sensing. Quorum-sensing bacteria make, release, and detect chemical signal molecules that increase in concentration in proportion to increasing cell population numbers. Cells respond to these chemicals with synchronous population-wide changes in behavior; community behavior allows bacteria to carry out tasks that could never be accomplished if a single bacterium acted alone. We suspect that the evolution of cell-cell communication in bacteria is one of the first steps in the development of multicellular organisms.

Vibrio harveyi is a bioluminescent gram-negative marine bacterium that regulates light production in response to two distinct chemical "words," or autoinducers. As a new professor, I wanted to answer a question that had baffled the field for several years: *Why does* V. harveyi *need two chemical signals for communication, when one should be sufficient?* The identity of one autoinducer, AI-1 (autoinducer-1), had been determined, but the other, AI-2, remained an enigma. Our lab cloned the gene that encodes the protein responsible for synthesizing AI-2, and sequenced the gene. At that time, there were no extensive databases of bacterial genome sequences available, only partial genomes for 40 or 50 different bacterial species. Nonetheless, we searched the incomplete database for a match to our sequence.

I recall sitting in front of the computer as the name of bacterium after bacterium scrolled up the screen. In the end, every single bacterium in the database contained a gene that closely matched our sequence! I realized at that moment that the bacteria were talking across species. The mysterious autoinducer AI-2 was, in fact, a chemical that enables different species to communicate with each other, a system that would obviously be very useful in natural settings where many different kinds of bacteria live together. This discovery changed the entire course of research in the field and led me to focus over the past decade on the mechanisms of interspecies communication.

—*Bonnie Bassler, on her discovery of interspecies quorum sensing*

As we discussed in Chapter 19, cells typically express just a subset of their genes at any given time. Some gene products are synthesized in large amounts, and many others in only small amounts or not at all, depending on the needs of the cell. For example, in bacteria, proteins directly involved in DNA replication and protein biosynthesis are required continuously during active growth, whereas proteins that mediate DNA repair or the metabolism of rare sugars may be needed only sporadically. Furthermore, requirements for many gene products change over time. The need for enzymes that participate in various metabolic pathways increases or decreases as nutrient types and levels change. The regulation of gene expression is essential for making optimal use of available energy and for enabling cells to adapt to a wide variety of environmental changes.

Much of what we know about gene regulation comes from studies that focused, at least initially, on bacterial systems. Microbes are masters at regulating gene expression, due to their need to adapt quickly to changing conditions. Thus they have provided investigators with many opportunities to discover fundamental mechanisms that, as it turns out, also characterize the gene regulatory pathways in humans, plants, and other eukaryotes. Recall that there are seven points in the flow of biological information where regulation can take place (see Figure 19-1). Not all of these occur in bacteria, however, due to the absence or rarity of certain of the processes, such as pre-mRNA splicing, in bacterial cells.

In this chapter, we focus on some of the central aspects of bacterial gene regulation by examining specific examples. Although much of the classic research in this field concentrated on the regulation of transcription, more recent investigations have revealed that other stages of gene expression, notably translation, provide bacterial cells with exquisite tools to fine-tune their protein levels. In addition to the roles of regulatory proteins in altering gene expression levels, regulatory RNAs are ubiquitous in controlling how, when, and where proteins are made. The multiple levels of gene regulation observed in the bacteriophage λ (λ phage) infection and replication cycle set the stage for explaining the kinds of complex regulatory interactions found in eukaryotes—the topic of Chapters 21 and 22.

20.1 TRANSCRIPTIONAL REGULATION

Cells need to maintain control of their growth and must be able to adapt quickly to a changing environment, but they also strive for energy efficiency. For this reason, transcription is a common site of regulation, because unnecessary downstream steps in gene expression can be avoided, thus conserving energy. Turning transcription off—or down—can alter protein levels without involving the cell's protein biosynthetic machinery at all. The control of transcription also permits the synchronized regulation of multiple genes encoding products with interdependent activities; for example, when their DNA is heavily damaged, bacterial cells require a coordinated increase in the levels of many DNA repair enzymes. Interactions between proteins and DNA are the key to transcriptional regulation, and much is now known about the specificity of transcriptional control.

As we discussed in Chapters 15 and 19, bacterial genes contain relatively simple promoters with sequences that allow RNA polymerase to bind and initiate transcription at specific sites. Variations in promoter sequences, or in the space between promoters and the gene(s) they control, affect transcriptional efficiency. In addition, bacteria have mechanisms for regulating groups of genes. For example, functionally related genes frequently cluster together in operons (see Figure 19-11), where they can be controlled by a single promoter. Sigma factors, which bind and regulate RNA polymerase, also contribute to the global control of transcription (see Table 15-2). In some cases, activator and repressor proteins confer further levels of regulation by altering transcription in response to metabolites. The activities of multiple activators and repressors can converge on a single promoter to fine-tune transcription in response to various stimuli.

In this section we discuss the regulation of two bacterial operons for which mechanistic details are particularly well established. The lactose (*lac*) and tryptophan (*trp*) operons both require multiple regulatory proteins, but the overall mechanisms of regulation are distinct. We then consider the SOS response in *E. coli*, illustrating how genes scattered throughout the genome can be coordinately regulated. Throughout this discussion we cover some of the experimental approaches that have provided insights into these regulatory pathways.

The *lac* Operon Is Subject to Negative Regulation

Many of the principles of bacterial gene expression were first discovered in studies of sugar metabolism in *Escherichia coli*. This bacterium can use a variety of sugars as an energy source, depending on what is available in its environment. Metabolism of each sugar type requires a unique set of enzymes. The genes encoding this set of enzymes are often grouped together into an operon, which allows the genes to be coordinately regulated. In the 1960s, two French scientists, François Jacob and Jacques Monod, examined the *E. coli* genes involved in metabolizing the sugar lactose. Through genetic experiments, they determined how expression of these genes is coordinately regulated in response to the presence or

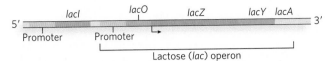

FIGURE 20-1 The lactose (lac) operon of E. coli. The three genes of the *lac* operon are transcribed as a single unit from a single promoter. The operator region regulates transcription through interaction with the Lac repressor protein, encoded by *lacI*. The repressor is transcribed separately from the operon (i.e., has a separate promoter) and is constitutively expressed.

absence of lactose. This work, for which they won the 1965 Nobel Prize in Physiology or Medicine, uncovered one of the central themes in molecular biology: some genes encode proteins with the sole function of regulating the expression of other genes.

The *lac* operon (**Figure 20-1**) includes the genes for β-galactosidase (*lacZ*), galactoside permease (*lacY*), and thiogalactoside transacetylase (*lacA*)—sometimes referred to collectively as the *lac* genes. Although the operon is transcribed as a single unit (i.e., the mRNA is polycistronic), the transcript contains three ribosome-binding sites, one preceding each open reading frame, that allow independent translation of each protein product. Each resulting protein functions in the metabolism of lactose. β-Galactosidase catalyzes cleavage of lactose into its components, glucose and galactose (**Figure 20-2**), which can then be metabolized further to generate ATP. The galactoside permease protein inserts into the bacterial plasma membrane and imports lactose into the cell. Thiogalactoside transacetylase modifies toxic galactosides that are imported along with lactose, facilitating their removal from the cell. When lactose is available, wild-type *E. coli* expresses these three genes. When lactose is unavailable, transcription from the *lac* operon is greatly reduced (**Figure 20-3a**).

Jacob and Monod isolated mutants of *E. coli* with defective regulation of the *lac* operon. They identified *lacI* and *lacO*, two DNA regions where mutations led to constitutive expression of the operon whether or not lactose was present (**Figure 20-3b**). To understand how *lacI* and *lacO* worked, the researchers performed merodiploid analysis, a procedure that essentially makes the bacterial cell diploid for the *lac* operon locus (see the How We Know section at the end of Chapter 5).

The first partial diploids they created combined wild-type strains with either the *lacI* or *lacO* mutants (**Figure 20-3c**, top and middle). The wild-type *lacI* allele was able to make up for (or "rescue") the defect in the *lacI* mutant; these partial diploids had normal regulation of both sets of lactose metabolism genes. Jacob and Monod hypothesized that the *lacI* locus produced a diffusible product that could act on any DNA molecule, not just the DNA from which it was generated. In this case,

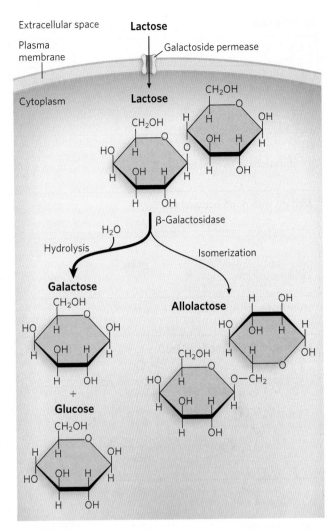

FIGURE 20-2 Lactose metabolism in E. coli. Galactoside permease, encoded by *lacY*, is a membrane protein that permits entry of lactose into the cell. β-Galactosidase, encoded by *lacZ*, converts lactose to galactose and glucose, and also converts a small amount of lactose to allolactose, the *lac* operon inducer.

the *lacI* gene product from the wild-type strain successfully regulated the *lac* operon DNA of the *lacI* mutant. However, a wild-type copy of the *lacO* regulatory region was *not* capable of rescuing the defect in a *lacO* mutant; these partial diploids still constitutively expressed the *lac* genes. Jacob and Monod hypothesized that *lacO* did not produce a diffusible substance: the *lac* operon DNA from the *lacO* mutant could not be correctly regulated, even in the presence of a wild-type copy of *lacO*.

Jacob had served in the military, and he likened the observations on the *lac* operon to the communication link between a bomber aircraft and a ground-based radio transmitter. If the transmitter on the ground were knocked out, a second transmitter could be used to direct the actions of the bomber. But if the receiver in the bomber were knocked out, neither a second transmitter nor a new bomber could direct the actions of the first bomber. Jacob

(a) **Wild type**

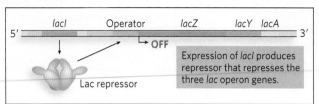

(b) **Mutations in normal (haploid) *E. coli***

(c) **Merodiploid analysis**

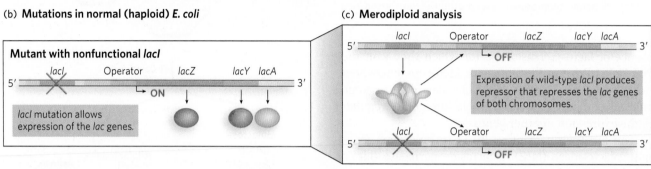

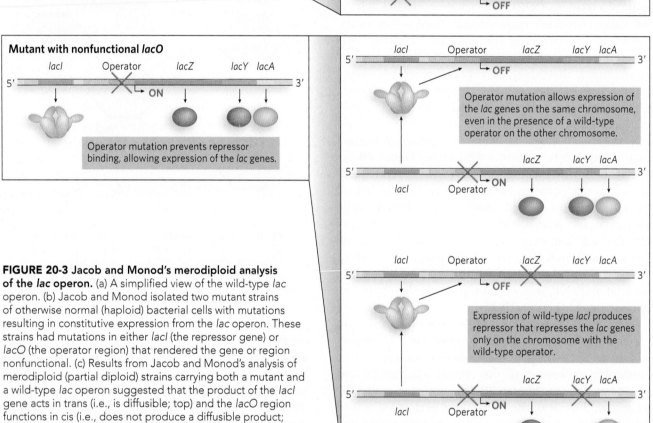

FIGURE 20-3 Jacob and Monod's merodiploid analysis of the *lac* operon. (a) A simplified view of the wild-type *lac* operon. (b) Jacob and Monod isolated two mutant strains of otherwise normal (haploid) bacterial cells with mutations resulting in constitutive expression from the *lac* operon. These strains had mutations in either *lacI* (the repressor gene) or *lacO* (the operator region) that rendered the gene or region nonfunctional. (c) Results from Jacob and Monod's analysis of merodiploid (partial diploid) strains carrying both a mutant and a wild-type *lac* operon suggested that the product of the *lacI* gene acts in trans (i.e., is diffusible; top) and the *lacO* region functions in cis (i.e., does not produce a diffusible product; middle). A double mutant analysis confirmed this hypothesis (bottom).

and Monod hypothesized that *lacI* functioned like the transmitter; if knocked out it could be replaced by a second transmitter. But *lacO* functioned like the receiver in the bomber: if a message could not be received, the action (transcription of the *lac* genes) could not be controlled.

With this hypothesis, the researchers tested a prediction (see **Figure 20-3c**, bottom). In a further

merodiploid analysis, they combined an operon carrying a *lacZ* mutation with an operon carrying mutations in *lacO* and *lacY*. The mutations in genes encoding different enzymes (*lacZ* or *lacY*) effectively "marked" the operons and allowed the researchers to determine which operon was giving rise to a particular gene product. They predicted that the *lacI* gene product (no matter which allele produced it) would not be capable of repressing the *lacZ* gene in an operon containing a *lacO* mutation. In the absence of lactose, this diploid construction produced β-galactosidase (the *lacZ* gene product)—exactly the result they predicted! These findings confirmed that the *lacI* gene encodes a diffusible molecule (i.e., acts in trans) that represses *lac* gene expression (whether on the same or, experimentally, on a different DNA), whereas *lacO* controls only the expression of *lac* operon genes to which it is connected (i.e., acts in cis).

From the experiments performed by Jacob and Monod, we know how the *lac* operon functions. Operon control consists of two main elements: a protein repressor (the Lac repressor, encoded by the *lacI* gene) and a DNA sequence called the operator (*lacO*) to which the Lac repressor binds. In the absence of lactose, the *lac* genes are not transcribed, because the Lac repressor binds to the operator sequence. The *lacI* gene is located near the *lac* operon, but it is transcribed from its own promoter, independent of the *lac* genes. The operator is adjacent to the *lac* operon promoter, and repressor binding to the operator prevents RNA polymerase from initiating transcription of the DNA (**Figure 20-4**). When lactose is present, a small amount of allolactose, an isomer of lactose, is produced (see Figure 20-2). Allolactose is a small effector molecule that functions as an **inducer** of the *lac* operon, binding to the Lac repressor and causing the repressor to lose affinity for and dissociate from the operator. On dissociation of the repressor, the operon becomes active, because RNA polymerase is able to initiate transcription and synthesize the polycistronic mRNA encoding the *lac* genes. It is important to note that Jacob and Monod's initial experiments examined *E. coli* grown solely in the presence of lactose, with no other sugars available for metabolism. As we will see shortly, bacteria preferentially metabolize some sugars over others and impose additional levels of regulation on the *lac* operon to shut down its transcription if a more highly preferred sugar source (such as glucose) is also available.

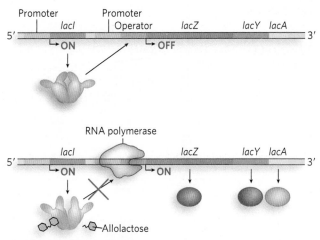

FIGURE 20-4 Negative regulation of the *lac* operon by the Lac repressor. In the absence of lactose (top), the Lac repressor binds the operator region and thus prevents RNA polymerase from leaving the promoter site and transcribing the operon. In the presence of lactose (bottom), its metabolite allolactose binds the Lac repressor, resulting in a conformational change that causes the repressor to dissociate from the operator. RNA polymerase can then initiate transcription. (Note that allolactose is enlarged for clarity; it is actually much smaller relative to the Lac repressor.)

Although it may seem simple today, the regulatory circuitry of the *lac* operon, the first to be discovered, was revealed only through powerful insight, prediction, intuitive reasoning, and creative thinking. Prior studies had focused on the fact that DNA encoded enzymes. But research on lactose metabolism revealed that some genes encode other kinds of proteins, such as DNA-binding proteins (e.g., the Lac repressor), and that some DNA sequences do not code for a gene product at all, but instead form genetic loci that affect cell function (e.g., *lacO*, the operator region).

The *lac* operon has been the subject of intense scrutiny by many laboratories since Jacob and Monod made their first observations. As research has progressed, new details of the regulatory machinery have come to light. We now know, for instance, that the operator region is more complex than suggested in Figure 20-1; in fact, there are three operator sequences.

The O_1 operator, to which the Lac repressor binds most tightly, abuts the *lac* operon's transcription start site (**Figure 20-5a**), but the operon also has two secondary binding sites for the repressor. One (O_2) is about 400 bp downstream, within the gene encoding β-galactosidase (*lacZ*); the other (O_3) is about 100 bp upstream, at the end of the *lacI* gene. As described in Chapter 19, most repressor proteins function as dimers, with each subunit binding to one half of an inverted repeat. The Lac repressor is unusual in that it functions as a tetramer of identical subunits, with two dimers tethered together at the end distant from the DNA-binding

(a)

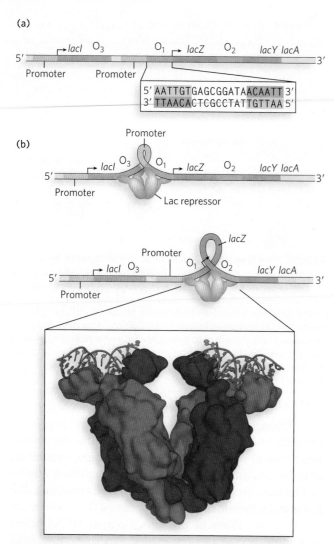

(b)

FIGURE 20-5 **Interaction of the Lac repressor and the operator region.** (a) The *lac* operon contains three operator sequences to which Lac repressor can bind. O₁ is adjacent to the *lac* operon promoter. The inverted repeat of the O₁ site is shown (sequence repeats are shaded orange). (b) The Lac repressor tetramer can bind to O₁ and O₂ or to O₁ and O₃. The intervening DNA is looped out. The molecular model shows the Lac repressor, a tetramer formed from two homodimers. Each homodimer can bind one operator sequence. *[Sources: (b) (bottom) PDB ID 1LBG and PDB ID 2PE5.]*

sites (**Figure 20-5b**). DNA recognition occurs in the major groove by means of a helix-turn-helix motif near the N-terminus of the repressor protein. An adjacent helix known as the hinge is important for positioning the helix-turn-helix and ensuring high-affinity DNA binding. As shown by crystal structures of the repressor, both alone and bound to inducers similar to allolactose, inducer binding causes disordering of the hinge and a resulting increase in flexibility of the DNA-binding region. DNA-binding affinity of inducer-bound Lac repressor is decreased, in a classic example of allosteric regulation (see Chapter 5).

When lactose levels are low, an *E. coli* cell contains about 20 tetramers of the Lac repressor. Each tethered dimer of the repressor tetramer separately binds to one of the three inverted-repeat operator sequences (see Figure 20-5b). To repress the operon, one dimer binds to the O₁ operator and the other dimer binds simultaneously to one of the two secondary sites, O₂ or O₃. The symmetry of the O₁ sequence corresponds to the twofold axis of symmetry of two paired Lac repressor subunits. The tetrameric Lac repressor binds to its operator sequences in vivo with very high affinity, with an estimated dissociation constant of about 10^{-10} M. The repressor discriminates between the operator and nonoperator sequences by a factor of about 10^6, so binding to just these few base pairs among the 4.6×10^6 bp of the *E. coli* chromosome is highly specific.

The simultaneous binding of the Lac repressor tetramer to O₁ and to O₂ or O₃ most likely results in a looped DNA structure, providing an effective steric block to transcription initiation by RNA polymerase. Because each dimer of the repressor binds to a separate region of DNA, and DNA looping enables two regulatory sites to be bound at the same time, the sensitivity of the system is enhanced by the cooperative nature of the binding. In other words, the affinity of one dimer for DNA is affected by the conformation (DNA-bound or not) of the other dimer. The process of working out how the Lac repressor functions required the development of techniques for detecting when and how proteins bind to specific sites in DNA (**Highlight 20-1** on p. 700). These methods are still widely used to analyze the properties of DNA-binding proteins in a variety of systems.

Despite formation of an elaborate complex, transcriptional repression of the *lac* operon by the Lac repressor is not absolute. Binding of the repressor reduces the rate of transcription initiation by a factor of 10^3. If the O₂ and O₃ sites are eliminated by deletion or mutation, the binding of repressor to O₁ alone reduces transcription by a factor of about 10^2. Even in the repressed state, each cell has a few molecules of β-galactosidase and galactoside permease, presumably synthesized on the rare occasions when the repressor transiently dissociates from the operators. This basal level of transcription is essential to operon regulation.

When cells are provided with lactose, the *lac* operon is induced. The few existing molecules of galactoside permease enable lactose from the medium to enter the cell, where it is converted by β-galactosidase to allolactose, a lactose isomer (see Figure 20-2). Allolactose binds to a specific site on the Lac repressor, causing a conformational change that results in dissociation of the repressor from the operator. Release of the operator, triggered as the repressor binds to the inducer allolactose, allows expression of the *lac* genes and a 10^3-fold increase in the concentration of β-galactosidase.

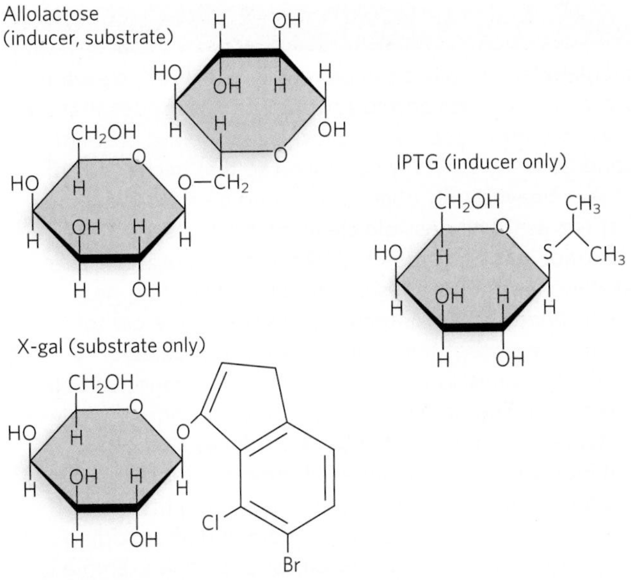

Allolactose
(inducer, substrate)

IPTG (inducer only)

X-gal (substrate only)

FIGURE 20-6 Chemical structures of some small-molecule effectors of the lac operon. Like allolactose, IPTG (isopropyl β-D-1-thiogalactopyranoside) can bind the Lac repressor and cause its dissociation from the operator, inducing transcription of the *lac* operon. However, IPTG is not a substrate for β-galactosidase. The β-galactoside X-gal (5-bromo-4-chloro-3-indolyl-β-D-galactopyranoside) does not induce expression of the *lac* operon, but it does serve as an experimentally useful substrate for β-galactosidase, producing a blue color when metabolized.

Several β-galactosides structurally related to allolactose are inducers of the *lac* operon but are not substrates for β-galactosidase; others are substrates but not inducers. One very effective and nonmetabolizable inducer of the *lac* operon often used experimentally is isopropyl β-D-1-thiogalactopyranoside (IPTG) (**Figure 20-6**). An inducer that cannot be metabolized lets researchers study the regulation of the *lac* operon without concern about the inducer being depleted. Equally useful as a tool in molecular biology is the noninducer substrate X-gal (5-bromo-4-chloro-3-indolyl-β-D-galactopyranoside), which consists of galactose linked to a substituted indole. β-Galactosidase cleaves X-gal to produce galactose and 5-bromo-4-chloro-3-hydroxyindole, which is oxidized to an insoluble blue compound, 5,5′-dibromo-4,4′-dichloro-indigo. Bacterial colonies grown on an agar medium containing X-gal and an inducer of β-galactosidase (usually IPTG) turn blue if they contain a functional *lacZ* gene, a useful marker in molecular cloning (see the How We Know section at the end of Chapter 5).

The *lac* Operon Also Undergoes Positive Regulation

The operator-repressor-inducer interactions affecting the *lac* operon provide an intuitively satisfying model for an on/off switch in the regulation of gene expression.

However, operon regulation is rarely that simple. A bacterium's environment is too complex for its genes to be controlled by one signal. Other factors besides lactose affect the expression of the *lac* genes, such as the availability of glucose. Glucose, metabolized directly by glycolysis, is *E. coli*'s preferred energy source. Other sugars can serve as the main or sole nutrient, but extra steps are required to prepare them for entry into glycolysis, necessitating the synthesis of additional enzymes. Clearly, expressing the genes for proteins that metabolize other sugars, such as lactose or arabinose, is wasteful when glucose is abundant. Only in the absence of glucose is it in the cell's best interest to increase the expression of genes that allow the use of alternative energy sources.

What happens to the expression of the *lac* operon when both glucose and lactose are available? A regulatory mechanism known as **catabolite repression** restricts expression of the genes required for metabolizing lactose, arabinose, or other sugars in the presence of glucose, even when these secondary sugars are also present. At first glance this may seem like another example of negative regulation, but it is a form of positive regulation for the *lac* operon. As we will see, the *lac* operon is activated for gene expression when glucose is absent.

The effect of glucose on expression of the *lac* operon is mediated by cyclic AMP (cAMP), a small-molecule effector, and by the activator cAMP receptor protein, or CRP, a homodimer with binding sites for DNA and cAMP (**Figure 20-7**). DNA binding is mediated by a helix-turn-helix motif within the protein's DNA-binding domain. When glucose is absent, CRP-cAMP binds to a site near the Lac promoter and stimulates RNA transcription fiftyfold. CRP-cAMP is therefore a positive regulatory factor responsive to glucose levels, whereas the Lac repressor

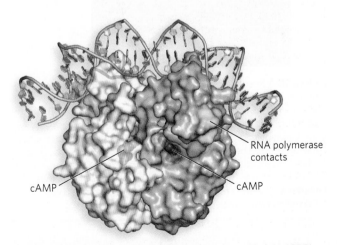

RNA polymerase contacts

cAMP

cAMP

FIGURE 20-7 Structure of the cAMP receptor protein (CRP) homodimer bound to DNA. In the absence of glucose, the CRP-cAMP complex binds near the Lac promoter and associates with RNA polymerase to induce transcription. [*Source: PDB ID 1RUN.*]

Classical Techniques in the Analysis of Gene Regulation

Many proteins affect gene expression, and some of them do so in part by interacting directly with DNA. Researchers often want to find out whether regulatory proteins bind DNA and, if so, what sequence or structure they recognize and how tight the interaction is. The electrophoretic mobility shift assay (EMSA) and DNA footprinting experiments (in various forms) are widely used to address these questions.

Both of these techniques test the ability of a protein to interact directly with DNA. In EMSA, fragments of DNA are incubated with the protein of interest and then analyzed on a nondenaturing polyacrylamide or agarose gel. The DNA used in the experiment is visualized either by staining with a dye or by covalently attaching a radioactive phosphate group at one end. Free DNA fragments migrate more quickly through the gel than DNA bound by protein. Thus, a shift in migration of DNA from fast to slow in the presence of protein indicates a direct binding interaction between the protein and DNA (Figure 1).

Once a direct DNA-binding interaction has been established, DNA footprinting can be used to map the exact nucleotide bases in contact with the bound protein (see Figure 15-12). Nucleases are incubated with DNA-protein complexes to cleave the DNA—often

radiolabeled so that it can be readily visualized; cleavage occurs at sites exposed to solvent, but not at sites that are physically protected by the presence of bound protein. Conditions are carefully controlled so that each piece of DNA is cleaved only once, generating a set of fragments that represent all possible cleavage products, with the exception of DNA protected by the bound protein. The resulting fragments are separated by denaturing gel electrophoresis and detected by exposing the gel to film or to a phosphorimager screen (which detects radioactive emissions from the ^{32}P-labeled bands of DNA fragments). The gap in the cleavage sites where the protein associates with the DNA produces a "footprint" that indicates the boundaries of the protein-binding site. The footprint can be identified by analyzing the sites of nuclease cleavage in the DNA before and after adding the protein (Figure 2a; the result for a specific example is shown in Figure 2b).

A related approach, called chemical protection footprinting, uses chemical reagents such as dimethyl sulfate to covalently modify DNA at nucleotide bases not protected by bound protein. This results in DNA containing methyl groups at sites outside the protein-binding site; the position of the protein creates a "footprint" marked by an absence of methylation sites. To locate the footprint, modified sites are detected using a DNA polymerase to copy the methylated template DNA by extending an annealed primer oligonucleotide that binds to a region just outside the DNA segment to be analyzed. The methylated bases lead to chemical fragmentation of the DNA, and the ends of the resulting DNA fragments are mapped by primer elongation. The products of the elongation reaction terminate at sites of methylated bases. These products are identified by analysis on a denaturing polyacrylamide gel, alongside the products of control primer-extension reactions conducted in parallel on unmodified DNA in the presence of dNTPs plus a small amount of A, C, G, or T dideoxynucleotide, which are chain-terminating nucleotide analogs. The products in these control reactions correspond to the positions of T, G, C, or A in the sequence, enabling exact determination of the sites of the protein footprint.

Chemical modification interference, a related approach, involves first reacting the DNA with a limiting amount of a chemical reagent to introduce nucleotide modifications randomly, at just one or a few sites. The resulting pool of modified DNA molecules, each containing a modification at a different site or sites, is then allowed to bind to protein, and free DNA is separated from DNA-protein complexes by nondenaturing gel electrophoresis, as in EMSA. Free and protein-bound DNA can then be

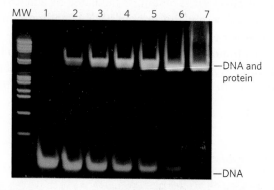

FIGURE 1 Results of an electrophoretic mobility shift assay demonstrate the shift in migration of Lac operator DNA (lane 1) with increasing concentrations of Lac repressor protein (lanes 2 through 7). The nondenaturing polyacrylamide gel preserves the Lac repressor in its folded state so that it can interact with the operator. The gel was stained with fluorophores that bind DNA (green) and protein (red). Yellow bands indicate protein-DNA complexes. (MW indicates the lane with molecular weight markers.) [Source: Photograph used courtesy of Thermo Fisher Scientific; copying prohibited.]

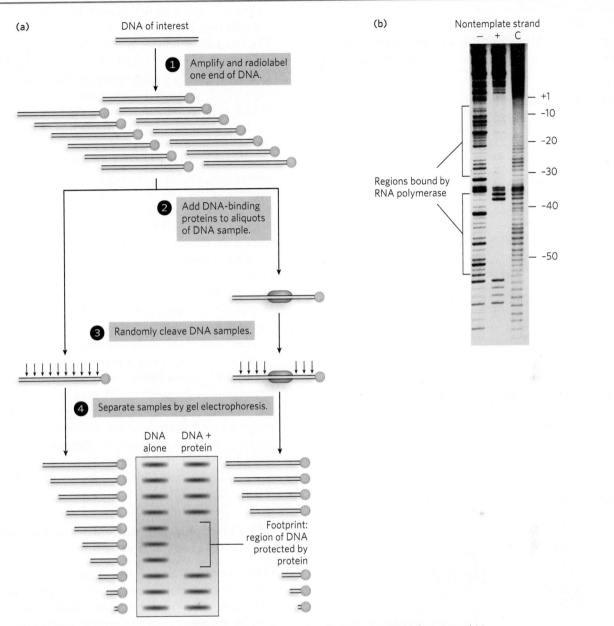

FIGURE 2 (a) DNA footprinting analysis reveals a protein's binding site on a DNA fragment. (b) In this example, the binding site of RNA polymerase at the Lac promoter is determined using DNase to digest the lac DNA wherever the polymerase is not directly binding it and protecting it. The lanes show no polymerase added (−), polymerase added (+), and the control reaction with no DNase added (C). The upstream sites are indicated to the right of the gel. [*Source: (b) Carol Gross, University of California, San Francisco, Department of Stomatology.*]

excised from the gel, eluted from the gel matrix, and analyzed by the primer-extension method. The DNA in the protein-bound sample contain chemical modifications only at sites that do not interfere with protein binding.

DNA molecules in the unbound sample—which migrated differently in the gel than the protein-bound DNA—contain chemical modifications primarily at sites that interfere with protein recognition.

is a negative regulatory factor responsive to lactose. The two act in concert.

CRP-cAMP has little effect on the *lac* operon when the Lac repressor is blocking transcription (**Figure 20-8a, b**), and dissociation of the repressor from the operator has

(a) Glucose high, cAMP low, lactose absent

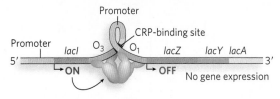

(b) Glucose low, cAMP high, lactose absent

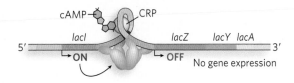

(c) Glucose high, cAMP low, lactose present

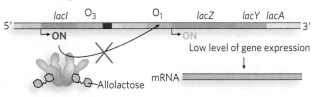

(d) Glucose low, cAMP high, lactose present

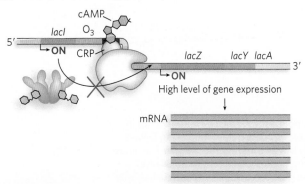

FIGURE 20-8 Positive regulation of the *lac* operon by CRP. (a), (b) When lactose is absent, the repressor binds the operator, blocking RNA polymerase and preventing transcription of the *lac* genes. It does not matter whether glucose is present or absent (and thus whether or not CRP-cAMP binds the operon). (c) When lactose is available, the repressor dissociates from the operator. However, if glucose is also available, low cAMP levels prevent CRP-cAMP formation and DNA binding. RNA polymerase may weakly bind the promoter and occasionally initiate transcription, leading to a very low level of *lac* operon expression. (d) Only when glucose levels are low, causing cAMP levels to rise and CRP-cAMP to bind the operon, and when lactose is present, causing repressor to dissociate, does the polymerase robustly bind and transcription proceed. (Note that cAMP is enlarged for clarity; it is actually much smaller relative to CRP; see Figure 20-7).

little effect unless CRP-cAMP is present to facilitate transcription (**Figure 20-8c**). When CRP-cAMP is not bound, the wild-type Lac promoter is a relatively weak promoter. The open RNA polymerase–promoter complex does not form readily unless CRP-cAMP is present. CRP interacts directly with RNA polymerase (at the region shown in Figure 20-7) through the polymerase's α subunit. Binding of CRP to the α subunit stimulates polymerase binding to the Lac promoter, triggering formation of the open polymerase-promoter complex.

The effect of glucose on CRP is mediated by the cAMP interaction. CRP binds to DNA most avidly when cAMP concentrations are high. When glucose is transported into the cell, the synthesis of cAMP is inhibited and efflux of cAMP from the cell is stimulated. As the cAMP concentration declines, CRP binding to DNA declines, thereby decreasing expression of the *lac* operon. Strong induction of the *lac* operon therefore requires both lactose (to inactivate the Lac repressor) and a lowered concentration of glucose (to trigger an increase in cAMP concentration and increased binding of cAMP to CRP) (**Figure 20-8d**).

CRP and cAMP are involved in the coordinated regulation of many operons, primarily those that encode enzymes for the metabolism of secondary sugars such as lactose and arabinose. Recall from Chapter 19 that a network of operons with a common regulator is known as a regulon. Other bacterial regulons include the heat shock gene system that responds to changes in temperature (see Chapter 15) and the genes induced in *E. coli* as part of the SOS response to DNA damage (described later in this section).

CRP Functions with Activators or Repressors to Control Gene Transcription

Other secondary sugars also trigger expression of their metabolic enzymes when present in the environment, and again, CRP provides a mechanism for activation only in the absence of the preferred sugar, glucose. For example, arabinose metabolism is regulated by CRP and the protein AraC, which acts as either an activator or a repressor of the arabinose (*ara*) operon, depending on whether arabinose is present. When arabinose is absent, AraC forms a dimeric structure in which one AraC monomer binds to the *ara* operon gene *araI₁* and the other binds a separate site much farther upstream called *araO₂* (**Figure 20-9a**). Similar to the effect of the Lac repressor, this mode of DNA binding causes the DNA to loop into a configuration that inhibits polymerase binding. When arabinose is present, it binds to AraC, causing AraC to adopt a different dimeric conformation that allows binding to two adjacent DNA half-sites, *araI₁* and *araI₂* (**Figure 20-9b**). This positions one monomer of AraC close to the promoter, where it can recruit RNA polymerase to activate transcription.

(a) AraC dimer without arabinose

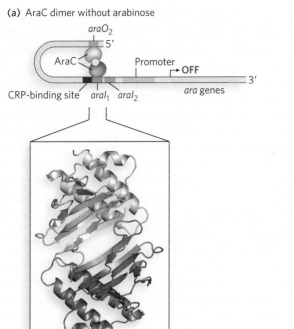

(b) AraC dimer with arabinose

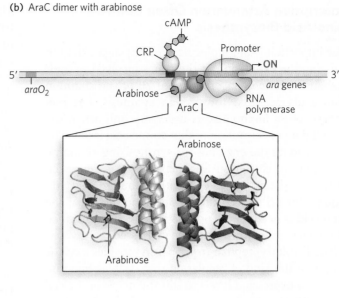

FIGURE 20-9 Regulation of the *ara* operon. (a) When arabinose is absent, AraC forms a dimer in which one monomer binds to *araO₂* and the other to *araI₁*, preventing RNA polymerase binding and transcription of the operon. (b) Activation of the *ara* operon occurs when AraC binds arabinose (its small-molecule effector) and CRP-cAMP (formed in the absence of glucose). The AraC dimer changes conformation such that one monomer binds *araI₁* and the other binds *araI₂*. The interaction with *araI₂* recruits RNA polymerase to the promoter and activates transcription of the *ara* operon. The molecular models show the AraC dimerization domain in the (a) absence and (b) presence of arabinose. *[Sources: (a) (bottom) PDB ID 2ARA. (b) (bottom) PDB ID 2ARC.]*

Cynthia Wolberger *[Source: Courtesy Cynthia Wolberger.]*

In determining the crystal structures of the AraC arabinose-binding and dimerization domains in the presence and absence of L-arabinose, Cynthia Wolberger found that arabinose binding changes the structure of the AraC dimerization domain. As long as glucose is absent, CRP-cAMP occupies a site on the DNA preceding the AraC-binding site and helps activate transcription of the *ara* operon.

In the case of the galactose (*gal*) operon, the Gal repressor inhibits transcription of the operon in the absence of galactose, and CRP-cAMP serves as the activator in the absence of glucose. The Gal repressor works differently from the Lac repressor in that it does not prevent RNA polymerase from binding to the Gal promoter. Instead, the Gal repressor probably prevents transition of the polymerase-promoter complex from the closed to the open form, thereby blocking formation of the elongation-competent form of RNA polymerase (**Figure 20-10**).

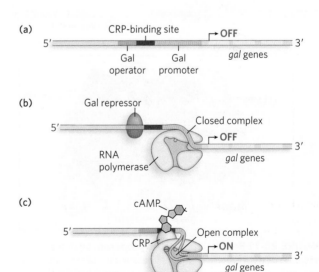

FIGURE 20-10 Regulation of the *gal* operon. (a) Structure of the *gal* operon. (b) The Gal repressor does not prevent RNA polymerase from binding the promoter; rather, it prevents formation of the open promoter-polymerase complex that is required for transcription initiation. (c) Similar to regulation of the *lac* and *ara* operons, transcription of the *gal* operon is increased only when glucose is absent, and thus CRP-cAMP binding is required.

Transcription Attenuation Often Controls Amino Acid Biosynthesis

Other important small molecules besides sugars help in regulating the expression of the genes involved in their metabolism. *E. coli* can produce all 20 of the common amino acids required for protein synthesis, but biosynthesis of an amino acid is necessary only when the intracellular concentration of that amino acid is low. The genes encoding the enzymes for synthesizing an amino acid generally cluster in an operon that is repressed whenever existing supplies of that amino acid are adequate for cellular requirements. When more of the amino acid is needed, the operon is actively transcribed and the biosynthetic enzymes are expressed.

The *E. coli* tryptophan (*trp*) operon provides a classic example of the kind of regulation that enables fine-tuning of gene expression levels to suit the needs of the cell (**Figure 20-11a, b**). The *trp* operon includes five genes encoding the enzymes required to synthesize tryptophan.

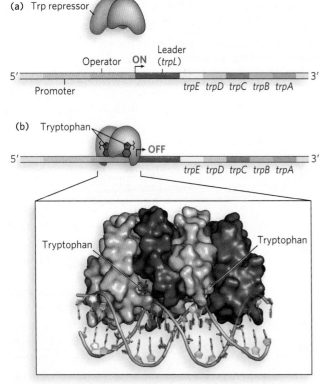

FIGURE 20-11 Regulation of the *trp* operon. (a) In the absence of tryptophan, the Trp repressor cannot bind the operator, and transcription of the *trp* operon is initiated. (b) When tryptophan is abundant, the protein products from the *trp* operon are no longer needed. Tryptophan serves as the effector molecule for the Trp repressor; their association causes the Trp repressor to bind the operator, blocking transcription. Notice the presence of the leader sequence; this is required for a second level of transcriptional control (see Figure 20-12). The molecular model shows the homodimeric *trp* repressor bound to DNA. [Source: (b) (bottom) PDB ID 1TRO.]

The short half-life (~3 minutes) of the mRNA transcribed from the *trp* operon allows the cell to respond rapidly to changing needs for tryptophan. A homodimeric repressor protein, the Trp repressor, regulates the operon. Tryptophan acts as a small-molecule effector for the Trp repressor (**Figure 20-11b**). When tryptophan is abundant, it binds the repressor and induces a conformational change that permits the repressor to bind the Trp operator and inhibit expression of the *trp* operon. The Trp operator site overlaps the promoter such that binding of the repressor blocks the binding of RNA polymerase. In this way, the *trp* operon is negatively regulated: a **corepressor** (in this case tryptophan) binds the repressor protein, rendering the repressor competent for DNA binding. This is distinct from the negative regulation of the *lac* operon, in which the Lac repressor binds the operator in the absence of inducer (allolactose), dissociating from DNA only when the small-molecule effector is present.

Once again, this simple on/off circuit mediated by a repressor protein and small effector molecule is only part of the regulatory story. Different cellular concentrations of tryptophan can alter the rate of synthesis of the biosynthetic enzymes over a 700-fold range. Repressor action accounts for only about a 70-fold difference in gene expression between the repressed and activated states of the operon. Once repression is lifted and transcription begins, the rate of transcription is modulated by a second regulatory process, **transcription attenuation**, in which transcription is initiated normally but is abruptly halted *before* the operon genes are transcribed. Attenuation provides a honing of gene expression that, when combined with repressor action, results in the 700-fold difference in expression of the tryptophan biosynthetic enzymes.

The frequency with which transcription of the *trp* operon is attenuated is regulated by the availability of tryptophan in the cell and relies on the very close coupling of transcription and translation in bacteria. This mode of regulation is necessarily unique to cells that lack a nucleus. In eukaryotic cells, where transcription and translation are physically and temporally separated, these processes cannot be coupled for the kind of attenuation described here.

The mechanism of attenuation in the *trp* operon relies on a 162-nucleotide region at the 5′ end of the mRNA, called the **leader sequence**, which precedes the initiation codon of the first gene (**Figure 20-12a**). Within the leader sequence are four regulatory regions, sequences 1 through 4. Sequences 3 and 4 can base-pair to form a **terminator**, a G≡C-rich stem-and-loop (hairpin) structure, closely followed by a series of U residues. Formation of the terminator causes RNA polymerase to terminate transcription prematurely and dissociate from the DNA before the operon genes can be transcribed. The termination mechanism involves polymerase slowing or

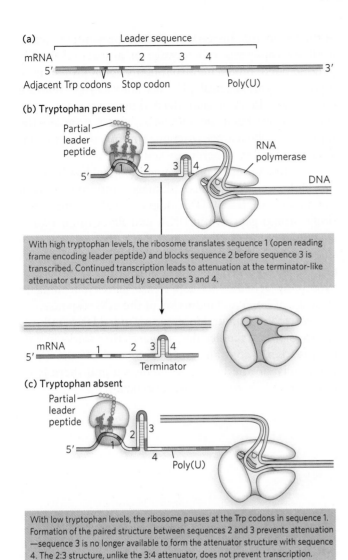

(a)

Leader sequence

mRNA

Adjacent Trp codons   Stop codon   Poly(U)

(b) Tryptophan present

Partial leader peptide

RNA polymerase

DNA

With high tryptophan levels, the ribosome translates sequence 1 (open reading frame encoding leader peptide) and blocks sequence 2 before sequence 3 is transcribed. Continued transcription leads to attenuation at the terminator-like attenuator structure formed by sequences 3 and 4.

mRNA

Terminator

(c) Tryptophan absent

Partial leader peptide

Poly(U)

With low tryptophan levels, the ribosome pauses at the Trp codons in sequence 1. Formation of the paired structure between sequences 2 and 3 prevents attenuation —sequence 3 is no longer available to form the attenuator structure with sequence 4. The 2:3 structure, unlike the 3:4 attenuator, does not prevent transcription.

FIGURE 20-12 Graded control of the *trp* operon through transcription attenuation. (a) The leader sequence of the *trp* mRNA. The transcript generated from the *trp* promoter includes a leader sequence at the 5′ end (containing four regulatory regions labeled 1 through 4). A portion of this sequence (sequence 1) is translated into the leader peptide, which has no known function other than to regulate the *trp* operon. (b) In the presence of tryptophan, the ribosome translates quickly through the Trp codons of sequence 1 and into sequence 2, allowing sequences 3 and 4 to associate to form a hairpin that stalls the RNA polymerase and terminates transcription. (c) In the absence of tryptophan, the ribosome stalls in sequence 1, allowing sequences 2 and 3 to associate. With sequence 3 unavailable to associate with sequence 4, the terminator structure is not formed and transcription can proceed. The amount of free tryptophan available for protein synthesis thus determines whether the *trp* operon is transcribed.

stalling when encountering the stable hairpin (terminator), then dissociating from the DNA as a result of relatively weak base pairing between the adjacent U-rich sequence and the complementary A-rich sequence in the DNA template. However, sequence 3 can also base-pair with sequence 2. When sequences 2 and 3 associate, the terminator cannot form, and uninterrupted transcription

continues into the *trp* genes. The loop formed by the pairing of sequences 2 and 3 does not block transcription.

How is hairpin choice determined? Regulatory sequence 1 and the availability of tryptophan are crucial for determining whether sequence 3 pairs with sequence 2 (letting transcription continue) or with sequence 4 (attenuating transcription). Formation of the terminator stem-and-loop structure depends on events that occur during *translation* of regulatory sequence 1. Sequence 1 encodes a **leader peptide** of 14 amino acids (**Figure 20-12b**), two of which are Trp residues. The leader peptide has no other known cellular function; its synthesis is simply an operon regulatory device. This peptide is translated immediately after the leader RNA is transcribed, by a ribosome on the nascent mRNA that follows closely behind RNA polymerase on the DNA as transcription proceeds.

When tryptophan concentrations are high, concentrations of Trp-tRNA^{Trp} are also high. This allows translation to proceed rapidly past the two Trp codons of sequence 1 and into sequence 2, before sequence 3 is transcribed by RNA polymerase. In this situation, sequence 2 is covered by the ribosome and unavailable for pairing to sequence 3 when sequence 3 is synthesized; the terminator structure (sequences 3 and 4) forms instead, and transcription halts (see Figure 20-12b). However, when tryptophan concentrations are low, the ribosome stalls at the two Trp codons in sequence 1, because Trp-tRNA^{Trp} is less available. Sequence 2 remains free while sequence 3 is transcribed, allowing these two sequences to base-pair. Sequence 3 is then unavailable for pairing with sequence 4, preventing formation of the terminator and letting transcription proceed (**Figure 20-12c**). In this way, the proportion of transcripts that are prematurely terminated declines as tryptophan concentration declines. Other bacteria also use multiple levels of regulation for control of tryptophan biosynthetic genes; an example of this is the TRAP system found in *Bacillus subtilis* (see the How We Know section at the end of this chapter).

In *E. coli*, other amino acid biosynthetic operons use a similar attenuation strategy to fine-tune enzyme production to meet the prevailing cellular requirements. For example, the 15-residue leader peptide produced by the *phe* operon contains seven Phe residues. The *leu* operon leader peptide has four contiguous Leu residues. The *his* operon leader peptide has seven contiguous His residues. In fact, in the *his* operon and several others, attenuation is sufficiently sensitive to be the only regulatory mechanism.

The SOS Response Leads to Coordinated Transcription of Many Genes

As described in Chapter 19, the many different genes that are required for a particular cell function are sometimes regulated together by a single transcription factor and/or small-molecule effector. This global regulation of transcription is an economical way for the cell to coordinate the expression of multiple genes that are needed at the same time.

An interesting example of this kind of genetic control is the cell's response to DNA damage. Extensive breakage or mutation of the bacterial chromosome triggers the expression of genes involved in DNA repair, which are located at different sites in the chromosome. This response is known as the **SOS response** and requires two key regulatory proteins: the RecA protein and the LexA repressor protein.

SOS genes encode proteins useful to cells with damaged DNA. These include Y-family polymerases, also known as translesion synthesis (TLS) polymerases, which have relaxed fidelity and can replicate DNA containing chemical lesions, such as UV-induced cross-links. The LexA repressor inhibits transcription of all the SOS genes by binding near their promoters, and induction of the SOS response requires the removal of LexA (**Figure 20-13a**). This is not a simple dissociation from DNA in response to the binding of a small molecule, as in the regulation

of the *lac* operon. Instead, the LexA repressor inactivates itself by catalyzing self-cleavage at a specific Ala–Gly peptide bond, producing two roughly equal protein fragments. At physiological pH, this autocleavage reaction requires the RecA protein. RecA is not a protease in the classical sense, but its interaction with LexA promotes the repressor's self-cleavage reaction. This function of RecA is sometimes called a coprotease activity.

The RecA protein provides a functional link between the biological signal (DNA damage) and activation of the SOS genes. Heavy DNA damage leads to numerous single-strand gaps in the DNA, and RecA binds tightly to single-stranded DNA. Only RecA that is bound to single-stranded DNA can facilitate cleavage of the LexA repressor (**Figure 20-13b**). Binding of RecA at the gaps eventually activates its coprotease activity, leading to cleavage of LexA and induction of the SOS response.

Active RecA, bound to single-stranded DNA, induces cleavage of LexA molecules, a system that works in part because LexA constantly cycles on and off the DNA. Eventually, all the LexA is proteolyzed and there is no intact LexA left to repress the SOS genes.

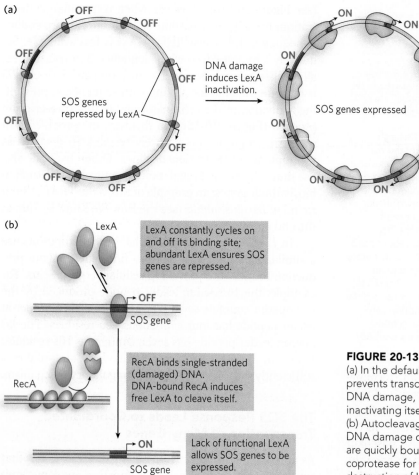

FIGURE 20-13 The global SOS response to DNA damage. (a) In the default state of the *E. coli* cell, the LexA repressor prevents transcription of the SOS genes. In response to DNA damage, LexA is stimulated to undergo autocleavage, inactivating itself and allowing transcription of the SOS genes. (b) Autocleavage of the LexA repressor requires RecA protein. DNA damage creates sites of single-stranded DNA, which are quickly bound by RecA. DNA-bound RecA becomes a coprotease for LexA, and their association facilitates the destruction of LexA and induction of the SOS response.

The SOS response is an example of how a single regulatory mechanism can coordinate the expression of related sets of genes. It also provides a remarkable illustration of evolutionary adaptation. During induction of the SOS response in a severely damaged bacterial cell, RecA also facilitates cleavage of repressors that allow the propagation of certain viruses present in the cell in a dormant, lysogenic state (a state described in Section 20.3). These repressors, like LexA, undergo self-cleavage at a specific Ala–Gly peptide bond. Induction of the SOS response permits replication of the virus and lysis of the cell, releasing new viral particles. Thus, bacteriophages have evolved to use the bacterial SOS system to their advantage, giving themselves the opportunity to make a hasty exit from a compromised bacterial host cell.

The bacterial SOS response is just one of the many ways in which cells control the expression of related genes. Another kind of mechanism involves the synthesis and detection of small molecules that can diffuse between cells in a process called **quorum sensing** (see this chapter's Moment of Discovery and How We Know section). Understanding how some kinds of pathogenic bacteria use quorum sensing (and other regulatory mechanisms) to control the genes necessary for rapid growth in infected individuals could offer new avenues for therapeutic intervention.

- Dissociation of a repressor from, or binding of an activator to, its target sequence to activate transcription can be triggered by a specific small molecule called an inducer. This was first elucidated in studies of the *lac* operon of *E. coli*. The Lac repressor dissociates from the Lac operator when the repressor binds its inducer, allolactose, activating expression of genes needed to metabolize lactose when this sugar is abundant.

- Catabolite repression is a mechanism of positive gene regulation in bacteria in which the presence of a preferred carbon source, such as glucose, prevents the activation of operons encoding enzymes required for metabolizing secondary sugars, such as lactose and arabinose. When glucose is depleted, cAMP concentrations increase and, in turn, increase the amount of CRP-cAMP complex, which stimulates transcription of these operons.

- When arabinose is present, CRP binds to the activator protein AraC, causing AraC to dimerize and bind to two DNA sites, activating the promoter of the *ara* operon. Alternative AraC-binding sites occupied by AraC in the absence of arabinose and CRP configure the promoter in an inactive state.

- Bacterial operons that produce the enzymes of amino acid synthesis use transcription attenuation, a regulatory process that involves a transcription termination site in the mRNA. Formation of the terminator is modulated by a mechanism that couples transcription and translation while responding to small changes in amino acid concentration.

- Many biosynthetic pathway operons, such as those encoding amino acid–synthesizing enzymes, are repressed by the end product of the pathway. In this way, amino acids inhibit their own production.

- In the SOS response, multiple genes throughout the chromosome, repressed by a single repressor protein, LexA, are activated simultaneously when DNA damage triggers RecA protein–facilitated autocatalytic proteolysis of LexA.

20.2 BEYOND TRANSCRIPTION: CONTROL OF OTHER STEPS IN THE GENE EXPRESSION PATHWAY

Genetic regulation in its simplest form is the process by which cells sense their metabolic needs and modulate the levels of certain gene products in response to those needs. Transcription, particularly at the initiation step, was for a long time the focus of research efforts to understand how bacteria change gene expression in response to various signals. But, as eventually became apparent, several key points after transcription have equally interesting mechanisms for controlling levels of active protein product. These points include mRNA stability, translation (protein synthesis), and protein modification and degradation. As noted earlier, some of the regulatory steps shown in Figure 19-1, such as certain aspects of mRNA processing and protein transport, are not relevant in bacteria, given the absence of a nucleus and lack of pre-mRNA splicing in these cells.

The control of gene expression at steps after transcription allows a rapid up- or down-regulation of protein synthesis in response to molecular signals. Such changes can occur more quickly at posttranscriptional levels because the cell does not need to wait for mRNA levels to change through increased or decreased transcription. We discuss here some of the ways in which bacteria regulate gene expression beyond transcription, as a means of altering the amount of protein produced from a given mRNA. As we will see, RNA transcripts themselves can be central players in these types of regulatory pathways.

RNA Sequences or Structures Can Control Gene Expression Levels

The most direct way of controlling gene expression at the mRNA level is for the RNA itself to have regulatory properties. These regulatory regions of mRNAs, called **riboswitches**, exist within the 5′ untranslated region (5′UTR) of the RNA; they bind to small-molecule metabolites with the affinity and specificity required for the precise regulation of gene expression. Riboswitches consist of a small-molecule-binding element connected to a regulatory region; binding of the small molecule (the ligand) triggers a conformational change in the regulatory region, such that the entire RNA molecule changes shape.

Different riboswitches have different downstream effects on gene expression; in most cases, ligand binding affects either transcription or translation (**Figure 20-14**). Depending on how well the ligand binds to the riboswitch, the regulation may be very sensitive to the presence of low ligand concentrations or may occur only when ligand concentrations rise to a significant level. Thus far, more than a dozen distinct classes of riboswitches have been identified, each class with common sequence and structural features, as well as distinct ligand-binding specificities, as shown in **Table 20-1**.

The discovery of riboswitches led to questions about how they translate binding events into changes in gene expression levels. Ronald Breaker and his colleagues at Yale University carried out elegant experiments, in vitro and in vivo, to analyze what happens to an mRNA containing a riboswitch on exposure to a ligand. The researchers used chemicals and enzymes that cleave single-stranded, but not double-stranded, regions of RNA to digest riboswitch-containing transcripts (referred to simply as "riboswitch RNA") in the presence or absence of ligand. By comparing the different patterns of RNA fragments generated with and without bound ligand, they found that riboswitches undergo conformational changes on binding to a favored ligand. Genetic experiments showed that mutations in

Ronald Breaker [Source: Courtesy Ronald Breaker.]

(a) **Transcription riboswitch**

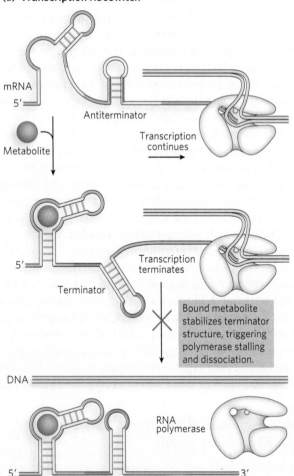

(b) **Translation riboswitch**

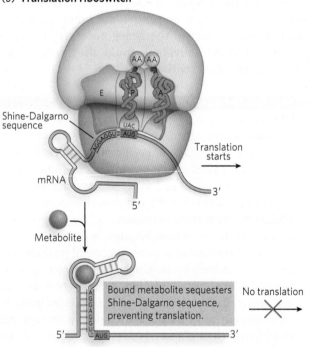

FIGURE 20-14 Riboswitches. Riboswitches are structures within an mRNA that bind metabolites (which act as ligands) and undergo a conformational change to regulate gene expression. Riboswitches exert their control at the level of (a) transcription or (b) translation.

TABLE 20-1		
Types of Riboswitches		
Riboswitch Class	*Ligand*	*Function of Regulated Genes*
FMN	Flavin mononucleotide (FMN)	Riboflavin biosynthesis and transport
THI box	Thiamine pyrophosphate (TPP)	Thiamine biosynthesis and transport
B12	Adenosylcobalamin	Vitamin B_{12} biosynthesis and transport
S box (SAM-I)	S-Adenosylmethionine (adoMet)	Methionine and adoMet biosynthesis and transport
SAM-II	S-Adenosylmethionine (adoMet)	Methionine and adoMet biosynthesis and transport
S_{MK} box (SAM-III)	S-Adenosylmethionine (adoMet)	Methionine and adoMet biosynthesis and transport
SAH	S-Adenosylhomocysteine (adoHcy)	Recycling of adoHcy, a metabolite of adoMet
L box	Lysine	Lysine metabolism and transport
Glycine	Glycine	Glycine metabolism
Purine	Guanine/adenine	Purine metabolism and transport
dG	2′-Deoxyguanosine	Deoxyribonucleotide biosynthesis
Cyclic di-GMP	Cyclic di-GMP	Virulence, motility, and biofilm formation
glmS	Glucosamine 6-phosphate	Glucosamine 6-phosphate biosynthesis
preQ1	7-Aminoethyl-7-deazaguanine (preQ1)	Synthesis of queuosine, a modified nucleotide in wobble position of some tRNAs
Mg^{2+}	Magnesium	Mg^{2+} transport
T box	Uncharged tRNAs	Aminoacyl-tRNA and amino acid biosynthesis

Source: Data from T. Henkin, *Genes Dev.* 22:3383–3390, 2008.

these regulatory regions prevented changes in gene expression in response to changing levels of ligand. This led to the idea that riboswitch RNAs undergo structural rearrangements on binding to their target ligands, resulting in either transcription termination or sequestration of a Shine-Dalgarno sequence, thus preventing translation. Although these mechanisms are established for some riboswitches, other riboswitches may have different kinds of downstream effects stemming from their ability to undergo conformational change on binding a specific ligand.

More detailed insights have come from structural studies of the riboswitch RNAs, particularly by x-ray crystallography. To use this approach, large amounts of homogeneous RNA corresponding to the riboswitch must be purified and concentrated slowly in the presence of salts that favor crystallization. Several riboswitch structures determined in this way, in the presence or absence of a bound ligand, have revealed the nature and magnitude of the structural changes that occur on ligand binding. For example, the thiamine pyrophosphate (TPP)–binding riboswitch, located in the 5′UTR region of mRNAs involved in vitamin B_1 (thiamine) biosynthesis, controls gene expression by inhibiting translation in the presence of abundant vitamin B_1. This regulatory RNA region, also known as the THI box or THI element, is one of the few examples of a riboswitch found in all organisms—archaea, eukaryotes, and bacteria.

Like other riboswitches, the THI box adopts a globular structure that encircles the TPP ligand (**Figure 20-15a**). In the absence of TPP, this structure is not energetically favored. Evidence for this conclusion comes from experiments in which the TPP riboswitch RNA is incubated in a buffered solution with or without TPP, then subjected to ribonuclease cleavage. When TPP is present, ribonuclease cleaves the riboswitch RNA at only a few positions in the nucleotide sequence, but in the absence of TPP, most of the RNA is susceptible to cleavage (**Figure 20-16** on p. 711). This result indicates that the three-dimensional structure of the riboswitch is stable only in the presence of the TPP ligand.

This discovery immediately suggested a mechanism by which the TPP riboswitch could regulate gene expression. Inspection of sequences containing the THI box showed that it occurs near the gene's Shine-Dalgarno sequence. When no TPP is present, the riboswitch structure does not form, or at least is unstable, and the Shine-Dalgarno sequence is accessible for ribosome binding. However, when TPP is abundant and the cell no longer needs to synthesize the enzymes required for vitamin B_1 production, the riboswitch structure forms, and the Shine-Dalgarno sequence is no longer available to the ribosome.

(a)

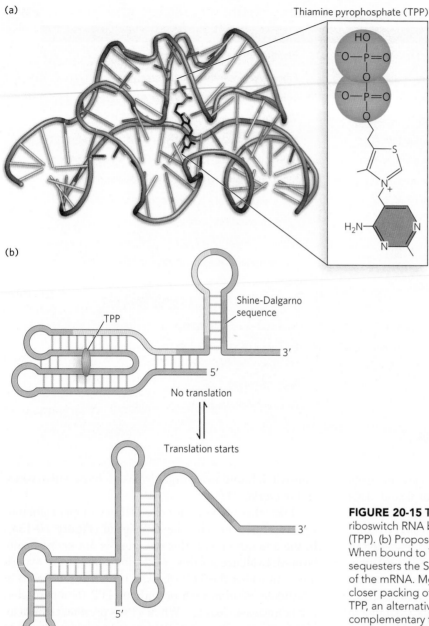

Thiamine pyrophosphate (TPP)

(b)

Shine-Dalgarno sequence

TPP

3′

5′

No translation

Translation starts

3′

5′

FIGURE 20-15 The TPP riboswitch. (a) Structure of the riboswitch RNA bound to its ligand, thiamine pyrophosphate (TPP). (b) Proposed mechanism of TPP riboswitch function. When bound to TPP, the riboswitch assumes a conformation that sequesters the Shine-Dalgarno sequence, preventing translation of the mRNA. Mg^{2+} ions (not shown) also play a role, allowing closer packing of the RNA helices and TPP. In the absence of TPP, an alternative secondary structure forms (yellow regions are complementary to and base-pair with each other), and translation can be initiated. [*Sources: (a) PDB ID 2GDI. (b) Data from A. Serganov et al., Nature 441:1167–1171, 2006, Fig. 4.*]

Examination of the TPP-bound form of the riboswitch shows how this works: TPP binds at the surface between two stem-and-loop structures in the RNA, creating a network of contacts between the small-molecule ligand (TPP) and the two RNA helices (**Figure 20-15b**). Nearby Mg^{2+} ions help neutralize the negative charges on the pyrophosphate group of the ligand, as well as the charges on the RNA phosphodiester backbone, allowing the TPP and the RNA helices to pack close together. This mode of binding, in which the structure of the ligand-recognition cavity forms only in the presence of its cognate ligand, is an example of induced fit (see Chapter 5).

To be useful as gene regulators, riboswitches must be able to distinguish between chemically related small molecules. Riboswitch RNAs prepared by in vitro transcription were found to bind much better to their natural ligand than to small molecules with similar but distinct chemical structures. Riboswitch RNAs can distinguish between molecules based on atomic charge, stereochemistry, and the presence or absence of particular functional groups.

Breaker and colleagues' analysis of an interesting bacterial riboswitch that controls the gene *glmS* gives us an example of the specificity of riboswitches. This gene encodes an enzyme that catalyzes the conversion of fructose 6-phosphate and glutamine to glucosamine 6-phosphate—a metabolite that down-regulates *glmS* expression. The riboswitch, in the 5′UTR of the *glmS* mRNA, binds glucosamine 6-phosphate to form a

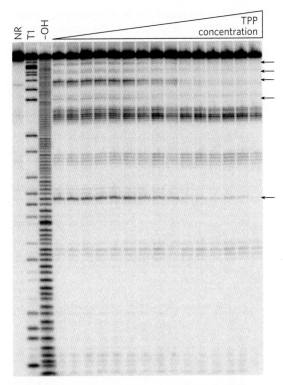

FIGURE 20-16 The effect of ligand binding on RNA susceptibility to spontaneous breakdown. In this experiment, a transcript containing just the TPP riboswitch aptamer sequence was incubated under slightly alkaline conditions in the presence of increasing concentrations of TPP. The leftmost lane (NR) of the polyacrylamide gel contains untreated RNA; the dark band indicates the size of the full-length RNA. The next two lanes show the fragments generated by partial digestion of the RNA with either riibonuclease T1 (T1) or strong alkali (-OH) in the absence of TPP. Many sites on the RNA are susceptible to cleavage when TPP is not present, indicating that the RNA does not form a stable three-dimensional structure. As the remaining lanes show, when the RNA is incubated for long periods of time under slightly alkaline conditions, fewer fragments are generated as the TPP concentration is increased (arrows). This demonstrates that TPP promotes formation of secondary and tertiary structures that protect certain residues from spontaneous cleavage. [Source: Welz, R., and Breaker, R., Ligand binding and gene control characteristics of tandem riboswitches in Bacillus anthracis. 2007. RNA 13:573–582. Figure 4a. © Cold Spring Harbor Laboratory Press.]

catalytically active structure—a ribozyme—that cleaves the *glmS* mRNA and thus leads to its degradation (**Figure 20-17a**). Glucosamine 6-phosphate is a cofactor for the catalytic reaction and participates directly in the reaction chemistry. Experiments with different, structurally related sugars showed that sugars with even single-atom differences relative to glucosamine 6-phosphate could not support *glmS* ribozyme cleavage (**Figure 20-17b, c**).

Because the *glmS* ribozyme is active only when bound to glucosamine 6-phosphate, this system affords a simple but effective mechanism for reducing *glmS* transcript levels when the gene product is not needed, thereby preventing translation. Although this is the only known example of a ribozyme that uses a small-molecule

cofactor as part of its catalytic mechanism, it suggests that ribozymes are inherently capable of such chemical collaborations and hence might be capable of more complex functions than have been discovered so far.

The *glmS* ribozyme has been crystallized in the precleaved and postcleaved states, as well as bound to its ligand. Unlike the TPP riboswitch, the *glmS* riboswitch was found to have essentially the same structure in all three states. Preformed catalytic and cofactor-binding sites may favor glucosamine 6-phosphate binding, enabling greater sensitivity to glucosamine 6-phosphate levels and hence to the metabolic state of the cell. Another notable feature of the *glmS* riboswitch structure is that the cofactor binds in an open, accessible pocket, perhaps further favoring ligand binding and ribozyme activation (see Figure 20-17a). How does glucosamine 6-phosphate participate in the ribozyme reaction chemistry? Although not yet confirmed, the crystal structures suggest that the amine group of the ligand may help to activate a specific 2′-hydroxyl group in the RNA backbone for nucleophilic attack.

Some riboswitches occur in similar types of genes and share common regulatory features. For example, levels of aminoacylated tRNAs in certain bacteria are controlled by RNA structures that respond to the concentration of specific amino acids in the cell (**Highlight 20-2** on p. 713).

Although examples of riboswitches have been identified in all three domains of life—bacteria, archaea, and eukaryotes—they seem to be most common in the bacterial world, where they are employed in the regulation of genes involved in the biosynthesis of vitamins, enzyme cofactors, and various amino acids. What might be the reason for this greater frequency of riboswitches in bacterial cells? The use of RNA as a metabolite detector affords cells the opportunity to regulate gene expression without requiring the presence or synthesis of additional regulatory proteins. For microbes that must rapidly adjust to changing environmental conditions and availability of vitamins, enzyme cofactors, amino acids, and other small molecules, riboswitch mechanisms seem to be a highly suitable mode of regulation. However, these mechanisms may lack the multiple layers of regulatory control that are typically found in eukaryotic organisms, and they may not be sufficient to enable the kind of coordination and fine-tuning of gene expression required for such complex processes as development and organ differentiation (processes discussed in Chapter 22).

Translation of Ribosomal Proteins Is Coordinated with rRNA Synthesis

In bacteria, an increased cellular demand for protein synthesis is met by increasing the number of ribosomes, rather than increasing the activity of individual ribosomes. In general, the number of ribosomes rises

(a)

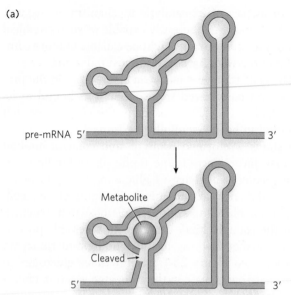

pre-mRNA 5′   3′

Metabolite

Cleaved

5′   3′

FIGURE 20-17 Ligand specificity in the *glmS* riboswitch. The *glmS* riboswitch is a ribozyme. (a) On binding of its small-molecule metabolite, glucosamine 6-phosphate, the *glmS* precursor mRNA (pre-mRNA) cleaves itself, promoting its own degradation. (b) The structure of glucosamine 6-phosphate (GlcN6P) and some close structural analogs. (c) Result of agarose gel electrophoresis of ribonuclease-cleaved samples. The *glmS* RNA was incubated with its known effector (GlcN6P) or with one of the other compounds shown in (b), for the times indicated. The leftmost lanes show untreated pre-mRNA (NR) and pre-mRNA incubated with no effector added (−). The top row of bands shows the full-length pre-mRNA (Pre); the bottom row, the ribonuclease cleavage product (Clv). Only GlcN6P induced cleavage of the pre-mRNA, indicating that the *glmS* riboswitch is highly specific for this ligand; the results also show that the riboswitch requires Mg^{2+}. *[Source: (c) Reprinted by permission from Macmillan Publishers Ltd: Winkler, W.C., et al. 2004. Nature 428: 281–286. (18 March 2004) doi:10.1038/nature02362.]*

(b)

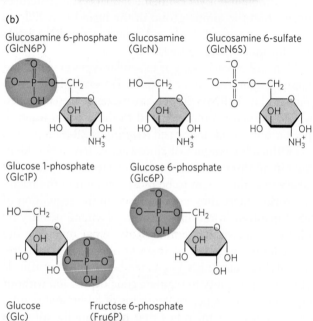

Glucosamine 6-phosphate (GlcN6P)

Glucosamine (GlcN)

Glucosamine 6-sulfate (GlcN6S)

Glucose 1-phosphate (Glc1P)

Glucose 6-phosphate (Glc6P)

Glucose (Glc)

Fructose 6-phosphate (Fru6P)

(c)

Effector	−	GlcN6P							GlcN	GlcN6S	Glc1P	Glc6P	Glc	Fru6P
Mg^{2+}		+	−	+	+	+	+	+	+	+	+	+	+	+
Time (seconds)	NR	60	60	0	15	30	45	60	60	60	60	60	60	60

Pre—

Clv—

HIGHLIGHT 20-2 ⎯ **A CLOSER LOOK**

T-Box Riboswitches

RNA molecules rather than proteins are sometimes employed by bacterial cells to detect the presence of small molecules. While studying how the soil bacterium *Bacillus subtilis* responds to environmental changes, Tina Henkin and her colleagues at the Ohio State University discovered that the *tyrS* gene transcript, which encodes tyrosyl-tRNA synthetase, includes an RNA structure in the noncoding leader region that monitors levels of tyrosine in the cell. Rather than the *tyrS* mRNA leader region binding directly to tyrosine, the limitation of tyrosine availability in the cell is detected by interaction of the leader region with uncharged tRNA^Tyr. The anticodon of the tRNA^Tyr base-pairs with a single detector "codon" in the mRNA leader, and the amino acid arm of the tRNA^Tyr makes a second interaction with the leader, promoting RNA polymerase read-through of a structure that would otherwise cause transcription termination (Figure 1). In this way, the depletion of tyrosine, leading to increased levels of uncharged tRNA^Tyr, enhances transcription of the mRNA encoding tyrosyl-tRNA synthetase, which in turn charges more tRNA^Tyr to Tyr-tRNA^Tyr. This mechanism is not unique to tyrosine: the leader regions of at least 18 aminoacyl-tRNA synthetase and amino acid biosynthetic gene transcripts in *B. subtilis*

Tina Henkin *[Source: Courtesy Tina Henkin.]*

and related gram-positive bacteria have conserved structural features similar to those of the *tyrS* gene. An example is regulation of tryptophan levels by the anti-TRAP protein (see the How We Know section at the end of this chapter). Collectively, these RNAs are known as the T-box family of regulators, or T-box riboswitches.

Using genetic methods, Henkin and her colleagues found that mutations in these regulatory regions of the mRNA prevent or change the ability of the genes to respond to specific amino acid levels. For example, a change in the detector codon of the *tyrS* mRNA leader from UAC (which base-pairs with tRNA^Tyr) to UUC (which base-pairs with tRNA^Phe) resulted in loss of transcription induction by low tyrosine levels and a switch to induction by low phenylalanine levels. In contrast, insertion of an extra nucleotide immediately before the detector codon did not affect regulation. Because such an insertion would change the reading frame of the ribosome, this finding indicates that the regulation results directly from RNA-mediated structural changes, rather than requiring production of a protein. Ultimately, Henkin demonstrated that *tyrS* antitermination can occur in a purified transcription system with no additional cellular factors, indicating that the leader RNA is sufficient for specific recognition of the cognate tRNA. These findings were among the first hints that RNA molecules play broader roles in gene regulation than previously expected.

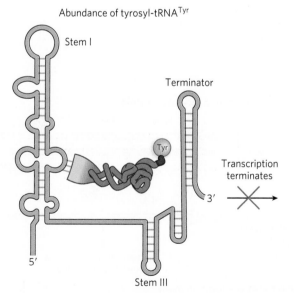

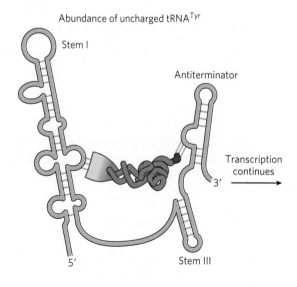

FIGURE 1 The *tyrS* riboswitch senses the cellular level of uncharged tRNA^Tyr. Both charged and uncharged tRNA^Tyr interact with the leader region of the mRNA at the detector codon, but only uncharged tRNA^Tyr is capable of a second interaction between its amino acid arm and the antiterminator. This second interaction stabilizes the antiterminator and promotes continued transcription of the tyrosyl-tRNA synthetase gene. *[Source: Data from T. M. Henkin and F. J. Grundy, Cold Spring Harb. Symp. Quant. Biol. 71:231–237, 2006, Fig. 2b.]*

as the cellular growth rate increases. At high growth rates, ribosomes make up about 45% of the bacterial cell's dry weight. The proportion of cellular resources devoted to making ribosomes is so large, and the function of ribosomes so important, that bacteria must coordinate the synthesis of the ribosomal components: the ribosomal proteins (r-proteins) and RNAs (rRNAs). This regulation occurs largely at the level of synthesis of r-proteins.

The 52 genes that encode the r-proteins occur in at least 20 operons, each containing 1 to 11 genes. Some of these operons also contain the genes for the subunits of DNA primase, RNA polymerase, and the elongation factors required for protein synthesis—revealing the close coupling of replication, transcription, and translation during bacterial cell growth.

The r-protein operons are regulated primarily through a translational feedback mechanism. One r-protein encoded by each operon also functions as a **translational repressor**, which binds the mRNA transcribed from that operon and blocks translation of all the genes encoded by the mRNA (**Figure 20-18**). In general, the r-protein that plays the role of repressor also binds directly to an rRNA. Each translational repressor r-protein binds with higher affinity to the appropriate rRNA than to its mRNA, so the mRNA is bound and translation is repressed only when the level of the r-protein exceeds that of the rRNA. This ensures that translation of the mRNAs encoding r-proteins is repressed only when synthesis of these r-proteins exceeds the level needed to make functional ribosomes. In this way, the rate of r-protein synthesis is kept in balance with rRNA availability.

The mRNA binding site for the translational repressor is near the start site of one gene in the r-protein operon, usually the first gene (see Figure 20-18). In other operons this would affect only that one gene, because in polycistronic mRNAs, most genes have independent translation signals. In r-protein operons, however, the translation of one gene depends on the translation of all the others. The mechanism of this translational coupling is not yet understood in detail. In some cases, the translation of multiple genes seems to be blocked by folding of the mRNA into an elaborate three-dimensional structure that is stabilized by both internal base pairing and binding of the translational repressor. When the repressor is absent, ribosome binding and translation of one or more genes disrupts the folded structure of the mRNA, allowing all the genes to be translated.

r-Protein operon

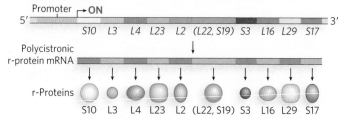

Overabundance of rRNA

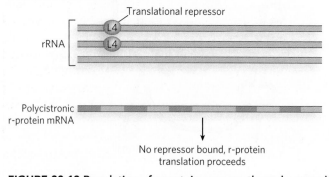

No repressor bound, r-protein translation proceeds

Overabundance of r-protein

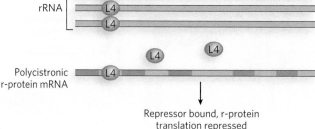

Repressor bound, r-protein translation repressed

FIGURE 20-18 Regulation of r-protein operons through a translational feedback loop. In many r-protein operons, such as the one shown here, one of the r-proteins produced by the operon (in this case, protein L4) also functions as a translational repressor in a mechanism that involves sensing the relative levels of r-proteins and rRNAs. L4 protein has a higher affinity for rRNA and binds it preferentially over the r-protein mRNA. When levels of r-protein are low relative to rRNA, few L4 protein molecules are available to bind the mRNA and translation proceeds, making more r-proteins. But when r-protein concentrations build up in excess of rRNA levels, the L4 protein binds to the mRNA generated from the operon and blocks production of additional r-proteins by preventing efficient initiation of translation.

Because the synthesis of r-proteins is coordinated with the availability of rRNA, ribosome production reflects the regulation of rRNA synthesis. In *E. coli*, rRNA synthesis from the seven rRNA operons responds to cellular growth rate and to changes in the availability of crucial nutrients, particularly amino acids. The regulation that is coordinated with amino acid concentrations is known as the **stringent response (Figure 20-19)**. It enables cells to direct resources away from growth and division and toward the biosynthesis of amino acids, to ensure survival until amino acid availability increases. The effects include down-regulation of rRNA and tRNA transcription. When amino acid concentrations are low, rRNA synthesis is halted. Amino acid starvation leads to the binding of uncharged tRNAs to the ribosomal A site, triggering a sequence of events that begins with the binding of an enzyme called **stringent factor** (RelA protein) to the ribosome. When bound to the ribosome, stringent factor catalyzes formation of the unusual nucleotide guanosine tetraphosphate (ppGpp). The enzyme adds pyrophosphate to the 3′ position of GTP, in the reaction:

$$\text{GTP} + \text{ATP} \rightarrow \text{pppGpp} + \text{AMP} \qquad (20\text{-}1)$$

A phosphohydrolase then cleaves off one phosphate to form ppGpp. The abrupt rise in ppGpp concentration in response to amino acid starvation results in greatly reduced rRNA synthesis; this is mediated, at least in part, by the binding of ppGpp to RNA polymerase, blocking transcription of the rRNA genes.

Like cAMP, ppGpp belongs to a class of modified nucleotides that act as cellular second messengers (see Chapter 6). In *E. coli*, ppGpp and cAMP serve as starvation signals; they cause large changes in cellular metabolism by increasing or decreasing the transcription of hundreds of genes. Similar nucleotide second messengers also have multiple regulatory functions in eukaryotic cells. The coordination of cellular metabolism with cell growth is highly complex, and further regulatory mechanisms undoubtedly remain to be discovered.

SECTION 20.2 SUMMARY

- Riboswitches regulate gene expression without the need for a separate regulatory protein, such as a repressor or an activator, to respond to signaling molecules. Direct binding of a riboswitch to a small-molecule ligand triggers a conformational change in the adjacent regulatory region that alters mRNA stability or translational efficiency.
- Ribosomal protein operons are regulated primarily through a translational feedback mechanism. One r-protein encoded by each operon functions as a translational repressor by binding to the mRNA transcribed from that operon and blocking translation of all the genes it encodes.
- In the stringent response, a starvation-induced pathway, the small signaling molecule ppGpp binds to RNA polymerase and reduces transcription of rRNA genes (and thus the number of ribosomes) and other genes needed for rapid growth.

20.3 CONTROL OF GENE EXPRESSION IN BACTERIOPHAGES

Our discussion of gene regulatory pathways up to this point has focused on bacteria. Historically, however, much of the early research on gene regulation was done on viruses—specifically, bacteriophages (phages), viruses that infect bacterial cells. Phage genomes are small, yet the genes they contain must be carefully regulated to enable efficient infection and viral reproduction. In addition, phage genes are frequently transferred to or from host cell chromosomes, thereby providing a critical means of transferring genetic information between organisms (i.e., from one host to another). Such horizontal gene transfer plays a significant role in driving the evolution of new bacterial traits, including resistance to drugs and toxins.

We focus here on the bacteriophage λ (λ phage) system, one of the best-studied systems of gene regulation in all of biology. Gene regulation in λ phage is intimately coupled to the state of the host cell, and we discuss some examples illustrating the principles involved. Gene regulation in λ phage involves a series of integrated pathways

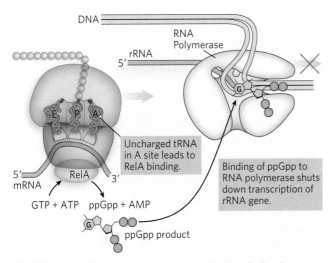

FIGURE 20-19 The stringent response in *E. coli.* Synthesis of rRNA from the seven rRNA operons is regulated by amino acid concentrations. When amino acid supplies are low, uncharged tRNAs can enter the ribosomal A site, triggering the stringent response and repressing rRNA synthesis. The formation of ppGpp occurs in two steps, as described in the text.

that exemplify the coordinated gene expression occurring in cells: some of the mechanisms used by λ phage have been found to govern gene expression in other systems. For example, animal cells take advantage of differential binding affinities and cooperative interactions of regulators to turn genes and gene networks on and off during development. Thus, insights from the λ phage system continue to guide our understanding of more complex organisms.

Phage Propagation Can Take One of Two Forms

Most organisms are susceptible to infection by viruses, which usurp the host cell machinery to produce more viral particles. One well-studied class of bacterial viruses, the lysogenic bacteriophages, uses two kinds of replication mechanism to ensure viral propagation and transmission. As we noted in Chapter 14, after introduction of its DNA into a host cell, a phage has the potential to enter one of two pathways for propagation (**Figure 20-20**). Most of the time, viruses use the **lytic pathway**, in which phage DNA is immediately replicated and viral proteins are synthesized to construct the viral coating around the DNA. Once many viral particles have been made, they burst from the cell, killing it, and enter the extracellular environment; they can now infect other cells. Occasionally, viruses enter the **lysogenic pathway**, whereby phage DNA integrates into the host cell chromosome and is replicated along with the chromosome as the cell divides. In this state, the bacteriophage is called a **prophage**, and the cell carrying it is a **lysogen**. Under normal conditions the prophage is stable within the lysogen, but if the host cell is stressed by DNA-damaging agents, nutritional limitations, or other conditions that threaten

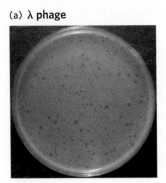

(a) λ phage

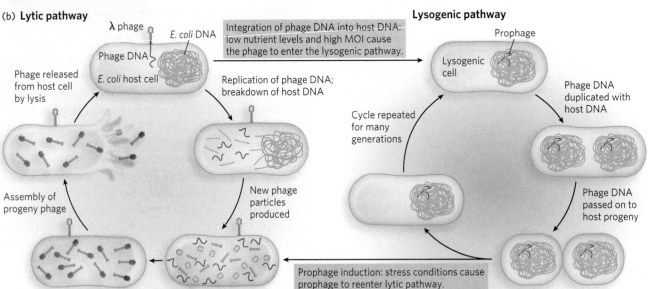

(b) **Lytic pathway**

FIGURE 20-20 The growth and life cycle phases of λ phage. (a) Bacteriophage λ infecting a lawn of *E. coli*. The viruses eventually kill their host cells, leaving cleared spots, or plaques, in the bacterial lawn. (b) The lytic and lysogenic pathways. Use of the lytic versus the lysogenic pathway is based on cellular nutrients and the viral multiplicity of infection (MOI). Under conditions of cellular stress, λ phages can exit the lysogenic pathway and enter the lytic pathway, in the process of prophage induction. [*Source: (a) Courtesy of Maria MacWilliams, University of Wisconsin–Parkside.*]

survival, the prophage can rapidly excise from the chromosome and enter the lytic pathway. The switch from lysogenic to lytic growth is referred to as **prophage induction**.

Many phages are capable of this kind of genetic switch, but the underlying mechanisms are best understood for λ phage, a virus of *E. coli* that consists of a double-stranded linear DNA encapsulated in a head structure and attached tail region made of virally encoded proteins. This phage has been used extensively as a model system for studying integrated gene regulatory networks, and it has also provided many tools useful in molecular biology research. For example, engineered versions of λ phage have been co-opted to transfer and express genes in *E. coli* and to inactivate host genes.

Two regulatory proteins, the λ repressor (also called cI) and Cro, govern the growth pathway of the virus. When the λ repressor protein is predominant, λ phage enters the lysogenic pathway; its DNA integrates into the chromosome, and only the λ repressor itself is expressed. When Cro predominates, however, λ phage enters the lytic pathway; most of the λ genes are expressed, and viral replication and packaging ensue. Eventually, the host cell is broken open by cell **lysis**, a process that releases the progeny phages.

Whether the phage enters lytic or lysogenic growth is determined early in a λ phage infection and depends largely on two other λ proteins: cII and cIII. The cII protein is a transcription activator that enhances transcription of the λ repressor gene and hence stimulates production of the repressor protein (cI). When cII is abundant and active, infection proceeds through the lysogenic pathway. But because cII is susceptible to degradation by bacterial proteases, it is often destroyed before it can trigger substantial production of the λ repressor. This tends to occur under nutrient-rich conditions, when proteases are present in high concentrations. It makes sense for the phage to enter the lytic pathway, triggered by low cII levels, when plenty of materials are available for viral reproduction and particle assembly.

The cIII protein stabilizes cII, probably by acting as a decoy, or alternative substrate, for protease molecules that would otherwise degrade cII. In this way, elevated cIII levels can trigger the switch to lysogen formation by enhancing cII concentrations, which increases production of the λ repressor.

Usually, λ phage infection leads to lytic growth. In addition to nutrient levels, the **multiplicity of infection (MOI)** also affects this growth pathway. This is because infections typically occur at a low MOI, when there are many more host cells than viral particles. In this situation, it is advantageous for viruses to grow lytically so that more progeny can be made and released to infect the available host cells. However, once there is an abundance of viral particles relative to host cells, multiple viruses begin to infect each host. In this circumstance, the high MOI leads to more frequent lysogen formation. The virus propagates silently within the chromosome and awaits future opportunities to enter the lytic pathway when host cells are again abundant. The mechanisms underlying this fascinating genetic switch have been elucidated over many years, using numerous genetic and biochemical methods.

Differential Activation of Promoters Regulates λ Phage Infection

Most of the ~50 genes of the λ phage genome are required for viral replication and packaging; a relatively small region of the genome is critical for the gene regulation necessary to induce lytic versus lysogenic growth (**Figure 20-21**). On initial infection of a host cell by λ DNA, viral transcription begins at two opposing promoters, P_L (leftward promoter) and P_R (rightward promoter), to produce the "immediate early" transcripts. This leads to the production of two proteins, N and Cro, which begin to increase in concentration. Two additional promoters, P_{RM} (promoter for *repressor maintenance*) and P_{RE} (promoter for *repressor establishment*), drive transcription of the *cI* gene, which encodes the λ repressor. However, unlike the strongly constitutive P_L and P_R, P_{RM} and P_{RE} are weak promoters and require activators to recruit RNA polymerase. Thus, at first, the *cI* gene is not transcribed and no λ repressor is produced.

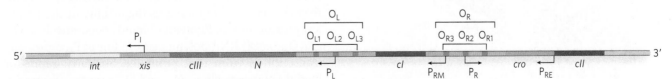

FIGURE 20-21 A partial map of λ phage. The genes and regulatory sites involved in establishing the lytic and lysogenic pathways are shown. Transcription begins at two opposing promoters, P_L and P_R, to produce the "immediate early" transcripts, which encode the N and Cro proteins. Two weak promoters, P_{RM} and P_{RE}, drive transcription of the *cI* gene, which encodes the λ repressor (cI). Overlapping the P_L, P_{RM}, and P_R promoters are two operator sites, O_L and O_R, each containing three binding sites for the λ repressor and Cro.

Overlapping the P_L, P_{RM}, and P_R promoters are two operator sites, O_L and O_R, each of which contains three binding sites for the λ repressor and Cro. Like the Lac repressor, both the λ repressor and Cro form homodimers in which two DNA-binding domains in the protein recognize DNA through a helix-turn-helix motif. In x-ray crystallographic structures of these proteins bound to DNA, researchers observed that each DNA-binding domain in the dimer binds to half of the 17 bp inverted repeat of the operator sequence (**Figure 20-22**). Each of the six operator binding sites, three sites in O_R and three in O_L (see Figure 20-21), can bind a λ repressor dimer or a Cro dimer. However, these sites have different affinities for the regulatory proteins, so they are not all occupied at once, or at random. In addition, operator-protein binding at one site influences binding at the other sites, an example of cooperative interaction.

During initial viral infection, once appreciable levels of Cro are expressed, this protein binds the O_{R3} site; because O_{R3} overlaps P_{RM}, Cro blocks the access of RNA polymerase to P_{RM}, and the *cI* gene is not transcribed. Cro does not bind as well to O_{R1} and O_{R2}, and thus these operator sites are not occupied by Cro. If this situation continues, lytic growth ensues (**Figure 20-23a**).

Lysogenic growth requires the activity of P_{RE} (**Figure 20-23b**). Initial transcription of the *cI* gene requires activation of P_{RE} by cII, another gene product expressed early in the infection process. The cII protein can activate the weak P_{RE} promoter by binding to a site upstream from the transcription start site (the -35 region; see Chapter 15) and recruiting RNA polymerase to P_{RE}. However, cII is a substrate of the host cell protease FtsH, which cleaves and inactivates the activator. At a low MOI, in which one viral particle at most has infected any one cell, the concentration of cII does not reach a level sufficient to activate P_{RE}. But under conditions of high MOI, in which multiple viral particles have infected the cell, more cII is produced because of the multiple copies of the *cII* gene present in the cell. Furthermore, the phage protein cIII (also expressed early in infection) helps stabilize cII by competing as a substrate for FtsH. In this situation, FtsH cannot keep up with cII production, and the increased levels of cII lead to activation of P_{RE}.

As a result of the activation of P_{RE}, the λ repressor protein is produced. The repressor activates P_{RM} by binding to O_{R1} and O_{R2}, leading to stable expression of the repressor (**Figure 20-23c**). The P_{RM} promoter is necessary to maintain ongoing production of the λ repressor because, when the repressor binds operator sites O_{R1} and O_{R2}, P_R is silenced and production of cII stops, leading to loss of the activator (cII) for P_{RE}. Under these conditions, lysogenic growth is favored.

Note that if Cro were to bind O_{R3} before cI (the repressor) bound O_{R1} and O_{R2}, the phage would have trouble establishing lysogeny. This interference by Cro probably never occurs, however, because when cII is highly active, cI production is sufficient to shut off the *cro* gene before enough Cro is made to turn off P_{RM}.

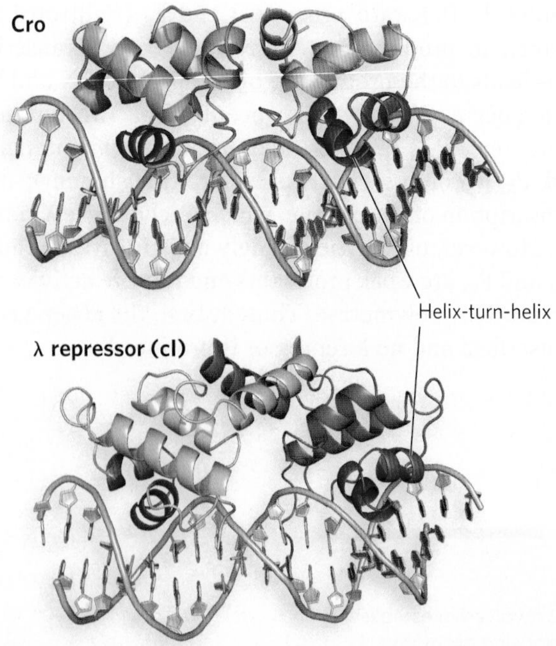

The λ Repressor Functions as Both an Activator and a Repressor

The λ repressor is capable of complex regulation, in part because it is a two-domain protein that forms a functional dimer. The N-terminal region of the protein contains the helix-turn-helix DNA-binding motif, and the C-terminal domain is responsible for dimerization. As we have seen, the λ repressor can bind to any of six operator binding sites in the O_L and O_R regions (see Figure 20-21). Despite its name, the repressor bound at O_{R2} actually *activates* transcription from P_{RM}, and this activation is critical to the lysogenic switch. However, the repressor has highest binding affinity for O_{R1}, and because binding is cooperative, O_{R1} recognition increases the affinity of the bound λ repressor for O_{R2}. Repressor bound cooperatively at O_{R1} and O_{R2} blocks RNA polymerase binding at P_R, repressing transcription from that promoter. Similarly, repressor bound cooperatively at O_{L1} and O_{L2} prevents transcription from P_L. Thus, the repressor can simultaneously activate transcription of its own gene from P_{RM} and repress transcription of the immediate early genes necessary for lytic growth from P_L and P_R.

Once the lytic-to-lysogenic switch has occurred, the phage DNA integrates into the host chromosome by a

FIGURE 20-22 The structures of Cro and λ repressor proteins. Both Cro and λ repressor form homodimers that bind the O_L and O_R operator regions. Both use helix-turn-helix motifs to associate with the DNA. [*Sources: PDB ID 3CRO (top) and PDB ID 1LMB (bottom).*]

Cro

λ repressor (cI)

Helix-turn-helix

(a) **Lytic pathway**

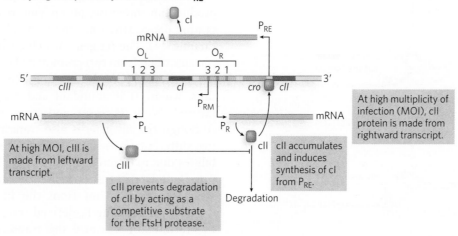

(b) **Lysogenic pathway: activation of P$_{RE}$**

At high MOI, cIII is made from leftward transcript.

cIII prevents degradation of cII by acting as a competitive substrate for the FtsH protease.

At high multiplicity of infection (MOI), cII protein is made from rightward transcript.

cII accumulates and induces synthesis of cI from P$_{RE}$.

Degradation

(c) **Lysogenic pathway: activation of P$_{RM}$**

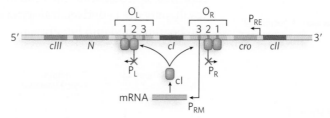

FIGURE 20-23 The λ phage genetic switch between lytic and lysogenic growth. A small region of the λ phage genome controls the genetic switch between lysis and lysogeny. Two critical proteins, Cro and λ repressor (cI), regulate the switch.

mechanism described below. The integrated viral DNA can be maintained and replicated stably as part of the host cell chromosome. But, as described earlier, when the host cell is exposed to agents or conditions that damage DNA, the prophage is rapidly excised, and lytic growth begins. This lysogenic-to-lytic switch comes about because the λ repressor protein resembles the bacterial protein LexA, which undergoes self-cleavage when stimulated by a second bacterial protein, RecA, that is a sensor of DNA damage in the bacterial cell (see Figure 20-13). When RecA is activated in response to DNA damage (the SOS response), the λ repressor protein, if present in the cell, also undergoes self-cleavage. This reaction clips off the C-terminal region of the repressor, removing its dimerization domain and thereby destroying its cooperative binding to O$_L$ and O$_R$, sites 1 and 2. As a result, the cleaved

λ repressor dissociates from the operator DNA, allowing transcription from P$_R$ and P$_L$, and thus lytic growth.

More Regulation Levels Are Invoked during the λ Phage Life Cycle

Switching between the activation and repression of promoters enables λ phage to alternate between lytic and lysogenic growth, as dictated by environmental conditions. But additional levels of regulation further control the expression of genes required to establish and maintain early steps in the infection cycle. Two λ proteins, the N and Q proteins, are known as antiterminators because they prevent RNA polymerase from prematurely stopping transcription of the genes they regulate. N protein binds to specific regions of the viral transcript called

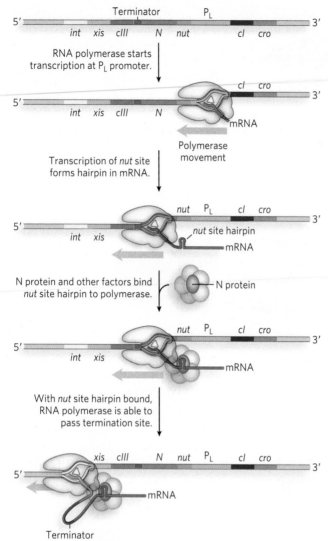

FIGURE 20-24 N protein–mediated antitermination of transcription. N protein is transcribed early in the infection process from the P_L promoter, but early transcription stops at the terminator just downstream from the *N* gene. N protein binds newly transcribed *nut* sites and, through interaction with several other factors, causes a change in RNA polymerase that allows it to read through terminator structures, generating longer mRNA transcripts.

nut (*N utilization*) sites (**Figure 20-24**). The resulting RNA-protein complexes assemble with additional proteins produced from the host cell genes *nusA, nusB, nusE,* and *nusG*. Although the functions of these proteins are not known in detail, they somehow help the phage-produced N protein bind to RNA polymerase and enhance its ability to bypass terminator structures downstream from the genes for N and Cro. In this way, elongation of the viral transcript is favored once initial infection is established.

One target of the N protein antitermination mechanism is the viral gene encoding the Q antiterminator protein. Unlike N, Q binds DNA sequences between the −10 and −35 regions of PR′, the promoter for genes that are active late in the infection process. In the absence

of Q, RNA polymerase pauses shortly after initiation of the transcript, then falls off the template when encountering a terminator hairpin structure, 200 bp further on. When Q is present, it first binds in the promoter region, then transfers to the polymerase when the enzyme pauses after initiation. With Q bound, the polymerase can transcribe through the terminator hairpin, thereby producing full-length transcripts; the mechanism of this antitermination process is not yet clear.

To establish a lysogenic infection, the virus must produce an integrase, an enzyme responsible for integrating phage DNA into the host cell chromosome (see Chapter 14). The *int* gene, which encodes the integrase, is transcribed from two promoters, P_I and P_L. The cII protein activates P_I, in addition to P_RE (as described earlier); thus, when cII is abundant and the lysogenic state is favored, both repressor and integrase are expressed. Although the *int* gene is also transcribed from P_L, the two different transcripts have inherently different stabilities due to different RNA structures at their 3′ ends. The P_I-derived transcript ends at a terminator hairpin structure downstream from the integrase-coding sequence, whereas the P_L-derived transcript extends past the terminator—because the transcript is made by a polymerase modified by the N protein (see Figure 20-24). This longer transcript forms an alternative hairpin structure that is a substrate for cellular ribonucleases. As a result, only the P_I-derived transcript is maintained, and it can be translated into integrase protein (**Figure 20-25**).

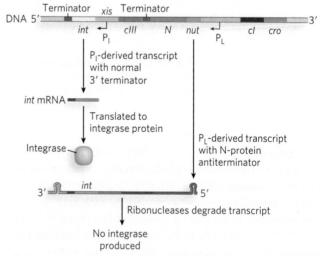

FIGURE 20-25 Regulation of *int* gene expression. Transcripts of the integrase gene (*int*) are generated from two promoters, P_I and P_L. The P_I-derived transcript is short, ending at a terminator structure that follows the coding sequence of the *int* gene. P_L-derived transcripts are longer and are generated as a result of N protein–mediated antitermination. The 3′ end of the P_L-derived transcript extends beyond the terminator. A structure at the 3′ end makes the transcript a substrate for cellular ribonucleases, and the transcript is degraded. Thus, more integrase protein is generated from the P_I-derived transcript than from the P_L-derived transcript.

In a twist on this regulatory mechanism, the λ DNA integrates into the host chromosome such that a small portion of the phage DNA is removed—the portion that, when transcribed, becomes the destabilizing RNA sequence at the end of the P_L-derived integrase transcript. Thus, when λ phage is in the lysogenic state and integrated into the host chromosome, integrase can be produced from transcripts derived from either P_I or P_L. Because integrase is also required to excise the prophage from the host genome, the ability to use either type of *int* gene transcript ensures that this step is not dependent on cII protein concentration.

SECTION 20.3 SUMMARY

- Bacteriophage gene activities can be controlled by regulatory networks that integrate multiple signals into a common gene regulatory response.

- Infection of a bacterial cell by λ phage can lead to either (1) a lytic pathway, in which host cells are lysed as new viral particles are assembled and released, or (2) a lysogenic pathway, in which the viral DNA integrates into the host chromosome and is propagated through chromosome replication and cell division, without immediate production and release of new viruses.

- In λ phage, two regulators, Cro and the λ repressor, control whether the phage propagates through the lytic pathway (when Cro dominates) or the lysogenic pathway (when the λ repressor dominates). Both repressors bind to similar phage DNA sequences, but with different binding affinities. Cro blocks λ repressor synthesis while allowing expression of other genes needed for lytic growth; the λ repressor blocks transcription of all phage genes except its own.

- In addition to Cro, the N protein, expressed early in the infection process, favors λ gene expression by binding *nut* sites in the viral transcripts. The resulting RNA-protein complexes favor transcription by assembling with host proteins that aid binding of N protein to RNA polymerase, enhancing the enzyme's ability to bypass terminator structures downstream from the N and Cro genes.

- N protein-mediated antitermination favors production of the Q protein, another antiterminator that binds DNA sites near the promoter of genes expressed later in the infection cycle. Binding of the Q protein allows it to transfer to paused RNA polymerase molecules and enhance their ability to traverse terminator structures in the regulated genes.

- To establish a lysogenic infection, the virus produces an integrase that integrates the phage DNA into the host chromosome. Integrase is also required to excise the prophage from the host genome.

? UNANSWERED QUESTIONS

The study of gene regulation continues to expand as new regulatory mechanisms are uncovered. RNA has emerged as a major player in controlling levels of protein production, and many fascinating aspects of its involvement remain to be deciphered.

1. How widespread are RNA-based gene regulatory mechanisms? Early work on the regulation of gene expression focused on proteins that have the sole role of controlling when and how much of a particular protein is synthesized. More recent research has revealed an increasing number of instances in which RNA plays this regulatory role. In addition to the use of riboswitches, bacteria produce many small noncoding RNAs that are important for modulating gene expression. Understanding how these work and how their mechanisms relate to those responsible for regulating gene expression in plants and animals (see Chapter 22) is a fascinating area of research.

2. How does a bacteriophage compete for a host cell's gene expression machinery? Studies of λ phage have provided an exciting introduction to the world of phage–host cell competition, but there are many different mechanisms by which viruses might take over a host cell's gene expression machinery and thus regulate the propagation of new viral particles. Some researchers estimate that Earth is home to more phage particles than cells! Phages thus provide an enormous pool of genes that are readily exchanged and introduced into new hosts, driving the evolution of new traits and viral defense mechanisms. A broader understanding of gene regulatory pathways in phages will offer new insights into bacterial gene regulation, and perhaps into the relationships between viral propagation and gene transfer.

3. How are gene regulatory networks integrated? Much of the research on gene regulation in bacteria has focused on individual genes or operons, giving the impression that just one or a few changes occur in response to signaling molecules. Studies using DNA microarrays and high-throughput sequencing, however, show that changes in gene expression in response to stresses or altered nutrient levels occur in hundreds of different genes. How these changes are coordinated and how different gene regulatory pathways integrate multiple signals at once are the subjects of active research. Investigators are using traditional genetic and biochemical methods, as well as approaches including high-throughput RNA sequencing (RNA-Seq) and bioinformatics. This area of research is referred to as systems biology, to indicate that gene expression operates not in isolation but as part of a system, as defined by the cell or organism.

HOW WE KNOW

TRAPped RNA Inhibits Expression of Tryptophan Biosynthetic Genes in *Bacillus subtilis*

Babitzke, P., and P. Gollnick. 2001. Posttranscription initiation control of tryptophan metabolism in *Bacillus subtilis* by the *trp* RNA-binding attenuation protein (TRAP), anti-TRAP, and RNA structure. *J. Bacteriol.* 183:5795–5802.

The TRAP system of tryptophan regulation in *B. subtilis* beautifully illustrates how the mechanistic details of multilevel gene regulation were eventually worked out using a combination of experimental methods. Similar to the process in other bacteria, the *trp* genes of *B. subtilis*, contained in the *trpEDCFBA* operon (each letter after *trp* standing for a gene in the operon), are transcribed only when tryptophan is in short supply in the cell. Genetic experiments revealed that transcription of the *trpEDCFBA* operon requires a regulatory protein called TRAP (*trp* RNA–binding *attenuation protein*). Using purified TRAP, RNA polymerase, nucleoside triphosphates, and a plasmid DNA with an inserted *trpEDCFBA* operon, researchers found that adding L-tryptophan to the mix caused transcription to stop in the leader sequence of the operon, upstream from the coding sequences. In the presence of L-tryptophan, TRAP bound the newly synthesized leader RNA and prevented formation of an antiterminator structure (Figure 1a). As a result, a competing RNA structure—a terminator—could form, blocking passage of the polymerase and causing premature transcription termination (see Chapter 15), as shown in Figure 1b.

How does TRAP respond to L-tryptophan? TRAP is composed of 11 subunits, each of 6 to 8 kDa. TRAP binds to 11 triplet repeats, primarily GAG and UAG, in the leader sequence that are separated from each other by two or three nonconserved nucleotides. The crystal structure of TRAP bound to a 53-nucleotide single-stranded *trp* leader RNA looks like a molecular spool in which the RNA wraps around the outer surface of the protein core (see Figure 1b). Each GAG or UAG triplet tucks into a binding pocket formed by one of the TRAP subunits, and tryptophan molecules are positioned between the subunits, where they presumably stabilize interactions required for high-affinity protein-RNA binding.

Subsequent experiments showed that TRAP can also bind its target sequence in the leader region of

mature mRNAs, blocking access of the ribosome to the Shine-Dalgarno sequence and thus preventing efficient translation initiation. Furthermore, a second protein, called anti-TRAP, induced by uncharged tRNA[Trp], can bind TRAP and prevent its binding to the *trp* leader RNA, allowing transcription of the *trp* operon to proceed. Through TRAP and anti-TRAP, *B. subtilis* senses the levels of both tryptophan and uncharged tRNA[Trp] in order to regulate tryptophan biosynthesis by changing the accessibility of the RNA to both RNA polymerase and the ribosome.

(a) Tryptophan absent

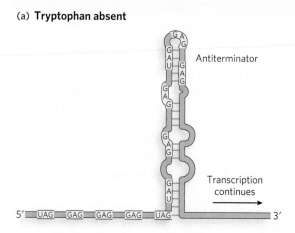

(b) Tryptophan present

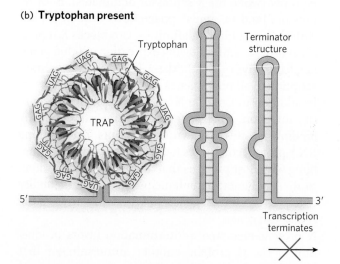

FIGURE 1 Regulation of the *B. subtilis trp* operon. (a) In the absence of tryptophan, the structure of the *trp* leader RNA allows continued transcription of the *trp* operon. (b) In the presence of tryptophan, the TRAP protein binds tryptophan and associates with the *trp* operon leader through interaction with 11 GAG and UAG triplets. This leads to formation of a terminator structure in the RNA, which halts transcription. [Sources: (a) Data from P. Babitzke and P. Gollnick, J. Bacteriol. 183:5795–5802, 2001, Fig. 2. (b) PDB ID 1C9S.]

Autoinducer Analysis Reveals Possibilities for Treating Cholera

Higgins, D.A., M.E. Pomianek, C.M. Kraml, R.K. Taylor, M.F. Semmelhack, and B.L. Bassler. 2007. The major *Vibrio cholerae* autoinducer and its role in virulence factor production. *Nature* 450:883–886.

Many single-celled organisms and multicellular tissues use cell-to-cell signaling to communicate information about population density, allowing cells to change gene expression levels in response to the group environment. In bacteria, such signaling molecules are secreted from one cell and detected or imported by neighboring cells, where the signal triggers complex changes in gene expression (Figure 2a). The ability to sense and respond to high population density, a type of cell-to-cell signaling referred to as quorum sensing, frequently involves secreted peptides known as autoinducers. In the pathogenic bacterium *Vibrio cholerae*, which causes cholera, autoinducers terminate rather than pro-mote virulence, and activation of quorum sensing by introducing an autoinducer could form the basis of a treatment for cholera.

Bonnie Bassler's laboratory at Princeton has investigated the molecular details of quorum sensing in *V. cholerae*. The bacterium uses quorum sensing to control its production of virulence factors and its ability to grow as a biofilm, a contiguous layer of cells. At low cell density, *V. cholerae* expresses virulence factors and forms a biofilm. As cell density increases, two autoinducers increase in concentration until they repress both virulence factor expression and biofilm formation (Figure 2b). Synthesized by autoinducer synthase enzymes, the small-molecule autoinducers are secreted from the bacterial cells and bind to receptors for import into neighboring cells.

Information from both types of autoinducers is transmitted through the protein LuxO, which in turn controls the level of HapR, a transcription factor that controls the expression of many other genes. At low cell density, in the absence of autoinducers, HapR is not produced, virulence factors are expressed, and biofilms form. Because HapR is required for the expression of genes that cause the cells to luminesce, the cells are not bioluminescent. This provides an easy way for researchers to determine whether HapR is turned on. At high cell density, autoinducers increase and bind to LuxO, leading to the production of HapR. HapR represses the genes for virulence factor production

and biofilm formation, while activating expression of the bioluminescence genes. The end result is that the production of autoinducers and activation of quorum sensing terminate virulence in *V. cholerae* cells growing in crowded conditions.

Bassler and colleagues cloned the autoinducer synthase genes and introduced them into *E. coli*. Because *E. coli* is not sensitive to the *V. cholerae* autoinducers, the signaling molecules were produced and secreted in large amounts without interfering with cell growth. The experimenters could then purify the autoinducers and determine their chemical structure by nuclear magnetic resonance spectroscopy (NMR). The NMR method enables exact determination of the chemical structure of the autoinducers, because it provides information about the chemical environment of each proton in the molecule. Bassler's group has also developed a method for chemically synthesizing these autoinducers—an exciting development that might allow researchers to "trick" cells into the quorum-sensing response even at low cell density. This could provide a clever way to expropriate the biology of gene regulation to prevent cholera infection, a strategy that could also be developed for other kinds of pathogenic bacteria.

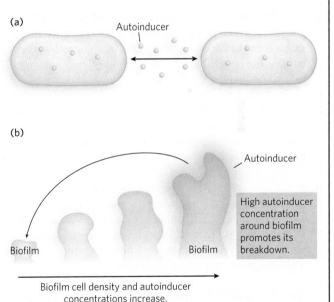

(a)

Autoinducer

(b)

Autoinducer

High autoinducer concentration around biofilm promotes its breakdown.

Biofilm   Biofilm

Biofilm cell density and autoinducer concentrations increase.

FIGURE 2 (a) In quorum sensing, bacterial cells sense and respond to high population density by sending and receiving small-molecule signals called autoinducers, which freely diffuse through the bacterial membranes. (b) In *V. cholerae*, autoinducers stimulate the breakdown of biofilms, reducing biofilm size as autoinducer concentration increases.

PROBLEMS

1. A researcher engineers a *lac* operon on a plasmid, but inactivates all parts of the Lac operator (*lacO*) and the Lac promoter, replacing them with the binding site for the LexA repressor (which acts in the SOS response) and a promoter regulated by LexA. The plasmid is introduced into *E. coli* cells that have a *lac* operon with an inactive *lacZ* gene. Under what conditions will these transformed cells produce β-galactosidase?

2. Describe the probable effects on the expression of *lac* genes of mutations that (a) relocate the Lac operator so that it is on the other side of the operon, (b) inactivate the binding site for CRP, and (c) alter the promoter sequence around position –10.

3. In the *ara* operon, the AraC protein can act as either an activator or a repressor. If AraC remains bound to the DNA in the absence of arabinose, why does the protein not always function as an activator?

4. *E. coli* cells are growing in a medium with glucose as the sole carbon source. Tryptophan is suddenly added. The cells continue to grow, and they divide every 30 minutes. Describe (qualitatively) how the levels of tryptophan synthase (an enzyme produced by the *trp* operon) change with time under the following conditions:

(a) The *trp* mRNA is stable (degraded slowly over many hours).
(b) The *trp* mRNA is degraded rapidly, but tryptophan synthase is stable.
(c) Both the *trp* mRNA and tryptophan synthase are degraded rapidly.

5. How would the SOS response in *E. coli* be affected by mutations in the *lexA* gene that (a) prevented autocatalytic cleavage of the LexA protein or (b) weakened the interaction of LexA with its normal binding site?

6. A typical bacterial repressor protein discriminates between its specific DNA binding site (operator) and nonspecific DNA by a factor of 10^4 to 10^6. About 10 molecules of repressor per cell are sufficient to ensure a high level of repression. Assume that a very similar repressor existed in a human cell, with a similar specificity for its binding site. How many copies of the repressor would be required to elicit a level of repression similar to that in the bacterial cell? (Hint: The *E. coli* genome contains about 4.6×10^6 bp; the human haploid genome has about 3.2×10^9 bp.)

7. The dissociation constant for a particular repressor-operator complex is very low, about 10^{-13} M. An *E. coli* cell (volume 2×10^{-12} mL) contains 10 copies of the repressor. Calculate the cellular concentration of the repressor protein. How does this value compare with the dissociation constant of the repressor-operator complex? What is the significance of this answer?

8. *E. coli* cells are growing in a medium containing lactose but no glucose. Indicate whether each of the following changes or conditions would increase, decrease, or not change expression of the *lac* operon. It may be helpful to draw a model depicting what is happening in each situation.

(a) Addition of a high concentration of glucose
(b) A mutation that prevents Lac repressor binding to the operator
(c) A mutation that completely inactivates β-galactosidase
(d) A mutation that completely inactivates galactoside permease
(e) A mutation that prevents binding of CRP to its binding site near the Lac promoter

9. How would transcription of the *E. coli trp* operon be affected by the following manipulations of the leader region of the *trp* mRNA?

(a) Increasing the distance (number of bases) between the leader peptide gene and sequence 2
(b) Increasing the distance between sequences 2 and 3
(c) Removing sequence 4
(d) Changing the two Trp codons in the leader peptide gene to His codons
(e) Eliminating the ribosome-binding site for the gene that encodes the leader peptide
(f) Changing several nucleotides in sequence 3 so that it can base-pair with sequence 4 but not with sequence 2

10. Many riboswitches have been characterized in bacteria, including one that binds to thiamine pyrophosphate (TPP) and another that binds to glucosamine 6-phosphate. Compare and contrast the mechanisms by which these two riboswitches inhibit translation of their RNAs.

11. A mutation is found in the gene encoding the translational repressor of an r-protein operon. The mutation increases the affinity of the repressor protein for mRNA and decreases its affinity for rRNA. What is the likely effect of such a mutation?

12. A λ phage lysogen (an *E. coli* cell with a λ prophage integrated into its genome) is largely immune to lysis by λ phages introduced into the cell later. Explain.

13. Mutant versions of CRP have been isolated that bind DNA normally but do not activate transcription. What does the existence of these mutants indicate about the mechanism of CRP-mediated transcription activation, and what phenotype would you expect to observe for cells expressing one of these mutant CRP alleles?

14. How does the organization of related genes into operons enable the coordinated production of proteins? Suggest a way that would allow different genes in an operon to be expressed at different levels.

15. Name one advantage and one disadvantage to using RNA structures such as riboswitches to regulate gene expression in response to small-molecule effectors.

16. What general principle of gene regulation is illustrated by transcription attenuation?

17. Would you expect the mechanism of transcription attenuation described for the *trp* operon to function similarly in eukaryotic cells? Why or why not?

18. Name three properties of riboswitch-regulated mRNAs.

19. How would a large increase in the intracellular concentration of ppGpp affect the growth of bacterial cells in a nutrient-rich medium?

DATA ANALYSIS PROBLEM

Oxender, D.L., G. Zurawski, and C. Yanofsky. 1979. Attenuation in the *Escherichia coli* tryptophan operon: Role of RNA secondary structure involving the tryptophan codon region. *Proc. Natl. Acad. Sci USA* 76:5524–5528.

20. Unraveling of the complicated regulatory mechanism of the *trp* operon proceeded in stages over more than a decade. The work was carried out mainly in the lab of Charles Yanofsky at Stanford University. Beginning with a study of tryptophan synthase in the early 1970s, Yanofsky's group gradually focused on regulation of the *trp* operon. They initially thought that regulation could be explained in terms of a *lac* operon–like repressor-operator interaction. However, they found a long leader region between the promoter and the first *trp* operon gene. Certain deletions in that leader region increased the expression of tryptophan synthase even when the repressor was present. The leader region was sequenced (laboriously, using the methods in existence before invention of the Sanger method), revealing a short open reading frame. When RNA was labeled in the cell, the researchers detected a large amount of truncated *trp* mRNA, terminated before the *trp* genes were transcribed. The sequence of this truncated mRNA is shown in Figure 1.

In the study published in 1979, Yanofsky and his colleagues explored the secondary structure in the leader region of the *trp* mRNA, in work that defined the outlines of the overall regulatory system. They used the enzyme RNase T1, a ribonuclease that cuts single-stranded RNA (unpaired linear regions or the unpaired loop ends of hairpins) much faster than double-stranded RNA. A *partial* digest was carried out on the labeled, attenuated RNA (so that the RNAs in the population were not cut at every single-stranded region). This produced the RNA species shown in the polyacrylamide gel (run under nondenaturing conditions) in Figure 2, with the three most prominent bands labeled A, B, and C. The RNAs in these three bands were separately isolated. When they were run on denaturing gels that would separate any paired RNAs, the A and C bands separated into multiple species; the major species are numbered in Figure 3 and are the focus of the questions below.

(a) The band A RNA (Figure 2) is approximately 140 nucleotides long. What species is this, and what can be said about the three bands that appear when it is separated on the denaturing gel (Figure 3)?

(b) Bands B and C are smaller than band A. What does this tell you?

(c) The band C RNA also separates into three species on the denaturing gel, but the band B RNA does not. What can you conclude from this?

The authors went on to identify the RNA species on the gels. Band B consisted of sequences from approximately position 108 to the 3′ end of the RNA. Band C1 was the segment from approximately position 51 to position 95, with C2 and C3 arising from an additional cleavage at around position 70.

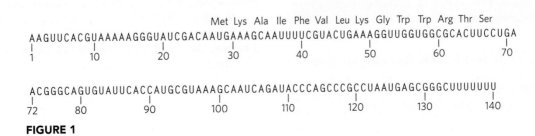

FIGURE 1

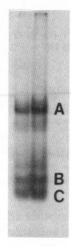

FIGURE 2 *[Source: Oxender, DL, Zurawski, G, and Yanofsky, C. (1979) Attenuation in the Escherichia coli tryptophan operon—role of RNA secondary structure involving the tryptophan codon region. Proc. Natl. Acad. Sci USA 76, 5524–5528. Fig. 1.]*

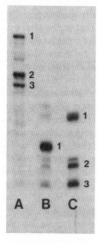

FIGURE 3 *[Source: Oxender, DL, Zurawski, G, and Yanofsky, C. (1979) Attenuation in the Escherichia coli tryptophan operon—role of RNA secondary structure involving the tryptophan codon region. Proc. Natl. Acad. Sci USA 76, 5524–5528. Fig. 2.]*

(d) From what you know about the attenuation mechanism, which regions of the RNA are present in band B?

(e) Which regions are present in band C?

(f) Are there any missing elements of secondary structure that are important to the regulatory mechanism? If so, what are they, and why might they be missing from the isolated RNA?

ADDITIONAL READING

General

Bassler, B.L., and R. Losick. 2006. Bacterially speaking. *Cell* 125:237–246.

Fang, F.C. 2005. Sigma cascades in prokaryotic regulatory networks. *Proc. Natl. Acad. Sci. USA* 102:4933–4934.

Gollnick, P., P. Babitzke, A. Antson, and C. Yanofsky. 2005. Complexity in regulation of tryptophan biosynthesis in *Bacillus subtilis. Annu. Rev. Genet.* 39:47–68.

Lewis, M. 2005. The *lac* repressor. *Crit. Rev. Biol.* 328:521–548.

von Hippel, P.H. 2007. From "simple" DNA-protein interactions to the macromolecular machines of gene expression. *Annu. Rev. Biophys. Biomol. Struct.* 36:79–105.

Wakeman, C.A., W.C. Winkler, and C.E. Dann III. 2007. Structural features of metabolite-sensing riboswitches. *Trends Biochem. Sci.* 32:415–424.

Transcriptional Regulation

Jacob, F., and J. Monod. 1961. Genetic regulatory mechanisms in the synthesis of proteins. *J. Mol. Biol.* 3:318–356.

Kolb, A., S. Busby, H. Buc, S. Garges, and S. Adhya. 1993. Transcriptional regulation by cAMP and its receptor protein. *Annu. Rev. Biochem.* 62:749–795.

Lawson, C.L., D. Swigon, K.S. Murakami, S.A. Darst, H.M. Berman, and R.H. Ebright. 2004. Catabolite activator protein: DNA binding and transcription activation. *Curr. Opin. Struct. Biol.* 14:10–20.

Yanofsky, C., K.V. Konan, and J.P. Sarsero. 1996. Some novel transcription attenuation mechanisms used by bacteria. *Biochimie* 78:1017–1024.

Beyond Transcription: Control of Other Steps in the Gene Expression Pathway

Barrick, J.E., and R.R. Breaker. 2007. The power of riboswitches: Discovering relics from a lost world run by RNA molecules may lead to modern tools for fighting disease. *Sci. Am.* 296(1):50–57.

Coppins, R.I., K.B. Hall, and E.A. Groisman. 2007. The intricate world of riboswitches. *Curr. Opin. Microbiol.* 10:176–181.

Gao, R., and A.M. Stock. 2010. Molecular strategies for phosphorylation-mediated regulation of response regulator activity. *Curr. Opin. Microbiol.* 13:160–167.

Magasanik, B. 2000. Global regulation of gene expression. *Proc. Natl. Acad. Sci. USA* 97:14,044–14,045.

Shapiro, L., H.H. McAdams, and R. Losick. 2009. Why and how bacteria localize proteins. *Science* 326:1225–1228.

Control of Gene Expression in Bacteriophages

Gottesmann, M., and R. Weisberg. 2004. Little lambda, who made thee? *Microbiol. Mol. Biol. Rev.* 68:796–813.

Hochschild, A. 2002. The switch: *cI* closes the gap in autoregulation. *Curr. Biol.* 12:R87–89.

Murray, N.E., and A. Gann. 2007. What has phage lambda ever done for us? *Curr. Biol.* 17:R305–312.

Oppenheim, A.B., O. Kobiler, J. Stavans, D.J. Court, and S. Adhya. 2005. Switches in bacteriophage lambda development. *Annu. Rev. Genet.* 39:409–429.

Ptashne, M. 1992. *A Genetic Switch.* Cambridge, MA: Cell Press.

21

The Transcriptional Regulation of Gene Expression in Eukaryotes

Tracy Johnson *[Source: Courtesy Tracy Johnson.]*

MOMENT OF DISCOVERY

Our lab has been trying to understand *how pre-mRNA splicing can occur cotranscriptionally*, which would neatly tie together the steps in producing functional mRNAs and provide cells with many interesting avenues for gene regulation along the way. Starting with a genetic screen in yeast in which nonessential transcription factors were mutated, we looked for mutations in a second gene that would cause cell death, an effect referred to as synthetic lethal. For a long time we found absolutely nothing interesting. But we kept working, and at last discovered that a deletion of the gene encoding the Gcn5 protein was synthetic lethal when combined with mutations in genes encoding parts of the U2 snRNP component of spliceosomes. Gcn5 is a histone acetyltransferase (HAT), a well-characterized enzyme that adds acetyl groups to histone proteins within nucleosomes, but it had no known connection to pre-mRNA splicing.

The real moment of surprise came when we found that when the *GCN5* gene is deleted, cotranscriptional splicing is completely messed up! This is because the U2 snRNP is no longer recruited to pre-mRNA splice sites. The splicing defect is specific to this HAT and requires the enzyme's catalytic activity, which is targeted toward promoter-bound histones. I never imagined there would be a link between chromatin structure and pre-mRNA splicing, as is implied by this finding. We envision that a specific pattern of histone acetylation leads to physical recruitment of proteins to acetylated histones within chromatin, which in turn recruits spliceosomes to newly transcribed pre-mRNAs. Because the HAT is very well conserved in mammals, it could be a general mechanism that affects which splice sites are chosen in pre-mRNAs, depending on the acetylation state of the histones associated with the parent gene.

—*Tracy Johnson, on discovering that pre-mRNA splicing requires specific histone acetylation*

Eukaryotic cells, like bacteria, express only a subset of their genes at any given time. We saw in Chapter 20 that bacteria, through gene regulation, are able to adapt to environmental changes and respond to signaling molecules and viral assaults. Eukaryotes, too, must respond to their environment and external stimuli. But in addition, multicellular eukaryotes must manage complex pathways of cell division and differentiation that give rise to the multitude of cell types required for organismal development. Developmental programs are extremely precise—it is critical that each protein influencing cellular differentiation is active at the right time and in the right place—and any deviation from the program can have drastic consequences. Many of the genes needed for development are so critical that if mutation renders them nonfunctional, the embryo dies before the organism is fully formed. Yet, even though the needs of a eukaryote are more complex than those of a bacterium, basic principles of gene regulation are still the key to all of these processes.

Recall that many bacterial genes and operons are regulated at the level of transcription initiation. This is true in eukaryotes as well, and as we will see, many eukaryotic regulatory mechanisms build on those used in bacteria. However, there is a fundamental difference between bacterial and eukaryotic regulation of transcription. The **transcriptional ground state**, the inherent activity of promoters and transcription machinery in vivo in the absence of regulatory mechanisms, is not the same in bacteria and eukaryotes. In bacteria, the transcriptional ground state is nonrestrictive; RNA polymerase generally has access to every promoter and can bind and initiate transcription at some level of efficiency in the absence of activators or repressors. In contrast, eukaryotic genes contain strong promoters that are generally inactive in the absence of regulatory proteins—the transcriptional ground state in eukaryotes is restrictive.

Crucial differences in DNA packaging and cell structure give rise to at least four important distinguishing features of regulation of gene expression in eukaryotes. First, access to eukaryotic promoters is restricted by the structure of chromatin, and transcriptional activation is associated with many changes in chromatin structure in the transcribed region. Second, although eukaryotic cells have both positive and negative regulatory mechanisms, positive mechanisms predominate in all systems investigated so far; given that the transcriptional ground state is restrictive, virtually every eukaryotic gene requires activation. Third, eukaryotic cells have larger, more complex, multiprotein regulatory networks than bacteria. And finally, transcription in the nucleus is separated from translation in the cytoplasm, in both space and time. As a result, posttranscriptional control plays a larger role in controlling gene expression in eukaryotes, as we will see in Chapter 22.

The complexity of regulatory circuits in eukaryotic cells is extraordinary, and we will not attempt comprehensive coverage of all aspects. This chapter and the next cover some of the guiding principles of eukaryotic gene regulation, drawing parallels to the mechanisms discussed for bacteria in Chapter 20, wherever applicable. The need to control the multitude of genes in a higher eukaryote requires an array of regulatory proteins for every single gene. We begin by discussing the basic logic of gene activation as used in essentially all eukaryotes. We take a brief look at experiments that first revealed the modular architecture of gene activators and highlight some of the regulatory networks that govern gene expression, from the simple system in yeast to the complex developmental controls typical in a multicellular eukaryote. The chapter concludes with a discussion of transcriptional control processes unique to eukaryotic gene expression, some of which are still far from understood.

21.1   BASIC MECHANISMS OF EUKARYOTIC TRANSCRIPTIONAL ACTIVATION

As in bacteria, the basal level of eukaryotic transcription is determined by the effect of regulatory sequences on the function of RNA polymerase and its associated transcription factors. As we discussed in Chapter 19, the nature of the eukaryotic genome lends itself to different regulatory strategies from those used in bacteria. The eukaryotic genome is packaged in chromatin, which presents a physical block to RNA polymerases, and therefore the majority of eukaryotic genes are repressed in their default (ground) state and require protein activators to stimulate expression. Because of the large size of eukaryotic genomes and the need to guard against nonspecific protein-DNA interactions, the binding of multiple protein regulators is required to activate each gene. As a result, eukaryotic promoters are more complex than their bacterial counterparts and contain many more regulatory protein-binding sites. In reality, though, the additional complexity in eukaryotes is handled in strategic ways that are not as complicated as we might expect, given the overwhelming difference between a bacterium and an animal.

Eukaryotic Transcription Is Regulated by Chromatin Structure

The genomic DNA of eukaryotes wraps around small basic proteins called histones to form nucleosomes, the building blocks of chromatin (see Figure 10-4). The transcription machinery must necessarily deal with chromatin structure in order to access particular genes. As a result, eukaryotic genes are generally expressed at low levels—or not at all—in the absence of regulatory proteins.

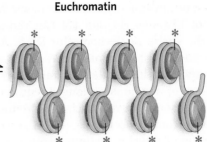

Heterochromatin

Nucleosome

DNA

Histone

Coactivator complexes
Loss of histone H1
Histone modifications (such as acetylation, phosphorylation, and methylation)

Histone deacetylation, dephosphorylation, and demethylation
Corepressor complexes

Euchromatin

FIGURE 21-1 Heterochromatin and euchromatin. Nucleosomes in heterochromatin are tightly packed together, and the DNA is transcriptionally silent. Nucleosomes in euchromatin are spaced farther apart, and the DNA can be decondensed by the loss of histone H1, thus becoming accessible to the transcription machinery. The two chromatin states are regulated by histone modifications (represented by red asterisks) and the binding of other factors (see Chapter 10).

Chromatin structure is controlled and altered by at least three interrelated mechanisms: ATP-dependent changes in nucleosome positioning on the DNA, posttranslational chemical modifications of histone proteins, and substitution of specialized histone variants into chromatin. These mechanisms were discussed in detail in Chapter 10, and we recap briefly here. Nucleosome remodeling complexes use ATP to shift nucleosomes along the DNA. Active promoters contain open regions with nucleosomes positioned away from the promoter region, allowing access to transcription factors. Some posttranslational modifications of histones, including acetylation by histone acetyltransferases (HATs), result in the decondensing of chromatin and provide access to DNA-binding factors; proteins containing a bromodomain bind acetylated histones and facilitate opening of the chromatin structure. Alternatively, histone modifications cause chromatin to become tightly closed to transcription. For example, methylated histones are bound by proteins containing chromodomains, and these proteins help condense the chromatin. Chromatin structure is also modulated by several histone variants. These proteins are homologous to the common histones and can take their place in nucleosomes, but they also contain amino acid extensions that have a variety of functional consequences.

In the eukaryotic cell cycle, interphase chromosomes appear to be dispersed and amorphous. However, chromosomes are not uniform structures, and several different forms of chromatin can be found along each chromosome. About 10% of the chromatin in a typical eukaryotic cell is in a much more condensed form than the rest of the chromatin. This form, **heterochromatin**, is transcriptionally inactive. Although heterochromatin does not contain any genes (which is why it is inactive), the heterochromatin structure itself is known to be repressive, because, experimentally, a gene is silenced when it is placed in heterochromatin. Heterochromatin is often associated with particular chromosome structures, including centromeres and telomeres.

The remaining, less condensed chromatin is called **euchromatin** (**Figure 21-1**).

Transcription of a eukaryotic gene is strongly repressed when its DNA is condensed within heterochromatin, but in euchromatin, some of the DNA is transcriptionally active. Regions of transcriptionally active DNA can be detected on the basis of their increased sensitivity to nuclease-mediated degradation. Nucleases such as DNase I tend to cleave the DNA of carefully isolated chromatin into fragments of multiples of about 200 bp, reflecting the regular repeating structure of the nucleosome (see Figure 10-1). However, in actively transcribed regions, the fragments produced by nuclease activity are smaller and more heterogeneous in size. Actively transcribed regions contain **hypersensitive sites**, sequences especially sensitive to DNase I, which are typically found in noncoding regions within 1,000 bp of the 5′ ends of transcribed genes. In some genes, hypersensitive sites are found farther from the 5′ end, or near the 3′ end, or even within the gene itself. The presence of hypersensitive sites suggests that DNA in that region is not packaged in the regular repeating nucleosomal structure.

Many hypersensitive sites correspond to binding sites for known regulatory proteins, and the relative absence of nucleosomes in these regions may allow the binding of these proteins. Nucleosomes are entirely absent in some regions that are very active in transcription, such as the rRNA genes. Transcriptionally active chromatin also tends to be deficient in histone H1, which binds the linker DNA between nucleosome particles.

Histones within transcriptionally active chromatin and heterochromatin also differ in their patterns of covalent modification. The C-terminal tails of the core histones are modified by the acetylation and methylation of Lys and Arg residues, phosphorylation of Ser or Thr residues, and ubiquitination or sumoylation (see Chapter 22). In particular, the acetylation-deacetylation of histones figures prominently in the processes that activate chromatin for transcription. The HAT-mediated acetylation of multiple Lys residues in the N-terminal domains of histones H3 and H4 can reduce the affinity of the entire nucleosome for DNA. Acetylation may also prevent or promote interactions with other proteins involved in regulating transcription. When transcription of a gene is no longer required, acetylation of nucleosomes in that vicinity is reduced by the activity of histone deacetylases (HDACs), resulting in condensation of the chromatin to reduce or inactivate gene transcription. HDACs often function through protein-protein interactions, such as by binding corepressors or acting as components of chromatin remodeling complexes.

A general model for deacetylation and gene inactivation is shown in **Figure 21-2**. In addition to the removal of certain acetyl groups, new covalent modifications of histones mark chromatin as transcriptionally inactive. For example, the Lys residue at position 9 in histone H3 is often methylated in heterochromatin.

Gene regulation through histone modifications is typically achieved through activation or inhibition of transcription initiation. However, an exciting finding has demonstrated that chromatin structure is also involved in the control of mRNA splicing for some genes (see this chapter's Moment of Discovery). This may seem surprising, given that histones do not bind RNA. Yet Gcn5, a well-studied transcription factor with HAT activity, is now known to also affect RNA processing: loss of Gcn5 HAT activity prevents proper pre-mRNA splicing in yeast, because components of the splicing machinery fail to properly bind the pre-mRNA splice sites (**Highlight 21-1** on p. 732). This example shows how different types of regulation (in this case, transcription and mRNA processing) can be interrelated. It seems likely that processes previously thought to be isolated and separable events may be interwoven in complex regulatory networks in the living cell.

Positive Regulation of Eukaryotic Promoters Involves Multiple Protein Activators

Each of the three eukaryotic RNA polymerases has little or no intrinsic affinity for its promoters. Instead, initiation of transcription almost always requires activator proteins. An important reason for the apparent predominance of positive regulation is clear from the earlier discussion: chromatin structure effectively renders most promoters inaccessible, so genes are normally silent in the absence of other regulation. The structure of chromatin affects access to some promoters more than others, but repressor binding to DNA to block access of RNA polymerase (negative regulation) would often be simply redundant. Other factors are also at play in the

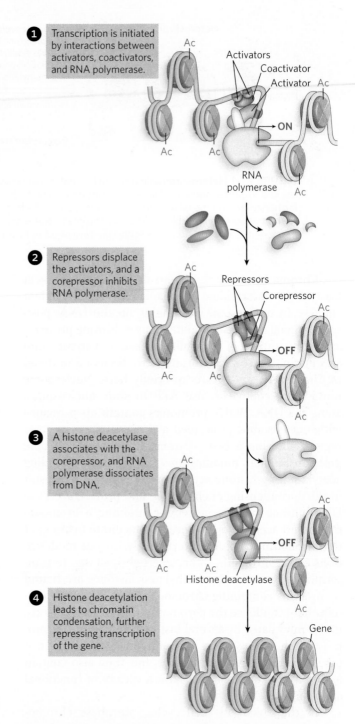

1 Transcription is initiated by interactions between activators, coactivators, and RNA polymerase.

2 Repressors displace the activators, and a corepressor inhibits RNA polymerase.

3 A histone deacetylase associates with the corepressor, and RNA polymerase dissociates from DNA.

4 Histone deacetylation leads to chromatin condensation, further repressing transcription of the gene.

FIGURE 21-2 Gene inactivation by histone deacetylation. Removal of acetyl groups from histones leads to dissociation of RNA polymerase and condensation of the chromatin. [Source: Data from T.M. Malavé and S.Y.R. Dent, Biochem. Cell Biol. 84:437–443, 2006, Fig. 1.]

use of positive regulation, however, and speculation generally centers on two: the large size of eukaryotic genomes and the greater efficiency of positive regulation.

Because eukaryotes have much larger genomes than bacteria, there is an increased likelihood that a specific

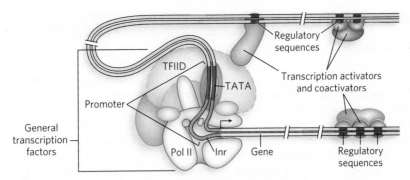

FIGURE 21-3 A typical eukaryotic promoter. General transcription factors and RNA polymerase II bind the promoter, assisted by transcription activators. Activator-binding sites (regulatory sequences) can be distant from the promoter and located either before or after the gene. Activators bind regulatory sequences in DNA directly, whereas coactivators bind activators instead of DNA. Activation of Pol II is mediated by coactivators binding to core subunits of the polymerase through DNA looping.

binding sequence for a regulatory protein will occur randomly in other regions of the DNA. Recall that a sequence of n nucleotides will occur randomly every 4^n bp. Thus, any single regulatory protein with a small binding site will probably bind nonspecifically to multiple places in the eukaryotic genome (see the How We Know section at the end of this chapter). Specific transcriptional activation of a gene through the binding of one regulatory protein to a small binding site, as often occurs in bacteria, would be ineffective in eukaryotes. In theory, specificity of regulator binding would be increased if the DNA sequences recognized by the proteins were longer. Yet eukaryotes did not evolve in this way: eukaryotic regulators do not bind longer DNA sequences than their bacterial counterparts. Instead, to achieve specific transcriptional activation of a gene, eukaryotes employ *multiple* regulatory proteins, or transcription factors, each of which binds a short sequence; successful gene activation occurs only when all the factors are bound at their individual sites. This "combination" of factors for activating one gene is used in combinatorial control (see Section 21.2).

To accommodate the binding of multiple transcription factors, eukaryotic promoters are necessarily more complicated than their bacterial counterparts (**Figure 21-3**). Take, for example, a typical promoter recognized by RNA polymerase II (Pol II), the enzyme responsible for mRNA synthesis. Many (but not all) Pol II promoters include the TATA box and Inr (initiator) sequences, with their standard spacing (see Figure 15-20). These sequences comprise the core promoter.

Eukaryotic genes also include regulatory sequences called **enhancers** in higher eukaryotes and **upstream activator sequences (UASs)** in yeast, to which transcription activators bind. These sequences cannot all be positioned adjacent to the promoter—there is simply not enough room to accommodate the binding of so

many regulatory proteins. The binding sites for multiple transcription factors must be able to act at a distance. In fact, they can be surprisingly far from the promoter. A typical enhancer may be hundreds or even thousands of base pairs upstream from the transcription start site, or downstream from the gene, or even within the gene itself. When bound by the appropriate regulatory proteins, an enhancer increases transcription at nearby promoters, regardless of its orientation in the DNA. Yeast UASs function in a similar way, although generally they must be positioned upstream and within a few hundred base pairs of the transcription start site. An average Pol II promoter may be affected by half a dozen regulatory sequences of this type, and many promoters are even more complex. In contrast, bacteria have very few genes that use a distantly bound transcription activator.

The more complex the eukaryotic organism, the more complex its promoters are likely to be. For example, mammalian promoters are generally much more complex than yeast promoters (**Figure 21-4**).

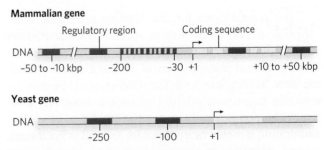

FIGURE 21-4 A comparison of mammalian and yeast promoter regions. The promoter regions of multicellular organisms, such as mammals, contain more control elements than those of unicellular eukaryotes, such as yeast. This reflects the need in higher eukaryotes for changes in gene expression during development and for intercellular communication. All regulatory regions are shown in dark blue, coding regions in yellow.

The Intertwining of Transcription and mRNA Splicing

Initiation is the most highly regulated step in transcription, an intricate process that requires, in eukaryotes, the coordinated action of numerous proteins. Transcription generates a pre-mRNA needing many modifications before it can be transported to the cytoplasm for translation. One of the most complex modifications en route to active mRNA is the removal of introns. The splicing machinery requires more than 100 proteins and five different splicing RNAs with complicated three-dimensional structures. Splicing is generally regarded as a separate step occurring after transcription initiation, or even after generation of the entire pre-mRNA, partly because of the complexity of the transcription and splicing processes. So it was surprising to discover that these two complicated processes—transcription and splicing—can happen simultaneously for some genes: transcription seems to deposit the U2 snRNP component of the spliceosome at specific sites in the pre-mRNA as it is synthesized. These sites correspond to branch points, the sites containing the 2'-OH nucleophile that initiates intron splicing and results in a branched, lariat-type structure when the intron RNA is excised (see Chapter 16). To explain why transcription and splicing would coordinate in this fashion, researchers have proposed that the rate of transcription elongation may be regulated by the spliceosome to help pick and choose alternative splice sites, thereby controlling the relative levels of different mRNAs produced from a pre-mRNA.

The true picture of what is going on is even more complicated, however, as revealed by recent work in Tracy Johnson's laboratory (see this chapter's Moment of Discovery). Johnson made the fascinating observation that Gcn5, a histone acetyltransferase (HAT), is an integral component in the coregulation of transcription and pre-mRNA splicing. The HAT activity of Gcn5, like other HATs, can alter chromatin structure, which is thought to be important in regulating transcription initiation. But results from Johnson's lab demonstrate that accurate mRNA splicing, too, requires the Gcn5 HAT activity.

How does a HAT help splicing? After all, RNA is not bound by histones, so what role does Gcn5 play in the splicing process? Johnson found that the coordination between transcription and splicing occurs even before the pre-mRNA is fully synthesized. The first evidence hinting at this conclusion came from genetic experiments showing that deletion of the gene encoding Gcn5 (and not other yeast lysine acetyltransferases that target histones) is lethal in yeast cells that also lack either of the genes encoding the U2 snRNP proteins Lea1 and Msl1. Neither Lea1 nor Msl1 is an essential protein in yeast, except when Gcn5 is missing.

Next, using the technique of chromatin immunoprecipitation (ChIP), Johnson's group showed that spliceosomal proteins are recruited directly to an intron branch point within the well-characterized *DBP2* gene. In the ChIP experiment, individual snRNP particles are formaldehyde cross-linked to the transcription complex or to the nascent RNA and immunoprecipitated (see Figure 10-21). When the associated DNA is amplified using specific PCR primer sets, the signal is enriched in regions of the gene where the snRNPs associate with the corresponding pre-mRNA. Johnson's results revealed that antibodies to Lea1, a component of the spliceosome, immunoprecipitated a relatively large amount of DNA corresponding to the branch point of the *DBP2* pre-mRNA (Figure 1a, b). Recruitment of Lea1 to the branch point depended on the presence and catalytic activity of Gcn5 (Figure 1c). A control experiment showed that the occupancy by Pol II of these regions of the *DPB2* gene was unaltered, whether or not Gcn5 was active (Figure 1d). Thus, the data indicate that Gcn5 sets the stage for the

---

The requirement for the binding of several transcription activators to several specific DNA sequences vastly reduces the probability of the random occurrence of a functional juxtaposition of all the necessary binding sites. In principle, a similar strategy could be used by multiple negative-regulatory elements. However, positive regulation is simply more efficient. From an energy standpoint, it makes more sense for the cell to synthesize several activators to promote transcription of the subset of genes needed at that time, rather than constantly synthesize one or more repressors for every gene in the genome to keep them turned off until needed. Positive regulation of transcription predominates in eukaryotes, although, as we will see, there are some examples of negative regulation.

To further conserve resources, differently regulated eukaryotic promoters often use some of the same protein activators, so diverse promoters can have some of the same binding sequences. However, only a specific combination of regulatory factors can unlock a given promoter and activate transcription of that gene. With this mechanism, the cell can achieve specificity of gene regulation with a smaller number of transcription activators than if each gene were regulated by a set of unique proteins (see Figure 19-14). Some regulatory proteins facilitate transcription at hundreds of promoters, whereas others are specific for only a few promoters. In addition, many transcription activators are sensitive to the binding of effector signal molecules, providing the capacity to activate

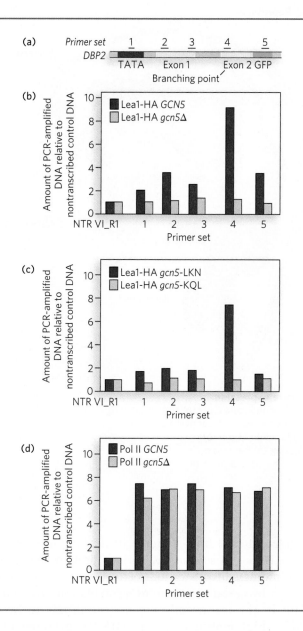

FIGURE 1 Gcn5 activity helps recruit spliceosomal components to *DBP2* pre-mRNA. (a) Numbers represent regions of DNA in the *DBP2* gene that are amplified in the ChIP analysis. (b) ChIP analysis of yeast cells expressing an engineered version of Lea1 tagged with a hemagglutinin (HA) peptide. Lea1-HA was immunoprecipitated with anti-HA antibodies, and Lea1 occupancy in the indicated regions of *DBP2* was compared with that of a nontranscribed region of DNA (NTR VI_R1). Sets of PCR primers corresponding to the regions indicated in (a) were used to amplify specific segments of chromatin after Lea1-HA immunoprecipitation. Dark blue bars are data for cells with wild-type Gcn (*GCN5*); light blue bars are data for cells with a Gcn5 deletion (*gcn5Δ*). (c) ChIP analysis as in (b), but the dark blue bars are results for cells with a point mutation in a nonessential region of Gcn5 (*gcn5*-LKN) and the light blue bars are results for cells with a point mutation in the Gcn5 active site (*gcn5*-KQL). (d) ChIP analysis as in (b), but this control experiment uses Pol II instead of Lea1. [*Source: Data from F. Q. Gunderson and T. L. Johnson, PLoS Genet. 5:1–12, 2009.*]

recruitment of spliceosomal components before the splice site junctions are even transcribed. Further experiments have demonstrated the same results for other genes.

Johnson also proposes other explanations for these observations. One possibility is that Gcn5 acetylates nonhistone proteins, perhaps even spliceosomal subunits. Another possibility is that hyperacetylation of histones at the promoter may facilitate recruitment of the splicing apparatus. Understanding the full details of coordinated regulation of transcription initiation, histone acetylation, and recruitment of the spliceosomal machinery will take considerably more time and work.

or deactivate transcription in response to a changing cellular environment.

Transcription Activators and Coactivators Help Assemble General Transcription Factors

Successful binding of active Pol II holoenzyme at one of its promoters usually requires the action of three types of regulatory proteins: general transcription factors, DNA-binding transcription activators, and coactivators. **General (basal) transcription factors** are required at every Pol II promoter; **DNA-binding transcription activators**, or **DNA-binding transactivators**, bind to enhancers or UASs to facilitate transcription; and **coactivators** act indirectly—by binding other

proteins rather than DNA—and are required for essential communication between the DNA-binding transactivators and the complex composed of Pol II and the general transcription factors (**Figure 21-5a**). Sometimes, a variety of repressor proteins can interfere with communication between Pol II and the DNA-binding transactivators, resulting in repression of transcription (**Figure 21-5b**). In fact, some proteins act as an activator or coactivator at one promoter and a repressor or corepressor at another promoter. Here we focus on the protein complexes shown in Figure 21-5a and how they interact to activate transcription.

For transcription to begin, the Pol II holoenzyme must be recruited to the promoter to form a preinitiation complex with the general transcription factors. Assembly

(a) **Activation**

(b) **Repression**

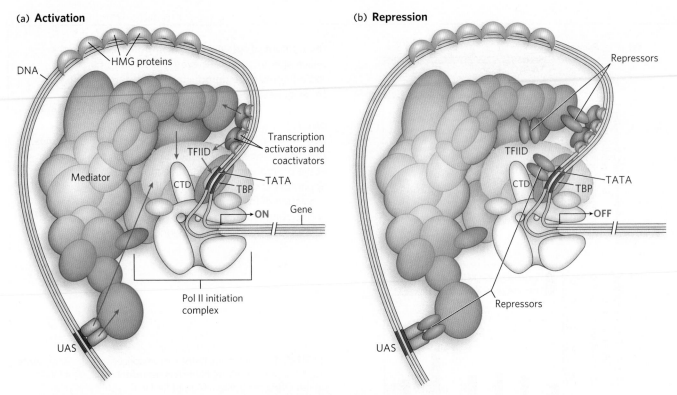

FIGURE 21-5 Mechanisms of activation and repression of eukaryotic gene expression.
(a) Transcription activators and coactivators bound to distant regulatory sites (enhancers and UASs) recruit components of the Pol II general (basal) transcription machinery to the promoter. Coactivators such as Mediator and TFIID are required at essentially all promoters. They function as a bridge between activators and the polymerase, and do not interact with DNA directly. (b) Repression is mediated by proteins that disrupt or prevent essential contacts between Pol II and activators or coactivators. See Figure 21-7 for greater detail.

of a preinitiation complex at a typical Pol II promoter begins with the binding of **TATA-binding protein (TBP)** to the TATA box. TBP, which is part of the larger transcription factor complex called TFIID, then recruits additional general transcription factors and Pol II (see Figure 15-24). The minimal preinitiation complex, however, is often insufficient for the initiation of transcription, and it generally does not form at all if the promoter is buried in chromatin. Positive regulation by transcription activators and coactivators is required. We now know that the basal Pol II machinery is not as uniform as originally thought; the individual components can vary with cell type. A well-documented example is muscle cells (see the How We Know section at the end of this chapter). Thus, different combinations of general transcription factors form a complex at a promoter and are acted on by specific activator and coactivator binding proteins, adding a further level of control to the regulation of that gene.

As noted above, binding sites for transcription activators are often located far from the promoters they

regulate. Recall from Chapter 19 that the intervening DNA is looped so that the various protein complexes can interact, directly or indirectly. DNA looping is promoted by certain nonhistone proteins that are abundant in chromatin and bind nonspecifically to DNA. These **high-mobility group (HMG) proteins** play an important structural role in chromatin remodeling and transcriptional activation ("high mobility" refers to the proteins' rapid electrophoretic mobility in polyacrylamide gels). A structure formed by an HMG-box domain in HMG proteins can bind directly to nucleosomes, leading to altered local chromatin structure. **Figure 21-6** shows the high degree to which DNA is bent by the HMG-box DNA-binding domain of the protein HMG-D of *Drosophila melanogaster*, one of many DNA-interactive protein structures determined in the laboratory of Mair Churchill.

In addition to transcription activators, most transcription requires coactivator protein complexes. Some major regulatory protein complexes that interact with Pol II have been defined both genetically and

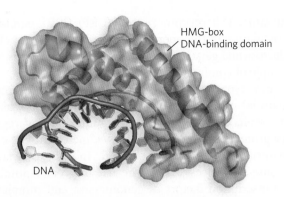

FIGURE 21-6 DNA looping facilitated by HMG proteins. HMG proteins bend DNA, helping form loops between enhancer and promoter elements. Binding is nonspecific. Shown here is the HMG-box DNA-binding domain of the protein HMG-D of *Drosophila*, bound to DNA. [Source: PDB ID 1QRV.]

biochemically. They act as intermediaries between the DNA-binding transactivators and the Pol II complex (Pol II and the general transcription factors). The best-characterized coactivator is TFIID (see Chapter 15). In eukaryotes, TFIID includes TBP and 10 or more TBP-associated factors (TAFs). Some TAFs resemble histones and may play a role in competing with and thus displacing nucleosomes during the activation of transcription. Many DNA-binding transactivators aid in transcription initiation by interacting with one or more TAFs.

Another important coactivator is the **Mediator complex** (see Figure 15-25), which consists of 20 core polypeptides that are highly conserved from fungi to humans. Mediator binds tightly to the C-terminal domain (CTD) of the largest Pol II subunit. The Mediator complex is required for both basal and regulated transcription at Pol II promoters, and it also stimulates phosphorylation of the Pol II CTD by the general transcription factor TFIIH. Phosphorylation of the CTD enhances the efficiency of Pol II. As with TFIID, some DNA-binding transactivators interact with one or more components of the Mediator complex. Some promoters require both Mediator and TFIID coactivators. The coactivator complexes function at or near the promoter's TATA box.

We can now begin to piece together the sequence of transcriptional activation events at a typical Pol II promoter. Crucial remodeling of the chromatin takes place in stages. Some DNA-binding transactivators have significant affinity for their binding sites even when the sites are within condensed chromatin. Binding of one transactivator may facilitate the binding of others, gradually displacing some of the nucleosomes that previously obscured the relevant DNA.

The bound transcription activators may have HAT activity or may recruit HATs or enzyme complexes

such as SWI/SNF, accelerating the remodeling of surrounding chromatin (**Figure 21-7**). In this way, transcription activator binding can lead to the stepwise assembly of components necessary for further chromatin remodeling, to permit the transcription of specific genes. The bound transactivators, acting through complexes such as TFIID or Mediator (or both), stabilize the binding of Pol II and its associated general transcription factors, greatly facilitating formation of the preinitiation complex. Complexity in these regulatory circuits is the rule rather than the exception, with multiple DNA-bound transactivators promoting transcription.

The script can change from one promoter to another, but most promoters seem to require a precisely ordered assembling of components to initiate transcription. The assembly process is not always fast: at some genes it may

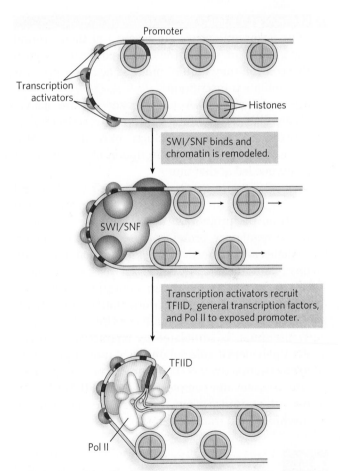

FIGURE 21-7 Transcription activator–mediated chromatin remodeling. Transcription activators can remodel chromatin structure by mobilizing nucleosomes; nucleosome repositioning is influenced by histone modifications. Some transcription activators have HAT activity or recruit enzyme complexes such as SWI/SNF, accelerating the remodeling of chromatin by relocating nucleosomes near a promoter. This leads to recruitment of the transcription machinery to newly exposed promoters, stimulating transcription.

take minutes, but it can take days at certain genes in higher eukaryotes.

Although rarer, some eukaryotic regulatory proteins that bind Pol II promoters can act as repressors, inhibiting the formation of active preinitiation complexes. Some transcription activators can adopt different conformations, enabling them to serve as activators or repressors. For example, some steroid hormone receptors function in the nucleus as DNA-binding transactivators, stimulating transcription of certain genes when a particular steroid hormone signal is present (see Section 21.3). When the hormone is absent, the receptor proteins revert to a repressor conformation, *preventing* formation of preinitiation complexes. In some cases this repression involves interaction with HDACs and other proteins that help restore the surrounding chromatin to its transcriptionally inactive state.

SECTION 21.1 SUMMARY

- Most eukaryotic genes are inactive in their ground state, as histones cover the DNA, and are under positive control; they require multiple activator proteins to stimulate transcription.
- The eukaryotic RNA polymerases require activator binding to promoter sequences to activate gene expression. The cell produces only the activator proteins necessary for transcription of the subset of genes needed at that time.
- Many Pol II promoters include the TATA box and Inr sequences, as well as other sequences located far from the promoter. When bound by the appropriate regulatory proteins, these distant regulatory sequences—enhancers in higher eukaryotes and upstream activator sequences in yeast—function at the promoter through DNA looping, increasing transcription regardless of their orientation in the DNA. The DNA bending is facilitated by HMG proteins.
- Transcription is stimulated by interactions between RNA polymerase core subunits and transcription activators (transactivators) bound to enhancer sequences. Often, coactivator complexes such as TFIID or Mediator act as bridges between the core transcription machinery and transactivators.

21.2 COMBINATORIAL CONTROL OF GENE EXPRESSION

The expression of eukaryotic genes is modulated by combinations of transcription factors, and when some of these factors are common to the regulation of multiple genes, the regulation is called **combinatorial control.** We learned in Chapter 20 that different bacterial genes driving sugar metabolism use a common transcription activator, cAMP receptor protein (CRP). CRP is employed in regulation of the operons involved in the metabolism of lactose and galactose, as well as other sugars. This is an example of combinatorial control.

Eukaryotes make much more extensive use of combinatorial control than do bacteria. First of all, as we have seen, eukaryotes generally require many regulatory proteins at any given promoter, increasing the combinatorial possibilities severalfold. Indeed, analysis of genome sequences reveals the use of greater numbers of transcription factors as genome size and complexity increase. For example, yeast are thought to use about 300 transcription factors, *Caenorhabditis elegans* and *D. melanogaster* more than 1,000, and humans more than 3,000. Although the number of transcription factors increases with the number of genes, there are still many fewer factors than there are genes to be regulated. Somehow, different genes must use the same transcription factors, but in different ways, to achieve activation. Given the increasing complexity of promoter sequences in more complex genomes and the greater number of transcription factors, combinatorial control allows higher eukaryotes to achieve exquisite specificity in gene regulation.

We begin with the relatively simple combinatorial control system that regulates the yeast *GAL* genes, driving the metabolism of galactose. The mechanism behind galactose metabolism is one of the best-understood systems (**Highlight 21-2** on p. 738). We then describe some increasingly complex mechanisms of combinatorial gene regulation.

Combinatorial Control of the Yeast *GAL* Genes Involves Positive and Negative Regulation

The enzymes required for importing and metabolizing galactose in yeast are encoded by *GAL* genes scattered over several chromosomes. Yeast cells have no operons like those in bacteria, and each of the *GAL* genes is transcribed separately. However, all the *GAL* genes have similar promoters and are regulated coordinately by a common set of proteins. The promoters for the *GAL* genes consist of the TATA box and an upstream activator sequence, which for each *GAL* gene is composed of one or more sequences denoted UAS_{GAL}. Each UAS_{GAL} site is recognized by a DNA-binding transactivator, the Gal4 protein (Gal4p). For example, the UAS of the gene *GAL1* is 118 bp long and contains four Gal4p-binding sites of 17 bp each (**Figure 21-8**).

Like the bacterial *lac* operon, the yeast *GAL* genes require more than just one protein (Gal4p) for activation. Control of gene expression by galactose depends on three proteins: the transcription activator Gal4p, the inhibitor Gal80p, and the ligand sensor Gal3p (**Figure 21-9**). Gal4p binds the 17mer UAS_{GAL} sites and, left to its own devices,

FIGURE 21-8 The GAL1 promoter. The promoters of the *GAL* genes of yeast each contain an upstream activator sequence (UAS), composed of one or more UAS_GAL sites. Each 17 bp UAS_GAL sequence is a binding site for the transcription activator Gal4p. The UAS of the *GAL1* gene has four UAS_GAL sites.

would activate gene expression at GAL promoters. However, at low galactose concentrations, Gal80p binds to Gal4p and blocks its transcription-activating region. When galactose is present, it binds Gal3p; Gal3p also binds ATP, and the Gal3p-galactose-ATP complex then interacts with Gal80p. This interaction causes a conformational change that relieves the inhibition of Gal4p and allows it to function as a transactivator at GAL promoters.

Glucose is the preferred carbon source for yeast, as it is for bacteria. When glucose is present, most of the *GAL* genes are repressed—whether galactose is available or not. The *GAL* gene regulatory system described above is effectively overridden by a global catabolite repression system. Global repression is achieved by the protein Mig1, which binds near the GAL promoter. Repression of the *GAL* genes also requires Tup1, a corepressor that binds Mig1 (**Figure 21-10**). Mig1 is regulated in a way that is not possible in bacteria—namely, through intracellular localization, which is regulated by phosphorylation. In the absence of glucose, Mig1 is phosphorylated and cannot enter the nucleus. Relegated to the cytoplasmic compartment, it is unable to bind DNA and repress the *GAL* genes. But when glucose is present, phosphorylation of Mig1 is blocked and the protein enters the nucleus, where it can bind DNA and associate with Tup1. Tup1 represses *GAL* gene expression by blocking transcription initiation, and possibly also by stimulating histone deacetylation at neighboring nucleosomes.

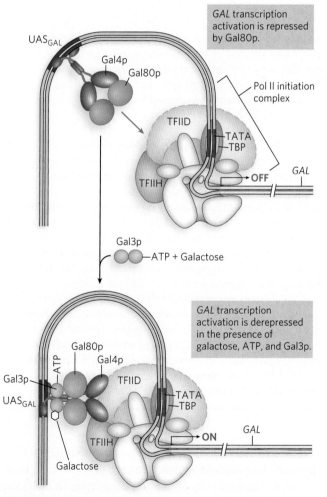

FIGURE 21-9 Regulation of *GAL* genes by the proteins Gal3p, Gal4p, and Gal80p. (a) Gal4p binds UAS_GAL, but Gal80p binds Gal4p and prevents its activation of Pol II and the general transcription factors. (b) Galactose is a small-molecule effector for Gal3p, causing it to bind Gal80p and alter Gal80p conformation, which frees Gal4p to activate transcription.

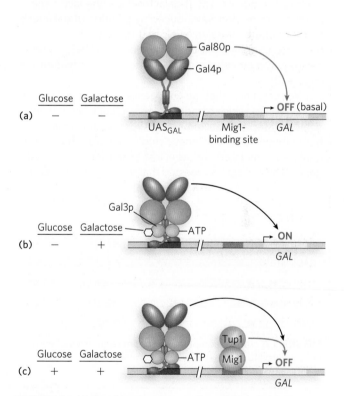

FIGURE 21-10 Combinatorial control in global repression of yeast *GAL* genes. Expression levels of a *GAL* gene are shown under three different growth conditions, with (+) or without (−) glucose or galactose. (a) In the absence of glucose and galactose, Gal4p occupies UAS_GAL, but the *GAL* gene is repressed by Gal80p. (b) In the presence of galactose and absence of glucose, Gal4p activates transcription of the *GAL* gene because Gal80p repression is relieved by binding of Gal3p. (c) In the presence of both glucose and galactose, glucose is the preferred carbon source; there is no transcription of the *GAL* gene because the Mig1-Tup1 complex represses its expression below basal levels.

Discovering and Analyzing DNA-Binding Proteins

Regulatory DNA sequences, such as the binding site for Gal4p in yeast, can be identified by sequence comparisons of genes that code for proteins of the same metabolic pathway. The Gal4p-binding site was one of the first eukaryotic activator-binding sites to be recognized. Genetic studies identified several genes in the pathway of galactose metabolism in yeast. In the presence of galactose, expression of the *GAL* genes increases as much as 1,000-fold. Clones containing the regulated *GAL* genes were sequenced, and comparison of the regions upstream from the TATA boxes revealed a common sequence, designated UAS$_{GAL}$ (Figure 1). The UAS$_{GAL}$ sequence is a 17mer, CGG(N)$_{11}$CCG, with a twofold axis of symmetry, indicating that the protein that binds it probably functions as a dimer. In vivo, mutation of UAS$_{GAL}$ sequences upstream from the *GAL* genes eliminated the usual activation in response to galactose. In a reporter gene assay in which the UAS$_{GAL}$ sequence was cloned into the upstream region of *lacZ*, β-galactosidase (the *lacZ* gene product) expression was induced by addition of galactose. Furthermore, expression levels of β-galactosidase depended on the number and sequence of UAS$_{GAL}$ sites, confirming their importance in transcription activation (Figure 2).

We now know that the *GAL* genes are activated by the protein Gal4p, which recognizes UAS$_{GAL}$. Early experiments demonstrated that Gal4p binds UAS$_{GAL}$ and functions as a transcription activator. Genetic studies revealed that a single gene, when mutated, results in loss of activation of all *GAL* genes. These results suggested that this single gene, *GAL4*,

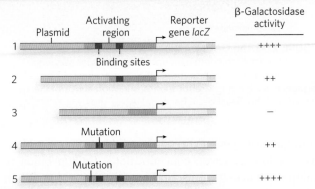

FIGURE 2 The function of UAS$_{GAL}$ sequences was confirmed in reporter gene assays in which promoter activity was determined by the activity of β-galactosidase (produced by the bacterial *lacZ* gene). As shown in these five assays (1 is the wild-type), deletion or mutation of UAS$_{GAL}$ elements, but not other areas close to the promoter, resulted in decreased promoter activity (β-galactosidase level).

was a master regulator, much like the bacterial CRP protein. *GAL4* was isolated by transforming a yeast genomic library into *GAL4*-mutant cells and selecting for colonies in which the *GAL* genes were again activated in the presence of galactose. *GAL4* was then cloned into an *E. coli* expression vector, and Gal4p was purified (see Chapter 7 for these cloning methods).

The technique of **deletion analysis** revealed the modular architecture of Gal4p, a structure now known to be common among many bacterial and eukaryotic transcription activators. In deletion analysis, nucleases or restriction enzymes are used to selectively delete pieces of DNA from a specified gene. The truncated protein product of this gene can be purified and tested for activity in vitro, or tested for function in vivo using a reporter assay. Studies such as these were performed with deletion constructs of Gal4p. DNA binding of the truncated proteins was measured in vitro with electrophoretic mobility shift assays, and the ability of the truncated proteins to activate transcription was tested in vivo with a reporter gene assay. In the reporter assay, deletion constructs of *GAL4* were transferred into *GAL4*-mutant yeast cells containing a plasmid with the bacterial *lacZ* reporter gene, driven by a typical GAL promoter with a UAS$_{GAL}$ sequence (Figure 3a). The ability of each

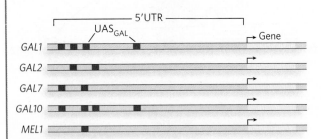

FIGURE 1 A comparison of the upstream sequences of the yeast *GAL* genes showed that they have common sequences, the UAS$_{GAL}$ sites, each 17 bp long (dark blue).

Combinatorial Control of Transcription Causes Mating-Type Switches in Yeast

Saccharomyces cerevisiae (baker's yeast) can grow as either diploid or haploid cells, both of which reproduce by mitosis (see the Model Organisms Appendix). The diploid cells contain two copies of each of the four yeast chromosomes, and haploid cells contain one copy of each. When

stressed by starvation, diploid cells can undergo meiosis to produce four haploid spores, two each of the mating types **a** and α. Haploid cells of the **a** mating type (**a** cells) can mate only with α haploids (α cells), and vice versa; thus, haploid cells display a simple sexual differentiation that is readily distinguishable when tested for mating ability.

Mating type is determined by the allele present at a single genetic locus, *MAT*. The identity of the allele

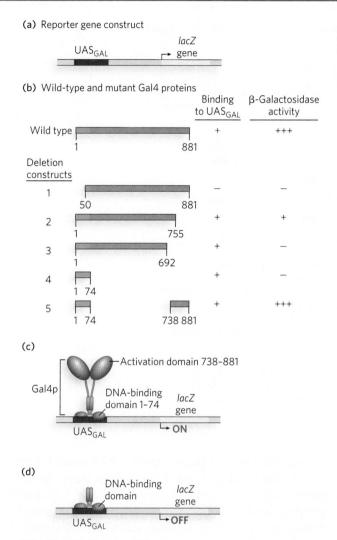

Gal4p-deletion construct to activate transcription of the *lacZ* gene was determined by measuring the activity of β-galactosidase (Figure 3b).

The in vitro DNA-binding activity of Gal4p was destroyed by a small deletion at the protein's N-terminus, but was not affected by small or large C-terminal deletions. Only the N-terminal 74 amino acid residues were needed for DNA-binding activity. Transcriptional activation required the DNA-binding region, as one would expect. Deletion of 60 residues from the C-terminus of Gal4p had little effect on gene activation. But deletion of 126 C-terminal residues reduced activation substantially, and a 191 residue C-terminal deletion completely eliminated activation. A large segment between these N- and C-terminal regions could be deleted without interfering with either activity.

The findings suggested that the two activities inherent in Gal4p require 260 or fewer residues: 74 at the N-terminus and 191 at the C-terminus. This result was surprising, given that the entire Gal4 protein is 881 amino acids long. To confirm the result, the researchers spliced together DNA for the 74 residue N-terminal DNA-binding domain and various lengths of the C-terminal transcription-activation region. They found that a 217 residue protein, missing 664 amino acids between the two regions, restored full activity in both DNA-binding and transcription-activation assays.

Clearly, the ability of Gal4p to activate transcription is the result of two distinct and separable domains. Similar results were obtained with other transcription activators from several different eukaryotes. Furthermore, examination of some transcription activators showed that the region between the two functional domains is highly sensitive to proteases, suggesting that the two domains are linked by sections of polypeptide that are open and flexible. These experiments gave rise to a model for some transcription activators, with two functional domains joined by a flexible linker (Figure 3c, d). The flexible region may help loosen the geometric constraints imposed by the DNA loop that forms between the transcription activator at an upstream binding site and the proteins it binds at the distant promoter. That the DNA-binding and transcription-activation domains of regulatory proteins can act independently has been demonstrated by "domain-swapping" experiments (see Figure 19-23).

FIGURE 3 (a) The reporter gene construct used for deletion analysis of Gal4p. Only constructs with functional Gal4p will bind UAS_GAL and drive expression of the reporter gene (*lacZ*). (b) Deletion analysis of Gal4p. Two activities were measured: in vitro DNA binding (indicated by + or − in the first column on the right) and in vivo transcriptional activation of the reporter gene construct (second column). (c) In this model of the Gal4 protein, derived from the deletion analysis, Gal4p has separable DNA-binding and transcription-activation domains joined by a flexible linker. (d) The DNA-binding domain, expressed alone, will bind DNA but will not activate transcription.

at the *MAT* locus can switch as often as every cell division cycle. The mating-type switch occurs through site-specific recombination (see Chapter 14), to express either the *MAT***a** allele or the *MAT*α allele. The *MAT***a** allele encodes the **a**1 protein, which directs transcription of **a**-specific genes, and the *MAT*α allele encodes the α1 and α2 proteins, which stimulate transcription of α-specific genes (**Figure 21-11**). After mating, the resulting diploid

cells contain two *MAT* loci, one with the *MAT***a** allele and the other with the *MAT*α allele; the presence of both the **a** and α gene products directs the diploid-specific transcriptional program, and haploid-specific gene expression is turned off.

The transcriptional activation and repression of genes in each mating type is an example of combinatorial control, because control is achieved by combinations of regulators,

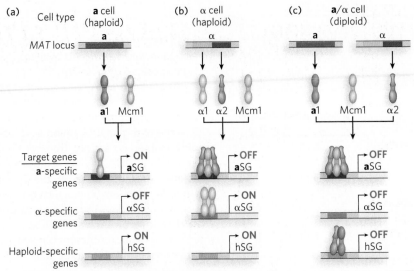

FIGURE 21-11 Combinatorial control of the yeast mating-type switch. In all *S. cerevisiae* cells, haploid and diploid, Mcm1 is expressed and is used in combinatorial control. (a) The haploid **a** cell expresses protein **a**1, but this protein is used only in diploid cells. Mcm1 alone turns on **a**-specific genes (**a**SG). Other haploid-specific genes (hSG) are also expressed. (b) The haploid α cell expresses the α1 activator and α2 repressor; α2 associates with Mcm1 to turn off α-specific genes, and α1 binds Mcm1 to turn on α-specific genes (αSG). Other haploid-specific genes are expressed. (c) Diploid cells express both **a**1 and α2. Each, in conjunction with Mcm1, represses transcription of a set of genes: **a**1-Mcm1 represses haploid-specific genes, and α2-Mcm1 represses **a**-specific genes. Because α1 is not expressed, α-specific genes are also not expressed.

at least one of which is common to the different cell types. In addition to the presence or absence of the **a**1, α1, and α2 proteins, specific activation and repression also involves Mcm1, expressed by both haploid cell types, as well as by diploid cells. In **a** cells, Mcm1 binds the promoters of **a**-specific genes and activates transcription. The genes specific to α cells are turned off in **a** cells, because the α1 activator is not present (see **Figure 21-11a**). In α cells, Mcm1 and α1 interact to activate α-specific gene transcription, while α2 (in association with Mcm1) represses transcription of **a**-specific genes (see **Figure 21-11b**).

There are also genes specific to both haploid states, but on mating to produce a diploid cell, the haploid-specific genes are turned off. Repression of genes specific to all haploid cells is made possible by the interactions of **a**1 and α2 repressor proteins, which are always expressed together in diploid cells (see **Figure 21-11c**).

Combinatorial Mixtures of Heterodimers Regulate Transcription

Like their bacterial counterparts, most eukaryotic transcription factors bind to DNA as homodimers. However, several types of eukaryotic transcription factors can form heterodimers of two different members of a family of similarly structured proteins, creating a larger number

of functional transcription factors from a smaller number of individual proteins. For example, three possible dimers can form from just two similarly structured proteins: two homodimers and one heterodimer; a hypothetical family of four different but structurally related proteins could form up to 10 different dimeric species (**Figure 21-12**).

An example of proteins that behave in this fashion is the mammalian AP-1 transcription activators. AP-1 activators can be either homodimers or heterodimers, formed from subunits that belong to the family of proteins that includes Fos, Jun, and ATF. Gene regulation by AP-1 homodimers and heterodimers occurs in response to a variety of external stimuli, including growth factors, cytokines, and factors involved in stress and infection. Thus, AP-1 transcription factors control such important processes as cell proliferation, differentiation, and programmed cell death. Indeed, some members of the Fos and Jun protein family are encoded by proto-oncogenes, which are genes that promote tumor formation when overexpressed. In other words, alterations in one or more of the subunits that make up AP-1 can be fatal for the cell, or even the entire organism.

The protein-dimerization and DNA-binding regions of AP-1 family members are of the basic leucine zipper type. The crystal structure of the dimerization and

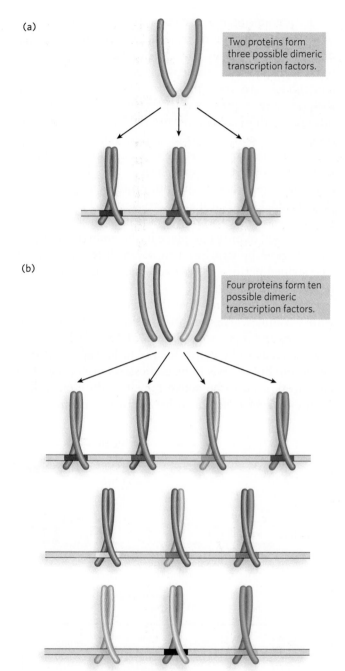

FIGURE 21-12 Combinatorial control by heterodimer formation. (a) Two regulatory proteins that form homodimers and a heterodimer could form 3 different structures, which could bind 3 different regulatory sites. (b) Four proteins have the potential to form 10 different structures and bind 10 regulatory sites. The possible combinations increase dramatically as the number of potential dimerization partners increases.

DNA-binding portions of a Fos-Jun heterodimer bound to DNA is shown in **Figure 21-13a**. AP-1 dimers activate genes containing an AP-1–binding site. AP-1 variants bind to AP-1–binding sites with different affinities and activate gene transcription to different extents, depending on the composition of that AP-1. **Figure 21-13b** shows the result of an electrophoretic mobility shift assay that

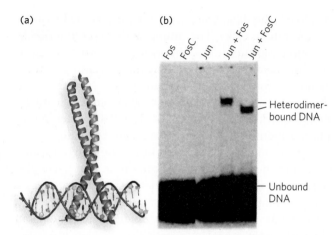

FIGURE 21-13 AP-1 transcription factors. (a) Structural model of the AP-1 heterodimer of Fos (purple) and Jun (green), bound to DNA. (b) Gel from an electrophoretic mobility shift assay using a ^{32}P-end-labeled DNA fragment containing the AP-1–binding site sequence. The DNA was mixed with Fos, FosC (a fragment of Fos), or Jun (all of which would form a homodimer), or with a mixture of Fos and Jun or a mixture of FosC and Jun (both of which would form the two types of homodimer and the heterodimer). Reactions were analyzed by polyacrylamide gel electrophoresis, then autoradiography. Binding of protein dimers to the DNA causes the complex to migrate more slowly through the gel, resulting in distinct bands. The radioactive signal at the bottom of the gel is unbound DNA. *[Sources: (a) PDB ID 1FOS. (b) Reprinted by permission from Macmillan Publishers Ltd: Kouzarides, T. and Ziff, E. (1988), Nature 336, 646–651 (15 December 1988). © 1988. doi:10.1038/336646a0.]*

examined DNA-binding affinity of Jun-Jun or Fos-Fos homodimers, as well as an AP-1 Fos-Jun heterodimer. A short DNA fragment containing an AP-1–binding site was end-labeled with ^{32}P, then mixed with Fos, Jun, or the Fos-Jun heterodimer. The experiment also examined the binding affinity of a subfragment of Fos (FosC) that contains the DNA-binding and dimerization elements. The resulting gel shows that Fos, FosC, and Jun do not bind appreciably to the AP-1–binding site on their own. However, the Fos-Jun and FosC-Jun heterodimers bind the AP-1 site much more tightly, such that they could be detected in this experiment.

This differential DNA binding, depending on the composition of the AP-1 transcriptional control complex, is another example of combinatorial control. Although many AP-1 variants contain transcription-activation domains, some lack them and instead function as transcription inhibitors. Thus, the effect of AP-1 can be varied by changing its composition, depending on the needs of the cell.

Differentiation Requires Extensive Use of Combinatorial Control

A more complicated example of combinatorial control can be seen in body plan development in the fruit fly, *D. melanogaster*. Before it is released to become fertilized, the developing oocyte is surrounded by cells called nurse cells. The nurse cells secrete mRNAs encoding various transcription factors into the egg at specific locations, establishing concentration gradients of mRNA for the different transcription factors within the egg. During early embryonic development the nuclei divide quickly, producing 3,000 to 6,000 nuclei before plasma membranes form to delineate individual cells. When plasma membranes do form, the newly formed cells trap the specific mRNAs present at that particular position in the embryo. Each new cell thus produces a unique complement of transcription factors that act in a combinatorial fashion to express different proteins in the early embryo.

An example of combinatorial control by these unevenly distributed transcription factors is regulation of the *eve* gene, which produces a protein called even-skipped. Even-skipped is expressed only in specific cells of the embryo, generating a pattern of seven stripes that can be visualized using a fluorescent antibody to even-skipped (**Figure 21-14a**). The *eve* gene is essential to development; the even-skipped product is a transcription activator that promotes further differentiation in the cells where it is expressed.

Expression of *eve* is controlled by the concentrations of four proteins translated from the original mRNAs deposited in the developing oocyte by the nurse cells. Two of these four proteins, Bicoid and Hunchback, are activators; the other two, Giant and Krüppel, are repressors. Different gradients of the mRNAs for these activators and repressors, established by the nurse cells, result in unique ratios of the four regulatory proteins in nearly every cell of the embryo. Expression of even-skipped occurs only in cells that have the proper ratio of the four proteins to activate *eve* (**Figure 21-14b**). But if *eve* were activated by only one particular ratio of protein concentrations, *eve* would be expressed in only one place in the embryo. How can the *eve* gene be expressed in seven different stripes? The striped pattern of *eve* expression is made possible by combinatorial control.

The *eve* gene has five different enhancers, each with a complex array of binding sites for transcription activators and repressors (**Figure 21-15a**). Only one enhancer needs to be active for *eve* to be expressed in a given cell. But if *eve* is to be expressed normally, all five enhancers need to be active (albeit in different cells). Each enhancer is activated by a different combination

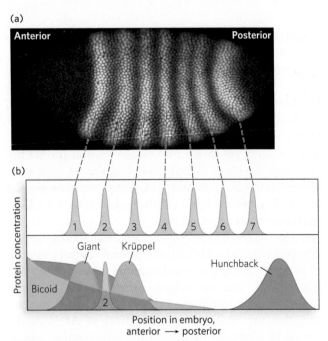

FIGURE 21-14 Combinatorial control of *eve* gene expression in fruit fly development. (a) A *Drosophila* embryo stained with fluorescent antibodies that recognize the protein even-skipped (product of *eve*), showing its striped pattern of expression. (b) The graphs represent the relative levels and positions along the length of the embryo of even-skipped (top) and four transcription factors that regulate its expression (bottom). Specific combinations of transcription factors activate the *eve* gene. [*Source: From In Silico Biol. 3, A. V. Spirov and D. M. Holloway, 89–100, 2003, Fig. 1. © 2003 with permission from IOS Press.*]

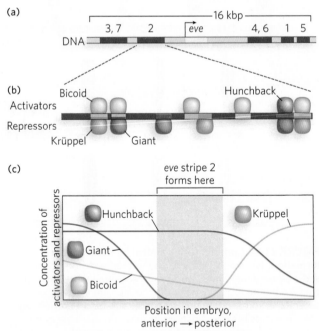

FIGURE 21-15 Five independently acting enhancers of the *eve* gene producing seven stripes of *eve* expression in the early embryo. (a) The *eve* gene and its upstream and downstream enhancers, any one of which can activate *eve* expression if bound by the correct combination of transcription factors. Numbers 1 through 7 indicate the stripe(s) activated by each enhancer (see Figure 21-14). (b) The binding sites in the stripe 2 enhancer for the Bicoid and Hunchback activators and the Krüppel and Giant repressors. (c) Changes in concentration of the four transcription factors along the length of the embryo in the region that expresses *eve* stripe 2.

Michael Levine
[Source: Courtesy of Michael Levine.]

of transcription factors. Some activator and repressor sites overlap and are controlled by competition, whereas some repressor sites are distinct from activator sites and repress the gene at a distance (**Figure 21-15b**). Seven stripes of even-skipped expression, each four cells wide, are formed because the local concentration of each activator and repressor is just right for activation of one of the five enhancers in particular cells along the length of the embryo. The same four transcription factors are used by the five different enhancers in different ways. Expression of the *eve* gene is an example of exquisite, complex combinatorial control.

The enhancer that activates *eve* expression in stripe 2 has been extensively studied in Michael Levine's laboratory. This enhancer is 500 bp long and contains binding sites for both repressors and activators (see Figure 21-15b). Both activators, Bicoid and Hunchback, must bind to their sites for gene expression to occur. Some binding sites for these activators overlap repressor-binding sites, and other activators bind DNA but are inactivated by repressors that bind within about 100 bp of the activators' binding sites. Increasing the distance between activator- and repressor-binding sites of this type prevents repressor function. Although the mechanism of repression is unclear, it might occur through covalent modification of the activator. The region of the embryo that expresses *eve* in stripe 2 is largely deficient in both repressors, yet contains both activators (**Figure 21-15c**). Hence, the particular cells that express *eve* in stripe 2 do so because this is the only location in the embryo where the condition for activator binding in the absence of repressors is met. In all other cells, the stripe 2 enhancer is silent. Combinatorial control also governs formation of the other stripes expressing *eve*. The other *eve* enhancers contain different combinations and arrangements of the repressor- and activator-binding sites, such that each enhancer is active in only a narrow region of the embryo.

These examples of combinatorial control of transcription illustrate a central mechanism by which eukaryotic cells govern gene expression. Through the use of a relatively small number of regulatory proteins in each case, many different genes can be regulated either in concert or differentially, depending on the immediate needs of the cell. In this way, cells can respond quickly and appropriately to changing environmental conditions or to developmental requirements, within the context of a tissue or an entire organism.

SECTION 21.2 SUMMARY

- Eukaryotic transcription activators such as Gal4p have DNA-binding and transcription-activation domains.
- Eukaryotes make greater use of combinatorial control of gene expression than do bacteria. In combinatorial control, the same transcription factor is used in the regulation of more than one gene.
- Combinatorial control can be achieved in a variety of ways. Some transcription factors are formed from combinations of two different subunits that form heterodimers, each of which has different strengths as an activator. Or a gene has several enhancers, each of which uses a different combination of transcription factors.
- Mating-type switching in yeast is a classic example of combinatorial control. Unique sets of genes are expressed specifically in the **a** and α haploid states, due to activation by specific regulatory factors. On mating to produce a diploid cell, the haploid-specific genes are repressed by the interactions of **a**1 and α2 repressor proteins, which are expressed together only in diploid cells.
- Body plan organization in *D. melanogaster* uses gradients of mRNAs for different transcription factors in the developing embryo. Different concentrations of transcription activators and repressors control where the gene *eve* is activated, producing seven stripes that influence cell differentiation.

21.3 TRANSCRIPTIONAL REGULATION MECHANISMS UNIQUE TO EUKARYOTES

Gene regulation is necessarily more complex in eukaryotes than in bacteria, as a consequence of some of the key differences between these two domains of organisms. The larger eukaryotic genomes entail more nonspecific DNA binding and more genes to regulate. And the multicellular nature of most eukaryotes requires mechanisms for development and intercellular communication, the formation and function of intracellular compartments, and the speedy control of gene expression as cells grow and change. We now turn to a discussion of some regulatory processes that are necessary to deal with the complexity of the eukaryotic genome. Such gene control mechanisms provide for situations that arise only in eukaryotes, such as the need to regulate gene dosage in diploid cells.

Insulators Separate Adjacent Genes in a Chromosome

The use of multiple transcription factors to control eukaryotic genes requires dispersed binding sites. Some enhancers are located well over 1,000 bp from the promoters they regulate, or they can be within the gene or at the noncoding 3′ end of the gene. This is quite different from the situation in bacteria, where control elements are almost always located close to, or overlap with, the promoter. As discussed earlier, DNA looping accommodates the large distances between enhancer and promoter elements in eukaryotes. Indeed, the large size of DNA loops provides the flexibility that enables enhancers to function in either orientation. However, this raises a new question: what prevents the enhancer for one gene forming a loop to interact with the promoter of another gene, thereby activating the wrong gene? In part, misregulation of this type is prevented by **insulators** (sometimes referred to as boundary elements), DNA sequences that form boundaries between genes or groups of genes in eukaryotes.

Insulators are relatively short sequences, sometimes fewer than 50 bp. Exactly how insulators function is still unknown, and it is likely that they have a variety of functions. An example of an insulator in T cells (T lymphocytes, white blood cells involved in the immune response) is shown in **Figure 21-16**. T cells must express either the α chain or the δ chain of the T-cell receptor, but not both. The promoters for these genes are adjacent in human cells, but an insulator sequence between them prevents the enhancer region for one gene activating transcription from the promoter of the other gene. Insulators can also prevent packaging of a gene into heterochromatin. When a gene is experimentally inserted into a heterochromatic region of a chromosome, it is typically repressed. But if the gene also contains an adjacent insulator, the gene remains active even after insertion into a heterochromatic region. An insulator can have a repressive effect if located between the promoter and enhancer of a gene.

All insulator sequences in higher eukaryotes require CTC-binding factor (CTCF) to function (see Figure 21-16). CTCF was first identified as a protein that binds to a site containing a 5′-CCCTC sequence, from which the protein derives its name. The binding site for CTCF is much longer than this sequence, but the flanking sequences are divergent and not as easy to recognize. CTCF contains 11 zinc fingers and binds a diverse set of DNA sequences up to 50 bp long. All insulators seem to require CTCF binding, although the mechanism of the insulators' function and the role of CTCF are currently unknown. CTCF might recruit other proteins to the insulator.

Some Activators Assemble into Enhanceosomes

Exquisite transcriptional regulation can be achieved through the interaction of multiple activators at a single gene. In some cases, cooperating activators form a stable, tightly folded nucleoprotein complex called an **enhanceosome**, which integrates regulatory information from multiple signaling cascades and generates a single transcriptional outcome at the target promoter. An example of an enhanceosome can be seen in the regulation of the gene for interferon beta (IFNβ). Interferons are produced in response to a viral infection and lead to programmed cell death, thereby halting further production of viral particles and infection of surrounding cells. At the IFNβ promoter, multiple activators can present their activation domains together to simultaneously interact with the cofactor protein complex CBP-p300 (**Figure 21-17**). Recruitment of this cofactor is most efficient only when all of the activators in the enhanceosome are present together. The placement of each activator-binding site in the DNA is critical, because of the three-dimensional

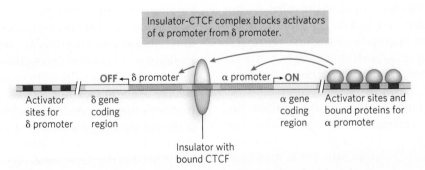

FIGURE 21-16 Insulator regulation of the expression of T-cell receptors.
With DNA looping over long distances, binding at enhancers could activate the wrong promoter. The insulator confines the action of enhancers to their matching promoter. In the regulatory regions of the genes shown here, for the α and δ chains of the T-cell receptor, the insulator prevents activation of the α promoter by the δ enhancer, and of the δ promoter by the α enhancer. CTC-binding factor (CTCF) binds to insulators, but how it functions is still uncertain.

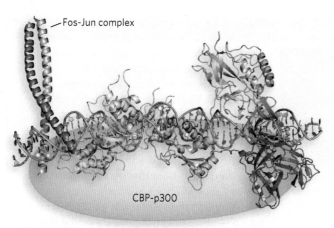

FIGURE 21-17 Hypothetical structure of the IFNβ enhanceosome. The IFNβ enhancer is referred to as an enhanceosome because, unlike other enhancers, it requires accurate spacing and helical phasing of the DNA between several protein-binding sites. This requirement indicates that a specific tertiary structure is formed with the various regulatory proteins. HMG proteins, while not part of the completed complex, facilitate enhanceosome formation by helping to bend the DNA (see Figure 21-6). Individual regulatory proteins in the complex are shown in different colors. [Sources: PDB ID 2O6G, PDB ID 2O6I, and PDB ID 1T2K.]

structure required for enhanceosome function. For example, the experimental insertion of 5 bp (i.e., a half-turn of the helix) between regulatory elements inactivates the function of the IFNβ enhanceosome.

Similar activator clusters can also function together to repress transcription, and it is possible that an enhanceosome can switch to a repressor function under different cellular conditions. Enhanceosomes tend to form at genes that need to be tightly regulated in pathways important to the organism's defense system, such as wound healing and antipathogen mechanisms.

Gene Silencing Can Inactivate Large Regions of Chromosomes

Thus far we have focused on the activation or repression of gene expression through the action of activator or repressor proteins at single promoters or enhancers. In some cases, however, the position of a gene within the chromatin, or its location on a particular chromosome, leads to almost complete repression of gene expression. This **gene silencing**, a powerful mode of regulation in eukaryotes, is the absence of gene expression due to the location of the gene, rather than its response to the presence or absence of a regulatory factor or complex. As a result, silencing can encompass relatively large segments of a chromosome, or an entire chromosome, thus controlling the expression of many genes at once.

Recall that chromatin is organized into heterochromatin and euchromatin. The loosely packed euchromatin is often transcriptionally active, whereas the densely packed heterochromatin is transcriptionally silent. Heterochromatin is often found at centromeres and telomeres, as well as other inactive parts of the genome. Experiments have shown that genes normally active in a region of euchromatin become transcriptionally silent when moved into heterochromatic regions. These observations led to the conclusion that a primary function of heterochromatin is to maintain certain parts of the genome in a transcriptionally inactive state by preventing access of the transcription machinery to these chromatin regions. Indeed, burying a gene within heterochromatin may be a preferred mechanism for long-term silencing.

The formation of heterochromatin requires many different factors, depending on the specific region of the chromosome. For example, a mechanism known as gene dosage compensation, occurring in the cells of female mammals, involves the formation of heterochromatin over an entire X chromosome, inactivating it (as we discuss in more detail below). Recent studies of heterochromatin in other regions of a chromosome show that small nuclear RNAs (snRNAs) are required for heterochromatin formation, along with certain proteins and histone modifications. **Highlight 21-3** describes studies of heterochromatin in the centromere region of yeast chromosomes that reveal a role of the silencing machinery mediated by small RNAs.

Imprinting Allows Selective Gene Expression from One Allele Only

In most diploid cells, both homologous genes (i.e., both alleles of a gene) are expressed equally. One allele may be dominant over the other, as Mendel found in his work on the garden pea, or the two alleles may deviate from Mendelian behavior and both may contribute to the phenotype. For example, we learned in Chapter 2 that both alleles of the genes responsible for human blood type, when expressed together, can lead to type AB (see Figure 2-5). But regardless of the phenotype, both alleles of a gene are usually expressed in the diploid cell. Some higher eukaryotes, however, have mechanisms to completely shut down the expression of an allele derived from one parent, in a process called **imprinting**. Because only one parental allele of an imprinted gene is repressed, the usual rule of equal expression of each allele in a diploid cell is violated.

Imprinting is typically restricted to mammals, although some examples have been found in flowering plants. About 80 genes in the human genome are currently known to be imprinted. Imprinting occurs during development of the gametes; a set of genes is imprinted during oocyte development, and a different set of genes is imprinted during sperm cell development. Thus, all

Gene Silencing by Small RNAs

Large tracts of DNA are encased in heterochromatin, a form of DNA so compact that its genes are silenced by the exclusion of RNA polymerase. The nucleosomes in heterochromatin have a distinctive histone modification pattern of methylation and hypoacetylation compared with actively transcribed regions of DNA (euchromatin). The epigenetic marks (marks that are inheritable but do not occur in the DNA) result in stable inheritance of the silent heterochromatin state.

Research from Danesh Moazed's laboratory shows that the formation of heterochromatin in fission yeast (*Schizosaccharomyces pombe*) requires the RNA silencing machinery, which localizes to heterochromatin nucleation sites. To study the process of heterochromatin formation, the researchers purified the RNA silencing complex of *S. pombe* called RITS (*RNA-induced transcriptional silencing*), and also identified a second complex that interacts and functions with RITS. The second complex is referred to as RDRC (*RNA-directed RNA polymerase complex*), because it contains an RNA-directed RNA polymerase, Rdp1. RDRC also contains a helicase known as Hrr1, and a putative poly(A) polymerase known as Cid12. A hypothesis for the process of heterochromatin formation at the centromere, based on the possible functions of these complexes, is shown in Figure 1.

The RITS complex contains siRNAs, derived from heterochromatic regions of the chromatin, and the protein Argonaute, which belongs to a family of proteins implicated in RNA-induced silencing pathways. RITS matches the siRNAs to complementary RNA sites known as cenRNA sequences, which are noncoding transcripts produced at repeat sequences (*cen* DNA) near the centromeres (see Figure 1, step 1). The RITS complex also contains a chromodomain protein, which binds methylated histones of heterochromatin and probably helps target RITS to heterochromatin. When RITS associates with a growing cenRNA transcript, Rdp1 within the RDRC complex uses the siRNA-cenRNA as a primer to synthesize double-stranded RNA (step 2). Dicer then chops the double-stranded RNA into multiple siRNAs (step 3). Moazed and his colleagues identified another complex, called ARC (*Argonaute chaperone*), that contains Argonaute and two other protein components. ARC binds individual siRNAs and is thought to chaperone them into the RITS complex (step 4). During this step, the siRNA base-pairs to complementary sequences in heterochromatin, recruiting RITS and completing a self-propagating cycle that expands the heterochromatic region.

Methylation of histone H3 at Lys^9, initiated by RNA silencing complexes, can help expand the heterochromatin.

Danesh Moazed [*Source: Steve Gilbert, Studioflex Productions.*]

The methylation is associated with recruitment of HP-1 protein, which is required for heterochromatin formation. The poly(A) polymerase Cid12 belongs to a family of proteins that target RNAs to the exosome, a large ribonuclease complex that catalyzes RNA degradation. Hence the Cid12 component of RDRC is thought to add another layer of RNA surveillance to ensure that heterochromatic regions are completely silenced.

Although this system was elucidated in a single-celled eukaryote, we know that the *S. pombe* proteins identified in the processes described here are conserved in higher eukaryotes. Perhaps a similar mechanism of siRNA-mediated heterochromatin formation occurs in mammals.

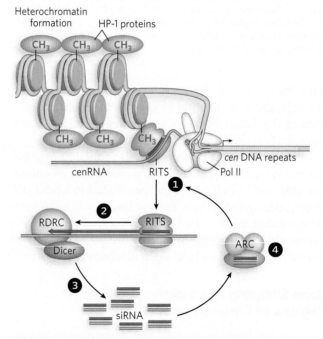

FIGURE 1 Proposed model for heterochromatin formation at the centromere in *S. pombe*. RITS is the RNA-induced transcriptional silencing complex, RDRC is the RNA-directed RNA polymerase complex, and ARC is an Argonaute chaperone complex that contains duplex siRNA. Synthesis of dsRNA and generation of siRNA occur in association with specific chromosome regions. The nascent transcript model proposes that the RITS complex mediates heterochromatin formation by associating with nascent transcripts via siRNA base pairing, and with methylated H3K9 (histone H3 with methylated Lys^9). RNA-directed association of RITS with specific transcripts (step 1) leads to RNA-dependent RNA copying (step 2) and processing (step 3) to increase the level of small RNAs that can bind to ARC and associate with RITS to continue the cycle (step 4). [*Source: Data from M. Buhler and D. Moazed, Nat. Struct. Mol. Biol. 14:1041–1048, 2007 (online, November 5, 2007), Fig. 3.*]

gametes from the mother have one set of genes imprinted, and all gametes from the father have a different set of genes imprinted. Nearly every cell of the offspring has the gene expression pattern dictated by the imprinted genes from the egg and sperm. The sole exception is in cells that will give rise to gametes. All imprinting information from the parents is lost during development of the germ cells. Developing gametes adopt the imprinting pattern specific to the sex of that individual organism.

Imprinting of a gene is not based on the DNA sequence; it is an epigenetic process (see Chapter 10). Epigenetic marks are created by nucleosome modification patterns and DNA methylation. **Figure 21-18** shows imprinting based on DNA methylation for the mammalian gene for insulin growth factor-2 (IGF-2). In this case, DNA near the paternally inherited *IGF2* allele becomes methylated (imprinted) and is active, but DNA near the maternally inherited allele is unmethylated and is not expressed in the offspring. Imprinting of this gene is important, because expression of both alleles tends to result in cancer. The mechanism of imprinting in this case involves an insulator sequence. The *IGF2* gene contains an insulator sequence between its promoter and an enhancer, rather than between the gene and an adjacent gene. Thus, when the insulator is active, *IGF2* is repressed because the activating effect of the enhancer is insulated from the promoter.

The biochemical explanation for imprinting in *IGF2* is thought to lie in 5-methylation of C residues of CpG sequences (see Figure 6-34a). Cytosine methylation in eukaryotic DNA is generally associated with gene repression, whereas transcriptionally active regions of DNA tend to be undermethylated. In the case of *IGF2*,

cytosines in CpG sequences recognized by CTCF in the insulator sequence that regulates the paternal *IGF2* gene become methylated when the gene is imprinted, thereby inactivating *IGF2*. When CpG sites are methylated, CTCF can no longer bind the insulator. Hence, when the insulator in *IGF2* is methylated during sperm development, the paternally inherited copy of *IGF2* is expressed because CTCF no longer binds the insulator and the enhancer activates the promoter. The insulator site is not methylated in the egg, so CTCF binds the insulator and represses the maternally inherited copy of *IGF2*.

Imprinting is essential for development in mammals. Studies in mice have shown that a genetically engineered egg containing two complete sets of maternally inherited chromosomes will not develop past the blastula stage. The same is observed for eggs containing two complete sets of chromosomes from sperm. In either of these situations, the alleles of the genes that are normally differentially marked will have identical imprinting patterns in the developing egg—leading to either no expression or double expression of those genes. For example, if an embryo were to develop from a fully diploid cell in which the two chromosome sets were derived from a female, both *IGF2* alleles would be inactive. This explains why mammals are not capable of **parthenogenesis**—the development of an embryo with a diploid genome that is entirely maternally derived.

It is possible that imprinting evolved by natural selection to increase the fitness of the organism. Imprinting in mammals is thought to confer certain behavioral traits that enhance fitness. One hypothesis proposes that imprinting reflects the different interests of the mother and father in the growth and development of offspring. The father is more interested in seeing that the offspring grow rapidly, regardless of whether this occurs at the expense of the mother. The mother is more interested in balancing her own survival with sufficient nourishment of the offspring and thus tends toward growth-limiting measures that conserve resources. In support of this "parental conflict" hypothesis, male-expressed imprinted genes tend to promote growth, and female-expressed imprinted genes tend to limit growth. The model is also supported by the lack of imprinting in animals, such as birds, that have lower requirements for raising offspring.

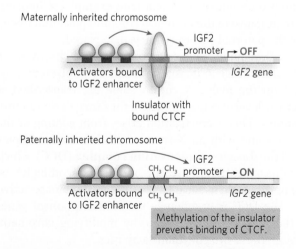

FIGURE 21-18 Imprinting of the mammalian *IGF2* gene. When CTC-binding factor (CTCF) binds the insulator of *IGF2* (top), this enables insulator function and turns off *IGF2*, because the insulator is located between the promoter and the enhancer. When the insulator sequence is methylated (bottom), CTCF can no longer bind the DNA, and *IGF2* can be activated by its enhancer.

Dosage Compensation Balances Gene Expression from Sex Chromosomes

Diploid organisms carry two copies of each autosomal chromosome, but the sex chromosomes are unequal in copy number in females and males. In mammals, females have two copies of the X chromosome, and males have one X chromosome and a Y chromosome. The

X chromosome carries genes that are required by both males and females, and the gene products are required in the same amounts in both sexes. **Dosage compensation** mechanisms have evolved to control the level of gene expression from these chromosomes so that the levels are similar in males and females. We can imagine three different ways in which gene dosage compensation could occur. (1) Total inactivation of one X chromosome in the female would make gene expression equal to that from the single X chromosome in the male. (2) Expression of the single X chromosome in the male could be doubled. (3) Expression of the two X chromosomes in the female could be halved. All three of these mechanisms are employed in one species or another (**Figure 21-19**).

In mammals, one X chromosome of female cells is inactivated, compacted into a tightly condensed structure called a **Barr body**. This process of **X chromosome inactivation** starts at the **X inactivation center (XIC)**, a region of about 10^6 bp near the middle of the X chromosome, which condenses into heterochromatin. Condensation spreads from this nucleation point in both directions until the entire X chromosome is compacted (**Figure 21-20**). The process involves XIST, an RNA produced from the XIC DNA in the inactivated chromosome. XIST is not translated into protein, but instead coats the chromosome non-sequence-specifically at the XIST locus, and then spreads in both directions. In addition to XIST, X chromosome inactivation in mammals also involves a histone variant, macroH2A (see Highlight 10-1).

In humans, inactivation occurs on only one of the two X chromosomes in each cell and thus seems to be a type of imprinting. However, the selection of which X chromosome becomes inactive is random; either the maternal or the paternal X chromosome is inactivated in any given cell. Therefore, this X chromosome inactivation is not

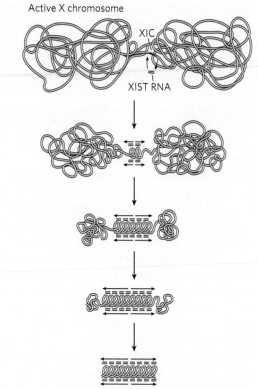

Active X chromosome

XIC

XIST RNA

Inactive X chromosome (Barr body)

FIGURE 21-20 X chromosome inactivation in mammals. Inactivation of one X chromosome in each cell of the female mammal begins at the XIC locus, which encodes XIST (a non-protein-coding RNA), and also requires the histone variant macroH2A and other chromatin regulatory factors. XIST coats the X chromosome, starting at the nucleation site and then spreading in both directions until the entire chromosome is coated, resulting in repression of nearly all the genes on the chromosome.

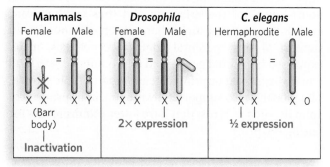

Mammals		Drosophila		C. elegans	
Female	Male	Female	Male	Hermaphrodite	Male
X X	X Y	X X	X Y	X X	X 0
(Barr body)			2× expression	½ expression	
Inactivation					

FIGURE 21-19 Different dosage compensation mechanisms control the level of gene expression from the X chromosomes. (a) In mammals, one X chromosome of the female (XX) is inactivated, forming a compact structure called a Barr body. (b) In *Drosophila*, the single X chromosome of the male (XY) is transcribed at twice the level of each X chromosome in the female (XX). (c) In *C. elegans*, the two X chromosomes of the hermaphrodite (XX) are transcribed at half the level of the single X chromosome of the male (X0).

strictly imprinting. There are some tissues in which X chromosome inactivation is a true example of imprinting in humans: in the umbilical cord and placental tissue, only the paternal chromosome is inactivated.

In *Drosophila*, male cells contain one X chromosome and solve the gene dosage problem in a different way: genes on the male's X chromosome are transcribed at twice the level of genes on each of the female's two X chromosomes. This overactivation arises from coating of the chromosome with an X-encoded RNA-protein complex called the **dosage compensation complex (DCC)**, which contains five different proteins and two noncoding RNAs. Two of the proteins have HAT and phosphokinase activities. The DCC may enhance the transcription of genes on the male's X chromosome by modifying chromatin through these two enzymatic activities.

The nematode *C. elegans* solves the gene dosage problem in the opposite way. The two sexes in *C. elegans* are the hermaphrodite (XX) and the male (X0). Expression from each of the two X chromosomes in the hermaphrodite is reduced by half relative to gene expression from the single X chromosome of the male. Like *Drosophila*, *C. elegans* has a dosage compensation complex, but here its function is to repress gene transcription instead of activating it. The DCC is expressed only in cells with two X chromosomes (i.e., in the hermaphrodite), where it coats both X chromosomes. The *C. elegans* DCC is composed of proteins that resemble the condensing complex of mitotic and meiotic chromosomes. Thus, evolution has modified and recruited these chromosome-condensing proteins to the task of gene dosage compensation. The DCC partially condenses both X chromosomes such that transcription is reduced by half, balancing the gene dosage with that of the male cell with its single X chromosome. Targeting of the *C. elegans* gene dosage complex to X chromosomes is accomplished by DNA sequence elements dispersed along the X chromosome that nucleate the complex. Nucleation is followed by spreading of the complex across the entire chromosome, maintaining the repressed epigenetic state throughout the life of the hermaphrodite.

Steroid Hormones Bind Nuclear Receptors That Regulate Gene Expression

Whereas dosage compensation provides for inherited control of gene expression levels, transient changes in transcription are often triggered in response to hormones. Intercellular communication is essential in multicellular organisms; tissues, organs, and organ systems need to work together and must be able to respond to external signals. One group of molecular signals is the steroid hormones, which operate in the nucleus to activate transcription of particular genes in response to tissue or system requirements.

The effects of steroid hormones (and thyroid and retinoid hormones, which have the same mode of action) provide well-studied examples of the modulation of eukaryotic regulatory proteins by direct interaction with molecular signals. Steroid hormones too hydrophobic to dissolve readily in the blood (e.g., estrogen, progesterone, and cortisol) travel on specific carrier proteins from their point of release to their target tissues, where the hormone enters cells by simple diffusion and binds to its specific nuclear receptor protein, which is a transcription activator.

There are two major types of steroid-binding nuclear receptors: those initially located in the cytoplasm (type I) and those always located in the nucleus, bound to DNA (type II). Examples of steroid hormones that bind type I nuclear receptors are estrogen, progesterone, androgens, and glucocorticoids. The action of type I nuclear receptors is shown in **Figure 21-21a**. The nuclear receptor (NR) is initially bound to a heat shock protein (Hsp70) in the cytoplasm, keeping the receptor in its monomeric state. On binding the steroid hormone, Hsp70 dissociates, the receptor dimerizes and exposes a nuclear import signal, and the receptor-hormone complex migrates into the nucleus, where it acts as a transcription factor.

Type II receptors also require binding of the hormone before activating transcription, but these receptors are already bound to the DNA, whether their molecular signal (steroid hormone) is present or not. In addition, type II receptors typically bind DNA as a heterodimer. Thyroid hormone receptor (TR) is an example of a type II nuclear receptor (**Figure 21-21b**). TR forms a heterodimer with a protein known as the retinoid X receptor (RXR, another type II receptor) and, in the absence of thyroid hormone, also binds a third protein, a corepressor. This complex does not activate transcription. When thyroid hormone enters the nucleus and binds to TR, this releases the corepressor and promotes binding of a coactivator, resulting in recruitment of Pol II. Expression of the thyroid hormone–induced genes produces proteins involved in metabolism and in regulation of heart rate.

Steroid hormone–nuclear receptor complexes act by binding to highly specific DNA sequences known as **hormone response elements (HREs)**. The bound hormone-receptor complex can either enhance or suppress the expression of adjacent genes. The HREs for the various steroid hormones are similar in length and organization in the genome, but they differ in sequence. Each receptor has a consensus HRE sequence to which the hormone-receptor complex binds well (**Figure 21-22 on p. 751**). The consensus sequence consists of two six-nucleotide sequences, either contiguous or separated by three nucleotides, in tandem or inverted with respect to each other. The steroid hormone receptors have a highly conserved DNA-binding domain with two zinc fingers. The hormone-receptor complex binds to the DNA as a dimer, and the zinc finger domains of each monomer recognize the six-nucleotide HRE sequences. The ability of a given hormone to act through its hormone-receptor complex to alter the expression of a specific gene depends on the exact sequence of the HRE, its position relative to the gene, and the number of HREs associated with the gene.

The ligand-binding region of the steroid hormone receptor protein—always located at the C-terminus—is specific to the particular receptor. For example, the ligand-binding region of the glucocorticoid receptor shares only 30% sequence similarity with the estrogen receptor, and only 17% similarity with the thyroid hormone receptor. The size of the ligand-binding region also varies dramatically: the vitamin D receptor has only 25 residues, whereas the mineralocorticoid receptor has 603 residues in this region. Mutations that change one amino acid in the ligand-binding region can result in loss of responsiveness to a specific hormone. In humans, medical conditions resulting from the inability to respond to cortisol, testosterone, vitamin D, or thyroid hormone are caused by mutations of this type.

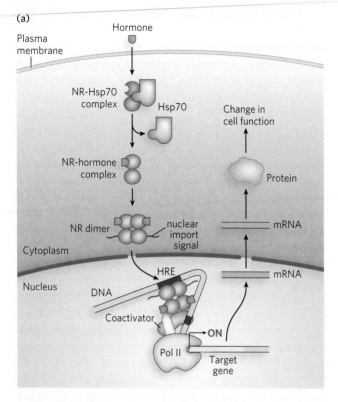

(a)

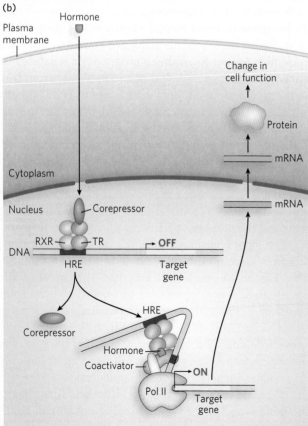

(b)

Responsiveness to a steroid hormone is tissue-specific. The specificity is due in part to transcriptional regulation of the gene encoding the hormone receptor. Cells that are not responsive to a particular steroid hormone do not seem to express its receptor. This can be seen experimentally by using radioactive steroid hormone to examine different tissues for accumulation of the hormone in the nucleus. For example, radioactive progesterone accumulates in the nuclei of endometrial cells, which prepare the uterus for pregnancy, but not in the nuclei of other tissues such as muscle.

Nonsteroid Hormones Control Gene Expression by Triggering Protein Phosphorylation

Nonsteroid hormones, grouped together as chemically distinct from steroid hormones, cannot cross the plasma membrane. Instead, they deliver their regulatory message via a cell surface receptor. We saw in Chapter 19 how the effects of insulin on gene expression are mediated by a series of steps leading ultimately to activation of a protein kinase that phosphorylates specific DNA-binding proteins. Phosphorylation alters the ability of the proteins to act as transcription factors (see Highlight 19-1). This general mechanism mediates the effects of many nonsteroid hormones on gene regulation.

A widely used mechanism of signal transduction for many nonsteroid hormones and other ligands involves the action of **G protein–coupled receptors (GPCRs)** that span the plasma membrane. In this pathway, a transmembrane GPCR binds the signal molecule on the outside of the cell, and binding activates a guanine nucleotide–binding protein (G protein) on the cytoplasmic side of the membrane. G proteins function as molecular switches: when bound to GTP they are active; on hydrolysis of the GTP to GDP they are inactive. When activated, G proteins promote the phosphorylation of proteins that activate gene transcription.

FIGURE 21-21 Steroid hormone receptor action. Steroid hormones diffuse across the plasma membrane and associate with a type I or type II nuclear receptor. (a) The type I nuclear receptor (NR), located in the cytoplasm, is complexed with a heat shock protein (Hsp70). Hormone binding releases Hsp70, and the NR dimerizes and exposes a nuclear import signal sequence. The NR-hormone complex then enters the nucleus and binds to a hormone response element (HRE) to activate transcription. (b) Type II nuclear receptors are bound to DNA whether or not the hormone signal is present. For example, the thyroid hormone receptor (TR) forms a heterodimer with the protein RXR to bind the HRE, but it is inactive without thyroid hormone. When the hormone enters the cell and the nucleus and binds the complex at the HRE site, it activates gene transcription.

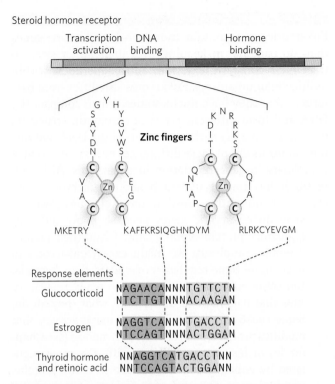

FIGURE 21-22 Structural organization of steroid hormone receptors and hormone response elements. Nuclear receptors are multidomain proteins (top, showing the three domains) that bind steroid hormones and DNA to activate transcription. As shown in the enlarged structure of the DNA-binding region (middle), two adjacent zinc fingers bind to the HRE in the DNA (bottom; the binding regions are indicated by dashed lines). The receptors bind the DNA as dimers. The HREs of several steroid hormone receptors are inverted repeats (highlighted in yellow and green).

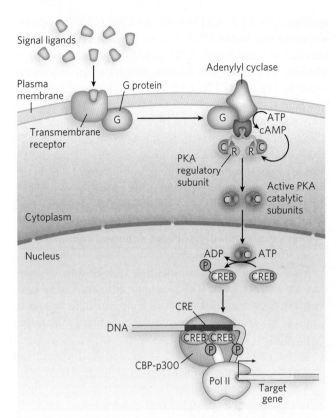

FIGURE 21-23 Gene expression regulated by protein phosphorylation and cAMP. Cyclic AMP–dependent protein kinase A (PKA) is repressed by a regulatory subunit of the adenylyl cyclase holoenzyme and becomes active only on binding of cAMP to this subunit. The cAMP is produced when a signal molecule binds a transmembrane receptor and induces it to activate adenylyl cyclase. (Several steps are omitted here.) Once active, PKA catalytic subunits enter the nucleus and phosphorylate various target proteins, such as CREB, which then recruits RNA polymerase to DNA.

The many different types of nonsteroid ligands that function through GPCRs include olfactory molecules such as odorants and pheromones; peptide hormones such as insulin, calcitonin, follicle-stimulating hormone, and oxytocin; neurotransmitters such as dopamine, epinephrine (adrenaline), and acetylcholine; glucagon; prostaglandins; and leukotrienes. Ligand binding to GPCRs triggers a wide variety of physiological processes. For example, serotonin and dopamine act through GPCRs in the mammalian brain to regulate mood and behavior. Glucagon and prostaglandins bind to GPCRs to trigger changes in metabolism and contraction of smooth muscle. The malfunctioning of GPCRs is associated with a range of human diseases, and they are the target of more than 25% of pharmaceuticals used in medicine.

A G protein-coupled pathway is shown in **Figure 21-23**. First, a signal ligand binds to and activates a surface receptor—the GPCR—that spans the plasma membrane. The signal is then transduced through a G protein to activate adenylyl cyclase, the enzyme that converts ATP to cyclic AMP (cAMP), leading to elevated levels of cytosolic cAMP. Recall from Chapter 20 that bacteria also use cAMP as a signal molecule; the cAMP binds directly to a transcription activator such as CRP. Eukaryotic cells use cAMP in a very different way. Instead of binding directly to a transcription factor, cAMP acts as a **second messenger** that carries a message received from outside the cell (from the first messenger) to proteins inside the cell. The target of cAMP is a kinase called cyclic AMP–dependent protein kinase A (PKA). It is bound in the cytoplasm by a regulatory subunit of the adenylyl cyclase holoenzyme that inhibits its kinase activity. When cAMP binds to the regulatory protein, the protein dissociates, releasing active PKA.

PKA acts on many different target proteins, leading to the activation or repression of various sets of genes. Figure 21-23 shows the activation of CREB (cAMP-responsive element-binding protein) by phosphorylation. CREB is a transcription activator that is inactive when unphosphorylated, but when phosphorylated and activated by PKA, CREB binds its CRE (cAMP-response element) site in the DNA. It then activates transcription through a coactivator, CBP (CREB-binding protein) of the CBP-p300 complex (shown as part of an enhanceosome in Figure 21-17). CBP is a coactivator for numerous genes, including genes encoding other transcription activators, and it functions in many organs. Most of its effects are still unknown. The most widely studied CREB functions are related to the brain, where CREB is implicated in the formation of long-term memories.

SECTION 21.3 SUMMARY

- Insulators are DNA sequences that prevent transcription factors bound at distant enhancers from activating the wrong promoters.

- Enhanceosomes are stable, tightly folded nucleoprotein complexes in which cooperating activators integrate regulatory information from multiple signals to produce a single transcriptional outcome at the target promoter.

- Some genes are blocked from active transcription within regions of densely packed heterochromatin. The formation of heterochromatin requires small RNAs, as well as proteins that condense the DNA.

- In imprinting, which occurs in the genes of some higher eukaryotes, the expression of an allele derived from one parent is shut down. Imprinting is an epigenetic process based on nucleosome modification patterns and DNA methylation.

- Gene dosage compensation, required because of the different number of X chromosomes in males and females, is achieved in one of three ways, depending on the organism. A protein-RNA complex covers the X chromosome(s) to inactivate one female X chromosome, double the expression of the single male X chromosome, or halve gene expression from each female X chromosome.

- Steroid hormones control the transcription of specific genes by interacting with intracellular receptors that are transcription activators. Hormone binding triggers interaction of receptor proteins with additional transcription factors. Hormone-receptor complexes bind hormone response elements in the DNA, altering gene expression.

- Nonsteroid hormones and other signal molecules regulate genes through binding to cell surface receptors, triggering phosphorylation of "second messenger" proteins that leads to modulation of gene expression.

? UNANSWERED QUESTIONS

Eukaryotic cells contain more DNA and more genes than do bacteria, in keeping with their larger size, intracellular compartmentation, and cooperation within multicellular organisms. Eukaryotes also differ from bacteria in that they have nucleosomes, which compact the DNA and form different types of chromatin structure, depending on epigenetic alterations of the DNA and histone subunits. All of these differences necessitate greater complexity in gene regulation in eukaryotes. Although research has taught us much about eukaryotic gene expression, numerous questions have yet to be answered.

1. Why do eukaryotic genes need so many different regulatory protein–binding sites? Given their greater genomic complexity, we might expect eukaryotes to need more gene-regulatory elements than bacteria. But some eukaryotic genes have so many regulatory sites that it is hard to understand what they all do. Some coactivators even have enzymatic activity that modifies proteins, such as RNA polymerase (see Chapter 16) or histones (see Chapter 10). For genes regulated by enhanceosomes, why are so many proteins required to come together to activate a single gene? An exciting area of study will be to gain a deeper understanding of how transcription modulators function.

2. Do different gene-regulatory processes intertwine? Transcription, mRNA processing, replication, recombination, and repair all occur in the nucleus. It seems possible that additional levels of gene regulation might be achieved by interconnections among these different processes. There is also evidence that transcription and mRNA splicing are coordinated. Future studies are likely to reveal increasingly complex regulatory networks among these diverse processes.

3. How is heterochromatin assembled and regulated? The role of RNA-mediated silencing machinery in assembling heterochromatic regions of a chromosome is a fascinating topic. Studies suggest that different areas of heterochromatin may have unique mechanisms of formation, depending on their location along the chromosome. Indeed, heterochromatin formation in X chromosome inactivation occurs by a different process than heterochromatin formation at a centromere. Understanding the generation of this important epigenetic silencing mechanism and how heterochromatin formation is regulated during differentiation are important avenues of future research.

HOW WE KNOW

Transcription Factors Bind Thousands of Sites in the Fruit Fly Genome

Li, X., S. MacArthur, R. Bourgon, D. Nix, D.A. Pollard, V.N. Iyer, A. Hechmer, L. Simirenko, M. Stapleton, C.L. Luengo Hendriks, et al. 2008. Transcription factors bind thousands of active and inactive regions in the *Drosophila* blastoderm. *PLoS Biol.* 6(2):e27, doi: 10.1371/journal.pbio.0060027.

Mark Biggin *[Source: Courtesy Mark Biggin.]*

Until recently, much of the research on transcription factor binding to DNA focused on experiments with purified proteins and short DNA sequences in vitro. Mark Biggin, of the Lawrence Berkeley National Laboratory, wondered how transcription factors might interact with DNA in living cells. Using *Drosophila* embryos (undergoing a transcriptionally controlled program of anterior-posterior segmentation) and the ChIP-Chip method (see Figure 10-21), Biggin and his colleagues set out to identify the binding sites for six transcription factors known to be active at this stage of fruit fly development. In the ChIP (chromatin immunoprecipitation) part of the experiment, chromatin from the embryos was chemically cross-linked to bound proteins, then purified by precipitation with antibodies to the six transcription factors. Next, in the Chip (DNA microarray chip analysis) part, the DNA in the immunoprecipitated samples was identified using microarray chips containing short DNA segments corresponding to every sequence in the fruit fly genome.

The results were surprising (Figure 1). The six transcription factors were bound to several thousand DNA segments located near half of all the protein-coding genes in the *Drosophila* genome! These binding sites corresponded to many more sequences, and many more genes, than the transcription factors were thought to regulate, based on DNA-binding preferences determined in vitro. However, only some of the in vivo binding sites showed up repeatedly in the data analysis, indicating that these sites are frequently occupied by the transcription factors. These high-occupancy sites correspond to DNA targets that are almost certainly regulatory, given their proximity to genes that are activated during fruit fly development. The remainder of the in vivo binding sites are less frequently bound, suggesting that they may not be used to regulate transcription. Instead, these may represent sites where transcription factors can bind nonproductively, perhaps as part of their search for higher-affinity sites along the chromatin. Biggin and coworkers' study should open the way for further investigation of the binding and regulatory roles of transcription factors in the context of chromatin-packaged DNA, as well as the myriad other proteins and regulatory factors found in vivo.

Transcription factors	mRNA localization in embryo	Number of sites bound by transcription factor
Bicoid		692
Caudal		1,392
Hunchback		1,789
Knirps		199
Giant		966
Krüppel		3,415

FIGURE 1 The patterns of mRNA expression for six transcription factors in the *Drosophila* embryo show that each factor is expressed in a unique subset of cells. The fruit fly embryos are shown with the anterior end to the left and the dorsal surface at the top. *[Source: Data from X. Li et al., PLoS Biol. 6(2):e27, doi: 10.1371/journal.pbio.0060027, 2008, Table 1 and Fig. 1. Courtesy Mark Biggin.]*

HOW WE KNOW

Muscle Tissue Differentiation Reveals Surprising Plasticity in the Basal Transcription Machinery

Deato, M.D.E., and R. Tjian. 2007. Switching of the core transcription machinery during myogenesis. *Genes Dev.* 21:2137–2149.

Hu, P., K.G. Geles, J.H. Paik, R.A. DePinho, and R. Tjian. 2008. Codependent activators direct myoblast-specific MyoD transcription. *Dev. Cell* 15:534–546.

Like many tissue-development processes, muscle differentiation begins with the development of progenitor cells into cells with more specialized functions. In mammalian muscle, myoblasts, the precursor cells, differentiate into myotubes, which subsequently form the muscle fibers of skeletal muscle tissue. The transformation of myoblasts to myotubes involves both selective gene silencing and gene activation pathways. Transcriptional regulation in these cells has long been known to require cell type–specific basic helix-loop-helix activator proteins, and many researchers suspected that these activators somehow modify the function of the basal transcription machinery in developing muscle.

To test this idea directly, Robert Tjian and his colleagues examined mouse myoblasts to determine which transcription factors are important for the differentiation process. The researchers used the Western blot method (see Figure 7-25), treating cell extracts with antibodies that recognize specific transcription factors, including the TATA-associated factors (TAFs) and TATA-binding protein (TBP) components of the TFIID general transcription factor complex, as well as TAFs present only in certain cell types. Because TFIID is part of the basal transcription machinery thought to be common to all cells, Tjian and coworkers expected to find it in muscle cells harvested at all stages of differentiation, along with variable levels of muscle-specific TAFs.

What they discovered instead was that an alternative form of general transcription factor complex, containing the activator proteins TAF3 and TRF3 in place of TFIID, initially coexists with the TFIID-containing core complex in myoblasts. As the cells differentiate into myotubes, however, TFIID decreases to undetectable levels, while TAF3 and TRF3 levels are maintained and eventually become dominant (Figure 2a). When Tjian and colleagues used short interfering RNAs (siRNAs) to reduce the amount of either TAF3 or TRF3 in myoblasts (in experiments not shown here), the expression of the muscle-specific protein MyoD also dropped, and muscle differentiation was compromised. These effects could be reversed by supplying fresh TAF3 and TRF3 to depleted myoblast cells.

These findings implicate TAF3-TRF3 complexes in the transcription of proteins central to the muscle cell differentiation pathway (Figure 2b). More importantly, they suggest that previously unexpected changes in the basal transcription machinery are required for the widespread changes in transcription patterns responsible for cellular differentiation in higher eukaryotes.

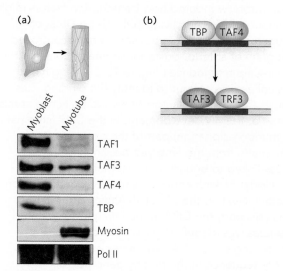

FIGURE 2 (a) Gels resulting from Western blot analysis of TFIID components (TAFs and TBP) involved in differentiation of myoblasts to myotubes in mouse muscle tissue. TFIID is represented by its component TBP; the TAF3-TRF3 complex is represented by TAF3. (b) A proposed model for cell differentiation from myoblast to myotube cells. A core transcription initiation complex including TAF3 and TRF3 functionally replaces the canonical TFIID complex in myotube cells, switching on the unique transcription pattern required during cell type–specific terminal differentiation. *[Source: (a) Data from Maria Divina E. Deato and Robert Tjian. Switching of the core transcription machine during myogenesis. (2007) Genes Dev.21: 2137–2149. Figure 1b. © Cold Spring Harbor Laboratory Press.]*

KEY TERMS

transcriptional ground
   state, 728
heterochromatin, 729
euchromatin, 729
hypersensitive site, 729
enhancer, 731
upstream activator
   sequence (UAS), 731
general (basal)
   transcription factor, 733
DNA-binding
   transactivator, 733
coactivator, 733
TATA-binding protein
   (TBP), 734
high-mobility group
   (HMG) protein, 734
Mediator complex, 735
combinatorial control,
   736

insulator, 744
enhanceosome, 744
gene silencing, 745
imprinting, 745
parthenogenesis, 747
dosage compensation,
   748
Barr body, 748
X chromosome
   inactivation, 748
X inactivation center
   (XIC), 748
dosage compensation
   complex (DCC), 748
hormone response
   element (HRE), 749
G protein–coupled
   receptor (GPCR), 750
second messenger, 751

PROBLEMS

1. In eukaryotes, most genes are normally turned off, and RNA polymerases do not function without activation. In bacteria, RNA polymerase can transcribe almost any gene in the absence of bound inhibitors. Suggest a few reasons for this difference between bacteria and eukaryotes.

2. The regulatory proteins in eukaryotes bind to DNA sequences of about the same length as those bound by bacterial regulatory proteins. However, the genomes of eukaryotes are generally orders of magnitude larger than those of bacteria. What effect does this have on the strategy of eukaryotes for regulating a particular gene?

3. A histone acetyltransferase (HAT) is activated, transferring acetyl groups to histones in a particular region of the genome. What amino acid residues in histones are generally modified by HATs? What is the likely effect of the modifications on the transcription levels of genes in that region? What enzymes reverse the effects of HATs?

4. Optimal activation of transcription of the *GAL* genes in yeast requires the function of two proteins: Gal4p and Gal11p. Elimination of either protein decreases activation of the GAL promoters. However, inactivation of Gal11p has the additional and dramatic effect of cell lethality. Why might elimination of Gal11p have a greater effect than elimination of Gal4p?

5. What is the phosphorylation state of the yeast protein Mig1 when: (a) glucose and galactose are absent; (b) galactose is present and glucose is absent; (c) glucose is present and galactose is absent; and (d) both glucose and galactose are present?

6. Perhaps 3,000 or more transcription factors participate in the activation of human genes. However, this is far fewer than the number of genes in the human genome (~20,000 to 25,000). Explain how specific gene activation is achieved when genes outnumber gene activators by 10:1.

7. If mice are engineered with a homozygous gene knockout (inactivation) for the gene encoding CTC-binding factor (CTCF), they exhibit an embryonic lethal phenotype. Explain.

8. Enhanceosomes consist of multiple transcription factors that activate transcription at particular genes. The enhanceosomes also often include HMG proteins (see Figure 21-6). Suggest a function for the HMG proteins.

9. Housekeeping genes are those that must be expressed at all times, providing a protein or RNA that is essential for general cellular metabolism. They are often expressed at a low but constant level. If an essential housekeeping gene were experimentally moved from euchromatin to a region of heterochromatin, what would be the likely effect on the cell?

10. Certain genes expressed on the X chromosome in mammals must be expressed at the same levels in males and females. However, there are two copies of each gene in females and only one copy in males. How is X-chromosome gene dosage controlled in mammals?

11. A scientist is studying the function of a type of nuclear steroid receptor protein in mouse cells. She introduces various mutations into the gene encoding the receptor protein and transfers the genes into mice. If mutations are introduced that (a) eliminate the nuclear import signal in the receptor protein or (b) alter the receptor protein surface so that the receptor can no longer interact with Hsp70 protein, how will the molecular pathway of hormone-receptor interaction be altered?

DATA ANALYSIS PROBLEM

Chung, J.H., M. Whiteley, and G. Felsenfeld. 1993. A 5′ element of the chicken β-globin domain serves as an insulator in human erythroid cells and protects against position effect in *Drosophila. Cell* 74:505–514.

12. By the early 1990s, a few examples of insulator sequences had been discovered and characterized in eukaryotic cells. These insulators were mostly found in *Drosophila*. To extend the work to vertebrates, Chung,

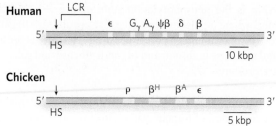

FIGURE 1

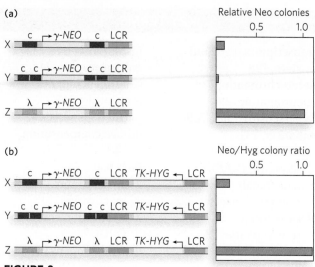

FIGURE 2

Whitely, and Felsenfeld focused on features of the β-globin gene cluster that were conserved in chickens, mice, and humans. They noted that locus control regions (LCRs) near the 5' end of the β-globin gene cluster served to attract enzymes that opened up the chromatin from the 5' to the 3' end of the cluster, a distance encompassing a few hundred thousand base pairs of the human genome. In principle, the LCR can serve that function—attracting enzymes to open up the chromatin to prepare it for transcription—in either direction. However, at the 5' end of the cluster and beyond, the chromatin remained condensed. Something was blocking the chromatin remodeling in that direction. The investigators focused on a prominent and constitutive nuclease-hypersensitive site (HS) at the 5' end of the cluster. The location of that site for the human and chicken systems is indicated by the black arrows in Figure 1, which shows the genes of the β-globin gene cluster in each organism.

(a) What is a nuclease-hypersensitive site in chromatin, and what is its significance?

The investigators constructed a series of vectors in which the constitutive hypersensitive region (denoted c in Figure 2) was placed at sites on either side of a gene conferring resistance to the antibiotic neomycin (γ-NEO in Figure 2). For eukaryotes, researchers generally use the related antibiotic geneticin, or G418, to kill cells. The γ-NEO gene confers resistance. The constructs are labeled X and Y in Figure 2. As a control, the investigators replaced the constitutive hypersensitive site (c) with a fairly random DNA segment of comparable length derived from λ phage (construct Z in Figure 2). They transfected human erythroleukemia cells with these constructs, isolated stably transfected cell lines (in which the construct had integrated at some random site in the genome), and counted the number of colonies produced when the cells were suspended in semisolid agar medium containing G418. As shown in the graph in Figure 2a, the number of G418-resistant colonies decreased when one or two of the chicken hypersensitive sites were inserted between the neomycin-resistance gene and the LCR.

(b) The LCR generally controls genes on its 3' side (to the right as shown in the figures). Why would it affect a neomycin-resistance gene placed on the 5' side?

(c) Why was it necessary to insert the hypersensitive sites on both sides of the neomycin-resistance gene?

(d) Was the chicken hypersensitive site effective in isolating the neomycin-resistance gene?

The investigators made a second series of constructs, shown in Figure 2b. Here, they added a gene for resistance to the antibiotic hygromycin (TK-HYG). This was set up in each construct so that the gene would be expressed constitutively. The constructs were again transfected into the same human cell line, and the numbers of colonies growing in media containing either hygromycin or G418 were tallied. The ratio again revealed that the number of neomycin-resistant colonies was greatly reduced when the neomycin-resistance gene was flanked by the chicken hypersensitive site.

(e) Why was this control experiment necessary?

(f) Given the results presented above, can you conclude that the constitutive hypersensitive site from the chicken is an insulator that affects gene transcription? Justify your answer.

(g) What other experiments would be needed to demonstrate or confirm that transcription was affected?

ADDITIONAL READING

General

D'Alessio, J.A., K.J. Wright, and R. Tjian. 2009. Shifting players and paradigms in cell-specific transcription. *Mol. Cell* 36:924–931.

Michel, D. 2010. How transcription factors can adjust the gene expression floodgates. *Prog. Biophys. Mol. Biol.* 102:16–37.

Basic Mechanisms of Eukaryotic Transcriptional Activation

Arnosti, D.N., and M.M. Kulkarin. 2005. Transcriptional enhancers: Intelligent enhanceosomes or flexible billboards? *J. Cell Biol.* 94:890–898.

Bernstein, E., and S.B. Hake. 2006. The nucleosome: A little variation goes a long way. *Biochem. Cell Biol.* 84:505-517.

Björklund, S., and C.M. Gustafsson. 2005. The yeast Mediator complex and its regulation. *Trends Biochem. Sci.* 30:240-244.

Segal, E., and J. Widom. 2009. What controls nucleosome positions? *Trends Genet.* 25:335-343.

Combinatorial Control of Gene Expression

Campbell, R.N., M.K. Leverentz, L.A. Ryan, and R.J. Reece. 2008. Metabolic control of transcription: Paradigms and lessons from *Saccharomyces cerevisiae*. *Biochem. J.* 414:177-187.

Davidson, E.H., and M.S. Levine. 2008. Properties of developmental gene regulatory networks. *Proc. Natl. Acad. Sci. USA* 105:20,063-20,066.

Juven-Gershon, T., J.Y. Hsu, J.W. Theisen, and J.T. Kadonaga. 2008. The RNA polymerase II core promoter: The gateway to transcription. *Curr. Opin. Cell Biol.* 20:253-259.

Transcriptional Regulation Mechanisms Unique to Eukaryotes

Gaszner, M., and G. Felsenfeld. 2006. Insulators: Exploiting transcriptional and epigenetic mechanisms. *Nat. Rev. Genet.* 7:703-713.

Grzechnik, P., S.M. Tan-Wong, and N.J. Proudfoot. 2014. Terminate and make a loop: Regulation of transcriptional directionality. *Trends Biochem. Sci.* 39:319-327.

Werner, M.H., and S.K. Burley. 1997. Architectural transcription factors: Proteins that remodel DNA. *Cell* 88:733-736.

Wood, A.J., and R.J. Oakey. 2006. Transcriptional control: Imprinting insulation. *Curr. Biol.* 10:R463-465.

22 The Posttranscriptional Regulation of Gene Expression in Eukaryotes

Judith Kimble [*Source: Courtesy Judith Kimble.*]

MOMENT OF DISCOVERY

One of the more thrilling moments in my career happened when our lab uncovered a previously unknown mechanism of gene control. I have long been interested in how stem cells are maintained and how they manage to generate different cell types. *To get at the molecular identity of stem cell regulators*, our lab started genetically and screened for mutations that would help find them. For these studies, we chose the worm *C. elegans*, because we could easily select for regulatory mutations. One particularly exciting group of mutations fell in a single gene and transformed cells beautifully from one fate to another, depending only on incubation temperature. Julie Ahringer, a student in my lab at the time, began to sequence the gene mutants, but found no molecular changes in the gene's open reading frame! This was really puzzling, but she pushed on into noncoding regions and discovered a single base-pair change in the part of the gene corresponding to the 3'UTR of the mRNA.

When Julie tested the effect of introducing wild-type or mutant 3'UTRs back into the worm, she confirmed the effect of that single base-pair change. She soon sequenced the same region in nine independently isolated mutations of the same class, and they all carried 3'UTR mutations falling within a five base-pair region.

This was a truly exhilarating discovery, because at the time, no one expected the noncoding bits of an mRNA to be so important for cell-fate regulation. That breakthrough paved the way for many subsequent studies of 3'UTR regulation, which we now know is a fundamental and conserved mechanism of gene control.

—*Judith Kimble, on the discovery that noncoding regions of mRNA regulate cell fate*

One of the most complex and fascinating processes in molecular biology is how different cell types arise during the development of a multicellular organism. For example, the adult human body contains about 50 trillion (50×10^{12}) cells that originated from a single fertilized egg cell. Almost all these cells contain the same DNA, yet the cells of each organ, and even those within an organ, have vastly different shapes and functions. The differences must reflect gene regulation.

In eukaryotic cells, gene transcription and pre-mRNA processing—splicing, 5′ capping, and 3′ polyadenylation—occur in the nucleus. Only after export to the cytoplasm are mature mRNAs recognized by ribosomes for translation into proteins. This physical and temporal separation of transcription and translation, distinct from the situation in bacterial cells, requires additional steps in the information pathways of eukaryotes. Although these extra steps take time (which is not always available), they provide unique opportunities to impose control.

The early research on eukaryotic gene regulation focused on transcription, particularly transcription initiation. It made practical sense that cells would control gene expression by regulating the first step, avoiding energy expenditure on unneeded transcripts. However, the experimental data have increasingly pointed to an abundance of regulatory mechanisms that occur *after* transcription. In humans and other multicellular organisms, for many genes, transcripts and even proteins are routinely produced that are not immediately used. Instead, the mRNAs and proteins are stored and used later, bypassing the time-consuming transcription and transport steps and thus allowing a more rapid response to cellular needs or metabolic signals.

To a large degree, the importance of posttranscriptional regulation parallels the complexity of the cellular processes that are regulated. Signal transmission in the brain, color patterns in flower petals, and that most complex of all biological processes, the development of a multicellular organism, are all governed by regulatory processes that take place after transcription. In this chapter, we discuss some of the predominant ways in which cells select which mRNAs are to be translated into protein and how much protein is to be made.

We begin with overviews of mechanisms that provide exquisite posttranscriptional control of gene expression levels in the nucleus and in the cytoplasm. We then turn to pathways for regulating groups of genes, including a discussion of the exciting and eminently exploitable discovery of small interfering RNAs (siRNAs) that alter gene expression through a process commonly called RNA interference (RNAi). Next, we discuss embryonic development, a process in which almost all the transcriptional and posttranscriptional regulatory mechanisms described in the last several chapters come

together. Most of the regulatory mechanisms that guide development are highly conserved in eukaryotes, from nematodes to humans. In addition to exemplifying mechanisms of gene regulation, elucidation of developmental pathways has taught us much about evolution and how it generates alterations in function and appearance in organisms. Thus, we end the book where we started—with a discussion of molecular biology from the perspective of evolution.

22.1 POSTTRANSCRIPTIONAL CONTROL INSIDE THE NUCLEUS

The multitude of posttranscriptional regulatory mechanisms in eukaryotes is usefully divided into those that occur in the nucleus and those that occur in the cytoplasm. In the nucleus, mRNA is modified in many ways and prepared for transport to the cytoplasm. In the cytoplasm, the focus shifts to translation. Control mechanisms determine which transcripts are translated, which are stored (and where they are stored), and how long each transcript is present in the cell. Additional controls are layered onto the translation process itself and the fate of the proteins thus produced (see Chapter 18).

Experiments to test which step in gene expression—transcription or translation—is regulated in cells often take advantage of molecular inhibitors that are specific for one step or the other. Two natural antibiotics made by bacteria of the genus *Streptomyces* have been particularly useful. Actinomycin D is a polypeptide that specifically blocks eukaryotic gene transcription by binding to DNA within the transcription initiation complex, preventing transcript elongation by RNA polymerase II (**Figure 22-1a**). Cycloheximide is a small molecule that interferes with the movement of tRNAs during polypeptide elongation on the ribosome, thus blocking protein synthesis (**Figure 22-1b**). These compounds are useful tools for the preliminary dissection of mechanisms of eukaryotic gene regulation. For example, Mark Ashe and Alan Sachs discovered that yeast cells that are starved for sugar rapidly shut down their protein synthesis. To determine whether this regulation is based in transcription or translation, yeast cells growing in a nutrient-rich medium were isolated by centrifugation and transferred to a medium containing no sugar. Protein production in these cells, as measured by the abundance of large ribosome-mRNA complexes, decreased to almost zero within a few minutes. On addition of sugar to the growth medium, protein synthesis was restored. Actinomycin D had no effect on either the inhibition or the reactivation of protein production, showing that no new transcription was needed for this kind of regulation.

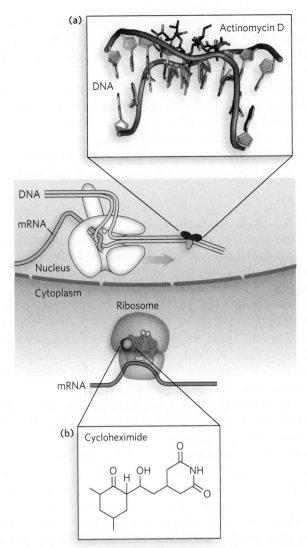

FIGURE 22-1 Compounds that selectively block transcription or translation. (a) Actinomycin D blocks transcription elongation by RNA polymerase II, by inserting itself into DNA. (See Figure 15-8 for the chemical structure of actinomycin D.) (b) Cycloheximide blocks translation elongation by the ribosome. [Source: (a) PDB ID 1DSC.]

In this section, we embark on detailed descriptions of the processes by which mRNAs are altered in the nucleus to control the temporal expression and structure of the final gene products. Regulation by the alteration of mRNA sequences can allow a single gene to yield more than one protein product, thus optimizing the efficiency of information storage in a genome. We focus on some examples of mRNA alterations that illustrate this potential, including mRNA splicing, 3'-end cleavage, and mRNA transport. These processes were introduced in Chapter 16 and are expanded on here. RNA editing, another process that makes important contributions to the coding potential of some eukaryotic mRNAs, is also described in Chapter 16.

Alternative Splicing Controls Sex Determination in Fruit Flies

The splicing of a new mRNA transcript does not always lead to the exclusive excision of all the introns. Instead, for some genes, one or more exons may also be deleted during maturation of a subset of the mRNAs. This process, called **alternative splicing**, creates new forms of the mature transcript that encode versions of the protein lacking one or more peptide segments. In this way, two or more different proteins can be produced from a single coding sequence.

Alternative splicing has a substantial effect on the coding capacity of eukaryotic genomes. It is relatively rare in single-celled eukaryotes such as yeast, but its frequency increases in multicellular organisms. Alternative splicing affects more than 95% of mammalian genes. Anywhere from a few to several thousand different mRNA variants may be generated from a single gene. The existence of alternative splicing implies a level of complexity in the splicing process that goes beyond the signal sequences at the exon-intron boundaries and the spliceosome proteins that bind to them, as described in Chapter 16. The process relies on an extra layer of mRNA sequences and the *trans*-acting proteins that bind to them to determine which splicing events occur in which cells and under what conditions. That extra layer of control can also be inferred from an examination of introns themselves. Whereas most exons fall within a range of 100 to 300 nucleotides in length, introns exhibit much more variation, ranging from about 50 to more than 100,000 nucleotides. In longer introns, canonical splicing signals that could lead to removal of parts of the intron are common, but are ignored by the cellular splicing systems.

The extra layer of control that guides alternative splicing and silences inappropriate splicing events is provided by mRNA sequences in the exons and introns, called exonic splicing enhancers and exonic splicing silencers (ESEs and ESSs, respectively) and intronic splicing enhancers and intronic splicing silencers (ISEs and ISSs). Whether a sequence is an enhancer or a silencer depends a great deal on the proteins that bind to it. A wide range of *trans*-acting RNA-binding proteins have been discovered, each one recognizing one or more of these sequences and thereby affecting the splicing process. There are two main families of alternative splicing regulatory proteins, the Ser/Arg-rich (SR) proteins and the heterogeneous nuclear ribonucleoproteins (hnRNPs). Most of these proteins interact with components of the eukaryotic spliceosome. Their function is illustrated in **Figure 22-2**.

In *Drosophila*, an interesting example of alternative pre-mRNA splicing dictates sexual fate during development. It is a complex system, but not unusually so in the

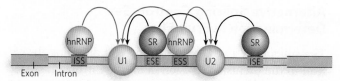

FIGURE 22-2 The regulation of splicing. Whether or not an exon is included in the final, mature mRNA transcript in a eukaryotic cell depends on a range of sequences in the primary transcript and the regulatory proteins that bind to them. The sequences that participate in splice site activation are exonic splicing enhancers (ESEs) or intronic splicing enhancers (ISEs), located in the exon or in the adjacent intron. The sequences that participate in splice site silencing are exonic splicing silencers (ESSs) and intronic splicing silencers (ISSs). The most common families of regulatory proteins that bind to these sequences are the serine/arginine-rich proteins (SR proteins) and the heterogeneous nuclear ribonucleoproteins (hnRNPs). [Source: Data from A. R. Kornblihtt et al. Nat. Rev. Cell. Mol. Biol. 14:153–165. 2013.]

world of alternative splicing. In both male and female flies, each cell contains two copies of the autosomes; female cells also have two X chromosomes, and male cells contain just one X chromosome (unlike in mammals, an X chromosome is not inactivated in fruit flies). The resulting difference in the ratio of X chromosomes to autosomes leads to different levels of expression of the X-chromosomally encoded SisA (*sisterless*) and SisB transcription activators. These regulatory proteins, along with an autosomally encoded repressor called Deadpan (Dpn), act on the *Sex-lethal* (*Sxl*) gene. The Sxl protein is an RNA-binding protein that regulates splicing. In females, with SisA and SisB in twice the amounts found in males, *Sxl* is activated; in males, *Sxl* is repressed.

Two promoters govern *Sxl* expression: P_e (establishment) and P_m (maintenance) (**Figure 22-3a**). The *Sxl* gene has one exon, L3, that contains an in-reading-frame UGA stop codon, and this exon must be removed if a full-length protein is to be produced. When *Sxl* is expressed from P_e, the primary transcript does not include the L1, L2, or L3 exons. Due to the action of regulatory components (e.g., ESS and ESE sites) not yet identified, the default splicing pathway of this transcript connects the exon E1 (encoding the N-terminal 20 amino acid residues of the protein) directly to L4. The splice junctions at the ends of exons L2 and L3 are not recognized by the splicing machinery and thus are deleted along with the neighboring introns. The mature transcript encodes an active Sxl protein with the N-terminal residues encoded by exon E1. When *Sxl* is expressed from P_m, the default splicing pathway eliminates all introns precisely; translation is initiated in L2 and halted at the UGA stop codon in L3, and inactive Sxl protein results.

In females, higher levels of SisA and SisB overcome Dpn repression and activate P_e early in development, generating active Sxl protein. P_m is switched on slightly later, and transcription from P_m is not dependent on SisA, SisB, or Dpn. The Sxl protein, already present at this point in development, is a splicing repressor that binds to the primary P_m-dependent transcript at particular ISSs and prevents the splicing machinery from recognizing the splice junctions on either side of the E1 or L3 exon. L1 is spliced directly to L2, and L2 to L4. In this way, the L3 exon is spliced out of the transcript as part of a larger segment effectively containing two introns, and production of functional Sxl protein continues (now with N-terminal residues encoded by L2).

The Sxl protein does not simply facilitate its own expression, however; it is also needed to regulate the expression of several additional genes downstream in the female developmental pathway (**Figure 22-3b**, left). Sxl regulates the splicing of additional transcripts, including that from the *transformer* (*tra*) gene. Sxl-mediated alternative splicing produces transcripts that encode functional Tra protein, another RNA-binding protein that regulates splicing. The Tra protein activates splicing of pre-mRNA from the *double sex* (*dsx*) gene, such that the truncated protein expressed from this spliced form represses the expression of male-specific genes.

In males, P_e is repressed by Dpn and never activated, due to the lower levels of SisA and SisB than in females. There is no early production of the Sxl protein. When, later, the expression of *Sxl* occurs from P_m, the lack of presynthesized Sxl protein leads to splicing of the P_m-produced transcript via its default pathway, with production of inactive Sxl fragments (see Figure 22-3a). Without functional Sxl protein to regulate splicing of the *tra* gene, no functional Tra protein is produced. In the absence of Tra protein, splicing of the *dsx* transcript produces mRNAs encoding an extended form of Dsx protein that represses transcription of female-specific genes. In the absence of genes enforcing the female developmental pathway, a male-specific pathway is initiated (see Figure 22-3b, right).

This interwoven regulatory cascade, with many steps involving alternative splicing, provides exquisite control over the sexual development of fruit flies by ensuring that sex-specific transcription patterns are maintained.

Multiple mRNA Cleavage Sites Allow the Production of Multiple Proteins

The **3′ cleavage** and polyadenylation of eukaryotic mRNAs is carried out by a protein complex recognizing a site defined, in part, by the sequence AAUAAA. This is a key event in the maturation of mRNAs (see Chapter 16). Many genes have multiple sites for 3′-end cleavage. By regulating which site is used, cells can produce different proteins from a single gene. Sequencing of the human

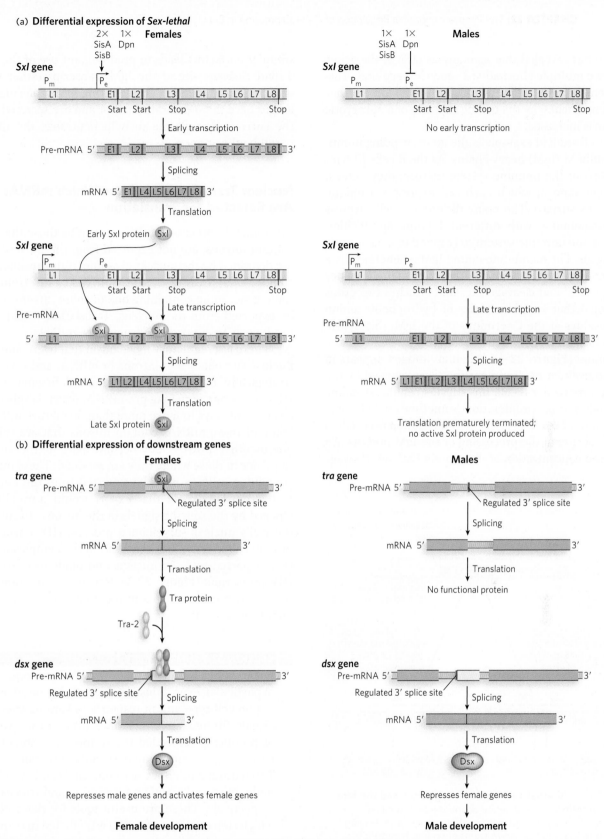

FIGURE 22-3 Alternative splicing of *Sxl* mRNA in male and female fruit flies. (a) The structure of the *Sxl* gene, showing the exons, introns, promoters, start codons, and stop codons, and its splicing patterns. In females, transcription from promoter P_e results in a characteristic splicing pattern and the production of a burst of active Sxl protein (top left). Later transcription from P_m, in the presence of Sxl, continues to produce active Sxl protein (bottom left). In males, transcription from P_e is repressed (top right). The absence of Sxl leads to a different splicing pattern of P_m transcripts and premature termination of Sxl (bottom right). (b) Sxl protein mediates differential expression of genes further downstream in the developmental pathway. In females, active Sxl protein produced from P_m leads to production of proteins that facilitate female development—first Tra protein, which then, with Tra-2, facilitates splicing of the mRNA for active Dsx protein (left). In males, the absence of Sxl leads to a different splicing pattern of *tra* transcripts; a different version of Dsx is produced that leads to male development (right).

genome has revealed that as many as 60% of the genes may have multiple alternative 3'-end cleavage sites. This represents another point of regulation and yet another mechanism whereby the coding capacity of a eukaryotic genome is increased.

A well-studied example is the gene encoding immunoglobulin M (IgM) heavy chains. As the B cells (B lymphocytes) of the immune system mature, they enter a quiescent state in which each cell expresses a unique IgM on its surface. The many different B cells express immunoglobulins with different binding specificities, enabling the immune system to respond to a wide array of antigens. The membrane-bound IgM is generated by a splicing event in the primary transcript that eliminates one of two 3'-end cleavage sites and attaches two exons (M exons) that encode a series of hydrophobic amino acid residues at the C-terminus of the IgM; this hydrophobic sequence serves to anchor the protein in the membrane (**Figure 22-4**). When an antigen appears in the extracellular environment that is recognized by the IgM of a particular B cell, this triggers cellular signaling that leads to rapid proliferation (sometimes called clonal expansion) of that B cell. As the proliferation proceeds, a change occurs in the processing of the IgM mRNAs. An increased concentration of the protein CstF-64 (*cleavage*

*st*imulation *f*actor) leads to predominant use of the first 3'-end cleavage site of the IgM transcript rather than its deletion by splicing. The resulting IgM molecules no longer have the membrane anchor and are secreted into the surrounding plasma to help neutralize the threat posed by the antigen.

Nuclear Transport Regulates Which mRNAs Are Selected for Translation

Incompletely processed mRNAs, such as those that still contain introns, are not exported from the nucleus and are degraded by the nuclear exosome. If this mechanism could be suppressed, various mRNAs made from the same gene, some containing one or more introns, could be exported from the nucleus. Translation would then produce different proteins from these mRNAs.

There is a regulatory mechanism that overcomes the nuclear surveillance of introns in mRNA, and it is used strategically by the human immunodeficiency virus. HIV makes one very long pre-mRNA, which is spliced in a variety of ways to make more than 30 mature mRNAs. Some of these mRNAs contain introns that are spliced out in other mRNAs but become part of the coding sequence in those where they are retained. These intron-containing mRNAs must be exported from the nucleus to be translated, and that export would normally be blocked by the splicing signals in the introns. To subvert the cell's nuclear surveillance process, HIV encodes a protein called Rev. After synthesis in the cytoplasm, Rev is transported into the nucleus and binds to introns in HIV transcripts (**Figure 22-5**). Rev also binds transport receptors, and thus escorts intron-containing viral RNAs out of the nucleus.

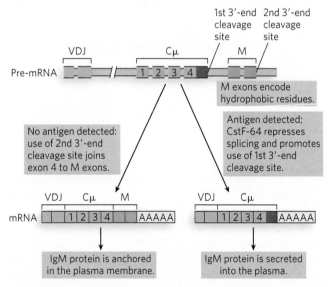

FIGURE 22-4 Alternative 3'-end cleavage sites and the fate of IgM proteins. In the absence of antigen (left, the first cleavage site is spliced out during mRNA processing, leaving the second site and the two M exons encoding hydrophobic C-terminal sequences that anchor the IgM in the membrane. When the cell encounters antigen (right), the splicing reaction that eliminates the first cleavage site is suppressed through the action of CstF-64; the mature mRNA produces an IgM that is secreted. Cμ exons encode the constant region of the IgM heavy chain. VDJ exons encode the variable, diversity, and joining segments of the IgM heavy chain. [*Source: Data from J. Zhao, L. Hyman, and C. Moore*, Microbiol. Mol. Biol. Rev. *63:405–445, 1999.*]

SECTION 22.1 SUMMARY

- Eukaryotic cells, and the cells of multicellular organisms in particular, express just a subset of proteins, depending on cell type and in response to various chemical signals. To maximize versatility, even at the expense of producing unneeded transcripts, cells sometimes regulate gene expression posttranscriptionally.

- Two natural antibiotics, actinomycin D and cycloheximide, block eukaryotic transcription and translation, respectively. These are useful tools for determining which step of gene expression is regulated in response to a particular stimulus.

- Alternative splicing greatly increases the coding capacity of eukaryotic genomes and is especially important in mammals.

- Alternative splicing is influenced by splicing enhancer and splicing silencing sites located in both exons and introns.

(a) **Early HIV infection**

(b) **Late HIV infection**

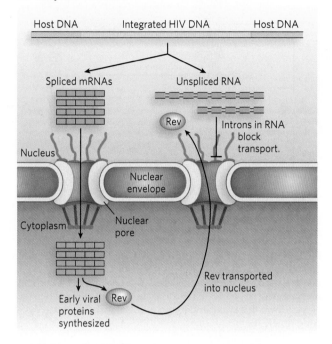
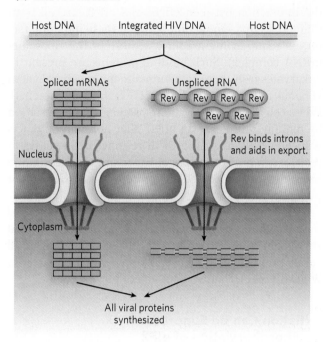

FIGURE 22-5 Regulation of nuclear export by the HIV Rev protein. (a) Early in HIV infection, only fully spliced viral mRNAs, containing the coding sequences for Rev and other proteins, are exported from the host cell nucleus and translated. (b) Once sufficient Rev protein has accumulated and been transported into the nucleus, unspliced viral RNAs can be exported into the cytoplasm. Many of these RNAs are translated into proteins. These and all other viral proteins are packaged into new viral particles.

- In fruit flies, alternative splicing of the pre-mRNA encoding the Sex-lethal (Sxl) protein determines sexual fate during development.
- Alternative selection of sites for 3′-end cleavage determines the fate of IgM molecules produced by B cells.
- HIV produces a protein (Rev) that suppresses cellular intron surveillance and allows export of intron-containing viral transcripts from the nucleus.

22.2 TRANSLATIONAL CONTROL IN THE CYTOPLASM

Regulation at the level of translation assumes a much more prominent role in eukaryotes than in bacteria and is observed in a range of cellular situations. In contrast to the tight coupling of transcription and translation in bacteria, transcripts generated in a eukaryotic nucleus must be processed and transported to the cytoplasm before translation. This can impose a significant delay on the availability of a protein function. When a rapid increase in protein production is needed, a translationally repressed mRNA already in the cytoplasm can be activated for translation without delay.

Translational control is responsible for activating the expression of proteins necessary for cell fate decisions. Translational regulation may play an especially important role in controlling the expression of certain very long eukaryotic genes (a few are measured in millions of base pairs!), for which transcription and mRNA processing can require many hours. Some genes are regulated at both the transcriptional and translational stages, with the latter playing a role in fine-tuning of cellular protein levels. In some non-nucleated cells, such as reticulocytes (immature red blood cells), transcriptional control is unavailable and translational control of stored mRNAs becomes essential. Translational controls are also essential during development, when the regulated translation of pre-positioned mRNAs creates a local gradient of the protein product (see Section 22.5).

Eukaryotes have at least three main mechanisms for regulating translation. First, various initiation factors are subject to phosphorylation by protein kinases. The phosphorylated forms are often less active and generally depress translation in the cell. Second, some proteins bind directly to mRNA and act as translational repressors. Many bind at specific sites in the 3′UTR and interact with other initiation factors bound to the mRNA, or they interact with the 40S ribosomal subunit to prevent initiation. Third, binding proteins can disrupt the interaction between eIF4E and eIF4G (recall from Chapter 18 that interaction between these initiation factors is required for proper assembly of the ribosome-mRNA complex). Such binding proteins are present in eukaryotes from yeast to mammals, and the mammalian versions are known as 4E-BPs (e*IF4E* *b*inding *p*roteins). When cell growth is slow, these proteins limit translation by binding to the site on eIF4E that normally interacts with eIF4G. When cell growth resumes or increases in response to growth factors or other stimuli, the binding proteins are inactivated by protein kinase–dependent phosphorylation.

The variety of translational regulation mechanisms in eukaryotes provides flexibility, allowing focused repression of a few mRNAs or global regulation of all cellular translation. Here we explore several examples in which these mechanisms come into play.

Initiation Can Be Suppressed by Phosphorylation of eIF2

Translation initiation in eukaryotes is a complex process involving multiple initiation factors that recruit ribosomes to mRNAs. Reversible phosphorylation of initiation factors plays a central role in regulating initiation. The phosphorylation is triggered by a wide range of cellular conditions, depending on cell type and function. One of the main phosphorylation pathways involves eIF2. In all eukaryotes, eIF2 is composed of three polypeptide subunits—eIF2α, eIF2β, and eIF2γ—that together bind the initiator tRNA and GTP. When Met-tRNAi^Met binds to the peptidyl (P) site on the 40S ribosomal subunit, GTP is hydrolyzed, and eIF2-GDP dissociates from the initiator tRNA. Recycling of eIF2-GDP to eIF2-GTP requires eIF2B (**Figure 22-6**). If eIF2 is phosphorylated, eIF2B binding to eIF2 is nearly irreversible and the GDP is not dislodged. Because the cell has less eIF2B than eIF2, only a little phosphorylated eIF2 is needed to sequester all the eIF2B, thereby shutting down protein synthesis.

This process has been well studied in mammalian reticulocytes. The maturation of reticulocytes into red blood cells includes destruction of the cell nucleus, leaving behind a hemoglobin-packed cell. Messenger RNAs deposited in the cytoplasm before loss of the nucleus allow for the replacement of hemoglobin. When reticulocytes become deficient in iron or heme, the translation of globin mRNAs is repressed. A protein kinase called HCR (*h*emin-*c*ontrolled *r*epressor) is activated, catalyzing phosphorylation of eIF2α, the smallest subunit of eIF2. When its eIF2α subunit is phosphorylated, eIF2 forms a stable complex with eIF2B, blocking dissociation of GDP after GTP hydrolysis and thus making these initiation

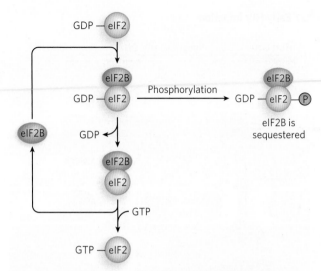

FIGURE 22-6 Halting the recycling of eIF2 by phosphorylation. The recycling of used eIF2 (eIF2-GDP), after it has served its role in translation initiation, is facilitated by a guanine nucleotide exchange factor, eIF2B. Phosphorylation of the eIF2 protein closes down this cycle, and thus controls translation rates, by tying up eIF2B.

factors unavailable for further rounds of translation. In this way, the reticulocyte coordinates the synthesis of globin with the availability of heme.

Phosphorylation of eIF2α regulates translation in other systems as well. For example, a double-stranded RNA–dependent protein kinase (PKR) phosphorylates eIF2α in some cells in response to viral infection. This helps block the translation of viral mRNAs and interferes with the viral life cycle. In yeast, activation of the kinase Gcn2 by nitrogen starvation leads to eIF2α phosphorylation and repression of most translation, until more nitrogen (and the amino acids that incorporate it) becomes available. In an interesting mechanistic twist, eIF2α phosphorylation in yeast induces translation of the transcription factor Gcn4 (which we discuss below).

The 3'UTR of Some mRNAs Controls Translational Efficiency

The 3' untranslated region of an mRNA communicates with the 5' end through protein-protein interactions between factors that bind specifically to the ends of a fully processed mRNA. This communication ensures that the mRNA is fully processed before translation begins. Circularization occurs when eIF4E bound at the 5' terminus and poly(A) binding protein (PABP) bound at the 3' poly(A) tail both interact with eIF4G (**Figure 22-7a**). Initiation of translation requires the recruitment of eIF4G by eIF4E, through a conserved motif in eIF4G that allows interaction with eIF4E. The same motif is used by other proteins to repress translation of certain mRNAs.

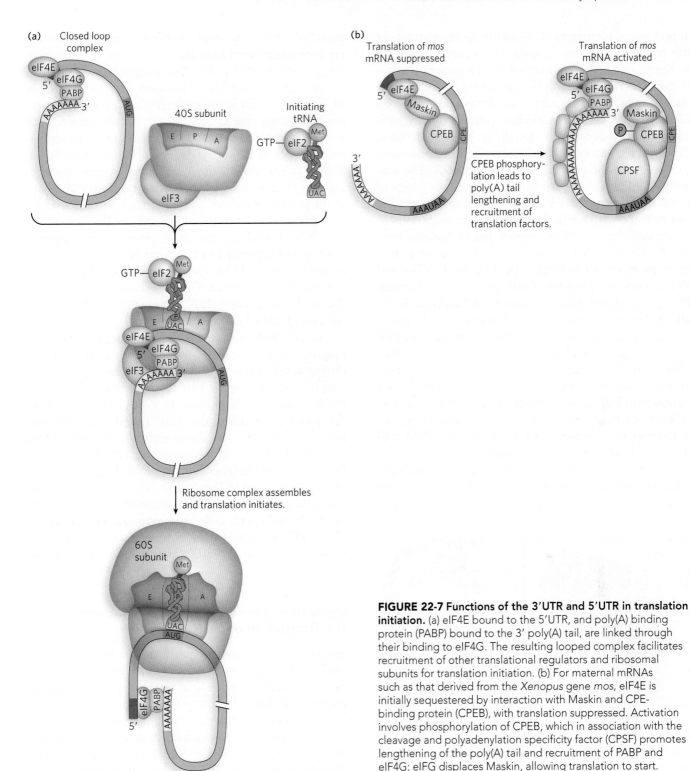

FIGURE 22-7 Functions of the 3′UTR and 5′UTR in translation initiation. (a) eIF4E bound to the 5′UTR, and poly(A) binding protein (PABP) bound to the 3′ poly(A) tail, are linked through their binding to eIF4G. The resulting looped complex facilitates recruitment of other translational regulators and ribosomal subunits for translation initiation. (b) For maternal mRNAs such as that derived from the *Xenopus* gene *mos*, eIF4E is initially sequestered by interaction with Maskin and CPE-binding protein (CPEB), with translation suppressed. Activation involves phosphorylation of CPEB, which in association with the cleavage and polyadenylation specificity factor (CPSF) promotes lengthening of the poly(A) tail and recruitment of PABP and eIF4G; eIFG displaces Maskin, allowing translation to start.

Regulation of translation by protein binding to the 3′UTR is especially important during the development of early embryos. Many maternal RNAs (deposited in the egg cytoplasm during oogenesis) have relatively short poly(A) tails (20 to 40 A residues), and their translation is suppressed until it is needed. Activation requires two sequences in the 3′UTR: the nuclear AAUAAA polyadenylation sequence and the cytoplasmic polyadenylation element (CPE, with consensus sequence UUUUUAUU). CPEs are bound by the protein CPEB (CPE-binding protein), which helps establish translational masking of maternal mRNAs by interacting with a protein known, appropriately, as Maskin. Maskin also interacts with eIF4E, functioning much like 4E-BPs in preventing binding of eIF4E to eIF4G. An example of this mechanism is the regulation of a maternal mRNA in *Xenopus* embryos, transcribed from a gene called *mos* (**Figure 22-7b**). To activate the *mos* mRNA, CPEB is first phosphorylated. This stimulates interaction between CPEB and the AAUAAA-binding protein CPSF (*cleavage and polyadenylation specificity factor*). In turn, CPSF recruits the cytoplasmic polyadenylation enzymes that lengthen the poly(A) tail on the mRNA. The longer poly(A) tails enable the binding of multiple copies of PABP. PABP recruits eIF4G, and eIF4G displaces Maskin and interacts with eIF4E. Translation then begins.

In addition to CPEBs, other types of proteins bind the 3'UTR of mRNA to regulate translation, including proteins of the **PUF family** (named for *Pumilio* and *FBF*, the first two proteins of this type to be discovered). PUF proteins are a highly conserved family of RNA-binding proteins associated with translational control in a variety of organisms, from yeast to mammals. PUF proteins typically include eight consecutive 40 residue repeat sequences, each of which contains characteristic aromatic and basic amino acids. The crescent-shaped structure of PUF proteins reveals two extended surfaces (**Figure 22-8**). Based on the location and effects of mutations, one surface probably binds to mRNA and the other to other regulatory proteins.

Each PUF protein is thought to regulate multiple mRNAs, because experiments have demonstrated the proteins' ability to bind multiple targets. Researchers have engineered fruit flies to express a molecularly tagged version of Pumilio. The tagged protein was purified with its bound RNA partners, and the identities of the bound RNAs were determined using microarray technology (see Chapter 7). The study showed that many of the bound mRNAs shared a short, characteristic sequence in their 3'UTRs. Furthermore, the mRNAs tended to encode functionally related proteins, suggesting that Pumilio acts by binding to sets of mRNAs. Many of the proteins regulated by Pumilio function in developmental pathways considered in Section 22.5.

Once bound by the PUF protein, the targeted mRNAs are typically blocked from efficient translation on the ribosome. Although the mechanism of repression is not completely known, PUF proteins seem to block initiation. In some cases, PUF proteins may also increase the rate of mRNA degradation.

In the nematode *Caenorhabditis elegans*, the hermaphrodite produces sperm and oocytes, successively, during development of the germ line. This cell fate decision is controlled by the PUF family protein FBF (*fem-3 binding factor*), which binds to a site in the 3'UTR of the *fem-3* gene called the PME (point mutation element) (see the How We Know section at the end of this chapter). This binding site was defined by gene mutants first isolated in the laboratory of Judith Kimble (see this chapter's Moment of Discovery). When FBF is not present in the germ-line cells, translation of *fem-3* transcripts proceeds and sperm cells are produced (**Figure 22-9**). When FBF protein is present, it binds the 3'UTR and blocks translation of *fem-3* transcripts, and oocytes are produced.

Upstream Open Reading Frames Control the Translation of *GCN4* mRNA

Some eukaryotic genes are controlled by short open reading frames located upstream from the gene's authentic start codon. These **upstream open reading frames (uORFs)** do not produce functional protein. Instead, they are a gene regulatory mechanism that generally decreases translation by diverting ribosomes, often making them halt and dissociate before reaching the AUG start codon. It might seem that uORFs, rather than regulating levels of gene expression, would simply repress genes under all conditions. However, the effectiveness of uORFs in terminating ribosome activity is altered by phosphorylation of eIF2α. An instance of this type of regulation is observed for the yeast *GCN4* gene (*general control nonderepressible*), encoding a transcription activator that regulates many other genes.

Although the phosphorylation of eIF2α typically down-regulates translation initiation, in *S. cerevisiae*,

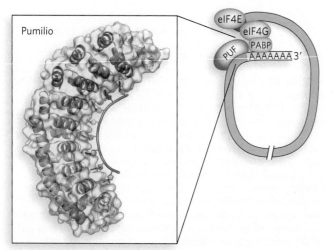

FIGURE 22-8 Structure and function of the PUF family protein Pumilio. The protein is crescent shaped, with 10 repeated α-helical segments, each consisting of about 40 amino acids. The 8 repeats in the middle (colored) are involved in mRNA binding. The 2 repeats on the ends (white) are important to Pumilio function, but do not directly participate in mRNA binding. Binding of a PUF family protein to the 3'UTR of an mRNA, as shown on the right, suppresses protein production either through interference with translation initiation by blocking assembly of translational factors or through promotion of mRNA degradation by recruiting RNA degradation enzymes. [*Source: PDB ID 1M8Y.*]

(a) Wild type: produces sperm, then oocytes

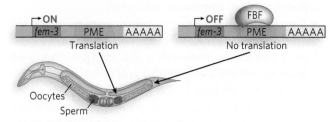

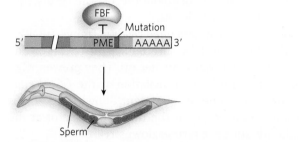

Oocytes
Sperm

(b) PME mutant: produces sperm only

FIGURE 22-9 The regulation of germ-line development in *C. elegans* hermaphrodites. (a) Hermaphrodites produce first sperm cells, then oocytes. The transition is regulated in part by interaction of the FBF protein with the transcript of the *fem-3* gene. When FBF is absent and the *fem-3* gene is activated, *fem-3* transcripts are translated and sperm are produced. When the FBF is present, *fem-3* translation is blocked regardless of the level of *fem-3* transcription, and oocytes are produced. (b) Mutations in the sequence PME (point mutation element) eliminate FBF binding. The mutant worms produce only sperm. The FBF protein is not drawn to scale; it spans a region much wider than the PME. [Source: Data from B. Zhang et al., Nature 390:477–484, 1997.]

low-level eIF2α phosphorylation in response to amino acid starvation induces expression of the transcription factor Gcn4. Because Gcn4 activates transcription of at least 40 genes encoding amino acid biosynthetic enzymes, its induction alleviates nutrient limitation that could otherwise trigger more extensive eIF2α phosphorylation and more general translational repression.

The mechanism of activation involves four short uORFs preceding the Gcn4-coding sequence of the mRNA. Located between 150 and 360 nucleotides upstream from the AUG start codon, the uORFs prevent ribosomes from initiating translation at the *GCN4* start site when nutrients are abundant. This occurs because ribosomes initiate efficiently at these "decoy" open reading frames instead of at the true protein-coding start site farther downstream (**Figure 22-10**). When eIF2α is phosphorylated, however, ribosomes are much less likely to initiate translation in general, and thus they have a greater propensity to continue scanning along a bound mRNA without forming an initiation complex. This circumstance favors initiation at the downstream *GCN4* start site, leading to expression of the Gcn4 protein.

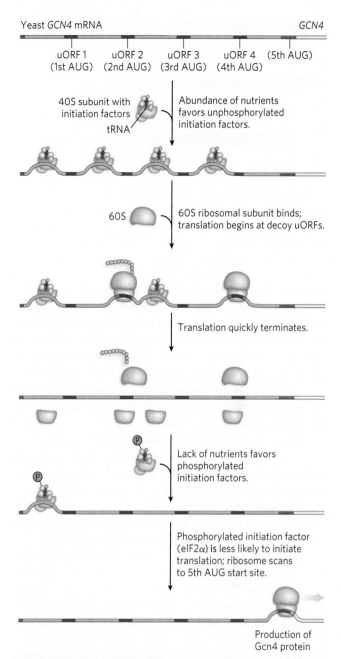

FIGURE 22-10 Translational regulation of *GCN4* in yeast by upstream ORFs. Unphosphorylated eIF2, present when nutrients are abundant, leads to frequent translation initiation at upstream ORFs (uORFS) and little initiation at the *GCN4* gene (first three steps). When eIF2α is phosphorylated, translation initiation in general, including at the uORFs, is reduced; initiation at *GCN4* is now more likely (last two steps). [Source: Data from M. Holcik and N. Sonenberg, Nat. Rev. Mol. Cell Biol. 6:318–327, 2005.]

mRNA Degradation Rates Can Control Translational Efficiency

Another important way in which translation is regulated in eukaryotic cells is by the degradation of mRNAs. Translation and degradation of particular mRNAs often show an inverse relationship. Those mRNAs that are efficiently translated tend to be stable in the cytoplasm,

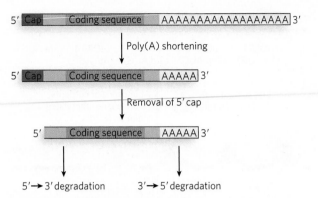

FIGURE 22-11 Eukaryotic mRNA decay. The initial steps involve removal of first the 3' poly(A) tail then the 5' cap. Exonuclease-catalyzed degradation of the remaining RNA is then rapid.

whereas those that are poorly translated tend to be degraded more quickly. The discovery of specific pathways of mRNA degradation has led to a better understanding of the close correlation between mRNA translation and turnover. Decay of mRNAs begins with removal of the 3' poly(A) tail (deadenylation), followed by removal of the 5' 7-meG cap (**Figure 22-11**). The mRNAs are then efficiently degraded by exonucleases, some in the 5'→3' direction, others in the 3'→5' direction.

Removal of the 5' cap involves the formation of macromolecular assemblies of translationally repressed mRNAs bound to decapping enzymes. Known as **processing bodies (P bodies)** (see Figure 16-26), these large cytoplasmic structures are easily seen by light microscopy in cells that have been starved or otherwise stressed to induce general translational repression.

Studies of the composition and formation of P bodies suggest that the rates of mRNA translation and degradation are influenced by the relative stability of the mRNA bound in polyribosomes and in P bodies. Regulatory proteins that bind sequences or structures within related groups of mRNAs serve either to recruit those mRNAs to P bodies or to stabilize their interaction with ribosomes. This is an exciting area of active research. P bodies are part of the conserved translational control machinery in eukaryotic cells, influencing patterns of gene expression in cells as diverse as oocytes and neurons.

SECTION 22.2 SUMMARY

- Eukaryotes have at least three mechanisms of translational regulation: phosphorylation-dependent repression of initiation factors, direct binding of mRNAs by translational repressors, and disruption of the required interaction between eIF4E and eIF4G.
- Reversible phosphorylation of initiation factors plays a central role in regulating translation initiation in response to cellular conditions. When the eIF2α subunit is phosphorylated, eIF2 forms a stable complex

with eIF2B, making these factors unavailable for further rounds of translation initiation.
- The CPE-binding protein binds sequences in the 3'UTR of maternal mRNAs and contributes both to suppressing translation (in conjunction with Maskin, a 4E-BP-like protein) and to activating translation (through interaction with CPSF).
- All PUF proteins have characteristic sequence and structural features that enable their binding to the 3'UTRs of mRNAs. Binding leads to translational repression and/or degradation of the transcript, reducing expression levels.
- In yeast cells, upstream open reading frames (uORFs) in the *GCN4* mRNA, preceding the Gcn4-coding sequence, act as ribosome decoys. They prevent ribosomes from initiating translation at the *GCN4* start site when nutrients are abundant, allowing ribosomes to initiate efficiently at the uORFs instead of at the authentic start site farther downstream.
- Efficiently translated mRNAs tend to be stable in the cytoplasm; those that are poorly translated tend to be rapidly degraded. The decay of an mRNA involves removal of the 3' poly(A) tail, decapping of the 5' end, then exonuclease-catalyzed degradation.
- Cytoplasmic foci, known as P bodies in yeast, are sites of mRNA decapping and degradation.

22.3 THE LARGE-SCALE REGULATION OF GROUPS OF GENES

Although early research on gene regulation often focused on the mechanism by which one specific gene could be controlled, it has become increasingly clear that cells regulate large numbers of their genes together, to bring about changes in cell fate. How this works is an area of active research and discovery. In fact, the new field of systems biology is devoted to understanding how whole networks of genes are controlled and how they function coordinately in cells and organisms. We present here three specific examples of multigene control in which posttranscriptional regulation is a key element.

Some Sets of Genes Are Regulated by Pre-mRNA Splicing in the Nucleus

Most eukaryotic genes, unlike bacterial genes, contain introns. Recent experiments conducted on *Saccharomyces cerevisiae* (budding yeast; referred to by nonscientists as baker's yeast) suggest that pre-mRNA splicing is regulated in response to alterations in nutrient availability. Yeast cells grown in a nutrient-rich medium then shifted to a medium lacking essential amino acids rapidly reduce their level of splicing of transcripts for ribosomal proteins. This was detected by isolating total RNA from the cells at different time points during growth and starvation, then using DNA microarrays to determine changes in the mRNA population. Notably, only the r-protein gene transcripts were affected by the nutrient shift; the splicing of other transcripts was maintained, or even enhanced (**Highlight 22-1**).

Although the mechanism of such targeted splicing regulation has yet to be worked out, it seems that, like transcription, splicing provides eukaryotic cells with a means of rapidly adjusting expression levels of sets of genes in response to environmental triggers. This may be an important reason that eukaryotic cells have acquired and maintained introns over evolutionary time.

5′UTRs and 3′UTRs Coordinate the Translation of Multiple mRNAs

Sequence or structural elements in the untranslated regions of mRNAs provide a mechanism by which sets of transcripts can be controlled. One of the best-studied examples involves the regulation of iron concentration in mammalian cells. Although an essential element for cellular function, iron is also highly toxic, and its intracellular concentration must be carefully regulated.

A controlled level of iron relative to other cellular constituents, or **iron homeostasis**, is achieved through the coordinated expression of proteins responsible for iron uptake, storage, utilization, and export. **Iron response**

HIGHLIGHT 22-1 — EVOLUTION

Regulation of Splicing in Response to Stress

Christine Guthrie [Source: Courtesy Christine Guthrie.]

The removal of introns from eukaryotic pre-mRNAs must occur before the mRNAs can be translated into protein. In higher eukaryotes, most genes have introns, and splicing regulates both when and where proteins are made. Furthermore, alternative splicing can produce different mRNAs and thus distinct proteins from a single initial transcript. In *S. cerevisiae*, however, only 5% of genes contain introns, and most of those genes have just one intron. All splice site sequences in yeast introns are very similar, and alternative splicing is rare. These observations led Christine Guthrie and her colleagues at the University of California, San Francisco, to wonder about the role of introns in the yeast genome: are introns maintained for a reason, perhaps to help cells respond to stressful situations?

To investigate this possibility, Guthrie and her students first examined the yeast genes that contain introns. They noticed that this set of genes includes many that code for metabolic regulators, such as enzymes that control the uptake and use of nutrients, and many ribosomal protein genes (RPGs). In fact, of the 139 RPGs in the yeast genome, 102 are interrupted by at least one intron. The RPGs are thus the largest functional category of intron-containing yeast genes.

Might these introns be playing a regulatory role in the expression of r-proteins? There were some hints that this might be the case. Previous research had shown that when yeast is starved for amino acids, which are essential for synthesizing new proteins, RPG transcription is suppressed and rRNA production and overall protein synthesis are reduced, whereas the production of enzymes required for

amino acid biosynthesis is increased. Guthrie hypothesized that the splicing of RPG transcripts might be regulated in response to amino acid starvation.

To test this idea, Guthrie's research team used DNA microarrays to examine the transcript-specific splicing changes resulting from exposure to two unrelated but environmentally relevant stresses: amino acid starvation and ethanol toxicity. RNA was purified from yeast cells at different times after exposure to each of these stresses and hybridized to microarrays containing DNA fragments representing the yeast genome. This analysis showed that splicing of the majority of RPGs is inhibited within minutes of induced amino acid starvation. By contrast, exposure to high levels of ethanol, which is not known to induce a global repression of translation, has little effect on RPG transcript splicing. Instead, in response to ethanol stress, the splicing of a different set of transcripts is reduced, and the splicing efficiency of a third group of transcripts is enhanced. The specificity of these responses and the speed of their onset—within minutes of stress induction—imply that splicing provides an important means of regulating gene expression in response to environmental stresses. This capacity for transcription-independent regulation could explain the evolutionary retention of introns in these yeast genes.

Since this initial study, several genome-wide surveys using microarray technology have shown distinct patterns of alternative splicing in various fruit fly and mammalian tissues. Furthermore, these experiments have identified RNA sequences that correspond to specific differential splicing events. These observations suggest that proteins controlling specific splicing events rely on an RNA code in their target transcripts and that this RNA code governs the correct set of differentially spliced mRNAs. Future research will focus on discovering the identities of these splicing regulators and understanding how they work together to bring about changes in pre-mRNA splicing in response to the cell's needs.

proteins **(IRPs)** bind to a hairpin structure called the **iron response element (IRE)** in the 5′UTR or 3′UTR of the mRNAs encoding the proteins for iron uptake (the transferrin receptor), storage (ferritin), utilization (aconitase), and export (Fpn1). IRP-IRE complexes formed in the 5′UTR of an mRNA inhibit translation, probably by physically blocking access to ribosomes. In contrast, IRP-IRE complexes formed in the 3′UTR, as is the case for the transferrin receptor, prevent mRNA degradation (**Figure 22-12**). The IRE-binding activity of IRPs is high in iron-deficient cells and low in iron-rich cells. This switch in binding affinity results from iron-sulfur clusters that assemble in the IRPs and thereby block IRE binding only when iron is abundant. In this way, groups of genes

in the iron-response pathway can be controlled together to bring about rapid changes in the relevant protein levels as cellular iron concentrations change.

Conserved AU-Rich Elements in 3′UTRs Control Global mRNA Stability for Some Genes

Large sets of genes can also be regulated by elements that affect the stability of RNA transcripts. Unlike bacterial mRNAs, many eukaryotic mRNAs are stable for hours or days. However, certain eukaryotic genes produce mRNAs with very rapid degradation rates. These genes are among those affected by a system for controlling mRNA stability that involves specific sequences in the 3′UTRs.

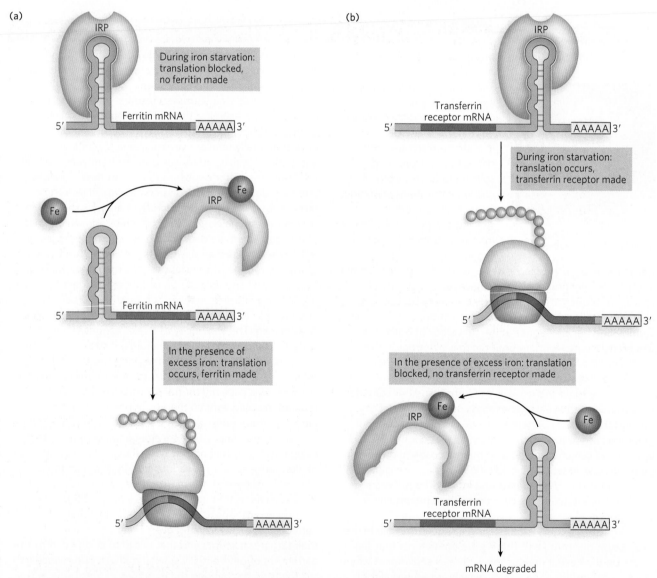

FIGURE 22-12 The function of iron response elements (IREs). In mammalian cells, IREs are bound by iron response proteins (IRPs) when iron levels are low, blocking the production of proteins that use and store iron (including ferritin), while promoting the production of proteins that facilitate iron uptake (the transferrin receptor). (a) IRP binding to the 5′UTR of ferritin mRNA inhibits translation (top). When iron is in excess (bottom), ferritin is produced. (b) IRP binding to the 3′UTR of the transferrin receptor mRNA stimulates translation (top); no transferrin receptor is made when iron is in excess (bottom).

In mammalian cells, mRNAs with sequence elements rich in A and U nucleotides, called **AU-rich elements (AREs)**, are targeted for rapid degradation. The most common ARE motif is AUUUA, and it is often part of a larger AU-rich region such as WWWUAUUUAUUUW, where W is A or U.

ARE-containing mRNAs typically encode proteins that regulate either cell growth or an organism's response to infection, inflammation, and environmental stimuli. For example, mRNAs transcribed from proto-oncogenes (many of which encode proteins that promote cell division) have a very short half-life in the cell. In resting or unstimulated cells, ARE-dependent degradation ensures there is only low-level expression of these potent proteins. When cells respond to particular signals, ARE-containing mRNAs are bound by certain RNA-binding proteins, especially those in the ELAV (*embryonic lethal abnormal visual*) family, first discovered in *Drosophila*. Thus bound, the mRNAs become more stable, and the expression levels of the proteins they encode increase. Notably, in cancerous or chronically inflamed human cells, this ARE-dependent regulation goes awry. This can be caused by increased levels of ELAV family proteins, such as HuR ("Hu" refers to an antibody, sometimes expressed in tumor-associated neurological disorders, that binds ELAV proteins). The increased mRNA stability conferred by the binding of HuR and related proteins allows increased expression of proteins that normally should not be present in cells, contributing to abnormal cell physiology and/or suppression of pathways that control cell growth.

One of the first demonstrations that an ARE sequence controls mRNA stability came from a study of genes for cytokines, cell proliferation factors produced principally by immune system cells that have very unstable mRNAs. In a comparison of the human and mouse gene sequences for a lymphokine (a type of cytokine) known as granulocyte-monocyte colony stimulating factor (GM-CSF), the most conserved sequence was outside the protein-coding region, in the 3′UTR. The protein-coding sequence was only 65% conserved, but a 51 nucleotide AT-rich sequence in the 3′UTR was 93% conserved, suggesting that the 3′UTR sequence plays a functional role. The sequence is also conserved in many other cytokine genes and proto-oncogenes. To test these 3′UTR sequence for function in mRNA stability, the AT-rich sequence in the 3′UTR of the GM-CSF gene was cloned into the 3′UTR of the human β-globin gene. The stability of the resultant mRNA was compared with that of the wild-type globin gene and that of a globin gene into which a GC-rich sequence (from the GM-CSF gene) was inserted. The AT-rich element in the 3′UTR caused a marked loss of mRNA stability (**Figure 22-13**).

With the human genome sequenced, we now know that 5% to 8% of human genes encode ARE-regulated mRNAs. Numerous ARE-binding proteins have been identified, in addition to the ELAV family proteins. As in

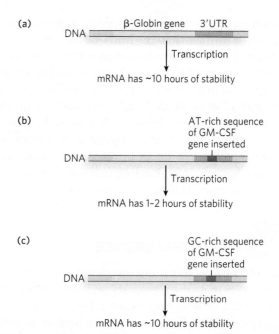

FIGURE 22-13 AU-rich elements (AREs) affect mRNA stability. AREs are binding sites for certain proteins that either stabilize or destabilize the bound mRNA. (a) mRNA transcribed from the β-globin gene is stable for about 10 hours. (b) When a single AT-rich ARE sequence is introduced into the globin gene, such as that derived from the GM-CSF gene, mRNA stability is greatly reduced. (c) When a GC-rich sequence is introduced into the gene, mRNA stability is restored.

the IRP-IRE system described above, these ARE-binding proteins can bind multiple mRNAs. In contrast to ELAV family proteins, which seem to increase mRNA stability, some ARE-binding proteins recruit bound mRNAs to the cell's 3′-deadenylation and 5′-decapping machinery, leading to rapid degradation of the transcript. When cells require increased amounts of the proteins encoded by ARE-containing mRNAs, the levels of destabilizing ARE-binding proteins decline and binding efficiencies drop. A concomitant increase in ELAV family proteins may occur, thereby stabilizing ARE-containing mRNAs and ensuring more translation from those transcripts.

SECTION 22.3 SUMMARY

- Cells often regulate large sets of genes together to bring about changes in cell fate.
- Pre-mRNA splicing provides one mechanism by which cells control the expression of sets of genes. In yeast, splicing of many r-protein RNA transcripts is regulated in response to changes in nutrient availability.
- Both mRNA stability and translational efficiency are controlled by elements in the untranslated regions of some mRNAs.
- In mammalian cells, the expression of proteins involved in iron homeostasis is controlled by the iron response element (IRE), a structure in the 5′UTR or 3′UTR of specific mRNAs. In the presence of iron, IREs bind iron response proteins (IRPs), triggering

increased or decreased expression, depending on the location of the IRE in the mRNA transcript.

- About 5% to 8% of mammalian mRNAs contain AU-rich elements (AREs) in their untranslated regions that bind ARE-binding proteins, leading to either stabilization (if ELAV family proteins) or rapid degradation (if destabilizing ARE-binding proteins) of the mRNA. When cells require a change in ARE-regulated protein levels in response to infection or other stimuli, ARE-binding activity is altered and expression levels increase or decrease.

22.4 RNA INTERFERENCE

Thus far, we have been discussing mechanisms of gene regulation that involve regulatory proteins. These proteins bind to DNA or RNA targets and bring about changes in the efficiency of transcription or translation in response to various stimuli. However, an entirely distinct mode of gene regulation became apparent in experiments initially carried out to examine petal color in petunias (**Figure 22-14**) and in experiments on nematodes. These studies revealed the existence of a regulatory mechanism involving small RNA molecules—**RNA interference (RNAi)**. Researchers discovered that when exogenous RNA was introduced into a eukaryotic cell, either experimentally or naturally, by infection with an RNA virus, the RNA was processed into 21- to 27-nucleotide species known as **short interfering RNAs (siRNAs)**; these mediated the silencing of certain genes. Although RNAi by siRNAs was the first RNA-mediated gene silencing pathway to be discovered, researchers soon found that silencing can also occur in pathways involving small RNAs called microRNAs

(miRNAs), encoded by the cells themselves. The two gene silencing mechanisms are very similar—siRNAs make use of cellular machinery associated with the endogenous miRNAs—but have some important differences. In both cases, though, these small RNAs function by base pairing with mRNAs, often in the 3′UTR, which results in mRNA degradation or translation inhibition. In some cases they can also repress transcription of the targeted mRNA.

RNA interference and related pathways control developmental timing in some organisms. They are also used as a mechanism to protect against invading RNA viruses (especially important in plants, which lack an immune system) and to control the activity of transposons. Small RNA molecules also play a critical, although still undefined, role in the formation of heterochromatin. In addition, RNAi has become a powerful tool for molecular biologists and has attracted attention as a potential therapeutic approach. In 2014, close to 150 trials of RNAi-based therapies were in progress. For example, an RNAi-based therapy has been developed for a rare disease called transthyretin-mediated amyloidosis, which is caused by a mutant liver protein that circulates in the blood and causes nerve damage, and even death. In an early trial, the new therapy reduced levels of the protein by as much as 96%.

Eukaryotic MicroRNAs Target mRNAs for Gene Silencing

In the 1980s and 1990s, numerous experiments were under way, on a variety of organisms, to use DNA or RNA oligonucleotides antisense (complementary) to an mRNA to block protein expression. The hypothesis was that base pairing between the antisense oligonucleotide and the target mRNA would prevent recognition by the translation machinery, or lead to degradation of the hybrid complex, or both. Experiments in plants, however, showed that many transgenic plants containing an artificial gene encoding an antisense RNA failed to suppress expression of the corresponding endogenous gene. Furthermore, in both plants and nematodes, "control" experiments in which the sense strand, rather than the antisense strand, of RNA was

FIGURE 22-14 RNAi pathway effects: petal color selection in petunias. In these petunia flowers, genes for pigmentation are silenced by RNAi. The flower on the left is the wild type; the other two flowers are from transgenic plants with inserted genes that produce siRNAs complementary in sequence to an endogenous gene required for the development of flower color. This leads to the unpigmented, white areas of the petals. [Source: (Part 1) Matzke MA, Matzke AJM. (2004). Planting the Seeds of a New Paradigm. PLoS Biol 2 (5):e133. doi:10.1371/journal.pbio.0020133. PMID 15138502.] Courtesy Dr. Jan M. Kooter. (Part 2) M. A. Matzke and A. J. M. Matzke, PLoS Biol 2(5):e133, 2004, doi: 10.1371/journal. pbio.0020133. PMID 15138502.] Courtesy Dr. Jan M. Kooter. (Part 3)M. A. Matzke and A. J. M. Matzke, PLoS Biol 2(5):e133, 2004, doi:10.1371/journal.pbio.0020133. PMID 15138502.] Courtesy Natlie Doestsch.]

Craig Mello [Source: Courtesy Craig Mello.]

Andrew Fire [Source: Getty Images.]

introduced into cells often showed just as much suppression of the targeted gene as did experiments using the antisense strand. Careful analysis of these phenomena in the nematode system by Craig Mello and Andrew Fire revealed a fascinating explanation for these puzzling observations. The observed RNAi in *C. elegans* resulted from the presence of small amounts of double-stranded RNA that contaminated the preparations of sense or antisense RNA injected into the worms. Researchers subsequently demonstrated in plants, worms, and other eukaryotes that the double-stranded siRNAs were much more efficient at inducing gene silencing than were single-stranded sense or antisense RNAs (**Figure 22-15**). Mello and Fire shared the 2006 Nobel Prize in Medicine or Physiology for this discovery.

Further experiments in plants, nematodes, fruit flies, and mammals revealed many endogenous, or naturally occurring, small RNAs—**microRNAs (miRNAs)**—that correspond to sequences in cellular mRNAs. In fact, hundreds of different miRNAs have been identified in higher eukaryotes (see the How We Know section at the end of this chapter). They are transcribed by Pol II, or in some cases by Pol III, as **primary miRNA transcripts (pri-miRNAs)** with one or more sets of internally complementary sequences that can fold to form hairpinlike structures. The pri-miRNAs are cleaved by the nuclear endonuclease Drosha, a member of the

ribonuclease III family of enzymes; Drosha cleaves RNA molecules that have significant double-stranded regions adjacent to single-stranded extensions of one or both strands. Drosha cleavage of its cellular substrates typically produces shortened hairpins—60 to 70 nucleotides long—with a 5′ phosphate and a two-nucleotide 3′ overhang (**Figure 22-16**). These partially processed

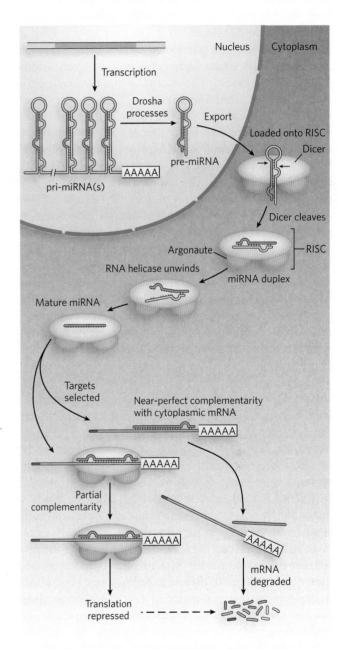

FIGURE 22-16 Drosha and Dicer process precursors to miRNAs. Processing of the transcribed pri-miRNA precursor by Drosha begins in the nucleus, generating a hairpin-shaped pre-miRNA that is exported to the cytoplasm. Dicer, as part of the larger RISC complex, continues the processing in the cytoplasm to generate a short double-stranded miRNA. One strand of the RNA is delivered to the target mRNA by the RISC-miRNA complex, leading to translational repression of the targeted mRNA. In plants, the targeted mRNA is often cleaved and degraded.

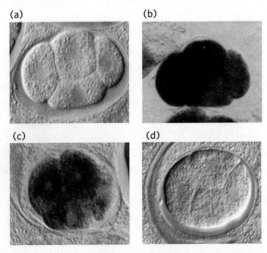

FIGURE 22-15 Silencing of gene expression by RNAi in a nematode. Mex-3 protein is a regulator that is expressed and functions early in nematode development. Four early embryos are shown at identical stages of development, with in situ hybridization used to detect the presence of *mex-3* transcripts. (a) No staining is seen in a control, with no hybridization probe added. (b) Staining reveals the normal pattern of *mex-3* expression in the embryo. (c) Injecting an embryo with an RNA complementary (antisense) to *mex-3* mRNA reduces gene expression somewhat. (d) Injecting an embryo with double-stranded RNA corresponding to *mex-3* reduces gene expression dramatically. [Source: Reprinted by permission from Macmillan Publishers Ltd: Andrew Fire, SiQun Xu, Mary K. Montgomery, Steven A. Kostas, Samuel E. Driver and Craig C. Mello, Nature 391, 806–811 (19 February 1998), Fig. 3. © 1998.]

precursor miRNAs (pre-miRNAs) then bind to export receptor proteins and are transported from the nucleus to the cytoplasm for further processing.

Once in the cytoplasm, pre-miRNAs are cleaved by Dicer, another member of the ribonuclease III family, to generate a 20 to 22 nucleotide miRNA paired with its complementary sequence. Some cells have multiple Dicer enzymes. Some of these enzymes simply cleave long double-stranded RNAs, while others are more specific for double-stranded RNAs that have a loop at one end and are of a particular length. Dicer is part of a larger complex that includes a protein from a family called Argonaute. The overall complex is called the **RNA-induced silencing complex (RISC)**. After cleavage, the miRNA is unwound and the unneeded strand is discarded. The strand complementary to the target is delivered to particular mRNAs. Note that complementarity between the miRNA and the targeted mRNA is typically imperfect, with one or more mismatched or unmatched bases within the duplex. These mismatches usually occur two to eight nucleotides downstream from the 5′ end of the miRNA. The nucleotides at the 5′ end are called the "seed" region and must be perfectly base-paired for efficient miRNA targeting. In animals, the resulting miRNA-mRNA-protein complex triggers RISC to inhibit translation of the bound mRNA, through a process that probably blocks translation initiation. In plants, miRNAs typically induce RISC-mediated cleavage of the targeted mRNA, leading to subsequent degradation.

Although the core components of RISC are conserved across organisms, some interesting differences have been noted. For example, in fruit flies and nematodes, RISC activation requires ATP hydrolysis, whereas in mammalian systems it does not. Also, some of the proteins found in purified RISC are unique to a specific system, suggesting that RISC may be fine-tuned for function in different situations. Also of interest is that Dicer and Argonaute proteins sometimes occur in multiple isoforms, depending on the organism. Humans have just one Dicer enzyme and four Argonaute proteins, whereas nematode worms have two Dicers and many (more than 20) different Argonautes. Plants have at least eight distinct forms of Dicer. Although there may be some functional redundancy among these different protein family members, they may also play discrete roles in the implementation of RNAi-mediated control of gene expression.

Our understanding of miRNA function continues to advance, and the importance of regulation by miRNAs becomes increasingly apparent. Genes for miRNAs represent at least 4% of the genes in the human genome, or more than 1,000 genes. A given miRNA may affect the regulation of up to 400 target genes. In all, the expression of at least 60% of all human protein-encoding genes is regulated in part by miRNAs. The pattern is similar for other vertebrates.

As you may have noticed, thus far we have not mentioned yeast in our discussion of RNAi. This is because *S. cerevisiae*, used so prominently in the study of many other aspects of eukaryotic gene regulation, does not seem to contain any of the enzymatic machinery required for RNAi. Furthermore, attempts to induce RNAi-mediated gene silencing in budding yeast have failed. However, other single-celled eukaryotes, such as fission yeast (*Schizosaccharomyces pombe*) and the human pathogen *Giardia intestinalis*, do express Dicer and Argonaute proteins. At least in fission yeast, small endogenous RNAs are important for silencing centromeric sequences through heterochromatin formation. Why *S. cerevisiae* lost the capacity to use RNAi as a gene regulatory mechanism, or never acquired it, remains a fascinating question.

Short Interfering RNAs Target mRNAs for Degradation

In addition to processing pre-miRNA transcripts exported to the cytoplasm, Dicer can also recognize and cleave long double-stranded RNAs. Double-stranded RNAs can arise naturally from viral infection, or they can be introduced by experimenters. The resulting diced products, 21 to 27 bp in length (depending on the organism) with two-nucleotide 3′ overhangs, are the short interfering RNAs (siRNAs) introduced earlier, and they can regulate gene expression by mechanisms similar to those described for miRNAs. In the cytoplasm, siRNAs can form perfect or imperfect base pairings with targeted mRNA or viral RNA. In either case, these RNA duplexes are contained within the RISC protein machinery that includes Argonaute and multiple additional factors. Perfect base pairing between the siRNA and its target triggers Argonaute-catalyzed cleavage of the RNA duplex, causing eventual degradation of the targeted RNA through the normal pathways, involving 3′ deadenylation, 5′ decapping, and subsequent exonucleolytic cleavage (**Figure 22-17**). Imperfect pairing leads to translation repression, as described for miRNAs, and sometimes to mRNA degradation. In some cases, siRNAs can enter the nucleus and induce heterochromatin formation at the promoters of targeted genes, providing yet a third mechanism of gene silencing.

Much of our current understanding of siRNA function comes from experiments in nematodes, in part because of the ease and efficiency of conducting such experiments in these worms. For example, they can be fed a diet of bacteria engineered to express specified

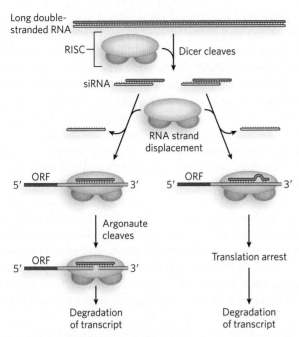

FIGURE 22-17 Alternative fates of siRNA-targeted mRNAs.
Once the mRNA has been targeted, the degree of siRNA-mRNA base pairing leads to either Argonaute-mediated degradation of the mRNA (left) or translational repression, followed by degradation of the mRNA (right).

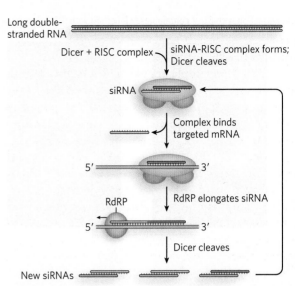

FIGURE 22-18 Amplification of siRNA-mediated gene regulation in nematodes by RNA-dependent RNA polymerases. The siRNAs target an mRNA by normal mechanisms. The annealed siRNA serves as a primer for RNA-dependent RNA polymerase (RdRP), which creates a longer double-stranded RNA. This is then used as a source of more siRNAs.

siRNA precursors, leading to the silencing of expression from siRNA-complementary mRNAs. This technique has been used in many investigations of the siRNA sequence, target site, and phenotypic consequences of siRNA-mediated gene silencing.

Another interesting aspect of RNAi in *C. elegans* is its extraordinary efficiency. Small amounts of siRNAs are sufficient to trigger almost complete silencing of target genes. RNA-dependent RNA polymerases have been implicated in the use of siRNAs to direct the synthesis of RNAs complementary to the targeted mRNAs, with increased numbers of siRNAs then resulting from Dicer-mediated cleavage of the new double-stranded RNAs (**Figure 22-18**). This amplification mechanism has not been detected in fruit flies or mammals, but does seem to occur in fission yeast (*S. pombe*). Furthermore, siRNA-mediated gene silencing in *C. elegans* can be inherited epigenetically. This means that worms in which one or more siRNAs have silenced the expression of a gene will produce offspring in which that gene remains silenced. Genetic and biochemical experiments have shown that an RNA-specific membrane channel is responsible for the transport of siRNAs from parent to fertilized oocyte, where the siRNAs are perhaps maintained by RNA-dependent RNA polymerase–mediated amplification. So far, this ability to pass siRNAs from one generation to the next has not been observed in other organisms.

RNAi Pathways Regulate Viral Gene Expression

Some lines of evidence suggest that RNAi originally evolved to suppress the replication of viruses and transposable elements that use double-stranded RNA as a replication intermediate. For example, plant viruses have acquired various mechanisms of suppressing the RNAi machinery in host cells. Most plant viruses have single-stranded RNA genomes that replicate through double-stranded RNA intermediates and thereby trigger RNA silencing in infected cells. On incorporation of viral-derived siRNAs into the cell's RISC machinery, the viral RNA is specifically targeted for degradation. Genetic experiments have identified numerous viral genes that can limit this effect. In tombusviruses, one of which infects carnations, a small viral protein called p19 was found to play a role in suppressing gene silencing. Traci Hall and her colleagues found that p19 binds specifically to siRNAs, based on their length and the presence of a two-nucleotide 3′ overhang. This p19-siRNA complex cannot assemble into RISCs to direct siRNA-mediated degradation of the viral RNA (**Figure 22-19**).

In an interesting twist on these plant viral siRNA-inhibitory mechanisms, experiments with hepatitis

Traci M. T. Hall [Source: Steven R. McCaw/Image Associates. Courtesy Traci Hall.]

(a)

Infection with CIRV containing
nonfunctional p19 protein

Infection with CIRV containing
wild-type p19 protein

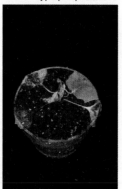

(b)

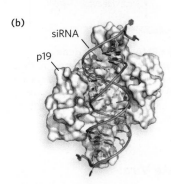

siRNA

p19

FIGURE 22-19 The binding of tombusvirus p19 to siRNAs and blocking of RISC assembly. (a) Plants infected with CIRV (a tombusvirus) that does not (left) or does (right) express the protein p19. (b) The structure of p19 bound to siRNA. [Sources: (a) Reprinted from Cell, Vol. 115, Jeffrey M Vargason, György Szittya, József Burgyán, Traci M. Tanaka Hall, Size Selective Recognition of siRNA by an RNA Silencing Suppressor, 799–811, Fig. 4, parts a and c. December 26, 2003, © 2003, with permission from Elsevier. (b) PDB ID 1RPU]

C virus (HCV) show that the virus takes advantage of a cellular miRNA to stimulate replication. HCV infects human liver cells, which express an endogenous miRNA known as miR-122. HCV is unable to replicate in cells that do not express miR-122. Moreover, HCV is unable to replicate efficiently in cultured liver cells pretreated with oligonucleotides having the antisense sequence to miR-122. Such engineered oligonucleotides, known as antagomirs, have been shown to repress viral replication in HCV-infected chimpanzees and could potentially be developed into antiviral therapeutics for humans (**Highlight 22-2**).

RNAi Provides a Useful Tool for Molecular Biologists

The discovery of RNA interference has an interesting and very useful practical side. If an investigator introduces into an organism a double-stranded RNA corresponding in sequence to virtually any mRNA, the Dicer endonuclease cleaves the duplex into siRNAs. These bind to the mRNA and silence it. Depending on

the gene and the site(s) selected for siRNA targeting, such expression "knockdown" results in a 50% to 90% reduction in the levels of protein produced from that gene (**Figure 22-20**). Removal of a protein is a powerful way to investigate that protein's cellular function. Compared with traditional genetic methods, RNAi provides a much easier way to alter gene expression for experimental or therapeutic purposes. The technique has rapidly become an important tool in ongoing efforts to study gene function, because it can disrupt functionality without creating a mutant organism.

In plants, virtually any gene can be effectively shut down in this way. In nematodes, simply introducing the double-stranded RNA into the worm's diet causes effective suppression of the target gene. The procedure can be applied to human cells as well. Laboratory-produced siRNAs have already been used to block HIV and poliovirus infections in cultured human cells for a week at a

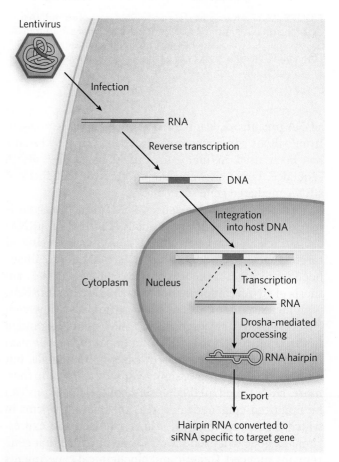

FIGURE 22-20 The siRNA-based knockdown of gene expression in human cells. Lentiviruses, RNA viruses that infect human cells, are used as vectors. A sequence encoding a hairpin homologous to the target gene is cloned into the lentiviral vector (red). In the host cell, the viral RNA is converted to double-stranded DNA by reverse transcriptase, and the DNA is integrated into the cellular genome. Transcription of the integrated gene produces a hairpin RNA that is converted to siRNA specific to the target gene.

Viral Takeover Using a Cell Type–Specific miRNA

Human tissues contain specific microRNAs that can base-pair with complementary sequences in mRNAs to trigger translational silencing and degradation. Even certain animal viruses that have RNA genomes encode miRNAs, indicating that viruses can use these small regulators during their life cycle. But how might viruses deal with, or even take advantage of, a host cell's miRNAs during an infection?

Peter Sarnow [Source: Courtesy Peter Sarnow.]

Many viruses infect and replicate in only certain tissue types, and hepatitis C virus (HCV) is a case in point. This dangerous human pathogen selectively targets liver cells, where it can cause cirrhosis and sometimes cancer. Analysis of cultured human liver cells shows that the liver specifically expresses an miRNA called miR-122; in fact, miR-122 constitutes about 70% of the total miRNA population in liver cells. Stanford University virologist Peter Sarnow and his colleagues wondered how miR-122 regulates mRNA function in both normal and virally infected cells. They first investigated miR-122 expression levels in various tissue and cell types. Using Northern blotting (see Figure 6-32), in which miR-122 was detected using radiolabeled complementary DNA oligonucleotides, the investigators detected miR-122 in cells from intact mouse and human liver and in cultured mouse and human liver (Huh7) cells, but not in human cervical carcinoma–derived HeLa cells or even in human liver–derived HepG2 cells.

Sarnow's research team next tried infecting these different tissues and cultured cells with HCV. Although both Huh7 and HepG2 cells are derived from human liver, HCV could replicate only in Huh7 cells. To find out whether this could be related to the presence of miR-122 in Huh7 cells, the researchers analyzed the 9,600 nucleotide HCV RNA genomic sequence for potential miR-122–binding sites that might enable a successful miRNA–target mRNA interaction. They found two sites in the noncoding regions of the viral RNA genome that are complementary to the miR-122 sequence and are conserved in many different isolates (genotypes) of HCV.

Mutation of those potential miR-122–binding sites in the viral RNA produced HCV variants that could not replicate efficiently in Huh7 cells. These mutant viruses could be "rescued" by introducing into the host Huh7 cells miR-122 variants that restored base-pair complementarity with the mutated viral RNA. Furthermore, the use of small complementary oligonucleotides to bind and sequester

miR-122 in the host Huh7 cells dramatically reduced the amount of HCV RNA replication in these cells (Figure 1). Sarnow concluded that HCV requires the host cell's miR-122 for efficient replication during infection. These findings suggest that miR-122 might be a good target for antiviral intervention.

To test this idea, Henrik Orum and his research team at Santaris Pharma in Denmark synthesized a short oligonucleotide with a sequence complementary to miR-122. To ensure that this oligonucleotide would not be rapidly destroyed in the body, they introduced chemical modifications into the sequence to block the action of degradatory enzymes. When introduced into chimpanzees chronically infected with HCV, the modified "anti–miR-122" oligonucleotide, called SPC3649, produced long-lasting suppression of HCV replication, with no evidence of viral resistance or side effects in the treated animals. This prolonged protection afforded by SPC3649 hints at a new antiviral strategy that takes aim at the underlying mechanism used by HCV to take over human liver cells.

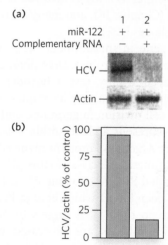

FIGURE 1 Hepatitis C virus replication is reduced in the absence of miR-122 RNA in the host cell. (a) When miR-122 was expressed in a cell harboring HCV, Northern blots showed production of HCV RNAs (lane 1). When an RNA complementary to miR-122 was also present, to sequester miR-122, HCV RNA production declined markedly (lane 2). Northern blots for a housekeeping gene that encodes actin, used as a control, are also shown. (b) Quantification of the results of the same experiments. [Source: (a) From Jopling et al., Modulation of Hepatitis C Virus RNA Abundance by a Liver-Specific MicroRNA, Science 2 September 2005: Vol. 309. no. 5740, pp. 1577–1581, 3c. Reprinted with permission from AAAS.]

time. The widespread use of this technology for medical applications has been limited by the difficulties inherent in getting the highly charged and ribonuclease-sensitive siRNAs across cellular membranes to their targets. New delivery technologies such as lipid nanoparticles and

specialized polymers have begun to solve that problem. Early efforts have focused on diseases of the liver, as this organ naturally takes up a range of material from the blood—as seen in the examples of transthyretin-mediated amyloidosis and HCV infection described earlier.

RNAs Regulate a Wide Range of Cellular Processes

An idea of the complexity and abundance of eukaryotic RNAs has begun to take shape, facilitated by new initiatives such as the ENCODE (*Encyclopedia of DNA elements*) project. Beginning with a pilot project in the early 2000s, and continuing as a much larger effort since 2007, ENCODE seeks to identify and map all functional elements in the human genome. A functional element is defined as any protein- or RNA-coding gene, transcription factor binding site, chromatin structure, or DNA methylation site. New information on RNA has been a highlight of this project. As reported in a 2012 update, more than 62% of the base pairs in the human genome are found in transcribed RNA molecules greater than 200 nucleotides in length. Regulation is a common theme, and the list of RNA-regulated processes keeps expanding. The list of small RNAs now includes the snRNAs involved in spliceosome function and the snoRNAs involved in rRNA processing (both discussed in Chapter 16), miRNAs (Chapter 16 and this chapter), RNAs that guide telomerase function (Chapter 11), RNAs involved in some types of RNA editing (Chapter 16), and many more. We cover two additional classes here.

A class of gene in multicellular eukaryotes called *piwi* (the name derives from a *Drosophila* gene called *P-element induced wimpy testis*) encodes regulatory proteins involved in maintaining incomplete differentiation and stable cell division in germ-line and stem cells. PIWI proteins represent a clade within a larger family that includes the Argonaute proteins involved in miRNA maturation and function. A class of short (21 to 35 nucleotides) RNAs called PIWI-interacting RNAs (piwi RNAs, or piRNAs) has been found, which bind to PIWI proteins and guide them to their RNA targets by base-pairing interactions. The piRNAs are processed from much longer, single-stranded transcripts derived from clusters of genes encoding piRNA precursors. Details of the biogenesis of piRNAs are still being worked out. In the germ line, piRNAs seem to play a variety of roles. These include the silencing of transposons (see Chapter 14) so that their movements around the genome, and the damage this would cause, are limited. The piRNAs also direct the placement of some epigenetic markers (particularly histone modifications and DNA methylation) at a variety of locations in the genome. During germ-line processes such as spermatogenesis, PIWI proteins and piRNAs trigger the opening of heterochromatin so that particular genes can be expressed, followed by piRNA-mediated reestablishment of the heterochromatin state. In mouse testes, there is a burst of piRNA expression in which piRNA production from more than 100 genes is increased about 6,000-fold during the pachytene phase of meiosis.

The piRNA/PIWI complexes carry out their functions by interacting with RNAs that are being transcribed (thus, for example, preventing translation of transposase mRNAs and perhaps triggering their degradation) and recruiting other proteins and enzymes to those locations (e.g., to modify histones or methylate DNA).

The broader screen of human genomic sequences has also revealed the presence of long noncoding RNAs (lncRNAs). These are defined as RNAs longer than 200 nucleotides that do not possess a significant open reading frame to encode a protein. Some of these are transcribed from long introns in protein-coding genes. Others are antisense transcripts, derived from the nontemplate (coding) DNA strand and often overlapping with the protein-coding exon sequences. Still others are intergenic, expressed from the large genomic regions between protein-coding genes. Some of these RNAs undergo splicing and even alternative splicing. Most of the lncRNAs analyzed thus far are found to play some role in differentiation and development. In some cases, elimination of an lncRNA results in a visible developmental defect; in others, the effects are subtle. In mice, for example, an lncRNA called *BC1* is derived from certain retrotransposons and is expressed at high levels in the brain. Elimination of this lncRNA does not result in physical deformity, but it does result in reduced environmental exploration and increased anxiety—behaviors likely to reduce survival in the wild. The investigation of RNA regulatory mechanisms is likely to be a major focus of molecular biology for decades to come.

SECTION 22.4 SUMMARY

- RNA interference (RNAi) silences gene expression by means of short interfering RNAs (siRNAs), using endogenous gene silencing pathways based on genome-encoded microRNAs (miRNAs).
- Endogenous to the cell, miRNAs are encoded by many eukaryotic genomes and cause translational inhibition of the mRNAs to which they bind. Introduced by viral infection or experimentally, exogenous siRNAs use cellular mechanisms associated with miRNAs to silence the expression of particular genes, by suppressing translation or causing mRNA degradation.
- Hundreds of different miRNAs in higher eukaryotes are synthesized by Pol II or Pol III as primary miRNA transcripts (pri-miRNAs), about 70 to 90 nucleotides long. After initial processing in the nucleus, miRNAs are exported to the cytoplasm, where further processing leads to assembly into RNA-induced silencing complexes (RISCs). In contrast, siRNAs are produced entirely in the cytoplasm; like miRNAs, they assemble into RISCs.
- RISCs facilitate base pairing between the small RNA and its complementary sequence in an mRNA,

triggering translational arrest (when the pairing is imperfect) or mRNA degradation (when the sequences are perfectly paired).

- RNAi probably evolved as a mechanism to suppress infection by RNA viruses. Molecular biologists and clinicians use RNAi to alter gene expression for experimental and therapeutic purposes.

- RNAs with regulatory functions are common in eukaryotic genomes and affect a wide range of processes.

22.5 PUTTING IT ALL TOGETHER: GENE REGULATION IN DEVELOPMENT

The patterns of gene regulation that bring about development from a zygote to a multicellular animal or plant are complex and intricately coordinated. Development requires transitions in protein composition and in morphology that depend on tightly coordinated changes in expression of the genome.

How is a complex organism produced, with its many tissues and organs and appendages, from a single cell? Some clues can be found in that single cell—the fertilized egg. More genes are expressed during early development than at any other stage of the life cycle. For example, in the sea urchin, an oocyte (an immature egg cell) has about 18,500 *different* mRNAs, compared with about 6,000 different mRNAs in the cells of a typical differentiated tissue. The mRNAs in the oocyte give rise to a cascade of events that regulate the expression of many genes across both space and time.

The regulatory mechanisms used in development encompass all of the regulatory processes discussed in Chapter 21 and in this chapter thus far. Transcriptional regulation occurs, but posttranscriptional regulatory processes are particularly important.

Development Depends on Asymmetric Cell Divisions and Cell-Cell Signaling

If all cells divided to produce two identical daughter cells, multicellular organisms could never be more than a ball of identical cells. Programmed asymmetric cell divisions are required for different cell fates. Cell-cell signaling also helps guide the eventual differentiation of tissues and organs with various functions. Asymmetry in the developing embryo is thus created in several ways (**Figure 22-21**).

Asymmetry within the cells themselves takes the form of gradients of mRNAs and proteins that define critical axes (posterior-anterior, dorsal-ventral). In the developing oocyte, some gradients are established by deposition of mRNAs at one end or the other. Active transport in the cell can also contribute to generating

(a) Distribution of cellular components (mRNA and protein) creates intracellular asymmetry

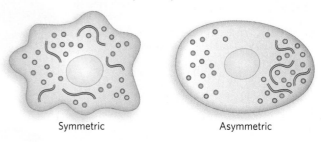

Symmetric   Asymmetric

(b) Cell-cell signaling creates extrinsic asymmetry

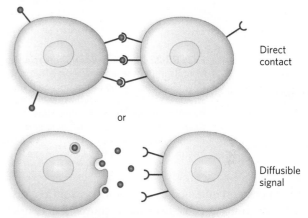

Direct contact

or

Diffusible signal

FIGURE 22-21 Several ways to generate asymmetry in a developing embryo. (a) Intrinsic asymmetry reflects the existing distribution of cellular components, especially mRNA and protein. The asymmetries are either inherent to the developing oocyte or created during fertilization. (b) Extrinsic asymmetry is generated by cell-cell signaling. Although asymmetry is not necessarily created in any given cell, the cell-cell signals alter the fate of a cell or a group of cells in the embryo, contributing to embryonic asymmetry. The signals can involve direct cell-cell contacts or the action of a secreted, diffusible signal.

a gradient. Fertilization can trigger events that create additional gradients in the fertilized egg (zygote). In many organisms, these gradients dictate different cell fates even for the daughter cells of the first cell division.

It is not enough to create a gradient in the cell, however. The mitotic spindle must also be aligned along the same axis as the gradient, so that the cell division occurs on an axis perpendicular to the gradient (see Chapter 2 for a reminder about mitotic cell division). Ensuring the proper alignment of the mitotic spindle in particular cell divisions is the function of some proteins critical to development.

In the developing embryo, cell-cell signaling generates additional asymmetry as development proceeds (see Figure 22-21). Direct contact between the lipids and glycoproteins on one cell surface and the receptors on another can guide changes in gene expression in the receptor-bearing cell. Some signals act at longer distances: diffusible molecules secreted by one cell or

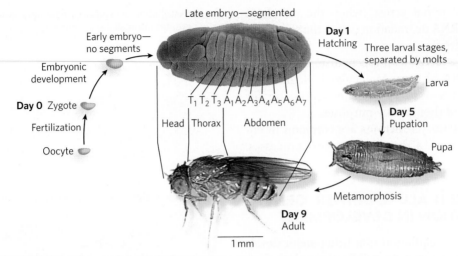

FIGURE 22-22 The fruit fly life cycle. The adult *Drosophila* is radically different in form from its immature stages, a transformation that requires extensive alterations during development. By the late embryonic stage, segments have formed, each containing specialized structures from which the various appendages and other features of the adult fly will develop. The segmented late embryo is enlarged compared with the other stages, to show detail. [*Sources: Top photo from F. R. Turner, Department of Biology, Indiana University, Bloomington. Other photos from Prof. Dr. Christian Klambt, Westfälische Wilhelms-Universität Münster, Institut für Neuro- und Verhaltensbiologie.*]

group of cells and detected by receptors on another, distant cell or group of cells. Ever more complex networks of signaling molecules and gene regulators are created as development proceeds.

Characteristic Stages of Development Several organisms became important model systems for the study of development because they are easy to maintain in a laboratory and have relatively short generation times. These include nematodes, fruit flies, zebra fish, mice, and the plant *Arabidopsis thaliana* (see the Model Organisms Appendix). The discussion here focuses on developmental pathways in the fruit fly. Our understanding of the molecular events during development of *Drosophila melanogaster* is especially well advanced and can be used to illustrate patterns and principles of general significance, and highlight the mechanisms of gene regulation that govern this complex process.

Multicellular eukaryotes develop in a process that begins with the union of an egg and a sperm cell by fertilization, to create a zygote. The egg cell has been preprogrammed by the deposition of maternal mRNAs in gradients, such that concentrations of certain maternal mRNAs vary greatly from one end of the oocyte to the other. On fertilization, cell division begins. Early in development, the fate of particular cells is determined by the concentration of maternal mRNAs, as well as by the actions of regulatory genes. As development proceeds, cascades of regulatory genes guide the various cell lineages as different tissue types develop. Although the regulatory genes are numerous, they generally fall into a

small number of classes that are highly conserved, from nematodes to fruit flies to humans. Signaling pathways and processes are also highly conserved.

The life cycle of the fruit fly is relatively complex, and the patterns are conserved in a wide range of multicellular eukaryotes. Complete metamorphosis occurs during progression from embryo to adult fly (**Figure 22-22**). The final structure of the adult is forecast by features that are evident in the embryo at a very early stage. One of the most important characteristics of the embryo is its **polarity**: the anterior and posterior ends of the animal are readily distinguished, as are its dorsal and ventral surfaces. The fly embryo also exhibits the key characteristic of **metamerism**, division of the body into serially repeating segments, each with characteristic features. During development, these segments become organized into head, thorax, and abdomen. Each segment of the adult thorax has a different set of appendages. The development of this complex pattern is genetically controlled, and pattern-regulating genes—almost all with close homologs, from nematodes to humans—dramatically affect the organization of the body.

The *Drosophila* egg, with its 15 nurse cells, is surrounded by a layer of follicle cells (**Figure 22-23**). As the oocyte matures (before fertilization), mRNAs and proteins originating in the nurse and follicle cells are deposited in the egg cell, where many play a crucial role in development. After the fertilized egg is laid, the nucleus divides and the nuclear descendants continue to divide in synchrony every 6 to 10 minutes. Plasma membranes are not formed around the nuclei, which are distributed

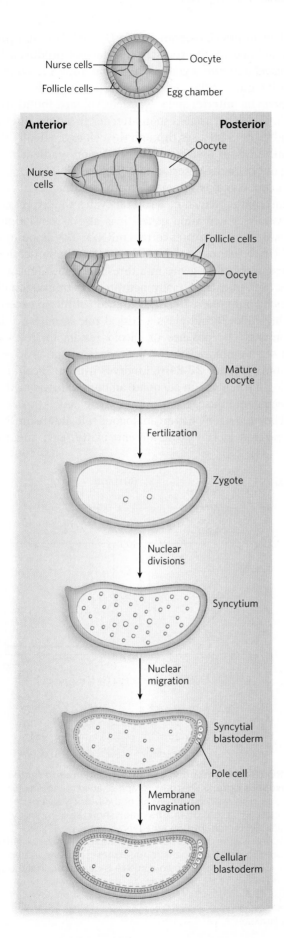

within the egg cytoplasm, forming a syncytium. During rounds 8 to 11 of nuclear division, the nuclei migrate to the egg's outer layer, forming a monolayer surrounding the common yolk-rich cytoplasm; this is the syncytial blastoderm. After a few additional divisions (producing up to 6,000 nuclei), membrane infoldings create a layer of cells, forming the cellular blastoderm. At this stage, the mitotic cycles in the cells lose their synchrony. The developmental fate of the cells is determined by the mRNAs and proteins originally deposited in the egg by the nurse and follicle cells.

Cascades of Regulatory Proteins in Development The role of key genes in development is to regulate other genes. Temporal and spatial regulation is critical to the gradual maturation of cells and tissues as cell divisions continue from embryo to adult. As each successive layer of regulatory genes is activated, the embryo acquires a finer specialization of cellular function.

Several types of RNAs and proteins in the early embryo, and proteins with essential roles in later stages of development, follow patterns widely conserved in multicellular eukaryotes. As defined by Christiane Nüsslein-Volhard, Edward B. Lewis, and Eric F. Wieschaus for *Drosophila*, three major classes of pattern-regulating genes—maternal, segmentation, and homeotic genes—function in successive developmental stages to specify the basic features of the fruit fly embryo body.

Maternal genes are expressed in the unfertilized egg, and the resulting **maternal mRNAs** remain dormant until fertilization. Maternal mRNAs provide most of the required proteins in the very early stages of development, and in fruit flies, this occurs until the cellular blastoderm forms. Some of the proteins encoded by maternal mRNAs direct the spatial organization of the developing embryo to establish its polarity. **Segmentation genes**, transcribed after fertilization, direct the formation of the proper number of body segments. In nematodes, similar genes guide the formation of specific tissues following completion of the earliest stages of embryogenesis. At least three subclasses of segmentation genes act at successive stages of *Drosophila* development. **Gap genes** divide the developing embryo into several

FIGURE 22-23 Early development in *Drosophila*. During oocyte development, maternal mRNAs and proteins are deposited in the oocyte by nurse cells and follicle cells. After fertilization, nuclear divisions occur in synchrony in the common cytoplasm (syncytium), and nuclei migrate to the periphery. Membrane invaginations surround the nuclei to create a monolayer of cells at the periphery; this is the cellular blastoderm stage. During the early nuclear divisions, several nuclei at the far posterior of the embryo become pole cells, which later become the germ-line cells.

broad regions, and **pair-rule genes**, along with **segment polarity genes**, define 14 stripes that become the 14 segments of a normal fly embryo. **Homeotic genes**, expressed at a later stage, specify the organs and appendages that will develop in particular body segments.

The many regulatory genes in these three classes direct the development of an adult organism, with a head, thorax, and abdomen, the proper number of segments, and the correct appendages on each segment. Although fruit fly embryogenesis takes about a day to complete, all these genes are activated during the first 4 hours. During this period, some mRNAs and proteins are present for only a few minutes at specific points in time. Some of the genes code for transcription factors that affect the expression of other genes in a kind of developmental cascade. Regulation at the level of translation also occurs, and many of the regulatory genes encode translational repressors, most of which bind to the 3′UTR of mRNAs. Because many mRNAs are deposited in the egg long before their translation is required, translational repression is especially important for regulation in developmental pathways.

Early Development Is Mediated by Maternal Genes

In invertebrates, a prescribed developmental path is evident from the very first embryonic cell division. The nonequivalence of the daughter cells of this first division implies a structural and functional asymmetry in the fertilized egg. The asymmetry is mediated by established gradients of molecules called morphogens—mRNAs and proteins produced by maternal genes.

In *Drosophila*, some maternal genes are expressed within the nurse and follicle cells, and some in the egg itself. In the unfertilized egg, the maternal gene products establish the critical anterior-posterior and dorsal-ventral axes, thereby defining which regions of the radially symmetric egg will develop into the head and abdomen and the top and bottom of the adult fly. A key event in very early development is establishing mRNA and protein gradients along the body axes. Some maternal mRNAs have protein products that diffuse through the cytoplasm, creating an asymmetric distribution in the egg. Various cells in the cellular blastoderm therefore inherit different amounts of these proteins, setting the cells on different developmental paths. The products of the maternal mRNAs include transcription activators or repressors as well as translational repressors, all regulating the expression of other pattern-regulating genes. Thus, the resulting gene expression sequences and patterns differ among cell lineages, ultimately orchestrating the development of each adult structure.

The anterior-posterior axis in *Drosophila* is also partially defined by the transcription factors produced by the *bicoid* and *nanos* genes. The *bicoid* mRNA is synthesized by nurse cells and deposited in the unfertilized egg near its anterior pole. Nüsslein-Volhard found that this mRNA is translated soon after fertilization, and the Bicoid protein diffuses through the cell to create, by the seventh nuclear division, a concentration gradient radiating out from the anterior pole (**Figure 22-24a**).

Bicoid contains a homeodomain (see Chapter 19). As a transcription activator, Bicoid activates the expression of several segmentation genes. It is also a translational repressor that inactivates certain mRNAs. The amounts of Bicoid in various parts of the embryo affect the subsequent expression of other genes in a threshold-dependent way. Genes are transcriptionally activated or translationally repressed only where the concentration of Bicoid exceeds the threshold. Bicoid plays a critical role in anterior development. The absence of Bicoid results in development of an embryo with two abdomens but no head and no thorax (**Figure 22-24b**). Embryos without Bicoid can develop normally if an adequate amount of *bicoid* mRNA is injected into the egg at the appropriate end.

The *nanos* gene has an analogous role, but its mRNA is deposited at the posterior end of the egg, and the

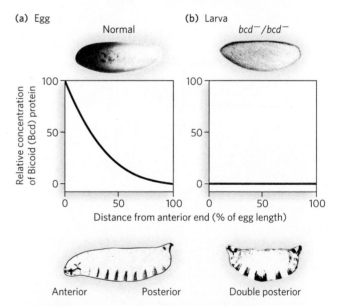

FIGURE 22-24 Distribution of a maternal gene product in a *Drosophila* egg. (a) The micrograph of an immunologically stained egg shows the distribution of the *bicoid* (*bcd*) gene product, the Bicoid protein. The graph shows stain intensity (protein concentration) along the length of the egg. This distribution is essential for normal development of the anterior structures. (b) If the *bcd* gene is not expressed by the mother (a *bcd⁻/bcd⁻* mutant), no *bicoid* mRNA is deposited in the egg, resulting in lack of Bicoid protein, as seen in the micrograph. The resulting embryo has two posteriors (and soon dies). [*Source: Wolfgang Driever and Christiane Nüsslein-Volhard, Max-Planck-Institut.*]

anterior-posterior Nanos protein gradient peaks at the posterior pole. Nanos is a translational repressor, conserved from worms to humans.

A broader view of the effects of maternal genes in *Drosophila* reveals a more precise picture of a developmental circuit. In addition to *bicoid* and *nanos* mRNAs, deposited in the egg asymmetrically, several other maternal mRNAs are deposited uniformly throughout the egg cytoplasm. Three of them encode the Pumilio, Hunchback, and Caudal proteins—all affected by Nanos and Bicoid (**Figure 22-25**).

Caudal and Pumilio are involved in the development of the fruit fly's posterior end. Caudal is a transcription activator with a homeodomain; Pumilio is a translational repressor from the PUF family of proteins (see Figure 22-28). Hunchback plays an important part in developing the anterior end; it is also a transcription factor for several genes, in some cases an activator and in others a repressor. Bicoid suppresses the translation of *caudal* mRNA at the anterior end and also acts as a

transcription activator of *hunchback* mRNA in the cellular blastoderm. Because *hunchback* is expressed through maternal mRNAs and from genes in the developing egg, it is considered a maternal as well as a segmentation gene. The result of Bicoid's activities is an increased concentration of Hunchback at the anterior end of the egg. Nanos and Pumilio act as translational repressors of *hunchback*, suppressing synthesis of Hunchback near the posterior end of the egg. Pumilio does not function in the absence of Nanos, and the gradient of *nanos* expression confines the activity of both proteins to the posterior region. Translational repression of the *hunchback* gene leads to degradation of *hunchback* mRNA near the posterior end. However, a lack of Bicoid in the posterior leads to expression of *caudal*. In this way, the Hunchback and Caudal proteins become asymmetrically distributed in the egg.

Segmentation Genes Specify the Development of Body Segments and Tissues

Segmentation genes are the zygotic genes that take over after maternal genes. Many operate at the level of transcriptional regulation. Gap genes, pair-rule genes, and segment polarity genes are activated in a cascadelike sequence at successive stages of embryonic development. The expression of gap genes is generally regulated by the products of one or more maternal genes. Gap genes activate the pair-rule genes, which in turn activate the segment polarity genes. This cascade of gene expression is accompanied by the gradual formation of 14 parasegments, then the true segments. Only a few cells (or nuclei) wide, parasegments are delimited by temporary grooves. Segments are offset from parasegments, so that each segment later encompasses the anterior part of a parasegment and the posterior part of the adjacent one. The anterior segments eventually fuse to form the head.

Pair-rule genes are expressed in alternating parasegments, and one well-characterized segmentation gene in the pair-rule subclass is *fushi tarazu* (*ftz*). When *ftz* is deleted, the embryo develops 7 segments instead of the normal 14, each segment twice the usual width. The Fushi-tarazu protein (Ftz) is a transcription activator with a homeodomain. The mRNAs and proteins derived from the *ftz* gene accumulate in a striking pattern of seven stripes that encircle the posterior two-thirds of the embryo (**Figure 22-26**). The stripes demarcate half of the parasegments; the development of alternating segments is compromised if *ftz* function is lost. The Ftz protein and a few similar regulatory proteins directly or indirectly regulate the expression of vast numbers of genes in the continuing developmental cascade.

In the stripes where *ftz* is repressed, repression is mediated in part by another pair-rule gene called *even-skipped* (*eve*) (see Figure 21-15). The expression of *eve*

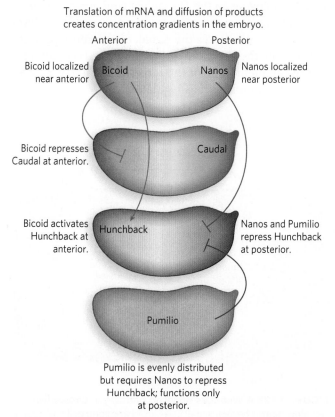

Translation of mRNA and diffusion of products creates concentration gradients in the embryo.

Anterior — Posterior

Bicoid localized near anterior — **Bicoid** — **Nanos** — Nanos localized near posterior

Bicoid represses Caudal at anterior. — **Caudal**

Bicoid activates Hunchback at anterior. — **Hunchback** — Nanos and Pumilio repress Hunchback at posterior.

Pumilio

Pumilio is evenly distributed but requires Nanos to repress Hunchback; functions only at posterior.

FIGURE 22-25 Regulatory circuits of the anterior-posterior axis in a *Drosophila* egg. The *bicoid* and *nanos* mRNAs are localized near the anterior and posterior poles of the egg, respectively. The *caudal, hunchback,* and *pumilio* mRNAs are distributed throughout the cytoplasm. Gradients of Bicoid and Nanos proteins lead to accumulation of Hunchback protein in the egg's anterior region and Caudal protein in its posterior. Because Pumilio requires Nanos for its activity as a translational repressor of *hunchback* mRNA, Pumilio functions only at the posterior end.

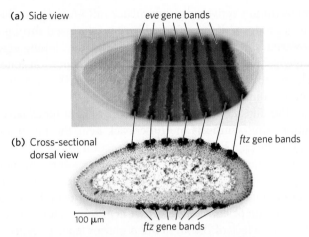

(a) Side view — eve gene bands

(b) Cross-sectional dorsal view — ftz gene bands

100 μm — ftz gene bands

FIGURE 22-26 Distribution of the *fushi tarazu (ftz)* and *even-skipped (eve)* gene products in early *Drosophila* embryos. (a) The *ftz* gene product can be detected in seven bands around the circumference of the embryo. These alternate with bands where the *eve* gene is expressed. (b) In a cross-sectional autoradiograph, the *ftz* bands appear as dark spots (generated by a radioactive label) and demarcate the anterior margins of the segments that will appear in the late embryo. [Sources: (a) Courtesy Stephen J. Small, Department of Biology, New York University. (b) Courtesy Phillip Ingham, Imperial Cancer Research Fund, Oxford University.]

is activated by Bicoid and Hunchback, and is highest in the parasegments where *ftz* gene expression is low. This gives rise to the alternating pattern of *ftz* and *eve* expression. The expression of *eve* is repressed by two other gap genes, *krüppel* and *giant*. The alternating pattern of *eve* and *ftz* expression is due to variations in gap gene expression from one parasegment to the next.

Gap and pair-rule genes operate during the first 2.5 hours of *Drosophila* development, when the embryo is still a syncytium. Virtually all these genes encode transcription factors, which have localized access to the nuclei in the syncytium. Segment polarity genes, the last group in the regulatory cascade, act at a stage when cells have formed. Some of the focus now shifts from transcription factors to cell-cell signaling pathways. The signaling helps reinforce the alternating boundaries between parasegments, then segments. Further, the function of adjacent segments is made interdependent by the pattern of segment polarity gene expression. A key example can be seen in the genes *wingless* (*wg*), *engrailed* (*en*), and *hedgehog* (*hh*). Wingless (Wg) and Engrailed (En) proteins are initially expressed in alternating cells, due to activation by pair-rule genes. However, the relevant pair-rule gene products recede after a few hours, and the continued expression of Wg and En becomes interdependent across opposite sides of parasegment boundaries (**Figure 22-27**).

The Wg protein is a signal of the Wnt class, and studies of the *wg* gene helped define the **Wnt-class signaling pathways**, which play key roles in development

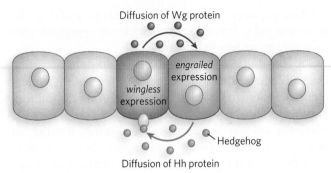

FIGURE 22-27 Interdependent signaling loops across segment boundaries in *Drosophila*. Part of the maintenance of parasegment, and later segment, boundaries involves cell-cell signaling. Wnt-class signaling with the Wg protein, the *wingless* (*wg*) gene product, induces expression of the *engrailed* (*en*) gene in recipient cells. The En protein triggers expression of *hedgehog*, part of a non–Wnt-class signaling pathway. The Hedgehog protein (Hh) promotes expression of Wg in the original cells and completes the closed signaling loop.

in eukaryotes, from nematodes to humans (the name "Wnt," *wingless type*, originated in the study of a mouse gene homologous to the *Drosophila* gene *wingless*). Wnt-class pathways generally consist of a secreted Wnt glycoprotein that constitutes the signal, one or more proteins involved in the secretion process, and a receptor protein in the membrane of the target cell (**Figure 22-28**). Additional proteins act as regulators.

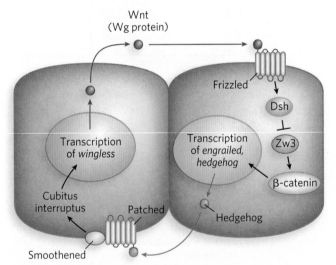

FIGURE 22-28 A Wnt-class signaling pathway in *Drosophila* development. The Wnt signal is secreted from one cell and interacts with a receptor on another cell. The signal in this case results in gene activation, via a pathway that uses a receptor (Frizzled) and signaling proteins Dsh, Zw3, and β-catenin. The secreted Hedgehog protein also works through a signaling pathway, involving a different receptor and different set of signaling proteins, as shown. Cubitus interruptus is a zinc finger transcription factor involved in Hedgehog signaling; Smoothened is a membrane receptor for Hedgehog; and Patched is a protein that regulates the expression and function of Smoothened.

Wnt proteins are highly homologous from one species to the next. They generally have a nearly invariant pattern of 23 Cys residues (some of which may form disulfide bonds needed for folding), an N-terminal signal sequence that helps guide secretion, and several N-glycosylation sites. They also have at least two lipid modifications: a palmitoyl group is added at a conserved Cys residue, and a palmitoleyl group at a conserved Ser. Wnt proteins are synthesized, and the lipid modifications are made, in the endoplasmic reticulum. The proteins move through a normal secretion pathway, from ER to Golgi complex, and are then transported to the cell surface in vesicles. Some of the secreted Wnt proteins are associated with lipoprotein particles.

The Wg protein is modified with lipids by an acyltransferase in the endoplasmic reticulum that is encoded by the gene *porcupine*. In nearby cells on the opposite side of the adjacent parasegment boundary, the secreted Wg protein interacts with a receptor that is the product of the gene *frizzled* (*fz*). In the recipient cell, the interaction triggers a signaling pathway that ultimately results in expression of the En protein. En is a transcription activator of the *hedgehog* gene. The Hedgehog (Hh) protein, part of a non–Wnt-class signaling pathway, is secreted and interacts with receptors on the Wg-producing cells. The resulting signal activates more Wg protein synthesis. The entire cycle is self-sustaining and self-reinforcing. Most of the other known segment polarity genes encode proteins that are part of either the *wingless* (Wnt) or the *hedgehog* (Hh) signaling pathway. In the alternating segments, the En protein and other transcription factors activate or repress a series of additional genes that now begin to give each segment a distinctive function. Many of these targets are homeotic genes.

Homeotic Genes Control the Development of Organs and Appendages

A set of 8 to 11 homeotic genes directs the formation of structures at specific locations in the body plan of most multicellular eukaryotes. Fewer homeotic genes are present in some simple eukaryotes. Even yeast has two homologs, regulators of mating-type switching (see Chapter 21). These genes are now more commonly referred to as **Hox genes** (from *homeobox*, the conserved gene sequence that encodes the homeodomain). However, these are not the only development-related proteins to include a homeodomain (as noted above, Bicoid has a homeodomain), and Hox is more a functional than a structural classification.

Hox genes are sometimes organized in genomic clusters. *Drosophila* has one such cluster, and mammals have four. The order of genes within the clusters is colinear with their targets of action, from the anterior to the

posterior of the developing embryo. In *Drosophila*, each Hox gene is expressed in a particular embryonic segment and controls the development of the corresponding part of the mature fly (**Figure 22-29a**). The terminology for describing Hox genes can be confusing. They have historical names in the fruit fly (e.g., *ultrabithorax*), whereas in mammals they are designated by two competing systems based on lettered (*A, B, C, D*) or numbered (*1, 2, 3, 4*) clusters (**Figure 22-29b**).

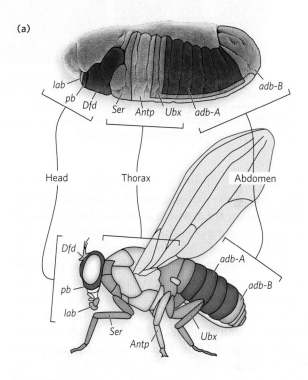

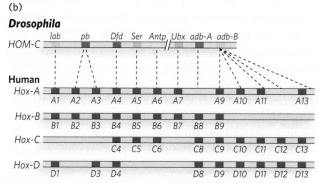

FIGURE 22-29 The Hox gene clusters and their effects on development. (a) Each Hox gene in the fruit fly directs the development of structures in a defined part of the body and is expressed in defined regions of the embryo (coded by color). (b) *Drosophila* has one Hox gene cluster (*HOM-C*). The genes are color-coded to match the fly segments in (a). The human genome has four Hox gene clusters (*Hox-A* through *Hox-D*). Many Hox genes are highly conserved in animals. Evolutionary relationships between genes in the *Drosophila* Hox gene cluster and those in the human (mammalian) Hox gene clusters are indicated by dashed lines. Similar relationships between the four sets of human Hox genes are indicated by vertical alignment. [*Source: (a) F. R. Turner, Department of Biology, Indiana University, Bloomington.*]

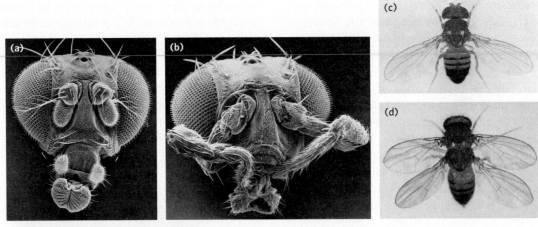

FIGURE 22-30 Effects of Hox gene mutations in *Drosophila*. (a) Normal head structure. (b) Homeotic mutant (*antennapedia*) in which antennae are replaced by legs. (c) Normal body structure. (d) Homeotic mutant (*bithorax*) in which a segment has developed incorrectly to produce an extra set of wings. [Sources: (a), (b) F. R. Turner, Department of Biology, Indiana University, Bloomington. (c), (d) Courtesy of the Archives, California Institute of Technology.]

The loss of Hox genes in fruit flies, by mutation or deletion, causes the appearance of a normal appendage or body structure at an inappropriate body position. An important example is the *ultrabithorax* (*ubx*) gene. When the Ubx protein function is lost, the first abdominal segment develops incorrectly, with the structure of the third thoracic segment. Other known homeotic mutations cause the formation of an extra set of wings, or two legs at the position in the head where the antennae are normally found (**Figure 22-30**). The Hox genes often span long regions of DNA. The *ubx* gene, for example, is 77,000 bp long. More than 73,000 bp are in introns, one of which is more than 50,000 bp long. Transcription of the *ubx* gene takes nearly an hour. The delay this imposes on *ubx* gene expression is believed to be a timing mechanism involved in the temporal regulation of subsequent steps in development. Many Hox genes are further regulated by miRNAs encoded by intergenic regions of the Hox gene clusters. All Hox gene products are themselves transcription factors that regulate the expression of an array of downstream genes.

The conservation of some Hox genes is extraordinary. For example, the products of the homeobox-containing *Hoxa-7* gene in mice and *antennapedia* gene in fruit flies differ in only one amino acid residue. Of course, although the molecular regulatory mechanisms may be similar, many of the ultimate developmental events are not conserved (humans do not have wings or antennae). The different outcomes are brought about by variance in the downstream target genes controlled by the Hox genes (see the How We Know section at the end of this chapter). The discovery of structural determinants with identifiable molecular functions is the first step in understanding the molecular events underlying development.

As more genes and their protein products are discovered, the biochemical side of this vast puzzle will be elucidated in increasingly rich detail.

Stem Cells Have Developmental Potential That Can Be Controlled

If we can understand development, and the mechanisms of gene regulation behind it, we can control it. An adult human has many different types of tissues. Many of the cells are terminally differentiated and no longer divide. If an organ malfunctions due to disease, or a limb is lost in an accident, the tissues are not readily replaced. Most cells, because of the regulatory processes that are in place, or even the loss of some or all genomic DNA, are not easily reprogrammed. Medical science has made organ transplants possible, but organ donors are a limited resource, and organ rejection remains a major medical problem. If humans could regenerate their own organs or limbs or nervous tissue, rejection would no longer be an issue. Real cures for kidney failure or neurodegenerative disorders could become reality.

The key to tissue regeneration lies in **stem cells**—cells that have retained the capacity to differentiate into various tissues. In humans, after an egg is fertilized, the first few cell divisions create a ball of **totipotent** cells (the morula), which have the capacity to differentiate individually into any tissue or even into a complete organism (**Figure 22-31**). Continued cell division produces a hollow ball, a blastocyst. The outer cells of the blastocyst eventually form the placenta. The inner layers form the germ layers of the developing fetus—the ectoderm, mesoderm, and endoderm. These cells are **pluripotent**: they can give rise to cells of all three germ layers and can be

Stem cells have two functions: to replenish themselves and, at the same time, to provide cells that can differentiate. These tasks are accomplished in multiple ways (**Figure 22-32a**). All or parts of the stem cell population can, in principle, be involved in replenishment, differentiation, or both.

Other types of stem cells can potentially be used for medical benefit. In the adult organism, **adult stem cells**, as products of additional differentiation, have a

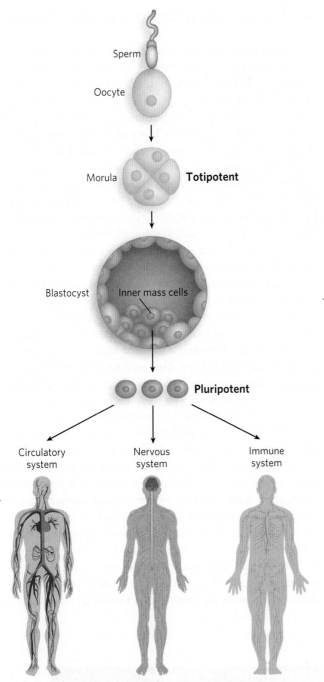

FIGURE 22-31 Totipotent and pluripotent stem cells. Cells of the morula stage are totipotent and have the capacity to differentiate into a complete organism. The source of pluripotent embryonic stem cells is the inner mass cells of the blastocyst. Pluripotent cells give rise to many tissue types, but they cannot form complete organisms. *[Source: (Circulatory system illustration) Matthew Cole/Shutterstock.]*

differentiated into many types of tissues. However, they cannot differentiate into a complete organism. Some of these cells are **unipotent**: they can develop into only one type of cell and/or tissue. It is the pluripotent cells of the blastocyst, the **embryonic stem cells**, that were originally used in embryonic stem cell research.

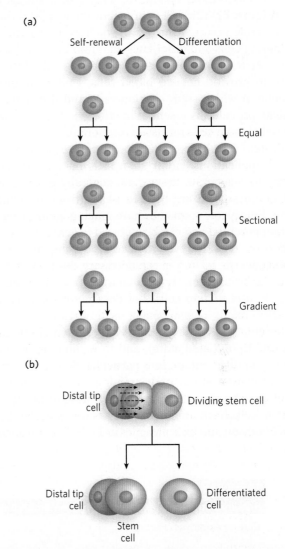

FIGURE 22-32 Stem cell proliferation versus differentiation and development. Stem cells must strike a balance between self-renewal and differentiation. (a) Some possible cell division patterns that allow for replenishment of stem cells and production of some differentiated cells. Each cell may produce one stem cell and one differentiated cell, or two differentiated cells or two stem cells in defined parts of the tissue or culture. Alternatively, a gradient of growth conditions can be established, with cell fates differing from one end of the gradient to the other. (b) Establishing a developmental niche through stem cell contact with a cell or group of cells. Molecular signals (dashed arrows) provided by niche cells (in this case, for plants, a distal tip cell) help orient the mitotic spindle for stem cell division and ensure that one daughter cell retains stem cell properties.

more limited potential for further development than do embryonic stem cells. For example, the hematopoietic stem cells of bone marrow can give rise to many types of blood cells, as well as to cells with the capacity to regenerate bone. They are referred to as **multipotent**. However, these cells cannot differentiate into a liver or kidney or neuron. Adult stem cells are often said to have a **niche**, a microenvironment that promotes stem cell maintenance while allowing differentiation of some daughter cells as replacements for cells in the tissue they serve (**Figure 22-32b**). Hematopoietic stem cells in the bone marrow occupy a niche in which signaling from neighboring cells and other cues maintain the stem cell lineage. At the same time, some daughter cells differentiate to provide needed blood cells. Understanding the niche in which stem cells operate, and the signals the niche provides, is essential in efforts to harness the potential of stem cells for tissue regeneration.

All stem cells have problems with respect to human medical applications. Adult stem cells have a limited capacity to regenerate tissues, are generally present in small numbers, and are hard to isolate from an adult human. Embryonic stem cells have much greater differentiation potential and can be cultured to generate large numbers of cells. However, their use is accompanied by ethical concerns related to the necessary destruction of human embryos. Identifying a source of plentiful and medically useful stem cells that does not raise ethical concerns remains a major goal of medical research.

Our ability to culture stem cells (i.e., maintain them in an undifferentiated state), and to manipulate them to grow and differentiate into particular tissues, is very much a function of our understanding of developmental biology. The identification and culturing of pluripotent stem cells from human blastocysts was reported by James Thomson and his colleagues in 1998. This advance led to the long-term availability of established cell lines for research.

In early work, mouse and human embryonic stem cells were used for most research. Although both types of stem cells are pluripotent, they require very different culture conditions, optimized to allow cell division indefinitely without differentiation. Mouse embryonic stem cells are grown on a layer of gelatin and require the presence of leukemia inhibitory factor (LIF). Human embryonic stem cells are grown on a feeder layer of mouse embryonic fibroblasts and require basic fibroblast growth factor (bFGF or FGF-2). The use of a feeder cell layer implies that the mouse cells are providing a diffusible product or some surface signal, not yet known, that is needed by human stem cells to either promote cell division or prevent differentiation. Recent research suggests that at least one of these diffusible products may be a Wnt-class protein. Some success has been achieved in directing the differentiation of human embryonic stem cells into particular tissue types; some of the progress in stimulating stem cell differentiation is summarized in **Table 22-1**. However, due to limited availability and ethical concerns, as noted above, embryonic stem cells have not been an ideal system on which to base continued research and potential medical applications.

The answer is to find another, more abundant and noncontroversial source of pluripotent stem cells. In 2007, researchers first reported success in reprogramming somatic cells to pluripotency. Skin cells—first from mice, then from humans—were reprogrammed to take on the characteristics of pluripotent stem cells. The reprogramming involves manipulations to get the cells to express some or all of four transcription factors involved in development: Oct4, Sox2, Krüppel-like factor 4 (KLF4), and cMyc—collectively known as OSKM factors. All of these are known to help maintain the stem cell–like state.

TABLE 22-1

Requirements for Differentiating Human Embryonic Stem Cells into Various Tissue Types	
Culture Additives/Protocol	*Differentiated Tissue*
Detach from mouse cell feeder layer Grow into neural-tube-like structures Grow in presence of FGF-2 Isolate neural precursors by treatment with dispase (a protease) Withdraw FGF-2; differentiation into neurons	Neural cell types (neurons, oligodendrocytes, and glial cells)
Co-culture with another cell line, derived from human embryonic stem cells treated with retinoic acid	Cardiomyocytes
Omit mouse cell feeder layer	Some insulin-producing pancreatic beta cells
Add ascorbate, β-glycerophosphate, and dexamethasone	Osteoblasts
Co-culture with bone marrow cell line Grow in presence of cytokines	Hematopoietic progenitors

The result is cells called induced pluripotent stem cells, or iPS cells. These, in turn, have been used to generate a range of tissue types. Gradual improvements in this technology are continuing. A new branch of medicine is slowly emerging, called regenerative medicine, which may eventually provide new approaches to the repair of damaged tissue after heart attacks and strokes and other traumas. The use of reprogrammed stem cells derived from the same patient may eliminate tissue rejection and provide a source of new, healthy tissue.

SECTION 22.5 SUMMARY

- Development of a multicellular organism presents the most complex regulatory challenge.
- The fate of cells in the early embryo is determined in part by establishment of anterior-posterior and dorsal-ventral gradients of proteins that act as transcription activators or translational repressors, regulating the genes required for the development of structures appropriate to a particular part of the organism.
- Sets of regulatory genes operate in temporal and spatial succession, transforming given areas of an egg cell into predictable structures in the adult organism.
- The developmental fate of cell lineages during development is also shaped by cell-cell signaling pathways in which signals from one cell lineage affect the fate of others. The Wnt-class signaling pathway is one well-studied example.
- In vertebrates, stem cells retain significant developmental potential. The differentiation of stem cells into functional tissues can be controlled by extracellular signals and conditions.

22.6 FINALE: MOLECULAR BIOLOGY, DEVELOPMENTAL BIOLOGY, AND EVOLUTION

Molecular biology is a story of biological information—the metabolism, maintenance, and transfer of that information from one generation to the next. As we go back almost unfathomable lengths of time, those generations link us to every living thing on our planet. Our genomic DNA makes our life possible. With its seemingly ragtag mix of piggy-backing transposons, integrated viruses, and genes, both borrowed and linearly evolved, our genome tells us about our past while at the same time linking us to a future rich with potential. Evolution continues.

Each topic in molecular biology has evolutionary significance. Errors or random events in DNA replication, recombination, and repair fuel genomic changes—some useful, many deleterious. Genomic changes are expressed, through transcription and translation, in the organismal phenotypes on which natural selection acts.

However, there are few areas where molecular biology meets evolution more dramatically than in the regulation of organismal development. Developmental biology thus provides a fitting final topic for our exploration of molecular biology.

The Interface of Evolutionary and Developmental Biology Defines a New Field

South America has several species of seed-eating finches, commonly known as grassquits. About 3 million years ago, a small group of a single species of grassquits took flight from the continent's Pacific coast. Perhaps driven by a storm, they lost sight of land and traveled nearly 1,000 km. Small birds such as these might easily have perished on such a journey, but the smallest of chances brought this group to a newly formed volcanic island in an archipelago later to be known as the Galápagos. It was a virgin landscape with untapped plant and insect food sources, and the newly arrived finches survived. Over many millennia, new islands formed and were colonized by new plants and insects—and by the finches. The birds exploited the new resources on the islands, and groups of birds gradually specialized and diverged into new species. By the time Charles Darwin stepped onto the islands in 1835, there were many different finch species in the archipelago, feeding on seeds, fruits, insects, pollen, and even blood. Some islands now have as many as 10 finch species, each adapted to a somewhat different lifestyle and food source.

The diversity of living creatures on our planet was a source of wonder for humans long before scientists sought to understand its origins. The extraordinary insight handed down to us by Darwin, inspired in part by his encounter with the Galápagos finches, provided a broad explanation for the existence of organisms with a vast array of appearances and characteristics. It also gave rise to many questions about the mechanisms underlying evolution. Answers to those questions have started to appear, first through the study of genomes and nucleic acid metabolism in the last half of the twentieth century, and more recently through an emerging field nicknamed **evo-devo**—a blend of evolutionary and developmental biology.

In its modern synthesis, the theory of evolution has two main elements: mutations in a population generate genetic diversity, and natural selection then acts on this diversity to favor individuals with more useful genomic tools and to disfavor others. Mutations occur at significant rates in every individual's genome, in every cell (see Chapters 3 and 12). Advantageous mutations in single-celled organisms or in the germ line of multicellular organisms can be inherited, and they are more likely to be inherited (i.e., are passed on to greater numbers of offspring) if they confer an advantage. It is a straight-forward scheme. But many have wondered whether that scheme is enough to explain, say, the many different beak shapes in the Galápagos finches, or the diversity of size and shape among mammals. Until recent decades, there were several widely held assumptions about the evolutionary process: that many mutations and new genes would be needed to bring about a new physical structure, that more-complex organisms would have larger genomes, and that very different species would have few genes in common and perhaps use very different patterns of gene regulation. All of these assumptions were wrong.

Modern genomics has revealed that the human genome contains fewer genes than expected—not many more than the fruit fly genome, and fewer than some amphibian genomes. The genomes of every mammal, from mouse to human, are surprisingly similar in the number, types, and chromosomal arrangement of genes. Meanwhile, evo-devo is telling us how complex and very different creatures can evolve within these genomic realities.

Small Genetic Differences Can Produce Dramatic Phenotypic Changes

The kinds of mutant organisms shown in Figure 22-30 were studied by the English biologist William Bateson in the late nineteenth century. Bateson used his observations to challenge the Darwinian notion that evolutionary change would have to be gradual. Recent studies of the genes that control organismal development have strongly supported Bateson's ideas. Subtle changes in regulatory patterns during development, reflecting just one or a few mutations, can result in startling physical changes and fuel surprisingly rapid evolution.

The Galápagos finches provide a wonderful example of the link between evolution and development. There are at least 14 species (some specialists list 15), and they are distinguished in large measure by their beak structure. The ground finches, for example, have broad, heavy beaks adapted to crushing hard, large seeds. The cactus finches have longer, slender beaks, ideal for probing cactus flowers and fruit (**Figure 22-33**).

Clifford Tabin and his colleagues carefully surveyed a set of genes expressed during avian craniofacial development. They identified a single gene, *Bmp4*, whose expression level correlated with the formation of the more robust beaks of the ground finches. More robust beaks were also formed in chicken embryos when high levels of the Bmp4 protein were artificially expressed in the appropriate tissues, confirming the importance of *Bmp4*. In a similar study, the formation of long, slender beaks was linked to the expression of the protein

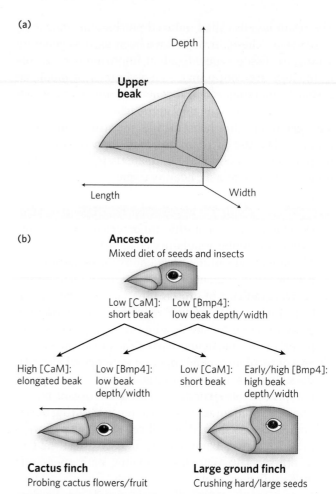

FIGURE 22-33 The evolution of new beak structures to exploit new food sources. Galápagos finches that feed on different, specialized food sources have different beak structures, as shown for the cactus finch and the large ground finch. (a) The beak structures can be varied along three dimensions. (b) The differences observed in the two finch species were produced largely through natural selection acting on a few mutations that altered the timing and level of expression of just two genes: those encoding Bmp4 and calmodulin (CaM).

calmodulin in particular tissues at appropriate developmental stages. Thus, major changes in the shape and function of the beak can be brought about by subtle changes in the expression of just two genes involved in developmental regulation. Very few mutations are required, and the needed mutations affect regulation. New genes are *not* required. Note that Bmp4 is a member of a family of signaling proteins, with roles in development similar to those of the Wnt and Hedgehog proteins. Like Wnt and Hedgehog, Bmp homologs are widely conserved in eukaryotes. And as in the other signaling pathways, alterations in Bmp signaling pathways can have large effects on development.

The system of regulatory genes that guides development is remarkably conserved among all vertebrates. Elevated expression of *Bmp4* in the right tissue at the right time leads to more robust jaw parts in zebra fish. The same gene plays a key role in tooth development in mammals. The development of eyes is triggered by the expression of a single gene, *Pax6*, in fruit flies and in mammals. The mouse *Pax6* gene will trigger the development of fruit fly eyes in the fruit fly, and the fruit fly *Pax6* gene will trigger the development of mouse eyes in the mouse. In each organism, these genes are part of the much larger regulatory cascade that ultimately creates the correct structures in the correct locations in each organism. The cascade is ancient; the Hox genes have been part of the developmental program of multicellular eukaryotes for more than 500 million years. Subtle changes in the cascade can have large effects on development, and thus on the ultimate appearance of the organism. These same subtle changes can promote rapid evolution. For example, the 400 to 500 described species of cichlids (spiny-finned fish) in Lake Malawi and Lake Victoria on the African continent are all derived from one or a few populations that colonized each lake over the past 100,000 to 200,000 years. The Galápagos finches simply followed a path of evolution and change that living creatures have been traveling for billions of years.

Our discussion of developmental regulation brings us full circle, back to a biochemical beginning—both figuratively and literally. Evolution appropriately provides a key backdrop for the first and last chapters in this book. If evolution is to generate the kind of changes in an organism that we associate with a different species, it is the developmental program that must be affected. Developmental and evolutionary processes are closely allied, each informing the other. Molecular biology ties the fields together, informs us about molecular mechanisms that underpin the changes, and provides the technology needed for new discovery. The continuing study of molecular biology has everything to do with enriching the future of humanity and understanding our origins.

SECTION 22.6 SUMMARY

- Developmental biology and evolutionary biology are closely related. The two fields inform each other, and molecular biology is intimately intertwined with both.
- Major changes in the appearance and/or function of multicellular eukaryotes can be effected by subtle changes in an organism's developmental program, involving mutations in the regulatory genes that guide the process.

? UNANSWERED QUESTIONS

Our understanding of gene regulation remains incomplete in many areas. Indeed, the recent discovery of RNA interference, which has proved to be a major mode of regulation, underscores the likelihood that more fundamentals remain to be elucidated.

1. How is alternative splicing regulated? Mounting evidence suggests that alternative splicing accounts for a much greater degree of protein complexity in higher eukaryotes than would be predicted by simply counting the number of open reading frames in a genome. How such alternative splicing is regulated is not well understood, nor have mechanisms of tissue-specific splicing been worked out.

2. How and when do miRNAs control gene expression in human cells? The human genome encodes several hundred miRNAs, yet the targets and functions of most of these are currently unknown. How do we harness these newly discovered regulatory mechanisms to provide new therapies for cancer and other diseases?

3. What other regulatory mechanisms have we just not uncovered yet? RNA interference is a fairly recent discovery. What are the functions of all the new RNAs being discovered in eukaryotic genomes?

4. Are transcriptional and posttranscriptional steps in gene expression coordinately regulated? What kinds of mechanisms might enable communication between the cytoplasm and the nucleus to adjust transcription, splicing, and mRNA transport rates in response to increased or decreased translation of a particular mRNA?

5. What are all of the signals that guide the action of regulatory proteins and the development of specific tissues? A much more detailed understanding is needed to completely unleash the potential of stem cell technologies, particularly with respect to the control of differentiation of induced pluripotent stem cells. That potential includes new cancer treatments, the regeneration of lost limbs, and the replacement of diseased tissues (e.g., heart, lung, kidney) without the danger of tissue rejection.

HOW WE KNOW

A Natural Collaboration Reveals a Binding Protein for a 3′UTR

Zhang, B., M. Gallegos, A. Puoti, E. Durkin,
S. Fields, J. Kimble, and M.P. Wickens.
1997. A conserved RNA-binding protein
that regulates sexual fates in the C.
elegans hermaphrodite germ line.
Nature 390:477–484.

Marvin Wickens *[Source: Courtesy Marvin Wickens.]*

Scientific collaborations come about in many ways, as the discovery of one gene-regulatory protein illustrates. The importance of the untranslated parts of an mRNA, particularly the 3′UTR, gradually became apparent over the course of the 1990s. In 1991, Judith Kimble's lab reported the discovery of a 3′UTR regulatory element in the *fem-3* gene of the nematode *C. elegans*, a sequence called PME (point mutation element) (see this chapter's Moment of Discovery). Single base-pair changes in this element had a dramatic effect on germ cell fate, specifically in the switch from sperm to eggs during germ-line development in the hermaphrodite. The PME sequence had to be interacting with something, but what? An RNA-binding protein seemed a likely candidate, but what protein? How could it be identified? For Kimble, the obvious approach was unusually close at hand.

In 1996, Marvin Wickens and his coworkers reported the invention of the three-hybrid method to identify proteins that bind to particular RNA sequences (see Figure 7-29). The problem posed by the Kimble group was a perfect test of this new technology. Happily, not only was the Wickens lab quite near the Kimble lab at the University of Wisconsin, but Wickens and Kimble were husband and wife. A new kind of collaboration (for them) was soon hatched.

The investigators initiated a screen of a cDNA library in which *C. elegans* genes were fused to the gene encoding the Gal4p activation domain. The new three-hybrid method worked as advertised; one clone was found that met all criteria for the study and activated expression of the reporter gene, a *his3-lacZ* fusion. Testing a wide array of RNA-binding substrates, they demonstrated that the protein encoded by the cloned nematode gene bound only to the target sequence in the *fem-3* mRNA (Figure 1). The protein was named

FBF-1 (*fem-3* binding factor). The *fbf-1* gene sequence was then used to search DNA databases for homologs. A second gene, *fbf-2*—91% identical to *fbf-1* and encoding another protein that bound to the *fem-3* 3′UTR—was identified in the *C. elegans* genome. Perhaps more significant was that both gene products, FBF-1 and FBF-2, were identified as members of the newly described PUF family of RNA-binding proteins, a group that includes the protein Pumilio.

RNAi experiments confirmed the role of FBF proteins in germ cell fate. Studies of the FBF proteins quickly became part of a still expanding effort to characterize the function of PUF family proteins.

(a) Three-hybrid system

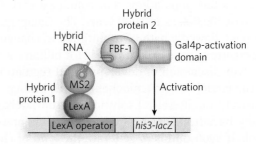

(b) Binding specificity

Hybrid RNA	β-Galactosidase filter assay	β-Galactosidase units
fem-3 wild type UCUUG		370
fem-3 ch8 AGAAC		3
fem-3 cq96 UCUUU		3
IRE		6
A₃₀		5
HIV-E		8

FIGURE 1 (a) The protein FBF-1 was identified in a three-hybrid screen. The hybrid RNA engineered to be the link between the binding protein MS2 and the (unknown) protein X–Gal4p fusion contained two PME sequences. (b) Activation of the *his3-lacZ* reporter gene led to production of β-galactosidase, which catalyzes the conversion of X-gal to a colored product. The color is seen only in cells expressing RNA containing PMEs of the proper sequence. The RNAs in other lanes of the gel provide controls that demonstrate specificity of binding. IRE is iron response element; A₃₀, a sequence of 30 A residues; and HIV-E, a 573-nucleotide RNA sequence derived from HIV. *[Source: Reprinted by permission from Macmillan Publishers Ltd: Nature, Zhang, B, et al. A conserved RNA-binding protein that regulates sexual fates in the C. elegans hermaphrodite germ line, Vol, 390, 477–484, © 1997.]*

Little RNAs Play a Big Role in Controlling Gene Expression

Lagos-Quintana, M., R. Rauhut, W. Lendeckel, and T. Tuschl. 2001. Identification of novel genes coding for small expressed RNAs. *Science* 294:853–858.

Lau, N.C., L.P. Lim, E.G. Weinstein, and D.P. Bartel. 2001. An abundant class of tiny RNAs with probable regulatory roles in *Caenorhabditis elegans*. *Science* 294:858–862.

Lee, R.C., and V. Ambros. 2001. An extensive class of small RNAs in *Caenorhabditis elegans*. *Science* 294:862–864.

Scientific discovery has a way of occurring in bursts of insight, often with input from multiple research teams whose ideas and experiments converge on a new line of thinking. In the field of RNA interference, such a conceptual breakthrough occurred in 2001 with the finding by three different labs that small regulatory RNA molecules are abundant in eukaryotic cells. Scientists had come to suspect that small RNAs might normally be produced in cells as a means of controlling gene expression. This suspicion was based on the discovery by Craig Mello and Andrew Fire that double-stranded RNA, when fed to *C. elegans*, could silence gene expression. Research teams led by Victor Ambros, David Bartel, and Thomas Tuschl set out to find evidence of small regulatory RNAs that might be produced naturally in cells.

Each team took a similar experimental approach, in which *C. elegans* or mammalian cells were grown in the laboratory and total cellular RNA was isolated. The total RNA was fractionated by size to enable purification of RNA molecules about 20 to 30 nucleotides long, the size of the molecules used in the Mello and Fire experiments. To identify these molecules, the RNAs were covalently linked at their 3' ends to oligonucleotide sequences that provided binding sites for a complementary oligonucleotide, which could be used to prime the reverse transcription of the RNA into DNA. The complementary strand of this DNA sequence could be produced in a similar fashion, by covalently attaching it to a second oligonucleotide of defined sequence at its 3' end. Once the small RNAs had been copied into double-stranded DNA, they were cloned into plasmids using standard techniques (see Chapter 7). The plasmids could then be propagated in bacterial cell culture, purified, and sequenced. The sizes of the small RNAs identified in *C. elegans* all fell within a narrow range (19 to 24 nucleotides), as opposed to the sizes of RNAs originating from *E. coli*, generated in a separate control experiment (Figure 2). These results suggested that the small RNAs cloned from *C. elegans* were indeed the product of transcription of the *C. elegans* genome, and not simply short degradation products.

The sequences of these small RNAs proved very exciting, because in many cases they were complementary to sequences found in the host genome. This finding suggested that the small RNAs were produced as part of a large regulatory pathway in which small RNA molecules, dubbed microRNAs (miRNAs), could base-pair with target sequences in mRNAs. Subsequent experiments verified that this mechanism occurs widely in eukaryotes.

Why were miRNAs overlooked by molecular biologists for so long? One reason is simply size: because they are so small, they tended to be ignored or were thought to be irrelevant degradation products rather than functional RNAs produced by the cells.

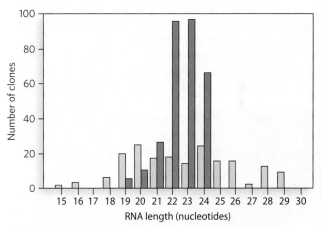

FIGURE 2 One class of short RNAs derived from *C. elegans* show a characteristically narrow length distribution (darker bars) compared with clones of *E. coli* RNA fragments (lighter bars) produced using the same protocol in an organism that lacks miRNAs. The short but uniform lengths of the *C. elegans* RNAs provided some of the evidence that the RNAs were functional, and not degradation products. [*Source: Data from N. C. Lau et al., Science 294:858–862, 2001, Fig. 2a.*]

HOW WE KNOW

[Source: Courtesy of the Sean Carroll Laboratory.]

Everything Old Is New Again: Beauty at the Turn of a Developmental Switch

Carroll, S.B., J. Gates, D.N. Keys, S.W. Paddock, G.E. Panganiban, J.E. Selegue, and J.A. Williams. 1994. Pattern formation and eyespot determination in butterfly wings. *Science* 265:109–114.

We might not notice the diminutive fruit fly, but butterflies rarely fail to inspire fascination. The bold colors and patterns in a butterfly wing that catch our eye—surely these are the product of an elaborate developmental program that is distinct from the program operative in fruit flies?

Sean Carroll's boyhood fascination with butterflies was eventually translated into research in a lab at the University of Wisconsin, where he studies insect development. In the early 1990s, it was already clear that many genes that control development are highly conserved, not just in insects, but in all higher eukaryotes. In setting out to decipher the development of butterfly wing patterns, Carroll decided that the genes known to affect the development of *Drosophila* wings were a good place to start.

Carroll's subject was the butterfly *Precis coenia*, also called the Buckeye, found each summer over much of the United States. His laboratory succeeded in cloning a series of *P. coenia* genes homologous to *Drosophila* genes known to control wing development, including the genes for signaling proteins called Wingless and Decapentaplegic, and transcription factors called Apterous, Invected, Scalloped, and Distal-less. From the cloned genes, Carroll and his colleagues made labeled DNA probes and developed in situ

Sean Carroll [Source: Courtesy Sean Carroll.]

hybridization methods to reveal the location of mRNAs in the butterfly wing at different stages of development. For months, the expression patterns looked identical to those already defined for the same genes in *Drosophila*. That changed when they got to *distal-less*, a gene that helps control appendage development in animals from insects to humans (Figure 3). Carroll describes it as one of his most thrilling moments in science.

One day, Carroll's student, Julie Gates, called him over because she saw a pattern of genes turned on in so-called eyespots, the concentric rings of pigment in butterfly wings that look like eyes and are used in both mate recognition and predator avoidance. Carroll was stunned to be staring at the developing spotted pattern of gene expression, realizing that the gene involved, *distal-less*, had been around for at least 500 million years and was used in other organisms for appendage patterning. In fruit flies, there is no counterpart to eyespot development, and it was suddenly clear that the ancient gene had been recruited to an entirely new function in butterflies. This turned out to be the first example of what became a major theme in understanding the evolution of animal form: old genes occasionally evolve to do something completely new.

FIGURE 3 Expression of the *distal-less* gene (dark coloring) is revealed by in situ hybridization at three different stages of upper wing development in *P. coenia*. (a) A wing bud. (b) Partially developed wing. Expression of *distal-less* is evident at the point where an eyespot will develop (arrow). (c) Fully developed wing with eyespot. [*Source: From Science 265, 109–114, Carroll, SB, Gates, J, Keys, DN, Paddock, SW, Panganiban, GE, Selegue, JE, and Williams, JA (1994) Pattern formation and eyespot determination in butterfly wings. Reprinted with permission from AAAS.*]

PROBLEMS

1. An investigator monitors the production of a particular mRNA in a mouse cell line. Expression of the mRNA is induced (i.e., its concentration in the cytoplasm increases) in response to the addition of a hormone. The observed increase in the mRNA concentration is blocked by actinomycin D and by cycloheximide. What does this tell you about the requirements for increased expression of this mRNA?

2. In female fruit fly embryos, the Sxl protein initially generated by transcription from the P_e promoter differs somewhat from that generated by later transcription from the P_m promoter. In what part of the protein does this difference arise?

3. As a researcher, you wish to scan the sequences of all the genes of the mouse genome to determine how many genes have multiple sites for 3'-end cleavage. What sequence would you scan for? Would a scan for a single sequence find all of the sites?

4. In eukaryotes, phosphorylation of the translation initiation factor eIF2α blocks translation of virtually all mRNAs. In a mammalian reticulocyte, a deficiency in iron or heme leads to eIF2α phosphorylation to block the translation of globin mRNAs. The phosphorylation of eIF2α does not create a problem for other cellular functions in reticulocytes. Suggest why.

5. What is the likely fate of an mRNA transcript containing the sequence (a) AAUAAA or (b) AUUUA?

6. As organisms become more complex, so do the numbers and structures of introns. Introns in vertebrates range up to 100,000 nucleotides in length. These long introns often include canonical splicing signals, but the signals are not recognized by the cellular spliceosome. How are these nonproductive splicing sites suppressed?

7. Suggest at least three cellular mechanisms that could establish a gradient of either a protein or an mRNA during maturation of an oocyte.

8. In *C. elegans*, the Pal-1 protein specifies some developmental fates in cells where it is highly expressed. Translation of the Pal-1 mRNA is suppressed by binding of the Mex-3 protein, concentrated in anterior cells of the embryo, to the 3'UTR of the Pal-1 mRNA. What would be the probable effect of mutations that eliminated the binding of Mex-3 to this 3'UTR?

9. A *Drosophila* female embryo that is bcd^-/bcd^- may develop normally, but the adult fruit fly will not be able to produce viable offspring. Explain why.

10. In the *Drosophila* ovary, a germ-line stem cell repeatedly divides. After each division, one daughter cell retains stem cell identity and the other begins to differentiate into an oocyte. The germ-line stem cell is associated with additional cells called cap cells. Describe how the asymmetric divisions might occur, and the possible role of the cap cells.

11. What is an induced pluripotent stem cell, and how does it differ from an embryonic stem cell?

12. The stem cell genes that regulate tissue regeneration tend to be highly conserved. *Planaria* (an aquatic flatworm) has an impressive capacity to regenerate its head and other structures when they are amputated, making this a favorite subject in grade school science labs. In the wild, *Planaria* eats smaller worms and eukaryotic organisms in its environment. In the lab, it can be fed clumps of bacteria mixed with pieces of liver and agar. As a biologist, you know that tissue regeneration mechanisms are likely to be conserved. You are interested in determining which *Planaria* genes are needed to guide head regeneration. Your reading tells you that regeneration depends on certain stem cells posterior to the animal's photoreceptors and excluded from its pharynx. Using methods described in this chapter, how would you go about discovering the key genes?

DATA ANALYSIS PROBLEM

Zhang, B., M. Gallegos, A. Puoti, E. Durkin, S. Fields, J. Kimble, and M.P. Wickens. 1997. A conserved

RNA-binding protein that regulates sexual fates in the *C. elegans* hermaphrodite germ line. *Nature* 390:477–484.

13. The collaborative effort by Judith Kimble, Marvin Wickens, and colleagues, as described in their 1997 paper, resulted in discovery of the FBF proteins that bind the 3'UTR of the mRNA from the *fem-3* developmental regulatory gene of nematodes. Compare the three-hybrid strategy used in this study (see Figure 1 in this chapter's How We Know section) with the three-hybrid method presented in Chapter 7.

   (a) How was the three-hybrid method modified in the Kimble and Wickens study?

   Six different RNA sequences were screened for FBF-1 binding. The normal PME sequence (UCUUG) gave a positive response. The other five sequences were two 5 nucleotide sequences, an iron response element, a segment containing 30 consecutive A residues, and a 573 nucleotide RNA sequence derived from HIV.

   (b) Of these controls, which would make the best case for specific binding of the PME by FBF-1?
   (c) Why might the other controls be useful?

   The investigators used immunofluorescence to detect expression of the FBF-1 protein in wild-type nematodes, as shown in the upper panel of Figure 1. All cells illuminated are in the germ line. The dark spots are cell nuclei.

   (d) What conclusion can you draw from the protein expression pattern in the upper panel of the figure?

   The lower panel in Figure 1 shows results for an animal treated with RNAi directed at the gene for FBF-1 (*fbf*). The expression of FBF-1 is essentially abolished.

   (e) Given the function of FBF-1 in the germ line, what is the likely effect of this RNAi treatment on the germ line of the treated animals?

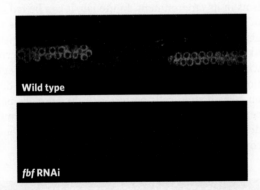

FIGURE 1 *[Source: Reprinted by permission from Macmillan Publishers Ltd: Nature, Zhang, B, et al. A conserved RNA-binding protein that regulates sexual fates in the C. elegans hermaphrodite germ line, Vol. 390, 477–484, © 1997.]*

ADDITIONAL READING

General

Carroll, S.B. 2005. *Endless Forms Most Beautiful: The New Science of Evo Devo and the Making of the Animal Kingdom.* New York: W. W. Norton & Company.

Granneman, S., and S.J. Baserga. 2006. Crosstalk in gene expression: Coupling and co-regulation of rDNA processing. *Curr. Opin. Cell Biol.* 17:281–286.

Rubin, G.M., and E.T. Lewis. 2000. A brief history of *Drosophila*'s contributions to genome research. *Science* 287:2216–2218.

Posttranscriptional Control inside the Nucleus

Bentley, D. 2005. Rules of engagement: Co-transcriptional recruitment of pre-mRNA processing factors. *Curr. Opin. Cell Biol.* 17:251–256.

Blencowe, B.J. 2006. Alternative splicing: New insights from global analyses. *Cell* 126:37–47.

Kornblihtt, A.R., I.E. Schor, M. Allõ, G. Dujardin, E. Petrillo, and M.J. Muñoz. 2013. Alternative splicing, a pivotal step between eukaryotic transcription and translation. *Nat. Rev. Mol. Cell Biol.* 14:153–165.

Neeman, Y., D. Dahary, and K. Nishikura. 2006. Editor meets silencer: Crosstalk between RNA editing and RNA interference. *Nat. Rev. Mol. Cell Biol.* 7:919–931.

Translational Control in the Cytoplasm

Fabian, M.R., N. Sonenberg, and W. Filipowicz. 2010. Regulation of mRNA translation and stability by micro-RNAs. *Annu. Rev. Biochem.* 79:351–379.

Gray, N.K., and M. Wickens. 1998. Control of translation initiation in animals. *Annu. Rev. Cell Dev. Biol.* 14: 399–458.

Hartmann-Petersen, R., M. Seeger, and C. Gordon. 2003. Transferring substrates to the 26S proteasome. *Trends Biochem. Sci.* 28:26–31.

Norbury, C.J. 2013. Cytoplasmic RNA: A case of the tail wagging the dog. *Nat. Rev. Cell Mol. Biol.* 14:643–653.

The Large-Scale Regulation of Groups of Genes

Goodrich, J.A., and J.F. Kugel. 2006. Non-coding RNA regulators of RNA polymerase II transcription. *Nat. Rev. Mol. Cell Biol.* 7:612–616.

Green, R., and J.A. Doudna. 2006. RNAs regulate biology. *ACS Chem. Biol.* 1:335–338.

Kapp, L.D., and J.R. Lorsch. 2004. The molecular mechanics of eukaryotic translation. *Annu. Rev. Biochem.* 73:657–704.

Liu, Q., and Z. Paroo. 2010. Biochemical principles of small RNA pathways. *Annu. Rev. Biochem.* 79:295–319.

RNA Interference

Cerutti, H. 2003. RNA interference: Traveling in the cell and gaining functions? *Trends Genet.* 19:9–46.

Crunkhorn, S. 2010. RNA interference: Clinical gene-silencing success. *Nat. Rev. Drug Discov.* 9:359.

ENCODE Project Consortium. 2012. An integrated encyclopedia of DNA elements in the human genome. *Nature* 489:57–74.

Kamath, R.S., A.G. Fraser, Y. Dong, G. Poulin, R. Durbin, M. Gotta, A. Kanapink, N. Le Bot, S. Moreno, M. Sohrmann, et al. 2003. Systematic functional analysis of the *Caenorhabditis elegans* genome using RNAi. *Nature* 421:231–237.

Luteijn, M.J., and R.F. Ketting. 2013. PIWI-interacting RNAs: From generational to trans-generational epigenetics. *Nat. Rev. Genet.* 14:523–534.

Morris, K.V., and J.S. Mattick. 2014. The rise of regulatory RNA. *Nat. Rev. Genet.* 15:423–437.

Putting It All Together: Gene Regulation in Development

Blow, N. 2008. In search of common ground. *Nature* 451:855–858.

Buganin, Y., D.A. Faddah, and R. Jaenisch. 2013. Mechanisms and models of somatic cell reprogramming. *Nat. Rev. Genet.* 14:427–439.

Farley, B.M., and S.P. Ryder. 2008. Regulation of maternal mRNAs in early development. *Crit. Rev. Biochem. Mol. Biol.* 43:135–162.

Gönczy, P. 2008. Mechanisms of asymmetric cell division: Flies and worms pave the way. *Nat. Rev. Mol. Cell Biol.* 9:355–366.

Lander, A.D. 2013. How cells know where they are. *Science* 339:923–927.

Morrison, S.J., and J. Kimble. 2006. Asymmetric and symmetric stem-cell divisions in development and cancer. *Nature* 441:1068–1074.

Nüsslein-Volhard, C., and E. Wieschaus. 1980. Mutations affecting segment number and polarity in *Drosophila*. *Nature* 287:795–801.

Passier, R., L.W. van Laake, and C.L. Mummery. 2008. Stem-cell-based therapy and lessons from the heart. *Nature* 453:322–329.

Pera, M.F. 2008. Stem cells: A new year and a new era. *Nature* 451:135–136.

Tajbakhsh, S., P. Rocheteau, and I. Le Roux. 2010. Asymmetric cell divisions and asymmetric cell fates. *Annu. Rev. Cell Dev. Biol.* 25:671–699.

Takai, Y., J. Miyoshi, W. Ikeda, and H. Ogita. 2008. Nectins and nectin-like molecules: Roles in contact inhibition of cell movement and proliferation. *Nat. Rev. Mol. Cell Biol.* 9:603–615.

Finale: Molecular Biology, Developmental Biology, and Evolution

Carroll, S.B. 2005. Evolution at two levels: On genes and form. *PLoS Biol.* 3:1159–1166.

Carroll, S.B., B. Prud'homme, and N. Gompel. 2008. Regulating evolution. *Sci. Am.* 298(5):60–67.

Haerty, W., and C.P. Ponting. 2014. No gene in the genome makes sense except in the light of evolution. *Annu. Rev. Genomics Hum. Genet.* 15:71–92.

Prud'homme, B., N. Gompel, and S.B. Carroll. 2007. Emerging principles of regulatory evolution. *Proc. Natl. Acad. Sci. USA* 104:8605–8612.

Raff, R.A. 2000. Evo-devo: The evolution of a new discipline. *Nat. Rev. Genet.* 1:74–79.

Model Organisms Appendix

Bacterium, *Escherichia coli*

Budding Yeast, *Saccharomyces cerevisiae*

Bread Mold, *Neurospora crassa*

Nematode, *Caenorhabditis elegans*

Mustard Weed, *Arabidopsis thaliana*

Fruit Fly, *Drosophila melanogaster*

House Mouse, *Mus musculus*

Humans differ from other organisms in their cognitive abilities and sense of wonder about their surroundings, and it is this curiosity that drives us to study life. What are we made of? How do we work? To understand humans and other living creatures, scientists have studied a wide variety of organisms, revealing a great deal, including the striking universal features shared by all living things. All organisms use the same amino acids, the same nucleotides, and essentially the same genetic code.

There is more to molecular biology than satisfying our curiosity about how life works. We also strive to understand the causes of disease and to apply our understanding to medicine, agriculture, and technology. This book points out numerous examples of how we have learned about human diseases—their causes and, in some cases, how to treat, cure, or prevent them. Scientists have discovered antibiotics to treat most bacterial infections, have developed vaccines for many types of viral infections, and now understand a great deal more about cancer and its treatments. The vast majority of these discoveries and developments came from studying model organisms.

A Few Organisms Are Models for Understanding Common Life Processes

When a particular species is chosen for intensive investigation by many laboratories, it is referred to as a model organism. This focus on one species by many labs allows the development of a large body of information that provides deep insights into that organism's living functions.

The organism is considered a model because researchers assume that what they learn about it will hold true for related organisms. The particular organism selected for study depends on the questions being asked. Throughout this book, we encounter the contributions of model organisms to our knowledge of molecular biology, and several of the most frequently used organisms are reviewed here.

We should note, however, that sometimes an organism that is "off the beaten track" is studied by only a few laboratories, purely out of curiosity—and these investigations can also have a profound effect on research. For example, *Thermus aquaticus* gave us Taq polymerase for the polymerase chain reaction (PCR; see the How We Know section at the end of Chapter 7). The study of *Tetrahymena thermophila* led to the discovery that RNAs can act as catalysts (i.e., ribozymes; see Chapter 16). And studies of some little-known insect viruses, the baculoviruses, gave rise to a recombinant protein expression system that is now in wide use (see Chapter 7).

Focusing on a handful of different organisms is important at a practical level: there are many more species than there are scientists. Indeed, developing an organism into a scientific tool of research is not easy. It requires many years of study to understand the organism and become familiar with its life cycle, proper nutrition, and optimum growth and storage conditions. Especially time-consuming is the development of genetic tools to manipulate the organism's genome. There are no "standard procedures" for this; genetic tools are largely specific to the organism and are often found by trial and error. This is why it is important that many laboratories work on the same organism and share their knowledge. A critical mass of interest in a model organism eventually leads to international conferences, online databases, and the formation of stock centers that maintain and distribute strains.

Of all the organisms in the world, why were certain ones chosen as models? The choices were often made with a healthy dose of serendipity. However, some common features underlie the utility of an organism as a model. Model organisms should have a rapid life cycle. They should produce many progeny, so that researchers can find and study rare genetic events. Size is important, too, because large organisms and their numerous large progeny would quickly exhaust the space of a typical laboratory. Model organisms should be easily propagated using a simple and inexpensive food source, and there should be a convenient method of long-term storage for accumulating strains for further study. **Table A-1** summarizes some features of the model organisms described here.

Studies on genetics and metabolic pathways in the early 1900s used complex multicellular organisms such as plants, fruit flies, rats, and mice. Later, researchers recognized that single-celled organisms are also amenable to fundamental studies of genes and cellular metabolism. In the 1940s, microbes such as *Escherichia coli*, yeast, and *Neurospora crassa* became the most useful models for understanding the basic chemistry of life. They also provided better starting material for biochemists than did animal tissues, because single-celled organisms are a uniform population of identical cells, whereas tissues are composed of different cell types.

No single model organism can answer all questions about life. Single-celled organisms continue to teach us about central aspects of fundamental life processes, such as chromosomal replication, DNA repair, recombination, gene expression, signal pathways, and control of the cell cycle. But single-celled organisms are insufficient for addressing questions about the development of multicellular organisms and most types of disease. Thus, the nematode worm and fruit fly are of enormous use in revealing how multicellular organisms are organized and the basics of how the animal body plan is determined. These organisms also provide insights about many types of disease. Similarly, the mustard weed was chosen as a model organism for plant development.

By far the most useful model of human disease is the mouse. It is, however, not the simplest of model organisms. For ease of growth and DNA manipulation, the mouse pales in comparison with the other model organisms. Genetic strains of mice are costly and time-consuming to construct. But one of the great advantages of the mouse is that 99% of its genes have homologs in the human, including the genes associated with human disease. So despite the difficulties, the mouse is an attractive model in which to study the diseases that afflict us.

We present here a brief overview of several model organisms in use today, including how they have contributed to, and continue to further, our understanding of life. As we have noted, many other organisms have also contributed greatly to our understanding of living processes, including bacteriophages and other viruses, *Tetrahymena thermophila* (a protozoan), *Schizosaccharomyces pombe* (fission yeast), *Xenopus laevis* (frog), and *Brachydanio rerio* (zebra fish). Before we launch into details of particular model organisms, we briefly describe a few highlights of how we learn about human disease from studying model organisms, in conjunction with genomics and cell culture.

Three Approaches Are Used to Study Human Disease

What causes heart disease, diabetes, neurodegenerative disorders, or cancer? How can these, and other diseases, be prevented, treated, or cured? The study of model organisms is usually the first step in understanding cellular processes that can be altered in human diseases. Using a homolog, we can study a human disease–causing mutation in a model organism and learn how the mutation disrupts the cellular process at the molecular, cellular, or organismal level. In addition, mouse models are often used to test treatments for diseases, as an early step in drug development. Model organisms are often the first avenue to understanding human disease, but human genomics and cell culture can provide an even deeper understanding.

The availability of the complete human genome sequence has been an enormous aid in our understanding of human disease genes at the molecular level, as well as in bioinformatics studies on human evolution and migrations (see Chapter 8). Our capacity for language and written history has played a large role in elucidating the genetics of human disease. In particular, people actively seek out medical and scientific advice for a disease, and often can recall a family pedigree stretching back generations that might provide information about how the disease is transmitted. We see examples of this throughout the book, including hemophilia in royalty (see Figure 2-27), sickle-cell anemia (see Highlight 2-1), and early-onset Alzheimer disease (see Figure 8-11). Identifying human disease genes is a difficult but important

TABLE A-1

Basic Information on Seven Common Model Organisms

	E. coli	S. cerevisiae	N. crassa	C. elegans	A. thaliana	D. melanogaster	M. musculus
Basic Facts							
Type of organism	Bacterium, single-celled	Eukaryote, single-celled ascomycete fungus	Eukaryote, multinucleate filamentous ascomycete fungus	Eukaryote, nematode worm	Eukaryote, angiosperm plant	Eukaryote, insect	Eukaryote, mammal
Natural habitat	Animal intestine	Plant surfaces	Dead vegetation	Rotting fruit	Global, temperate climate	Rotting fruit	House, fields
Size	1–2 μm	4–6 μm	Irregular	1 mm	10–20 cm	2.5 mm	17 cm
Reproduction	Asexual fission; some conjugation	Asexual, budding (mitosis); sexual, sporulation (meiosis)	Sexual, sporulation	Sexual, self- and cross-fertilization; some asexual	Sexual, self- and cross-fertilization	Sexual, mating	Sexual, mating
Generation time	20 min	70–90 min	2–4 weeks	50 h	6 weeks	10–12 days	9 weeks
Growth conditions	Petri plates or liquid culture	Petri plates or liquid culture	Petri plates, race tubes, or liquid culture	Petri plates or liquid culture	Petri plates or small horticultural trays	Bottles or vials	Cages
Food source	Yeast extract and tryptone broth; defined media possible	Yeast extract, peptone; defined media possible	Complete or defined media (sugar, inorganic salts, biotin, and nitrogen)	Bacteria (E. coli)	Light, water, source of nitrogen and other minerals (i.e., fertilizer)	Yeast and molasses, fruit	Plant matter
Storage	Frozen glycerol stocks; lyophilized cells	Frozen glycerol stocks; lyophilized cells	Frozen glycerol stocks, silica stocks	Frozen at 280°C	Seeds	Live propagation	Frozen embryos

(continued)

TABLE A-1

Basic Information on Seven Common Model Organisms (continued)

	E. coli	S. cerevisiae	N. crassa	C. elegans	A. thaliana	D. melanogaster	M. musculus
Genetic Statistics							
Genome size	4,639,675 bp	12,070,898 bp	42,900,000 bp	100,269,917 bp	119,186,997 bp	166,600,000 bp	2,729,273,687 bp
Chromosome number	1 (haploid)	16 (haploid); 32 (diploid)	7 (haploid); 14 (diploid)	11 (diploid, male), 12 (diploid, hermaphrodite)	10 (diploid)	8 (diploid)	40 (diploid)
Gene number	4,410	6,000	9,826	21,035	25,540	14,065	29,083
Genes similar to human	8%	30%	6%	36%	18%	50%	99%
Gene naming convention	dnaA	GCN4	arg-1	fem-3	AAO	ry	Cdc20
Protein naming convention	DnaA	Gcn4	ARG-1	FEM-3	AAO	Ry	Cdc20
Genome website	www.ecocyc.org	www.yeastgenome.org	www.broadinstitute.org/annotation/genome/neurospora/MultiHome.html	www.wormbase.org	www.arabidopsis.org	www.flybase.org	www.jax.org

task and is being accomplished at a rapid and accelerating rate. Knowledge of the genetics involved in transmitting a disease may help couples plan their families and cope with the possible maladies that may be passed on to their sons and daughters. For scientists, knowledge of a disease gene can help devise a treatment or cure.

A third way we study ourselves is by culturing individual human cells in vitro. Cells taken directly from the body and then grown in culture typically die within 40 (or fewer) generations. But cells taken from cancer tissue have altered growth control and can often be grown through countless generations; they are referred to as "immortalized." Cells can sometimes even be removed from normal tissue and then immortalized in tissue culture by infection with particular viruses. Through these

and other means, many different types of human tissue cells are grown and maintained in culture, including hepatocytes (liver), renal cells (kidney), fibroblasts (skin), glial cells (nerve), lymphocytes (blood), and myocytes (muscle). By investigating cancer cells and transformation agents, we have also learned a great deal about the genes involved in cancer (see Highlight 12-1). Studies of human and other primate cells in tissue culture have provided important information about surface receptors, protein trafficking, viral entry, and cellular reproduction. Human tissue cells can even be grown in quantities suitable for biochemical studies (see Chapter 7). Recent advances in stem cell research hold promise for the treatment of many diseases and for developing replacement tissue (see Chapter 22).

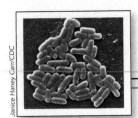
Janice Haney Carr/CDC

Bacterium, *Escherichia coli*

Escherichia coli is a single-celled bacterium. It is a natural occupant of the animal gut and is typically harmless, although unusual toxic varieties exist. *E. coli* is small (microscopic) and reproduces rapidly by fission to form clonal colonies on agar plates, with tens of millions of cells produced in a day. It can also be grown in liquid media to produce astounding numbers of progeny—an important characteristic for studies of rare genetic events. *E. coli* contains a single circular chromosome, and studying mutants is much easier in a haploid organism than in a diploid organism, as there are no dominant genes to mask a mutation. Common mutant phenotypes are resistance to bacteriophages, ability to grow on certain carbon sources, antibiotic resistance, and colony size and shape. *E. coli* also offers a plentiful, uniform source of starting material for biochemical studies. Thus, the study of *E. coli* combines the power of genetics with the power of biochemistry to understand life processes at the molecular level.

E. coli is not a eukaryote and thus has limited homology to humans. This restricts its usefulness in studying all but the most basic cellular processes. For example, *E. coli* lacks nucleosomes, and most of its proteins are unmodified. Eukaryotic protein expression in transformed *E. coli* sometimes does not work, because it requires posttranslational modifications that the bacterium cannot carry out.

Early Studies of *E. coli* as a Model Organism

The marriage of biochemistry and genetics made *E. coli* a rich source for the discovery of new information. The famous "fluctuation analysis" by Salvadore Luria and Max Delbruck in 1943 demonstrated that *E. coli* mutates spontaneously to become resistant to bacteriophage and passes the phage-resistance trait to new progeny—thus establishing *E. coli* as a model system for understanding the nature and function of genes. The discovery by Joshua Lederberg and Edward Tatum that *E. coli* undergoes a type of mating and crossing over (DNA conjugation mediated by the F plasmid) made possible DNA exchange and complementation tests, firmly rooting

E. coli as a model for genetics. In 1952, Alfred Hershey and Martha Chase used *E. coli* to identify DNA as the genetic material of T2 phage (see the How We Know section at the end of Chapter 2). In 1958, Matthew Meselson and Franklin Stahl selected *E. coli* for their studies to demonstrate that DNA replication occurs by a semiconservative mechanism (see Figure 11-1). In the same year, Arthur Kornberg and his colleagues published their results on the purification and characterization of the first-discovered DNA polymerase (see the How We Know section at the end of Chapter 11).

Francis Crick and Sydney Brenner used *E. coli* genetics to establish the triplet nature of the genetic code in 1961. By 1966, the genetic code had been cracked by Marshall Nirenberg, Heinrich Matthaei, Gobind Khorana, and their colleagues, using synthetic oligonucleotides and *E. coli* extracts (see Chapter 17). A completely new era of gene regulation research was opened up by the studies of François Jacob and Jacques Monod on the detailed workings of the *lac* operon (see Chapter 20 and the How We Know section at the end of Chapter 5). The identification of plasmids in *E. coli*, and methods to handle them, ushered in the era of recombinant DNA technology. Plasmids continue to be wonderful biotech tools, valuable beyond compare (see Chapter 7).

Life Cycle

Like most bacteria, *E. coli* reproduces by binary fission (**Figure A-1**). As the cell grows, it senses when it has reached the correct size to begin replication. The duplicate chromosomes partition to opposite poles of the cell. Cytokinesis splits the cell in half to produce two daughter cells, each containing an identical chromosome. The entire process requires only about 20 minutes at 37°C, the average mammalian body temperature. Thus, a large, visible colony of millions of identical bacteria form within only 24 hours on a Petri plate of rich medium. Huge quantities of *E. coli* can also be grown in liquid culture within a day.

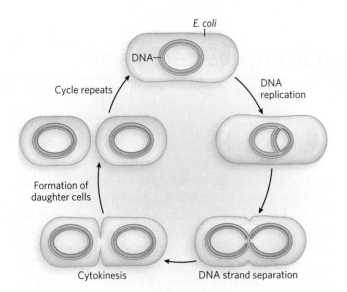

FIGURE A-1 The life cycle of *Escherichia coli*.

Genetic Techniques

For a complete description of many of these techniques, see Chapter 7.

Mutagenesis Mutation frequency in *E. coli* can be enhanced by chemicals or UV light. Mutant cells can be selected by phenotype or screened by replica plating on different media.

Plasmids *E. coli* has several natural plasmids that can harbor foreign DNA in the cell. Researchers use these plasmids for protein expression, as shuttle vectors, or for building and sequencing large genomes.

Introduction of DNA *E. coli* can be made to accept foreign DNA by chemical transformation or electroporation. Cells transformed with plasmid DNA containing an antibiotic-resistance gene can be selected and maintained by adding that antibiotic to the growth medium.

Gene Knockouts Plasmids without a replication origin but containing an antibiotic-resistance gene can be selected for integration into the bacterial chromosome. Integration is typically achieved by homologous recombination, convenient for gene deletion, or knockout.

Complementation This technique is often used to prove that a mutation is in a particular gene by adding a wild-type gene to restore function. Complementation analysis of mutant cells can be performed with plasmids that contain a wild-type copy of the gene of interest. Complementation analysis of unknown mutations often makes use of a genomic plasmid library. Positive clones (cells that survive under conditions in which the mutant cannot live) can then be sequenced to identify the complementing gene.

Recombinant Protein Expression Rapid and inexpensive growth of *E. coli* makes it the most widely used vehicle for the expression and purification of foreign proteins. Foreign proteins are typically expressed from genes located on high-copy-number plasmids. Expression is induced using regulatory elements derived from well-studied *E. coli* and phage promoters.

E. coli as a Model Organism Today

Multiprotein Machines Biochemical and structural studies are now illuminating the once unknown atomic details of structures and processes of information transfer. These include the structure and function of the DNA replication apparatus and the DNA repair and recombination proteins, regulation of RNA polymerase, and the structure and chemistry of the ribosome. These multiprotein structures and processes occur in bacteria and eukaryotes alike, but they are streamlined in the relatively simple *E. coli* cell.

Cytologic Studies Processes can be visualized in living *E. coli* cells through the use of fluorescent fusion proteins. For example, recent studies of this sort have allowed detection of the proteins involved in chromosomal replication and their movements along the chromosome during replication.

Recombinant DNA Technology *E. coli* holds the central position in recombinant DNA technology. Its plasmids are widely used as shuttle vectors to amplify and maintain the DNA libraries of other organisms. *E. coli* is also still one of the most widely used vehicles for the expression and purification of foreign proteins.

Systems Biology Its relatively small genome makes *E. coli* an attractive system in which to pioneer approaches to systems biology. DNA chips have been constructed to examine the response of every transcript to a wide variety of conditions. The functions of about 35% of *E. coli* genes are still unknown; a genome-wide collection of *E. coli* gene knockouts is helping to assign function to these genes and to identify new gene interaction networks.

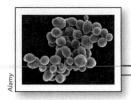
Alamy

Budding Yeast, *Saccharomyces cerevisiae*

Saccharomyces cerevisiae is a microbial ascomycete fungus, used by humans for centuries to make bread, beer, and wine. It is generally referred to by scientists as budding yeast, but is also called yeast, baker's yeast, and brewer's yeast. Yeast has several features that make it an attractive model organism. It is a single-celled haploid organism, reproduces rapidly in liquid media, forms clonal colonies on agar plates, and can be stored frozen. As is the case for *E. coli*, these features make yeast an excellent model organism for studying the genetics and biochemistry of fundamental life processes, such as mechanisms of replication, transcription, and translation. However, unlike *E. coli*, yeast is a eukaryote with a full complement of intracellular organelles, including mitochondria and a nucleus, with DNA packaged into chromatin, and with a cytoskeletal structure. *S. cerevisiae* is thus an apt model for mechanisms such as cell cycle control, regulation by protein phosphorylation, chromosome structure, and mitochondrial function.

The *S. cerevisiae* genome is only ~12 Mbp long, with ~6,000 genes. Yet it contains homologs to several human disease genes, making it a model for understanding the basic function of gene products involved in some forms of disease, as well as indicating that some human diseases result from the disruption of relatively simple and fundamental life processes. However, the single-celled yeast does not serve as a model of development, and the many human diseases arising from mutations that cause developmental abnormalities cannot be studied in yeast, limiting its usefulness in medical research.

The haploid nature of yeast allows convenient mutational studies. Yeast is also easily transformed with exogenous DNA. This DNA frequently integrates into the chromosome by homologous recombination, making possible the construction of gene knockout strains. The ease of genetics, combined with the ease of growth, makes *S. cerevisiae*, like *E. coli*, a model organism that beautifully combines the power of genetics with that of biochemical studies.

Early Studies of Yeast as a Model Organism

Yeast has been a model organism for more than 100 years, and it essentially spawned the field of modern biochemistry. Louis Pasteur proved that yeast ferments sugar to alcohol and CO_2; he declared fermentation a vital life force that could not be separated from the living organism. In 1897, Eduard Buchner accidentally observed fermentation in a yeast cell extract, and called the fermenting substance "zymase." Later studies showed that zymase was actually a mixture of many different enzymes. The suffix *ase* is used for many enzyme names to this day.

Life Cycle

Budding yeast can exist in either the haploid or the diploid state (**Figure A-2**). Both haploid and diploid cells can reproduce asexually by budding. Sex in *S. cerevisiae* is determined by the gene products of the mating type (MAT) locus, for which there are two alleles, **a** and α (see Figure 13-22). Haploids of different mating types can be mated to form diploids by plating them together. Diploid **a**/α yeast can be converted to the haploid state in media that induce them to sporulate, undergoing meiosis to produce four haploid spores contained in an ascus. The spores can be germinated, and each haploid phenotype can be analyzed after growing the spores into colonies.

Genetic Techniques

Mutagenesis Mutational studies in *S. cerevisiae* are simplified by its ability to grow as a haploid. This allows the study of mutant phenotypes detectable as failure to grow under diverse conditions (e.g., mutants requiring specific nutritional conditions, or high or low temperature). In the case of gene knockouts, it is very useful that particular nutritional requirements can be made to report the presence or absence of a given knockout.

Complementation Haploid yeast can be mated to produce diploid cells for complementation analysis of mutant haploids. Mutations in essential genes can be maintained by harboring them in the diploid state. On sporulation of a diploid with a mutation in an essential gene, two of the four spores will not be viable. Isolation of all four products of meiosis in an ascus is useful for the analysis of recombination during meiosis.

Introduction of DNA DNA can be introduced into *S. cerevisiae* by chemical or physical abrasion of the thick cell

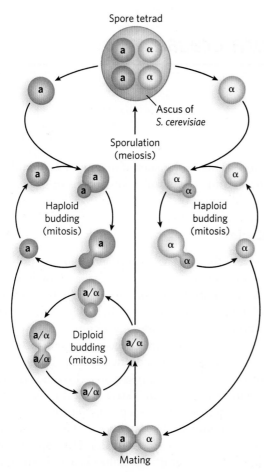

Spore tetrad

Ascus of
S. cerevisiae

Sporulation
(meiosis)

Haploid
budding
(mitosis)

Haploid
budding
(mitosis)

Diploid
budding
(mitosis)

Mating

FIGURE A-2 The life cycle of *Saccharomyces cerevisiae*.

wall. Linear DNA, with a selectable marker and ends homologous to the yeast chromosome, readily integrates into the genome by homologous recombination. This configuration permits the ends to recombine with their homologous sequence in the yeast chromosome, replacing the gene of interest with the selectable marker and effectively knocking out the gene.

Selectable Markers Selection in yeast is mostly performed with auxotrophs, strains that lack a functional gene for an enzyme in an amino acid biosynthetic pathway and therefore require that amino acid to grow. Cells are grown on defined media lacking the now essential amino acid. Only cells that acquire the marker and thus are capable of synthesizing the amino acid from raw materials will survive. Drug-resistance markers can also be used for selection.

Plasmids DNA can also be maintained in yeast as a circular plasmid, using one of the many chromosomal origins known as an ARS (autonomously replicating sequence). YACs (yeast artificial chromosomes) can maintain very

large DNA inserts (300 kbp to 1 Mbp), which are useful in genome sequencing projects (see Figure 7-7). *S. cerevisae* also contains a natural 2 micron (2μ) plasmid, which can be used to maintain exogenously derived DNA inside the cell.

Yeast as a Model Organism Today

Fundamental Mechanisms of Information Flow The fact that yeast are microbes, the ability to grow them in large quantities for biochemical studies, and the ease of genetic manipulation, all make yeast ideal for studying detailed mechanisms of fundamental processes, including replication, gene regulation, DNA repair, transcription, translation, and recombination.

Cell Division Cycle When yeast divides, the bud is visible in the microscope, as are the size changes at different phases of the cell cycle. Accumulation of mutant cells stuck in a budded or unbudded state has been used as a phenotype by which to identify numerous cell division cycle (*cdc*) mutants, useful for genetic and biochemical dissection of the cell cycle. Humans have homologs to many of the cell cycle genes discovered in yeast.

Signaling Pathways Yeast is a model for signal transduction pathways. For example, yeast pheromone signaling is a good model for 7-TM (transmembrane) G protein-coupled receptor (GPCR) and MAPK signaling.

Meiotic Recombination Because *S. cerevisiae* produces an ascus with four spores derived from a single cell, recovering all four products permits detection of more complex genetic events than simple crossovers. For example, study of gene conversion in yeast led to an understanding of the role of double-strand breaks in meiotic recombination.

Systems Biology It is now within the scope of systems biology to understand most of the genetic, physical, and molecular interactions of the 6,000 yeast genes. Pairwise examinations of genetic interactions between these gene knockouts are leading to the identification of new pathways. Proteomic approaches have been pioneered in yeast. Intracellular copy numbers of almost all yeast proteins, proteome-wide, have been determined. Protein-protein and protein-RNA interactions have also been assessed on a global scale, using yeast two-hybrid and three-hybrid techniques (see Chapter 7) and mass spectroscopy of protein complexes. Cell biological studies are facilitated by a library of fluorescent protein tags for every open reading frame in the yeast genome. In addition, yeast is used as a protein expression system for the study of foreign genes.

Bread Mold, *Neurospora crassa*

The bread mold, *Neurospora crassa*, is a filamentous asco-mycete fungus. It was one of the first eukaryotic microorganisms adopted for genetic studies. About 75% of all fungi are ascomycetes, and the remainder are basidiomycetes (including mushrooms). About 90% of ascomycetes are filamentous (and often with multinucleate hyphae, as further explained below), and the rest are single-celled yeasts. The genome of *N. crassa* (often referred to simply as *Neurospora*) has ~43 Mbp and 9,826 genes, making it comparable to the fruit fly in genetic complexity.

The war between filamentous fungi and bacteria has yielded important antibiotics, including penicillin. These complex organic molecules are found nowhere else in nature, and accordingly, more than 40% of *Neurospora* genes have no identifiable counterpart in any other organism studied. The ecological niche filled by filamentous fungi includes the decay of many types of biological material. This may explain why *Neurospora* can grow on highly unusual carbon and nitrogen sources—which is probably reflected in other uncommon genes.

A unique feature of *Neurospora* is the arrangement in the spore case (ascus) of the spores produced from meiosis of a single diploid nucleus. The spores are linearly arranged in the order in which they are produced by the meiotic divisions. With the ability to precisely separate spores and study them individually, researchers have an ideal model for the analysis of meiotic recombination. *Neurospora* is also attractive for studying several gene silencing mechanisms, some of which also exist in multicellular eukaryotes.

Early Studies of *Neurospora* as a Model Organism

Neurospora achieved its early standing as a model organism because of its haploid state, rapid growth, production of millions of asexual spores per colony, and ease of culture on simple, defined media. The capacity to synthesize all its essential biomolecules from media with known ingredients propelled *Neurospora* into the spotlight in the 1940s as an ideal model in which to study the genetics of metabolic pathways that are similar in all cells. Of particular historical note was the use of *Neurospora* by George Beadle and Edward Tatum in studies that formed the basis for the "one gene, one polypeptide"

hypothesis. This work ushered in a new era of molecular genetics and the use of microorganisms in genetic studies (see Figure 2-23).

Life Cycle

The asexual life cycle of *Neurospora* starts with haploid spores released from microconidia and macroconidia (**Figure A-3**, top). On germination of the haploid spore, a tubular projection elongates by tip growth to form a filamentous, multinucleate mycelium, which contains branching hyphae. Growth occurs through the replication and division of haploid nuclei, accompanied by the formation of incomplete cross walls. Growth is rapid, up to 10 cm in a single day. On Petri plates, the compact network of filaments produces a colony. The network of hyphae in a colony is essentially one large, continuous single cell with many haploid nuclei. The haploid colony sporulates by forming specialized aerial structures, termed conidiophores, which produce numerous asexual spores, termed conidia. One colony can produce spores that number in the millions.

Two different alleles at the mating type locus, *mat A* and *mat a*, control the *Neurospora* sexual cycle. Spores contain either the *MATA* or *MATa* allele. *Neurospora* is hermaphroditic, and both mating types produce female reproductive structures (protoperithecia). However, fertilization can occur only between reproductive structures of opposite mating type (protoperithecia and conidia/microconidia of opposite mating type). Fertilization results in formation of hyphae containing both mating type nuclei (a dikaryon). Nuclear fusion in the dikaryon is followed by meiotic cell divisions, each diploid nucleus yielding four haploid spores (ascospores) in an ascus (see Figure A-3, bottom). The haploid spores undergo one further mitotic division, forming eight spores arranged according to their cell of origin and stacked in a linear row within the long thin ascus. Numerous asci, each from an independent meiotic event, are held in a structure called a perithecium. Each ascospore in the perithecium is haploid and can form another *Neurospora* colony. Approximately 2 to 4 weeks elapse in proceeding from spore to haploid hyphae, to diploid nuclei, and back to the production of spores.

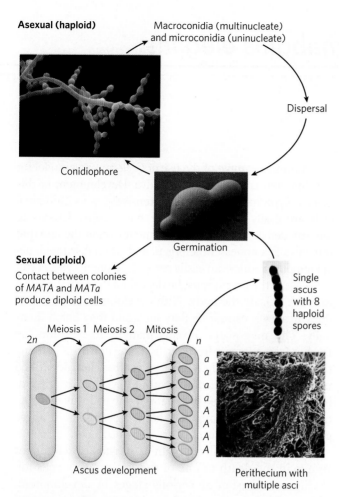

Asexual (haploid)

Macroconidia (multinucleate) and microconidia (uninucleate)

Dispersal

Conidiophore

Germination

Sexual (diploid)

Contact between colonies of *MATA* and *MATa* produce diploid cells

Single ascus with 8 haploid spores

Meiosis 1   Meiosis 2   Mitosis

2n

n

a
a
a
a
A
A
A
A

Ascus development

Perithecium with multiple asci

FIGURE A-3 The life cycle of *Neurospora* crassa.

[Sources: (Conidiophore) From Neurospora: Contributions of a Model Organism by Roland Davis © 2000 by Oxford University Press, Inc. Photo courtesy of Matthew L. Springer, University of California, San Francisco. (Germination) Roca, M.G., Arlt, J., Jeffree, C.E., Read, N.D. 2005. Cell biology of conidial anastomosis tubes in Neurospora crassa. Eukaryotic Cell 4: 911–919. Reproduced with permission from American Society for Microbiology. (Single ascus) Courtesy of Namboori B. Raju, Stanford University. (Perithecium) Courtesy of Louise Glass.]

Genetic Techniques

Mutagenesis *Neurospora* asexual spores can be mutagenized by treatment with ionizing irradiation. Irradiated spores are germinated in complete media and allowed to sporulate, forming stocks of mutant lines. Spores from these lines can then be screened for phenotypes of interest, and mutations can be mapped by genetic crosses.

Introduction of DNA *Neurospora* readily takes up exogenous DNA by transformation. The DNA must integrate into the genome to be stably inherited. The antibiotic hygromycin can be used to select for *Neurospora* with a transgene. Unlike in yeast, DNA insertion rarely occurs through homologous recombination in *Neurospora*; insertion is more often random and untargeted.

Gene Knockouts There is an unusual way of generating a specific gene knockout in *Neurospora*. A cell with a duplicate gene (constructed either by transformation or by crosses using a translocation strain), when crossed, undergoes a process called RIP (repeat-induced point mutation). RIP occurs in the haploid nuclei of premitotic cells; the repeated DNA is searched out and destroyed by littering it with numerous $G{\equiv}C$-to-$A{=}T$ transition mutations, in both copies of the duplicate gene. This process essentially results in a knockout strain. The discovery that the frequency of homologous integration in *Neurospora* can be increased to nearly 100% by using mutants containing lesions in *ku70/ku80* homologs (which disrupt nonhomologous end joining) led to the generation of a publicly available, full-genome deletion strain set for this organism (www.fgsc.net/ncrassa.html).

Sporulation As *Neurospora* chromosomes segregate during meiosis, the daughter cells become arranged linearly in a long axis. The frequency of crossing over can be used to map the distance of a mutant allele locus from the centromere.

Neurospora as a Model Organism Today

Meiotic Recombination As we have seen, *Neurospora* packages its ascospores from a single meiotic event in an arrangement (in the ascus) that reflects the order of chromosomal segregation during meiosis. This feature makes *Neurospora* an attractive system for studies of crossover recombination during meiosis.

Circadian Rhythms *Neurospora* has a daily circadian rhythm of conidial spore formation. Conidial spores are conspicuous to the naked eye, so the phenotype of rhythmic behavior can be examined on Petri plates or in race tubes—long tubes in which the rapid growth of hyphae can be measured. Why this rhythm exists in *Neurospora* is not clear, but it serves as a convenient model for understanding the molecular basis of rhythmic behavior.

Silencing Mechanisms *Neurospora* contains at least three different silencing mechanisms that are thought to have evolved as genomic defenses against invading DNA, such as transposons and viruses. Meiotic silencing by unpaired DNA (MSUD) is a process in which any unpaired gene that is detected during meiosis is silenced. Quelling occurs in haploid cells and detects duplicate sequences. RIP detects duplicated sequences just before meiosis, resulting in methylation and $G{\equiv}C$-to-$A{=}T$ mutations in both copies of a duplicated sequence. MSUD and quelling are related to RNA interference, while RIP is a consequence of cytosine methylation in duplicated sequences.

Nematode, *Caenorhabditis elegans*

Small single-celled microbes such as *E. coli* and yeast are overwhelmingly successful models for studying the fundamental chemistry of life processes at the cellular level, but multicellular models are needed to investigate the complexities of development and the nervous system. In the 1960s, Sidney Brenner recognized that it was time to ask questions about the nervous system and how a multicellular organism develops from a single cell. Brenner decided on the nematode worm, a small, translucent metazoan, as a model for pursuing these new questions.

Caenorhabditis elegans is a member of the Rhabditidae family of nematodes, but unlike other family members, it is not pathogenic or parasitic to other animals. *C. elegans* is an appropriate choice as a model for animal development because it grows rapidly and, depending on temperature, can develop from an egg to a sexually mature adult in 2½ to 5½ days. Easy to grow on agar plates with *E. coli* as a food source, the worms can even be frozen for long-term storage. *C. elegans* hermaphrodites can self-fertilize, a property that allows quick recovery of homozygous mutants from a population of mutagenized worms. Males, produced by self-fertilizing hermaphrodites at a low frequency, can mate with hermaphrodites to create strains with new mutant combinations, an essential aspect of any genetic model organism. In addition, *C. elegans* is transparent, allowing every cell to be visualized during development. Although the hermaphrodite contains only 959 somatic cells, the nematode is highly complex, with a nervous system, muscles, and digestive and reproductive systems, and exhibits a variety of behaviors for neurological genetic studies. Each worm produces hundreds of progeny, allowing mutant screens for all types of anatomic and behavioral changes.

Early Studies of *C. elegans* as a Model Organism

To lay a foundation for studies on how a multicellular organism develops, Brenner and others mapped the location of all 959 cells in the hermaphrodite worm by image reconstruction of tissue slices. During development, each migratory cell travels to a precise position in the animal. Further studies led researchers to the discovery that about 12% of cells consistently die during development and that cell death is genetically programmed. We now know that programmed cell death is important to the development of higher organisms, including humans, and that defects in this process can lead to cancer.

A prime example of the use of *C. elegans* as a model for development is studies on the vulva. Development of this 22-cell organ has been studied intensively, as its individual cells are easily visualized in the microscope. Screens to identify defects in vulval development exploit the fact that the vulva is a nonessential organ, and when disrupted, development produces an easily recognized phenotype. Fertilized eggs in the uterus must be deposited through the vulva to hatch outside the body. With developmental defects in the vulva, eggs cannot be deposited and they hatch internally. The microscope reveals many worms inside the mother (sometimes referred to as a "bag of worms"). Using this obvious mutant phenotype, researchers have identified numerous genes that control vulval development, including genes in a conserved phosphorylation pathway that controls cell growth. Mammalian homologs of some of these genes encode tumor suppressors and oncogenes.

Programmed cell death in *C. elegans* development results in the death of one of the two progeny cells following cell division at various stages of development. This mechanism is essential for the proper developmental path of the surviving progeny cell. Many examples of programmed cell death are documented in *C. elegans*, several of which occur during development of the nervous system. Programmed cell death is also important in human embryonic development, such as in the removal of tissue between the fingers and toes.

Life Cycle

There are two sexes in *C. elegans*, hermaphrodite and male. Hermaphrodites contain two copies of the X chromosome. A low percentage of hermaphrodite germ cells (0.05% to 1.0%) undergo nondisjunction of the X chromosomes during meiosis, to produce so-called null-X gametes. When a null-X gamete fertilizes a normal (X) gamete, the resulting zygote has only one X chromosome (X0) and develops as a male. Hermaphrodites produce 250 to 300 self-sperm, limiting self-progeny to 250 to 300 embryos; the fertilized eggs follow a multistage life cycle. Hermaphrodites crossed with a male preferentially use the male's sperm for fertilization and can produce hundreds of additional cross-progeny embryos that, other than the sex ratio, are identical to self-progeny and follow a multistage life cycle (**Figure A-4**). After embryogenesis, which takes about 12 hours at 25°C, eggs hatch to produce larvae 80 µm long. The initial larva contains

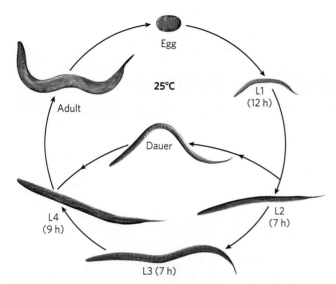

FIGURE A-4 The life cycle of *Caenorhabditis elegans*.
[Sources: (Egg) Reprinted with permission from Zeynep Altun and David Hall, Wormatlas (www.wormatlas.org). (Adult) Courtesy Ian Chin-Sang, Ph.D., Queen's University, Kingston, Ontario, Canada.]

558 cells. Development proceeds through four larval stages (L1 through L4) separated by molts. The final molt results in the sexually competent adult worm. The overall process takes about 52 hours at 25°C. Adults live for approximately 15 days. Under stressful conditions, L2 larvae may enter a dormant stage, the dauer larva, which can persist for several months, waiting for conditions to improve.

Genetic Techniques

Mutagenesis Worms can be mutagenized with chemical treatment or irradiation. Mutant worms can be screened for phenotypic alterations by direct observation with the microscope. Phenotypes include aberrant behavior or development, the inability of certain cells to undergo programmed cell death, altered life span, and inability of larvae to enter the dauer stage.

Self-Fertilization and Cross-Fertilization Self-fertilization produces about 250 to 300 progeny from one hermaphrodite and allows rapid recovery of recessive mutant alleles from individual mutant worms. Although males are produced only rarely, they provide opportunities to construct new genetic stocks by crossing with a hermaphrodite.

Introduction of DNA In transformation studies, linear recombinant DNA is injected directly into the gonad before the eggs are formed. Injected DNA from long tandem arrays that behave like artificial chromosomes is inefficiently transmitted through the germ line. This DNA can be induced to integrate into the genome by gamma irradiation. Rare integration events lead to stable inheritance of transgene arrays. Integration rarely occurs by homologous recombination.

Gene Knockouts Disruption by a transposon was a common method to destroy the function of a *C. elegans* gene. Transposon mutagenesis is rarely performed any more. It has been replaced by random mutagenesis and whole genome sequencing or, more recently, by site-directed cleavage and nonhomologous end joining to create deletions or homologous recombination to engineer mutants and transgenes into the genome.

RNA Interference RNA interference, originally discovered in the nematode, can be used to silence gene function by introducing double-stranded RNA homologous to the gene under study (see Chapter 22).

C. elegans as a Model Organism Today

Signaling Pathways Programmed cell death and vulval development use signaling pathways that are apt models for human signaling pathways. Studies of *C. elegans* development continue to elucidate important features of these processes and related pathways.

Human Disease *C. elegans* has many genes that are homologous to human disease genes, including those in the insulin-signaling pathway, as well as genes involved in heart, kidney, and neurological diseases. Study of these disease genes may illuminate the basis for human diseases.

Aging Genetic studies of the dauer larva have identified a set of genes that, when switched on in the adult, dramatically extend the worm's life span. The presence of homologs of these genes in other animals has obvious implications for the study of aging.

RNA-Based Control of Gene Expression Studies of RNA interference, first discovered in the nematode system, have identified miRNAs involved in gene expression. Indeed, miRNAs are now known to be involved in gene regulation in all plants and animals. The sensitivity of worms to ingested dsRNA (environmental RNAi) enables rapid, whole genome screens for genes important in specific processes.

Neurodevelopment The *C. elegans* nervous system is the animal's most complex organ, comprising over one-third of its cells (302 neurons and 56 glial cells). Unlike the highly branched neuronal connections of vertebrates, the connectivity of neurons in *C. elegans* is relatively simple, with about 5,000 chemical synapses and 2,000 neuromuscular junctions. Behavioral abnormalities are easily observed and can be mapped with precision to particular neuronal networks. Knowledge of the complete neuronal connectivity enables researchers to study how axon growth is guided and how synapses form. In addition, *C. elegans* has several different classes of neurons, enabling genetic studies of neuronal differentiation.

Mustard Weed, *Arabidopsis thaliana*

Arabidopsis thaliana is an angiosperm, a dicot of the mustard family (Brassicaceae), which includes the more familiar broccoli, cabbage, and radish. *Arabidopsis* is generally regarded as a weed, but what it lacks in economic importance it makes up for in its special features as an experimental model. *Arabidopsis* is relatively small, allowing many plants to be grown in a confined space. It has a short life cycle, about 5 to 6 weeks from seed to flower, and each plant is capable of self-pollination, producing thousands of seeds for large genetic mapping studies. In addition, the *Arabidopsis* genome is less than 5% the size of the maize genome (2,500 Mbp) and less than 1% that of wheat (16,000 Mbp). Numerous labs worldwide study *Arabidopsis*, and several public stock centers maintain seed stocks of mutant lines and natural *Arabidopsis* variants called ecotypes, or accessions, as well as genomic resources.

Arabidopsis is a model for the physiology, development, genetics, and structure of all plants. It is also a model for how plants interact with the environment and sense day length, cold, drought, and salt, and how they respond to pathogens. We can expect many differences between plants and animals in the genetics of developmental programs. Nonetheless, studies on development of the flower whorl reveal a deep interconnection with animals. The flower whorl consists of four concentric rings (**Figure A-5**). The outer ring (whorl 1) consists of four sepals; whorl 2, four petals; whorl 3, six anthers; and the inner whorl, two carpels that fuse to form a pistil. Studies of plants with homeotic mutations reveal altered whorl identities. For example, in one mutant, carpels are found in whorl 1 in place of sepals. This genetic behavior is reminiscent of homeotic gene mutations in *Drosophila*, in which limbs protrude from incorrect body segments.

Early Studies of *Arabidopsis* as a Model Organism

Arabidopsis is a relatively recent addition to the select group of model organisms, developed into a robust model in the 1980s. In 1907, Friedrich Laibach identified the number of chromosomes in *Arabidopsis*, and in

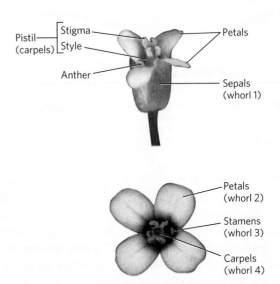

FIGURE A-5 *Arabidopsis thaliana* flower whorl anatomy. [Source: Courtesy Prof. Dr. Sabine Zachgo.]

the 1940s he proposed the use of this plant as a model organism. With the help of Albert Kranz, Laibach collected and organized a large variety of ecotypes, which contributed to the current collection of 750 accessions of *Arabidopsis*. The beginnings of an *Arabidopsis* research community became evident with publication of a newsletter in the early 1960s, and the first International *Arabidopsis* Conference was held in 1965. By the 1980s, *Arabidopsis* was one of several plant models that also included the well-established genetic model, maize. With the development of T-DNA–mediated transformation (described shortly) in 1986, *Arabidopsis* rapidly became the predominant model for plant research.

Life Cycle

Arabidopsis has a common plant life cycle (**Figure A-6**). As in many angiosperms, both male and female germ cells are present in the same flower, and self-fertilization (self-pollination) is readily accomplished. Plants can also be cross-fertilized (cross-pollinated). Each plant produces many flowers, which together can produce more

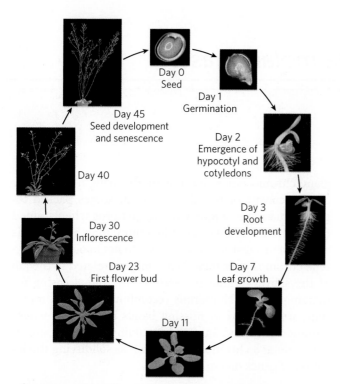

FIGURE A-6 The life cycle of *Arabidopsis thaliana*.
[Source: Douglas C. Boyes, Adel M. Zayed, Robert Ascenzi, Amy J. McCaskill, Neil E. Hoffman, Keith R. Davis1 and Jörn Görlach, Growth Stage—Based Phenotypic Analysis of Arabidopsis The Plant Cell, Vol. 13, 1499–1510, July 2001, Copyright © 2001, American Society of Plant Biologists].

than 10,000 seeds from a single plant. The life cycle from seed germination to a flowering plant and then a new crop of seeds takes 5 to 6 weeks.

Genetic Techniques

Mutagenesis Plants can be mutagenized by treating the seeds with ionizing radiation or chemical mutagens. Plants that are homozygous for recessive mutant genes are easily obtained by self-fertilization.

Complementation *Arabidopsis* in bloom is self-fertile or can be crossed with other plants by techniques similar to those used by Gregor Mendel, in which anthers are excised, thus permitting control over pollination and genetic cross (see Chapter 2).

Introduction of DNA Transformation with recombinant DNA is mediated by *Agrobacterium tumefaciens* T-DNA (transfer DNA), which typically inserts into chromosomes randomly. *A. tumefaciens* is a gram-negative bacterium that causes tumors in plants. It contains a 200 kbp tumor-inducing (Ti) plasmid, which includes genes that facilitate replicative transfer of the DNA into a plant cell. Once inside the plant cell, the plasmid integrates a particular segment of its DNA, the T-DNA, into the plant genome. When recombinant

DNA is present in the T-DNA, this is also inserted into the genome.

Gene Knockouts Homologous recombination of transgenic DNA, producing a gene knockout, occurs only rarely in plants and is not typically used in plant studies. However, large collections of sequenced *Arabidopsis* mutants are available, and specific insertions in individual genes can be ordered from stock centers.

Interference Gene function in *Arabidopsis* can be effectively knocked out using RNAi (see Chapter 22).

Arabidopsis as a Model Organism Today

Plant Evolution The 750 or so different natural accessions of *Arabidopsis* vary considerably in development and form (e.g., leaf shape, flowering time, hairiness, resistance to disease), and these are being studied to explain the evolution of traits and plant responses to the environment.

Light Sensing Plants have a variety of responses to light, and *Arabidopsis* is a genetic model for some of these. One such response is the switch from leaf to flower production. *Arabidopsis* flowers in response to an increase in day length and is a model for day-length sensing. Also, limiting light during seed germination results in altered growth through a developmentally programmed switch that produces tall but spindly plants with a limited root system. *Arabidopsis* is being used as a model to study the genetics of how light shapes this early development. Another light-sensing process under study in *Arabidopsis* is shade avoidance.

Circadian Rhythms As we might expect for organisms with an immobile lifestyle and a dependence on light, plants display strong circadian rhythms. *Arabidopsis* provides an excellent model for exploring the genetic basis of circadian rhythms in plants and the nature of the mysterious "rhythmic oscillator."

Plant Resistance to Pathogens Plants have a wide variety of strategies to survive stressful conditions, including invading pathogens, and *Arabidopsis* is a model for pathogen detection and defense. Among these defenses are antimicrobial molecules, development of physical barriers, and triggering of programmed cell death.

Genetics of Plant Development As a multicellular organism, *Arabidopsis* has a variety of organs and tissue types, each with its own genetics of development. Areas of developmental study include leaf growth, formation of flower whorls, seeds, and roots, development of vascular tissue, embryogenesis, and development of the flower body plan.

Fruit Fly, *Drosophila melanogaster*

We are all familiar with the fruit fly as a cosmopolitan nuisance, but *Drosophila melanogaster* has a 100-year history as an important model organism for studying genetics and development. The fruit fly's body is divided into several segments that form three major sections: the head, thorax, and abdomen. The body sections are encased in a hard chitin cuticle, secreted by underlying epidermal cells, and is rich in anatomic details (indentations, hairs) that serve as phenotypic landmarks for genetic studies.

Drosophila is small (about 2.5 mm long); it is easily and inexpensively grown in the laboratory, in bottles containing a layer of cornmeal, molasses, and yeast. The fruit fly has a reasonably rapid generation time (12 days), is simple to cross by mating, and produces hundreds of progeny in each generation. The main inconvenience for researchers is its ability to fly away. Although *Drosophila* was originally used to study the basic mechanisms of transmission genetics, many new genetic tools can now scrutinize the developmental basis of embryogenesis and the body plan.

Early Studies of *Drosophila* as a Model Organism

In 1908, Thomas Hunt Morgan looked for a suitable organism in which to study animal genetics. With little funding for science available at that time, he eventually settled on *Drosophila* because it was cheap and inexpensive to grow. In 1910, after two years of fruitless studies, Morgan found a male white-eyed spontaneous mutant (the flies normally have red eyes). Elegant and detailed studies of this single mutant revealed that genes are located on chromosomes, that each gene has two alleles, that alleles assort independently during meiosis, and that genes located on different chromosomes assort independently. These and other important findings supported Mendel's laws in the physical context of genes located on chromosomes (see Chapter 2).

The fruit fly is also the first organism for which a genetic map was constructed. Alfred Sturtevant, a student in Morgan's laboratory, mapped the relative distance between genes along chromosomes, based on their frequencies of crossover recombination. Calvin B. Bridges carried the studies further by identifying the exact positions of genes within polytene chromosomes. These giant chromosomes, located in the fruit fly's salivary glands, consist of bundles of chromosomes packed together and have unique banding patterns when stained (**Figure A-7**). Bridges identified more than 5,000 bands arranged in a distinct pattern. We do not know why polytene chromosomes form, but it may be related to the job of the salivary gland in excreting the pupal casing for metamorphosis. Numerous recombination-based mutations were traced to missing bands or to spots where chromosomes had inverted, enabling the mapping of genes along a chromosome, as well as solidifying the location of genes on chromosomes.

Life Cycle

Flies have a short (12 day) diploid life cycle (**Figure A-8**). Sex in flies is determined by X chromosome copy number, not by the Y chromosome (XX is female, XY and the rare X0 are male), although the Y chromosome is required for the production of sperm. About a day after mating, the female begins to lay hundreds of eggs. Nuclear divisions in the embryo are the most rapid of any multicellular organism, and eggs hatch in about one day. The maggot proceeds through three larval instar stages, separated by molts, that take about 5 days. Larvae contain imaginal discs of tissue that are destined to become each of the appendages of the adult fly (e.g., eyes, antennae, legs, wings, halteres (modified wings acting as flight stabilizers), and mouthparts). The pupal case is derived from the larval cuticle, within which metamorphosis occurs in 3½ to 4½ days to yield the adult fly. Flies live approximately 30 days.

FIGURE A-7 An insect polytene chromosome.
[Source: S. F. Werle, *Canadian Journal of Zoology*, 82:118–129. © 2008 Canadian Science Publishing or its licensors. Reproduced with permission.]

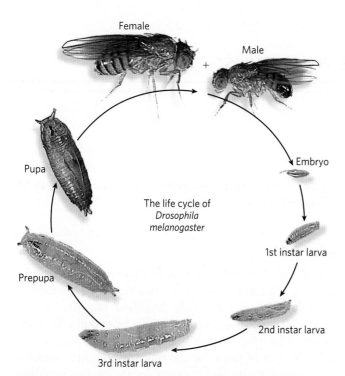

FIGURE A-8 The life cycle of *Drosophila melanogaster*.
[*Source: Prof. Dr. Christian Klambt, Westfälische Wilhelms-Universität Münster, Institut für Neuro- und Verhaltensbiologie.*]

Genetic Techniques

Mutagenesis Flies can be mutagenized by exposure to ionizing radiation or by feeding with mutagenic chemicals. Mutations that result in changes in eye color, wing shape, or body parts can be identified by visual screening.

Introduction of DNA A process known as P-element transformation is used to insert recombinant DNA, and one application of this approach is to study the effects of protein expression on transcriptional control elements. The P element is a 3 kbp transposon that can carry a section of recombinant DNA between its terminal repeats in place of the self-encoded transposase and repressor (see Chapter 14). The recombinant P-element DNA is injected into the fertilized egg, along with a transposase-encoding plasmid. Insertion is random, and the plasmid encoding the transposase gene is lost during cell divisions, thereby preventing reinsertion of the transposon elsewhere in the genome.

Balancer Chromosomes Recessive lethal alleles can be stably maintained in fruit flies when paired with a balancer chromosome—a chromosome that cannot recombine with its homologous pair, due to the presence of many internal inversions that prevent complete alignment during meiosis. Balancer chromosomes are homozygous lethal, so the only progeny that survive are heterozygous for the recessive lethal gene and the balancer.

Genetic Mosaics Genetic mosaics, patches of genetically altered tissue, can be formed in adult or developing flies by

x-ray irradiation to induce mitotic recombination. Mosaics are particularly useful in the study of lethal genes. Genetic mosaics of lethal genes can also be formed through the use of a heat-inducible yeast recombinase (called FLP) engineered into the genome. Heat induction results in a high frequency of mitotic recombination to form genetic mosaics. Although the genetic alterations may be lethal to embryos, they do not necessarily kill the adult, given the localized expression in only some cells.

Gene Knockouts and Gene Replacements Gene knockouts are obtained by homologous recombination, which occurs infrequently in *Drosophila*. The experimental strategies to edit genes in the fly genome depend on homologous recombination. Researchers inject into the fly embryo a modified copy of the targeted gene. A double-strand break in the gene is generated, and during its repair by homologous recombination, the wild-type gene is replaced by the modified copy. More recently, researchers have demonstrated that the CRISPR/Cas system can be used to directly edit endogenous genes, eliminating the need to go through this two-step approach.

RNA Interference RNAi can be used to effectively knock down specific gene products in lieu of a true gene knockout. RNAi can also be employed to knock down a gene product in a specific tissue at a specific time.

Drosophila as a Model Organism Today

Human Disease The ~170 Mbp *Drosophila* genome is about one-twentieth the size of the mouse and human genomes, yet it encodes nearly the same number of gene families. About 60% of the genes known to be involved in human disease have homologs in the fruit fly. For example, studies of embryonic lethal genes in *Drosophila* have helped explain the genetic basis of human birth defects. Other disease models include immunological disorders, diabetes, cancer, Huntington disease, Alzheimer disease, and Parkinson disease.

Body Plan Development Localization of maternal mRNA in eggs results in localized gene expression that sets up the anterior-posterior and dorsal-ventral axes in *Drosophila*. These genes control the body plan and have counterparts in more complex animals. Thus, the fly serves as a relatively simple system to understand body plan formation (see Figure 16-24; see also Chapters 21 and 22).

Behavior *Drosophila* provides a model for understanding the cellular and molecular basis for certain types of behavior. Fruit fly behavioral abnormalities include changes in learning and memory, foraging, resting behavior, sleep, and alcohol consumption, among other behaviors.

House Mouse, *Mus musculus*

The house mouse, *Mus musculus*, has been collected and bred by mouse "fanciers" for hundreds of years, and some of the purebreds developed by fanciers are used today as standards in scientific studies. The house mouse is the leading mammalian model, and it is more like a human than we may care to admit (see Figure 1-12). The mouse genome is almost the size of the human genome and encodes essentially the same number of genes, 99% of which have homologs in the human. In fact, much of the mouse's genome is syntenic with ours, meaning that whole blocks of genes occur in the same order in both species (see Figure 8-4).

Compared with other model organisms, mice are more cumbersome to work with in every way. They are larger, of course, but they also have a generation time of about 8 to 10 weeks and produce, on average, only 3 to 14 pups per litter. These statistics are attractive when compared with other mammals, but pale in comparison with other model organisms. Colonies of mice are also costly to maintain, and they simply cannot be dealt with in the numbers needed to perform large genetic screens, as with other model organisms. However, unlike the nematode and fruit fly, mice have biological systems that have no parallel in lower animal models, such as the immune and skeletal systems, or are simply better models for studies of complex systems such as the cardiovascular system, endocrine system, and many others. The mouse is a model for human disease, including cancer, in virtually all of these systems.

Early Studies of the Mouse as a Model Organism

Genetic studies in mice began in the early 1900s, when selection and breeding were the main methods of obtaining progeny with the desired traits. These early studies produced a general model that explained coat coloring in all other fur-bearing mammals. Mice and rats also have a long history in nutritional studies, especially in the identification of vitamins and the symptoms caused by vitamin deficiencies in the diet. Use of mice as a human disease model was pioneered by Clarence Cook Little. In the 1920s he developed an inbred mouse strain, C57BL/6 (commonly known as Black6), which eventually became the mouse strain used to determine the genome sequence. Little also founded the Jackson Laboratory in Bar Harbor, Maine, a center for mouse genetics that also serves as a public repository of mouse models for scientific research.

Life Cycle

The X and Y chromosomes determine sex in mice, as in humans. Fertilization gives rise to a blastocyst containing some undifferentiated cells bunched together in the inner cell mass, the source of embryonic stem cells. Gestation is complete within 19 to 21 days and gives rise to a litter of 3 to 14 pups. Sexual maturity requires about 6 weeks for females and 8 weeks for males, but breeding can take place in as short a time as 35 days. Mice live for 1 to 2 years, and females can produce about 5 to 10 litters, mainly in their first year of life.

Genetic Techniques

Mutagenesis Inbreeding over many generations has produced many useful strains of mutant mice. Adding mutagenic chemicals to the food supply also facilitates development of mutant strains.

Introduction of DNA Foreign DNA can be injected directly into the nucleus of fertilized eggs, followed by implantation of the eggs in the oviduct of the female recipient. Random integration occurs with high frequency. The recombinant DNA used typically has a mouse promoter that directs expression of a reporter gene, such as *lacZ* or GFP (green fluorescent protein), so that expression of the transgene can be followed during development. About half of the transgenic mice contain recombinant DNA in the germ line and therefore pass on the recombinant gene to future generations.

Gene Knockouts Targeted knockouts for mouse disease models are constructed in embryonic stem cells (**Figure A-9**). The stem cells are extracted from the inner cell mass of the blastocyst and grown in culture. Cultured stem cells are then transformed with linear DNA containing a mutated copy of the gene under study, along with genes for neomycin resistance (*neo^r*) and thymidine kinase (*tk*). Homologous DNA flanks the mutated gene (*gene^mut* in Figure A-9) and the *neo^r* gene, such that homologous integration replaces the wild-type gene with

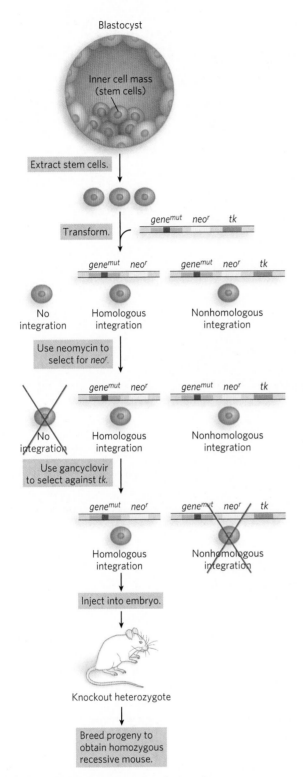

FIGURE A-9 The construction of a knockout mouse.

against *tk*, using the antiviral gancyclovir, kills cells with DNA integrated by nonhomologous recombination, because these cells contain *tk* and thymidine kinase converts gancyclovir to a toxin that kills these cells. Only cells that contain gene knockouts produced by homologous recombination survive both selections. Engineered stem cells are injected into a host wild-type blastocyst-stage embryo. This results in formation of an embryo that is a chimera of host wild-type and donor engineered cells. Resulting chimeras are bred for germ-line transmission of the genetic modification. The F$_1$ (first-generation) heterozygous mice are then crossed to obtain F$_2$ wild-type, heterozygous, and homozygous offspring, in the expected Mendelian 1:2:1 ratio. Selective breeding results in homozygous knockout mice. Methods have been worked out to preserve valuable, and hard to obtain, mouse strains by cryopreservation of sperm or egg cells.

More recently, flippase–flippase recognition target (FLP-FRT) and/or Cre-lox technology have been developed in mice and allow tissue-specific and cell type-specific genome editing, instead of whole body genome alterations. The mouse has also been used for more recent gene inactivation methodologies that include RNAi and novel techniques such as transcription activator–like effector nucleases (TALENs) and CRISPR.

The Mouse as a Model Organism Today

Human Disease The mouse is an important model for studying human disease and aging. Disease models can be derived by inbreeding or by producing knockouts of known disease genes. Mouse models of human disease include cancer, atherosclerosis, hypertension, metabolic diseases, immune disorders, type 1 and type 2 diabetes, Down syndrome, Alzheimer disease, glaucoma, osteoporosis, obesity, epilepsy, Lou Gehrig disease (amyotrophic lateral sclerosis), Huntington disease, blood disorders, and many others.

Mapping of Mutant Genes Mutant genes can be identified more quickly in the mouse than can mutant genes in humans. Closely related strains of mice (e.g., *Mus musculus* and *Mus spretus*) can be crossed to produce hybrids that usually contain different sequences at polymorphic positions, enabling researchers to use linkage analysis to develop detailed genetic maps and locate a disease gene.

Behavior Mouse models exist for certain types of behavior, including alcoholism, drug addiction, anxiety disorders, and aggressive behavior.

Mammalian Development Transgenic mice are used to study the location and timing of expression of particular genes at various stages of development. In addition, models exist for studying certain human developmental disorders, including cleft lip and cleft palate.

the mutant gene plus the *neo^r* gene. On the other hand, DNA that inserts randomly results in integration of the entire DNA fragment, including *tk*.

To select for cells with the desired gene knockout, two steps are required. Selection for *neo^r*, using neomycin, kills all cells that fail to integrate transformed DNA. Selection

Solutions to Problems

Chapter 2

1. **(a)** Plants with round seeds, 100%. **(b)** Plants with round seeds, 75%; plants with wrinkled seeds, 25%. **(c)** Plants with round seeds, 50%; plants with wrinkled seeds, 50%.

2. At least one of the parents was *RR*. The other parent was *RR*, *Rr*, or *rr*.

3. One parent was *RR*, and the other was *Rr*. In the F_1 generation, half of the plants were *RR* and half were *Rr*. In the F_1 crosses, all crosses involving *RR* plants (three-fourths of the total) produced progeny with round seeds. For the $Rr \times Rr$ crosses (one-fourth of the total), three-fourths of the progeny had round seeds and one-fourth had wrinkled seeds. The wrinkled seeds are thus found in $0.25 \times 0.25 = 0.0625$ (~8/129) of the total plants.

4. Recall that this eye-color gene is on the X chromosome. The F_1 generation will have $X^W Y$ males (red eyes) and $X^w X^W$ females (red eyes). The F_2 generation will have $X^W Y$ and $X^w Y$ males in a 50:50 mix, and $X^w X^W$ and $X^W X^W$ females in a 50:50 mix, so half of the males but no females will have white eyes. Some white-eyed females will appear in the F_3 generation, from crosses between $X^w Y$ males and $X^w X^W$ females.

5. The F_1 generation will have $X^W Y$ males (all red-eyed) and $X^w X^W$ females (all red-eyed). Half of the males in the F_2 generation will have white eyes (a result found by Morgan).

6. The gene producing the black color is on the Y chromosome. It is not present in any females, regardless of the generation. There are no other Y chromosomes in the population, so all males have the black trait.

7. All F_1 offspring are *RrLl* plants, with red flowers and large leaves. The F_2 generation is shown in the Punnett square.

	RL	**Rl**	**rL**	**rl**
RL	RRLL	RRLl	RrLL	RrLl
Rl	RRlL	RRll	RrlL	Rrll
rL	rRLL	rRLl	rrLL	rrLl
rl	rRlL	rRll	rrlL	rrll

9/16, or 56.25%, have red flowers and large leaves; 3/16, or 18.75%, have red flowers and small leaves; 3/16, or 18.75%, have white flowers and large leaves; 1/16, or 6.25%, have white flowers and small leaves.

8. 100% of the F_1 generation plants would have pink flowers. In the F_2 generation, 50% would have pink flowers.

9. 12.5% of the F_2 males will have red eyes and normal wings; these are marked in the Punnett square by an underline. They all have genotypes with at least one X^W gene and one *V* gene.

	$X^w V$	$X^W V$	$X^w v$	$X^W v$
$X^w V$	$X^w X^w VV$	$\underline{X^w X^W VV}$	$X^w X^w Vv$	$\underline{X^w X^W Vv}$
$X^w v$	$X^w X^w vV$	$\underline{X^w X^W vV}$	$X^w X^w vv$	$X^w X^W vv$
YV	$YX^w VV$	$\underline{YX^W VV}$	$YX^w Vv$	$\underline{YX^W Vv}$
Yv	$YX^w vV$	$\underline{YX^W vV}$	$YX^w vv$	$YX^W vv$

10. The *G* and *S* genes are linked: they are found on the same chromosome and are sufficiently close together that crossovers between them are rare. Thus, the *G* and *S* alleles are not independently assorted.

11. In mitosis, the paired chromosomes (in the sister chromatid pair) are separated, and one chromatid passes to each daughter cell. In meiosis, the sister chromatids are not separated but instead remain paired in the daughter cells produced by the first division.

12. Genes *A*, *B*, *C*, and *E* are on one chromosome, unlinked to genes *D* and *F*. Relative distances between genes are indicated by numbers under each chromosome.

Chromosome 1:

$$\underset{3 \quad\quad 6 \quad\quad\quad 4}{\underline{\text{A} \quad\quad \text{C} \quad\quad\quad\quad \text{B} \quad\quad \text{E}}}$$

Chromosome 2:

$$\underset{16}{\underline{\text{D} \quad\quad\quad\quad\quad\quad\quad \text{F}}}$$

13. <u>M O N</u> or <u>N M O</u>

14. Characteristics of rRNAs: only a few types in a cell; located in ribosomes; sequence highly conserved in different life forms; are functional RNAs (do not encode proteins). Characteristics of tRNAs: small; fold into a characteristic L shape; have many modified bases; can link to an amino acid; are functional RNAs. Characteristics of mRNAs: the most diverse of the RNAs in sequence—each cell type makes >1,000 different

mRNAs; can be very long; encode the sequences of proteins.

15. Hershey and Chase could have done the work with a single batch of phage with both radioactive labels. After the blending step, the ^{35}S-labeled protein heads would have ended up in the supernatant, and the ^{32}P-labeled DNA in the pellet containing the bacteria.

16. Genes *trpD* and *trpE* are involved in step 1, *trpC* in step 2, and *trpA* and *trpB* in step 3. The results suggest that the enzymes catalyzing steps 1 and 3 each consist of at least two polypeptides encoded by separate genes.

17. In mitotic cell division, the sister chromatid pairs are separated following replication: A1 and A1* go to different daughter cells, and A2 and A2* go to different daughter cells. In the first cell division of meiosis, the two sets of sister chromatids stay together, with one set going to each of the two daughter cells: A1/A1* goes to one cell, and A2/A2* goes to the other.

18. There are many possible answers; these suggestions focus on answers that might come from a student with a limited background in molecular biology. Some of the RNAs could have catalytic functions. Others could be cofactors for protein enzymes. Some could pair with RNA or DNA to regulate processes involving these nucleic acids. Some could be associated with RNA viruses.

Chapter 3

1. The O=O bond is stronger; its atoms are closer together.

2. The enantiomers have the same chemical formula, so the answer in each case is *yes*.

3. (d) It lowers the activation energy for a reaction.

4. pH 5

5. No. Peptides have directionality, established by the N-terminus and C-terminus. The order in which the amino acids are connected in a polypeptide confers unique properties on the polymer.

6. Activation energy is larger in the reverse direction.

7. The number of moles of NaCl remains unchanged; the *concentration* of NaCl is reduced by 50%.

8. The second law of thermodynamics, which states that in any chemical or physical process, the entropy of the universe tends to increase.

9. A and B are of equal length. Resonance structures, by definition, represent the shared distribution of electrons between sets of bonding atoms, and hence the average bond lengths are the same.

10. (b) Measuring rates.

11. (a) 4.76. (b) 9.19. (c) 4.0. (d) 4.82.

Details: $pH = -\log [H^+]$

(a) $-\log [1.75 \times 10^{-5}] = 4.76$

(b) $-\log [6.50 \times 10^{-10}] = 9.19$

(c) $-\log [1.0 \times 10^{-4}] = 4.0$

(d) $-\log [1.5 \times 10^{-5}] = 4.82$

12. (a) 1.5×10^{-4} M. (b) 3.0×10^{-7} M.
(c) 7.8×10^{-12} M.

Details: $[H^+] = 10^{-pH}$

(a) $[H^+] = 10^{-3.82} = 1.5 \times 10^{-4}$ mol/L

(b) $[H^+] = 10^{-6.53} = 3.0 \times 10^{-7}$ mol/L

(c) $[H^+] = 10^{-11.11} = 7.8 \times 10^{-12}$ mol/L

13. (a) 5.00. (b) 4.22. (c) 5.40. (d) 4.70. (e) 3.70.

Details: $pH = pK_a + \log$ [acetate]/[acetic acid]

(a) $pH = 4.70 + \log (2/1) = 5.00$

(b) $pH = 4.70 + \log (1/3) = 4.22$

(c) $pH = 4.70 + \log (5/1) = 5.40$

(d) $pH = 4.70 + \log 1 = 4.70$

(e) $pH = 4.70 + \log (1/10) = 3.70$

14. (a) 4.3. (b) pH decrease of 0.12. (c) pH decrease of 4.4.

Details:

(a) $pH = pK_a + \log$ [lactate]/[lactic acid] $= 3.60 + \log (0.05/0.01) = 3.60 + 0.69897 = 4.3$

(b) Strong acids ionize completely:

$$0.005 \text{ L} \times 0.5 \text{ mol/L} = 0.0025 \text{ mol of } H^+ \text{ added}$$

The added acid converts some of the salt form (lactate) to the acid form (lactic acid):

$$[H^+] = 3.60 + \log \frac{0.05 - 0.0025}{0.01 + 0.0025}$$

$$= 3.6 + \log 3.8 = 3.6 + 0.58 = 4.18$$

The change in pH $= 4.30 - 4.18 = 0.12$.

(c) $pH = -\log [H^+]$

$$[H^+] = 0.0025 \text{ mol}/1.005 \text{ L} \approx 0.0025 \text{ mol/L}$$

$$pH = -\log 0.0025 = 2.6$$

The pH of pure water is 7.0, so change in pH $= 7.0 - 2.6 = 4.4$.

15. pK_a 7.2

Details: At pH 2.0, 50% of the group with $pK_a = 2.0$ is ionized ($pK_a = pH$). Amount of base added $= 0.075$ L $\times 0.1$ mol/L $= 0.0075$ mol. Amount of compound $= 0.1$ L $\times 0.1$ mol/L $= 0.01$ mol. A pH increase to 6.72 completely titrates the remainder of the group with the low pK_a (because pH $>>$ pK_a). So, 50% $\times$ 0.01 mol of compound requires 0.005 mol of base to titrate the rest of that group. Thus, 0.0075 mol of base added $-$ 0.005 mol of base used $= 0.0025$ mol of base remaining to titrate the second group.

$$6.72 = pK_a + \log \frac{0.0025}{0.01 - 0.0025}$$

$$= pK_a + \log 0.333 = pK_a - 0.477$$

$$7.2 = pK_a$$

16. The ring is flat/planar. The conjugated double bonds in the ring produce considerable resonance, such that all bonds in the ring have a partial double-bond character and all lie in the same plane.

17. Free rotation is restricted about the bond with angle ω. Due to resonance between the N–C bond and the C=O bond, the N–C bond has a partial double-bond character.

18.

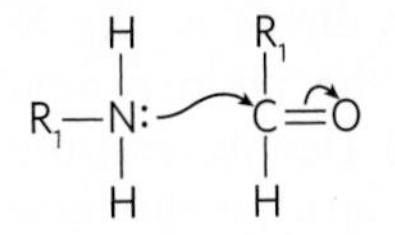

(a)

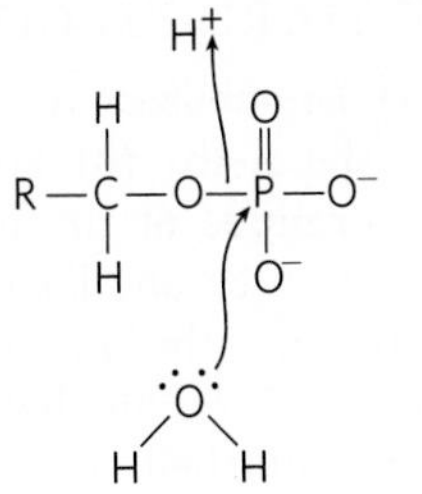

(b)

19. The molecule is L-alanine (see Figure 3-23). This is an example of an L-amino acid that is dextrorotatory.

20. The biochemist must add 13 mL of 1.0 M NaOH.

Details: pH = pK_a + log [base] / [acid]. The solution contains 0.01 mol of histidine. The beginning pH is 1.82, identical to one of the pK_a values. Since log [base] / [acid] = 0, [base] / [acid] = 1; the concentrations of base and acid are the same. For the ionizable group (α-carboxyl), there are effectively 0.005 mol of the acid form and 0.005 mol of the basic form. This group is titrated first by the NaOH, requiring 0.005 mol of NaOH (5 mL of 1 M solution) for complete titration. To reach pH 6.60, part of the next ionizable group (side-chain imidazole, pK_a 6.00) must be titrated. At the beginning of the titration, 100% (or very close) is in the protonated (acidic) form. Using the Henderson-Hasselbalch equation: 6.00 = 6.60 + (−0.60); thus, log [base] / [acid] = −0.60, and [base] / [acid] = 0.25, or 1/4. In the final solution, 80% of the His side chain (4 of 5 molecules) must be in the basic form, or 0.008 mol total. An additional 8 mL of the 1 M NaOH is needed to do this. Thus, the total 1 M NaOH needed to titrate the solution to pH 6.60 is 5 mL + 8 mL = 13 mL.

Chapter 4

1. (a)

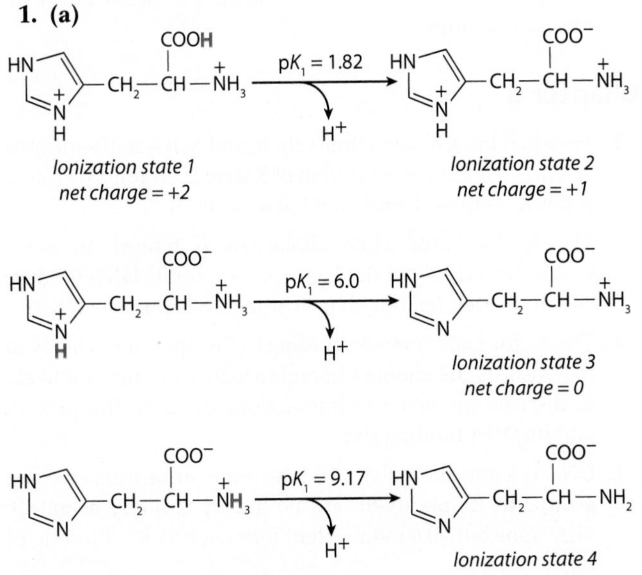

Ionization state 1
net charge = +2

pK_1 = 1.82

Ionization state 2
net charge = +1

pK_1 = 6.0

Ionization state 3
net charge = 0

pK_1 = 9.17

Ionization state 4
net charge = −1

(b)

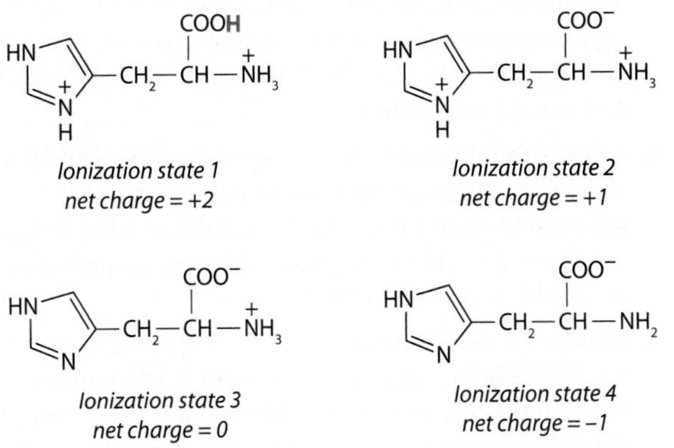

Ionization state 1
net charge = +2

Ionization state 2
net charge = +1

Ionization state 3
net charge = 0

Ionization state 4
net charge = −1

(c)

pH	Ionization state	Net charge	Migrates toward:
1	1	+2	Cathode
4	2	+1	Cathode
8	3	0	Does not migrate
12	4	−1	Anode

2. (a) Minimum M_r 32,000. Remember that the molecular weight of a Trp residue is not the same as the molecular weight of the free amino acid. **(b)** 2 Trp residues.

3. (a) At pH 3, +2; at pH 8, 0; at pH 11, −2. **(b)** pI 7.

Details:

(a) This peptide has five ionizable groups: (1) the α-amino group of E (pK_a 9.67); (2) the side chain of E (pK_a 4.25); (3) the side chain of H (pK_a 6.0); (4) the side chain of R (pK_a 12.48); and (5) the α-carboxyl group of G (pK_a 2.34). *At pH 3*, (1), (3), and (4) are protonated and positively charged, giving a net charge for these groups of +3. The pH of 3 is between the two pK_a values for (2) and (5) (one mostly protonated, the other mostly unprotonated), giving a net charge of about −1 for these two groups. This yields a net charge for the peptide of about +2. *At pH 8*, (2) and (5) are unprotonated, contributing a charge of −2; (3) is mostly unprotonated and neutral; (1) and (4) are mostly protonated, contributing a charge of nearly +2. The net charge is near zero. *At pH 11*, (2), (3), and (5) are unprotonated, with a charge of −2; (1) and (4) are mostly unprotonated and uncharged. The net charge is close to −2.

(b) The pI can be estimated by determining the pH at which the net charge is zero. In this peptide, (1), (3), and (4) can contribute + charges when protonated, and (2) and (5) can contribute − charges when unprotonated. Hence, a net charge of zero can occur at the pH where (1), (3), and (4) together contribute a net +2 charge to exactly balance the net −2 contributed by the two carboxyl groups of (2) and (5). This will occur about halfway between the pK_a values of the H and R side chains: the fraction of R side chains that are unprotonated is balanced exactly by the fraction of H side chains that are protonated. Thus, the pI is approximately 7.

4. **(a)** A1, R5, K20, R28. **(b)** D2, E4, D27, E29, T30. **(c)** C7, C23.

5. The sheets are likely to be antiparallel, because the linkers that connect one β strand to the next are too short to connect parallel strands. The linkers are four residues long and could form β turns.

6. Alternating R groups are on opposite sides of the β sheet structure. Therefore, in a continuous layer of β sheet, with alternating polar and nonpolar residues in the β strands, the sheet is likely to fold into a β barrel, sequestering hydrophobic residues inside.

7. Residues 1 and 3 are in β sheets, residue 2 is in a right-handed α helix, and residue 4 is in a left-handed α helix. Residue 5 is an outlier, and the electron density map should be examined to see whether the angles are correctly assigned. If residue 5 were Gly, it might be in an acceptable position in the plot, because Gly residues are more flexible than other residues and can adopt a greater range of acceptable bond angles.

8. AIPRKKR⌐EFICRFGAIR⌐PNT. The P3 and P18 residues disfavor helix formation, limiting the region favorable for helix formation to the sequence between residues 4 and 17. The several positively charged residues, 4 through 7, are not likely to initiate a helix, because they would repel one another and interact unfavorably with the helix dipole, which is positively charged at the N-terminus. In contrast, E8 will interact favorably with the helix dipole. Thus, the most likely α-helical region is residues 8 through 17 (boxed). Within this helix, the stabilizing interactions are: N-terminal positive dipole, stabilized by E8; C-terminal negative dipole, stabilized by R17; hydrophobic interactions between two F residues, spaced four residues (one turn) apart; ion-pair interaction between E8 and R12, spaced four residues (one turn) apart.

9. **(a)** and **(d)**. Both contain the consensus sequence for an ATP/GTP-binding site: (G/A)XXGXGK(T/S), where X is any amino acid.

10. As an α helix, 210 Å ([1.5 Å/residue] × 140 residues). As a β strand, 490 Å ([3.5 Å/residue] × 140 residues).

11. There are many possible answers; any of the following will suffice. NMR uses magnets and radiofrequency irradiation; crystallography uses x rays. NMR is performed on proteins in solution; x-ray crystallography requires a protein crystal. NMR measures a nuclear event; x-ray crystallography measures events in the electron shell. In NMR, irradiated proteins emit radiowaves; in x-ray crystallography, irradiated proteins emit x rays. NMR makes great use of protons; protons are largely ignored in x-ray crystallography. NMR can be applied only to small proteins; x-ray crystallography can solve large proteins and complexes. Both methods irradiate the protein sample with electromagnetic radiation (photons). Both methods yield structures with atomic resolution. Both methods make heavy use of computations.

12. **(a)**, **(b)**, **(c)** one domain; **(d)**, **(e)** two domains. When a protein reaches a size of about 150 to 200 residues (M_r ~20,000), the polypeptide chain usually folds into two domains.

13. Completely buried residues are likely to be hydrophobic: L2, F4, I6, V8, V12, L13, L18, and L19 fit this description. Highly polar residues, or at least their polar groups, are likely to be on the surface, exposed to water: D1, K3, T5, S7, T14, R15, E16, Q17, and E20 fit this description.

14. **(a)** Destabilizes the positive dipole at the N-terminus of the helix. **(b)** Stabilizes the positive dipole at the N-terminus of the helix. **(c)** Destabilizes; the ion pair between R2 and E5, one turn apart, is eliminated. **(d)** No difference; the ion pair between the residues is maintained. **(e)** Stabilizes; the hydrophobic interaction is on the same side of the helix, one turn apart. **(f)** Destabilizes; P (a helix breaker) destabilizes the helix.

15. **(a)** The N-terminus is the lower-right end; the C-terminus is the upper-right end; the β turns are the U turns at the lower left and upper right. **(b)** The more hydrophobic surface is likely to be on the right side of the β sheet.

16.

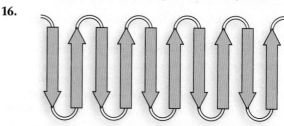

17. **(a)** Bovine serum albumin. **(b)** Green fluorescent protein.

    Details: Much of the structure of bovine serum albumin is in the form of an α helix, while green fluorescent protein is predominantly in the β conformation. The corresponding Ramachandran plots exhibit many residues with torsion angles characteristic of the predominant secondary structures.

18. The conversion of trans peptide bonds to cis peptide bonds is catalyzed by an enzyme, peptide prolyl cis-trans isomerase.

19. No correct answer is possible at this stage of the book. However, the answer is *yes*: virtually all of these complexes, no matter how large, are held together by weak noncovalent interactions.

Chapter 5

1. Protein B has a higher affinity for ligand X; it is half-saturated at a much lower concentration of X than is protein A. Protein A has $K_a = 10^6$ M^{-1}; protein B has $K_a = 10^9$ M^{-1}.

2. The K_d increases when allolactose is bound. In other words, the affinity of the Lac repressor for its DNA binding site decreases, leading to its dissociation from the DNA.

3. The K_d for Lac repressor binding to the operator is likely to increase, as the changes in amino acid sequence are likely to disrupt one or more interactions between the protein and its DNA binding site.

4. DNA is a polyelectrolyte, and the many negatively charged groups in its backbone are bound by cations, primarily Mg^{2+} ions but also monovalent ions such as K$^+$. Binding of

a protein to DNA involves displacement of some or all of the ions at the DNA site where binding occurs.

5. Nonspecific DNA binding generally involves interactions with the invariant parts of DNA structure—the phosphate and deoxyribose groups in the DNA backbone—and hydrophobic interactions with the nucleotide bases. Binding to a specific DNA sequence requires substantial interaction with groups in the bases that distinguish one nucleotide from another. These features of each base are largely accessible in the major and minor grooves of the DNA.

6. All of these situations give rise to real or apparent negative cooperativity. Apparent negative cooperativity in ligand binding can be caused by the presence of two or more different types of ligand-binding sites with different affinities for the same ligand on the same or different protein molecules in the same solution. Apparent negative cooperativity can also be observed in heterogeneous protein preparations. There are few well-documented cases of true negative cooperativity.

7. 2.4×10^{-6} M.

   Details: The volume of a cylinder is given by $V = \pi r^2 h$. For a cylinder of $r = 0.000050$ cm and $h = 0.00020$ cm, $V = 1.57 \times 10^{-12}$ cc (mL), or 1.57×10^{-15} L. The solution concentration in the cell is 1.20 g/mL, and the protein concentration is 20% of this, or 0.24 g/mL. Thus the cell contains 3.77×10^{-13} g of protein. If the cell contains 1,000 proteins of the same molecular weight, it contains 3.77×10^{-16} g of a given protein, or 0.24 g per liter of cytosol. Dividing 0.24 g/L by the 100,000 g/L of a 1 M solution gives a concentration of 2.4×10^{-6} M. This corresponds to just over 2,200 copies of each protein in the cell.

8. **(b)**, **(e)**, and **(g)**.

9. We now know that to catalyze a reaction, an enzyme active site must be complementary (in shape and charge) not to the substrate but to the transition state of the reaction that is catalyzed.

10. The inactivated enzyme will no longer have a measurable k_{cat} and K_m.

11. **(a)** $[S] = 1.7 \times 10^{-3}$ M. **(b)** $0.33V_{max}$, $0.67V_{max}$, and $0.91V_{max}$. **(c)** The upper (red) curve corresponds to enzyme B ($[X] > K_m$ for this enzyme); the lower (black) curve, enzyme A.

12. **(a)** $k_{cat} = 400$ s^{-1}. **(b)** $K_m = 10$ μM. **(c)** $\alpha = 2$, $\alpha' = 3$. **(d)** ANGER is a mixed inhibitor.

13. **(a)** $[E_t] = 24$ ηM. **(b)** $[A] = 4$ μM (V_0 is exactly $\frac{1}{2}V_{max}$, so $[A] = K_m$). **(c)** $[A] = 40$ μM (V_0 is exactly $\frac{1}{2}V_{max}$, so $[A] = 10K_m$ in the presence of inhibitor).

14. $V_{max} \approx 140$ mM min^{-1}; $K_m \approx 1 \times 10^{-5}$ M.

15. Movement along the DNA requires ATP hydrolysis. A normal RuvB subunit can still hydrolyze ATP, even if it is adjacent to a mutant subunit in a heterohexameric complex. However, movement along DNA requires cooperation between adjacent subunits, which cannot occur if one of the subunits is mutated.

16. In principle, ATM and ATR could be enzymes that covalently modify other proteins. In fact, ATM and ATR are the most common type of such proteins: they are protein kinases that add phosphoryl groups to hundreds of cellular protein targets.

17. It is likely that the helicase is present in one fraction, and another protein or macromolecule needed to activate the helicase is present in the other fraction.

Chapter 6

1. An enzyme (RNA or protein) must (1) increase the rate of a chemical reaction and (2) remain unchanged on completion of a catalytic cycle.

2. N-3 and N-7.

3. Within experimental error, the number of purines (A + G) equals the number of pyrimidines (C + T); the fractional amounts of A and T are the same, and the fractional amounts of G and C are the same; the relative ratios of the bases do not vary from one tissue to another.

4. **(a)** 5′-TACCAGCCTTAGAATTTAACTAAGGCTGTAATC-3′. Note that nucleic acid sequences are always written in the 5′→3′ direction, and the two strands of DNA are antiparallel. **(b)** Yes; 5′-CAGCCTTAG-3′ and 5′-CTAAGGCTG-3′ form an inverted repeat, so the strand has the potential to form a hairpin. The duplex can assume a cruciform structure.

5. Higher; RNA has greater thermal stability than DNA.

6. The DNA has a backbone with deoxyribose and will take up a B-form helix. The RNA has a backbone with ribose and will take up an A-form helix.

7. The presence of T rather than U as one of the primary pyrimidine nucleotides in RNA is a likely mechanism by which cells monitor mutations in DNA. Uracil is regularly produced in DNA largely by the slow, nonenzymatic hydrolytic deamination of cytosine; having thymine as the base in DNA allows the efficient detection and repair of C-to-U mutations.

8. **(a)** 5′ terminus. **(b)** GTC. **(c)** RNA. Note that the identity of a nucleic acid as RNA or DNA depends only on the ribose variant in its backbone. Even though thymine is relatively rare in RNA, an oligo- or polynucleotide containing D-ribose is RNA, and an oligo- or polynucleotide containing deoxy-D-ribose is DNA, regardless of whether thymine or uracil is present.

9. 5′-ATTGCATCCGCGCGTGCGCGCGCGATCCCGTTACTTTCCG-3′

10. The double helix is the most thermodynamically stable structure. It places the hydrophobic bases in the interior of the molecule, where they interact with one another through base stacking, and the charged phosphate groups on the outside, where they can interact with water and ions.

11. The phosphate groups between the sugars (deoxyribose or ribose) in the sugar–phosphate backbone are highly acidic, giving the nucleic acids an overall negative charge.

12. Cytosine deamination to form uracil is a slow but constant reaction in all cells. In many eukaryotes, hundreds of

C residues are converted to U residues every day, in every cell, creating G—U base pairs that are "seen" by the repair system as G—T pairs. Because the G is correct and the U is the damaged base, repairing G—T to G≡C restores the correct genetic information. Repair to A=T would cause a mutation.

13. The three-dimensional structure of a tDNA would probably be similar to that of a tRNA, in that the base pairing would make it fold into a similar cloverleaf. This is supported by experiments. However, the 2′-hydroxyl groups of ribose contribute significantly to tRNA folding, so subtle structural differences would exist. Enzymes specific for tRNAs modify certain of their bases, and many of these enzymes include the 2′-hydroxyl group in their recognition mechanism. This would prevent many, if not all, of the key base modifications from occurring. A tDNA generated in a cell would thus be unlikely to function as a tRNA substitute.

14. Both the G and C content and the length of the DNA influence the strength of association between the two strands in the double helix. G≡C pairs contribute more than A=T pairs, due to their stacking properties. The longer the DNA, the greater the number of base pairs and the greater the energy (i.e., the higher the temperature) required to break the hydrogen bonding between them.

15. An abundance of purines, especially A residues, which play an important role in the three-dimensional folding of RNA; also, multiple short segments capable of base pairing with adjacent or distant regions of the RNA molecule, particularly if these short segments are conserved in related organisms.

16. The regular repeating properties of the double helix produce characteristic x-ray diffraction patterns for DNA fibers, used in the earliest studies of DNA structure. However, these diffraction patterns result from the *averaged* properties of the DNA helices in a fiber. Determination of the properties of individual DNA sequences required single crystals containing a single, homogeneous form of the DNA molecule, arranged in a three-dimensional array. The x-ray diffraction patterns produced from single crystals could be used to determine the electron density map of the DNA in the crystal, providing an exact, rather than an averaged, image of the molecular structure.

17. DMT is a blocking group, preventing unwanted reactions at the 5′-hydroxyl group of the nucleotide.

18. (a) There is no sulfur in DNA, so proteins are uniquely labeled by ^{35}S. There is little or no phosphate in proteins (at least in bacteria), so DNA is uniquely labeled by ^{32}P. (b) ^{14}C or ^{3}H would have labeled both the DNA and the protein, permitting no differentiation. (c) The intact phages, the T2 ghosts, and the DNA are all insoluble in acid, and all are removed from solution by centrifugation. (d) The nucleotides liberated by DNase treatment are soluble in acid. Osmotic shock releases the T2 DNA into solution, where it is degraded by the DNase. The unplasmolyzed T2 phages contain DNA, but it is protected from the DNase, within the phage protein coat. (e) Both the intact viruses and the T2 ghosts adsorb to the bacteria. The components needed for attachment of T2 to the bacteria are located uniquely in the protein coat. (f) The antibodies recognize the T2 protein coat. In both the control and plasmolyzed samples, the protein coats are immunoprecipitated by the antisera, but in the plasmolyzed sample, the DNA is left behind in solution. (g) The material released by osmotic shock is entirely or almost entirely DNA. The T2 ghosts are almost entirely protein. Little or no protein is released from the phages with the DNA. The DNA does not adsorb to phage-susceptible bacteria on its own. The ghosts are protein coats that surround the DNA of the intact phage particles. These coats react with antibodies and protect the DNA within from DNase. They also are responsible for attaching phages to a bacterial host. (h) The centrifugation and resuspension remove unadsorbed phages from the solution, which otherwise would have added to the background signal. (i) About 80% of the ^{35}S-labeled phage heads are stripped from the cells by the blender, with only about 16% found in the supernatant without the blender treatment. The amount in the supernatant without blender treatment increases when the multiplicity of infection increases, as a result of some kind of displacement of phages by other phages attached to the same cells. (j) The bulk of the ^{35}S is removed from the cells by blender treatment, whereas a relatively small amount of the ^{32}P is removed. The capacity of the cells to survive and continue with the infection process is not affected by the treatment. The results indicate that the bulk of the protein remained in the protein coats at the cell surface during infection, while the bulk of the DNA entered the cells.

Chapter 7

1. (a)

5′ - - - G-3′	5′-AATTC - - - 3′
3′ - - - CTTAA-5′	3′-G - - - 5′

(b)

5′ - - - GAATT-3′	5′-AATTC - - - 3′
3′ - - - CTTAA-5′	3′-TTAAG - - - 5′

(c)

5′ - - - GAATTAATTC - - - 3′
3′ - - - CTTAATTAAG - - - 5′

(d)

5′ - - - G-3′	5′-C - - - 3′
3′ - - - C-5′	3′-G - - - 5′

(e)

5′ - - - GAATTC - - - 3′
3′ - - - CTTAAG - - - 5′

(f)

5′ - - - CAG-3′	5′-CTG - - - 3′
3′ - - - GTC -5′	3′-GAC - - - 5′

(g)

5'---CAGAATTC---3'   or   5'---GAATTCTG---3'
3'---GTCTTAAG---5'   3'---CTTAAGAC---5'

(h) *Method 1:* Cut the DNA with EcoRI as in (a). Then treat the DNA as in (b) or (d), and ligate a synthetic DNA fragment containing the BamHI recognition site between the two resulting blunt ends. *Method 2* (more efficient): Synthesize a DNA fragment with the structure:

5'-AATTGGATCC
3'-CCTAGGTTAA

This would ligate efficiently to the sticky ends generated by EcoRI cleavage, would introduce a BamHI site, but would not regenerate the EcoRI site.

(i) The four fragments (with N = any nucleotide), in order of discussion in the problem, are:

5'-AATTCNNNNCTGCA-3'
3'-GNNNNG-5'

5'-AATTCNNNNGTGCA-3'
3'-GNNNNC-5'

5'-AATTGNNNNCTGCA-3'
3'-CNNNNG-5'

5'-AATTGNNNNGTGCA-3'
3'-CNNNNC-5'

2.

5'-GAAAGTCCGCGTTATAGGCATG-3'
3'-ACGTCTTTCAGGCGCAATATCCGTACTTAA-5'

3. A YAC vector is not stably maintained as a yeast chromosome during mitosis unless it carries an insert of more than 100,000 bp.

4. (a) Some original pBR322 plasmids will be present, regenerated without insertion of a foreign DNA fragment. Also, two or more pBR322 plasmids might be ligated, with or without insertion of a segment of foreign DNA. All of these would retain resistance to ampicillin. **(b)** The clones in lanes 1 and 2 each have one DNA fragment, inserted in different orientations. The clone in lane 3 has incorporated two of the DNA fragments, ligated such that the ends closest to the EcoRI sites are joined.

5. The sequence will appear about once every $4^8 = 65,536$ bp. If the G + C content is greater than the A + T content (or vice versa), the frequency of occurrence of the restriction site will decrease.

6. In a large DNA molecule with a random sequence, a BamHI site will occur, on average, every 4,096 bp (assuming all four nucleotides are present in equal proportions). Cleavage of all BamHI sites in the DNA would produce fragments much smaller than the 100,000 to 300,000 bp needed for a BAC library.

7.

Primer 1: CCTCGAGTCAATCGATGCTG
Primer 2: CGCGCACATCAGACGAACCA

Recall that all DNA sequences are written in the 5'→3' direction, left to right; that the two strands of a DNA molecule are antiparallel; and that both PCR primers must target the end sequences so that their 3' ends are oriented toward the segment to be amplified.

8.

Primer 1: GAATTCCCTCGAGTCAATCGATGCTG
Primer 2: GAATTCCGCGCACATCAGACGAACCA

9. The test requires DNA primers, a heat-stable DNA polymerase, deoxynucleoside triphosphates, and a PCR machine. The primers are designed to amplify a DNA segment encompassing the CAG repeat. The DNA strand shown in the problem is the coding strand, oriented 5' → 3', left to right. The primer targeted to the DNA to the left of the CAG repeat should be identical to any 25-nucleotide sequence in the region to the left of the repeat. The primer on the right side must be *complementary and antiparallel* to a 25-nucleotide sequence to the right of the CAG repeat. With these primers, an investigator would use PCR to amplify the DNA including the CAG repeat, and then determine its size by comparison with size markers on electrophoresis. The length of the DNA reflects the length of the CAG repeat, providing a simple test for the disease.

10. The researcher could design PCR primers complementary to DNA in the deleted segment that will direct DNA synthesis away from each other. No PCR product will be generated unless the ends of the deleted segment are joined to create a circle.

11.

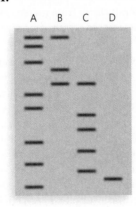

12. No. The orientation of the cloned gene is very important, because the information specifying the protein is contained in only one of the two DNA strands. The promoter specifies not only where RNA polymerase binds the DNA but also the direction in which it travels and the DNA strand that it uses as template for RNA synthesis. When the correct DNA strand is used as template, a functional protein results. If the gene is inverted, the opposite DNA strand will become the template for synthesis of RNA, with a much different nucleotide sequence. The resulting protein will be completely different, unrelated to the normal gene product, and most likely nonfunctional.

13. (a) The bacterial *recA* gene is readily cloned by using the bacterial plasmid. **(b)** For the mammalian DNA polymerase, the baculovirus system may have a better chance of generating an active protein.

14. The production of labeled antibodies is difficult and expensive. The labeling of every antibody to every protein target would be impractical. By labeling one antibody preparation for binding to all antibodies of a particular class, the same labeled antibody preparation can be used in many different Western blot experiments.

15.

```
     9    1       5     3     7   4        2      6   8
     ├────┼───────┼─────┼─────┼───┼────────┼──────┼───┤
     ├──────────────A───────────┤
             ├───────────────B──────────────────────┤
     ├───────────C────────────┤
               ├───────────D─────────┤
     ├──────E────────┤
                   ├──────────F────────────┤
```

16. Express the protein in yeast strain 1 as a fusion protein with one of the domains of Gal4p, such as the DNA-binding domain. Using yeast strain 2, make a library in which essentially every protein of the fungus is expressed as a fusion protein with the interaction domain of Gal4p. Mate strain 1 with the strain 2 library, and look for colonies that are colored due to expression of the reporter gene. These colonies will generally arise from mated cells containing fusion protein that interacts with your target protein.

17. Cover spot 4, add solution containing activated T, irradiate, wash. Result: 1. A–T; 2. G–T; 3. A–T; 4. G–C. Cover spots 2 and 4, add solution containing activated G, irradiate, wash. Result: 1. A–T–G; 2. G–T; 3. A–T–G; 4. G–C. Cover spot 3, add solution containing activated C, irradiate, wash. Result: 1. A–T–G–C; 2. G–T–C; 3. A–T–G; 4. G–C–C. Cover spots 1, 3, and 4, add solution containing activated C, irradiate, wash. Result: 1. A–T–G–C; 2. G–T–C–C; 3. A–T–G; 4. G–C–C. Cover spots 1 and 2, add solution containing activated G, irradiate, wash. Result: 1. A–T–G–C; 2. G–T–C–C; 3. A–T–G–G; 4. G–C–C–G.

18. In long repetitive DNA sequences, the fragments can be overlapped (see Figure 7-15), but there are many ways to overlap two fragments that contain such repeats. It is impossible to determine exactly how long the repeated region is unless all the repeats are contained in one continuous sequenced fragment. The read lengths for many next-generation sequencing technologies are insufficient to capture an entire region of this kind in one continuous sequence.

19. The sgRNA is required to pair the CRISPR/Cas9 with its intended target site in the DNA and to activate the CRISPR nuclease domains for cleavage.

20. (a) R6-5, at least 11; pSC101, 1; pSC102, 3. **(b)** Each of the observed bands in a given lane represents a DNA fragment, and each fragment is present in the same concentration (total molecules). However, the fragments to the left are longer than the ones to the right and thus take up more of the fluorescent stain. **(c)** Two EcoRI fragments derived from R6-5 are very nearly the same size, and they migrate together at this position. Thus, there are 12 fragments, derived from cleavage of R6-5 at its 12 EcoRI recognition sites. **(d)** The plasmid pSC102 is made up of these three EcoRI fragments from R6-5. **(e)** The larger fragment on the left, which comigrates with pSC101, is the only possible source of a tetracycline-resistance gene in the parent plasmids. **(f)** The smaller fragment on the right, which comigrates with one of the fragments of pSC102, is the only possible source of a kanamycin-resistance gene in the parent plasmids. **(g)** 7,000 bp. **(h)** Four phosphodiester bonds; two fragments were ligated, with two new phosphodiester bonds created at each of the two ligation sites. **(i)** The original ligation mixture included a wide range of combinations of DNA fragments. When the mixture was used to transform *E. coli* cells, only cells that took up a combination of fragments that allowed survival on the selection media would grow. Evidently, these two fragments of pSC101 did not include a gene for resistance to tetracycline or kanamycin, the antibiotics used for selection. The pSC101 plasmid that included the tetracycline-resistance gene also had a replication origin, so no new replication origin was required. The joining of three fragments into a circle by ligation is considerably less probable than the ligation of two fragments. In effect, the selection generated the simplest possible recombinant plasmid from the available fragments.

Chapter 8

1.

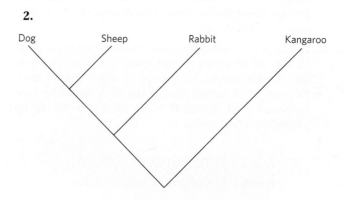

2.

3. The genomic DNA fragments are cloned into plasmid vectors. Although the sequence of the cloned DNA is not known, the plasmid sequences immediately adjacent to a DNA fragment are known. A single primer is used, targeted to the plasmid sequence near one end of the cloning site, and sequencing of each clone is initiated from that point.

4. ATSAAG**W**DEWEGGKV**L**IHL**D**GKLQNRGALLELDIGAV

5. (a) Modern sequencing of the cDNAs generated in RNA-Seq produces many short sequence reads; each read is linked to a particular gene by its sequence, and the number of reads from each gene provides a measure of the relative numbers of RNAs derived from that gene. (b) In most cells, the majority of the RNA is rRNA, and the rRNA will generate a very high background of sequence information if it is not removed before the analysis.

6. The possibilities include gene duplication, horizontal gene transfer, and transposon insertion.

7. The primers can be used to probe libraries containing long genomic clones to identify contig ends that lie close to each other. If the contigs are close enough, the primers can be used in PCR to directly amplify the intervening DNA separating the contigs, which can then be cloned and sequenced.

8. If the same procedure were used in both dimensions, all the proteins would form a single diagonal line in the final gel, and much of the potential for separation would be wasted. Using different protein properties in the two electrophoresis steps effectively spreads the proteins over the entire gel.

9. Proteins may precipitate nonspecifically for many reasons. In many cases, the precipitation is simply a property of the proteins and the conditions used. The interaction might also be indirect, with both the target protein and the nonspecific precipitating protein interacting in common with some other macromolecule, such as another protein or nucleic acid. Researchers focus on the precipitating proteins that are most abundant and confirm the putative interaction with additional experiments.

10. The same disease condition can be caused by defects in two or more genes, which are on different chromosomes.

11. If gene *X* has no relationship to any other gene in species B, it may have arisen by horizontal gene transfer. If gene *X* is homologous to gene 2, it may have arisen by gene duplication.

12. The pattern of haplotypes in the Aleut and Eskimo populations suggests that their ancestors migrated into the American Arctic regions in a separate migration from the one that led to the populating of the rest of North and South America.

13. Yes. Mitochondrial Eve lived thousands of years before Y chromosome Adam. Given that all modern humans (male and female) inherited their mitochondrial DNA from Eve, Adam must have had that DNA as well.

14. Interbreeding between Denisovans and *Homo sapiens* must have occurred in Asia, at some time in the many millennia during which humans migrated from Africa to Asia and then to Australia and Melanesia.

15. (a) Coronavirus genomes are composed of a single strand of RNA. At least one round of synthesis by reverse transcriptase is needed if the genome is to be amplified by PCR. (b) In all PCR experiments, researchers design primers that are unique and sufficiently long that they are unlikely to anneal well to other areas of a genome. In an experiment like this, one would also search for sequences that are highly conserved among similar genomes (e.g., sequences encoding parts of an enzyme that are critical to the enzyme's function), thereby maximizing the chance that the sequence will be present in an unchanged state in the target genome. (c) HEV and BCoV: 6; BCoV and SARS: 146; TGEV and SARS: 168. The coronaviruses BCoV and HEV are most closely related, with few differences between them. The differences among sequences are consistent with the phylogenetic tree.

Chapter 9

1. The length of chromosome 2 would be 0.34 nm/bp $\times$ 243 million bp = 82.62 million nm = 82,620 μm, or about 13,770 times longer than the diameter of the nucleus.

2. Lk_0 = (4,200 bp) / (10.5 bp/turn) = 400. From Equation 9-1, $\Delta Lk = Lk - Lk_0 = 374 - 400 = -26$. Substituting the values for ΔLk and Lk_0 into Equation 9-2: $\sigma = \Delta Lk/Lk_0 = -26/400 = -0.065$. The superhelical density is negative, so the DNA molecule is negatively supercoiled. When the same molecule has an Lk of 412, $\Delta Lk = 412 - 400 = 12$, and $\sigma = 12/400 = 0.03$. The superhelical density is positive, so the molecule is positively supercoiled.

3. (a) The DNA has ~171,000 bp; at 0.34 nm/bp, the DNA length is 58,140 nm. The DNA is almost 600 times longer than the JS98 head. (b) 170,523 bp.

4. The content of A does not equal the content of T. The simplest explanation is that the DNA is single-stranded.

5. The DNA has a molecular weight of 580,070 bp $\times$ 650/bp = 377,045,500. The contour length is 197,224 nm. Lk_0 = (580,070 bp) / (10.5 bp/turn) = 55,245. If $\sigma = -0.06$, Lk = 55,245 - (55,246 $\times$ 0.06) = 51,930.

6. The DNA has ~5,250 bp. (a) In the absence of strand breakage and resealing, Lk is unchanged; a positive supercoil must form elsewhere in the DNA to compensate. (b) Lk is undefined. (c) Lk decreases. (d) No change.

7. Lk remains unchanged because the topoisomerase introduces the same number of positive and negative supercoils.

8. Lk_0 = (13,800 bp) / (10.5 bp/turn) = 1,314. $\sigma = (Lk - Lk_0)/Lk_0$ = 292/1,314 = 20.07. Superhelical density is the same for the cellular chromosome and the plasmid, so the probability of infection is >70%.

9. (a) Lk undefined. (b) Lk = 500. (c) No effect. (d) Lk = 484. (e) Lk = 488. (f) Lk = 484.

10. Z-DNA is a left-handed double helix. Underwinding of the right-handed B-form helix will make a left-handed helix easier to form.

11. (a) The DNA must be unbroken and topologically constrained so that $Lk < Lk_0$. (b) Strand separation, formation

of hairpins and cruciforms, and formation of Z-DNA are all more favorable in negatively supercoiled DNA. **(c)** DNA gyrase introduces negative supercoils into DNA, with the aid of ATP. **(d)** The mechanism involves creation of a double-strand break, passage of an unbroken DNA segment through the break, followed by strand resealing. Transient phosphotyrosyl-DNA intermediates form, and the conformational changes are coupled to hydrolysis of ATP.

12. The DNA must include origins of replication, required for DNA replication; a centromere, for proper segregation of the chromosome at cell division; and telomeres, to protect the chromosomal ends.

13. **(a)** The lower, faster-migrating band is negatively supercoiled plasmid DNA. The upper band is nicked, relaxed DNA. **(b)** DNA topoisomerase I relaxes the supercoiled DNA. The lower band will disappear, and all of the DNA will converge on the upper band. **(c)** DNA ligase produces little change in the pattern. Some minor additional bands may appear near the upper band, due to the trapping of topoisomers not quite perfectly relaxed by the ligation reaction. **(d)** The upper band will disappear, and all of the DNA will be in the lower band. The supercoiled DNA in the lower band may become even more supercoiled and migrate somewhat faster.

14. **(a)** When DNA ends are sealed to create a relaxed, closed circle, some DNA species are completely relaxed but others are trapped in slightly under- or overwound states. This gives rise to a distribution of topoisomers centered on the most relaxed species. **(b)** Positively supercoiled. **(c)** The DNA that is relaxed despite the addition of dye is DNA with one or both strands broken. DNA isolation procedures inevitably introduce small numbers of strand breaks in some of the closed-circular molecules. **(d)** $\sigma \approx$ 20.05. This is determined by comparing native DNA with samples of known σ. In both gels, the native DNA migrates most closely with the sample of $\sigma = -0.049$.

15. Form I DNA was negatively supercoiled. When spread on an electron microscope grid, the DNA would tend to fold onto itself, creating DNA crossings or nodes. In form II DNA, the circles are relaxed.

16. The pattern in lane 2 is produced by DNA gyrase; that in lane 3 by DNA topoisomerase III (a type I topoisomerase).

17. The superhelical density would increase. The new mutations occur in the two genes encoding subunits of DNA polymerase II (gyrase), such that the activity of gyrase declines.

18. **(a)** 25 nodes. **(b)** Removal of 25 of 667 DNA turns would correspond to a σ of -0.037. **(c)** $\Delta Lk = 25/(-0.89) = -28$; thus, $\sigma = -0.042$. **(d)** No.

Chapter 10

1. Histones have an unusually high concentration of positively charged amino acid residues on their surface compared with most other proteins. Although many SDS molecules bind each protein and give it an overall negative charge, the SDS does not eliminate the positive charges on a protein, it just overwhelms them. For histones, the abundance of positive charges prevents the full effect of SDS on the charge of the proteins, and this manifests as a slower histone migration during electrophoresis compared with most other types of protein.

2. Approximately 150 bp are associated with each nucleosome in eukaryotic chromatin, with another 50 bp as linker, bringing the total to ~200 bp per nucleosome.

3. Phosphorylation adds negative charges that can alter the net charge. Acetylation of Lys residues can alter the net charge by creating an amide linkage. Methylation of Lys generally does not remove the positive charge of the terminal amino moiety, although it can alter the pK_a of the group.

4. Most histone modifications occur near the N-terminus. This part of the histone molecule is a relatively unstructured tail that extends out from the nucleosome core.

5. The bacterial chromosome is divided into topologically constrained loops, defined by bound proteins at their boundaries. When the DNA in one loop is relaxed, the DNA in other loops remains supercoiled.

6. Transcription will decrease. H2A and H2B are core histones and are closely paired in the nucleosome structure. H1 is generally bound in linker regions, between the core histones, and its level can be varied independent of the core histones. An increase in H1 will lead to greater compaction of the DNA and thus decreased transcription.

7. Histone H1 is in the center of the filament, along with the linker DNA. The nucleosomes are stacked along the outside of the filament.

8. Bacteria generally divide much more rapidly than eukaryotic cells. Stable protein-rich structures would impede the required replication and segregation of chromosomes at cell division.

9. Transcriptionally active genes are characterized by a decrease in histone H1, an absence of bound nucleosomes at the promoter regions, the presence of specialized chromatin remodeling complexes, and the presence of histone variants such as H2AZ and H3.3.

10. "Epigenetic inheritance" refers to chromatin modifications (particularly histone modifications) that are retained in the chromatin after cell division and affect gene transcription. Such modifications are not encoded in the DNA and thus are not subject to Mendelian inheritance.

11. **(c)**

12. 62×10^6 H2A molecules. ("Genome" refers to the haploid genetic content of the cell; the cell is diploid, so the number of nucleosomes is doubled.)

    Details: $[(3.1 \times 10^9 \text{ bp}) / (200 \text{ bp/nucleosome})] \times 2$ H2A/nucleosome $\times 2$ [for diploid cell] $= 62 \times 10^6$ H2A. The 62 million would double on replication.

13. Instead of observing eight different complexes, Kornberg would have observed five: H3, H4, H3-H3, H3-H4, and H3-H3-H4.

14. (a) 220 bp is the approximate spacing of adjacent nucleosomes in chromatin. **(b)** The excess of DNA sequences allowed the investigators to select for sequences that bound tightly and to eliminate weaker binders. **(c)** The salt interfered with protein-DNA interactions and ensured that only the most tightly binding DNA remained bound to the nucleosomes. **(d)** Isolation of the DNA-nucleosome complexes reduced the total amount of DNA in each cycle. The PCR step allowed the DNA levels to be increased again. However, only the bound DNA sequences, the "winners," were amplified; in each cycle, the solution was enriched in DNA sequences binding more tightly to the nucleosomes.

Chapter 11

1. The plasmid replicates unidirectionally. Molecules (c) and (d) are inverted relative to (a) and (b). Molecule (a) identifies the position of the origin relative to one end. The observation that (b), (c), and (d) have one forked end of similar size and the other forked end differing in size reveals that a single replication fork moves first through the short arm of (a) and then proceeds around the circular plasmid. The order of replication time is (a), (b), (d), (c).

2. The primer is a preexisting strand of RNA or DNA to which new nucleotides are added at the 3′ end. The template is a longer strand, paired with the primer, that determines the identity of each new nucleotide added via base pairing.

3. (a) No. In the absence of any one dNTP, the polymerase would stop incorporating the other three dNTPs as soon as it encountered a template residue that should pair with the missing dNTP, and incorporation of ^{32}P would be undetectable. **(b)** No. DNA synthesis releases the β and γ phosphates of dNTPs as pyrophosphate.

4. Possible answers: Pol I is slow in DNA synthesis compared with the rate of replication in *E. coli*. Pol I can be mutated and the cells still survive. Pol I is not highly processive.

5. The DNA polymerase contains a 3′→5′ exonuclease that degrades DNA to produce [^{32}P]dNMPs. The activity is not a 5′→3′ exonuclease, because addition of dNTPs inhibits [^{32}P]dNMP production: the polymerase extends radioactive 3′ termini by adding nonradioactive dNTPs, protecting the radioactive portion of DNA from the 3′→5′ exonuclease. This would *not* protect the 5′ terminus of radioactive DNA from a 5′→3′ exonuclease. Adding pyrophosphate would result in production of [^{32}P]dNTPs through reversal of the polymerase reaction.

6. Ligase will not seal a nick in which the 5′-terminal nucleotide is a ribonucleotide. Sealing is delayed until all the RNA has been removed.

7. The DnaA protein, assisted by the protein HU, creates a bubble of unwound DNA at the origin and also targets DnaB protein to that location. The DnaC protein loads the DnaB helicase onto the DNA at the origin.

8. (a) Either any combination of three A sites is sufficient for origin function, or three particular A sites are required. Construct four plasmids, each with a different mutant A site. Transfer the mutant plasmids into the host organism, and plate each transformed product on a medium containing the appropriate antibiotic. Use an unmutated plasmid and a plasmid without A sites as controls. If a particular A site is essential, the mutant plasmid will not form a colony. **(b)** Either the B sites are not essential, or one B site is needed but either one suffices. Construct a plasmid containing mutations in both B sites. If a particular B site is essential, a colony will not appear after transformation. Use an unmutated plasmid and a plasmid without B sites as controls.

9. The 3′→5′ exonuclease is a proofreading function, eliminating mismatched nucleotides incorrectly inserted into the growing DNA strand. DNA polymerases IV and V function in translesion DNA synthesis during DNA repair, a situation where the accuracy of DNA polymerization becomes less important (as described in Chapter 12).

10. A number of proteins and complexes, including the origin recognition complex (ORC), Cdc6, Cdt1, and the replicative helicase Mcm2-7, are loaded onto origins of replication in G_1 phase.

11. The preRC forms only in G_1, not in other phases of the cell cycle. Cyclin kinases produced only in S phase are needed to assemble the remaining proteins to produce active replication forks. Origins do not fire a second time, because new preRC complexes cannot form until the cell completes its cycle and returns to G_1.

12. The τ subunits link together the leading- and lagging-strand core polymerases, one τ linked to each core, and both connected to DnaB. **(a)** Two. **(b)** Zero. The core polymerase, in conjunction with a β sliding clamp, is capable of processive synthesis of a new DNA strand on a single-stranded DNA template without any other subunits being present. This is analogous to leading-strand synthesis without lagging-strand synthesis. **(c)** Three. Two of the three τ subunits, and their accompanying polymerase core subunits, act on the lagging strand and may permit more efficient DNA synthesis on the lagging strand.

13. DnaA: ATP hydrolysis inactivates the DnaA for replication initiation. DnaC: ATP hydrolysis helps release DnaB helicase as it is loaded onto the DNA. Pol III γ or τ subunits: ATP hydrolysis allows the β subunit (sliding clamp) to close around the DNA as it is loaded.

14. The two replication forks would never meet, and part of the chromosome near the terminus would remain unreplicated.

15. The RNA template for telomere synthesis is telomerase RNA (RT), which is bound tightly to the telomerase reverse transcriptase (TERT) to form the active telomerase holoenzyme. A very short segment of the TR, equivalent to an organism's telomere repeat sequence (typically 6 or 7 nucleotides), is used over and over again as a template in telomere DNA synthesis.

16. (a) As the DNA strands are linked together, the singly bonded phosphoryl groups are converted to phosphodiester bonds (doubly bonded phosphoryl groups) that are no longer susceptible to alkaline phosphatase. **(b)** The substrate for the reaction is a DNA strand break in one strand of double-stranded DNA. Ligation of single strands is not observed. **(c)** The reaction halts only because the enzyme runs out of substrate. Addition of poly(dA) creates more of the correct substrate and the reaction can continue. **(d)** The DNA ligase of *E. coli* uses NAD+ as cofactor rather than ATP.

Chapter 12

1. The cross-linked pyrimidine dimer causes a distortion in the DNA that prevents base pairing in the active site of the DNA polymerase.

2. **(a)**

    AACOTGCAC
    TTGTACGTG

    where O represents O^6-meG.

    (b)

    AACGTGCAC
    TTGTACGTG

    (c)

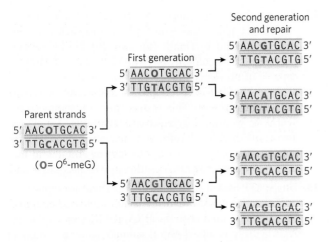

3. Any three of the following: defects in repair; UV light; TLS error-prone bypass; oxidative damage; spontaneous hydrolysis. For these lesions to become mutations, replication must occur (before repair), producing a naturally occurring base pair that is different from the original base pair.

4. Serine auxotrophs can grow only on serine-containing medium; when plated on a serine-free medium, the cells will die—unless they have undergone a reversion mutation to restore the serine-synthesizing pathway. Treatment with a genotoxic agent, even though it kills most cells by producing mutations in vital genes, produces some of these particular reversion mutations.

5. The survivors arise from spontaneous reversion mutations caused by spontaneous hydrolysis, oxidative damage, or natural sources of irradiation.

6. **(a)** The colonies arose because background (spontaneous) mutations included reversion mutations that reversed the cells' histidine dependency. **(b)** 2-Aminoanthracene causes DNA damage; some of the lesions escape repair before DNA replication, thus forming a mutation. Some of these are reversion mutations in the histidine-pathway gene, allowing cells to survive in a histidine-free medium. **(c)** 2-Aminoanthracene is a potential carcinogen, because it causes mutations in the Ames test.

7. **(a)** G≡C to A=T. **(b)** C≡C to T=A. **(c)** A=T to G≡C

8. **(a)** XP mutant cells usually contain a mutation in a gene required for NER, the main repair pathway for UV lesions in humans. Single-strand DNA breaks are produced during NER, accounting for the short ssDNA fragments observed in normal cells after irradiation. However, a defective NER system in the XPG mutant cells prevents formation of single-strand breaks. **(b)** XPG cells are defective at a step preceding the first strand incision of NER, which is needed to produce the fragmented ssDNA.

9. The initiation step. In global NER, the XPC protein recognizes the lesion. In TCR, RNA polymerase recognizes the lesion, by stalling at the lesion.

10. Both processes cleave the phosphodiester backbone, remove the pentose phosphate, then insert a correct dNTP and ligate the nick. In BER, a specific DNA glycosylase cleaves a damaged base from the pentose to form the abasic (AP) site.

11. The most common DNA lesions leading to G–T mismatches are insertion of T opposite an O^6-methylguanine lesion during replication and deamination of a 5-methylcytosine. In both cases, it is the T that is incorrect and potentially mutagenic if not repaired.

12. Frameshift mutations would lead to alteration of many amino acid residues in the protein product and could be caused by template slippage of DNA polymerase in the region with the repeated A residues.

13. Reactive oxygen species generated during normal aerobic metabolism are a major source of DNA damage, because they form free radicals that react with the DNA.

14. With only one good copy of a key DNA repair gene, there is a good chance that the gene will be inactivated in one or more cells by normal mutagenic processes such as unrepaired oxidative damage. Whenever this happens, a DNA repair system is no longer present in that cell, and the mutation rate increases substantially. Eventually, mutations occur in—and inactivate—one or more genes that normally control cell division, and a cancer cell is born.

15. About 1,700 phosphodiester bonds derived from dNTPs are expended in the repair: 850 in the DNA degraded between the mismatch and the GATC sequence, and another 850 in the DNA synthesis needed to fill the resulting gap. ATP is hydrolyzed by the MutL-MutS complex and by the UvrD helicase.

16. NER and BER occur only in double-stranded DNA. Both processes excise the damaged base or bases from the damaged strand, leaving a gap that can be filled in—but only if an undamaged complementary strand is present.

17. Each molecule of O^6-methylguanine methyltransferase is used only once and is degraded after the repair reaction. Thus, the reaction expends all the energy needed to synthesize the protein, along with all the energy used to mark the protein for degradation and to carry out that degradation.

18. **(a)** There may have been a trace contaminant of Pol III or Pol II in the preparation of UmuD′ and UmuC. **(b)** Pol III is the main replicative DNA polymerase in the cell. A strain with a completely inactivated Pol III would be dead. The temperature sensitivity allows the cells to be grown at a permissive temperature, but Pol III can be inactivated, when needed, by increasing the temperature. **(c)** Fraction 56 contains both Pol III and Pol V. The Pol V can replicate over the lesion, and Pol III can then extend the DNA. **(d)** The mutant Pol III is inactivated at 47°C. **(e)** Fraction 64 contains UmuC (and UmuD′, not shown) almost exclusively. The lack of temperature sensitivity provides evidence that UmuC and/or UmuD′ have a DNA polymerization activity.

Chapter 13

1. (1) The fork may bypass the lesion. If the DNA backbone is intact, the fork may either (2) stall or (3) leave the lesion behind in a single-strand gap. (4) If the fork encounters a break in a template strand, one arm of the fork is lost and the fork collapses.

2. **(a)**

(b)

3.

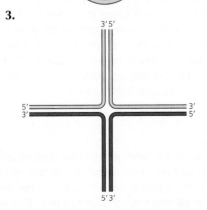

4. **(a)** RecB. **(b)** RecB. **(c)** RecD. **(d)** RecC. **(e)** RecC.

5. (1) Formation of a double-strand break, sometimes occurring spontaneously and sometimes created by an enzyme. (2) Degradation of the 5′-ending strand to generate a 3′extension. (3) Recombinase-promoted strand invasion, pairing the single-stranded region with a complementary strand in the duplex and displacing the other strand of the duplex to create a D-loop.

6.

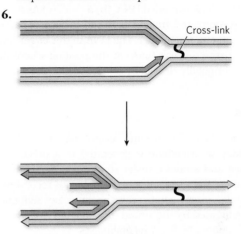

7. During normal growth, forks collapse at sites where there is a break in the template strand. The absence of RecBCD in the mutants will curtail the repair of such double-strand breaks, leading to the increase in linearized chromosomes.

8. The Rad51 recombinase is used in all recombinational processes in eukaryotes. The Dmc1 protein is used only during meiosis.

9. If gene conversion occurs across the region where the sequence difference between *A* and *a* is located, this will create a heteroduplex intermediate that has an *A*-containing strand paired with an *a*-containing complement. The mismatch at this gene locus will be resolved one way or the other by mismatch repair, resulting in the loss or gain of information.

10. NHEJ involves degradation by nucleases and processing of the DNA ends, leading to some loss of base pairs.

11. The information at *HMLα* would be subject to change. Whatever information was present in *MAT* would be transferred to *HMLα*.

12. The polyacrylamide gel separates proteins on the basis of molecular weight. For a protein covalently linked to DNA, the molecular weight is that of the combined protein and DNA. If the size of the linked DNA varies, the protein-DNA complex will have a highly variable molecular weight and will appear as a smear in the gel; detecting a single protein band will be impossible. Nuclease digestion eliminates this variability, allowing Spo11 to be detected as a discrete band.

13. **(a)** Points Y. **(b)** Points X.

14. During desiccation, DNA repair is impossible because of its requirement for metabolic energy in the form of ATP and dNTPs. Without a functioning cellular metabolism,

ATP cannot form. However, DNA degradation by nucleases requires no ATP or other cofactors. Thus, chromosomes with double-strand breaks formed during desiccation are rapidly degraded, eventually destroying the genome, unless the DNA ends at the break are protected by proteins such as DdrA.

15. **(a)** The single-stranded DNA substrate is degraded before agarose gel electrophoresis. In addition, restriction enzymes do not cleave single-stranded DNA. **(b)** $5' \rightarrow 3'$. **(c)** As strand exchange is completed, fragments 1b and 1a become part of a larger fragment, fragment 1. The DNA in this larger fragment has a strand break (nick) at the position where the duplex was originally cleaved by restriction enzyme A.

Chapter 14

1. In tyrosine-class site-specific recombination systems, the Holliday intermediate is generated in a precise set of cleavage and strand-transfer steps, all occurring at a single, unique DNA sequence. In homologous recombination, Holliday intermediates can appear at any sequence and are generated by the strand invasion and branch migration promoted by RecA recombinases.

2. Homologous recombination occurs only where two DNA molecules have identical or very similar sequences over a significant region; any sequence is permitted. Site-specific recombination occurs only at particular DNA sequences that are recognized, bound, and recombined by the recombinases. Transposition, with a few exceptions, can occur at almost any sequence.

3. The DNA between the sites would undergo both deletions and inversions.

4. Four active sites are present; only two are in the conformation required to catalyze a reaction.

5.

6.

(a) Recombination at the X sites inverts the intervening sequence because the X sites are in opposite orientations. **(b)** Recombination at the Y sites deletes the intervening sequence because the Y sites are in the same orientation,

leaving only one copy of the X site and one of the Y site. **(c)** If the X sites react first, the orientation of one Y site changes, leading to an inversion in the later reaction with the Y sites. **(d)** If the Y sites react first, deletion occurs. Reaction of the (deleted) X site on the circle with the (remaining) X site on the original DNA yields the same product as in (c).

7. Only **(b)** is an appropriate substrate. The Hin-*hix* system works only when the two *hix* sites are in the opposite orientation and on the same supercoiled DNA molecule.

8. Replicative transposition generates a cointegrate intermediate that joins the donor and target DNAs together. The cointegrate is resolved (the DNAs are separated) by the transposon-encoded site-specific recombinase (the Tn3 resolvase).

9. The λ site-specific system requires additional proteins, besides the Int recombinase, to promote the reaction. To adapt this system to eukaryotic cells, these additional proteins, as well as the recombinase, would have to be expressed. The λ *attP* site is also more complex, and is a larger segment of DNA to clone, than the FRT and lox sites.

10.

```
          ↓
5' ...ATGCAGGCTAATGGCTACCTGA...
3' ...TACGTCCGATTACCGATGGACT...
                    ↑
```

11. The tRNA, carried by the virus from one host to another, provides the primer for DNA synthesis by reverse transcriptase.

12. Only one active site is required to create one of the double-strand breaks. Three phosphoryl group transfer reactions are catalyzed in succession to bring this about.

13. The reverse transcriptase of TP (non-LTR) retrotransposons makes use of a 3' end derived from the target DNA to prime viral DNA synthesis. The endonuclease cleaves a phosphodiester bond in the target, exposing the needed primer terminus.

14. Exons are the coding regions of genes. Insertion of a transposon of any type into an exon would almost certainly disrupt the activity of the protein encoded by the gene.

15. **(a)** If Cre were used for the insertion, the cassette itself could be altered during the insertion process. **(b)** When the Flp reaction is complete, intact FRT sites flank the cassette. Each FRT could be a target for a new insertion event. **(c)** The core sequence (see Figure 14-2a) must be modified to prevent recombination between the different Cre sites. **(d)** Using R, Y, and C to denote RFP, YFP, and CFP, the possible combinations are: RRR, YYY, CCC, RRY, RRC, RYY, RCC, YCC, YYC, and RYC.

Chapter 15

1. The promoter is a site for RNA polymerase loading and initiation of transcription. To accomplish this, the RNA polymerase must form an open complex in which the two DNA strands are separated over a short distance. Due to

the effects of AT-rich versus GC-rich DNA on the overall thermal stability of DNA, the strand separation is more readily accomplished in sequences that are AT-rich (see Chapter 4).

2. The deletion would move the -35 sequence closer to the -10 sequence by half a helical turn of the DNA, putting the two elements on opposite faces of the DNA duplex. This would dramatically reduce binding of sigma factor to the promoter, thereby decreasing transcription efficiency.

3. 5′-AUGACCAUGAUUACG. The sequence reported for a gene is, by convention, that of the coding strand, and sequences are always written in the 5′→3′ direction.

4. At 50 to 90 nucleotides per second, the enzyme would take 34 to 61 seconds to transcribe the gene.

5. Assuming the two genes use similar transcription initiation modes, *tesA* will be more efficiently transcribed. The *tesB* -10 sequence deviates more from the consensus sequence, and its higher G≡C content will be more difficult to melt.

6. The number of transcripts would increase, because Pol III typically generates many more transcripts than Pol II. tRNAs, which are synthesized by Pol III, are required in much greater quantities in the cell than are most mRNAs, which are made by Pol II.

7. Initiation (including abortive initiation), the slowest step; elongation, the fastest step but punctuated by pauses; and termination, requiring stalling and subsequent dissociation of the polymerase from the DNA.

8. About 8 to 10 phosphodiester bonds can be formed in the initiation phase. RNA synthesis always begins at a unique location (defined as $+1$) relative to the promoter sequence. Elongation can be prevented by controlling the sequence of the DNA template strand and the rNTPs added to the reaction, and the reaction products are defined by the same parameters. For example, if the first G residue in the DNA template strand does not appear until position $+6$, and you add ATP, UTP, and GTP, but no CTP, RNA synthesis will be limited to the first five nucleotides, and the dominant reaction product is likely to be pentanucleotides. This strategy has been used in research on RNA polymerase. (See, for example, W. R. McClure, C. L. Cech, and D. E. Johnston, *J. Biol. Chem.* 253:8941–8948, 1978.)

9. Many RNAs are present in the body, synthesized before ingestion of the α-amanitin. The mRNAs support cell and tissue function until they are degraded, and typically last for about two days.

10. Intercalation means that the planar portion of a small molecule inserts into the double-helical DNA between successive base pairs, deforming the DNA and preventing movement of RNA polymerase along the template. Actinomycin D and acridine act in a similar way.

11. This is difficult to do experimentally. The TATA box and the Inr sequence are often, but not always, found upstream from genes transcribed by Pol II. A significant minority of eukaryotic promoters lack a well-defined TATA box—the so-called TATA-less promoters—and this can confound identification by computer algorithms alone.

12. No. The two strands of a DNA molecule are antiparallel and complementary (not identical). When the promoter is inverted, it will direct RNA synthesis that uses what was originally the coding strand as the template strand. The mRNA sequence, derived from a different DNA strand and synthesized in the opposite direction, would be very different from the mRNA produced by the original gene and might not even contain an open reading frame.

13. Errors in the genetic information in actively transcribed genes are of greater immediate importance to the cell than errors in silent genes. Transcription-coupled DNA repair focuses repair on those DNA sequences most heavily utilized by the cell.

13. There may be a palindromic sequence that allows formation of a hairpin structure during transcription, creating a ρ-independent terminator that is not perfectly efficient, or an imperfect rut sequence to guide the loading of the ρ protein. In both cases, the sequences would have to be imperfect so that some transcripts could be elongated through gene *B*.

14. (a) 10–13. (b) 5–8. (c) 2–13. (d) 5–8. (e) Transcription from the P_{HS} promoter requires a factor present in fractions 5–8. (f) Transcription from P1 and P2 uniquely requires the RNA polymerase reconstituted with the 32 kDa protein, and the designation "σ^{32}" is probably warranted. (g) The use of an alternative sigma factor is an efficient way to coordinate regulation of transcription of a group of genes that express products not always required by the cell.

Chapter 16

1. Inactivation of the polymerase would lead to incomplete pre-mRNA processing, including 3′ end formation, splicing, editing, and transport—and would certainly be lethal.

2. Two; one to cleave the 5′ exon-intron junction and one to join the two exons, with release of the intron.

3. Group I and group II introns are generally self-splicing, with only the RNA backbone of the intron required for the reaction in vitro. The nucleophile in the first step (cleavage of the 5′ splice site) for group I introns is a guanine nucleotide or nucleoside, and for group II introns is the 2′-OH of an internal A residue. The second step uses the liberated 3′-OH at the 5′ splice site as a nucleophile to attack the phosphodiester bond at the 3′ splice site, joining the exons. Site specificity is aided by guide sequences that are part of the intron structures. Introns removed with spliceosomes follow a mechanism similar to that for group II introns, but also rely on the spliceosome ribonucleoprotein complex to catalyze their excision.

4. Self-splicing catalyzes phosphodiester exchange reactions with no net loss or gain of energy. Bonds are broken and re-formed with different nucleotides, and there is no change in the number of phosphodiester bonds, just in the covalent bonding partners. Thus, there is no net change in free energy from reactants to products.

5. Incubate the total RNA in the presence of an appropriate buffer to ensure RNA folding. Add a fluorescently labeled

guanine nucleotide analog to label the 5′ ends of any RNA that uses this nucleotide as a nucleophile in the first step of splicing. After reaction, fractionate the RNA (by size) on a polyacrylamide gel and look for fluorescently labeled RNAs. Controls could include incubation with other labeled nucleotides or with just the fluorescent dye, in parallel reactions—which are not expected to result in labeled RNAs, based on the group I intron reaction mechanism.

6. Over many generations, the wild-type bacteria would win out and the mutant strains would disappear. The normal rRNA modifications add subtle but important thermodynamic stability to the ribosome structure, enabling more robust protein synthesis in the wild-type bacteria.

7. Sometimes. By definition, enzymes remain unchanged after catalysis, which is not the case for self-splicing and self-cleaving RNAs. They are catalysts because they enhance the rate of bond cleavage and/or joining, but in their natural form they have only one reaction turnover. However, some of these RNAs can be engineered to bind to separate substrate molecules and catalyze reactions on multiple such molecules, so they are inherently capable of functioning as enzymes. Some, such as RNase P, do so in their natural state.

8. The binding and hydrolysis of GTP by Ran ensures a cycle in which mRNA is bound in the nucleus and released in the cytoplasm, and not the reverse.

9. The lifetime would increase, because mRNAs in higher eukaryotes are degraded by 5′→3′ exonucleolytic digestion.

10. The most commonly altered bases are cytosine and adenine. The alterations generally involve deamination reactions that convert these bases to uracil and inosine, respectively. Uracil pairs with A rather than G. Inosine pairs with C rather than T or U. Both alterations can change the amino acid encoded by the mRNA codon where the deamination reaction occurred.

11. In each case, the nucleotide is modified by removing an exocyclic amine from the six-membered ring and replacing it with a keto oxygen. The change converts C to U and A to I.

12. The most important properties of living systems are the capacity to catalyze reactions and the capacity to store information that defines the structure of the catalysts. RNA has both of these properties.

13. 5′-AAUAAA and a GU-rich sequence. The 5′-AAUAAA is retained in the mature mRNA.

14. The 2′, 3′, and 5′ hydroxyls of ribose. The 3′-OH and 5′-OH remain linked to the nucleotides they were bonded to before the reaction. The 2′-OH is linked to the G residue on the 5′ end of the intron.

15. The function of Dicer is to cleave pre-miRNAs to generate mature miRNAs.

16. Inactivation of tRNA nucleotidyltransferase. The CCA-3′ would not be added to the tRNA, so amino acids could not attach.

17. Many naturally occurring ribozymes catalyze reactions involving other RNAs, mainly cleavage or splicing reactions. Ribozymes also catalyze the formation of peptide bonds on ribosomes.

18. (a) α-Amanitin inhibits Pol II. The rRNA precursors that the researchers wanted to isolate are synthesized by Pol I, and addition of α-amanitin eliminated a lot of background RNA synthesis. (b) Some very stable or tightly bound protein may have remained and catalyzed the intron-splicing reaction. (c) The reaction requires a guanine nucleoside or nucleotide with a 2′-OH (RNA form) and a free 3′-OH. (d) This experiment is a compelling demonstration of self-splicing and RNA catalysis. There are no introns in bacterial rRNA. Expressing the rRNA segment with the use of bacterial enzymes eliminated the possibility that contaminating intron-splicing enzymes from *Tetrahymena* were present. The *Tetrahymena* rRNA gene is the only *Tetrahymena* macromolecule in the reaction mixture and was deproteinized before the experiment.

Chapter 17

1. There are two possible codons: AUA and UAU; only two amino acids can be incorporated into a polymer, and the only polymer produced is poly(Tyr-Ile).

2. The codon for methionine is AUG, so an RNA that produces poly(Met) must be a repeating sequence of AUG. There are three reading frames for (AUG)$_n$. The (AUG-AUG-AUG)$_n$ frame produces poly(Met); UGA is a termination codon, so no polypeptide forms; and (GAU-GAU-GAU)$_n$ produces poly(Asp).

3. Start and stop codons are in red:

5′-CCG-AUG-CCA-UGG-CAG-CUC-GGU-GUU-ACA-AGG-CUU-GCA-UCA-GUA-CCA-GUU-UGA-AUCC-3′

Met-Pro-Trp-Gln-Leu-Gly-Val-Thr-Arg-Leu-Ala-Ser-Val-Pro-Val-(stop)

4. Four possible RNA sequences can encode Met-Asn-Trp-Tyr (the variations in codons are in red):

ATG-**AAU**-UGG-**UAU**
ATG-**AAC**-UGG-**UAU**
ATG-**AAU**-UGG-**UAC**
ATG-**AAC**-UGG-**UAC**

Addition of a Leu residue, which has 6 possible codons, increases the number of possible RNA sequences to 24: any of the 6 codons—UUA, UUG, CUU, CUC, CUA, or CUG—can be added to the end of each of the four sequences above.

5. Met-Tyr-Gln. These are the first three amino acids, beginning at the second AUG (initiation) codon: AUG-UAU-CAG. The first reading frame, AUG-UGU-UGA, ends in a stop codon after only two amino acids.

6.

5′-AUG-GGU-CGU-GAG-UCA-UCG-UUA-AUU-GUA-GCU-GGA-GGG-GAG-GAA-UGA-3′
Met-Gly-Arg-Glu-Ser-Ser-Leu-Ile-Val-Ala-Gly-Gly-Glu-Glu-(stop)

Ten tRNAs are needed for these 14 amino acids (two to encode the three Gly; one to encode both Ser; one to encode the three Glu; and one for each of the other residues).

7.

5'-AUG-GGU-CGU-GAG-UCA-UCG-UUA-AUU-GUA-GCU-GGA-GGG-GAG-GAA-UGA-3'

Met-Gly-Arg-Glu-Ser-Ser-Leu-Ile-Val-Ala-Gly-Gly-Glu-Glu-Trp

The peptide is 15 amino acids long, instead of 14; 10 tRNAs are needed, one for each different amino acid.

8.

Met-Ile-Leu-Leu-Ser-Trp-Thr
5'-AUG-AUA-UUG-CUA-UCU-UGG-ACU-3'

Transitions: C A U G   C   C Six positions
Transversions: C/U A/G   A/G Three positions

9.

5'-ACC-AUA-UUG-CUC-UCU-UCG-3'

Thr - Ile - Leu- Leu- Ser - Ser

Pro - Leu- Met - Phe- Pro - Pro

Ala - Val - Val - Ile - Thr- Thr

Ser - Thr- Ser - Val - Ala - Ala

Ile - Lys- Trp - Pro - Phe- Leu

Asn - Arg- Phe - His - Tyr - Trp

Met-(stop) Arg-Cys-(stop)

10. Met-Pro-Ala-Glu-Val. A tRNA^{Tyr} with an anticodon mutation (e.g., 5'-AUA to 5'-UUA) will suppress the first stop codon, resulting in Met-Pro-Ala-Glu-Val-Tyr-Ser-Glu-Ala.

11. **(a)** Poly(Arg-Glu). **(b)** Poly(Val-Cys). **(c)** Poly(Glu), poly(Thr), and poly(Asn). **(d)** Poly(Lys), poly(Glu), and poly(Arg). **(e)** Poly(Leu-Leu-Thr-Tyr).

12. Leu: U$_2$C and UC$_2$ (observed proportion 22.2; calculated proportion 20). Phe: UUU and U$_2$C (obs. 100; calc. 100). Pro: UC$_2$ and CCC (obs. 5.1; calc. 4). Ser: U$_2$C and UC$_2$ (obs. 23.6; calc. 20).

13. The anticodon contains an inosine: 5'-ICC.

14. Incorporation of a Sec residue occurs only at a UGA codon that has an adjacent sequence called a SECIS element; it also requires a specialized tRNA (SelC, or tRNA^{Sec}); an enzyme, RelA, that catalyzes conversion of the Ser that is initially used to charge tRNA^{Sec} to Sec, to form Sec-tRNA^{Sec}; and a specialized elongation factor called SelB that recognizes the SECIS element.

15. The maximum number for a four amino acid change is 11 nucleotides between the insertion (X) and the deletion:

CAT-XCA-TCA-TCA-TCA(omit T)-CAT

The minimum number is 7 nucleotides:

CAT-CAX-TCA-TCA-T(omit C)AT-CAT

16.

A-CGU-CGA-GUA-GCA-GUA-UCG-AUU-GAG-CUC-UUA-GAU-AAG-AUC-GC

The other reading frames would encode stop codons (red) in the middle of a protein:

ACG-UCG-AGU-AGC-AGU-AUC-GAU-**UGA**-GCU-CUU-AGA-**UAA**-GAU-CGC
AC-GUC-GAG-**UAG**-CAG-UAU-CGA-UUG-AGC-UCU-**UAG**-AUA-AGA-UCG-C

17. Only one set of answers is shown.

Amino Acid	Codon	Anticodon
Phe	UUU	GAA
	UUC	
Leu	UUA	UAA
	UUG	
	CUU	GAG
	CUC	
	CUA	UAG
	CUG	
Ile	AUU	IAU
	AUC	
	AUA	
Met	AUG	CAU

18. **(a)** Three different codons are present in the three reading frames of this oligonucleotide, and thus a maximum of three different aminoacyl-tRNAs might bind. **(b)** With limited knowledge of the genetic code, possible explanations included the presence of a four-base code rather than a triplet code, or a relaxed specificity due to reaction conditions. **(c)** AAG was assigned to Lys. **(d)** AAA had already been assigned to Lys in the earliest code-cracking experiments that used homopolymeric poly(A). **(e)** GAA was assigned to Glu. **(f)** Increased Mg^{2+} concentrations and lower temperatures relaxed the specificity of the binding. **(g)** AAG = Lys; GAA = Glu; AGA = Arg. **(h)** GAC and GAU encode Asp. The Asp-tRNA^{Asp} may have elicited a positive signal in some experiments, because coding specificity is weakest in the third (wobble) position of the codon.

Chapter 18

1. 1,600 (4 × 400) phosphoanhydride bonds. Typically, four ATP/GTP molecules are hydrolyzed per amino acid incorporated into protein.

2. **(a)** Amide or peptide. **(b)** Ester. **(c)** Phosphodiester. **(d)** Hydrogen bonds. **(e)** Noncovalent bonds, including hydrogen bonds and van der Waals forces.

3. The 15,000 ribosomes in an *E. coli* cell can synthesize more than 29,000 proteins in 20 minutes.

4. Using polysomes, the cell can produce several protein molecules on a single mRNA molecule. Because mRNAs have an average lifetime of just a few minutes in the cell, polysomes maximize the number of proteins that can be made per unit time.

5. The fMet-tRNA^{fMet} binds directly to the P site during initiation of protein synthesis, positioned there to make the first peptide bond with a second aminoacylated tRNA that is positioned in the A site.

6. Proline is incorporated during protein synthesis; post-translational processing adds the hydroxyl group.

7. Isoleucyl-tRNA synthetase sometimes catalyzes the addition of valine to tRNA^{Ile}; valine is similar to but smaller than isoleucine and can readily fit into the synthetase active site. Histidine has no close structural analogs among the amino acids, greatly lowering the chance that tRNA^{His} will be charged with the incorrect amino acid.

8. Yes. In polypeptide synthesis, removal of the last amino acid added (by hydrolytic cleavage of the last peptide bond to form) would sever the covalent link between the polypeptide and the tRNA in the ribosomal P site. This would terminate polypeptide synthesis.

9. IF-2: the 70S ribosome would form, but initiation factors would not be released and elongation could not start. EF-Tu: the second aminoacyl-tRNA would bind to the ribosomal A site, but no peptide bond would form. EF-G: the first peptide bond would form, but the ribosome would not move along the mRNA to vacate the A site for binding of a new EF-Tu–tRNA.

10. (a) The synthetase recognizes the G–U base pair (between G3 and U70) in the amino acid arm of tRNA^{Ala}. (b) The mutant tRNA would insert Pro residues at codons that specify Ala. (c) A mutation in tRNA^{Pro} that allowed it to be recognized and aminoacylated by Ala-tRNA synthetase would have similar effects. (d) Such changes would insert Pro residues at many inappropriate sites in polypeptides, inactivating many proteins, and thus would be lethal.

11. A tmRNA (transfer-messenger RNA) binds to the A site of the stalled ribosome. Part of its structure acts as an aminoacylated tRNA^{Ala}, leading to transfer of an Ala residue to the C-terminus of the partially synthesized protein. Another part of the tmRNA structure functions as an mRNA, permitting the addition of 10 additional amino acids to the protein before a stop codon is reached. After the polypeptide is released from the ribosome, the 10 amino acid residues incorporated at its C-terminal end act as a signal for degradation by cellular proteases.

12. An unknown location, but *not* the nucleus. The protein will be bound by SRP and transported as it is synthesized into the ER lumen. Its fate from there would depend on other signals. The NLS would not be used, because it would not be accessible to the cytosolic proteins that normally bind it.

13. Chloramphenicol inhibits bacterial protein synthesis *and* mitochondrial protein synthesis. The effects on mitochondrial ribosomes give rise to the human toxicity.

14. (a) Poly(UG) generates polypeptides containing Phe, Leu, Val, Cys, Trp, and Gly. (b) Ala is not normally present in polypeptides synthesized in response to poly(UG). (c) Yes. In both cases, about one-third of the added label is incorporated into the product polypeptide, although the yield is slightly reduced in the Raney nickel-treated preparation. (d) This experiment was a control to confirm that Raney nickel treatment did not have a general deleterious effect on tRNA. (e) In most scientific studies, it is useful to confirm a

result by two or more different methods. The extra experiment rendered the final conclusion unambiguous.

15. (a) The 50S subunit contains the peptidyl transferase activity of the ribosome. It includes the parts of the P and A sites that interact with the 3′ ends of charged tRNAs. (b) The CCA sequence at the 3′ end of the hexanucleotide is present at the 3′ end of every tRNA and is required for specific binding to the 50S subunit. (c) The fMet-oligonucleotide must bind in the P site, and the puromycin in the A site. (d) The reaction is much simpler and does not rely on added protein factors such as initiation and elongation factors, which would be inactivated by the protein removal treatments. (e) Given the capacity of *T. aquaticus* to grow at high temperatures, the rRNA in the 50S subunits might be particularly stable and able to withstand protein removal. (f) The result in Figure 2 shows that chloramphenicol and carbomycin strongly inhibit the reaction, so the reaction is indeed catalyzed by the peptidyl transferase activity of the 50S subunit. In addition, the RNA component of the subunit is essential for activity. (g) The protein extraction procedures were thorough, but there was a possibility that small amounts of protein remained associated with the 23S rRNA.

Chapter 19

1. Screening of a genomic library to identify interacting partners of a "bait" protein is one of the common uses of the two-hybrid assay. To complete the screening system, you make a genetic library of "prey" fusion proteins by splicing a cDNA library containing random genes from the organism (the source of the bait protein) with a gene encoding a transcription-activation domain (e.g., the activation domain of Gal4p). You transfer the prey plasmid library into yeast, along with the bait fusion plasmid. Expression of β-galactosidase (as reporter gene) in the transformed yeast could be measured by the presence of blue colonies on plates containing X-gal (see Chapter 7); this will occur only in cells where the prey fusion protein interacts with the bait fusion protein. You can then sequence the prey gene to identify the interacting partner.

2. Because the 18 bp site is a near palindrome, the activator probably functions as a dimer. Given that both protein A and protein B are required for activating gene *X*, they may form a heterodimer that has specificity for the binding site. This could be tested in an electrophoretic mobility shift assay or any other assay that measures DNA binding. Alternatively, the 18 bp site might be bound by a third, unidentified protein, and proteins A and B might bind different DNA sites. To test this, the site could be used in a functional DNA-binding assay to follow the binding protein during purification, allowing identification of the correct protein. Footprinting could be used as an assay (see Chapter 20), and the DNA sequence could be used to make an affinity chromatography resin to aid purification (see Chapter 7). A final possibility is that protein A and/or protein B interact with a different protein to bind the 18 bp site. This could be tested by purification using a

functional assay such as footprinting. The purified active protein would reveal the additional protein.

3. There are more hydrogen-bond donor and acceptor groups on the nucleotide bases in the major groove than on those in the minor groove, providing much better discrimination between bases.

4. Heterochromatin is highly condensed and transcriptionally inert, because the histone proteins make promoters inaccessible. The less-condensed euchromatin has undergone a structural remodeling, allowing some regions to be transcribed. The alterations include covalent modification (such as acetylation) of histones and displacement of nucleosomes, creating exposed regions of DNA that are probably binding sites for regulatory proteins.

5. The primary transcript of an miRNA that is an stRNA is about 70 nucleotides long, with self-complementary internal sequences that form hairpin structures. The precursor is cleaved into 20- to 25-nucleotide partial duplexes, one strand of which can bind complementary stretches in cellular mRNAs. This binding can inhibit gene expression by blocking translation or facilitating mRNA degradation.

6. Positive regulation. Positive and negative regulation are defined in terms of the type of protein involved in the regulation. Regulation by an activator is positive regulation; regulation by a repressor is negative regulation.

7. A leucine zipper motif. The motif contains Leu (L) residues at every seventh position. It often functions in transcription factors to form a dimer interface by forming a coiled-coil.

8. The regulon may be subject to combinatorial control, in which subsets of the regulon genes are needed in certain circumstances. The 13 genes may be subject to regulation by another regulatory protein—either another repressor that needs to be removed or an activator that needs to be present for transcription to occur.

9. Most repressors with helix-turn-helix motifs function as oligomers, many as homodimers. When the plasmid-encoded mutant repressor is synthesized at high levels in the cell, most of the wild-type repressor molecules synthesized are incorporated into less functional heterodimers with a mutant repressor subunit.

10. One of the best-characterized transcription-activation motifs in eukaryotes is a string of acidic amino acid residues. The acidic peptide segments fused to LexA functioned as activation motifs or domains. The LexA fusion proteins could bind to the correct site adjacent to gene *X* via the LexA operator and activate transcription of gene *X* by means of the acidic fused peptides.

11. Regulatory proteins with helix-turn-helix, helix-loop-helix, or homeodomain motifs generally function as dimers and bind to sequences with inverted repeats. Proteins with zinc finger motifs can function as monomers and have no constraint to bind inverted repeats. Thus, strings of zinc finger motifs can be linked together to bind almost any sequence, whether it contains repeated elements or not.

12. The process is called RNA interference. The enzyme Dicer cleaves the double-stranded RNA into small interfering RNAs (siRNAs), which can bind to the mRNA and prevent its translation.

13. On binding a hormone molecule, the steroid hormone receptor dimerizes and the hormone-receptor complex is transported into the nucleus.

14. There are several possible explanations. Many eukaryotic genes are regulated by more than one activator protein, and another activator may be needed. Many eukaryotic genes are encapsulated (and silenced) in heterochromatin, and remodeling of the chromatin may be required in the region where the gene is located to allow activation. The protein may need to be modified, and the modifying enzyme (e.g., kinase) may not be present in the cell. Finally, perhaps the activator cannot be transported into the nucleus.

15. (a) The regulatory sequences for the chromosomal *GAL1* gene were known to respond only to Gal4p, and the DNA-binding elements of Gal4p had been removed in the fusion protein. (b) Given the finding that the fusion protein functioned as a repressor in *E. coli*, the researchers knew that the DNA-binding elements were properly folded and bound to the normal LexA-binding sites. (c) The LexA-Gal4 fusion protein is functional and stimulates gene expression to a level similar to that stimulated by Gal4p at UAS_G. (d) LexA by itself does not activate gene expression in this system. (e) Positioning of the LexA operator 577 bp rather than 178 bp away from the transcription start site lowers transcription activation by about 25%. (f) When the upstream sequences contain UAS_G or the 17mer, expression depends on the cellular Gal4p, which in turn is expressed only in the presence of galactose. (g) The LexA protein itself does not activate transcription; instead, a segment of Gal4p (that does not include the DNA-binding elements) is required. Thus the LexA part of the LexA-Gal4 fusion protein is not altering the DNA structure in any way that facilitates transcription. Instead, the Gal4p portion must be directly interacting with RNA polymerase.

Chapter 20

1. The *E. coli* cells will produce β-galactosidase when they are subjected to high levels of a DNA-damaging agent, such as UV light. Under such conditions, RecA binds to single-stranded chromosomal DNA and catalyzes cleavage of LexA, releasing LexA from its binding site and allowing transcription of downstream genes.

2. (a) With the Lac operator on the other side of the *lac* operon, the *lac* genes would no longer be subject to repression by the Lac repressor. (b) Inactivation of the binding site for CRP would reduce or eliminate expression of the *lac* genes under all conditions. (c) Alterations in the promoter sequence could either increase or decrease expression of the *lac* genes, depending on the particular alteration.

3. The conformation of AraC is altered by arabinose binding, and the protein binds different sites on the DNA when arabinose is absent versus present. In the absence

of arabinose, AraC binding blocks RNA polymerase access to the promoter.

4. (a) With high tryptophan concentration (added to the medium), tryptophan synthase levels drop due to attenuation of mRNA translation, even if the mRNA is stable. **(b)** Tryptophan synthase levels remain high for a considerable time. **(c)** Tryptophan synthase levels decline rapidly.

5. (a) An uncleavable LexA protein would permanently repress the SOS genes and block induction of the SOS response. **(b)** Weakened LexA binding would lead to constitutive expression of the SOS response genes.

6. ~7,000 copies. (10 copies/ *E. coli* cell) × (3.2 × 10⁹ bp/ haploid human cell)/(4.6 × 10⁶ bp/*E. coli* cell) ≈ 7,000 copies/human cell.

7. Repressor concentration is ~8×10^{-9} M. The number of moles of repressor = 10 molecules/6.02×10^{23} molecules/mol = 1.66×10^{-23} mol. Dividing by the volume of the cell in liters gives the concentration: $(1.66 \times 10^{-23}$ mol$)/(2 \times 10^{-15}$ L$) = 8 \times 10^{-9}$ mol/L. This is about five orders of magnitude greater than the dissociation constant for the repressor; as a result, the operator site will almost always be bound by repressor.

8. (a), (c), (d), (e) Operon expression decreases. **(b)** Operon expression remains unchanged, as it is already running at maximum expression.

9. (a) Decreased attenuation and thus increased transcription. If the ribosome translating sequence 1 did not block sequence 2, sequences 2 and 3 would pair more often. **(b)** The efficiency of pairing between sequences 2 and 3 could decrease, leading to increased attenuation and thus decreased transcription. **(c)** No attenuation, and thus transcription would proceed at the maximum level. **(d)** The attenuation system would respond more to histidine than to tryptophan concentration. **(e)** Decreased attenuation and thus increased transcription, because sequences 2 and 3 would pair more often. **(f)** Increased attenuation and thus decreased transcription, because pairing of sequences 3 and 4 would almost always occur.

10. The TPP-binding riboswitch alters mRNA conformation when TPP is bound, making the Shine-Dalgarno sequence inaccessible to ribosome binding. The glucosamine 6-phosphate–binding riboswitch also changes the conformation of its mRNA, but in this case the change activates a ribozyme ribonuclease function that cleaves and inactivates the mRNA.

11. The repressor protein would probably block translation of the mRNA even when the protein's binding sites on the ribosome were available. This would tend to slow the assembly of active ribosomes.

12. The λ prophage produces the λ repressor (cI) and cII proteins, which will bind to any λ phages entering the cell and prevent induction of a lytic pathway.

13. These mutants indicate that activation is not caused by CRP changing the local DNA structure near a promoter; the activator must do more than simply bind to DNA. Because the mutant CRP cannot activate transcription

in response to low glucose levels, cells with this mutation will not grow well on lactose or other secondary sugars for which the metabolizing-enzyme genes use CRP as transcription activator.

14. Because genes in an operon are transcribed together from one promoter as a polycistronic mRNA, they can be regulated by one set of activators and/or repressors. In this way, enzymes required for a common pathway can be synthesized together. Translational regulation could be used to express the operon genes at different levels: by alteration of the ribosome-binding sites for each gene or binding of translational repressors to one or a few of the genes.

15. An advantage is that the signal sensor is contained within the mRNA itself, and thus regulation does not require a separate sensor molecule to be synthesized or maintained. A disadvantage is that coordinated gene expression through integration of different cellular signals is difficult.

16. Conserving energy is important, but careful modulation of gene expression in response to changing growth conditions is paramount. It is preferable to "waste" energy in synthesizing partial transcripts that will remain unused unless antitermination is triggered, so that the cell is primed to respond quickly to a sudden need for enhanced gene expression.

17. No. Eukaryotic transcription occurs in the nucleus, and translation in the cytoplasm. The spatial and temporal separation prevents the coupling of transcriptional and translational regulation that can occur in bacterial operons such as the *trp* operon.

18. (1) These mRNAs tend to have unusually long sequences upstream from the translation start site that are necessary to form the three-dimensional structure of the riboswitch. (2) The upstream sequence is conserved in the same mRNA in different bacterial species, and sometimes in archaea, plants, and fungi. (3) These RNAs do not have protein binding partners, consistent with their ability to function in directly regulating gene expression.

19. Cell growth would decline, even in the presence of abundant nutrients, due to the inhibitory effects of ppGpp on RNA polymerase.

20. (a) Band A is, in part, the undigested attenuated mRNA, and this undigested RNA accounts for one of the three A bands in Figure 3. The two additional A bands in the denaturing gel indicate the presence of a good cleavage site for RNase T1 near the middle of the RNA. Because the two parts of the RNA separate only on the denaturing gel, sequences on either side of the T1 cleavage site must be paired and thus migrate together in the first gel (Figure 2). **(b)** Bands B and C are segments of the larger attenuated mRNA. **(c)** Both band B and band C have paired RNA sequences that protect the RNA from RNase T1. Because band C is sometimes cleaved in two (the gel in Figure 3 has one band of undigested RNA and two bands derived from the cleavage), there must be a loop of significant size near the end of the paired sequences in this segment. **(d)** The band B RNA includes sequences 3 and 4. The loop at the end of this hairpin is small enough to limit

T1 cleavage. **(e)** The band C RNA derives from paired regions in sequences 1 and 2. **(f)** The pairing of sequences 2 and 3 is not present in this analysis. This hairpin must be less stable than the pairing of sequences 3 and 4 and of sequences 1 and 2.

Chapter 21

1. There are many more genes in a eukaryote, and most are not organized into operons. A huge number of repressors would be needed at all times if negative regulation predominated. Eukaryotic DNA is packaged into chromatin, which effectively silences most genes, and special mechanisms are needed to open up the chromatin when a gene must be expressed. Chromatin is not present in bacteria.

2. Eukaryotes often use multiple regulatory proteins to activate transcription of a single gene. Given the large size of eukaryotic genomes, the chance of nonspecific binding of a given regulator to DNA sequences unrelated to that particular gene is too great for the cell to rely on a single regulatory protein.

3. HATs generally modify Lys residues in the C-terminal tails of histones. Histone acetylation reduces the affinity of nucleosomes for DNA and can increase the transcriptional activity of a chromosomal region. The acetyl groups are removed by histone deacetylases (HDACs).

4. Gal4p is a transcription factor that largely functions as a transcription activator of the *GAL* genes. Gal11p is probably involved in regulation of a much larger set of genes. (In fact, Gal11p is part of the yeast Mediator coactivator complex.)

5. **(a)** Phosphorylated. **(b)** Phosphorylated. **(c)** Unphosphorylated. **(d)** Unphosphorylated.

6. First, most genes are regulated by multiple transcription factors (activators), and different (often unique) combinations of factors are used at different genes. Second, families of activators form heterodimers, such that a family of four related proteins can make a total of 10 different dimeric species that can recognize 10 different DNA sequences.

7. CTCF is a key component of the gene insulator system in eukaryotes; it binds to sequences that prevent inappropriate activation of certain genes during development and in the mature organism. Its loss would be lethal.

8. HMG proteins bind DNA and facilitate DNA bending. The presence of HMG protein–binding sites in the DNA that interacts with an enhanceosome may reflect HMG-mediated facilitation of DNA wrapping around the enhanceosome structure.

9. Gene expression is generally silenced in regions of heterochromatin. If the gene were essential, as most housekeeping genes are, moving it into heterochromatin would be lethal for the cell.

10. In mammalian females, one X chromosome is inactivated by condensation into a structure called a Barr body. Condensation begins at the X inactivation center, near the center of the X chromosome.

11. **(a)** With elimination of the nuclear import signal, the receptor could bind the hormone in the cytoplasm, but the complex would not be imported into the nucleus to activate gene expression. **(b)** With elimination of the interaction with Hsp70, most of the receptor molecules would be in the nucleus, where they would not have access to the hormone signal.

12. **(a)** A nuclease-hypersensitive site is a region of uncondensed (or less condensed) chromatin where the DNA is accessible to nuclease action. The decondensed state signals a site for the binding of proteins that function in genome maintenance and/or gene regulation. **(b)** The LCR triggers chromatin remodeling on both sides and can help activate transcription of genes on both sides. **(c)** The constructs are integrated at random locations in the chromosome. Some may accidentally be integrated near an LCR that could activate transcription from a site on the 5′ side. **(d)** Yes. Many more colonies are observed when the λ DNA is flanking the neomycin-resistance gene than when the chicken hypersensitive site is flanking this gene. **(e)** Because the constructs were integrated at random locations in the genome, they might have been subjected to the differential effects of sequences on either side of the integration sites that could affect expression of the neomycin-resistance gene. Sometimes, multiple copies of the constructs could have been integrated. If these effects were present, the hygromycin-resistance gene provided a way to normalize the results. **(f)** No. The effect could be at the level of posttranscriptional modification or translation. **(g)** To show that the effect was at the level of transcription, researchers would have to directly measure the production of mRNA from the genes in stably transfected cell lines (which the authors did).

Chapter 22

1. The effect of actinomycin D reflects simply a blockage of RNA synthesis. The effect of cycloheximide may reflect the requirement for a newly synthesized protein factor in the signaling pathway for induction of the gene encoding this mRNA.

2. The N-terminal amino acid residues of Sxl expressed from P_e (encoded by exon E1) differ from the N-terminal residues of Sxl expressed from P_m (encoded by exon L2).

3. The RNA-binding site for proteins that carry out 3′-end cleavage and polyadenylation is defined most reliably by the sequence AAUAAA. In the DNA, the sequence is AATAAA, located downstream from the final gene exon. Because AATAAA is a consensus sequence, scanning only for this sequence would not find all the 3′-end cleavage sites; some will vary slightly from the consensus.

4. Reticulocytes are the precursors of red blood cells, filled with hemoglobin and highly specialized for oxygen transport; their nucleus is destroyed during maturation. Almost all the mRNA deposited in a reticulocyte before destruction of the nucleus encodes hemoglobin. There are essentially no other translation-dependent cellular functions to disrupt.

5. **(a)** AAUAAA is a signal for 3′-end cleavage and polyadenylation. **(b)** AUUUA is an ARE motif that limits mRNA stability.

6. Sequences called intronic splicing silencers, or ISSs, are bound by proteins that suppress splicing at these sites.

7. (1) mRNAs could be deposited at one end of the oocyte, such as by *Drosophila* nurse cells. (2) Proteins could be deposited at one end of the oocyte. (3) mRNAs or proteins could be actively transported from one part of the oocyte to another. (4) A set of mRNAs or proteins could be subjected to differential stability by the introduction, at one end of the oocyte, of factors leading to mRNA or protein degradation.

8. Anterior cells of the embryo, where Mex-3 is concentrated and normally suppresses Pal-1 production, would take on fates similar to those of cells at the posterior end.

9. The *bcd* mRNA needed for development is contributed to the egg by the mother. The fertilized egg develops normally, even if its genotype is bcd^-/bcd^-, as long as the mother has one wild-type bcd allele (and thus contributed the mRNA to the oocyte) and the bcd^- allele is recessive. However, an adult bcd^-/bcd^- female will be sterile, because she cannot generate *bcd* mRNA for her oocytes.

10. The cap cells may create a niche for the germ-line stem cell, providing extrinsic signals that orient the cell division to ensure that one daughter retains the stem cell identity. In this case, the cap cells supply protein ligands that activate a Bmp-family signaling pathway in the stem cell that represses at least one gene required for differentiation.

11. An embryonic stem cell is a cell, derived from a mammalian embryo, that has the potential to differentiate into almost any kind of tissue. An induced pluripotent stem (iPS) cell is a somatic cell, such as a skin cell, that has been reprogrammed to have pluripotent potential to differentiate into different tissue types.

12. The ability to feed bacteria to the worms allows an RNAi approach. Given knowledge of the genome sequence and the types of genes that might be involved (e.g., Wnt-class signaling genes, homeotic genes), you could devise an RNAi screen. Short double-stranded RNAs complementary to worm genes are expressed in bacteria, using gene segments cloned between opposing promoters on a bacterial plasmid, and the bacteria are fed to worms. The heads of the fed worms are cut off, and the RNAi clones that affect regeneration are determined. Such an experiment has been done, with the finding that one homeobox gene of the *piwi* family plays a key role. (See P. W. Reddien et al., *Dev. Cell* 8:635–649, 2005; P. W. Reddien et al., *Science* 310:1327–1330, 2005.)

13. **(a)** As described in Chapter 7, the three-hybrid method is designed to screen for RNA sequences that bind to a particular protein. The method used here was modified to screen for unknown proteins that bind to a particular RNA sequence. **(b)** The second control from the top shows that a single nucleotide change in UCUUU abolishes binding. **(c)** FBF-1 may be binding to some other sequence in the engineered RNA, and the various control sequences help eliminate that possibility. **(d)** The FBF-1 not only localizes to the germ line but is predominantly present in the cytoplasm. **(e)** The animals should (and most do) produce only sperm.

Glossary

AAA+ proteins: A family of proteins with ATPase activity that share a common structural domain called the AAA domain, which includes Walker A and Walker B motifs. AAA stands for *ATPases associated with diverse cellular activities*.

abasic site: A position in an intact DNA backbone that is missing the base. Also called an AP (apurinic or apyrimidinic) site.

abortive initiation: Release of an 8 to 10 base pair RNA transcript from the bacterial RNA polymerase initiation complex before it clears the promoter and enters the elongation stage.

accommodation: A process of checking for appropriate codon-anticodon pairing prior to rotation of an incoming aminoacyl-tRNA into position for peptidyl transfer.

achiral: Describes a molecule that can be superimposed on its mirror image.

acid dissociation constant (K_a): The dissociation constant of an acid, HA, describing its dissociation into its conjugate base, A^-, and a proton. $K_a = [A^-][H_3O^+] / [HA]$.

acridine: A planar heterocyclic molecule isolated from coal tar that intercalates between successive $G \equiv C$ base pairs, deforming the DNA. Acridine inhibits transcription by preventing movement of the RNA polymerase along the DNA template. Acridine is also a mutagen, causing DNA polymerase to insert an extra base during replication.

actinomycin D: A peptide antibiotic that inhibits transcription elongation by RNA polymerase in bacteria, eukaryotes, and cell extracts. The molecule has a planar heterocyclic region that intercalates between successive $G \equiv C$ base pairs, deforming the DNA and preventing movement of the RNA polymerase along the template.

activation: The positive regulation of the expression of a gene or genes.

activation energy ($\Delta G^{\ddagger}$): The difference in free energy between the ground state of a reacting substance and the transition state.

activator: (1) A DNA-binding protein that positively regulates the expression of one or more genes; that is, transcription rates increase when an activator is bound to the DNA. (2) A positive modulator of an allosteric enzyme.

active site: The region of an enzyme surface that binds the substrate molecule and catalytically transforms it. Also called a catalytic site.

ADAR: *See* adenosine deaminase acting on RNA.

adenine (A): A purine base that is a component of DNA and RNA.

adenosine 3′,5′-cyclic monophosphate: *See* cyclic AMP.

adenosine deaminase acting on RNA (ADAR): An enzyme that catalyzes the conversion of adenosine to inosine by removal of the amino group at C-6 on the adenine ring.

adenylylation step: The first (activation) step in the attachment of an amino acid to a tRNA. The aminoacyl-tRNA synthetase reacts with the α-phosphoryl group of ATP, displacing pyrophosphate and forming a 5′-aminoacyl adenylate intermediate. *See also* tRNA-charging step.

A-DNA (A-form DNA): Conformation of double-stranded DNA observed under certain nonaqueous solvent conditions. The molecule assumes a right-handed helix with 11 base pairs per turn and a rise of 2.6 Å per base pair. *Compare* B-DNA *and* Z-DNA.

adult stem cells: The cells in adult mammals that retain the ability to divide and differentiate into other cell types. *Compare* embryonic stem cells.

affinity chromatography: A type of column chromatography in which molecules are separated based on their binding affinity for chemical groups present on the stationary phase.

A-form DNA: *See* A-DNA.

alkylation: The transfer of an alkyl group from one molecule to another.

allele: A variant form of a gene at a specific locus.

allopatric speciation: Geographic isolation of a group of individuals followed by evolution to form a distinct species that no longer can interbreed with the original one.

allosteric enzyme: A regulatory enzyme with catalytic activity modulated by the noncovalent binding of a specific metabolite at a site other than the active site.

allosteric modulator: A metabolite that, when bound to the allosteric site of an enzyme, alters its kinetic characteristics.

allosteric protein: A protein (generally with multiple subunits) with multiple ligand-binding sites, such that ligand binding at one site affects ligand binding at another.

α-amanitin: A cyclic polypeptide antibiotic that inhibits transcription in eukaryotic cells by binding Pol II and blocking its ability to translocate along the DNA template. At high concentrations it also binds and inhibits Pol III.

α/β barrel: Common protein domain architecture consisting of eight hydrogen-bonded β strands surrounded by eight α helices. The domain is formed by a series of β-α-β motifs.

α carbon (C_α): The first carbon atom attached to a functional group. In amino acids, the α carbon is the central carbon to which the amino, carboxyl, and R groups are bonded.

α helix: A helical conformation of a polypeptide chain, usually right-handed, with maximal intrachain hydrogen bonding; the most common secondary structure in proteins.

alternative splicing: The splicing of exons from a single gene in various combinations to produce different mRNAs and thus different polypeptides.

Ames test: A simple bacterial test for carcinogenicity, based on the assumption that carcinogens are mutagens.

amino acids: α-Amino–substituted carboxylic acids, the building blocks of proteins.

aminoacyl-tRNA: An aminoacyl ester of a tRNA; the tRNA is charged with an amino acid.

aminoacyl-tRNA synthetases: Enzymes that catalyze synthesis of an aminoacyl-tRNA at the expense of ATP energy.

amino terminus (N-terminus): The end of a polypeptide chain with a free α-amino group.

amphipathic helix: An α helix with both polar and nonpolar segments.

analyte: A molecule to be analyzed by mass spectrometry.

anaphase: The third stage of mitosis (M phase). Sister chromatid pairs held together at the centromere separate, and the two homologous chromosomes move toward opposite spindle poles.

annealing: Process in which single strands of nucleic acid in solution spontaneously rewind or renature with strands of complementary base sequence to form duplex structures.

anticodon: A specific sequence of three nucleotides in a tRNA, complementary to a codon for an amino acid in an mRNA.

antiparallel: Describes two linear polymers that are opposite in polarity or orientation.

antiparallel β sheet: *See* β sheet.

AP endonucleases: Enzymes that cleave the DNA backbone at an AP (apurinic or apyrimidinic; abasic) site as part of the base excision repair pathway.

apoenzyme: The protein portion of an enzyme, exclusive of any organic or inorganic cofactors or prosthetic groups that might be required for catalytic activity.

apoprotein: The protein portion of a protein, exclusive of any organic or inorganic cofactors or prosthetic groups that might be required for activity.

aqueous solution: Solution in which the solvent is water.

archaea: One of the three main groups of living organisms. Like bacteria, archaea are unicellular and contain no internal organelles or nucleus; however, archaea are more closely related to eukaryotes with respect to some genes and metabolic pathways. Archaea include many species that thrive in extreme environments of high ionic strength, high temperature, or low pH.

ARE: *See* AU-rich element.

A site: The site in a ribosome where the aminoacyl-tRNA binds.

association constant (K_a): An equilibrium constant for the association of a complex of two or more biomolecules from its components; for example, association of a substrate with an enzyme. K_a is the reciprocal of the dissociation constant, K_d. *Compare* dissociation constant (K_d).

atomic orbital: Mathematical function that describes the behavior of an electron in an atom.

ATP-coupling stoichiometry: A property of helicases and other motor proteins that describes the number of ATP molecules consumed per distance traveled or other defined work units.

AU-rich element (ARE): Sequences in mRNA with 5 to 13 residues of A and U, which target the mRNA for rapid degradation.

autoinhibition: The reduction or elimination of a molecule's activity by one of its own segments or domains.

autoradiograph: An image on an x-ray film or on certain photographic plates that is produced by decay emissions of a radioactive substance.

autosome: Any chromosome that is not a sex chromosome. *Compare* sex chromosome.

auxotrophic mutant (auxotroph): A mutant organism defective in the synthesis of a particular biomolecule, which must therefore be supplied for the organism's growth.

BAC: *See* bacterial artificial chromosome.

bacmid: A large circular DNA that includes the entire baculovirus genome and sequences that allow replication of the bacmid in *Escherichia coli*; a baculovirus shuttle vector.

bacteria: One of the three main groups of living organisms; bacteria have a plasma membrane but no internal organelles or nucleus.

bacterial artificial chromosome (BAC): A plasmid designed as a cloning vector for large segments of DNA. A BAC typically includes cloning sites, one or more selectable markers, and a stable origin of replication.

bacterial transduction: The transfer of genetic information from one bacterial cell to another by means of a viral vector.

Barr body: In the cells of female mammals, the inactivated X chromosome, which is compacted into a dense chromatin particle.

basal transcription factor: In eukaryotic cells, a protein required at every Pol II promoter. Also called a general transcription factor.

base excision repair (BER): A DNA repair pathway that involves excision of a damaged base by DNA glycosylase, followed by cleavage of the DNA backbone adjacent to the site by an AP endonuclease. Nick translation, DNA polymerization, and ligation complete the repair.

base pair: Two nucleotides in nucleic acid chains that are paired by hydrogen bonding of their bases; for example, A with T or U, and G with C.

base pairing: The weak binding of nucleotide bases with each other, via complementary hydrogen bonding, within a nucleic acid with at least two associated strands.

base stacking: A property of adjacent bases in a DNA strand or of base pairs in a DNA double helix that describes their orientation relative to each other. Parallel orientation of the hydrophobic planar rings minimizes their association with water and contributes to the stability of the B-form (Watson-Crick) double helix. Also called hydrophobic stacking.

basic helix-loop-helix motif: A protein secondary structural motif typical of transcription activators. It consists of two amphipathic α helices joined by a loop of variable length. Two such motifs dimerize through one pair of α helices. The other α helices have a series of basic amino acid residues along one side through which they bind DNA.

basic leucine zipper motif: A leucine zipper motif in which one side of the recognition helix has a series of basic residues, which facilitates DNA binding.

B-DNA (B-form DNA): Standard Watson-Crick conformation of double-stranded DNA. The molecule assumes a right-handed helix with 10.5 nucleotide residues per turn and a rise of 3.4 Å per base pair. *Compare* A-DNA *and* Z-DNA.

BER: *See* base excision repair.

β-α-β motif: A protein secondary structural motif in which two parallel β strands are connected by an α helix.

β barrel: A protein structural domain in which a β sheet of eight or more strands with one hydrophobic surface forms a cylinder in which the first β strand hydrogen-bonds with the last β strand.

β conformation: An extended conformation of a polypeptide chain, usually stabilized by interchain hydrogen bonding with adjacent polypeptide segments in the same conformation to form a sheet-like structure; the second most common secondary structure in proteins.

β hairpin: A protein structural motif in which two antiparallel β strands are connected, usually by an α, β, or γ turn.

β sheet: A common protein secondary structure in which a polypeptide chain assumes an extended, zigzag arrangement with extensive hydrogen bonding between adjacent segments or strands. In parallel β sheets, the strands are aligned with the same polarity. In antiparallel β sheets, adjacent strands have opposite polarity.

β sliding clamp: A component of the *E. coli* DNA polymerase III holoenzyme. The ring-shaped homodimer encircles and slides along the duplex DNA ahead of the Pol III core to which it is attached, greatly enhancing the processivity of DNA synthesis.

β turn: A type of protein secondary structure consisting of four amino acid residues arranged in a tight turn so that the polypeptide turns back on itself.

B-form DNA: *See* B-DNA.

binding energy (ΔG_B): The energy derived from noncovalent interactions between enzyme and substrate or receptor and ligand.

binding site: The crevice or pocket on a protein in which a ligand binds.

biochemical standard free-energy change ($\Delta G'^\circ$): The free-energy change for a reaction occurring under a set of standard conditions: temperature, 298 K; partial pressure of each gas, 1 atm or 101.3 kPa; all solutes at 1 M concentration, pH 7.0, in 55.5 M water.

biological information: Information required for cellular growth and metabolism, inherited from one generation of an organism to the next. Primarily imprinted within the sequences of nucleic acids, information may also be embedded within nucleic acid modifications and in the patterns of modification in certain proteins bound to nucleic acids; *see* epigenetic inheritance.

blunt ends: The product of restriction endonuclease action on double-stranded DNA that leaves no unpaired bases at the cleavage site.

bond angle: The angle between two adjacent bonds to the same atom.

branch migration: Movement of the branch point in a branched DNA formed from two DNA molecules with identical sequences. *See also* Holliday intermediate.

branch point: An internal A residue just upstream of the 3′ splice site of an intron that attacks the phosphate at the 5′ splice site, forming the loop of the intron lariat.

BRCA1: A vertebrate protein involved in DNA damage sensing and repair (primarily double-strand break repair). Mutations in the gene encoding BRCA1 confer a predisposition to breast and ovarian cancer.

BRCA2: A vertebrate recombination mediator protein involved in the repair of double-strand breaks. Mutations in the gene encoding BRCA2 confer a predisposition to breast and ovarian cancer.

bromodomain: A protein structural domain that recognizes and binds to certain acetylated Lys residues in proteins.

buffering capacity: Quantitative measure of the ability of a buffer solution to resist changes in pH.

buffer solution: A system capable of resisting changes in pH, consisting of a conjugate acid-base pair in which the ratio of proton acceptor to proton donor is near unity.

C_α: *See* α carbon.

cAMP: *See* cyclic AMP.

cAMP receptor protein (CRP): In bacteria, a specific regulatory protein that controls initiation of transcription of the genes that produce the enzymes required for the cell to use some other nutrient when glucose is lacking. Also called catabolite gene activator protein (CAP).

CAP: *See* cAMP receptor protein.

cap-binding complex (CBC): A protein complex that recruits capped mRNAs to the ribosome to initiate translation.

carboxyl terminus (C-terminus): The end of a polypeptide chain with a free α-carboxyl group.

carcinogen: A substance directly involved in causing cancer.

catabolite repression: The inhibition of the expression of genes required for the metabolism of other sugars in the presence of glucose.

catalysis: An increase in the rate of a chemical reaction caused by a substance that is not consumed by the reaction.

catalyst: A substance that increases the rate of a chemical reaction without being consumed by the reaction.

catalytic RNA: *See* ribozyme.

catenane: Two or more circular polymeric molecules interlinked by one or more noncovalent topological links, resembling the links of a chain.

CBC: *See* cap-binding complex.

cDNA: *See* complementary DNA.

cDNA library: DNA library consisting entirely of cloned cDNAs from a particular organism or cell type.

cell: Membrane-bounded structure that is the smallest unit of life.

cell cycle: The process by which cells replicate and divide. The bacterial cell cycle involves binary fission; the eukaryotic cell cycle has four phases, including mitosis.

cell theory: The theory proposed by Theodor Schwann in 1839 that cells are the basic units of all living things.

cellular function (of a gene product): The metabolic processes in which a gene product participates and the interactions of that gene product with other proteins or RNAs in the cell. *Compare* molecular function *and* phenotypic function.

central dogma: The organizing principle of molecular biology: genetic information flows from DNA to RNA to protein. The pathways of information flow have been expanded to include RNA to DNA, and RNA to RNA, and are now established. They are no longer constituted dogma.

centromere: A specialized site in a chromosome, serving as the attachment point for the mitotic or meiotic spindle.

centrosome: An organelle that serves as the microtubule organizing center. Two centrosomes are responsible for the creation of the spindle apparatus that moves chromosomes to opposite poles of the cell during mitosis and meiosis.

cGMP: *See* cyclic GMP.

chain topology diagram: A method of illustrating in two dimensions the topology of the polypeptide chain in supersecondary structures.

change in free energy (ΔG): *See* free-energy change (ΔG).

chaperone: Any protein that interacts with partially folded or improperly folded polypeptides, facilitating the correct folding pathway or providing a microenvironment where proper folding can occur.

chaperonins: A class of chaperones that form a large, barrel-like structure, inside which certain cellular proteins fold.

Chargaff's rules: A set of quantitative observations about the nucleotide content of DNA from many organisms and species that helped to lay the groundwork for the discovery of the structure of DNA.

chemical bond: An attractive force that holds atoms to each other in a molecule or crystal.

chemical reaction: A process that changes the structure or energy content of atoms in a molecule, but not their nuclei.

chemical shift: Variations of nuclear magnetic resonance frequencies, relative to a standard of the same kind of nucleus, caused by variations in the electron distribution within a molecule.

chi: The sequence 5′-GCTGGTGG-3′, which alters the endonuclease activity of bound RecC in the RecBCD complex so that it preferentially degrades the 5′ end of the molecule.

chiasma (*pl.* chiasmata): A cross-shaped junction that represents physical recombination between chromosomes.

ChIP-Chip: Chromatin immunoprecipitation followed by hybridization of the precipitated DNA to a genomic microarray (chip).

ChIP-Seq: Chromatin immunoprecipitation followed by DNA sequencing.

chiral: Describes a compound that contains an asymmetric center (chiral atom or chiral center) and thus can occur in two nonsuperimposable, mirror-image forms (enantiomers).

chiral center: An atom with substituents arranged so that the molecule is not superimposable on its mirror image.

chloramphenicol: An antibiotic that inhibits protein synthesis by bacterial, mitochondrial, and chloroplast ribosomes by blocking peptidyl transfer.

chromatin: A filamentous complex of DNA, histones, and other proteins, constituting the eukaryotic chromosome.

chromatin remodeling complex: A protein complex with ATPase activity that translocates nucleosomes along the DNA, making certain regions of DNA more or less accessible to transcription factors.

chromatography: A process in which complex mixtures of molecules are separated by many repeated partitionings between a flowing (mobile) phase and a stationary phase. The stationary phase may be packed into a tube (column chromatography) or planar (thin-layer chromatography).

chromodomain: A protein structural motif that recognizes and binds certain methylated Lys residues in proteins.

chromosomal scaffold: Proteinaceous residue after extraction of histones from chromosomes, consisting mainly of SMC proteins.

chromosome: A single large DNA molecule and its associated proteins, containing many genes; stores and transmits genetic information.

chromosome theory of inheritance: The hypothesis proposed by Walter Sutton in 1903 that genes are located on chromosomes.

clamp loader: The portion of the *E. coli* DNA polymerase III holoenzyme that assembles the β sliding clamps onto the DNA.

clone: An identical copy.

cloning: The production of large numbers of identical DNA molecules, cells, or organisms from a single ancestral DNA molecule, cell, or organism.

cloning vector: A DNA molecule known to replicate autonomously in a host cell, to which a segment of DNA may be spliced to allow its replication; for example, a plasmid or an artificial chromosome.

closed-circular DNA: A continuous double-stranded DNA molecule with no free 3′ or 5′ ends.

closed complex: A complex of the RNA polymerase bound to a promoter, in which the DNA is intact and double-stranded. *Compare* open complex.

closed form: The conformation assumed by *E. coli* DNA polymerase I when a primed template and the correct dNTP are both bound to the active site.

CMG complex: A complex of the proteins Cdc24, MCM helicase, and GINS proposed to function in the eukaryotic replisome.

coactivator: A protein that stimulates transcription by binding both the RNA polymerase and an activator or activators, without binding the DNA directly. *Compare* corepressor *and* DNA-binding transcription activator.

coalescent theory: Retrospective analysis of population genetics data (mutation rates, selection, genetic drift, and other factors) to trace a polymorphism back to the original ancestor in which it appeared.

coding strand: The strand of a double-stranded DNA that has the same sequence as the RNA transcript (with T in place of U) and is complementary to the template strand. Also called the nontemplate strand.

codominance: Non-Mendelian behavior in which two alleles of a gene produce distinct functional products, neither of which is dominant to the other. *Compare* incomplete dominance.

codon: In a nucleic acid, a sequence of three adjacent nucleotides that codes for a specific amino acid.

codon bias: The use of certain degenerate codons more than others to code for a given amino acid.

codon family: A set of multiple codons that specify the same amino acid.

coenzyme: An organic cofactor required for the action of certain enzymes; usually derived from a vitamin.

cofactor: An inorganic ion or a coenzyme required for enzyme activity.

cohesins: SMC proteins that link sister chromatids immediately after chromosomal replication and keep them together as the chromosomes condense to metaphase.

coiled-coil motif: Protein motif in which two α helices twist around each other in a left-handed supercoil, interacting through hydrophobic contacts.

cointegrate: An intermediate in the migration of certain DNA transposons in which the donor DNA and target DNA are covalently attached.

column chromatography: A process in which complex mixtures of molecules are separated by many repeated partitionings between a flowing (mobile) phase and a stationary phase packed into a column.

combinatorial control: The use of specific combinations of a limited number of regulatory proteins to exert fine control over gene expression.

comparative genomics: The study of genome structure, function, and evolution by comparison across different species.

competitive inhibitor: A molecule that competes with the normal substrate or ligand for a protein's binding site.

complementary: Having a molecular surface with chemical groups arranged to interact specifically with chemical groups on another molecule. Because of complementarity, if the nucleotide sequence of one strand of a double-stranded nucleic acid is known, the sequence of the opposite strand can be deduced.

complementary DNA (cDNA): A duplex DNA with one strand identical to a specific mRNA (with T residues generally substituting for the U residues in the mRNA), used in DNA cloning; usually made by reverse transcriptase.

complex transposon: A viruslike transposon with a large genome including genes not required for transposition.

composite transposon: A transposon that consists of two insertion elements flanking one or more genes not required for transposition, such as antibiotic-resistance genes.

condensins: SMC proteins that facilitate chromosomal condensation.

configuration: An arrangement of bonded atoms that can be changed to a different configuration only by breaking and re-forming one or more covalent bonds.

conformation: An arrangement of bonded atoms that can be changed to a different conformation without breaking and re-forming a covalent bond, such as by rotation about one or more single bonds.

consensus sequence: A DNA or amino acid sequence consisting of the residues that most commonly occur at each position in a set of similar sequences.

constitutive gene expression: The continual expression of a gene. *Compare* regulated gene expression.

constructive interference: Phenomenon in which waves in the same phase add to create waves of larger amplitude.

contig: A series of overlapping clones or a continuous sequence defining an uninterrupted section of a chromosome.

cooperativity: The characteristic of an enzyme or other protein in which binding of the first molecule of a ligand changes the affinity for the second molecule. In positive cooperativity, the affinity for the second ligand molecule increases; in negative cooperativity, it decreases.

core histones: The four histone proteins (H2A, H2B, H3, and H4) that form the octameric core of the most common type of nucleosome.

corepressor: A protein that inhibits transcription by binding both the RNA polymerase and a repressor or repressors, without binding the DNA directly. *Compare* coactivator.

core promoter: In eukaryotic cells, the DNA sequence elements common to promoters used by Pol II. The TATA box and initiator sequence (Inr) are required elements of a core promoter; a TFIIB recognition element (BRE) and downstream promoter element (DPE) may also be involved in transcription initiation from some core promoters.

correlation spectroscopy (COSY): A type of two-dimensional nuclear magnetic resonance spectroscopy in which atoms that are near to one another and connected through covalent bonds can be identified.

COSY: *See* correlation spectroscopy.

covalent bond: A chemical bond that involves sharing of electron pairs.

covalent modification: The addition, dissociation, or rearrangement of an atom or functional group covalently bonded in a molecule. In biological systems, common modifying groups include acetyl, adenylyl, amide, carboxyl, hydroxyl, methyl, myristoyl, palmitoyl, phosphoryl, prenyl, sulfate, and uridylyl groups.

CpG sequence: A DNA sequence (cytosine, guanine) that is a frequent substrate for cytosine methylation.

Cre-lox: A bacteriophage-encoded site-specific recombination system that promotes circularization of the phage P1 genome and aids in proper segregation at cell division of phage plasmids in the lysogenic state.

crossing over: The reciprocal exchange of DNA between paired homologous chromosomes during meiosis. Also called recombination.

cross-linking: The use of a small chemical agent with two reactive groups to covalently link molecules that are in close proximity.

crossover: *See* genetic crossover.

CRP: *See* cAMP receptor protein.

cruciform: A secondary structure in double-stranded RNA or DNA in which the double helix is denatured at palindromic repeat sequences in each strand, and each separated strand is paired internally to form opposing hairpin structures. *See also* hairpin.

C-terminus (carboxyl terminus): *See* carboxyl terminus.

cyclic AMP (cAMP): A second messenger, adenosine $3',5'$-cyclic monophosphate; its formation in a cell by adenylyl cyclase is stimulated by certain hormones or other molecular signals.

cyclic GMP (cGMP): A second messenger, guanosine $3',5'$-cyclic monophosphate; its formation in a cell by guanylyl cyclase is stimulated by certain hormones or other molecular signals.

cyclobutane ring: A structure formed by the condensation of two double-bonded $C_5=C_6$ atoms on adjacent pyrimidine bases in DNA.

cycloheximide: An antibiotic that inhibits protein synthesis by eukaryotic ribosomes by blocking peptidyl transfer.

cytogenetics: The study of chromosomes and their role in heredity.

cytokinesis: The final separation of daughter cells following mitosis.

cytology: The study of cells and cellular structures.

cytoplasmic membrane: The exterior membrane surrounding the cytoplasm of a cell. Also called the plasma membrane.

cytosine (C): A pyrimidine base that is a component of DNA and RNA.

cytotoxic: Deadly to cells.

Dam methylase (DNA adenine methyltransferase): An enzyme of *E. coli* that methylates adenine residues in the palindromic sequence GATC on both strands of the DNA. Transient hemimethylation of a DNA duplex following replication distinguishes the parental strand from the daughter strand.

DCC: *See* dosage compensation complex.

DDE motif: A protein secondary structure in which the amino acid residues D, D, and E (two aspartate residues and a glutamine residue) form the catalytic core in the active site of phosphoryltransferase enzymes such as integrases and transposases.

deamination: The enzymatic removal of amino groups from biomolecules such as amino acids or nucleotides.

deep sequencing: Extensive genomic sequencing designed to produce multiple (sometimes hundredfold or greater) coverage of all targeted sequences; used to detect sequence variants within a population.

degenerate code: A code in which a single element in one language is specified by more than one element in a second language. The genetic code is degenerate because some amino acids are specified by more than one codon.

deletion analysis: A method for assessing the functional importance of various regions of a protein by engineering a series of constructs with different parts of the gene deleted. The proteins expressed by these constructs can then be assayed for functionality.

deletion mutation: A mutation resulting from the deletion of one or more nucleotides from a gene or chromosome. *Compare* insertion mutation.

denaturation: The partial or complete unfolding of the specific native conformation of a polypeptide chain, protein, or nucleic acid such that the function of the molecule is lost. In the case of nucleic acids, also called melting.

deoxyribonucleic acid: *See* DNA.

deoxyribonucleotide: A nucleotide containing 2-deoxy-D-ribose as the pentose component. Also called a deoxynucleotide.

depurination: The enzymatic removal of a purine base from a nucleotide.

Dicer: An endonuclease in eukaryotic cells that catalyzes the hydrolysis of double-stranded RNAs, producing siRNAs or processing pre-miRNAs to mature miRNAs. Dicer also plays a role in the creation of RNA-induced silencing complexes.

diffraction pattern: The interference pattern that results when a wave or series of waves is diffracted by an object with a regular structure, such as a crystal.

dimer: A molecule with two subunits.

diphtheria toxin: A bacterial toxin that catalyzes the ADP-ribosylation of a diphthamide (a modified histidine) residue of elongation factor eEF2, thereby inactivating it and inhibiting protein synthesis by the eukaryotic ribosome.

diploid: Having two sets of genetic information; describes a cell with two chromosomes of each type. *Compare* haploid.

directionality: The direction in which a process or enzyme proceeds along an asymmetric molecule. For example, certain endonucleases act on DNA only in a 5′ to 3′ direction.

dissociation constant (K_d): An equilibrium constant for the dissociation of a complex of two biomolecules into its components; for example, dissociation of a substrate from an enzyme. K_d is the reciprocal of the association constant, K_a. *Compare* association constant (K_a)

distributive synthesis: The enzymatic synthesis of a biological polymer in which the enzyme dissociates from the substrate after the addition of each monomeric unit. *Compare* processive synthesis.

disulfide bond: A covalent bond involving the oxidative linkage of the sulfhydryl groups of two Cys residues, in the same or different polypeptide chains.

Dmc1: A eukaryotic recombinase structurally and functionally homologous to the RecA protein of *E. coli*. *See also* Rad51.

DNA (deoxyribonucleic acid): A polynucleotide with a specific sequence of deoxyribonucleotide units covalently joined through 3′,5′-phosphodiester bonds; serves as the carrier of genetic information.

DNA adenine methyltransferase: *See* Dam methylase.

DNA-binding transactivator: *See* DNA-binding transcription activator.

DNA-binding transcription activator: In eukaryotic cells, a protein that binds to enhancers or UASs to facilitate transcription. Also called a DNA-binding transactivator. *Compare* coactivator.

DNA cloning: *See* cloning.

DNA genotyping: The process of defining particular genomic sequences associated with an individual. Also called DNA fingerprinting or DNA profiling.

DNA glycosylase: An enzyme that hydrolyzes the *N*-β-glycosyl bond between a nucleotide base and pentose, creating an abasic site in the DNA.

DNA helicase: *See* helicase.

DNA library: A collection of cloned DNA fragments.

DNA ligase: An enzyme that creates a phosphodiester bond between the 3′ end of one DNA segment and the 5′ end of another.

DNA looping: The interaction of proteins bound at distant sites on a DNA molecule so that the intervening DNA forms a loop.

DNA microarray: A collection of DNA sequences immobilized on a solid surface, with individual sequences laid out in patterned arrays that can be probed by hybridization. Also called a DNA chip.

DNA nuclease: *See* nucleases.

DNA overwinding: The condition in which a closed-circular DNA has more helical turns than would be expected of B-form DNA. Its linking number, *Lk*, is increased relative to that of B-form DNA, and the molecule is positively supercoiled.

DNA photolyase: A flavoprotein enzyme that becomes an electron donor when activated by visible light. DNA photolyases can repair pyrimidine dimers and other lesions caused by ultraviolet light.

DNA polymerase: An enzyme that catalyzes template-dependent synthesis of DNA from its deoxyribonucleoside 5′-triphosphate precursors.

DNA polymerase I: A bacterial DNA polymerase engaged in DNA replication associated with DNA repair and processing of Okazaki fragments.

DNA polymerase II: A bacterial DNA polymerase engaged in translesion DNA synthesis.

DNA polymerase III: The primary replicative DNA polymerase in bacteria.

DNA polymerase IV: A bacterial DNA polymerase engaged in translesion DNA synthesis.

DNA polymerase V: A bacterial DNA polymerase engaged in translesion DNA synthesis.

DNA polymerase α (Pol α): A eukaryotic DNA polymerase with both primase and error-prone DNA polymerase activities. The enzyme synthesizes an RNA primer on a DNA template and then extends it with DNA.

DNA polymerase δ (Pol δ): A eukaryotic chromosomal replicase with both DNA polymerase and 3′→5′ exonuclease activities. It acts on the lagging strand of the replication fork.

DNA polymerase ε (Pol ε): A eukaryotic chromosomal replicase with both DNA polymerase and 3′→5′ exonuclease activities. It acts on the leading strand of the replication fork.

DNA replication: The synthesis of daughter DNA molecules identical to the parental DNA.

DNA strand invasion: The pairing of a single-stranded extension of a DNA molecule with a homologous region of another DNA molecule, with displacement of one strand of the recipient molecule by the invading strand.

DNA supercoiling: The coiling of DNA upon itself, generally as a result of bending, underwinding, or overwinding of the DNA helix.

DNA topology: The properties of DNA that do not change under continuous deformations such as twisting, bending, stretching, or binding other molecules.

DNA underwinding: The condition in which a closed-circular DNA has fewer helical turns than would be expected of B-form DNA. Its linking number, *Lk*, is reduced relative to that of B-form DNA, and the molecule is negatively supercoiled.

domain: A distinct structural unit of a polypeptide; domains may have separate functions and may fold as independent, compact units.

dominant: Describes the allele that determines the phenotype in a heterozygous individual. *Compare* recessive.

donor site: The location on a chromosome of a transposon before it moves to a target site. *Compare* target site.

dosage compensation: The control of gene expression from sex chromosomes to ensure that male and female cells express similar levels of each gene product.

dosage compensation complex (DCC): A ribonucleoprotein complex encoded by the X chromosome in *Drosophila*. The complex coats the single X chromosome in male cells, hyperstimulating transcription from its genes to compensate for the lack of a second X chromosome.

double bond: A bond between two elements that involves four electrons instead of two.

double-strand break (DSB): A break in the phosphodiester backbone of both strands of a double-stranded nucleic acid.

double-strand break repair (DSBR): A method for repairing double-strand breaks that creates two Holliday intermediates, which must be cleaved by resolvases. The genes flanking the repair site may be unchanged or may undergo a reciprocal exchange, depending on how the crossovers are resolved.

Drosha: An endonuclease in eukaryotic cells that cleaves the hairpin of primary miRNA transcripts to produce pre-miRNAs.

DSB: *See* double-strand break.

DSBR: *See* double-strand break repair.

duplication mutation: The duplication of a large tract of DNA, leading to an increased dosage of genes in the affected area.

editosome: A protein complex that catalyzes the insertion or deletion of nucleotide residues during the process of RNA editing.

eEF1α: In eukaryotic protein synthesis, an elongation factor that delivers aminoacyl-tRNAs to the A site of the elongation complex with the concomitant hydrolysis of bound GTP.

eEF1βγ: In eukaryotic protein synthesis, an elongation factor that uses bound GTP to regenerate eEF1α-GTP from eEF1α-GDP.

eEF2: In eukaryotic protein synthesis, an elongation factor with GTPase activity. GTP hydrolysis provides the energy for the ribosome to translocate along the mRNA to the next codon. Also called a translocase.

effector: A small molecule that binds a transcription activator or repressor, causing a conformational change in the regulatory protein that results in an increase or decrease in transcription from the gene.

EF-G: In bacterial protein synthesis, an elongation factor with GTPase activity. GTP hydrolysis provides the energy for the ribosome to translocate along the mRNA to the next codon. Also called a translocase.

EF-Ts: In bacterial protein synthesis, an elongation factor that uses bound GTP to regenerate EF-Tu-GTP from EF-Tu-GDP.

EF-Tu: In bacterial protein synthesis, an elongation factor that delivers aminoacyl-tRNAs to the A site of the elongation complex with the concomitant hydrolysis of bound GTP.

EJC: *See* exon junction complex.

electric dipole moment: A measure of the electrical polarity of a bond or molecule. It is equal to the magnitude of the charge times the distance separating the charges.

electron density map: A three-dimensional description of the electron density in a crystal, derived from x-ray diffraction data.

electronegative atoms: Atoms with a tendency to gain electrons.

electronegativity: The propensity of an atom to attract electrons to itself.

electrophoresis: *See* gel electrophoresis.

electroporation: Introduction of macromolecules into cells after rendering the cells transiently permeable by the application of a high-voltage pulse.

electropositive atoms: Atoms with a tendency to lose electrons.

elongation: (1) The second of three stages of RNA synthesis, in which ribonucleotides are added to the 3′ end of the growing RNA molecule. (2) The second of three stages of protein synthesis, in which amino acids are added to the C-terminal end of the growing peptide chain.

elongation complex: The complex of proteins required for efficient synthesis of the RNA transcript after the RNA polymerase has moved beyond the promoter.

elongation factors: (1) Proteins required in the elongation phase of eukaryotic transcription. (2) Proteins required in the elongation phase of protein synthesis. *See also* eEF1α, eEF1βγ, eEF2, EF-G, EF-Ts, *and* EF-Tu.

embryonic stem cells: The cells in a mammalian embryo that retain the ability to divide and differentiate into other cell types. *Compare* adult stem cells.

enantiomers: Stereoisomers that are nonsuperimposable mirror images of each other.

endonuclease: An enzyme that hydrolyzes the interior phosphodiester bonds of a nucleic acid; that is, it acts at bonds other than the terminal bonds.

end replication problem: The inability of DNA polymerases to replicate the final segment of DNA at the 3′ end of the lagging strand where there is no primer to provide a free 3′-OH group.

enhanceosome: A nucleoprotein complex of cooperating activators, which integrates regulatory information from multiple signals and generates a single transcriptional outcome at the target promoter.

enhancer: A DNA sequence that facilitates the expression of a given gene; it may be located a few hundred, or even thousand, base pairs away from the gene. In yeast, enhancers are called upstream activator sequences (UASs).

entropy (S): The extent of randomness or disorder in a system.

enzyme: A biomolecule, either protein or RNA, that catalyzes a specific chemical reaction. It does not affect the equilibrium of the catalyzed reaction; it enhances the rate of the reaction by providing a reaction path with lower activation energy.

enzyme kinetics: The study of the rates of reactions catalyzed by enzymes.

epigenetic inheritance: The inheritance of characteristics acquired by means that do not involve the nucleotide sequence of the parental chromosomes; for example, covalent modifications of histones.

epitope tag: A protein sequence or domain bound by some well-characterized antibody.

equilibrium expression: A mathematical expression for the equilibrium constant of a chemical reaction, expressed as the product of the molar concentrations of each reaction product, raised to its coefficient in the balanced reaction, over the product of the molar concentrations of each reactant, raised to its coefficient in the balanced reaction.

E site: The site in a ribosome occupied by the tRNA molecule released after the growing polypeptide chain is transferred to the aminoacyl-tRNA. Also called the exit site.

EST: *See* expressed sequence tag.

euchromatin: The regions of interphase chromosomes that stain diffusely, as opposed to the more condensed, heavily staining, heterochromatin. These are often regions in which genes are being actively expressed.

eukaryotes: One of the three main groups of living organisms; eukaryotes are unicellular or multicellular organisms with cells having a membrane-bounded nucleus, multiple chromosomes, and internal organelles.

evo-devo: The field of evolutionary development, which demonstrates that dramatic phenotypic differences between species can be accounted for by changes in the temporal expression of shared or homologous genes and regulatory networks.

evolution: A process in which the population of a species changes over time. Genetic variation occurs in the populations due to mutation; competitive pressures in the environment lead to the natural selection of individuals whose genetic makeup gives them a reproductive advantage. Over time, the genetic makeup of the surviving population shifts, sometimes creating new species.

excinuclease: An enzyme that cleaves a phosphodiester bond in the DNA on either side of a bulky lesion in DNA. Also called an excision endonuclease.

exon: The segment of a eukaryotic gene that encodes a portion of the final product of the gene; a segment of RNA that remains after posttranscriptional processing and is transcribed into a protein or incorporated into the structure of an RNA. *Compare* intron.

exon junction complex (EJC): A complex of proteins deposited on an mRNA by the spliceosome 20 to 24 nucleotides upstream of exon-exon junctions.

exonuclease: An enzyme that hydrolyzes only those phosphodiester bonds that are in the terminal positions of a nucleic acid.

exosome: A complex of $3' \rightarrow 5'$ exonucleases in eukaryotic cells that processes the $3'$ ends of rRNAs and tRNAs and is responsible for RNA degradation in higher eukaryotes.

exothermic reaction: A chemical reaction that releases heat (that is, for which ΔH is negative).

exportin: A protein receptor responsible for transporting RNAs from the nucleus, through a nuclear pore, into the cytoplasm.

expressed sequence tag (EST): A specific type of sequence-tagged site in DNA representing a gene that is expressed.

expression vector: A cloning vector with the transcription and translation signals needed for the regulated expression of a cloned gene. *See also* cloning vector.

extrachromosomally primed (EP) retrotransposon: A retrotransposon that moves via a double-stranded cDNA copy of its mRNA transcript. The cDNA inserts itself into the target site in a reaction catalyzed by a recombinase or integrase. *See also* target-primed (TP) retrotransposon.

first law of thermodynamics: The law stating that, in all processes, the total energy of the universe remains constant.

$5'$ cap: A residue of 7-methylguanosine (7-meG) linked to the $5'$-terminal residue of an mRNA through a $5',5'$-triphosphate linkage, which protects the mRNA from exoribonucleases.

F_1 generation: The first filial generation, the hybrid offspring in a genetic cross.

fork regression: Backward movement of the replication fork, which can occur when a replication fork encounters a lesion and stalls. Fork regression allows the parental strands to reanneal until the lesion is repaired.

four-helix bundle: A supersecondary protein structure in which four α helices associate through hydrophobic interactions.

frameshift mutation: A mutation caused by insertion or deletion of one or more paired nucleotides, changing the reading frame of codons during protein synthesis; the polypeptide product has an altered amino acid sequence beginning at the mutated codon.

free energy (G): The component of the total energy of a system that can do work at constant temperature and pressure.

free-energy change (ΔG): The amount of free energy released (negative ΔG) or absorbed (positive ΔG) in a reaction at constant temperature and pressure.

F_2 generation: The second filial generation, the offspring of crossing the F_1 generation.

functional RNA: An RNA molecule that is a functional end product, as distinct from messenger RNA (mRNA), which serves as a transient intermediary between DNA and a protein product it encodes.

fusion gene: A hybrid gene formed when chromosomal DNA is rearranged by deletion, duplication, insertion, or transposition.

fusion protein: The protein product of a gene created by the fusion of two distinct genes or portions of genes.

gamete cell: A reproductive cell with a haploid gene content; a sperm or egg cell.

gap genes: A subclass of the segmentation genes involved in dividing the developing *Drosophila* embryo into broad regions. Gap genes are expressed before the pair-rule genes.

gap repair: A process for repairing gaps left when the replication fork bypasses a lesion.

gel electrophoresis: A technique for separating mixtures of large charged molecules such as proteins or nucleic acids by causing them to move through a gel matrix in an applied electric field.

gel-exclusion chromatography: A type of column chromatography in which molecules are separated by size, based on the capacity of porous polymers to exclude solutes above a certain size.

gene: A chromosomal segment that codes for a single functional polypeptide chain or RNA molecule.

gene conversion: A nonreciprocal transfer of genetic information as an outcome of DNA repair, especially during meiosis.

general rate constant (k_{cat}): The constant defined by the limiting rate of an enzyme-catalyzed reaction. It describes the number of substrate molecules converted to product by a single molecule of enzyme at saturating levels of substrate. The constant has units of reciprocal time. *See also* turnover number.

general transcription factor: In eukaryotic cells, a protein required at every Pol II promoter. Also called a basal transcription factor.

gene silencing: (1) The suppression of gene expression by incorporation of the gene into transcriptionally inactive heterochromatin. (2) The suppression of gene expression by short interfering RNAs, which bind mRNAs and target them for degradation.

genetic code: The set of triplet code words in DNA (or mRNA) coding for the amino acids of proteins.

genetic crossover: Any redistribution of genes between two homologous chromosomes that results from a chromosomal crossover.

genetic drift: The change in frequency of an allele in a population due to random sampling, rather than selective pressure. Genetic drift is affected by such variables as the number of reproducing individuals in a population and the number of offspring generated.

genetic engineering: Manipulation of an organism's genome in the laboratory.

genetics: The science of heredity and the variation of inherited characteristics.

genome: One copy of all the genetic information encoded in a cell or virus. In a eukaryote, this generally constitutes one copy of all the genetic information in the nucleus. Separate genomes are found in certain organelles, particularly mitochondria and chloroplasts.

genome annotation: Information about the location and function of genes and other regulatory and functional sequences in a genome.

genome editing: The precise introduction, alteration, or removal of sequences in a genome using a variety of sequence-targeted methods.

genomic library: A DNA library containing DNA segments that represent all (or most) of the sequences in an organism's genome.

genomics: Broadly, the study of genomes. Genomics embraces sequencing, mapping, and annotating genomes; organizing databases to archive genomic data; developing computational tools to analyze the data; and application of genomic data to other fields, such as medicine.

genotoxic: Causing damage to the genomic DNA.

genotype: The genetic constitution of an organism, as distinct from its physical characteristics, or phenotype.

genotyping: *See* DNA genotyping.

GFP: *See* green fluorescent protein.

glycosidic bonds: Bonds formed between a sugar and another molecule (typically an alcohol, purine, pyrimidine, or sugar) through an intervening oxygen.

G_1 phase: The first gap phase of the eukaryotic cell cycle, in which the cell is diploid. G_1, part of interphase, occurs before the S (synthesis) phase, in which the DNA is replicated.

GPCR: *See* G protein–coupled receptor.

G protein–coupled receptor (GPCR): Any of a large family of membrane receptor proteins with seven transmembrane helical segments, often associating with G proteins to transduce an extracellular signal into a change in cellular metabolism.

Greek key motif: Supersecondary protein motif in which four antiparallel β strands combine in a pattern seen on ancient Greek pottery.

green fluorescent protein (GFP): A small protein that produces a bright fluorescence in the green region of the visible spectrum. Fusion proteins with GFP are commonly used to determine the subcellular location of the fused protein by fluorescence microscopy. Variants that produce other colors, such as red fluorescent protein (RFP) and cyan fluorescent protein (CFP), have also been produced.

group I intron: A large, self-splicing ribozyme that catalyzes its own excision from an mRNA, tRNA, or rRNA transcript in a reaction that requires a guanosine nucleotide or nucleoside to initiate the reaction.

group II intron: A large, self-splicing ribozyme that catalyzes its own excision from an mRNA transcript as a lariat structure.

G tetraplex: A four-stranded DNA structure that can form from G-rich segments of DNA.

G_2 phase: The second gap phase of the eukaryotic cell cycle, in which the cell is tetraploid. G_2, part of interphase, occurs between the S (synthesis) phase and the M (mitosis) phase.

guanine (G): A pyrimidine base that is a component of DNA and RNA.

guanosine 3′,5′-cyclic monophosphate: *See* cyclic GMP.

hairpin: A secondary structure in single-stranded RNA or DNA, in which complementary parts of a palindromic repeat fold back and are paired to form an antiparallel duplex helix that is closed at one end.

haploid: Having a single set of genetic information; describes a cell with one chromosome of each type. *Compare* diploid.

haplotype: (1) In genetics, a group of alleles on a chromosome that are nearly always inherited together. (2) In genomics, a set of single-nucleotide polymorphisms that are nearly always inherited together.

HAT: *See* histone acetyltransferase.

helicase: An enzyme that catalyzes the separation of strands in a nucleic acid molecule in a reaction coupled to the hydrolysis of ATP.

helix-turn-helix motif: A supersecondary protein motif consisting of two α helices separated by a β turn. This motif is crucial to the interaction of many bacterial regulatory proteins with DNA.

heterochromatin: The condensed, heavily staining portions of chromosomes that are not transcriptionally active, including centromeres, telomeres, some repetitive DNA sequences, and mitotic chromosomes.

heterocyclic compound: A molecule incorporating one or more rings that incorporate atoms of different elements.

heterooliogmer: A multisubunit molecule (oligomer) with nonidentical subunits.

heterotropic: Describes an allosteric modulator that is distinct from the normal ligand or an allosteric enzyme requiring a modulator other than its substrate.

heterozygous: Having different alleles at a specific genetic locus.

hierarchical model: A model for protein folding that proposes that local regions of secondary structure form first, followed by longer-range interactions, continuing until complete domains form and the entire polypeptide is folded. *Compare* molten globule model.

high-mobility group (HMG) proteins: Three families of chromosomal proteins that bind DNA nonspecifically, promoting chromatin remodeling and DNA looping for regulating DNA transcription.

histone acetyltransferase (HAT): Any of a family of enzymes that transfer an acetyl group from acetyl-CoA to the ε-amino group of specific Lys residues on histone tails.

histone chaperones: Acidic proteins required for the assembly of histone octamers on DNA.

histone code: A hypothetical code in which successive covalent modifications of histone tails and DNA trigger chromatin remodeling and transcriptional activation events.

histone-fold motif: A protein structural motif formed from three α helices connected by two loops. Histone-fold dimers are instrumental in the tight wrapping of the DNA helix around the histone core in nucleosomes.

histone modifying enzymes: A class of enzymes that covalently modify the N-terminal tails of histones.

histone octamer: The complex of two copies of each of the four core histones that forms the histone core of the nucleosome.

histones: The family of basic proteins that associate tightly with DNA in the chromosomes of all eukaryotic cells.

histone tails: The flexible, disordered N-terminal ends of the histone proteins that comprise the histone core. These ends protrude from the nucleosome and contact adjacent nucleosomes.

HMG proteins: *See* high-mobility group (HMG) proteins.

Holliday intermediate: An intermediate in genetic recombination in which two double-stranded DNA molecules are joined by a reciprocal crossover involving one strand of each molecule to form a junction with four DNA branches.

Holliday junction resolvase: A nuclease that specifically binds to and cleaves Holliday intermediates.

holoenzyme: A catalytically active enzyme, including all necessary subunits, prosthetic groups, and cofactors.

homeodomain motif: A conserved 60 amino acid sequence motif in transcription activators encoded by genes that regulate body pattern development.

homeotic genes: Genes that regulate development of the pattern of segments in the *Drosophila* body plan; similar genes are found in most vertebrates. Homeotic genes are expressed after the segmentation genes.

homing endonucleases: Intron-encoded restriction endonucleases that recognize and cleave an asymmetric sequence of 12 to 40 base pairs in the cellular DNA. Repair of the break results in insertion of a copy of the intron via homologous recombination.

homologous chromosomes: In diploid organisms, a pair of chromosomes, one inherited from each parent, that are of similar length, structure, and gene sequence. Also called homologs.

homologous recombination: Recombination between two DNA molecules of similar sequence, occurring in all cells; takes place during meiosis and mitosis in eukaryotes and during the repair of double-strand breaks in all organisms.

homologs: Genes or proteins with sequence similarity. Also shorthand for homologous chromosomes.

homooligomer: Multisubunit molecule (oligomer) with identical subunits.

homotropic: Describes an allosteric modulator that is identical to the normal ligand or an allosteric enzyme that uses its substrate as a modulator.

homozygous: Having identical alleles at a specific genetic locus.

Hoogsteen pairing: Non-Watson-Crick pairing of a pyrimidine base to a purine base that is already participating in a Watson-Crick base pair with another pyrimidine. The arrangement allows the formation of triplex DNA.

Hoogsteen position: The atoms in a purine base that participate in Hoogsteen pairs (non-Watson-Crick hydrogen bonding) with a pyrimidine base.

horizontal gene transfer: The process by which an organism receives genetic information from another organism of which it is not a descendant.

hormone response element (HRE): A short (12 to 20 bp) DNA sequence that binds receptors for steroid, retinoid, thyroid, and vitamin D hormones, altering the expression of the contiguous genes. Each hormone has a consensus sequence preferred by the cognate receptor.

housekeeping gene: A gene that must be expressed continually for the cell to survive.

Hox genes: A major class of homeotic genes.

HRE: *See* hormone response element.

Hsp70: A family of heat-shock proteins with $M_r \approx 70,000$ that constitute a class of molecular chaperones.

hybrid: The offspring of a cross between genetically nonidentical individuals.

hybrid duplex: A duplex experimentally reconstituted from single-stranded DNA (or RNA) from different sources.

hydrogen bond: A weak electrostatic attraction between one electronegative atom (such as oxygen or nitrogen) and a hydrogen atom covalently linked to a second electronegative atom.

hydrolysis: Cleavage of a bond, such as an anhydride or peptide bond, by the addition of the elements of water, yielding two or more products.

hydrophobic effect: The association of nonpolar groups or compounds with each other in aqueous systems, driven by the tendency of the surrounding water molecules to seek their most stable (disordered) state.

hydrophobic interactions: *See* hydrophobic effect.

hydrophobic stacking: *See* base stacking.

hyperchromic effect: The large increase in light absorption at 260 nm occurring as a double-helical DNA unwinds (melts). *Compare* hypochromic effect.

hypersensitive site: A DNA sequence within a chromosome that is especially sensitive to cleavage by DNase I and other nucleases due to the relative absence of binding proteins such as histones. These sites typically precede active promoters and may be binding sites for proteins regulating expression from the downstream gene.

hypochromic effect: The large decrease in light absorption at 260 nm occurring as single strands of DNA anneal to form double-helical DNA. *Compare* hyperchromic effect.

hypothesis: A proposal that provides a reasonable explanation for observations, but has not yet been substantiated by sufficient experimental evidence to stand up to rigorous critical examination.

IF-1: In bacterial protein synthesis, an initiation factor that binds the ribosomal A site and blocks tRNA binding.

IF-3: In bacterial protein synthesis, an initiation factor that prevents premature addition of the 50S ribosomal subunit to the assembling initiation complex.

IF-2: In bacterial protein synthesis, an initiation factor that directs the initiating tRNA to the P site of the 30S subunit.

When the 50S subunit binds to the complex, it hydrolyzes the GTP bound to IF-2, releasing IF-2 and allowing the 70S subunit to form.

immunofluorescence: The labeling of antibodies with a fluorescent dye to visualize or quantify an antigen in a biological, biochemical, or histological preparation.

immunoprecipitation: The use of antibodies against an epitope on a protein of interest (often with secondary antibodies against those primary antibodies) to precipitate the protein from a complex mixture.

importin: A protein receptor responsible for transporting noncoding RNAs processed in the cytoplasm into the nucleus through a nuclear pore.

imprinting: An epigenetic method of regulating gene expression based on the parental origin of the gene.

incomplete dominance: A condition in which alleles at a specific locus are neither dominant nor recessive, and the progeny express a phenotype intermediate between those of the two parents. *Compare* codominance.

indel: A collective term for insertion and deletion mutations.

induced fit: A change in the conformation of an enzyme in response to substrate binding that renders the enzyme catalytically active; also used to denote changes in the conformation of any macromolecule in response to ligand binding such that the binding site of the macromolecule better conforms to the shape of the ligand.

inducer: A signal molecule that, when bound to a regulatory protein, produces an increase in the expression of a given gene.

initial model: A protein structure derived from an electron density map before further refinements are made.

initial velocity (V_0): The velocity of a reaction while the concentration of substrate is saturating and can be regarded as constant relative to the enzyme concentration.

initiation: (1) The first of three stages in the synthesis of DNA, in which the DNA polymerase binds to the origin of replication. (2) The first of three stages in the synthesis of RNA, in which the RNA polymerase binds to the promoter sequence on the DNA. (3) The first of three stages in the synthesis of a protein, in which the ribosome binds to the mRNA and initiator aminoacyl-tRNA.

initiation codon: AUG (sometimes GUG or, even more rarely, UUG in bacteria and archaea); codes for the first amino acid in a polypeptide sequence: *N*-formylmethionine in bacteria; methionine in archaea and eukaryotes. Also called a start codon.

initiation complex: A complex of a ribosome with an mRNA and the initiating Met-tRNAi^{Met} or fMet-tRNA^{fMet}, ready for the elongation steps.

initiation factors: Three protein factors required to assemble the ribosomal subunits and initiator tRNA in preparation for protein synthesis in bacteria. *See also* IF-1, IF-2, *and* IF-3.

initiator protein: A protein that binds specific sites in an origin of replication and serves as a nucleation site for the assembly of other protein complexes necessary to initiate replication; for example, DnaA in *E. coli*, and ORC in eukaryotes.

insertion mutation: A mutation caused by insertion of one or more extra bases between successive bases in DNA. *Compare* deletion mutation.

insertion sequence: A specific base sequence at either end of a transposable segment of DNA.

insertion site: A site within the active site of a DNA polymerase where the template nucleotide and incoming dNTP are positioned. *Compare* postinsertion site.

insulator: A short sequence of DNA that prevents inappropriate cross-signaling between regulatory elements for different genes. Also called a boundary element.

integrase: An enzyme that catalyzes the insertion of a retrovirus or retrotransposon into its target site.

internal ribosome entry site (IRES): A site on the 5′ side of the start codon in some viral and eukaryotic mRNAs where a eukaryotic ribosome can bind in the absence of a 5′ cap.

interphase: The portion of the cell cycle that does not include mitosis. Subdivided into three phases: G_1 phase, S phase, and G_2 phase.

intervening sequence: *See* intron.

intrinsically unstructured protein or protein segment: A protein or segment of a protein that does not fold into an identifiable stable structure in solution. Intrinsically unstructured protein segments often take up a particular structure when they interact with another macromolecule.

intron: A sequence of nucleotides in a gene that is transcribed but excised before the gene is translated. Also called an intervening sequence. *Compare* exon.

inversion mutation: A mutation that results from the inversion of a large segment of DNA in a chromosome.

inverted repeat: A sequence that is the reversed complement of a downstream sequence.

ion-exchange chromatography: A type of column chromatography in which molecules are separated by charge, using a stationary phase that contains fixed charged groups.

ionic bond: A chemical bond in which the electrons of one atom are transferred to another, creating positive and negative ions that attract each other.

ion torrent: A method for rapid DNA sequencing.

IRE: *See* iron response element.

IRES: *See* internal ribosome entry site.

iron homeostasis: The maintenance of a dynamic steady-state concentration of cellular iron by regulatory mechanisms that compensate for changes in external circumstances.

iron response element (IRE): A hairpin structure in the 3′ or 5′ untranslated region of the mRNAs for proteins involved in iron homeostasis. Binding of the iron response protein (IRP) to a 5′ IRE inhibits translation of the mRNA; binding of IRP to a 3′ IRE inhibits degradation of the mRNA.

iron response protein (IRP): A protein that binds the iron response element (IRE) in mRNAs for proteins involved in

iron homeostasis, inhibiting their translation or degradation in response to the cell's need for iron. Iron-sulfur centers required for efficient binding of IRPs to IREs form only when iron is plentiful in the cell and therefore serve as a sensor of the cellular level of iron.

IRP: *See* iron response protein.

irreversible inhibitor: A molecule that either forms a stable noncovalent association with an enzyme or binds the enzyme covalently, destroying a functional group necessary for its catalytic activity.

isoenergetic: Describes a chemical reaction in which the reactants and products have the same or very similar free energy and therefore exist at similar concentrations at equilibrium.

K_a: *See* association constant.

karyopherins: A family of nuclear transport receptors including importins and exportins.

K_d: *See* dissociation constant.

kinetic proofreading: A mechanism for error correction in complex biological processes that maximizes the speed of correct reactions while stalling and allowing incorrect reactions to reverse.

kinetics: The study of reaction rates.

K_m: *See* Michaelis-Menten constant.

Kozak sequence: A sequence around the start codon in eukaryotic mRNA that enhances its translation. The Kozak sequence has a purine nucleotide three residues before, and a G residue immediately after, the start codon.

lagging strand: The DNA strand that, during replication, must be synthesized in the direction opposite to that in which the replication fork moves.

last universal common ancestor: *See* LUCA.

law of independent assortment: In the formation of gametes there is an independent assortment of alleles for different genes. Also known as Mendel's second law.

law of segregation: In the formation of gametes there is an equal segregation of alleles. In other words, a haploid gamete contains one copy of each gene. Also known as Mendel's first law.

leader peptide: A short sequence near the amino terminus of a protein that has a specialized targeting or regulatory function.

leader sequence: A short sequence near the 5′ end of an RNA that has a specialized targeting or regulatory function.

leading strand: The DNA strand that, during replication, is synthesized in the same direction in which the replication fork moves.

leucine zipper motif: A protein structural motif involved in protein-protein interactions in many eukaryotic regulatory proteins; consists of two interacting α helices in which Leu residues in every seventh position are a prominent feature of the interacting surfaces.

ligand: Any molecule, small or large, that is specifically bound by a protein without altering that bound molecule; for example, a hormone is the ligand for its specific protein receptor.

linkage analysis: The use of bioinformatics to analyze the statistical association between inheritance of a gene and the presence of specific single-nucleotide polymorphisms, with the goal of mapping the gene to a specific location on a chromosome.

linked genes: Genes that are close together on a chromosome and the alleles of which therefore assort together during meiosis, in contradiction to Mendel's second law.

linker: A synthetic DNA fragment inserted into a cloning vector, usually to provide a specific desired sequence, such as a restriction endonuclease recognition sequence.

linker histone: The histone protein H1, which binds to the linker DNA adjacent to the nucleosome.

linking number (*Lk*): The number of times one closed circular DNA strand is wound about another; the number of topological links holding the circles together.

Lk: *See* linking number.

LUCA (last universal common ancestor): The single-celled organism that gave rise to all life currently existing on Earth.

lysis: Destruction of a plasma membrane or (in bacteria) cell wall, releasing the cellular contents and killing the cell.

lysogen: A bacterial cell infected with a prophage.

lysogenic pathway: Bacteriophage infection in which the DNA is incorporated into the host chromosome or as an autonomously replicating plasmid with most of its genes repressed. *Compare* lytic pathway.

lytic pathway: Parasitic bacteriophage infection in which the DNA is replicated and packaged into phage heads, and the host cell is destroyed by lysis to disperse the progeny. *Compare* lysogenic pathway.

major groove: The wider of two grooves that wind around the outside of a DNA double helix.

mass spectrometry (MS): An analytical technique for determining the mass of a molecule, thus providing a clue to its identity, by measuring the charge-to-mass ratio of gaseous ions formed from the molecule as the ions pass through an electromagnetic field in a vacuum.

maternal genes: Genes expressed in the unfertilized egg that are required for development of the early embryo.

maternal mRNAs: Transcripts of maternal genes that are generated in the egg during oogenesis and remain dormant until fertilization.

mating type: In yeast, one of the two haploid forms, **a** and α, that can mate only with a haploid cell of the opposite type to form a diploid cell.

maximum velocity: *See* V_{max}.

MCM complex: A ring-shaped eukaryotic helicase complex that acts at the replication fork. It is composed of six homologous, but nonidentical, AAA+ proteins and interacts with two other proteins to form the CMG complex.

Mediator complex: A large, multiprotein complex in eukaryotic cells that serves as the mediator between the Pol II transcription complex and any upstream transcription activators or enhancers regulating Pol II–catalyzed transcription.

meiosis: A type of cell division in which diploid cells give rise to haploid cells destined to become gametes.

melting: The denaturation or unwinding of a double-stranded polynucleotide to form single-stranded polynucleotides.

melting point (*T*$_m$): The temperature at which a specific double-stranded polynucleotide separates into single strands.

messenger RNA (mRNA): A class of RNA molecules, each of which is complementary to one strand of DNA, that carry the genetic message from the chromosome to the ribosomes.

metagenomics: The structural and functional analysis of the collective genome of an environmental population of microorganisms rather than a pure population derived from a single cultured cell.

metamerism: The division of the body into a series of repeating segments, such as in insects.

metaphase: The second stage of mitosis (M phase). The spindle apparatus directs condensed sister chromatid pairs to align along the metaphase plate.

metaphase plate: The equatorial plane in a dividing cell along which chromosomes align during metaphase.

Michaelis-Menten constant (*K*$_m$): The substrate concentration at which an enzyme-catalyzed reaction proceeds at one-half its maximum velocity.

Michaelis-Menten equation: The equation describing the hyperbolic dependence of the initial reaction velocity, V_0, on substrate concentration, [S], in many enzyme-catalyzed reactions.

microprocessor complex: A nuclear complex responsible for the early stages of miRNA and siRNA processing in eukaryotic cells. It consists of a primary miRNA transcript, an miRNA recognition protein, and the endonuclease Drosha.

microRNA (miRNA): A class of small RNA molecules (21 to 23 nucleotides after processing is complete) involved in gene silencing by inhibiting translation and/or promoting the degradation of particular mRNAs.

migration: (1) The movement of a population to a new geographic location. (2) The movement of cells to a new location within an organism or tissue.

minor groove: The narrower of two grooves that wind around the outside of a DNA double helix.

miRNA: *See* microRNA.

mirror repeat: A segment of duplex DNA in which the base sequences exhibit symmetry on each single strand.

mismatch repair (MMR): An enzymatic system for repairing base mismatches (non-Watson-Crick pairs) in DNA.

missense mutation: A single-nucleotide change in a gene that results in an amino acid change in the protein product.

mitosis: In eukaryotic cells, the multistep process that results in the segregation of replicated cellular chromosomes and cell division.

mixed inhibitor: An inhibitor molecule that can bind to either the free enzyme or the enzyme-substrate complex (not necessarily with the same affinity).

MMR: *See* mismatch repair.

modulator: *See* allosteric modulator.

MOI: *See* multiplicity of infection.

mole: One gram molecular weight of a compound. A mole of any compound contains 6.02×10^{23} molecules.

molecular biology: The study of essential cellular macromolecules, including DNA, RNA, and proteins, and the biological pathways that link their biosynthesis.

molecular function (of a gene product): The precise biochemical activity of a protein or an RNA, such as the reactions an enzyme catalyzes, the ligands a receptor binds, or the complex formed between a specific RNA and a protein. *Compare* cellular function *and* phenotypic function.

molecular genetics: The study of the structure and function of genes at the molecular level.

molecular orbital model: A mathematical function describing the wavelike behavior of electrons in a molecule.

molten globule model: A model for protein folding in which the hydrophobic residues of a polypeptide chain rapidly collapse into a condensed, partially ordered state, which limits the conformations available to the rest of the molecule. As subdomains with tertiary structure develop, alternative conformations become increasingly limited, and the molecule achieves its native conformation. *Compare* hierarchical model.

motif: *See* sequence motif; structural motif.

motor protein: A protein that uses energy (typically from the hydrolysis of ATP) to undergo a cyclic conformational change that creates a unified, directional force.

M phase: The phase of the eukaryotic cell cycle during which mitosis, or cell division, occurs. The M phase follows the G_2 phase and precedes the G_1 phase.

mRNA: *See* messenger RNA.

MS: *See* mass spectrometry.

multiplicity of infection (MOI): The ratio of infectious particles to target cells.

multipotent: Describes stem cells that can differentiate into a number of types of closely related cells.

mutation: An inheritable change in the nucleotide sequence of a chromosome.

mutation rate: The frequency of new mutations (in a gene or in an organism) in each cellular generation.

natural selection: The process by which traits (phenotypes) become more prevalent in a population because those individuals best adapted to exploit the prevailing resources are the ones most likely to survive and reproduce, passing on their advantageous traits.

negative regulation: The decreased expression of a gene due to the binding of a repressor protein. *Compare* positive regulation.

negative supercoiling: The twisting of a helical (coiled) molecule on itself to form a right-handed supercoil.

NER: *See* nucleotide excision repair.

N-formylmethionyl-tRNA^{fmet}: The charged tRNA used to initiate protein synthesis in bacteria.

NHEJ: *See* nonhomologous end joining.

niche: In cellular differentiation, a microenvironment that allows the maintenance of both multipotent adult stem cells and their differentiated progeny.

nick translation: A concerted process of $5' \rightarrow 3'$ excision and DNA polymerization that shifts a discontinuity in the phosphodiester backbone between the $3'$ hydroxyl of one nucleotide and the $5'$ phosphate of the adjacent nucleotide along a DNA strand.

NLS: *See* nuclear localization sequence.

NMD: *See* nonsense-mediated mRNA decay.

NMR: *See* nuclear magnetic resonance spectroscopy.

NOESY: *See* nuclear Overhauser effect spectroscopy.

nondisjunction: The failure of paired chromosomes or sister chromatids to segregate during mitotic or meiotic cell division.

nonhomologous end joining (NHEJ): A method for repairing double-strand breaks by joining nonhomologous DNA ends in a process that does not conserve the original sequence.

nonpolar: Hydrophobic; describes molecules or groups that have no effective dipole moment and are therefore poorly soluble in water.

nonsense-mediated mRNA decay (NMD): A pathway for degradation of mRNA molecules with a premature stop codon, triggered by the presence of an exon junction complex on a transcript that has been translated. *Compare* non-stop mRNA decay.

nonsense mutation: A mutation that results in the premature termination of a polypeptide chain.

non-stop mRNA decay: A pathway for degradation of mRNA molecules lacking a stop codon, triggered by release of the ribosome from the $3'$ end of the message. *Compare* nonsense-mediated mRNA decay.

nontemplate strand: *See* coding strand.

Northern blotting: A nucleic acid hybridization procedure in which one or more specific RNA fragments are detected in a larger population by hybridization to a complementary, labeled DNA probe.

N-terminus (amino terminus): *See* amino terminus.

nuclear localization sequence (NLS): An amino acid sequence that targets a protein for transport to the nucleus.

nuclear magnetic resonance (NMR) spectroscopy: A technique that utilizes certain quantum mechanical properties of atomic nuclei to study the structure and dynamics of the molecules of which they are a part.

nuclear Overhauser effect spectroscopy (NOESY): A type of two-dimensional nuclear magnetic resonance spectroscopy in which atoms that are near to one another in space, but not necessarily nearby in the primary structure, can be identified.

nucleases: Enzymes that hydrolyze the internucleotide (phosphodiester) linkages of nucleic acids.

nucleic acids: Biologically occurring polynucleotides in which the nucleotide residues are linked in a specific sequence by phosphodiester bonds; DNA and RNA.

nucleoid: In bacteria, the nuclear zone that contains the chromosome but has no surrounding membrane.

nucleolytic proofreading: A pathway for the correction of errors in an RNA transcript in which the RNA polymerase reverses direction by one or a few nucleotides on the template, and its endonuclease activity hydrolyzes the phosphodiester backbone of the transcript proximal to the mismatched base.

nucleophile: An electron-rich group with a strong tendency to donate electrons to an electron-deficient nucleus (electrophile); the entering reactant in a bimolecular substitution reaction.

nucleoside: A compound consisting of a purine or pyrimidine base covalently linked to a pentose.

nucleosome: In eukaryotes, the structural unit for packaging chromatin; consists of a DNA strand wound around a histone core.

nucleotide: A nucleoside phosphorylated at one or more of its pentose hydroxyl groups.

nucleotide excision repair (NER): A DNA repair pathway that involves excinuclease-catalyzed cleavage of the phosphodiester bond on either side of a bulky DNA lesion such as a pyrimidine dimer or base adduct, followed by removal of the segment containing the lesion, then DNA polymerization and ligation to fill the gap.

Okazaki fragment: A short segment of DNA synthesized on the lagging strand during DNA replication.

oligomer: A short polymer, usually of amino acids, sugars, or nucleotides; the definition of "short" is somewhat arbitrary, but usually fewer than 50 subunits.

oligomeric state: The number of identical polypeptide subunits in a particular form of a protein. For example, monomer, dimer, and trimer are different oligomeric states.

oligonucleotide: A short polymer of nucleotides (usually fewer than 50).

oligonucleotide-directed mutagenesis: A method for creating a mutation in a cloned gene. Two short, complementary synthetic DNA strands, each with the desired base change, are annealed to opposite strands of the cloned gene within a suitable vector. The two annealed oligonucleotides prime DNA synthesis, creating two complementary strands with the mutation.

oncogene: A gene that, when subjected to particular mutations or introduced as part of the genome of particular viruses, causes cells to exhibit rapid, uncontrolled proliferation leading to cancer.

open complex: (1) A complex of the RNA polymerase bound to a promoter, in which the bound DNA is partially unwound. Transcription initiation occurs in the open complex. *Compare* closed complex. (2) A complex assembled on the *E. coli* origin of replication, *oriC*, at an early stage of replication initiation. It includes an oligomer of the AAA+ protein DnaA, ATP, and the histonelike protein HU.

open form: The conformation assumed by *E. coli* DNA polymerase I when a primed template is bound to the active site, but the correct dNTP is not.

open reading frame (ORF): A group of contiguous nonoverlapping nucleotide codons in a DNA or RNA molecule that does not include a termination codon.

operator: A region of DNA that interacts with a repressor protein to control the expression of a group of genes organized in an operon.

operon: A unit of genetic expression consisting of one or more cotranscribed genes and the operator and promoter sequences that regulate their transcription.

optically active: Able to rotate the plane of plane-polarized light.

ORC: *See* origin recognition complex.

ORF: *See* open reading frame.

organelles: Membrane-bounded structures found in eukaryotic cells; contain enzymes and other components required for specialized cell functions.

ori: *See* origin of replication.

origin of replication (ori): The nucleotide sequence or site in DNA where DNA replication is initiated.

origin recognition complex (ORC): A eukaryotic initiator protein complex that assembles at an origin to initiate replication.

orthologs: Genes in different organisms that possess a clear sequence and functional relationship to each other. *Compare* paralogs.

outgroup: A taxon outside the group of interest in a phylogenetic tree.

pair-rule genes: A subclass of the segmentation genes expressed in alternate body segments of the developing *Drosophila* embryo, after the gap genes and before the segment polarity genes.

palindrome: A segment of duplex DNA in which the base sequences of the two strands exhibit twofold rotational symmetry about an axis.

parallel β sheet: *See* β sheet.

paralogs: Genes within a species that possess a clear sequence and functional relationship to each other and probably arose as the result of a gene duplication. *Compare* orthologs.

parthenogenesis: Reproduction by the growth and development of an unfertilized egg.

P body: *See* processing body.

PCNA: Proliferating cell nuclear antigen, the eukaryotic sliding clamp protein that tethers DNA polymerase to the DNA at the replication fork.

PCR: *See* polymerase chain reaction.

PDB: *See* Protein Data Bank.

peptide bond: A substituted amide linkage between the α-amino group of one amino acid and the α-carboxyl group of another, with elimination of the elements of water.

peptide prolyl cis-trans isomerase: An enzyme that catalyzes the interconversion of the cis and trans isomers of proline peptide bonds.

peptide translocation complex: A complex in the endoplasmic reticulum (ER) that catalyzes the translocation into the ER lumen of a growing polypeptide chain containing an N-terminal signal sequence.

peptidyl transferase reaction: The reaction that synthesizes the peptide bonds of proteins—nucleophilic attack of the α-amino group of the ribosomal A-site aminoacyl-tRNA on the carbonyl carbon of the ester bond linking the fMet (or the growing peptide chain) to the P-site tRNA. The reaction is catalyzed by a ribozyme, part of the rRNA of the large ribosomal subunit.

P generation: The parental generation in a genetic cross.

pH: The negative logarithm of the hydronium ion concentration of an aqueous solution.

phase variation: The expression of alternative primary cell surface antigens used by some pathogenic bacteria and parasitic protists as a means of eluding a host's immune system.

phenotype: The observable characteristics of an organism.

phenotypic function (of a gene product): The effect of a gene product on the entire organism. *Compare* cellular function *and* molecular function.

phosphatases: Enzymes that hydrolyze a phosphate ester or anhydride, releasing inorganic phosphate, P_i.

phosphodiester bond: A chemical grouping that contains two alcohols esterified to one molecule of phosphoric acid, which thus serves as a bridge between them.

photoreactivation: The repair of a cyclobutane pyrimidine dimer by electron transfer from a DNA photolyase.

phylogenetic profiling: A bioinformatic technique used to discover structure-function relationships by searching for genes that consistently appear together across many genomes.

phylogenetics: The study of the evolutionary relationships among organisms.

phylogeny: The evolutionary relationships among organisms.

pi stacking: The attractive, noncovalent interactions between aromatic rings; important in base stacking in nucleic acids.

pK_a: The negative logarithm of an acid dissociation constant.

plasmid: An extrachromosomal, independently replicating, small circular DNA molecule; commonly used in genetic engineering.

plectonemic supercoiling: A structure in a molecular polymer that has a net twisting of strands about each other in some simple and regular way.

pluripotent: Describes stem cells that can differentiate into cells derived from any of the three germ layers.

pOH: The negative logarithm of the hydroxyl ion concentration of an aqueous solution.

point mutation: A mutation consisting of a single base-pair change.

polar covalent: Describes a type of covalent bond between atoms of different electronegativities, such that the electrons are shared unequally between the atoms.

polar: Hydrophilic, or "water-loving"; describes molecules or groups that have a dipole moment and are therefore soluble in water.

polarity: In a developing embryo, the distinct areas that will become the anterior, posterior, dorsal, and ventral parts of the adult organism.

Pol I: *See* DNA polymerase I; RNA polymerase I.

Pol II: *See* DNA polymerase II; RNA polymerase II.

Pol III: *See* DNA polymerase III; RNA polymerase III.

Pol III core: A complex of the α, ε, and θ subunits of the *E. coli* DNA polymerase III with polymerase and $5' \rightarrow 3'$ proofreading exonuclease activities.

Pol III holoenzyme: The 17-subunit *E. coli* DNA polymerase III complex, responsible for chromosomal replication. It includes two Pol III, two sliding clamps, and a clamp-loading complex. *Compare* RNA polymerase holoenzyme *and* RNA polymerase III.

poly(A) addition site: The site where an mRNA is cleaved by a specific endonuclease to generate the free 3′ hydroxyl to which A residues are added. The site is marked by a highly conserved 5′-AAUAAA sequence 10 to 30 nucleotides on the 5′ side, and a G- and U-rich region 20 to 40 nucleotides on the 3′ side.

poly(A) site choice: The existence of more than one site in an mRNA that may be cleaved to generate the free 3′ hydroxyl to which A residues are added, which can generate diverse transcripts from a single gene.

poly(A) tail: *See* 3′ poly(A) tail.

polycistronic mRNA: A contiguous mRNA with more than two genes that can be translated into proteins.

polyglutamine (polyQ) disease: A triplet expansion disease caused by the insertion of many additional glutamine codons in a gene. Fragile X syndrome and Huntington disease are polyglutamine diseases.

polylinker: A short, synthetic fragment of DNA containing recognition sequences for several restriction endonucleases that is inserted into a cloning vector.

polymerase chain reaction (PCR): A repetitive laboratory procedure that results in a geometric amplification of a specific DNA sequence.

polynucleotide: A covalently linked sequence of nucleotides in which the 3′ hydroxyl of the pentose of one nucleotide residue is joined by a phosphodiester bond to the 5′ hydroxyl of the pentose of the next residue.

polypeptide chain: A long chain of amino acids linked by peptide bonds; the molecular weight is generally less than 10,000. Also called a polypeptide.

polyQ disease: *See* polyglutamine disease.

polyribosome: *See* polysome.

polysome: A complex of an mRNA molecule and two or more ribosomes. Also called a polyribosome.

positive regulation: The increased expression of a gene by the binding of an activator protein. *Compare* negative regulation.

positive supercoiling: The twisting of a helical (coiled) molecule on itself to form a left-handed supercoil.

postinsertion site: A site within the active site of a DNA polymerase where the primer 3′-terminal base pair is positioned. *Compare* insertion site.

posttranslational modification: The enzymatic processing of a polypeptide chain after translation from its mRNA.

postulate of objectivity: The only assumption made by scientists—that basic forces and laws in the universe are not subject to change and can thus be studied and defined by scientific inquiry. The term was introduced by Jacques Monod.

precursor miRNA (pre-miRNA): A partially processed RNA intermediate that is transported from the nucleus to the cytoplasm for final processing into an miRNA by the endonuclease Dicer.

preinitiation complex: A eukaryotic nucleoprotein complex consisting of intact, double-stranded promoter DNA, Pol II, and various transcription factors. *Compare* closed complex (in bacteria).

pre-miRNA: *See* precursor miRNA.

prepriming complex: The complex of proteins assembled at *oriC* in *E. coli* at an early stage of replication fork assembly. The complex includes a DnaA oligomer bound to the DNA and DnaB helicases stabilizing the single strands of DNA in the replication "bubble."

preRC: *See* prereplication complex.

prereplication complex (preRC): The complex of proteins, including the origin recognition complex (ORC) and MCMs, that assembles during the G₁ phase of the eukaryotic cell cycle, thereby marking the origin for replication during S phase.

preribosomal RNA (pre-rRNA): The primary transcript of ribosomal RNAs in bacterial and eukaryotic cells, which is processed into mature ribosomal RNAs (and transfer RNAs in bacteria).

pre-rRNA: *See* preribosomal RNA.

pre–steady state: The time immediately after an enzyme is mixed with its substrate, before the free enzyme and its intermediates have reached their steady-state concentrations.

primary miRNA transcript (pri-miRNA): An RNA transcript that can fold into an extensive hairpin structure, which becomes a substrate for cleavage by the nuclear endonuclease Drosha into a smaller hairpin structure called pre-miRNA.

primary structure: A description of the covalent backbone of a polymer (macromolecule), including the sequence of monomeric subunits and any interchain and intrachain covalent bonds.

primary transcript: The immediate RNA product of transcription before any posttranscriptional processing reactions.

primase: An enzyme that catalyzes the formation of RNA oligonucleotides used as primers by DNA polymerases.

primed template: A template nucleic acid strand annealed to an RNA or DNA primer.

primer strand: A strand of nucleic acid with a free 3′-OH group to which a DNA polymerase can add nucleotides.

primer terminus: The end of a primer to which monomeric subunits are added.

pri-miRNA: *See* primary miRNA transcript.

prion: A misfolded protein in the nervous tissue of mammals that acts as an infectious agent, causing other proteins to misfold and accumulate, leading to the development of spongiform encephalopathy.

probe: A labeled fragment of nucleic acid containing a nucleotide sequence complementary to a genomic sequence that one wishes to detect in a hybridization experiment.

processing body (P body): An area in the cytoplasm of a eukaryotic cell where mRNAs that are not being translated are sequestered, possibly for degradation.

processive synthesis: The enzymatic synthesis of a biological polymer in which the enzyme adds multiple subunits without dissociating from the substrate. *Compare* distributive synthesis.

processivity: For any enzyme that catalyzes the synthesis of a biological polymer, the property of adding multiple subunits to the polymer without dissociating from the substrate.

product: A molecule formed in a chemical reaction.

proenzyme: A precursor form of an enzyme, before it is cleaved into its active form.

promoter: A DNA sequence at which RNA polymerase may bind, leading to initiation of transcription.

promoter clearance: Movement of the transcription complex away from the promoter, which marks the beginning of the elongation stage of transcription.

proofreading: The correction of errors in the synthesis of an information-containing biopolymer by removing incorrect monomeric subunits after they have been covalently added to the growing polymer.

prophage: A bacteriophage genome incorporated into the host DNA or as an autonomously replicating plasmid, with most of its genes repressed; a lysogenized bacteriophage genome.

prophage induction: The process in which a lysogen switches from lysogenic growth to lytic growth.

prophase: The first stage of mitosis (M phase). Chromosomes duplicated in S phase begin to condense and become visible in the light microscope.

proprotein: A precursor form of a protein, before it is cleaved into its functional form.

prosthetic group: A metal ion or an organic compound (other than an amino acid) that is covalently bound to a protein and is essential to its activity.

proteasome: A large assembly of enzymatic complexes that function in the degradation of damaged or unneeded cellular proteins. In eukaryotes, also called a 26S proteasome.

Protein Data Bank (PDB): An international database (www.rcsb.org/pdb) that archives data describing the three-dimensional structure of nearly all macromolecules for which structures have been published.

protein disulfide isomerase: An enzyme that catalyzes the breakage and formation of disulfide cross-links in a protein.

protein family: A group of evolutionarily related proteins with similar primary sequence and function.

protein folding: The process by which a polypeptide chain attains its biologically active conformation.

protein kinases: Enzymes that transfer the terminal phosphoryl group of ATP or another nucleoside triphosphate to a Ser, Thr, Tyr, Asp, or His side chain in a target protein, thereby regulating the activity or other properties of that protein.

protein phosphatases: Enzymes that hydrolyze a phosphate ester or anhydride on specific amino acid residues of a protein, releasing inorganic phosphate, P_i.

proteolytic cleavage: The enzyme-catalyzed breakage of peptide bonds in proteins.

proteome: The full complement of proteins expressed in a given cell under a given set of conditions, or the complete complement of proteins that can be expressed by a given genome.

proteomics: Broadly, the study of the protein complement of a cell or organism.

protomer: A general term describing any repeated unit of one or more stably associated protein subunits in a larger protein structure. If a protomer has multiple subunits, the subunits may be identical or different.

P site: The site in a ribosome occupied by the peptidyl-tRNA.

PUF family: In eukaryotic cells, a family of proteins that bind the 3′ untranslated region of mRNAs to suppress their translation.

pulsed field gel electrophoresis: A variation on the technique of gel electrophoresis in which the direction of the current passed through the gel is altered at regular intervals. This allows the separation of larger molecules of DNA than is possible with conventional gel electrophoresis. *See also* gel electrophoresis.

Punnett square: A matrix for displaying the genes involved in a cross and the possible combinations of alleles in the progeny. The gamete genotypes are written along the top and sides of the square; the possible combinations of alleles are shown in the matrix.

purebred: Describes an individual homozygous for a given trait or set of traits.

purine: A nitrogenous heterocyclic base found in nucleotides and nucleic acids; contains fused pyrimidine and imidazole rings.

puromycin: An antibiotic that inhibits polypeptide synthesis by being incorporated into a growing polypeptide chain, causing its premature termination.

pyrimidine: A nitrogenous heterocyclic base found in nucleotides and nucleic acids.

pyrimidine dimer: A covalently joined dimer of two adjacent pyrimidine residues in DNA, induced by absorption of UV light; most commonly derived from two adjacent thymines (a thymine dimer).

pyrophosphorolysis: The reverse of a nucleotide polymerization reaction, in which pyrophosphate reacts with the 3′-nucleotide monophosphate of an oligonucleotide, releasing the corresponding nucleotide triphosphate.

pyrosequencing: A method for rapid DNA sequencing in which nucleotide additions are detected with flashes of light.

qPCR: *See* quantitative PCR.

quantitative PCR (qPCR): A polymerase chain reaction (PCR) protocol that allows the simultaneous amplification and detection of a sequence through use of a fluorescent probe. *See also* polymerase chain reaction.

quaternary structure: The three-dimensional structure of a multisubunit protein, particularly the manner in which the subunits fit together.

quorum sensing: The regulation of gene expression in response to fluctuations in cell population density, assessed by the detection of small, diffusible signaling molecules secreted by the cells.

Rad51: A eukaryotic recombinase structurally and functionally homologous to the RecA protein of *E. coli*. *See also* Dmc1.

Ramachandran plot: A graphical representation of the φ and ψ angles of the amino acid residues in a polypeptide.

rate constant: The proportionality constant that relates the velocity of a chemical reaction to the concentration(s) of the reactant(s).

rate-limiting step: (1) Generally, the step in an enzymatic reaction with the greatest activation energy or the transition state of highest free energy. (2) The slowest step in a metabolic pathway.

reactant: A starting material in a chemical reaction.

reaction intermediate: Any chemical species in a reaction pathway that has a finite chemical lifetime (greater than a molecular vibration, or 10^{-13} seconds).

reaction kinetics: *See* kinetics.

reaction mechanism: The sequence of individual steps that take place during the conversion of reactants to products in a chemical reaction.

reading frame: A contiguous, nonoverlapping set of three-nucleotide codons in DNA or RNA.

RecA protein: A non-site-specific bacterial recombinase that binds single-stranded DNA and promotes homologous recombination. RecA protein also has co-protease activity in the autocatalytic cleavage of some transcription repressors.

RecBCD: A bacterial protein complex that prepares DNA at a double-strand break for repair. The complex has helicase activity to unwind the DNA, an endonuclease activity that creates 3′ single-stranded overhangs, and an activity that loads RecA protein on the 3′-ending single strand.

recessive: Describes an allele that manifests in a homozygous individual but is masked by the dominant allele in heterozygotes. *Compare* dominant.

RecFOR: A bacterial recombination mediator that loads RecA protein on single-strand gaps in need of repair.

recognition helix: The α helix in a DNA regulatory protein that recognizes and binds to the DNA regulatory site.

recognition sequence: A specific nucleotide sequence in a double-stranded DNA molecule that is recognized by a restriction endonuclease as a substrate.

recombinant: Any DNA or chromosome in which segments from different sources are combined.

recombinant DNA: DNA formed by joining DNA molecules, sometimes from different species, in new combinations.

recombinant DNA technology: Laboratory methods used for genetic engineering.

recombinase: An enzyme that catalyzes genetic recombination by the reciprocal exchange of short pieces of DNA between longer DNA molecules.

recombination: (1) The reciprocal exchange of alleles between chromosomes. Also called crossing over. (2) At the molecular level, any enzymatic process by which the linear arrangement of nucleic acid sequences in a chromosome or plasmid is altered by cleavage and rejoining.

recombinational DNA repair: Recombinational processes directed at the repair of DNA strand breaks or cross-links, especially at inactivated replication forks.

recombination mapping: The process of determining the relative distance between genes on a chromosome based on the frequency of recombination of alleles during meiosis.

refinement: In x-ray crystallography, an iterative stage in which the computed diffraction pattern of a predicted model of a three-dimensional structure is compared with the diffraction pattern obtained from the crystal in an experiment.

reflection spot: In x-ray crystallography, an area on a film or x-ray detector created by the constructive interference of x rays diffracted from the atoms in a unit cell of a crystal.

regulated gene expression: The conditional expression of a gene based on the cellular need for the gene product, achieved by the presence or absence of activators, repressors, enhancers, and other regulatory factors. *Compare* constitutive gene expression.

regulatory enzyme: An enzyme with a regulatory function, through its capacity to undergo a change in catalytic activity by allosteric mechanisms or by covalent modification.

regulatory sequence: A DNA sequence involved in regulating the expression of a gene; for example, a promoter or operator. Also called a regulatory site.

regulatory site: *See* regulatory sequence.

regulon: A group of genes or operons that are coordinately regulated even though some, or all, may be spatially distant in the chromosome or genome.

relaxed DNA: Any DNA that exists in its most stable and unstrained structure, typically the B form under most cellular conditions.

release factors: Protein factors required for the release of a completed polypeptide chain from a ribosome. Also called termination factors. *See also* RF-1, RF-2, *and* RF-3.

replicase: A general term describing any polymerase enzyme that duplicates chromosomes. Also called a chromosomal replicase.

replication factor C (RFC): The eukaryotic clamp loading complex.

replication fork: The Y-shaped structure generally found at the point where DNA is being synthesized.

replication protein A (RPA): A eukaryotic single-stranded DNA-binding protein; its bacterial homolog is SSB.

replicon: The length of DNA replicated from a single origin.

replisome: The multiprotein complex that promotes DNA synthesis at the replication fork.

replisome progression complex: A large protein assemblage that is part of the eukaryotic replication machinery; includes proteins that move with the replication fork, proteins that participate in affixing Pol α in the replisome, and several nonessential proteins thought to control the rate of replication during times of cellular stress.

repression: A decrease in the expression of a gene in response to a change in the activity of a regulatory protein.

repressor: A protein that binds to the regulatory sequence or operator for a gene, blocking its transcription.

resonance: A conceptual view of the delocalized electrons in the bonding structure of a molecule that can only be described as the average of two or more Lewis structures.

resonance hybrid: A molecule that exists in an average of two possible resonance forms. *See also* resonance.

restriction endonucleases: Site-specific endodeoxyribo-nucleases that cleave both strands of DNA at points in or near the specific site recognized by the enzyme; important tools in genetic engineering.

retrohoming: The integration of a mobile group II intron into DNA by the reverse splicing of its RNA transcript into a target site (catalyzed by an encoded endonuclease), followed by DNA synthesis (catalyzed by an encoded reverse transcriptase).

retrotransposable element: *See* retrotransposon.

retrotransposon (retrotransposable element): A transposon that moves via an RNA intermediate that is converted back to DNA by reverse transcriptase.

retrovirus: An RNA virus containing a reverse transcriptase.

reverse transcriptase: An RNA-directed DNA polymerase in retroviruses; capable of making DNA complementary to an RNA.

reverse transcriptase PCR (RT-PCR): A polymerase chain reaction (PCR) protocol for amplifying an RNA sequence by first using reverse transcriptase to create a DNA copy. *See also* polymerase chain reaction *and* reverse transcriptase.

reverse turn: A segment of a polypeptide chain in a folded protein or domain that connects two regions of secondary structure with reversed N-to-C directions.

reversible inhibition: Inhibition by a molecule that binds reversibly to the enzyme, such that the enzyme activity returns when the inhibitor is no longer present.

reversible terminator sequencing: A method used in many protocols for rapid DNA sequencing.

reversion mutation: A mutation in a gene that reverses a previous mutation. Also called a back mutation. A true reversion restores the original gene sequence; a second-site reversion restores the functionality (phenotype).

R factor: The residual error in a model of the three-dimensional structure of a molecule obtained by x-ray crystallography. The R factor is the difference between the calculated diffraction pattern based on the model and the actual diffraction pattern obtained from the crystal.

RFC: *See* replication factor C.

RF-1: A bacterial class I release factor that recognizes the stop codons UAG and UAA and induces peptidyl transferase to transfer the growing polypeptide to water.

RF-3: A bacterial class II release factor with GTPase activity that catalyzes the dissociation of RF-1 and RF-2 from the ribosome.

RF-2: A bacterial class I release factor that recognizes the stop codons UGA and UAA and induces peptidyl transferase to transfer the growing polypeptide to water.

R group: (1) Formally, an abbreviation denoting any alkyl group. (2) Occasionally, used in a more general sense to denote virtually any organic substituent (the R groups of amino acids, for example).

ribonuclease: A nuclease that catalyzes the hydrolysis of certain internucleotide linkages of RNA.

ribonucleic acid: *See* RNA.

ribonucleoprotein (RNP): A molecular complex of RNA and protein, such as the ribosome.

ribonucleoside 3′-monophosphates: Metabolites produced during the enzymatic or alkaline hydrolysis of RNA.

ribonucleoside 2′-monophosphates: Metabolites produced during the enzymatic or alkaline hydrolysis of RNA.

ribonucleoside 2′,3′-cyclic monophosphates: Metabolites produced during the enzymatic or alkaline hydrolysis of RNA.

ribonucleotide: A nucleotide containing D-ribose as its pentose component.

ribosomal protein (r-protein): A protein serving as a component of ribosomes.

ribosomal RNA (rRNA): A class of RNA molecules serving as components of ribosomes.

ribosome: A macromolecular complex of rRNAs and r-proteins; the site of protein synthesis.

ribosome binding site: A sequence in an mRNA that is required for binding bacterial ribosomes. Also called the Shine-Dalgarno sequence.

ribosome recycling: The disassembly of a translated mRNA, deacylated tRNAs, and ribosomal subunits in preparation for new rounds of translation.

ribosome recycling factor: A bacterial factor involved in ribosome recycling that binds to the empty ribosomal A site and recruits EF-G to stimulate release of the deacylated tRNAs in the P and E sites.

riboswitch: A structured segment of an mRNA that binds to a specific ligand and affects the translation or processing of the mRNA.

ribozyme (catalytic RNA): A ribonucleic acid molecule with catalytic activity; an RNA enzyme.

ricin: An extremely toxic protein of the castor bean that inactivates the 60S subunit of eukaryotic ribosomes by depurinating a specific adenosine in 23S rRNA.

rifampicin: An antibiotic that acts as a bacterial transcription inhibitor by binding to the β subunit of bacterial RNA polymerase, preventing promoter clearance.

RISC: *See* RNA-induced silencing complex.

RNA (ribonucleic acid): A polyribonucleotide of a specific sequence linked by successive 3′,5′-phosphodiester bonds.

RNA degradation: The complete hydrolysis of RNA into its component ribonucleotides, usually catalyzed by ribonucleases or the exosome.

RNA editing: The posttranscriptional modification of an mRNA that alters one or more codons prior to translation.

RNAi: *See* RNA interference.

RNA-induced silencing complex (RISC): A cytoplasmic complex, including the endonucleases Dicer and Argonaute, that incorporates an miRNA or an siRNA, delivers it to its complementary mRNA target, and then cleaves the mRNA.

RNA interference (RNAi): Methods of gene silencing, mediated by short interfering RNAs (siRNAs) or microRNAs (miRNAs), which bind to and silence the mRNA transcript, usually by targeting it for degradation. MicroRNAs are endogenous to the cell, produced by nucleases from longer transcripts encoded in the genome; siRNAs are generated by the same cellular enzymes from exogenous double-stranded RNA, introduced into the cell by viral infection or experimental manipulation.

RNA polymerase: An enzyme that catalyzes the formation of RNA from ribonucleoside 5′-triphosphates, using a strand of DNA as a template. Some RNA polymerases, primarily in viruses, use RNA as a template and are called RNA-dependent RNA polymerases (RDRPs).

RNA polymerase I (Pol I): One of the three eukaryotic RNA polymerases; Pol I transcribes genes encoding large rRNA precursors.

RNA polymerase II (Pol II): One of the three eukaryotic RNA polymerases; Pol II transcribes most of the protein-coding genes.

RNA polymerase III (Pol III): One of the three eukaryotic RNA polymerases; Pol III transcribes genes encoding tRNAs, some snRNAs, 5S ribosomal RNA, and other small functional RNAs.

RNA polymerase core: The *E. coli* RNA polymerase complex of subunits $\alpha_2\beta\beta'\omega$, exclusive of the σ subunit. *Compare* Pol III core.

RNA polymerase holoenzyme: The complete *E. coli* RNA polymerase complex, including the $\alpha_2\beta\beta'\omega$ core and the σ subunit. *Compare* Pol III holoenzyme.

RNA secondary structure: The local spatial arrangement of an RNA strand, including a description of any intrachain base pairing.

RNaseH: An endonuclease that cleaves the 3′-O-P bond of RNA in an RNA-DNA duplex.

RNA-Seq: A strategy for determining the level of expression of all the genes in a cell or tissue by sequencing the transcriptome.

RNA splicing: The removal of introns and joining of exons in a primary transcript.

RNA world hypothesis: The hypothesis that in an early stage of evolution, a living system was based on RNA. In this system, RNA enzymes could catalyze the synthesis of all the molecules required for life from simpler molecules available in the environment.

RNP: *See* ribonucleoprotein.

Rossmann fold: A protein supersecondary structure with alternating β strands and α helices (β-α-β-α-β motif), common in nucleotide-binding proteins.

RPA: *See* replication protein A.

r-protein: *See* ribosomal protein.

RRF: *See* ribosome recycling factor.

rRNA: *See* ribosomal RNA.

RT-PCR: *See* reverse transcriptase PCR.

Sanger sequencing method: Refers to two methods developed by Frederick Sanger, (1) for determining the sequence of a polypeptide and (2) for determining the base sequence of a DNA molecule. The DNA sequencing method uses dideoxynucleotides to terminate synthesis; products are analyzed on a gel.

scanning: The process by which a partially assembled eukaryotic initiation complex slides along the mRNA until it comes to a start codon.

scientific method: A variety of approaches for generating new knowledge, all focused exclusively on the natural world. In a common variant, a hypothesis is generated, tested, and then strengthened or discarded, depending on the outcome of the test.

scientific theory: An idea or principle that has been verified in numerous independent experiments over a significant period of time and that is used as the basis for generating new hypotheses.

SCOP database: *See* Structural Classification of Proteins (SCOP) database.

screenable marker: A gene introduced into a cell that allows any colony expressing it to be readily identifiable by its color or fluorescence. *Compare* selectable marker.

SDSA: *See* synthesis-dependent strand annealing.

SDS-PAGE: *See* sodium dodecyl sulfate–polyacrylamide gel electrophoresis.

secondary structure: The local spatial arrangement of the main-chain atoms in a segment of a polypeptide chain; the term is also applied to polynucleotide structure. *See also* RNA secondary structure.

second law of thermodynamics: The law stating that, in any chemical or physical process, the entropy of the universe tends to increase.

second messenger: An effector molecule synthesized in a cell in response to an external signal (first messenger) such as a hormone.

segmentation genes: A group of genes involved in pattern formation in the *Drosophila* embryo. Segmentation genes are divided into three subclasses: gap genes, pair-rule genes, and segment polarity genes.

segment polarity genes: A subclass of the segmentation genes, expressed after the pair-rule genes, that determine the anterior-posterior polarity of the developing *Drosophila* embryo.

selectable marker: A gene introduced into a cell that either permits the growth of the cell or kills it under a defined set of conditions. *Compare* screenable marker.

semiconservative: Describes a mode of DNA replication in which the daughter duplex has one intact parental strand and one newly synthesized strand.

semidiscontinuous: Describes a mode of DNA replication in which one strand, the leading strand, is replicated continuously, but the opposite strand, the lagging strand, is replicated in shorter, discontinuous segments.

sequence polymorphisms: Any alterations in genomic sequence (base-pair changes, insertions, deletions, rearrangements) that help distinguish subsets of individuals in a population or distinguish one species from another.

sequence motif: Any nucleotide or amino acid sequence that appears in multiple nucleic acids or proteins and has a demonstrable molecular function.

sequence tagged site (STS): Any known sequence that has been mapped in a chromosome and/or clones derived from it.

sequencing depth: The number of times, on average, that any given sequence among the targeted genomic sequences appears in the sequencing dataset. For example, sequencing to 100-fold coverage indicates that each targeted sequence is represented an average of 100 times in the dataset.

sex chromosome: A chromosome that determines the male or female sex of an organism. *Compare* autosome.

shelterin: A set of proteins that bind to and protect telomere sequences at the ends of eukaryotic chromosomes.

Shine-Dalgarno sequence: A sequence in an mRNA that is required for binding bacterial ribosomes. Also called the ribosome binding site (RBS).

short interfering RNA (siRNA): A short (~21 to 27 nucleotide) double-stranded RNA with 3′ overhangs, created from exogenous double-stranded RNA by the endonuclease Dicer, that participates in the RNAi gene silencing pathway.

short tandem repeat (STR): A short (typically 3 to 6 bp) DNA sequence repeated multiple times in tandem at a particular location in a chromosome.

shuttle vector: A recombinant DNA vector that can be replicated in two or more different host species. *See also* cloning vector.

sigma factor (σ): A transient subunit of the bacterial RNA polymerase that directs the enzyme to the promoter. Different sigma factors are specific for different promoters. The core RNA polymerase plus the sigma factor constitutes the RNA polymerase holoenzyme.

signal integration: The control of gene expression by the net effect of multiple, sometimes conflicting, regulatory signals.

signal recognition particle (SRP): A protein-RNA complex that recognizes and binds the signal sequence in a nascent polypeptide and delivers the ribosome to a peptide translocation complex in the endoplasmic reticulum.

signal sequence: An amino acid sequence, often at the amino terminus, that signals the cellular fate or destination of a newly synthesized protein.

silent mutation: A mutation in a gene that causes no detectable change in the peptide sequence of the gene product.

simple-sequence repeats (SSRs): Highly repeated, nontranslated segments of DNA in eukaryotic chromosomes, often associated with the centromere and telomere, but not restricted to these regions. Their function is unknown.

single bond: A bond between two elements that involves two electrons.

single molecule real time (SMRT) sequencing: A method for rapidly sequencing a long segment of a single strand of DNA.

single nucleotide polymorphism (SNP): A genomic base-pair change that helps distinguish one species from another or one subset of individuals in a population.

single-stranded DNA–binding protein (SSB): A bacterial protein that binds single-stranded DNA in a sequence-independent fashion.

siRNA: *See* short interfering RNA.

sister chromatid pair: Duplicate chromosomes, attached at the centromere, produced during the S phase of the eukaryotic cell cycle. Sister chromatids separate into individual chromosomes during anaphase of mitosis or anaphase II of meiosis, when the centromere divides.

site-directed mutagenesis: A set of methods used to create specific alterations in the sequence of a gene.

site-specific recombination: A type of genetic recombination that occurs only at specific sequences.

6-4 photoproduct: A pyrimidine dimer in which the C-6 and C-4 atoms of adjacent pyrimidines are linked.

small nuclear ribonucleoprotein (snRNP): A protein and snRNA complex, found in the nucleus and a component of the spliceosome.

small nuclear RNA (snRNA): A class of short, noncoding RNAs, typically 100 to 200 nucleotides long, found in the nucleus and involved in the splicing of eukaryotic mRNAs.

small nucleolar ribonucleoprotein (snoRNP): A protein and snoRNA complex that guides the modification of rRNAs in the nucleolus.

small nucleolar RNA (snoRNA): A class of short, noncoding RNAs, generally 60 to 300 nucleotides long, found in the nucleolus and involved in the modification of rRNAs.

SMC proteins: A family of ATPases that modulate the structure and organization of chromosomes.

snoRNA: *See* small nucleolar RNA.

snoRNP: *See* small nucleolar ribonucleoprotein.

SNP: *See* single nucleotide polymorphism.

snRNA: *See* small nuclear RNA.

snRNP: *See* small nuclear ribonucleoprotein.

sodium dodecyl sulfate–polyacrylamide gel electrophoresis (SDS-PAGE): A type of electrophoresis used to separate proteins on the basis of size. Proteins in experimental samples are denatured and bound along their length by negatively charged molecules of the detergent SDS, giving them a charge proportional to their length. They move through a gel matrix of cross-linked polyacrylamide under an electric field at a rate proportional to their size. *See also* gel electrophoresis.

solenoidal supercoiling: The wrapping of a helical molecule to form a coiled superstructure.

solenoid model: A model for the arrangement of nucleosomes in the 30 nm filament in which the nucleosome array assumes a spiral shape, with the flat sides of adjacent nucleosomes next to each other. *Compare* zigzag model.

somatic cells: All cells in a multicellular organism except the germ-line cells.

SOS response: In bacteria, a coordinated induction of a variety of genes in response to high levels of DNA damage.

Southern blotting: A DNA hybridization procedure in which one or more specific DNA fragments are detected in a larger population by hybridization to a complementary, labeled nucleic acid probe.

S phase: The phase of the cell cycle during which the DNA is replicated. The S phase occurs between the G_1 and G_2 phases.

spliceosome: A ribonucleoprotein complex that splices mRNAs in eukaryotic cells.

splice site: A nucleotide sequence within an intron at the intron-exon border, where a primary mRNA transcript may be spliced.

Spo11: A eukaryotic protein that creates double-strand breaks in DNA to initiate recombination in meiotic prophase I.

spongiform encephalopathies: Transmissible, progressive neurodegenerative diseases caused by prions.

SRP: *See* signal recognition particle.

SSB: *See* single-stranded DNA-binding protein.

SSR: *See* simple-sequence repeats.

standard (Gibbs) free-energy change ($\Delta G'$): The free-energy change for a reaction occurring under a set of standard conditions: temperature, 298 K; pressure, 1 atm or 101.3 kPa; and all solutes at 1 M concentration. $\Delta G'^\circ$ denotes the standard free-energy change at pH 7.0 in 55.5 M water.

start codon: *See* initiation codon.

steady state: A nonequilibrium state of a system through which matter is flowing and in which all components remain at a constant concentration.

steady-state kinetics: The study of reaction rates under steady-state conditions. *See also* steady state.

stem cells: The cells in multicellular organisms that retain the ability to divide and differentiate into other cell types.

step size: The average number of subunits over which a motor protein moves for each ATP molecule hydrolyzed.

stereochemistry: The spatial arrangement of the atoms within a molecule.

stereoisomers: Compounds that have the same composition and the same order of atomic connections but different molecular arrangements.

sticky ends: Two DNA ends in the same DNA molecule, or in different molecules, with short, overhanging, single-stranded segments that are complementary to each other, facilitating ligation of the ends.

stop codons: *See* termination codons.

STR: *See* short tandem repeat.

streptomycin: An aminoglycoside antibiotic that disrupts or inhibits bacterial protein synthesis. At low concentrations it causes misreading of the genetic code; at higher concentrations it inhibits translation initiation by preventing fMet-tRNA^{fMet} from binding to the ribosome.

stringent factor: A bacterial protein (RelA) recruited to a ribosome when an uncharged tRNA binds. Stringent factor catalyzes the formation of guanosine tetraphosphate (ppGpp), which binds RNA polymerase, reducing transcription from rRNA and tRNA genes and increasing transcription from biosynthetic genes.

stringent response: A mechanism for coordinating transcriptional activity in bacteria with the levels of amino acids available in the cell. Triggered by the binding of uncharged tRNAs to the ribosome, the stringent response directs the cellular machinery toward amino acid synthesis rather than growth and reproduction.

Structural Classification of Proteins (SCOP) database: A database (http://scop.berkeley.edu) of the structural and evolutionary relationships among all proteins for which the structures are known.

structural motif: Any distinct folding pattern for elements of secondary structure, observed in one or more proteins. A motif can be simple or complex and can represent all or just a small part of a polypeptide chain. Also called a fold or a supersecondary structure.

STS: *See* sequence-tagged site.

substrate: A molecule that undergoes an enzyme-catalyzed reaction.

supercoiled DNA: DNA that twists upon itself because it is underwound or overwound (and thereby strained) relative to B-form DNA.

superfamily: A structural classification that includes protein families with little sequence similarity but with the same supersecondary structural motif and functional similarities.

superhelical density (σ): In a helical molecule such as DNA, the number of supercoils (superhelical turns) relative to the number of coils (turns) in the relaxed molecule.

supersecondary structure: *See* structural motif.

suppressor tRNA: A mutant tRNA that binds to a termination codon but carries an amino acyl residue that can be incorporated into the growing amino acid chain, suppressing the termination signal.

surroundings: All matter in the universe that is outside the system being considered. *See also* system.

synteny: Conserved gene order along the chromosomes of different species.

synthesis-dependent strand annealing (SDSA): A pathway for repairing double-strand breaks that ends with the invading strands dissociating and reannealing and with the homologous DNA molecule intact. *See also* double-strand break repair.

system: An isolated collection of matter; all other matter in the universe apart from the system is called the surroundings. *See also* surroundings.

systems biology: In biochemistry or molecular biology, the study of complex systems, integrating the functions of the macromolecules in a cell (RNA, DNA, proteins).

tag: A peptide or protein that binds a simple, stable ligand with high affinity and specificity.

tandem affinity purification (TAP) tags: Two tags (such as Protein A and calmodulin-binding peptide) fused to the same target protein to allow sequential, highly specific affinity purification steps of the expressed protein.

tandem MS: Two mass spectrometry procedures carried out in succession to define the mass of peptides generated by proteolytic cleavage of a protein.

TAP tags: *See* tandem affinity purification tags.

target-primed (TP) retrotransposon: A retrotransposon that moves via a cDNA copy of its mRNA transcript and is synthesized as a direct extension of the 3′ end created in the target site by a retrotransposon-encoded endonuclease. *See also* extrachromosomally primed (EP) retrotransposon.

target site: The location on a chromosome where a transposon inserts itself. *Compare* donor site.

TATA-binding protein (TBP): A eukaryotic transcription factor that binds all three RNA polymerases as well as the AT-rich region of many promoters known as the TATA box.

τ (tau) complex: A complex of four different proteins that functions in bacteria to load the processivity clamp onto DNA in the bacterial replisome.

taxon: A grouping of organisms in a phylogenetic tree.

TBP: *See* TATA-binding protein.

TCR: *See* transcription-coupled repair.

telomerase: A ribonucleoprotein enzyme that has reverse transcriptase activity and a short RNA sequence that primes the addition of nucleotide repeats to the 3′ ends of DNA. Telomerase activity ensures that no unique sequence information is lost as a result of the end replication problem.

telomerase reverse transcriptase (TERT): The protein component of telomerase.

telomerase RNA (TR): The RNA component of telomerase.

telomere: A specialized nucleic acid structure found at the ends of linear eukaryotic chromosomes.

telomere loop: *See* t-loop.

telophase: The final stage of mitosis (M phase), in which the two sets of homologous chromosomes reach opposite spindle poles and begin to decondense. The cell physically divides in the process of cytokinesis, resulting in two daughter cells.

template strand: A strand of nucleic acid used by a polymerase as a template to synthesize a complementary strand.

termination: (1) The third of three stages of RNA synthesis, in which the RNA polymerase and the RNA product are released from the DNA template. (2) The third of three stages of protein synthesis, in which the ribosome and the peptide product are released from the mRNA template.

termination codons: UAA, UAG, and UGA; in protein synthesis, these codons signal the termination of a polypeptide chain. Also called stop codons.

termination factors: *See* release factors.

termination sequence: A DNA sequence, at the end of a transcriptional unit, that signals the end of transcription.

terminator: Broadly, a place where transcription is halted. This can occur at the end of a gene (where some termination sequences are also called terminators) or in regulatory sequences preceding some operons (as occurs in transcription attenuation). *See also* termination sequence.

Ter site: A 23 bp sequence that serves as a DNA replication termination site in *E. coli.*

TERT: *See* telomerase reverse transcriptase.

tertiary structure: The three-dimensional conformation of a polymer in its native folded state.

testcross: Genetic cross of an F₁ hybrid with a homozygous recessive strain to determine the genotype of the F₁ individual. A testcross also reveals linked genes.

tetracyclines: A family of antibiotics that inhibit bacterial protein synthesis by occupying the ribosomal A site, thereby preventing the binding of aminoacyl-tRNAs.

tetrad: A structure formed in meiotic prophase I by the association of two homologous sister chromatid pairs.

thin-layer chromatography: A process in which complex mixtures of molecules are separated by many repeated partitionings between a flowing (mobile) phase and a stationary phase coated onto a planar surface. Molecules separate based on the rate of their migration across the stationary phase as the mobile phase is drawn through it by capillary action.

30 nm filament: A higher-order organization of nucleosomes seen in condensed chromosomes.

3′ cleavage: The cleavage of nascent mRNA molecules in eukaryotes, prior to the addition of poly(A) tails.

3′ poly(A) tail: A length of adenosine residues (typically 80 to 250) added to the 3′ end of many mRNAs in eukaryotes (and sometimes in bacteria), which serves as a binding site for proteins that protect the mRNA from exonucleases.

thymine (T): A pyrimidine base that is a component of DNA, but not RNA.

t-loop (telomere loop): A looped structure seen in mammalian telomeres, in which the single-stranded 3′ end of the chromosome folds back and hybridizes to a duplex portion of the telomere.

TLS: *See* translesion synthesis.

T_m: *See* melting point.

tmRNA: A bacterial RNA that has the properties of a tRNA at its 5′ end and the properties of an mRNA, including a stop codon, at its 3′ end. When aminoacylated, the 5′ end can bind in the A site of a ribosome stalled on a truncated mRNA, and the 3′ end can serve as a template for continued translation through a termination codon that recruits the termination factors required for proper termination and ribosome recycling. Also called transfer-messenger RNA.

topoisomerase: An enzyme that catalyzes alterations in DNA topology, introducing and/or removing positive or negative supercoils in closed, circular duplex DNA.

topoisomers: Different forms of a covalently closed, circular DNA molecule that differ only in their linking number.

totipotent: Describes stem cells from the first few divisions of the fertilized egg that can differentiate into a complete, viable organism.

TR: *See* telomerase RNA.

transcription: The enzymatic process whereby the genetic information contained in one strand of DNA is used to specify a complementary sequence of bases in an RNA strand.

transcriptional ground state: The inherent activity of promoters and transcription machinery in vivo in the absence of regulatory mechanisms.

transcription attenuation: A process for the regulation of expression of a bacterial operon in which transcription begins but is halted before transcription of the operon genes.

transcription-coupled repair (TCR): A nucleotide excision repair pathway in eukaryotes that is triggered when RNA polymerase encounters a lesion in the DNA and stalls.

transcription factor: In eukaryotes, a protein that affects the regulation and transcription initiation of a gene by binding to a regulatory sequence near or within the gene and interacting with RNA polymerase and/or other transcription factors.

transcriptome: The entire complement of RNA transcripts present in a given cell or tissue under specific conditions.

transcriptomics: The study of transcriptomes.

transfection: Incorporation of exogenous DNA into a eukaryotic cellular genome by any of several methods.

transfer-messenger RNA: *See* tmRNA.

transfer RNA (tRNA): A class of RNA molecules (M_r 25,000 to 30,000), each of which combines covalently with a specific amino acid for use in protein synthesis.

transformation: (1) Introduction of an exogenous DNA into a bacterial cell, causing the cell to acquire a new phenotype. (2) The conversion of a cell in a multicellular eukaryote into a cancer cell.

transition mutation: A point mutation resulting in the exchange of one purine-pyrimidine base pair for another purine-pyrimidine pair. *Compare* transversion mutation.

transition state: An activated form of a molecule in which the molecule has undergone a partial chemical reaction; the highest point on the reaction coordinate.

translation: The process in which the genetic information present in an mRNA molecule specifies the sequence of amino acids during protein synthesis.

translational repressor: A repressor that binds to an mRNA, blocking translation.

translesion synthesis (TLS): A pathway for replicating DNA across lesions that occur in unwound DNA at the replication fork. The pathway uses a TLS polymerase that lacks a proofreading exonuclease and has a less-selective active site. Although this polymerase may introduce a mutation, it allows replication to proceed.

translocase: (1) An enzyme that catalyzes membrane transport. (2) An enzyme that causes movement, such as movement along a double-stranded nucleic acid or the movement of a ribosome along an mRNA.

translocation: (1) Enzyme-catalyzed movement across a membrane. (2) Movement along a double-stranded nucleic acid without strand separation. (3) Movement of a ribosome by one codon along the mRNA.

translocation mutation: A mutation that results from the exchange of large segments of DNA between nonhomologous chromosomes.

transposable element: *See* transposon.

transposases: Transposon-encoded enzymes that catalyze the reactions required for the transposon to excise itself from the donor site and insert itself into the target site. These reactions typically include hydrolysis of a specific phosphodiester bond and transesterification involving attack of the liberated 3′ hydroxyl on another phosphodiester bond.

transposition: The movement of a gene or set of genes from one site in the genome to another.

transposon (transposable element): A segment of DNA that can move from one position in the genome to another.

***trans*-splicing:** A process in nematode worms in which a short leader sequence is spliced to the 5′ end of a primary transcript from a separate RNA molecule.

transversion mutation: A point mutation resulting in the exchange of a purine-pyrimidine base pair for a pyrimidine-purine pair, or vice versa. *Compare* transition mutation.

triplet expansion disease: A disease caused by the insertion of many additional copies of a repeated codon triplet into a gene due to template slippage in the DNA replication process.

triplex DNA: A DNA structure involving three polynucleotide strands bonded through non-Watson-Crick interactions called Hoogsteen pairings.

tRNA: *See* transfer RNA.

tRNA-charging step: The second step in the attachment of an amino acid to a tRNA, in which the aminoacyl-tRNA synthetase transfers the bound aminoacyl-AMP to the 2′-OH or 3′-OH of the 3′-terminal adenosine residue of the tRNA. *See also* adenylylation step.

trombone model: A description of DNA replication on the lagging strand, with its repeated cycles of loop growth and disassembly, by analogy with the movement of a slide on a trombone.

tumor suppressor genes: A class of genes that encode proteins that normally suppress the division of cells. When defective, the normal gene becomes a tumor suppressor gene, and when both copies are defective, the cell is allowed to continue dividing without limitation; it becomes a tumor cell.

tunicamycin: An antibiotic that inhibits the *N*-glycosylation of proteins in eukaryotic cells.

turnover number: The number of times an enzyme molecule transforms a substrate molecule per unit time, under conditions giving maximal activity at substrate concentrations that are saturating. *See also* general rate constant.

26S proteasome: In eukaryotes, a large assembly of enzymatic complexes that function in the degradation of damaged or unneeded cellular proteins.

twist (*Tw*): The net number of helical turns in a DNA molecule. *Compare* writhe (*Wr*).

two-dimensional gel electrophoresis: A technique that separates the components of a sample on the basis of two properties, in successive steps. For example, a protein sample may be separated on the basis of isoelectric point in one dimension, followed by separation on the basis of relative molecular mass in the other dimension.

two-dimensional NMR: A type of nuclear magnetic resonance spectroscopy in which different pulses of an electromagnetic field provide two qualitatively distinct signals.

type II restriction endonucleases: Enzymes that cleave DNA at a specific site within a short recognition sequence, with no requirement for a nucleotide triphosphate cofactor.

type I topoisomerase: An enzyme that introduces positive or negative supercoils in closed, circular duplex DNA by cleaving one of the two DNA strands, passing the intact strand through the break, and ligating the broken ends. Type I topoisomerases change *Lk* in increments of 1.

type II topoisomerase: An enzyme that introduces positive or negative supercoils in closed, circular duplex DNA by cleaving both DNA strands, passing an intact segment of DNA through the break, then religating the broken ends. Type II topoisomerases change *Lk* in increments of 2.

UAS: *See* upstream activator sequence.

uncompetitive inhibitor: An inhibitor molecule that can bind to the enzyme-substrate complex but not to the free enzyme.

unipotent: Describes mammalian cells that can reproduce to form more of the same kind of differentiated cell.

unit cell: The smallest regularly repeating unit in a crystal.

unwinding: The separation of paired strands of a nucleic acid.

uORF: *See* upstream open reading frame.

UP element: *See* upstream promoter element.

upstream activator sequence (UAS): A regulatory sequence in yeast DNA to which transcription activators bind. *See also* enhancer.

upstream open reading frame (uORF): A short open reading frame upstream of a gene's start codon that serves as a decoy to divert ribosomes, thereby down-regulating expression.

upstream promoter (UP) element: An AT-rich sequence between positions -40 and -60 in the promoters of some highly expressed bacterial genes. The sequence is bound by an α subunit of RNA polymerase, improving the efficiency of transcription initiation for that gene.

uracil (U): A pyrimidine base that is a component of RNA but not DNA.

valence: The number of covalent bonds formed by atoms of a particular element.

valence bond model: A model that proposes that chemical bonds form when half-filled valence atomic orbitals from two atoms overlap.

van der Waals interactions: Weak intermolecular forces between molecules as a result of each molecule inducing polarization in the other.

van der Waals radius: Half the distance between two atoms of an element that are as close to each other as possible without being formally bonded. The van der Waals radius defines an imaginary sphere that represents the size of an atom in models.

V_{max}: The maximum velocity of an enzymatic reaction when the binding site is saturated with substrate.

V_0: *See* initial velocity.

weak chemical bonds: Chemical interactions such as van der Waals interactions, hydrophobic interactions, and hydrogen bonds that are weaker than formal ionic or covalent bonds.

Western blotting: A technique that uses antibodies to detect the presence of a protein in a biological sample, after the proteins from the sample have been separated in a gel. Also called immunoblotting.

whole-genome shotgun sequencing: A strategy for sequencing a genome in which random segments of DNA are sequenced and the segments are ordered by the computerized identification of sequence overlaps.

wild type: The allele or phenotype that appears with the greatest frequency in a natural population of a species.

Wnt-class signaling pathway: A type of cell-cell signaling during development that does not require cell-cell contact. Wnt-class genes and proteins are involved in the synthesis, secretion, and reception of glycoprotein signals in developing embryos.

wobble base: The base at the 5′ end of an anticodon, which pairs loosely and can form mispairs with the base at the 3′ end of the codon.

wobble hypothesis: A hypothesis proposed by Francis Crick in 1966 to describe how some anticodons can recognize more than one codon.

wobble position: The first position of the anticodon, at the 5′ end, which may contain a wobble base.

writhe (*Wr*): The net number of supercoils in a DNA molecule. *Compare* twist (*Tw*).

X chromosome inactivation: A method of dosage compensation in mammals that involves inactivating one of the two X chromosomes in females by compacting it into heterochromatin.

XIC: *See* X inactivation center.

X inactivation center (XIC): A point on the mammalian X chromosome that is the nucleation center for heterochromatin formation when the chromosome is inactivated.

x-ray crystallography: The analysis of x-ray diffraction patterns of a crystalline compound, used to determine the molecule's three-dimensional structure.

YAC: *See* yeast artificial chromosome.

yeast artificial chromosome (YAC): An expression vector for cloning eukaryotic genes in yeast. YAC vectors have a yeast origin of replication, two selectable markers, and telomere and centromere sequences for maintaining chromosome integrity.

yeast three-hybrid analysis: A method for defining protein-RNA interactions in vivo. A series of engineered proteins and a randomized library of RNA sequences are set up such that expression of a reported gene occurs only when the RNA sequence bound by the target protein is present.

yeast two-hybrid analysis: A method for defining protein-protein interactions. Target protein interaction activates transcription of a reporter gene.

Z-DNA (Z-form DNA): A conformation of double-stranded DNA observed in solvents with a high salt concentration or with sequences rich in G≡C base pairs. The molecule assumes a left-handed helical conformation with 12 base pairs per turn and a rise of 3.7 Å per base pair. The backbone of the helix has a zigzag structure. *Compare* A-DNA *and* B-DNA.

Z-form DNA: *See* Z-DNA.

zigzag model: A model for the arrangement of nucleosomes in the 30 nm filament in which zigzag histone pairs stack on each other and twist about a central axis. *Compare* solenoid model.

zinc finger motif: A protein structural motif involved in DNA recognition by some DNA-binding proteins; characterized by a single atom of zinc coordinated to four Cys residues or to two His and two Cys residues.

Index

Note: Page numbers followed by f, t, and b indicate figures, tables, and boxed material, respectively.
Page numbers preceded by A refer to the Model Organisms Appendix.